CENTRES OF PLANT DIVERSITY

A Guide and Strategy for their Conservation

WWF

The World Wide Fund For Nature (WWF) – founded in 1961 – is the largest international, private nature conservation organization in the world. Based in Switzerland, WWF has national affiliates and associate organizations on five continents. WWF works to conserve the natural environment and ecological processes essential to life on Earth.

WWF aims to create awareness of threats to the natural environment, to generate and attract on a worldwide basis the strongest moral and financial support for safeguarding the living world and to convert such support into action based on scientific priorities. Since 1961, WWF has channelled over US$130 million into more than 5000 projects in some 130 countries which have saved animals and plants from extinction and helped to conserve natural areas all over the world. It has served as a catalyst for conservation action, and has brought its influence to bear on critical conservation needs by working with and influencing governments, non-governmental organizations, scientists, industry and the general public.

IUCN – THE WORLD CONSERVATION UNION

IUCN – The World Conservation Union – founded in 1948, is a membership organization comprising governments, non-governmental organizations, research institutions and conservation agencies in 120 countries. The Union promotes the protection and utilization of living resources.

Several thousand scientists and experts from all continents form part of a network supporting the work of its six Commissions: threatened species, protected areas, ecology, sustainable development, environmental law and environmental education and training. Its thematic programmes include tropical forests, wetlands, marine ecosystems, plants, the Sahel, Antarctica, population and natural resources, and women in conservation. These activities enable IUCN to develop sound policies and programmes for the conservation of biological diversity and sustainable development of natural resources.

NATIONAL MUSEUM OF NATURAL HISTORY, SMITHSONIAN INSTITUTION

The Smithsonian Institution, founded in 1846 by an Act of Congress, counts among its most important missions the discovery, identification and understanding of the world around us. Personnel and research support for the North America, Middle America and South America sections of *Centres of Plant Diversity* were provided by the Department of Botany, National Museum of Natural History, Smithsonian Institution with partial support from other sources. Its cadre of research scientists, which includes a large number of biologists specializing in taxonomy, systematics and evolutionary studies, is contributing to the global inventory of species and to the revision of our view of natural systems. The production of the data for this volume is but one of the efforts to provide accurate information for the use of resource managers and planners and as an educational tool for future generations.

Availability of further information from WCMC

Background data and more general information on the sites documented in this volume have been donated by WWF and IUCN to the World Conservation Monitoring Centre (WCMC), 219 Huntingdon Road, Cambridge CB3 0DL, U.K., where they form part of WCMC's Biodiversity Information Service and where they are available for consultation. Both map and text information are also available from WCMC in electronic format. Readers may be interested to note that the maps (or modified versions of the maps) of Centres of Plant Diversity in this volume can be overlain with other geographic information, such as protected area boundaries and vegetation cover, through the Geographical Information System (Biodiversity Map Library) at WCMC.

CENTRES OF PLANT DIVERSITY

A Guide and Strategy for their Conservation

Project Director: V.H. HEYWOOD
Project Coordinator: S.D. DAVIS

VOLUME 3
THE AMERICAS

edited by

**S.D. DAVIS, V.H. HEYWOOD, O. HERRERA-MACBRYDE, J. VILLA-LOBOS
AND A.C. HAMILTON**

published by
The World Wide Fund For Nature (WWF)
and IUCN – The World Conservation Union

with financial support from the
European Commission (EC) and the
U.K. Overseas Development Administration (ODA)

1997

Citation: WWF and IUCN (1994–1997). Centres of plant diversity. A guide and strategy for their conservation. 3 volumes. IUCN Publications Unit, Cambridge, U.K.

British Library Cataloguing-in-Publication Data
A catalogue record for this book is available from the British Library.
ISBN 2-8317-0199-6

Designed and produced by the Nature Conservation Bureau Limited, Newbury, Berkshire, U.K.

Printed by Information Press, Oxford, U.K.

TABLE OF CONTENTS: VOLUME 3

CONTENTS OF VOLUMES 1 AND 2

**Volume 1: EUROPE, AFRICA, SOUTH WEST ASIA
AND THE MIDDLE EAST**

Volume 2: ASIA, AUSTRALASIA AND THE PACIFIC

PREFACE

The balance between people and their environment is being upset. Escalating human numbers and increasing demands for material resources are leading to the transformation and degradation of ecosystems worldwide, with consequent loss of genetic diversity and an inevitable rise in the extinction of species. More and more land is being converted to intensive production of food, timber and other plant products, much of the world's pasture land is overgrazed, and soil erosion and salinization are reducing the fertility of farmlands, especially in semi-arid regions. Wild species are being displaced or overcropped, and wild ecosystems can no longer be taken for granted as reservoirs of genetic diversity and regulators of the cycles of the elements.

Plants are a central component of this threatened nature. The loss of plants is very significant, for they stand at the base of food webs and provide habitats for other organisms. Although people in some modern societies may be mentally distanced from the reality of nature and are unaware of the services it provides, even city inhabitants are part of wider ecosystems, based on wild plants and natural vegetation. For example, genes from the wild can play major roles in the breeding of new varieties of food crops and other cultivated plants. New medicines continue to be derived from wild species. In many countries, natural and semi-natural ecosystems provide many plant products essential for human welfare, including fuelwood, timber, fibre, medicinal plants, fruits and nuts. Forests protect catchments and thus help regulate the flow of water for drinking, hydropower and irrigation. And wild nature – not the least wild plants – contribute to the natural beauty of the world that we rightly cherish.

In Rio de Janeiro in June 1992 over a hundred Heads of States and Governments signed a new International Convention on the Conservation of Biological Diversity – that is, of the world's rich variety of genes, species and ecosystems. They did so not out of altruism, but because they recognized that these resources were economically valuable and important to the future of humanity. Under the Convention – which entered into legal force at the end of 1993 – each country accepts a responsibility to safeguard its own natural diversity and to cooperate internationally, especially to help poorer countries enjoy the benefits of their living resources.

All ecosystems contribute to the biological wealth of the planet, but some have a much higher diversity than others. Tropical forest contains a much richer flora than tundra, and some areas of tropical forest have many more plant species than others. Even agricultural land varies in the diversity it supports, from the richness of traditional agriculture with its many local varieties of crops, to the ecological deserts of industrial farming with high-yielding monocultures. To be effective, conservation of biodiversity must depend in part on surveys to identify how species and ecosystems are distributed and to identify key sites where diversity is greatest.

The three volumes comprising *Centres of Plant Diversity* are the product of information received from hundreds of botanists from many countries. They have worked together to identify some of the most important sites for plants worldwide. These are priority areas for conservation. These volumes are offered to national conservation authorities and to global conservation organizations as an aid in their work, especially in implementing the obligations of States under the Convention.

A publication like this can only be a beginning. Strategies to conserve plant diversity are needed at all geographical levels – global, national and local – and should form an integral part of all land development plans. And strategies have to be turned into action. Many of the sites described in these volumes are subject to threats and pressures and solutions will depend on finding a balance between different interests, site by site. The solutions are likely to be almost as diverse as the sites, since the environmental features of the latter and the nature of the threats to them vary so widely.

People are at the heart of conservation as well as the source of many threats. Virtually all terrestrial ecosystems, even those that appear to be more pristine, have long included people as components. For example, there are very few, if any, areas of tropical forest which do not provide local people with products for their use, and which have not been altered as a result of interaction with the users over the centuries. And, although local people have been components of natural ecosystems, sometimes for tens of thousands of years, their present day interactions with their environment are not necessarily harmonious.

The conservation aim is to conserve the diversity of nature: clearly this must be achieved within the context of cultural diversity. Conservationists must work with local communities, understanding how they use and manage nature, examining whether current practices are sustainable (both for particular species and in terms of wider impacts) and, if necessary, searching for alternative approaches. Moreover, cultures, whether of local people or professionals, including botanists, land managers and legislators, are not static. The search for appropriate practices, which will conserve plant diversity, perhaps while using it and other environmental resources sustainably, will be a continuing process.

It is our hope that these volumes will stimulate such activities throughout the world, not only at the key localities identified here. Without urgent, informed, practical action, the marvellous plant wealth of our planet will not be conserved, and future generations will be the poorer.

Claude Martin, Director General, WWF International (World Wide Fund For Nature).

Martin Holdgate, former Director General, IUCN – The World Conservation Union.

This book is dedicated to the memory of Dr Alwyn H. Gentry,
who contributed so much to neotropical botany.
He died in a plane crash in Ecuador at a time when he was
preparing the South America overview for this volume.

It is very appropriate to dedicate this book to Alwyn H. Gentry, and I am honoured to be able to give him a brief tribute here.

Al's knowledge of the tropics made him indispensable in the preparation of this CPD volume. The editors received advice, direction and extremely valuable information from him about sites to include, their richness, ecological features, and conservation status. Volume 3 is richer, and its conservation effects will be better, because of his broad and specific contributions.

Al and I shared 20 years of productive botanical interactions. We received our doctoral degrees in the early 1970s. We met during the autumn 1972 Systematics Symposium at the Missouri Botanical Garden in St. Louis, where Peter Raven suggested that we start a botanical venture on the Chocó region of Colombia. The Chocó was botanically one of the least known places on Earth. Not more than 4000 collections from the region were available in 1973 in herbaria in Colombia and the United States. We submitted separate grant proposals to the U.S. National Science Foundation (NSF) and the Colombian National Research Fund (COLCIENCIAS), and received separate but coordinated funding through the Colombia-U.S. International Program of NSF and COLCIENCIAS. At the end of a 12-year programme of exploration in Colombia's Chocó Department, we had gathered over 17,000 collection numbers and more than 80,000 herbarium specimens, provided field training for many Colombian and U.S. students, gotten the scientific community excited about the region (CPD Site SA39), and developed possible explanations for its species richness. In the process, Al and I developed a close friendship. Al later expanded his work to the whole Chocó biogeographic region from Panama to Ecuador. We collaborated in a number of other projects, shared responsibilities within the Flora Neotropica Organization for several years, jointly advised graduate students in Colombia and the U.S.A., published a few papers together including the Checklist of the Flora of Chocó, and worked together at the Missouri Botanical Garden during my five years there.

Al's work in Chocó was only one of the many exciting things he did in his life. He was a tireless traveler. During his short life he visited all continents and worked in many countries. At one point, in or around 1990, he made it a priority to visit places like China and several other Asian and African countries he had never visited, to study the flora. He gathered support from the U.S. National Geographic Society and other sources to fulfill this wish, and was able to travel to remote places all over the globe to collect plants and set up his well-known and now famous plots. Al was one of the most "fundable" botanists I have known. The excellence of his research was recognized by the Pew Charitable Trust with a prestigious Pew Fellowship.

He was extremely hard working. One could find him literally in the middle of the forest, at his home or a friend's apartment, at any airport, or very late at night or early in the morning (he didn't sleep much) writing his own papers, correcting a student's doctoral dissertation, dictating letters into his dictaphone, or even reading a novel!

He was an excellent teacher and was always willing to share his knowledge with anyone willing to listen and learn. This is clearly shown not only in the many courses he taught throughout Latin America and in the U.S.A. but also in several of his books that were aimed at field botanists and students, with easy to use keys and descriptions always illustrated with simple but useful drawings. He made a point of writing about his research findings and as a result, his publication record is truly outstanding.

Al had been trained as a plant taxonomist, and taxonomy is mainly what he did most of his life. During the last few years, however, he openly recognized that he had become more of an ecologist and less of a taxonomist – and it seems to me that it got harder and harder for him to sit down and finish his monographic work on Bignoniaceae. He did publish two parts of a three-part treatment of the family for *Flora Neotropica*, and he also prepared treatments for many national and regional Floras.

His contributions to our understanding of tropical ecosystems are many and of considerable importance. He was at one point the sole person in the world who could talk intelligently, based on his own first-hand experience, about forests located anywhere, and who was, therefore, able to establish comparisons and parallels and to draw conclusions about differences and similarities better than anybody else. He was forever looking for the "richest place on Earth" in terms of numbers of species. It seemed at times that every new place he visited had the honour of being the "most species-rich in the world", even if that distinction only lasted until the next place was sought out and visited. The Chocó was "it" for a while, then the Western Amazon region, and so on. Because of his amazing knowledge of plants in the field, he was one of the botanists in the rapid assessment team put together by Conservation International to recognize endangered habitats throughout the tropics. At the same time he became one of the most sought-after botanical advisors for the New World tropics. Recognizing biogeographic patterns constituted one of his research interests.

Members of his generation will remember him for his devotion to his work, his scientific excellence, his inquiring mind, and his strong opinions about certain botanical subjects. Members of younger generations who knew him will cherish the excellence of his teaching, his warm attitude towards his students, his willingness to share information with them, and his ability to make them love their work.

Enrique Forero
Director
Instituto de Ciencias Naturales
Universidad Nacional
Bogotá, Colombia

ACKNOWLEDGEMENTS

The *Centres of Plant Diversity (CPD)* project, a major international collaborative exercise, has involved over 400 botanists, conservationists and resource managers worldwide, together with over 100 collaborating institutions and organizations. Without this enormous amount of help and support, the project would not have been possible. We wish to thank all those who contributed to the text.

A full list of contributors and collaborating organizations is given below, and authors and contributors to individual Data Sheets are given at the end of each sheet. Here, we would like in particular to acknowledge the major contributions to the project provided by Dr Dennis Adams, Dr John R. Akeroyd, Professor Peter S. Ashton, Dr Henk J. Beentje, Dr Robert W. Boden, Professor Loutfy Boulos, Dr Alwyn Gentry, Dr David R. Given, Juan Carlos Godoy, Luis Diego Gómez, Dr Alan C. Hamilton, Professor K. Iwatsuki, Professor R.J. Johns, Dr Ruth Kiew, Professor Valentin A. Krassilov, Dr Domingo Madulid, Dr Robert R. Mill, Dr Tony Miller, Professor P. van Royen, Dr Jerzy Rzedowski, Dr B.D. Sharma, Dr Sy Sohmer, Dr Peter F. Stevens, Dr Wendy Strahm, Dr Víctor Toledo, Professor Wang Xianpu and Professor Yang Zhouhuai, who undertook the preparation of regional texts and/or contributed a number of site Data Sheets, and (with others) helped in the selection of sites for inclusion in the project.

The CPD project has been co-ordinated by the IUCN Plant Conservation Office at Kew, U.K. Throughout this period, the project benefited enormously from the valuable help, advice and information provided by colleagues and staff of the Herbarium of the Royal Botanic Gardens, Kew, as well as from use of Kew's extensive library facilities. We thank all those members of Kew's staff who contributed to individual Data Sheets, in particular Professor Robert J. Johns, who provided extensive help with writing the section on New Guinea, and to Dr John Dransfield for checking sheets for Malaysia and Indonesia. Special thanks are also extended to Milan Svanderlik and colleagues in Media Resources for their help in producing a set of posters on the project which were used throughout the world at CPD Workshops. Particular thanks are extended to the Director, Professor Ghillean Prance, and to the Keeper of the Herbarium, Professor Gren Ll. Lucas, for their support. Indeed, the initial concept of identifying centres of plant diversity and endemism owes much to Gren Lucas and Hugh Synge (then of the IUCN Threatened Plants Unit). Acknowledgement must also be made of the role played by the IUCN/WWF Plant Advisory Group, initially under the chairmanship of Dr Peter Raven, and subsequently Professor Arturo Gómez-Pompa, in planning and developing the CPD project.

Much of the co-ordinating work for the Americas was undertaken by colleagues at the Department of Botany at the National Museum of Natural History, Smithsonian Institution, Washington, D.C., U.S.A. In particular, we would like to thank Olga Herrera-MacBryde for her dedication in compiling and writing many of the accounts for South America and Middle America, and for editing the Latin America part of this volume on the Americas (Volume 3 in the series). Additional editors for the Americas volume were Dr Bruce MacBryde, and also Jane MacKnight and Dr Wayt Thomas.

Sadly, during the latter stages of the project, Dr Alwyn Gentry, who had substantially prepared the Regional Overview for South America, died in a plane crash whilst carrying out a forest survey in Ecuador. The text on South America owes much to the extensive knowledge on the botany of the region which he had accumulated. We acknowledge the help of Dr Carlos B. Villamil, Dr Otto Huber and Olga Herrera-MacBryde in completing the South America overview, and of Jane Villa-Lobos for help with North America and Middle America. Shirley L. Maina and Dr Robert A. DeFilipps are thanked for their contributions for North America.

We are indebted to colleagues at the Royal Botanic Garden, Edinburgh, who undertook (with Professor Loutfy Boulos) much of the co-ordinating work for South West Asia and the Middle East. We particularly thank the Deputy Regius Keeper, Dr David G. Mann, and make special mention of the valuable contributions to the project provided by Dr Tony Miller and Dr Robert Mill.

Grateful thanks are also extended to the members of the IUCN Australasian Plant Specialist Group (co-ordinated by Dr Robert W. Boden): John Benson, Stephen Harris, Frank Ingwersen, Dr John Leigh, Dr Ian Lunt, Dr Bob Parsons and Neville Scarlett. Drs Garry Werren, Geoff Tracey, Stephen Goosem and Peter Stanton also provided much valuable advice and contributions for the Australian section. Similarly, we would like to thank the Secretary for the Environment (Government of India), the Botanical Survey of India and the IUCN Plant Specialist Groups for China and Lower Plants.

For Africa, we would particularly like to thank members of AETFAT (the Association pour l'Etude Taxonomique de la Flore d'Afrique Tropicale) for much helpful advice and valuable contributions.

At the World Conservation Monitoring Centre (WCMC), Cambridge, U.K., the following are thanked for their help: Dr Kerry Walter (Threatened Plants Unit) for providing country plant biodiversity tables; Dr Mark Collins, Dr Richard Luxmoore, Mary Edwards and Clare Billington (Habitats Data Unit) for advice on mapwork and for producing regional maps which formed the basis for those used in all three volumes; and to Jerry Harrison, James Paine, Michael Green and Harriet Gillett (Protected Areas Data Unit) for checking protected areas information. Andrew McCarthy provided much valuable assistance with information on Indonesian protected areas. Dr Tim Johnson is thanked for co-ordinating WCMC's input to the project.

CPD benefited from close collaboration with BirdLife International (formerly known as the International Council for Bird Preservation). We would like to make special mention of contributions of bird data provided by Alison Stattersfield, Mike Crosby, Adrian Long and David Wege. The data on birds will be published more fully by BirdLife International in the *Global directory of Endemic Bird Areas* (Stattersfield *et al.*, in prep.), in which the distributions of all restricted-range bird species will be analysed.

Thanks are expressed to colleagues at Botanic Gardens Conservation International (BGCI), with whom the CPD project shared office space and equipment, and benefited from computer and secretarial support. In particular, we would like to thank Diane Wyse Jackson for technical computer support and Nicky Powell, Erika Keiss and Christine Allen for secretarial help and typing manuscripts. Ros Coles, at WWF, is also thanked for her secretarial support. Kevin McPaul (Computer Unit, Royal Botanic Gardens, Kew) provided valuable assistance in converting several incoming computer diskettes into a readable form. For translations, we thank Sally Horan, Doreen Abeledo, Jennifer Moog and Barbara Windisch, and for map drawing Cecilia Andrade-Herrera, Raúl Puente-Martínez, Alice Tangerini, Carlos Bazán, Jeff Edwards and Martin Walters.

Finally, we give particular thanks to the organizations who provided financial support for the CPD project and without which none of this work would have been possible, namely the Commission of the European Communities (EC), the U.K. Overseas Development Administration (ODA), the World Wide Fund For Nature (WWF), IUCN – The World Conservation Union and, in the U.S.A., Conservation International, the Smithsonian Institution and the Wildcat Foundation. We are most grateful for their support and encouragement throughout the project.

Stephen D. Davis.   Vernon H. Heywood.

The successful completion of this major project has in no small measure been due to the outstanding efforts and commitment of Stephen Davis who has worked unflaggingly and consistently over the whole period of its preparation.

V.H.H.

The opportunity has been taken with production of this volume to make further corrections and additions to the introductory material which precedes the regional and data sheet accounts, these being in addition to those already made in Volume 2.

This work has proved much more difficult to produce than originally envisaged. Credit for formulating the project is due to Professor Vernon Heywood, formerly of IUCN, and Hugh Synge, formerly of WWF. WWF wishes to acknowledge the efforts of the many contributors and especially the dedication of Stephen Davis, Olga Herrera-MacBryde, Bruce MacBryde and Jane Villa-Lobos. The assistance of Botanic Gardens Conservation International (BGCI) (Peter Wyse Jackson) and of the Department of Botany, National Museum of Natural History, Smithsonian Institution for accommodating some of those working on the project is gratefully acknowledged. Here at WWF, Ros Coles has put in monumental efforts. Also at WWF, I would like to show my appreciation to Clive Wicks, Peter Newborne, Peter Ramshaw and Michael Pimbert for their encouragement and support.

Special thanks are due to the patience, creativity and understanding of the staff of the Nature Conservation Bureau for their task of laying-out and overseeing the printing of the books. Special contributions have been made by Peter Creed, Charlotte Matthews and Joe Little.

Alan Hamilton
Plants Conservation Officer
WWF International.

The Mexico, Central America and South America sections of this volume would not have been possible or substantial without the professional dedication of my colleague and friend, Olga Herrera-MacBryde. Words do not express the gratitude she deserves for her inexhaustible efforts, and her cooperative spirit in working with our many generous authors – including through the last two years when she volunteered her time to complete the manuscript and to painstakingly edit the first-stage and final proofs. Very special thanks should also be given to Olga's husband, Bruce MacBryde, for his wise council and dedicated professional assistance in enhancing the Latin America texts and maps, and his careful scrutiny in proof-reading as well.

Much of the remaining natural legacy we have been given from evolutionary time can be all of ours for the indefinite future, if we choose to notice, and to act individually, and together. We fervently hope that the perspective, information, generosity, and will that have created this volume on the Americas will be taken up and extended, to truly conserve the natural areas and their plants throughout this botanically richest portion of the Earth, with its wonderful landscapes and so remarkable peoples.

Jane Villa-Lobos
Latin American Plants Program
National Museum of Natural History
Smithsonian Institution.

LIST OF CONTRIBUTORS

Regional co-ordinators

Volume One:
Europe, Africa, South West Asia and the Middle East

Europe:
    Dr John R. Akeroyd and
    Professor Vernon H. Heywood
Atlantic Ocean Islands:
    Dr Alan C. Hamilton
Africa:
    Dr Henk J. Beentje and
    Stephen D. Davis
Indian Ocean Islands:
    Wendy Strahm
South West Asia and the Middle East:
    Professor Loutfy Boulos,
    Dr Tony Miller and
    Dr Robert R. Mill

Volume Two:
Asia, Australasia and the Pacific

Central and Northern Asia:
    Professor Valentin A. Krassilov
Indian Subcontinent:
    Dr B.D. Sharma,
    Mr A.K. Narayanan and
    Dr Robert R. Mill
China and East Asia:
    Professors Wang Xianpu and Yang Zhouhuai (China),
    Professor Kunio Iwatsuki (Japan) and
    Dr Vu Van Dung (Vietnam)
South East Asia (Malesia):
    Stephen D. Davis
Australia and New Zealand:
    Dr Robert W. Boden and
    Dr David R. Given
Pacific Ocean Islands:
    Professor P. van Royen, Dr Sy Sohmer and
    Stephen D. Davis

Volume Three:
The Americas

North America:
    Dr Robert DeFilipps and
    Shirley L. Maina
Middle America:
    Olga Herrera-MacBryde
Caribbean Islands:
    Dr Dennis Adams
South America:
    Olga Herrera-MacBryde

Major collaborating organizations and institutions

Africa

Association pour l'Etude Taxonomique de la Flore d'Afrique
    Tropicale (AETFAT)

Argentina

Universidad Nacional de Córdoba, Centro de Ecología y
    Recursos Naturales Renovables
Universidad Nacional del Sur, Departamento de Biología,
    Bahía Blanca

Australia

A.C.T. Parks and Conservation Service
Australian National Parks and Wildlife Service, New South
    Wales
Conservation Commission of the Northern Territory
Queensland Herbarium

Belgium

Nationale Plantentuin van België, Meise

Bolivia

Centro de Investigaciones de la Capacidad de Uso Mayor
    de la Tierra (CUMAT), La Paz
Universidad Mayor de San Andrés, Centro de Datos para la
    Conservación, Herbario Nacional de Bolivia, Instituto
    de Ecología, La Paz

Brazil

Centro Nacional de Pesquisas de Recursos Genéticos e
    Biotecnologia (CENARGEN), Brasília
Centro de Pesquisas do Cacau (CEPEC), Bahía
Fundação Estadual de Engenharia do Meio Ambiente
    (FEEMA), Centro de Botânica do Rio de Janeiro
Instituto Brasileiro de Geografia e Estatística (IBGE), Brasília
Instituto Nacional de Pesquisas da Amazônia (INPA), Manaus
Jardim Botânico do Rio de Janeiro
Museu Paraense Emilio Goeldi, Departamento de Ecologia,
    Belém
Secretaria de Estado do Meio Ambiente, Instituto de Botânica,
    São Paulo
Universidade Estadual de Campinas, Departamento de Botânica
Universidade Estadual Paulista, Departamento de Botânica,
    Rio Claro
Universidade Federal Rural do Rio de Janeiro, Seropédica
Universidade de São Paulo, Instituto de Biociências,
    Departamento de Botânica

Chile

Comité Nacional Pro Defensa de la Fauna y Flora (CODEFF), Santiago
Corporación Nacional Forestal (CONAF), Santiago
Fundación Claudio Gay, Santiago
Pontificia Universidad Católica de Chile, Departamento de Biología Ambiental y de Poblaciones, Santiago
Universidad de Chile, Departamento de Biología, Santiago

China

Commission for Integrated Survey of Natural Resources

Colombia

Corporación Colombiana para la Amazonia Araracuara (COA), Santafé de Bogotá
Fundación Pro-Sierra Nevada de Santa Marta, Santafé de Bogotá
Fundación Tropenbos-Colombia, Santafé de Bogotá
Instituto Nacional de los Recursos Naturales y del Ambiente (INDERENA), Santafé de Bogotá
Universidad Nacional de Colombia, Instituto de Ciencias Naturales, Santafé de Bogotá

Costa Rica

Fundación Neotrópica, San José
Instituto Nacional de Biodiversidad (INBio), Heredia
Las Cruces Botanical Garden, San Vito, Coto Brus
Organization for Tropical Studies, Moravia
Universidad Nacional, Escuela de Ciencias Ambientales, Heredia

Cuba

Jardín Botánico Nacional, La Habana

Denmark

University of Aarhus, Botanical Institute
University of Copenhagen, Botanical Museum and Herbarium

Ecuador

Fundación Ecuatoriana de Estudios Ecológicos (ECOCIENCIA), Quito
Ministerio de Agricultura y Ganadería, Dirección de Desarrollo Forestal, Quito
Museo Nacional de Ciencias Naturales, Herbario Nacional, Quito
Pontificia Universidad Católica del Ecuador, Instituto de Ciencias Naturales, Quito
Río Palenque Science Center, Santo Domingo de los Colorados
Universidad de Guayaquil, Facultad de Ciencias Naturales

France

Centre ORSTOM, Nouméa, New Caledonia
Laboratoire de Phanérogamie, Muséum National d'Histoire Naturelle, Paris

French Guiana

Centre ORSTOM, Cayenne

Germany

Universität Hamburg, Institut für Allgemeine Botanik
University of Kassel

Guatemala

Fundación Defensores de la Naturaleza, Guatemala City
Universidad de San Carlos, Centro de Estudios Conservacionistas (CECON), Guatemala City
Universidad del Valle de Guatemala, Departamento de Biología, Guatemala City

Honduras

Mosquitia Pawisa (MOPAWI), Tegucigalpa
Universidad Nacional Autónoma de Honduras, Departamento de Biología, Tegucigalpa

Hungary

Eszterhazy Teachers' College

India

Botanical Survey of India
Government of India, Department of Forests

Indonesia

Southeast Asian Regional Centre for Tropical Biology (BIOTROP)
WWF Representation in Indonesia

Italy

International Plant Genetic Resources Institute (IPGRI), formerly International Board for Plant Genetic Resources (IBPGR)

Japan

Botanical Gardens, University of Tokyo

Kuwait

University of Kuwait

Malaysia

Malaysian Nature Society
Sabah Parks
WWF Malaysia

Mexico

Centro Interdisciplinario de Investigación para el Desarrollo Integral Regional, Instituto Politécnico Nacional (CIIDIR-IPN), Durango
Centro de Investigaciones Biológicas de Baja California Sur, La Paz

Instituto de Ecología, Centro Regional del Bajío, Pátzcuaro
Instituto de Ecología, Xalapa, Veracruz
Secretaría de Desarrollo Urbano y Ecología (SEDUE), Mexico City
Universidad Autónoma Agraria Antonio Narro, Saltillo
Universidad de Guadalajara, Instituto Manantlán de Ecología y Conservación de la Biodiversidad
Universidad Nacional Autónoma de México (UNAM), Instituto de Biología, Departamento de Botánica and Jardín Botánico; Laboratorio de Plantas Vasculares; and Centro de Ecología, Mexico City

Netherlands

Rijksherbarium, Leiden
University of Amsterdam, Hugo de Vries Laboratorium
Wageningen Agricultural University

New Zealand

Invermay Agricultural Centre, Crop and Food Research, Mosgiel

Oman

Office for Conservation of the Environment

Panama

Asociación Nacional para la Conservación de la Naturaleza (ANCON), Panama City
Universidad de Panamá, Departamento de Botánica

Paraguay

Centro de Estudios y Colecciones Biológicas para la Conservación, Asunción
Museo Nacional de Historia Natural del Paraguay, Asunción

Peru

Fundación Peruana para la Conservación de la Naturaleza, Lima
Universidad Nacional Mayor de San Marcos, Museo de Historia Natural, Lima
Universidad Nacional Agraria La Molina, Centro de Datos para la Conservación, Lima
Universidad Nacional de la Amazonia Peruana, Iquitos

Portugal

Universidade dos Açores, Depart. Ciencias Agrárias

Russia

Institute of Nature Conservation and Reserves
Research Institute of Nature Protection

Saudi Arabia

National Herbarium, Riyadh

South Africa

National Botanical Institute, Kirstenbosch

Rhodes University, Department of Botany
University of Cape Town
University of Pretoria

Spain

Instituto Pirenaico de Ecología, Jaca
Jardín Botánico, Universidad de Valencia
Jardín Botánico "Viera y Clavijo", Las Palmas de Gran Canaria
Real Jardín Botánico, Consejo Superior de Investigaciones Científicas, Madrid
Universidad de Granada

Sri Lanka

University of Peradeniya, Department of Botany

Sweden

University of Uppsala, Department of Systematic Botany

Switzerland

Conservatoire et Jardin botaniques de la Ville de Genève, Geneva

Taiwan

National Taiwan University, Department of Forestry

Turkey

Istanbul Üniversitesi Eczacilik Fakültesi
Society for the Protection of Nature (DHKD)

U.K.

BirdLife International
Botanic Gardens Conservation International
Royal Botanic Garden, Edinburgh
Royal Botanic Gardens, Kew
The Natural History Museum, London
World Conservation Monitoring Centre
WWF-U.K.

U.S.A.

Arizona State University, Department of Botany, Tempe
California Native Plant Society, Sacramento
Conservation International, Washington, D.C.
Drylands Institute, Tuscon
Field Museum of Natural History, Department of Botany, Chicago
Harvard University Herbaria, Cambridge
Louisiana State University, Department of Plant Biology, Baton Rouge
Missouri Botanical Garden, St Louis
New York Botanical Garden, Bronx, New York
Smithsonian Institution, National Museum of Natural History, Department of Botany, Washington, D.C.
The Nature Conservancy – Latin America Program, Arlington
University of California at Riverside, Department of Botany and Plant Sciences

University of Colorado, Department of Geography, Boulder
University of Florida, Department of Wildlife and Range Sciences, Gainesville
University of Hawaii at Manoa, Honolulu
University of Maryland Baltimore County, Department of Geography, Baltimore
University of Oregon, Department of Biology, Eugene
University of South Florida, Department of Biology, Tampa
U.S. Agency for International Development, Washington, D.C.
U.S. Fish and Wildlife Service, Washington, D.C.
U.S. National Park Service, Washington, D.C.
USDA Agricultural Research Service, Beltsville
WWF U.S., Washington, D.C.

Vietnam

Forest Inventory and Planning Institute, Ministry of Forestry

Zaïre

Association des Botanistes du Zaïre (ASBOZA)

Individual contributors

The following kindly contributed information to the CPD project:

B. Adams, C.D. Adams, D. Aeschimann, M. Aguilar, J. Aguirre, M. Ahmedullah, L. Aké Assi, J.R. Akeroyd, R. Alfaro, K. Alpinar, J. Aranda, D.S.D. Araújo, A. Arévalo, G. Argent, H. Arnal, P.S. Ashton, A. de Avila, M.M.J. van Balgooy, J. Balmer, P. Bamps, J. Beaman, J.S. Beard, S. Beck, H.J. Beentje, D. Benson, J. Benson, B.F. Benz, R. Berazaín, J.B. Besong, M. Bingham, J. Black-Maldonado, R.W. Boden, Bo-Myeong Woo, J. Bosser, L. Boulos, D. Bramwell, F.J. Breteler, D. Brummitt, D. Brunner, T.M. Butynski, R. Bye, A. Byfield, A. Cano, J. Cardiel, L.G. Carrasquilla, A.M.V. de Carvalho, H. Centeno, P. Chai, D. Chamberlain, J.D. Chapman, J. Charles, Shankat Chaudhary, A.S. Chauhan, A. Chaverri, J. Chávez-Salas, M. Cheek, Chin See Chung, M. Cifuentes, A. Cleef, S. Collenette, A. Contreras, M. Costa, I. Cordeiro, M. Correa, I. Cowie, R. Cowling, The Earl of Cranbrook, P.J. Cribb, M. Crosby, H. Cuadros, D. Cuartas, J. Cuatrecasas, R. Cuevas-Guzmán, A. Cunningham, J. Cunningham, R. Daly, P. Dávila, S.D. Davis, G.W.H. Davison, R.A. DeFilipps, C. Dendaletche, L.V. Denisova, E. Dias, N. Diego, M.O. Dillon, R. Dirzo, M.A. Dix, C. Dodson, C. Doumenge, C. Downer, F. Dowsett-Lemaire, J. Dransfield, S.J.M. Droop, G.R.F. Drucker, O.A. Druzhinina, J.F. Duivenvoorden, B. Eastwood, G. Echeverría, D.S. Edwards, I. Edwards, S. Elliott, J.L. Ellis, J. Estrada, T. Ju. Fedorovskaya, R. Felger, R. Ferreyra, T.S. Filgueiras, P. Fisher, R.M. Fonseca, A. Forbes, E. Forero, F.R. Fosberg, R. Foster, J.E.D. Fox, H. Freitag, F. Friedmann, I. Friis, J. Fuertes, F.M. Galera, S.M. Gan III, C. García-Kirkbride, S. Gartlan, A. Garzón, L. Gautier, A.H. Gentry, A.M. Giulietti, D.R. Given, D. Glick, J.C. Godoy, L.D. Gómez, A. Gómez-Pompa, J.A. González, S. González-Elizondo, R.B. Good, S. Goosem, R. Gopalan, J.-J. de Granville, G. Green, M.J.B. Green, P.S. Green, P. Gregerson, L. Guarino, R. Guedes-Bruni, N. Gunatilleke, R. Guzmán, P.K. Hajra, T. Hallingbäck, S. Halloy, O. Hamann,

A.C. Hamilton, B. Hammel, D. Harder, R.M. Harley, S. Harris, He Shan-an, I. Hedberg, O. Hedberg, I. Hedge, A.N. Henry, F.N. Hepper, P. Herlihy, A. Hernández, J. Hernández-Camacho, B. Herrera, O. Herrera-MacBryde, H. Hewson, V.H. Heywood, C. Hilton-Taylor, R. Hnatiuk, A. Hoffmann, Horng Jye-Su, V.B. Hosagoudar, R.A. Howard, Huang Shiman, O. Huber, B.J. Huntley, K. Hurlbert, Indraneil Das, F. Ingwersen, S. Iremonger, S. Iversen, K. Iwatsuki, N. Jacobsen, T. Jaffré, D. Janzen, J. Jaramillo, E.J. Jardel-Peláez, C. Jeffrey, J. Jérémie, C. Jermy, V. Jiménez, R.J. Johns, M.C. Johnston, M. Jørgensen, W.S. Judd, N. Jürgens, C. Kabuye, M.T. Kalin Arroyo, M. Kappelle, S. Keel, R. Kiew, T. Killeen, D.J.B. Killick, Kim Yong Shik, J. Kirkbride Jr., K. Kitayama, C. Kofron, J. Kokwaro, V.A. Krassilov, A.N. Kuliev, A. Lamb, M. Lamotte, A. Lara, R. Lara, P.K. Latz, Y. Laumonier, A. Lehnhoff, J. Leigh, H.F. Leitão Filho, A. Leiva, J. Lejoly, D. Lellinger, B. León, J.L. León de la Luz, J. Léonard, H.C. de Lima, Li Zhiji, E. Lleras, M. Lock, G.A. Lomakina, A. Long, F. Lorea, E. Lott, J. Lovett, P.P. Lowry II, P. Lowy, L. Lozada, C.A. Lubini, I. Lunt, J. Luteyn, B. MacBryde, K. MacKinnon, L. Madrigal, D.A. Madulid, G. Maggs, S.L. Maina, B. Makinson, F. Malaisse, T. Maldonado, M.C.H. Mamede, M.A. Mandango, S. Manktelow, W. Mantovani, M. Marconi, G. Martin, J.F. Maxwell, N.F. McCarten, A.J. McCarthy, W.J.F. McDonald, I. McLeish, J.A. McNeely, R.A. Medellín, W. Meijer, P. Mena, T. Messick, R.R. Mill, A.G. Miller, L.P. de Molas, J. Molero-Mesa, E. Moll, L. Monroy, J. Moore, M. Moraes, Ph. Morat, L.P.C. Morellato, J. Morello, S. Mori, H. Moss, M. Mössmer, C. Muñoz, G.P. Nabhan, J. Nais, J.C. Navarro, D. Neill, B.W. Nelson, C. Nelson, Ngui Siew Kong, S.V. Nikitina, H.P. Nooteboom, R.M. Nowak, P. Núñez, H. Ohashi, J.C. Okafor, B. Ollgaard, C. Ormazábal, C.I. Orozco, B. Orr, A. Ortega, R. Ortiz, R. Ortiz-Quijano, G. Palacios, W. Palacios, Pan Borong, B. Parsons, A.L. Peixoto, R. Petocz, Phan Ke Loc, A. Phillipps, P.B. Phillipson, D.J. Pinkava, M.J. Pires-O'Brien, B. Pitts, M. Plotkin, T. Pócs, R. Polhill, D. Poore, G.T. Prance, J.R. Press, J. Proctor, Qiu Xuezhong, N. Quansah, T.P. Ramamoorthy, L. Ramella, O. Rangel-Ch., T. Ravisankar, A. Rebelo, C. Reynel, M. Ríos, D. Roguet, I. Rojas, M. Roos, P. Rosales, L. Rossi, R. Rowe, P. van Royen, J. Russell-Smith, J. Rzedowski, C. Sáenz, N. Salazar, J.G. Saldarriaga, K.A. Salim, Samhan Nyawa, H. Sánchez, M. Sánchez S., M.J.S. Sands, T. Santisuk, N. Scarlett, C. Schnell, B.D. Sharma, C. Sharpe, D. Sheil, T. Shimizu, P. Silverstone-Sopkin, D.K. Singh, M.W. Skinner, D. Smith, S.H. Sohmer, J. Solomon, V.J. Sosa, J.P. Stanton, A. Stattersfield, P.F. Stevens, B. Stewart-Cox, J. Steyermark, W. Strahm, T.F. Stuessy, Su Zhixian, W.R Sykes, H. Synge, M. Syphan Ouk, Tae Wook Kim, A. Telesca, D. Thomas, W. Thomas, K. Thomsen, M. Thulin, J. Timberlake, V. Toledo, L. Torres, R. Torres, A. Touw, J.G. Tracey, C. Ulloa, L.E. Urrego, E. Vajravelu, J. Valdés-Reyna, O. Valdéz-Rodas, F.M. Valverde, T. Veblen, J.-M. Veillon, J. Vermuelen, H. Verscheure, J. Vidal, J. Villa-Lobos, L.M. Villarreal de Puga, J.A. Villarreal-Quintanilla, C.B. Villamil, L. Villar, J.-F. Villiers, W. Vink, K. Vollesen, L.I. Vorontsova, Vu Van Dung, D.H. Wagner, W.L. Wagner, K. Walter, Wang Xianpu, D. Wege, T. Wendt, M.J.A. Werger, G.L. Werren, J. Whinam, W.A. Whistler, A. Whitten, J.J.F.E. de Wilde, B. Wilson, P. Windisch, B. Woodley, R.P. Wunderlin, J. Wurdack, A.E. van Wyk, G. Yeoman, Yang Zhouhuai, K.R. Young, D. Yuck Beld, T.A. Zanoni, E. Zardini and Zhou Yilian.

INTRODUCTION

Vernon H. Heywood and Stephen D. Davis

The primary tactic in conservation must be to locate the world's hot spots and to protect the entire environment they contain.
Edward O. Wilson, *The Diversity of Life* (1992).

The importance of plant diversity

The diversity of plant life is an essential underpinning of most of our terrestrial ecosystems. Humans and most other animals are almost totally dependent on plants, directly or indirectly, as a source of energy through their ability to convert the sun's energy through photosynthesis. Worldwide tens of thousands of species of higher plants, and several hundred lower plants, are currently used by humans for a wide diversity of purposes – as food, fuel, fibre, oil, herbs, spices, industrial crops and as forage and fodder for domesticated animals. In the tropics alone it has been estimated that 25,000–30,000 species are in use (Heywood 1992) and up to 25,000 species have been used in traditional medicines. In addition, many thousands of species are grown as ornamentals in parks, public and private gardens, as street trees and for shade and shelter.

Very few of these species enter into world trade and only 20–30 of them are staple crops that supply most of human nutrition. A recent study by Prescott-Allen and Prescott-Allen (1990) indicates that 103 species contribute 90% of the national per capita supplies of food plants. The vast majority of the species used by humans do not form part of recorded trade and, therefore, do not appear in official trade statistics. They form a significant part of what is called the hidden economy.

Another important role of plant life is the provision of ecosystem services – the protection of watersheds, stabilization of slopes, improvement of soils, moderation of climate and the provision of a habitat for much of our wild fauna. It is impossible to attach a precise value to such ecosystem services except by counting the costs of failing to maintain them, and of repairing the consequent damage, such as soil erosion and deforestation.

While it is generally accepted today that the conservation of all biodiversity should be our goal, especially through the preservation and sustainable use of natural habitats, this is an ideal that is unlikely to be achieved and there are convincing scientific, economic and sociological reasons for giving priority to the conservation of the major centres of plant diversity throughout the world, especially as this will very often also lead to the conservation of much animal and micro-organism diversity as well.

Recently, BirdLife International (formerly the International Council for Bird Preservation) has published a survey of Endemic Bird Areas (EBAs) as hotspots for biodiversity (ICBP 1992; Stattersfield *et al.*, in prep.) and has suggested that birds can make a unique contribution to determining priorities for the conservation of global biodiversity because (1) they have dispersed to, and diversified in, all regions of the world and (2) they occur in virtually all habitat types and altitudinal zones. These features apply equally well, or with even more force, to plants. However, a third factor, namely that avian taxonomy and geographical distribution of individual bird species are sufficiently well known to permit a comprehensive and rigorous global review and analysis, cannot be claimed for plants. On the other hand, the bird survey covers only 2609 species of birds – those that have had in historical times a global breeding range below 50,000 km² – while there are an estimated 250,000 species of higher plants, the taxonomy and detailed distribution of most of which are poorly known. On the other hand, as we have noted, plants contribute the background habitat for vast numbers of other species, provide many of them with a food source and interact with many of them in pollination and fruit and seed dispersal, so that their significance as determinants of conservation priorities is unrivalled.

Historical background to the Centres of Plant Diversity project

The idea of preparing a world survey of the centres of plant diversity had its origins in an informal meeting convened by G.Ll. Lucas of botanists from the Threatened Plants Unit of the IUCN Conservation Monitoring Centre and staff from the Herbarium of the Royal Botanic Gardens, Kew in 1982, inspired in part by the work of Haffer (1969) on South American bird species. The suggestion was incorporated into the Plants Programme which was being developed by IUCN and WWF in 1984 and, in the first published draft of this Programme (IUCN/WWF 1984), one of the key themes was "Promoting plant conservation in selected countries". The choice of countries was based on an options paper prepared in 1982 by Hugh Synge based on data from the Threatened Plants Unit of the Conservation Monitoring Centre. The countries were:

Africa and Madagascar
Côte d'Ivoire, Liberia, Madagascar, Mauritius, Morocco, Niger, Tanzania;
Asia
India, Indonesia, Malaysia, Nepal, New Caledonia, Sri Lanka;
Central and South America
Brazil, Chile: Juan Fernández, Costa Rica, Ecuador: Galápagos, Honduras, Peru;
Europe
Atlantic Islands, Greece.

Within each country, project sites for conservation action were proposed, based on biological, operational, political and socio-economic considerations.

As part of the process of plant conservation in selected countries, the Threatened Plants Unit of CMC decided to prepare a "Plant Sites Directory" (Davis 1986), later known as "The Plant Sites Red Data Book" (IUCN 1986), that would include accounts of about 150 areas around the world which botanists consider to be of top priority for plant conservation, building on the areas previously selected as noted above. At a meeting in 1987 the Joint IUCN-WWF Plant Advisory Group (PAG) which had been established in 1984 by the Directors General of IUCN and WWF to advise on the overall content and direction of the Plants Programme gave approval to the preparation of a Plant Sites Red Data Book and agreed the concepts and criteria for site selection.

Subsequently, the PAG made a thorough review of the Plant Sites Red Data Book concept and decided to broaden the concept by not just paying attention to sites whose conservation would ensure the survival of most species, but by aiming at a listing of all the major botanical sites and vegetation types considered to be of international importance for the conservation of plant diversity. It was also envisaged that the work would document the many benefits, economic and scientific, that conservation of those areas would bring, outline the potential of each for sustainable development in line with the principles of the World Conservation Strategy (IUCN 1980) and provide an outline strategy for the effective conservation of each centre. The project was renamed *Centres of Plant Diversity: A Guide and Strategy for their Conservation*. Details were given in a brochure published in 1988 (IUCN 1988).

Although work started on the preparation of the project in 1987 and some sample Data Sheets were prepared, major finance for the project was not received until 1989 when WWF International provided a grant and negotiated an arrangement with the UK Overseas Development Administration (ODA) for matching funding and, subsequently, obtained a grant from the European Commission (EC) to fund the project over a three-year period. WWF contracted IUCN to arrange for the implementation of the project and Professor Vernon Heywood, then IUCN's Chief Scientist, Plant Conservation, was nominated to organize and supervise the work. Stephen Davis, then a member of the Threatened Plants Unit of the World Conservation Monitoring Centre, and who had been closely involved in the development of the project concept, was appointed full-time Project Co-ordinator. Olga Herrera-MacBryde, Smithsonian Institution, Washington, D.C., worked full-time on the Latin American section of the project from 1989–1994 under the IUCN-SI Latin American Plants Project. Networks of regional contributors and advisers were subsequently established.

The concept of identifying centres of diversity and endemism

The idea of seeking out high concentrations of diversity among plants, animals or both has a long history in biogeography in one form or another. Attention has frequently been paid to the floristic or faunistic richness of certain areas, such as the tropics of Asia, Africa and the Americas, the Mediterranean climatic regions, such as the Cape of Good Hope, and the concentrations of species on islands, such as Madagascar, Cuba and the islands of Indonesia. Particular emphasis has been given to the large numbers of species that are endemic to such areas, most often with an emphasis on animals, particularly large vertebrates. Another focus has been on particular areas that have been identified as the centres of origin and diversity of crop plants – the so-called Vavilov Centres of Crop Genetic Diversity (Hawkes 1983).

More recently, the concept of sites or centres of high diversity has attracted the attention of conservationists, both as a tool for helping determine which areas should receive priority attention, and also as a challenge as to how to undertake the conservation action necessary, especially as the areas of high diversity are most often found in developing countries which usually have limited human and financial resources available for this purpose.

Much attention has also been directed during the past two or three decades at the large numbers of species that are threatened with extinction at some time in the coming decades (Myers 1986, 1988b; Simberloff 1986; Raven 1987, 1990; Wilson 1988, 1992); the World Conservation Strategy (IUCN 1980) suggested giving conservation priority to those areas where a number of threatened species occur together so as to maximize the benefit from conservation efforts and to reduce the risk of losing large numbers of species if particular areas are not conserved.

Such efforts to seek out areas of high priority for conservation have acquired increased urgency in the light of the accelerating losses throughout the world of natural habitats and the biodiversity they contain, as a result of human action and the growth of the world's population. In particular, attention has been directed at the plight of the world's tropical rain forests which are believed to contain the majority of living organisms but which are being destroyed at an alarmingly high rate (FAO 1990a, b; Whitmore and Sayer 1992).

Determining priority areas for plants

The problem of determining priority areas can be approached at different geographical scales – global, regional, national or local. At a global level, Raven (1987) developed an approach based on analysis of the size of floras that are threatened, and highlighted the fact that about 170,000 of the world's estimated total of 250,000 species of angiosperms grow in tropical regions of the world, with an estimated 85,000 in Latin America, 35,000 in tropical and subtropical Africa (excluding the Cape), and at least 50,000 in tropical and subtropical Asia. He drew attention to the remarkable fact that more than 40,000 plant species – about a quarter of total tropical diversity – occur in Colombia, Ecuador and Peru. Also, the flora of Brazil should be highlighted since it has been estimated to contain between 40,000 and 80,000 species.

Some of these regional figures have been modified subsequently: for example, the count for tropical Africa has been reduced from 35,000 to 21,000 in the light of more accurate assessments (A.L. Stork, pers. comm. to P. Raven 1991) and the figure for tropical Asia appears to have been under-estimated. The total number of single country endemics (excluding Brazil, Paraguay and Papua New Guinea), recorded by the World Conservation Monitoring Centre (see Table 1), and updated by information arising from the present study, is a remarkable 175,976 species, an estimate which casts doubt on the generally accepted global total of about 250,000 species (see also below, p. 7). What is remarkable too is the fact that for many countries of the tropics it is still not possible to provide more than a very rough estimate of the number of species of plants (or of most other groups of organisms) and,

2

for the majority of these countries, the inventory is neither accurate nor complete.

Raven (1987) also singled out areas such as Madagascar, lowland Western Ecuador and the Atlantic forests of Brazil as deserving of critical attention. Each of these areas houses about 10,000 higher plant species and, in each, forest has been reduced to less than 10% of the area which it occupied 50 years ago.

This analysis was developed further by Myers (1988a) who identified 10 tropical forest "hotspots" (defined as areas that feature exceptional concentrations of species with high levels of endemism and face exceptional threats of destruction). Two further hotspots are in the developed world (Hawai'i and Queensland, Australia). Together, these hotspots total about 3.5% of the remaining primary tropical forest, occupy only 0.2% of the land surface of the planet, but contain around 13.8% of the world's plant species. He later extended this analysis by adding another eight areas, four of them in tropical forests and four in Mediterranean type vegetation zones (Myers 1990).

The five areas of Mediterranean-type vegetation – around the Mediterranean itself, in south-western Western Australia, California, central Chile and the Cape region of South Africa – house some 45,000–80,000 higher plant species, depending on how narrowly or widely "Mediterranean" is defined (Heywood 1994b). Of these, an estimated 27,000–35,000 are estimated to be endemic to the areas concerned (data from various sources including Quézel 1985; Cowling et al. 1989; Myers 1990; Greuter 1991; Heywood 1991, 1994b). Southern Africa with some 21,000 species of plants (Cowling et al. 1989), of which 80% are endemic, presents a special case and has the highest species/area ratio in the world (Huntley 1988). All these areas are subjected to a high degree of human disturbance; consequently, the flora and vegetation are significantly threatened.

The selection of sites

The analyses of floristic richness and endemism described above, while providing useful general indications as to which areas might be considered for priority action, have severe limitations in that they are based essentially on species richness and endemism in selected areas, irrespective of the nature, relationships and values of the species concerned, the ecological diversity of the areas and socio-economic factors. Nonetheless they give useful pointers.

In the last 10 years, much effort has been put into considering how habitats of conservation importance should be chosen and which of them should be given priority. These problems have been addressed on many occasions, such as the IVth World Congress on National Parks and Protected Areas (McNeely 1993) and they are reviewed in the WRI-IUCN-UNEP *Global Biodiversity Strategy* (WRI, IUCN and UNEP 1992). The Convention on Biological Diversity, which was agreed at the UNCED at Rio de Janeiro in June 1992, and which came into effect in December 1993, also stresses (in Annex I) the importance of identifying ecosystems and habitats containing high diversity, large numbers of endemic or threatened species and those of social, economic, cultural or scientific importance.

Some authors believe that less time should be spent worrying about the persistence of particular species and more time spent on maintaining the nature and diversity of ecosystem processes. Yet it is species that take part in ecosystem processes

and the fact is that species conservation cannot be separated from that of the habitats in which they occur (Heywood 1994a).

A further criticism of the use of ecological or taxonomic hotspots or mega-diversity regions or countries (Mittermeier and Werner 1988; Myers 1988a) to establish priorities to determine the most important areas to conserve comes from authors such as Dinerstein and Wikramanayake (1993), Pressey et al. (1993) and Williams, Vane-Wright and Humphries (1993). They regard such methods as arbitrary, unsystematic and lacking a paradigm. Pressey et al. (1993) propose three principles for selecting priority regions and regional reserves for the conservation of biodiversity: complementarity, flexibility and irreplaceability and suggest that they can be applied in practice at different scales. At the global level they advocate the use of the WORLDMAP computer program (Vane-Wright, Humphries and Williams 1991) which identifies key regions for conserving the biodiversity of one or more groups at global and national scales. Biodiversity is measured in this case as a combination of the number of species or higher taxa in a region and the taxonomic differences between them, although measures of endemism are also supported. They note that a critical aspect of the system is the implementation of the principle of complementarity which is used to find a priority sequence of regions to represent all taxa by identifying the maximum increment of unrepresented biodiversity possible at each step.

Once a priority region has been identified, there remains the problem of identifying a network of reserves that is able to represent all the features considered as requiring protection. Pressey et al. (1993) give examples of the application of the principles of complementarity and flexibility such as the CODA (Conservation Options and Decisions Analysis) procedure (Bedward, Pressey and Keith 1992) which has been applied to the south-eastern forests of New South Wales to find a network of sites which represent a minimum percentage area of all environments as well as occurrences of rare species and other important features.

Dinerstein and Wikramanayake (1993) present a new approach to conservation planning which they call a Conservation Potential/Threat Index. This index forecasts "how deforestation during the coming decade will affect conservation or establishment of forest reserves." It compares biological richness with reserve size, size of protected area, size of remaining forest cover and deforestation rate and is used to identify conservation potentials, threats and strategies for the 23 Indo-Pacific countries.

The above, and other approaches that will undoubtedly be developed, are to be welcomed. They reflect a growing concern that current approaches to biodiversity conservation worldwide are largely serendipitous, poorly co-ordinated, often ineffective and leave many major problems unsolved. What none of them addresses adequately is the great range of perceptions of biodiversity and priorities from a broad array of different interest groups, be they land use planners, conservation biologists, taxonomists, sociologists, economists, genetic resource agencies or politicians. Any top-down approach, no matter how sophisticated the science, is liable to fail unless full cognizance is taken of the detailed needs, perceptions, aspirations and political realities of the countries and regions concerned. As noted below, in the preparation of this book we adopted from the beginning a principle of involving local experts and, wherever possible, national governmental and non-governmental conservation bodies.

TABLE 1. SPECIES RICHNESS AND ENDEMISM

The following table provides a world list of vascular plant floras arranged alphabetically by region. The data are based on those provided by the World Conservation Monitoring Centre (WCMC) and updated with statistics arising from the CPD project. It should be emphasized that many of the figures are estimates. The reader is referred to the Regional Overviews which provide, in many cases, more detailed flora statistics and information sources. Note that the figures given below for the number of vascular plants have been rounded to the nearest 50–100 species for some regions. A note has been added where this is the case.

	Native vascular plant species	Endemic species	% Species endemism		Native vascular plant species	Endemic species	% Species endemism
AFRICA				Aruba, Bonaire and Curaçao	460	25	5.4
Algeria	3164	250	7.9	Bahamas	1129	118	10.5
Angola	5185	1260	24.3	Barbados	572	3	0.5
Benin	2201	0	0	Bermuda	166	15	9.0
Botswana	2015	17	0.8	Cayman Islands	539	19	3.6
Burkina Faso	>1100	0	0	Cuba	6505	3224	49.6
Burundi	>2500	?	?	Dominica	1227	12	1.0
Cameroon	8260	156	1.9	Grenada	875	4	0.5
Central African Republic	>3600	100	2.8	Grenadines	473	0	0
Chad	>1600	?	?	Guadeloupe	1672	23	1.4
Congo	6000	1200	20.0	Hispaniola	5135	1445	28.1
Côte d'Ivoire	3660	62	1.7	Jamaica	3304	923	27.9
Djibouti	641	2	0.3	Martinique	1505	24	1.6
Egypt	2076	70	3.4	Montserrat	670	2	0.3
Equatorial Guinea	3250	66	2.0	Nevis	260	1	0.4
Ethiopia	>6100	>600	>9.8	Puerto Rico	2492	236	9.5
Gabon	7151	1573	22.0	Saint Kitts	659	1	0.2
Gambia	974	0	0	Saint Lucia	1028	11	1.1
Ghana	3725	43	1.2	Saint Vincent	1134	20	1.8
Guinea	>3000	88	2.9	Trinidad and Tobago	2259	236	10.5
Guinea-Bissau	>1000	12	1.2	Turks and Caicos	448	9	2.0
Kenya	6506	265	4.1				
Lesotho	1591	2	0.1	**CENTRAL AND NORTHERN ASIA**			
Liberia	>2200	103	4.7	Estimate for whole region,			
Libya	1825	134	7.3	comprising Asiatic part of			
Malawi	3765	49	1.3	the former U.S.S.R.	17,500	2500	14.3
Mali	>1741	11	0.6				
Mauritania	1100	?	?	**CHINA AND EAST ASIA**			
Morocco	3675	625	17.0	Cambodia/Laos/Vietnam	>12,800	?	10.0
Mozambique	5692	219	3.8	China	27,100	10,000	36.9
Namibia	3174	?	?	Hong Kong	1984	25	1.3
Niger	1178	0	0	Japan (main islands)	5565	>222	>4.0
Nigeria	4715	205	4.3	Korean Peninsula	2898	407	14.0
Rwanda	>2288	26	1.1	Mongolia	2272	229	10.1
São Tomé and Príncipe	895	134[1]	15.0	Taiwan	3577	1075	30.1
Senegal	2086	26	1.2	Thailand	12,000	?	10.0
Sierra Leone	>1700	74	3.5	Vietnam	8000	>800	>10.0
Somalia	3028	500	16.5	(Note that the statistics for China and Indochina are very			
South Africa	23,420	>16,500	>70.0	approximate.)			
Sudan	>3132	50	1.6				
Swaziland	2715	4	0.2	**EUROPE**			
Tanzania	>10,000	1122	11.2	Albania	3000	24	0.8
Togo	2501	0	0	Austria	3100	35	1.1
Tunisia	2196	?	?	Belgium	1550	1	0.1
Uganda	5406	30	0.6	Bulgaria	3600	320	8.8
Western Sahara	>330	?	?	Cyprus	1650	88	5.3
Zaïre	11,000	1100	10.0	Czech Republic and Slovakia	2600	62	2.4
Zambia	4747	211	4.4	Denmark	1450	1	0.1
Zimbabwe	4440	95	2.1	Faroes	250	1	0.4
				Finland	1045	?	?
ATLANTIC OCEAN ISLANDS				France	4650	133	2.9
Ascension	25	11	44.0	Germany	2700	6	0.2
Azores	300	81	27.0	Greece	5000	742	14.9
Cape Verde	740	92	12.4	Hungary	2200	38	1.7
Canary Islands	1200	500	41.6	Ireland	950	?	?
Iceland	378	1	0.3	Italy	5600	712	12.7
Madeira	1119	106	9.7	Liechtenstein	1400	?	?
Saint Helena	60	50	83.3	Luxembourg	1200	0	0
Tristan da Cunha	40	?	?	Malta	914	5	0.5
				Netherlands	1200	?	?
AUSTRALIA AND NEW ZEALAND				Norway	1600	1	0.1
Australia (mainland)[2]	15,638	14,290	95.4	Poland	2450	3	0.1
Chatham Islands[3]	320	40	12.5	Portugal (mainland)	2600	3	0.1
Lord Howe Island	228	93	40.8	Romania	3400	41	1.4
New Zealand	2400	1942	80.9	Spain (mainland)	5050	941	18.6
Norfolk Island	165	50	30.3	Sweden	1750	1	0.1
Subantarctic Islands	?	35	?	Switzerland	3000	1	0.03
				United Kingdom	1550	16	1.0
CARIBBEAN ISLANDS				Yugoslavia (former territory of)	5350	137	2.6
Anguilla	321	1	3.1	(The figures for vascular plants are rounded to the nearest			
Antigua	845	0	0	50 species.)			

4

TABLE 1. SPECIES RICHNESS AND ENDEMISM ...continued

	Native vascular plant species	Endemic species	% Species endemism		Native vascular plant species	Endemic species	% Species endemism
INDIAN OCEAN ISLANDS				Tuvalu	44	0	0
British Indian Ocean Territory				Vanuatu	870	150	17.2
(Chagos Archipelago)	100	0	0	Wallis and Futuna	475	7	1.5
Comoros	416	136	33.0	Western Samoa	894	134	15.0
Madagascar	10,000	8000	80.0				
Maldives	277	5	1.8	**SOUTH AMERICA**			
Mauritius[4]	685	311	45.4	Argentina	9370	1100	11.7
Réunion[4]	546	189	34.6	Bolivia	17,350	4000	23.0
Rodrigues[4]	134	47	35.1	Brazil	56,000	?	?
Seychelles	250	87	34.8	Chile	5215	2698	51.7
				Colombia	51,000	1500	2.9
INDIAN SUBCONTINENT				Ecuador	17,600–21,100	4000	19.0–22.7
Andaman and Nicobar Islands	2270	225	9.9	French Guiana	4000	>150	>3.8
Bangladesh	5000	?	?	Guyana	6400	?	?
Bhutan	5500	100	1.8	Paraguay	7000–8000	?	?
India	17,000	6800–7650	40.0–45.0	Peru	18,000–20,000	5356	26.8–29.7
Lakshadweep	348	0	0	Suriname	5000	?	?
Maldives	277	5	1.8	Uruguay	2270	40	1.8
Myanmā (Burma)	14,000	1700	12.1	Venezuela	21,070	8000	38.0
Nepal	7000	350	5.0	(The figures for South America are revised from those presented in			
Pakistan	5100	400	7.8	Volumes 1 and 2, and are quite preliminary estimates for some			
Sri Lanka	3370	902	24.5	countries.)			
MIDDLE AMERICA				**SOUTH EAST ASIA (MALESIA)**			
Belize	4423	53	1.2	Borneo	20,000–25,000	6000–7000	30.0
Costa Rica	9500–10,500	600	5.7–6.3	Brunei	6000	?	?
El Salvador	2500	17	0.7	D'Entrecasteaux	>2500	?	?
Guatemala	8000	1171	14.6	Indonesia (available estimates):			
Honduras	6000	148	2.5	Java	4598	230	5.0
Mexico	18,000–30,000	10,000–15,000	50.5–55.5	Sulawesi	5000	?	?
Nicaragua	7000	60	0.9	Sumatra	>10,000	>1200	>10.0
Panama	8500–9000	1230	13.7–14.5	Louisiade Archipelago	>3000	?	?
(The figures for Middle America are revised from those presented in				Malaysia (available estimates):			
Volumes 1 and 2.)				Peninsular Malaysia	>9000	2700–4500	30.0–50.0
				Sabah and Sarawak	>10,000	?	?
NORTH AMERICA				New Guinea	15,000–20,000	10,500–16,000	70.0–80.0
Canada	3269	147	4.5	Philippines	8931	3500	39.2
Greenland	497	15	3.0	Singapore[6]	1293	2	0.2
U.S.A. (continental)[5]	18,000	4036	20.2	Solomon Islands	3172	30	0.9
PACIFIC OCEAN ISLANDS				**SOUTH WEST ASIA AND THE MIDDLE EAST**			
American Samoa	471	15	3.2	Afghanistan	4000	800	20.0
Bonin (Ogasawara) Islands	400	150	37.5	Bahrain	248	0	0
Cook Islands	284	33	11.6	Iran	8000	1400	17.5
Fed. Micronesia	1194	293	24.5	Iraq	3000	190	6.3
Fiji	1628	812	49.9	Israel	2225	165	7.4
French Polynesia				Jordan	2100	145	7.3
(incl. Marquesas)	959	560	58.3	Kuwait	282	0	0
Galápagos	541	224	41.4	Lebanon	2600	311	12.0
Guam	330	69	20.9	Oman	1200	73	6.1
Hawaiian Islands	1200	1000	83.3	Qatar	306	0	0
Juan Fernández	210	127	60.4	Saudi Arabia	2028	34	1.7
Kiribati	22	2	9.1	Syria	3100	395	13.0
Marquesas	318	132	41.5	Turkey	8650	2675	30.9
Marshall Islands	100	5	5.0	United Arab Emirates	340	0	0
Nauru	54	1	1.9	Yemen[7] (N)	1650	58	3.5
New Caledonia	3322	2551	76.8	Yemen[7] (S)	1180	77	6.5
Niue	178	1	0.6	Socotra (Yemen)	815	>230	>28.2
North Marianas	221	81	36.7				
Palau	175	?	?	**[ANTARCTIC, SOUTH ATLANTIC AND SOUTHERN OCEAN]**			
Pitcairn Islands	76	14	18.4	Antarctic continent	2	0	0
Tokelau Islands	32	0	0	Falkland Islands (Malvinas)	165	14	8.5
Tonga	463	25	5.4	French Southern Territories	50	11	22.0
				(The above region is included here for completeness.)			

Notes:

[1] Refers to single-island endemics.
[2] Latest count for Australia, but an estimated 3000–5000 vascular plant taxa yet to be named.
[3] Figures refer to vascular plant taxa (i.e. species, subspecies and varieties).
[4] Figures refer to flowering plant taxa (i.e. species, subspecies and varieties).
[5] Figures refer to flowering plant species.
[6] Figures for gymnosperms and dicots only.
[7] Yemen was unified in 1991; figures for North Yemen (Yemen Arab Republic), South Yemen (People's Democratic Republic of Yemen) and Socotra are kept separate for convenience.

The objectives of the *Centres of Plant Diversity (CPD)* project are:

❖ to identify which areas around the world, if conserved, would safeguard the greatest number of plant species;
❖ to document the many benefits, economic and scientific, that conservation of those areas would bring to society and to outline the potential value of each for sustainable development;
❖ to outline a strategy for the conservation of the areas selected.

These objectives are fully consonant with the *Convention on Biological Diversity*.

As a consequence of the very extensive involvement of local specialists, and well over 100 government and non-governmental agencies and other conservation bodies in the preparation of the project, including the holding of workshops in several parts of the world, we believe that the resultant three volumes comprising *Centres of Plant Diversity* will provide a unique global and regional review of the nature and distribution of the main concentrations of plant diversity in the world, and a guide to the most practical and cost-effective ways of conserving as much of this diversity as possible, together with its sustainable use. It is our aspiration that CPD will provide not only its sponsors, IUCN, WWF, ODA and the EC, but also governments, aid agencies, development banks and conservation organizations, with a considered overview of the status of plant diversity worldwide and clear guidance on which areas are global and regional priorities for its conservation.

Although the original intention was to select between 150 and 200 sites of global priority, the total number finally chosen greatly exceeds these figures. In addition to the 234 priority sites selected for Data Sheet treatment, many more sites are treated in summary paragraphs in the Regional Overviews. By extending the Regional Overviews in this way, we overcame the problem of making an arbitrary selection of a single site where several potential sites occur in a particular area. In some cases, the Regional Overviews also contain a selection of important botanical areas which do not meet the criteria for selection as CPD sites but which are, nonetheless, areas of important conservation concern because, for example, they may contain the best surviving examples of certain vegetation types.

The criteria and methodology used for selecting sites

The criteria adopted for the selection of sites and vegetation types was based principally on a requirement that each must have one or both of the following two characteristics:

❖ the area is evidently species-rich, even though the number of species present may not be accurately known;
❖ the area is known to contain a large number of species endemic to it.

The following characteristics were also considered in the selection:

❖ the site contains an important genepool of plants of value to humans or that are potentially useful;
❖ the site contains a diverse range of habitat types;

❖ the site contains a significant proportion of species adapted to special edaphic conditions;
❖ the site is threatened or under imminent threat of large-scale devastation.

It has to be stressed that *Centres of Plant Diversity* is concerned with "first order" sites that are of *global* botanical importance. As a consequence some countries have had a number of sites selected, while others have none that qualify for inclusion when viewed on a world basis. If viewed from a national perspective, the selection of sites would have been different. We hope, however, that the publication of these volumes will serve as a stimulus for national programmes aimed at identifying plant sites that are important at a more local level.

The selection process has involved extensive consultations with experts in all the major regions of the world. Networks of individuals and collaborating institutions were established and their advice sought on the listing of sites for the Regional Overviews and for selecting those sites which merit Data Sheet treatment. In Africa, China, India, North America and South America, this involved the holding of workshops within these regions at which data on proposed CPD sites were reviewed and the final selection of Data Sheet sites made. Sometimes this led to some sites being rejected and replaced by others. This is especially true for South America, where the Regional Workshop held in Quito, Ecuador, in 1991 to review the site selection led to a total revision of the phytogeographical divisions of South America and to the major revision of the list of sites previously chosen.

To qualify for Data Sheet treatment, most mainland sites have (or are believed to have) in excess of 1000 vascular plant species, of which at least 100 (i.e. 10%) are endemic either to the site (strictly endemic) or to the phytogeographical region in which the site occurs. In many cases, the number of regional endemics is very much higher than 10% of the flora. In all cases the sites have at least some strict endemics.

The criteria for the selection of islands treated as Data Sheets were somewhat different from those used for mainland sites. Many islands have depauperate floras compared with continental areas, but the level of endemism is often very high. For example, Saint Helena has a flora of only 60 vascular plant species but 50 of these are endemic. Clearly, the high concentration of endemics is of considerable conservation importance, and to restrict the selection of sites to those with floras of, or in excess of, 1000 species would lead to the omission of such important areas.

To qualify for Data Sheet treatment, an island flora must contain at least 50 endemic species or at least 10% of the flora must be endemic. Even so, data were not available for some islands that perhaps warranted inclusion. Comoros, in the Indian Ocean, is a good example of such a situation: the only available figures (416 vascular plant species, of which 136 are endemic) are based on an assessment of the flora made in 1917 which is almost certainly incomplete.

For some islands, the Data Sheet covers the whole island (e.g. Socotra) or an archipelago (e.g. the Canary Islands). South East Asia presented a somewhat different problem: practically the whole region is insular and floristically very rich with high levels of endemism on many of the larger islands. In this case, a number of sites and vegetation types were selected throughout the archipelago, with some islands being represented by more than one Data Sheet to cover a range of vegetation and community types.

There was, in some cases, an element of subjectivity involved in the selection process, and other criteria were used (such as degree of threat, diversity of habitats or soil types, presence of scientifically or economically important species) in deciding which out of a number of similar sites, would be chosen as Data Sheets. But the main criteria and principles outlined above were used to make the initial selection, and always with the help and advice of regional and local expertise.

Content of the Regional Overviews and Data Sheets

Within the constraints imposed by the enormous diversity of the areas covered, a standard format has been adopted for each of the Regional Overviews and Data Sheets. A summary table is provided in each case. There is a certain amount of variation in the headings adopted in the Regional Overviews, but for nearly all the site Data Sheets the following sections are included: Geography, Vegetation, Flora, Useful Plants, Social and Environmental Values, Threats, Conservation and References. An Economic Assessment section is included where data are available.

Analysis of the information on the sites

(i) Regional distribution of Data Sheets

A total of 234 sites are selected for Data Sheet treatment. Their general location is shown in the map on page 11. The distribution of the sites in the different regions is given in Table 2.

TABLE 2. DISTRIBUTION OF DATA SHEETS BY REGION

EUROPE, AFRICA, SOUTH WEST ASIA AND THE MIDDLE EAST

Europe	9
Atlantic Ocean Islands	4
Africa	30
Indian Ocean Islands	3
South West Asia and the Middle East	11
Total	**57**

ASIA, AUSTRALASIA AND THE PACIFIC

Central and Northern Asia	5
Indian Subcontinent	13
China and East Asia	21
South East Asia (Malesia)	41
Australia and New Zealand	14
Pacific Ocean Islands	8
Total	**102**

THE AMERICAS

North America	6
Middle America	20
South America	46
Caribbean Islands	3
Total	**75**
Grand Total	**234**

When considering the figures in Table 2, it must be remembered that the sites vary enormously in size, from extensive mountain systems, such as the Alps, to island complexes, such as the Hawaiian Islands, and to much smaller areas, such as the Sinharaja forest of Sri Lanka. This makes direct comparisons between sites impossible. Attention should also be drawn to the Regional Overviews which place the sites selected in a much wider context and provide a comparative assessment of the biodiversity and conservation status of the different areas within the regions.

(ii) Floristic diversity

It has proved difficult to obtain accurate data on the number of species in some of the sites and even for some areas. Even estimates for individual countries are imprecise in some instances, such as Bolivia, Indonesia and Thailand. The global figures for species richness and endemism can, however, be summarized as follows:

TABLE 3. FLORISTIC DIVERSITY AND ENDEMISM BY REGION

	Species	Endemics	% Endemism
EUROPE, AFRICA, SOUTH WEST ASIA AND THE MIDDLE EAST			
Europe	12,500	3500	28
Atlantic Ocean Islands	2650	785	29.5
Africa	40–45,000	35,000	77–87.5
Indian Ocean Islands	11,000	9000	82
South West Asia and the Middle East	23,000	7100	31
Totals	**89–94,150**	**55,385**	
ASIA, AUSTRALASIA AND THE PACIFIC			
Central and Northern Asia	17,500	2500	14
Indian Subcontinent	25,000	12,000*	48
China and East Asia	45,000*	18,650*	41.5*
South East Asia (Malesia)	42–50,000	29–40,000	70–80
Australia and New Zealand	17,580	16,202	81–90
Pacific Ocean Islands	11–12,000	7000	58–63
Totals	**156–167,080**	**81–96,352**	
THE AMERICAS**			
North America	20,000	4198	21
Middle America	21–33,500*	12–18,500*	55–57
South America	83,000*	>50,000*	>60
Caribbean Islands	13,000	6555	50
Totals	**137–149,500**	**72–79,253**	
Grand Totals	**382–410,730**	**209–230,990**	

* very approximate estimate based on available data.
** Note: the figures for the Americas differ from those given in Volumes 1 and 2.

Considerable caution needs to be used in interpreting the figures in this table. It has been exceedingly difficult to obtain accurate figures for many countries. This reflects the relatively poor state of our floristic knowledge, especially in the tropics and subtropics, as has been repeatedly pointed out elsewhere. Effective conservation of biodiversity in the face of such ignorance will be difficult and this is a problem which has to be faced as a matter of urgency. It is notable too that in several regions, such as Africa, Asia, Australia and Latin America, tens of thousands of species are believed still to be undescribed.

The numbers of endemic species given in Table 3 can be totalled meaningfully but the totals of all species in each of the regions, and the sum total of these, are only indicative, as no account is made of the species that are shared between two or more regions. The remarkable total of 230,990 endemic species recorded from the information given in the regional reviews, which themselves reflect the data given in the individual Data Sheets, is surprisingly high when one considers that the *total* number of flowering plants and ferns known today is generally accepted as being around 250,000. It is also substantially higher than the figure of 175,976 endemic species derived from WCMC data. The implications are that, if these data can be confirmed, the total number of flowering plant and fern species must be at least of the order of 300,000–350,000 if allowance is made for those species that occur in two or more regions, in more than one continent or are even more widespread.

It is notable too that the total numbers of endemic species (and therefore by implication the total flora) recorded for Asia, Australasia and the Pacific are appreciably higher than those of Latin America and the Caribbean combined. If one takes into account the probability that narrower species concepts are applied in Latin America than in parts, at least, of Asia and the Pacific, as several authors have suggested (e.g. Gentry 1990), the difference will be even greater. This would suggest that conservation agencies might take a closer look at the comparative richness of Latin America/Caribbean and Asia/Pacific in assessing priorities and balance of investment.

If the numbers of Data Sheets for each of the three volumes are compared (Table 4), it will be seen that there is a very broad consistency of treatment.

TABLE 4. COMPARISON OF FLORISTIC RICHNESS (AS MEASURED BY ENDEMICS) AND NUMBERS OF DATA SHEETS FOR REGIONS IN EACH VOLUME OF CPD

Region	Total endemics	Number of Data Sheets
EUROPE, AFRICA, SOUTH WEST ASIA AND THE MIDDLE EAST	55,385	57
ASIA, AUSTRALASIA AND THE PACIFIC	96,352	102
THE AMERICAS	79,253	75

(iii) Conservation status

As one reads through the paragraphs in the Data Sheets on the conservation status of the sites, two worrying trends may be discerned. On the one hand, many sites are not legally protected, or are only protected in part. On the other hand, a considerable proportion of those sites that are officially protected are not effectively managed, being subject to various forms of degradation, ranging from logging to gathering of fuelwood. This emphasizes the need for extending and strengthening the protected area systems of the countries concerned, if the strategy of protecting biodiversity through setting aside and protecting areas of high floristic and ecological richness is to succeed.

Table 5 gives a breakdown by region of the number of sites included partly or fully within legally designated protected areas. From Table 5 and the information presented in the Summary Table (see final section in Introduction), it is encouraging that many CPD sites are represented (at least in part) in existing protected areas, or are proposed for inclusion. Australia is perhaps the best illustration of this. Here, most CPD sites are well represented within the National Park system, and some areas have additional protection as World Heritage Sites. Of particular note is the Wet Tropics World Heritage Area which includes virtually all of the remaining tropical rain forest in Australia (see the Data Sheet on the Wet Tropics of Queensland – CPD Site Au10, in Volume 2). In other cases, such as Kakadu-Alligator Rivers region and the Western Tasmanian Wilderness (CPD Sites Au4 and Au9, respectively, also covered in Data Sheets in Volume 2), there are important areas outside the existing reserve boundaries which need to be brought into protection to conserve significant elements of the flora.

Closer examination of the current protection of the CPD sites, reveals that the present coverage of legally protected areas is often inadequate to "capture" the full range of vegetation types and areas of greatest floristic richness within a region. Worldwide, less than one in four of the Data Sheet sites (21%) are legally protected in full, and only about one-third of the selected sites (35%) have more than 50% of their areas occurring within existing protected areas.

The largest number of totally unprotected sites occurs in Africa (8 sites) and South East Asia (6 sites); many more sites (17) in South East Asia have less than 50% of their areas occurring within legally designated protected areas. Also of significance is the large number of South American sites (32) which have only a relatively small part of their areas covered by any form of legal protection. A priority must be to extend the coverage of the existing network of protected areas in these regions, and particularly to safeguard remaining plant-rich lowland forests.

Better protection of the lowlands is also a priority for areas with Mediterranean-type climates. In the South-West Botanical Province of Western Australia (CPD Site Au7, see Data Sheet in Volume 2), for example, over a third of the state's Rare and Endangered plants are not protected within the existing reserve system. In the California Floristic Province of U.S.A. (CPD Site NA16, see Data Sheet in Volume 3), the lowlands are particularly under-protected. The lowland flora is also under-represented in the present reserve system in the Cape Floristic Province of South Africa (CPD Site Af53, see Data Sheet in Volume 1), which in terms of the number of species per unit area is probably the richest plant area in the world.

In Central and Northern Asia, the Indian Subcontinent, and South West Asia and the Middle East, few CPD sites have more than 50% of their areas protected within existing reserves. In the case of Central and Northern Asia, none of the CPD sites have more than 50% of their areas protected and, of those which are protected in part, the proportion of the area protected is relatively small. Some of the sites in this region are, however, very large. Again, a priority should be to extend the coverage of the reserve system.

In Europe, Australia and New Zealand, and North America, virtually all CPD sites selected for Data Sheet treatment are included (at least in part) within the existing reserve systems of the regions concerned. Mention has already been made of the protection afforded to CPD sites in Australia; in New Zealand the situation is rather different in that all the sites selected for Data Sheet treatment, whilst being partly included within existing reserves, are nevertheless under threat in at least part of their range. One site in Europe (the Massifs of Gudar and Javalambre in Spain – CPD Site Eu7, see Data Sheet in Volume 1) has no current legal protection.

Even when areas are designated for protection, many are nevertheless threatened, some severely so. This can be seen by comparing figures in Table 5 with those in Table 6, which provides a summary of the degree of threat to CPD sites in each region.

Only 33 sites (15% of the total number of Data Sheet sites) are considered to be safe, or reasonably safe, whereas 119 sites (c. 50% of the total) are either threatened or severely threatened. A further 41 sites worldwide are considered to be vulnerable or at risk, and the remainder (40 sites) are partly threatened (for example where encroachment is occurring in part of an area or around its perimeter).

Of particular note is the degree of threat to many sites in the tropics. For example, none of the sites in Middle America are assessed as "safe" or "reasonably safe". Over half the sites in the region are threatened or severely threatened, irrespective of their designation, in whole or in part, as protected areas.

Some of the most seriously threatened floras are those on islands. For example, all of the CPD Data Sheet sites for the Indian Ocean and the Pacific Ocean are assessed as either at risk or threatened. Of the 8 CPD sites selected for Data Sheet treatment for the Pacific, 5 are classified as severely threatened, including the Hawaiian Islands, which include some of the most endangered floras in the world.

In the moist tropics, the main threats to sites are from encroaching agriculture, including slash and burn agriculture, logging and road building. Road building is often followed by unplanned colonization. In Middle and South America, cattle ranching is also a serious threat to a number of sites. An additional threat in some cases is the quest for oil and gas, together with the associated problems of pollution. In South East Asia, many CPD sites are already National Parks, but their protection status is still not secure; most suffer from encroachment from slash and burn cultivators, and from logging or conversion to oil palm plantations. In some cases in the moist tropics, forest clearance has occurred up to the park boundaries.

Almost all the tropical sites which are already protected (at least on paper) suffer from a lack of adequate funding and trained manpower. This affects the security of the areas. Boundaries may not be properly marked or policed, resulting in valuable resources, such as rattans and timber trees, being illegally exploited in an uncontrolled manner, perhaps not sustainable in the long-term. For some areas there are no effective management plans to protect or utilize important genetic resources, and there is mostly no out-reach programme to involve local people in the management and protection of the forests, and to ensure that utilization of plant resources is undertaken on a sustainable basis. In many cases, land zonation within the protected area and its vicinity is needed to ensure that conservation of biodiversity is balanced against other demands on natural resources.

Effective conservation of the CPD sites depends, therefore, on adequate funding and the political will to establish more protected areas where this is necessary and to ensure that all protected areas are effectively managed.

While it is encouraging to note that there are still many opportunities available for protecting and conserving a large proportion of the Earth's wild plant resources *in situ* (supplemented by *ex situ* procedures as necessary), it is alarming that very few of the areas identified as global priorities are assessed as "safe". With the ever-increasing threats to many sites, the opportunities for conservation are disappearing rapidly and the need for emergency action becomes more frequent. There is therefore an urgent need for action from governments, aid agencies and conservation organizations to achieve the conservation of the areas selected as CPD sites before it is too late.

(iv) Useful plants, social, economic and environmental values

A remarkable diversity of uses of plants is noted in the Regional Overviews, Data Sheets and Summary Tables which follow. Almost all the sites contain important timber trees, fruit trees or medicinal plants. Even in Europe, the native flora contains over 200 crop relatives which represent important genetic resources. For most of the tropical sites detailed information on the uses of plants is not available and a vast amount of research still needs to be undertaken.

TABLE 5. ANALYSIS OF THE CONSERVATION STATUS OF CPD SITES

Region	% area of CPD site within protected area(s)				Total no. of sites	% of sites with >50% or 100% protection	
	0	>0–50	>50–<100	100		>50	100
Africa	8	7	8	7	30	50	23
Atlantic Ocean Islands	-	3	1	-	4	25	0
Australia/New Zealand	-	8	4	2	14	43	14
Caribbean Islands	1	-	1	1	3	67	33
Central and Northern Asia	1	4	-	-	5	0	0
China and East Asia	1	7	4	9	21	62	43
Europe	1	8	-	-	9	0	0
Indian Ocean Islands	-	3	-	-	3	0	0
Indian Subcontinent	2	9	-	2	13	15	15
Middle America	3	8	5	4	20	45	20
North America	-	6	-	-	6	0	0
Pacific	-	5	1	2	8	38	25
South America **	2	32	6	5	45	24	11
South East Asia	6	17	2	16	41	44	39
South West Asia and the Middle East	2	8	-	1	11	9	9
Totals	**27**	**125**	**32**	**49**	**233**	**35**	**21**

Notes:

** The figures for South America exclude one Data Sheet (SA34), for which data are not available.

For CPD sites having Forest Reserve status, a judgement had to be made whether such a designation confers any degree of protection on the flora. If it is clear from the text of the Data Sheet that no protection is afforded, and especially if the area is under imminent threat of large-scale logging, the percentage area of the CPD site protected has been adjusted accordingly.

TABLE 6. DEGREE OF THREAT TO CPD SITES

The table shows the number of CPD Data Sheet sites classified under broad categories of threat. For details on each site, see the Summary Table on page 12, and the individual site Data Sheets.

Region	Safe or reasonably safe	Partly safe but some areas threatened	Vulnerable or at risk	Threatened	Severely threatened
Africa	7	5	7	8	3
Atlantic Ocean Islands	1	-	-	1	2
Australia/ New Zealand	9	2	1	2	-
Caribbean Islands	-	1	1	-	1
Central and Northern Asia	-	-	4	1	-
China and East Asia	2	5	5	5	4
Europe	2	2	-	5	-
Indian Ocean Islands	-	-	1	1	1
Indian Subcontinent	2	6	-	3	2
Middle America	-	5	2	10	3
North America	-	2	-	3	1
Pacific	-	-	1	2	5
South America **	7	5	3	14	16
South East Asia	3	7	12	15	4
South West Asia and the Middle East	-	-	4	4	3
Totals	**33**	**40**	**41**	**74**	**45**

Notes:

** The figures for South America exclude one Data Sheet (SA34), for which data are not available.

Many of the areas attract large numbers of tourists. Often, it is intact natural vegetation which provides the scenic backdrop, or is the actual focus, for tourism in the areas concerned. If properly planned and controlled to prevent visitor pressure and tourist developments becoming threats to the sites, such visitation could provide substantial long-term financial and employment benefits to the immediate area, as well as to national economies. In a few cases, a partial and preliminary economic assessment of the value of plant resources is given in the Regional Overviews and Data Sheets. This is usually based upon the number of visitors to a protected area, or the amount of wild plant resources gathered from a site.

Mention must also be made of the valuable environmental services provided by the sites, such as the prevention of soil erosion and flooding of downslope agricultural and settled areas, as well as safeguarding watersheds and contributing to climatic stability. It has been impossible to place monetary values on these services within the context of this present work.

The information presented in the Data Sheets and Regional Overviews demonstrates all too clearly the enormous value to humanity of the vast range of riches and economic potential of plant resources. This poses a dramatic challenge to the nations of the world and to the international agencies that are called upon to provide the necessary aid, support, and strategic and practical advice to ensure the survival into the future of this treasure house for humanity.

Summary information on sites selected for Data Sheet treatment

The Summary Table below includes a world list of the 234 sites which have been selected for detailed treatment as Data Sheets in the three volumes comprising *Centres of Plant Diversity*. Any updates to the information given in the Summary Table will be noted in subsequent volumes.

The following notes indicate the sorts of data presented under each column in the table:

(i) Type

A letter code categorizing each of the areas selected for Data Sheet treatment. The following codes are used:

S   Site; where the area is a discrete geographical unit, and where the whole area needs to be conserved.
F   Floristic province, often covering a very wide area, or CPD site covering a whole region. Effective conservation of the flora of such areas often requires a network of reserves to be established, as in many cases it would be impractical to protect the entire province or region.
V   Vegetation type. As in "F", effective conservation often requires representative samples to be protected.

(ii) Area

From information given in the Data Sheets; usually to nearest 1 km² for individual sites, sometimes an estimate to nearest 100 km² or 1000 km² for large regions.

(iii) Altitude

Altitude range in metres.

(iv) Flora

Unless otherwise stated, numbers refer to indigenous vascular plant species, or an estimate based on current botanical knowledge of that (or similar) sites, usually to the nearest 100 species, or to nearest 1000 species for some large tropical sites. An asterisk (*) denotes the exact number of plant species present or so far recorded for a site.

taxa – refers to the number of species, subspecies and varieties.
angiosp. – angiosperms.

(v) Examples of Useful Plants

Important plants or major groups are listed, including names of some commodity groups.

(vi) Vegetation

The major vegetation formations are listed. Information has been summarized from the Data Sheets. The use of some local names for vegetation types has been retained.

(vii) Protected Areas

Categories of protected areas are given where a CPD site is fully or partially protected. In most cases, the area of the protected site is given after the category, or a percentage figure is given for that part of the site which is protected. If a site is fully protected, the entry will just show the category of protection.

Abbreviations used in this volume:

BG	Botanic Garden	PF	Protection/Protected
BR	Biosphere Reserve		Forest
FoR	Forest Reserve	PFA	Protected Forest Area
GR	Game Reserve	RF	Reserved Forest
GS	Game Sanctuary	SF	State Forest
NM	Nature Monument	SFoRP	State Forest Park
NNP	Natural National Park	SP	State Park
NP	National Park	SpNR	Special Nature Reserve
NR	Nature/Natural	VJR	Virgin Jungle Reserve
	Reserve	WA	Wilderness Area
NS	Nature Sanctuary	WHS	World Heritage Site
NatM	National Monument	WMA	Wilderness
NatP	Natural Park		Management Area
NatS	National Sanctuary	WR	Wildlife Reserve
PA	Protected Area	WS	Wildlife Sanctuary

In many cases, the IUCN Management Category of each protected area is given in the Data Sheet. The definitions of the IUCN Management Categories are given in a separate Appendix.

(viii) Threats

Only the main threats, with the most important ones first, are listed.

(ix) Assessment

A summary of the conservation status of the area, including whether the area is safe, reasonably safe, at risk, threatened or severely threatened. An analysis of the threats and conservation status of the CPD sites is given in the Introduction to this volume.

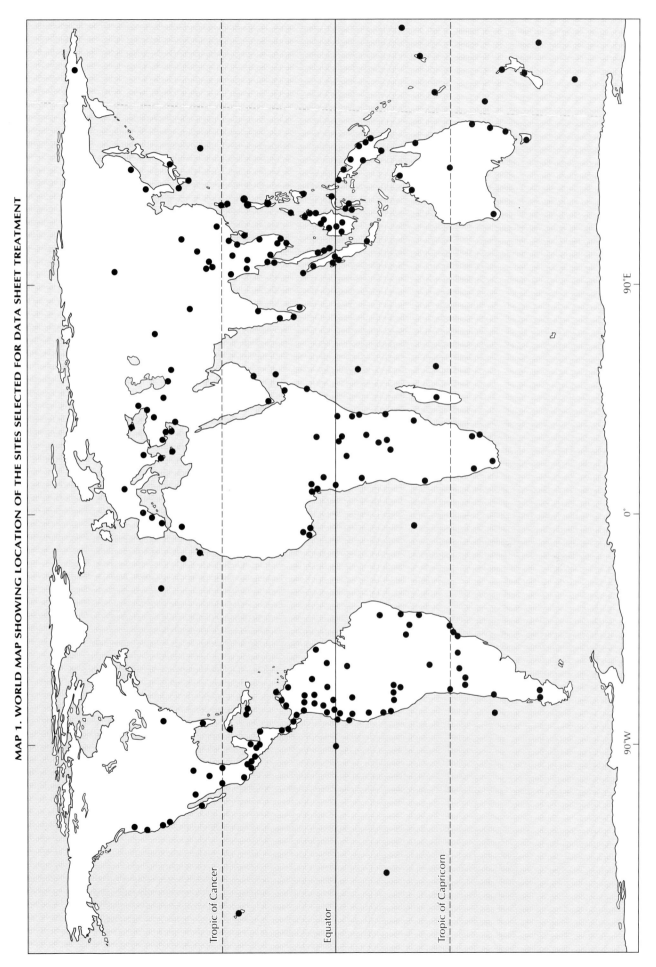

MAP 1. WORLD MAP SHOWING LOCATION OF THE SITES SELECTED FOR DATA SHEET TREATMENT

Tropic of Cancer

Equator

Tropic of Capricorn

90°W

0°

90°E

CENTRES OF PLANT DIVERSITY: SUMMARY INFORMATION ON SITES SELECTED FOR DATA SHEET TREATMENT

Code	Site	Type	Size (km²)	Altitude	Flora	Examples of Useful Plants	Vegetation	Protected Areas	Threats	Assessment
AFRICA										
Af81	Afroalpine region	F	3500	3500–5890 m	350	Grazing, browsing, fuel	Moist tree composite woodland, scrub, tussock grassland, bogs	Considerable areas fall within existing NPs and reserves	Overgrazing, fuelwood collection, fire, tourism	Some areas safe, but others severely threatened; management plans needed
Cameroon										
Af11	Forest zone, River Dja region	S	8100	200–500 m	2000	Timber trees, essential oils, medicinal plants	Tropical evergreen rain forest, semi-deciduous forest, swamps, secondary forests	FoR and Fauna Reserve, BR, and proposed NP (5000 km²)	Some encroachment on boundaries from cocoa, coffee and subsistence plots	Reasonably safe; threats not as serious as elsewhere
Af12	Korup NP	S	1259	100–1075 m	3500	Medicinal plants, palm canes, chewing sticks, fruit trees	Lowland tropical evergreen rain forest	NP	Over-collecting of bush mango and palm canes, some encroachment	Reasonably secure
Af13	Mount Cameroon	S	1100	0–4095 m	3500	Timber trees (especially African mahogany), medicinal plants	Lowland evergreen rain forest to Afromontane forest, scrub, subalpine and montane grasslands, coastal mangroves	6 FoRs (700 km²), mostly in foothills; proposed Etinde Reserve (360 km²) and Mabeta-Moliwe Reserve (36 km²); Limbe BG.	Agricultural encroachment, fire, logging; potentially grazing	Threatened; most seriously on lower slopes of east and north; montane forest at risk from fire
Congo/Cabinda/Zaïre										
Af16	Mayombe	S	2500	150–350 m	>1100	Timber trees, medicinal plants, food plants, plants of cultural value	Tropical lowland semi-evergreen forest, Zambezian savanna, wetlands	Proposed BR	Logging, charcoal production, clearance for agriculture	Threatened
Côte d'Ivoire										
Af2	Taï NP	S	3500	80–623 m	1300	Timber trees, many plants used locally for a variety of purposes	Tropical evergreen rain forest	NP, WHS (all of area), BR (3300 km²)	Logging, agricultural encroachment, gold mining, population pressure	Threatened around perimeter
Gabon										
Af18	Cristal Mountains	S	9000	0–911 m	>3000	Timber trees (e.g. okoumé), semi-wild oil palms, raphia, medicinal plants	Mainly tropical lowland and hill rain forest	None	No serious threats at present; logging is a potential threat	Not protected; at risk
Guinea/Côte d'Ivoire/Liberia										
Af4	Mont Nimba	S	480	450–1752 m	>2000	Timber trees, oil palm	Lowland and transitional rain forest, grasslands	Strict NR & WHS (Guinea: 130 km², Côte d'Ivoire: 50 km²), BR (Guinea: 171 km²), proposed NR (Liberian part)	Mining of iron ore (Liberia/Guinea), some clearance for agriculture	Liberian and Guinean part greatly damaged by mining and severely threatened
Kenya										
Af62	Mount Kenya	S	1500	1600–5199 m	800	Timber trees, fruit trees, medicinal plants	Montane moist and dry forest, bamboo, woodland, giant heaths, moorland	Above 3100 m, NP (715 km²); BR (718 km²); FoR covers 1421 km² of lower slopes	Logging, monocultures on the lower slopes, visitor pressure at higher altitudes	Lower slopes severely threatened

Code	Site	Type	Size (km²)	Altitude	Flora	Examples of Useful Plants	Vegetation	Protected Areas	Threats	Assessment
Liberia										
Af7	Sapo NP	S	1307	100–400 m		Timber trees, medicinal plants, wide range of plants used for making artefacts, tools	Lowland tropical rain forest	NP	Potential logging, illegal hunting	At risk
Malawi										
Af64	Mount Mulanje	S	500	750–3002 m	>800	Mulanje cedar, Brachystegia, bamboos, fruit trees	Woodland, evergreen forest, montane grassland, high-altitude scrub, rupicolous communities	FoR	Deforestation, illegal logging of cedars, uncontrolled collection of fuelwood, invasive pines	Severely threatened; action urgently needed
Morocco										
Af84	High Atlas	F	7000	1000–4165 m	1000	Timber trees (e.g. walnut, pine, juniper, relict stands of Atlantic cedar), fodder plants	Cedar, juniper, pine and holm oak forests, scrub, alpine meadows, pseudo-steppe, alpine scree communities	2 NPs (860 km², incl. 1 WS of 8 km²), 1 NR (12 km²); c. 12% of total area	Population pressure leading to clearance for cultivation, overgrazing, fuelwood cutting, timber cutting	Protected area reasonably secure but management needs strengthening; areas outside threatened
Namibia/Angola										
Af50	Kaokoveld	F	70,000	0–2000 m	952*	Medicinal plants, plants of cultural value, fodder plants	Desert/escarpment vegetation, mopane savanna	Angola: most in Iona NP (15,150 km²) and a Partial Reserve; Namibia: small area in NP, rest as Game Conservation Area	Timber cutting, overgrazing, over-collecting of food and medicinal plants	Protection needs strengthening; some parts threatened
Nigeria										
Af24	Cross River NP	S	4227	150–1700 m	>400 trees*	Timber trees, rattans, edible fruits, medicinal plants	Tropical lowland rain forest, freshwater swamp forest, montane forest, grassland	NP	Potential logging, agricultural encroachment	Reasonably secure
Somalia										
Af42	Cal Madow	S	9600	0–2400 m	1000	Frankincense, myrrh, carob relative, timber trees	Dry montane forest, evergreen to deciduous woodland, bushland, semi-desert	Daalo FoR is proposed as NP	Logging, grazing	No protection in practice, but threats to flora probably not severe
Af44	Hobyo	S	3000	0–440 m	<1000	Medicinal plants, plants used for construction	Deciduous bushland and woodland, dune vegetation	None; proposed GR for part of area	Overgrazing, fuelwood collection	No protection, but threats to flora probably not severe
South Africa										
Af53	Cape Floristic Region	F	90,000	0–2325 m	8600	Ornamentals (e.g. bulbs, succulents, proteas, ericas)	Fynbos, shrubland, Afromontane forest	Montane areas well protected in many reserves; <3% of lowlands protected	Agriculture, urbanization, fire, invasive species, high population growth rate	Most mountain fynbos effectively protected; lowlands much less so; 44% of remaining fynbos area protected
South Africa/Lesotho										
Af82	Drakensberg Alpine Region	F	40,000	1800–3482 m	>1750	Forage grasses, grasses and sedges for thatching, rope, hats, fuelwood	Subalpine and alpine grassland and shrubland, scrub, savanna, wetlands	NPs, NRs, WAs cover 2194 km²; Drakensberg/Maluti Ecosystem Conservation Area is a proposed WHS	Overgrazing, soil erosion, arable agriculture, invasive plants	Severely threatened in places; protected areas safe, but coverage inadequate

Code	Site	Type	Size (km²)	Altitude	Flora	Examples of Useful Plants	Vegetation	Protected Areas	Threats	Assessment
South Africa/Swaziland/Mozambique										
Af59	Maputaland-Pondoland Region	F	201,640	0–1800 m	6000–7000	Ornamental plants (e.g. *Agapanthus*, *Gladiolus*), cowpea relatives, 900 medicinal plant species	Grassland, Afromontane forest, coastal, sand and swamp forests, valley bushveld, semi-evergreen bushland/thicket, palmveld, aquatic communities	7.5% in conservation areas, but 93% of conserved area is in northern savanna zone which is not rich in plant endemics	Rapid population growth, slash and burn agriculture, plantation crops, afforestation, urbanization, invasive species, mining	Existing reserve system inadequate to protect plant-rich vegetation types, which remain threatened
South Africa/Namibia										
Af51	Western Cape Domain (Succulent Karoo)	F	111,212	0–1907 m	5000	Ornamentals (e.g. bulbs, succulents), food and medicinal plants	Succulent shrubland (veld) with associated annuals	NP (270 km²) and a few other reserves covering >2% of the region in total	Overgrazing, agriculture, mining, plant collecting, invasive species, urban development	Inadequate coverage of protected areas; threatened
Tanzania										
Af71	East Usambara Mountains	S	280	150–1506 m	1921 taxa*	Ornamental plants (incl. African violet, *Streptocarpus*), timber and pole species	Lowland semi-deciduous and evergreen submontane forests	25 Catchment FoRs, proposed NRs	Logging, pole-cutting, clearance of forest for agriculture, invasive species	Reserves intact; more extensive forest outside threatened and declining
Af33	Mahale-Karobwa Hills	F	24,000	773–2496 m	>2000	Timber trees, plants used for honey and wax collection	Miombo woodland, riverine, lowland, submontane and montane forests, dambos	Mahale Mountain NP (1613 km²)	Shifting cultivation, influx of refugees, proposed road	Threatened
Af57	Rondo Plateau	F	250	300–700 m	800	Timber trees	Dry semi-deciduous lowland forest, woodland and thicket	All remaining forests protected in Rondo FoR	Logging, pole cutting, clearance for farming, burning, inadequate staffing	Threatened
Uganda										
Af25	Bwindi (Impenetrable) Forest	S	321	1160–2607 m	1000	Timber trees, bamboos, medicinal plants	Moist evergreen submontane and montane forests	NP	Logging, over-collection of forest products (e.g. removal of timber and fuelwood)	Isolated forest very threatened despite protection status and conservation projects
Zaïre										
Af49	Garamba NP and surrounding Domaines de Chasses	S	56,727	710–1061 m	1000	Timber trees, papyrus, medicinal plants	Sudanian woodland, savanna, papyrus swamps, riverine forests	NP and WHS (49,200 km²), 3 Domaines de Chasses (7527 km²)	Uncontrolled fires	NP/WHS area safe
Af35	Kundelungu	S	9000	1500–2000 m		Medicinal plants, timber trees, plants of cultural importance	Miombo woodland, grassland, gallery forest	NP (7600 km²)	Fires, charcoal production, hunting	At risk
Af29	Maiko NP	S	10,830	1000–1200 m		Timber trees	Tropical evergreen rain forest, Afromontane forest	NP	Minor threats from conversion to savanna, gold mining	Reasonably safe due to difficulties of access, but more management resources needed
Af30	Salonga NP	S	36,560	350–700 m	1500–2000	Timber trees, medicinal plants	Tropical evergreen rain forest, swamp and riverine forests, grasslands	NP (whole area), WHS (36,000 km²)	Local population pressure, timber cutting, fuelwood cutting, fire, over-collection of medicinal plants	At risk; inadequate trained staff, infrastructure and management planning

Code	Site	Type	Size (km²)	Altitude	Flora	Examples of Useful Plants	Vegetation	Protected Areas	Threats	Assessment
Af37	Upemba NP	S	11,730	350–1100 m	>2400	Fruit trees, fibres, some timber species	Miombo woodland, dry evergreen forest, wooded grassland, swamps	NP	Illegal commercial logging, fire	At risk; inadequate management resources
Zambia										
Af39	Zambezi Source Area	F	1700	1200–1490 m	>1000	Timber trees, melliferous plants	Riverine, swamp and dry evergreen forest; miombo woodland, bushland, savanna	9% gazetted as PFA	Clearance for agriculture, refugees	At risk from inappropriate land management practices
ATLANTIC OCEAN ISLANDS										
AO1	Azores	F	2304	0–1351 m	300	Timber trees (logged out), fodder plants, fruits, dyes, medicinal plants	Evergreen (laurisilva) forest, montane cloud forest, grasslands, seral communities on volcanic rocks	12 NRs, but only 3 over 10 km²	Clearance for pastures, exotic forestry plantations, invasive plants	Inadequate coverage of protected areas and poor management; threatened
AO2	Canary Islands	F	7542	0–3717 m	1200	Ornamental plants, medicinal plants, timber trees, dry zone pasture grasses	Evergreen (laurisilva and pine) forests, montane and coastal scrub, woodland	4 NPs (273.5 km²), WHS (39.9 km²), BR (5.1 km²), 98 other protected landscapes and parks	Tourist and residential developments, overgrazing, off-road vehicles, invasive plants, fire	Lowlands inadequately protected; seriously threatened
AO3	Madeira/Salvage Islands	F	728 (M) 3 (S)	0–1861 m 0–153 m	1191	Timber trees (now depleted), medicinal plants, Madeiran bilberry, potential ornamental plants	Dry & humid evergreen forests (incl. laurel forest), coastal herb and shrub communities, cliff vegetation	Madeira: NatP (567 km², incl. 6 fully protected areas), Natural Reserve (14 km², land area); Salvage Is: Natural Reserve (whole area)	Introduced plants, tourism, rats, former clearance for agriculture, fire	Reasonably well protected, but control on tourism and eradication of invasive plants needed
AO4	St Helena	F	122	0–823 m	60	Threatened timber trees, incl. endemic ebony, redwood, gumwood	Tree fern thicket, semi-desert, scrub, woodland, severely degraded	c. 10% protected as forest, whole island proposed as BR & WHS	Invasive plants, overgrazing by domestic and feral animals	Rescue and rehabilitation programme underway, but remaining native vegetation still severely threatened
AUSTRALIA/NEW ZEALAND										
Australia										
Au1	Australian Alps	F	30,000	200–2228 m	780	Timber trees (especially eucalypts)	Grassland, woodland, shrubland, forest, alpine vegetation	NPs and SPs cover 17,000 km²	Tourism, grazing, hydroelectric facilities, fire, feral animals, exotic plants	Generally well protected; reasonably secure
Au2	Border Ranges	F	600	0–1360 m	>1200	Timber trees (e.g. red cedar, rose mahogany), macadamia nuts, ornamental plants	Various types of rain forest, eucalypt forest, woodland, shrubland, heath	NP (majority of area), SF, WHS (NSW part)	Clearance for grazing, rural development, visitor pressure, fire, invasive plants	Generally well protected; proposal to add Queensland part to WHS will improve protection
Au3	Central Australian Mountain Ranges	F	168,000	500–1531 m	1300	Food plants (140 spp. previously used), medicinal plants (70 spp.), timber for artefacts	Hummock grasslands, shrublands, woodlands, open rock and cliff vegetation	23 NPs and reserves cover 2.2% of area	Increasing tourist pressure, exotic weeds, grazing, mining, fire	More or less intact, but Petermann Ranges degraded by grazing. No reserves in Petermann and Musgrave areas

Code	Site	Type	Size (km²)	Altitude	Flora	Examples of Useful Plants	Vegetation	Protected Areas	Threats	Assessment
Au4	Kakadu-Alligator Rivers Region	F	30,000	0–370 m	1400	Many plants used by aborigines	Tropical sclerophyll forest, woodland, rain forest, swamp forest, mangrove, saltmarsh, grassland, sedgeland	NP and WHS covers 19,804 km² (c. 66% of region)	Tourism, invasive weeds	WHS area well protected
Au5	Norfolk and Lord Howe Islands	F	39 (NI) 15 (LH)	0–319 m 0–875 m	392*	Ornamental plants (e.g. Kentia palms, Norfolk Island pine, also a timber tree)	Evergreen rain forest, palm forest, "mossy" forest, pine and hardwood forests	WHS covers all Lord Howe; NP of 4.6 km² on Norfolk Island	Tourism, cattle grazing, introduced plants on Norfolk Island; no major threats on Lord Howe	Lord Howe well protected; both areas secure
Au6	North Kimberley Region	F	99,100	0–854 m	1476*	Pasture grasses, wild fruits and roots formerly eaten by aborigines, medicinal plants	Mostly high-grass savanna woodland, some eucalypt woodland, rain forest patches, mangroves	13,855 km² as NP and NRs, much of rest is virtual wilderness	Grazing pressure in parts, feral animals including cattle, donkeys	Safe due to remoteness, terrain and small permanent population
Au7	South-west Botanical Province	F	309,840	0–400 m	5500	Timber trees, ornamental plants	Eucalypt forest, woodland, mallee, scrub	NPs & NRs cover 25,969 km² (8.4% of area); SForPs cover 17,459 km² (5.6% of area)	Root rot fungus, wild fires	Reserves well protected, but coverage inadequate in agricultural belt
Au8	Sydney Sandstone Region	F	24,000	0–1300 m	2200	Australian red cedar and other hardwoods, ornamental plants	Forest, woodland, shrubland, grassland, coastal dune and swamp, mangrove	NPs and reserves cover 11,709 km² (c. 49% of region)	Urban and industrial development, tourism	Generally well protected
Au9	Western Tasmanian Wilderness	F	14,050	0–1617 m	800	Timbers, especially Huon pine, King Billy pine, ornamental plants	Cool temperate rain and eucalypt forests, alpine vegetation, sub-alpine scrub, moorland, grassland	WHS (13,800 km²) incl. NPs, BR; most of WHS has NP status	Tourism, fire, Phytophthora root fungus	WHS well protected. North-west forests unprotected and at risk, but recommended as reserve
Au10	Wet Tropics of Queensland	F	11,000	0–1622 m	>3400	Timber trees, ornamental plants, native food plants, potential medicinal plants	Various types of rain forest, woodlands, shrublands, mangroves	WHS (8990 km²), c. 50 NPs (2073 km²), remainder mainly Crown Land	Tourism, lowland forest clearance, hydro-electric scheme, telecommunication facilities, feral animals, local grazing and logging	Well protected now as WHS, but some areas still at risk, particularly from increasing tourism and illegal clearance
Australia/New Zealand										
Au17	Subantarctic Islands	F	949	0–668 m	35 endemic taxa	Macquarie island cabbage, ornamental plants	Grasslands, fellfields and herbaceous communities, wetlands, forests, coastal vegetation	NRs cover all islands, Macquarie Island has BR status	Introduced plants; introduced cats, rabbits and wekas a threat on Macquarie Island	High level of protection; reasonably safe
New Zealand										
Au16	Chatham Islands	F	965	0–300 m	320	Many species used in Maori culture, ornamental plants	Evergreen cool-temperate forest, peatlands, lagoons, coastal communities	Network of reserves, mostly small in size, more recommended	Clearance for agriculture, feral animals, invasive plants, fire	Threatened; degradation persists with many important sites unprotected
Au14	Northland	F	14,000	0–776 m	620	Timber trees, ornamental plants, many species used in traditional culture	Moist temperate evergreen lowland forest, swamps, coastal communities	>200 protected areas, incl. Waipoua State Forest Sanctuary (91 km²)	Clearance for agriculture, feral animals, weeds, tourism, plantations	Despite many protected areas, many endemics insufficiently protected and at risk

Code	Site	Type	Size (km²)	Altitude	Flora	Examples of Useful Plants	Vegetation	Protected Areas	Threats	Assessment
Au15	North-west Nelson	F	9500	0–1875 m	1200* angiosp.	Timber trees, ornamental plants, plants used in traditional culture	Moist temperate to montane forest, wetlands, alpine vegetation, grassland	NP (225 km², >40 reserves, proposed NP for much of Crown Land	Logging, mining, agriculture, introduced plants and animals	Threatened
CARIBBEAN ISLANDS										
Cuba										
Cb3†	Cajálbana Tableland/Preluda Mountain region	S	100	0–464 m	353*	Pines (especially *Pinus caribaea*) used as sources of timber and resins, palms used for roofing	Pine forest, xerophytic thorn scrub, riverine forests	Partly in Mil Cumbres Integrated Management Area (166 km²); whole area traditionally managed as "Forestry Patrimony"	Fire, clearance for agriculture, logging, invasive species, tourism	At risk
Jamaica										
Cb10	Blue and John Crow Mountains	S	782	380–2256 m	>600	Timber trees, medicinal plants, ornamental plants	Montane rain forests, scrub, savanna, cliff vegetation	NP	Clearance for subsistence farming and commercial crops, fire, invasive species, plant collecting	Southern slopes of Blue Mountains severely threatened; rest of area at risk, regulations need to be enforced
Cb11	Cockpit Country	F	430	300–746 m	1500	Few remaining timber trees, potential ornamental plants, yam relatives, several medicinal plants	Evergreen seasonal subtropical forest, mesic limestone forest, scrub thicket	None, although much has FoR status	Clearance for agriculture, fire, road building, illegal timber cutting, fuelwood collection	Severely threatened
CENTRAL AND NORTHERN ASIA										
Armenia/Azerbaijan/Georgia/Russia										
CA2	Caucasus	F	440,000	0–5642 m	6000	c. 300 food plants, 300 medicinal plants, timber trees, ornamental plants	Broadleaved and coniferous forests, montane steppe, subalpine meadows, semi-desert	37 NRs cover 8982 km² (c. 2% of region)	Logging, overgrazing, plant collecting, visitor pressure	Inadequate coverage of protected areas; threatened
Kazakhstan/Kirghizia/Tadzhikistan/Turkmenistan/Uzbekistan										
CA3	Mountains of Middle Asia	F	550,000	100–7495 m	5500	c. 1000 species of useful plants, incl. timbers, medicinal plants, fruit and nut trees, ornamentals	Semi-desert, steppes, broadleaved and coniferous forests, juniper and pistachio woodlands, meadows	18 reserves cover 5720 km² (c. 1% of region)	Industrial/agricultural development in lowlands, overgrazing, tourism, over-collection of medicinal plants	Inadequate coverage of reserves; at risk
Russia										
CA4	Chukotskiy Peninsula	F	117,000	0–2300 m	939*	c. 40 species of food and medicinal plants, c. 400 fodder species	Tundra, steppe, shrub communities, rare hot spring communities	Proposed International Ethno-ecological Park to cover c. 50,000 km²	Mining, overgrazing, waste disposal, visitor pressure	Not protected at present; vulnerable
CA5	Primorye	F	165,900	0–1933 m	1850	Medicinal plants, honey and food plants, timbers, ornamentals	Mixed coniferous-forest, oak-pine broadleaved montane woodlands, wetlands	6.5% of region protected in 5 NRs (5299 km²), 8 Refuges (2032 km²), BR (3402 km²)	Logging, pollution, fire, visitor pressure, over-collecting	Some vegetation types and threatened species not included in existing reserves; vulnerable
Russia/Kazakhstan										
CA1	Altai-Sayan	F	1,100,000	300–4506 m	2500	Timber trees, medicinal plants, food plants, ornamentals	Coniferous forests, steppe, subalpine and alpine vegetation, tundra	1.5% protected in 5 NRs (15,595 km²), c. 300 NMs, small floristic refuges	Forest clearance, fire, overgrazing, plant collecting	Coverage of existing reserves inadequate, more being planned

† Please note that the statistics for site Cb3 have been revised since Volumes 1 and 2 were published.

Code	Site	Type	Size (km²)	Altitude	Flora	Examples of Useful Plants	Vegetation	Protected Areas	Threats	Assessment
CHINA AND EAST ASIA										
China										
EA1	Changbai Mts region, Jilin	F	30,000	300–2691 m	2000–2500	>80 medicinal spp.; timber trees, fruits, fodder plants	Coniferous and broadleaved forests, meadows, tundra, alpine vegetation	NR/BR (1906 km²)	Logging, tourism, over-collection of medicinal plants	Area lacks effective management; at risk
EA26	Gaoligong Mts, Nu Jiang River and Biluo Snow Mts, Yunnan	F	10,000	1090–5128 m	2000	Timber trees, medicinal plants, ornamental plants	Evergreen broadleaved and sub-alpine coniferous forests, alpine vegetation	2 NRs (4994 km²)	Clearance for cultivation	Part safe; part severely threatened
EA27	Tropical forests of Hainan Island	V	33,920	0–1867 m	4200–4500	>2900 species used locally; timber trees, medicinal plants, rattans, wild litchi	Tropical lowland seasonal rain and monsoon forests, montane seasonal rain forest, mangroves, savanna	51 NRs cover c. 1500 km² (including marine reserves)	Clearance for cultivation, illegal logging, over-collection of medicinal plants	Severely threatened; reserve coverage adequate, but lack of funds and trained personnel
EA40	High Mountain and Deep Gorge region, Hengduan Mts and Min Jiang River basin, Sichuan	F	4000	600–6250 m	>4000	Medicinal plants, timber trees, oil and starch plants, bamboos	Evergreen and semi-deciduous broadleaved forests, coniferous forests, alpine scrub and meadows, savanna	BR (2072 km²) including NR (2000 km²)	Logging, visitor pressure, road building, over-collection of medicinal plants	Area lacks effective management; at risk
EA31	Limestone region, Zhuang Autonomous Region	F	20,000	100–1300 m	2500–3000	>1000 species used locally; timbers, medicines, bamboo, rattan, ornamentals	Seasonal rain forest, montane evergreen broadleaved and limestone forests	11 NRs (4000 km²)	Illegal timber cutting, lack of conservation awareness by local people	Protected areas need effective management; severely threatened
EA13	Nanling Mt Range	F	30,000	200–2142 m	>3000	>800 spp. used locally; medicinal plants, timber trees, fruits, oil, fibres, dyes, wild vegetables	Evergreen broadleaved forest, montane semi-deciduous broadleaved mixed forests	17 NRs (c. 5000 km²)	Clearance for cultivation, over-collection of medicinal plants	Part safe; part severely threatened
EA6	Taibai Mt region of Qinling Mts, Shaanxi	F	2779	500–3767 m	1900	Medicinal plants, oils, fibres, tannins, fodder plants	Broadleaved deciduous and mixed deciduous evergreen forests, sub-alpine coniferous forests, alpine vegetation	NR (540 km²)	Logging, illegal collection of medicinal plants, tourist pressure	Part safe; part severely threatened
EA29	Xishuangbanna region, Yunnan	F	19,690	500–2429 m	4000–4500	>800 medicinal species, 128 timber trees, bamboos, rattans, fruits	Tropical lowland and montane seasonal rain forests, evergreen dipterocarp forest, monsoon forest, montane evergreen broadleaved forest	5 NRs (2416 km²)	Slash and burn agriculture, clearance for plantation crops, illegal logging, colonization	Severely threatened; reserves cover major plant-rich sites, but threats continue and management inadequate
Japan										
EA49	Mount Hakusan	S	480	1700–2702 m	1300	Timber trees, ornamental plants	Montane deciduous forests, subalpine coniferous forests, alpine vegetation	BR (whole area), NP (477 km²)	Tourism, plant collecting, introduced plants	Management needs strengthening; at risk

Code	Site	Type	Size (km²)	Altitude	Flora	Examples of Useful Plants	Vegetation	Protected Areas	Threats	Assessment
EA50	Yakushima	S	503	0–1935 m	1343*	Genetic resources of the timber tree Cryptomeria japonica	Evergreen broadleaved and laurel forests, Cryptomeria forest, montane scrub	NP covers almost all of island, BR (190 km²)	Tourism	Urgent conservation measures needed; threatened
Korea										
EA44	Mount Halla (Cheju Do)	S	151	800–1950 m	1453	Medicinal plants, fibres, timber trees, ornamental plants	Deciduous broadleaved forest, coniferous forest, scrub, secondary grassland	NP (151 km²), mostly including NR (91 km²)	Visitor pressure, over-collection of medicinal and ornamental plants	Reasonably safe, but summit vegetation at risk from trampling and measures to protect rare species need to be implemented
Taiwan										
EA41	Kenting NP	S	177 (land)	0–526 m	1350	Timber trees, medicinal plants, legumes, ornamental plants	Evergreen broadleaved rain forest, semi-deciduous and littoral forests, grassland, scrub	NP	Tourism, grazing, plant collecting, military facilities	Generally well protected, some parts under threat
Thailand										
EA53	Doi Chiang Dao WS	S	521	<800–2175 m	1200	Teak (logged out) edible fruits	Lowland mixed evergreen-deciduous and dipterocarp-oak forests, montane forest	WS	Fire	Vulnerable
EA55	Doi Suthep-pui NP	S	261	360–1685 m	2063* taxa	Timber trees, bamboos, fruit trees, medicinal and ornamental plants, mushrooms	Monsoon forests, incl. lowland dipterocarp-oak and mixed evergreen-deciduous forests, pine forests	NP, BR (whole area)	Encroachment, tourism, road building, tree cutting, fire, over-collecting of ornamental plants	Lack of resources for conservation measures to be implemented fully; threatened
EA57	Khao Yai NP	S	2168	100–1351 m	2000–2500	Medicinal plants, rattans	Moist evergreen forest, dry evergreen and mixed deciduous forests, grassland	NP	Encroachment of agriculture, illegal logging, over-collection of forest products, dam construction	Threatened around perimeter; conservation and development programme underway
EA59	Thung Yai-Huai Kha Khaeng WHS	S	6427	200–1811 m	>2500	Crop relatives, ornamental plants	Lowland evergreen and deciduous forests, evergreen montane forest	WS/WHS	Encroachment of agriculture, logging, fire, potential dam construction	Management under-resourced; threatened
Vietnam										
EA62	Bach Ma-Hai Van	S	600	0–1450 m	2500	200 timber trees, 108 medicinals, ornamentals, fibres, rattans, edible fruits	Lowland evergreen forest, tropical montane evergreen forest	NP (220 km²), 2 Environment Protection Forests	High local population density, felling of timber trees, fuelwood cutting	Threatened
EA63	Cat Tien BR	S	1372	60–754 m	2500	200 timber trees, 120 medicinals, rattans, bamboos, orchids	Lowland evergreen and semi-deciduous forests, freshwater swamps, bamboos	BR includes NP (379 km²), Rhino Sanctuary (360 km²)	Illegal logging, over-exploitation of rattans and resins	Reasonably secure
EA64	Cuc Phuong NP	S	300	200–636 m	1980*	Timber trees, medicinal plants, bamboos, ornamental plants	Lowland evergreen forest, incl. limestone forest, semi-deciduous forest	NP (222 km²)	Illegal felling of timber trees, fuelwood cutting	Reasonably secure

Code	Site	Type	Size (km²)	Altitude	Flora	Examples of Useful Plants	Vegetation	Protected Areas	Threats	Assessment
EA65	Langbian-Dalat Highland	S	4000	1400–2289 m	2000	100 medicinal plants, ornamental plants (especially orchids), resin trees, rattans	Pine forest, tropical montane evergreen forest, subtropical montane forest	3 NRs (535 km²)	Logging, slash and burn cultivation, fire	Severely threatened
EA66	Yok Don NP	S	650	200–482 m	1500	150 timber trees, tannins, resin trees, edible fruits, ornamental plants	Dry dipterocarp forest, lowland semi-evergreen forest, riverine forest	NP (580 km²)	Illegal logging, hunting, forest fires	Threatened
EUROPE										
Bulgaria/Greece/Serbia										
Eu14	Balkan and Rhodope Massifs	F	10,000	400–2900 m	3000	Medicinal plants, timber trees, ornamental plants	Oak and mixed deciduous forests, fir/spruce forests, shiblyak, grasslands, montane and alpine vegetation	Several NPs; 4 BRs (140 km²)	Tourism, agriculture, overgrazing, fire, political conflicts, changes in land tenure	Extensive and expanding areas under protection; some threats arising from new political order
Cyprus										
Eu18	Troodos Mts	S	1800	1000–1960 m	1650	Timber trees, medicinal herbs, ornamentals	Evergreen scrub, some evergreen oak and coniferous forests, rock and cliff communities	Some areas of forest protected	Tourism and skiing development, road building, fire	Conflict between tourism and conservation, but forests well protected
France/Spain/Andorra										
Eu10	Pyrenees	F	30,000	0–3404 m	3500	c. 20 timber spp., 600 medicinal plant spp., honey and food plants, ornamentals	Evergreen forests, semi-deciduous and deciduous forests, montane and subalpine vegetation, coniferous forests	c. 50 protected areas, including 3 NPs, 7 NatPs, numerous NRs	Tourism, fires, overgrazing leading to soil erosion	Some areas well protected; others threatened
Germany/Austria/France/Italy/Liechtenstein/Slovenia/Switzerland										
Eu11	Alps	F	200,000	100–4800 m	5500	Medicinal plants, food plants, timber trees; 0–20% of the flora is traditionally used	Broadleaved and coniferous forests, meadows, marshes, alpine grasslands, scree/rock vegetation	9 NPs, many NRs; national and regional legislation	Tourism and associated development, intensive land management, hydroelectric schemes, possibly global warming	Threatened; conflict between tourism and conservation. Legislation needs to be enforced
Greece										
Eu17	Crete	F	8700	0–2456 m	1600	Crop relatives, culinary herbs, ornamentals	Evergreen scrub, fragments of evergreen oak, pine and cypress forest, rock and cliff vegetation	1 NP (48.5 km²), a few other small reserves	Tourism, agricultural and industrial developments, plant collecting	Protected area coverage inadequate, but threats less acute than in many other areas
Eu16	Mountains of Southern and Central Greece	F	18,000	1000–2495 m	4000	Timber trees, culinary herbs, chickpea relative	Coniferous and deciduous forests; rock, cliff and scree vegetation	4 NPs (173 km²)	Fire, overgrazing, tourism, mining, plant collecting	Threatened
Spain										
Eu4	Baetic and Sub-Baetic Mts	F	18,000	0–3481 m	>3000	Lavender, thyme and other perfumes, medicinal plants, pines, cork oak, wild olives, ornamentals	Evergreen oak forest, deciduous/coniferous woodlands, pine-juniper scrub, pastures, scree vegetation	13 NatPs (7700 km²)	Overgrazing, urbanisation, tourism, fires, agriculture, plant collecting	Threatened: coastal regions, particularly by tourism, montane areas by overgrazing

Code	Site	Type	Size (km²)	Altitude	Flora	Examples of Useful Plants	Vegetation	Protected Areas	Threats	Assessment
Eu7	Massifs of Gudar and Javalambre	F	445	1200–2020 m	>1500	Timber trees (especially pines), medicinal plants	Juniper and pine woodlands, grassland, rock communities	None; until recently traditional land management conserved area well	Tourist developments (e.g. ski resorts)	Until recently well conserved; now threatened
	Ukraine/Russia									
Eu21	South Crimean Mountains and Novorossia	F	80,500	0–1200 m	2200	Fruits, medicinal plants, aromatic plants, ornamentals	Maquis, shiblyak; oak, pine, pine-beech woodland, grassland	Crimea: 3 Reserves (160.8 km²), Game Preserve (150 km²), 5 Refuges; Novorossia: 15 Refuges for selected biotopes	Forest clearance, pollution, over-collection of ornamental and medicinal plants	Inadequate reserve coverage; threatened
INDIAN OCEAN ISLANDS										
IO1	**Madagascar**	F	587,000	0–2876 m	9345	Medicinal plants, fibres, timber trees, edible plants, crop relatives, ornamentals	Rain forest, dry deciduous forest, montane forests, tapia forest, bushland, thicket, mangroves	38 protected areas covering c. 0.7% of land area, 2 BRs, 1 WHS (1520 km²); protected forests cover 4.6% of land area	Clearance for agriculture, fire, overgrazing, mining, tree felling, plant collecting, invasive species	Many important plant sites included in existing protected areas but level of protection inadequate; threatened
	Mauritius/Reunion									
IO2	Mascarene Islands	F	4481	0–3069 m	>955 angiosp. taxa	Timber trees (e.g. black ebony) virtually exhausted, palm hearts, coffee relatives, ornamentals	Tropical/subtropical coastal and dry lowland forests to moist montane forests, high-altitude heath	Mauritius: NP & numerous NRs cover 2.4% of land area, BR (36 km²); Rodrigues: 3 NRs (58 ha); Réunion: 1 NR (68 ha), more proposed	Invasive plants, introduced animals, development pressure, population pressure	Réunion: inadequate reserve system; Mauritius and Rodrigues: good reserve network but flora severely threatened
IO3	**Seychelles** (granitic islands)	F	230	0–905 m	200 angiosp.	Timber trees (over-exploited in past), medicinal plants	Lowland rain forest, hygrophile peak forest, coastal forest, mangroves	2 NPs (37 km²), 8 NRs, 2 SpNRs (351 km²), 2 PAs, 2 WHS	Invasive plants, tourist pressure	Good reserve coverage, but some important areas not covered, management needed; at risk
INDIAN SUBCONTINENT										
	India									
IS7	Agastyamalai Hills	F	2000	67–1868 m	2000	Medicinal herbs, timber trees, bamboos, rattans, crop relatives	Tropical dry to wet forests	3 Sanctuaries and a number of RFs protect c. 919 km²; proposed BR	Clearance for hydro-electric projects, plantations, tourism, fire, grazing	Some areas threatened
IS17	Andaman and Nicobar Islands	F	8249	0–726 m	2270	Rice and pepper relatives, timber trees, rattans	Tropical evergreen, semi-evergreen and moist deciduous forests, beach forest, bamboo scrub, mangroves	c. 500 km² land protected in 6 parks, 94 Sanctuaries (mostly small islets); part of Great Nicobar proposed as BR, proposed North Andaman BR	Population pressure, logging, hydroelectric schemes, clearance for agriculture	Some areas severely threatened; some areas reasonably safe at present; protected area coverage inadequate
IS9	**Nallamalai Hills**	F	7640	300–939 m	758*	Medicinal plants, rice and pepper relatives, bamboos, timbers	Dry and moist deciduous forests, dry evergreen forest, scrub	WS covers 357 km²	Forest clearance, bamboo cutting for paper, fire, fuelwood cutting	Severely threatened

Code	Site	Type	Size (km²)	Altitude	Flora	Examples of Useful Plants	Vegetation	Protected Areas	Threats	Assessment
IS4	Namdapha	S	7000	200–4578 m	5000	Wild relatives of banana, citrus, pepper; timber trees, ornamental plants	Tropical evergreen and semi-evergreen rain forests, alpine vegetation	NP (1985 km²); whole area proposed as BR with core of c. 2500 km²	Shifting cultivation, refugees and other settlers, illegal timber felling	Core area not threatened; perimeter areas under threat
IS2	Nanda Devi	S	2000	1000–7817 m	900* angiosp.	Medicinal plants, ornamental plants, edible plants	Coniferous, birch and rhododendron forests, alpine vegetation	NP and WHS (630 km²), proposed as BR	No major threats; potential dam construction	Effectively secure
IS8	Nilgiri Hills	F	5520	250–2500 m	3240	Medicinal plants, timber trees, fruit tree relatives	Tropical evergreen rain forest to tropical dry thorn forest, montane shola forest and grassland	Several NPs (incl. Silent Valley NP – 89.5 km²), Sanctuaries and RFs; proposed BR	Forest clearance for timber, plantations, roads, development projects	Some areas threatened
	Myanmā (Burma)									
IS14	Bago (Pegu) Yomas	F	40,000	100–821 m	2000	Timber trees (especially teak)	Wet evergreen dipterocarp forest, moist and dry teak forests, bamboo scrub	Proposed NP (146 km²), some RFs	Slash and burn cultivation, road building, dam construction	Protection needs strengthening; threatened
IS5	Natma Taung (Mt Victoria) and Rongklang Range (Chin Hills)	F	25,000	500–3053 m	2500	Gingers, peppers, ornamental plants (e.g. orchids)	Tropical and temperate semi-evergreen forests, subtropical evergreen forests, savanna, alpine vegetation	Proposed NP (364 km²)	Slash and burn agriculture	Threatened
IS6	North Myanmā	F	115,712	150–5881 m	6000	Medicinal plants, bamboos, rice	Lowland tropical evergreen rain forest to cool temperate rain forest, pine–oak forest, subalpine scrub	GS (705 km²), WS (215 km²); important plant sites not protected	Slash and burn agriculture	Threatened; inadequate coverage of protected areas
IS16	Taninthayi (Tenasserim)	F	73,845	0–2275 m	3000	Timber trees (especially teak, rosewood), dyes, vegetables, fibres, medicinal plants	Wet evergreen dipterocarp forests to montane rain forest, bamboo scrub, mangroves	WS (49 km²), GS (139 km²), proposed NR (259 km²)	Logging, resin tapping, fuelwood cutting	Severely threatened in north, inadequate coverage of protected areas
	Sri Lanka									
IS11	Knuckles	S	182	1068–1906 m	>1000	Timbers, medicinal plants, bamboos, fruit tree relatives, spices, ornamental plants	Lowland dry semi-evergreen to montane evergreen forests, grasslands	No legal protection	Clearance for cardamom, settlements and agriculture, mining, fuelwood cutting	Area below 1200 m under threat
IS12	Peak Wilderness and Horton Plains	S	224/32	700–2238 m/ 1800–2389 m	>1000	Timbers, medicinal plants, bamboos, fruit tree relatives, spices, ornamental plants	Lowland, submontane and montane wet evergreen forests, montane grassland	Sanctuary (Peak Wilderness), NP (Horton Plains)	Religious tourism, fuelwood and timber cutting, mining, fire, invasive grasses	Reasonably secure, but management plan required; some rich forests in upper regions should also be included in protected areas
IS13	Sinharaja	S	112	210–1170 m	700 angiosp.	Timber trees, rattans, fruit tree relatives, medicinal plants, ornamentals	Lowland and submontane wet tropical evergreen rain forests, grasslands	National Heritage Wilderness Area and WHS	Population pressure, encroachment of agriculture, over-collection of medicinal plants, potential hydroelectric scheme	Southern part still under threat, otherwise threats declining and reasonably secure

Code	Site	Type	Size (km²)	Altitude	Flora	Examples of Useful Plants	Vegetation	Protected Areas	Threats	Assessment
MIDDLE AMERICA										
Costa Rica										
MA16	Braulio Carrillo-La Selva region	S	500	35–2906 m	4000–6000	Timber trees, ornamental plants (incl. *Monstera*), vanilla, edible palms	Tropical wet forest to montane rain forests	NP, Biological Station, BR	Ecological isolation, local population pressure, illegal logging, agriculture, grazing	Reasonably secure, but at risk
MA18†	Osa Peninsula and Corcovado NP	F	2330	0–745 m	4000–5000	Timber trees, medicinal plants, fruit trees, fibres	Mostly tropical wet forest; also premontane and cloud forests, swamp forest, mangroves	NP (572 km²), FoR (592 km²), small Amerindian Reserve, regional Conservation Area	Mining, logging, road building, colonization	Threatened
Costa Rica/Panama										
MA17	La Amistad BR	S	>10,000	0–3819 m	10,000	Timber trees, medicinal plants, dyes, ornamental plants	Lowland humid forest to subalpine rain páramo	Costa Rica: BR and WHS (6126 km²); Panama: 3 NPs (2345 km²), FoR (200 km²), PF (1250 km²), planned BR	Colonization, agriculture, cattle ranching, fire, oil pipelines, potential mining, road building	Threatened
Guatemala										
MA13	Petén region and Maya BR	F	36,000	10–800 m	3000	Mahogany, edible palms, parlour palms, medicinal plants	Subtropical semi-deciduous moist forest, savanna, wetlands	Maya BR (c. 15,000 km²)	Logging, road building, colonization, grazing, slash and burn agriculture, oil exploration	Region under threat
MA14	Sierra de las Minas region and BR	F	4374	150–3015 m	>2000	Timber trees (e.g. pines), tree ferns, bamboos, medicinal and edible plants	Tropical dry forest and thorn scrub, tropical wet forest to premontane dry forest	BR (2363 km²)	Clearance for agriculture, colonization, logging, road building	BR reasonably safe, other areas threatened
Honduras										
MA15	N.E. Honduras and Río Plátano BR	S	5250	0–1500 m	>2000	Timber trees (especially mahogany and tropical cedar), many plants used locally	Mostly humid tropical and subtropical forests; also mangroves, savanna	BR and WHS (whole area)	Logging, shifting agriculture, cattle grazing, road building, colonization, mining	Threatened
Mexico										
MA5	Canyon of the Zopilote region	F	4383	600–3550 m	>2000	Softwood and hardwood timber trees, medicinal plants, plants of cultural value	Deciduous tropical forest, oak and coniferous forests, montane forests	Ecological SP (36 km²)	Logging, agricultural encroachment, coffee plantations, colonization, potential road building	Inadequate coverage of protected areas; threatened
MA12†	Central region of Baja California Peninsula	F	36,000	0–1985 m	>500	Food and fodder plants, medicinal plants, ornamental plants, fuelwood species	Desert vegetation, lagoon communities	Portion of BR covers 15,000 km² (c. 42%) of region	Overgrazing, expansion of agriculture, salinization, road building, mining, oil and gas exploration	Severely threatened
MA10	Cuatro Ciénegas region	F	2000	740–3000 m	860*	Timber trees, forage plants, medicinal plants	Grasslands, dune communities, desert scrub, oak-pine woodlands, montane coniferous forests	None	Mining, clearance for agriculture, overgrazing, plant collecting (especially of cacti)	Threatened

† Please note that protected area information for sites MA18 and MA12 has been revised since Volumes 1 and 2 were published.

Code	Site	Type	Size (km²)	Altitude	Flora	Examples of Useful Plants	Vegetation	Protected Areas	Threats	Assessment
MA9	Gómez Farías region and El Cielo BR	F	2400	200–2200 m	>1000	Temperate and tropical timber trees, fibres, medicinal plants, ornamental plants	Tropical dry forest, tropical semi-deciduous forest, cloud forest, oak and pine forests, scrub	El Cielo BR (1445 km²)	Logging, grazing, agriculture (incl. shifting cultivation), fire, expanding settlements	Regulations to protect BR need to be enforced; threatened, although some areas remote and safe
MA1†	Lacandon Rain Forest region	F	6000	80–1750 m	4000	Timber, fruit and spice trees, chicle gum, ornamental palms	Tropical lowland to montane rain forests, semi-deciduous forest, cloud forest, savanna, seasonally-inundated forest, wetlands	1 BR (3312 km²), 1 national BR (619 km²), 2 NMs (70 km²), Flora & Fauna Reserve (122 km²)	Road building, logging, colonization, clearance for agriculture, cattle grazing, oil drilling	>50% of forest lost, much of rest going at an accelerating rate; severely threatened
MA7†	Pacific lowlands, Jalisco: Chamela Biological Station and Cumbres de Cuixmala Reserve	F	350	0–500 m	1120*	Valuable timber trees (e.g. rosewood, lignum vitae), potential ornamentals	Tropical deciduous and semi-deciduous forests	1 Biological Station (3.46 km²), 1 Reserve (7 km²)	Clearance for agriculture, resort development, logging	Reserves reasonably safe, but inadequate coverage; no protection of coastal palm forest; threatened
MA3	Sierra de Juárez	F	1700	500–3250 m	2000	Medicinal plants, timber trees, ornamental plants	Mainly montane cloud forest; also tropical evergreen forest, pine and pine-oak forests	None	Logging, grazing, colonization, fire, clearance for agriculture, plant collecting, potential dam construction	Severely threatened
MA6	Sierra de Manantlán region and BR	S	1396	400–2860 m	2800	>500 spp. have been used traditionally; wild perennial maize, wild beans, oaks, pines	Tropical deciduous and semi-deciduous forests to cloud forest, pine and pine-oak forests	BR	Logging, fire, agriculture, cattle ranching, fuelwood cutting	About 30% in good condition; threatened, although some areas safe
MA4	Tehuacán-Cuicatlán region	F	9000	600–2200 m	2700	Cacti and other ornamental plants, medicinal plants, fibres	Sclerophyll scrub, thorn scrub, early-deciduous forest	BG (1 km²)	Agriculture, overgrazing by goats, salinization, plant collecting (especially of cacti)	Threatened
MA8	Upper Mezquital River region, Sierra Madre Occidental	F	4600	800–3350 m	2900	>450 species used by local people, including food and medicinal plants; timber trees	Conifer, pine-oak and oak forests, some tropical semi-deciduous forest	BR (700 km²)	Logging, overgrazing, road building	Threatened
MA2†	Uxpanapa-Chimalapa region	F	7700	80–2250 m	3500	Timber trees (including mahogany), ornamental palms, fruit trees	Evergreen and semi-evergreen rain forest, montane rain forest	None	Logging, colonization, agriculture, grazing, construction of dams and roads	Threatened
Mexico/U.S.A.										
MA11	Apachian/Madrean region	F	180,000	500–3500 m	4000	180 wild relatives of crop plants, >300 food plants, >450 medicinal plants	Coniferous, oak-coniferous and tropical deciduous forests, savanna, chapparal, thorn scrub	<10% of region protected in Mexico; most of region in U.S.A. has some form of protection	Logging, erosion, agricultural encroachment, overgrazing	Part safe, but majority of region not formally protected; areas in northern Mexico severely threatened
Panama										
MA19	Cerro Azul-Cerro Jefe region	S	53	300–1007 m	934*	Timber trees, ornamental plants, potential medicinal plants	Tropical premontane wet and rain forests, tropical wet forest	Whole area within Chagres NP	Grazing, agricultural encroachment, fire	At risk

† Please note that the protected area figures for sites MA1 and MA7 and altitude for site MA2 have been revised since Volumes 1 and 2 were published.

Code	Site	Type	Size (km²)	Altitude	Flora	Examples of Useful Plants	Vegetation	Protected Areas	Threats	Assessment
MA20	Darién Province and Darién NP	F	16,671	0–1875 m	2440*	Cativo and other valuable timbers, medicinal plants	Tropical lowland dry, moist and wet forests, swamps, premontane and montane forests, cloud forest	NP (also a WHS and BR) covers 5790 km²	Logging, mining, agriculture, colonization, proposed Pan-American Highway	Threatened
NORTH AMERICA										
U.S.A.										
NA16*	California Floristic Province	F	324,000	0–4418 m	3500	Timber trees and other forest products, wild grape, ornamental plants, medicinal plants	Coniferous and mixed evergreen forests, oak woodlands, chaparral, coastal scrub, grassland	Parks and reserves protect c. 11% of land area, mostly montane	Population pressure, urban and agricultural development, mining, dams, overgrazing, exotic weeds, off-road vehicle recreation	Highlands reasonably well protected; lowland habitats threatened and inadequately protected
NA29	Central Highlands, Florida	F	10,000	20–94 m		Few timber trees, food plants, potential insecticides	Xerophytic pine and oak scrub, palmetto flatwoods, wetlands	State, federal and private reserves, state and local parks	Citrus plantations, urban, tourist and recreational developments	Inadequate protected area coverage; threatened
NA32	Edwards Plateau, Texas	F	100,000	100–1000 m	2300	Forage species, juniper oil, few timber trees	Semi-arid temperate semi-evergreen forest, grassland, semi-desert scrub	State and municipal parks, several reserves cover <0.05% of region	Clearance for agriculture, grazing, dams, urbanization, introduced species	Protected area coverage very inadequate; threatened
NA16c†	Klamath-Siskiyou region	F	55,000	0–2280 m	3500	Timber trees, beargrass, medicinal plants, including *Taxus brevifolia* for cancer drug	Coniferous and mixed evergreen forests, prairies, savanna, subalpine meadows	WAs, NatMs, federal, state and private reserves	Logging, agriculture, urbanization, mining, dams, tourism	Most is relatively safe due to remoteness, but logging and mining are considerable threats
U.S.A./Canada										
NA25*	Serpentine flora	V	4550	0–2900 m	2000 taxa	Flax and sunflower relatives adapted to low nutrient soils	Grasslands, chaparral, montane woodland	Frenzel Creek Research Natural Area (2.7 km²), TNC Ring Mountain Preserve (25 ha)	Mining, logging, off-road vehicle recreation, urbanization	Inadequate protected area coverage; threatened
U.S.A./Mexico										
NA16g	Vernal pools (California and Baja California)	V	20,000	0–600 m (–2500 m)	>200	*Limnanthes* is potential sperm whale oil substitute, ornamental plants	Annual herbs, wet grassland, aquatics, oak and pine forests	Numerous small pools protected, but more reserves needed	Agriculture, grazing, urban development, mining	Inadequate protected area coverage; severely threatened
PACIFIC OCEAN ISLANDS†										
Chile										
PO8†	Juan Fernández	F	100	0–1319 m	210*	Some potential horticultural species	Mainly forested	NP	Feral animals, invasive plants	Flora acutely threatened; WWF-sponsored rescue programme underway
Ecuador										
PO7†	Galápagos Islands	F	7900	0–1707 m	541*	Tomato and cotton relatives, native timber trees	Wide range of generally arid vegetation types, *Scalesia* forests (much reduced on higher islands)	NP: 7278 km² (96.7% of land area); WHS and BR (7665 km²)	Feral animals, invasive plants, over-exploitation of native woody species, increasing human population, fire	Many plants still at risk despite conservation measures

† Please note that the Data Sheet site codes for the Pacific Ocean Islands have been revised since Volume 1 was published; the codes in Volumes 2 and 3 are the correct ones. The altitude and flora statistics for sites NA16, NA16c and NA25 have been revised since Volumes 1 and 2 were published.

Code	Site	Type	Size (km²)	Altitude	Flora	Examples of Useful Plants	Vegetation	Protected Areas	Threats	Assessment
PO3†	**Fiji**	F	18,270	0–1324 m	1628	Timber trees, medicinal plants, culturally important plants	Tropical rain, dry and montane forests, grassland, scrub, coastal vegetation	16 protected areas (65 km²) incl. 8 NRs (62 km², mainly rain forest), 1 NP (2.4 km², coastal), several other small forest parks	Logging, clearance for agriculture, feral animals, invasive plants, population pressure, tourism, mining	Threatened; reserves and funding inadequate, numerous proposals never implemented. Need involvement of local population
	French Polynesia									
PO5†	Marquesas	F	1275	0–1260 m	318*	Ornamental species, food plants	Lowland forest, remnants of dry forest, grasslands, xerophytic scrub	4 reserves, but inadequate coverage	Overgrazing, introduced species, fire, many species reduced to small populations	Severely threatened
	New Caledonia									
PO2†	Grande Terre	F	16,890	0–1628 m	3322	Timber trees, ornamental plants, medicinal plants, aromatic plants	Humid evergreen and sclerophyll forests, maquis, mangroves, marshes	9% covered by protected areas, incl. 13 Special Botanical Reserves (155 km²), 1 of which has Strict Nature Reserve status, and 4 Provincial Parks (105 km²)	Fire, mining, clearance for agriculture and grazing, urbanization, invasive plants	Coverage of protected areas inadequate. Sclerophyll and calcareous forests particularly threatened, along with many point endemics
	Japan									
PO1†	Bonin (Ogasawara) Islands	F	73	0–462 m	456*	Ornamental plants (orchids)	Broadleaved evergreen forest, but mostly destroyed	NP (61 km²)	Population pressure, grazing, introduced plants	NP, but much degraded and highly endangered
	U.S.A.									
PO6†	Hawaiian Islands	F	16,641	0–4205 m	1200	Native plants used by indigenous people for all their needs; ornamentals, hardwood trees	Wide range of vegetation types: dry to mesic lowland to subalpine forests, shrublands, grasslands, herblands	2029 km² (12% of land area) protected in many state, federal and private reserves, but not all plant-rich communities covered	Development, introduced plants and animals, fire	Possibly the most endangered island flora in the world; severely threatened
	Western Samoa/American Samoa									
PO4†	Samoan Islands	F	3114	0–1857 m	775	Timber trees, >100 medicinal plants, incl. possible cancer & AIDS treatments	Coastal, lowland and montane rain forests, cloud forest, upland and volcanic scrub, mangroves, marshes	Proposed NPs on Tutuila and Ta'u, American Samoa; 2 village-level reserves on Western Samoa	Increasing population pressure, shifting cultivation, cash crops, logging, invasive plants, hydroelectric scheme	Severely threatened; lowland forests almost eliminated, montane forests increasingly threatened
SOUTH AMERICA										
	Argentina									
SA35	Anconquija region	F	6000	400–5550 m	2000	Timber trees, crop relatives (e.g. tomato relatives), medicinal and ornamental plants	Amazonian winter-dry rain forest to temperate cloud forest, Andean páramo grassland, spiny shrubland	Several protected areas covering c. 370 km², proposed NP of 3000 km²	Logging, clearance for agriculture, grazing, plant collecting, road building, dams, potential mining	Varying degrees of threat; some protected areas reasonably secure, remote areas self-protected; other areas severely threatened
	Argentina/Chile									
SA34†	Altoandina	F	?	500–5900 m	?	Forage and medicinals	Grass-steppe, chamaephytes, bog	No information	No information	No information
SA45	Temperate rain forest of Chile	V	110,000	0–2000 m	450	Timber trees (e.g. Araucaria, Fitz-roya)	Evergreen and deciduous temperate rain forests	25 NPs, 14 NRs, 1 NatM	Logging, agriculture, grazing, fire, plantation forestry	Coverage and staffing of existing reserves inadequate; severely threatened

† Please note that the Data Sheet site codes for the Pacific Ocean Islands have been revised since Volume 1 was published; the codes in Volumes 2 and 3 are the correct ones. The information for site SA34 has been revised since Volumes 1 and 2 were published.

Code	Site	Type	Size (km²)	Altitude	Flora	Examples of Useful Plants	Vegetation	Protected Areas	Threats	Assessment
SA46	Patagonia	F	500,000	0–2000 m	1200	Forage plants, many species used traditionally for food and medicine	Steppe shrublands and grasslands	Lower portions of 3 NPs/NRs, all of 1 NP/NR (112 km²), 1 NatM (100 km²)	Overgrazing, desertification	Inadequate coverage of protected areas; some areas threatened
Argentina/Paraguay/Brazil/Bolivia										
SA22†	Gran Chaco	F	1,010,000	100–2795 m	1200	Timber trees, medicinal plants, ornamentals, fibres	Xerophytic deciduous to semi-evergreen forests, palm woodlands, savanna, steppes, wetlands	6 NPs and 1 National NR cover 12,720 km² (c. 1%) of region	Logging, agricultural development, oil and gas exploration, road building, colonization, fire	Inadequate protected area coverage; some areas threatened
Bolivia										
SA24	Llanos de Mojos region	F	270,000	130–235 m	5000	Forage grasses, legumes, rubber, Brazil nut, palm hearts, pineapple relatives	Mosaic of savanna and various evergreen forest types, wetlands	BR (1350 km²), FoRs, Biological Station, Amerindian territories, regional park, private reserves	Overgrazing, fire, road building, logging	Inadequate protected area coverage; key forest areas severely threatened
SA36†	Madidi-Apolo region	F	30,000	250–2000 m	>5000	Timber trees (e.g. mahogany), quinine, medicinal plants, ornamental plants, palm oils	Mainly humid evergreen tropical montane forest, dry forest, cloud forest, grasslands	New NP recommended	Logging, road building, colonization, oil exploration	Mostly intact, with new NP; threatened
SA23†	South-eastern Santa Cruz	F	70,000	350–1290 m	2000–2500	Timber trees, forage grasses, Copernicia palms for telephone poles, forage plants	Dry forest, savanna (cerrado and campo rupestre), wetlands, thorn scrub	National Historical Park (c. 170 km²), 1 NP (24,000 km² recommended), proposed FoRs, 3 proposed Biological Reserves	Cattle ranching, agriculture, mining, gas pipeline, road building	Inadequate coverage of protected areas; threatened
Brazil										
SA12	Atlantic moist forest of southern Bahia	V	3500	0–1000 m		Brazil-wood, rosewood, and other valuable timbers	Tropical Atlantic rain forest, semi- and dry deciduous forests, littoral forest	<300 km² or <0.1% of original wet forests protected in 6 reserves	Logging, clearance for cattle grazing, cash crops and subsistence agriculture, pulpwood plantations	Inadequate coverage of protected areas, even existing reserves inadequately protected; severely threatened
SA14	Cabo Frio region	F	1500	0–500 m	1500–2200	Brazil-wood, medicinal plants	Coastal evergreen scrub, xeromorphic forest, mangroves, submontane rain forest	Protected areas cover c. 10% of region	Land development, tourism, clearance for cattle grazing, sugar cane plantations	Inadequate coverage of protected areas, even existing reserves inadequately protected; severely threatened
SA19†	Caatinga of north-eastern Brazil	V	1,000,000	0–1000 m		Forage plants, fruits, timber trees, palms	Xerophytic deciduous forest to sparse scrub, savanna, gallery and montane forests, cerrado, grassland	3 NPs (1048 km²), 5 Biological Reserves (132 km²), 5 Ecological Stations (1452 km²), 1 National Forest (383 km²), 2 Environmental Protection Zones	Overgrazing, timber and fuelwood extraction, agriculture, salinization, plant collecting	Inadequate coverage of protected areas; severely threatened
SA21	Distrito Federal	F	5814	750–1336 m	>3000	Timbers, fruit trees, medicinal plants, ornamental plants	Gallery, mesophytic and cerrado forests, wetlands, savannas	Protected areas cover only 3.2% of region and do not protect full range of vegetation types	Population pressure, fires, invasive species, potential dam construction	Inadequate funding and coverage of existing reserves; threatened

† Please note that protected area information for sites SA36, SA23 and SA19 and assessments for sites SA22, SA36 and SA23 have been revised since Volumes 1 and 2 were published.

Code	Site	Type	Size (km²)	Altitude	Flora	Examples of Useful Plants	Vegetation	Protected Areas	Threats	Assessment
SA20	Espinhaço Range region	F	7000	1000–2107 m	>4000	Fruit trees, ornamental plants, medicinal plants	Mainly campo rupestre, with cerrado, marshes, montane to dry deciduous forests	2 NPs (1858 km²), 2 SPs (430 km²), Ecological Station, 3 Environmental Protection Zones	Grazing, erosion, fire, charcoal production, plant collecting, hydroelectric schemes, road building, mining	Inadequate coverage of protected areas; threatened
SA17	Juréia-Itatins Ecological Station	S	792	0–800 m	>500	Timber trees, edible palms, medicinal plants, ornamental plants	Tropical Atlantic rain forest, restinga forest and scrub, littoral forest, mangroves, grassland	Ecological Station	Illegal logging, over-collection of palm hearts	Isolated and relatively secure
SA5	Manaus region	F	71,000	16–130 m	>1000 tree spp.	Timber trees, rosewood, Brazil nut, fruit trees, edible palms	Terra-firme rain forest, permanently and seasonally inundated forests, caatinga forest, scrub, grasslands	Numerous protected areas	Clearance for cattle ranching, oil palm plantations, urban expansion, road building	Inadequate reserve coverage, no várzea (inundated) forests protected nearby
SA15	Mountain ranges of Rio de Janeiro	F	7000	60–2800 m	5000–6000	Timber trees (e.g. Brazil-wood), palmito palm hearts, medicinal plants, ornamental plants	Tropical Atlantic rain forests, high-altitude fields	20 conservation areas cover c. 3220 km² (46%) of region	Clearance for agriculture, settlements, over-collecting of palms and ornamentals, logging	Protected areas lack management plans; forests outside protected areas severely threatened
SA16†	Serra do Japi	S	354	800–1300 m	300* tree spp.	Timber trees, fruit trees, medicinal and ornamental plants	Semi-deciduous forests, semi-arid vegetation on rock outcrops	"Historical patrimony" (191 km²), Environmental Protection Zone, small municipal reserve, part of BR	Population pressure leading to clearance for settlements, industrial developments, logging, air pollution, fire, mining, tourism	Protection status needs strengthening; severely threatened
SA13	Tabuleiro forests, N. Espírito Santo	V	484	28–90 m	637* tree spp.	Timber trees, resins, oils, medicinal plants	Tropical Atlantic rain forest, várzea forest, savanna	FoR (220 km²), 2 Biological Reserves (264 km²)	Fire, logging	Protected areas reasonably secure and protect almost all the remaining forests
Brazil/Colombia/Venezuela										
SA6†	Upper Rio Negro region	F	>250,000	<100–1000 m	>15,000	Centre of diversity for many useful plants, incl. rubber, rosewood, Brazil nut	Amazon caatinga forest and scrub, flood forest (igapó), submontane forest	Well protected in 5 NPs, 2 NNRs (c. 34% of region), 1 FoR, Amerindian Reserves	None; potential threats: gold mining, coca plantations, logging	Relatively safe; further areas proposed as NPs and reserves
Brazil/Guyana/Surinam										
SA4†	Transverse Dry Belt of Brazil	F	500,000	5–600 m		Timber trees, Brazil nut, rubber and cocoa relatives, vanilla, curare, rosewood	Lowland (terra firme) semi-deciduous and deciduous forests, permanently and seasonally inundated swamp forests, various savannas	1 Biological Reserve (38,500 km²), 1 National Forest (3150 km²), 2 Ecological Stations (5771 km²), Amerindian Reserves	Mining, pollution, logging	Relatively inaccessible but development beginning; protection status needs strengthening
Chile										
SA43	Lomas formations of the Atacama Desert	V	>5000	0–1100 m	550	Ornamental plants, potential genetic resources	Desert vegetation of annual, short-lived perennials and woody scrub	NP (438 km²), Nature Reserve (30 km²)	Urbanization, mining, fuelwood cutting, grazing	Inadequate coverage of protected areas; threatened

† Please note that area, flora and protected area information for sites SA16, SA6 and SA4 have been revised since Volumes 1 and 2 were published.

Code	Site	Type	Size (km²)	Altitude	Flora	Examples of Useful Plants	Vegetation	Protected Areas	Threats	Assessment
SA44†	Mediterranean region and La Campana NP	F	5000	300–2222 m	1800–2400	Timber trees (e.g. *Nothofagus*), fruit trees, Chilean honey palm, dyes, medicinal plants, ornamentals	*Nothofagus* forest, submontane to alpine vegetation, palm forest, bamboo thickets, matorral	2 NPs (117 km²), 5 National Reserves (723 km²), 1 BR, 1 NS (116 km²), proposed reserve (c. 5000 km²)	Agriculture, cattle ranching, urbanization, road building, mining, fire, fuelwood cutting	Partially protected, but areas outside protected areas are severely threatened
Colombia										
SA7†	Chiribiquete-Araracuara-Cahuinarí region	F	50,000	150–700 m	12,000	Timber trees, rubber, cocoa and hot pepper relatives, fruit trees, medicinal plants	Tropical lowland rain forest, evergreen sclerophyllous scrub, grassland	2 NNPs (18,550 km²), 3 Amerindian Reserves (3470 km²)	Potential unplanned colonization, burning, road building, mining	Reasonably safe at present due to remoteness; management plan for protected areas yet to be developed
SA39†	Colombian Pacific Coast region (Chocó)	F/V	130,000	0–1000 m	8000–9000	Timber trees, fruit trees, medicinal plants, ornamentals	Lowland and premontane tropical pluvial to moist forests, mangroves	4 NNPs (2013 km²) and lower portions of 2 NNPs	Logging, colonization, agriculture, grazing, mining	Present reserve system inadequate; large areas of forest under threat
SA29†	Colombian Central Massif	F	2000	1000–5000 m	1200	Medicinal plants, fodder plants, fuelwood	Páramo, montane and mid-altitude forests, tropical deciduous forest	NNP (830 km²), BR (in part), private reserve, Amerindian Reserves	Mining, road building, cattle ranching, fire, tree felling, draining of wetlands	Inadequate coverage of protected areas; threatened
SA28†	Los Nevados Natural National Park region	S	12,200	300–4600 m	1250	Medicinal plants, dyes, timbers, edible species	Lowland dry forests, oak and laurel forests, páramo, super-páramo	NNP (583 km²), 3 other protected areas (115 km²)	Logging, fire, volcanic eruptions, cattle grazing	Region at mid-altitudes severely threatened
SA27†	Páramo de Sumapaz	F	15,000	300–4250 m		Plants used for medicine, agriculture, construction	Lowland to montane forests, dry forests, páramo	1 NNP (1540 km²)	Road construction, colonization, cattle ranching, logging, mining	Increasingly at risk from population growth and needs; more reserves needed
SA25†	Sierra Nevada de Santa Marta	F	12,232	0–5776 m	>1800	Medicinal plants, dyes, cultural species, edible species, timber trees	Humid tropical forest, montane tropical rain forest, cloud forest, páramo	2 NNPs (3980 km²), Amerindian Reserves	Colonization, erosion, agriculture, cattle ranching	Severely threatened
SA26†	Sierra Nevada del Cocuy-Guantiva	F	4260	500–5493 m		Medicinal plants, spices, plants for construction	Dry xerophytic scrub, Andean forests, páramo	1 NNP (3060 km²), including Amerindian Reserve (216 km²)	Excessive sheep grazing, agricultural spread, tourism	Improved regulation; an additional reserve and management plans needed
Colombia/Ecuador										
SA30	Volcanoes of Nariñense Plateau	F	1400	3000–4500 m	450	Medicinal plants, edible plants, ornamental plants	Páramo grassland and peat bogs, scrub, forests with *Miconia*	Fauna and Flora Sanctuary (176 km²)	Volcanic eruptions, agriculture, cattle ranching, plantation forestry using exotics	Inadequate protection; severely threatened
Ecuador										
SA40	Ecuadorian Pacific Coast mesic forests	V	3700	0–900 m	5300	Timber trees, ivory-nut palm, ornamental plants	Lowland tropical moist, wet and pluvial forests	FoR (1300 km²), Ecological Reserve (2040 km²), several other small reserves (c. 5 km²)	Logging, colonization, agriculture, grazing	Threatened; some areas severely threatened
SA38†	Gran Sumaco and Upper Napo River region	F	9000	300–3732 m	6000	Medicinal plants, crop relatives, culturally important plants	Lowland tropical forest to wet páramo	Biological Station (15 km²); whole area proposed as BR; NP (2052 km²)	Road building, subsistence agriculture, colonization, cattle ranching, potential mining	Inadequately protected, especially lowlands; threatened

† Please note that the protected area information for sites SA44, SA7, SA39, SA29, SA28, SA27, SA25, SA26 and SA38, and altitude figures for site SA29, have been revised since Volumes 1 and 2 were published.

Code	Site	Type	Size (km²)	Altitude	Flora	Examples of Useful Plants	Vegetation	Protected Areas	Threats	Assessment
SA31	Páramo and Andean forests of Sangay NP	V/S	5717	1000–5319 m	>3000	Food, fodder and medicinal plants, fibres, ornamental plants	Páramo, Andean and sub-Andean forests	Included within NP and WHS	Road building, logging, colonization, mining, fire, overgrazing, tourism	Reasonably safe, but management plan needs revision and implementation
SA8	Yasuní NP and Waorani Ethnic Reserve	S	15,920	200–350 m	4000	Wild rubber, vegetable ivory palm, valuable hardwoods, medicinal plants	Tropical moist forest	BR: NP (9820 km²), Ethnic Reserve (6100 km²)	Oil extraction, logging, road building, colonization	Severely threatened
Ecuador/Peru										
SA32	Huancabamba region	F	29,000	1000–4000 m	2000–2500	Timber trees, medicinal plants, ornamental plants	Montane cloud forest, dry forest, páramo	NatS (295 km²)	Logging, agriculture; 75% of original humid forest destroyed	Inadequate protection; severely threatened
French Guiana										
SA3†	Saül region	S	>1340	200–762 m	2000	Timber trees, edible palms, rosewood oil, medicinal plants	Mostly lowland moist forest, swamp forest, submontane forest	2 potential NRs (790 km²); proposed NP (1336 km²)	Slash-and-burn agriculture, fuelwood cutting, gold mining, potential road	Seriously threatened at present; repeated efforts to protect area since 1975
Paraguay										
SA18	Mbaracayú Reserve	S	600	140–450 m		Valuable timbers, fruit trees, edible palms, Paraguayan tea, medicinal plants	Semi-evergreen subtropical moist forest, savanna, lowland bogs	Private reserve	Agriculture, logging, cultivation of narcotic plants	Inadequately protected; region threatened
Peru										
SA41	Cerros de Amotape NP region	S	2314	100–1618 m	>500	Timber trees, forage plants, medicinal plants, ornamental plants, craft plants	Dry forest, matorral	NP (913 km²) forming core area of a BR of 2314 km²	Fuelwood and timber cutting, fire, overgrazing, soil erosion, desertification	At risk
SA37†	Eastern slopes of Peruvian Andes	F	250,000	400–3500 m	7000–10,000	Timber trees, ornamental plants	Dry, wet and pluvial tropical and subtropical lowland, premontane and montane forests	3 NPs protect c. 18,300 km², 2 NatSs, 2 Reserved Zones, 1 PF, 1 National Forest cover 6000 km²	Deforestation from colonization, roads, agriculture, logging, cultivation of narcotics	NPs reasonably safe due to inaccessibility; protection of other reserves inadequate; severe threats in places
SA42†	Lomas formations	V	>2000	0–1000 m	600	Potential genetic resources of crop plants (e.g. tomato)	Desert vegetation of annual, short-lived perennial and woody scrub	2 National Reserves (1221 km²), 1 NatS (7 km²), but mostly covering aquatic territory	Urbanization, grazing, mining, fuelwood cutting	Coastal areas severely threatened, more protected areas and conservation measures needed
SA11	Lowlands of Manu NP: Cocha Cashu Biological Station	S	7500	300–400 m	1900	Spanish cedar, mahogany, cocoa relatives, edible palms and fruits	Mostly evergreen tropical forest	Biological Station (10 km²) within NP of 15,328 km², forming part of a BR of 18,812 km²	Potential road building, oil exploration and mining	Inadequate funding resulting in lack of trained staff; threatened
SA33	Peruvian puna	V	230,000	3300–5000 m	1000–1500	Potato relatives, many traditional Andean crops and medicinal plants, spices	Grasslands, scrub, wetlands, tropical alpine vegetation	9500 km² (4% of area) within 3 NPs, 5 National Reserves, 3 NatSs, 3 Historical Sanctuaries	Overgrazing, fire, soil erosion, mining, fuelwood cutting	Inadequate funding resulting in lack of trained staff; inadequate coverage of protected areas; threatened
SA10	Tambopata region	F	15,000	250–3000 m	2500–3000	Timber trees, Brazil nut, rubber, fruit trees	Subtropical premontane/montane wet forests, tropical moist forest, swamp forest, savanna	NS (1021 km²), Reserved Zone (14,000 km²), Tambopata Reserve (5.5 km²)	Cattle ranching, subsistence agriculture, gold mining, local population pressure	Conservation measures need implementing in Reserved Zone; threatened

† Please note that the protected area information for sites SA3, SA37 and SA42 have been revised since Volumes 1 and 2 were published.

Code	Site	Type	Size (km²)	Altitude	Flora	Examples of Useful Plants	Vegetation	Protected Areas	Threats	Assessment
Peru/Colombia										
SA9	Iquitos region	F	80,000	105–140 m	>2265	Timber trees, medicinal plants (incl. curare), fibres, >120 spp. of edible fruits	Evergreen Amazonian moist forests, swamps	Peru: National Reserve (in part), Communal Reserve (3225 km²); Colombia: National Natural Park (1700 km²)	Clearance for settlements and agriculture	Inadequate coverage of protected areas; some habitats threatened
Venezuela										
SA1	Coastal Cordillera	F	45,000	0–2765 m	5000	Quinine and other medicinal plants, quality hardwoods	Montane/submontane semi-deciduous and evergreen forests, hill savanna, cloud forest, coastal and upper montane scrub, mangroves	11 NPs (6640 km²) and 5 NMs	Population pressure, deforestation, colonization, fire, agriculture, roads	Protected areas reasonably secure, but coverage inadequate
Venezuela/Brazil/Guyana										
SA2†	Pantepui region of Venezuela	S/F	7000	1300–3015 m	3000	Ornamentals (e.g. bromeliads, orchids, carnivorous plants), undoubtedly rich in potentially useful species	Montane forests, tepui scrub, pioneer communities on cliffs and rocky areas	NPs, NMs and a BR cover entire region	No serious threats, but potential threats from increased tourism and over-collecting	Safe at present but potentially at risk without more effective controls
SOUTH EAST ASIA (MALESIA)										
Brunei Darussalam										
SEA13	Batu Apoi FoR, Ulu Temburong	S	488	50–1850 m	3000	Timber trees (especially dipterocarps, Agathis), medicinal plants, ornamentals	Lowland dipterocarp rain forest, lower and upper montane forests	FoR, planned NP	Logging, potential dam construction	At risk
Indonesia (Irian Jaya)										
SEA68	Arfak Mountains	F	2200	100–3100 m	3000–4000	Timber trees, rattans, fruit tree relatives, ornamental plants (especially rhododendrons)	Lowland, hill and lower montane rain forests, grassland/heath communities, lake vegetation	NR (450 km²); proposal to extend to 653 km² as Nature Conservation Area	Population pressure, resettlement schemes, agriculture and logging, road building	Lowlands threatened in particular
SEA69	Gunung Lorentz	S	21,500	0–4884 m	3000–4000	Fruits, vegetables, fibres, building materials	Lowland to montane rain forests, mangroves, bogs, swamps, heaths, grasslands, alpine vegetation	NR; proposals for NP, BR & WHS status	Mining, logging, petroleum exploitation, road building, tourism, colonization	Threatened; high-altitude vegetation vulnerable to trampling
SEA70	Mamberamo-Pegunungan Jayawijaya	F	23,244	0–4680 m	2000–3000	Timber trees, especially southern beech, podocarps, conifers	Lowland to montane rain forest, lowland swamp forest, mangroves	Proposed NP/WHS (14,425 km²), GR (8000 km²), proposed GR (819 km²)	Petroleum exploitation, logging at lower altitudes	Lowlands particularly at risk
SEA71	Waigeo	S	14,784	0–999 m		Timber trees, wild sugar cane	Lowland to lower montane rain forest, riverine forest, mangroves, limestone and ultramafic vegetation	NR (1530 km²), marine reserve covers offshore islets and reefs	Potential nickel mining	At risk if mining goes ahead

† Please note that the protected area information for site SA2 has been revised since Volumes 1 and 2 were published.

31

Code	Site	Type	Size (km²)	Altitude	Flora	Examples of Useful Plants	Vegetation	Protected Areas	Threats	Assessment
Indonesia (Java)										
SEA64	Gunung Gede-Pangrango NP	S	150	1000–3019 m	>1000	Timber trees, medicinal plants, ornamental plants	Mostly montane and submontane rain forests, grass plains	NP (whole areae), BR (140 km²)	Timber and fuelwood cutting, agricultural encroachment, visitor pressure, plant collecting	Encroachment around boundaries; at risk
Indonesia (Kalimantan)										
SEA15	Bukit Raya and Bukit Baka	S	7705	100–2278 m	2000–4000	Timber trees (especially dipterocarps), fruit trees, illipe nuts, rattans	Lowland tropical rain forest, swamp forest, lower and upper montane forests, ericaceous scrub	NP (1811 km²)	Logging, road construction, shifting cultivation	Encroachment in west; at risk
SEA17	Gunung Palung	S	900	0–1160 m		Timber trees, fruit trees, ornamental plants	Dipterocarp rain forests, montane forests, swamp forests, beach forest, mangroves	NP	Logging, shifting cultivation	Buffer zones needed; most of area safe, but some parts at risk
SEA19	Sungai Kayan-Sungai Mentarang NR	S	29,000	100–2556 m	2000	Timber trees, fruit trees, gingers, rattans	Lowland and hill dipterocarp rain forests, montane forests, riverine, swamp and heath forests	NR (16,000 km²); proposed extensions (13,000 km²)	Logging, mining, shifting cultivation	Boundaries at risk
Indonesia (Sulawesi)										
SEA46	Dumoga-Bone NP	S	3000	200–1968 m		Timber trees, rattans	Tropical lowland semi-evergreen rain forest, riverine rain forest, montane forest, some limestone forest	NP	Over-collection of forest products, shifting cultivation, potential mining, road building	Boundaries threatened but demarcation and zoning being implemented
SEA47	Limestone flora of Sulawesi	V		150–1000 m		Sugar palm, fruit tree relatives, plants for degraded land	Forest over limestone, scrub, lithophytic vegetation	c. 70% of all outcrops unprotected	Quarrying, firewood cutting, clearance for agriculture, fire	Threatened
SEA50	Pegunungan Latimojong	S	580	1000–3455 m		Ornamental plants	Lower and upper montane forests, hill forest, montane grassland, subalpine vegetation	PF; proposed NR	Clearance of lower slopes for agriculture	Lowlands threatened, but most of area not threatened
SEA48	Ultramafic flora of Sulawesi	V	12,000			Plants of potential value for rehabilitating degraded areas	Ultramafic facies of lowland forest, scrub, some montane vegetation	Mostly unprotected	Mostly intact, but agricultural development planned	Threatened
Indonesia (Sumatra)										
SEA41	Gunung Leuser	S	>9000	0–3466 m	2000–3000	Timber trees (especially dipterocarps), fruit trees, medicinal plants, ornamentals	Lowland dipterocarp rain forest, montane and subalpine forests, freshwater swamp forest, marshes	NP (7927 km²), BR (9464 km²)	Encroachment of settlements, agriculture, illegal logging, over-collection of rattans	Threatened, especially in the lowlands; inadequate funding
SEA42	Kerinci-Seblat NP	S	1485	200–3805 m	2000–3000	Fruit trees, timbers (e.g. dipterocarps, *Agathis*), medicinal plants, rattans	Lowland and hill dipterocarp rain forests, montane forests, montane swamp forest	NP	Encroachment of settlements, agriculture, illegal logging, over-collection of rattans	Threatened, especially in the lowlands

Code	Site	Type	Size (km²)	Altitude	Flora	Examples of Useful Plants	Vegetation	Protected Areas	Threats	Assessment
SEA43	Limestone flora of Sumatra	V	5000	150–1500 m	1500–2000	Ornamental plants, fruit tree relatives	Forest over limestone, scrub, lithophytic vegetation	Very few outcrops protected	Quarrying	Threatened
SEA45	Tigapuluh Mountains	S	2000	150–800 m	2000–3000	Timber trees	Tropical lowland evergreen and hill dipterocarp rain forests	None	Logging, conversion to forestry and rubber plantations, shifting cultivation	Eastern lowlands severely threatened
Indonesia/Malaysia										
SEA16	Lanjak-Entimau WS, Batang Ai NP, Gunung Bentuang dan Karimun	S	10,112	500–1284 m		Timber trees, especially dipterocarps, illipe nuts, rattans, fruit tree relatives	Lowland and hill evergreen rain forests, heath, swamp and montane forests	WS (1688 km²), proposed extensions (184 km²), NP (240 km²), NR (8000 km²)	Agricultural encroachment, logging	Boundaries and lower slopes at risk
SEA18	Limestone flora of Borneo	V		0–1710 m		Ornamental plants (especially orchids, gesneriads, begonias, balsams, ferns)	Lowland to upper montane forest on limestone, lithophytic vegetation	Some major areas in Sarawak in NP, few in Sabah in FoRs, proposed NPs in Kalimantan, other sites unprotected	Quarrying, fire, clearance of surrounding forests for agriculture, tourism	Some major areas safe, some relatively safe but unprotected, most others threatened
Malaysia (Peninsular Malaysia)										
SEA2	Endau-Rompin State Parks	S	500	100–1000 m		Timber trees (including dipterocarps), rattans, fruit trees, wild banana relative, medicinal herbs	Mainly tropical lowland rain and hill dipterocarp forest, hill swamp forest, riparian communities	FoR; proposed State Parks	Logging, clearance for development schemes, tourist facilities, commercial collection of ornamental plants	Protection needs strengthening; at risk
SEA3	Limestone flora of Peninsular Malaysia	V	260	0–713 m	>1300	Ornamental plants, especially orchids, begonias, palms, gesneriads	Limestone forest, scrub, lithophytic vegetation	A few outcrops protected in Taman Negara (NP); some occur in FoRs; some are Temple Reserves (no protection to flora)	Quarrying, mining, encroachment from agriculture, fire, tourism, plant collecting	Many outcrops severely threatened, some at risk, a few safe
SEA4	Montane flora of Peninsular Malaysia	V	2180	810–2188 m	>3000	Ornamental plants (especially orchids, pitcher plants, rhododendrons)	Lower and upper montane forests	Most peaks fall within FoRs, G. Tahan is in NP, Cameron Highlands is in WS, G. Kajang is in WR	Large-scale resort development, agriculture/horticulture, road building, plant collecting	Most areas outside NP severely threatened
SEA5	Pulau Tioman	S	72	0–1038 m	1500	Ornamental plants, especially slipper orchids, Rafflesia used medicinally	Coastal forest, hill and upper montane forests, some mangroves	WR	Large-scale resort development, airstrip construction, over-collecting of orchids and Rafflesia	Threatened; protected status not enforced
SEA8	Taman Negara	S	4343	75–2188 m	>3000	Timber and fruit trees, rattans, ornamental plants (e.g. orchids), potential medicinal plants	Lowland, hill and montane rain forests, "padang" vegetation, limestone and quartzite vegetation, riparian communities	NP	Logging, hydroelectric dams, lack of buffer zone, some tourist developments	Safe at present, but frequently threatened

Code	Site	Type	Size (km²)	Altitude	Flora	Examples of Useful Plants	Vegetation	Protected Areas	Threats	Assessment
SEA10	Trengganu Hills	S	150	60–920 m	1500	Timber trees, rattans, ornamental plants	Lowland and hill rain forest	FoR, 3 small VJRs	Logging, land clearance for cultivation	At risk
Malaysia (Sabah)										
SEA22	East Sabah lowland/hill dipterocarp forests	V	20,000	0–1298 m	5000–6000	Timber trees (seraya, keruing, kapur), rattans, fruit trees, ornamentals	Tropical evergreen lowland and hill dipterocarp rain forests	4077 km² protected in conservation areas	Conversion to agriculture, human settlement, tree plantations, unsustainable logging	Threatened, some areas seriously threatened; protection of reserves needs strengthening
SEA24	Kinabalu Park	S	754	150–4101 m	4500	Timber trees, ornamental plants (e.g. pitcher plants, orchids)	Mostly montane rain forest, some tropical lowland rain forest, ultramafic forest, alpine vegetation	SP	Clearance for cultivation, illegal logging, mining, tourism	Boundaries threatened
SEA27	North-east Borneo ultramafic flora	V	3500	0–3000 m		Timber trees, potential ornamental plants, nickel- and manganese-tolerant species	Ultramafic facies of tropical lowland evergreen rain forest, lower and upper montane forests	Kinabalu SP only area of formal protection; small areas in Danum Valley Conservation Area; Mt Silam is a protected watershed	Clearance for cultivation and golf course, logging, fire, dam construction, road building	Inadequate coverage of protected areas; threatened
Malaysia (Sarawak)										
SEA34	Lambir Hills	S	69	30–457 m	1500	Timber trees (incl. 69 dipterocarp spp.), mango and durian relatives, rattans	Lowland mixed dipterocarp forest, heath forest, scrub	NP	Logging, clearance for agriculture	Encroachment around boundaries; threatened
Malaysia/Brunei										
SEA33	Gunung Mulu NP/ Medalam PF/ Labi Hills/Bukit Teraja/Ulu Ingei/ Sungei Ingei	S	1521	30–2376 m	3500	Timber trees, especially dipterocarps, fruit and nut trees, sago palm, rattans, medicinal plants	Lowland mixed dipterocarp to montane forests on sandstones, limestones and shales, heath forests, peat swamp forest	Malaysia: NP, PF; Brunei: PF and Conservation Area within FoR	Logging around perimeter of Mulu NP, shifting cultivation along some rivers, potential threat from road construction	Mostly safe at present, but would be at risk if proposed road went ahead; buffer zones to Mulu NP need to be implemented
Papua New Guinea										
SEA89	Bismarck Falls–Mt Wilhelm–Mt Otto-Schrader Range-Mt Hellwig-Gahavisuka	S	9754	250–4499 m	5000–6000	Traditional food and medicinal plants, timbers, fibres, plants of cultural value	Lowland swamp and rain forest, montane forests, alpine vegetation	Proposed NP (Mt Wilhelm), small Provincial Park (Gahavisuka); region proposed as WHS	Population pressure, logging, agriculture, coffee and cardamom plantations	Protected area coverage inadequate; at risk
SEA91	Huon Peninsula (Mt Bangeta-Rawlinson Range; Cromwell Ranges-Sialum Terraces)	S	3415	0–4120 m	4000–5000	Fruit trees, vegetables, fibres, potential timber species	Lowland tropical rain forest to subalpine forest, grasslands, mangroves	No formal protection	Logging, road building	At risk
SEA98	Menyamya-Aseki-Amungwiwa-Bowutu Mts-Lasanga Island	F	6695	0–3278 m	1500–3000	Fruits, vegetables, fibres, building materials, potential timber species	Lowland rain forests to upper montane forests, ultramafic vegetation	NP (21 km²), local support for conservation area in Bowutu Mts, several reserves throughout region needed	Logging, road building, local population pressure	Severely threatened in places, protected area coverage inadequate

Code	Site	Type	Size (km²)	Altitude	Flora	Examples of Useful Plants	Vegetation	Protected Areas	Threats	Assessment
SEA86	Mt Giluwe-Tari Gap-Doma Peaks	S	3346	1000–4368 m	>3000	Traditional food and medicinal plants, fibres, ornamental plants	Montane and subalpine forests, grasslands, alpine communities	Local reserves in Tari Gap area	Logging, road building, clearance for agricultural plantations, dieback of *Nothofagus*	Protected area coverage inadequate; at risk
SEA92	Southern Fly Platform	F	18,644	0–30 m	>2000	Edible palms, traditional food and medicinal plants	Monsoon and savanna vegetation, mangroves, lowland swamps	Wildlife Management Area (5900 km²)	No major threats, some grazing pressure from introduced deer, potential threat from mining	Reasonably safe, but protected area coverage inadequate
Philippines										
SEA51	Batan Islands	S	209	0–1008 m	>500	Timbers, fibres, medicinal plants, food plants	Lowland evergreen to mid-montane rain forest, grassland, secondary vegetation	Proposed as Protected Landscape and 2 "critical watersheds" under National Integrated Protected Area System	Typhoons, clearance for grazing, crops, shifting cultivation, over-collection of forest products	Vulnerable, but growing conservation awareness among local people
SEA52	Mt Apo	S	769	500–2954 m	>800	Ornamental plants (e.g. orchids, aroids, begonias), timber trees	Lowland rain forest (mostly cleared), montane forests, "elfin woodland", scrub, grassland	NP	Construction of geothermal plant, clearance for agriculture, illegal logging, shifting cultivation	Severely threatened
SEA57	Mt Pulog	S	115	1000–2929 m	800	Timber tree provenances (especially pines), ornamental plants	Montane forest, pine forest, grassland	NP	Conversion of forest to vegetable and cut-flower gardens, fire	Threatened
SEA59	Palanan Wilderness Area	S	2168	0–1672 m	1500	Timber trees, rattans	Lowland and hill dipterocarp forest, lower montane forest, ultramafic and limestone forests	Wilderness Area; proposed as NP	Illegal logging, shifting cultivation, over-collection of forest products, potential large-scale logging	Mostly safe at present, but could become severely threatened
SEA60	Palawan	F	14,896	0–2085 m	>2000	Timber trees, rattans, almaciga resin, fruit trees, orchids, nipa palm	Lowland evergreen dipterocarp and semi-deciduous forests, ultramafic and limestone forests, mangroves	BR (11,508 km²), NP (39 km²), various other protected areas covering 3.4% of land area	Logging, mining, shifting cultivation, tourism, over-collection of forest products	Inadequately protected; threatened
SEA61	Sibuyan Island	S	445	0–2052 m	700	Timber trees, almaciga resin, ornamental plants	Lowland dipterocarp forest, montane forest, grassland, mangroves	Proposed NR	Logging, slash and burn agriculture, fire, over-exploitation of rattan	Threatened
SOUTH WEST ASIA AND THE MIDDLE EAST										
Iran										
SWA10	Touran Protected Area BR	S	18,604	690–2281 m	1000	Medicinal herbs, food and fodder plants	Semi-desert, psammophytic and halophytic vegetation	Protected Area divided into WR (5650 km²) and Protected Area (12,954 km²); BR covers 10,000 km²	Overgrazing, fuelwood cutting	At risk; protection measures need to be enforced
Iran/Azerbaijan										
SWA18	Hyrcanian forests	V	50,000	0–2500 m		Timber trees, ornamental plants	Broadleaved deciduous forest, shrubland	Several protected areas; no information available on current status	Clearance for agriculture, logging, invasive plants in some coastal areas	Threatened

Code	Site	Type	Size (km²)	Altitude	Flora	Examples of Useful Plants	Vegetation	Protected Areas	Threats	Assessment
Oman/Yemen										
SWA1	Dhofar Fog Oasis	F	30,000	0–2100 m	900	Frankincense, traditional food, fibre and medicinal plants	Mainly dry deciduous shrubland, montane evergreen shrubland, semi-desert grassland	1 small Bird Sanctuary, otherwise none	Population pressure, overgrazing, cutting of wood for fuel and timber	Severely threatened; IUCN proposals for reserves in Oman not yet implemented
Saudi Arabia/Yemen										
SWA5	Highlands of South-western Arabia	F	70,000	200–3760 m	2000	Qat (stimulant), coffee, barley, wheat and sorghum relatives, myrrh	Deciduous and evergreen bushland and thicket, juniper woodland	Asir NP (4150 km²) in Saudi Arabia, proposed areas in Yemen	Uncontrolled cutting of wood for fuel, timber and charcoal; overgrazing, erosion	No effective protection over most of area, remaining woodland severely threatened
Turkey										
SWA12	Anti-Taurus Mts/ Upper Euphrates	F	60,000	700–3734 m	3200	Walnut, medicinal plants, dyes, cereal crop relatives	Oak forest, steppe, montane steppe	NP (428 km²)	Dam construction, re-afforestation projects, rock climbing leading to erosion	Urgent action needed to protect more of the region; threatened
SWA15	Isaurian, Lycaonian and Cilician Taurus	F	45,120	0–3524 m	>2500	Fig, pomegranate, nuts, dune stabilizers	Cilician fir and cedar forests, scrub, thorn-cushion plants, scree vegetation, dune communities	NP (very small), Bird Sanctuary, GR (30 km²)	Road building, export of wild bulbs, overgrazing, tourism, clearance for agriculture	Many important ecosystems unprotected; threatened
SWA19	N.E. Anatolia	F	33,200	0–3932 m	>2460	Cherries, hazelnuts, timber trees	Coastal humid subtropical forest to fir forest, rhododendron scrub, scree vegetation	1 NP	Illegal logging, clearance for agriculture, export of wild bulbs	None of the Little Caucasus is protected; at risk but not as seriously threatened as other areas
SWA16	S.W. Anatolia	F	75,680	0–3070 m	>3365	Cypress, cedar, oriental sweet gum, medicinal herbs	Mediterranean forest and scrub, cedar forest	Several NPs and Protected Zones, 5 Biogenetic Reserves (264.5 km²)	Tourism, overgrazing by goats, export of wild bulbs	Coastal areas severely threatened
Turkey/Iran/Iraq										
SWA14	Mountains of S.E. Turkey, N.W. Iran, N. Iraq	F	147,332	1400–4168 m	2500	Pears, almonds, hawthorns, gum tragacanth	Oak forest, alpine thorn cushion scrub, montane steppe, alpine grassland, scree vegetation	Iran: NP (4636 km²); Turkey: 1 Forest Recreation Area, 1 GR	Influx of refugees, re-afforestation projects, potential dam and irrigation schemes	Inadequate coverage of reserve system, but threats to flora less severe than elsewhere
Turkey/Syria/Lebanon/Israel/Jordan										
SWA17	Levantine Uplands	F	96,675	0–3083 m	4160	Timber trees (e.g. Cilician fir, cedar of Lebanon), fodder plants, spices, edible oils, root vegetables	Oak, pine, cypress, fir and cedar of Lebanon forests; juniper scrub, maquis, garigue, geophytic communities, alpine vegetation	Israel: 7 small reserves; Syria: 2 Protected Areas (6320 km²)	Logging of remaining forests, clearance for agriculture, tourism, urbanization, over-exploitation of essential oils	Threatened; inadequate coverage of protected areas
Yemen										
SWA4	Socotra	F	3625	0–1519 m	815*	Dragon's blood, other resins, gums, aloes	Semi-desert, dry deciduous shrubland, montane semi-evergreen thicket, secondary grassland	None	Overgrazing by goats, fuelwood cutting, potential new development projects	No protection but traditional practices have prevented serious exploitation so far

36

References

Bedward, M., Pressey, R.L. and Keith, D.A. (1992). A new approach for selecting fully representative reserve networks: addressing efficiency, reserve design and land suitability with an iterative analysis. *Biological Conservation* 62: 115–125.

Cowling, R.M., Gibbs Russell, G.E., Hoffman, M.T and Hilton-Taylor, C. (1989). Patterns of species diversity in Southern Africa. In Huntley, B.J. (ed.), *Biotic diversity in Southern Africa: concepts and conservation*. Oxford University Press, Cape Town. Pp. 19–50.

Davis, S. (1986). Plant Sites Directory. *Threatened Plants Newsletter* No. 16: 17.

Dinerstein, E. and Wikramanayake, E.D. (1993). Beyond "hotspots": how to prioritize investments to conserve biodiversity in the Indo-Pacific region. *Conservation Biology* 7: 530–565.

FAO (1990a). *TFAP Independent Review Report 1990. Appendix 3, section 3.3.* FAO, Rome.

FAO (1990b). *Interim report on Forest Resources Assessment 1990 Project.* Committee on Forestry 10th Session. COFO–90/8(a).

Gentry, A.H. (1990). Herbarium taxonomy versus field knowledge. *Flora Malesiana Bulletin Special Volume* 1: 31–35.

Greuter, W. (1991). Botanical diversity, endemism, rarity, and extinction in the Mediterranean area: an analysis based on the published volumes of Med-Checklist. *Bot. Chron.* 10: 63–79.

Haffer, J. (1969). Speciation in Amazonian forest birds. *Science* 165: 131–137.

Hawkes, J.G. (1983). *The diversity of crop plants.* Harvard University Press, Cambridge, Massachusetts.

Heywood, V.H. (1991). Assessment of the state of the flora of the west Mediterranean basin. In Rejdali, M. and Heywood, V.H. (eds), *Conservation des Ressources Végétales*. Actes Editions, Institut Agronomique et Vétérinaire Hassan II, Rabat. Pp. 9–17.

Heywood, V.H. (1992). Conservation of germplasm of wild species. In Sandlund, O.T., Hindar, K. and Brown, A.H.D. (eds), *Conservation of biodiversity for sustainable development*. Scandinavian University Press, Oslo. Pp. 189–203.

Heywood, V.H. (1994a). The measurement of biodiversity and the politics of implementation. In Forey, P.L., Humphries, C.J. and Vane-Wright, R.I. (eds), *Systematics and conservation evaluation*. Oxford University Press. Pp. 15–22.

Heywood, V.H. (1994b). The Mediterranean flora in the context of world biodiversity. In Olivier, L. and Muracciole, M. (eds), *Connaissance et Conservation de la Flore des Iles de la Méditerranée*. (In press.)

Huntley, B. J. (1988). Conserving and monitoring biotic diversity. Some South African examples. In Wilson, E.O. (ed.), *Biodiversity*. National Academy Press, Washington, D.C. Pp. 248–260.

ICBP (1992). *Putting biodiversity on the map: priority areas for global conservation*. International Council for Bird Preservation, Cambridge, U.K. 90 pp.

IUCN (1980). *World Conservation Strategy: living resource conservation for sustainable development*. IUCN, UNEP and WWF, Gland, Switzerland.

IUCN (1986). *The Plant Sites Red Data Book: a botanists' view of the places that matter*. IUCN, Richmond, U.K. 48 pp.

IUCN (1988). *Centres of Plant Diversity. A guide and strategy for their conservation*. IUCN, Richmond, U.K. 40 pp.

IUCN (1990). *Centres of Plant Diversity. An introduction to the project with guidelines for collaborators*. IUCN, Richmond, U.K. 31 pp.

IUCN/WWF (1984). *The IUCN/WWF Plants Conservation Programme 1984–85*. IUCN/WWF, Gland, Switzerland. 29 pp.

McNeely, J.A. (ed.) (1993). *Parks for life*. Report of the IVth World Congress on National Parks and Protected Areas. IUCN, Gland, Switzerland.

Miller, K.R. (1984). Selecting terrestrial habitats for conservation. In Hall, A.V. (ed.), *Conservation of threatened natural habitats*. South African National Scientific Programmes Report No. 92, Pretoria. Pp. 95–108.

Mittermeier, R.A. and Werner, T.B. (1988). Wealth of plants and animals unites "megadiversity" countries. *Tropicus* 4: 1, 4–5.

Myers, N. (1986). Tackling mass extinctions of species: a great creative challenge. The Horace M. Albright Lectureship in Conservation. University of California, Berkeley.

Myers, N. (1988a). Threatened biotas: "hot-spots" in tropical forests. *Environmentalist* 8: 187–208.

Myers, N. (1988b). Tropical forest and their species. Going, going...? In Wilson, E.O. (ed.), *Biodiversity*. National Academy Press, Washington, D.C. Pp. 28–35.

Myers, N. (1990). The biological challenge extended: extended hot-spots analysis. *Environmentalist* 10: 243–256.

Prescott-Allen, R. and Prescott-Allen, C. (1990). How many plants feed the world? *Conservation Biology* 4: 365–374.

Pressey, R.L., Humphries, C.J., Margules, C.R., Vane-Wright, R.I. and Williams, P.H. (1993). Beyond opportunism: key principles for systematic reserve selection. *Trends in Ecology and Evolution* 8: 124–128.

Quézel, P. (1985). Definition of the Mediterranean region and the origin of its flora. In Gómez Campo, C. (ed.), *Plant conservation in the Mediterranean area*. Junk, Dordrecht.

Raven, P.H. (1987). The scope of the plant conservation problem world-wide. In Bramwell, D., Hamann, O., Heywood, V. and Synge, H. (eds), *Botanic gardens and the World Conservation Strategy*. Academic Press, London. Pp. 10–19.

Raven, P.H. (1990). The politics of preserving biodiversity. *BioScience* 40: 769.

Simberloff, D. (1986). Are we on the verge of a mass extinction in tropical rain forests? In Elliott, D.K. (ed.), *Dynamics of extinction*. Wiley, New York. Pp. 165–180.

Stattersfield, A.J., Crosby, M.J., Long, A.J. and Wege, D.C. (in prep.). *Global directory of Endemic Bird Areas*. BirdLife International, Cambridge, U.K.

Vane-Wright, R.I., Humphries, C.J. and Williams, P.H. (1991). What to protect and the agony of choice. *Biological Conservation* 55: 235–254.

Whitmore, T.C. and Sayer, J.A. (eds) (1992). *Tropical deforestation and species extinction*. IUCN, Gland, Switzerland and Cambridge, U.K., and Chapman and Hall, London. 153 pp.

Williams, P.H., Vane-Wright, R.I. and Humphries, C.J. (1993). Measuring biodiversity for choosing conservation areas. In LaSalle, J. and Gauls, I.D. (eds), *Hymenoptera and biodiversity*. CABI. Pp. 309–328.

Wilson, E.O. (1988). The current state of biodiversity. In Wilson, E.O. (ed.), *Biodiversity*. National Academy Press, Washington, D.C. Pp. 3–18.

Wilson, E.O. (1992). *The diversity of life*. Belknap Press, Harvard University Press, Cambridge, Massachusetts.

WRI, IUCN and UNEP (1992). *Global biodiversity strategy. Guidelines for action to save, study and use the Earth's biotic wealth sustainably and equitably*. WRI, IUCN and UNEP. 244 pp.

REGIONAL OVERVIEW: NORTH AMERICA

Total land area: 21,500,000 km^2.

Population (1994): 287,713,408[a].

Maximum altitude: 6194 m (summit of Mount McKinley, Alaska).

Natural vegetation: Major types include: coniferous forest, in Alaska, Canada, around the Great Lakes and on the upper slopes of the higher eastern and western mountains, but mainly on Pacific Coast Ranges, Cascades, Sierra Nevada, and Rocky Mountains south to Mexico; tundra, across north of continent and on high mountains; temperate deciduous forest, in east from central Minnesota to Quebec, south to Texas; grassland (prairie), in mid-continent; desert, mainly in south-west; subtropical vegetation, including mangroves, in Florida.

Number of vascular plants: c. 20,000 species.

Number of endemic species: c. 4198 (est. 20%).

Number of genera: 2350.

Number of endemic genera: 900 (or nearly endemic).

Number of vascular plant families: 210.

Number of endemic families: 2.

Important plant families: Gramineae, Leguminosae, Pinaceae, Rosaceae, Ericaceae, Cucurbitaceae.

Source: [a] Famighetti (1995); flora statistics based on Thorne (1993); N. Moran 1995, pers. comm.

Introduction

The area covered in this region includes the U.S.A. (excluding Hawaii), Canada and Greenland.

U.S.A.

The U.S.A. (including Hawaii; see Volume 2 for Hawaiian Islands, CPD Site PO6), is one of the largest countries in the world. As a whole, it extends over more than 50 degrees of latitude, from north of the Arctic Circle in northern Alaska to southernmost Hawaii, and over 120 degrees of longitude, from the east coast of Maine to the westernmost part of the Aleutian Islands. The geographic centre of the 48 contiguous states is at about 37°N, 96°W (in Kansas). To the north, these 48 states are bordered by Canada. The east coast extends from Maine to Florida along the Atlantic Ocean, then west along the north coast of the Gulf of Mexico to the Mexican border at the Rio Grande. The west coast extends from Washington through California along the Pacific Ocean to the Mexican border. Continental U.S.A. consists of a vast central plain, with high mountains in the west and hills and low mountains in the east. The highest point is Mount McKinley in Alaska, with an elevation of 6194 m; in the conterminous states it is Mount Whitney in California, with an altitude of 4418 m. The lowest point is in Death Valley, California at 86 m below sea-level.

Alaska, the largest state, is separated from the rest of the U.S.A. by Canada, which borders it to the south and east; it lies about 800 km north-west of the state of Washington. The Arctic Ocean lies to the north, the Chukchi and Bering Seas to the west, and the Pacific Ocean to the south. Alaska is mountainous and dominated by the Brooks Range to the north and the Alaska and Coastal ranges to the south.

Canada

Canada comprises the northern half of the continent (except Alaska) and the arctic islands. It is a confederation of 10 provinces and 2 territories. Covering a land area of 9,976,186 km^2, Canada is the largest country in the Western Hemisphere. It is bounded on the north by the Arctic Ocean, on the south by the U.S.A., on the east by the Atlantic Ocean, and on the west by Alaska and the Pacific Ocean. The highest point is Mount Logan, at an elevation of 6050 m, in southwestern Yukon Territory. The Hudson Bay, an inland sea about 1370 km long by 965 km wide, is a major feature of eastern Canada. Continental Canada consists of interior and arctic plains and lowlands with the immense Canadian Shield to the east and the Cordilleran mountain system to the west. Cold temperate boreal climates dominate over much of Canada, with polar and subpolar climates to the north and cool temperate ones in the south and west. An oceanic climate prevails off the west coast from the Queen Charlotte Islands to Vancouver Island (Brouillet and Whetstone 1993).

Greenland

Greenland is the world's largest island, with an area of 2,175,600 km^2. Most of it lies within the Arctic Circle. It is located east of Canada, separated by Baffin Bay and the Davis Strait. The coast is intersected by deep fjords and is much influenced by sea-ice. The country is largely mountainous (highest elevation 3700 m), but there are some areas of gentle relief. Although the climate is primarily Arctic, it is sub-Arctic in some sheltered valleys in the far south.

Population

The U.S.A. (excluding Hawaii) has a population of 259,542,408 (mid-1994 estimate). Since 1990, the population has increased by about 2.8 million people per year. The population density is 26 people per km². Regarding distribution, 79.5% of the population live in metropolitan areas, more than half of these live in just one of 41 metropolitan areas with populations of at least 1 million. Slightly over half of the population (56.4%) lives in the South and West. Between 1990 and 1993, the population in the West grew by 6.2% and in the South by 4.7%. The five most populous states are California (31.2 million), New York (18.2 million), Texas (18 million), Florida (13.7 million) and Pennsylvania (12 million). The five states with the lowest populations are Delaware (700,000), North Dakota (635,000), Alaska (599,000), Vermont (576,000) and Wyoming (470,000) (Famighetti 1995).

Canada has a population of about 28,114,000 (mid-1994 estimate). The population density is 2 people per km². About 77% of the population is urban. The most populous provinces are Ontario (10 million) and Quebec (nearly 7 million).

Greenland, long a possession of Denmark, became independent in 1979. At that time Greenlandic place names came into official use. The correct official name for Greenland is Kalaallit Nunaat. The island has a population of some 57,000 people (mid-1994 estimate), with a natural rate of increase of 1.3% per annum (IUCN 1992).

Geology

North America is the product of a very long geologic evolution which began early in the Archean (3750–3400 Ma). Plate-tectonic processes have largely dictated this evolution (Bally, Scotese and Ross 1989), including major plate movements during the Mesozoic and Tertiary. According to Bond, Nickerson and Kominz (1984), after the breakup of Pangea and Gondwana during the Middle Jurassic (175 Ma) the continents drifted away and were dispersed during the lower Palaeozoic (62.5–55.5 Ma). It is not known with certainty which continents originally bordered North America. However, the Palaeozoic folded belts of North America are continued in North Africa, Europe and Siberia. Much of North America and Greenland is underlain by Precambrian (3750–570 Ma) crust. The diversity and complexity of North American geology is illustrated in Thornbury (1965), Hunt (1967, 1974), Graf (1987), Bally and Palmer (1989) and Brouillet and Whetstone (1993).

Climate

The climate of North America is extremely diverse. Two geographical features have major effects on North American climates: the western Cordillera, trending north-south; and the Interior Plains to the east. The former constitutes a major obstacle to westerlies and trade winds, while the latter provides an interrupted path for the flow of arctic and tropical air masses. The major climatic regions of North America are: polar and subpolar; cold temperate boreal; cool temperate; warm temperate and subtropical; and tropical. See Brouillet and Whetstone (1993) for a detailed treatment.

Physiographic provinces

The division of North America into physiographic units has been detailed by various authors (Atwood 1940; Hunt 1967, 1974; Brouillet and Whetstone 1993). They include:

❖ **Atlantic Coastal Plain**: a lowland area bordering the ocean into which it slopes and extending from Cape Cod south along the Atlantic Ocean and then west along the Gulf of Mexico into Yucatán. It is an area of generally low relief (altitudes less than 150 m), with some of the major hills and ridges rising only some 60 to 90 m. The surface of the Coastal Plain has been extensively reworked by coastal and fluvial processes over the last 2–3 million years. The Mississippi alluvial plain divides this province into the Atlantic and eastern Gulf coasts, the Mississippi flood plain, and the coastal band of Texas.

❖ **Appalachian Highlands**: extending from the Maritime provinces of eastern Canada south-westward through the eastern U.S.A. It consists of an association of 3000 km long mountain ranges, plateaux and rolling uplands with altitudes ranging from 150 to 2300 m. The Highlands include several physiographic units: Maritime and New England, Adirondack Mountains, Piedmont Plateau, Blue Ridge, Valley and Ridge, and Appalachian Plateau.

❖ **Canadian Shield**: geologically ancient area occupying much of Canada, extending from Baffin Island in the Arctic Ocean, south to northern Wisconsin and east to Labrador. Altitudes range up to 1500 m, but there is little local relief. To the north is found the Arctic Lowlands and Coastal Plain province.

❖ **Greenland Shield**: an ice-cap (2500 km long, 1000 km wide and up to 3 km thick/avg. 300 m) covers 84% of the total land area, and an ice-free zone (generally very narrow, but sometimes broadening to 200–300 km) borders the coast (IUCN 1992).

❖ **St Lawrence Lowlands**: between the Canadian Shield and the Appalachians. Characterized by flat limestone and sandstone outcrops. Elevations are sea level to 150 m.

❖ **Central Lowlands**: gently rolling prairies forming a vast central plain extending from Ohio north-west to central Saskatchewan and south to the Gulf Coastal Plain. Elevations are 150 to 600 m.

❖ **The Great Plains Province**: semi-arid western extension of the Central Lowlands, rising westward from 600 to 1500 m, and forming a broad belt extending from Rio Grande north almost to the Arctic Ocean, and from the Rocky Mountains in the west to the Central Lowlands in the east. Although appearing relatively flat, it slopes to the east.

❖ **Interior Highlands**: between the southern end of the Central Lowlands and the Gulf Coastal Plain, south of glacial limits. It includes the Ozark Plateau, a rolling upland mostly above 300 m, and the Ouachita province, with altitudes of 150–800 m, and the Interior Low Plateau east of the Mississippi River, mostly below 300 m. It is an area composed mostly of sedimentary rock.

❖ **Rocky Mountains:** Alaska, from the Brooks Range southward, to north-western Wyoming, where the mountain ranges are interrupted by a large plateau (not part of the province), and then continuing south into Mexico (the Sierra Madre ranges). The mountains of this province originated in the early Tertiary, primarily through anticlinal folding. The southern mountain ranges have altitudes of 1500–4265 m; the semi-arid Wyoming Basin has altitudes mostly of 1200–3000 m; the middle Rocky Mountains have altitudes mostly of 1525–2135 m and contain some semi-arid inter-montane basins (major ranges were glaciated, and snow and ice still cover most of the mountains above 2400 m); the Northern Rockies are highly irregular granitic mountains with altitudes mostly between 1220 and 2135 m.

❖ **Cordilleran or Intermontane Plateaux:** a disjunctive province comprised of a series of plateaux extending from central Alaska southward into south-central Mexico. The Colorado Plateau section has the highest plateaux in the U.S.A., with its surface rising to 3355 m. It has numerous canyons and is semi-arid. The Basin and Range section, with its blocky mountains separated by desert basins, has altitudes from below sea-level to over 3660 m. However, relief between its mountains and basins is usually no more than 1525 m. The Columbia Plateau is primarily composed of lava flows. Altitudes are mostly below 1525 m. Although semi-arid, two major rivers, the Columbia and the Snake, traverse it.

❖ **The Pacific Border System:** extends from western Alaska to Baja California, fronting the Pacific Coast. The ranges of the Cascade-Sierra Nevada province are north-south trending. The Cascades are a series of volcanoes, while the Sierra Nevada is composed of blocky granite. Altitudes reach 4270 m; western slopes are humid and eastern slopes semi-arid. The Pacific Border province consists of coastal ranges with altitudes mostly below 600 m. These ranges are separated from the Cascade-Sierra Nevada province by troughs 155 m or lower in altitude. The Lower California province is comprised of the northern end of a granitic ridge forming the Baja California Peninsula. In Alaska a belt of mountains forms the South Central Alaska province, leading into the Alaska Peninsula and Aleutian Islands province.

Vegetation

The following simplified classification provides an overview of the major types of vegetation in North America. Major references on the vegetation of the U.S.A. are Gleason and Cronquist (1964), Küchler (1964), Bailey (1976, 1978), Barbour and Billings (1988) and Barbour and Christensen (1993). Regional accounts for the U.S.A. may be found in Oosting (1956), Braun (1964), Waggoner (1975) and Benson (1979). More detailed accounts of Canada's vegetation may be found in such publications as Macoun and Malte (1917), Halliday (1937) and Taylor and Ludwig (1966). For North America as a whole, see Harshberger (1911), Weaver and Clements (1938) and Barbour, Burk and Pitts (1987).

Tundra

Tundra forms a broad band across the top of North America beyond the northern limit of trees, including the ice-free parts of the Arctic islands and coastal margins of Greenland, and above the timberline on the high mountains south into Mexico. Average temperatures are low, and drying winds are ever present. At treeline tundra is in contact with either boreal or subalpine forest.

Forest

Boreal forest, or taiga, is the most widespread and typical vegetation type of Canada. It extends across the continent as a broad band, with its northern boundary running from the Mackenzie delta to the Hudson Bay, and reaches the Atlantic Coast in Newfoundland. From Cook Inlet (Alaska), the southern boundary trends south-eastward to Saskatchewan, eastward to Lake Winnipeg and then into northern New Brunswick. Small pockets occur on the highest mountains of eastern North America, south to the Great Smoky Mountains. Conifers (*Abies, Picea* and their associates) are dominants, but they are variously mixed with *Populus* (aspens) and *Betula* (birches) throughout, and replaced by them in some areas. Although the climate is less severe in this region than in the tundra region, plants here must still tolerate extremes of environment and short growing season.

Pacific coastal coniferous forest occurs from southern Alaska and British Columbia into northern California. It is characterized by high rainfall and low evaporation. *Thuja-Tsuga* (cedar-hemlock) coniferous forest is most extensive in Washington and British Columbia, while *Larix-Pinus* (larch-pine) forest is found in the subalpine zone across the mountains of British Columbia and northern Washington. (Similar vegetation occurs on west-facing slopes in Idaho and north-western Montana as far as the Continental Divide). *Sequoia sempervirens* (redwood) occurs in southern Oregon and northern California.

Subalpine forest (usually at 3000–3500 m) of *Abies-Picea* (fir-spruce) and associated species occupies the upper slopes of high ranges from southern Alaska and adjacent British Columbia through California into Mexico. There is a short growing season, relatively high precipitation (mostly snow), and wide diurnal and seasonal ranges of temperature. A long winter with high winds is characteristic.

Western montane forest, the most extensive of the western forests, stretches from the eastern slopes of the Rocky Mountains and the mountains of western Texas to the Pacific Coast, where it extends from the mountains of western British Columbia to those of Baja California and mainland Mexico. Dominants are *Pinus, Abies* and *Pseudotsuga menziesii* (pine, fir, Douglas fir). The latitudinal range is great, as is the corresponding range of mean annual rainfall (500–1500 mm) and temperature (7–16°C).

Lake forest, with a *Pinus-Tsuga* (pine-hemlock) association, occurs notably around the Great Lakes. Precipitation is 640–1140 mm annually, and temperatures range from -10° to 41°C over the course of the year. The growing season averages four months.

Deciduous forest is dominant in eastern North America and comprises an array of forest types and woody species, with broad transitions to boreal forest to the north, pine and broadleaved evergreen forests to the south, and grassland to the west. It reaches its northern limits in Minnesota, Ontario, and southern Quebec and alternates or is intermixed with boreal forest in New Brunswick and Nova Scotia. From southern Maine it extends southward to central Georgia, southern Louisiana and eastern Texas. The climate of the deciduous forest region is temperate, with definite summer and winter, a growing season varying over the region from less than 150 to more than 280 days, mean annual temperatures of 8–19°C, and mean annual precipitation of 625–2000 mm; frost to some degree is common to the region. The deciduous forest is the best developed in the Great Smoky Mountains.

The South-eastern coastal plain pine and broadleaved forests are upland pine forests that include the northern pine barrens and mesic pine communities, upland hardwood forests with trees such as *Quercus virginiana*, *Magnolia grandiflora* or *Fagus grandifolia* prominent in the overstorey, and tropical hardwood hammocks.

Grassland

Grassland is perhaps the most extensive and varied vegetation formation of the North American continent. Dominating the centre of the continent from south-eastern Alberta, southern Saskatchewan, and south-western Manitoba to Texas and from Indiana to the Rocky Mountains, with significant prairie islands throughout the intermountain west and in California, grassland once covered 50 million ha in Canada (Barbour and Billings 1988) and almost 40% (300 million ha) of the area of the U.S.A. (Küchler 1964). This was upwards of 25% of the landmass north of Mexico. Even today, despite long human activity, grassland still covers more than 125 million ha in the U.S.A. Climate is a major determinant of grassland and grassland type, it being characterized usually by a wet-season/dry-season regime and temperature and precipitation extremes. The three major types of prairies are tall-grass, mixed-grass, and short-grass, the latter being the steppe grasslands of the high plains. All are dominated by grass and grass-like species, but also have a rich assemblage of forbs associated with them. Risser (1985) estimates that North American grasslands contain 7500 plant species. The Great Plains flora contains about 3000 taxa (Great Plains Flora Association 1986).

Grassland comprises several major associations:

❖ The tall-grass prairie is the most mesic of the grasslands and is characterized by the greatest north-south species diversity and the largest number of dominants. It extends from south-eastern Manitoba and Minnesota to eastern Oklahoma and Texas and from Indiana to western Nebraska. Most of the tall-grass prairie is now in cultivation.

❖ The mixed-grass prairie includes both short and tall grasses and associated forbs and forms an ecotone between the tall-grass prairie to the east and the short-grass prairie to the west. In the northern Great Plains, it is found in the western Dakotas, north-eastern Wyoming, eastern Montana and the southern parts of the central Canadian provinces. It is scattered in Nebraska, being interrupted by the sand hills, and occurs south-west from south-western Nebraska and western Kansas through Oklahoma and central Texas.

❖ The short-grass prairie extends from Montana, south-eastern Wyoming, and the Nebraska panhandle southward through eastern Colorado and western Kansas to the high plains of Oklahoma, Texas and New Mexico. Arizona has similar grasslands. Generally less than 50% of the ground is covered by vegetation. Dominants include several species of *Bouteloua*.

❖ Desert grassland extends from the short-grass area in Texas and New Mexico through the south-west into northern Mexico. In pristine times it was dominated by short bunchgrasses but today these have been replaced by desert shrubs.

❖ The original dominants of the California grasslands or Pacific prairie, which once covered over 5 million ha, were cool-season, perennial bunchgrasses (Weaver and Clements 1938). The Pacific prairie has been replaced by farmland and urban sprawl, with non-cultivated areas being dominated by weedy annual species of grasses and forbs.

❖ The Palouse prairie originally extended throughout south-western Canada, eastern Washington and Oregon, and from south-western Idaho into western Montana. Eighty percent of the flora consisted of cool-season grasses. Excessive grazing and cultivation (as well as fire) have resulted in shrub-steppe grassland.

Woodland and scrub

Pine-juniper woodland (*Pinus-Juniperus*), a formation composed of small trees, is essentially south-western, xeric and subtropical. It occurs discontinuously from western Texas through northern Mexico to southern California, extending northward into New Mexico and Colorado to south-western Wyoming and westward through Utah and Nevada to northern California.

Chaparral is characterized by a dense vegetation of broadleaved evergreen, sclerophyllous shrubs. Dominants include *Adenostoma fasciculatum* (chamise), *Arctostaphylos* (manzanita) and *Ceanothus* (California lilac). It occurs in the west largely in the southern Rocky Mountains, mountain ranges in Utah and Arizona, and in the Sierra Nevada, California Coast ranges and Cascade Mountains northward to southern Oregon. Climatically, it occupies xeric zones intermediate between grassland and forest.

Other shrub formations: sagebrush (*Artemisia*) and desert-scrub. The former is found from the Black Hills to southern British Columbia, south-eastern California and northern Arizona. Precipitation (annually 130–500 mm) is lowest in summer and heaviest (as snow) during the four-month winter. Desert scrub, distinguished by its more open structure, is found in the Gila-Sonoran, Colorado, Mohave and Death Valley deserts. It is the most xerophytic type, with annual rainfall of 85–150 mm.

Flora

Due to the basically temperate nature of the flora of North America, the floristic richness of this large area is considerably less than other regions in the Americas, except for the Caribbean. According to Reveal and Pringle's (1993) in-depth history of North American taxonomy and floristics, "the modern history of systematic botany and floristics in North America began when the first Europeans landed and began to collect objects of curiosity". However, native Americans who had arrived millenia earlier had developed their own systems of classification and nomenclature. The origins of the North American flora are diverse. Takhtajan (1986) divides North America north of Mexico into two floristic kingdoms, the Holarctic and Neotropical (see Map 2). The Holarctic is represented by two subkingdoms and four regions with ten provinces. The Neotropical Kingdom is

MAP 2. FLORISTIC REGIONS OF NORTH AMERICA

Source: Regional schema adapted from A.L. Takhtajan (1986), with permission from University of California Press as published in Thorne (1993).

KEY:
A. HOLARCTIC KINGDOM
(Boreal Subkingdom)
1. Circumboreal Region
   1a. Arctic Province
   1b. Canadian Province
3. North American Atlantic Region
   3a. Appalachian Province
   3b. Atlantic and Gulf Coastal Plain
       Province
   3c. North American Prairies
       Province
4. Rocky Mountain Region
   4a. Vancouverian Province
   4b. Rocky Mountain Province
(Madrean or Sonoran Subkingdom)
9. Madrean Region
   9a. Great Basin Province
   9b. Californian Province
   9c. Sonoran Province
       9c1. Mojavean Subprovince
       9c2. Sonoran Subprovince
       9c3. Chihuahuan Subprovince
       9c4. Tamaulipan Subprovince
   9d. Mexican Highlands Province
B. NEOTROPICAL KINGDOM
23. Caribbean Region
   23a. Central American Province
   23b. West Indian Province

represented by the West Indian province of the Caribbean region, which includes the southern third of the Florida peninsula and the adjacent Florida Keys. For more information on phytogeography see Thorne (1993).

The flora of North America north of Mexico consists of approximately 20,000 vascular plant species in approximately 2350 genera (Morin, unpublished data) and 210 families (Thorne 1993). Two families, Leitneriaceae and Limnanthaceae are entirely endemic to the area; another two, Simmondsiaceae and Fouquieriaceae, also extend into Baja California and Sonora in Mexico. About 900 genera are endemic or nearly endemic to this area (some extend into Mexico) (Takhtajan 1986). It is estimated that approximately 4198 (est. 20%) plant species are endemic to North America

(N. Morin 1995, pers. comm.). In the U.S.A., California is the state with the greatest diversity of native plants, with over 5500 species (see Table 12).

Canada has 3269 native vascular plant species (Scoggan 1978–1979). Except for a few relictual areas, most of the flora has reoccupied the land since the Pleistocene glaciations. British Columbia, Ontario and Quebec have the highest number of native vascular plant species (see Table 12).

Greenland has 497 species of vascular plants and only 15 endemic species (Böcher *et al.* 1978). Most of Greenland is covered with permanent ice. The vegetation of the ice-free coastal strip is composed of arctic/alpine and boreal elements. Patchy mats and herbaceous or shrubby heaths are formed, depending on the environmental conditions.

The flora of North America has been documented in many regional floras (e.g. Small 1933; Correll *et al.* 1970; Cronquist *et al.* 1972; Hickman 1993). Recently, two efforts have advanced the knowledge of the North America flora. The Flora of North America Project, a collaboration of botanists in the U.S. and Canada, with its organizational centre at the Missouri Botanical Garden, is producing books and an on-line database on the vascular plants and bryophytes growing outside of cultivation in North America north of Mexico. Two volumes were published in 1993. Kartesz (1994) published a synonymized checklist of the vascular plants of the U.S.A., Canada and Greenland.

Useful plants

North America, as a temperate region, is relatively impoverished with regard to its gene pools and potential to provide major crops, in contrast to tropical areas, which exhibit great diversity. Native North American crops include plants such as cranberries, blueberries, strawberries, pecans and sunflowers (Myers 1983).

In industrial timber harvest, however, North America exceeds all other continents. Some 687 million m³ were harvested annually (five-sixths softwood) from approximately 700 million ha of forest north of Mexico. About three-quarters of the timber harvest comes from the U.S.A. and a quarter from Canada. Table 7 shows annual wood production in the U.S.A. and Canada from 1989–1991 (World Resources Institute 1994).

TABLE 7. ANNUAL PRODUCTION OF WOOD AND WOOD PRODUCTS IN U.S.A. AND CANADA (1989–1991)

Country	Fuelwood and charcoal (10³ m³)	Industrial roundwood (10³ m³)	Sawnwood and panels (10³ m³)	Paper (10³ metric tons)
U.S.A.	90,300	417,900	137,832	71,401
Canada	6834	172,170	61,674	16,527

Source: World Resources Institute (1994).

Table 8, adapted from Schery (1972), shows the principal trees used for timber in the U.S.A. They are listed in order of volume of timber produced, with Douglas fir being the most exploited. It should be noted that western species occur in significantly greater volume than eastern ones.

Prescott-Allen and Prescott-Allen (1986) list and discuss the many North American plants used for lumber and paper products, medicine, food and industrial products, new domesticates, wild genetic resources, pollution control and pest control. They present numerous tables, listing all of the plants used for these purposes in North America, as well as quantitative information concerning their economic values and a valuable bibliography of the literature on the subject (see also Roecklein and Leung 1987).

For North America, including Greenland and northern Mexico, Prescott-Allen and Prescott-Allen (1986) list the average annual value (in U.S.$ million, for the years 1976–1980) of terrestrial wild plant resources produced (or imported) by the U.S.A. (see Table 9).

TABLE 8. PRINCIPAL TIMBER SPECIES IN U.S. FORESTS

Scientific name	Common name	Area
Pseudotsuga menziesii	Douglas fir	primarily in the Pacific Northwest
Tsuga heterophylla	western hemlock	Pacific Northwest
Pinus ponderosa	Ponderosa (western yellow) pine	all western forests
Abies spp.	firs	various western forests
Pinus palustris	longleaf pine	south-eastern states
Pinus elliottii	slash pine	south-eastern states
Pinus echinata	shortleaf pine	south-eastern states
Pinus taeda	loblolly pine	south-eastern states
Quercus rubra and other *Quercus* spp.	red oaks	mainly central and upper south
Picea engelmannii and other *Picea* spp.	Engelmann and related spruces	Rocky Mountain belt
Quercus alba and other *Quercus* spp.	white oaks	mainly central and upper south
Picea sitchensis	Sitka spruce	Pacific Northwest
Pinus contorta	lodgepole pine	Rocky Mountain belt
Pinus monticola *Pinus lambertiana*	western white pine and sugar pine	northern Rocky Mountains and Pacific Southwest
Sequoia sempervirens	redwood	California
Carya spp.	hickories	central and upper south
Liquidambar styraciflua	sweetgum	central and upper south
Nyssa sylvatica *Nyssa aquatica*	black tupelo and cotton gum	south-central
Acer spp.	hard maples	deciduous hardwood belt
Pinus strobus *Pinus resinosa*	eastern white pine and red pine	northern coniferous belt

TABLE 9. AVERAGE ANNUAL VALUE OF TERRESTRIAL WILD PLANT RESOURCES FOR NORTH AMERICA

Biome	Average annual value (US$ million) (1976–1980) Subtotals	Biome total	Biome	Average annual value (US$ million) (1976–1980) Subtotals	Biome total
1. Boreal forest and taiga		1405.1	**6. Mediterranean type**		502.3
a. timber from Canada	1402.1		timber from U.S.A.	502.3	
b. food and industrial products			**7. Prairie**		65.0
blueberry (*Vaccinium* spp.)	2.1		a. timber from U.S.A.	51.9	
wild rice (*Zizania aquatica*)	0.9		b. medicine: American ginseng	0.4	
2. Temperate mixed forest		2193.9	c. food and industrial products: pecan	12.7	
a. timber from U.S.A.	470.0		d. new domesticates:		
b. timber from Canada	1688.9		wheatgrass (*Agropyron* spp.; part of		
c. medicine:			total US$1.9 million with 2 other biomes)		
American ginseng			**8. Steppe**		0.3
(*Panax quinquefolius*)	0.4		a. food and industrial products:		
d. food and industrial products:			oregano (*Origanum* spp.) and *Lippia* spp.	0.3	
sugar maple (*Acer saccharum*)	21.4		b. new domesticates:		
blueberry	10.6		wheatgrass (*Agropyron* spp.; part of		
wild rice	2.3		total US$1.9 million with 2 other biomes)		
cedar leaf oil from			**9. Desert**		3.3
Thuja occidentalis (white cedar)	0.3		a. food and industrial products:		
e. new domestications:			tampico, made from *Agave lechequilla*		
highbush blueberry (half of US$44.8 million)			and *Yucca carnerosa*	2.4	
and wild rice (US$2.8 million)			and candelilla wax, from several		
3. Temperate broadleaved forest		1377.6	species of *Euphorbia* and		
a. timber from U.S.A.	1353.4		*Pedilanthus*/source – primarily Mexico	0.6	
b. timber from Canada	15.0		oregano	0.3	
c. medicine: American ginseng	7.7		**10. Mountains**		2719.4
d. food and industrial:			a. timber from U.S.A.	1230.5	
sugar maple products	1.5		b. timber from Canada	1485.5	
e. highbush blueberry			c. food and industrial products:		
(other half of US$44.8 million)			tampico	2.4	
4. Temperate maritime forest		8353.7	candelilla wax	0.7	
a. timber from U.S.A.	6638.3		oregano	0.3	
b. timber from Canada	1715.4		d. new domesticates:		
5. Subtropical forest		6176.7	wheatgrass (*Agropyron* spp.; part of total		
a. timber from U.S.A.	6167.5		US$1.9 million with 2 other biomes)		
b. food and industrial products:			**Total**		**22,797.3**
pecan (*Carya illinoensis*)	9.2				

Source: Prescott-Allen and Prescott-Allen (1986).

TABLE 10. U.S. TERRESTRIAL WILD PLANT FOOD PRODUCTION RESOURCES WITH AVERAGE ANNUAL VALUE OF MORE THAN US$1 MILLION

Plant	Product	% product from wild	Producer states and %
Carya illinoensis	pecan	18	Texas 50, Louisiana 26, Oklahoma 16, Mississippi 5, Arkansas 3
Vaccinium spp.	blueberry	15	Maine 99; Oregon, Michigan, New Jersey, Washington 1
Acer saccharum	maple syrup and sugar	100	Vermont 36, New York 25, Wisconsin 9, Michigan 8, New Hampshire and Ohio 7, Others 15
Zizania aquatica	wild rice	32	Minnesota 95, Wisconsin (the rest)

Source: Prescott-Allen and Prescott-Allen (1986).

With regard to domestic production of food in the U.S.A. from wild or wild and cultivated sources, Table 10 shows plants which have an average annual value of more than $1 million (Prescott-Allen and Prescott-Allen 1986).

Long before other cultures arrived on the North American continent, American Indians made much use of native plants for food, medicine, dyes, shelter, tools, charms and in the decorative arts. According to Kavasch (1984), almost 80% of all medicinal plants of North America are currently harvested east of the Mississippi River. Numerous publications have been written on economic plant use by native Americans, such as Scully (1970), Moerman (1986) and Duke (1992).

TABLE 11. ECONOMIC PLANTS NATIVE TO NORTH AMERICA

Medicinal plants

Apocynum androsaemifolium	bitteroot	Impatiens pallida	pale touch-me-not
Apocynum cannabinum	Indian hemp	Lithocarpus densiflora	tan oak
Artemisia tridentata	sagebrush	Monarda didyma	Oswego tea or bee balm
Asclepias syriaca	common milkweed	Monarda fistulosa	wild bergamot
Asclepias tuberosa	butterfly weed	Monarda punctata	horsemint
Castanea pumila	chinquapin	Montia perfoliata	winter purslane
Crambe abyssinica	crambe	Panax quinquefolius	American ginseng
Fraxinus spp.	ashes	Quercus alba	white oak
Gossypium hirsutum	American upland cotton	Quercus lobata	California white oak
Hamamelis virginiana	witch hazel	Salix spp.	willows
Ilex vomitoria	Yaupon holly	Sambucus canadensis	American elderberry
Impatiens capensis	jewelweed	Veratrum viride	American hellebore

Food plants

Acer saccharum	sugar maple (sap)	Parthenocissus quinquefolia	Virginia creeper
Amelanchier canadensis	juneberry (fruit)		(stalk and sap next to bark)
Amphicarpaea bracteata	wild bean; hog-peanut (root)	Populus tremuloides	aspen (sap)
Asarum canadense	wild ginger (root)	Prunus americana	chokecherry (twigs)
Asimina triloba	wooly pawpaw (fruit	Prunus serotina	wild cherry (twigs)
Asclepias syriaca	common milkweed (flowers)	Prunus virginiana	chokecherry (twigs)
Aster spp.	asters (leaves)	Pycnanthemum virginianum	mountain mint (flowers and buds)
Cornus canadensis	bunchberry (fruit)	Quercus macrocarpa	bur oak (fruit – acorns)
Corylus americana	American hazelnut (nut)	Ribes triste	red currant (fruit)
Crataegus spp.	hawthorns; thornapples (fruit)	Ribes spp.	wild currants (fruit)
Diospyros virginiana	common persimmon (fruit)	Rubus idaeus var. strigosus	red raspberry (fruit)
Fragaria virginiana	wild strawberry (fruit)	Rubus pensilvanicus	Pennsylvania blackberry (fruit)
Gaultheria hispidula	creeping snowberry (leaves)	and other Rubus spp.	blackberries (fruit)
Gaultheria procumbens	wintergreen (leaves)	Sambucus canadensis	American elderberry (fruit)
Gaylussacia baccata	huckleberry (fruit)	Tilia americana	basswood (sap next to bark)
Juglans cinerea	butternut (nut)	Tsuga canadensis	hemlock (leaves)
Juglans lindsii	California black walnut (nut)	Vaccinium angustifolium	wild blueberry (fruit)
Ledum groenlandicum	Labrador tea (leaves)	Vaccinium macrocarpon	cranberry (fruit)
Lycopus asper	rough bugleweed (root)	Viburnum edule	squashberry (fruit)
Oryzopsis cuspidata	Indian rice	Vitis labrusca	fox grape (fruit)
Oryzopsis hymenoides	Indian rice	Zizania palustris	Indian rice (fruit)

Dyes

Acer spp.	maples	Lithospermum carolinense	puccoon
Betula papyrifera	white or paper birch	Prunus americana	chokecherry
Coptis trifolia	goldthread	Quercus macrocarpa	bur oak
Cornus stolonifera	red-osier dogwood	Rhus glabra	smooth sumac
Corylus americana	American hazelnut	Sanguinaria canadensis	bloodroot
Juglans cinerea	butternut	Tsuga canadensis	hemlock
Juniperus virginiana	cedar		

Oil plants

Gaultheria procumbens	wintergreen	Limnanthes bakeri	Baker's meadowfoam
Lesquerella fendleri	Fendler bladderpod	Limnanthes douglasii	Douglas' meadowfoam

Forage, pasture and feed grain plants

Agropyron spicatum	bluebench wheatgrass	Elymus cinereus	Basin wildrye
Andropogon gerardii	big bluestem	Elymus condensatus	giant wildrye
Bouteloua eriopoda	black grama	Sporobolus airoides	alkali sacaton
Bouteloua gracilis	blue grama	Stipa comata	needle and thread grass
Buchloe dactyloides	buffalograss	Stipa viridula	green needlegrass
Elymus canadensis	Canada wildrye		

Source: Densmore (1974); Kavasch (1984); Roecklein and Leung (1987).

Some of the most important indigenous economic plants are listed in Table 11.

Factors causing loss of biodiversity

The primary cause of loss of plant diversity in North America is the habitat destruction resulting, ultimately, from the continuously increasing human population. Even publicly owned, "protected" areas (federal, state, local) are constantly being subjected to various uses that are detrimental to the conservation of biodiversity. Residential, industrial and agricultural development, with the consequent alteration and loss of habitat, is the primary threat associated with rising population. Additional threats include invasive plant and animal species, fires and ecologically unsound fire suppression, logging, mining, pollution (including pesticides), recreational land use, alteration of drainages, filling in of wetlands, road building, maintenance of utility rights-of-way, plant collecting for the domestic and international trade (especially cacti, carnivorous plants, ginseng and orchids) and insufficient or faulty management practices.

Impact on North American vegetation formations

Boreal forests
According to Elliott-Fisk (1988), human disruption of this forest type generally results in occupancy by successional communities similar to those associated with fire. Sometimes disturbances alter the substrate. The water budget is easily altered by mining, hydropower development and destruction of surface vegetation. In turn, this can lead to destruction of permafrost and rapid landscape degradation in northern regions. In certain cases, trees may fail to regenerate following logging, and a more or less permanent tundra subclimax may result.

Tundra
Inuit and Eskimos use plants of the tundra to some extent. Some areas are now being affected by livestock grazing. Exploitation of mineral resources, especially oil, may negatively impact the tundra (Barbour and Christensen 1993).

Rocky Mountain forests
As stated by Peet (1988), nearly all these forests are in some stage of recovery from significant prior disturbance by fire, wind, insect outbreaks, extreme weather, volcanism, ungulate browsing and events associated directly with people. Fire has historically been the most significant natural disturbance. In prehistoric times, the activities of the native Americans involved periodic burning, but the early Europeans caused a major increase in the frequency of fires. Fire suppression practices have reduced fire destruction significantly in this century, but the ecological consequences of this have not always been positive; long-term, unnecessary fire control may result in loss of biodiversity.

Pacific Northwest forests
Clear-cutting is the most common disturbance in these forests, with consequent erosion and nutrient loss as well as forest destruction. Also, rates and paths of succession are altered by planting and the elimination of non-arboreal species (Franklin 1988). Less than 10% of old-growth rain forests survive, scattered in small fragments throughout the Pacific Northwest.

California upland forest and woodlands
Air pollutants, primarily those from automobiles, play a significant role here. The ozone, hydrocarbons, nitrogen oxides and sulphur dioxide are major phytotoxicants, the last two also being components of acid rain (Barbour 1988).

Intermountain deserts, shrub steppes and woodlands
As stated by West (1988), livestock grazing over the last 100 years has been a significant cause of vegetation change in the Great Basin sagebrush, as well as in other vegetation types. One result is loss of dominance by perennial grasses. Undesirable plants, especially introduced annuals, have increased in abundance. Two examples of these are *Halogeton glomeratus* and *Ceratocephala testiculata*, both poisonous to livestock. The conversion from perennial to annual dominants seems to be a permanent trend, and succession is leading to tree dominance in many areas. Also, in recent times, erosion has become more severe because of reduced soil and litter cover.

Warm deserts
Activities associated with increasing urban development have been affecting the desert vegetation, often with long-term deleterious effects. Cattle grazing and recreation, especially involving off-road vehicles, also have serious consequences for the vegetation (altering its composition by, for example, reducing the diversity of annuals) and the soil. Among the desert plants being destroyed are such especially valuable plants as *Simmondsia chinensis* (jojoba; its oil can substitute for whale oil) and *Parthenium argentatum* (guayule; a rubber substitute can be produced from it) (McKell 1988).

Grasslands
The grasslands that once dominated central North America have been greatly reduced through extensive overgrazing, agricultural development and fire suppression. Fire has played a significant role in the development and maintenance of grasslands by, for example, suppressing the encroachment of trees and shrubs and by reducing competition from invaders such as *Poa pratensis* and *Bromus inermis*. It also improves the palatability and nutritional value of forage. Although they harbour relatively few endemics and lack high species diversity, the grasslands are very important for their value as watershed, forage and habitat for large numbers of domestic and wild animals and are the home of species of realized, or potential food crops (Sims 1988). Grasslands are the principal agricultural lands of North America; only 1% of the original prairie ecosystem is still intact (Barbour and Christensen 1993).

Deciduous forests
Greller (1988) states that, in general, the distribution of tree species in the east has not changed since pre-colonial times, but tree stature was significantly larger in the past. Much of the forest that once occupied 95% of the land has been cut over one or more times or has been completely destroyed by agriculture, urbanization and many other types of development and exploitation. Human use and activity continue to pose many threats to the present-day forests. The Carolinian forest in Ontario, Canada has been largely eliminated (McAllister 1994).

South-eastern coastal plain
Prehistorically, Indians are believed to have occupied this land around 12,000 BP. From 8–10,000 BP, land management by the Archaic Indian culture in the Coastal Plain was intensive, involving the use of fire. The first European colonists in temperate North America settled in the Coastal Plain. Their impacts varied with their location and cultural traditions, but all quickly learned to use fire to manipulate the landscape (Christensen 1988). Present-day threats are much the same as everywhere else in North America – the consequences of relentless development.

Wetlands
Wetlands are not a specific vegetation formation but they are of considerable conservation concern. Only half of the original wetlands in the U.S.A. remain, and 1300 km² of wetlands are being lost through agriculture and construction every year (N. Morin 1995, pers. comm.). Draining of wetlands is one of the most severe threats to biodiversity on the continent, endangering species and threatening the existence of some vegetation types.

TABLE 12. ESTIMATED NUMBER OF NATIVE AND GLOBALLY RARE VASCULAR SPECIES IN NORTH AMERICA

State/province	Number of native species [a]	Number rare (G1–G3) [b]	% rare per state/province
U.S.A.			
Alabama	2870	273	9.5
Alaska	1300	125	9.6
Arizona	3510	616	17.5
Arkansas	2140	86	4.0
California	5510	1762	32.0
Colorado	2630	313	11.9
Connecticut	1820	46	2.5
Delaware	1610	52	3.2
Florida	3080	452	14.7
Georgia	2980	333	11.2
Idaho	2360	187	7.9
Illinois	2220	53	2.4
Indiana	2020	49	2.4
Iowa	1600	24	1.5
Kansas	1730	34	2.0
Kentucky	2000	84	4.2
Louisiana	2400	108	4.5
Maine	1680	54	3.2
Maryland	2210	80	3.6
Massachusetts	2000	65	3.3
Michigan	2040	58	2.8
Minnesota	1790	34	1.9
Mississippi	2300	116	5.0
Missouri	2100	74	3.5
Montana	2160	115	5.3
Nebraska	1520	21	1.4
Nevada	2850	495	17.4
New Hampshire	1610	37	2.3
New Jersey	2130	71	3.3
New Mexico	3290	399	12.1
New York	2360	90	3.8
North Carolina	2770	261	9.4
North Dakota	1170	18	1.5
Ohio	2010	42	2.1
Oklahoma	2320	77	3.3
Oregon	3130	367	11.7
Pennsylvania	2210	66	3.0
Rhode Island	1430	27	1.9
South Carolina	2500	204	8.2
South Dakota	1460	14	1.0
Tennessee	2350	161	6.9
Texas	4500	439	9.8
Utah	2960	516	17.4
Vermont	1620	43	2.7
Virginia	2580	151	5.9
Washington	2380	198	8.3
West Virginia	1840	65	3.5
Wisconsin	1860	43	2.3
Wyoming	2240	175	7.8
CANADA			
Alberta	1710	50	2.9
British Columbia	2360	138	5.8
Franklin (N.W.T.)	340	18	5.3
Keewatin (N.W.T.)	520	12	2.3
Labrador	710	16	2.3
Mackenzie (N.W.T)	1030	45	4.4
Manitoba	1510	27	1.8
New Brunswick	1110	27	2.4
Newfoundland (insular)	1000	38	3.8
Nova Scotia	1150	35	3.0
Ontario	2330	60	2.6
Prince Edward Island	720	11	1.5
Quebec	1990	78	3.9
Saskatchewan	1420	24	1.7
Yukon	1090	78	7.2

Note: List includes native full species only.

Sources:

[a] J.T. Kartesz, 1995, pers. comm.
[b] The Network of Natural Heritage Programs and Conservation Data Centers and The Nature Conservancy.

Rare plant species

About 10% of the species in the flora of the U.S.A. (Elias 1977) and 31% of the flora of Canada (Argus and Pryer 1990) are considered to be endangered or rare. About 90 plant species became extinct in North America between 1800 and 1950 (Yatskievych and Spellenberg 1993). A survey conducted by the Center for Plant Conservation suggests that as many as an additional 475 continental U.S.A. taxa may become extinct by the year 1998. This is five times as many extinctions in one-third the time compared with the 1800–1950 period.

Figures on the number of native vascular plants and number of rare (G1–G3) species per state have been calculated by the Network of Natural Heritage Programs and Conservation Data Centers and The Nature Conservancy (Table 12). Data on numbers of native plant species were obtained from a phytogeographical summary in preparation by Kartesz (1995, pers. comm.). The states with the highest number of globally threatened and rare species are California (1762); Arizona (616); Utah (516); Nevada (495); Texas (439); New Mexico (399); and Oregon (367). In Canada, British Columbia (138); Quebec (78) and the Yukon (78) are the areas with the largest number of globally threatened species.

In 1990 Argus and Pryer published a study on the rare vascular plants in Canada. Since nearly 70% of the Canadian population is located in a narrow band along the border with the contiguous U.S.A., the natural vegetation has been drastically reduced. This border area cuts through the northern edge of four major floristic provinces (Argus and Pryer 1990). As a consequence, a large number of plant species that have a small range in Canada are regarded as nationally rare. The concept of rarity is based on a phytogeographical concept, not a threat concept. The purpose of the list was not only to promote the conservation of individual species, but to emphasize the need to preserve what is left of their habitats and ecosystems. Over 1000 taxa, or 31% of the native flora, are considered rare in Canada (G. Argus 1995, pers. comm.). Table 13 gives an estimate of the number of rare species in each Canadian province/territory. British Columbia and Ontario have the highest number of rare plants, with 426 and 355 respectively. In British Columbia, many of the rare species occur on the Queen Charlotte Islands (CPD Site NA9). In Ontario, the

TABLE 13. ESTIMATED NUMBER OF RARE SPECIES IN CANADA

Region	Rare endemics	Rare plants [a]	Rare in Canada [b]
Alberta	16	307	125
British Columbia	47	816	426
Manitoba	1	291	52
New Brunswick	5	207	25
Newfoundland	17	271	40
Northwest Territories	18	206	62
Nova Scotia	3	219	45
Ontario	15	600	355
Prince Edward Island	3	191	6
Quebec	22	408	106
Saskatchewan	11	300	77
Yukon	28	313	91

Notes:

[a] Rare in province.
[b] Rare in province that are rare throughout Canada.

Source: G. Argus 1995, pers. comm.

Carolinian Forest of the Eastern Deciduous Forest province is a floristically diverse vegetation that not only has a limited distribution in Canada, but is severely threatened by agriculture and urbanization.

Conservation

Federal, state, and provincial legislation exists to protect plant diversity and endangered species to some degree in both Canada and the U.S.A. Of these two countries, the U.S.A. has the best developed national legislation for plant protection. The federal Endangered Species Act, passed in 1973, provides legal protection for listed taxa on federal lands. Approximately one-third of the land (2,994,688 km²) in the U.S.A. is owned and administered, with at least nominal protection, by federal government agencies. The vast majority of this land lies in the western third of the nation (IUCN 1992).

Federal agencies concerned with conservation and land management in the U.S.A. include the United States Fish and Wildlife Service (responsible for enforcing the Endangered Species Act and for habitat conservation, especially of wetlands), the United States Forest Service, the Bureau of Land Management and the National Park Service. There is no comprehensive inventory of U.S.A. native plant species or community diversity on which to base management and land-use decisions. Therefore, in 1993, the National Biological Survey was established to undertake research aimed at biological and ecosystem monitoring.

At the state level, conservation activities are undertaken by a network of state Natural Heritage Programs and by various conservation, natural resource, fish and game, forestry or other departments, and by native plant societies. While somewhat more than half of the states have rare plant laws, virtually all have rare plant lists. Some major private organizations in the U.S.A. concerned with conservation are The Nature Conservancy, Natural Resources Defense Council and the Center for Plant Conservation.

In 1988, Natural Heritage inventories were initiated in Canada (starting in Quebec) by The Nature Conservancy of Canada, as had previously been done in the U.S.A. Today, conservation data centres exist in Quebec, Ontario, Manitoba, Saskatchewan and British Columbia. The Canadian Government has jurisdiction in the territories and provides information and coordination for the country as a whole, but the provinces initiate and have the final responsibility for their own floras. Manitoba, Ontario, Quebec and New Brunswick have Endangered Species Acts. The provinces of Alberta, British Columbia, New Brunswick, Saskatchewan and Quebec have Ecological Reserves or Wilderness Areas, with conservation as one of their purposes, but protection is limited. The provinces have provincial parks or Crown Lands that provide some species and habitat protection. The federal government is establishing ecological reserves in the Northwest Territories. Under the federal Green Plan, Parks Canada has taken a more active role in the conservation of biodiversity in its holdings and has set up policies to this effect (Argus 1977a, b; see also Maini and Carlisle 1974; Argus and Pryer 1990).

In 1992, Canada's federal, provincial and territorial ministers of the environment, parks, wildlife and forestry signed "A Statement of Commitment to Complete Canada's Network of Protected Areas" (Canadian Museum of Nature 1994). This plan proposes to expand the National Park system with the ultimate goal of preserving about 2.8% of Canada's lands.

Greenland has no national legislation specifically designed to protect plants. The country, however, contains the largest National Park in the world, North East Greenland National Park, encompassing 700,000 km² (IUCN 1992), and by its size alone protects significant portions of the country's biological diversity.

A review of the protected areas system and conservation legislation throughout the region is given in IUCN (1992, 1994). Table 14 provides a summary of protected area coverage.

TABLE 14. NATIONAL PROTECTION SYSTEMS* IN NORTH AMERICA

Country	Number of protected sites >10 km²	Combined area protected (km²)	% of country protected
Canada	426	825,454	8.3
Greenland	2	982,500	45.2
U.S.A.	1355	1,041,724	11.4

Note:

* The figures presented here refer only to those areas which cover **10 km² or more** and are in IUCN Management Categories I–V. It is very important to note that, in some cases, figures include sites which are only partially protected. The combined area protected and percentage of country protected are summations of individual areas and do not take into account areas which may have overlapping boundaries. In some cases, therefore, the figures imply a greater area protected than is really the case. A further consideration is that many protected areas have been established primarily for their fauna and do not necessarily protect the best areas for plants.

Source: IUCN (1994).

Centres of plant diversity and endemism

Sites have been selected for Data Sheet treatment in the U.S.A. and adjacent parts of Mexico over those in Canada and Greenland because of their relatively higher levels of plant diversity and endemism. Some of the important centres of plant diversity and endemism, although lying predominantly in the U.S.A., extend into Canada. This explains in part the relatively larger number of endemics in southern as compared to northern Canada.

CANADA

The Wisconsin glaciation destroyed much of the Canadian flora, resulting, overall, in a relatively young flora. There are, however, a number of local centres of plant diversity and endemism, as listed below. Most of these centres probably represent refugia. Some are also active centres of evolution which need protection. Centres of evolution are important to the study of evolutionary biology and biogeography and are gene reservoirs for plant and animal breeding. They are defined as geographical areas in which biological populations are actually or potentially undergoing evolutionary change, especially where this

MAP 3. CENTRES OF PLANT DIVERSITY AND ENDEMISM: NORTH AMERICA
The map shows the location of the CPD Data Sheet sites for North America.

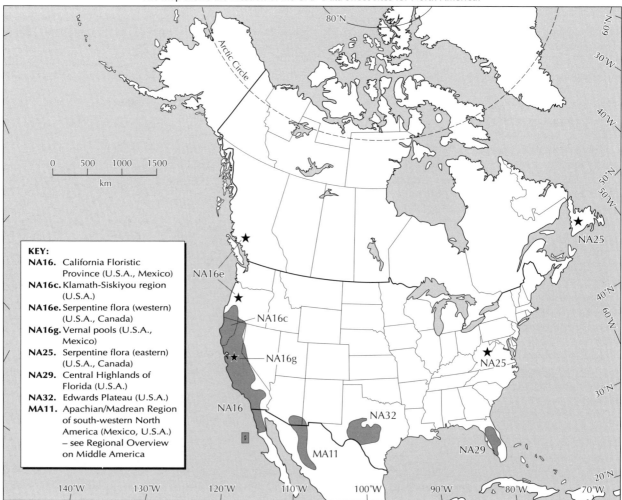

KEY:
NA16. California Floristic Province (U.S.A., Mexico)
NA16c. Klamath-Siskiyou region (U.S.A.)
NA16e. Serpentine flora (western) (U.S.A., Canada)
NA16g. Vernal pools (U.S.A., Mexico)
NA25. Serpentine flora (eastern) (U.S.A., Canada)
NA29. Central Highlands of Florida (U.S.A.)
NA32. Edwards Plateau (U.S.A.)
MA11. Apachian/Madrean Region of south-western North America (Mexico, U.S.A.) – see Regional Overview on Middle America

TABLE 15. NORTH AMERICAN SITES IDENTIFIED AS CENTRES OF PLANT DIVERSITY AND ENDEMISM

The list of sites is arranged below according to the sequence adopted in the Regional Overview. Sites selected for Data Sheet treatment appear in bold.

NA1. Ellesmere Island (Canada)
NA2. The Arctic Islands (Canada)
NA3. Baffin Island (Canada)
NA4. Central Yukon Plateau (Canada)
NA5. Mackenzie Mountains (Canada)
NA6. Lake Athabasca, Saskatchewan, sand-dune region (Canada)
NA7. Western Newfoundland (Canada)
NA8. Torngat Mountains (Canada)
NA9. The Queen Charlotte Islands (Canada)
NA10. Rocky Mountains (Canada)
NA11. Gulf of St Lawrence (Canada)
NA12. Owyhee region, Oregon to Idaho (U.S.A.)
NA13. Wenatchee Mountains, Washington (U.S.A.)
NA14. Olympic Mountains, Washington (U.S.A.)
NA15. Southern British Columbia, primarily the Queen Charlotte Islands (Canada)
NA16. **California Floristic Province** (U.S.A., Mexico)
NA16a. Big Bear Valley and Baldwin Lake area (San Bernandino National Forest) (U.S.A.)
NA16b. Guadalupe Island (Mexico)
NA16c. **Klamath-Siskiyou region** (U.S.A.)
NA16d. Santa Lucia range and Monterey Peninsula (U.S.A.)
NA16e. **Serpentine flora (western)** (U.S.A., Canada)
NA16f. Kern Plateau and southern Sierra Nevada (U.S.A.)
NA16g. **Vernal pools** (U.S.A., Mexico)
NA17. Inyo region of California and Nevada (U.S.A.)
NA18. White Mountains (U.S.A.)

NA19. Death Valley (U.S.A.)
NA20. Colorado Plateau division (U.S.A.)
NA20a. Canyon Lands section (U.S.A.)
NA20b. Utah Plateaux section (U.S.A.)
NA21. Central Rocky Mountains (U.S.A.)
NA21a. Uinta Mountains (U.S.A.)
NA21b. Wasatch Mountains (U.S.A.)
NA22. Shale barrens (U.S.A.)
NA23. Great Smoky Mountains (U.S.A.)
NA24. Piedmont rock outcrops in Georgia, South Carolina and Alabama (U.S.A.)
NA25. **Serpentine flora (eastern)** (U.S.A., Canada)
NA26. Cedar glades (U.S.A.)
NA27. Mesic savannas (U.S.A.)
NA28. Apalachicola River drainage of north-western Florida (panhandle) and adjacent Georgia (U.S.A.)
NA29. **Central Highlands of Florida** (U.S.A.)
NA30. Miami Ridge rocklands (U.S.A.)
NA31. Atlantic Coastal Plain (U.S.A.)
NA32. **Edwards Plateau** (U.S.A.)
NA33. Trans-Pecos region, including Big Bend National Park (U.S.A.)
NA34. Gulf or coastal prairie (U.S.A.)
NA35. Chihuahuan Desert (U.S.A., Mexico)
NA36. Great Basin Desert (U.S.A.)
NA37. Mojave Desert (U.S.A.)
NA38. Sonoran Desert, including Baja California (U.S.A., Mexico)

change is of greater than average rate or extent (Argus and McNeill 1974). They also can contain high endemism or morphologically and physiologically distinct populations in a restricted area. Areas in British Columbia and Ontario merit protection due to their high plant diversity and greatest number of both rare and endemic taxa.

Centres of plant diversity and endemism

Centres of plant diversity and endemism in Canada include:

NA1. Ellesmere Island

NA2. The Arctic Islands (Brassard 1971)

NA3. Baffin Island (Brassard 1971)

NA4. Central Yukon Plateau (Cody 1971)

NA5. Mackenzie Mountains (Cody 1971)

NA6. Lake Athabasca, Saskatchewan, sand-dune region (Argus and McNeill 1974)

NA7. Western Newfoundland (Fernald 1926; Argus and McNeill 1974)

NA8. Torngat Mountains (Abbe 1936)

NA9. The Queen Charlotte Islands (Calder and Taylor 1968)

NA10. Rocky Mountains (Packer 1971)

NA11. Gulf of St Lawrence (Morisset 1971; Argus and McNeill 1974)

PACIFIC NORTHWEST

The Pacific Northwest of North America should be considered a single floristic unit, due largely to the uniform moist climate (Brooks 1987). It encompasses Oregon, Washington and northern California in the U.S.A. and southern British Columbia in Canada, and corresponds roughly to the Vancouverian province of Takhtajan (1986). Whittaker (1960, 1961), Franklin and Dyrness (1988) and Kruckeberg (1969, 1984a–c, 1992) have described the geology, floristics and ecology of the region. Detling (1968) covers the historical background of the flora. The Pacific Northwest is rich in centres of plant endemism, with some of these centres also extending into Idaho and California (Siddall, Chambers and Wagner 1979). It is also known for its areas of serpentine-associated endemism. These may be found at: Bralorne and the upper Tulameen River in British Columbia; Double Eagle Lakes and the Sumas, Twin Sisters and Wenatchee Mountains in Washington; and Fields Creek, Baldy Mountain and the Klamath/Siskiyou Mountains in Oregon. In total, perhaps 500 to 600 species are restricted or nearly restricted to the Pacific Northwest (Thorne 1993).

Centres of plant diversity and endemism

U.S.A.

NA12. Owyhee region, Oregon to Idaho

Leslie Gulch (in Oregon) is a drainage of about 90 km² at the centre of the Owyhee region with a high number of unusual, rare and endemic plants. It is an area from which normal vegetation is excluded. This entire region has a complex stratigraphy of volcanic rocks and sediments, with ashtuff as the typical substrate. It is thought likely that many endemics that occur here are pioneers adapted to recently exposed habitats and might be competitively excluded from more normal substrates.

NA13. Wenatchee Mountains, Washington

The Wenatchee Mountains form a series of spurs along the eastern side of the central Washington Cascades. Portions of these mountains provided refugia along the southern margins of the Pleistocene continental ice sheet. Many mountain tops and some valleys not only escaped alpine glaciation, but were presumably sufficiently isolated for speciation to occur. Additionally, these mountains have the most extensive ultramafic (primarily serpentine and peridotite) rock outcrops in Washington (Sheehan and Schuller 1981). In combination, these factors resulted in the highest concentration of rare and endemic plant taxa in the state.

NA14. Olympic Mountains, Washington

The Olympic Mountains in north-west Washington are isolated from other mountains of comparable size. Although most of this range was glaciated during the Pleistocene, the few peaks and ridges that escaped glaciation provided refugia for some constituents of older floras. These former refugia are responsible, at least partly, for relatively high levels of endemism (Sheehan and Schuller 1981). About 20 species are restricted to the Olympic Peninsula (Jones 1936).

Canada

NA15. Southern British Columbia, primarily the Queen Charlotte Islands

The vegetation largely belongs to the Cordilleran Forest province of the U.S.A., which reaches a short distance into Canada. Of the 47 taxa endemic to British Columbia, only 14 are restricted to the province itself. These occur primarily on the Queen Charlotte Islands (and also on north-western Vancouver Island) of southern British Columbia, which provided refugia during the Pleistocene and thus possess a relictual flora (Argus and Pryer 1990).

CALIFORNIA

For additional information on the flora of California and the California Floristic Province, see Abrams (1925, 1926), Campbell and Wiggins (1947), Howell (1957), Stebbins and Major (1965), Ornduff (1974), Raven (1977), Raven and Axelrod (1978), Barbour (1988), Barbour and Christensen (1993), Hickman (1993), Thorne (1993) and Yatskievych and Spellenberg (1993).

Centres of plant diversity and endemism

U.S.A., Mexico

NA16. California Floristic Province

– see Data Sheet.

U.S.A.

NA16a. Big Bear Valley and Baldwin Lake area (San Bernardino National Forest)

– see Data Sheet on California Floristic Province. This area contains relicts of a mesic coniferous and deciduous forest which was widespread throughout the Northern Hemisphere in the middle of the Tertiary period (Axelrod 1958).

Mexico

NA16b. Guadalupe Island

– see Data Sheet on California Floristic Province. Located 265 km off the mainland of Baja California, Guadalupe Island is known for its high degree of endemism. The flora consists of 164 native species, of which 31 are endemic, including two endemic genera (*Baeriopsis* and the extinct *Hesperelaea*) (Raven and Axelrod 1978).

U.S.A.

NA16c. Klamath-Siskiyou region

– see Data Sheet.

NA16d. Santa Lucia range and Monterey Peninsula

– see Data Sheet on California Floristic Province. This area contains high endemism.

U.S.A., Canada

NA16e. Serpentine flora (western)

– see Data Sheet on North American serpentine flora.

U.S.A.

NA16f. Kern Plateau and southern Sierra Nevada

– see Data Sheet on California Floristic Province.

U.S.A., Mexico

NA16g. Vernal Pools

– see Data Sheet.

U.S.A.

NA17. Inyo region of California and Nevada

This is considered both the floristically richest region in transmontane California and the most important centre of endemism east of the mountains in California. Of the 454 species occurring only within this region in California, at least 199 are not found elsewhere outside of the state. Also, there are at least 43 endemic species within the Inyo region and over 25 other endemic species that range a short distance into Nevada (Raven 1977).

NA18. White Mountains

The White Mountains of California and Nevada are of exceptional botanical interest due to the presence of the oldest known living plant (*Pinus longaeva*). The vegetation is distinctly Great Basin in character consisting of desert scrub, pinyon woodland, subalpine forest and alpine zones. Within these zones nearly 1100 vascular plant taxa are known to occur (Morefield 1988). The desert ranges such as those in the southern White Mountains of eastern California have edaphic endemics confined almost entirely to limestone soils derived from dolomite (Lloyd and Mitchell 1973).

NA19. Death Valley

The extremely dry Death Valley, with temperatures frequently approaching 52°C, is part of the Mojave Desert. The Mojave covers the southern tip of Nevada and the eastern edge of California. Its unique appearance comes from the characteristic regular spacing of the creosote bush. Death Valley itself contains a rich flora, while endemism on limestone in the deep canyons flanking the valley is another important feature. In addition, many genera occur in California only within this region.

INTERMOUNTAIN REGION

This area, as described in Cronquist *et al.* (1972), covers the dryland region between the Sierra Nevada on the west and the Rocky Mountains on the east, and between the moister country of the Pacific Northwest on the north and warmer drylands to the south. It includes all of Utah, most of Nevada (see Tidestrom 1923) and portions of Idaho, Oregon, Arizona and California. It includes the hydrographic Great Basin and a considerable portion of the geologic Colorado Plateau. The region coincides to a large extent with the Great Basin Floristic province, as defined by Gleason and Cronquist (1964). Although the Intermountain region is mountainous, its mountains are discontinuous and surrounded by desert.

Although the region occurs in the arid to semi-arid part of the U.S.A., the flora shows great diversity, because the many mountain ranges contain habitats humid enough to support woodland and forest vegetation. Thus, the flora ranges from alpine to xeric to subtropic. The region as a whole has at least 215 endemic species and 58 endemic varieties and subspecies (Cronquist *et al.* 1972). For more detailed information on the Intermountain region and its endemic taxa see also Tidestrom (1923), Graham (1937), Kearney and Peebles (1960), Franklin and Dyrness (1988), Welsh, Atwood and Reveal (1975), Welsh (1979), Mozingo and Williams (1980), Steele *et al.* (1981), Washington Natural Heritage Program (1990), Skinner and Pavlik (1994), Rutman (1990), Barbour and Christensen (1993), Thorne (1993) and Yatskievych and Spellenberg (1993).

The area encompassed in the Intermountain Floristic region includes all or parts of four major physiographic provinces or subprovinces: Colorado Plateau, Central Rocky Mountains, Great Basin (forms nearly three-quarters of the total Intermountain region) and the Snake River Plains. Each of the following divisions has at least one section containing a number of endemic taxa: Great Basin division (the Calcareous Mountains section is the richest area in this division for plant endemism); Wasatch Mountains division (only one section, by the same name, has some endemic taxa); the Uinta Mountains division (comprising only the Uinta Mountain section); and the Colorado Plateau division (Canyon Lands and Utah Plateaux sections).

Centres of plant diversity and endemism

U.S.A.

NA20. Colorado Plateau division

A large system of plateaux and intervening valleys encompassing an area of around 1,341,620 km². This system contains some of the world's most scenic river canyons. The areas of high endemism in this division include the following two sections. These three areas, in Utah alone, have 120 endemics (Shultz 1993).

NA20a. Canyon Lands section

This section lies almost entirely in south-eastern Utah, covering an area of 61,515 km². It is characterized by a broad desert plain broken by deep canyons. The Colorado River cuts diagonally across the section in a deep, rugged canyon. It is the richest section in the Intermountain region for endemic taxa (approximately 90 species).

NA20b. Utah Plateaux section

This area consists of nine fault-blocks or individual plateaux, lined up in three series separated by flat-bottomed valleys,

occupying 31,080 km². It is the second richest section in the Intermountain region for endemic taxa (approximately 40 species).

NA21. Central Rocky Mountains

The branch of the Laramide uplift that trends southward from Idaho into Utah and Colorado and comprises the central Rocky Mountain province. Considerable work has been done recently at the University of Wyoming to survey the flora of Wyoming and the adjacent Middle Rocky Mountain region. The following ecological regions have been found to have high levels of endemism: Absaroka/Owl Creek (9); Bighorn/Pryor Mountains (6); Colorado Front Range/Laramie Range/Medecine Bow Mountains (4); Uinta Mountains foothills (9); Wyoming Basin (17).

NA21a. Uinta Mountains

This range, located mainly in north-eastern Utah along the Wyoming boundary, is the largest east-west trending mountain range in the Western Hemisphere (Fenneman 1938). The Uinta Mountains section covers an area of approximately 12,175 km². Many peaks are over 3660 m in elevation and are composed mainly of quartzites. The upper portions of the range were extensively glaciated. The portion of this basin in Utah has 25 endemics.

NA21b. Wasatch Mountains section

For nearly 320 km, the Wasatch Mountain range, with peaks up to 3660 m in elevation, trends north-south in Utah and south-eastern Idaho. Various rivers and creeks have dissected the steep western escarpment of these mountains into a number of U-shaped canyons. Also several areas along the range were glaciated. Fifteen taxa are known only from this section.

APPALACHIAN MOUNTAIN REGION

The Appalachian mountain region, composed primarily of crystalline rocks, is considered one of the most important areas of plant endemism in the eastern U.S.A. It is part of the Piedmont Upland geological province, a belt of metamorphic rock extending from Trenton, New Jersey to Alabama (Brooks 1987). This, in turn, forms part of a very long belt of ultramafics extending through New York state, the Gaspé Peninsula, Newfoundland and the northern tip of Quebec in the Ungava.

The Southern Appalachians are unique in that they represent the oldest land mass in eastern North America thought to have been relatively unaffected by marine waters or continental glaciation. They are also thought to have served as "...a source of inoculum for the development of the floras of the south-eastern coastal plains and the revegetation of the glaciated terrain to the north" (Sharp 1970). Holt (1970) covers the distributional history of the flora. Wood (1970) states that "of the approximately 100 genera endemic to eastern North America, 52 are represented in the southern Appalachian region...". Harper (1947) listed 188 species of vascular plants found in the southern Appalachians. About half of them are definitely narrow Appalachian endemics, while some of the others range westward to the sandstone of the Cumberland Plateau in Kentucky and Tennessee. Many of the taxa on this list are endemic to shale barrens or to limestone or dolomitic valleys and cliffs.

Centres of plant diversity and endemism

U.S.A.

NA22. Shale barrens

The term was first used by Steele (1911) to designate a peculiar type of plant habitat found in the mid-Appalachians and extending northward along the Virginia-West Virginia border, across Maryland and into central Pennsylvania. These barrens developed on areas with outcrops of certain types of hard shaly rock. The barrens are particularly well-developed from central Pennsylvania to south-western Virginia and adjacent West Virginia, where they occur at elevations of 305–610 m (Keener 1970). Their characteristics include a generally southern exposure, usually a steep slope on a low hill, a stream undercutting the base and sparse vegetation growing on a mantle of thin weather-resistant rock.

A scrubby growth of pines, oaks and other woody plants, with herbaceous species scattered beneath, typically occupies these barrens (Henry 1954). The peculiar flora includes a number of endemic and near-endemic taxa (Keener 1970).

See Platt (1951) for a complete list of shale barren plants, including 97 angiosperm taxa. See also Henry (1954) for another treatment of floristic aspects of the endemic flora. Localities of the Virginia-West Virginia barrens, primarily along a narrow strip of shale outcropping along both sides of the states' border (where a considerable number of endemics is found), are covered by Core (1940); for locations of the Pennsylvania barrens, see Henry (1954).

NA23. Great Smoky Mountains

Several peaks with elevations of over 1830 m occur in the Blue Ridge and Smoky Mountains of North Carolina and easternmost Tennessee. Many of the numerous southern Appalachian endemics and near-endemics have patchy distributions and may occur at fairly high altitudes (Reed 1986). This area is now well protected as the Great Smoky Mountains National Park, established in 1934 and covering 2092 km².

NA24. Piedmont rock outcrops in Georgia, South Carolina and Alabama

Outcrops of ancient granitic rock in the Piedmont, known as "flat-rocks" or "cedar rocks," are "without parallel in eastern North America" and support an unusual flora with one-third (17) of the outcrops' 44 most characteristic taxa being endemic, an unusually high percentage for such a small area (Braun 1955). The escarpment region, owing to its location at the edge of the Piedmont, has served as a pathway for migrations during the numerous periods of climatic change that have occurred in the Tertiary and Pleistocene. Its gorges and their surrounding areas now have a significant number of southern Appalachian endemics.

U.S.A., Canada

NA25. Serpentine flora (eastern)

– see Data Sheet on North American serpentine flora (NA16e and NA25).

INTERIOR LOW PLATEAUX PROVINCE

The Interior Low Plateaux province includes southernmost Illinois and Indiana, south-easternmost Ohio, central and western Kentucky and central Tennessee and extends into northernmost Alabama. The general geological structure is similar to the Appalachian Plateau, except that altitudes are lower and relief is more subdued. The northernmost portion of the province underwent glaciation. There are two major elevations or domes, one centred in Kentucky and the other in Tennessee. Owing to the erosion of these domes, hills have been formed surrounding them. These are composed largely of sandstone and chert-rich layers of limestone. The centre of the Nashville dome is eroded into a basin. In a number of areas, limestone is exposed or, where moderately flat, is covered with only a very thin soil. In the Nashville area, such places are called barrens, while elsewhere in the province they are called glades. Rivers tend to be subterranean in limestone areas.

Centres of plant diversity and endemism

U.S.A.

NA26. Cedar glades

One of the most unusual areas of rock-outcrop vegetation in the eastern U.S.A. is the cedar glades. Red cedar, *Juniperus virginiana*, is one of the most common woody species growing on the deeper soils in and around these openings. Neo- and palaeoendemics, as well as disjuncts and eastern species of western genera, are among the plant taxa characteristic of the glades. Although originally known from and predominantly found in central Tennessee, the glades have also been described from a number of localities in the eastern deciduous forest. These areas include eastern Tennessee, northern Alabama, north-western Georgia, the Ozark region of Missouri, and extreme south-western Virginia and Kentucky. As determined by Baskin and Baskin (1988) and others, the cedar glade flora includes some 30 endemic and near endemic taxa. Baskin and Baskin state that the unusually high number of endemics, combined with the presence of disjuncts, is indicative of the sites having served as centres of speciation for a long time.

SOUTH-EASTERN COASTAL PLAIN

The south-eastern Coastal Plain, of low relief and dipping seaward, extends some 3540 km from Cape Cod south to Florida and west to Texas and for an additional 1609 km along the Gulf coast of Mexico (Shimer 1972). The underwater seaward extension forms the continental shelf. Overall, the plain is composed of poorly consolidated sediments of Cretaceous to Recent age. Elevations of hills and ridges reach only 60–90 m. The most recently formed features, which lie along the coast, include cliffs, sand-dunes, beaches, sandspits and sandbars. The area from Cape Cod to Long Island was glaciated. The vegetation can be classified as the southern mixed hardwood region of the deciduous forest. The flora includes several hundred endemic species, several endemic genera, and one endemic family, the monotypic Leitneriaceae (Thorne 1993). The states with important centres of plant diversity and endemism are North and South Carolina, Florida and Texas.

Florida contains a large number of endemic plant species due to three main factors: (1) most of Florida is a peninsula, relatively isolated from other land masses; (2) it projects southward from the continent, and there is no nearby land mass with climates similar to those of Florida's southern portion (except the Bahamas, which are relatively small and have very different soil from most of Florida); and (3) Florida has more sandy soils than any other state, as well as considerable areas of limestone in its central and southern portions (Harper 1914). Also, past geological factors have played a major role in determining the centres (Woodson 1947; Thorne 1949; James 1961). For example, the existence of an island refugium, which could have functioned as a centre of dispersal during the Tertiary, has been postulated. Howard (1954) indicated the areas within which these endemics are most concentrated. Muller *et al.* (1989) have provided current data on the centres and their endemic taxa. They list some 235 endemic and an additional 40 near-endemic vascular plant taxa.

Texas displays wide ranges in rainfall, soil type and elevation, resulting in great differences in habitat and vegetation from area to area. The extreme variation and isolation that occurs within Texas has produced both a large, very diverse flora and a high degree of endemism. Some 5480 known taxa of flowering plants and ferns occur here, including half of the grass species indigenous to the U.S.A. Furthermore, Texas is the only state in which are found, not only Rocky Mountain and eastern plants, especially oaks and pines, but also, in the state's southernmost section, many subtropical species.

Of the flowering plant and fern taxa, 464 are considered endemic to Texas, about 8.3% of the total (Correll and Johnston 1970). Important centres of plant endemism occur in five of the state's ten major vegetational areas: East Texas pinewoods; coastal prairie (Gulf prairies and marshes); Rio Grande plains (South Texas plains); Edwards Plateau; and Trans-Pecos region. The last two areas are especially important centres of plant diversity and endemism.

Centres of plant diversity and endemism

U.S.A.

NA27. Mesic savannas of North and South Carolina

In these savannas, graminoid diversity and herb diversity are generally very high. Common graminoids are *Sporobolus* *teretifolius*, *Muhlenbergia expansa*, *Ctenium aromaticum*, *Andropogon* spp. and *Rhynchospora plumosa*. Other indicative species are *Lycopodium carolinianum*, *Lachnocaulon anceps* and *Xyris smalliana*. However, the most distinctive taxa occurring here are insectivorous plants, including *Drosera* spp. (sundews), *Pinguicula* spp. (butterworts), *Sarracenia* spp. (pitcher plants) and the endemic Venus flytrap, *Dionaea muscipula*. The Venus flytrap is found only in savannas of the outer Coastal Plain of south-eastern North Carolina and at a few localities in South Carolina. Serious threats include human alteration and destruction of habitat, fire suppression and collecting and trade in insectivorous plants (Christensen 1988).

NA28. Apalachicola River drainage of north-western Florida (panhandle) and adjacent Georgia

The east side of the Apalachicola River is one of the classic areas of both endemics and rare plants, such as *Torreya taxifolia*, with its nearest relative in California, and the associated herb *Croomia pauciflora*, a member of a family (Croomiaceae) not elsewhere known outside of Asia. The flora contains many endemics and Tertiary relicts. The endemics occur primarily in the cool wet flatlands (savannas, seepage slopes and flatwoods).

NA29. Central highlands of Florida

– see Data Sheet.

NA30. Miami Ridge rocklands

Found in Dade and Monroe counties of Florida, the Miami Ridge rocklands contain a large number of endemic taxa, and support an endemic community. Dade County is considered to have a larger number of endemic plants than any other Florida county, with estimates ranging from 55–65 taxa (Muller *et al.* 1989).

NA31. Atlantic Coastal Plain

The area from south-eastern North Carolina south to north-eastern Florida between the coast and St John's River is an important centre of plant diversity. Many now feel that coastal North Carolina-Florida should be considered a separate region since numerous endemic plants occur in its habitats, including coastal hammocks, dunes, shell mounds, marshes and flatwoods. There are 73 species endemic to northern Florida.

NA32. Edwards Plateau

– see Data Sheet.

NA33. Trans-Pecos region, including Big Bend National Park

The Trans-Pecos region, an area of about 7,695,000 ha west of the Pecos River, consists of mountains and arid valleys. Some consider the extreme south-eastern portion to be part of the Edwards Plateau. The primary representation of the Chihuahuan Desert in the U.S.A. lies in this region (see

NA35). About one out of every twelve species in the Texas flora occurs in the Trans-Pecos and nowhere else in Texas (Correll and Johnston 1970). Owing to the variety of available habitats, the flora is both rich and diverse; the primary vegetation types include creosote-tarbush desert shrub, grama grassland, yucca and juniper savannas, pinyon pine, oak forests and (more locally) ponderosa pine.

NA34. Gulf or coastal prairie

This is a nearly level plain of 3,847,500 ha lying along the coast of Texas. With an elevation of less than 45 m, it has numerous sluggish rivers, creeks, bayous and sloughs. It includes low flat woodlands, freshwater marshes, salt meadows and salt marshes. The natural vegetation consists of tall-grass prairie or post oak savanna. Cattle graze much of the marsh, with rangelands and farms in the uplands. Owing to human disturbance, much of the area has been invaded by certain trees, shrubs and cacti. This area contains many threatened plants.

DESERT AREAS

As stated in MacMahon (1979), low elevation "warm" deserts occupy the south-western portion of the U.S.A. and the northern quarter of Mexico: the Mojave and Chihuahuan deserts are warm temperate, while the Sonoran Desert is classified as subtropical. In contrast, the Great Basin Desert to the north, with 60% of its precipitation in the form of snow, is known as a cold desert. Except for a piece of the Great Basin Desert that overlaps onto the Columbia Plateau province and Baja California, considered a province in itself, all of the North American deserts are contained in the physiographic area referred to as the Basin and Range province.

Desert elevations range from -86 m in the Mojave to 1525 m in the southern Chihuahuan in Mexico; latitudes range from 22°N to 36.5°N. Consequently, these deserts collectively possess a wide range of vegetation types. In fact, most American deserts have a much greater diversity of species than do many temperate forests. All of these deserts possess endemic species. For general coverage of North American deserts, see MacMahon (1979) and Bender (1982).

Table 16 shows the approximate areas of North American deserts.

TABLE 16. APPROXIMATE AREAS OF NORTH AMERICAN DESERTS

Unit	Area (km²)	Portion of North American desert area (%)
Great Basin Desert	409,000	32.0
Sonoran Desert	275,000	21.5
Mojave Desert	140,000	11.0
Chihuahuan Desert	453,000	35.5
Total desert	**1,277,000**	**100.0**
Warm deserts	868,000	68.0
Cold deserts	409,000	32.0

Source: MacMahon (1979).

Centres of plant diversity and endemism

U.S.A., Mexico

NA35. Chihuahuan Desert

The Chihuahuan Desert is of generally high elevations, with most sites between 1100 and 1500 m. It is the largest North American desert with a total land area of 450,000 km². Its lowest elevations, along the Rio Grande River, are near 400 m, while its highest elevations, in the south, reach up to 2000 m. Much of this desert is dominated by limestone or gypsum which often have endemic plants (MacMahon 1988). The mean annual temperature is 18.6°C, and precipitation ranges from 150 to 400 mm. It consists of three regions: the northernmost, Trans-Pecos region (see CPD site NA33), encompassing approximately 40% of the desert and including all of the sections in the U.S.A. and more than half of the desert areas of Chihuahua (Mexico); the Mapimian, or middle region, including Coahuila and parts of Chihuahua and Durango in Mexico; and the Saladan, or most southern region, including primarily the states of Zacatecas and San Luis Potosí in Mexico. The vegetation may be divided into eight primary subdivisions of desert scrub and woodlands: Chihuahuan Desert scrub, lechuguilla scrub, yucca woodland, *Prosopis-Atriplex* scrub, alkali scrub, gypsophilous scrub, cactus scrub and riparian woodland (Henrickson and Johnston 1986). Grasses are conspicuous in most areas of this desert, especially at higher elevations; common genera are *Sporobolus*, *Muhlenbergia* and *Bouteloua*. There are also many species of cacti; *Opuntia phaeacantha* and *Echinocactus horizonthalonius* cover large areas in dense stands.

U.S.A.

NA36. Great Basin Desert

This desert covers most of Nevada and Utah and south-western Colorado to northern New Mexico and Arizona. The part that extends north of 42°N latitude has been described as shrub steppe and is considered part of the Intermountain region. Its latitude and elevation (mostly above 1000 m) result in a temperate climate. Because of these factors, it is not a desert in the generally accepted sense, and should really be classified as a semi-desert with a shrub-steppe landscape. *Artemisia* spp. (sagebrush) are the most characteristic and widespread dominants constituting more than 70% of the cover (West 1988).

NA37. Mojave Desert

North of the Sonoran Desert is the Mojave, the smallest of the four deserts. It includes portions of southern California, southern Nevada, south-western Utah and north-western Arizona. The vegetation differs from that of the other deserts in that it is generally dominated by low-growing, usually widely spaced perennial shrubs, representing few species. Also, although cacti are present, they are generally of low stature. Five general vegetation types are found: (1)

creosote bush in which the most common dominants are *Larrea tridentata* and *Ambrosia dumosa*, covering 70% of the desert; (2) shadscale (*Atriplex confertifolia*); (3) saltbush (*Atriplex polycarpa*); (4) blackbrush (*Coleogyne ramosissima*); and (5) Joshua tree (the usual dominant is *Yucca brevifolia*, with a total cover of 10.15%), associated with variable quantities of perennials at different elevations (Vasek and Barbour 1988). About 25% of this desert's plant species are endemic (nearly 80% of the approximately 250 annuals). Stebbins and Major (1965) listed 138 endemics.

U.S.A., Mexico

NA38. Sonoran Desert, including Baja California

The Sonoran Desert covers part of south-western Arizona and northern Mexico, as well as the Baja California peninsula. It is considered perhaps the richest of all North American deserts. Kearney *et al.* (1960) list endemic plant species found in the portion of the desert lying in Arizona and Bowers (1981) has an excellent annotated bibliography of local floras. See also Shreve and Wiggins (1964) for more information on the vegetation of the Sonoran Desert. (Part of this area is included in CPD Data Sheet MA11, Apachian/Madrean region of south-western North America.)

References

Abbe, E.C. (1936). Botanical results of the Grenfell-Forbes northern Labrador expedition 1931. *Rhodora* 38: 102–161.

Abrams, L. (1925). The origin and geographical affinities of the flora of California. *Ecology* 6: 1–6.

Abrams, L. (1926). *Endemism and its significance in the California flora*. Proceedings of the International Congress of Plant Science 2. Pp. 1520–1523.

Argus, G.W. (1977a). The conservation of Canadian rare and endangered plants. In Mosquin, T. and Suchal, C. (eds), *Canada's threatened species and habitats*. Proceedings of the Symposium on Canada's Threatened Species and Habitats. Canadian Nature Federation, Ottawa. Pp. 139–143.

Argus, G.W. (1977b). Canada. In Prance, G.T. and Elias, T.S. (eds), *Extinction is forever: threatened and endangered species of plants in the Americas and their significance in ecosystems today and in the future*. New York Botanical Garden, New York. Pp. 17–27.

Argus, G.W. and McNeill, J. (1974). Conservation of evolutionary centres in Canada. In Maini, J.S. and Carlisle, A. (eds), *Conservation in Canada, a conspectus*. Department of the Environment, Canadian Forestry Service Publication 1340. Pp. 131–141.

Argus, G.W. and Pryer, K.M. (1990). *Rare vascular plants in Canada: our natural heritage*. Rare and Endangered Plants Project. Canadian Museum of Nature, Ottawa. 191 pp.

Atwood, W.W. (1940). *The physiographic provinces of North America*. Ginn, Boston. 535 pp.

Axelrod, D.I. (1958). Evolution of the Madro-Tertiary geoflora. *Bot. Rev.* 24: 433–509.

Bailey, R.G. (1976). *Ecoregions of the United States*. United States Forest Service, Ogden. (Map)

Bailey, R.G. (1978). *Description of the ecoregions of the United States* (Manual). United States Forest Service, Ogden. 77 pp.

Bally, A.W. and Palmer, A.R. (eds) (1989). *The geology of North America – an overview*. The Geographical Society of America, Boulder. 619 pp.

Bally, A.W., Scotese, C.R. and Ross, M.I. (1989). North America: plate-tectonic setting and tectonic elements. In Bally, A.W. and Palmer. A.R. (eds), *The geology of North America – an overview*. The Geological Society of America, Boulder. Pp. 1–17.

Barbour, M.G. (1988). California upland forests and woodlands. In Barbour, M.G. and Billings, W.D. (eds), *North American terrestrial vegetation*. Cambridge University Press, Cambridge. Pp. 131–164.

Barbour, M.G. and Billings, W.D. (eds) (1988). *North American terrestrial vegetation*. Cambridge University Press, Cambridge. 434 pp.

Barbour, M.G., Burk, J.H. and Pitts, W.D. (1987). *Terrestrial plant ecology*. (2nd ed.) Benjamin Cummings Publishing Company, Menlo Park. 634 pp.

Barbour, M.G. and Christensen, N.L. (1993). Vegetation. In Flora of North America Editorial Committee (eds), *Flora of North America north of Mexico. Vol. 1. Introduction*. Oxford University Press, New York. Pp. 97–131.

Barbour, M.G. and Major, J. (eds) (1988). *Terrestrial vegetation of California*. California Native Plant Society, Sacramento. 1002 pp.

Baskin, J.M. and Baskin, C. (1988). Endemism in rock outcrop plant communities of the unglaciated eastern United States: an evaluation of the roles of edaphic, genetic and light factors. *J. Biogeography* 15: 829–840.

Bender, G.L. (ed.) (1982). *Reference handbook on the deserts of North America*. Greenwood Press, Westport. 594 pp.

Benson, L. (1979). *Plant classification*. 2nd ed. Heath, Boston. 901 pp.

Böcher, T.W., Fredskild, B., Holmen, K. and Jacobsen, K. (1978). *Grönlands Flora*. 3rd ed. Hasse, Köbenhaven. 326 pp. (trans. T.T. Elkington and M.C. Lewis, from Danish 2nd ed.).

Bond, G.C., Nickerson, P.A. and Kominz, M.A. (1984). Breakup of a supercontinent between 650 Ma and 555 Ma; new evidence and implications for continental histories. *Earth and Planetary Science Letters* 70: 325–345.

Bowers, J.E. (1981). Local floras of Arizona: an annotated bibliography. *Madroño* 28: 193–209.

Brassard, G.R. (1971). Endemism in the flora of the Canadian high Arctic. *Naturaliste Canadien* 98: 159–166.

Braun, E.L. (1955). Phytogeography of the unglaciated eastern United States and its interpretation. *Bot. Rev.* 21: 297–375.

Braun, E.L. (1964). *Deciduous forests of eastern North America*. Hafner, New York. 596 pp.

Brooks, R.R. (1987). *Serpentine vegetation, a multi-disciplinary approach*. Ecology, Phytogeography and Physiology Series. Vol. I. Dioscorides Press, Portland. 454 pp.

Brouillet, L. and Whetstone, R.D. (1993). Climate and physiography. In Flora of North America Editorial Committee (eds), *Flora of North America north of Mexico. Vol. 1. Introduction*. Oxford University Press, New York. Pp. 15–46.

Calder, J.A. and Taylor, R.L. (1968). Flora of the Queen Charlotte Islands. Part 1. Systematics of the vascular plants. *Can. Dep. Agric. Res. Branch Monograph* 4(1): 1–659 pp.

Campbell, D.H. and Wiggins, I.L. (1947). Origins of the flora of California. *Stanford University Publications in Biological Science* 10: 1–20.

Canadian Museum of Nature (1994). *Canadian biodiversity strategy. Canada's response to the Convention on Biological Diversity*. Report of the Biodiversity Working Group, Ottawa. 61 pp.

Christensen, N.L. (1988). Vegetation of the southeastern Coastal Plain. In Barbour, M.G. and Billings, W.D. (eds), *North American terrestrial vegetation*. Cambridge University Press, Cambridge. Pp. 317–364.

Cody, W.J. (1971). A phytogeographic study of the floras of the continental Northwest Territories and Yukon. *Naturaliste Canadien* 98: 145–158.

Core, E.L. (1940). The shale barren flora of West Virginia. *Proc. W. Va. Acad. Sci.* 14: 27–36.

Correll, D.S., Johnston, M.C. and collaborators (1970). *Manual of the vascular plants of Texas*. Texas Research Foundation, Renner. 1881 pp.

Cronquist, A., Holmgren, A.H., Holmgren, N.H. and Reveal, J.L. (1972). *Intermountain flora: vascular plants of the Intermountain West, U.S.A. Vol. 1. Geological and botanical history of the region, its plant geography and a glossary. The vascular cryptogams and the gymnosperms*. Hafner Publishing Co., New York. 270 pp.

Densmore, F. (1974). *How Indians use wild plants – for food, medicine and crafts*. Dover Publications, New York. 397 pp.

Detling, L.E. (1968). Historical background of the flora of the Pacific Northwest. *University of Oregon Museum of Natural History Bulletin* 13: 1–57.

Dick-Peddie, W.A. (1993). *New Mexico vegetation – past, present, and future*. University of New Mexico Press, Albuquerque. 244 pp.

Duke, J.A. (1992). *Handbook of edible weeds*. CRC Press, Boca Raton. 246 pp.

Elias, T. (1977). Threatened and endangered species problems in North America. In Prance, G. and Elias, T. (eds), *Extinction is forever*. New York Botanical Garden, Bronx. Pp. 13–16.

Elliott-Fisk, D.L. (1988). The boreal forest. In Barbour, M.G. and Billings, W.D. (eds), *North American terrestrial vegetation*. Cambridge University Press, Cambridge. Pp. 33–62.

Famighetti, R. (1995). *The world almanac and book of facts 1995*. Funk and Wagnalls Corp., Mahwah, New Jersey. 976 pp.

Fenneman, N.M. (1938). *Physiography of the western United States*. McGraw-Hill, New York. 714 pp.

Fernald, M.L. (1926). Two summers of botanizing in Newfoundland. III. Noteworthy plants collected in Newfoundland, 1924 and 1925. *Rhodora* 28: 145–155, 161–178, 181–204, 210–225, 234–241.

Franklin, J.F. (1988). Pacific Northwest forests. In Barbour, M.G. and Billings, W.D. (eds), *North American terrestrial vegetation*. Cambridge University Press, Cambridge. Pp. 103–130.

Franklin, J.F. and Dyrness, C.T. (1988). *Natural vegetation of Oregon and Washington*. Oregon State University Press, Eugene. 452 pp.

Gentry, A.W. (1986). Endemism in tropical versus temperate plant communities. In Soulé, M. (ed.), *Conservation biology: the science of scarcity and diversity*. Sinauer, Sunderland. Pp. 153–181.

Gleason, H.A. and Cronquist, A. (1964). *The natural geography of plants*. Columbia University Press, New York. 420 pp.

Graf, W.L. (ed.) (1987). *Geomorphic systems of North America*. Geological Society of America, Boulder. Centennial Special Vol. 2. 643 pp.

Graham, E.H. (1937). Botanical studies in the Uinta Basin of Utah and Colorado. *Annals of Carnegie Museum* 26: 1–432.

Great Plains Flora Association (1986). *Flora of the Great Plains*. University of Kansas, Lawrence. 1392 pp.

Greller, A.M. (1988). Deciduous forest. In Barbour, M.G. and Billings, W.D. (eds), *North American terrestrial vegetation*. Cambridge University Press, Cambridge. Pp. 287–316.

Halliday, W.E.D. (1937). A forest classification for Canada. *Can. For. Serv. Bull.* 89: 1–50.

Harper, R.M. (1914). Geography and vegetation of northern Florida. *Ann. Re. Fla. Geol. Surv.* 6: 163–416.

Harper, R.M. (1947). Preliminary list of southern Appalachian endemics. *Castanea* 12: 100–112.

Harshberger, J.W. (1911). *Phytogeographic survey of North America*. G.E. Stechert, New York. 790 pp.

Henrickson, J. and Johnston, M.C. (1986). Vegetation and community types of the Chihuahuan Desert. In Barlow, J.C., Powell, A.M. and Timmermann, B.N. (eds), *Chihuahuan Desert – U.S. and Mexico*. Vol. 11. Chihuahuan Desert Research Institute, Sul Ross State University, Alpine. Pp. 20–39.

Henry, L.K. (1954). Shale-barren flora in Pennsylvania. *Proceedings of the Pennsylvania Academy of Science* 28: 65–68.

Hickman, J. (ed.) (1993). *The Jepson manual: higher plants of California*. University of California Press, Berkeley. 1400 pp.

Holt, P.C. (1970). *The distributional history of the biota of the southern Appalachians. Part 2. Flora*. Virginia Polytechnical Institute and State University, Blacksburg. 414 pp.

Howard, R.A. (1954). *Contribution of the Caribbean flora to the southeastern coastal plain*. (Paper addressed to the Systematic Section of the AIBS in Gainesville, Florida in 1954.)

Howell, J.T. (1957). The California flora and its provenance. *Leaflets of Western Botany* 8: 133–138.

Hunt, C.B. (1967). *Physiography of the United States*. Freeman, San Francisco. 480 pp.

Hunt, C.B. (1974). *Natural regions of the United States and Canada*. Freeman, San Francisco. 725 pp.

IUCN (1992). *Protected areas of the world: a review of national systems. Vol. 4. Nearctic and Neotropical*. IUCN, Gland, Switzerland and Cambridge, U.K. 460 pp. (Prepared by World Conservation Monitoring Centre.)

IUCN (1994). *1993 United Nations list of national parks and protected areas*. IUCN, Gland, Switzerland. 313 pp. (Prepared by the World Conservation Monitoring Centre and the IUCN Commission on National Parks and Protected Areas.)

James, C.W. (1961). Endemism in Florida. *Brittonia* 13: 225–244.

Jones, G.N. (1936). A botanical survey of the Olympic Peninsula. *Univ. Wash. Publ. Biol.* 5: 1–286.

Kartesz, J.T. (1994). *A synonymized checklist of the vascular flora of the United States, Canada, and Greenland*. Timber Press, Portland. 2 vols.

Kartesz, J.T. (1995). Personal communication to The Nature Conservancy. March 1995.

Kavasch, E.B. (1984). *Medicinal plants in American Indian life*. Smithsonian Institution Traveling Exhibition Service (SITES) booklet, Washington. 24 pp.

Kearney, T.H., Peebles, R.H. and collaborators. (1960). *The Arizona flora*. 2nd ed. University of California Press, Berkeley. 1085 pp.

Keener, C.S. (1970). The natural history of the mid-Appalachian shale barren flora. In Holt, P.C. (ed.), *The distributional history of the biota of the southern Appalachians. Part 2. Flora*. Virginia Polytechnic Institute and State University, Blacksburg. Pp. 215–248.

Kruckeberg, A.R. (1969). Soil diversity and the distribution of plants, with examples from western North America. *Madroño* 20: 129–154.

Kruckeberg, A.R. (1984a). California serpentines: flora, vegetation, geology, soils and management problems. *Univ. Calif. Publs. in Botany* 78: 1–180.

Kruckeberg, A.R. (1984b). California's serpentine. *Fremontia* 11(4): 11–17.

Kruckeberg, A.R. (1984c). The flora on California's serpentine. *Fremontia* 11(5): 3–10.

Kruckeberg, A.R. (1992). Plant life of western North America. In Roberts, B. and Proctor, J. (eds), *The ecology of areas with serpentinized rocks. A world view*. Kluwer Academic Publishers, Dordrecht. Pp. 31–73.

Küchler, A.W. (1964). *Potential natural vegetation of the conterminous United States*. American Geographical Society Special Publication No. 36, New York. 116 pp.

Lloyd, R.M. and Mitchell, R.S. (1973). *A flora of the White Mountains of California and Nevada*. University of California Press, Berkeley. 202 pp.

MacMahon, J.A. (1979). North American deserts: their floral and faunal components. In Goodall, D. and Perry, R.A. (eds), *Arid-land ecosystems: structure, functioning and management*. Vol. 1. Cambridge University Press, Cambridge. Pp. 21–82.

MacMahon, J.A. (1988). Warm deserts. In Barbour, M.G. and Billings, W.D. (eds), *North American terrestrial vegetation*. Cambridge University Press, Cambridge. Pp. 232–264.

Macoun, J.M. and Malte, M.O. (1917). *The flora of Canada*. National Museum of Canada Mus. Bull. No. 26. Canadian Department of Mines, Ottawa. 14 pp.

Maini, J.S. and Carlisle, A. (eds) (1974). *Conservation in Canada: a conspectus*. Department of Environment, Canadian Forestry Service, Publication No. 1340, Ottawa.

McAllister, D. (1994). Why is biodiversity being lost? *Different Drummer* 1(3): 32–34.

McKell, A.M. (1988). Deciduous forest. In Barbour, M.G. and Billings, W.D. (eds), *North American terrestrial vegetation*. Cambridge University Press, Cambridge. Pp. 287–316.

Moerman, D.E. (1986). *Medicinal plants of native America*. Vol. 1–2. University of Michigan Museum of Anthropology, Tech. Rep. No. 19. Ann Arbor. 910 pp.

Morefield, J. (1988). Floristic habitats of the White Mountains, California and Nevada: a local approach to plant communities. In Hall, C. and Doyle-Jones, V. (eds), *Plant biology of eastern California*. University of California, Los Angeles. Pp. 1–19.

Morisset, P. (1971). Endemism in the vascular plants of the Gulf of St Lawrence region. *Naturaliste Canadien* 98: 167–177.

Mozingo, H.N. and Williams, M. (1980). *Threatened and endangered plants of Nevada: an illustrated manual*. United States Fish and Wildlife Service, Portland and Bureau of Land Management, Reno. 268 pp.

Muller, J.W., Hardin, E.D., Jackson, D.R., Gatewood, S.E. and Caire, N. (1989). *Summary report on the vascular plants, animals and plant communities endemic to Florida*. Florida Game and Fresh Water Fish Commission, Nongame Wildlife Program Technical Report No. 7, Tallahassee. 113 pp.

Myers, N. (1983). *A wealth of wild species: storehouse for human welfare*. Westview Press, Boulder. 74 pp.

Oosting, H.J. (1956). *The study of plant communities*. 2nd ed. Freeman, San Francisco. 440 pp.

Or; nduff, R. (1974). *Introduction to California plant life*. University of California Press, Berkeley. 152 pp.

Packer, J.G. (1971). Endemism in the flora of western Canada. *Naturaliste Canadien* 98: 131–144.

Peet, R.K. (1988). Forests of the Rocky Mountains. In Barbour, M.G. and Billings, W.D. (eds), *North American terrestrial vegetation*. Cambridge University Press, Cambridge. Pp. 63–102.

Platt, R.B. (1951). An ecological study of the mid-Appalachian shale barrens and the plants endemic to them. *Ecol. Monogr.* 21: 269–300.

Prescott-Allen, C. and Prescott-Allen, R. (1986). *The first resource: wild species in the North American economy*. Yale University Press, New Haven. 529 pp.

Raven, P.H. (1977). The California flora. In Barbour, M.G. and Major, J. (eds), *Terrestrial vegetation of California*. California Native Plant Society, Sacramento. Pp. 109–137.

Raven, P.H. and Axelrod, D.I. (1978). Origin and relationships of the California flora. *University of California Publications in Botany* 72: 1–134.

Reed, C.F. (1986). *Flora of the serpentine formations in eastern North America; with descriptions of geomorphology and mineralogy of the formations*. Reed Herbarium, Baltimore. 858 pp.

Reveal, J. and Pringle, J. (1993). Taxonomic botany and floristics. In Flora of North American Editorial Committee (eds), *Flora of North America north of Mexico. Volume 1. Introduction*. Oxford University Press, New York. Pp. 157–192.

Risser, P.G. (1985). Grasslands. In Chabot, B. and Mooney, H. (eds), *Physiological ecology of North American plant communities*. Chapman and Hall, New York. Pp. 232–256.

Roecklein, J.C. and Leung, P. (1987). *A profile of economic plants*. Transaction, Inc., New Brunswick. 623 pp.

Rutman, S. (1990). *Handbook of federally endangered, threatened and candidate plants of Arizona*. United States Fish and Wildlife Service, Phoenix. 34 pp.

Schery, R.W. (1972). *Plants for man*. 2nd ed. Prentice-Hall, Englewood Cliffs. 657 pp.

Scoggan, H.J. (1978-1979). *The flora of Canada*. National Museum of Natural Sciences, Publications in Botany 7, Ottawa. 4 vols.

Scully, V. (1970). *A treasury of American Indian herbs*. Crown Publishers, New York. 306 pp.

Sharp, A.J. (1970). Epilogue. In Holt, P.C. (ed.), *The distributional history of the biota of the southern Appalachians. Part II: Flora*. Virginia Polytechnic Institute and State University, Blacksburg. Pp. 405–410.

Sheehan, M. and Schuller, R. (1981). *An illustrated guide to endangered, threatened and sensitive vascular plants in Washington*. Washington Natural Area Heritage Program, Olympia. 328 pp.

Shimer, J.A. (1972). *Field guide to the landforms in the United States*. MacMillan, New York. 272 pp.

Shreve, F. and Wiggins, I. (1964). *Vegetation and flora of the Sonoran Desert*. Vol. 1. Stanford University Press, Stanford. 840 pp.

Shultz, L. (1993). Patterns of endemism in the Utah Flora. In Sivinski, R. and Lightfoot, K. (eds), *Proceedings of the Southwestern rare and endangered plant conference*. New Mexico Forestry and Resources Conservation Division Misc. Publ. 2. Pp. 249–263.

Siddall, J., Chambers, K. and Wagner, D. (1979). *Rare, threatened and endangered vascular plants in Oregon – an interim report*. Oregon Natural Area Preserves Advisory Committee to the State Land Board, Salem. 109 pp.

Sims, P.L. (1988). Grasslands. In Barbour, M.G. and Billings, W.D. (eds), *North American terrestrial vegetation*. Cambridge University Press, Cambridge. Pp. 265–286.

Skinner, M. and Pavlik, B. (eds) (1994). *Inventory of rare and endangered vascular plants of California*. California Native Plant Society, Sacramento. 338 pp.

Small, J.K. (1933). *Manual of the southeastern flora*. Published by the author, New York. 1554 pp.

Stebbins, G.L. and Major, J. (1965). Endemism and speciation in the California flora. *Ecol. Monogr.* 35: 1–35.

Steele, E.S. (1911). New or noteworthy plants from the eastern United States. *Contrib. U. S. Nat. Herb.* 13: 359–374.

Steele, R., Brunsfeld, S., Henderson, D., Holte, K., Johnson, F. and Packard, P. (1981). *Vascular plant species of concern in Idaho*. University of Idaho Forest, Wildlife and Range Experiment, Moscow. Contribution No. 210. 161 pp.

Takhtajan, A.L. (1986). *Floristic regions of the world*. University of California Press, Berkeley. 522 pp.

Taylor, R.L. and Ludwig, R.L. (eds) (1966). *The evolution of Canada's flora*. University of Toronto Press, Toronto. 137 pp.

Thornbury, W.D. (1965). *Regional geomorphology of the United States*. John Wiley, New York. 609 pp.

Thorne, R.F. (1949). Inland plants on the Gulf Coastal Plain of Georgia. *Castanea* 14: 88–97.

Thorne, R.F. (1993). Phytogeography. In Flora of North America Editorial Committee (eds), *Flora of North America north of Mexico. Volume 1. Introduction*. Oxford University Press, New York. Pp. 132–153.

Tidestrom, I. (1923). Flora of Utah and Nevada. *Contrib. U. S. Nat. Herb.* 25: 1–665.

Vasek, F.C. and Barbour, M.G. (1988). Mojave desert scrub vegetation. In Barbour, M.G. and Major, J. (eds), *Terrestrial vegetation of California*. California Native Plant Society, Sacramento. Pp. 835–867.

Waggoner, G.S. (1975). *Eastern deciduous forest. Vol. I. Southeastern evergreen and oak-pine region*. United States National Park Service, Washington. 206 pp.

Washington Natural Heritage Program (1990). *Endangered, threatened and sensitive vascular plants of Washington*. Washington State Department of Natural Resources, Olympia. 52 pp.

Weaver, J.E. and Clements, F.E. (1938). *Plant ecology*. McGraw-Hill, New York. 520 pp.

Welsh, S.L. (1979). *Illustrated manual of proposed endangered and threatened plants of Utah*. United States Fish and Wildlife Service, Denver. 318 pp.

Welsh, S.L., Atwood, N.E. and Reveal, R.L. (1975). Endangered, threatened, extinct, endemic and rare or restricted Utah vascular plants. *Great Basin Naturalist* 35: 327–376.

West, N.E. (1988). Intermountain deserts. In Barbour, M.G. and Billings, W.D. (eds), *North American terrestrial vegetation*. Cambridge University Press, Cambridge. Pp. 209–231.

Whittaker, R.H. (1960). Vegetation of the Siskiyou Mountains, Oregon and California. *Ecol. Monogr.* 30: 279–338.

Whittaker, R.H. (1961). Vegetation history of the Pacific coast states and the "central" significance of the Klamath Region. *Madroño* 16: 5–23.

Wood, C.E., Jr. (1970). Some floristic relationships between the southern Appalachians and western North America. In Holt, P.C. (ed.), *The distributional history of the biota of the southern Appalachians. Part II: Flora*. Virginia Polytechnic Institute and State University, Blacksburg. Pp. 331–404.

Woodson, R.E., Jr. (1947). Notes on the "historical factor" in plant geography. *Contrib. Gray Herbarium* 165: 12–25.

World Conservation Monitoring Centre (1992). *Global biodiversity: status of the earth's living resources*. Chapman and Hall, London. 594 pp.

World Resources Institute (1994). *World resources 1994–95: a guide to the global environment*. Oxford University Press, New York. 400 pp.

Yatskievych, G. and Spellenberg, R.W. (1993). Plant Conservation. In Flora of North American Editorial Committee (eds), *Flora of North America north of Mexico. Volume 1. Introduction*. Oxford University Press, New York. Pp. 207–226.

Authors

This overview was written by Shirley L. Maina and Jane Villa-Lobos (Smithsonian Institution, Department of Botany, NHB 166, Washington, DC 20560, U.S.A.).

Acknowledgements

The preparation of this text has involved a number of individuals. Particular thanks go to Dr Stanwyn Shetler (Smithsonian Institution, Department of Botany, NHB 166, Washington, DC 20560, U.S.A.) and Dr Nancy Morin (Missouri

Botanical Garden, P.O. Box 299, St Louis, MO 63166, U.S.A.) who reviewed an earlier draft and provided additional information. Valuable comments and information were also provided by the following: Raymond Angelo (New England Botanical Club Herbarium, Harvard University, 22 Divinity Avenue, Cambridge, MA 02138, U.S.A.), Dr George Argus (Canadian Museum of Nature, National Herbarium of Canada, P.O. Box 3443, Sta. "D", Ottawa, Ontario, Canada K1P 6P4), Dr Theodore Barkley (Kansas State University, Division of Biology, Ackert Hall, Manhattan, Kansas 66506, U.S.A.), Dr Luc Brouillet (Institut botanique, Université de Montréal, 4101, rue Sherbrooke est, Montréal, Québec H1X 2B2, Canada), Dr Ronald Hartman (University of Wyoming, 3165 University Station, Laramie, WY 82701-3165, U.S.A.), Dr Marshall Johnston (University of Texas, Plant Resources Center, Botany Department, Austin, TX 78713-7640, U.S.A.), Dr Brien Meilleur (Center for Plant Conservation, Missouri Botanical Garden, P.O. Box 299, St Louis, MO 63166, U.S.A.), Dr Leila Shultz (Harvard University, 22 Divinity Avenue, Cambridge, MA 02138, U.S.A.), Dr Richard Spellenberg (New Mexico State University, Biology Department, P.O. Box 3AF, Las Cruces, NM 88003, U.S.A.), and Anukriti Sud (Center for Plant Conservation, Missouri Botanical Garden, P.O. Box 299, St Louis, MO 63166, U.S.A.). Information on rare plants was provided by the Network of Natural Heritage Programs and Conservation Data Centers and The Nature Conservancy and compiled by Lynn Kutner (The Nature Conservancy, 1815 N. Lynn St., Arlington, VA 22209, U.S.A.). Data on numbers of native vascular plant species were prepared by J. T. Kartesz (Biota of North America Program, North Carolina Botanical Garden, Chapel Hill, NC 27599-3375, U.S.A.).

CALIFORNIA FLORISTIC PROVINCE
California and Oregon, U.S.A.
and Baja California, Mexico

Location: California west of the deserts and small areas in Oregon and Baja California, at latitudes c. 30°–42°N and longitudes 115°–124°W.

Area: 324,000 km².

Altitude: 0–4418 m.

Vegetation: Montane coniferous and mixed evergreen forests, oak woodlands, chaparral, coastal scrubs, annual grassland; more locally other types.

Flora: 3488 native species; of which 2124 are endemic; 565 taxa within the Californian part of the area are threatened or endangered.

Useful plants: Timber trees, agricultural rootstocks (e.g. *Vitis californicus* for grape vines), ornamentals, medicinals.

Threats: Population growth, fuelling many activities threatening the flora, including conversion of land for agricultural or urban uses, dam construction, excessive groundwater abstraction, recreation and mining. The deliberate starting of fires is a serious problem.

Conservation: Many protected areas, including 29,000 km² devoted primarily to biodiversity protection (but many threats to the flora remain).

Geography

The California Floristic Province (CFP) lies on the west coast of North America, mostly within the state of California (U.S.A.); however, small portions of south-western Oregon (U.S.A.) and north-western Baja California (Mexico) are included on the basis of similarities in flora and climate (Map 4).

In California the CFP includes all areas west of the Sierra Nevada-Cascade mountain crests and the islands off the shore of southern California. In Oregon the CFP includes the coastal mountains south of Cape Blanco and most of the Rogue River watershed. In Baja California the CFP includes the forest and chaparral belts of the Sierra Juarez and the Sierra San Pedro Martir (but not their desert slopes to the east), coastal areas south to about El Rosario, and Guadalupe Island. The CFP is approximately 324,000 km² in area: 285,000 km² in California, 25,000 km² in Oregon and 14,000 km² in Baja California (Raven and Axelrod 1978).

Major physiographic regions within the CFP include all or portions of four coastal mountain ranges, a large interior valley and two interior mountain ranges. These regions are: (1) the Klamath Mountains (in south-east Oregon and north-west California) (see Data Sheet NA16c); (2) the Coast ranges (from about Trinidad Head to Point Arguello); (3) the Transverse ranges (from Point Arguello and San Miguel Island east into the deserts); (4) the Peninsular ranges (from the Los Angeles Basin and San Nicolas Island south into Baja California); (5) the Central (or Great) Valley (the Sacramento Valley east of the North Coast ranges and the San Joaquin Valley east of the South Coast ranges); (6) the Cascade range (mostly volcanic mountains from Mt Lassen north into Oregon); and (7) the Sierra Nevada (mostly granitic mountains from near Mt Lassen south to the Mojave Desert). Physiographic regions east of the CFP are the Great Basin, Mojave and Colorado deserts (Norris and Webb 1990).

The coastal slopes of the CFP drain directly to the Pacific Ocean by many perennial and seasonal rivers and streams. The interior slopes of the North and South Coast ranges and the west slopes of the Sierra Nevada and southern Cascade mountains drain by numerous tributary rivers to the Sacramento and San Joaquin rivers, through the Sacramento and San Joaquin valleys to the inland delta where these two rivers meet, and finally to the ocean through San Francisco Bay.

The climate throughout the CFP is Mediterranean, characterized by cool, wet winters and warm-to-hot, dry summers. Local climatic conditions within the CFP vary on several axes: (1) from relatively wet and temperate at the northern end to relatively dry and subtropical at the southern end; (2) from relatively moderate daily and annual temperature variations with cooling summer fogs along the coast to more extreme daily and annual temperature variations with intense summer heat in the interior; and (3) from long, hot summers and short, mild winters in the interior lowlands to short, mild summers and long, frigid winters in the high mountains (Major 1977; Barbour *et al.* 1991).

Geology and soils are extremely varied within the CFP. Granitic, volcanic and marine sedimentary rocks occupy large portions of the region. Serpentinite, limestone, several unusual rock types, and soils ranging from very young to very old and from highly acidic to highly alkaline contribute to habitat diversity (Stebbins and Major 1965; Stebbins 1978a).

Diversity in topography, climate, geology and soils are the main factors leading to the unusual diversity of habitats and plants in the CFP. The CFP exceeds the other four Mediterranean climate regions of the world in diversity of soil types and available moisture regimes (Stebbins 1978a). The newest manual for the California flora recognizes 28 non-overlapping floristic subregions within the CFP of California (Hickman 1989). The CFP also includes portions of 17 different horticultural zones (Sunset Books 1986).

MAP 4. CALIFORNIA FLORISTIC PROVINCE (CPD SITE NA16)

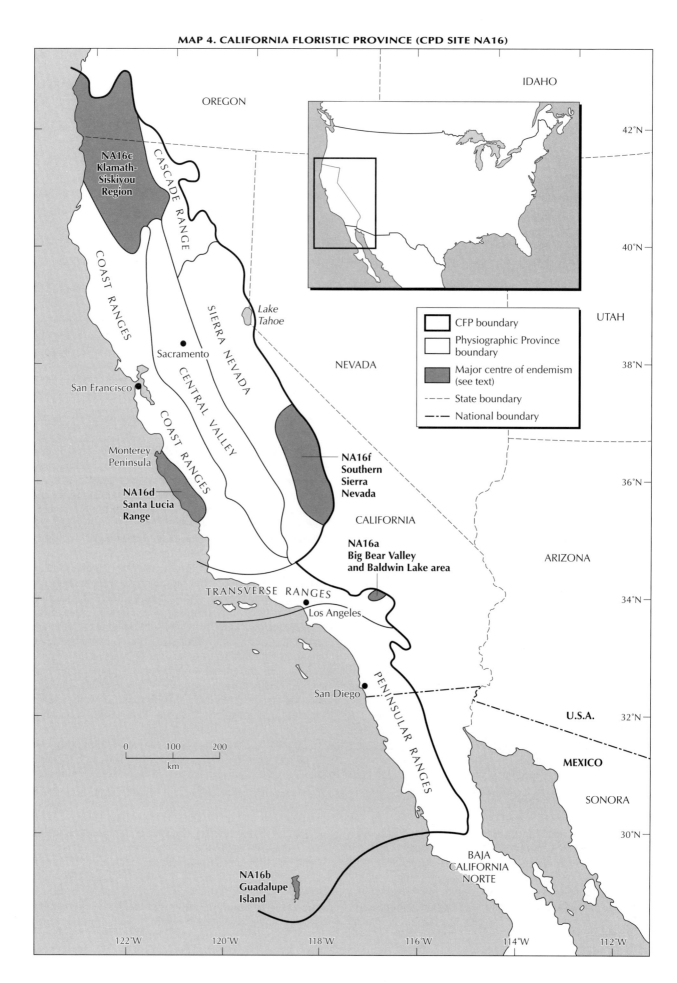

Vegetation

Numerous vegetation classifications have been proposed in the CFP (Munz and Keck 1959; Cheatham and Haller 1975; Kuchler 1977; Matyas and Parker 1980; Barry 1985; Holland 1986), but no single system is widely accepted. One of the most detailed recognizes over 210 separate plant communities in the California portion of the CFP (Holland 1986). General vegetation types covering extensive areas of the CFP and their most common dominant species are listed below.

1. Montane coniferous forests (5,937,000 ha): *Pinus ponderosa, P. contorta* subsp. *murrayana, P. jeffreyi, P. coulteri, Abies concolor, A. magnifica, Pseudotsuga menziesii, Calocedrus decurrens, Juniperus occidentalis* and others (Rundel, Parsons and Gordon 1977; Thorne 1977).

2. Oak woodlands (3,809,000 ha): *Quercus agrifolia, Q. wislizenii, Q. douglasii, Q. lobata, Q. garryana, Q. engelmanii* and *Pinus sabiniana* (Griffin 1977).

3. Mixed evergreen forests (829,000 ha): *Pseudotsuga menziesii, Quercus kelloggii, Q. chrysolepis, Lithocarpus densiflorus, Arbutus menziesii, Heteromeles arbutifolia* and others (Sawyer, Thronburgh and Griffin 1977).

4. Chaparral (3,141,000 ha): *Adenostoma fasciculatum, Arctostaphylos* spp., *Ceanothus* spp., *Quercus dumosa, Q. durata, Eriogonum* spp., *Toxicodendron diversilobum* and others (Hanes 1977).

5. Montane chaparral (420,500 ha): *Arctostaphylos patula, Castanopsis sempervirens, Quercus vaccinifolia* and *Ceanothus velutinus* (Hanes 1977).

6. Northern and southern coastal scrubs (1,015,000 ha): Northern coastal scrub: *Baccharis pilularis, Mimulus (=Diplacus) aurantiacus, Ceanothus thyrsiflorus, Gaultheria shallon, Lupinus arboreus, Eriophyllum stoechadifolium, Rubus* spp., *Pteridium aquilinum, Castilleja latifolia, Heracleum lanatum* and others. Southern coastal scrub: *Artemisia californica, Salvia mellifera, S. leucophylla, Eriogonum fasciculatum, Eriophyllum confertiflorum, Encelia californica, Haplopappus* spp. and others (Heady *et al.* 1977; Mooney 1977).

7. Annual grassland (3,502,000 ha): *Bromus mollis, B. diandrus, B. rubens, Avena fatua, A. barbata, Lolium multiflorum, Hordeum* spp., *Vulpia* spp., *Taeniatherum asperum, Erodium botrys, E. cicutarium, Centaurea solstitialis, Medicago* spp., *Trifolium* spp. and other non-native species (Heady 1977).

Several less extensive vegetation and habitat types that contribute significantly to diversity and endemism within the CFP include:

1. Valley and montane riparian habitats (20,000 and 35,000 ha, respectively): *Quercus lobata, Acer negundo, Fraxinus latifolia, Alnus rhombifolia, A. rubra, Cephalanthus occidentalis, Platanus racemosa, Populus fremontii, P. trichocarpa, Juglans hindsii, Salix* spp., *Rubus* spp. and others (Holstein 1984).

2. Closed-cone pine and cypress forests (32,000 ha, combined): The pines are *Pinus attenuata, P. muricata, P. remorata, P. contorta* subsp. *contorta, P. contorta* subsp. *bolanderi, P. radiata* and *P. torreyana.* Common cypresses are *Cupressus macnabiana* and *C. sargentii;*

TABLE 17. APPROXIMATE PRISTINE AND PRESENT AREA (HECTARES) AND PERCENTAGE OF CHANGE FOR MAJOR VEGETATION TYPES IN CALIFORNIA[a]

	Pristine [b] (c. 1800)	Present [c] (c. 1990)	% change
Forests, woodlands, and scrub			
Douglas-fir (*Pseudotsuga*)	815,900	717,100	-12.1
Mixed conifer	5,4478,000	3,751,000	-32.1
Redwood (*Sequoia, Sequoiadendron*)	938,900	635,400	-32.3
Ponderosa pine (*Pinus ponderosa*) - shrub	686,000	1,073,000	+56.4
Oak	3,867,000	3,809,000	-1.5
Riparian (Central Valley only)	373,000	41,300	-88.9
Chaparral	3,440,000	3,141,000	-8.7
Montane chaparral	232,000	420,500	+81.3
Herbaceous			
North coastal bunchgrass	356,000	36,400	-90.0
Needlegrass (*Stipa*) steppe	5,350,900	400	-99.9
Annual grassland	400	3,502,000	+8653.0
Alpine meadow	302,000	96,300	-68.2
Coastal wetlands	102,400	20,600	-80
Inland wetlands	1,600,000	101,000	-94
Tule (*Scirpus*) marsh	752,300	233,000	-69.0
Urban and agriculture (excluding grazing)	400	6,155,800	+15,211

Sources:

[a] Primary source: Barbour *et al.* (1991). Figures are for vegetation types throughout the state of California. They have not been recalculated to exclude desert areas east of the CFP and include the Oregon and Baja California portions of the CFP. For the vegetation types listed, percentages of change for California are probably similar to those for the CFP. Figures originally given in acres were converted to ha with the same number of significant digits.

[b] Figures taken from Barbour and Major (1977), except: riparian from Katibah (1984); annual grassland and urban from Barbour *et al.* (1991); wetlands from Jones and Stokes Associates (1987).

[c] Figures taken from Forest and Rangeland Resources Assessment Program (1988), except: riparian from Katibah (1984); needlegrass from Barbour *et al.* (1991); wetlands from Jones and Stokes Associates (1987).

rare cypresses are *C. forbesii*, *C. stephansonii*, *C. macrocarpa*, *C. goveniana*, *C. pygmaea*, *C. abramsiana*, *C. nevadensis* and *C. bakeri* (Vogl *et al.* 1977).

3. Alkali sink, grassland and scrub communities (area undetermined): *Allenrolfea occidentalis*, *Distichlis spicata*, *Sporobolus airoides*, *Suaeda fruticosa*, *Atriplex spinifera* and *A. polycarpa* (Griggs and Zaninovich 1984; Coates, Showers and Pavlik 1989).

4. Coastal salt marsh (20,600 ha): *Salicornia virginica*, *Spartina foliosa*, *Frankenia grandifolia*, *Distichlis spicata*, *Atriplex patula*, *Batis maritima*, *Scirpus robustus*, *Grindelia* spp. and others (MacDonald 1977).

5. Coastal dunes (area undetermined): Native dominants: *Abronia latifolia*, *Ambrosia chamissonis*, *Calystegia soldanella*, *Camissonia cheiranthifolia*, *Elymus mollis* and *Poa douglasii* (foredune habitats); *Artemisia pycnocephala*, *Baccharis pilularis*, *Croton californicus*, *Ericameria ericoides*, *Eriogonum latifolium*, *Haplopappus venetus*, *Lathyrus littoralis*, *Lupinus arboreus*, *L. chamissonis* and *Scrophularia californica* (backdune habitats). Introduced dominants: *Ammophila arenaria*, *Cakile maritima* and *Mesembryanthemum* spp. (Holland 1986; Barbour and Johnson 1977).

Two habitat types especially notable for their high levels of endemism in the CFP, **serpentine soils** and **vernal pools**, are described in separate Data Sheets (NA16e and NA16g).

The extent and character of natural vegetation throughout the CFP have changed dramatically in the last two centuries (Table 17). The most serious losses of habitat important to plant diversity include a 91% loss of wetlands (due to filling and draining for urban and agricultural growth), an 89% loss of riparian woodlands (due to agricultural and urban development, river channelization and diversions), a >90% loss of native perennial grasslands (through urbanization, conversion to agriculture and a conversion to non-native annual grassland mediated by overgrazing) and a >30% loss of the most diverse conifer forests (through conversion to less diverse forest-scrub, montane chaparral and clear-cut habitats by logging).

Flora

Size and degree of endemism

The flora of the CFP is well documented, but remains fertile ground for research and discovery. Major Floras have been published for California (Jepson 1925; Munz and Keck 1959; Munz 1968), southern California (Munz 1974), Oregon (Peck 1961) and Baja California (Wiggins 1980). A completely new manual of the California flora was published in 1993 (Hickman 1993). New taxa continue to be described, especially from the southern Sierra Nevada. Between 1968 and 1986, 219 taxa (116 species and 103 lower taxa) were described or renamed in the CFP (Shevock and Taylor 1986; J. Shevock, pers. comm.).

As of 1978, the flora of the CFP was estimated to include 794 genera and 4452 species; about 3488 species (78.3%) are native. Fifty-two genera (6.5%) (Table 18) and 2124

TABLE 18. GENERA ENDEMIC TO THE CALIFORNIA FLORISTIC PROVINCE

Family/Genus	Number of species	State
Apiaceae		
Oreonana	3	California
Asteraceae		
Achyrachaena	1	Oregon, California
Adenothamnus	1	Baja California
Baeriopsis	1	Baja California
Benitoa	1	California
Blepharizonia	1	California
Calycadenia	12	Oregon, California, Baja California
Corethrogyne	3	Oregon, California, Baja California
Eastwoodia	1	California
Holocarpha	4	California
Holozonia	1	California
Lembertia	1	California
Monolopia	4	California
Orochaenactis	1	California
Pentachaeta	6	California, Baja California
Phalacroseris	1	California
Pseudobahia	3	California
Tracyina	1	California
Venegasia	1	California, Baja California
Whitneya	1	California
Brassicaceae		
Heterodraba	1	Oregon, California, Baja California
Cactaceae		
Bergerocactus	1	California, Baja California
Crassulaceae		
Parvisedum	4	California
Ericaceae		
Ornithostaphylos	1	California, Baja California
Sarcodes	1	California, Baja California
Fabaceae		
Pickeringia	1	California, Baja California
Hydrophyllaceae		
Draperia	1	California
Lemmonia	1	California, Baja California
Lamiaceae		
Acanthomintha	3	California, Baja California
Pogogyne	5	Oregon, California, Baja California
Lauraceae		
Umbellularia	1	Oregon, California
Liliaceae		
Bloomeria	2	California, Baja California
Chlorogalum	5	Oregon, California, Baja California
Odontostomum	1	California
Oleaceae		
Hesperelea	1	Baja California
Onagraceae		
Heterogaura	1	Oregon, California
Papaveraceae		
Dendromecon	2	California, Baja California
Hesperomecon	1	Oregon, California, Baja California
Romneya	2	California, Baja California
Stylomecon	1	California, Baja California
Poaceae		
Neostaphia	1	California
Orcuttia	6	California, Baja California
Tuctoria	2	California
Polygonaceae		
Hollisteria	1	California
Rosaceae		
Chamaebatia	2	California, Baja California
Lyonothamnus	1	California
Saxifragaceae		
Bensoniella	1	Oregon, California
Carpenteria	1	California
Jepsonia	3	California, Baja California
Scrophulariaceae		
Ophiocephalus	1	Baja California
Taxodiaceae		
Sequoia	1	Oregon, California
Sequoiadendron	1	California

Sources: Raven and Axelrod (1978), Howell (1957).

species (47.7%) are endemic (Raven and Axelrod 1978; Howell 1957). This flora is nearly as large as that of the entire central and north-eastern U.S.A. and adjacent Canada (5523 species, of which 4425, 80.0%, are native) (Fernald 1970), an area over 6 times as large as the CFP. The largest families in the CFP (in order of decreasing size) are: Asteraceae, Poaceae, Fabaceae, Scrophulariaceae, Brassicaceae, Cyperaceae, Polygonaceae, Polemoniaceae, Boraginaceae and Hydrophyllaceae. The largest genera are: *Carex, Astragalus, Phacelia, Lupinus, Eriogonum* and *Mimulus* (Smith and Noldeke 1960). Of the 964 or so introduced species, about 672 (15.1% of the flora) are weedy (Raven and Axelrod 1978).

Floristic affinities

Evergreen sclerophyll vegetation in the CFP (mixed evergreen forest, live-oak woodland, chaparral, closed-cone pine and cypress forests, and the soft-leaved coastal scrub) are derived from Eocene floras of the south-western U.S.A. and north-western Mexico which lived in a warm, summer-moist climate. Conifer forests of the Sierra Nevada and more southerly ranges are derived from Miocene forests of the Great Basin and Columbia Plateau, where the climate was also warm and moist. Conifer forests of the north coast are more closely related to forests of the Miocene north-west, which flourished in a cooler climate (Raven and Axelrod 1978; Axelrod 1977).

Increasing topographic relief, climatic fluctuations and climatic heterogeneity during the Miocene and Pliocene promoted migration, mixing, segregation and differentiation among these floras throughout western North America. Coastal and interior floras became distinct, and the mosaic of lowlands and mountains became complex. As the Sierra Nevada uplifted and the Mediterranean climate developed fully during the Pleistocene, the rich Tertiary floras became progressively impoverished, but new species evolved from Tertiary ancestors to occupy new habitats. The Xerothermic period of the early Holocene, warmer and drier than the present, established many of the plant distributional patterns seen today (Raven and Axelrod 1978; Axelrod 1977).

Thus, the CFP emerged as a recognizable entity during the Quaternary, enriched by both old and young species with both northern and southern affinities. Numerous Tertiary relicts (mostly trees and some shrubs) have survived in several coastal and montane mild-climate refugia. Many plants (especially herbs) that became isolated in topographic, edaphic, or other ecological islands in the CFP have diverged from their closest relatives in other physiographic or floristic provinces to become local endemics. Many of the most notable narrow endemics evolved by adapting to unusual combinations of soil and climate that were too physiologically stressful or disturbance-prone for many more common plants to tolerate (Raven and Axelrod 1978; Axelrod 1977).

Numbers of rare and endangered species

Approximately 565 taxa within the CFP of California are recognized by conservation organizations as threatened or endangered. About 32 are believed to be extinct.

Approximately 453 more are taxa of limited distribution and 122 are taxa of uncertain but potentially threatened status (Smith and Berg 1988; M. Skinner, pers. comm.). In the California portion of the CFP, 24 plants are listed as threatened or endangered under the federal Endangered Species Act; 191 are listed under the California Endangered Species Act (California Department of Fish and Game 1991). More species are proposed and listed each year. Several additional rare plants occur in the Oregon portion of the Klamath region (see Klamath-Siskiyou region Data Sheet NA16c). Approximately 150 Baja California endemics occur in the CFP of Mexico (R. Thorne, pers. comm.), of which many are rare or threatened.

Centres of diversity and endemism

Floristic diversity in the CFP is highest in the Sierra Nevada and Transverse ranges and nearly as high in the Klamath Mountains and Coast ranges (Stebbins 1978a; Richerson and Lum 1980). Endemism is highest regionally in the Transverse and Peninsular ranges, Coast ranges and Klamath Mountains (Stebbins and Major 1965), although local centres of endemism are widely scattered throughout the CFP. Major centres of endemism within the CFP are listed below, together with their locally endemic plants. Endemism associated with the Klamath-Siskiyou region, serpentine soils and vernal pools are listed in separate Data Sheets (NA16c, NA16e and NA16g).

NA16a – Big Bear Valley and Baldwin Lake area
Neighbouring sites in the eastern Transverse ranges with 11 local endemics on Pleistocene clay deposits: *Arabis parishii, Arenaria ursina, Castilleja cinerea, Eriogonum kennedyi* subsp. *austromontanum, Linanthus killipii, Mimulus exiguus, Pyrrocoma uniflora* subsp. *gossypina, Senecio bernardinus, Sidalcea pedata, Taraxacum californicum* and *Thelypodium stenopetalum*; and 4 local endemics on limestone formations: *Astragalus albens, Lesquerella kingii* subsp. *bernardina, Erigeron parishii* and *Eriogonum ovalifolium* var. *vimineum* (Krantz 1983, 1988). Plants endemic to the pebble plain and adjacent habitats are relatively well protected in private and state preserves, although populations on national forest lands are not adequately protected. Plants endemic to the limestone habitats are seriously threatened by mining and lack of attention by the local national forest authorities. Five of these plants were proposed for endangered status under the federal Endangered Species Act in November 1991. If these plants are listed as endangered, they will receive significant legal protection on federal lands.

NA16b – Guadalupe Island
A mountainous Pacific island 265 km off Baja California at 29°N, 118°17'W, with 17 local endemics (*Baeriopsis guadalupensis, Castilleja guadalupensis, Dudleya virens, Eriogonum zapotense, Erysimum moranii, Erythea edulis, Euphorbia pondii, Hemizonia palmeri, Lavatera lindsayi, Lupinus niveus, Mirabilis heimerlii, Phacelia phyllomanica, Senecio palmeri, Sphaeralcea palmeri, S. sulphurea, Stephanomeria guadalupensis* and *Talinum guadalupense*) (Wiggins 1980). Guadalupe Island is entirely unprotected and is probably the most seriously threatened centre of diversity in the CFP. The island is sparsely settled and mostly

undeveloped, but is severely overgrazed by feral goats. The grazing allows little or no successful reproduction among many native woody plants. If goat populations are not reduced and sensitive sites fenced before existing mature plants die, numerous extinctions are likely.

NA16d - Santa Lucia Range and Monterey Peninsula

A large, rugged and vegetationally diverse coastal mountain range near the middle of the CFP, with the coastal scrub, dunes and forests of the Monterey Peninsula at its northern end. At least 50 plants are locally endemic, including *Abies bracteata, Allium hickmanii, Arctostaphylos* (10 taxa), *Bloomeria humilis, Brodiaea lutea* var. *cookii, Carex obispoensis, Ceanothus maritimus, C. rigidus, Centrostegia vortriedei, Chlorogalum purpureum, Chorizanthe breweri, Cirsium fontinale* var. *obispoense, Clarkia jolonensis, Collinsia antonina, Delphinium huthchinsoniae, D. umbraculorum, Dudleya bettinae, D. abramsii* var. *murina, Eriogonum butterworthianum, Galium californicum* subsp. *luciense, G. clementis, G. hardhamae, Hemizonia paniculata* subsp. *cruzensis, Layia jonesii, Lupinus abramsii, L. cervinus, Malacothamnus jonesii, M. palmeri* (3 varieties), *Malacothrix saxatalis* (2 varieties), *Mimulus guttatus* subsp. *arenicola, Monardella antonina, M. palmeri, M. subglabrata, Orthocarpus densiflorus* var. *obispoensis, Pogogyne clareana, Ribes menziesii* var. *hystrix, R. sericeum* and *Sidalcea hickmanii*. At least 7 more are endemic to Monterey Peninsula: *Astragalus tener* var. *titi, Brodiaea versicolor, Cupressus macrocarpa, C. goveniana, Haplopappus*

eastwoodiae, Lupinus tidestromii var. *tidestromii, Trifolium polyodon* and *T. trichocalyx* (Rogers 1991). Approximately 8% of this area is protected in State Parks and National Forest Wilderness Areas. Another 8% is on non-wilderness national forest land. Many endemic plants occur in these areas, but important sites on the Monterey Peninsula and at two large military reservations are threatened by urban growth and military training activities.

NA16f - Kern Plateau and Southern Sierra Nevada

Rugged and diverse mountains with the highest local floristic diversity in the CFP (approximately 2200 species on the Kern Plateau) and over 75 local endemics, including *Abronia alpina, Allium shevockii, Angelica callii, Arabis pygmaea, Astragalus ertterae, A. ravenii, A. shevockii, A. subvestitus, Brodiaea insignis, Calochortus westonii, Calyptridium pulchellum, Camissonia kernensis, C. sierrae* subsp. *alticola, Carpenteria californica, Castilleja praeterita, Ceanothus fresnensis, C. pinetorum, Clarkia springvillensis, Cordylanthus eremicus* subsp. *kernensis, Cupressus nevadensis, Delphinium purpusii, Dicentra nevadensis, Draba cruciata* (2 varieties), *Dudleya calcicola, D. cymosa* subsp. *costafolia, Erigeron aequifolius, E. multiceps, Eriogonum* (9 taxa), *Eriophyllum lanatum* var. *obovatum, Erythronium pluriflorum, E. pusaterii, Eschscholzia procera, Frasera tubulosa, Fritillaria brandegei, F. striata, Galium angustifolium* subsp. *onycense, Horkelia tularensis, Ivesia purpurascens, Linanthus oblanceolatus, L. serrulatus, Lomatium rigidum, L. shevockii, Lupinus citrinus, L. culbertsonii, L. padre-crowleyi, Mimulus norisii,*

CPD Site NA16d: Santa Lucia Range and Monterey Peninsula. Scrub and woodland near Point Sur. Photo: Tiffany Passmore.

M. pictus, M. shevockii, Monardella beneolens, Nemacladus twisselmannii, Oreonana clementis, O. purpurascens, Phacelia novenmillensis, P. orogenes, Ribes tularense, Sidalcea keckii, Silene aperta, Streptanthus cordatus var. *piutensis, S. fenestratus, S. gracilis* and *Trifolium bolanderi* (J. Shevock, pers. comm.; Twisselman 1972). Approximately 60% of this area is protected in National Parks and National Forest Wilderness Areas. Most of the rest is on non-wilderness forest land. Overall protection is high, but some endemic plants are threatened by grazing, development and other land uses.

Lesser centres of endemism in the CFP include Pine Hill, the Pitkin-Bodega area, Suisun Bay-west delta area, Ione chaparral, Diablo range, Santa Cruz Mountains, Pajaro dunes, southern San Joaquin Valley and Carrizo plains, Channel Islands and Sierra San Pedro Martir (Emory 1989; Stebbins and Major 1965; Hoover 1937). In addition to its local endemics, each centre of endemism hosts several other rare CFP endemics.

Useful plants

Plants native or endemic to the CFP have been used for wood products, agricultural rootstocks, ornamental plants and sources of medicine. The Monterey pine (*Pinus radiata*), endemic to 3 small areas on the central coast of California, has become a major component of the forest products industries in New Zealand and Australia. Sugar pine (*P. lambertiana*) and giant sequoia (*Sequoiadendron giganteum*, a CFP endemic), were once major suppliers of wood products. The endemic coast redwood (*Sequoia sempervirens*) and other more widespread native conifers (especially *Abies, Pinus* and *Pseudotsuga* spp.) are the base of a major forest products industry.

California wild grape (*Vitis californicus*) has been used in developing disease-resistant rootstocks for European grapes that support a major wine industry in California. California black walnut (*Juglans hindsii*) is widely used as a disease-resistant rootstock for locally grown European walnuts.

The horticultural industry in California and throughout the world has been enriched by many contributions from the CFP. Prized horticultural subjects include: members of several endemic genera (especially *Carpenteria, Dendromecon, Lyonothamnus, Romneya, Sequoia, Sequoiadendron* and *Umbellularia*); many species of *Arctostaphylos, Ceanothus, Clarkia, Fremontodendron, Iris, Lupinus, Penstemon, Rhamnus, Salvia* and *Zauchneria*; most other native or endemic trees; numerous bulbs (especially *Brodiaea, Calochortus, Erythronium* and *Fritillaria*); many annuals; and others (Schmidt 1980). Many native grasses, shrubs and trees are valued for use in erosion control, revegetation and wildlife habitat improvement.

Native Americans in the CFP used many local plants as remedies for illness. Among the most important were yerba santa (*Eriodictyon californicum*), yerba mansa (*Anemopsis californica*) and cascara (*Rhamnus purshiana*) (Balls 1962). Modern medicine has scarcely explored the medicinal potential of the CFP flora, but Pacific yew (*Taxus brevifolia*), native to the northern CFP and Pacific Northwest, is the source of a potential new anticancer drug called taxol. Vast quantities of bark are being collected for research and, although the species will stump-sprout, it grows very slowly and some populations are threatened (Scher 1991).

Social and environmental values

Individuals, businesses and industries throughout the CFP enjoy substantial social and environmental benefits from its native vegetation and diversity of habitats. Economically and socially important benefits include moderation of local climate, protection of watersheds and water resources, flood abatement, nonpoint source pollution abatement, habitat for wild pollinators of food crops, habitat for a diverse non-game fauna that includes many rare and endangered species, critical links in the Pacific flyway, commercial and recreational fisheries, large and small game for hunters, many forms of passive recreation and opportunities for education and research.

Economic assessment

California is the largest agricultural producer (200 crops worth $1690 million in 1988) and forest products producer (about 4450 million board feet annually) in the U.S.A. (Jones and Stokes Associates 1992). Although made possible by the region's abundance of water, fertile soils and forests, not all of this production can be considered sustainable. For example, the water that irrigates up to 75% of California's cropland is supplied through state and federal water projects that are heavily subsidized by taxpayers, promote agriculture on marginal lands, and have caused many kinds of impacts on native biological resources (El-Ashry and Gibbons 1986).

Amenity values of natural habitats in the CFP are extremely high to residents of, and visitors to, this very populous and mostly affluent region. The CFP contains one of the most visited National Parks in the U.S.A. (Yosemite) and the national forest most used for recreation (San Bernardino) (Jones and Stokes Associates 1992). California's state parks and recreation areas (not counting National Parks) received 10% of all park use in the U.S.A. in 1988. The tourism industry was valued at $3990 million in 1989. During the late 1980s, approximately 2.4 million fishing licenses and 420,000 hunting licenses were purchased annually in California (Jones and Stokes Associates 1992).

The monetary value of the social and environmental values described above is undoubtedly astronomical, but few efforts at quantification have been made. One recent survey of households to assess the "existence value" of wildlands indicated that Californians would be willing to pay $1.5 billion annually to maintain wetlands and $2.4 billion annually for ecological restoration of wetland habitats (Jones and Stokes Associates 1990).

Threats

California is the most populous State in the U.S.A. (about 29,976,000 in 1990), with over 10% of the U.S. population. Most of these people (over 29,000,000) live in the CFP portion of the state. Approximately 182,000 and 800,000–900,000 people live in the Oregon and Baja California portions of the CFP, respectively. The highest population densities occur along the south coast (Santa Barbara to Tijuana), central coast (San Francisco Bay and Monterey Bay areas) and the California state capital area (Sacramento and vicinity). California's population is projected to grow

1–2% per year to about 39,000,000 by 2005 and over 44,000,000 by 2035 (California Department of Finance 1991).

Population growth is the driving force behind nearly all the direct causes of habitat loss, fragmentation and degradation in the CFP today. Threats to plant diversity in the CFP include habitat conversion, water use and water projects, unsustainable natural resource harvest and extraction, abusive recreation and vandalism, biological invasions, and others (Jensen, Torn and Harte 1990). Habitats on undeveloped lands are being widely converted to agricultural and urban uses. An estimated 1.7 million acres will be converted to agricultural uses and 1 million acres to urban and rural uses between about 1986 and 2000. Water projects have disrupted flow, degraded water quality and flooded canyons on most rivers in the CFP. Over 100 major and 1200 minor dams exist in California and over 30 more major dams are proposed. Groundwater overdraft has caused and continues to cause wetland and riparian loss throughout the region and land subsidence in the Sacramento-San Joaquin Delta. Marginally productive soils are destroyed through heavy irrigation and fertilization leading to salinization. Off-road vehicle activity degrades many sand-dune and serpentine barren habitats. Although much vegetation in the CFP is fire-adapted, fires caused by arson and accident are unnaturally damaging to biodiversity where many years of fire suppression have led to excessive fuel accumulations. Forests, especially old growth, are being cut faster than they will regrow and are being converted to types less supportive of indigenous biodiversity. Livestock overgrazing damages range and riparian vegetation. Mining of precious metals devastates many serpentine areas. Sand and gravel mining has degraded many riparian and coastal dune areas. Invasive exotic plants (especially *Bromus, Cortaderia, Cytisus, Ilex, Mesymbryanthemum, Carpobrotus* and *Senecio mikanioides*) outcompete much native vegetation, and some animals present in large numbers (especially introduced feral pigs) cause much direct damage to vegetation.

Conservation

Status of land protection

Plants and other elements of natural diversity are protected to varying degrees in dozens of federal, state, regional and private parks and preserves ranging in size from a fraction of a hectare to over 300,000 hectares. Lands devoted primarily to biodiversity protection include: National Parks, Monuments, seashores and recreation areas; National Forest Research Natural Areas, Special Interest Areas and Wilderness Areas; California ecological reserves and state preserves; University of California natural reserves; and Nature Conservancy preserves. Lands managed for mixed uses (recreation, hunting, etc.) in addition to biodiversity protection include National Wildlife Refuges, State Wildlife Areas, and State Parks, beaches and recreation areas.

No database has been compiled that summarizes land protection in the CFP separately from other parts of the state of California, or by vegetation type or bioregion within the CFP. Summary data from Jensen, Torn and Harte (1990), revised to exclude several large sites in the California deserts and including a Wilderness Area in Oregon and two National Parks in northern Baja California, result in the following

A range of vegetation types is protected in the state of California including 11% of its redwood forests.
Photo: Beryl Hulbert.

figures: approximately 2.9 million ha in sites devoted primarily to biodiversity protection and approximately 0.7 million ha protected in mixed-management sites (together, about 3.6 million ha, or 11% of the CFP). Areas of selected vegetation types protected in the state of California include: 83,970 ha of subalpine conifer forests (91% of the habitat in California); 252,600 ha of lodgepole pine forest (83%); 377,900 ha of red fir forest (49%); 1655 ha of closed-cone pine and cypress forests (13%); 494,620 ha of mixed conifer forests (12%); 69,890 ha of redwood forests (11%); 60,870 ha of coastal scrub (6%); 119,200 ha of valley-foothill hardwoods (4%); and 390 ha of valley riparian (2%) (Jensen, Torn and Harte 1990; Forest and Rangeland Resources Assessment Program 1988).

Barriers to progress

A legal basis for biodiversity protection exists in California and Oregon through a variety of federal, state, and local laws and regulations, including endangered species acts (both California and federal), the National Environmental Policy Act, the National Forest Management Act, California Environmental Quality Act, state fish and game department codes, and the open space and conservation elements of county and city general plans (Jones and Stokes Associates 1992). Although significant protection has been achieved through these laws and regulations, protection measures are too often uncoordinated, slow, overly bureaucratic, poorly or unevenly implemented, underfunded and weakened by loopholes.

While public interest in biodiversity protection is growing in the CFP, losses of habitat and diversity continue to accelerate. Jensen, Torn and Harte (1990) identified a variety of legal, social, and scientific barriers that limit progress in biodiversity protection in the state of California:

❖ California lacks a state policy mandating conservation of biodiversity;

❖ legislative protection focuses on species, not ecosystems;

❖ conservation interests are under-represented in resource policy-making bodies in California state government, while industries and special interest groups that extract goods and values from biological resources are over-represented;

❖ responses to threats are mostly local and reactive, not regional and anticipatory;

❖ some diverse ecosystems are seriously under-represented on public lands and in protected areas;

❖ serious gaps remain in knowledge of the status, ecological requirements, and management needs of plants and other biodiversity in the CFP;

❖ environmental and scientific illiteracy among many private landowners, responsible officials, and elected representatives hinders wise resource stewardship.

In Baja California, relatively few laws are available to support biodiversity protection and the two designated National Parks are not fully protected from grazing and timber extraction. The Mexican government is encouraging many people to move to Baja California from other parts of Mexico and establish new settlements. These settlements and the agriculture that supports them are eliminating much of the species-rich coastal scrub in north-western Baja California (Thorne and Moran, pers. comm.).

Current actions and recommendations for the future

Existing and emerging efforts to improve biodiversity protection in the CFP are too numerous and diverse to describe here. Most are occurring in the state of California and are the result of combined governmental and private efforts. Some of the more promising programmes and mechanisms include informal management agreements, memoranda of understanding, transfer of development rights, habitat conservation plans, mitigation banking, open space and conservation easements (Jones and Stokes Associates 1992). Memoranda of understanding are frequently used by the California Department of Fish and Game to enforce implementation of protection and mitigation measures by landowners whose activities impact state-listed threatened or endangered plants.

Ecological and social-economic guidelines have been identified that would improve site selection and management planning for biodiversity protection (Jones and Stokes Associates 1992; J. Shevock, pers. comm.). General ecological guidelines include the following: (1) emphasize contiguous blocks of habitat area; (2) select habitat blocks that are close together; (3) protect blocks of habitat large enough to contain large populations of species; (4) connect habitat blocks that support similar habitats; (5) maintain a broad distribution of protected areas; and (6) ensure adequate representation of all major plant community types in protected areas. Rare plants with narrow distributions and very specific habitat requirements will often need more focused efforts.

Two general social and economic guidelines are: (1) incorporate and maximize compatibility with current patterns of ownership and land use, projected future land use needs and the management objectives of private landowners, public land managers and regulatory agencies and (2) identify the roles and responsibilities of land owners, managers and regulatory agencies early in the planning process. New initiatives for biodiversity conservation must be increasingly strategic (rather than opportunistic), because of competing interests, multiple uses and complex ownership and use patterns on most of the lands now targeted for protection and enhancement.

Further recommendations for improving protection of all biodiversity in California focus on immediate actions, institutional changes and anticipatory actions (Jensen, Torn and Harte 1990): (1) establish a state policy mandating protection of biodiversity; (2) establish a habitat protection act; (3) establish a biodiversity conservation board (a state governmental agency) to coordinate, advocate, administer and enforce biodiversity protection laws and policies; (4) establish a biodiversity research institute; (5) ensure implementation and enforcement of existing laws; (6) acquire significant natural areas and provide additional funding for parkland operations and maintenance; (7) reduce environmental illiteracy by increasing requirements for high school and college-level courses in biology and environmental issues, and increase training of teachers in these fields; (8) broaden representation on state policy-making bodies; (9) close loopholes in the California Endangered Species Act, including those that permit destruction of endangered plants and their habitats; and (10) protect biological diversity in the face of global atmospheric change by mitigating impacts and providing leadership in slowing the rate of atmospheric change. Little action has been taken yet on any of these recommendations.

References

Axelrod, D. I. (1977). Outline history of California vegetation. In Barbour, M. and Major, J. (eds), *Terrestrial vegetation of California*. John Wiley and Sons, New York. Pp. 139–194.

Balls, E. K. (1962). *Early uses of California plants*. University of California Press, Berkeley. 103 pp.

Barbour, M. G. and Johnson, A. F. (1977). Beach and dune. In Barbour, M. and Major, J. (eds), *Terrestrial vegetation of California*. John Wiley and Sons, New York. Pp. 223–262.

Barbour, M. and Major, J. (1977). Introduction. In Barbour, M. and Major, J. (eds), *Terrestrial vegetation of California*. John Wiley and Sons, New York. Pp. 3–10.

Barbour, M., Pavlik, B., Drysdale, F. and Lindstrom, S. (1991). California vegetation: diversity and change. *Fremontia* 19(1): 3–12.

Barry, W. J. (1985). *A hierarchical classification system with emphasis on California plant communities*. California Department of Parks and Recreation, Sacramento. 130 pp.

California Department of Finance, Demographic Research Unit (1989). *Total population projections, 2020–2035.* (Report 89 P-1E). Sacramento.

California Department of Finance, Demographic Research Unit (1991). *Interim population projections for California state and counties, 1990–2005.* (Report 91 P-1). Sacramento.

California Department of Fish and Game, Endangered Plant Program (1991). Designated endangered, threatened, or rare plants. Unpublished. 5 pp.

Cheatham, N. H. and Haller, R. (1975). *An annotated list of California habitat types.* University of California Natural Reserves System, Berkeley. 80 pp.

Coates, R., Showers, M. A. and Pavlik, B. (1989). The Springtown alkali sink: an endangered ecosystem. *Fremontia* 17(1): 20–23.

Dahl, T. E. (1990). *Wetland losses in the United States: 1780's to 1980's.* U.S. Department of the Interior, Fish and Wildlife Service, Washington, D.C. 21 pp.

El-Ashry, M. T. and Gibbons, D. C. (1986). *Troubled waters: new policies for managing water in the American West.* World Resources Institute, Washington, D.C. 89 pp.

Emory, J. (1989). The Carrizo Plain: a thin slice of California's past. *Fremontia* 17(3): 3–10.

Fernald, M. L. (1970). *Gray's manual of botany.* Van Nostrand, New York. 1632 pp.

Forest and Rangeland Resources Assessment Program (1988). California's forests and rangelands: growing conflict over changing uses. California Department of Forestry and Fire Protection, Sacramento. 348 pp.

Griffin, J. R. (1977). Oak woodland. In Barbour, M. and Major, J. (eds), *Terrestrial vegetation of California.* John Wiley and Sons, New York. Pp. 383–416.

Griggs, F. T. and Zaninovich, J. M. (1984). Definitions of Tulare Basin plant associations (unpublished). The Nature Conservancy, San Francisco. 10 pp.

Hanes, T. L. (1977). Chaparral. In Barbour, M. and Major, J. (eds), *Terrestrial vegetation of California.* John Wiley and Sons, New York. Pp. 417–470.

Heady, H. F. (1977). Valley grassland. In Barbour, M. and Major, J. (eds), *Terrestrial vegetation of California.* John Wiley and Sons, New York. Pp. 491–514.

Heady H. F., Foin, T. C., Hektner, M. M., Taylor, D. W., Barbour, M. G. and Barry, W. J. (1977). Coastal prairie and northern coastal scrub. In Barbour, M. and Major, J. (eds), *Terrestrial vegetation of California.* John Wiley and Sons, New York. Pp. 733–757.

Hickman, J. (ed.) (1989). *Introduction to the Jepson manual.* Jepson Herbarium and Library, University of California, Berkeley. 74 pp.

Hickman, J. (ed.) (1993). *The Jepson manual: higher plants of California.* University of California Press. Berkeley. 1400 pp.

Holland, R. (1986). *Preliminary descriptions of the natural communities of California.* California Department of Fish and Game, Sacramento. 156 pp.

Holstein, G. (1984). California riparian forests: deciduous islands in an evergreen sea. In Warner, R.E. and Hendrix, K.M. (eds), *California riparian systems: ecology, conservation, and productive management.* University of California Press, Berkeley. Pp. 2–22.

Hoover, R. F. (1937). *Endemism in the flora of the Great Central Valley of California.* Ph.D. Dissertation, University of California, Berkeley. 76 pp.

Howell, J. T. (1955). A tabulation of California endemics. *Leaflets of Western Botany* 7(11): 257–264.

Howell, J. T. (1957). The California Floral Province and its endemic genera. *Leaflets of Western Botany* 8(5): 138–141.

Jensen, D. B. (1991). A strategy for the future of California's flora. *Fremontia* 19(2): 3–9.

Jensen, D. B., Torn, M. and Harte, M. (1990). *In our own hands: a strategy for conserving biological diversity in California.* California Policy Seminar, University of California, Berkeley. 184 pp.

Jepson, W. L. (1925). *A manual of the flowering plants of California.* University of California Press, Berkeley. 1238 pp.

Jones and Stokes Associates (1987). *Sliding toward extinction: the state of California's natural heritage.* Prepared for the California Nature Conservancy. Sacramento. 105 pp.

Jones and Stokes Associates (1990). *Environmental benefits study of San Joaquin Valley's fish and wildlife resources.* Prepared for the Federal-State San Joaquin Valley Drainage Program, Sacramento.

Jones and Stokes Associates (1992). *Reassembling the pieces: a strategy for maintaining biological diversity in California.* Prepared for the California Nature Conservancy, California Department of Forestry and Fire Protection, and California Department of Fish and Game, Sacramento.

Katibah, E. F. (1984). A brief history of riparian forests in the Central Valley of California. In Warner, R.E. and Hendrix, K.M. (eds), *California riparian systems: ecology, conservation, and productive management.* University of California Press, Berkeley. Pp. 23–29.

Krantz, T. (1983). The pebbleplains of Baldwin Lake. *Fremontia* 10(4): 9–13.

Krantz, T. (1988). Limestone endemics of Big Bear Valley. *Fremontia* 16(1): 20–21.

Kuchler, A. W. (1977). The map of the natural vegetation of California. In Barbour, M. and Major, J. (eds), *Terrestrial vegetation of California*. John Wiley and Sons, New York. Pp. 909–938.

MacDonald, K. B. (1977). Coastal salt marsh. In Barbour, M. and Major, J. (eds), *Terrestrial vegetation of California*. John Wiley and Sons, New York. Pp. 263–294.

Major, J. (1977). California climate in relation to vegetation. In Barbour, M. and Major, J. (eds), *Terrestrial vegetation of California*. John Wiley and Sons, New York. Pp. 11–74.

Matyas, W. J. and Parker, I. (1980). *Calveg, mosaic of existing vegetation of California, 1979*. California Department of Parks and Recreation, Sacramento. 27 pp.

Mooney, H. A. (1977). Southern coastal scrub. In Barbour, M. and Major, J. (eds), *Terrestrial vegetation of California*. John Wiley and Sons, New York. Pp. 471–490.

Munz, P. A. (1968). *Supplement to a California flora*. University of California Press, Berkeley. 224 pp.

Munz, P. A. (1974). *A flora of southern California*. University of California Press, Berkeley. 1086 pp.

Munz, P. A. and Keck, D. D. (1959). *A California flora*. University of California Press, Berkeley. 1681 pp.

Norris, R. M. and Webb. R. W. (1990). *Geology of California*. 2nd ed. John Wiley and Sons, New York. 541 pp.

Peck, M. E. (1961). *A manual of the higher plants of Oregon*. 2nd ed. Binfords and Mort, Portland. 936 pp.

Raven, P. (1977). The California flora. In Barbour, M. and Major, J. (eds), *Terrestrial vegetation of California*. John Wiley and Sons, New York. Pp. 109–137.

Raven, P. and Axelrod, D. I. (1978). Origin and relationships of the California flora. *University of California Publications in Botany* 72: 1–134.

Richerson, P. J. and Lum, K. (1980). Patterns of plant species diversity in California: relation to weather and topography. *The American Naturalist* 116(4): 504–536.

Rogers, D. (1991). The Santa Lucia Mountains: diversity, endemism, and austere beauty. *Fremontia* 19(4): 3–11.

Rundel, P. W., Parsons, D. J. and Gordon, D. T. (1977). Montane and subalpine vegetation of the Sierra Nevada and Cascade Ranges. In Barbour, M. and Major, J. (eds), *Terrestrial vegetation of California*. John Wiley and Sons, New York. Pp. 559–600.

Sawyer, J. O., Thornburgh, D. A. and Griffin, J. R. (1977). Mixed evergreen forest. In Barbour, M. and Major, J. (eds), *Terrestrial vegetation of California*. John Wiley and Sons, New York. Pp. 359–382.

Scher, S. (1991). Conserving the Pacific yew. *Fremontia* 19(4): 15–18.

Schmidt, M. G. (1980). *Growing California native plants*. University of California Press, Berkeley. 366 pp.

Shevock, J. R. and Taylor, D. W. (1986). Plant explorations in California: the frontier is still here. In Elias, T.S. and Nelson, J. (eds), *Conservation and management of rare and endangered plants: proceedings of a California conference on the conservation and management of rare and endangered plants*. California Native Plant Society, Sacramento. Pp. 91–98.

Smith, G. L. and Noldeke, A. M. (1960). A statistical report on a California flora. *Leaflets of Western Botany* 9(8): 117–132.

Smith, J. P., Jr. and Berg, K. (1988). *Inventory of rare and endangered vascular plants of California*. Special Publication No. 1, 4th ed. California Native Plant Society, Sacramento. 168 pp.

Stebbins, G. L. (1978a). Why are there so many rare plants in California? I. Environmental factors. *Fremontia* 5(4): 6–10.

Stebbins, G. L. (1978b). Why are there so many rare plants in California? II. Youth and age of species. *Fremontia* 6(1): 17–20.

Stebbins, G. L. and Major, J. (1965). Endemism and speciation in the California flora. *Ecological Monographs* 35(1): 1–36.

Sunset Books (1986). *The western garden book*. Lane Publishing Company, Menlo Park. 512 pp.

Thorne, R. F. (1977). Montane and subalpine forests of the Transverse and Peninsular Ranges. In Barbour, M. and Major, J. (eds), *Terrestrial vegetation of California*. John Wiley and Sons, New York. Pp. 537–558.

Twisselman, E. (1972). The Kern Plateau. *California Native Plant Society Newsletter* 8(2): 4–13.

Vogl, R. J., Armstrong, W. P., White, K. L. and Cole, K. L. (1977). The closed-cone pines and cypresses. In Barbour, M. and Major, J. (eds), *Terrestrial vegetation of California*. John Wiley and Sons, New York. Pp. 295–358.

Wiggins, I. L. (1980). *Flora of Baja California*. Stanford University Press, Stanford. 1025 pp.

Author

This Data Sheet was written by Tim Messick (Jones and Stokes Associates, 2600 V Street, Suite 100, Sacramento, CA 95818, U.S.A.).

Acknowledgements

Mr Messick thanks Reid Moran, Jim Shevock, Mark Skinner, Robert Thorne and Roger Trott for providing information and comments.

KLAMATH-SISKIYOU REGION
California and Oregon, U.S.A.

Location: South-western Oregon and north-western California at latitudes c. 41°–43°N and longitudes 121°–124°W.

Area: 55,000 km².

Altitude: 600–2280 m. Highest peak is Mount Ashland in Oregon.

Vegetation: Mixed evergreen, montane and subalpine forests; also serpentine vegetation, redwood forest and other types.

Flora: A rich flora, estimated 3500 vascular plant species, with representatives of many floristic elements; nearly 280 endemic taxa, but few palaeoendemics.

Useful plants: Timber species, medicinal species (e.g. *Taxus brevifolia*), craft and horticultural species.

Other values: Scenery, wilderness experience, white-water boating.

Threats: Logging, mining, agriculture and grazing; potentially hydroelectric dams.

Conservation: Large areas designated Wilderness Areas (although mining permitted); small areas of selected habitats protected by Federal Research Natural Area Program.

Geography

The Klamath-Siskiyou region of south-western Oregon and north-western California is an area of old and geologically complex mountains (see Map 4, California Floristic Province Data Sheet NA16). It forms a bridge between the Cascade-Sierra axis and the Pacific Ocean, and also bridges the Coast ranges of Oregon and California. It covers an area of 55,000 km² (Smith and Sawyer 1988). The land surface rose above sea-level in the Mesozoic, mainly as islands, but then as part of the continental land mass (since the beginning of the Tertiary). The Coast ranges to the north and south are much younger and the volcanoes of the Cascades only a few million years old, or less.

The topography is rugged, with numerous fault systems. Elevation ranges from 600 m in the western portion to as high as 2280 m at Mount Ashland (Oregon). The main rivers are the Rogue and Klamath rivers. The oldest rocks are schists, presumably originating in the Palaeozoic (but their exact age is uncertain due to faulting and extensive metamorphism in the late Jurassic). Thick beds of metasedimentary and volcanic rocks are of Triassic age. Granitic intrusions and folding occurred during the Nevadan Orogeny in the late Jurassic-early Cretaceous. It was preceded by intrusions of pyroxenite and peridotite which have been altered to various serpentine minerals. Deposits of sedimentary and volcanic rocks were emplaced at later times. Being the leading point of a continental plate, compression, folding, uplift and erosion have continuously affected the region.

The Klamath region is still affected by earth movements today. As a result of the complex geological history, a mosaic of parent materials is concentrated in a relatively small area, giving rise to a variety of soil types.

The climate of the region is Mediterranean, that is with mild, moist winters and a period of drought in the summer.

Near the coast, annual precipitation may exceed 2500 mm, while in interior valleys it can be less than 250 mm.

Frosts at any time are rare at low altitudes in the southern part of the region; the summit areas of higher peaks receive enough snow in the winter to support ski facilities. Severe storms with high winds almost never occur.

Vegetation

The Klamath-Siskiyou region is notable for its complex mosaic of vegetation types. It represents the contact between the Pacific Northwest Floristic Province and the Californian Province. In addition to a mixing of formations from the north and south, it contains a substantial complement of elements from the east. The presence of extensive deposits of serpentine substrates adds much to the complexity of vegetation patterns. The classic treatment of the vegetation of the area is that of Whittaker (1960) which describes three main types of vegetation on normal soils:

1. **Mixed evergreen forests**: These occur at lower elevations, inland from the coast. The only ubiquitous conifer is *Pseudotsuga menziesii*, while *Chamaecyparis lawsoniana* and *Pinus* spp. are frequent. The type is characterized by prominence of evergreen hardwood trees, *Arbutus*, *Lithocarpus*, *Castanopsis*, *Umbellularia* and *Quercus* spp. being typical. Other broadleaved trees, such as *Acer macrophyllum* and *Alnus rubra*, are equally widespread on more mesic sites (Sawyer, Thornburgh and Griffin 1977). Mixed hardwood forests and oak woodlands occur along the southern fringe of the region.

2. **Montane forests**: Above 600–1300 m are conifer-dominated forests best defined by the presence of *Abies*

concolor (Sawyer and Thornburgh 1977). *Pseudotsuga, Pinus* spp. and *Chamaecyparis* are prominent, as well as *Calocedrus*. Evergreen hardwoods are found at lower altitudes. The shrub and herb layer is rich in species. Extensive shrub fields are common on drier sites.

3. Subalpine forests: Above 1400 m there are three forest zones dominated by various conifers. The lowest is dominated by *Abies magnifica* or *A. procera,* the next by *Tsuga mertensiana* and the highest by *Pinus albicaulis.* Open meadows are found at sites where soils are too thin to support trees. The upper tree-line is defined by edaphic, rather than climatic, limits.

Serpentine soils support a distinctive vegetation, which is usually quite sparse, with the trees widely spaced and much bare ground exposed. Within these areas are pine woodlands (*Pinus jeffreyi*) at lower elevations, a forest-shrub complex with *Pseudotsuga, Chamaecyparis lawsoniana* and *Pinus monticola,* and subalpine communities with *Pinus balfouriana* (Whittaker 1960; Sawyer and Thornburgh 1977).

The Klamath-Siskiyou region, as defined here, also includes a portion of the magnificent redwood forest near the coast (Zinke 1977), coastal prairie and coastal scrub (Franklin and Dyrness 1973; Heady *et al.* 1977), Interior Valley grasslands (Franklin and Dyrness 1973), as well as fragments of typical forests of the *Tsuga heterophylla* zone (Franklin and Dyrness 1973) and Californian chapparal (Hanes 1977).

Flora

Although several botanists had visited the area earlier, the exploration of Thomas Jefferson Howell in the latter part of the 19th Century first brought to light the richness of the endemic flora of the region. He found the serpentine region around Waldo, in south-western Oregon, to be particularly rich and many new discoveries were made there. The endemic flora has recently been enumerated by Smith and Sawyer (1988). Their paper lists nearly 280 endemic taxa, many of which are infraspecific taxa. Although there are a number of species which can be considered relicts – *Picea breweriana* and *Kalmiopsis leachiana* are the best known examples – most of the endemic taxa are not palaeoendemics. Endemism above the rank of species is limited, the monotypic *Bensoniella* and *Tracyina* being the only endemic genera. Smith and Sawyer (1988) contend that the high degree of endemism at the infraspecific level reflects a pattern of recent diversification of the flora into the region's heterogeneous habitats. The open habitats of serpentine materials are particularly rich in endemics.

The area has served as a refugium for species that moved into the area in the past and which have been stranded in small sites of favourable relief and climate. The diversity of relictual species in such refugia is notable. For example there may be as many as 12 different species of conifers in one small valley (D. Axelrod, pers. comm.). At various times in the past, beginning early in the Tertiary, when the climate was warmer or cooler, wetter or drier, species moved into the area. They have come from all directions and now occur mixed together in complex plant communities. For this reason, Whittaker (1961) suggested that the area has a "central" significance.

There is no floristic treatment that covers the whole region well. The California portion of the region is covered by the new Jepson Manual (Hickman 1993), and is estimated to contain over 3500 taxa of vascular plants (Smith and Sawyer 1988). The most useful references for the area north of the California border, in addition to the Jepson Manual, are Abrams and Ferris (1923–1960), Munz and Keck (1959), Munz (1968) and Peck (1961).

Useful plants

The region, as much of the Pacific Northwest, has been a major source of timber. As the more productive forests further to the north have begun to show signs of being depleted, pressure on logging in the Klamath area has increased. The old-growth forests near the coast, in particular, have been extensively logged during the 1960s–1990s. Decorative foliage is a significant product; boughs for Christmas wreaths are exported to Europe. The bark of the western yew, *Taxus brevifolia,* is being stripped at a rapid rate for the production of taxol, a new anti-cancer drug of great promise. There is an annual harvest of beargrass (*Xerophyllum tenax)* by native Americans for trade. It is used to make fine baskets. Commercial mushroom picking has increased dramatically in recent years. Grasslands have value as livestock forage; grazing permits are issued in suitable areas. Many species are of horticultural value; several nurseries specialize in rare plants of the area.

Social and environmental values

As there are no roads throughout much of the region, its remoteness provides significant amounts of land for people seeking wilderness experiences. The Rogue and Klamath rivers are famous for spectacular scenery and exciting white water, and are used for kayak, canoe and rafting expeditions. The pressure is so intense that a permit system is in effect on both of these rivers during the most popular months. Other forms of outdoor recreation are practised extensively.

Economic assessment

The bulk of revenue generated from forestry comes from four forests in the area. In fiscal year 1990, nearly US$80,000,000 was received from cut timber. The Bureau of Land Management also manages forests in the area, but data are not available. Lesser amounts (less than $50,000 per item per forest) are received from grazing fees, campground fees, power generation, land use fees, mineral leases and other recreation fees. Mining has also been an important activity in the area.

Threats

Old-growth stands of trees have been much reduced in extent, with most outside the Wilderness Areas or parks scheduled for clear-cutting. Placer mining has disturbed many of the valley gravel deposits in the past. There are plans for utilizing the weathered soils for extraction of chromium and other rare metals; this process could have significant impacts on large areas, since only the overburden is desired. Over-collecting by commercial plant dealers has reduced some populations of rare plants. Road building has had a similar effect on the periphery of the area. Off-road vehicles damage open slopes where the substrate is unstable. Agriculture and urbanization has eliminated almost all of the valley bottom grasslands, while the grassy areas on hillsides are grazed by livestock. Many of the watersheds on the eastern side of the region, especially in their upper reaches, have been severely impacted by nearly a century of overgrazing. There is pressure to develop the hydroelectric potential of some of the rivers. Conflicts have led to halting construction on a partially completed dam in Oregon.

Conservation

Much of the area is protected by its remoteness and the rugged nature of the terrain. Large areas have been designated Wilderness Areas (although mining is still permitted in such areas). The central and western Siskiyous of Oregon are within the Siskiyou National Forest and the eastern Siskiyou in the Rogue River National Forest (Whittaker 1960). Extensive Wilderness Areas have been set aside in the Yolla Bolly (2891 ha), Salmon and Trinity Alps (1871 ha), and the Marble Mountains (97,831 ha) (IUCN 1992); and small areas of the coastal redwood forests are protected as parks. Several conservation groups have proposed a Siskiyou National Park, but the likelihood of this proposal being implemented seems remote at present. The federal Endangered Species Act is having significant impact on conservation, especially in areas that are classified as old-growth forests. Both Oregon and California have state endangered species legislation. Oregon has very strong land use laws that specify conservation of natural values, although they are not always carefully enforced in the less-populated counties. The Federal Research Natural Area Program is important for protecting small areas of selected habitats. State programs are less successful in this area; private groups such as The Nature Conservancy are more effective. Conservation organizations such as the Sierra Club, the Audubon Society, the Wilderness Society and the Oregon Natural Resources Council are important for promoting conservation issues in the political arena.

References

Abrams, L. and Ferris, R.S. (1923-1960). *Illustrated flora of the Pacific states. Vols. 1–4.* Stanford University Press, Stanford.

Franklin, J.F. and Dyrness, C.T. (1973). *Natural vegetation of Oregon and Washington.* USDA Forest Service Gen. Tech. Rep. PNW-80. Portland.

Hanes, T.L. (1977). California chaparral. In Barbour, M.G. and Major, J. (eds), *Terrestrial vegetation of California.* John Wiley and Sons, New York. Pp. 417–469.

Heady, H.F., Foin, T.C., Hektner, M.M., Taylor, D.W., Barbour, M.B. and James Barry, W. (1977). Coastal prairie and northern coastal scrub. In Barbour, M.G. and Major, J. (eds), *Terrestrial vegetation of California.* John Wiley and Sons, New York. Pp. 733–760.

Hickman, J. (ed.) 1993. *The Jepson manual: higher plants of California.* University of California Press, Berkeley. 1400 pp.

IUCN (1992). *Protected areas of the world. A review of national systems. Vol. 4: Nearctic and Neotropical.* IUCN, Gland, Switzerland and Cambridge, U.K. (Compiled by the World Conservation Monitoring Centre.) 459 pp.

Munz, P.A. (1968). *Supplement to a California flora.* University of California Press, Berkeley. 224 pp.

Munz, P.A. and Keck, D.D. (1959). *A California flora.* University of California Press, Berkeley. 1681 pp.

Peck, M.E. (1961). *A manual of the higher plants of Oregon.* 2nd ed. Binsford and Mort, Portland.

Sawyer, J.O. and Thornburgh, D.A. (1977). Montane and subalpine vegetation of the Klamath Mountains. In Barbour, M.G. and Major, J. (eds), *Terrestrial vegetation of California.* John Wiley and Sons, New York. Pp. 699–732.

Sawyer, J.O., Thornburgh, D.A. and Griffin, J.R. (1977). Mixed evergreen forest. In Barbour, M.G. and Major, J. (eds), *Terrestrial vegetation of California.* John Wiley and Sons, New York. Pp. 359–381.

Smith, J.P. and Sawyer, J.O. Jr. (1988). Endemic vascular plants of northwestern California and southwestern Oregon. *Madroño* 35(1): 54–69.

Whittaker, R.H. (1960). Vegetation of the Siskiyou Mountains, Oregon and California. *Ecological Monographs* 30(3): 279–338.

Whittaker, R.H. (1961). Vegetation history of the Pacific coast states and the "central" significance of the Klamath region. *Madroño* 16(1): 5–23.

Zinke, P.J. (1977). The redwood forest and associated north coast forests. In Barbour, M.G. and Major, J. (eds), *Terrestrial vegetation of California.* John Wiley and Sons, New York. Pp. 679–698.

Author

This Data Sheet was written by Dr David H. Wagner (Biology Department, University of Oregon, Eugene, Oregon 97403, U.S.A.).

NORTH AMERICAN SERPENTINE FLORA
U.S.A. and Canada

Location: NA16e – Western North America: California, Oregon, Washington and British Columbia; NA25 – Eastern North America: Alabama, Georgia, Maryland, Massachusetts, New Jersey, New York, North Carolina, Pennsylvania, South Carolina, Vermont, Virginia, Quebec and Newfoundland. Found scattered throughout North America between latitudes 35°–70°N and longitudes 55°–135°W.

Area: California (2860 km²); Oregon (1170 km²); Washington (520 km²); Newfoundland (3000 km²); other areas undetermined.

Altitude: Western North America: 0–2900 m. Eastern North America: 0–1500 m.

Vegetation: Grasslands, chaparral, woodlands and forest.

Flora: High species diversity (over 2000 taxa); high species endemism, including neo- and palaeoendemics.

Useful plants: Genetic resources for agricultural relatives for adaption to low nutrient soils and soils contaminated with heavy metals; drought tolerant.

Other values: Refuges for native plants with low levels of non-invasive plants.

Threats: Mining, urbanization, off-road vehicles, logging, road-building.

Conservation: Frenzel Creek Research Natural Area (270 ha); Butterfly Valley Botanical Area in Plumas National Forest; Harrison Grade Ecological Reserve in Sonoma County; Ring Mountain Preserve in Marin County (25 ha); Cedars Area of Critical Environmental Concern in Lake and Napa counties.

Geography

Serpentine soil habitats occur in areas that have been geologically active, particularly where plate tectonics has occurred and uplifting of altered peridotites has exposed these formations (Brooks 1987; Coleman and Jove 1992). These formations are known throughout the world and often form linear zones that indicate the margins of old continents (Coleman and Jove 1992). These linear zones are very apparent in North America where two major zones occur, one on the western side of the continent from mid California, U.S.A., north to British Columbia, Canada, and another on the eastern part of the continent from the state of Alabama, U.S.A., north-east to Quebec, Canada.

In western North America (CPD Site NA16e), serpentine soil habitats occur along fault zones in the Central and North Coast and Cascade ranges near the subduction contact of the Pacific plate with the North American plate. Western Sierra Nevadan serpentine habitats are associated with the uplifting of the Sierran batholith and regions of vulcanism. The geographic area of serpentine habitats in the western U.S.A. includes California (2860 km²), Oregon (1170 km²) and Washington (520 km²), and British Columbia in western Canada (Map 5). The serpentine habitats in western North America occur from sea-level to an elevation of 2900 m. Precipitation in western North American areas falls primarily between October and March; mean annual rainfall increases from 600 mm in the south to 2000 mm in the north-west. Some of the precipitation falls in the form of snow.

In eastern North America (CPD Site NA25), serpentine habitats occur in the U.S.A. in the states of Alabama, Georgia, South and North Carolina, Virginia, Maryland, New Jersey, Pennsylvania, New York, Massachusetts and Vermont (Reed 1986; Brooks 1987), and in Quebec (Brooks 1987) and Newfoundland (3000 km²) (Roberts 1992) in Canada (Map 5). The serpentine areas of eastern North America occur at elevations from near sea-level to 1500 m. Precipitation occurs primarily from October to April with substantial amounts in the form of snow in the northern parts. In the southern areas, such as Alabama and Georgia, summer monsoons (July through September) may contribute to the annual precipitation.

Ecological factors

Serpentine soil habitats are distinct due to a variety of influencing factors not controlled by local climate or elevational conditions. Specifically, the soil chemical and physical characteristics make these habitats very poor in terms of nutrients, sometimes toxic due to the presence of heavy metals, and potentially having lower soil moisture availability. The basic nutrient limitations of low nitrogen and phosphorus exist in serpentine soils. In addition, there is an imbalance between calcium and magnesium, where calcium is relatively low compared to what are often high concentrations of magnesium (Kruckeberg 1984; Brooks 1987). The heavy metals of chromium and nickel can occur in concentrations high enough to invoke toxic conditions for most plant species (Brooks 1987). In addition, the mineralogical conditions and equilibrium conditions of them has made serpentine rocks highly erodible under normal atmospheric conditions. The montmorillonite clays formed by the minerals adsorb more water than many other clay surfaces, thus reducing the available water to plants. All these factors reduce the ability of plants to adapt to serpentine soil habitats and has led to a highly specialized flora.

Vegetation

Vegetation types occurring in serpentine soil habitats include grasslands, chaparral, woodlands and forest. Also, in some areas there are extensive serpentine habitats referred to as serpentine barrens which are sparsely vegetated by annual and perennial herbaceous plant species.

In the western U.S.A. serpentine grasslands are relatively uncommon and mostly occur near the Pacific coast such as in the San Francisco Bay region of California (McCarten 1986, 1987). There are approximately 5000 ha of serpentine grassland in the western U.S.A. In California, grassland dominants include the native perennial bunchgrasses; *Melica imperfecta, M. torreyana, Poa scabrella, Sitanion hystrix,* and *Stipa pulchra.* In disturbed grassland areas, especially where extensive cattle grazing occurs, non-native grasses such as *Lolium multiflorum, Bromus mollis,* and *B. rubens* can be dominants. Locally, the grassland habitats can have a dominance of non-grass herbaceous plant species including the genera *Allium, Gilia, Layia, Lasthenia, Microceris, Phacelia, Plantago* and *Zygadenus* (Kruckeberg 1984; McCarten 1986, 1988a).

Serpentine chaparral vegetation is also restricted to the western U.S.A., primarily in California. This Mediterranean vegetation type is the more common serpentine vegetation in

California and collectively covers over 1000 km². The dominant plant species include *Quercus durata* and *Arctostaphylos viscida.* These two species often occur in nearly pure stands; however, mixtures of these and other species occur in habitat gradients where other *Arctostaphylos* species may occur with *Garrya congdonii, Rhamnus californica* and several species of *Ceanothus* (McCarten and Rogers 1991).

Serpentine woodland vegetation occurs in a variéty of types both in western and eastern North America. In western North America, serpentine woodlands include evergreen oak woodlands, often dominated by *Quercus chrysolepis,* and pine woodlands that are generally dominated by *Pinus sabiniana.* In eastern North America pines are also a dominant in woodland vegetation.

Forests on serpentine soils are extremely uncommon due to the low nutrient levels in the soil. However, some areas do have a higher density vegetation particularly montane areas with higher rainfall such as the Cascade Mountains in northern California and southern Oregon. In those areas *Pinus jeffreyi* and *P. sabiniana* form patchy forested areas. These forests are often interrupted by open areas of serpentine barrens, or as a function of slope the steeper areas may support chaparral or woodlands.

The serpentine barrens are recognized as a major habitat in serpentine areas throughout the world (Kruckeberg 1984; Brooks 1987). The areas are often exposed areas with dramatic changes in topography such as steep, high erodible slopes. The serpentine barrens are often areas where high concentrations of toxic heavy metals occur in the serpentine

MAP 5. SERPENTINE AREAS OF NORTH AMERICA

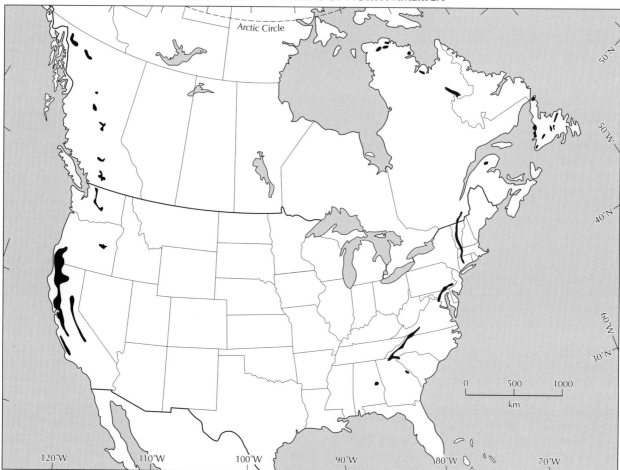

minerals. Extensive studies have been conducted on plants adapted to these serpentine barrens where heavy metals occur (Kruckeberg 1984; Brooks 1987; Baker, Proctor and Reeves 1992).

In general, the vegetation types on serpentine soils are representative of the vegetation one might expect based on climate and elevation. Two notable differences are that for some forest and woodland species their elevational zonation may be several hundred metres lower than in non-serpentine areas having the same climate and exposure (Kruckeberg 1984); the other difference is the lack of deciduous species which are predominately only found in very moist sites such as along creeks and watercourses.

Flora

The serpentine flora is recognized throughout the world for its high level of diversity (Brooks 1987; Baker, Proctor and Reeves 1992). In North America mostly local floristic studies have been conducted (McCarten 1986, 1988a), with the exception of Reed (1986) who has developed a detailed flora for serpentine areas in the eastern U.S.A. Reed's flora (1986) includes over 1600 taxa of plants that occur in serpentine areas of the eastern U.S.A. From that flora it is not clear how many may be endemic to serpentines and what proportion of the native endemic flora of the regions or states covered is represented. Reed's finding of a very high number of taxa of plants growing in serpentine habitats is consistent with other serpentine floras, including western North America where the flora may exceed 2000 taxa (Kruckeberg 1984; Brooks 1987; McCarten 1986, 1988a; McCarten and Rogers 1991).

The most significant aspect of the serpentine floras of North America, and elsewhere in the world, is the diversity which is primarily due to the high percentage of serpentine endemics. In California, Kruckeberg (1984) estimated that the serpentine endemic plant species represents 10% of the 2300 species of plants that are endemic to the California Floristic Province. Other estimates indicate that the serpentine endemics may be as high as 15% of the endemic flora (McCarten 1987). In Newfoundland, Canada, Roberts (1992) identified 190 plant taxa as endemic to the serpentine flora of that area.

Many of these serpentine endemics are recognized as rare and endangered plant species (Kruckeberg 1984; McCarten 1986, 1987, 1988a; McCarten and Rogers 1991; Callizo 1992). Over 25% of the 1300 plant taxa considered to be rare in California are represented by taxa that occur on serpentine soils (McCarten 1988a, 1995). Important genera that are broadly recognized as having speciated into and with the serpentine soil habitats in western North America include *Allium, Brodiea, Calochortus, Calycadenia, Cordylanthus, Delphinium, Eriogonum, Hesperolinon, Phacelia*, and *Streptanthus* to name a few (Kruckeberg 1984; McCarten 1988a).

In relation to the plant species diversity that occurs in serpentine soil habitats the low number of non-native plants, such as introduced weeds, is important. Due to the extreme habitat conditions, non-native species only occur in areas where the serpentine soils have been disturbed or altered such as through the addition of fertilizer. While many vegetation types have been invaded by non-native species, particularly in the Mediterranean climate areas of western North America, the serpentine areas have been resistant to this invasion. In undisturbed areas the local flora may be represented by over 95% native plant species. This is significant since many other areas of non-serpentine grassland and chaparral have as much as 50% or more of the flora non-native species.

Useful plants

Two examples of useful plants are the genus *Hesperolinon* (Linaceae) which is related to the flax of commerce (*Linum usitatissimum*), and *Helianthus exilis*, which is related to sunflower (*Helianthus annuus*). *Helianthus exilis* represents a novel genetic resource of potential agronomic value in which some of the characters of interest are its high linoleic acid content, cold tolerance and genetic variation in germinability (Jain, Kesseli and Olivieri 1992).

Social and environmental values

The significance of serpentine habitats and the vegetation that occur in them include all the reasons for recognizing values for plant conservation in general. Recognized social and environmental values include: (1) economic value of plants as resources for human uses, (2) role of plants in maintaining environmental stability, (3) the scientific value as a tool for understanding ecological processes, and (4) cultural and inspirational values (Given 1994).

The social values of specific serpentine plants is reflected in the few species that were mentioned as being useful plants, especially for their potential use in agriculture (Jain, Kesseli and Olivieri 1992). Other examples involve the use of serpentine adapted plants that have been identified as hyperaccumulators of heavy metals, such as nickel, that have been used in bioremediation of mine tailings (Kruckeberg 1984; Brooks 1987). The specificity of particular plant groups and species as endemics on serpentine soils having high concentrations of nickel has been suggested to be a means of identifying areas that produce these economically important minerals (Reeves 1992).

The environmental values of North American serpentine habitats follow closely from the high degree of plant species diversity and endemism in these areas. Serpentine endemics often occur in very localized areas and contribute significantly to local floral diversity. A significant number of serpentine endemics, or species that predominately grow on serpentine soils are either neo-endemics or palaeo-endemics (Raven and Axelrod 1978; Kruckeberg 1984). Representation of the primitive flora and the new rapidly speciating flora is critical to studies on plant evolutionary processes. Further, the serpentine floras in relatively undisturbed areas are generally represented by a higher percentage of native species (McCarten 1986, 1988a). The low nutrient conditions of the serpentine soils often limits the ability of non-native weedy plants to invade these habitats. Therefore, serpentine habitats are one of the few remaining areas where a significant level of invasion by non-native weedy plants has not occurred. Serpentine native bunchgrass areas are considered to potentially represent examples of non-serpentine native grassland plant communities prior to the invasion of non-native Mediterranean grasses.

Threats

Threats have been from industrial, urban, recreational, and occasionally agricultural development and other activities. Historically, mining for heavy metals including chromium, nickel, and mercury, have had a significant negative impact on serpentine habitats (Kruckeberg 1984; Brooks 1987; McCarten 1987, 1988a). Mining for heavy metals has been reduced. More recently, however, gold mining has begun to have an impact by processing large quantities of rock that are often associated with serpentine habitats. Gold mining activities often have both indirect and direct effects on serpentine areas by physically altering the surface and subsurface areas to gain access to the gold-bearing rocks. Indirect, but possibly more destructive with regard to the vegetation, is alteration of surface and ground water hydrology by changing the direction of natural precipitation run-off and adding water to some areas. Attempts to restore mine tailings have not always recognized the unique ecological conditions of serpentine habitats and fertilizers have been used in an attempt to grow plants not adapted to serpentine soils. In those cases the restoration experiments have failed.

Urbanization from housing developments and the development of infrastructural systems, such as major highways and roads, have often utilized the apparent poorly vegetated areas of serpentine habitat. These urban forms of development have directly reduced the amount of serpentine area particularly in the San Francisco Bay region of California (McCarten 1986, 1987). Indirect impacts from urban development have been the alteration of surface and ground water hydrology that altered the vegetation of any area, sometimes affecting endangered plant species populations (McCarten 1987). In addition, use of water, fertilizer, and pesticides has reduced the native endemic plant species ability to grow in some areas and allowed non-native weedy plants to invade serpentine habitats.

Recreational activities, particularly by off-road vehicles, have damaged often sensitive serpentine habitats. Specifically, areas that are perceived as unvegetated barrens, which often support endangered herbaceous plant species, are used for off-road vehicle use. These activities directly damage the plant species, increase the erosion rates of the soils, and otherwise physically damage the rare or unique conditions that occur in these areas.

Agricultural activities rarely affect serpentine areas due to the low nutrient levels, occurrence of toxic levels of heavy metals, and overall poor soil conditions. However, through liming and the use of fertilizers some lands have been converted to low quality arable lands. The more destructive agriculturally-related practise has been forestry. Even though serpentine areas are generally low in forest productivity these lands have been cut for timber resulting in accelerated erosion of the shallow soil. Also, related activities, such as road building and development of heavy equipment staging areas, has also directly damaged the serpentine soil habitat. Natural recolonization of disturbed serpentine soils is generally slow often taking decades for vegetation to become established.

Conservation

Relatively little of the serpentine habitats in North America are preserved despite the numerous unique features and rare and endangered plant species (Kruckeberg 1984, 1992; McCarten 1987, 1995). Some preserves are managed by U.S. federal and state agencies, as well as some private conservation organizations.

Conservation areas in California: the U.S. Department of Agriculture's National Forest Service maintains some research natural areas (RNAs) including the Frenzel Creek RNA in the Mendocino National Forest, Colusa County. The Frenzel Creek RNA (270 ha) includes seven rare plant species and the plant communities of serpentine barrens, extensive chaparral, and a corridor of riparian vegetation including MacNab cypress (*Cupressus macnabiana*). Other RNAs may exist that include serpentine habitats, but the preserves are not specifically designed to protect these habitats. The Forest Service also maintains areas referred to as Botanical Areas (BAs) that have been identified as important for botanical resources such as the presence of rare and endangered plant species. The Butterfly Valley BA, in the Plumas National Forest, Plumas County, includes a wetland area known as a fen which flows over serpentine formations. The serpentine fen supports a range of wetland plants including the insectivorous cobra lily (*Darlingtonia californica*).

The U.S. Bureau of Land Management (BLM) maintains an Area of Critical Environmental Concern (ACEC) that includes serpentine habitat (McCarten 1988a). The Cedars ACEC, Lake and Napa counties, supports several plant communities including MacNab cypress woodland and extensive chaparral.

The California State agency, Department of Fish and Game, manages a serpentine area known as the Harrison Grade Ecological Reserve, in Sonoma County. This reserve includes Sargent cypress (*Cupressus sargentii*) woodland, serpentine chaparral, and serpentine seeps supporting a distinct wetland vegetation (McCarten 1988b). In addition, this reserve is habitat for a rare plant, Pennel's birds-beak (*Cordylanthes tenuis* subsp. *capillaris*).

The California Nature Conservancy manages the Ring Mountain Preserve in Marin County. This preserve, covering approximately 25 ha of serpentine habitat, is the only reserve protecting native serpentine bunchgrass habitat (McCarten 1986). In addition, four rare plants are known from the preserve including two endangered species, Tiburon Mariposa lily (*Calochortus tiburonensis*) and the Marin western flax (*Hesperolinon congestum*).

Numerous other sites have been identified in the North Coast ranges of California as areas needing preservation due to their unique floras and habitats associated with serpentine (McCarten 1988a, 1995). Currently, a serpentine ecosystem study of California is being conducted to identify areas needing preservation in the state (McCarten 1995).

In eastern North America, several sites of serpentine habitat are being preserved and managed by The Nature Conservancy. Details of the location, size, and flora were not available.

Overall, conservation efforts for protecting serpentine habitats and the associated flora have been poor even though these areas are widely recognized for their ecological value and floristic diversity. Regional efforts are needed to identify plant communities and floristic associations that should be protected.

References

Baker, A., Proctor, J. and Reeves, R. (eds) (1992). *Vegetation of ultramafic (serpentine) soils: proceedings of the first international conference on serpentine ecology.* Intercept Limited, U.K. 509 pp.

Brooks, R. (1987). *Serpentine and its vegetation.* Dioscorides Press, Portland. 454 pp.

Callizo, J. (1992). Serpentine habitats for the rare plants of Lake, Napa, and Yolo counties, California. In Baker, A., Proctor, J. and Reeves, R. (eds), *The vegetation of ultramafic (serpentine) soils: proceedings of the first international conference on serpentine ecology.* Intercept Limited, U.K. Pp. 35–53.

Coleman, R. and Jove, C. (1992). Geologic origin of serpentinites. In Baker, A., Proctor, J. and Reeves, R. (eds), *The vegetation of ultramafic (serpentine) soils: proceedings of the first international conference on serpentine ecology.* Intercept Limited, U.K. Pp. 1–18.

Given, D. (1994). *Principles and practice of plant conservation.* Chapman and Hall, London. 292 pp.

Jain, S., Kesseli, R. and Olivieri, A. (1992). Biosystematic status of the serpentine sunflower, *Helianthus exilis* Gray. In Baker, A., Proctor, J. and Reeves, R. (eds), *The vegetation of ultramafic (serpentine) soils: proceedings of the first international conference on serpentine ecology.* Intercept Limited, U.K. Pp. 391–408.

Kruckeberg, A. (1984). *California serpentines: flora, vegetation, geology, soils, and management problems.* University of California Press, Berkeley. 180 pp.

Kruckeberg, A. (1992). Plant life of western North American ultramafics. In Roberts, B. and Proctor, J. (eds), *The ecology of areas with serpentinized rocks: a world view.* Kluwer, Dordrecht. Pp. 31–74.

McCarten, N. (1986). *Serpentine flora of the San Francisco Bay region.* California Endangered Plant Program, California Department of Fish and Game, Sacramento. 93 pp.

McCarten, N. (1987). Serpentine plant communities of the San Francisco Bay region. In Elias, T. (ed.), *Conservation and management of rare and endangered plants: proceedings from a conference of the California Native Plant Society.* California Native Plant Society, Sacramento. Pp. 335–340.

McCarten, N. (1988a). *Rare and endemic plants of Lake County serpentine soil habitats.* California Endangered Plant Program, California Department of Fish and Game, Sacramento. 137 pp.

McCarten, N. (1988b). *Habitat management plan for the Harrison Grade Ecological Reserve.* Technical report for the California Department of Fish and Game, Sacramento. 35 pp.

McCarten, N. (1995). Conservation strategies for an ecosystem level approach to preserving plant biodiversity in serpentine soil habitats. *Conserv. Biol.* (in review).

McCarten, N. and Rogers, C. (1991). *Habitat management study of rare plants and communities associated with serpentine soil habitats in the Mendocino National Forest.* U.S. Department of Agriculture, Mendocino National Forest, Colusa. 70 pp.

Raven, P. and Axelrod, D. (1978). Origin and relationships of the California flora. *Univ. of California Publ. in Botany* 72: 1–134.

Reed, C. (1986). *Floras of the serpentinite formations of eastern North America.* Contributions to the Reed Herbarium No. 30, Baltimore. 858 pp.

Reeves, R. (1992). The hyperaccumulation of nickel by serpentine plants. In Baker, A., Proctor, J. and Reeves, R. (eds), *The vegetation of ultramafic (serpentine) soils: proceedings of the first international conference on serpentine ecology.* Intercept Limited, U.K. Pp. 253–278.

Roberts, B. (1992). The serpentinized areas of Newfoundland, Canada: a brief review of their soils and vegetation. In Baker, A., Proctor, J. and Reeves, R. (eds), *The vegetation of ultramafic (serpentine) soils: proceedings of the first international conference on serpentine ecology.* Intercept Limited, U.K. Pp. 53–66.

Author

This Data Sheet was written by Dr Niall F. McCarten (CH$_2$ Hill, P.O. Box 12681, Oakland, CA 94604-2681, U.S.A. and University of California Herbarium, University of California, Berkeley, California 94720, U.S.A.).

VERNAL POOLS
California, U.S.A. and Baja California, Mexico

Location: California Floristic Province (CFP), concentrated in Great Valley grasslands, especially on terrace soils along the eastern edge, less abundantly on old mesas along south coast of California and Baja California, and in northern San Francisco Bay Area; also less commonly in mountain valleys. Found between latitudes 30°–42°N and longitudes 116°–124°W.

Area: Vernal pool habitat covers c. 20,000 km² within the CFP.

Altitude: Mostly 0–600 m, ranging to 2500 m.

Vegetation: Ephemeral pools, formed by winter rains of California's Mediterranean climate, support a distinct vegetation type containing a large number of showy annuals that flower in decreasing concentric rings as the pools dry.

Flora: Diverse for a limited habitat type, with over 200 vascular plant species in California; very high degree of endemism: 69% restricted to the CFP; about one-third of the vernal pool flora is endangered.

Useful plants: Mostly unknown; artificial selection of *Limnanthes floccosa* var. *californica* oils in progress; *Orcuttia* exudates may prove interesting; some use of annual ornamentals.

Other values: Aesthetic values unsurpassed: concentric rings of brightly coloured flowers a powerful incentive to preserve grassland and other habitat in California; obligate pollination relationships with andrenid bees are of great ecological and evolutionary interest; pools contain endangered crustaceans and amphibians, and provide essential habitat for migrating and overwintering birds and endemic insects.

Threats: Agriculture, heavy grazing, gold mining and urban development have destroyed 90–95% of vernal pool habitat in California.

Conservation: Acquisition of existing vernal pool habitat urgently needed; very few are protected adequately; creation of artificial vernal pools are a threat to preservation of existing pools.

Geography

Vernal pools form in small depressions underlain by a subsurface layer that impedes drainage, so that water stands sufficiently long during the winter rainy season to prohibit terrestrial vegetation from developing. Many vernal pool plants flower in showy concentric rings as pools dry out. Standing water during the winter, followed by total desiccation during the dry Mediterranean summer, are essential for the development of vernal pool vegetation and for exclusion of strictly aquatic or strictly dryland species (Holland 1976).

Soil texture, moisture holding capacity, salinity and microrelief are important determinants of vernal pool vegetation (Holland and Jain 1988). Many vernal pool soils are very old (600,000–4,000,000 years BP) (Harden 1987) and prolonged profile development has led to formation of aquatards (commonly called duripans, hardpans or claypans) which create locally perched water-tables (Holland and Jain 1988). As the water evaporates, ionic concentrations rise in the pools and in surrounding soils, resulting in saline, alkaline or acidic conditions (Holland and Jain 1988).

Conditions for vernal pool formation occur in much of the CFP where various impediments to rapid drainage allow pooling and subsequent evaporation. Holland (1978) estimates that 16,800 km² of vernal pool habitat originally occurred in the Great or Central Valley, which had the vast majority of such habitat within the CFP. Vernal pool vegetation is most developed on terrace soils along the east side of the Great Valley at the base of the Sierra foothills (Hoover 1937), but there were once abundant pools on terraces along the western valley margin and on the valley floor. Old marine terraces in San Diego County and adjacent Baja California support pools with endemic plant species (Holland 1976) (Map 6). Well developed pools also occur on soils derived from volcanic mudflows in the northern Great Valley, on lava-capped mesas in the foothills of the northern Sierra Nevada in Butte County, on the Modoc Plateau, and in the Peninsular ranges in western Riverside County (Holland 1976). Pools are rarer in Sierran foothill valleys and in the Coast Ranges from San Luis Obispo County north to Lake County.

Ephemeral pools supporting endemic-rich floras are found in other parts of the world, especially where a Mediterranean climate of winter rains and summer drought occurs (Zedler 1987). Perhaps the closest analogues to California's pools are found in Spain, western Australia, western South America and South Africa. Genera such as *Eryngium, Lythrum, Isoetes, Myosurus* and *Callitriche* are found in common between vernal pools in Spain and California (Zedler 1987), but these plants are cosmopolitan and broadly aquatic rather than limited to vernal pools. Stronger floristic affinities occur with vernal pools in Chile and Argentina, and involve the amphitropic distributions of *Lasthenia, Downingia, Navarretia* and *Legenere*, all genera more distinctively associated with California vernal pools.

Vegetation

The characteristic feature of ephemeral pools is a unique vegetation type distinguished by a preponderance of showy annuals that flower in decreasing concentric rings as the pools dry (Holland and Jain 1988). Growth form ranges from diminutive plants 3–5 cm tall in *Myosurus, Crassula* (*=Tillaea*) and *Psilocarphus* to 30–50 cm tall *Eryngium, Navarretia* and *Pogogyne* (Jain 1976).

Vernal pools are most common and conspicuous in open grassland, but occur under a variety of tree canopies ranging from *Quercus lobata* (valley oak) or *Q. douglasii* (blue oak) savannah to *Pinus ponderosa* (yellow pine) forests (Holland and Jain 1988). Pools are also known from understorey herbaceous cover in the Coast, Peninsular and Transverse ranges, from sagebrush steppe (R.F. Holland, pers. comm.) and as vernal ponds in dune slacks (Holland and Jain 1988).

Vernal pools vary in size and have been given many names, but the smallest are often called vernal swales and retain standing water only briefly or not at all; intermediate depressions are typically called vernal pools, and the largest (over 20 ha) are frequently referred to as vernal lakes (Crampton 1976; Holland 1976). Taxa found in the bottom of smaller, shallower pools often occur along the margins of deeper pools (Holland 1976), whereas other species, particularly members of the unusual grass tribe Orcuttieae (Griggs 1976; Stone *et al.* 1988), are most typically found only in the deepest pools or vernal lakes.

Flora

Over 200 species of vascular plants (Holland 1976) in about 56 genera (Jain 1976) are known from vernal pools in California. About half of these plants are known primarily from vernal pools (Holland and Jain 1988) and more than 70% are native annuals that reproduce by seed, mostly through selfing (Zedler 1987). Introduced annuals comprise less than 7% of the species, and most of these are from South America (Holland and Jain 1988). The only perennial genera that typically occur in vernal pools in California are *Isoetes, Marsilea, Eleocharis, Eryngium* and *Lythrum* (Jain 1976). Ferns are uncommon and represent less than 3% of the species total. Most sizeable pools contain 15–25 vascular plant species (Holland 1976), but never the same ones, so no two vernal pools are alike in species composition. A list of some important genera in vernal pools is given in Table 19.

Vernal pools contain a remarkable part of the Californian flora that has resisted the invasion of primarily Mediterranean exotic plants better than most other Californian vegetation types (Holland 1976). In fact, 91% of vascular plants currently found in California vernal pools are natives (Holland 1976). This contrasts favourably with surrounding grasslands which frequently comprise 35–40% exotic annual grass and herbaceous species, in which 99% of annual biomass is contributed by exotics (R.F. Holland, pers. comm.). Undoubtedly the sharp contrasts in soil hydrology caused by cycles of inundation and desiccation promote specific adaptations that are lacking in most exotic plants, hence their inability to survive in vernal pools. Vernal pools are thus "too wet for exotics, too dry for aquatics" (R.F. Holland, pers. comm.).

The families Asteraceae, Poaceae, Campanulaceae (Lobelioideae), Limnanthaceae, Scrophulariaceae and Marsileaceae are well-represented in California's vernal pool flora (Jain 1976). Endemism is exceedingly high among California vernal pool flora; about 69% occur only in the CFP

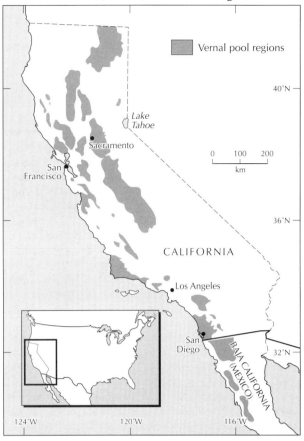

MAP 6. VERNAL POOLS, CALIFORNIA AND BAJA CALIFORNIA (CPD SITE NA16g)

TABLE 19. IMPORTANT GENERA AND THEIR FAMILIES OF CALIFORNIA VERNAL POOL PLANTS

Genus	Family
Atriplex	Chenopodiaceae
Bergia	Elatinaceae
Blennosperma	Asteraceae
Boisduvalia	Onagraceae
Brodiaea	Liliaceae
Callitriche	Callitrichaceae
Crassula (*=Tillaea*)	Crassulaceae
Damasonium	Alismataceae
Deschampsia	Poaceae
Downingia	Campanulaceae
Elatine	Elatinaceae
Eryngium	Apiaceae
Gnaphalium	Asteraceae
Gratiola	Scrophulariaceae
Isoetes	Isoetaceae
Juncus	Juncaceae
Lasthenia	Asteraceae
Legenere	Campanulaceae
Limnanthes	Limnanthaceae
Limosella	Scrophulariaceae
Marsilea	Marsileaceae
Myosurus	Ranunculaceae
Navarretia	Polemoniaceae
Neostapfia	Poaceae
Orcuttia	Poaceae
Pilularia	Marsileaceae
Plagiobothrys	Boraginaceae
Pleuropogon	Poaceae
Pogogyne	Lamiaceae
Porterella	Campanulaceae
Psilocarphus	Asteraceae
Ranunculus	Ranunculaceae
Rorippa	Brassicaceae
Sidalcea	Malvaceae
Tuctoria	Poaceae
Veronica	Scrophulariaceae

Source: Jain 1976; Thorne 1981

and 55% are found only in California itself (Holland 1976). For example, eight of nine members of the grass tribe Orcuttieae are restricted to vernal pools within the CFP; all are rare and endangered (Skinner and Pavlik 1994). Most species of *Downingia* and *Limnanthes* are endemic to vernal pools within the CFP.

Centres of endemism within California's vernal pool flora lie in Sonoma County and adjacent Lake and Mendocino counties. They support several endemic and endangered taxa, including *Blennosperma bakeri*, *Eryngium constancei*, *Limnanthes bakeri*, *L. vinculans*, *Lasthenia burkei*, *Navarretia leucocephala* subsp. *pauciflora*, *N. leucocephala* subsp. *pleiantha*, *Parvisedum leiocarpum* and *Pogogyne douglasii* subsp. *parviflora*. The Great Valley is isolated by the Coast Ranges and contains the bulk of vernal pool endemics, including most members of the grass tribe Orcuttieae, and endemics in *Chamaesyce*, *Downingia*, *Eryngium*, *Gratiola*, *Juncus*, *Limnanthes*, *Navarretia* and *Orthocarpus*, among others. San Diego Mesa vernal pools and adjacent Baja California are home to endangered endemics such as *Eryngium aristulatum* var. *parishii*, *Muilla clevelandii*, *Navarretia fossalis*, *Orcuttia californica*, *Pogogyne abramsii* and *P. nudiuscula*.

California's vernal pool flora is highly endangered: 73 vernal pool taxa, about one-third of the total, are listed in the California Native Plant Society's *Inventory of Rare and Endangered Vascular Plants of California* (Skinner and Pavlik 1994). Two of these are extinct, 53 are "rare, threatened, or endangered in California, and elsewhere", four are "rare, threatened, or endangered in California, but more common elsewhere", five are poorly known and more information about them is needed, and nine are plants of limited distribution.

Useful plants

Practical uses for vernal pool plants are virtually unknown. Artificial selection of Shippee meadowfoam (*Limnanthes floccosa* var. *californica*) oils is in progress; this oil possesses some of the lubricating properties of sperm whale oil and remains liquid at high temperatures (Gentry and Miller 1965). *Orcuttia* exudates may prove useful for insect pest deterrence; genes for this feature could eventually be transferred to important cereal crops (Griggs 1981). *Downingia*, *Limnanthes* and *Navarretia* are used occasionally as ornamentals, and there is similar potential in other groups, including *Sidalcea* and several genera of Asteraceae.

Social and environmental values

Vernal pools are associations of native plants in ecological island habitats that have a high proportion of endemics and native species. Springtime vernal pool floral displays are spectacularly beautiful and stimulate great interest in the preservation of vernal pools, their component species and surrounding habitats. Several vernal pools genera, such as *Limnanthes*, *Blennosperma*, *Orcuttia* and *Downingia*, have radiated recently, some in concert with specialized or obligate bee pollinators, particularly in the genera *Andrena* and *Panurginus* (Andrenidae) (Thorp 1976). Vernal pools also support perhaps 25 endemic invertebrates, including several endangered fairy shrimps (Anostraca: *Branchinecta* and

Linderiella) and various clam shrimps (Conchostraca) and tadpole shrimps (Notostraca) (Martin 1990).

Economic assessment

Vernal pool habitat is widely used for cattle and sheep grazing; under less intensive use, monetary yields are about $300/acre/year (R.F. Holland, pers. comm.). This represents tens of millions of dollars per year within the CFP, but precise quantification is difficult. Vernal pools aid flood control and also provide invaluable habitat for migrating and wintering waterfowl, but the monetary values of these, too, are unknown. Vernal pools in San Diego County, the North Bay (Napa and Sonoma counties) and much of the Great Valley occur on extremely valuable real estate which is under relentless development pressure.

Threats

More than 90% of vernal pool habitat in California has been destroyed or degraded by agricultural conversion, heavy livestock grazing, unregulated gold mining and urbanization, with lesser impacts from various hydrological alterations, brush clearing for fire control and off-road vehicle activity. As much as 40% of the soils of the Great Valley and associated terraces had the potential to support vernal pools (Holland 1976), and it is thought that vernal pools were once common in most of these areas, being most numerous on younger terrace soils. Agriculture and urban development have eliminated about 95% of these (extrapolated from Holland 1978) and the best remaining pools are now found on higher, older terraces (Holland and Jain 1988).

Extensive vernal pools containing several rare endemic plants (*Blennosperma bakeri*, *Lasthenia burkei*, and *Limnanthes vinculans*) previously occurred in Sonoma County north of San Francisco. The U.S. Fish and Wildlife Service (1990) estimates 90% of the primary habitat for these plants has been altered or destroyed by flood control, urbanization, agricultural development or livestock grazing. In southern California, losses are estimated at 97% for San Diego Mesa vernal pool communities (Bauder 1986; Gustafson 1990).

Surprisingly, light cattle or horse grazing does not appear to be a significant threat, despite trampling (Zedler 1987). Vernal pools also have a surprising ability to recover from partial disruption such as discing (light ploughing) over the course of a few decades, though deep vehicle ruts through pool areas remain unfilled by earth for 10–15 years (Zedler 1987).

Conservation

Vernal pools are among the most severely threatened vegetation types within the CFP (Wickenheiser 1990). Most of the remaining sites have no legal protection. An integrated conservation strategy for vernal pools will require several different approaches. Many taxa are quite localized geographically and different types of vernal pools occur in different areas, so preserves must be scattered throughout the CFP, that is, from southern Oregon to northern Baja California. Furthermore, the genetic diversity of a single species is exhibited over a large geographical scale and is, in part, a consequence of different environmental conditions in different areas (Griggs 1981). Therefore, preserves should encompass the maximum environmental variation experienced by a given taxon.

Vernal pools are essentially habitat islands within other communities (Jain 1976; Wickenheiser 1990), so island biogeographic theory predicts high rates of extinction in particular pools. The conservation implication of this is that preserves must contain many pools to serve as sources of propagules for colonization. Because proper floristic development hinges on precise hydrologic relationships, surrounding watershed areas for pools need to be preserved along with the pools themselves. Large unfragmented areas preserving adjacent habitats are required to support pollinators, and will incidentally protect many other rare non-pool plant species that occur in the vicinity of vernal pools (Zedler 1987).

Some of the exceedingly localized vernal pool species have already been targeted and protected by conservationists. For example, *Tuctoria mucronata* is known from only one set of pools in Solano County, but is protected in The Nature Conservancy's (TNC) Jepson Prairie Preserve along with a carabid beetle (*Elaphrus viridis*) found nowhere else. TNC's Boggs Lake Preserve protects *Navarretia leucocephala* subsp. *pleiantha* and *Orcuttia tenuis*, among other rare plants. On the other hand, *Lasthenia conjugens* is now known from only two of its seven original counties and most remaining sites are threatened by agricultural or urban development.

Some examples of different types of vernal pools are currently protected. The California Department of Fish and Game (DFG) recently acquired Table Mountain in Butte County, which is one of the finest Northern Basalt Flow vernal pool sites and supports the endangered *Juncus leiospermus* var. *leiospermus*. In 1992, the DFG, California Department of Parks and Recreation, U.S. Bureau of Land Management and TNC purchased Big Table Mountain in Fresno County, another important Northern Basalt Flow vernal pool site protecting endangered *Castilleja campestris* subsp. *succulenta*. The DFG already owns Loch Lomond in Lake County, a Northern Volcanic Ash Fall vernal pool supporting the narrow endemics *Eryngium constancei* and *Navarretia leucocephala* subsp. *pleiantha*, and also Phoenix Field in Sacramento County, a Northern Hardpan vernal pool site with endangered *Orcuttia viscida*. Much of the Southern Interior Basalt Flow vernal pool habitat on the Santa Rosa Plateau in western Riverside County is protected by TNC. Several rare plants including three Orcuttieae are protected in the Northern Volcanic Mudflow vernal pools at TNC's Vina Plains Preserve in Butte County. Unfortunately, because of limited resources, management is inadequate at many of these preserves protecting vernal pool species.

Perhaps the greatest preservation priorities for California's remaining vernal pools are Sonoma County's rapidly dwindling Northern vernal pools, the nearly extinct San Diego Mesa vernal pools and, especially, the Northern vernal pools and their habitat within the Great Valley and associated terraces. Protection and regional natural resource planning in these areas is currently promoted by recent formal listings of many vernal pool plants under either the federal Endangered Species Act or its California counterpart. Twenty-nine California vernal pool plants now or will shortly benefit from such

protection, including all eight California Orcuttieae and many members of *Limnanthes, Navarretia*, and *Pogogyne*.

Artificial creation of new vernal pools to compensate for destruction of existing pools during development is of concern, since there is no evidence that artificial pools retain their vernal pool plants over a long period of time. Frequently, artificial pools are established without regard for essential associated species, such as specialized pollinators. While it appears that artificial pools may initially be able to retain vernal pool plants, visual appearance and functional values, such as food chain support, do not approximate to conditions in naturally occurring pools (Ferren and Gevirtz 1990). Moreover, created pools are often intermixed with naturally occurring pools. This misguided mitigation may engender outbreeding depression (Dole and Sun 1992) or alteration of natural hydrology, and promote subsequent degradation of both natural pools and the landscape.

References

Bauder, E.T. (1986). *San Diego vernal pools: recent and projected losses; their condition; and threats to their existence*. Prepared for Endangered Plant Program, California Department of Fish and Game.

Crampton, B. (1976). A historical perspective on the botany of the vernal pools in California. In Jain, S. (ed.), *Vernal pools, their ecology and conservation*. Institute of Ecology Publication No. 9: 5–11. (University of California, Berkeley.)

Dole, J. and Sun, M. (1992). Field and genetic survey of the endangered Butte County meadowfoam – *Limnanthes floccosa* subsp. *californica* (Limnanthaceae). *Conservation Biology* 6(4): 549–558.

Ferren, W.R. and Gevirtz, E.M. (1990). Restoration and creation of vernal pools: cookbook recipes or complex science. In Ikeda, D.H. and Schlising, R.A. (eds), *Vernal pool plants – their habitat and biology*. Studies from the Herbarium No. 8: 147–178. (California State University, Chico.)

Gentry, H.S. and Miller, R.W. (1965). The search for new industrial crops. IV: prospectus of *Limnanthes. Economic Botany* 19: 25–32.

Griggs, F.T. (1976). Life history strategies of the genus Orcuttia (Gramineae). In Jain, S. (ed.), *Vernal pools, their ecology and conservation*. Institute of Ecology Publication No. 9: 57–63. (University of California, Davis.)

Griggs, F.T. (1981). A strategy for the conservation of the genus *Orcuttia*. In Jain, S.K. and Moyle, P.B. (eds), *Vernal pools and intermittent streams*. Institute of Ecology Publication No. 28: 255–262. (University of California, Davis.)

Gustafson, S.S. (1990). Ephemeral edens. *Pacific Discovery*, Spring 1990: 23–32.

Harden, J.W. (1987). Soils developed in granitic alluvium near Merced, California. In Harden, J.W. (ed.), *A series of soil chronosequences in the western United States*. United States Geological Survey Bulletin 1590-A.

Holland, R.F. (1976). The vegetation of vernal pools: a survey. In Jain, S. (ed.), *Vernal pools, their ecology and conservation*. Institute of Ecology Publication No. 9: 11–15. (University of California, Davis.)

Holland, R.F. (1978). *Geographic and edaphic distribution of vernal pools in the Great Central Valley, California*. California Native Plant Society Special Publication No 4. Berkeley, California.

Holland, R.F. and Jain, S.K. (1988). Vernal pools. In Barbour, M.G. and Major, J. (eds), *Terrestrial vegetation of California*. California Native Plant Society Special Publication No. 9: 515–531. Sacramento.

Hoover, R.F. (1937). Endemism in the flora of the Great Valley of California. Ph.D. dissertation, University of California, Berkeley.

Jain, S. (1976). Some biogeographic aspects of plant communities in vernal pools. In Jain, S. (ed.), *Vernal pools, their ecology and conservation*. Institute of Ecology Publication No. 9: 15–22. (University of California, Davis.)

Martin, G. (1990). Spring fever. *Discover*, March 1990: 70–74.

Skinner, M.W. and Pavlik, B.M. (1994). *Inventory of rare and endangered vascular plants of California*. California Native Plant Society Special Publication No. 1, 5th Edition. Sacramento. 338 pp. + vi.

Stone, R.D., Davilla, W.B., Taylor, D.W., Clifton, G.L. and Stebbins, J.C. (1988). *Status survey of the grass tribe Orcuttieae and Chamaesyce hooveri in the Central Valley of California*. Prepared for the U.S. Fish and Wildlife Survey, Biosystems Analysis, Santa Cruz.

Thorne, R.F. (1981). Are California's vernal pools unique? In Jain, S.K. and Moyle, P.B. (eds), *Vernal pools and intermittent streams*. Institute of Ecology Publication No. 28: 1–8. (University of California, Davis.)

Thorp, R.W. (1976). Insect pollination of vernal pool flowers. In Jain, S. (ed.), *Vernal pools, their ecology and conservation*. Institute of Ecology Publication No. 9: 36–40. (University of California, Davis.)

U.S. Fish and Wildlife Service (1990). Endangered and threatened wildlife and plants; proposed endangered status for the plants *Blennosperma bakeri* (Baker's sticky seed), *Limnanthes vinculans* (Sebastopol meadowfoam), and *Lasthenia burkei* (Burke's goldfields). *Federal Register* 55(109): 23109–23115.

Wickenheiser, L.P. (1990). Vernal pools – springtime treasure of California. *Outdoor California*, March–April: 11–13.

Zedler, P.H. (1987). *The ecology of southern California vernal pools: a community profile*. U.S. Fish and Wildlife Service Biological Report 85 (7.11).

Author

This Data Sheet was written by Dr Mark W. Skinner (California Native Plant Society, 1722 J Street, Suite 17, Sacramento, California 95814, U.S.A.).

Acknowledgements

Dr Skinner thanks Ms Roxanne Bittman, Mr Tim Messick and especially Dr Robert F. Holland for their helpful comments.

CENTRAL HIGHLANDS OF FLORIDA
U.S.A.

Location: Physiographic province of the Atlantic Coastal Plain, central peninsular Florida, between latitudes 27°–29°N and longitudes 81°–82°W.

Area: 10,000 km².

Altitude: c. 20–94 m.

Vegetation: Xerophytic scrub dominated by sand pine, several species of sclerophyllous oaks and rosemary, interspersed with xeric sandhills, palmetto flatwoods, and various aquatic and wetland systems.

Flora: High endemism, many endangered and threatened species, species with eastern and south-western U.S. affinities, centre of diversity for woody *Dicerandra* species.

Useful plants: Timber and food plants; plants containing chemicals used as insecticides and herbicides.

Other values: Citrus production, residential, tourist attraction.

Threats: Increasing conversion of land for citrus production, residential development, commercial use, recreation.

Conservation: Archbold Biological Station (315 ha); Saddle Blanket Lakes Preservation (30 ha); Lake Arbuckle State Park (300 ha); proposed program by state, federal, and national agencies to protect 13,000 ha of scrub habitat.

Geography

The Central Highlands Region of peninsular Florida consists of a series of rather localized high grounds, comprising near parallel north-south ridges that are remnants of beach and sand-dune systems associated with Miocene, Pliocene or Early Pleistocene shorelines. The region consists of xeric residual sandhills, beach ridges and dune fields, the whole of which is interspersed with numerous sinkhole lakes and basins caused by erosion of the underlying limestone bedrock. The main axis of the Central Highlands is the Central Ridge, extending from south-eastern Lake County in the north to southern Highlands County in the south. This comprises the Lake Wales Ridge, Winter Haven Ridge, Lake Henry Ridge and Bombing Range Ridge. This is the oldest of the ridge systems. An outlying ridge system to the north-west and extending from Gilchrist County to Polk County comprises the Bell Ridge, Brooksville Ridge, Cotton Plant Ridge and Lakeland Ridge. Slightly to the north-east and ranging from Marion and Putnam counties south to Osceola County are the Mount Dora Ridge, Crescent City Ridge, DeLand Ridge and Orlando Ridge. To the north and extending to the Georgia border is the Trail Ridge (Map 7). To the far east, beginning in the south-eastern tip of Duval County and extending southward along the coast is the Atlantic Coastal Ridge. The latter, the youngest of the ridge systems, is not included as part of the Central Highlands, although it does have some phytogeographic affinities with the Central Highlands. These ridge systems rise above the Polk, Lane, Sumter and Marion Uplands, which have less relief. The general area also encloses large lowlands – the Central and Western Valleys, and the St Johns River Valley. The Lake Wales Ridge, one of the most prominent physiographic features of peninsular Florida, is only a few kilometres wide but more than 150 km long.

Vegetation

The predominant vegetation of the xeric uplands is scrub, consisting of sand pine (*Pinus clausa*) and sclerophyllous oaks (*Quercus inopina, Q. geminata, Q. chapmanii, Q. myrtifolia*). Along a hydrological gradient, scrubs vary from low scrubs, also called scrubby flatwoods, which are dominated by *Serenoa repens, Quercus* spp., *Lyonia* spp. and *Ilex glabra*, to very xeric rosemary balds dominated by *Ceratiola ericoides*. Interspersed is the sandhill or wiregrass community and the more xeric turkey oak barrens, the latter dominated by *Quercus laevis*. In the lowlands are numerous wet swales, ponds and lakes, which are often of sinkhole origin. Natural or anthropogenic fires help maintain these xeric uplands.

Flora

Florida has more federally listed endangered and threatened species than any state east of the Mississippi River. Thirty-eight plant species occurring in Florida are listed with the U.S. Fish and Wildlife Service (USFWS). Of these, 16 occur in the Central Highlands Region (Table 20). Nowhere else in the state is there a higher concentration of federally endangered or threatened plant taxa.

The Central Highlands has a high number of endemic or near endemic (>90% of range in Florida) species, a large majority occurring in the scrub habitat. The Central Ridge has 27 families containing endemic species. It is also the centre of speciation for woody species of *Dicerandra* (Lamiaceae). Forty-three species have their centre of distribution in the Central Ridge Region (Table 21).

Within the last ten years, five species of plants new to science have been described from the Central Highlands Region: *Conradina etonia* (Kral and McCartney 1991),

MAP 7. CENTRAL HIGHLANDS OF FLORIDA, U.S.A. (CPD SITE NA29)

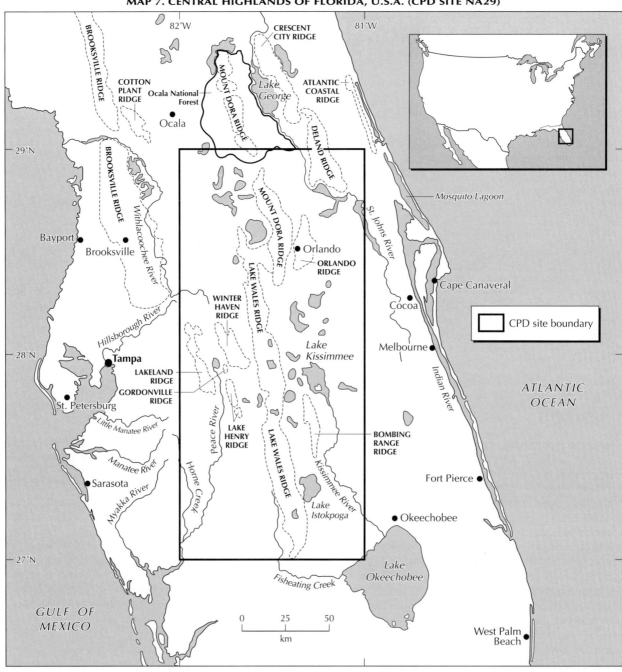

TABLE 20. FEDERALLY LISTED PLANT SPECIES IN FLORIDA

Scientific name	Status	Scientific name	Status	Scientific name	Status
Amorpha crenulata	E	Dicerandra frutescens*	E	Polygala smallii	E
Asimina tetramera	E	Dicerandra immaculata	E	Polygonella ciliata var. basiramia*	
Bonamia grandiflora*	T	Eryngium cuneifolium*	E	(= Polygonella basiramia)	E
Campanula robinsiae*	E	Euphorbia deltoidea	E	Prunus geniculata*	E
Cereus eriophorus var. fragrans	E	Euphorbia garberi	T	Rhododendron chapmanii	E
Cereus robinii	E	Galactia smallii	E	Ribes echinellum	T
Chionanthus pygmaeus*	E	Harperocallis flava	E	Silene polypetala	E
Chrysopsis floridana	E	Hypericum cumulicola*	E	Spigelia gentianoides	E
Deeringothamnus pulchellus	E	Justicia cooleyi	E	Thalictrum cooleyi	E
Deeringothamnus rugellii	E	Liatris ohlingerae*	E	Torreya taxifolia	E
Dennstaedtia bipinnata	E	Lindera melissifolia	E	Warea amplexifolia*	E
Dicerandra christmannii*	E	Lupinus aridorum*	E	Warea carteri*	E
Dicerandra cornutissima*	E	Paronychia chartacea*	T	Ziziphus celata*	E

* indicates species is found in the Central Highlands region.
Total: 38(16); Endangered (E): 34 (11); Threatened (T): 4(2).
Numbers in parentheses are those in the Central Highlands Region.

TABLE 21. SPECIES ENDEMIC OR NEARLY ENDEMIC (>90% WITHIN STATE) AND WHOSE CENTRE OF DISTRIBUTION IS THE CENTRAL RIDGE REGION

Asclepias curtissii	Garberia heterophylla	Persea borbonia var. humilis
Asimina obovata	Hartwrightia floridana	(= Persea humilis)
Bonamia grandiflora	Hypericum cumulicola*	Polygala lewtonii
Calamintha ashei	Hypericum edisonianum	Polygonella ciliata var. basiramia
Carya floridana	Ilex opaca var. arenicola	(= Polygonella basiramia)*
Chapmannia floridana	Lechea cernua	Polygonella myriophylla
Chionanthus pygmaeus	Lechea divaricata	Prunus geniculata
Clitoria fragrans	Liatris ohlingerae*	Quercus inopina
Crataegus lepida	Lupinus aridorum	Sabal etonia
Crotalaia avonensis*	Osmanthus megacarpus	Schizachyrium niveum
Dicerandra christmanii*	Palafoxia feayii	Sisyrinchium xerophyllum
Dicerandra frutescens*	Paronychia chartacea	Stylisma abdita
Eryngium cuneifolium*	Nolina brittoniana	Tradescantia roseolens
Eriogonum longifolium var. gnaphalifolium	Panicum abscissum	Warrea amplexifolia
(= Eriogonum floridanum)	Paronychia chartacea	Warrea carteri
		Ziziphus celata*

* indicates species which are restricted to the Central Highlands Region.

Crotalaria avonensis (DeLaney and Wunderlin 1989), *Dicerandra christmanii* (Huck *et al.* 1989), *Lupinus aridorum* (Beckner 1982), and *Ziziphus celata* (DeLaney, Wunderlin and Hansen 1989). Three of these (*Crotalaria, Dicerandra* and *Ziziphus)* are restricted to the Lake Wales Ridge.

Christman (1988) considers 114 plant species to be characteristic of scrubs in peninsular Florida. A further 202 plant species are known to occur at least occasionally in peninsular Florida scrubs. There are 41 vascular plant species restricted to the Central Highlands scrubs, of which 19 are restricted to the Central Ridge Region.

Although most of peninsular Florida was inundated in the past, particularly during the Pliocene and Pleistocene interglacial periods, the southern portion of the Lake Wales Ridge has apparently been emergent and suitable for plant habitation since the Late Miocene or the Early Pliocene (Huck *et al.* 1989). It is probable that the Lake Wales Ridge served as a refugium for plants during times of higher sea-levels. An equally important factor in the development of high endemism are the deep, dry, sandy soils of the palaeo-dune fields and ridges of the Lake Wales Ridge. A large number of taxa have become adapted to this xeric environment. More recently, other ridge systems in the Central Highlands apparently served a similar role.

The endemic plant species of the Central Highlands are undoubtedly of various origins and ages (Huck *et al.* 1989). Many appear to be neoendemics that share a recent common ancestor or are derived from a wide ranging eastern North American taxon (e.g. *Osmanthus megacarpus, O. americanus, Chionanthus pygmaeus, C. virginicus*), while others appear to have a south-western affinity (e.g. *Bonamia grandiflora, Carya floridana, Liatris ohlingerae, Palafoxia feayi, Ziziphus celata*). There is little, if any, Caribbean influence in the region.

Although a precise inventory of plant species has never been completed for any of the ridges, species diversity is high. For example, the Archbold Biological Station, a 1590 ha preserve located at the southern end of the Lake Wales Ridge and with excellent examples of scrub vegetation, contains 535 species of vascular plants (Abrahamson *et al.* 1984).

Useful plants

Although little is known of economically important plant species other than a few timber and food plants, several scientists are currently investigating the natural chemical defences of scrub plants for compounds which may prove useful as insecticides, insect repellents (Eisner *et al.* 1990; Eisner 1994) or herbicides. One example is *Diceranda frutescens*, an endangered species found in the Archbold Biological Station, which contains trans-pulegol, a potent insect repellent (Eisner 1994).

Social and environmental values

The many lakes and white sandy beaches, often bordered by citrus groves, are highly attractive for residential and recreational development. The unique habitat and geological history also contribute to high endemism in the fauna. Three species of vertebrates with distributions largely in the Central Highlands are listed by the USFWS as threatened. Arthropods endemic to the Florida scrub habitat are concentrated on the Lake Wales Ridge (Deyrup 1990).

Economic assessment

The deep sands, highly favoured for citrus cultivation, have resulted in the Central Highlands Region being a major citrus producing region in the state, a $1.4 billion industry.

Threats

Like many other parts of Florida, the Central Highlands is undergoing rapid and intense development. The major threat to the Central Highlands natural environment is conversion of scrub habitat to citrus groves. Other major threats are conversion to residential, commercial and recreational uses. About 80% of the upland habitats on the southern Lake Wales Ridge have been converted to citrus groves or residential developments (Peroni and Abrahamson 1985a, 1985b). Although no data are currently available, similar amounts of conversion are expected for other areas of the Central Highlands. Of major impact is Walt Disney World and its satellite developments of hotels, restaurants, homes and other theme parks on the Orlando and Mount Dora ridges. Some property owners faced with existing or pending developmental restrictions if their property is found to contain endangered or threatened plant or animal species have wilfully destroyed scrub habitat before proper plant and animal surveys could be conducted.

Conservation

Because of the high level of endemism in the Central Highlands Region, considerable interest has been generated by environmentalists in preserving this fragile habitat and its flora. Recovery plans are completed for all 13 plant species occurring on the Central Ridge that are listed by the USFWS as endangered or threatened species.

Several natural area preserves on the Central Ridge provide protection for some state endemics (Christman and Judd 1990). The Archbold Biological Station in southern Highlands County protects about 315 ha of scrub and 13 endemic species (Abrahamson *et al.* 1984). The Nature Conservancy's (TNC) Saddle Blanket Lakes Preservation covers about 30 ha in Polk County and offers protection for 11 plant species. The proposed Lake Arbuckle State Park will include about 300 ha and 10 plant species. The TNC Tiger Creek Preserve does not contain any true scrub, but does have natural turkey oak barrens with five endemic plant species. At Catfish Creek in Polk County, TNC has recently acquired about 120 ha, including scrub with five endemic plant species. Two endemics are protected at Highlands Hammock State Park. Two or three endemic plant species are protected at Turkey Lake, Lake Cain and Lake Marsha parks, and several small parks in Orange County. Four species of Central Ridge endemics are not protected anywhere, and 12 have fewer than five protected populations.

A conference "Biological Priorities for a Network of Scrub Preserves on the Lake Wales Ridge" was held at the Archbold Biological Station, 29–30 November 1989. As a result, there is now an active programme at the local, state and national level to preserve scrub habitat on the Lake Wales Ridge. Nearly 13,000 ha of scrub habitat have been proposed for purchase by the State through its 1992 Conservation and Recreational Lands (CARL) programme, which is part of the ten year $3 billion Preservation 2000 programme. Efforts are also underway by the USFWS to establish the first U.S. refuge specifically for endangered plant species.

Through the state's wetland protection laws, three endemic plant species of wetland habitats are protected on the Central Ridge. They are *Hartwrightia floridana*, *Hypericum edisonianum* and *Panicum abscissum*.

Bok Tower Gardens, Lake Wales, operating under the National Collection of Endangered Plants programme of the Center for Plant Conservation (CPC), maintains a living collection of 19 species of Central Highlands plants.

References

Abrahamson, W.G., Johnson, A.F., Layne, J.N. and Peroni, P.A. (1984). Vegetation of the Archbold Biological Station, Florida: an example of the southern Lake Wales Ridge. *Florida Scientist* 47: 209–250.

Beckner, J. (1982). *Lupinus aridorum* J.B. McFarlin ex Beckner (Fabaceae), a new species from central Florida. *Phytologia* 50(3): 209–211.

Christman, S.P. (1988). *Endemism and Florida's interior scrub*. Final Report to Florida Game and Fresh Water Fish Commission. Tallahassee. Contract No. GFC–84–101. 247 pp.

Christman, S.P. and Judd, W.S. (1990). Notes on plants endemic to Florida scrub. *Florida Scientist* 53: 52–73.

DeLaney, K.R. and Wunderlin, R.P. (1989). A new species of *Crotalaria* (Fabaceae) from the Florida Central Ridge. *Sida* 13(3): 315–324.

DeLaney, K.R., Wunderlin, R.P. and Hansen, B.F. (1989). Rediscovery of *Ziziphus celata* (Rhamnaceae). *Sida* 13(3): 325–330.

Deyrup, M. (1990). Arthropod footprints in the sands of time. *Florida Entomologist* 73: 529–538.

Eisner, T. (1994). Chemical prospecting: a global imperative. *Proceedings of the American Philosophical Society* 148(3): 385–393.

Eisner, T., McCormick, K.D., Sakaino, M., Eisner, M., Smedley, S.R., Aneshansley, D.J., Deyrup, M., Myers, R.L. and Meinwald, J. (1990). Chemical defense of a rare mint plant. *Chemoecology* 1: 30–37.

Florida Natural Areas Inventory (1990). *Matrix of habitats and distribution by county of rare/endangered species in Florida*. Tallahassee.

Huck, R.B., Judd, W.S., Whitten, W.M., Skean, J.D., Wunderlin, R.P. and DeLaney, K.R. (1989). A new *Dicerandra* (Labiatae) from the Lake Wales Ridge of Florida, with a cladistic analysis and discussion of endemism. *Syst. Bot.* 14: 197–213.

Kral, R. and McCartney, R.B. (1991). A new species of *Conradina* (Lamiaceae) from northeastern peninsular Florida. *Sida* 14(3): 391–398.

Myers, R.L. and Ewel, J.J. (eds) (1990). *Ecosystems of Florida*. University of Central Florida Press, Orlando. 765 pp.

Peroni, P.A. and Abrahamson, W.G. (1985a). Vegetation loss on the southern Lake Wales Ridge. *Palmetto* 5: 6–7.

Peroni, P.A. and Abrahamson, W.G. (1985b). A rapid method for determining loss of native vegetation. *Natural Areas J.* 5: 20–24.

White, W.A. (1970). *The geomorphology of the Florida peninsula*. Geological Bull. 51. Bureau of Geology, Florida Department of Natural Resources, Tallahassee. 164 pp.

Wunderlin, R.P. (1982). *Guide to the vascular plants of Central Florida*. University of South Florida Press, Tampa. 472 pp.

Zona, S. and Judd, W.S. (1986). *Sabal etonia* (Palmae): systematics, distribution, ecology and comparisons to other Florida scrub endemics. *Sida* 11(4): 417–427.

Author

This Data Sheet was written by Richard P. Wunderlin (Department of Biology, University of South Florida, Tampa, FL 33620–5150, U.S.A.).

NORTH AMERICA REGIONAL CENTRE OF ENDEMISM: CPD SITE NA32

EDWARDS PLATEAU
Texas, U.S.A.

Location: Central-western Texas, between latitudes 29°–32°N and longitudes 97°30'–102°30'W.

Area: c. 100,000 km².

Altitude: 100–1000 m.

Vegetation: Low semi-arid temperate semi-evergreen forest, grassland, semi-desert scrub.

Flora: c. 2300 indigenous species, with moderate species endemism; many taxa at limits of their ranges from all direction; threatened and endangered species.

Useful plants: Many range and forage species, a few timber species, one arboreal species exploited for aromatic oil.

Other values: Ranching, tourism and retirement urbanization, wildlife, watershed protection, farming.

Threats: Exploitation, clearing to maximize agricultural value, dams, grazing and browsing, roads, housing developments, industrial development.

Conservation: A few scattered reserves, and a few parks heavily used for recreational purposes, c. 0.02% of total area; a few concerned private landowners with c. 0.02% total area.

Geography

The Edwards Plateau, lying east of the continental divide, is a plateau sloping gently eastward, dropping on average about 180 cm per km, but steeper at the western margin. The southern and eastern margins, and to a lesser extent the other parts, are much dissected by the following rivers: Colorado of Texas, Guadalupe, Nueces, Rio Grande/Pecos, and tributaries thereof (Map 8). This is an ancient evolutionary arena. Most of the land surface has been exposed continuously for occupation by terrestrial biota for at least 65,000,000 years. The plateau consists of three sub-regions:

1. The north-western sub-region, with little macrorelief (i.e., flat or gently undulating), a true plateau geomorphically, comprising about half the Edwards Plateau. Elevation 700–1000 m. Soil is a dark stony clay loam where present, but much of the surface consists of outcrops of Cretaceous limestone. The part of the Edwards Plateau west of the Pecos River is sometimes called the Stockton Plateau.

2. The southern and eastern margins are dissected by canyons separated by flat or undulating divides. The dissected part is locally known as the Texas Hill Country; roughly half of the total area of the Edwards Plateau. Virtually the entire surface consists of outcrops of Cretaceous limestone. Elevation: 250–800 m, or as low as 100 m in the bottom of the canyon mouths.

3. Most geomorphologists also include in this broad category a relatively small area of about 6000 km², known as the Granitic Central Basin, or Central Mineral Region, in the north-central margin. In this area the Cretaceous limestone has been removed by erosion. A variety of rocks crop out, ranging from Archean through upper Palaeozoic in age. The soil, where present, is thin and stony. The exposed rocks are largely granitic and gneissic, with some sandstones and limestones. Elevation: 500–800 m.

Average annual rainfall varies from about 800 mm on the eastern encanyonated margin to 500 mm at the western margin. Rainfall is erratic on the eastern margin and its dependability declines even further westward. Virtually all rain runs off rapidly in the eastern and southern parts. In the north-western quarter of the Plateau, water accumulates for some days after rains, gradually permeating the underlying strata. Some years are virtually rainless. In relatively "normal" years, rainfall is low in winter and spring through April, but a peak of rainfall occurs in May/June followed by a summer slump and then another peak in September, tapering off in October. Rainfall usually occurs as local, moving, showers or storms, originating from cumulus clouds. Rare flooding is associated with the "hurricane season", May through October. Winter snow is rare, scant and usually melts within 24 hours.

Average annual temperature for the Edwards Plateau is approximately 21°C. Monthly average temperatures vary from around 11°C in January to 30°C in August. Temperatures are lowest at the western, higher parts of the Edwards Plateau, and higher at the eastern, lower parts. But, as in the case of precipitation records, extremes are in several ways more important statistics for the biologist. July afternoon highs often range to 41° or 42°C and January night-time lows often approximate -10°C. The lowest night-time temperature ever recorded in the Plateau was about -23°C. The average last spring frost is usually in late March and the earliest frost in autumn is in mid-November.

Vegetation

The "original" (c. 1800) vegetation was grassland or, more commonly, a type of open savanna, with shrubs and low trees along rocky slopes (Correll and Johnston 1970; Stanford 1976; Hatch, Gandhi and Brown 1990). "Tall" prairie grasses (about 1 m tall in late September, the month of maximum height) are still common on level or nearly level rocky outcrops and protected areas having good soil moisture. Shallow or more xeric, exposed sites support "midgrasses" (maximum heights usually 20–50 cm) and those areas with much grazing have a predominance of "shortgrasses" not more than about 15 cm in height. The vegetation in 1800 was therefore not as dense or lush as would be expected in a climate with a mean annual precipitation of 500 to 800 mm, but this is because of the unpredictable rainfall. In the early 1800s and even up into the 1840s, 1850s and later, the open, grassy nature of the landscape was associated with recurrent fires which suppressed woody growth. Fires are well known to be propagated on relatively level ground, but not on steep slopes. Thus the woody brush was confined to the steeper slopes and canyon walls.

Presently not only the rocky slopes but many of the undulating uplands, especially in the eastern and southern halves, carry a dense growth, 3–8 m tall, of shrubs and small trees, mostly oaks (*Quercus fusiformis* and other species) and juniper (*Juniperus ashei*). This invasion of woody plants into former grasslands is attributed to the suppression of fires in historical times, which allowed the woody invaders to displace some of the grassland. Some writers invoke the working of cattle and other livestock on the landscape to help explain these historical trends. At the north-western margin of the Edwards Plateau, the vegetation grades into that of the "mesquite-tobosa country" of the

Rolling Plains, also a short-grass savanna with mesquite (*Prosopis glandulosa*) dotting the short tobosa grass (*Hilaria mutica*) which forms vast almost mono-specific stands on flatlands with slow run-off.

Flora

The earliest professional botanical collections in the Edwards Plateau were those of Jean Louis Berlandier, who accompanied a bison-hunting party in the Hill Country late in 1828. Another notable Texas botanist, Ferdinand Lindheimer, settled at the mouth of a Hill Country canyon in 1843 and collected in the area for some years. Many of the species of the region carry his name. A few other botanists collected in the area in the nineteenth century and early twentieth centuries. From the 1920s onward, many botanists have collected here. The Edwards Plateau is well known botanically. Approximately 2300 species of native vascular plants have been recorded from the Edwards Plateau (Correll and Johnston 1970; Stanford 1976; Hatch, Gandhi and Brown 1990). An additional 200 species are introduced. A study of the distributions of the species outside of the Edwards Plateau indicates that about 500 of the gramineous and herbaceous, especially the prairie-type, species tend to be those of wide distribution in the grasslands of North America, many extending south into north-central Mexico and north into the Great Plains and even into southern Canada. Of the shrubs and small trees, most reach their northern and/or eastern limits in the Edwards Plateau, and thus can be thought of as having

MAP 8. EDWARDS PLATEAU, TEXAS, U.S.A. (CPD SITE NA32)

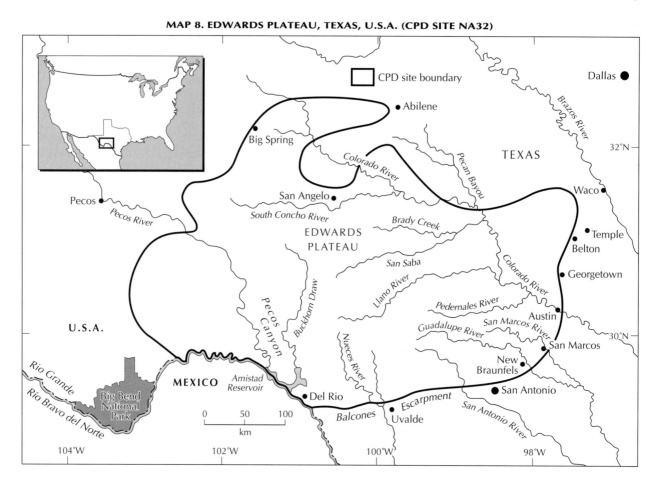

floristic relationships more with the calcareous mountains and plateaux of northern Mexico (the Sierra Madre Oriental, broadly construed). Many of these plants seem to "spill over" to some extent onto the edaphically similar limestone uplands just east of the Edwards Plateau, for example, the upper Cretaceous Austin Chalk outcrop, or farther south onto the calcareous hills of the Rio Grande plain, especially the Bordas Scarp.

Only a few species (about 10% of the Plateau flora) are strictly endemic, that is, they are not known to occur at all outside of the Edwards Plateau. But the list of strictly endemic species includes some truly fascinating threatened and endangered species. In the listings below a careful attempt has been made to list the rarest species first, and less and less rare species in sequence. Topping the list are the beautiful and endangered *Styrax texana* and the threatened *Styrax platanifolia*, known from a few canyons in the Hill Country (Gonsoulin 1974). The rare, beautiful and probably endangered *Salvia penstemonoides* deserves an early listing, followed by: *Dalea sabinalis* (Barneby 1977), *Streptanthus bracteatus*, *Crataegus secreta* (Phipps 1990), *Philadelphus ernestii*, *P. texanus*, *Penstemon triflorus*, *Carex edwardsensis* (Bridges and Orzell 1989), *Seymeria texana* (Turner 1982), *Tridens buckleyanus* (Gould 1975), *Anemone edwardsiana*, *Penstemon helleri*, *Matelea edwardsensis*, *Amsonia tharpii*, *Ancistrocactus tobuschii*, *Onosmodium helleri*, *Erigeron mimegletes*, *Tragia nigricans*, *Berberis swaseyi*, *Amorpha texana* (Wilbur 1975), *Hesperaloë parviflora*, *Galactia texana*, *Opuntia edwardsensis* (Grant and Grant 1979, 1982), *Kuhnia leptophylla* (Turner 1989), *Perityle lindheimeri* (Powell 1974), *Tradescantia edwardsiana*, *Chaetopappa effusa* (Nesom 1988), *C. bellidifolia* (Nesom 1988), *Quercus laceyi*, *Vitis monticola* (Moore 1991), *Buddleja racemosa*, *Garrya lindheimeri* (Dahling 1978) and *Verbesina lindheimeri*.

One special case is that of *Muhlenbergia involuta* which appears to be a series of sterile first-generation hybrids between *M. lindheimeri* and *M. reverchonii*. These hybrids are known only from the Edwards Plateau, which is the overlap-area of the ranges of the two putative parental species.

Examples of more widespread species of which endemic varieties occur in the Edwards Plateau are few. Some examples are *Samolus ebracteatus* var. *cuneatus*, a variety of a very widespread American tropical complex (Henrickson 1983) and *Aesculus pavia* var. *flavescens*, a local variety of a species widespread in the woodlands of south-eastern U.S.A. Likewise, *Vitis aestivalis* var. *lincecumii* is the Edwards Plateau race of a grape widespread in eastern North America (Moore 1991). *Pediomelum hypogaeum* var. *scaposum* on the other hand is a variety of a species widespread in the southern Great Plains and Prairie States (Grimes 1988). *Croton alabamensis* is a special case, with one extremely rare variety endemic to northern Alabama and an even rarer variety endemic to the eastern Edwards Plateau (Ginzbarg 1991, 1992).

If we extend our consideration of endemism to limestone uplands generally in northern Mexico and southern, central and western Texas, we find many species that occur in the Edwards Plateau and range perhaps 100 to 300 km outside of the strictly defined Edwards Plateau. Among the species that are not strictly endemic to the Edwards Plateau but also occur on the edaphically similar limestone uplands of

central and southern Texas (including the Austin chalk) are *Euphorbia jejuna*, *Brickellia dentata*, *Agalinis edwardsiana*, *Physostegia correllii* (Cantino 1982), *Clematis texensis*, *Penstemon guadalupensis*, *P. brevibarbatus*, *Yucca rupicola*, *Muhlenbergia reverchonii*, *Vernonia larsenii* (King and Jones 1975), *Argythamnia simulans*, *Dichromena nivea*, *Thelesperma curvicarpum*, *Mirabilis lindheimeri*, *Desmanthus reticulatus*, *Pediomelum cyphocalyx* (Grimes 1988), *Desmanthus velutinus*, *Hedeoma acinoides*, *Panicum pedicellatum*, *Salvia roemeriana*, *Tetragonotheca texana*, *Lespedeza texana*, *Salvia dolichantha*, *Salvia texana* and *Quercus buckleyi* (Dorr and Nixon 1985).

Species common to the flora of limestone uplands of western Nuevo Leon and Coahuila (Mexico) and also the Edwards Plateau include: *Colubrina stricta*, *Hesperaloë funifera*, *Pinus remota* (Bailey and Hawksworth 1979), *Bouteloua uniflora*, *Pistacia texana*, *Pavonia lasiopetala*, *Muhlenbergia lindheimeri*, *Hunzikeria texana* (Hunziker and Subils 1979), *Antiphytum heliotropioides*, *Salvia engelmannii*, *Lythrum ovalifolium*, *Passiflora affinis*, *Penstemon baccharifolius*, *Scutellaria microphylla*, *Forestiera reticulata*, *Rhus virens*, *Thelesperma longipes*, *T. simplicifolium*, *Chaptalia texana*, *Chrysactinia mexicana*, *Bernardia myricifolia*, *Galphimia angustifolia*, *Croton fruticulosus*, *Acacia roemeriana*, *Stillingia texana*, *Indigofera lindheimeri* and *Cassia lindheimeriana*.

Many other species of the Edwards Plateau could be listed as widely distributed in the south-western U.S.A. and northern Mexico. Examples are *Thamnosma texanum* and *Pinaropappus roseus*. A compilation of species common to both the desertic and montane floras of trans-Pecos Texas and adjacent areas and the Edwards Plateau will probably total 300 species. A few species are known to be endemic to non-limestone substrate in the Central Mineral Region, e.g. *Campanula reverchonii* and *Valerianella texana* (Mahler 1981). Two widespread subtropical ferns are found in Texas only on Enchanted Rock, one of the granitic knobs of the Central Mineral Region: *Blechnum occidentale* (Seigler and Lockwood 1975) and *Cheilanthes kaulfussii*.

Useful plants

The region abounds, during years of normal rainfall, with grasses and herbs useful for foraging by domestic stock. In comparison to this role in ranching, the direct human values of these and other plants of the Edwards Plateau are miniscule. An industry of moderate proportion derives fence posts from the abundant *Juniperus ashei* of the region. These posts have the reputation of high decay-resistance, when in contact with soil, as compared to any other readily available natural posts. A small industry in the region derives, through steam-distillation, an aromatic oil from the roots, trunks and branches of the same *Juniperus ashei*. The aromatic oil is incorporated into germicidal bathroom cleaners, all of which have the word "pine" in the name. Exploitation of trees for timber and lumber is of negligible importance. A small industry derives firewood for sale to householders and restaurateurs, mainly from *Prosopis glandulosa* and *Quercus fusiformis*. A couple of endemic species have been taken into the horticultural trade, namely *Clematis texana* and more importantly *Hesperaloë parviflora*.

Social and environmental values

The region comprises mainly small to large ranches with a mixture of domestic stock including many cattle and slightly fewer goats. Sheep are infrequent, followed by even fewer pigs. Hunting privileges, contracted months or even years in advance of the annual season, provide significant revenues to landowners. The major species hunted is the white tailed deer. Harvests of Rio Grande turkey, wild boar, javalina (collared peccary), bobwhite and scaled quail, and mourning doves are of lesser value. Some ranchers stock exotic species, such as various African and Indian antelopes, for sport hunting. Others stock ostriches and emus, these large birds being valued principally for their skins used in making "cowboy boots". Non-hunting recreation and tourism brings in some revenue. Eco-tourism, specifically ornitho-tourism brings thousands to catch a glimpse of the increasingly uncommon endemic black-capped vireo (*Vireo atricapillus*) and the golden-cheeked warbler (*Dendroica chrysopareia*). Hill Country rivers, clear, relatively cool and lined with stately *Taxodium distichum* trees, attract thousands of visitors, principally from the even hotter and more humid coastal cities such as Houston. In the last 20 years the Hill Country has blossomed as a sun-belt retirement area with relatively low land values and other low living costs, compared to the traditional overcrowded areas in California, Arizona and Florida. The population approximately doubled in the decade 1981 through 1990, from 1,000,000 to 2,000,000

(these estimates do not include the nearby cities east and south of the Hill Country).

Economic assessment

The current value of resources and services that the Edwards Plateau provides to the national economy on a sustainable basis is difficult to ascertain. Minor forest products are not only minor, but negligible in importance.

The Edwards Plateau serves as an important watershed and ground-water recharge zone for several of the most important cities of the region (Waco, Temple, Belton, Georgetown, Austin, San Marcos, New Braunfels, San Antonio, Uvalde, Del Rio), lined up along its eastern and southern margins where rivers disgorge onto the coastal plain. All these cities rely for their water supplies on the streams flowing out of the Edwards Plateau and the large springs at its eastern and southern edges (which are marked by the Balcones Fault Zone). Water percolates through the permeable limestone to the water-table, thence to the springs and wells. A number of streams have been dammed in the last 50 years to ensure the continuity of water-supplies and to support aquatic recreation. Watershed protection has not been, and still is not, a priority with governmental agencies of the region. As a result of the rapid population growth, the quality of the surficial and subterranean water has deteriorated in recent years. Nevertheless, the Edwards Plateau's function as a water

CPD Site NA32: Edwards Plateau, Texas. Dense growth of shrubs and small trees on a rocky slope near Fredricksburg, in the Texas Hill Country of the southern and eastern margins of the Edwards Plateau. Photo: John Shaw/NHPA.

recharge and supply zone will in the future become more and more its major economic value to the human population of the region. Vegetational integrity and maturity are known to correlate positively with the quality of underground and surficial water.

Threats

The plant resources of the region have been threatened for hundreds of years by an over-abundance of non-native and native animals. Some species have survived the intensive over-browsing only as a few individuals. For example, probably fewer than 15 individuals of *Styrax texana* exist in the wild. This is an extreme case of poor survival and perhaps indicates the likelihood that some species became extinct prior to botanical exploration. The threat continues and in some areas is further exacerbated by high deer populations which are no longer controlled by natural predation. The major acute new threat, accelerating in recent years, is the development of land for residential and light industrial purposes, especially in the Hill Country near the largest cities (Austin, San Antonio, Waco). Diffuse urbanization, being the result of literally millions of independent decisions to buy small parcels and settle in the area, appears to be immune from governmental regulation without a massive public awakening.

Conservation

The only lasting and worthwhile effort to minimize loss of biotic and genetic diversity will be to establish large biological reserves selected for their relatively undisturbed state. While this concept is not new to existing policymakers and those with financial resources, at the present time only a miniscule area of the region has been set aside. It should be noted that, in contrast to other western states which initially were carved out of federal territorial land (and which therefore presented more opportunities for establishment of federal parks and forest reserves), Texas initially had virtually no federal lands. State and municipal parks in the Edwards Plateau total only c. 30 km² or possibly 35 km². There is no National Park, save one of c. 1 km² devoted mainly to historical interpretation. While no true wilderness presently survives in the Edwards Plateau, tracts of several hundred to several thousand hectares still exist that could be set aside. It is suggested here that a minimum 2% criterion be applied, that is that concerned governmental and private initiatives work together to set aside appropriate tracts with the minimum goal being 2% of the land area, or about 2000 km². That will represent an almost hundred-fold increase over present holdings. Only then can any reasonable expectation that the biotic richness of the region will be passed on to succeeding generations.

References

Bailey, D.K. and Hawksworth, F.G. (1979). Pinyons of the Chihuahuan Desert Region. *Phytologia* 44: 129–133.

Barneby, R. (1977). Daleae imagines. *Mem. N.Y. Bot. Gard.* 27: 1–891.

Bridges, E.L. and Orzell, S.L. (1989). A new species of *Carex* (sect. Oligocarpae) from the Edwards Plateau of Texas. *Phytologia* 67: 148–154.

Cantino, P.D. (1982). A monograph of the genus *Physostegia* (Labiatae). *Contr. Gray Herb.* 211: 1–105.

Correll, D.S., Johnston, M.C. and collaborators (1970). *Manual of the vascular plants of Texas.* Texas Research Foundation, Renner. 1881 pp.

Dahling, G.V. (1978). Systematics and evolution of *Garrya. Contr. Gray Herb.* 20: 3–104.

Dorr, L.J. and Nixon, K. (1985). Typification of the oak (*Quercus*) taxa described by S.B. Buckley (1809–1884). *Taxon* 34: 211–228.

Ginzbarg, S. (1991). *Croton alabamensis* Chapm. (Euphorbiaceae), disjunct populations in Texas. Master's Degree Report, The University of Texas at Austin. 38 pp.

Ginzbarg, S. (1992). A new disjunct variety of *Croton alabamensis* (Euphorbiaceae) from Texas. *Sida* 15(1): 41–52.

Gonsoulin, G.J. (1974). A revision of *Styrax* (Styracaceae) in North America, Central America, and the Caribbean. *Sida* 5: 191–258.

Gould, F.W. (1975). *The grasses of Texas.* Texas A and M University Press, College Station. 654 pp.

Grant, V. and Grant, K.A. (1979). Systematics of the *Opuntia phaeacantha* group in central Texas. *Bot. Gaz.* 140: 205–214.

Grant, V. and Grant, K.A. (1982). Natural pentaploids in the *Opuntia lindheimeri-phaeacantha* group in Texas. *Bot. Gaz.* 143: 117–120.

Grimes, J. (1988). Systematics of the New World Psoraleae (Leguminosae: Papilionoideae). Doctoral dissertation, Univ. of Texas at Austin. 307 pp.

Hatch, S.L., Gandhi, K.N. and Brown, L.E. (1990). *Checklist of the vascular plants of Texas.* Texas Agricultural Experiment Station, College Station, MP-1655. 158 pp.

Henrickson, J. (1983). A revision of *Samolus* (Primulaceae). *S.W. Nat.* 28: 303–314.

Hunziker, A.T. and Subils, R. (1979). *Salpiglossis, Leptoglossis* and *Reyesia* (Solanaceae): a synoptical approach. *Bot. Mus. Leafl. Harvard* 27 (1–2): 1–44.

Johnston, M. (1990). *The vascular plants of Texas – a list, up-dating the manual of the vascular plants of Texas.* Second edition, published by the author. Austin, 108 pp.

King, B. and Jones, S.B. (1975). The *Vernonia lindheimeri* complex (Compositae). *Brittonia* 27: 74–86.

Mahler, W.F. (1981). Field studies in Texas endemics. *Sida* 9: 176–181.

Moore, M.O. (1991). Classification and systematics of eastern North American *Vitis* L. (Vitaceae) north of Mexico. *Sida* 14: 339–367.

Nesom, G. (1988). Synopsis of *Chaetopappa* (Compositae – Astereae) with new species and the inclusion of *Leucelene*. *Phytologia* 64: 448–456.

Phipps, J.B. (1990). *Crataegus secreta* (Rosaceae), a new species of hawthorn from the Edwards Plateau, Texas. *Sida* 14: 13–19.

Powell, A.M. (1974). Taxonomy of *Perityle* section *Perityle* (Compositae – Peritylinae). *Rhodora* 76: 229–306.

Seigler, D.F. and Lockwood, T.F. (1975). *Blechnum occidentale*, new to Texas. *American Fern Journal* 65: 96.

Stanford, J.W. (1976). *Keys to the vascular plants of the Texas Edwards Plateau and adjacent areas.* Published by the author. Brownwood, Texas. 366 pp.

Turner, B.L. (1982). Revisional treatment of Mexican species of *Seymeria* (Scrophulariaceae). *Phytologia* 51: 403–422.

Turner, B.L. (1989). An overview of the *Brickellia (Kuhnia) eupatorioides* (Asteraceae, Eupatorieae) complex. *Phytologia* 67: 121–131.

Wilbur, R.L. (1975). Revision of North American *Amorpha* (Fabaceae). *Rhodora* 77: 337–409.

Author

This Data Sheet was written by Dr Marshall Johnston (The University of Texas, Botany Department, Austin, Texas 78713-7640, U.S.A.).

REGIONAL OVERVIEW: MIDDLE AMERICA

MEXICO

Total land area: 1,972,546 km².

Population (1990): 81,000,000.

Maximum altitude: c. 5700 m.

Vegetation: Seven main vegetation types – primarily xerophyllous scrublands and grasslands, conifer and oak forests, and tropical deciduous forests.

Number of vascular plants: 18,000–30,000.

Number of endemic species: 10,000–15,000.

Number of genera: 2410.

Number of endemic genera: 283.

Vascular plant families: 220.

Number of endemic families: 1.

CENTRAL AMERICA

Total land area: 509,640 km².

Population (1991): 29,160,000.

Maximum altitude: 4211 m; more than 80 volcanoes.

Vegetation: Many types in c. 20 life zones; especially coastal swamp and mangroves, semi-arid scrub, pine (*Pinus*) savanna, conifer forest, tropical deciduous and semi-deciduous forests, evergreen rain forest, cloud forest, páramo.

Number of vascular plant species: 16,000–18,000.

Number of endemic species: 3300.

Number of endemic genera: c. 100.

Number of endemic families: 0.

MEXICO

Introduction

Mexico, with a territory of close to 2 million km², is a country of major biodiversity – it is home to between 10% and 12% of all living organisms on the planet (cf. Gómez-Pompa *et al.* 1994). This is evident in groups such as flowering plants, mammals, birds, reptiles, butterflies and bees (Toledo and Ordóñez 1993). Some authors (e.g. Mittermeier 1988) rank Mexico third in biological richness after Brazil and Colombia, and ahead of Indonesia, Australia, Zaïre and Madagascar. This wealth of biodiversity results from the size of the territory and the great variety of natural habitats that result from its complex orography and geologic history, with its latitudinal location at the intersection of two biogeographic realms: the Nearctic and the Neotropical.

Geology

Mexico is very mountainous, with over half of the territory higher than 900 m above sea-level. There is evidence of past volcanic activity throughout most of the area. The most spectacular volcanic feature is the great Trans-Mexican Volcanic Belt (17°30'–20°25'N), which crosses the entire country at about the latitude of Mexico City. The landscape is characterized by thousands of old cinder cones and dozens of tall volcanic peaks. Volcanism is continuing, with many active or merely dormant volcanoes. Earthquakes are common, mostly along the Pacific coast and Gulf of California; they are also frequent in the Trans-Mexican Volcanic Belt – often causing considerable damage in this heavily populated region.

Mexico can be divided into five general physical regions based on landforms: Baja California and the Buried Ranges of north-west Mexico; the Central Plateau and the bordering Sierras Madre; the Gulf Coast Plain and Yucatán Peninsula; the Trans-Mexican Volcanic Belt; and the Highlands of Southern Mexico.

1. Baja California continues the California Coast Range of the U.S.A. as a mountain-dominated peninsula about 1300 km long and 50–240 km wide. Geologically it has the form of a tilted fault block, with its crest close to the Gulf of California. The Buried Ranges of north-west Mexico rise 600–1500 m above sea-level and form a continuation of

the Basin and Range region from southern Arizona, U.S.A. These rugged mountains have been partly buried by immense outwashes of debris westward from the much higher and wetter Sierra Madre Occidental. Geologically the Buried Ranges are very complex, with large deposits of minerals.

2. The Central Plateau or Altiplano extends southward from the U.S.A. to the latitude of Mexico City. Geologically it is a continuation of the Basin and Range region of the U.S. intermountain west, and consists of a series of basins separated by small scattered mountain ranges. It is the most heavily mineralized area in Mexico and one of the great mining zones of the world. The Sierra Madre Occidental and Sierra Madre Oriental form dissected borders respectively on the western and eastern edges of the plateau.

   The broad crest of the Sierra Madre Occidental rises to well over 3000 m. The upper portion of the range is covered with thick layers of lava; the western slope forms rugged canyons and narrow ridges descending to the Pacific Coastal Plain. The Sierra Madre Oriental rises to a sharper crest on the eastern rim of the Central Plateau, with elevations to 4000 m. In the north this sierra is composed of several irregular ridges separated by basins descending gradually to the Gulf Coast Plain.

3. The Gulf Coast Plain and Yucatán Peninsula are the two largest lowland areas of the country and are quite distinct. The Gulf Coast Plain is located mostly in the states of Tamaulipas, Veracruz and Tabasco; the low land is very flat and bordered by offshore barrier beaches and lagoons. The Yucatán Peninsula is a limestone platform geologically similar to western Cuba and peninsular Florida, U.S.A. The surface is stony and pitted with sinkholes, and has little surface drainage.

4. The Trans-Mexican Volcanic Belt forms a major geological break with the Central Plateau. Hundreds of volcanic peaks, cinder cones, lava flows, hot springs and ash deposits provide evidence of past and present volcanic activity. The region is bordered on the north by a series of high basins; to the south, the land drops sharply into the Balsas Depression. Included in this belt are Mexico's highest and best known peaks: Pico de Orizaba or Citlaltépetl (c. 5700 m), Popocatépetl (5452 m), Ixtaccíhuatl (5286 m) and Nevado de Toluca or Zinantécatl (4392 m).

5. The Highlands of Southern Mexico are a geologically complex region separated into two sections by the Isthmus of Tehuantepec: the Sierra Madre del Sur in the west and the Chiapas Highlands in the east. The Sierra Madre del Sur is a much dissected mountain system with narrow valleys, a discontinuous Pacific Coastal Plain and a few highland basins. The south-eastern highlands are dominated by the Chiapas Highlands, a plateau rising to 2500 m.

The geological composition of Mexico is thus varied and complex. Nearly 7000 bibliographic references deal with Mexico's geology. The territory was divided into eleven morphotectonic provinces in the recent synthesis by Ferrusquía-Villafranca (1993). This make-up provides a heterogeneous physical-geographical stage for one of the world's most diverse biotas. The regional diversity of Mexico as a result of its complex topography, geography, soils and climate has produced complex mosaics of natural ecosystems and types of vegetation.

Climate

Probably the determining factors for the most significant features of Mexico's climatic diversity are the wide range of elevations (mostly 0–5000 m), its location straddling the Tropic of Cancer, and the oceanic influences due to the narrow breadth of its continental mass. The Tropic of Cancer is not only a significant thermal demarcation, it also approximately marks the transitional strip between arid and semi-arid climates of anticyclone high pressures to the north, and humid and semi-humid climates under the influence of trade winds and cyclones to the south.

The complex topography, together with the differences determined by latitude and altitude, result in a climatic mosaic with very many variations (García 1973, 1989).

Temperature

The great diversity of thermal conditions in Mexico is shown by the fact that, although the subtropical line of the Tropic of Cancer spans its territory, some of its mountains have glaciers and permanent snow-caps. The range of most frequently recorded temperatures varies between 10°–28°C. The lowest known value (-6°C) is at the summit of Pico de Orizaba (c. 5700 m) in Veracruz; the average maximum temperatures recorded (28°–30°C) are in the low-lying regions of the Balsas Depression and adjacent Pacific coastal zones.

Precipitation

The precipitation also presents notable contrasts, from annual average values less than 50 mm with no wet season (e.g. in parts of Baja California), to more than 5500 mm with no dry season (e.g. in parts of Tabasco and Chiapas). The region with more humid continuity extends from south-east of San Luis Potosí through most of the territory of the states of Veracruz and Tabasco to the Yucatán Peninsula, including also northern Chiapas and parts of Oaxaca, Puebla and Hidalgo states. In these areas the most abundant precipitation, with annual values exceeding 4000 mm, is on the windward (eastern) slopes of the Sierra Madre Oriental and on the hills north of Oaxaca and the Central Massif of Chiapas.

Average annual precipitation below 500 mm is found north of the 20° parallel, except for a small enclave in the Tehuacán-Cuicatlán Valley of Puebla and Oaxaca with possibly a minimal area of Veracruz (CPD Site MA4). The Sierra Madre Occidental separates the two principal dry zones of Mexico: the Chihuahuan Desert to the east and the Sonoran Desert to the west. The Chihuahuan Desert corresponds to the major portion of the Central Plateau (Altiplano) from west of Hidalgo, north of Guanajuato and Aguascalientes, to the border with the U.S.A., extending somewhat to the north-east coastal plain in the extreme north of Tamaulipas and areas adjacent to Nuevo León. In this region the annual rainfall generally averages between 200–500 mm; a few small areas register less than 200 mm.

In the Sonora Desert, comprising the Sonora Coastal Plain and the major part of Baja California, the precipitation is less, particularly on the peninsula, where almost throughout its length the average rainfall is less than 200 mm and in some areas less than 50 mm. For a phytogeographic review of the arid zones of Mexico, see Rzedowski (1973).

The distribution of rainfall during the year constitutes a factor of the greatest importance for the plants, particularly where moisture is scarce, which is the case for most of the country's territory. In general the months of June through September are the rainy season, and May and October can also be moist. On the Atlantic slope and in large areas of northern Mexico, 5–18% of the precipitation occurs as "winter" rainfall, as a consequence of the incursion of polar air masses. In contrast, on the Pacific slope from Sinaloa to Chiapas, the months of November to April usually are absolutely dry.

Winds and cyclones

In broad terms, the majority of Mexico is under the influence of trade winds bringing moisture from the east and the north. During the colder season of the year, dry winds from the north-west and west prevail in the north, the west and the centre of the country. Along most of the Pacific coast, at least between Nayarit and Chiapas, there is a monsoon-type regimen, with humid air currents flowing toward the mainland during half the year, and dry air currents flowing seaward during the following six months. The Atlantic and Pacific coasts (except for Sonora and most of Baja California) are in the pathway of tropical cyclones that originate on the high seas during June–October, travelling great distances and many times penetrating onto the continental mass. In the vicinity of their centres, hurricane-type winds are generated that can cause great destruction, not only in coastal areas, but also on the windward slopes of the mountain ranges. Together with their direct devastating effect, the cyclones carry great quantities of moisture, producing copious precipitation in vast areas and frequently affecting extensive parts of the Altiplano.

Population

Mexico is one of the most culturally diverse countries in the New World. In addition to Spanish, there are 54 indigenous languages spoken, in five main language groups: Náhuatl, Maya, Zapotec, Otomi and Mixtec. The population has grown steadily, from 28 million people in 1950 to over 81 million in 1990; more than 40% are under the age of 15. Population density is 73 people per km², with 72% living in urban areas. It is projected that the population will grow just 1.8% between 1995 and 2000, due to population control policies (World Resources Institute 1992).

Vegetation and ecological zones

Mexico has a great diversity of vegetation types, comparable only to India and Peru. Although from studies of physiognomy and floristic composition in Mexico up to 70 different units of vegetation have been distinguished, it is possible to differentiate several principal types of vegetation (Rzedowski 1978) that are conceptually equivalent to the biome category some authors distinguish. This allows for a synthetic profile of the vegetation, and differentiates large natural (terrestrial) habitats or ecological zones.

The zonal habitats in this Regional Overview are defined very broadly (rather than in the usual strict sense, using criteria to define a habitat as determined by the distinctive nature of the species under study). The terrestrial as well as marine territory of Mexico should be classified into units that make sense from a biological point of view and are valid for the broad spectrum of organisms that make up this biotic universe. Mexico may be subdivided relatively easily based on the distribution of vegetation and climate.

Figure 1 shows the distribution of the main vegetation types of Mexico as a function of two main climatic gradients: precipitation and temperature. The first gradient is represented by the average annual precipitation, which ranges from 0–5000 mm; the second gradient is represented by the altitudinal belts, which range from 0–5700 m. The use of certain biogeographic and ecological criteria in this bioclimatic synthesis of the vegetation determines the zonal habitat types or ecological zones in the Mexican territory. Six natural terrestrial habitats, defined by climate, biogeography, ecology and vegetation, are then recognized (Map 9). The detailed methodology used to define these six zonal habitats is given in Toledo (1995a) and Toledo and Ordóñez (1993). This subdivision of the territory coincides in general with those adopted by most authors studying biogeographic and bioclimatic aspects of Mexico (West 1964; Rzedowski 1978, 1993; Flores-Villela 1993; Flores-Villela and Geréz 1988; García 1989).

1. The first of these terrestrial zonal habitats is the **humid (and warm) tropical zone**, which is characterized by the highest thermal systems and precipitation, and by an original cover of medium and high tropical forests and (very rarely) savannas. This zone extends over nine southern and south-eastern states and covers 200,000 km².

2. The **subhumid (and warm) tropical zone** extends over 320,000 km² and 21 states. This zonal habitat covers important portions of western and southern Mexico including the Pacific Coastal Plain, the Yucatán Peninsula and as well central Veracruz and southern Tamaulipas. It is characterized by a hot climate with an annual dry period of five to nine months, and by tropical deciduous forests.

3. Cloud forests, with a floristic composition including both boreal and tropical elements, characterize the **humid temperate zone**. Located at 600–2500 m, this zonal habitat occupies very restricted sites, especially on slopes of the Gulf of Mexico where it extends from Tamaulipas to Chiapas. Distributed in 21 states, it covers 10,000 km².

4. The **subhumid temperate zone** covers the greatest portion of the mountainous areas of Mexico. This zonal habitat, which is represented by pine, oak and mixed forests, is distributed through 20 states and covers 330,000 km². This vegetation is most extensive in the states of Chihuahua, Durango, Michoacán and Oaxaca.

5. The **arid and semi-arid zone** is the most extensive zonal habitat, with an area almost equal to half the country – 900,000 km². With a very low annual precipitation, this

FIGURE 1. CLIMATIC DISTRIBUTION OF MAIN VEGETATION TYPES OF MEXICO

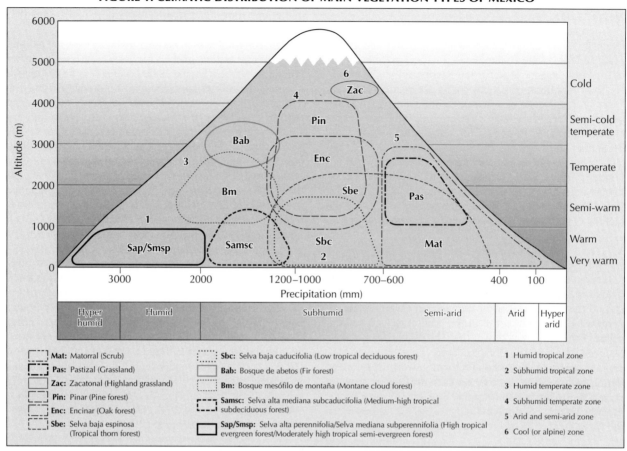

Mat: Matorral (Scrub)
Pas: Pastizal (Grassland)
Zac: Zacatonal (Highland grassland)
Pin: Pinar (Pine forest)
Enc: Encinar (Oak forest)
Sbe: Selva baja espinosa (Tropical thorn forest)

Sbc: Selva baja caducifolia (Low tropical deciduous forest)
Bab: Bosque de abetos (Fir forest)
Bm: Bosque mesófilo de montaña (Montane cloud forest)
Samsc: Selva alta mediana subcaducifolia (Medium-high tropical subdeciduous forest)
Sap/Smsp: Selva alta perennifolia/Selva mediana subperennifolia (High tropical evergreen forest/Moderately high tropical semi-evergreen forest)

1 Humid tropical zone
2 Subhumid tropical zone
3 Humid temperate zone
4 Subhumid temperate zone
5 Arid and semi-arid zone
6 Cool (or alpine) zone

MAP 9. MAIN VEGETATION TYPES OF MEXICO

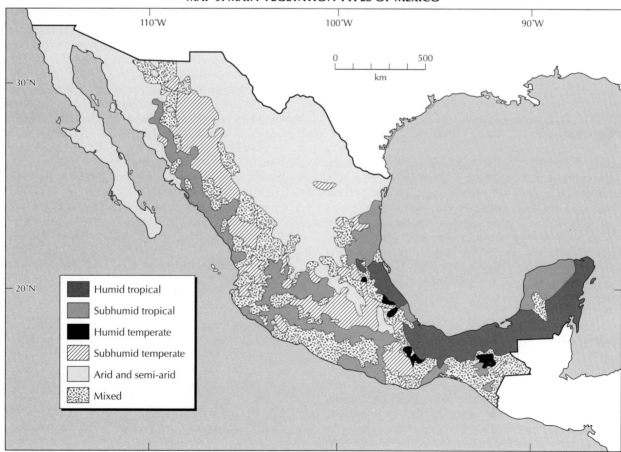

Humid tropical
Subhumid tropical
Humid temperate
Subhumid temperate
Arid and semi-arid
Mixed

TABLE 22. NUMBER OF PLANT SPECIES IN SIX DIVERSE ECOLOGICAL ZONES OF MEXICO (based on Toledo 1995c)

Region	Species	Biogeographic province [a]	Ecological zone [b]
Baja California	2705	California/Baja California	Arid
Nueva Galicia	7500 [c]	Pacific coast/ Southern serranías/ Central Plateau	Temperate, and subhumid tropical
Veracruz	7400	Gulf of Mexico coast/ Sierra Madre Oriental	Humid and subhumid tropical
Tehuacán-Cuicatlán	2700	Tehuacán-Cuicatlán Valley	Arid
Yucatán Peninsula	1936	Yucatán Peninsula	Subhumid to humid tropical
Chiapas	8248	Trans-isthmian serranías/ Soconusco	Humid and subhumid, tropical and temperate

Notes:
[a] Source: Rzedowski 1978.
[b] Source: Toledo *et al.* 1989.
[c] Estimate.

zonal habitat includes two large phytoclimatic zones: the arid zone with scrub vegetation, an annual precipitation of 400 mm or less and 8 to 12 dry months; and the semi-arid zone partly covered with temperate grasslands, with an annual precipitation of 400–700 mm and 6 to 8 dry months.

6. The **cool (or alpine) zone**, located above treeline at 4000 m on the 12 highest mountains of Mexico, represents the sixth zonal habitat. It is dominated by "zacatonales" or páramos, which are high-altitude grasslands.

7. In addition to these six zonal habitats, at least one more terrestrial ecological zone should be recognized: the **wetlands zone**, which includes both interior or continental water bodies (e.g. rivers, lakes, swamps) and coastal waters (lagoons).

Flora

The floristic knowledge of Mexico is still incomplete. However, since c. 1975 there has been strong development of Mexican botanical institutions, collections and researchers; a consortium of 40 institutions – the National Council of the Flora of Mexico – has been created to carry out the ambitious project to inventory the flora. As a result, almost 20 main regional Floras have either been completed or are in progress (Map 10), as well as numerous Florulas and checklists (Rzedowski 1991; Toledo 1985, 1995c). The collections have grown from 567,000 plant specimens and 18 herbaria in 1974 to 2,100,000 specimens and 40 herbaria in 1990 (Toledo and Sosa 1993).

Floristic richness

It has long been known that Mexico together with Central America constitutes one of the regions of greatest plant diversity. However, attempts to quantify this diversity in Mexico have been hindered not only by the lack of a comprehensive inventory of the known species, but also because a significant number have yet to be discovered or described for science. Based on the total number of species (30,489) in six regional Mexican Floras with different

– even contrasting – biogeographic and ecological conditions (Table 22), and considering the high proportions of regional endemism in Mexico, Toledo (1995c) estimated that the entire flora of Mexico might have over 30,000 species of vascular plants. (This rough approximation recognized that some species are found in several of these six regions, but that other Mexican species are in none of these particular regional floras.)

Calculations using different methods indicate a moderately rich flora. Dirzo and Gómez (1996) estimated the total vascular flora to be 19,500 by counting the number of species ascribed to Mexico in *Index Kewensis* (named 1753–1885) and in the Gray Herbarium Card Index (named 1886–1988). Rzedowski (1993) figured floristic richness based on the observation that in latitudes close to Mexico where Compositae are prominent, the ratio of the number of species to the number of genera (s/g coefficient) in this family is similar to the s/g ratio of the entire phanerogamic flora. As a result, Rzedowski extrapolated that there are c. 18,000 species of flowering plants in the country. If undescribed species are added, the possible total reaches c. 21,600, and adding the 1200 species of pteridophytes, brings the total vascular plant species to 22,800. Gómez-Pompa *et al.* (1994) used an estimate of 26,000 species, which placed Mexico fourth in ranking countries with the most species of vascular plants – after Brazil, Colombia and China.

Families with the largest number of species are listed in Table 23. Six families make up c. 40% of the flora; their relative importance varies from region to region. Compositae, Gramineae and Cactaceae are best represented in the northern and central parts of the country, whereas Orchidaceae and Rubiaceae are more diverse in the southern half. Leguminosae become more abundant in warmer

TABLE 23. BEST-REPRESENTED FAMILIES IN MEXICAN PHANEROGAMIC FLORA

Family	Number of genera (approx.)	Number of species (approx.)
Compositae	314	2400
Leguminosae	130	1800
Gramineae	170	950
Orchidaceae	140	920
Cactaceae	70	900
Rubiaceae	80	510
Total	**904**	**7480**

climates. The Orchidaceae may prove to be richer than the Gramineae, as there are many orchid species in Mexico yet to be discovered or scientifically described (Soto-Arenas 1988), whereas the grasses are relatively well known (Valdés-Reyna and Cabral-Cordero 1993).

Map 10 also shows what is known or estimated regarding floristic richness of various states and regions in Mexico. The states with the highest concentrations of species are Chiapas, Oaxaca and Veracruz. A significantly lower number of species occurs on the Yucatán Peninsula, which includes the states of Yucatán, Quintana Roo and Campeche. It is interesting to compare the floristic richness of some areas to their size. For instance, the Baja California Peninsula (two states), with 73,475 km², has 2705 species, whereas the State of Tabasco, which is one-third the size, has one and a half times to perhaps twice as many species.

Endemism

The flora of Mexico is notable for its richness and also its large number of endemics. Ramamoorthy and Lorence (1987) compiled a list of 283 endemic genera of vascular plants, which represents 12% of the presently estimated total of 2410 genera – one of the highest percentages of generic endemism in the world. Rzedowski (1993) estimated 52% endemism at the species level in flowering plants. Thus with a flora of 20,000 to 30,000 species, 10,000 to 15,000 may live only in Mexico. Villa-Lobos (in prep.) has a database with over 3200 of the species considered to be Mexican endemics. According to Rzedowski (1991), plant endemism at the species level is

particularly high in temperate and subhumid montane highlands (70%) and arid and semi-arid areas (60%), moderately high (40%) in subhumid tropical areas, moderate (30%) in humid temperate areas and low (5%) in humid tropical lowlands.

Table 24 shows the richness and endemism of the six terrestrial zonal habitats in Mexico. The reason for this significant wealth of endemic organisms is the existence of a number of regions that function as ecological islands and peninsulas (some extending over large portions of the country), and events and environmental conditions of the geologic past. Particularly, during much of the Cenozoic no terrestrial connection existed with South America. Mexico was a peninsula, much like South Africa today, which experienced sharply varying climatic conditions in contrast with those prevailing on the wider part of the continent.

A remarkable correlation can be observed between high endemism at the levels of genera and species and strong climatic aridity. At the species level, temperate and semi-humid areas are equally endowed with endemics. However in warm humid regions, endemism is poor. In general, endemic genera are better represented in the northern half of the country, and endemic species are more numerous on the Pacific slopes than the Atlantic slopes.

A large assemblage of palaeoendemics can be distinguished, partly concentrated in areas that acted as refugia during epochs of changing climates in the Tertiary and Quaternary. Gypsophytes, which bear a long evolutionary history, stand out among edaphic endemics. An important proportion of very local and/or rare endemic species can be

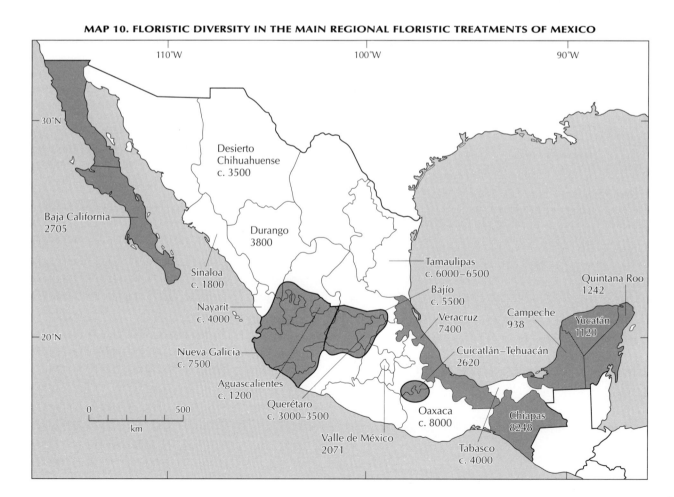

MAP 10. FLORISTIC DIVERSITY IN THE MAIN REGIONAL FLORISTIC TREATMENTS OF MEXICO

TABLE 24. PLANT DIVERSITY AND ENDEMISM OF MEXICO'S TERRESTRIAL HABITATS

Ecological zone	Number of species (approx.)	Number of endemic species (approx.)
1 Humid tropical	5000	250
2 Subhumid tropical	6000	2400
3 Humid temperate	3000	900
4 Subhumid temperate	7000	4900
5 Arid/Semi-arid	6000	3600
7 Aquatic	1000	150
Others	2000	400

TABLE 25. APPROXIMATE PERCENTAGE OF MEXICAN ENDEMISM IN SOME FAMILIES OF FLOWERING PLANTS

Family	Genera %	Species %
Cactaceae[a]	36	72
Rubiaceae[a]	14	69
Compositae[b]	14	66
Orchidaceae[c]	8	35
Gramineae[d]	6	30
Malvaceae[e]	5	48
Leguminosae[f]	2	52
Burseraceae	20	89
Bignoniaceae	0	1
Hernandiaceae	0	0
Taxaceae	0	100
Lacandoniaceae	100	100

Sources:

[a] Values obtained by sampling.
[b] Turner and Nesom, pers. comm.
[c] Soto-Arenas 1988.
[d] Valdés-Reyna and Cabral-Cordero 1993.
[e] Fryxell 1988.
[f] M. Sousa, pers. comm.

recognized; however, the majority of endemics do not belong to this group – many of the most common and characteristic plants of Mexico are endemic taxa, including a large number of weeds and some cultivars. A rough estimate indicates that Mexican endemism is highest among shrubs and perennial terrestrial herbs, whereas lianas and aquatic plants show the lowest incidence. Among the larger families, Cactaceae, Rubiaceae and Compositae stand out with an average species endemism of 69%, whereas Orchidaceae and Gramineae have respectively 35% and 30% (Table 25).

Evolution of plant lineages in Mexico

The grand profusion and high degree of endemism of Mexico's flora, associated with its remarkable diversity, indicate that the country has been the place of origin and/or development for a great number of plant groups and life forms (Rzedowski 1993).

For example the Cactaceae, albeit of South American origin, have reached their maximum diversity, abundance and importance in Mexico – with c. 900 species, over 95% restricted to Mexico (or sometimes including nearby areas of the U.S.A., i.e. Mega-Mexico 1 of Rzedowski 1991, 1993). It has been known for a century or more that Mexico harbours more genera and species of Compositae than any other country (Turner and Nesom 1993). Within the Mexican Gramineae, after *Muhlenbergia* (see below), *Bouteloua* is particularly outstanding. This genus is now widely distributed on the American continent, but its diversity (c. 40 species) is concentrated almost entirely in Mexico and its origin, like that

of eight derived satellite genera, is probably in Mexico (Rzedowski 1993; Stebbins 1975). Similar cases are presented for example by *Cucurbita* (Cucurbitaceae), *Dalea* and *Marina* (Leguminosae), *Lopezia* (Onagraceae), *Karwinskia* (Rhamnaceae), *Lamourouxia* (Scrophulariaceae), *Achimenes* (Gesneriaceae) and *Bouvardia* (Rubiaceae).

The Fouquieriaceae, endemic to Mexico and some portions of south-western U.S.A. and probably having originated in that region, is notable for its growth forms, which are unusual even among xerophytes. The variations offered by the species of *Agave*, which at present is not restricted to Mexico, are no less remarkable. The genus has diversified morphologically and taxonomically in Mexico and in all probability originated there. *Yucca*, *Dasylirion*, *Nolina*, *Krameria* and other taxa present similar examples. In the hot parts of the Pacific drainage area the outstanding feature is the diversity and importance of the some 60 species of *Bursera*, almost all endemic to Mexico and some portion of Central America, followed closely by many endemics in *Acacia*, *Euphorbia*, *Ipomoea* and the Malvaceae.

Other zones of the country also have been active centres of speciation. The montane cool semi-humid areas deserve special emphasis, as a surprisingly rich flora has evolved, not only of herbaceous plants but also shrubs and trees. Examples of this abundance are *Pinus* (c. 48 species), *Quercus* (c. 165 species), *Sedum* (c. 60 species), *Eryngium* (c. 50 species), *Salvia* (c. 300 species – Ramamoorthy 1984; Ramamoorthy and Elliott 1993), *Castilleja* (c. 50 species), *Eupatorium sensu lato* (c. 220 species), *Senecio sensu lato* (c. 180 species), *Stevia* (c. 70 species) and *Muhlenbergia* (c. 100 species). The cloud forests, which cover less than 1% of the territory, have been an important theatre of speciation for epiphytes – the most diverse and prolific have been *Epidendrum*, *Peperomia* and *Tillandsia*.

· Mexico is also an important centre of evolution for weeds. Unlike countries such as Canada, U.S.A., Uruguay and Argentina, in which almost all of the weedy flora is composed of introduced species, it is remarkable that in most of Mexico native weeds strongly prevail. A great number of them have preserved their endemic character. Several representatives of genera such as *Argemone*, *Sicyos*, *Euphorbia*, *Physalis*, *Solanum*, *Bidens* and *Melampodium* are weedy and in an active state of evolution.

Ethnobotany

Mexico has been a notable laboratory for the interaction of plants and human cultures for thousands of years, dating back to the development of pre-Columbian civilizations in

TABLE 26. PLANT KNOWLEDGE AMONG SELECTED INDIGENOUS GROUPS OF MEXICO

Ethnic group	Area	Number of species used	Ecological zone
Seri	Sonora	516	Arid
Tarahumara	Chihuahua	398	Subhumid temperate
Pu'rhépecha	Michoacán	230	Subhumid temperate
Tzeltal	Chiapas	1040	Subhumid and humid temperate
Maya	Yucatán	909	Subhumid tropical
Huasteco	San Luis Potosí	657	Humid tropical

Mesoamerica (Bye 1993; Hernández-Xolocotzi 1993). In 1990 there were an estimated 50 indigenous cultures, with a population near 10 million. These ethnic groups occupy almost all of the main ecological zones or zonal habitats recognized in the territory, and have a profound knowledge of local and regional plants (Table 26). Mexico is considered one of the world's main centres of genetic diversification of cultivated plants, and a centre of origin for agriculture – 116 plant species have been domesticated (Hernández-Xolocotzi 1993). The large number of Mexican and foreign researchers interested in ethnobotany in Mexico has stimulated Mexican scientists to initiate their own research (Toledo 1995b). By May 1991 there were 116 studies published on the knowledge and uses of plants among 26 different indigenous groups in the country (Toledo and Cortés, unpublished), plus numerous studies on useful wild and cultivated plants.

The long interaction between indigenous cultures and the flora in Mexico is likely to yield a considerable number of plants with some degree of modern utility. Recent studies include a catalogue of more than 700 species of edible wild plants (J. Arellano, pers. comm.); a database with more than 3000 medicinal species (Argueta 1992); and a list of 1500 useful plants of the humid tropical flora (Toledo *et al.* 1992). So far over 5000 useful vascular plants have been documented for Mexico.

Factors causing loss of biodiversity

Mexico's natural plant communities and their botanical richness have undergone great modification, as in other countries in Latin America. The main cause of species loss is deforestation. Mexico has lost a large portion of its original forest cover, particularly during the 1960s, and especially in the tropical lowlands. As in the rest of the Latin American countries, expansion of cattle-ranching has been the leading factor by far in the loss of tropical and temperate forests (Toledo 1992). Other factors causing deforestation are forest fires, agriculture, over-exploitation of the forests and urban expansion.

Although there are no definitive statistics on deforestation rates in Mexico, a recent review on this problem offers the best source of information (Masera, Ordóñez and Dirzo 1992, 1995). Slightly more than 8000 km² of closed forests were lost each year during the mid 1980s, leading to an overall deforestation rate of 1.56% per year (Table 27 and Table 28). This ranks Mexico as the country with the third highest rate of deforestation (after Brazil and Indonesia) – a dramatically accelerating phenomenon in tropical nations during the 1980s (World Resources Institute 1992). Gone are 2450 km² of temperate forests and 5590 km² of tropical

TABLE 27. DEFORESTATION RATE BY TYPE OF CLOSED FOREST

Forest type	Estimated area (km²)	Deforestation (km²/yr)	Deforestation rate (%/yr)
Temperate coniferous	169,000	1630	0.96
Temperate broadleaved	88,000	820	0.93
Tropical evergreen	97,000	2370	2.44
Tropical deciduous	161,000	3220	2.00
Total	**515,000**	**8040**	**1.56**

Source: after Masera, Ordóñez and Dirzo (1995).

TABLE 28. TYPES OF CLOSED FOREST AND CAUSES OF DEFORESTATION (%)

Forest type	Cause/Result				
	Pasture	Agriculture	Timber extraction	Fire	Other *
Temperate coniferous	28	16	5	49	3
Temperate broadleaved	28	17	5	47	3
Tropical evergreen	58	10	2	22	7
Tropical deciduous	57	14	5	7	16
All forests	49	13	4	24	10

* Other causes include forest loss by erosion, road building, etc.
Source: Masera, Ordóñez and Dirzo (1995).

TABLE 29. THREATENED PLANT SPECIES OF MEXICO

Plant group	Ex	Ex/V	E	E/V	V	R	I	Total
Lichens	1		1					2
Ferns	1		1		4	19	4	29
Gymnosperms	2		18		8	18	1	47
Monocotyledons	2		42	1	126	169	29	369
Dicotyledons	8	4	82	2	179	471	54	800
Total	**14**	**4**	**144**	**3**	**317**	**677**	**88**	**1247**

Ex = Extinct, E = Endangered, V = Vulnerable, R = Rare, I = Indeterminate (E, V or R).
Source: World Conservation Monitoring Centre (WCMC) 1993.

forests, affecting extensive areas of four of the main zonal habitats or ecological zones and their floras.

According to the World Conservation Monitoring Centre's plants database, over 1200 species in Mexico are threatened (Table 29). The families with the most threatened species recorded are Cactaceae, Orchidaceae and Palmae. The status of these species results from diverse actions, mainly destruction of habitat through deforestation, illegal commerce of plants, and industrial and urban pollution (especially of aquatic plants).

Conservation

Gómez-Pompa *et al.* (1994) report that more than 385 areas are protected in Mexico, and can be grouped into seven broad management categories: (1) Protection Areas; (2) Forest Protection Zones (for: sierras and their forests; watersheds; streams, rivers and lakes; reservoirs; national irrigation systems; farms and ranches; cities; and reforestation); (3) Reserves, variously as: Forest Reserves, Ecological Reserves, UNESCO-MAB or solely national-level Biosphere and Special Biosphere Reserves; (4) Parks, with their many variations; (5) Natural Monuments; (6) Refuges; and (7) Biological Stations. They provide a list of the c. 340 protected natural areas that have been created by federal decree, as well as the biologically most important areas created by State decree or private initiative. Forty-nine areas are identified as having priority for conservation; most of them are administered under the responsibility of the Secretaría de Medio Ambiente, Recursos Naturales y Pesca (SEMARNAP) [formerly the Secretaría de Desarrollo Social (SEDESOL), Subsecretaría de Ecología].

Although Mexico has protected some natural areas since the turn of the century (Vargas 1984), establishment of a national protection system for nature is recent. In 1982 the government created the Secretaría de Desarrollo Urbano y Ecología (SEDUE), followed by founding the Sistema Nacional de Areas Protegidas (SINAP) which is now under SEMARNAP. SINAP is the organization specifically in charge of administration of the country's natural reserves. SINAP recognizes nine different categories of protected areas and by 1989 administered 66 natural areas totalling 57,301 km², representing 2.9% of the national territory. The major protected areas under SINAP in 1989 were 9 UNESCO-MAB Biosphere Reserves (44,739 km²), 43 National Parks (6890 km²) and 10 Flora and Fauna Protected Areas (4864 km²) (Table 30).

An analysis of the distribution of land area officially protected in relation to the main natural zonal habitats or ecological zones was made by overlaying the distribution of the protected areas on the Toledo *et al.* 1989 map of ecological zones (Table 31).

The zonal habitat most under protection is the arid and semi-arid zone, with more than 30,000 km² distributed in 15 reserves. This sizeable coverage is the result of recent establishment of two large Biosphere Reserves: in Baja California (Ojo de Liebre BR) and in Sonora (El Pinacate y Gran Desierto de Altar BR). Nevertheless, there is still not enough protection provided for the xerophytic flora in other parts of Mexico, such as in Tamaulipas, Querétaro, the Mezquital Valley, Veracruz, and especially the Tehuacán-Cuicatlán Valley (Puebla and Oaxaca) (CPD Site MA4) – which is

endowed with the most exclusive flora in Mexico (Dávila *et al.* 1991). It is important to protect enough areas of the arid and semi-arid ecological zone, which has a series of well-defined phytogeographic patterns and richness in endemism and unusual biological forms (Rzedowski 1973).

The humid tropical zone is the second most protected zone in Mexico, with 12,500 km² in six areas. Two Biosphere Reserves include almost the totality of the protected forest: Sian Ka'an (5280 km²) and Calakmul (7230 km²), both on the Yucatán Peninsula. Considering the profound ecological transformation that this zone has endured in Mexico, attention is drawn to the need to protect two other areas of importance: Los Tuxtlas region in Veracruz (Dirzo 1991) and the Uxpanapa-Chimalapa region on the border of Veracruz, Oaxaca and Chiapas (CPD Site MA2). If these two regions are protected and considered together with the Montes Azules BR which includes medium to high forests (see Lacandon Data Sheet, CPD Site MA1), this zone will be well represented in Mexico's system of protected natural areas.

The subhumid temperate zone, dominated by pine and/ or oak forests, is third with 3250 km² and 22 areas protected. In relative terms, these areas protect a minimum portion of this zonal habitat, which is considered from a floristic point of view (by numbers of species and endemics) to be the most important in Mexico (Rzedowski 1991). The protected areas cover only a little more than 1% of this zone, perhaps less, because with the exception of La Michilía BR in Durango (see Upper Mezquital River Region Data Sheet, CPD Site MA8), the remainder is in many small National Parks, which tend to be poorly managed (Vargas 1984).

Although the areas that fall within the humid temperate zone (with mesophyllous mountain forests and firs) are only represented by El Triunfo BR (c. 300 km²) in Chiapas, it is likely that this zonal habitat type will be better represented – if El Cielo BR (1445 km²) in Tamaulipas (in CPD Site MA9) and the Sierra de Manantlán (1396 km²) in Jalisco (CPD Site MA6) are included under SINAP. But even if 1000 km² (i.e. 10% of the total) of this zonal habitat can be designated, the only way to provide adequate protection for the flora – some 3000 species – would be to integrate a network of reserves through the country, since the species are distributed in an ecologically insular pattern (which explains the high degree of endemism). At present there are more than a dozen studies on the flora of different sites in Mexico's humid temperate zonal habitat (see Puig, Bracho and Sosa 1983; Luna-Vega, Almeida and Llorente 1989), from which the criteria for its conservation can be generated.

Almost unprotected is Mexico's subhumid tropical zone (primarily lowland deciduous forest), home to an estimated 6000 plant species, with 40% endemism – 2400 species. Presently, SINAP protects 1100 km², which corresponds roughly to two Flora and Fauna Protection Areas, both on the Yucatán Peninsula: Río Celestún (590 km²) and Río Lagartos (480 km²). The lowland forests on the Pacific slope, which have been much altered by agriculture and cattle-raising, lack much protection. The same can be said about the Balsas Depression, with an exceptional endemic flora, and the dry zone in the centre of Veracruz.

The last region distinguished is composed of multizonal areas constituting 8460 km², where there is no clear predominance of an ecological zone, but a mosaic of habitats. Basically this regional complexity is represented by three Biosphere Reserves: Sierra de Manantlán BR and El Cielo BR

TABLE 30. STATUS OF PROTECTED AREAS IN MEXICO (1989)

Management category	Number	Surface area (km²)	% of national total
National Park	43	6890	0.35
Biosphere Reserve	9	44,739	2.27
Special Biosphere Reserve	2	496	0.025
Flora and Fauna Protected Area	10	4864	0.25
Protected Forest Reserve and Animal Refuge	1	286	0.015
Natural Monument	1	26	0.0001
Total	**66**	**57,301**	**2.9**

Source:
Secretaría de Desarrollo Urbano y Ecología (SEDUE) 1989.

TABLE 31. PROTECTED AREAS IN TERRESTRIAL ECOLOGICAL ZONES

Ecological zone	Number of areas	Surface area (km²)	% of ecological zone	% of national total
Humid tropical	6	12,465	6.10	0.63
Subhumid tropical	6	1109	0.34	0.06
Humid temperate	1	300	5.30	0.015
Subhumid temperate	22	3250	1.2	0.16
Arid and semi-arid	15	31,723	3.7	1.61
Multizonal areas	16	8454	2.4	0.43
Total	**66**	**57,301**		**2.90**

(where mesophyllous mountain forest and pine and oak forests dominate) and Montes Azules BR (which mainly covers a tropical humid area, with ecological islands of temperate forest and lower deciduous forest).

Centres of plant diversity and endemism

Eleven exclusively Mexican centres of plant diversity and one binational centre have been selected. Two identified plant centres encompass primarily north-western Mexico as well as the borderlands of south-western U.S.A. (within Rzedowski's Mega-Mexico 1): the Sonoran Desert, including Baja California (Mexico) – where the sectoral CPD centre has been selected (MA12), and the Apachian/Madrean

region of south-western North America. The location of each of the 12 areas is shown on the regional map of Middle America (Map 11). These selected centres of plant diversity that involve Mexico are:

MA1. Lacandon Rain Forest Region

– see Data Sheet.

MA2. Uxpanapa-Chimalapa Region

– see Data Sheet.

MA3. Sierra de Juárez, Oaxaca

– see Data Sheet.

MAP 11. CENTRES OF PLANT DIVERSITY AND ENDEMISM: MIDDLE AMERICA
The map shows the locations of the CPD Data Sheet sites for Mexico and Central America.

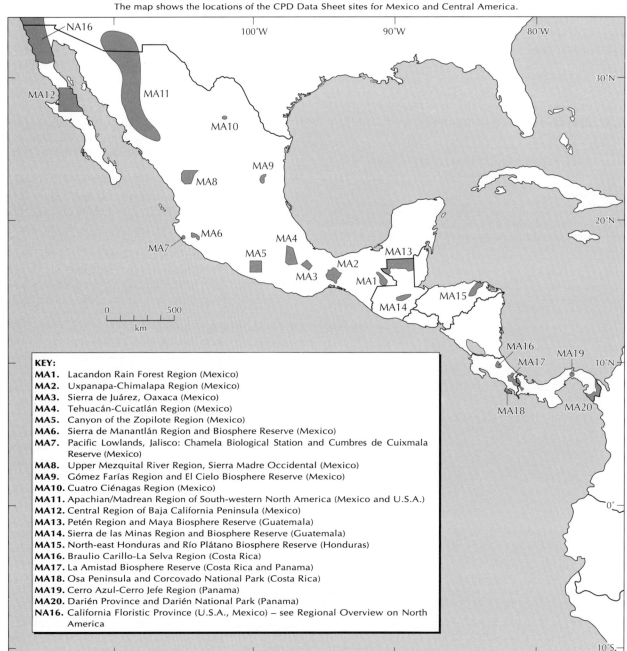

KEY:
MA1. Lacandon Rain Forest Region (Mexico)
MA2. Uxpanapa-Chimalapa Region (Mexico)
MA3. Sierra de Juárez, Oaxaca (Mexico)
MA4. Tehuacán-Cuicatlán Region (Mexico)
MA5. Canyon of the Zopilote Region (Mexico)
MA6. Sierra de Manantlán Region and Biosphere Reserve (Mexico)
MA7. Pacific Lowlands, Jalisco: Chamela Biological Station and Cumbres de Cuixmala Reserve (Mexico)
MA8. Upper Mezquital River Region, Sierra Madre Occidental (Mexico)
MA9. Gómez Farías Region and El Cielo Biosphere Reserve (Mexico)
MA10. Cuatro Ciénagas Region (Mexico)
MA11. Apachian/Madrean Region of South-western North America (Mexico and U.S.A.)
MA12. Central Region of Baja California Peninsula (Mexico)
MA13. Petén Region and Maya Biosphere Reserve (Guatemala)
MA14. Sierra de las Minas Region and Biosphere Reserve (Guatemala)
MA15. North-east Honduras and Río Plátano Biosphere Reserve (Honduras)
MA16. Braulio Carillo-La Selva Region (Costa Rica)
MA17. La Amistad Biosphere Reserve (Costa Rica and Panama)
MA18. Osa Peninsula and Corcovado National Park (Costa Rica)
MA19. Cerro Azul-Cerro Jefe Region (Panama)
MA20. Darién Province and Darién National Park (Panama)
NA16. California Floristic Province (U.S.A., Mexico) – see Regional Overview on North America

MA4. Tehuacán-Cuicatlán Region

– see Data Sheet.

MA5. Canyon of the Zopilote Region

– see Data Sheet.

MA6. Sierra de Manantlán Region and Biosphere Reserve

– see Data Sheet.

MA7. Pacific Lowlands, Jalisco: Chamela Biological Station and Cumbres de Cuixmala Reserve

– see Data Sheet.

MA8. Upper Mezquital River Region, Sierra Madre Occidental

– see Data Sheet.

MA9. Gómez Farías Region and El Cielo Biosphere Reserve

– see Data Sheet.

MA10. Cuatro Ciénagas Region

– see Data Sheet.

MA11. Apachian/Madrean Region of south-western North America

– see Data Sheet.

MA12. Central Region of Baja California Peninsula

– see Data Sheet.

These sites cover five of the principal terrestrial ecological zones of the country. More than one-third of Mexico's flora is represented in them, including an important percentage of endemic species. Additionally, the primarily U.S. California Floristic Province (CPD Site NA16), including the vegetation type of vernal pools (CPD Site NA16g), extends into north-western Baja California. Portions of Mexico's cold (alpine) zone have not been covered (cf. McDonald 1993) – several National Parks contain samples, nor have grasslands (cf. Valdés-Reyna and Cabral-Cordero 1993) or the wetlands zone (cf. Lot-H., Novelo and Ramírez-García 1993).

There are 65 main wetland sites in Mexico designated to be of special importance and meriting international attention (Scott and Carbonell 1986). This list is headed by the Usumacinta-Grijalva Delta, a vast marsh-mangrove wetland in southern Mexico with 10,000 km² of distributory riverine channels, freshwater lagoons, swamps and seasonally inundated marshes. In addition, the coastal zone of Mexico, extending along 10,000 km, contains over 125 lagoons (Lankford 1977; Contreras 1985). Tropical coastal lagoons are now considered to be among the most biologically important ecosystems on Earth – within them a wide variety of nutrients are concentrated that form essential support for a great diversity of plants and animals (e.g. fishes, crustaceans, molluscs). In addition, most of the tropical wetlands of Mexico are extremely important for the breeding, passage and/or wintering of a rich diversity of waterfowl.

Some of the 12 sites listed above now function as reserves, and others are in the process of becoming protected areas under SINAP. When the areas are provided adequate protection, they will constitute much of the basic network of protected areas where all the phytogeographic regions of Mexico are represented. However, it is essential to identify additional sites in parts of the country that have special natural conditions and concentrations of endemic plants. For example, Gómez-Pompa *et al.* (1994) list 62 natural areas recommended for protection by Mexican experts. The 12 areas that have been recognized as centres of plant diversity and the other areas specifically mentioned do not represent all of the important plant centres in Mexico. Comprehensive coverage will require considerably more work.

CENTRAL AMERICA

Introduction

The Central American region is a land bridge connecting North America and South America, extending diagonally from north-west to south-east and separating the Pacific Ocean from the Caribbean Sea. The region extends from c. 7°–18°N and comprises 533,726 km² – near the size of Thailand or continental France. At its widest (more or less at the Honduras-Nicaragua border), Central America is c. 480 km across, whereas in the area of the Panama Canal the isthmus narrows to 65 km (Leonard 1987). The region is composed of seven relatively small countries (Table 32) which base their economies on agriculture and use of natural resources.

Natural richness is evident considering that this small tropical region has 20 life zones ranging from semi-desert to cloud forest, with 8% of the world's known plant species and 10% of its vertebrates. Central America "is a delicate strip of land, a region of extremes. It has a distinct, heterogeneous nature with extremely steep terrains, ample variety of climate and perhaps a higher propensity for natural disasters than any other territory on the planet" (Leonard 1987).

Central America is one of the world's richer regions in culture and tradition. It is pluri-cultural and multi-lingual, with more than 43 distinct indigenous-linguistic groups, making up a total indigenous population of 4–6 million people (de Souza 1992). The region is inhabited by Maya groups (e.g. Quiché, Mam, Kaqchiquel, Kekchí) and various others –

TABLE 32. CENTRAL AMERICAN COUNTRIES AND THEIR TERRITORIAL COVERAGE

Country	Area (km²)	Land area (km²)
Guatemala	108,889	108,430
Belize	22,965	22,800
Honduras	112,088	111,890
El Salvador	20,877	20,720
Nicaragua	140,746	118,750
Costa Rica	51,101	51,060
Panama	77,060	75,990

e.g. Miskito, Guaymí, Lenca, Kuna, Emberá, Sumu, Cabécar, Tol (Xicaque) and Bribri. The largest concentration of indigenous peoples is in Guatemala, where 3–4.5 million Amerindians of 23 different ethnic-linguistic groups comprise 30–50% of the population. Spanish is the official language of Central America except in Belize, where English is primary.

Geology

The region is relatively young, having formed c. 5 million years ago. Only a western archipelago consisting of a series of volcanoes existed during the Jurassic and Cretaceous. The Talamanca Range was formed during the Miocene. During the Pliocene, five tectonic plates converged (the American, Pacific, Cocos, Nazca and Caribbean plates), forming the isthmus of Central America (Heckadon 1992). Only Belize has geologic stability, since it is situated on a portion of a platform of bedrock, which as well underlies the Gulf of Mexico.

The zone of convergence between the Cocos Plate and the Caribbean Plate gave rise to a volcanic chain that borders the Pacific, extending from Guatemala to Panama. Sixty-eight terrestrial volcanoes are included within this chain – three are active in Guatemala, three in El Salvador, more than 15 in Nicaragua and four in Costa Rica. In the last 30 years at least four major earthquakes have affected Guatemala, El Salvador, Nicaragua and Costa Rica, causing thousands of deaths and millions of dollars in damage.

There are four principal geological faults: (1) Motagua, extending across southern Guatemala, at the fracture zone between the North American and Caribbean plates; (2) the highlands of the Pacific, which are characterized by much geological activity related to the shifting of the Cocos Plate; (3) the Nicaragua Depression, which forms Managua and Nicaragua lakes and the San Juan River Basin; and (4) the Chagres River Basin in Panama. These last two fault zones represent the final points at which plates joined to form the Central American land bridge.

The formation of the isthmus of Central America influenced the climatic evolution of the whole planet. It altered marine currents by closing circulation between the Atlantic and the Pacific oceans, and resulted in the formation of a current (the Gulf Stream) flowing north-eastward from the Gulf of Mexico. Another consequence was the major biotic exchange made possible between North America and South America.

Physiography

The convergence of the tectonic plates has resulted in a region of sharp topography, with varied landforms from plains near sea-level to peaks rising over 4000 m (Leonard 1987), which tend to be more precipitous on the Pacific slope. Hillsides and highlands comprise 77% of Central America.

The highland landforms consist of many mountain ranges, volcanoes, valleys and high plateaux. Adding the Maya Mountains which are highest in central Belize, mountainous regions are continuous from Guatemala and northern El Salvador through Honduras and Nicaragua. They are interrupted in the area of the Nicaragua Depression at the San Juan River, which forms part of the border with Costa Rica. The chain of volcanoes begins with the highest Central American peaks, Guatemala's western volcanoes Tacaná (c. 4093 m) and

Tajumulco (4211 m), and ends in the Cordillera de Talamanca, which begins near central Costa Rica and extends into western Panama. Several serranías in eastern Panama are near the north-western extension of the Andes of South America.

Narrow plains border the Pacific Coast from Mexico to eastern Panama. Interior lowlands known as the Nicaragua Depression (Limón Basin) extend diagonally southward from the Gulf of Fonseca across the Central American isthmus to the coastal plain of north-eastern Costa Rica. There are lowlands on the Pacific side between the isthmian spine and Costa Rica's Nicoya and Osa peninsulas and western Panama's Azuero Peninsula; in eastern Panama there are lowlands inland from the Gulf of San Miguel in the Darién near Colombia. On the Atlantic side, low limestone tableland sloping north-eastward as the Yucatán Peninsula forms the Guatemalan Petén and northern Belize, which has a broad coastal zone to the Gulf of Honduras. From north-eastern Honduras the northerly much broader flatlands known as the Mosquitia region extend to central Panama (the broad La Mosquitia narrowing southward to the Costa de Mosquitos). The Caribbean coast continues east of the Panama Canal as a rather narrow strip which broadens again in Colombia.

Climate

Central America is influenced notably by its location between two oceanic climates, its chain of high mountains and its highly heterogeneous physiography. Generalizations can be made about the climate based upon average temperatures, annual precipitation and duration of the dry season (Leonard 1987). The lowlands are warmer on both coasts, with gradual change in ascending to the cool and pleasant climate in the temperate zones of the interior.

The total average precipitation tends to increase from north to south. Local climates range from semi-desert areas which receive only 400 mm of precipitation a year to areas of cloud forest with annual precipitation up to 7500 mm. Two seasons exist – the dry or summer season which begins in November, and the rainy season which begins in May. Each – in general – lasts about six months, but the lengths of the seasons vary greatly. There are places where it rains year round, principally in Costa Rica. In contrast, a more accentuated dry season, with less precipitation over intermediate times and annually, is found in some areas on the Pacific side. Some areas are prone to droughts.

The amount of annual precipitation and the duration of the rainy season are the two factors that determine the three principal climatic zones in Central America: (1) the tropical lowlands of the Caribbean region, which are hot and humid, receiving some rain all year; (2) the interior uplands of the isthmus where the temperate climate is cool and damp in the intermontane valleys and plateaux, to cold and cloudy higher on the mountains; and (3) the lower Pacific slope and coastal plains, which have hot and dry conditions, except for intermittent periods of torrential rain between May and October (Leonard 1987).

Population

The population of Central America has grown rapidly over the last few decades; it presently has the world's second highest growth rate – c. 2.8%. The population reached 29,160,000 in 1991, with a projected total by the year 2025 of c. 62 million (Table 33). Sixty percent of the people are in Guatemala, El Salvador and Honduras. In 1991 El Salvador had the highest population density with 261 people per km², whereas Belize had 10 people per km². Nicaragua, Guatemala and Honduras have the highest growth rates. Forty-four percent of the region's population is under 15 years of age.

TABLE 33. SIZE AND GROWTH OF POPULATION IN CENTRAL AMERICA

Country	Population (millions)		Average annual population change (%) 2000–2005 [b]
	1991 [a]	2025 [b] (projected)	
Guatemala	9.26	21.67	2.68
Belize	0.22	0.29	1.46
Honduras	4.94	11.51	2.48
El Salvador	5.41	9.74	2.00
Nicaragua	3.75	9.08	2.78
Costa Rica	3.11	5.61	1.84
Panama	2.47	3.86	1.48
Total	**29.16**	**61.76**	**2.10**

Sources:
[a] Hoffman 1993.
[b] WRI, UNEP and UNDP 1994.

From 1940 to 1990, the population of Central America more than tripled. During that period an estimated two-thirds of the region's tropical rain forests were cut, burned and converted to agriculture (de Souza 1992; Ypsilantis 1992). Most of the population (c. 90%) remains concentrated in capital cities or along the Pacific slope, where soils are suitable for agriculture and the climate is pleasant. Colonization in recent years has brought about expansion of the agricultural frontier from the crowded Pacific coast and interior regions, by penetrating the forests of the Caribbean coastal plain, which had suffered little destruction prior to the 1940s. New technologies and motorized machinery such as chain saws and bulldozers facilitate building roads and clearing the land.

The natural environments and their vegetation

The Holdridge life-zone system has been applied to all of the countries in Central America except the eastern part of Nicaragua (Leonard 1987; Campbell and Hammond 1989). At least 20 bioclimatically distinct life zones are represented. Most of the region east of the continental divide and most of Costa Rica and Panama are in the moist or wet-forest life zones. Montane forest and alpine-like vegetation occur on the highest peaks of Guatemala and Costa Rica. On the western slope, from north-western Costa Rica northward, most of the natural vegetation is in dry-forest life zones. Semi-desert thorny scrub occurs in east-central Nicaragua and Guatemala's Oriente Region.

In considering vegetation types, the published Floras (e.g. dealing with Guatemala, Costa Rica and Panama) have overviews of varying geobotanical scope, but their introductions lack an underlying and uniform system of classification (cf. Campbell and Hammond 1989). Consequently they are helpful in the context of each country but of limited applicability when considering the whole isthmus as interlinked biotic units. The life zones broadly coincide with simplified natural vegetational regimes. For example, high humid forest, oak and deciduous forest, and conifer forest tend to occur in the pluvial, wet, moist and montane forest life zones, whereas low and medium forests and savanna prevail in the dry-forest life zones (Leonard 1987).

Main vegetation biotopes

Gómez (1985) recognized 27 major biotopes in a biogeographical concept for the Neotropical Realm between the boundaries of U.S.A./Mexico and Panama/Colombia (excluding a few areas of continental U.S.A. and the Caribbean Islands). These broad biotopes incorporate the analysis of vegetation with faunal distributions and the climatic patterns. A summary of the intra-isthmian connections is in Gómez (1986); also see Herrera and Gómez (1992). The biotopes are characterized in the following sketch mainly by vegetation in a north to south order by country and generally from west to east. The major vegetational types are based on Gómez's system of biotic units.

1. Guatemala
The country has 12 biotopes, all shared with Mexico:

❖ 1.1. Coastal swamp and mangroves, along the Pacific coast without *Pelliciera rhizophorae*. A highly endangered and reduced zonal habitat, most of which is degraded.

❖ 1.2. Southern Sonoran elements in the semi-arid Pacific coastal lowlands. Highly degraded.

❖ 1.3. Deciduous and semi-deciduous forests of one or two strata. The dominant arborescent elements to 15 m tall, in semi-open woodland with a high percentage of thorn and microphyllous scrub. Also found isolated in areas of Guatemala's Petén, but with a large number of changes induced by people.

❖ 1.4. Exclusively on the Pacific coast, semi-deciduous forest with "cardonales" (succulents and similar plants – e.g. Cactaceae and Agavaceae) and a lower percentage of microphyllous scrub.

❖ 1.5. Intermediate vegetation, from approximately 800 m to 1500 m on the Pacific slope and 500 m to 1500 m on the Caribbean side. High percentage of Mexican elements on both slopes. This biotope in Mexico corresponds to the Central American element recognized by Rzedowski, which is described in the *Flora of Chiapas* (Breedlove 1981).

❖ 1.6. Oak and other broadleaved forests, on both slopes. The understorey has abundant representatives of Cactaceae and Agavaceae (forming cardonales).

❖ 1.7. Oak and conifer forest. No fewer than seven species of *Pinus* in various combinations with species of *Quercus*, Juglandaceae, *Liquidambar* and *Carpinus*. With a high percentage of boreal elements which also occur in the Mexican highlands. On the Caribbean slope there is an increasing representation of southern Central American species.

❖ 1.8. Mexican-Guatemalan conifer forest. With *Pinus, Abies, Juniperus, Cupressus* and *Taxus* – essentially the southern limit for some of these genera.

❖ 1.9. Guatemalan highland meadow. Referred to in literature as "páramo", but Gómez (1985) prefers the concept highland meadow. Rich in boreal elements, this type of vegetation is limited in extent, and confined in Guatemala to the highest volcanic peaks. This zonal habitat is also the northernmost extension of some South American páramo plants, particularly in Umbelliferae and Compositae and for some vascular cryptogams. Little known floristically.

❖ 1.10. Petén vegetation. With fewer Mexican-Yucatán elements, this biotope is nevertheless diverse for edaphic reasons, having swamps of fresh or brackish water depending on distance from the coast, as well as limestone outcrops with varying percentages of thick-stemmed and thorny plants (cf. Gentry 1942; Standley and Record 1936). There is a conspicuous presence of regionally endemic cycads (*Zamia, Ceratozamia*).

❖ 1.11. Caribbean lowland rain forest. A rich mixture of broadleaved forests of diverse origins, mostly with Mexican affinities (cf. Standley 1931; Gómez 1986).

❖ 1.12. Caribbean lowland pine savanna. Extending from south of the Petén vegetation but including some of its elements. This biotope is characterized by the presence of *Pinus caribaea* var. *hondurensis* and *Quercus oleoides*. It extends, with a significant loss of Mexican elements (particularly *Quercus*), to the Nicaraguan Mosquitia or Mosquito Coast.

2. Belize

The vegetation of Belize, which is well described by Standley and Record (1936), constitutes a mixture of Yucatán-Petén and Caribbean elements (cf. Standley 1930; Gómez 1986). Belize does not have exclusive biotopes, sharing those presented above for Guatemala in 1.1 and 1.10–1.12. Perhaps the most singular vegetational type in Belize is the halophyte reef community. Belizean vegetation is highly degraded and depauperate as a result of intense human activity since pre-Columbian times.

Amazonian elements are poorly represented in the flora of Belize, in contrast with the other Caribbean coastal wet areas of Central America, where those elements are pervasive and conspicuous. This is partly due to the singular geological configuration and origin of Belize, which is further emphasized by its unusual mammalian and herpetological affinities with the Greater Antilles.

❖ 2.1. Halophyte reef community. These communities find their highest diversity in the Belizean keys off the coast. Not only are these communities an indispensable link in the trophic chain that supports an extensive reef system and Caribbean fisheries (both for Belize and Caribbean islands), but there is nothing comparable elsewhere along the entire east coast of Central America. This is an area in urgent need of protection.

❖ 2.2. Coastal and lagoon swamps. Populated by *Rhizophora, Conocarpus, Pelliciera, Prioria* and the palms *Manicaria* and *Raphia*.

❖ 2.3. Petén vegetation, as described for Guatemala (1.10 above).

❖ 2.4. Caribbean lowland rain forest (1.11) (cf. Standley and Record 1936).

❖ 2.5. Pine savanna (Miskito savanna). An open woodland dominated mainly by *Pinus caribaea* var. *hondurensis*, and also connected by corridors to the higher elevation pine forests with *Pinus ayacahuite, P. oocarpa* and *P. rudis*, interspersed with broadleaved species. The savanna extends south, with interruptions, to just north of Bluefields, Nicaragua (5.9 below). Even though this zonal habitat has been much degraded by human activities, natural patches remain of this diverse biotope.

3. Honduras

Basically composed of the same vegetation types that occur in Guatemala, except without those peculiar to the Pacific coast. Inland valleys show a remarkable Pacific climatic influence where there is a predominance of *Quercus-Pinus* forest, with isolated patches of broadleaved forests that are rich in endemics. The conifer forest is less diverse, for example with *Pinus, Liquidambar* and *Carpinus* and no Cupressaceae. Gómez (1985) recognized eight major vegetational types in Honduras:

❖ 3.1. Coastal swamp and mangroves. Identical in species composition to those of Belize (2.2), but not as extensive.

❖ 3.2. Semi-deciduous broadleaved forest on karst with microphyllous elements in the understorey. A much reduced biotope.

❖ 3.3. Lowland pine savannas. In Honduras these may be divided into two subtypes depending on the presence or absence of *Quercus oleoides*, which reaches its southern limit here on the east coast. Savanna vegetation without *Quercus oleoides* continues south into the Nicaraguan Mosquitia or Mosquito Coast (5.9).

❖ 3.4. The Aguán Valley represents the southern limit of the Yucatán-type vegetation (1.10) within isthmian Central America; for a floristic analysis, see Standley (1931).

❖ 3.5. Intermediate area. Situated at 500–1500 m, this zone is a composite of floristic constituents of Mexico and northern South America, with a Mexican predominance. Where soils permit, this zone includes the Central American conifers.

❖ 3.6. Oak-pine forest. Occurring at elevations of 900 m and above, this biotope corresponds with the Nuclear Central America region defined by geologists, and marks the southern limit of boreal floristic influence.

❖ 3.7. Semi-arid and deciduous inner-valley vegetation. Occurs in the valleys of Comayagua and Tegucigalpa, with many representatives of the Sonoran to isthmian progression. Also in the Choluteca Valley in the south and on the Salvadoran coastal plains (4.2).

❖ 3.8. Broadleaved montane forests, well represented on the upper reaches of the Uyuca and San Juancito mountains and on La Tigra. This biotope compares well with forests at similar elevations in Nicaragua (5.7) and northern Costa Rica (6.9).

4. El Salvador

El Salvador has the least national area and the lowest biological diversity in the region. Among the landscapes of Central America, it is also the most altered. Many factors have contributed, but over-population is perhaps the most significant (Pacheco and Martínez 1949). Floristically, the vegetation types are an extension of those mentioned for Honduras, except without those peculiar to the Caribbean lowlands but with the types as found on the Pacific coast in Guatemala.

❖ 4.1. Coastal swamp and mangroves, without *Pelliciera*. As described for Guatemala (1.1) but less expansive. Exploitation of mangroves for fuelwood and tannins has depleted this biotope.
❖ 4.2. Southern Sonoran elements on semi-arid coastal lowlands (1.2).
❖ 4.3. Deciduous and semi-deciduous forests of one or two strata. Fewer cardonales than in the Guatemalan area (1.4). Occurring on higher and drier slopes in the Trifinio area (where El Salvador borders Guatemala and Honduras) and in the Gulf of Fonseca area. Within historical times there were extensive almost pure stands of *Myroxylon balsamum* var. *pereirae* (5.3), which is now very scarce. The narrow riverine valleys contain a larger number of broadleaved evergreen species. Some ravines and canyons eroded from Pliocene diatomite have an interesting calciphilous flora – still evident a decade ago.
❖ 4.4. Intermediate area. The natural vegetation much reduced with development of extensive coffee plantations, particularly towards the interior and on volcanic slopes. A few privately protected areas remain, but they are too fragmented to be a potentially viable mosaic for the spontaneous regeneration of natural cover.
❖ 4.5. Oak-pine mixed forests (1.7). Forests severely depleted by extraction for fuelwood, and by urban development, preventing their regeneration.
❖ 4.6. Conifer forests dominated by *Pinus*, but with fewer species than the ecologically equivalent areas in Guatemala (1.8) and Honduras.

5. Nicaragua

North of 12°30'N, Nicaragua may be geologically considered part of Nuclear Central America. Biologically, Nicaragua south of this latitude has closer affinities to the younger territories of Costa Rica and Panama. Based on its geological origins, Nicaragua has three major morphotectonic units: (a) Pacific Coastal Plains, mostly a mixture of Quaternary volcanism over remnant older (Cretaceous) outcrops, the Guanarivas Formation and serpentines; (b) Central Highlands, which are plateaux and valleys of tectonic origin; and (c) Caribbean Plains, the northern part being very similar in origin and vegetation to south-eastern Honduras.

Floristic inventories have been conducted, and a comprehensive *Flora de Nicaragua* is nearing completion. The following vegetation types are identified:

❖ 5.1. Coastal swamp and mangroves of the Pacific coast, mostly isolated at the mouths of larger rivers having estuarine conditions. As in other northern localities without *Pelliciera*, but with it occurring sporadically just south of León. Threats are exploitation of the mangrove wood (and collection of shells).
❖ 5.2. Sonoran elements on semi-arid coastal lowlands. *Quercus oleoides* occurs in patches. There are isolated "vernal" pools with a high proportion of seasonal aquatics toward the south-western part of the country.

Occasionally interrupted by volcanic peaks of Quaternary origin with slopes covered by seasonal yet evergreen broadleaved vegetation.
❖ 5.3. Deciduous and semi-deciduous forests of one or two strata, with an understorey much reduced in numbers of microphyllous, succulent and thick-stemmed species. North of the Gulf of Fonseca much of this biotope once contained large stands of *Myroxylon balsamum* var. *pereirae* (4.3), which has been cut for use in rural construction. *Erythroxylum* appears from León southward. The vegetation type is also found, isolated by recent volcanic events, in inland areas closer to Managua Lake than to the coast (i.e. Santiago and Masaya volcanoes).
❖ 5.4. Intermediate zone, from 500 m to nearly 1000 m. Enriched by having South American elements mixed with the Mexican and Central American elements. This biotope is more extensive and less degraded along Caribbean slopes. In 1991 Nicaragua established the Bosawas Natural Reserve (8000 km²), the largest reserve in the country, protecting a great diversity of ecosystems.
❖ 5.5. Oak-pine forest. As described for Honduras (3.6) and El Salvador (4.5), with fewer species of *Pinus* – including *P. caribaea* var. *hondurensis*, *P. chiapensis*, *P. oocarpa* subsp. *oocarpa*, *P. patula* subsp. *tecunumanii* and *P. pseudostrobus* var. *oaxacana*; the southern limit of the Central American pines is 12°N. Unusual examples of this vegetation type are on calcareous outcrops north of Estelí and on an isolated gently sloping piedmont overlooking the Gulf of Fonseca (and reaching the forest types of 5.3).
❖ 5.6. Conifer forest. Some *Liquidambar* and *Carpinus* mixed in stands of *Pinus* (see 5.5). Now limited toward the central northern part of Nicaragua and much reduced by human encroachment and exploitation.
❖ 5.7. Montane broadleaved forest. Surrounding the oak-pine forest and conifer forest on both slopes, this biotope finds its maximum development along the Caribbean side. In Guatemala presented in 1.6; rich in diversity and with a high number of endemics.
❖ 5.8. Caribbean lowland rain forest. Two regional subtypes: (a) in the north-east near the Coco River and border with Honduras, where less moist than (b) in the very wet south-east, south of the Siuna River (and extending into north-eastern Costa Rica). The wetter southern formation is found on deltaic deposits basically from Bluefields south, and follows the course of the San Juan River westward to Lake Nicaragua.
❖ 5.9. Caribbean lowland pine savanna. On sandy loams around Puerto Cabezas and on gleys north and west of Bluefields Lagoon. As described for Honduras (3.3) and Belize (2.5), but more patchy and reduced in extent, and less diverse. Much degraded.
❖ 5.10. Coastal swamp and mangroves of the Caribbean, with relictual *Pelliciera*. Sometimes with extensive outer rings of palm swamp, mostly *Raphia* and *Manicaria*. Also inland just south of Castillo with relictual Yucatán-Antillean elements, such as the palm *Acoelorraphe* on permanent and seasonal waterlogged flat ground.

❖ 5.11. Lacustrine communities. Extensive areas dominated by *Eichhornia* and *Pistia*. A rich assemblage of hydrophytes, most important for the restoration of the manatee populations (*Trichechus manatus*) found in the lakes and riverine system that flows into the San Juan River.

6. Costa Rica

This country, together with Guatemala, has received most of the attention in botanical studies. Gómez (1986) discussed in detail the vegetational types of Costa Rica and produced a series of maps that indicated 44 macro-types. Herrera and Gómez (1992) refined the maps and brought the number of biotopes to 56. Several biotopes are being analysed in phytosociological detail by European researchers – high montane oak forest (Kappelle 1995) and páramo (see La Amistad Data Sheet, CPD Site MA17). Costa Rica has designated 27% of its territory with some category of protected status.

For this Regional Overview, these more than 50 biotic units have been grouped to conform to the format of the other six countries as follows:

❖ 6.1. Coastal swamp and mangroves. With *Rhizophora*, *Conocarpus* and *Laguncularia* on both coasts, and *Pelliciera* on the Osa Peninsula and other isolated spots along the Pacific, but not extending north of Bahía Salinas. On the Caribbean shore in estuarine conditions, the mangrove forest may include *Prioria*. The Tortuguero lagoons and channels of the Caribbean coast constitute an extensive system of aquatic plant communities where many South American hydrophytes find their northernmost distribution. On both coasts these natural resources suffer from over-exploitation.

❖ 6.2. Halophyte community. Mainly *Thalassia* beds, which are shared with Caribbean Nicaragua from just south of Bluefields to Punta Mono (near Panama), and also occur northward in Honduras, Guatemala and Belize (2.1). Although not of significant extent, they are very important in the life-cycles of green and leatherback sea-turtles.

❖ 6.3. Santa Elena Peninsula serpentine vegetation, in north-western Costa Rica. A continuation of the contingent in Nicaragua, but rather different even though very sparse and restricted in area. The calciphilous vegetation on the Guanarivas Formation reappears in Costa Rica mostly in the inner arc of the Nicoya Peninsula, representing an isolated Pleistocene relict.

❖ 6.4. Deciduous and semi-deciduous forests of Pacific coast. Very similar floristically to those sketched for the rest of Central America (e.g. 1.3, 4.3, 5.3), but the percentage of succulents and thick-stemmed species is negligible. One singular occurrence is *Melocactus curvispinus* (*M. maxonii* var. *sanctae-rosae*), found not only on marine sediments but often on ultramafic layers along the coast. *Myroxylon* and *Erythroxylum* are also present and together with the conspicuous cactus *Stenocereus aragonii*, mark the penetration of this type of vegetation inland and to the west of San José.

❖ 6.5. Pacific lowland savanna, with *Quercus oleoides* (5.2). Vegetation identical to that in the area of influence of the Gulf of Fonseca and the Choluteca area. *Quercus oleoides* extends up the slopes to c. 800 m,

marking the limit of this biotope and the beginning of the intermediate zone.

❖ 6.6. Seasonal yet evergreen broadleaved forest. Found in isolated areas along the Pacific coastal lowlands and mid-elevations (see 5.2), often in association with riverine conditions. Extends to the western portion of the Central Valley and in the General Térraba area, surrounded by the Cordillera Costera to the west and the Cordillera de Talamanca to the north and east.

❖ 6.7. Lowland rain forest. Below 800 m on the Pacific side and 500 m on the Atlantic slope (cf. 5.8). The Caribbean lowlands are remarkably homogeneous, whereas the Pacific lowlands are less so. The Osa Peninsula shares c. 90% of its arborescent flora with areas in north-eastern Costa Rica (e.g. La Selva – see Data Sheet, CPD Site MA16), but also has an unusual number of disjuncts shared with the Guayanan flora. The forests between the mouths of the Grande de Térraba and Grande de Tárcoles rivers in the south-west are remarkable in having a high number of endemics that have not been found in similarly structured counterpart forests elsewhere.

❖ 6.8. Intermediate zone (1.5, 3.5, 5.4). Along the Pacific slope from 800–1500 m and on the Atlantic slope from 500 m, this is the richest and most diverse of the biotopes. It is also one of the more endangered, having climatic and edaphic conditions that promote urban development as well as agricultural activities.

❖ 6.9. Montane forest. Found in the Guanacaste, Central Volcanic and Talamanca cordilleras. In Costa Rica including the *Quercus* forests noted for more northern areas, as well as broadleaved forests without *Quercus*. The Costa Rican highlands do not have native species of *Pinus*, whose southern limit is in Nicaragua (see 5.5); the only conifers are *Podocarpus* and *Prumnopitys*.

❖ 6.10. Mid-elevation and highland savannas of the Talamancas. Three subtypes occur: the hyper-humid savanna of Esperanza, on the western slopes above the Grande de Térraba River; *Myrica*-dominated savanna, which extends south of Dúrika Massif above 2000 m; and at 3000 m, the "punoid" dry savanna Los Leones, which is mostly composed of bunch grasses.

❖ 6.11. Subpáramo and páramo (1.9). Crowning the highest peaks of the Cordillera de Talamanca starting with Buenavista and Cuerici and forming isolated ecological islands above 3000 m.

7. Panama

Panama's location on an east-west axis, instead of the basically north-south orientation of the rest of Central America, accounts for somewhat different climatic patterns. Geologically this is the youngest part of the Central American isthmus. Its connection to the South American land mass nearly 5 million years ago and its unusual topography with highlands at both ends of the country and rather flat areas in the middle, and with the isthmus itself only c. 65 km wide from coast to coast, have contributed to its rich flora. In keeping with the rest of the overview, the vegetation types are:

❖ 7.1. Coastal swamp and mangroves. As on both coasts of Costa Rica (6.1), with *Pelliciera* also found on the Caribbean side.

❖ 7.2. Deciduous and semi-deciduous forests. Along the Pacific coast from Punta Burica eastward. *Quercus oleoides* supposedly reached as far south as Panama City, an assertion for which there is no known herbarium voucher nor mention in the literature. Briefly described by Standley (1928).

❖ 7.3. Pacific lowland savanna. Similar to Costa Rican counterpart (6.5), it extends to the west of San Francisco Veraguas where a marked South American llano influence is obvious, particularly in the diversity of vascular cryptogams and aquatic phanerogams.

❖ 7.4. Seasonal yet evergreen broadleaved forest (5.2, 6.6), well described in the *Flora of the Panama Canal Zone* (Standley 1928) and the *Flora of Barro Colorado Island* (Croat 1978).

❖ 7.5. Lowland rain forest. In western to central Panama very similar to Costa Rica (6.7). Based on studies so far, eastern Panama has a higher number of Guayanan-Amazonian elements. Partly because of its relative isolation, botanical explorations have been limited.

❖ 7.6. Intermediate zone. As in Costa Rica (6.8); partly included in the *Flora of Barro Colorado Island*. Standley (1928) pointed out that the Caribbean slope is richer. The flora is different in far eastern Panama, where both slopes have higher diversity and affinities to the Colombian Chocó.

❖ 7.7. Montane forest. In two major regions: western Panama on the extension of the Talamanca Range from Costa Rica (6.9); and in the Darién, where species share affinities with neighbouring Colombia (cf. Lamb 1953; Meyers 1969). La Amistad International Park (CPD Site MA17) provides ample protection for these western formations; however, the Pacific slope of the park is badly degraded on the Panamanian side.

❖ 7.8. Subpáramo and páramo vegetation (6.11). Richly expressed on Fábrega and Barú massifs. This biotope has many plants occurring especially in Andean South America.

Flora

Floristically, Central America is the least known part of the world, yet it is extremely rich in numbers of plant species. Of the world's c. 250,000 species of flowering plants, an estimated 15,000–17,000 species live in Central America (Gentry 1978; cf. Davidse, Sousa-S. and Chater 1994). With continued plant exploration and taxonomic studies, these figures could rise, since in the neotropics as many as 10,000 species of vascular plants may be undescribed (Gentry 1982). Floristic knowledge of Central America is still quite limited, although every country has either some checklist or Flora (Campbell and Hammond 1989). Several extensive

Flora projects have been finished (*Flora of Guatemala* by Standley *et al.* 1946–1977 and *Flora of Panama* by Woodson *et al.* 1943–1980), but these works are incomplete and now out of date. Some major publications describing the floras of the region are given in the References section below.

Work began in 1980 on a major project, *Flora Mesoamericana*, undertaken by the Universidad Nacional Autónoma de México's Instituto de Biología, the Missouri Botanical Garden and The Natural History Museum (London). This project covers the region from the Isthmus of Tehuantepec in Mexico through Central America, and will produce seven volumes treating an estimated 18,000–19,000 native, naturalized and cultivated species in 225 families. A first volume which covers 40% of the monocotyledons has been published (Davidse, Sousa-S. and Chater 1994), and the volume on ferns and fern allies was published in 1995.

For each of the countries in Central America, Table 34 provides estimated numbers of vascular plant species, recorded endemics of the country and threatened species, based on information from specialists and literature gathered for *Threatened Plants of Middle America* (Villa-Lobos, in prep.). These figures sometimes update those in Table 1 of the CPD Introduction, at least in earlier volumes.

Endemism

A large portion of the flora of Central America is composed of widely distributed species; nevertheless, studies have shown this region also contains a high number of endemic species. Gentry (1982) calculated that the Central American region has 14% endemism, based on estimates of the number of species and the percentage of endemism by predominant habitat group, using available Floras and monographs for geographical distributions.

The greatest concentrations of endemic plants occur high on the mountains. In the early 1970s it was estimated that 70% of the vascular plants of the high mountains of Mexico and Guatemala are endemic, with the high mountains of Costa Rica and Chiriquí in Panama surpassing 50% endemism (D'Arcy 1977). Table 34 gives the percentage of endemism based on recorded endemics in each country; the region's total endemism is c. 19%.

The concept of endemism is relative, and often based on artificial limits such as political boundaries; this is well illustrated in Central America. For example, the family Rubiaceae in Costa Rica has some 405 species, including cultivated, disjunct and possibly occurring species. Removing those three categories of plants, it may be assumed that the Rubiaceae is represented in Costa Rica by 345 species, of which 66 (19%) are endemic to Costa Rica. But, if we instead consider the natural region of the Talamanca-Tabasara ranges and include the data of adjacent western Panama, the number of endemics increases to 109 (32%). Of the 345 species,

TABLE 34. FLORISTIC DIVERSITY IN CENTRAL AMERICA

Country	Land area (km²)	Vascular plants	Endemics	Endemism (%)	Threatened species
Guatemala	108,430	8000	1171	14.6	477
Belize	22,800	4423	53	1.2	49
Honduras	111,890	6000	148	2.5	84
El Salvador	20,720	2500	17	0.7	86
Nicaragua	118,750	7000	60	0.9	134
Costa Rica	51,060	9500–10,500	600	5.7– 6.3	799
Panama	75,990	8500–9000	1230	13.7–14.5	838

83 (24%) range from just Nicaragua to Panama, resulting in a total of 192 endemic species (56%) in that region (W.C. Burger 1993, pers. comm.). Similar perspectives about endemics of narrow and broader geographical ranges are being discovered for many other groups of organisms.

Endemic genera

There are c. 100 genera endemic to Central America, and another 65 genera that are essentially Central American with only one of the several species occurring outside this region (D'Arcy 1977). Intensive collecting in Costa Rica has resulted in the discovery of many monotypic genera – e.g. *Gamanthera* and *Povedadaphne* (Lauraceae), and *Panamanthus* and *Gaiadendron* (Loranthaceae).

Endemic families

Discovery of new families of flowering plants is uncommon. However, Ticodendraceae was described in 1991 as a result of research conducted in Costa Rica. It is now known to be endemic to Middle America, occurring in mid-elevation cloud forests on both sides of the continental divide through Mexico and Central America.

Useful plants

The number of economic plants in Central America is very large. The forests and other vegetation are important for example for foods, flavourings, medicinals, oils, dyes, natural pesticides, fibres, wood, energy and ornamentals (Table 35). Major compilations have been published on useful plants in Central America (Morton 1981; Williams 1981), including information on those species that are more or less endemic to Middle America or Central America, or either of these regions and also the Caribbean and/or northern South America.

Many of the timber species are commercially extracted for their valuable wood. Several timber species have been placed on CITES appendices, which regulates their international trade. One of the best known tropical hardwoods is genuine mahogany, represented by a genus of three species in tropical America: *Swietenia humilis*, which is distributed along the west coast from south-western Mexico to Costa Rica; *Swietenia macrophylla*, the most widespread of the three, which occurs from southern Mexico through Central America, south to Brazil and Bolivia; and *Swietenia mahagoni* (the original "genuine" mahogany), restricted in distribution from the

TABLE 35. SELECTED ECONOMIC PLANTS OF CENTRAL AMERICA

Timber species

Abies guatemalensis	Guatemalan fir, pinabete	*Nectandra* spp.	aguacatillo, quizarrá
Alfaroa spp.	gualín, gavilán (colorado)	*Ocotea* spp.	
Aspidosperma megalocarpon	chichique	*Oreomunnea pterocarpa*	palo colorado
Calophyllum brasiliense		*Pachira quinata*	
var. *rekoi*	Santa María, jacareuba	(= *Bombacopsis quinatum*)	pochote
Campnosperma panamensis	sajo, orey	*Pinus ayacahuite*	Mexican white pine, pachá
Carapa guianensis	crabwood, andiroba	*Pinus caribaea*	
Carapa nicaraguensis	royal mahogany	var. *hondurensis*	Caribbean pine, pino colorado
Caryocar costaricense	ajo, ají	*Pinus chiapensis*	
Chaetoptelea mexicana	duraznillo	*Pinus oocarpa*	pino de ocote
Cordia dodecandra	ziricote	*Pinus pseudostrobus*	pino blanco
Cupressus lusitanica	Mexican cypress	*Pithecellobium dulce*	Manila tamarind, jaguay
Cynometra hemitomophylla	guapinol negro	*Platymiscium pleiostachyum*	Cristóbal
Dalbergia retusa	cocobolo	*Podocarpus guatemalensis*	podo
Dalbergia stevensonii	Honduras rosewood	*Quercus copeyensis*	roble, Copey oak
Dalbergia tucurensis	granadilla, ronrón	*Swietenia humilis*	Pacific Coast mahogany
Guaiacum sanctum	lignum vitae, guayacán blanco	*Tabebuia donnell-smithii*	primavera
Juglans olanchana	nogal	*Tachigali versicolor*	caña fístula
Karwinskia calderonii	huilihuiste	*Vantanea barbourii*	caracolillo
Macrohasseltia macroterantha	areno amarillo	*Virola koschnyi*	drago
Myroxylon balsamum		*Vochysia guatemalensis*	
var. *pereirae*	bálsamo		

Food species

Acacia spadicigera	pico de gorrión, subín (blanco)	*Manilkara zapota*	sapodilla, chicozapote
Acrocomia mexicana	coyol, coyoli palm	*Pachyrhizus erosus*	yam-bean, jícama
Anacardium excelsum	marañón	*Parmentiera aculeata*	cuajilote
Annona liebmanniana		*Passiflora quadrangularis*	giant granadilla
(syn. *Annona scleroderma*)	poshte	*Persea americana*	avocado, aguacate
Annona purpurea	soncoya, custard-apple	*Persea nubigena*	
Beilschmiedia anay	anay	var. *guatemalensis*	Guatemalan avocado
Capsicum frutescens	chili pepper, chile	*Persea schiedeana*	chinene, chucte
Carica papaya	papaya	*Pimenta dioica*	allspice
Casimiroa edulis	white sapote, zapote blanco	*Pouteria campechiana*	yellow sapote, zapote amarillo
Chamaedorea tepejilote	tepejilote, pacaya	*Pouteria sapota*	zapote mamey
Chrysophyllum mexicanum	caimito silvestre	*Pouteria viridis*	green sapote, zapote injerto
Couepia polyandra	zapote bolo	*Prunus salicifolia*	capulín
Crotalaria longirostrata	chipilín de comer	*Pseudolmedia oxyphyllaria*	ojoche
Dioon mejiae	palma teosinte	*Psidium friedrichsthalianum*	Costa Rican guava, cas
Diospyros digyna	black sapote	*Sideroxylon capiri*	
Fernaldia spp.	loroco	subsp. *tempisque*	tempisque
Gonolobus edulis	guayote, guayato	*Spondias radlkoferi*	
Inga edulis	ingá-cipó	*Talisia floresii*	poloc
Licania platypus	sunza	*Talisia oliviformis*	tinalujo, tapajocote
Litsea spp.	laurel	*Theobroma angustifolium*	cacao silvestre
Malpighia emarginata	acerola, Barbados cherry		
(*Malpighia glabra* misapplied)			

Bahamas through some Caribbean islands including Cuba, Jamaica and the Cayman Islands.

In 1992, Costa Rica and the U.S.A. proposed adding the latter two American mahoganies to *Swietenia humilis* in Appendix II of CITES in order to avoid loss of their wild populations and consequent ecological extinction, by reducing international trade in the woods to sustainable levels. After much controversy in a few countries, *S. macrophylla*, by far the most heavily traded species, was not listed under this treaty. *Swietenia mahagoni* was listed, but it was already ecologically compromised and commercially extinct. In 1994, The Netherlands proposed including *Swietenia macrophylla* in Appendix II, and the CITES countries voted 50 to 33 in favour of doing so, which was just six votes short of the two-thirds of voting countries needed. It was put in Appendix III by Costa Rica in 1995.

Studies are being conducted to inventory and document the numerous species being utilized for medicinal purposes in local cultures. For example, Duke (1985) provided a list of neotropical anti-cancer plants, Balick's floristic checklist for Belize (in prep.) records plant usage, and the world's largest pharmaceutical company (Merck and Co.) made a major investment and agreement in 1991 for bio-prospecting with Costa Rica's National Institute of Biodiversity (INBio) (Booth 1993; Reid *et al.* 1993).

Various families endemic to or richer in the tropics are important ornamentally, such as Orchidaceae, Bromeliaceae, Palmae and Zamiaceae. Wild-collection of some of these species and/or their cultivation in the region is undertaken for personal use and also for local, national and international markets.

The orchids of Central America are relatively well documented and show considerable diversity (e.g. Hamer 1988, 1990; Williams 1956). They are popular in the region as ornamentals, and a resource for recreation and tourism. Guatemala and Honduras export large numbers of beautiful species. All Orchidaceae are regulated by CITES, to differentiate wild from propagated plants and curtail excessive exportation of the wild plants.

The bromeliad genus *Tillandsia* has become popular as house-plants and for use as a non-viable decorative in novelties, with grey-leaved species being particularly favoured. The main exporting countries are Guatemala and Honduras, and the main importer is Europe. Data from

TABLE 35. SELECTED ECONOMIC PLANTS OF CENTRAL AMERICA ...continued

Medicinal species

Bocconia arborea	quiebra muelas, chicalote	*Myroxylon balsamum*	
Calea zacatechichi	zacatechichi	var. *pereirae*	Peruvian balsam, chuchupate
Casimiroa edulis	chapote	*Passiflora ornithoura*	bejuco de calzoncillo largo
Ceanothus coeruleus	chaquirilla, tlaxistle	*Passiflora trinifolia*	bejuco de calzoncillo corto
Cecropia peltata	guarumo	*Picramnia antidesma*	Macary bitter, cáscara de Honduras
Cestrum lanatum	zorillo blanco	*Siparuna nicaraguensis*	limoncillo, cerbatana, kex
Chrysothemis		*Solanum hartwegii*	flor de pajalkish, pajl
friedrichsthaliana	desbaratador	*Solanum nudum*	sauco, yerba de barrer
Croton guatemalensis	copalchillo, zicche	*Stemmadenia donnell-smithii*	cojón
Dicliptera unguiculata	panalitos	*Strychnos panamensis*	guaco, snale seed
Eryngium carlinae	achicoria cimarrona	*Styrax argenteus*	tepeaguacate
Exostema mexicanum	melena de león	*Tagetes lucida*	pericón
Gonolobus barbatus	cuayote	*Talauma mexicana*	flor del corazón, cocte
Ipomoea purga	jalapa	*Thevetia plumeriifolia*	chilindrón blanco
Jatropha curcas	physic nut, piñoncillo	*Tournefortia petiolaris*	pie de guaca
Juglans steyermarkii	nogal	*Trichilia arborea*	choben-che
Justicia spicigera	mohintle, muicle	*Vernonia patens*	tuete blanco
Justicia tinctoria	sacatinta	*Vismia mexicana*	guayabón de montaña
Lacmellea panamensis	espinudo	*Zuelania guidonia*	atamte, tepecacao

Species with other uses

Agave letonae	Salvador henequén	*Omphalea oleifera*	chirán
Anthurium scherzerianum	lengua del diablo, anthurium	*Orbignya cohune*	cohune palm
Asplundia utilis	cuajiote	*Otoba novogranatensis*	
(syn. *Carludovica costaricensis*)		*Pinus caribaea*	
Bambusa paniculata	tarro	var. *hondurensis*	Caribbean pine, pino colorado
(= *Guadua paniculata*)		*Pinus oocarpa*	pino de ocote
Brosimum alicastrum		*Pithecellobium albicans*	chimay, huisache
var. *alicastrum*	ramón blanco	*Plumeria rubra*	frangipani
Bursera simaruba	palo mulato	*Prosopis juliflora*	mesquite, algarroba
Calliandra calothyrsus	calliandra	*Protium copal*	copal
Calycophyllum candidissimum	salamo, degame, lemonwood	*Reinhardtia koschnyana*	window palm
Castilla elastica var. *elastica*	Central American rubber tree, balata, hule	*Rollinia membranacea* (syn. *Rollinia rensoniana*)	churumuyo
Ceiba aesculifolia	algodón de monte, pochote de pelota	*Scheelea lundellii*	corozo
Chamaedorea spp.	parlour- or house-palms	*Scheelea preussii*	corozo, manaca
Chusquea longifolia	cañito	*Schippia concolor*	pimento palm
Cyperus canus	tule	*Sideroxylon stevensonii*	zapote faisán
Cyphomandra costaricensis	contra gallinazo	(= *Dipholis stevensonii*)	
Geonoma hoffmanniana		*Simarouba glauca*	aceituno
Hirtella americana		*Simira salvadorensis*	campeche, John Crow redwood
Hura polyandra	javillo, jabillo	*Solandra grandiflora*	copa de oro
Inga paterno	paterno	*Styrax argenteus*	resino
Licania arborea	alcornoque	*Trophis racemosa* subsp. *ramon*	ramón colorado, ojushte
Luehea candida	algodoncillo	*Virola guatemalensis*	palo de seba, chucul
Manilkara chicle	crown gum, níspero	*Xylosma flexuosum*	pepenance, sweetwood
Monstera deliciosa	ceriman, harpón	*Zamia* spp.	cocalito, chacuhua
Neonicholsonia watsonii		*Zea mexicana*	teosinte

World Wildlife Fund - Germany indicate that from January to March 1988, c. 6 million plants were exported from Guatemala, primarily to Germany and the Netherlands (Rauh 1992). These were a mixture of species, and included both wild-collected plants and plants produced by the developing local horticultural industry.

Chamaedorea palms and their products are used extensively in the floricultural and horticultural industries. Cut leaves of several species are a staple item in the florist trade of the U.S.A. and large quantities of leaves are also imported by Europe. Most of the leaves (both wild-collected and cultivated) appear to originate in Mexico, but some come from Guatemala, Honduras and Costa Rica. In 1986, an estimated nearly 360 million leaves were imported by the U.S.A. – 314,419,000 from Mexico, 40,179,000 from Guatemala and 4,145,000 from Costa Rica (Hodel 1992).

Prior to the 1950s, cycads (Zamiaceae: *Ceratozamia, Dioon, Zamia*) had been exploited in the region for their wood, for making laundry starch, alcohol, poison, fertilizer and various medicines, and as well for food. They are popular as ornamentals due to their beauty, rarity and toughness. Although many cycad species are exported by Mexico, countries such as Guatemala, Nicaragua, Costa Rica and Panama also have trade in their various native species (Jones 1993). Many of the Central American species are commercially exploited due to the proximity of the U.S. horticultural market. All cycads are regulated by CITES, which has encouraged their horticultural production.

Factors causing loss of biodiversity

Central America has experienced alarming deforestation, especially since 1950. The region originally had extensive forests, mainly species-rich rain forests but also rich dry forests; about one-third of them remain. The causes of deforestation are numerous and vary somewhat from country to country (Nations and Leonard 1986; Leonard 1987; Utting 1991; Butterfield 1994).

Unsustainable agricultural and cattle-raising practices have resulted in previously productive land becoming unproductive, causing a demand for frontier land. In 1977, 10% of Central America was considered to be unproductive lands, but by 1987, 24% of the land area had become unproductive (Ypsilantis 1992). For example, the amount of land eroded or degraded seriously has reached 30% in Guatemala and 50% in El Salvador.

Much of the land deforested in the region has been converted in three stages: road construction, colonization and cattle-ranching. In the past few decades and years the pace of deforestation has accelerated, from the plains of the Petén in Guatemala to the Darién rain forest in Panama.

Road construction

Rain forests are being bulldozed to build roads for a variety of reasons, such as national development, military control and oil exploration (Leonard 1987), but mostly for logging commercially valuable timber (cf. MacKerron 1993). Unfortunately, even though only several prized species may be sought in an area, many other trees and the surrounding vegetation are affected during the logging operations. Areas also are cleared for feeder roads, staging and stacking, so the whole operation is quite destructive to the forest. After the loggers leave, the roads through the forest give access to poor farmers in search of land, who establish new settlements.

Colonization

Direct population growth is a factor that brings colonists into the rain forests. With the Central American population growing almost as fast as Africa's, farmers are clearing forest to grow basic food crops. Over the next 20 years, more than 90% of the region's population growth will take place in areas that presently are covered – or recently were covered – by rain forest (Nations 1987).

Roads into forest provide access for landless farmers to clear the forest to raise subsistence crops, such as maize, beans and manioc, as well as cash crops such as coffee, cacao and chili peppers. After 3–4 years of subsistence harvests, the soil is usually unproductive and farmers move on and repeat the cycle in areas newly disturbed. This shifting cultivation has more impact than it did a few generations ago, because with increasing populations the land is more extensively and intensively exploited, rather than being left to natural processes of recovery for many years. Some of the itinerant farmers seed their old plots with grasses and sell the plots to cattle ranchers, who consolidate them into large holdings.

The land in Central America is unequally distributed, which contributes to this second stage of forest loss. In 1975–1976, the 175,000 km² of farmland were 70% in large estates (of more than 200 ha) and 20% in medium-sized estates (10–200 ha), with just 10% of the area but 80% of the farms smaller than 10 ha (Heckadon 1992; cf. Leonard 1987). For example, the growing of bananas and cotton by major estates (including corporations) caused much conversion of forests in Guatemala, Honduras and Costa Rica.

Cattle production

Colonization often facilitates or leads to the third stage of tropical deforestation, which is expansion of commercial export crops into the rain-forest zone. The main factor in the elimination of Central America's tropical forests may be the raising of cattle. In the early decades, most of the forests along the Pacific coast were cleared by settlers to grow food crops and export products (e.g. cacao, indigo, coffee, bananas, cotton, sugar). With increased internal urban and export demand for beef from the 1960s and 1970s, rain forests on the Caribbean side were cleared and burned to establish pasturage. After 7–10 years, overgrazed pastures often turn into eroded wastelands, and ranchers must move to new areas. Cattle-ranching has thus contributed to the destruction of more than half of Central America's jungle since the 19th century, with most of the conversion taking place since 1950 (Nations and Komer 1984; Nations and Leonard 1986; Leonard 1987; Heckadon 1992).

Fuelwood

Gathering fuelwood also has caused serious forest degradation in certain countries in Central America. Most affected have been areas of high population density and agricultural areas in the Pacific coastal regions. Half of all the energy consumed in Central America comes from fuelwood, and the demand is increasing along with the number of small and medium-sized rural industries; less than 10% of the whole region's energy is produced hydroelectrically. Nearly three-quarters of the households use fuelwood, with the consumption per person c. 1 m^3 each year (Heckadon 1992). In El Salvador, where 75% of the homes use fuelwood, gathering the wood constitutes the principal cause of forest degradation and deforestation (Utting 1991). With only 5% of its territory now forested, El Salvador has to import fuelwood from Honduras and Guatemala. Similar problems are occurring in Costa Rica, the most prosperous country in the region, where half the population still uses fuelwood. The same problem is arising in Guatemala (cf. Leonard 1987).

These several forces destructive to the tropical forests are causing major changes in land use. Two estimates of the recent extent of forest cover and deforestation are provided in Tables 36 and 37 (WRI, UNEP and UNDP 1994). The data in Table 36 are from the U.N. Food and Agriculture Organization's (FAO) Forest Resources Division, which defines total forest to consist of closed forest and open forest, including mixed forest/grassland with at least 10% tree cover and a continuous grass layer. The data in Table 37 are based on land-use data provided to FAO by national governments, and refer to closed, degraded and planted forest and woodland (but not cropland, e.g. with coffee or cacao). In Table 36 vs. Table 37, substantially larger amounts of forested land are estimated for Belize, Honduras and Nicaragua but somewhat smaller amounts for Costa Rica and Panama. According to data gathered by FAO and others (e.g. Collins 1990; Ypsilantis 1992), Central America's area in forest and woodland has decreased from c. 56% of the total land area in the first half of the 1960s to c. 33% in 1989–1991.

Tourism

Badly managed tourism has led to degradation of natural and cultural resources in Central America. Many protected areas that are not well managed are affected by constant pillaging from both local and foreign tourists. Throughout the region, one can observe the exploitation of bromeliads, orchids and ferns; in Belize, Honduras and Costa Rica coral-reef specimens are taken; in Guatemala and Belize archaeological objects are looted from National Parks. Tourism also has affected indigenous cultural practices, for example by influencing the design and colours of textiles and the items offered. In some cases the standard of living has increased in communities, but in others tourism has led to an increase in begging.

Most sites visited for their natural values do not have management plans, or evaluations of environmental impact or carrying capacity for visitors, which contributes to the ecosystems becoming highly altered. Nevertheless, there are programmes such as Mundo Maya (Maya World), Paseo Pantera (Path of the Panther) and specific management plans for some protected areas that are striving to diminish the negative impacts of tourism on the environment.

Conservation

In recent years, the countries of Central America have agreed to several regional endeavours to help balance environmental concerns and development, with creation of a Comisión Centroamericana de Ambiente y Desarrollo (Central American Commission for Environment and Development) (CCAD) in 1989, which entered into effect 14/6/90, and was followed by a *Central American Agenda on Environment and Development* (1992, 26 pp.) and an Alianza para el Desarrollo Sostenible de Centroamérica (Alliance for Sustainable Development of Central America),

TABLE 36. ESTIMATES OF CLOSED AND OPEN FOREST RESOURCES AND CONVERSION IN CENTRAL AMERICA
(after FAO, *Forest resources assessment 1990: tropical countries*, 1993)

Country	Closed and open forest area 1990 (km²)	% of land area	% deforested since 1980	Average annual deforestation (km²) 1980–1990
Guatemala	42,250	39	16.1	813
Belize	19,960	88	2.4	50
Honduras	46,050	41	19.5	1115
El Salvador	1230	6	20.6	32
Nicaragua	60,130	51	17.1	1241
Costa Rica	14,280	28	25.7	495
Panama	31,170	41	17.1	644

TABLE 37. ESTIMATES OF FOREST AND WOODLAND LAND USE AND CONVERSION IN CENTRAL AMERICA
(after FAO, *Agrostat-PC*, 1993)

Country	Forest and woodland area 1989–1991 (km²)	% of land area	% deforested since 1979–1981	Average annual deforestation (km²) 1979–1981 to 1989–1991
Guatemala	37,500	35	17.6	801
Belize [a]	10,120 [a]	44 [a]	0.2	2.1
Honduras	32,600	29	18.8	755
El Salvador	1040	5	25.7	36
Nicaragua	33,800	28	24.7	1109
Costa Rica	16,400	32	9.9	180
Panama	33,000	43	20.4	846

Note: [a] The 88% estimate in Table 36 is closer to being correct (L. Nicolait, 1995 pers. comm.).

12/10/94 (13 pp.). The region also has agreed to a Convenio para la Conservación de la Biodiversidad y Protección de Areas Silvestres Prioritarias en América Central (Convention for the Conservation of Biodiversity and Protection of Priority Wild Areas in Central America), 1992. The Global Environment Facility (GEF) agreed in 1994 to provide an initial US$50 million to the Central American Environmental Fund in support of this Alliance.

Creation of protected areas

The origins of modern protected areas date from the 19th century and beginning of this century. Registrations began with establishment of the first Municipal Forests in 1870 in Guatemala – natural forests put under a special system of management for production of forest products – and with the creation of the first forestry laws in the region between 1905 and 1940 (Ugalde and Godoy 1993).

In 1923 Barro Colorado Island (15.4 km²), isolated as an island by the waters of the Panama Canal, was declared a Biological Reserve. In 1928 the colonial administration of British Honduras (Belize) declared Half Moon Cay (39 km²) a Crown Reserve. In 1939 some of the volcanic craters in Costa Rica were declared National Parks, which in the 1970s were ratified as protected areas. In Honduras in 1952 the Forest Reserve of San Juancito was declared, which in 1980 became La Tıgra National Park (76 km²). In 1955 Guatemala declared its first ten National Parks, whereas 1958 marked the creation of the first National Park in Nicaragua.

In the 1970s, the first regional meetings to discuss the development of protected areas took place. During that decade agencies or institutions for the administration of National Parks were established. During the 1980s, ideas related to national systems of protected areas were developed as well as a regional system – the Central American System of Protected Areas (SICAP). The most significant changes toward strengthening the protected areas in the region began to take place during the 1980s.

In 1981 Belize passed its Law of Protected Areas. Between 1983 and 1985, Costa Rica strengthened its system by declaring Refuges for Wild Fauna. During this same decade and the beginning of the 1990s, Honduras declared its most important protected areas, such as Río Plátano Biosphere Reserve (CPD Site MA15), Ríos de Cuero y Salado Wildlife Refuge (85 km²) and reserves for 37 cloud forests. El Salvador declared its first legally protected area in 1987 – Montecristo National Park (39 km²). In 1989 Guatemala proclaimed its Law of Protected Areas which in turn stimulated the creation of the two largest Biosphere Reserves in the country, Maya Biosphere Reserve and Sierra de las Minas BR (CPD Sites MA13 and MA14). Between 1980 and 1988, 14 of the 20 protected areas in Panama were declared.

Recently, an important factor in the creation of protected areas has been the inclusion in regional and national political agendas of the necessity to halt the general environmental degradation which is occurring and the severe loss of biodiversity. For example in October 1991, Nicaragua declared the two largest reserves in the country (SIAPAZ and Bosawas), equivalent to more than half the total area of the country's system of protected areas, and also declared more than 40 natural areas (IRENA 1993). In November 1991, Belize declared the National Parks of Chiquibul (c. 2659 km²) and Laughing Bird Cay (c. 14 km²).

Coverage in the system

Despite the fact that more than 300 conservation units and more than 91 proposed areas have been identified within the region, covering 100,000 km² and 19% of Central America, as of mid-1991 only 173 protected areas covering a little more than than 56,000 km² and 10.5% of the Central American territory were recognized under IUCN categories I–V of protected areas, for which there are management objectives emphasizing environmental preservation (WCMC 1992).

These 173 protected areas, plus the territory declared in the region that comprises 71 Forest Reserves, Indigenous Reserves, Protected Zones and Areas of Multiple Use, bring the total to 244 declared units, covering c. 88,000 km² or 16.5% of the Central American region.

Panama, Nicaragua and Costa Rica have the most national territory under the highest protective categories of management (12–13%). Recent data indicate that the protected areas of Panama, Nicaragua and Guatemala make up 75% of the protected territory under SICAP. Nicaragua, for example, has 17,500 km² (14.7%) of its land surface in 71 protected areas that have been legally decreed (IRENA 1993). El Salvador and Belize are the countries with the least amount of land under SICAP (2.76%).

In addition to considering the amount of land designated for protected areas and the classification of the areas into categories of management objectives and use, it is essential to consider the effectiveness of administration and enforcement for protected areas. Although differing considerably, all the countries of Central America need to significantly augment their capacities to protect their designated areas (Barzetti 1993; Ugalde and Godoy 1993). For example, in Nicaragua 600 km² in protected areas are actively managed, 2500 km² are minimally managed, and 13,000 km² that are legally protected are without planning or management (IRENA 1993).

Gaps in coverage in the system

The majority of protected areas in the region have an emphasis on protecting mountain ecosystems (such as the peaks and volcanoes with cloud forests) and low tropical rain forests. Many areas of endemics and unique ecosystems are not well represented in SICAP. Examples are dry or semi-arid zones, humid mountain forests of cold altiplanos, zones with nearctic vegetation (oaks, pines, *Liquidambar*, etc.) and rare vegetation associations.

Consequently, many important additions need to be made to SICAP, including the following natural areas: the Maya Mountains in the south of Belize; in Guatemala, the region of Morazán in the semi-arid zone and the cold altiplano of the Cuchumatanes; in Honduras, the pine forests on Guanaja Island and the Tawahka Mountain Reserve zone; Los Morrales de Chalatenango in El Salvador; and the Arenal Cordillera in Costa Rica.

The protected areas located in the most populated zones of the region (mid-elevation plateaux and zones of the Pacific coastal plain) are the smallest and their biotopes the most threatened, because of the length of time human populations have been there and current social pressures. Urgent protection is needed for these small vegetational preserves to create areas which, through ecological restoration, will provide a return of the ecological goods and services required for wise development in these zones (e.g. north-western Costa Rica's Guanacaste Conservation Area – see Janzen 1986; Tenenbaum 1994).

Nicaragua and Guatemala offer the greatest potential within Central America for the creation of new units of conservation that would protect both ecosystems in the mountain zones and the lowland rain forest of the Atlantic coast. IRENA (1993) identified 1320 km² for the possible addition of protected areas in Nicaragua. Studies carried out in Costa Rica show the necessity to modify some of the boundaries of protected areas, with objectives (among others) of improving the representation of vegetation communities, enlarging key habitats, protecting endemic species and uniting strategic ecosystems (including remnants of natural resources, and incorporating marine resources).

Other ecosystems, such as mangrove swamps and humid coastal zones, need greater protection and more effective management throughout the region. National Parks, such as Corcovado (CPD Site MA18) and Tortuguero in Costa Rica, have official marine portions that need better management.

The region lacks development in management and conservation of marine resources. Very few wetland coastal parks and marine parks exist – but there are the Ríos de Cuero y Salado refuge in Honduras (85 km²), Cayos Miskitos (c. 5027 km²) in Nicaragua, and Islas del Coco National Park (24 km²) and Caño Negro Faunal Refuge (c. 100 km²) in Costa Rica.

Suitable additions to the Central American system of coastal and marine protected areas are multiple bays and reefs through Belize; some areas of mangrove swamp in southern Belize; Punta de Manabique - La Graciosa and Manchón in Guatemala; Humedales de Caratasca, Laguna de Guaymoreto and Islas del Cisne in Honduras; Los Cóbanos in El Salvador; the Golfo de Fonseca as a tri-national micro-region of El Salvador, Honduras and Nicaragua; Río Grande de Matagalpa Delta, Tapamlaya and Kukulaya in Nicaragua; and Isla de Coiba in Panama.

The desirability of creating on frontier international borders both terrestrial and as well coastal marine protected areas has been recognized since 1974, but not until recent years have real interest and action been demonstrated. The best known projects include the establishment of La Amistad International Park between Costa Rica and Panama in 1982/1986 (CPD Site MA17); Trifinio or La Fraternidad Biosphere Reserve between Guatemala, El Salvador and Honduras in 1987; and the International System of Protected Areas for Peace (SIAPAZ) between Nicaragua and Costa Rica in 1989. These three areas plus the Gulf of Fonseca, the Gulf of Honduras and the Miskito Cays are six of the eleven areas given priority under the region's 1992 Convenio para la Conservación de la Biodiversidad y Protección de Areas Silvestres Prioritarias en América Central.

Many cooperative initiatives between Central American countries demonstrate the high priority being given to conserving natural resources. Examples include the Honduras and Nicaragua project for an ecological corridor of Río Plátano-Tawahka-Río Coco-Bosawas; the Guatemala and Belize project for the Chiquibul-Maya Mountain area; and the System of Protected Areas of the Gran Petén (SIAP)

which includes Calakmul in Mexico, El Mirador-Río Azul in Guatemala and Río Bravo-Lamanai in Belize.

Centres of plant diversity and endemism

The Central American region has widely distributed plant species, and notable areas with high levels of endemism and diversity, eight of which are covered as Data Sheets (see Map 11 for their locations). Further research is likely to highlight some additional areas of major importance.

Guatemala

MA13. Petén Region and Maya Biosphere Reserve

– see Data Sheet.

MA14. Sierra de las Minas Region and Biosphere Reserve

– see Data Sheet.

Honduras

MA15. North-east Honduras and Río Plátano Biosphere Reserve

– see Data Sheet.

Costa Rica

MA16. Braulio Carrillo-La Selva Region

– see Data Sheet.

MA18. Osa Peninsula and Corcovado National Park

– see Data Sheet.

Costa Rica, Panama

MA17. La Amistad Biosphere Reserve

– see Data Sheet.

Panama

MA19. Cerro Azul-Cerro Jefe Region

– see Data Sheet.

MA20. Darién Province and Darién National Park

– see Data Sheet.

References

Allen, P.H (1956). *The rain forests of Golfo Dulce*. University of Florida Press, Gainesville. 417 pp.

Ames, O. and Correll, D.S. (1952-1953). *Orchids of Guatemala. Fieldiana, Bot.* 26: 1–727.

Argueta, A. (1992). El banco de plantas medicinales del Instituto Nacional Indigenista de México. Presented at the Congreso Etnobotánica 92, Córdoba, Spain.

Balick, M.J. (in prep.). Checklist of the plants of Belize, with annotations on common names and uses. New York Botanical Garden, Bronx.

Barzetti, V. (ed.) (1993). *Parques y progreso: áreas protegidas y desarrollo económico en América Latina y el Caribe.* IUCN, Washington, D.C. 258 pp.

Booth, W. (1993). U.S. drug firm signs up to farm tropical forests. In Place, S.E. (ed.), *Tropical rainforests: Latin American nature and society in transition.* Scholarly Resources, Wilmington, Delaware, U.S.A. Pp. 211–213.

Breedlove, D.E. (1981). Introduction to the *Flora of Chiapas*. In Breedlove, D.E. (ed.), *Flora of Chiapas*. Part 1. California Academy of Sciences, San Francisco. 35 pp.

Burger, W.C. (ed.) (1971-). *Flora Costaricensis. Fieldiana, Bot.* 35, 40, and *Fieldiana, Bot. New Series* 4, 13, 18, 35.

Butterfield, R.P. (1994). Forestry in Costa Rica: status, research priorities, and the role of La Selva Biological Station. In McDade, L.A., Bawa, K.S., Hespenheide, H.A. and Hartshorn, G.S. (eds), *La Selva: ecology and natural history of a neotropical rainforest.* University of Chicago Press, Chicago. Pp. 317–328.

Bye, R. (1993). The role of humans in the diversification of plants in Mexico. In Ramamoorthy, T.P., Bye, R., Lot, A. and Fa, J.E. (eds), *Biological diversity of Mexico: origins and distribution.* Oxford University Press, New York. Pp. 707–731.

Calderón, S. and Standley, P.C. (1944). *Lista preliminar de plantas de El Salvador*, 2nd edition. San Salvador. 450 pp.

Campbell, D.G. and Hammond, H.D. (eds) (1989). *Floristic inventory of tropical countries: the status of plant systematics, collections, and vegetation, plus recommendations for the future.* Central America. New York Botanical Garden, Bronx. Pp. 281–312.

Catling, P. and Catling, V. (1988). An annotated checklist of the orchids of Belize. *Orquídea (Méx.)* 11: 85–102.

Collins, M. (ed.) (1990). *The last rain forests: a world conservation atlas.* Oxford University Press, New York. 200 pp.

Contreras, F. (1985). *Las lagunas costeras mexicanas.* Centro de Ecodesarrollo, Secretaría de Pesca, Mexico, D.F. 81 pp.

Correll, D.S. (1965). Supplement to the orchids of Guatemala and British Honduras. *Fieldiana, Bot.* 31(7): 177–221.

Croat, T.B. (1978). *Flora of Barro Colorado Island*. Stanford University Press, Stanford, California. 943 pp.

D'Arcy, W.G. (1977). Endangered landscapes in Panama and Central America: the threat to plant species. In Prance, G.T. and Elias, T.S. (eds), *Extinction is forever: the status of threatened and endangered plants of the Americas.* New York Botanical Garden, Bronx. Pp. 89–102.

D'Arcy, W.G. (1987). *Flora of Panama: checklist and index.* Monogr. Syst. Bot., Missouri Bot. Gard. 17 and 18: 1–1000.

Davidse, G., Sousa-S., M. and Chater, A.O. (eds) (1994). *Flora Mesoamericana*. Vol. 6, Alismataceae a Cyperaceae. Universidad Nacional Autónoma de México, Mexico City. 543 pp.

Dávila, P., Medina, R., Ramírez, A., Salinas, A. and Tenorio, P. (1991). Análisis de la flora del Valle de Tehuacán: endemismo y diversidad. In *Resúmenes del Simposio Sobre Plantas en Peligro de Extinción*. UNAM and SEDUE, Mexico, D.F. Pg. 26.

de Souza, A.R. (ed.) (1992). The coexistence of indigenous peoples and the natural environment in Central America. *Research and Exploration (Washington, D.C.)* 8(2): map supplement.

Dirzo, R. (1991). Rescate y restauración ecológica de la selva de Los Tuxtlas. *Ciencia y Desarrollo* 17: 33–45.

Dirzo, R. and Gómez, G. (1996). Ritmos temporales de la investigación taxonómica de plantas vasculares en México y una estimación del número de especies conocidas. *Ann. Missouri Bot. Gard.* 83(3): 396–403.

Duke, J.A. (1985). Neotropical anticancer plants. In D'Arcy, W.G. and Correa-A., M.D. (eds), *The botany and natural history of Panama.* Monogr. Syst. Bot. 10, Missouri Botanical Garden, St. Louis. Pp. 299–304.

Dwyer, J.D. and Spellman, D.L. (1981). A list of the Dicotyledoneae of Belize. *Rhodora* 83: 161–236.

FAO (1993). *Agrostat-PC*. FAO, Rome. Computer diskette.

FAO (1993). *Forest resources assessment 1990: tropical countries.* FAO Forestry Paper No. 112, FAO, Rome. 112 pp.

Ferrusquía-Villafranca, I. (1993). Geology of Mexico: a synopsis. In Ramamoorthy, T.P., Bye, R., Lot, A. and Fa, J.E. (eds), *Biological diversity of Mexico: origins and distribution.* Oxford University Press, New York. Pp. 3–107.

Flores-Villela, O.A. (1993). Herpetofauna of Mexico: distribution and endemism. In Ramamoorthy, T.P., Bye, R., Lot, A. and Fa, J.E. (eds), *Biological diversity of Mexico: origins and distribution.* Oxford University Press, New York. Pp. 253–280.

Flores-Villela, O.A. and Geréz, P. (1988). *Conservación en México: síntesis sobre vertebrados terrestres, vegetación y uso del suelo.* INIREB and Conservation International, Mexico, D.F. 302 pp.

Fryxell, P.A. (1988). *Malvaceae of Mexico.* Syst. Bot. Monogr. 25. 522 pp.

García, E. (1973). *Modificaciones al sistema de clasificación climática de Köppen (para adaptarlo a las condiciones de la república mexicana),* 2nd edition. Instituto de Geografía, UNAM, Mexico, D.F. 246 pp.

García, E. (1989). *Diversidad climática vegetal en México.* Presented at the Simposio Sobre la Diversidad Biológica de México, Oaxtepec, Morelos. 27 pp.

Gentry, A.H. (1978). Floristic knowledge and needs in Pacific tropical America. *Brittonia* 30: 134–153.

Gentry, A.H. (1982). Neotropical floristic diversity: phytogeographical connections between Central and South America, Pleistocene climatic fluctuations, or an accident of the Andean orogeny? *Ann. Missouri Bot. Gard.* 69: 557–593.

Gentry, H.S. (1942). *Río Mayo plants: a study of the flora and vegetation of the valley of Río Mayo, Sonora.* Carnegie Institution of Washington Public. No. 527. 328 pp.

Gómez-P., L.D. (1985). *Biotic units of neotropical region.* The Nature Conservancy, International Program, Science Division. Washington, D.C. 55 pp. + 3 maps. Unpublished.

Gómez-P., L.D. (1986). *Vegetación de Costa Rica: apuntes para una biogeografía costarricense.* Editorial Universidad Estatal a Distancia, San José. 327 pp.

Gómez-Pompa, A. and Dirzo, R. with Kaus, A., Noguerón-Chang, C.R. and Ordoñez, M. de J. (1994). *Las áreas naturales protegidas de México de la Secretaría de Desarrollo Social.* SEDESOL, Mexico, D.F. 331 pp. Unpublished.

Hamer, F. (1988). *Orchids of Central America: an illustrated field guide. A-L. Selbyana* 10 (Supplement): 1–422.

Hamer, F. (1990). *Orchids of Central America: an illustrated field guide. M-Z. Selbyana* 10 (Supplement): 423–860.

Heckadon, S.P. (1992). Central America: tropical land of mountains and volcanoes. In Barzetti, V. and Rovinski, Y. (eds), *Toward a green Central America: integrating conservation and development.* Kumarian Press, Hartford, Connecticut, U.S.A. Pp. 5–20.

Hernández-Xolocotzi, E. (1993). Aspects of plant domestication in Mexico: a personal view. In Ramamoorthy, T.P., Bye, R., Lot, A. and Fa, J.E. (eds), *Biological diversity of Mexico: origins and distribution.* Oxford University Press, New York. Pp. 733–753.

Herrera, W. and Gómez-P., L.D. (1992). *Mapa de unidades bióticas de Costa Rica.* Incafo, Madrid.

Hodel, D. (1992). Chamaedorea *palms. The species and their cultivation.* Allen Press, Lawrence, Kansas, U.S.A. 338 pp.

Hoffman, M. (ed.) (1993). *The world almanac and book of facts 1993.* Pharos Books, New York. 960 pp.

Holdridge, L.R. and Poveda, L.J. (1975). *Arboles de Costa Rica,* Vol. 1. Centro Científico Tropical, San José. 546 pp.

IRENA (1993). *Plan de acción ambiental de Nicaragua – PAANIC: sumario ejecutivo y matrices de planificación.* Ministerio de Economía y Desarollo, Instituto Nicaraguense de Recursos Naturales y del Ambiente (IRENA), Managua. 19 pp.

Janzen, D.H. (1986). *Guanacaste National Park: tropical ecology and cultural restoration.* Editorial Universidad Estatal a Distancia, San José. 103 pp.

Jiménez, Q. and Poveda, L.J. (1991). *Arboles maderables nativos de Costa Rica.* Contr. Depto. Hist. Nat. 5, Museo Nacional de Costa Rica, San José. 32 pp.

Jones, D.L. (1993). *Cycads of the world.* Smithsonian Institution Press, Washington, D.C. 312 pp.

Kappelle, M. (1995). *Ecology of mature and recovering Talamancan montane* Quercus *forests, Costa Rica.* Universiteit van Amsterdam, Amsterdam. 273 pp.

Lamb, F.B. (1953). The forests of Darién. *Caribbean Forest.* 14: 128–135.

Lankford, R.R. (1977). Coastal lagoons of Mexico: their origin and classification. In Wiley, M.L. (ed.), *Estuarine processes.* Academic Press, San Diego, California. Pp. 182–215.

Leonard, H.J. (1987). *Natural resources and economic development in Central America: a regional environmental profile.* International Institute for Environment and Development, Washington, D.C. Transaction Books, New Brunswick, New Jersey, U.S.A. and Oxford, U.K. 269 pp.

Lot-H., A., Novelo, A. and Ramírez-García, P. (1993). Diversity of Mexican aquatic vascular plant flora. In Ramamoorthy, T.P., Bye, R., Lot, A. and Fa, J.E. (eds), *Biological diversity of Mexico: origins and distribution.* Oxford University Press, New York. Pp. 577–591.

Luna-Vega, I., Almeida, L. and Llorente, J. (1989). Florística y fitogeografía del bosque mesófilo de montaña de las cañadas de Ocuilán, Morelos. *Ann. Inst. Biol. Univ. Nac. Autón. México, Ser. Bot.* 59: 63–87.

MacKerron, C.B. (1993). *Business in the rain forests: corporations, deforestation and sustainability* (Cogan, D.G., ed.). Investor Responsibility Research Center, Washington, D.C. 239 pp.

Masera, O., Ordóñez, M. de J. and Dirzo, R. (1992). Emisiones de carbono a partir de la deforestación en México. *Ciencia* 43 (número especial): 151–153.

Masera, O., Ordóñez, M. de J. and Dirzo, R. (1995). Carbon emissions from deforestation in Mexico: current situation and long-term scenarios. *Global Environmental Change*: in press.

McDonald, J.A. (1993). Phytogeography and history of the alpine-subalpine flora of northeastern Mexico. In Ramamoorthy, T.P., Bye, R., Lot, A. and Fa, J.E. (eds), *Biological diversity of Mexico: origins and distribution*. Oxford University Press, New York. Pp. 681–703.

Meyers, C.W. (1969). The ecological geography of cloud forests in Panama. *Amer. Mus. Novitates* 2396: 1–52.

Mittermeier, R.A. (1988). Primate diversity and the tropical forest: case studies from Brazil and Madagascar and the importance of the megadiversity countries. In Wilson, E.O. and Peter, F.M. (eds), *Biodiversity*. National Academy Press, Washington, D.C. Pp. 145–154.

Molina, A. (1975). Enumeración de las plantas de Honduras. *Ceiba* 19(1): 1–118.

Morton, J.F. (1981). *Atlas of medicinal plants of Middle America. Bahamas to Yucatan*. Charles Thomas, Springfield, Illinois, U.S.A. 1420 pp.

Nations, J.D. (1987). Mesoamerica's tropical rainforests: conflicts and conservation. *Tulane Stud. Zool. Bot.* 26: 59–76.

Nations, J.D. and Komer, D.I. (1984). Chewing up the jungle. *Int. Wildlife* 14(5): 14–16.

Nations, J.D. and Leonard, H.J. (1986). Grounds of conflict in Central America. In Maguire, A. and Brown, J.W. (eds), *Bordering on trouble: resources and politics in Latin America*. Adler and Adler, Bethesda, Maryland, U.S.A. Pp. 55–98.

Nelson, C. (1986). *Plantas comunes de Honduras*, Vols. I and II. Editorial Universidad Nacional Autónoma de Honduras, Tegucigalpa. 922 pp.

Pacheco, M. and Martínez, A. (1949). Population of El Salvador and its natural resources. *Proceedings of the Inter-American Conference on Conservation of Renewable Natural Resources, Denver, Colorado, September 7–20, 1948*. U.S. Dept. of State Public. 3382: 125–133.

Puig, H., Bracho, R. and Sosa, V.J. (1983). Composición florística y estructura del bosque mesófilo de Gómez-Farías, Tamaulipas. *Biótica* 8: 339–359.

Ramamoorthy, T.P. (1984). Notes on the genus *Salvia* (Lamiaceae) in Mexico with three new species. *J. Arnold Arbor.* 65: 135–143.

Ramamoorthy, T.P. and Elliott, M. (1993). Mexican Lamiaceae: diversity, distribution, endemism and evolution. In Ramamoorthy, T.P., Bye, R., Lot, A. and Fa, J.E. (eds), *Biological diversity of Mexico: origins and distribution*. Oxford University Press, New York. Pp. 513–539.

Ramamoorthy, T.P. and Lorence, D.H. (1987). Species vicariance in Mexican flora and a new species of *Salvia* from Mexico. *Adansonia* 2: 167–175.

Ramírez, M. (1909, 1911). *Flora nicaraguense*. Compañía Tipográfica Internacional, Managua.

Rauh, W. (1992). Are tillandsias endangered plants? *Selbyana* 13: 138–139.

Reid, W.V., Laird, S.A., Meyer, C.A., Gámez, R., Sittenfeld, A., Janzen, D.H., Gollin, M.A. and Juma, C. (1993). *Biodiversity prospecting: using genetic resources for sustainable development*. World Resources Institute, Washington, D.C. 341 pp.

Rzedowski, J. (1973). Geographical relationships of the flora of Mexican dry regions. In Graham, A. (ed.), *Vegetation and vegetational history of northern Latin America*. Elsevier, New York. Pp. 61–72.

Rzedowski, J. (1978). *Vegetación de México*. Editorial Limusa, Mexico, D.F. 432 pp.

Rzedowski, J. (1991). El endemismo en la flora fanerogámica mexicana: una apreciación analítica preliminar. *Acta Bot. Mex.* 15: 47–64.

Rzedowski, J. (1993). Diversity and origins of the phanerogamic flora of Mexico. In Ramamoorthy, T.P., Bye, R., Lot, A. and Fa, J.E. (eds), *Biological diversity of Mexico: origins and distribution*. Oxford University Press, New York. Pp. 129–144.

Scott, D. and Carbonell, M. (eds) (1986). *A directory of neotropical wetlands*. IUCN, Cambridge, U.K. 684 pp.

Seymour, F.C. (1980). *A checklist of the vascular plants of Nicaragua*. Phytologia Memoirs 1: 1–314.

Soto-Arenas, M.A. (1988). Updated list of the orchids of Mexico. *Orquídea (Méx.)* 11: 273–276.

Spellman, D.L., Dwyer, J.D. and Davidse, G. (1975). A list of the Monocotyledoneae of Belize including a historical introduction to plant collection in Belize. *Rhodora* 77: 105–140.

Standley, P.C. (1928). *Flora of the Panama Canal Zone. Contr. U.S. Natl. Herb.* 27: 1–416.

Standley, P.C. (1930). *Flora of Yucatan. Field Mus. Nat. Hist., Bot.* 3(3): 157–492.

Standley, P.C. (1931). *Flora of the Lancetilla Valley, Honduras. Field Mus. Nat. Hist., Bot.* 10: 1–418.

Standley, P.C. (1937-1938). *Flora of Costa Rica. Field Mus. Nat. Hist., Bot.* 18: 1–1616.

Standley, P.C. and Record, S.J. (1936). *The forests and flora of British Honduras. Fieldiana, Bot.* 12: 1–432.

Standley, P.C., Steyermark, J.A. and Williams, L.O. (1946-1977). *Flora of Guatemala. Fieldiana, Bot.* 24(1–13).

Stebbins, G.L. (1975). The role of polyploid complexes in the evolution of North American grasslands. *Taxon* 24: 91–106.

Stolze, R.S. (1976). *Ferns and fern allies of Guatemala.* Part I. *Fieldiana, Bot.* 39: 1–130.

Stolze, R.S. (1981). *Ferns and fern allies of Guatemala.* Part II. *Fieldiana, Bot. New Series* 6: 1–522.

Stolze, R.S. (1983). *Ferns and fern allies of Guatemala.* Part III. *Fieldiana, Bot. New Series* 12: 1–91.

Tenenbaum, D. (1994). The Guanacaste idea. *Amer. Forests* 100 (11/12): 28–31, 58–59.

Toledo, V.M. (1985). *A critical evaluation of the floristic knowledge in Latin America and the Caribbean.* A report presented to The Nature Conservancy International Program. Washington, D.C. 108 pp.

Toledo, V.M. (1992). Bio-economic costs. In Downing, T.E., Hecht, S.B., Pearson, H.A. and García-Downing, C. (eds), *Development or destruction: the conversion of tropical forest to pasture in Latin America.* Westview Press, Boulder, Colorado. Pp. 67–93.

Toledo, V.M. (1995a). Los habitats naturales de México: una visión sintética. In Ceballos, G. and Navarro, D. (eds), *Fauna mexicana en peligro de extinción.* UNAM, Mexico, D.F. In press.

Toledo, V.M. (1995b). New paradigms for a new ethnobotany: reflections of the Mexican case. In Schultes, R.E. and von Reis, S. (eds), *Ethnobotany today.* Dioscorides Press. Pp. 125–134.

Toledo, V.M. (1995c). La riqueza florística de México: un análisis para conservacionistas. In Guevara, S. and Moreno, P. (eds), *La botánica mexicana hacia el fin del milenio.* Instituto de Ecología. In press.

Toledo, V.M. and Ordóñez, M. de J. (1993). The biodiversity scenarios of Mexico: an analysis of terrestrial habitats. In Ramamoorthy, T.P., Bye, R., Lot, A. and Fa, J.E. (eds), *Biological diversity of Mexico: origins and distribution.* Oxford University Press, New York. Pp. 757–777.

Toledo, V.M. and Sosa, V. (1993). Floristics in Latin America and the Caribbean: an evaluation of plant collections and botanists. *Taxon* 42: 355–364.

Toledo, V.M., Batis, A., Becerra, R., Martínez, E. and Ramos, C.H. (1992). Products from the tropical rain forests of Mexico: an ethnoecological approach. In Plotkin, M. and Famolare, L. (eds), *Sustainable harvest and marketing of rain forest products.* Island Press, Washington, D.C. Pp. 99–109.

Toledo, V.M., Carabias, J., Toledo, C. and González-Pacheco, C. (1989). *La producción rural en México: alternativas ecológicas.* Fundación Universo Veintiuno, Mexico, D.F. 365 pp.

Turner, B.L. and Nesom, G.L. (1993). Biogeography, diversity, and endangered or threatened status of Mexican Asteraceae. In Ramamoorthy, T.P., Bye, R., Lot, A. and Fa, J.E. (eds), *Biological diversity of Mexico: origins and distribution.* Oxford University Press, New York. Pp. 559–575.

Ugalde, A. and Godoy, J.C. (1993). *La situación de las áreas protegidas en Centroamérica.* Informe al IV Congreso Mundial de Parques, Caracas, Venezuela, Febrero de 1992. IUCN, San José, Costa Rica and Guatemala City, Guatemala. 80 pp.

Utting, P. (1991). *The social origins and impact of deforestation in Central America.* United Nations Research Institute for Social Development. Geneva, Switzerland. 43 pp.

Valdés-Reyna, J. and Cabral-Cordero, I. (1993). Chorology of Mexican grasses. In Ramamoorthy, T.P., Bye, R., Lot, A. and Fa, J.E. (eds), *Biological diversity of Mexico: origins and distribution.* Oxford University Press, New York. Pp. 439–446.

Vargas, F. (1984). *Los parques nacionales de México.* Instituto de Investigaciones Ecónomicas, UNAM, Mexico, D.F. 145 pp.

Villa-Lobos, J. (in prep.). *Threatened plants of Middle America.* Smithsonian Institution Press, Washington, D.C.

WCMC (1992). *Protected areas of the world. A review of national systems. Vol. 4. Nearctic and Neotropical.* IUCN, Gland, Switzerland and Cambridge, U.K. 459 pp.

West, R.C. (1964). The natural regions of Middle America. In Wauchope, R. (ed.), *Handbook of Middle American Indians,* Vol. 1. West, R.C. (ed.), *Natural environment and early cultures.* University of Texas Press, Austin. Pp. 363–383.

Williams, L.O. (1956). An enumeration of the Orchidaceae of Central America, British Honduras, and Panama. *Ceiba* 5: 1–256.

Williams, L.O. (1981). Useful plants of Central America. *Ceiba* 24: 1–381.

Woodson, R.E., Schery, R.W. *et al.* (1943-1980). Flora of Panama. In *Ann. Missouri Bot. Gard.* Vols. 30–67.

World Resources Institute (1992). *World resources 1992–93. A guide to the global environment.* Oxford University Press, New York. 385 pp.

WRI (World Resources Institute), UNEP (United Nations Environment Programme) and UNDP (United Nations Development Programme) (1994). *World resources 1994–95. A guide to the global environment.* Oxford University Press, New York. 403 pp.

Ypsilantis, J. (1992). Fragile isthmus under pressure. *People and Planet* 1(3): 19–22.

Yuncker, T.G. (1938). A contribution to the Flora of Honduras. *Field Mus. Nat. Hist., Bot.* 17(4): 287–407.

Authors

The overview for Mexico was written by Dr Víctor M. Toledo [Universidad Nacional Autónoma de México (UNAM), Centro de Ecología, Apdo. Postal 70-275, 04510 Mexico, D.F., Mexico], Dr Jerzy Rzedowski (Instituto de Ecología, Centro Regional del Bajío, Apdo. Postal 386, 61600 Pátzcuaro, Michoacán, Mexico) and Jane Villa-Lobos (Smithsonian Institution, Department of Botany, NHB-166, Washington, DC 20560, U.S.A.).

The overview for Central America was written by Luis Diego Gómez (Las Cruces Botanical Garden, P.O. Box 73, San Vito, Coto Brus, Costa Rica), Juan Carlos Godoy (IUCN Protected Areas and Biodiversity, Apdo. 112-I Zona 7, Guatemala City, Guatemala), Olga Herrera-MacBryde and Jane Villa-Lobos (Smithsonian Institution, Department of Botany, NHB-166, Washington, DC 20560, U.S.A.).

LACANDON RAIN FOREST REGION
Mexico

Location: In eastern Chiapas between latitudes 16°05'–17°45'N and longitudes 90°25'–91°45'W, between the Usumacinta River and the Perlas and Lacantún rivers.

Area: Region c. 6000 km²; reserves 4122 km².

Altitude: Region 80–1750 m; reserves 80–1400 m.

Vegetation: Tropical and montane rain forests, cloud forest, semi-deciduous tropical forest, savanna, pine-oak forest, seasonally flooded forest, gallery forest, open wetlands.

Flora: High diversity: c. 4000 species of vascular plants, some endemism; threatened species.

Useful plants: Reserve for timber, fruit and gum trees; heavy use of palm leaves for local roofing or export as ornamentals.

Other values: Watershed protection, Amerindian lands, archaeological sites, high biodiversity, large fauna refuge, potential tourism.

Threats: Road construction, logging, colonization, fire, agriculture, cattle-raising, oil exploration.

Conservation: 1 UNESCO-MAB Biosphere Reserve, 1 national-level Biosphere Reserve, 2 Natural Monuments, 1 Flora and Fauna Reserve, 1 Protection Forest.

Geography

The tropical rain forests of southern and central Mexico are the northernmost extension of this ecosystem in the Americas, and once covered c. 11% of the country – forming a continuous corridor in the states of the Yucatán Peninsula, Chiapas, Tabasco, Oaxaca, Veracruz and Puebla, and occurring as well in Hidalgo and San Luis Potosí (Rzedowski 1978; Wendt 1993). However, since 1950 more than half has been destroyed for crops, lumber, cattle-ranching and oil exploration. What remains is in six large and various small ecological islands (Estrada and Coates-Estrada 1983; Dirzo and Miranda 1991).

The largest remaining segment of this northern tropical rain forest is along the eastern edge of the highlands of Chiapas, Mexico's southernmost state. The region known as Selva Lacandona is delimited to the north by the Lower Usumacinta River and the savannas and wetlands of eastern Tabasco; to the east the limits are the Usumacinta and Salinas rivers and the Guatemalan Petén (CPD Site MA13, see Data Sheet). The region continues south into Guatemala to a small extent; to the west it is delimited by the central highlands of Chiapas. The surviving Lacandon forest covers 6000 km² in the Jataté and Upper Usumacinta river basins. The Usumacinta Basin and Grijalva Basin (farther west) contain one-third of the freshwater resources of Mexico.

The region is within the Sierra Madre de Chiapas morphotectonic province, in the Northern Folded Ranges and Plateaux subprovince (Ferrusquía-Villafranca 1993). The Chiapas Highlands, from 400–2200 m, are mostly limestone with some sandstone and volcanic extrusions. The western half of the Selva Lacandona is primarily mountain ranges (to 1750 m) trending from north-west to south-east, and narrow intermontane valleys draining south-easterly. The eastern half of the region is primarily alluvial plains at low elevations, with isolated hills and valleys as major features; in general its altitude ranges from 80 m to 500 m. Much of the regional geology is based on marine and transitional sedimentary rock types from the Middle Cretaceous to Early Tertiary.

About one-fifth of the original Selva Lacandona is in the Montes Azules Biosphere Reserve (16°06'–16°49'N, c. 90°45'–91°30'W), which encompasses 3312 km² (c. 100 km × 15–55 km) including the south-central section of the forest with the lakes Ocotal, Lacanjá and Miramar (79 km²). In 1992 Mexico reserved an adjacent 663 km² to the north-east in the Lacan-tún Biosphere Reserve (619 km²) and Bonampak Natural Monument (44 km²). Montes Azules BR encompasses c. 70% of the remaining original vegetation, and has most of the vegetation types and weather found in the Lacandon region (cf. Orellana 1978).

The climate is warm and humid, with the median annual temperature above 22°C and the coolest month averaging above 18°C. The median annual precipitation is over 2500 mm, with a summer rainy season which experiences monsoons (Gómez-Pompa *et al.* 1994). Generally, winds from the Chiapas Highlands predominate.

Vegetation

According to Breedlove (1981), in Chiapas true tropical rain forest occurs only in a few locations in the flat valleys of the upper drainage of the Usumacinta River, and is surrounded by more common lower montane rain forest. Rzedowski (1978) included both in evergreen tropical forest. This true tropical rain forest is multistoreyed. Some of the most common canopy trees (50–60 m tall) are *Aspidosperma megalocarpon, Brosimum alicastrum* var. *alicastrum, Dialium guianense, Erblichia odorata* var. *odorata, Guatteria anomala,*

Licania platypus, Manilkara zapota, Poulsenia armata, Swietenia macrophylla and *Terminalia amazonia* ("canshán"). The most abundant trees of the continuous second-storey canopy (25–40 m high) and trees of the understorey (10–20 m) are *Alchornea latifolia, Alibertia edulis, Trichospermum grewiifolium, Bumelia persimilis, Bursera simaruba, Cassia grandis, Blepharidium guatemalense (B. mexicanum), Guarea glabra, Pleuranthodendron lindenii, Licaria peckii, Orthion subsessile, Pithecellobium arboreum, Quararibea funebris, Simira salvadorensis, Wimmeria bartlettii* and *Zuelania guidonia*. Epiphytes are only in the upper stories; small shrubs and herbaceous cover are sparse.

The lower montane rain forest contains nearly all species found in the tropical rain forest; it is well adapted to the well-drained conditions of the region's calcareous mountains. Most of the highlands of Lacandona are covered with this formation. Its canopy is 25–45 m high, lianas are much more common, epiphytes occur throughout, and the shrub layer is dense and well developed. Additional canopy trees include *Trichospermum* sp., *Calophyllum brasiliense* var. *rekoi, Ocotea rubriflora, Ocotea* sp. (*Nectandra sinuata*), *Quercus oleoides* and *Talauma mexicana* (Miranda 1961; Breedlove 1981).

Higher ridges with montane rain forest and other types of montane forest occur to the west (cf. Calzada and Valdivia 1979). In the Lower Lacantún River Basin there is considerable seasonally flooded forest. Marshes or seasonally flooded areas have special formations, such as almost pure stands of *Haematoxylum campechianum* surrounding many lakes in the forest (Nations and Nigh 1980). The south-eastern Marqués de Comillas area includes drier, semi-deciduous lowland forest, and areas of savanna occur in the east (see Breedlove 1981; Gómez-Pompa *et al.* 1994).

Flora

There are an estimated 4000 species of vascular plants in the Selva Lacandona (Medellín 1991). Ongoing research in the Lacandon region by the Universidad Nacional Autónoma de México (UNAM) has resulted in many new collections. Studies are urgently needed to recognize threatened species, which may include the cycads *Ceratozamia matudae, Dioon merolae* and *Zamia splendens*.

The tropical rain forests of Mexico are species-rich for northern areas, particularly in the State of Chiapas where c. 1500 species of trees have been found. However, Mexican rain forests show low species diversity compared to those of Central America and South America – the diversity of trees increases in a southward gradient (Toledo 1982). For example, there are more than three times as many tree species in the Lacandon forest as in the far north-western rain forests of Huichihuayán (San Luis Potosí) and Misantla (Veracruz) (cf. Dirzo and Miranda 1991).

A few of the non-endemic canopy tree species in Mexico are found only in the Lacandon forest (e.g. *Luehea seemannii, Orthion subsessile, Pourouma guianense*), due to a mixture of floristic rain-forest elements of the Yucatán Peninsula and Gulf of Mexico (Wendt 1993). Floristic associations in the Lacandon forest are continuous with those in the Guatemalan Petén (Breedlove 1981) (CPD Site MA13); their compositions are relatively poorly known. The floras of the Chiapan highlands and Petén share some endemic taxa (Rzedowski 1962; Breedlove 1981), but the majority of their species are

shared with Central America. Although the known endemic species restricted to the Mexican Lacandon forest are few (e.g. *Yucca lacandonica*), they include the startling new endemic family Lacandoniaceae, represented by *Lacandonia schismatica* (Martínez and Ramos 1989; Mestel 1995).

Toledo (1982) tentatively identified the Lacandon forest as one of two primary Pleistocene refugia for Mexican tropical rain forest, characterized by high rainfall (over 3000–4000 mm a year), high temperature (annual average over 25°C), and a high concentration of species. Zoogeographic data on the Lacandon region also have given evidence of isolation and protected conditions during the Pleistocene and other unstable periods (Smith 1949; Barrera 1962).

Nevertheless, Wendt (1989, 1993) found little floristic evidence of a past refuge restricted to the Lacandon region; high diversity may not be a good indicator of former refugia. The endemic taxa are mostly shared with the Petén and other areas of nearby Guatemala and Belize. The larger area (Lacandona, eastern Guatemala and southern Belize) is relatively important for rain-forest species endemism. Furthermore, in the extreme northern part of the Lacandon region the average annual precipitation can exceed 3000 mm; this is the eastern end of another endemic-rich area of rain forest known as "the crescent area". It terminates to the west in the Uxpanapa area of southern Veracruz (CPD Site MA2, see Data Sheet), which may have been a refuge (Wendt 1989).

Useful plants

The Lacandon forest contains important reserves of timber, such as *Calophyllum brasiliense* var. *rekoi, Cedrela odorata, Cordia* spp., *Dialium guianense, Lonchocarpus castilloi, Swietenia macrophylla, Tabebuia guayacan, Talisia oliviformis* and *Trema micranthum*. Other species of economic importance are *Manilkara zapota* – the "sapodilla" tree from which chicle gum is extracted; *Castilla elastica* var. *elastica* – its latex was the source of rubber ("caoutchouc") for the aborigines of Mexico; *Cymbopetalum penduliflorum* – the flowers are used among the Maya Amerindians for flavouring and medicine (Wagner 1964); and species that bear edible fruits, such as (among trees) *Manilkara zapota, Pimenta dioica* (allspice), *Poulsenia armata, Pouteria mammosum* and many others (Miranda 1961). Peters and Pardo-Tejeda (1982) have evaluated the promising economic potential of *Brosimum alicastrum* ("ramón") (by using fruits, seeds, leaves, wood, latex, bark).

Several species of palms (e.g. *Geonoma oxycarpa, Scheelea liebmannii*) are used by the local inhabitants for roofing. Additionally, seeds, seedlings and leaves of some small palms called "xate" (e.g. *Chamaedorea tepejilote, C. oblongata, C. elegans*) are being removed from the forest by the millions and with increasing frequency. They are dispatched from peripheral cities by refrigerated truck to Texas, U.S.A., where they are commercialized (Marshall 1989).

Social and environmental values

The Selva Lacandona is important for management of its large watersheds and soil conservation, and encompasses a significant portion of Mexican biodiversity. About 20–25% of the Mexican species of many groups are present in the Montes

Azules BR, which is only 0.16% of the Mexican territory. Among the best represented groups are birds – 345 species, over 33% of the Mexican species, including harpy eagle, black-and-white hawk-eagle (*Spizastur melanoleucus*), macaws and parrots; mammals – 112 species, c. 25%; and diurnal butterflies – 800 species, 44% (Medellín 1991, 1994). The forest is one of the few areas in Middle America large enough for viable populations of animals such as jaguars, ocelots, howler and spider monkeys, Baird's tapirs, white-lipped peccaries, kinkajous and caimans (cf. Vega-Rivera 1990).

The region is variously inhabited by five Amerindian groups (Lacandons, Tojolobals, Chols, Tzeltals, Tzotzils), including descendants of the pre-Columbian Maya civilization that flourished for nearly ten centuries, practising a highly diverse, long-term system of food production that showed sustained use of the tropical forest ecosystem. Among the few Mayan groups that continue using elements of this knowledge are the c. 300–450 Lacandons (Nations 1984, 1985), whose techniques of farming (a milpa style mimicking the forest's dynamics) allow the forest to regenerate without significant loss. Aspects of the Lacandon Maya subsistence and forest-management systems are being examined and applied by some institutions to demonstrate that the practices that serve them – and served the Classic Maya (250–950 AD) – are helpful for the modern development of sustained-yield use of tropical forest ecosystems (Gómez-Pompa 1987; Nations 1988).

Economic assessment

The Selva Lacandona seems not to be integrated into the economic processes of Mexico or the State of Chiapas, but the situation suggests that the overall region is barely able to maintain its present population of over 150,000. About 9800 people (1990 census) live within the Montes Azules BR, especially in the south-western portion. From c. 1960, Tzeltals and Tzotzils migrated to the region from central Chiapas. Immigration of mestizo people is increasing from various parts of southern and central Mexico.

The inhabitants produce maize, cacao, rice and other crops mostly for themselves or regional use. Only crops such as the jalapeño pepper (*Capsicum annuum* var. *annuum*) represent a relatively important source of income for a few farmers, who can afford the high costs of intensive human labour and frequent applications of pesticides; these crops are shipped to Mexico City. In general, farmers invest their money in cattle, which are maintained in extensive grasslands at high environmental cost.

Important quantities of valuable woods had been obtained c. 1850–1948; most of the good trees accessible by river were gone by 1949, when the Mexican Government prohibited export of unprocessed trunks (Medellín 1991). Logging roads began to open up the region in 1965. The increased rate of timber extraction is far from sustainable.

Alternatives that may provide sustainable economic development are scarcely explored or implemented. Several studies are beginning to examine the possibilities of sustained use with newly developed technologies, such as farming of butterflies and iguanas, fisheries with native species, and production of xate palms (*Chamaedorea* spp.) and orchids.

The region has considerable potential for tourism, especially for nature (CI 1993) and archaeological sites. The Selva Lacandona is a part of the large area dominated by the ancient Mayan culture and shared with Guatemala, Belize, Honduras and El Salvador. Major archaeological sites include Yaxchilán and Bonampak, and many still unstudied sites occur in the Lacandona.

Threats

Accelerating rates of loss of the Lacandon forest threaten its flora and fauna and the survival of its indigenous peoples. In 1943, the region was covered by more than 13,000 km² of tropical rain forest, but over half has been cleared and burned, and a significant part of the remainder is being converted by modern agricultural colonization, logging and cattle-ranching. Recently, oil exploration and road construction have significantly increased the rate of destruction of the lowland forest. Through the past 30 years, annually over 150 km² have been converted. The 1990 census found over 9800 people living within the Montes Azules Biosphere Reserve, with c. 40% concentrated in the south-western portion in 14 locales.

Planned development also threatens the future of the Selva Lacandona. One section of the plan intends to expand colonization, extending agricultural activity over 2000 km² farther into the Marqués de Comillas zone (the south-eastern area next to Guatemala). A road would be built south of the Lacantún River and parallel to the Usumacinta River and Guatemalan border to link the area with San Cristóbal de las Casas and Palenque. The Mexican forest also became home for 70,000 Guatemalans displaced by guerrilla activity, creating additional strains on forest management. The region has been threatened by a proposed binational hydroelectric project on the Usumacinta River (Wilkerson 1985), which was halted in 1992 by Mexico's president (GCTM 1992).

Conservation

The Mexican scientific community and government have become increasingly concerned about social and ecological costs of economic growth in the State of Chiapas. A 1971 presidential decree set aside over 6140 km² as a Forest Reserve for the sole property and home of the Lacandons. In the middle 1970s, several institutions and agencies undertook a programme of intensive investigation in the Lacandon forest. The study recommended the creation of a Natural Reserve.

The Montes Azules Biosphere Reserve (3312 km²) was established by federal decree in 1977, and approved under UNESCO's Man and the Biosphere Programme in 1979. Mexico at the same time included the area in a Protection Forest of 26,123 km² which was established for the upper basin of the Usumacinta River and the Tulijah River Basin. The reserve is the responsibility of the Instituto Nacional de Ecología of the Secretaría de Medio Ambiente, Recursos Naturales y Pesca (SEMARNAP) (formerly in SEDESOL). A management plan was produced in November 1992. It proposes new zoning in buffer areas to regulate land uses under a realistic view for conservation and sustainable development. The reserve's exact boundaries were being established and marked in 1993. In August 1992, establishment of the following four areas protected an additional c. 810 km² in the region (Map 12): Lacan-tún Biosphere Reserve (recognized nationally) (619 km²), Chan Kin Flora and Fauna Reserve (122 km²) and the Natural Monuments Bonampak (44 km²) and Yaxchilán (26 km²). The Sierra de la Cojolita Communal Reserve, which was established by the Lacandons, is the unique forested connection between Montes Azules BR and the protected areas in Guatemala (Gómez-Pompa *et al.* 1994). Studies are underway to ensure preservation of the ecological corridor between the Selva Lacandona and Guatemala's Maya Biosphere Reserve to the north-east.

The first Mexican debt-for-nature swap was signed in 1991 by the federal government and Conservation International (CI) (*Excelsior* 1991). These financial resources are being used to protect this forest by considering sustainable economic alternatives and providing logistical support for researchers working in the region (CI 1993). A recently restored research facility (Chajul Scientific Station) is managed by UNAM and SEMARNAP (instead of CI), and will help generate the information needed on the region and take early steps toward sustainable development and long-term conservation of the reserves. Other Mexican and U.S. institutions are among those carrying out research and conservation projects.

Scientists have a preliminary biological perspective on the region. A coalition of agencies and institutions needs to continue working to carry out a plan of action combining conservation with sound development (cf. Medellín 1991; CI 1993). The intent is to include environmental education programmes and a coordinated policy of eco-development to solve economic difficulties of the inhabitants through creative programmes without having to continue to lose large areas of natural forest. Integrated and growing activities during the past few years suggest that much of the remaining Selva Lacandona may still be saved.

References

Barrera, A. (1962). La península de Yucatán como provincia biótica. *Rev. Soc. Mex. Hist. Nat.* 23: 71–105.

Breedlove, D.E. (1981). Introduction to the *Flora of Chiapas*. In Breedlove, D.E. (ed.), *Flora of Chiapas*. Part 1. California Academy of Sciences, San Francisco. 35 pp.

Calzada, I. and Valdivia, E. (1979). Introducción al estudio de la vegetación de la Selva Lacandona, México. *Biótica* 4: 149–162.

MAP 12. LACANDON RAIN FOREST REGION, MEXICO (CPD SITE MA1)

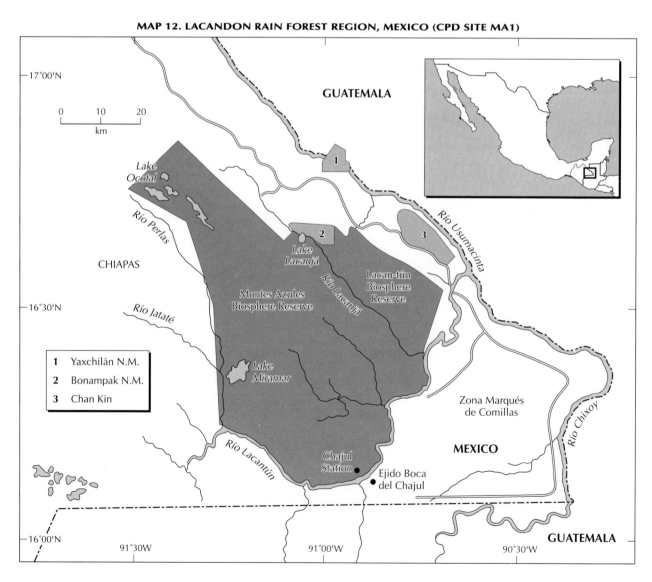

CI (1993). *Selva Lacandona/Montes Azules Biosphere Reserve: a priority ecosystem.* Conservation International (CI), Washington, D.C. 7 pp.

Dirzo, R. and Miranda, A. (1991). El límite boreal de la selva tropical húmeda en el continente americano: contracción de la vegetación, y solución de una controversia. *Interciencia* 16: 240–247.

Estrada, A. and Coates-Estrada, R. (1983). Rain forest in Mexico: research and conservation at Los Tuxtlas. *Oryx* 17: 201–204.

Excelsior (1991). No perderá México su soberanía por el canje de deuda externa por inversión ecológica. *Excelsior*, 23/2/1991.

Ferrusquía-Villafranca, I. (1993). Geology of Mexico: a synopsis. In Ramamoorthy, T.P., Bye, R., Lot, A. and Fa, J.E. (eds), *Biological diversity of Mexico: origins and distribution.* Oxford University Press, New York. Pp. 3–107.

GCTM (Group for the Conservation of the Tropics in Mexico) (1992). *Commitments to the Mexican tropics. El Ocote Reserve, Usumacinta River, and Campesinos Reserve.* Mexico, D.F. 31 pp.

Gómez-Pompa, A. (1987). On Maya silviculture. *Mexican Studies* 3: 1–17.

Gómez-Pompa, A. and Dirzo, R. with Kaus, A., Noguerón-Chang, C.R. and Ordoñez, M. de J. (1994). *Las áreas naturales protegidas de México de la Secretaría de Desarrollo Social.* SEDESOL, Mexico, D.F. 331 pp. Unpublished.

Marshall, N.T. (1989). Parlor palms. Increasing popularity threatens Central American species. *TRAFFIC USA* 9(3): 1–3.

Martínez, E. and Ramos, C.H. (1989). Lacandoniaceae (Triuridales): una nueva familia de México. *Ann. Missouri Bot. Gard.* 76: 128–135.

Medellín, R.A. (1991). The Selva Lacandona: an overview. *TCD (Tropical Conservation and Development Program) Newsletter (Gainesville, Florida)* 24: 1–5.

Medellín, R.A. (1994). Mammal diversity and conservation in the Selva Lacandona, Chiapas, Mexico. *Conserv. Biol.* 8: 780–799.

Mestel, R. (1995). Is that a pistil in your pocket? *Discover* 16(1): 88–89.

Miranda, F. (1961). Tres estudios botánicos en la Selva Lacandona, Chiapas, México. *Bol. Soc. Bot. Méx.* 26: 133–176.

Nations, J.D. (1984). The Lacandones, Gertrude Blom, and the Selva Lacandona. In Harris, A. and Sartor, M. (eds), *Gertrude Blom bearing witness.* University of North Carolina Press, Chapel Hill and London. Pp. 27–41.

Nations, J.D. (1985). Bearing witness. *Natural History* 94(3): 50–59.

Nations, J.D. (1988). The Lacandon Maya. In Denslow, J.S. and Padoch, C. (eds), *People of the tropical rain forest.* University of California Press, Berkeley. Pp. 86–88.

Nations, J.D. and Nigh, R.B. (1980). The evolutionary potential of Lacandon Maya sustained-yield tropical forest agriculture. *J. Anthrop. Res.* 36: 1–30.

Orellana, L. (1978). *Relación clima-vegetación en la región lacandona, Chiapas.* Tesis de Licenciatura. UNAM, Mexico, D.F.

Peters, C.M. and Pardo-Tejeda, E. (1982). *Brosimum alicastrum* (Moraceae): uses and potential in Mexico. *Economic Bot.* 36: 166–175.

Rzedowski, J. (1962). Contribuciones a la fitogeografía florística e histórica de México. *Bol. Soc. Bot. Méx.* 27: 52–65.

Rzedowski, J. (1978). *Vegetación de México.* Editorial Limusa, Mexico, D.F. 432 pp.

Smith, H.M. (1949). Herpetology in Mexico and Guatemala. *Ann. Assoc. Amer. Geogr.* 39: 219–238.

Toledo, V.M. (1982). Pleistocene changes of vegetation in tropical Mexico. In Prance, G.T. (ed.), *Biological diversification in the tropics.* Columbia University Press, New York. Pp. 93–111.

Vega-Rivera, J.H. (1990). Situación actual del conocimiento faunístico de la Reserva Montes Azules. In Camarillo, J.L. and Rivera, F. (eds), *Areas naturales protegidas en México y especies en extinción.* UNAM, Escuela Nacional de Estudios Profesionales (ENEP) Iztacala.

Wagner, P.L. (1964). Natural vegetation of Middle America. In Wauchope, R. (ed.), *Handbook of Middle American Indians*, Vol. 1. West, R.C. (ed.), *Natural environment and early cultures.* University Texas Press, Austin. Pp. 216–264.

Wendt, T. (1989). Las selvas de Uxpanapa, Veracruz-Oaxaca, México: evidencia de refugios florísticos cenozoicos. *Anales Inst. Biol. Univ. Nac. Méx., Ser. Bot.* 58: 29–54.

Wendt, T. (1993). Composition, floristic affinities, and origins of the canopy tree flora of the Mexican Atlantic slope rain forests. In Ramamoorthy, T.P., Bye, R., Lot, A. and Fa, J.E. (eds), *Biological diversity of Mexico: origins and distribution.* Oxford University Press, London. Pp. 595–680.

Wilkerson, J.K. (1985). The Usumacinta River: troubles on a wild frontier. *National Geogr.* 168(4): 514–543.

Authors

This Data Sheet was written by Olga Herrera-MacBryde (Smithsonian Institution, Department of Botany, NHB-166, Washington, DC 20560, U.S.A.) and Dr Rodrigo A. Medellín [Universidad Nacional Autónoma de México (UNAM), Centro de Ecología, Apdo. Postal 70-275, 04510 Mexico, D.F., Mexico].

UXPANAPA-CHIMALAPA REGION
Mexico

Location: South-eastern Veracruz and eastern Oaxaca, between latitudes 16°27'–17°30'N and longitudes 93°52'–94°56'W.

Area: c. 7700 km².

Altitude: c. 80–2250 m.

Vegetation: Evergreen, semi-evergreen and semi-deciduous tropical rain forests, montane rain forest, pine and pine-oak forests, xeric vegetation.

Flora: 3500 species of vascular plants estimated; high species endemism, many taxa disjunct or at northern limit of their ranges; threatened species.

Useful plants: Timber trees, fruit trees, ornamental palms.

Other values: Wilderness, fauna refuge, genetic resources, watershed protection, scenery, tourism.

Threats: Logging, colonization, agriculture, grazing, construction of roads and dams.

Conservation: None.

Geography

The Uxpanapa-Chimalapa region covers c. 7700 km² in extreme south-eastern Veracruz and eastern Oaxaca states (Map 13). The region includes the Atlantic slope of the eastern portion of the Isthmus of Tehuantepec, mostly in the drainage of the Coatzacoalcos River and its major tributary the Uxpanapa River, and extends from the Gulf of Mexico lowlands southward to the highlands of the continental divide; a part of eastern Chimalapa is in the Grijalva River drainage.

The Uxpanapa section (c. 2600 km²) includes the parts in Veracruz and in the municipality of Matías Romero (the "Colonia Cuauhtemoc") in Oaxaca. The Chimalapa section (c. 5100 km²) encompasses essentially all of the municipality of Santa María Chimalapa and the gulf slope and highlands of the municipality of San Miguel Chimalapa, both in Oaxaca; the southern boundary of the region is the southern boundary of the montane forests of the Sierra Atravesada.

Seven physiographic regions may be distinguished (cf. Ferrusquía-Villafranca 1993):

1. A karst area in northern Uxpanapa, with relatively little macrorelief but numerous small to large limestone outcrops. Average elevation 100–150 m.

2. An east-west strip in central Uxpanapa, with deep, flat alluvial soils formed by north-flowing rivers and alluvial outwash from the mountains to the south. Average elevation 100–130 m.

3. In southern Uxpanapa – northern Chimalapa, an area of steep hilly terrain fringing the Sierra de Tres Picos, with very deep soils, and the relief becoming gradually higher toward the sierra.

4. The Sierra de Tres Picos, a small granitic system with a number of steep-sided peaks reaching 1450 m, in north-central Chimalapa.

5. The east-west Río del Corte Valley (Upper Coatzacoalcos River) in central Chimalapa. The river is mostly at an elevation of 80–250 m, its tributaries going through a complexly dissected central area with no major flats in the montane areas.

6. The Sierra Atravesada (or Sierra Niltepec) of southern Chimalapa, a major east-west range of mostly granitic and metamorphic substrates, which forms the continental divide and includes peaks such as Cerro Azul (the highest – 2250 m) and Cerro Baúl (2050 m). Its eastern part is drained by the Negro River, a tributary of the Grijalva River. This range along with the Sierra de Tres Picos forms the north-western end of the Sierra Madre de Chiapas (Wendt 1983).

7. The limestone Sierra Espinazo del Diablo of north-eastern Chimalapa and a small part of extreme south-eastern Uxpanapa, reaching 1350 m.

Moist gulf winds moving south across the Isthmus of Tehuantepec produce one of the highest rainfalls in lowland Mexico in the Uxpanapa area. Average annual precipitation is 2800 mm in the west and 4400 mm in the east (Comisión del Papaloapan, unpublished). Even higher rainfalls probably occur in the relatively unknown Sierra de Tres Picos. To the south (leeward) of this range in Chimalapa, the precipitation in the Río del Corte lowlands is less than in Uxpanapa, probably mostly 2000–3000 mm or less in some areas, and greatest in the lower (western) part of the valley. The higher portions of the Sierra Atravesada and Sierra Espinazo del Diablo are clearly very wet (but without records).

Rain is seasonal, with a marked dry season from March through May, when monthly rainfall averages 60–100 mm in Uxpanapa. The wettest months are June to October. The Uxpanapa area forms the western end of a roughly crescent-shaped lowland of high precipitation ("the crescent area") that extends eastward in Mexico to southern Tabasco and northern Chiapas (Wendt 1983, 1989, 1993).

Average annual temperature in the lowlands is 24°–25°C; monthly average temperatures vary from around 20°C in January (and as low as 8°–10°C at night), to above 28°C in May. Daytime temperatures can exceed 41°C (Comisión del Papaloapan, unpublished). The highlands are cooler, but frosts appear to be rare and snow unknown.

Vegetation

The Uxpanapa-Chimalapa region includes the only major remaining well-developed rain forest in Mexico apart from the larger but much more disturbed Lacandon forest of Chiapas (Ewell and Poleman 1980) (CPD Site MA1, see Data Sheet). On the lowland and hill areas of Uxpanapa and northern Chimalapa are evergreen and semi-evergreen tropical rain forests, which vary according to the substrate (Wendt, unpublished):

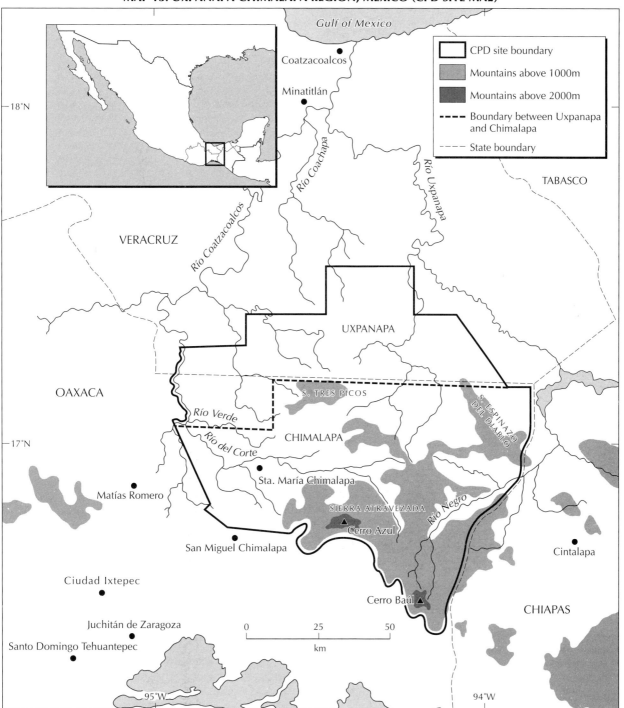

MAP 13. UXPANAPA-CHIMALAPA REGION, MEXICO (CPD SITE MA2)

CPD Site MA2: Uxpanapa-Chimalapa region, Mexico. Cerro Blanco, limestone ridge with karst forest, eastern Uxpanapa, State of Veracruz. Photo: Tom Wendt.

1. **Karst area forests** – with irregular canopies c. 25–35 m high, and many gaps due to tree falls and irregular substrate. In the drier western areas, a notable portion of the trees shed their leaves in the dry season. Common canopy trees include *Bernoullia flammea, Brosimum alicastrum* var. *alicastrum, Bursera simaruba, Cedrela odorata, Chione chiapasensis, Dendropanax arboreus, Dialium guianense, Guarea glabra, Lonchocarpus guatemalensis, Omphalea oleifera* and *Spondias radlkoferi*.

2. **Hill forest** – evergreen, on the deep soils of the hill area, with a continuous canopy c. 30–40 m high. Common or characteristic canopy species include *Brosimum guianense, B. lactescens, Calophyllum brasiliense* var. *rekoi, Cordia megalantha, Dialium guianense, Elaeagia uxpanapensis, Enterolobium schomburgkii, Eschweilera mexicana, Hirtella triandra* subsp. *media, Licania hypoleuca* var. *hypoleuca, L. sparsipilis, Pouteria neglecta, Sloanea meianthera, Spondias radlkoferi, Sterculia* new sp. Wendt & E. Taylor, *Tapirira* new sp. Wendt and *Terminalia amazonia*. In many areas *Sloanea tuerckheimii* is the most common subcanopy tree.

3. **Forest of the central Uxpanapa alluvial plain** – almost completely destroyed, its original composition is not well known. It appears to have been more similar to the hill forest, with *Ceiba pentandra, Dialium guianense, Terminalia amazonia* and *Vochysia guatemalensis* important.

4. **Riparian forest** – the most common tall trees are *Ficus insipida* and *Ocotea uxpanapana*.

The Chimalapa area is notable for its very large tracts of undisturbed and little disturbed lowland and montane forests; especially, there are large transects of undisturbed lowland rain forest to montane cloud forest. The hill and riparian forests of Uxpanapa extend into the Chimalapa area. Hill forest is widespread in the lower Río del Corte Valley and areas near the Sierra de Tres Picos and Sierra Atravesada. The lower elevations of the dissected topography of the central Río del Corte Valley support a complex mixture of lowland oak forest (*Quercus oleoides* and other oaks), pine forest (*Pinus oocarpa*), tropical rain forest, and semi-deciduous tropical forest, with some elements of montane mesic forests (e.g. *Liquidambar styraciflua, Pinus chiapensis*); this mixture, all at less than 600 m, is perhaps unique in Mexico. Very small areas of rock outcrop near the river support unique xeric vegetation including species of *Agave, Beaucarnea* and *Yucca*.

Montane areas of the Sierra de Tres Picos and Sierra Atravesada support diverse cloud forests, which may represent the largest area of undisturbed cloud forest in Mexico and Central America. These are isolated from other cloud forests of Oaxaca, Veracruz and Chiapas by lower and/or drier intervening areas. Among the genera of canopy trees with one or more common species are *Alfaroa, Billia, Cedrela, Clethra, Genipa, Inga, Liquidambar, Magnolia, Matayba, Oreomunnea, Pinus, Podocarpus, Quercus, Ticodendron* and *Weinmannia*, along with numerous species of Lauraceae. Elfin forest occurs along the crest of the Sierra de Tres Picos and on some peaks of the Sierra Atravesada. Extensive pine forests with *Pinus oocarpa* and other pines are found in the lower, eastern Sierra Atravesada. The montane forests of the Sierra Atravesada become drier south of the continental divide and give way to tropical deciduous forest at lower elevations on the Pacific slope (beyond our region).

Flora

The first collections in the Uxpanapa area were those in the early 1970s by the *Flora de Veracruz* project (cf. Márquez-Ramírez, Gómez-Pompa and Vázquez-Torres 1981), except for some in the 1930s along the Coatzacoalcos River in the west. Since 1980, the Colegio de Postgraduados, Chapingo, has had an intensive collecting programme (MacDougall 1971; Myers 1980; cf. Wendt 1987). The Uxpanapa lowlands are now relatively well known; over 200 species of canopy trees have been recorded. The entire region is estimated to have 3500 vascular plant species.

Wendt (1989) reported on the high endemism of the Uxpanapa flora (including northern Chimalapa), and proposed the area as a refuge for rain-forest species during cycles of climatic deterioration during the Pleistocene and earlier. His partial lists include at least 36 species apparently endemic to the Uxpanapa area and 11 more endemic to the high-precipitation crescent area. Many other species are known in Mexico only from this area. Wendt (1993) has shown that the Uxpanapa area is perhaps the most important centre for endemism in canopy trees of the Mexican rain forest.

Several genera and one family are known in Mexico only from Uxpanapa-Chimalapa (Wendt 1988, 1989), and more genera are known only from the crescent area (cf. Wendt 1993). Noteworthy rain-forest trees include the monotypic genus *Chiangiodendron*, abundant only in Uxpanapa (although also known from Chiapas) and the only New World member of its tribe (Wendt 1988); *Eschweilera mexicana* (endemic to Uxpanapa-Chimalapa), the only representative of the Lecythidaceae in Mexico (Wendt, Mori and Prance 1985); many other endemic canopy tree species including *Ocotea uxpanapana*, *Sterculia* new sp. and *Tapirira chimalapana*; two new flagelliflorous understorey species of Annonaceae, one of which appears to be a new genus (Schatz and Wendt, unpublished); and a new understorey genus of Rutaceae (Chiang, unpublished). All of the canopy trees just mentioned are very common to co-dominant elements of the rain forests of the region. A number of other abundant canopy trees are more widespread, but restricted in Mexico to Uxpanapa-Chimalapa (e.g. *Pouteria torta* subsp. *tuberculata*).

The Chimalapa area had been only sporadically collected (MacDougall 1971) until the Colegio de Postgraduados, Chapingo, began intensive collecting in 1984. However, due to the very remote nature of most of the area and its great vegetational diversity, it remains poorly known floristically. Its rain forests are mostly similar to the Uxpanapa hill forests and share many of their unique species. The montane forests include many species also known from the Sierra de Juárez of northern Oaxaca (CPD Site MA3, see Data Sheet) and the cloud forests of Chiapas, but already have yielded a number of new species in both the tall forests and elfin forests. The *Oreomunnea* forests are probably better developed in Chimalapa than in any other part of Mexico, and as well as less disturbed. Preliminary studies of the xerophytic outcrop vegetation near El Corte River indicate high endemism in its small flora.

Useful plants

The region contains important timber resources, including not only such widespread high quality tropical woods as *Cedrela odorata* (tropical red-cedar), *Calophyllum brasiliense* var. *rekoi* (Santa María) and *Swietenia macrophylla* (bigleaf mahogany), but important endemics such as the new, undescribed species of *Sterculia*, which has been much used locally in the manufacture of fine plywood. Furthermore, a large native population of the important fruit tree *Pouteria sapota* (the "zapote mamey") is present; there are few such populations elsewhere. Several non-timber montane species are important in the local economy, especially "palmita" (*Chamaedorea* sp.), the leaves of which are carefully harvested without killing the plant, and sold for ornamental purposes. Caballero *et al.* (1978) emphasize the usefulness to local peoples of a large proportion of the native Uxpanapa species. The endemic species still include many that have never been investigated for possible uses, since the region was virtually uninhabited until the past few decades.

Social and environmental values

The Uxpanapa-Chimalapa region is one of the most important remaining natural areas in Mexico for wild animals. There are large populations of such threatened species as jaguar and Baird's tapir, together with spider monkeys, tayra, agouti, kinkajou and many others.

Birds include the northernmost population of quetzal, large populations of curassow, and many others; harpy eagles have been reported in the Chimalapa area. There are no Endemic Bird Areas (EBAs) in the region, but the whole range of the highly localized and threatened Nava's wren (*Hylorchilus navai*) is restricted to karst forest in Uxpanapa and immediately adjacent Chiapas; the entire genus *Hylorchilus* is endemic to the Mexican karst rain forest (Atkinson *et al.* 1993).

The montane forests of the Sierra Atravesada contain a biogeographically distinctive herpetofauna with a significant endemic component; Campbell (1984) states that this range represents "a pivotal point of major importance in the distribution of Middle American cloud-forest reptiles and amphibians" – yet most of the range remains herpetologically unknown.

Uxpanapa and Chimalapa form the headwaters of the Coatzacoalcos River drainage, which has important downstream cities (Coatzacoalcos, Minatitlán) where flooding has increased since development of the Uxpanapa area began. Adequate forest and rational management are essential for control of soil erosion, flooding and water quality, especially considering the steep fragile soils of southern Uxpanapa and northern and eastern Chimalapa.

The region has potential for tourism, especially with regard to the animals, the beautiful rock formations and flora of the karst zone, and the true wilderness quality of much of Chimalapa.

Threats

Uxpanapa did not support a large indigenous population, and except for logging operations and other development near the Coatzacoalcos River (Río del Corte) in the west, it was almost completely undeveloped and unsettled until the early 1970s, when the federal government (Comisión del Papaloapan) started a major project in the Veracruz portion of Uxpanapa to resettle indigenous Chinanteco people displaced by the Cerro de Oro dam. Huge areas were cleared mechanically, mostly on the deep flat soils of the central part, but including some karst areas completely unsuited for agriculture or livestock. Early large-scale agricultural schemes (notably with rice) were conspicuous failures (Ewell and Poleman 1980), although many areas proved suitable for smaller scale subsistence agriculture. Large areas of closer hill forest were burned by escaped "milpa" fires during spring; some was converted into pasture. There also have been incursions into the bordering Chimalapa forests.

By 1980, perhaps one-half of the Uxpanapa area still supported primary forest. During the 1980s, the attrition rate in the Veracruz part of the area was slower, due to the limited number of settlers allowed, a partially enforced statewide ban on cutting of primary forest, and the fact that the prime agricultural lands had already been cleared (cf. Wendt 1993). However, logging in the Colonia Cuauhtemoc area in Oaxaca has intensified, often including part of the Chimalapa area and advancing toward the Sierra de Tres Picos. Large tracts of forest are subsequently cleared for agriculture, grazing and colonization.

The Comisión del Papaloapan in the 1980s developed a master plan for the Veracruz part; included were areas designated for forestry and for biotic reserves. Many karst areas and hill areas with extreme slopes were not to be developed. However, the Veracruz part went completely into State responsibility in 1985; its future is unclear. The karst forests are less threatened at present due to the difficulty of logging and the low agricultural potential of their soils. However, the rich hill forests in both Veracruz and Oaxaca have, as of 1995, almost vanished.

The history of the Chimalapa area has been quite different. Since pre-Hispanic times the Chima (a small Zoque population) have been there, centred in the area of the present town of Santa María Chimalapa. The Chimas communally own all of the municipality of Santa María Chimalapa, but the right to use parcels of the land may not only be inherited but bought, including by outsiders who are accepted as "comuneros". A similar situation pertains in the south-eastern portion of the area for some communal lands of San Miguel Chimalapa.

Before the 1970s, the only significant impact was near the town of Santa María and a few small areas westward, and along the Chiapas border to the east. Since then, logging, agriculture and/or settlements have increased westward and along the borders with Veracruz and Chiapas. As of 1995, only the vast heart of the Chimalapa area – Sierra de Tres Picos, upper Río del Corte Valley, much of the Sierra Atravesada – remains little disturbed and completely without roads. However, several large-scale developments have been proposed, including major dams on El Corte River and its tributaries as part of an irrigation scheme for the southern dry side of the Isthmus of Tehuantepec, and also major internationally financed forestry projects.

A proposed second trans-isthmian highway to connect the area of Acayucán, Veracruz with Cintalapa, Chiapas would cut diagonally across Uxpanapa and north-eastern Chimalapa, further threatening the region due to concomitant colonization.

Conservation

For the Uxpanapa area, no official plan for conservation exists. Before the dissolution of the Comisión del Papaloapan, the agency's field personnel in Uxpanapa were much interested in the possibility of parks and tourism to boost the local economy and protect the watersheds. Nevertheless, presently there appears to be no effort within the government to establish parks or to preserve part of the area.

There has, however, been much interest in the future of the Chimalapa area within the now vigorous Mexican conservation community, including the formation in 1991 of a National Committee for the Defense of the Chimalapas. Interest thus focused has led to increased appreciation by local people for the unique nature of the area, international support for studies of the zone, an attempt to solve boundary disputes and encroachments, and a government proposal for a Biosphere Reserve, which however is opposed by the local populace and some conservation groups. Although details are still sketchy, present plans made with greater local input call for a "campesino"-administered Biological Reserve.

References

Atkinson, P.W., Whittingham, M.J., de Silva Garza, H., Kent, A.M. and Maier, R.T. (1993). Notes on the ecology, conservation and taxonomic status of *Hylorchilus* wrens. *Bird Conserv. Intl.* 3: 75–85.

Caballero, J., Toledo, V.M., **Argueta, A., Aguirre, E., Rojas, P. and Viccon, J.** (1978). Estudio botánico y ecológico de la región del río Uxpanapa, Veracruz. Núm. 8. Flora útil o el uso tradicional de las plantas. *Biótica* 3: 103–144.

Campbell, J.A. (1984). A new species of *Abronia* (Sauria: Anguidae) with comments on the herpetogeography of the highlands of southern Mexico. *Herpetologica* 40: 373–381.

Ewell, P.T. and Poleman, T.T. (1980). *Uxpanapa: reacomodo desarrollo agrícola del trópico mexicano.* Instituto Nacional de Investigaciones sobre Recursos Bióticos (INIREB), Xalapa, Veracruz.

Ferrusquía-Villafranca, I. (1993). Geology of Mexico: a synopsis. In Ramamoorthy, T.P., Bye, R., Lot, A. and Fa, J.E. (eds), *Biological diversity of Mexico: origins and distribution.* Oxford University Press, New York. Pp. 3–107.

MacDougall, T. (1971). The Chima wilderness. *Explorers J.* 49: 86–103.

Márquez-Ramírez, W., Gómez-Pompa, A. and Vázquez-Torres, M. (1981). Estudio botánico y ecológico de la región del río Uxpanapa, Veracruz. Núm. 10. La vegetación y la flora. *Biótica* 6: 181–217.

Myers, N. (1980). *Conversion of tropical moist forest.* National Academy of Sciences, Washington, D.C. 205 pp.

Wendt, T. (1983). Plantae Uxpanapae I. *Colubrina johnstonii* sp. nov. (Rhamnaceae). *Bol. Soc. Bot. Méx.* 44: 81–90.

Wendt, T. (1987). Plantae Uxpanapae III. A new species of *Biophytum* (Oxalidaceae) and five genera new for the Mexican flora. *Brittonia* 39: 133–138.

Wendt, T. (1988). *Chiangiodendron* (Flacourtiaceae: Pangieae), a new genus from southeastern Mexico representing a new tribe for the New World flora. *Syst. Bot.* 13: 435–441.

Wendt, T. (1989). Las selvas de Uxpanapa, Veracruz-Oaxaca, México: evidencia de refugios florísticos cenozoicos. *Anales Inst. Biol. Univ. Nac. Méx., Ser. Bot.* 58: 29–54.

Wendt, T. (1993). Composition, floristic affinities, and origins of the canopy tree flora of the Mexican Atlantic slope rain forests. In Ramamoorthy, T.P., Bye, R., Lot, A. and Fa, J.E. (eds), *Biological diversity of Mexico: origins and distribution.* Oxford University Press, New York. Pp. 595–680.

Wendt, T., Mori, S.A. and Prance, G.T. (1985). *Eschweilera mexicana* (Lecythidaceae): a new family for the flora of Mexico. *Brittonia* 37: 347–351.

Author

This Data Sheet was written by Dr Tom Wendt (Louisiana State University, Department of Plant Biology, Herbarium, Baton Rouge, Louisiana 70803-1705, U.S.A.).

MEXICO: CPD SITE MA3

SIERRA DE JUAREZ, OAXACA
Mexico

Location: Southern Mexico in north-eastern Oaxaca State, north of the city of Oaxaca between about latitudes 17°20'–17°50'N and longitudes 96°15'–97°00'W.

Area: c. 1700 km².

Altitude: c. 500–3250 m.

Vegetation: Mainly montane cloud forest; also tropical evergreen forest and forests of pine, pine-oak and oak.

Flora: c. 2000 species, high endemism; threatened species.

Useful plants: Extensive uses by several Amerindian peoples, including as medicinals; also timber trees, ornamentals.

Other values: Watershed protection, genetic resources, Amerindian lands, rich in fauna including high butterfly diversity, tourist potential.

Threats: Logging, agriculture, colonization, erosion, fire, over-collection of ornamental species; potential hydroelectric dams.

Conservation: No reserves.

Geography

In north-eastern Oaxaca between Tuxtepec and Ixtlán de Juárez (Map 14), the Sierra de Juárez is separated from the Sierra de Zongólica to its north by the Santo Domingo River with Tecomavaca Canyon, and extends south-eastward to the Cajones River and Sierra de Villa Alta, which connects to the Sierra de Los Mixes (Paray 1951; Lorence and García-Mendoza 1989). These mountain chains are part of the Sierra Madre de Oaxaca in the Oaxaca-Puebla Uplands subprovince of the Sierra Madre del Sur morphotectonic province (cf. Ferrusquía-Villafranca 1993).

North of the Trans-Mexican Volcanic Belt is the Sierra Madre Oriental, and just south of the belt begins the Sierra Madre de Oaxaca. Mountains thus front the Gulf of Mexico lowlands from the belt's Pico de Orizaba (Veracruz) to the Isthmus of Tehuantepec, where they meet the Sierra Madre del Sur.

The region is characterized by large deep ravines (barrancas). The Sierra de Juárez geological subprovince is composed of folded sedimentary rocks with series of younger granitic intrusions dating from the Palaeozoic to Cenozoic, with the majority being Mesozoic, but the region is complex and not well known (Lorence and García-Mendoza 1989; Ferrusquía-Villafranca 1993). Various watercourses (e.g. the Valle Nacional River) have their origin in the Sierra de Juárez and eventually form the Papaloapan River which empties into the Gulf of Mexico.

The climate ranges from subtropical to mostly temperate and subhumid – above 1000 m. The average temperature varies between 16°–20°C, with regular frost in the high mountains. Average annual precipitation varies locally from 700 mm to 2000–4000 mm or more, often as moisture removed from trade winds coming off the Caribbean Sea – which in some years reach gale force.

Vegetation

The Sierra de Juárez has a great variety of habitats due to variations in topography, altitude, geological substrates and climate. These environmental characteristics determine a mosaic of formations and communities that always include some evergreen vegetation. Although there are many factors that affect seasonality within these vegetation types, humidity seems to limit distribution in the mountain areas in comparison to the vegetation that extends to the lower, warmer areas (Lorence and García-Mendoza 1989).

Five general formations are distinguished in the Sierra de Juárez, of which broadleaved montane cloud forest is predominant (while extensive on these mountains, it covers less than 1% of Mexico).

1. **Tropical evergreen forest** (800–1000 m) is dominated by evergreen trees 30–40 m tall. There are abundant lianas and epiphytes of tropical affinity. The characteristic or dominant trees are *Acosmium panamense*, *Andira* sp., *Brosimum alicastrum* var. *alicastrum*, *Calophyllum brasiliense* var. *rekoi*, *Dialium guianense*, *Dussia mexicana*, *Ormosia isthmensis* and *Robinsonella* sp. (cf. Wendt 1993).

2. **Montane cloud forest** (bosque mesófilo de montaña – Rzedowski 1978) forms a band between (1000–) 1400–2250 m along the northern and eastern slopes of the Sierra de Juárez and Sierra de Los Mixes. The climate is cool (20°–14°C) and humid, with a mean annual precipitation exceeding 2000 mm and probably reaching 6000 mm in places (e.g. at Vista Hermosa) (Rzedowski and Palacios-Chávez 1977). Dominant trees average 20–30 m tall. Evergreen and deciduous species bearing many epiphytes occur together with palms, tree ferns, Ericaceae shrubs, vines, and moisture-loving herbs (Paray 1951; Lorence and García-Mendoza 1989; Martin and de Avila-B. 1990).

Floristically this formation is a mixture of both neo-tropical and holarctic elements, including affinities with South America and Asia. Rzedowski and Palacios-Chávez (1977) studied a community at Vista Hermosa, where they reported *Oreomunnea mexicana*, *Weinmannia pinnata* and *Liquidambar styraciflua* to be dominant. Additional characteristic species were *Magnolia schiedeana*, *Brunellia mexicana*, *Nyssa sylvatica* and species of *Alnus*, *Carpinus*, *Cedrela*, *Clethra*, *Ilex*, *Ocotea*, *Phoebe* and *Podocarpus*. This vegetation type in the region remains little known.

3. **Pine forest**, on basaltic substrate at 1600–2600 m. The evergreen trees are 25–40 m tall. Dominant are *Pinus ayacahuite*, *P. cornuta*, *P. lawsonii*, *P. chiapensis*, *P. devoniana* (*P. michoacana*) and *P. pseudostrobus* var. *oaxacana*. Grasses dominate the lower stratum.

4. **Pine-oak forest** (2000–2800 m). Predominant are *Pinus rudis*, *P. devoniana* (*P. michoacana*), *P. lawsonti* and *P. montezumae* together with *Quercus laurina* and *Q. rugosa*. Some individuals of the rare *Abies guatemalensis* and also *A. oaxacana* are associated

MAP 14. SIERRA DE JUAREZ, OAXACA, MEXICO (CPD SITE MA3)

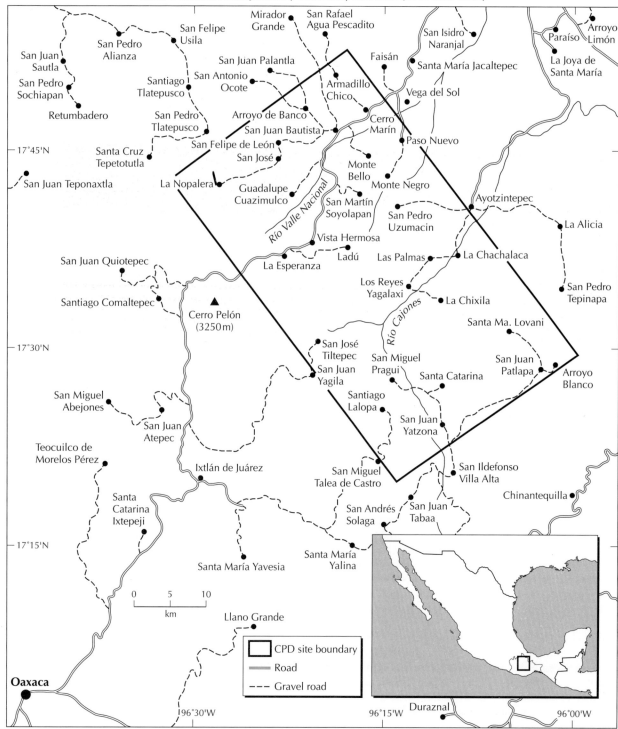

with the *Pinus*, mainly in ravines and above 2700 m. *Plecosorus speciosissimus* and *Dryopteris wallichiana* are exceptional among the terrestrial ferns (at 2800 m) (Riba 1993).

5. Oak forest (2000–2500 m). Precipitation is relatively low, with a summer dry season. The dominants are *Quercus crassifolia*, *Q. castanea*, *Q. crassipes*, *Q. rugosa* and *Q. laurina*. This formation occurs westward (inland) to the Grande River Basin, where it changes to matorral and/ or low forest.

Flora

Located in the Mixteca-Oaxaqueña Floristic Province, the Sierra de Juárez is one of Oaxaca State's wettest and floristically richest areas, with probably 2000 of the 8000 or more species in Oaxaca. The northernmost limit of many Mesoamerican taxa occurs in the Sierra de Juárez (Lorence and García-Mendoza 1989), and higher areas in Oaxaca may be ancient centres of diversity (Hunt 1993). The Instituto de Biología of the Universidad Nacional Autónoma de México (UNAM) is conducting floristic inventories in the region. Although knowledge of the sierra's flora is limited, the number of endemics is expected to be significant.

For example, the Sierra de Juárez is one of five Oaxacan centres of endemism for the Leguminosae, and 6–7% of Mexican legume species are endemic to Oaxaca (Lorence and García-Mendoza 1989). Estimates of species endemic to Oaxaca for several other families occurring in the Sierra de Juárez are Rubiaceae (18%) and Monimiaceae (40%). Most of these endemics seem to be restricted to the sierra or its vicinity. As in some other families, Oaxaca seems to be a refugium for relict species of Commelinaceae – e.g. in temperate forest, the generically monotypic endemics *Gibasoides laxiflora* and *Matudanthus nanus* (Hunt 1993). Oaxaca also has many endemic Labiatae (Ramamoorthy and Elliott 1993) and Gramineae (31%) (Valdés-Reyna and Cabral-Cordero 1993). Some endemics in other families are: *Siparuna scandens*, *Carica cnidoscoloides*, *Rondeletia ginettei*, *Anthurium cerroplonense*, *A. subovatum*, *A. yetlense* and *Syngonium sagittatum*.

The probable high endemism in the Sierra de Juárez supports Rzedowski's (1991) finding that 60% of the species occurring in the montane cloud forests of MegaMexico (south-western U.S.A., Mexico and northern Central America) are endemic species.

Useful plants

In the Sierra Norte of Oaxaca, which includes the Sierra de Juárez, several indigenous peoples (Chinantec, Mixe and Mixtec) have extensive knowledge of and uses for the flora, which have been receiving thorough ethnobotanical study (Martin and de Avila-B. 1990; Martin 1992). In the 1970s there were large-scale collections of *Dioscorea* tubers, used in the synthesis of birth-control pills. The Sierra de Juárez contains rich timber resources such as *Abies*, *Pinus*, *Liquidambar* and *Quercus*. Among the region's various ornamental species are tree ferns, cycads, pipers, aroids, bromeliads and orchids (Paray 1951).

CPD Site MA3: Sierra de Juárez, Oaxaca, Mexico. Two Chinantec women from Santiago Comaltepec collecting sap from *Agave salmiana* for preparation of puque, a fermented beverage. This agave is among the hundreds of plant species used by local people for medicine, food, construction, fuel and other purposes. Photo: Gary J. Martin.

Social and environmental values

A major highway from Tuxtepec south-westward to the city of Oaxaca crosses the Sierra de Juárez north of Cerro Pelón (3250 m), and unpaved roads provide access to many of the indigenous communities and forests in this rugged region.

The sierra harbours various threatened mammals such as jaguar, ocelot and brocket deer. There are a number of species of birds from the Central Mexican Highlands Endemic Bird Area (EBA A11) and the Balsas Drainage and Interior Oaxaca EBA (A27). Among them is the threatened dwarf jay (*Cyanloyca nana*), which occurs only in central Oaxaca and western Veracruz in pine-oak forest. The Sierra de Juárez is the richest region in Mexico for butterflies and is significant for endemic butterflies (Llorente-Bousquets and Luis-Martínez 1993).

Threats

The flora is threatened by logging (wood, pulp and cellulose), agriculture and grazing (crops, cattle and sheep), colonization, commercial collecting of those species with ornamental value, and potentially threatened by construction of hydroelectric dams. Over one-third of the *Pinus* forest of the Sierra Norte may have been destroyed from 1956–1981, with much of the remainder being degraded (G.J. Martin, pers. comm.).

Conservation

Abies guatemalensis is in Appendix I of CITES. There is general interest within academia for protection of some areas in the Sierra de Juárez, but there is no official plan to establish nature reserves. Several foundations have supported ethnobotanical research and conservation projects in collaboration with SERBO (the Society for the Study of Biotic Resources of Oaxaca/Sociedad para el Estudio de Recursos Bióticos de Oaxaca), which hopes to establish an outreach centre of information in the city of Oaxaca (Martin and de Avila-B. 1990). This sierra may harbour the best remaining occurrence of the highly threatened Mexican montane cloud forest; a substantial portion needs to be saved.

References

Ferrusquía-Villafranca, I. (1993). Geology of Mexico: a synopsis. In Ramamoorthy, T.P., Bye, R., Lot, A. and Fa, J.E. (eds), *Biological diversity of Mexico: origins and distribution*. Oxford University Press, New York. Pp. 3–107.

Hunt, D.R. (1993). The Commelinaceae of Mexico. In Ramamoorthy, T.P., Bye, R., Lot, A. and Fa, J.E. (eds), *Biological diversity of Mexico: origins and distribution*. Oxford University Press, New York. Pp. 421–437.

Llorente-Bousquets, J. and Luis-Martínez, A. (1993). Conservation-oriented analysis of Mexican butterflies: Papilionidae (Lepidoptera, Papilionoidea). In Ramamoorthy, T.P., Bye, R., Lot, A. and Fa, J.E. (eds), *Biological diversity of Mexico: origins and distribution*. Oxford University Press, New York. Pp. 147–177.

Lorence, D.H. and García-Mendoza, A. (1989). Oaxaca, Mexico. In Campbell, D.G. and Hammond, H.D. (eds), *Floristic inventory of tropical countries: the status of plant systematics, collections, and vegetation, plus recommendations for the future*. New York Botanical Garden, Bronx. Pp. 253–269.

Martin, G.J. (1992). Searching for plants in peasant marketplaces. In Plotkin, M. and Famolare, L. (eds), *Sustainable harvest and marketing of rain forest products*. Island Press, Washington, D.C. Pp. 212–223.

Martin, G.J. and de Avila-B., A. (1990). Exploring the cloud forests of Oaxaca, Mexico / Explorando el bosque nuboso de Oaxaca, México. *WWF Reports* (Oct.–Dec.): 8–11/ 11–14.

Paray, L. (1951). Exploraciones en la Sierra de Juárez. *Bol. Soc. Bot. Méx.* 13: 4–10.

Ramamoorthy, T.P. and Elliott, M. (1993). Mexican Lamiaceae: diversity, distribution, endemism and evolution. In Ramamoorthy, T.P., Bye, R., Lot, A. and Fa, J.E. (eds), *Biological diversity of Mexico: origins and distribution*. Oxford University Press, New York. Pp. 513–539.

Riba, R. (1993). Mexican pteridophytes: distribution and endemism. In Ramamoorthy, T.P., Bye, R., Lot, A. and Fa, J.E. (eds), *Biological diversity of Mexico: origins and distribution*. Oxford University Press, New York. Pp. 379–395.

Rzedowski, J. (1978). *Vegetación de México*. Editorial Limusa, Mexico, D.F. 432 pp.

Rzedowski, J. (1991). Diversidad y orígenes de la flora fanerogámica de México. *Acta Bot. Mex.* 14: 3–21.

Rzedowski, J. and Palacios-Chávez, R. (1977). El bosque de *Engelhardtia* (*Oreomunnea*) *mexicana* en la región de la Chinantla (Oaxaca, México) – una reliquia del cenozoico. *Bol. Soc. Bot. Méx.* 36: 93–123.

Valdés-Reyna, J. and Cabral-Cordero, I. (1993). Chorology of Mexican grasses. In Ramamoorthy, T.P., Bye, R., Lot, A. and Fa, J.E. (eds), *Biological diversity of Mexico: origins and distribution*. Oxford University Press, New York. Pp. 439–446.

Wendt, T. (1993). Composition, floristic affinities, and origins of the canopy tree flora of the Mexican Atlantic slope rain forests. In Ramamoorthy, T.P., Bye, R., Lot, A. and Fa, J.E. (eds), *Biological diversity of Mexico: origins and distribution*. Oxford University Press, New York. Pp. 595–680.

Authors

This Data Sheet was written by Dra. Patricia Dávila and Leticia Torres and Rafael Torres [Universidad Nacional Autónoma de México (UNAM), Instituto de Biología, Departamento de Botánica, Apartado Postal 70-367, Mexico 04510, D.F., Mexico] and Olga Herrera-MacBryde (Smithsonian Institution, Department of Botany, NHB-166, Washington, DC 20560, U.S.A.).

TEHUACAN-CUICATLAN REGION
Mexico

Location: Southern central Mexico, south-east of Mexico City in south-eastern Puebla and northern Oaxaca states, between latitudes 17°39'–18°53'N and longitudes 96°55'–97°44'W.

Area: c. 9000 km²; the broad upper Tehuacán Valley over 2000 km².

Altitude: Floor of valley from 2200 m in north-west to c. 600 m in south-east; dryland below 1800 m.

Vegetation: Several dryland scrub formations with many species succulent, spiny or thorny, or forming rosettes; low, early deciduous forest.

Flora: c. 2700 species; high diversity – c. 910 genera of seed plants, high endemism – c. 30%; threatened species.

Useful plants: Ornamentals (e.g. cacti, *Dasylirion*, *Nolina*, *Beaucarnea*, *Agave*, bromeliads); traditional medicinals, and for food (e.g. cactus fruits), ceremonial uses, fibre and fuelwood.

Other values: Extensive benchmark research on ecodevelopment from 10,000 BC onwards; freshwater springs, caves.

Threats: Agricultural and pastoral activities, soil salinization and hardpan formation (caliche), depletion of groundwater, overgrazing by goats; road improvements; collection of cacti, other succulent and thick-stemmed species.

Conservation: Botanical Garden (1 km²).

Geography

The Tehuacán-Cuicatlán region begins c. 150 km south-east of Mexico City approximately between the cities of Puebla and Orizaba in south-eastern Puebla State, and extends into northern Oaxaca. The region is dominated by a rift valley, La Cañada or Tehuacán-Cuicatlán-Quiotepec depression (Map 15), in the Oaxaca-Puebla Uplands subprovince of the Sierra Madre del Sur morphotectonic province (Ferrusquía-Villafranca 1993). Tamayo-L. (1962) considered the region within a Mixteca-Oaxaqueña physiographic province.

The predominating Tehuacán Valley opens to the north-west with low borders c. 15 km apart and declines gradually over some 100 km to the south-east, where it narrows to Tomellín Canyon. The north-eastern to eastern boundary is formed by a part of the Sierra Madre de Oaxaca known as the Sierra de Zongólica, which reaches 2500–3000 m or more (its northern portion is the Sierra de Tecamachalco, its south-eastern portion the Sierra Mazateca). The valley is bounded to the south and west by the Sierra de Zapotitlán, a part of the Sierras Mixtecas (to 2500 m high). A major valley adjoining from the west is Zapotitlán Valley.

Topography in the Tehuacán Valley is varied. Over much of its upper expanse, wide plains slope gradually to the south. There are many gently rolling hills, outliers from the parallel mountain borders. The northern portion across the valley floor lies at an elevation of 1500–1700 m and is drained southward, forming the Salado River. Permanent surface water begins more or less near Coxcatlán where several watercourses converge, including the Zapotitlán River, which is the valley's major seasonal source of water from rainfall in the bordering highlands. The more narrow southern portion of the valley is drained northward by the

Grande River. Altitude in the Tehuacán Valley declines in gradual steps to the confluence of these rivers at an elevation of less than 600 m.

Early in the Quaternary, the Tehuacán-Cuicatlán region was transformed from a closed basin with a large lake at 1700 m into the dry eroding valleys – which are still hydrostatically draining their Quaternary groundwater reserves – by the upstream erosion of the Santo Domingo River with creation of Tecomavaca Canyon. The region thus forms part of the watershed of the Papaloapan River, which empties eastward into the Gulf of Mexico.

López-Ramos (1981) considered the region part of the Tlaxiaco geologic province. The Tehuacán Valley has a mosaic of outcrops of different geologic ages and composition. In general the area presents Early Tertiary sedimentation, followed by some volcanic activity in the Pliocene and Pleistocene. Centrally, near the regional capital Tehuacán (which is at 1676 m) are limestones, sandstones and gypsum outcrops from the Late and Early Cretaceous. From Tehuacán south-east to Teotitlán del Camino are Precambrian outcrops with undifferentiated metamorphic rock, as well as Early Jurassic mudstones and sandstones. Part of the Sierra de Zongólica has outcrops of metamorphic rocks from the Palaeozoic. In the lower parts are outcrops of sediments from the Tertiary and sandstones and conglomerates from the Quaternary.

The mean annual temperature in the Tehuacán Valley varies between 18°–22°C, rising to 24.5°C at Cuicatlán; the mean January temperature in the city Tehuacán is 15°C. The arid climate is largely controlled by the Sierra de Zongólica, which is between the valley and the Gulf of Mexico. These mountains intercept much of the moisture from trade winds blowing westward off the Caribbean Sea, resulting in a rain-shadow enclave. The average annual rainfall in the valley region varies from 250–500 mm, mainly occurring from May to October, with more precipitation in June and September (Smith 1965a, 1965b; Enge and Whiteford 1989).

During the Pleistocene the climate had less differentiated seasons, might have been slightly more moist and was cooler with regular winter frosts. Probably there was extensive grassland, much less thorn forest, and many of the cacti, *Agave* and other desert plants could not have lived in the valley. The transition to the Recent climate occurred between 7800–7400 BC (MacNeish, Peterson and Neely 1972). The generally grey soils are either intrazonal, ranging from halomorphic alkaline and saline to calcimorphic rendzina soils, or zonal sierozem, characteristic of arid and semi-arid regions.

Vegetation

The Tehuacán-Cuicatlán region has a great variety of habitats related to the variations in topography, altitude, geological substrates and climate. There are many vegetation formations and associations (Miranda 1948; Rzedowski 1978; Zavala-H. 1982; Dávila-Aranda 1983; Jaramillo-Luque and González-Medrano 1983). Some are edaphically determined, such as communities of gypsophytes and halophytes. Communities are strongly affected by the marked seasonality of precipitation – with six essentially dry months, as well as frequent droughts; some types of vegetation lose their leaves very rapidly. Six vegetation formations are recognized, along a descending altitudinal gradient:

MAP 15. TEHUACAN-CUICATLAN REGION, MEXICO (CPD SITE MA4)

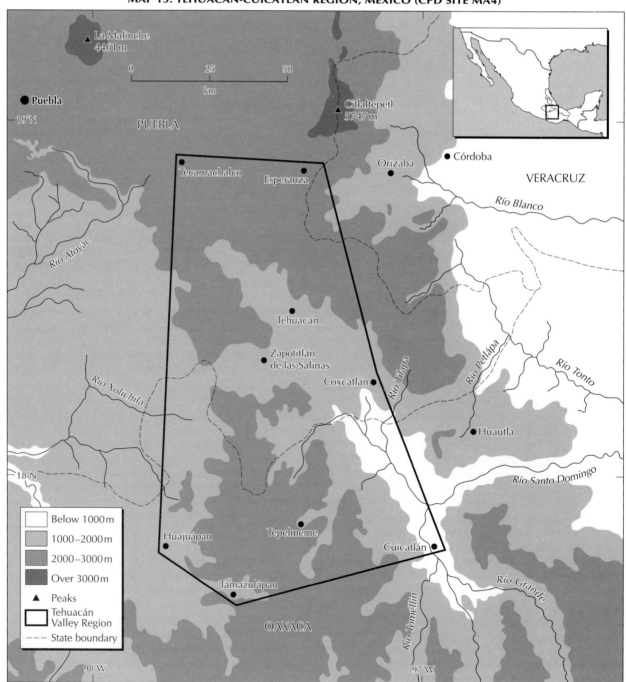

1. **Sclerophyllous scrub or chaparral** is located above 1900 m on limestone. It is a dense community 1–2.5 m high, composed of evergreen or briefly deciduous shrubs with leathery leaves. Common genera include *Ceanothus*, *Cercocarpus*, *Condalia*, *Ephedra*, *Quercus*, *Rhus*, *Sophora* and *Vauquelinia*.

2. **Spiny rosette scrub** is located at 1500–1800 m on limestone. It is characterized by rosette-formed shrubs and plants with somewhat thick, spiny leaves. The dominant shrubs are 50–60 cm tall, and 1–3 m tall plants also occur. Typical genera are *Agave*, *Dasylirion*, *Nolina*, *Beaucarnea* and *Yucca* – all of which have the majority of their species and perhaps their origins in Mexico.

3. **High, scarcely thorny scrub** is a complex, variable formation with great species diversity. It is dominated by shrubs over 2 m tall, some with thorns or spines but most unarmed or nearly so. Low trees are sometimes interspersed. Some species are deciduous during the dry season. Frequent genera are *Acacia*, *Forestiera*, *Fraxinus*, *Gochnatia*, *Leucaena*, *Ptelea* and *Schaefferia*.

4. **High thorn scrub** is characteristic of alluvial soils in the lower portion of valleys. Thorny shrubs 2–3 m tall are dominant, and trees 5–6 m tall are frequently interspersed. Common genera are *Acacia*, *Agonandra*, *Bumelia*, *Celtis*, *Cercidium*, *Lysiloma*, *Myrtillocactus*, *Opuntia*, *Stenocereus*, *Parkinsonia*, *Prosopis* and *Ximenia*.

5. **High, spiny thick-stemmed scrub** is dominated by cacti – which characterize the Tehuacán region (Meyrán-García 1980). Species may be unbranched, or slightly to strongly branched. Outstanding is *Neobuxbaumia tetetzo*, a few-branched columnar cactus 10–15 m tall, which sometimes densely dominates in communities ("tetecheras"). Other associated species are *Cephalocereus columna-trajani* and *Neobuxbaumia mezcalaensis*, and sometimes *Beaucarnea gracilis* – a notable rosette plant with distinctive thick stems. A strikingly dense stand of arborescent cacti is near Calipan, south-east of the city of Tehuacán. This type of scrub grows on latosol soils and often on mudstone. Other important genera are *Escontria*, *Pachycereus*, *Pilosocereus*, *Polaskia*, *Stenocereus*, *Fouquieria*, *Hesperothamnus* and *Lasiocarpus*.

6. **Low early deciduous forest** is very rich in number of species, and notable for the mostly quickly deciduous trees 8–10 m tall with exfoliating bark. This type of vegetation is characteristic of the warmer and moderately more humid areas in the lower portion of valleys and canyons, on varied soils ranging from latosols and conglomerate to mudstone. Often represented are cacti in *Cephalocereus* and *Neobuxbaumia*, as well as species of *Beaucarnea*. Also important are *Amphipterygium*, *Bursera*, *Ceiba*, *Cercidium*, *Cyrtocarpa*, *Lasiocarpus*, *Lysiloma*, *Pseudobombax* and *Pseudosmodingium*.

Flora

This arid region has been recognized as the Tehuacán-Cuicatlán Floristic Province (Rzedowski 1978; cf. Delgadillo-

M. and Zander 1984). The flora is very rich (Dávila-Aranda 1983), with c. 910 genera of seed plants, including 8 gymnosperm genera; there are c. 2700 species of vascular plants. Endemism is estimated to be 30%. The southern Tehuacán Valley is considered one of three Mexican centres rich in *Agave* taxa. Of the c. 250 genera endemic to Mexico, c. 10% occur in this region (Jaramillo-Luque and González-Medrano 1983; Rzedowski 1978, 1993). Local endemics include several plants recognized near or at the generic level, such as *Gypsacanthus*, *Mammillaria pectinifera* (*Solisia*), *Oaxacania*, *Pringleochloa*, *Salvia* sect. *Conzattiana* (*S. aspera*) – which may be relictual (Ramamoorthy and Lorence 1987), and *Trichostema* sect. *Rhodanthum* (*T. purpusii*). Rare endemics include *Agave lurida*, *A. peacockii*, *A. titanota* (perhaps), *Mammillaria pectinifera* (*Solisia pectinata*) and *Salvia aspera*.

The surrounding mountains are part of an interrupted chain from the U.S.A. into Central America. From the north, the Sierra Madre Oriental separates Puebla and Veracruz states and ends at the Trans-Mexican Volcanic Belt, which extends to the Atlantic coast; south of the belt is the Sierra de Zongólica, and then the Sierra de Juárez (CPD Site MA3, see Data Sheet). These extended mountains have been a bridge for the dispersal of plants and animals for millennia. South of the Tehuacán Valley are other dry valleys and mountains that may serve as a pathway for many plants that require xeric habitats (Smith 1965b; López-Ramos 1981).

Phytogeographically, the region's flora is related to the floras of other arid and semi-arid regions, such as the Balsas Depression nearby to the west; to the north-west in Hidalgo State, a small desert enclave – the only other locale for *Lepechinia mexicana* (*Sphacele mexicana*); and farther away (250 km) to the north, the Chihuahuan Desert and as well southern Texas. The valley has floristic affinities not only with the arid regions; there also is a marked influence of tropical elements, especially those characteristic of Mexico's Pacific slope (Smith 1965b; Rzedowski 1978; Villaseñor, Dávila and Chiang 1990). The Tehuacán-Cuicatlán region is geologically ancient, and probably was more climatically isolated in the past – which helps to explain the high endemism (Smith 1965b).

Useful plants

Native species are used traditionally for example as medicinals, food (e.g. cactus fruits), fibre (e.g. basket-making), fuelwood, living fences and in ceremonials; some are sold in local markets (Meyrán-García 1980). Various species of considerable horticultural value have attracted national and international markets, and some are being readily produced in cultivation elsewhere. For example among hobbyists and landscapers, interest is strong for some cacti, *Nolina*, *Beaucarnea*, *Dasylirion*, *Agave*, *Hechtia* and *Tillandsia* (Dávila-Aranda 1983).

Social and environmental values

The Tehuacán-Cuicatlán region has localities (e.g. San Juan Raya and near Tepeji de Rodríguez) with important Mesozoic fossils. The region is one of five truly desert biotic provinces in Mexico. The Tehuacán-Cuicatlán Valley is within the arid intermontane areas of the Balsas drainage and interior Oaxaca, which is recognized as an Endemic Bird Area (EBA A27). A number of restricted-range species occur, including dusky hummingbird (*Cynanthus sordidus*), grey-breasted woodpecker (*Melanerpes hypopolius*), pileated flycatcher (*Xenotriccus mexicanus*), Boucard's wren (*Campylorhynchus jocosus*), bridled sparrow (*Aimophila mystacalis*) and white-throated towhee (*Pipilo albicollis*).

Archaeological research in the Tehuacán Valley has contributed greatly to understanding processes on the origin and spread of plant cultivation and domestication (MacNeish 1972; Bye 1993; Hernández-Xolocotzi 1993). Comprehensive research by the Tehuacan Archaeological-Botanical Project has reconstructed the 11,500-year prehistoric chronology of the inhabitants' patterns of subsistence and development of agriculture, including the domestication of some plants (Byers 1967; MacNeish, Peterson and Neely 1972). Remains include up to 15 races of maize (*Zea mays* subsp. *mays*) – the earliest dated 5600 years ago (Long *et al.* 1989); several kinds of squashes and gourds (*Cucurbita* spp., *Lagenaria siceraria*, *Crescentia cujete*, *Apodanthera* sp.); species and races of beans (*Phaseolus* spp., *Canavalia* sp.); amaranths (*Amaranthus* spp.); cassava (*Manihot esculenta*); chili pepper (*Capsicum annuum*); avocado (*Persea americana*); guava (*Psidium guajava*); and tempesquite (*Mastichodendron*).

Economic assessment

The Tehuacán Valley has been irrigated for over two millennia. Local communities have transformed areas by means of an irrigation system that has grown elaborate and very extensive. Construction between 1944–1969 greatly expanded the system and has resulted in irrigation of 166 km², mostly throughout the year. The water-capture technology includes not only use of surface water in a lengthy network of canals, but also tunnels from the New World's only well-known development of qanats – laterally hand-dug tunnels ("galerías") that tap distant aquifers, creating gradual flow of their groundwater into wells (Enge and Whiteford 1989).

Amerindian bands apparently developed into loosely organized agricultural communities that were chiefdoms, independent small villages or city-states prior to conquest by the Aztecs at the beginning of the 16th century (Flannery 1983). The present regional inhabitants tend to be ethnically associated with local communities and speak dialects of Náhuatl and/or Spanish. Population density in the productive valley is c. 100 persons per km².

Agriculture in the Tehuacán Valley is of two main types: upland farming on good areas above 1800 m in the oak-pine zone of the surrounding mountains, and intensive valley farming. About 410 km² of the valley were in cultivation in 1981: on the limited alluvial areas of canyons and watercourses, c. 40% of the total as irrigated croplands on the valley floor and gentle hillsides, and dry farming on the marginal land of hillsides. Crops are diverse, e.g. maize, beans, squash, tomatillo (*Physalis philadelphica*), tomato, garlic, dry peppers, peas, sugarcane, saffron, alfalfa and sorghum. Between 1973 and 1981, many of the region's farmers converted from mainly subsistence cultivation to almost completely cash-crop agriculture.

The Tehuacán-Cuicatlán region connects the city of Oaxaca with Mexico's heavily populated Central Plateau. The Mexico City-Oaxaca railroad (constructed in 1892) and a major highway pass through the valley. Another main highway links the cities of Tehuacán and Orizaba (67 km away) and continues east to important markets in the Veracruz lowlands.

Threats

The surrounding mountains have been heavily deforested, with the inevitable results of decreased water infiltration for aquifer recharge, and rapid runoff with erosion and flash floods. The groundwater level is decreasing in the Tehuacán Valley's upper aquifer stratum (which is between 30–50 m subsurface). Natural springs around the city Tehuacán decreased c. 12% in flow rate from 1976 to 1983; galería systems measured from 1968 or 1976 to 1983 decreased c. 38%. Demand is intensifying to drill more deep wells to obtain fossil water from a separate stratum at 114–195 m – an essentially non-renewable supply already also showing decline. Yet increasing the use of water might expand agriculture.

The valley's flora is severely threatened by agricultural and pastoral activities, especially overgrazing by goats. About 500 km² in the Tehuacán Valley are used for grazing. In the broad lower part of the valley, the vegetation is affected most by the agricultural activities. Road improvements cause additional losses of natural vegetation. Also, some of the native species and populations are threatened by trade and commercialization of wild specimens, especially those with ornamental value such as some of the cacti and other succulent or thick-stemmed (pachycaul) species.

Conservation

Although there is general interest in protection of representative areas within the region, there is no official plan to establish nature reserves. So far, just 1 km² near Zapotitlán de las Salinas has been set aside, as the Helia Bravo Hollis Botanical Garden. The necessity for conservation is clear from the biological standpoint, and because the region's hydrology is critically important to support an increasing human population (Enge and Whiteford 1989).

References

Bye, R. (1993). The role of humans in the diversification of plants in Mexico. In Ramamoorthy, T.P., Bye, R., Lot, A. and Fa, J.E. (eds), *Biological diversity of Mexico: origins and distribution*. Oxford University Press, New York. Pp. 707–731.

Byers, D.S. (ed.) (1967). *The prehistory of the Tehuacan Valley*, Vol. 1: *Environment and subsistence*. University of Texas Press, Austin. 331 pp.

Dávila-Aranda, P.D. (1983). *Flora genérica del Valle de Tehuacán-Cuicatlán*. M.S. thesis. Universidad Nacional Autónoma de México, Mexico, D.F. 694 pp.

Delgadillo-M., C. and Zander, R.H. (1984). The mosses of the Tehuacan Valley, Mexico, and notes on their distribution. *Bryologist* 87: 319–322.

Enge, K.I. and Whiteford, S. (1989). *The keepers of water and earth: Mexican rural social organization and irrigation.* University of Texas Press, Austin. 222 pp.

Ferrusquía-Villafranca, I. (1993). Geology of Mexico: a synopsis. In Ramamoorthy, T.P., Bye, R., Lot, A. and Fa, J.E. (eds), *Biological diversity of Mexico: origins and distribution.* Oxford University Press, New York. Pp. 3–107.

Flannery, K.V. (1983). Pre-Columbian farming in the valleys of Oaxaca, Nochixtlán, Tehuacán, and Cuicatlán: a comparative study. In Flannery, K.V. and Marcus, J. (eds), *The cloud people: divergent evolution of the Zapotec and Mixtec civilizations.* Academic Press, New York.

Hernández-Xolocotzi, E. (1993). Aspects of plant domestication in Mexico: a personal view. In Ramamoorthy, T.P., Bye, R., Lot, A. and Fa, J.E. (eds), *Biological diversity of Mexico: origins and distribution.* Oxford University Press, New York. Pp. 733–753.

Jaramillo-Luque, V. and González-Medrano, F. (1983). Análisis de la vegetación arbórea en la provincia florística de Tehuacán-Cuicatlán. *Bol. Soc. Bot. Méx.* 45: 49–64.

Long, A., Benz, B.F., Donahue, D.J., Jull, A.T. and Toolin, L.J. (1989). First direct AMS dates on early maize from Tehuacán, Mexico. *Radiocarbon* 31: 1035–1040.

López-Ramos, E. (1981). *Geología de México*, Tomo III. Publ. particular autorizada, Mexico, D.F. 446 pp.

MacNeish, R.S. (1972). Summary of the cultural sequence and its implications in the Tehuacan Valley. In MacNeish, R.S. *et al.*, *The prehistory of the Tehuacan Valley*, Vol. 5: *Excavations and reconnaissance.* University of Texas Press, Austin. Pp. 496–504.

MacNeish, R.S., Peterson, F.A. and Neely, J.A. (1972). The archaeological reconnaissance: [introduction]. In MacNeish, R.S. *et al.*, *The prehistory of the Tehuacan Valley*, Vol. 5: *Excavations and reconnaissance.* University of Texas Press, Austin. Pp. 341–360.

Meyrán-García, J. (1980). *Guía botánica de cactáceas y otras suculentas del Valle de Tehuacán*, 2nd edition. Soc. Mexicana de Cactología, Mexico, D.F. 52 pp.

Miranda, F. (1948). Datos sobre la vegetación en la cuenca alta del Papaloapan. *Anales Inst. Biol. Univ. Nac. Méx.* 19: 333–364.

Ramamoorthy, T.P. and Lorence, D.H. (1987). Species vicariance in the Mexican flora and description of a new species of *Salvia* (Lamiaceae). *Adansonia* 2: 167–175.

Rzedowski, J. (1978). *Vegetación de México.* Editorial Limusa, Mexico, D.F. 432 pp.

Rzedowski, J. (1993). Diversity and origins of the phanerogamic flora of Mexico. In Ramamoorthy, T.P., Bye, R., Lot, A. and Fa, J.E. (eds), *Biological diversity of Mexico: origins and distribution.* Oxford University Press, New York. Pp. 129–144.

Smith, C.E., Jr. (1965a). Agriculture, Tehuacan Valley. *Fieldiana, Bot.* 31: 49–100.

Smith, C.E., Jr. (1965b). Flora, Tehuacán Valley. *Fieldiana, Bot.* 31: 101–143.

Tamayo-L., J. (1962). *Geografía general de México*, 2nd edition, Vol. 1. Inst. Mex. Invest. Econ., Mexico, D.F. 562 pp.

Villaseñor, J.L., Dávila, P. and Chiang, F. (1990). Fitogeografía del Valle de Tehuacán-Cuicatlán. *Bol. Soc. Bot. Méx.* 50: 135–149.

Zavala-H., J.A. (1982). Estudios ecológicos en el valle semiárido de Zapotitlán, Pue. I. Clasificación numérica de la vegetación basada en atributos binarios de presencia o ausencia de las especies. *Biótica* 7: 99–120.

Authors

This Data Sheet was written by Dra. Patricia D. Dávila [Universidad Nacional Autónoma de México (UNAM), Instituto de Biología, Departamento de Botánica, Apdo. Postal 70-367, Mexico 04510, D.F., Mexico] and Olga Herrera-MacBryde (Smithsonian Institution, Department of Botany, NHB-166, Washington, DC 20560, U.S.A.).

CANYON OF THE ZOPILOTE REGION
Mexico

Location: South of Mexico City in central Guerrero State, between latitudes 17°20'–17°50'N and longitudes 99°30'–100°15'W.

Area: 4383 km².

Altitude: c. 600–3550 m.

Vegetation: Xerophilous scrub, deciduous and subdeciduous tropical forests, mesophyllous montane forest, oak, pine and fir forests.

Flora: Over 2000 species, high diversity, endemic genera and species; threatened species.

Useful plants: Timber trees; medicinals, ceremonial species.

Other values: Watershed protection, genetic reserve, refuge for fauna – including many endemics; archaeological sites, scenery, tourism.

Threats: Logging, agriculture, colonization; potential roads development and railroad extension.

Conservation: Omitelmi Ecological State Park; its expansion, and three other areas for reserves suggested.

Geography

The Canyon of the Zopilote River and neighbouring area southward beyond the state capital Chilpancingo and westward to the Sierra El Plateado cover 4383 km², situated between Mexico City and Acapulco in Guerrero, a state along the Pacific coast of south-western Mexico. The region is part of seven municipalities: Zumpango del Río, Chilpancingo, Chichihualco, Tlacotepec, San Miguel Totolapan, Atoyac de Alvarez and Coyuca de Benítez. It is in the Sierra Madre del Sur morphotectonic province and mainly in the Pacific Ranges and Cuestas subprovince (including the Sierra de Igualatlaco), with the northern lowlands in the Balsas Depression subprovince (Ferrusquía-Villafranca 1993).

The topography is generally mountainous, with c. 40% above 2000 m, and drainage mostly to the Balsas River in the north. The minimum elevation in the north-east is 600 m, in the north-west 2000 m. Important peaks ("cerros") include Yextla (2950 m) centrally, and in the south-west Teotepec (3550 m), Jilguero (2850–2900 m) and Tlacotepec (3330 m). Minimum elevations in the south-west and south-east are 800 m and 1600 m, within watersheds draining directly to the coast.

The region's geology is very complex (Ferrusquía-Villafranca 1993). Sedimentary rocks are the most prevalent, principally Mesozoic (Cretaceous) marine and transitional limestones. Igneous rocks are volcanic – extruded during the Cenozoic (e.g. basalts, andesites, tufas, breccias), or intrusive (e.g. granites, granodiorites).

The local weather is considerably influenced by the region's diverse topography. The general climate changes with increasing elevation from hot and dry in the lowland interior to warm to temperate and subhumid, to cool on the highest peaks. Generally there is a dry season during winter and spring, with precipitation mostly in summer through autumn. In the semi-arid area near Chilpancingo above Zopilote Canyon, the annual rainfall of 800 mm occurs mainly in May–November, and the mean annual temperature is 24°C with a 4.5°C oscillation and the extremes 5°C and 40°C; in May the diurnal fluctuation is 14°C (Rzedowski 1978). Higher in the mountains, precipitation may be over 1600 mm and temperatures become cooler, but the region includes many different restricted habitats, for example related to rain shadows and the full compass of slope aspects.

Vegetation

Of the ten principal vegetation types in Mexico, seven occur in the region (Rzedowski and Vela 1966; Rzedowski 1978; Fonseca and Lorea 1980). The varied topography and climate have resulted in a diverse mosaic. Along a north-east to south-west transect from Mezcala over the crest to Paraíso, the main formations are:

1. **Deciduous tropical forest** (500 m at Mezcala ascending to Xochipala). Many species of *Bursera* are characteristic, such as *B. bonetii*, *B. longipes*, *B. morelensis*, and columnar cacti especially in the canyon, such as *Neobuxbaumia mezcalaensis* and *Pachycereus weberi* (*P. gigas*).

2. With increasing altitude develops a **transitional forest** very rich in species, e.g. *Ostrya virginiana*, *Cercocarpus macrophyllus*, *Juniperus flaccida* and *Actinocheita potentillifolia* mixed with *Quercus* spp. from higher elevations. East of the transect near more arid Chilpancingo (1275 m) is xerophilous scrub, where *Quercus magnoliifolia* has become shrub-like (Miranda 1947), and *Agave cupreata* occurs on warmer slopes.

3. **Oak forest** (1500–2000 m), where the extreme minimum temperature may fall below 0°C. Predominant are *Quercus glaucoides*, *Q. resinosa* and *Q. magnoliifolia*.

4. **Pine-oak forest** (2000–2400 m or more) comprised of *Pinus devoniana* (*P. michoacana*), *P. teocote* and *P. leiophylla*, mainly with *Quercus uxoris*, *Q. laurina*, *Q. acutifolia*, *Q. glaucescens* and *Q. crassipes*.

5. **Pine forest** (above 2400 m) in less humid locales, with *Pinus herrerae*, *P. leiophylla*, *P. ayacahuite* and *P. pseudostrobus* var. *oaxacana*. The pine forest in some lower and warmer areas (not along this transect) is an association of just *P. oocarpa* and *P. pringlei*.

6. In more humid locales, an exuberant **mesophyllous montane forest** of several strata, with abundant climbers and epiphytes. Characteristic trees 30 m or more tall include *Chiranthodendron pentadactylon*, *Chaetoptelea mexicana*, *Abies guatemalensis* and *Pinus ayacahuite*. Shorter trees are *Ostrya virginiana*, *Clethra mexicana*, *Styrax ramirezii*, *S. argenteus*, *Tilia occidentalis*, *Saurauia serrata*, *Viburnum ciliatum* and *Meliosma dentata*. This community covers 40% of Omiltemi Ecological State Park.

7. **Fir forest** (2500–3000 m), with individuals of *Abies religiosa* and *A. hickelii* to 30 m tall. On the highest cerro Teotepec there is also a low forest of stout *Pinus hartwegii*; in rocky places grow *Juniperus monticola* var. *monticola* and rosette and cushion species of the páramo.

8. Descending the **Pacific slope** (below c. 2450 m) is a **mesophyllous montane forest** with a shrubby stratum rich in Melastomataceae, and many epiphytes – orchids, *Peperomia* and ferns. Characteristic trees include *Podocarpus matudae*, *Pinus chiapensis*, *P. maximinoi*, *Chaetoptelea mexicana*, *Saurauia angustifolia*, *Hedyosmum mexicanum*, *Oreopanax obtusifolius*, *Dendropanax arboreus*, *Persea schiedeana*, *Drimys granadensis*, *Sloanea medusula* and *Magnolia schiedeana*.

9. From 1570 m to 800 m toward Paraíso there is a very diverse **subdeciduous tropical forest** with species such as *Licaria peckii*, *Persea schiedeana*, *Phoebe ehrenbergii*, *Hibiscus uncinellus*, *Synardisia venosa*, *Guarea glabra*, *Trophis chiapensis* and *Ardisia compressa*.

Flora

The northern basically lowland portion of the Canyon of the Zopilote region is within the Balsas Depression Floristic Province (Rzedowski 1978), a centre important for endemic species and the spectacular diversification of *Bursera* species (Toledo-Manzur 1982). Most of the region is in the Middle Serranías Floristic Province, one of the major centres for endemics in Mexico, which includes the genera *Silviella*, *Omiltemia*, *Microspermum*, *Peyritschia* and *Hintonella*, and species such as *Arracacia ovata*, *Coaxana bambusoides* and *Donnellsmithia ampulliformis*. Endemism tends to be associated with high and humid locales, which function as isolated ecological islands.

The flora of this region and the whole state are not thoroughly known. Inventories suggest 30% of the 7000–7500 species in Guerrero are present. Some taxa have been more completely collected and studied, e.g. pteridophytes, Pinaceae, Lauraceae, Fagaceae, Melastomataceae, Araliaceae,

Apiaceae, Rubiaceae and Orchidaceae. Arborescent ferns (e.g. *Cyathea divergens*, *Lophosoria quadripinnata*) are widely distributed in the general area. *Peltogyne mexicana* is a pre-Cenozoic tropical relict found in a few populations, which is also known from Panama and Colombia (Sousa-S. and Delgado-S. 1993). The taxonomic inventories illustrate the region's biological richness – including 195 of the 207 estimated orchid species in Guerrero and half of the 351 collected species of pteridophytes (Lorea 1990). In the Canyon of the Zopilote occur 20 of the 64 species of Burseraceae reported for all of Mexico.

Useful plants

The region is very rich in timber resources. For general purposes including fuelwood, most used are the pines and oaks *Pinus ayacahuite*, *P. devoniana* (*P. michoacana*), *P. chiapensis*, *P. herrerae*, *Quercus uxoris* and *Q. laurina*, and as well *Abies religiosa* and *A. guatemalensis*; for particular construction, the palm *Brahea dulcis*, *Cordia elaeagnoides* and *Pithecellobium dulce*; and for artisanry and carvings, the preceding two hardwoods and *Actinocheita potentillifolia*.

Some species are used in local ceremonies, such as *Bursera copallifera* ("copal") and *Solandra* spp. ("copa de oro"). Among medicinals are *Ternstroemia pringlei* ("té de tila"), *Juniperus flaccida*, *Magnolia schiedeana* and *Chiranthodendron pentadactylon* ("flor de la manita") – which is now cultivated in Europe and U.S.A.

Social and environmental values

The Canyon of the Zopilote region harbours abundant diverse fauna, also not well known. A study of birds nearby to the south-west in the Sierra de Atoyac de Alvarez found 161 species, with 21 endemics (Navarro-Sigüenza 1986). Another study in that sierra reported 339 species of butterflies, 76% of the species known in Guerrero (Vargas 1990). The canyon is one of nine Mexican areas rich in endemic butterflies (Llorente-Bousquets and Luis-Martínez 1993). In Omiltemi Ecological State Park there are 37 species of amphibians and reptiles, with 13 endemics (Muñoz-Alonso 1988).

The Canyon of the Zopilote River is within the Sierra Madre del Sur of Guerrero and Oaxaca Endemic Bird Area (EBA A12), which has nine bird species of restricted range. Three of the four threatened birds within the EBA have significant populations in the canyon – the white-tailed hummingbird (*Eupherusa poliocerca*), short-crested coquette (*Lophornis brachylopha*) and white-throated jay (*Cyanolyca mirabilis*).

Proper management of the mesophyllous montane forests is essential to maintain their catchment of rainfall. For example, the Omiltemi forest captures over half the water for the city of Chilpancingo. The number of indigenous people inhabiting the region is small, in dispersed communities. They grow subsistence crops, and their impact on the vegetation is minimal.

The region also has archaeological importance. Xochipala was a pre-Hispanic ceremonial centre in 965 AD (Schmidt 1986). The northern area between Mezcala and Tetela del Río has evidence of human settlements of the Preclassic, Protoclassic, Classic and Postclassic periods (Rodríguez

1986). The regional variety thus offers potential for tourism, for example from Acapulco, Chilpancingo and Mexico City.

Economic assessment

The region is accessible by all the major means of transportation. The general area's much-used major highway from the interior to the coast extends through the canyon and near the eastern edge of the region, and a railroad from the north terminates near the region's northern border at the Balsas River. From the south-west coast, a major road extends as far as Puerto del Gallo and connects by minor road with the principal road west of Chichihualco.

The Instituto Nacional de Investigaciones Forestales (INIF 1972) has pointed out the region's importance for forestry, with the highest timber volume in the state – estimated at 126–346 m³ per ha of coniferous forests with broadleaved trees (hardwoods) present, 91–245 m³/ha of coniferous forests with hardwoods codominant, and 110–233 m³/ha of hardwood forests with conifers present. By 1971, Guerrero may have provided up to 10.9% of Mexico's annual yield of saw timber (Styles 1993).

The state is fifth in Mexican coffee production, and the Canyon of the Zopilote region produces c. 70% of Guerrero's coffee. In the extreme south-west, the natural forest canopy has been left almost intact to shade the coffee plantations.

Threats

Farmers using the forest to shade their coffee plantations selectively cut mainly species of *Pinus*, *Quercus* and *Ficus*. With the population increasing, probably the road from Puerto del Gallo will be improved into a principal highway across the region and the railroad will be extended across

MAP 16. CANYON OF THE ZOPILOTE REGION, MEXICO (CPD SITE MA5), SHOWING PRIORITY SITES FOR CONSERVATION OF PLANT COMMUNITIES

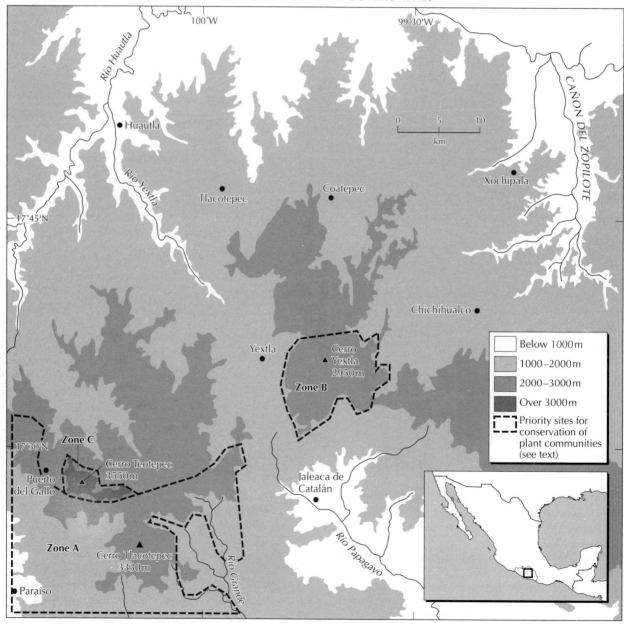

to the coast. However, the greatest threat to the region is deforestation subsequent to commercial logging of coniferous and hardwood trees for the lumber and paper industries, without reforestation, and as well there is clandestine use by individuals. Many endemic species of flora and fauna are threatened by the deforestation.

Conservation

This region is one of the few large and highly diverse areas remaining in Mexico that maintains relatively undisturbed native flora and fauna. Many plant communities still occur that elsewhere have been greatly altered or destroyed. Already the region has become a refuge for many threatened species.

Omiltemi Ecological State Park (17°31'–17°35'N and 99°30'–99°44'W) is 27 km west of Chilpancingo, including 36 km² of the western portion of the Sierra de Igualatlaco. The reserve was established in 1984 particularly for watershed management, through an agreement of the state governor and the Universidad Nacional Autónoma de México (UNAM) Facultad de Ciencias. Commercial export of *Abies guatemalensis* is prohibited by Appendix I of CITES.

Botanists from the UNAM Department of Biology's Vascular Plants Laboratory, who have been studying forests in Guerrero State for several years, suggest several communities that need to be protected: (1) the *Abies* forest (30 km²) on cerros Teotepec and Zacatonal (Map 16, zone C); (2) the adjacent area of mesophyllous montane and subdeciduous tropical forests (100 km²) between Puerto del Gallo and Paraíso (Map 16, zone A); (3) around Cerro Yextla (Map 16, zone B), by expanding Omiltemi park (30 km²) between Puerto Soleares and Cruz de Ocote – the park harbours c. 1.4% of Guerrero's mesophyllous forest, yet 11% could be reserved in zones A and B; and (4) deciduous tropical forest in the Canyon of the Zopilote River.

References

Ferrusquía-Villafranca, I. (1993). Geology of Mexico: a synopsis. In Ramamoorthy, T.P., Bye, R., Lot, A. and Fa, J.E. (eds), *Biological diversity of Mexico: origins and distribution.* Oxford University Press, New York. Pp. 3–107.

Fonseca, R.M. and Lorea, F. (1980). *Recursos bióticos de la cuenca del Río Zopilote, area Filo de Caballo.* Universidad Nacional Autónoma de México (UNAM), Fac. Ciencias, Archivo de la Comisión de Biologías de Campo, Mexico, D.F. 42 pp. Unpublished report.

INIF (1972). *Inventario forestal del estado de Guerrero.* Secretaría de Agricultura y Ganadería (SAG), Subsecretaría Forestal y de la Fauna, Instituto Nacional de Investigaciones Forestales (INIF) Publ. No. 24. Mexico, D.F. 66 pp.

Llorente-Bousquets, J. and Luis-Martínez, A. (1993). Conservation-oriented analysis of Mexican butterflies: Papilionidae (Lepidoptera, Papilionoidea). In Ramamoorthy, T.P., Bye, R., Lot, A. and Fa, J.E. (eds), *Biological diversity of Mexico: origins and distribution.* Oxford University Press, New York. Pp. 147–177.

Lorea, F. (1990). *Estudios pteridológicos en el estado de Guerrero, México (Diversidad, distribución y relaciones fitogeográficas de la pteridoflora).* Thesis. UNAM, Mexico, D.F. 43 pp.

Miranda, F. (1947). Estudios sobre la vegetación de México. V. Rasgos de la vegetación en la cuenca del Río de las Balsas. *Rev. Soc. Mex. Hist. Nat.* 8: 95–114.

Muñoz-Alonso, L.A. (1988). *Estudio herpetofaunístico del Parque Ecológico Estatal de Omiltemi, Mpio. de Chilpancingo de los Bravo, Guerrero.* Thesis. UNAM, Mexico, D.F. 111 pp.

Navarro-Sigüenza, A.G. (1986). *Distribución altitudinal de las aves en la Sierra de Atoyac, Guerrero.* Thesis. UNAM, Mexico, D.F. 79 pp.

Rodríguez, F. (1986). Desarrollo cultural en la región de Mezcala-Tetela del Río. In *Primer Coloquio de Arqueología y Etnohistoria del Estado de Guerrero.* Instituto Nacional de Antropología e Historia (INAH), Mexico, D.F. Pp. 155–170.

Rzedowski, J. (1978). *Vegetación de México.* Editorial Limusa, Mexico, D.F. 432 pp.

Rzedowski, J. and Vela, L. (1966). *Pinus strobus* var. *chiapensis* en la Sierra Madre del Sur de México. *Ciencia (México)* 24: 211–216.

Schmidt, P. (1986). Secuencia arqueológica de Xochipala. In *Primer Coloquio de Arqueología e Historia del Estado de Guerrero.* INAH, Mexico, D.F.

Sousa-S., M. and Delgado-S., A. (1993). Mexican Leguminosae: phytogeography, endemism, and origins. In Ramamoorthy, T.P., Bye, R., Lot, A. and Fa, J.E. (eds), *Biological diversity of Mexico: origins and distribution.* Oxford University Press, New York. Pp. 459–511.

Styles, B.T. (1993). Genus *Pinus*: a Mexican purview. In Ramamoorthy, T.P., Bye, R., Lot, A. and Fa, J.E. (eds), *Biological diversity of Mexico: origins and distribution.* Oxford University Press, New York. Pp. 397–420.

Toledo-Manzur, C.A. (1982). *El género Bursera (Burseraceae) en el estado de Guerrero.* Thesis. UNAM, Mexico, D.F. 182 pp.

Vargas, I. (1990). *Listado lepidopterofaunístico de la Sierra de Atoyac de Alvarez en el estado de Guerrero. Nota acerca de su distribución local y estacional (Rhopalocera: Papilionoidea).* Thesis. UNAM, Mexico, D.F. 149 pp.

Authors

This Data Sheet was written by Nelly Diego, Rosa M. Fonseca, Francisco Lorea, Lucio Lozada and Lino Monroy [Universidad Nacional Autónoma de México (UNAM), Laboratorio de Plantas Vasculares, Departamento de Biología, Facultad de Ciencias, Circuito exterior, Ciudad Universitaria, Mexico 04510, D.F., Mexico].

SIERRA DE MANANTLAN REGION AND BIOSPHERE RESERVE
Mexico

Location: In south-western Jalisco and north-eastern Colima, between about latitudes 19°27'–19°42'N and longitudes 103°51'–104°27'W.

Area: 1396 km².

Altitude: 400–2860 m.

Vegetation: Tropical dry forest, tropical subdeciduous forest, mesophyllous mountain forest, oak, pine-oak and pine forests, fir (*Abies*) forest.

Flora: c. 2800 vascular plant species; endemic and rare species; mix of neotropical and holarctic species.

Useful plants: Germplasm of wild taxa for important crop and tree species (*Zea, Phaseolus, Pinus, Abies*); over 500 species used traditionally.

Other values: Watershed protection, base for biological research; diverse fauna, threatened species.

Threats: Logging, fuelwood-gathering, agriculture, extensive cattle-ranching, erosion, fires.

Conservation: Biosphere Reserve, Biological Research Station, broad management plan.

Geography

The Sierra de Manantlán is a north-western portion of the Sierra Madre del Sur c. 50 km inland from the Pacific Ocean (Map 17), and ranges from c. 400 m at Casimiro Castillo (La Resolana) to 2860 m centrally at Cerro La Bandera (El Muñeco). The Sierra de Manantlán lies in two hydrographic regions and contributes to the Armería, Marabasco and Purificación watersheds. The western and central portions of the range are of igneous origin, from the Tertiary (extrusive) and Cretaceous (intrusive); the eastern portion is of sedimentary Late Cretaceous origin (Jardel-P. 1992).

The region's climate (in the Köppen system) ranges from subhumid warm to semi-warm, to subhumid temperate. Mean temperatures range from 27°–12°C depending on elevation, with some frost. The median annual rainfall is c. 800 mm in the driest northern lowlands, to 1800 mm at higher elevations. The rainy season is in summer, and fog is frequent above 1500 m (Jardel-P. 1992).

Vegetation

The Sierra de Manantlán is in a transition zone between the Neotropical and Holarctic biogeographic regions (Rzedowski and McVaugh 1966). Additionally, lowland tropical forests change to temperate forests at higher elevations. The region's complex topography, geologic substrates and climate are combined with the history of human influences into an intricate mosaic of extraordinarily different vegetation types within a relatively limited area. Transitions between vegetation types generally are gradual, but can be abrupt (Vázquez-G. *et al.* 1990; Jardel-P. 1992).

Tropical dry forest

Tropical dry forest occurs at 600–1200 m, and is composed of non-spiny tree species 8–15 m tall which are deciduous during a lengthy dry season (Rzedowski and McVaugh 1966; McVaugh 1984). These forests are very diverse floristically. Some families (e.g. Leguminosae) are represented by many species. Dominant species include *Lysiloma microphyllum*, *L. acapulcense, Jacaratia mexicana* and *Ceiba aesculifolia*. Principal threats are conversion to cattle pastures and clearance to grow maize, and as well collection of fuelwood.

Tropical subdeciduous forest

This vegetation type is located at 400–1200 m and composed of tree species 15–35 m tall, some of which are deciduous during the short dry season (Rzedowski and McVaugh 1966). Among the dominants are *Brosimum alicastrum*, *Hura polyandra* and *Enterolobium cyclocarpum*. *Bursera arborea* is very rare and threatened.

Several economically valuable species are extracted from this type of forest, including *Cedrela odorata* and *Swietenia humilis*; the inhabitants gather edible and medicinal plants (Robles-H. *et al.*, in press). Principal threats are the logging, and conversion for agriculture and pasturage.

Oak forest

Two types of oak forest are differentiated (Guzmán-Mejía 1985; Vázquez-G. *et al.* 1990; Jardel-P. 1991) from the 27 *Quercus* species recognized in the region by Trelease (cf. Nixon 1993). **Dry oak forest** (400–1500 m) is characterized by trees reaching 5–15 m, which are deciduous for the short period of the driest season. Prominent are *Quercus castanea*, *Q. glaucescens, Q. magnoliifolia* and *Q. rugosa*. **Subdeciduous oak forest** (above 1500 m) is characterized by trees reaching 20–35 m that are deciduous for very short periods, but the different species do not simultaneously lose their leaves. Among the main species are *Quercus crassipes, Q. candicans,*

Q. acutifolia and *Q. laurina*. The principal threat is over-exploitation for lumber and fuelwood.

Pine forest

Pine occurs in forests from 800 m, with climax pine forest above 2100 m; tree height is 10–35 m. Prominent species include *Pinus oocarpa*, *P. douglasiana*, *P. herrerae* and *P. pseudostrobus* (Cuevas-Guzmán and Núñez-L. 1988). These species have been the most exploited commercially, and also are threatened by fire.

Pine-oak forest

Pine-oak forest has a mix of species dominated by *Pinus* and *Quercus*; located at 1000–2500 m, it is characterized by species 10–40 m tall with persistent needle-like or deciduous leathery leaves. Threats are logging, cattle-ranching and fire.

The pine-oak forest has been greatly affected by human activities and its composition reflects this impact. The pine forests are mostly patches of second growth. The climax vegetation in sites currently occupied by the pines corresponds to pine-oak forest in dry environments or cloud forest in humid areas (Jardel-P. 1991). Cloud-forest tree species and some shade-tolerant oaks colonize patches of land dominated by the pines, and can replace them successionally (Sánchez-Velásquez 1988; Saldaña-A. and Jardel-P. 1992).

Mesophyllous mountain forest (cloud forest)

The distribution of the mesophyllous mountain forest is very irregular in the sierra, and now very limited in Jalisco and all of Mexico. It occurs at 700–2600 m in gorges and ravines, sometimes forming a definite fringe that abruptly gives way to other communities (such as those with conifers). The trees are 20–35 m tall and epiphytes are very common (Santiago-P. 1992). Many of the tree species have holarctic affinities. Although floristically quite variable at different sites (Muñoz-M. 1992), fairly consistently dominant are *Ilex brandegeana*, *Cornus disciflora*, *Carpinus tropicalis* and *Ostrya virginiana*. Other trees include *Quercus* spp., *Meliosma dentata*, *Dendropanax arboreus*, *Magnolia iltisiana* (recently discovered), *Rapanea juergensenii* and *Persea hintonii*. The understorey and ground layer have more neotropical affinities.

This type of vegetation has many species used for many purposes, but it is threatened principally by logging (e.g. for *Fraxinus uhdei*, *Magnolia iltisiana*, *Juglans major*, *Tilia mexicana*) and cattle-ranching.

Fir forest

Fir forest has a relatively homogenous floristic composition dominated by *Abies*. These forests are located at 2200–2800 m, with trees c. 30–40 m tall. Representative species are *Abies religiosa* var. *religiosa*, *A. religiosa* var. *emarginata*, *Pinus pseudostrobus* and *Quercus laurina* (Figueroa-R. 1991).

Flora

The discovery of *Zea diploperennis* in 1977, a disease-resistant perennial relative of maize (Iltis *et al.* 1979), sparked intensive study of the plants in the Sierra de Manantlán. Currently, 85–90% of the vascular flora in the c. 1400 km² of the sierra may have been recorded – 2500 species (Vázquez-G. *et al.* 1990).

The Sierra de Manantlán is part of a western zone considered a centre of diversification for several families, including Malvaceae, Compositae and possibly Euphorbiaceae

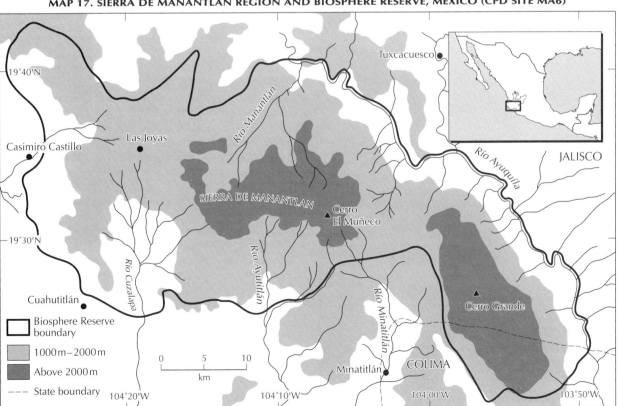

MAP 17. SIERRA DE MANANTLAN REGION AND BIOSPHERE RESERVE, MEXICO (CPD SITE MA6)

and Malpighiaceae (Vázquez-G. *et al.* 1990; Fryxell 1988; McVaugh 1984). Generic diversification includes *Periptera*, *Jaliscoa*, *Tripsacum* and *Zea*. Endemics reported from the Sierra de Manantlán include *Zea diploperennis*, *Agrostis novogaliciana*, *Populus guzmanantlensis* ("álamo"), *Croton wilburii*, *Cnidoscolus autlanensis* and *Vernonia pugana*.

The tropical dry forest of western Mexico may be the most diverse in woody species in the world. The montane mesophyllous forest is considered relictual of the Tertiary humid forests in northern North America. The region is also important as part of a transitional area which has a mix of species with boreal and tropical affinities.

Some of the species threatened due to selective exploitation are *Cedrela odorata*, *Swietenia humilis*, *Fraxinus uhdei*, *Juglans major*, *Tilia mexicana*, *Abies religiosa*, *Guaiacum coulteri*, *Talauma* sp. and *Magnolia iltisiana*.

Useful plants

Preliminary data from current studies indicate that local rural communities have a profound knowledge of the utility of the flora in the Sierra de Manantlán (Benz *et al.* 1994; Robles-H. *et al.*, in press). Based on interviews with the local inhabitants, over 500 species are utilized – yet only 179 are used in more than one of the ten communities in the study area. About 284 species have more than one use; 300 are used for traditional

CPD Site MA6: Sierra de Manantlán region and Biosphere Reserve, Mexico, showing *Pinus durangensis* forest on Cerro de las Capillas. Photo: Enrique Jardel.

medicinal purposes, 226 for timber, 170 for food and 100 for forage. Fruits of *Stenocereus queretaroensis*, *Jacaratia mexicana*, *Pithecellobium dulce* and *Ternstroemia lineata* are among those gathered for marketing.

The Sierra de Manantlán contains wild relatives of maize – the "teosintes" *Zea diploperennis* and *Z. mays* subsp. *parviglumis*, and of beans – *Phaseolus coccineus* and *P. vulgaris* (Benz 1988; Gepts 1988). The *Zea* germplasm is already considered to be of monumental economic value for the world's agriculture (e.g. Vietmeyer 1979; Nault and Findley 1982). For example, *Zea diploperennis* ("milpilla") is resistant to various viruses that have infected cultivated maize.

The region's cultivated maize (*Zea mays* subsp. *mays*) includes two of the most primitive traditional varieties in Mexico, one of which is only found in small isolated towns, so this land race perhaps is being lost. Wild relatives of crops may survive and evolve with traditional agricultural practices of slash-and-burn cultivation ("coamil") and cattle-ranching. The traditional agriculture practised in the region may be responsible for the genetic amplitude of the teosintes and very probably the wild beans (Sánchez-G. and Ordaz-S. 1987; Benz, Sánchez-Velásquez and Santana-Michel 1990). These plants demonstrate the necessity to conserve traditional agricultural systems (Guzmán-M. and Iltis 1991).

Preliminary surveys indicate that the Sierra de Manantlán may contain c. 40 widespread tree species of economic value, primarily to the silvicultural industry. Examples include *Pinus douglasiana*, *P. durangensis*, *P. herrerae*, *P. oocarpa*, *P. pseudostrobus* and some *Quercus* spp.

Social and environmental values

Water is one of the most important resources of the region. About 400,000 people rely on the sierra's water catchment for industry, agriculture and other purposes (Jardel-P. 1992). The agricultural crops are grown in the valleys, for local and export markets. None of the land in the Sierra de Manantlán Biosphere Reserve has been purchased by governmental authorities – it represents a vast zoning experiment, with 20% in indigenous communities, 40% community-owned ("ejido") lands and 40% privately owned. About 9000 people inhabit the BR, and a biological research field station is located within its western area (at 19°35'N, 104°16'W). Cooperation among the reserve's various parties is crucial (Sheean-Stone 1989).

The Sierra de Manantlán is an important refuge for animals (Sheean-Stone 1989; Jardel-P. 1992), including threatened species such as the jaguar. So far, 108 species of mammals are known from the region – 25% of Mexico's mammals; at least 12 of the species are endemic to montane areas of western Mexico and 2 subspecies to the reserve. The bird tally for the general area is 336 species – 30% of Mexico's birds, including the locally rare lilac-crowned amazon or "cotorra" (*Amazona finschi*), military macaw or "guacamaya verde" (*Ara militaris*) and golden eagle or "águila real" (*Aquila chrysaetos*); 30% of the species are totally or partially migratory, and 32 species are endemic to western and central Mexico. Lists have been compiled for the species of reptiles (60) and amphibians (20). The fishes (16 species) have a high proportion of endemics. There are 180 or more families of insects represented in the reserve.

In the mountains of central Mexico (Endemic Bird Area A11) are seven bird species of restricted range; four occur in the Sierra de Manantlán: long-tailed wood-partridge (*Dendrortyx macroura*), grey-barred wren (*Campylorhynchus megalopterus*), green-striped bush-finch (*Atlapetes virenticeps*) and collared towhee (*Pipilo ocai*). The threatened black-capped vireo (*Vireo atricapillus*), which breeds in south-eastern U.S.A. and Coahuila, Mexico, winters in the highlands of western Mexico and probably within this sierra.

Threats

Timber has been one of the most coveted resources in the Sierra de Manantlán. Between 1940–1985, the timber industry used intensive poorly planned extraction methods, resulting in a network of logging roads; the agricultural communities, which are the forest owners, received marginal benefits (Jardel-P. *et al.* 1989). Currently logging has been suspended, but private timber companies exert pressure to resume. The management plan for the reserve permits sustainable forest utilization in the buffer zone, under control of the agrarian communities that own the land (Jardel-P. 1992).

Cattle-grazing, goat-browsing and human-induced fires are other major causes of habitat degradation (Jardel-P. 1992). Only c. 30% of the region remains in good condition, and forest destruction was occurring at a rate of 20 km² per year. However, fairly recently the Jalisco Government has increased protection.

Conservation

Discovery of the rare *Zea diploperennis* brought the Sierra de Manantlán to the attention of national and international scientific communities and the public worldwide – a famous, indeed classic example of the need to preserve genetic diversity *in situ* (Iltis 1983; Halffter 1987). Further ecological studies have indicated that the sierra is significant not only because of the diploid wild maize, but also for the sierra's generally distinctive flora and fauna and its forestry potential. *Swietenia humilis* (Pacific Coast mahogany) is in Appendix II of CITES.

An initial protective step for the region was taken in 1984 when the State of Jalisco purchased 12.5 km² at Las Joyas, a private locality with a large population of *Zea diploperennis*. In 1985 the Universidad de Guadalajara created the Laboratorio Natural Las Joyas de la Sierra de Manantlán (LNLJ). Efforts were directed toward protection of the larger area. The Reserva de la Biósfera de la Sierra de Manantlán was established in 1987 by Presidential decree, created under the auspices of the Jalisco Government, with support from the Consejo Nacional de Ciencia y Tecnología (CONACYT), and was recognized by UNESCO's Man and the Biosphere Programme in 1988 (LNLJ 1989; Santana-C., Guzmán-M. and Jardel-P. 1989).

The reserve encompasses 1396 km² of the volcanic Sierra de Manantlán and neighbouring calcareous Cerro Grande to the east. The 345 km² of tropical dry forest in the reserve is the largest area of this formation protected in Mexico, and the 50 km² of cloud forest is the largest protected on Mexico's Pacific slope. The reserve and plan (Jardel-P. 1992) have presented a splendid opportunity to create a model of wise conservation and management of natural resources, with the assistance of the Universidad de Guadalajara's Instituto Manantlán de Ecología y Conservación de la Biodiversidad, which conducts regional multidisciplinary studies through facilities such as the LNLJ, and the full involvement of the local inhabitants (mostly indigenous people) and the nearby communities. For example, it is planned to continue the coamil system in a few restricted areas within the BR so that the milpilla populations can survive.

References

Benz, B.F. (1988). *In situ* conservation of the genus *Zea* in the Sierra de Manantlán Biosphere Reserve. In CIMMYT, *Recent advances in the conservation and utilization of genetic resources: proceedings of the Global Maize Germplasm Workshop.* Centro Internacional del Mejoramiento de Maíz y Trigo (CIMMYT), Mexico, D.F. Pp. 59–69.

Benz, B.F., Sánchez-Velásquez, L.R. and Santana-Michel, F.J. (1990). Ecology and ethnobotany of *Zea diploperennis*: preliminary results. *Maydica* 35: 85–94.

Benz, B.F., Santana-M., F.J., Pineda-L., R., Cevallos-E., J., Robles-H., L. and DeNiz-L., D. (1994). Characterization of mestizo plant use in the Sierra de Manantlán, Jalisco-Colima, Mexico. *J. Ethnobiology* 14: 23–41.

Cuevas-Guzmán, R. and Núñez-L., N. (1988). *Taxonomía de los pinos de la Sierra de Manantlán, Jalisco.* Prof. thesis. Facultad de Agronomía, Universidad de Guadalajara, Jalisco, Mexico. 104 pp.

Figueroa-R., B.L. (1991). *Estructura y distribución de las poblaciones de* Abies *sp. en Cerro Grande, municipios de Tolimán, Jalisco y Minatitlán, Colima.* Prof. thesis. Facultad de Ciencias Biológicas, Universidad de Guadalajara, Jalisco, Mexico. 83 pp.

Fryxell, P.A. (1988). Malvaceae of Mexico. *Systematic Botany Monogr.* 25: 1–522.

Gepts, P.L. (ed.) (1988). *Genetic resources of* Phaseolus *beans: their maintenance, domestication, evolution, and utilization.* Kluwer Academic Publishers, Dordrecht, The Netherlands. 613 pp.

Guzmán-Mejía, R. (1985). Reserva de la Biósfera de la Sierra de Manantlán, Jalisco: estudio descriptivo. *Tiempos de Ciencia, Univ. Guadalajara, Jalisco, Méx.* 1: 10–26.

Guzmán-M., R. and Iltis, H.H. (1991). Biosphere reserve established in Mexico to protect rare maize relative. *Diversity* 7(1–2): 82–84.

Halffter, G. (1987). La Reserva de la Biósfera de Manantlán y la conservación *in situ* de los recursos bióticos. *Rev. Soc. Mex. Hist. Nat.* 39: 27–34.

Iltis, H.H. (1983). *The 3rd University of Wisconsin – University of Guadalajara Teosinte Expedition to the Sierra de Manantlán, Jalisco, Mexico: December 28, 1979 to January 21, 1980. Contr. Univ. Wisconsin Herbarium* 1(1), 2nd revised edition. University of Wisconsin, Madison. 78 pp. Unpublished.

Iltis, H.H., Doebley, J.F., Guzmán-M., R. and Pazy, B. (1979). *Zea diploperennis* (Gramineae): a new teosinte from Mexico. *Science* 203: 186–188.

Jardel-P., E.J. (1991). Perturbaciones naturales y antropogénicas y su influencia en la dinámica sucesional de los bosques de Las Joyas, Sierra de Manantlán, Jalisco. *Tiempos de Ciencia, Univ. Guadalajara, Jalisco, Méx.* 22: 9–26.

Jardel-P., E.J. (ed.) (1992). *Estrategia para la conservación de la Reserva de la Biósfera Sierra de Manantlán.* Editorial Universidad de Guadalajara, Guadalajara, Jalisco, Mexico. 312 pp.

Jardel-P., E.J., Cuevas-G., R., León-C., P., León-C., M.A., Mariscal-L., G., Pineda-L., R., Saldaña-A., M.A., Sánchez-V., L.R. and Téllez-L., J. (1989). Conservación y aprovechamiento de los recursos forestales de la Reserva de la Biósfera Sierra de Manantlán. *Tiempos de Ciencia, Univ. Guadalajara, Jalisco, Méx.* 16: 18–24.

LNLJ (1989). *Plan operativo 1989–1990. Reserva de la Biósfera Sierra de Manantlán.* Universidad de Guadalajara, Laboratorio Natural Las Joyas de la Sierra de Manantlán (LNLJ), El Grullo, Jalisco, Mexico. Pp. 4–11.

McVaugh, R. (1984). Compositae. *Flora Novo-Galiciana* 12: 1–1157.

Muñoz-M., M.E. (1992). *Distribución de especies arbóreas del bosque mesófilo de montaña en la Reserva de la Biósfera Sierra de Manantlán.* Prof. thesis. Universidad de Guadalajara, Jalisco, Mexico. 102 pp.

Nault, L.R. and Findley, W.R. (1982). *Zea diploperennis*: a primitive relative offers new traits to improve corn. *Desert Plants* 3: 202–205.

Nixon, K.C. (1993). The genus *Quercus* in Mexico. In Ramamoorthy, T.P., Bye, R., Lot, A. and Fa, J.E. (eds), *Biological diversity of Mexico: origins and distribution.* Oxford University Press, New York. Pp. 447–458.

Robles-H., L., DeNiz-L., D., Benz, B.F., Santana-M., F.J., Pineda-L., R. and Anaya-C., M. (in press). Flora útil de la Reserva de la Biósfera Sierra de Manantlán. In Bye, R. (ed.), *Memoria del primer encuentro sobre bosque tropical caducifolio.* Universidad Nacional Autónoma de México, Mexico, D.F.

Rzedowski, J. and McVaugh, R. (1966). La vegetación de Nueva Galicia. *Contr. Univ. Michigan Herb.* 9: 1–123.

Saldaña-A., M.A. and Jardel-P., E.J. (1992). Regeneración natural de especies arbóreas en los bosques subtropicales de montaña de la Sierra de Manantlán. *Biotam, Univ. Autón. Tamaulipas, Méx.* 3(3): 36–50.

Sánchez-G., J.J. and Ordaz-S., L. (1987). *El teocintle en México.* Systematics and ecogeographic studies in crop genepools. Vol. 2. International Board for Plant Genetic Resources (IBPGR), Rome. 98 pp.

Sánchez-Velásquez, L.R. (1988). *Sucesión forestal en la Sierra de Manantlán, Jalisco, México.* M.S. thesis. Colegio de Postgraduados, Centro de Botánica, Chapingo, Mexico. 54 pp.

Santana-C., E., Guzmán-M., R. and Jardel-P., E.J. (1989). The Sierra de Manantlán Biosphere Reserve: the difficult task of becoming a catalyst for regional sustained development. In Gregg Jr., W.P., Krugman, S.L. and Wood Jr., J.D. (eds), *Proceedings of the symposium on biosphere reserves.* 4th World Wilderness Congress. U.S. National Park Service, Washington, D.C. Pp. 212–222.

Santiago-P., A.L. (1992). *Estudio fitosociológico del bosque mesófilo de montaña de la Sierra de Manantlán.* Prof. thesis. Facultad de Ciencias Biológicas, Universidad de Guadalajara, Jalisco, Mexico. 120 pp.

Sheean-Stone, O. (1989). Mexico's wonder weed. *WWF Reports (Gland, Switzerland)* (Aug.–Sept.): 9–12.

Vázquez-G., J.A., Cuevas-G., R., Cochrane, T.S. and Iltis, H.H. (1990). *Flora de la Reserva de la Biósfera Sierra de Manantlán, Jalisco-Colima, México.* Universidad de Guadalajara, LNLJ, Public. Especial No. 1, El Grullo, Jalisco, Mexico and Contr. Univ. Wisconsin Herb. No. 9, Madison. 166 pp.

Vietmeyer, N.D. (1979). A wild relative may give corn perennial genes. *Smithsonian* 10(9): 68–76.

Authors

This Data Sheet was written by Ramón Cuevas-Guzmán, Dr Bruce F. Benz and Enrique J. Jardel-Peláez (Universidad de Guadalajara, Instituto Manantlán de Ecología y Conservación de la Biodiversidad, Apdo. Postal 1-3933, Guadalajara, Jalisco C.P. 44100, Mexico) and Olga Herrera-MacBryde (Smithsonian Institution, Department of Botany, NHB-166, Washington, DC 20560, U.S.A.).

PACIFIC LOWLANDS, JALISCO: CHAMELA BIOLOGICAL STATION AND CUMBRES DE CUIXMALA RESERVE
Mexico

Location: Coastal Jalisco south-west of Guadalajara, between latitudes 19°22'–19°39'N and longitudes 104°56'–105°10'W.

Area: Region c. 350 km²; 131 km² in national-level Biosphere Reserve, 10.46 km² fully conserved.

Altitude: Region from sea-level to c. 500 m.

Vegetation: Fully conserved reserves mainly tropical deciduous forest, with some tropical semi-deciduous forest and riparian wetland.

Flora: High diversity, especially of woody plants; 1120 vascular plant species in 544 genera of 124 families known from region; c. 16% of station's species are regional endemics; threatened species.

Useful plants: Locally marketed fine or speciality woods (*Cordia, Dalbergia, Guaiacum, Platymiscium, Swietenia*); potential ornamentals.

Other values: Wild fauna, including many threatened species; scientific benchmark, scenery.

Threats: Agriculture, resort development, selective logging.

Conservation: National-level Biosphere Reserve, including Chamela Biological Station owned and protected by Universidad Nacional Autónoma de México (UNAM), Cumbres de Cuixmala Reserve owned and protected by Fundación Ecológica de Cuixmala.

Geography

This region of coastal Jalisco State is south of Puerto Vallarta near Chamela Bay and includes 350 km² in the municipality La Huerta. It is bounded to the west by the Pacific Ocean, to the north by the San Nicolás River, to the east (c. 10 km inland) by an arbitrary limit at about 500 m elevation and the Juan Gil-Rancho San Borja road, and to the south by the Cuitzmala River (which begins in the Sierra de Cacoma) (Map 18). Two adjacent forested reserves that are managed for full protection have been established in the region: Chamela Biological Station (3.46 km²) and Cumbres de Cuixmala Reserve (7 km²) (Lott 1993).

The region is within the Sierra Madre del Sur morphotectonic province, in the Pacific Coastal Plain (0–200 m) and the Pacific Ranges and Cuestas subprovinces (Ferrusquía-Villafranca 1993). Soils are volcanic; those of the hillsides are usually derived either from basalt or rhyolites (Bullock 1986). The region has hilly eroded plains with many small seasonal drainages and a few larger river valleys with permanent water. Only the San Nicolás and Cuitzmala rivers are major watercourses; the Chamela River has significant subterranean flow, but has surface flow only in occasional very wet years.

The mean annual precipitation was 707 mm in the period 1977–1983 (Bullock 1988; cf. Bullock and Solís-Magallanes 1990). The rainy season is only four months long, and the dry season is severe. Most rains (80%) fall between early July and early November, in some years with occasional December or January storms ("cabañuelas"). The mean annual temperature was 24.9°C over an 8-year period (1977–1984) (Bullock 1986), and monthly average temperatures varied from about 22.3°C in March to 27.3°C in May. Fire has not been important in this ecosystem.

Vegetation

The region's vegetation is dominated by tropical deciduous forest, with tropical semi-deciduous forest along drainages (arroyos). According to Rzedowski (1978), tropical deciduous forest originally occupied c. 8% of Mexico. His vegetation map designated a part of coastal Jalisco as thorn forest, which might be better classified as a secondary stage of tropical deciduous forest. The thorn forest covered a large part of coastal Sonora and Sinaloa and in isolated patches extended to the Balsas River Basin and the Isthmus of Tehuantepec, occupying c. 5% of Mexico (Rzedowski 1978).

The tropical deciduous forest is characteristic of the Pacific slope from the states of extreme southern Baja California Sur, Sonora and Chihuahua to Chiapas and onward into Central America. On Mexico's Atlantic slope, it is found mainly in three limited patches: in southern Tamaulipas and south-eastern San Luis Potosí, in central Veracruz, and on the Yucatán Peninsula. However, the Atlantic coast tropical deciduous forest is very poorly explored floristically, its affinities are uncertain, and it may be more physiognomically than floristically similar.

The Chamela station ecosystem includes solely tropical deciduous and semi-deciduous forests, and in this region the dry forest has its greatest diversity (Lott, Bullock and Solís-Magallanes 1987). The structure of the forest on hillsides is mostly a closed canopy of trees 4–15 m tall, with an understorey of shrubs and herbaceous species; in arroyos the trees are 8–25 m tall, in two strata (cf. Gómez-Pompa *et al*. 1994). There is high diversity in trees, shrubs and vines; very few herbaceous species occur within the closed forest and true annuals are rare (Lott 1993). There are few naturally occurring open areas except for ecotones. Many of the grasses and herbs found in disturbed areas may not be local natives.

A few significant plants carry on considerable photosynthesis during the long dry season – the abundance of four species of arborescent cacti [e.g. *Pilosocereus (Cephalocereus) purpusii, Stenocereus chrysocarpus*] is only apparent during the forest's leafless periods (cf. Bullock and Solís-Magallanes 1990). Visitors frequently remark on the preponderance of white tree trunks which are especially notable on some slopes in the dry season, when bark shedders in several families are also conspicuous (*Brongniartia, Bursera, Caesalpinia, Celaenodendron, Jatropha, Psidium*).

The Cumbres de Cuixmala Reserve includes a small area of riparian wetlands as well as the tropical deciduous and semi-deciduous forests (Rothschild, Lott and Sanders 1992; Lott 1993; cf. Gómez-Pompa *et al.* 1994). The heterogeneous riparian woodlands are 5–40 m high.

MAP 18. PACIFIC LOWLANDS, JALISCO: CHAMELA BIOLOGICAL STATION AND CUMBRES DE CUIXMALA RESERVE, MEXICO (CPD SITE MA7)

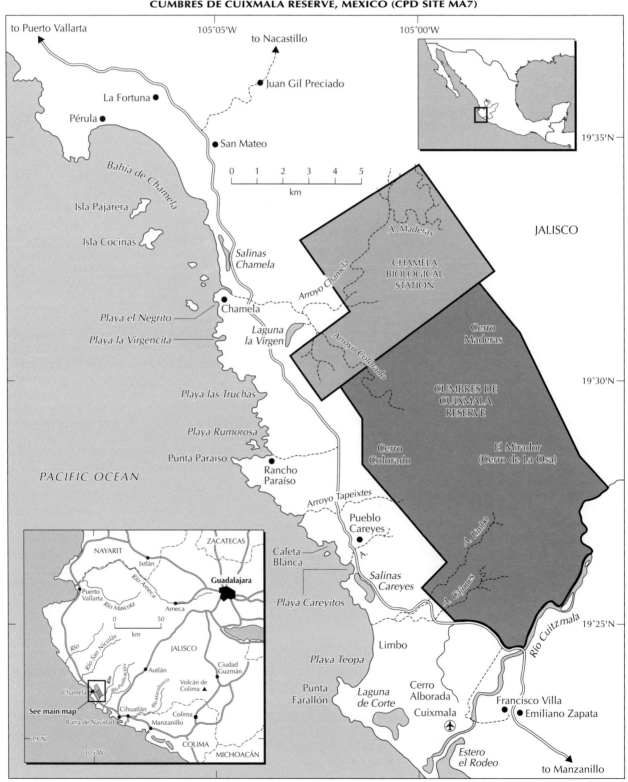

Flora

The region is in the Middle Serranías Floristic Province (Rzedowski 1978). A checklist of the vascular plants of the Chamela Biological Station has been prepared (Lott 1985) and a floristic survey of the Cuixmala reserve carried out (Rothschild, Lott and Sanders 1992). An annotated checklist of the Chamela Bay region includes both reserves and other habitats (Lott 1993). A Florula with keys and illustrations is planned. The most frequent families are Leguminosae, Euphorbiaceae, Compositae, Gramineae, Convolvulaceae and Malvaceae. The region is rich in species, with 1120 vascular plant species known, and supports a high level of endemism.

About 10% of the species may be endemic to coastal Jalisco and Colima or with some also in Michoacán. There is c. 16% regional endemism at the station (Lott 1985), since it does not include the marine coast or true riparian habitats where most of the more widely distributed species occur. Nearly 40% of the species in the flora are endemic to the dry forest of the Pacific slope, and 28% are restricted to Mexico (Lott 1993). Four genera are endemic to the Pacific coast: *Chalema*, *Dieterlea*, *Celaenodendron* and *Mexacanthus*. *Celaenodendron mexicanum* (Euphorbiaceae), the only species in its genus, has a very patchy coastal distribution and dominates a distinctive forest association that intergrades with the tropical deciduous forest in large patches of both reserves, which are important in protecting this species.

Useful plants

Because the region has only been easily accessible to settlement since a coast highway opened in 1972, there is no local tradition of useful plants. Many residents are from upland Jalisco or the neighbouring states (mostly Colima and Michoacán), and have brought common names and uses from their home areas for application to these sometimes different species; over 100 are used. Examples include a decoction of *Plumeria rubra* leaves used to relieve earache, *Spondias purpurea* fruits used in a drink, *Stenocereus chrysocarpus* fruits eaten as a delicacy and *Hura polyandra* used for wood, even though it can cause dermatitis. *Hintonia latiflora* and *Physodium adenodes* var. *adenodes* have ornamental potential.

Although the region's forests are not suitable for large-scale commercial logging, several timber species occur and their fine or regular woods are locally marketed: *Cordia alliodora*, *C. dentata*, *C. elaeagnoides*, *C. seleriana*, *Dalbergia congestiflora*, *Guaiacum coulteri*, *Platymiscium lasiocarpum* and *Swietenia humilis*. Already *Celaenodendron mexicanum* is a locally choice timber for roof beams and building posts.

Social and environmental values

These two fully protected forest reserves conserve representatives of a vegetation type highly threatened in Mexico by conversion to pastoral and agricultural uses (cattle, maize). The Mexican lands occupied by tropical deciduous forest have been considered poor, but are rapidly being deforested now because of the increasing human population. The conversion is highly destructive, since long-term sustainable production is not possible due to thin soils and lack of water except along river-valley bottoms, where the native vegetation (tropical semi-deciduous forest) already is almost totally gone.

There is tourist interest because the region has a spectacular coastline with beautiful beaches and is still relatively unspoiled, with interesting animals (particularly birds and sea-turtles) and plants. Controlled ecotourism offers one of the few long-term potentials. Typical tourism is of questionable viability because of competition from more established areas with easier access and greater availability of water. Several marginal or failed tourist centres and small real-estate developments exist in the area, and some have altered or drained salt marshes and mangrove swamps.

Of the 270 species of birds in the region, 40% are migratory and 60% resident. The Chamela Bay region is at the southern end of the North-west Mexican Pacific Slope Endemic Bird Area (EBA A05), which has seven species of restricted range. Two threatened species possibly occur within this CPD site: the yellow-headed parrot (*Amazona oratrix*), recorded from Chamela, and the Mexican woodnymph (*Thalurania ridgwayi*), known from only a handful of humid barrancas in western Mexico.

Many scientific papers and student theses have been produced at Chamela Biological Station, and long-term ecological studies are being carried out there (e.g. Bullock and Solís-Magallanes 1990) and at the Cumbres de Cuixmala Reserve. There are manuals of the mammals (Ceballos and Miranda 1986) and reptiles and amphibians (García and Ceballos 1994; Ramírez-Bautista 1994), a checklist of the avifauna (Arizmendi-A. *et al.* 1990) and lists, taxonomic treatments or natural histories of various insect groups (Atkinson and Equihua 1986; Morón 1988). Research on reforestation at the Cuixmala reserve, which had a 3-km² deforested area, was initiated in 1992. The proceedings of a 1991 symposium at the station on the ecology of tropical dry forests are forthcoming (Bullock, Mooney and Medina 1995).

Threats

Threats include burning and cutting, with conversion of neighbouring lands to pastoral and agricultural uses (although they are not very productive), and ecological isolation due to destruction of vegetation with the building of resort hotels, real-estate developments, polo fields and marinas.

Invasions and property disputes occur. Lack of perimeter access to monitor and prevent invasions has made the situation at Chamela more difficult, and is expected at Cuixmala. The selective hardwood exploitation may be causing or compounding problems for certain species: e.g. under Mexican law, *Dalbergia congestiflora* and *Platymiscium lasiocarpum* are considered endangered species and *Guaiacum coulteri* is a species subject to special protection (SEDESOL 1994); *Celaenodendron mexicanum* could become endangered from continuing use and further coastal development for tourism; and *Swietenia humilis* is in Appendix II of CITES.

Conservation

Since 1971 the area of the Chamela Biological Station has been managed and protected by UNAM. The management has succeeded in resolving a long-term dispute with invaders and is committed to maintaining the station's integrity. The Cumbres de Cuixmala Reserve began in 1987 and has established an official status with the Mexican Government.

At the end of 1993, a Chamela-Cuixmala Biosphere Reserve at the national level (covering 131 km²) was established by Presidential decree (Gómez-Pompa *et al.* 1994). In 1986 Playa de Cuixmala was decreed a Reserved Zone and Wildlife Refuge, but no coastal areas of the region are included in a fully protected Nature Reserve. It is highly desirable to fully conserve a portion of the remaining coastal palm forest, of *Orbignya guacuyule*. The status for protection of the Chamela Bay islands is unclear.

References

Arizmendi-A., M. del C., Berlanga, H., Márquez-Valdemar, L., Navarijo, L. and Ornelas, F. (1990). *Avifauna de la región de Chamela, Jalisco*. Universidad Nacional Autónoma de México (UNAM), Cuadernos del Instituto de Biología 4.

Atkinson, T.H. and Equihua, A. (1986). Biology of the Scolytidae and Platypodidae (Coleoptera) in a tropical deciduous forest at Chamela, Jalisco, Mexico. *Florida Entomol.* 69: 303–310.

Bullock, S.H. (1986). Climate of Chamela, Jalisco, and trends in the south coastal region of Mexico. *Arch. Meteorol. Geophys. Bioklimatol. Ser. B* 36: 297–316.

Bullock, S.H. (1988). Rasgos del ambiente físico y biológico de Chamela, Jalisco, México. In Morón, M.A. (ed.), *La entomofauna de Chamela, Jalisco, México. Folia Entomol. Mex.* Vol. 77. Pp. 5–17.

Bullock, S.H. and Solís-Magallanes, J.A. (1990). Phenology of canopy trees of a tropical deciduous forest in México. *Biotropica* 22: 22–35.

Bullock, S.H., Mooney, H.A. and Medina, E. (eds) (1995). *Seasonally dry tropical forests*. Cambridge University Press, Cambridge, U.K. 521 pp.

Ceballos, G. and Miranda, A. (1986). *Los mamíferos de Chamela, Jalisco; manual de campo*. UNAM, Inst. Biol., Mexico, D.F.

Ferrusquía-Villafranca, I. (1993). Geology of Mexico: a synopsis. In Ramamoorthy, T.P., Bye, R., Lot, A. and Fa, J.E. (eds), *Biological diversity of Mexico: origins and distribution*. Oxford University Press, New York. Pp. 3–107.

García, A. and Ceballos, G. (1994). *Guía de campo de los reptiles y anfibios de la costa de Jalisco, México/Field guide to the reptiles and amphibians of the coast of Jalisco, Mexico*. Fundación Ecológica de Cuixmala and UNAM, Inst. Biol., Mexico, D.F. 184 pp.

Gómez-Pompa, A. and Dirzo, R. with Kaus, A., Noguerón-Chang, C.R. and Ordoñez, M. de J. (1994). *Las áreas naturales protegidas de México de la Secretaría de Desarrollo Social*. SEDESOL, Mexico, D.F. 331 pp. Unpublished.

Lott, E.J. (1985). *Listados florísticos de México, III. La Estación de Biología Chamela, Jalisco*. UNAM, Inst. Biol., Mexico, D.F. 47 pp.

Lott, E.J. (1993). *Annotated checklist of the vascular flora of the Chamela Bay region, Jalisco, Mexico. Occas. Pap. Calif. Acad. Sci.* No. 148. 60 pp.

Lott, E.J., Bullock, S.H. and Solís-Magallanes, J.A. (1987). Floristic diversity and structure of upland and arroyo forests in coastal Jalisco. *Biotropica* 19: 228–235.

Morón, M.A. (ed.) (1988). *La entomofauna de Chamela, Jalisco, México. Folia Entomol. Mex.* Vol. 77. 525 pp.

Ramírez-Bautista, A. (1994). *Manual y claves ilustradas de los anfibios y reptiles de la región de Chamela, Jalisco*. UNAM, Cuadernos del Instituto de Biología 23.

Rothschild, B.M., Lott, E.J. and Sanders, A.C. (1992). *A report to the Fundación Ecológica de Cuixmala on the floristic surveys of 1990–1991 of the Cuixmala-Cumbres and El Jabalí reserves in Mexico*. University of California, Riverside, U.S.A. and IUCN, Richmond, U.K. 134 pp.

Rzedowski, J. (1978). *Vegetación de México*. Editorial Limusa, Mexico, D.F. 432 pp.

SEDESOL (Secretaría de Desarrollo Social) (1994). Norma Oficial Mexicana NOM-059-ECOL-1994, que determina las especies y subespecies de flora y fauna silvestres terrestres y acuáticas en peligro de extinción, amenazadas, raras y las sujetas a protección especial, y que establece especificaciones para su protección. *Diario Oficial de la Federación* 16/05/94: 2–60.

Author

This Data Sheet was written by Emily J. Lott (University of California, Department of Botany and Plant Sciences, Riverside, CA 92521, U.S.A.).

Acknowledgement

Beth M. Rothschild (Botanic Gardens Conservation International, Kew, U.K.) served as liaison with the Cumbres de Cuixmala Reserve.

UPPER MEZQUITAL RIVER REGION, SIERRA MADRE OCCIDENTAL
Mexico

Location: Western Sierra Madre mountains in the south of Durango State, between latitudes 22°50'–23°40'N and longitudes 104°05'–105°30'W.

Area: c. 4600 km².

Altitude: 800–3350 m.

Vegetation: Principally conifer, pine-oak and oak forests; tropical dry forests, and patches of tropical subdeciduous forest.

Flora: High diversity – c. 2900 species of vascular plants; relatively high endemism.

Useful plants: More than 450 wild species used for medicinal, food and other purposes by local people; many timber species extracted.

Other values: Water resources, archaeological sites, refuge of Tepehuan Amerindians, germplasm reserve for timber species and wild relatives of cultivated plants, habitats for wild animals, threatened species.

Threats: Logging, erosion due to overgrazing and road construction, loss of traditional knowledge.

Conservation: La Michilía Biosphere Reserve (700 km², 70 km² as core). Part of temperate forests in area of the Tepehuan could be declared special reserve.

Geography

The Upper Mezquital River region of the Sierra Madre Occidental morphotectonic province is in north-central Mexico's State of Durango in the southern part, south of the city of Durango (Map 19). The Tropic of Cancer crosses the region's northern third. This region has a complex physiography (Ferrusquía-Villafranca 1993; Gómez-Pompa *et al.* 1994); the highest peak is Cerro Gordo. The western flank of the sierra is extremely steep; its rivers form deep gorges, in some sites with walls more than 2000 m high and up to 10 km in extent. Most prominent is the gorge of the Mezquital River, the only river which originates on the eastern side and crosses the sierra – it becomes the San Pedro River and flows into the Pacific Ocean in Nayarit State.

The region has a wide range of climates, from subhumid warm and semi-warm with a pronounced winter dry season at lower elevations on the western slope of the sierra, to various temperate and semi-cold climates at higher elevations. The annual precipitation ranges between 800 mm and 1400 mm.

The formation of the Sierra Madre Occidental began in the Eocene, but its upper layers are more recent, derived from a period of intense volcanic activity during the Pliocene and Pleistocene. Volcanic rocks 34 to 27 million years old predominate in the region, principally ignimbrites, rhyolites and rhyodacites as well as tuffs; very small outcrops of basalt also occur. The upper layer of the Sierra Madre Occidental is the Earth's most continuous ignimbritic layer.

Vegetation

The region is a mosaic comprised of a great variety of ecosystems because of its transitional location between the Holarctic and Neotropical realms, and the complexity of its physiography and climates. In the higher areas, temperate pine and/or oak forests predominate, with small enclaves of fir and pine-fir forests. In the gorges toward the west, tropical dry and subdeciduous forests are found (Rzedowski 1978; S. González 1983; Gómez-Pompa *et al.* 1994; González, González and Cortés, in press).

Fir and pine-fir forests
The fir and pine-fir forests occupy humid ravines and hillsides with northern exposure at altitudes above 2700 m. In the most extensive of these coniferous forests *Pseudotsuga menziesii* (Douglas fir) predominates. There are smaller areas of *Abies durangensis* and *Picea chihuahuana* – which is restricted to just a few sites in the Sierra Madre Occidental.

Pine forests
At elevations above 2500 m on hillsides and in broad ravines occur diverse combinations of pine species, principally *Pinus durangensis*, *P. teocote* and *P. arizonica* var. *stormiae*, and in more humid sites *P. leiophylla* and *P. ayacahuite*.

Pinus cooperi alone is dominant in broad valleys with deep soils, as well as on slight hillsides at 2400–2650 m. On the western slope of the sierra at 1350–1700 m, there are hillside communities dominated by *Pinus douglasiana*.

Pine-oak forests
The pine-oak forests occur in very diverse associations, depending upon the altitude and physiography:

Above 2500 m, *Pinus durangensis* and *Quercus sideroxyla* associate with *P. teocote*, *P. cooperi*, *Q. crassifolia*, *Q. rugosa* and *Arbutus madrensis*, which in disturbed areas mingle with *Alnus, Arbutus arizonica* and *A. tessellata*. This mixture is probably the most representative of the forests of the Sierra Madre Occidental in Durango. In sites with more humidity, *Pinus leiophylla* becomes dominant.

At 2300–2500 m, *Pinus oocarpa* subsp. *trifoliata* is the dominant tree on mesas and hillsides that are not very steep, and frequently mixes with *P. lumholtzii, Quercus urbanii* and some *P. engelmannii*, or in areas of more tropical influence with more *P. engelmannii* and *P. douglasiana, Q. viminea* and *Q. fulva*.

In semi-dry temperate areas, diverse species of *Quercus* may combine with *Pinus engelmannii, P. chihuahuana* and madrones – principally *Arbutus arizonica* and *A. tessellata*. In the drier forests, *Pinus cembroides* and *Quercus grisea* or *Q. eduardii* predominate (Passini 1985).

In areas between 1800–1900 m, *Quercus coccolobifolia* mixes with *Pinus chihuahuana, Q. resinosa, Q. crassifolia* and *Q. viminea*. On the western slope of the sierra at 1350–1700 m are communities dominated by *Pinus herrerae, P. douglasiana* and/or *P. oocarpa* along with *Quercus* and *Arbutus*.

In areas of shallow soil or strong outcrops of very unsheltered bedrock, forests of *Pinus lumholtzii* ("pino triste") mixed with oaks (*Quercus urbanii* and *Q. crassifolia*) are prominent. Isolated shrub-like elements include manzanita (*Arctostaphylos pungens*) and *Juniperus durangensis*.

Oak forests

Oak forests are common on the eastern slope of the Sierra Madre Occidental at its lower elevations (1900–2400 m), where the following associations are predominant:

In areas with a semi-dry temperate climate there are low open forests, in which the dominants are white oaks such as *Quercus grisea, Q. arizonica* and *Q. laeta*, red oaks such as *Q. eduardii* and *Q. durifolia*, and other oaks' (e.g. *Q. crassifolia*).

Where there is a semi-humid climate, forests of *Quercus rugosa* occur on mesas and hillsides, sometimes up to 2800 m. In areas with deeper soil, *Q. sideroxyla* can also be found.

Tropical dry forest

The dry hillsides with a subtropical climate between 800–1900 m have open forests, in which the dominant species lose their leaves during the dry season. Among them are *Acacia pennatula, Bursera* spp., *Ipomoea murucoides, Lysiloma divaricata* and *Plumeria rubra*. Arborescent cacti are also dominant.

Tropical subdeciduous forest

This is restricted to sites with greater humidity, especially in gorges and along permanent streams; many of the principal species have foliage throughout the year. Common species are *Pithecellobium dulce, Ceiba acuminata, Ficus* spp., *Brosimum alicastrum, Bursera* spp. and *Enterolobium cyclocarpum*.

MAP 19. UPPER MEZQUITAL RIVER REGION, SIERRA MADRE OCCIDENTAL, MEXICO (CPD SITE MA8)

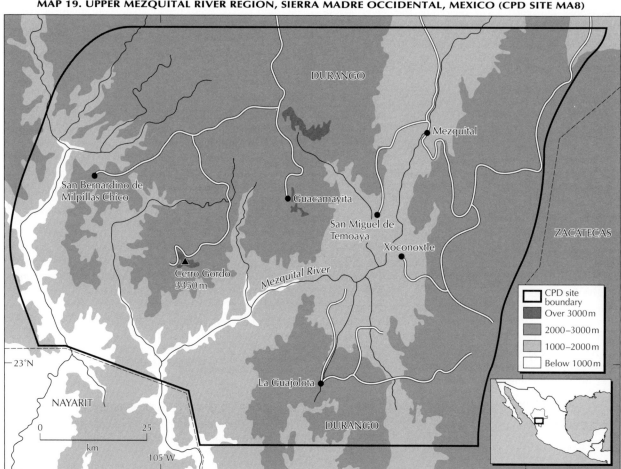

Flora

Of the 3800 phanerogamous species known to occur in Durango (González, González and Herrera 1991), c. 2900 are estimated to be present in the Upper Mezquital River region. This floristic diversity is high considering the latitude, and is enriched by the confluence of elements from the Holarctic and Neotropical realms and three floristic provinces: Sierra Madre Occidental, Altiplano and Pacific Coast (Rzedowski 1978, 1991), as well as the region's physiographic and climatic complexities.

The great gorge of the Mezquital River functions as a biological corridor that provides entrance of tropical elements to the eastern side of the Sierra Madre Occidental, but also functions as the major north-south geographic barrier to various plant species of the sierra. The region's location makes it an important zone of confluence for boreal and equatorial genera, whereas at the species level it is an area where Mexican endemics predominate. Probably because of its physiography and location, it is a zone favourable to speciation – many plant genera are in active evolution. The zone is consequently a refuge for many endemic species of western Mexico and at least sixteen local endemics – e.g. *Eleocharis svensoniana*, *Muhlenbergia durangensis*, *M. michisiensis*, *Axiniphyllum durangense*, *Sabazia gonzalezae*, *Senecio gesneriifolius*, *S. gonzalezae*, *Tridax durangensis*, *Verbesina durangensis* and *Pavonia durangensis*.

Floristic inventories have been carried out through the Flora of Durango project of the Centro Interdisciplinario de Investigación para el Desarrollo Integral Regional, Unidad Durango – Instituto Politécnico Nacional (CIIDIR-IPN). Endemic species have been found and described in the last ten years but a good part of the region is still unexplored. It is highly probable that there are many taxa not yet discovered, and others that have been lost forever because of the intense exploitation in some areas.

Useful plants

Among the species with economic value, predominant are *Pinus durangensis*, *P. cooperi*, *P. teocote* and *P. ayacahuite*, which are extracted for their wood. Diverse species of *Quercus* have economic value for charcoal, and lumber potential as well. In addition to using various species for timber, there is a great potential for utilization of the region's flora for non-timber products.

Over 450 wild plant species are used by the indigenous and mestizo populations in the region and elsewhere in Durango for medicinal, food, construction, forage and handicraft purposes. Some of these species have possibilities at the industrial level. A few species, such as *Laelia speciosa*, are collected to sell in nearby cities as ornamentals, whereas others such as *Senecio sessilifolius* ("peyotillo") and *S. albolutescens* ("matarique") are sold for medicinal purposes in local markets (cf. M. González 1984).

The northern part of the region is home to the mestizo population, whereas the south is inhabited by the Tepehuan indigenous people. The ethnobotany of the region's mestizos and Tepehuans has been studied by M. González (1984, 1991) and González and Galván (1992), and several additional works on the useful plants and agro-ecological practices are in preparation.

Social and environmental values

Archaeological sites are found in the region, including the core area (Cerro Blanco) of the Biosphere Reserve (Gómez-Pompa *et al.* 1994). Because of its inaccessibility, the region has been a refuge for indigenous peoples, especially Tepehuans, who were displaced from the lower lands that they originally had occupied. The main Tepehuan community, Santa María Ocotán y Xoconoste, is the principal political and religious centre. Its 13,500 inhabitants are distributed in four towns and 22 dependencies. Other important communities are Taxicaringa and San Bernardino de Milpillas. In lower areas, including within Nayarit State, the Mezquital watershed also sustains the indigenous towns of Coras and Huicholes.

The region's water resources are utilized for agricultural areas of the lower lands of Durango and in the coastal plain of Nayarit. The watershed's vegetation protects the soil from erosion and helps avoid flooding at lower elevations – thus its conservation is of vital importance for the equilibrium of a diversity of ecosystems.

The populations of timber species and wild relatives of cultivated plants are reserves for germplasm. Examples of such crop relatives are found in *Phaseolus* and *Solanum*. The region also helps to sustain a diverse wild fauna (Halffter 1978), including the threatened thick-billed parrot (*Rhynchopsitta pachyrhyncha*) and perhaps the imperial woodpecker (*Campephilus imperialis*) – if it is not already extinct. The Instituto de Ecología has a research station (Piedra Herrada) in the Cerro Blanco area.

Economic assessment

During the last 25 years the region has been exploited intensively by a private timber company. Currently, a complex socio-economic problem has allowed a profound deterioration of the forests to take place in the southern area inhabited by the Tepehuan. Considering their low standard of living and that these forests constitute not only their patrimony but their source of sustenance and protection for their culture, it has become urgent to reach a solution to the deteriorating situation (González and González 1992).

Threats

Many populations of the plants and animals have been affected by destruction of their natural habitats. The main problem is deforestation resulting from the intense exploitation of timber trees, which is being carried out without any conservation or management programmes. Along with extraction of the pines, it is common that the oaks are girdled or eliminated in other ways to comply with guidelines of SARH (Mexico's Secretaría de Agricultura y Recursos Hidráulicos), which has stated that a volume of oaks must be extracted equal to half that of the pines extracted.

Furthermore, overgrazing and the introduction of sheep and goats, seasonal utilization of steep hillsides as agricultural areas and the construction of a road south-westward to link the cities of Durango and Tepic have caused erosion and reduced the regeneration of plants (Carrillo-S. 1982) and the recuperation of the forests.

The unregulated exploitation of the resources and road construction are affecting the cultural identity of the indigenous people. Symptoms of social maladjustment have become clearly manifest, and their traditional knowledge is in danger of disappearing.

Conservation

An area reaching from 2000–2985 m on the eastern flank of the Sierra Madre Occidental in 1979 was officially declared a Biosphere Reserve (La Michilía) and a Zone of Forest Protection, but within the sierra's expansive range through Durango there is no protected area for its ecosystems. La Michilía BR is located between 23°15'–23°30'N and 104°05'–104°20'W (Halffter 1978; González, González and Cortés, in press); it is administered by the federal government's Instituto de Ecología. The reserve's core (Cerro Blanco) includes 70 km², and a buffer zone increases the area to (350–) 700 km². This large buffer zone has public grazing pastures and private properties in which grazing and occasional extraction of wood take place.

To achieve an optimal and sustained management of the natural resources of the Tepehuan area in the region's south, it is essential to address the socio-economic problems of the indigenous community and the problem of logging.

Among technical recommendations for management of the region are the following: Exclude from utilization the steep hillsides, and strips 100 m wide along permanent streams, as well as isolated areas of 1 km² and all areas with *Picea* and *Abies*. Control grazing of goats and sheep, which damage regeneration of the forests. Revegetate with native species those areas that are severely damaged.

References

Carrillo-S., A. (1982). *Producción primaria neta aérea del estrato herbáceo y efecto del ganado sobre su composición florística en la Reserva de la Biósfera "La Michilía", Dgo.* Professional Thesis. Facultad de Ciencias, UNAM, Mexico, D.F. 187 pp.

Ferrusquía-Villafranca, I. (1993). Geology of Mexico: a synopsis. In Ramamoorthy, T.P., Bye, R., Lot, A. and Fa, J.E. (eds), *Biological diversity of Mexico: origins and distribution.* Oxford University Press, New York. Pp. 3–107.

Gómez-Pompa, A. and Dirzo, R. with Kaus, A., Noguerón-Chang, C.R. and Ordoñez, M. de J. (1994). *Las áreas naturales protegidas de México de la Secretaría de Desarrollo Social.* SEDESOL, Mexico, D.F. 331 pp. Unpublished.

González, M. (1984). *Las plantas medicinales de Durango.* Centro Interdisciplinario de Investigación para el Desarrollo Integral Regional, Instituto Politécnico Nacional (CIIDIR-IPN), Durango, Mexico. *Cuad. Inv. Tecnol. (Durango)* 1(2): 1–117.

González, M. (1991). Ethnobotany of the Southern Tepehuan of Durango, Mexico: I. Edible mushrooms. *J. Ethnobiol.* 11: 165–173.

González, M. and Galván, R. (1992). El maguey (*Agave* spp.) y los Tepehuanes de Durango. *Cact. Suc. Mex.* 37: 3–11.

González, M. and González, S. (1992). The plight of Tepehuan forest land. *Abstracts of the III International Congress of Ethnobiology*, Mexico, D.F.

González, M., González, S. and Herrera, Y. (1991). *Listados florísticos de México. IX. Flora de Durango.* Instituto de Biología, Universidad Nacional Autónoma de México, Mexico, D.F. 167 pp.

González, S. (1983). *La vegetación de Durango.* CIIDIR-IPN, Durango, Mexico. *Cuad. Inv. Tecnol. (Durango)* 1(1): 1–114.

González, S., González, M. and Cortés, A. (in press). Vegetación de la Reserva de la Biósfera La Michilía. *Acta Bot. Mex.* [140 pp.]

Halffter, G. (ed.) (1978). *Reservas de la Biósfera en el estado de Durango.* Instituto de Ecología, Mexico, D.F. 198 pp.

Passini, M.F. (1985). Les fôrets de *Pinus cembroides* Zucc. de la Sierra de Urica. Rèserve de la Biosphère "La Michilía" (Estado de Durango, Mexique). *Bull. Ecol.* 16: 161–166.

Rzedowski, J. (1978). *Vegetación de México.* Editorial Limusa, Mexico, D.F. 432 pp.

Rzedowski, J. (1991). Diversidad y orígenes de la flora fanerogámica de México. *Acta Bot. Mex.* 14: 3–21.

Author

This Data Sheet was written by Dra. Socorro González-Elizondo (CIIDIR and COFAA, Instituto Politécnico Nacional, Apdo. Postal 738, Durango, Dgo., Mexico).

GOMEZ FARIAS REGION AND EL CIELO BIOSPHERE RESERVE
Mexico

Location: North-east Mexico in south-western Tamaulipas State, between latitudes 22°55'–23°30'N and longitudes 99°02'–99°30'W.

Area: c. 2400 km².

Altitude: 200–2200 m.

Vegetation: Tropical dry forest; tropical semi-deciduous forest; cloud forest; oak, pine and mixed oak and pine forests; desert scrubs or brushlands; riparian vegetation.

Flora: Over 1000 vascular plant species; transition zone with tropical and temperate, lowland to highland, and humid to arid elements; threatened species.

Useful plants: Temperate and tropical timbers; medicinal, ornamental, edible, fodder and fibre species.

Other values: Watershed protection for large region; refuge for wild fauna; extensive biological research base; archaeological sites; scenery and ecotourism.

Threats: Fire, cattle and goats, agriculture, irregular settlements, road construction, potential logging, over-collection of some species (for ornamentals and fuelwood).

Conservation: 60% of region in El Cielo Biosphere Reserve (1445 km²).

Geography

The Gómez Farías region is in Mexico's north-eastern State of Tamaulipas (Map 20) on the eastern slope of the Sierra Madre Oriental, known locally as the Sierra de Guatemala, and includes portions of the municipalities Jaumave, Llera de Canales, Gómez Farías and Ocampo (Sosa 1987). This region is part of the eastern sector of the Sierra Madre Oriental morphotectonic province, in the subprovince of Closely Spaced Ridges (Ferrusquía-Villafranca 1993). The region is formed by several mountain ranges trending north-south, of which the most prominent are the Sierra Cucharas, Sierra Chiquita and Sierra Los Nogales.

The mountains rise abruptly from 200 m in the east to 2200 m westward, and form two hilly plateaux – at 900–1200 m, and beginning above 1900 m. About 98% of the region is slopes steeper than 20%, c. 0.5% has slopes of 5–10% and c. 1% consists of alluvial plains with slopes of 1–2% (Castro-Meza, Plácido de la Cruz and Almaguer 1989). Geologically the general region is characterized by secondary limestone masses of sedimentary origin from the Early Cretaceous (Puig 1976), which have formed a karstic topography, with many rocky outcrops, sinkholes and caves.

The dominant soils are lithosols and rendzinas less than 40 cm deep. In the bottom of valleys and ravines and on gentle slopes have developed luvisols, acrisols, vertisols, phaeozems, regosols and cambisols. On these soils slash-and-burn agriculture and grasslands are sometimes found. Aridisols have developed in the driest inland portion of the region (Bracho and Sosa 1987; Contreras 1991).

There are several climatic zones. The highlands are cooler than the lowlands and generally more humid. The Sierra Madre Oriental is a barrier against the humid winds from the Gulf of Mexico, causing considerable regular precipitation on the eastern slopes. This rain and mist feed numerous streams that flow into the south-eastern Sabinas and Frío rivers – tributaries for the Guayalejo River, which bounds the region in the north-east. Leeward slopes of the Gómez Farías region, which are inland over the mountain crests, and the uppermost elevations, are drier than lower and mid-elevation windward slopes due to the loss of moisture with altitude and then rain-shadow effects. The area including El Cielo Biosphere Reserve is subjected to frequent cyclones from the Caribbean Sea (145 km distant), and as well to incursions of polar air masses ("nortes") that may cause freezing conditions – usually only in the highlands – or bring some rainfall, from November to January (Puig and Bracho 1987).

The tropical deciduous and tropical semi-deciduous forests have a semi-warm humid climate with rains from May to October; the mean annual temperature is 22.8°C and total annual rainfall reaches 1852 mm. The mid-elevation cloud forest has a temperate humid climate with a mean annual temperature of 13.8°C and mean annual rainfall of 2527 mm – mist is always present, the relative humidity is over 90%, and the wettest months are May to October. Highlands variously covered with oak, pine and mixed pine and oak forests or chaparral, have a generally temperate subhumid climate, with a cool summer – the annual temperature averages 16°C, the annual rainfall 889 mm. In the north-western portion of the region, the desert scrublands have a semi-warm dry climate, with summer rains; the mean annual temperature is 21.7°C and mean annual rainfall 505 mm (Puig and Bracho 1987; Contreras 1991).

Vegetation

The mosaic of vegetation types in the Gómez Farías region is diverse (Wood 1975a; cf. Rzedowski 1978), and includes the northernmost tropical forests on the Atlantic slope. Cloud forest, a formation with a much reduced distribution – in Mexico on less than 1% of the area – is exuberant in this region. The region's principal vegetation types are:

1. **Tropical dry forest (dry deciduous forest)**, which originally included broad flat areas below 300 m at the base of the Sierra Madre Oriental. Most of those areas have been transformed into croplands. Characteristic forest species

are *Croton cortesianus*, *Beaucarnea inermis*, *Acacia coulteri*, *Guazuma ulmifolia*, *Cassia emarginata*, *Ficus cotinifolia*, *Pseudobombax ellipticum*, *Bursera simaruba* and *Neobuxbaumia euphorbioides* (Martin 1958; Valiente-Banuet 1984).

2. **Tropical semi-deciduous forest (moist semi-deciduous forest)** occurs at 300–800 m on portions of Sierra Chiquita, and the eastern slope of Sierra Cucharas east and south of the Gómez Farías region. The continuous canopy is c. 20 m high, and at least half of the species shed their leaves during the dry season (December to April). The dominant trees are *Brosimum alicastrum*,

MAP 20. GOMEZ FARIAS REGION AND EL CIELO BIOSPHERE RESERVE, MEXICO (CPD SITE MA9)

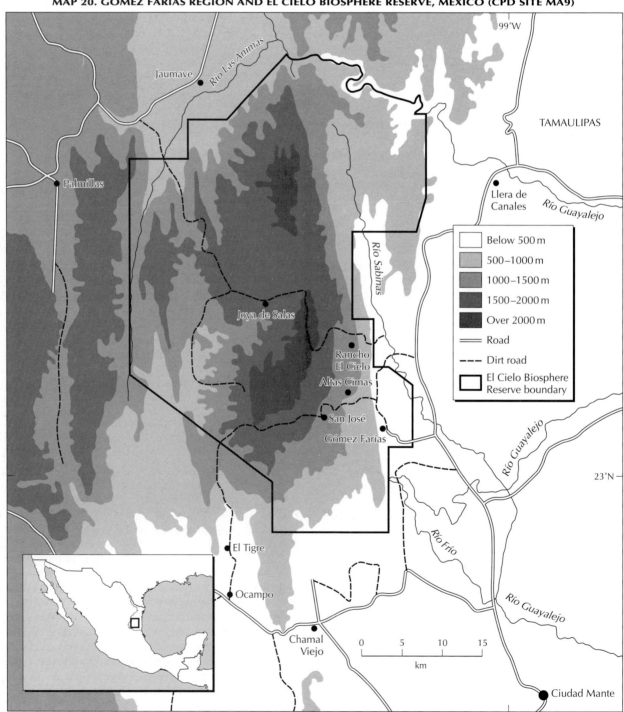

Bursera simaruba, Croton niveus, Drypetes lateriflora, Guazuma ulmifolia, Aphananthe monoica (Mirandaceltis monoica) and two *Lysiloma* species. The most common species in the dense shrub layer are *Acalypha schiedeana, Psychotria erythrocarpa, Randia laetevirens* and *Savia sessiflora.* Lianas and epiphytes are scarce (Valiente-Banuet 1984; Sosa 1987).

3. **Cloud forest** occurs at 800–1400 m. It has three tree storeys, the highest in climax stands reaching 25–30 m. Epiphytes and lianas are abundant. The dominant trees are *Liquidambar styraciflua, Quercus sartorii, Q. germana, Clethra pringlei, Magnolia schiedeana, Podocarpus reichei, Cercis canadensis* and *Acer skutchii* (Puig, Bracho and Sosa 1983). El Cielo Biosphere Reserve was established to protect this area, which is one of the most important sites along the Sierra Madre Oriental.

4. **Pure and mixed oak forests** cover the largest area of the region between 1500–2200 m. Humid oak forests thrive in ravines and other depressions of the eastern slope, and dry oak forests thrive on the north-facing hills. The common species are *Quercus sartorii, Q. affinis, Q. germana, Q. laeta* and *Q. glabrescens.* There are many lianas of *Antigonon, Dioscorea, Serjania* and *Smilax.*

5. **Pine forest** has a discontinuous distribution on summits above 1800 m. The canopy reaches 25 m. The dominant species are *Pinus patula, P. pseudostrobus, P. teocote* and *P. oocarpa.* The shrub layer is dominated by *Eupatorium, Gaultheria, Vaccinium, Myrica* and *Staphylea.*

6. **Mixed pine-oak forest and oak-pine forest** are the most extensive vegetation types in the region, occurring at 1500–1900 m in an inland south-west to north-west band. Characteristic are *Quercus glabrescens, Q. affinis, Q. mexicana, Q. polymorpha* and *Q. castanea.*

7. In the driest areas of the region between 900–1600 m, two types of **desert scrub** are present. Shrubs 2–3 m in height dominate, with some arborescent emergents 4–5 m tall and a stratum of low shrubs and herbs less than 1 m in height. About 50% of the plants bear thorns or spines. Some conspicuous species are *Acacia anisophylla, A. berlandieri, Cordia boissieri, Dasylirion acrotriche, Helietta parvifolia, Neopringlea integrifolia, Yucca treculeana, Bursera fagaroides, Cassia greggii, Forestiera angustifolia* and *Randia laetevirens. Agave lechuguilla* and *Hechtia glomerata* are localized. Some species of oak and cactus also occur (Chimal *et al.* 1989).

8. In addition to these principal vegetation types, there are smaller plant communities, e.g. enclaves of chaparral within the eastern dry oak-pine forest, a marshy pond and a shallow lake at Joya de Salas and lowland riparian vegetation along the major rivers (with species such as *Ficus, Taxodium, Platanus, Salix*).

Flora and fungi

The flora of the Gómez Farías region is one of the best known in Mexico, having been studied since 1950 directly or indirectly. Over 1000 species have been recorded, including 95 pteridophytes (cf. Lof 1980) and 21 gymnosperms; the flowering plants are represented by 968 species in 547 genera of 132 families (Johnston *et al.* 1989). The macroscopic species of fungi are represented by 149 basidiomycetes, 19 ascomycetes and 5 myxomycetes (Heredia 1989; Valenzuela and Chacón-Jiménez 1991).

In the region, particularly in the cloud forest, plants of diverse phytogeographic affinities coexist. Cloud forest habitat harbours one of the richest floras in Mexico. This region's cloud forest is notably unique because taxa (families, genera, species) from so many varied geographic distributions and origins come together (Hernández-Xolocotzi *et al.* 1951). Of the vascular plant families represented, 56% are tropical, 20% temperate, 19% cosmopolitan and 5% with other distributions. Representative examples among species include: broadly neotropical *Bursera simaruba, Callicarpa acuminata, Cedrela odorata, Celtis iguanea* and *Nectandra sanguinea*; Mesoamerican neotropical *Aphananthe monoica (Mirandaceltis monoica), Bomarea acutifolia, Cupania glabra* and *Rapanea myricoides*; Mexican neotropical *Gymnanthes longipes* and *Oyedaea ovalifolia* and montane neotropical *Beilschmiedia mexicana, Clethra pringlei, Magnolia schiedeana* and *Meliosma oaxacana*; and holarctic *Acer skutchii, Carpinus caroliniana, Carya ovata, Cercis canadensis* and *Taxus globosa* (Puig, Bracho and Sosa 1987).

Generic phytogeographic affinities or lineages considered in a broad perspective add to the northern North American and/or South American genera some representatives from Asia – e.g. *Rhus, Dendropanax, Berberis, Cedrela, Bocconia, Meliosma, Ternstroemia* and *Aphananthe*; and as well from Africa – e.g. *Bursera, Dioscorea* and *Trichilia* (cf. Puig, Bracho and Sosa 1987; Rzedowski 1993).

There are also broad to narrow endemics in the region. *Quercus germana, Q. xalapensis* and *Ternstroemia sylvatica* are endemic to the Sierra Madre Oriental, whereas *Wimmeria concolor, Bernardia interrupta* and *Decatropis bicolor* are endemic to north-eastern Mexico (Puig, Bracho and Sosa 1987). The following are endemic to the Sierra de Guatemala: *Louteridium tamaulipense, Eupatorium richardsonii, Senecio richardsonii, Verbesina richardsonii, Omphalodes richardsonii, Comarostaphylis sharpii, Acalypha tamaulipensis, Phyllanthus barbarae* and *Abutilon procerum* (Johnston *et al.* 1989).

Various species are considered threatened overall or in the region, such as the cacti *Ariocarpus retusus, A. trigonus, Ferocactus hamatacanthus, Obregonia denegrii* and *Pelecyphora aselliformis*; the cycads *Ceratozamia kusteriana* and *Dioon edule*; the orchids *Laelia speciosa* and *Lycaste deppei*; and *Abies vejari, Acalypha tamaulipensis, Diospyros riojae, Magnolia schiedeana, Taxus globosa* and *Tillandsia ionantha* (Lucas and Synge 1978; Vovides 1981; Chimal *et al.* 1989; SEDUE 1991).

Useful plants

From a report on useful plants of Tamaulipas (Hernández-Sandoval, González and González-Medrano 1991), the many species and uses that follow were derived for the

Gómez Farías region: 167 medicinals, 98 edible, 11 fodder, 5 energy sources, 84 timbers, 16 industrial usage and 69 ornamentals. Several native species are grown in family gardens to provide spices or medicinals. In the past, cloud forest and oak and pine forests were exploited for their timber, and they are now recovering; these forests harbour at least 14 oak and four pine species.

In the tropical forests are several species with commercial value, such as *Bursera simaruba* ("gumbo limbo") – wood, resin, incense; *Enterolobium cyclocarpum* ("guanacaste") – timber, fodder; *Cedrela odorata* ("cedro") – cabinet wood; *Tabebuia pentaphylla* ("cinco hojas") – timber; *Brosimum alicastrum* ("ojite" or "ramón") – wood, food (edible seeds, potable latex); and *Lysiloma divaricata* ("rajador") – timber.

The desert scrublands have various species of local importance, such as *Helietta parvifolia* – timber; *Acacia berlandieri*, *Gochnatia hypoleuca*, *Opuntia* spp. – fodder; *Dasylirion* spp. – edible, alcohol fermentation, ornamental; *Brahea berlandieri* – house-building; *Agave* spp. (e.g. *A. lechuguilla*), *Yucca carnerosana* – fibre; *Quercus* spp., *Rhus microphylla*, *R. virens*, *Krameria ramosissima* – tannin; and *Turnera diffusa*, *Chrysactinia mexicana*, *Hesperozygis marifolia*, *Jatropha dioica*, *Larrea tridentata* – medicinal (Chimal *et al.* 1989).

Leaves of the small *Chamaedorea* palms ("camedor") (e.g. *C. radicalis*, "palmilla") are collected in the area's tropical and cloud forests and exported to the U.S.A. for floral arrangements – a frequent activity in the Gómez Farías region, providing income for the local inhabitants (cf. Marshall 1989). On a much smaller scale, *Magnolia schiedeana* flowers and *Ternstroemia sylvatica* ("trompillo") fruits are occasionally collected.

Social and environmental values

The Gómez Farías region provides water for villages and the sugar industry in the lowlands, and is the source for irrigation of a vast agricultural zone. The mountain vegetation prevents soil erosion downstream by moderating large volumes of water (from rainfall and mist condensation) that steadily flow into the Guayalejo river system. The Guayalejo River is a tributary of the Pánuco River, which has one of the five most important river basins in the country in surface and draining volume (Contreras 1991).

Because of its environmental heterogeneity and transitional location between the Neotropical and Nearctic biogeographical regions, the Gómez Farías fauna is diverse and unique for Mexico. The terrestrial vertebrate species consist of 24 amphibians, 65 reptiles, 182 resident and seasonal birds and more than 80 mammals (Wood 1975b; Hernández-M. 1989). Two endemic rodents live in the area (*Peromyscus ochraventer*, *Neotoma angustapalata*) (Hooper 1953). There are populations of taxa threatened overall or in the region – such as black bear, jaguar, margay, ocelot, jaguarundi, great curassow, military macaw, white-crowned parrot and boa constrictor.

The Gómez Farías region is included in two Endemic Bird Areas (EBAs); it is the extreme southern portion of the Highland Sierra Madre Oriental EBA (A06) and on the western edge of the Lowland Gulf Slope EBA (A07). The region is particularly important for parrots. There are

historical records for the maroon-fronted parrot (*Rhynchopsitta terrisi*), which is a threatened species of mixed conifer forests between 1800–3500 m in a narrow (300 × 60 km) section of the Sierra Madre Oriental. In the lowlands occur the restricted-range green-cheeked or red-crowned parrot (*Amazona viridigenalis*) and threatened yellow-headed parrot (*A. oratrix*). Other restricted-range species in the lower elevations of Gómez Farías are the crimson-collared grosbeak (*Rhodothraupis celaeno*) and Tamaulipas crow (*Corvus imparatus*).

El Cielo Biosphere Reserve protects cultural concerns as well as the biological diversity. At the time of Spanish conquest, nomadic peoples inhabited the region's lowlands (Pisones, Janambres, Pames, Siquillones). Maps from the 16th and 17th centuries note the names of Huasteco settlements in the Guayalejo-Tamesi Basin. Artefacts and sites from Amerindian cultures have been found through the Sierra Chiquita cloud forest and lowlands. Archaeological studies in southern Tamaulipas caves reveal that early cultures practised agriculture with squash, beans and chili peppers (MacNeish 1964).

In more recent times, with the expansion of timber companies into the region, settlers from western Mexico's Michoacán State moved into the mountains to work the forests. El Cielo BR has a research station, and is an excellent base for conducting cultural or social as well as biological research, and for environmental education.

Bird watching has been a long-standing activity in the region. Throughout the general area, spectacular landscapes and caves can be enjoyed. With appropriate management, the Gómez Farías region could sustain small-scale ecotourism (cf. Wood 1975a, 1975b).

Economic assessment

About 3500 people live in the Gómez Farías region (1990 census), in one village and 26 small rural settlements. The region is peripheral to the economic activities of Tampico, Ciudad Mante and the state capital Ciudad Victoria. During colonial times the forests of the Sierra Madre Oriental supplied raw materials to towns and mines. On the flat terrain of the adjacent Gulf Coast Plain developed especially the sugarcane industry and cattle-ranching. In the 19th century railroads helped integrate the area with broader markets. During the 1930s, with government investment and private capital, settlements proliferated and agricultural production diversified. Many important crops are now produced in the region such as maize, sorghum, soybeans, rye, tomatoes, other vegetables and cotton (Pepin-Lehalleur 1986).

Threats

In lowlands throughout the region, the remaining environmental diversity and biological richness are threatened by modern agricultural and industrial development, such as the highly technological systems of less diversification and mass production. Changes in the landscape are caused by hydrological works and the conversion to vast croplands. Clearing is almost total in the Sierra Madre Occidental valleys and on gentle slopes, and in the adjacent plains.

In some areas, environmental degradation is associated with the typical expansion of settlements; various problems are proportional to the settlement size, such as land clearing, improper soil use, landscape modification, water and air pollution and solid-waste disposal. The pollution of water is mainly from by-products of sugarcane industrialization.

The construction of a road between the villages Gómez Farías and El Azteca is expected, which will separate the tropical semi-deciduous forest from the montane vegetation types – with unknown ecological consequences. The tropical semi-deciduous forest is already under the most pressure, because it is closer to the major populated areas – it urgently needs conservation action.

In the highlands, deforestation is a possible threat as a result of shifting agriculture as well as logging. The subsistence economy of the mountain residents is based on cultivating a plot of 3–5 ha for food, gathering fuelwood, logging for home construction and selectively harvesting forest products for income. Intensive timber extraction in the region took place from 1952 to 1973, but logging has diminished and become local. Timber companies persistently seek permits to log the buffer zone of El Cielo Biosphere Reserve.

Livestock impact vegetation and soil, such as the free-roaming cattle in the forests. Browsing by goats is changing the vegetation structure and floristic composition of desert scrubs.

Despite federal and state laws regulating conservation of wild fauna and flora, over-exploitation of plants and animals occurs in El Cielo Biosphere Reserve. Particular regulations for the use of species in El Cielo BR need to be enacted and enforced. There are problems for example with collection of orchids and cacti, extraction of palm leaves and gathering of fuelwood.

Of basic concern is the scarce attention from government agencies and managers for the reserve. Budgets for maintenance have been reduced, the protection of identified threatened species has diminished, and support is lacking for environmental education and research programmes.

Conservation

Rancho del Cielo began as the remote homestead of the Canadian J.W.F. Harrison (1901–1966), a self-taught botanist and horticulturist (growing tuberous begonias and amaryllises) who deeded his land to a non-profit Mexican corporation with the stipulation that the cloud forest be preserved. Beginning in 1964, the area has served as a biological research station (at 1140 m) for Texas Southmost College, of Brownsville in U.S.A. Some of its funding has come from visiting bird-watching enthusiasts (Wood 1975b).

In 1985 the State of Tamaulipas declared 1445.3 km² as El Cielo Biosphere Reserve, which helps to protect c. 60% of the Gómez Farías region. In 1987 the reserve was officially recognized within the UNESCO-MAB reserve network. The reserve has two core zones: the larger area (287 km²) includes a broad portion of cloud forest and a strip through the region's altitudinal gradient; the smaller area (78 km²) includes especially tropical forests. The region's brushlands are least represented within the reserve.

Because of the rugged terrain, the northern, western and southern areas of the Gómez Farías region are less accessible and less threatened by development. The region remained nearly undisturbed until the 1950s, when the major logging began. Cycads, cacti and orchids are regulated by CITES, with the genera *Ceratozamia, Ariocarpus, Obregonia* and *Pelecyphora* in Appendix I. The difficult access and low human population within the reserve have kept this area relatively pristine, despite limited enforcement.

References

Bracho, R. and Sosa, V.J. (1987). Edafología. In Puig, H. and Bracho, R. (eds), *El bosque mesófilo de montaña de Tamaulipas*. Instituto de Ecología, Public. 21, Mexico, D.F. Pp. 29–37.

Castro-Meza, B.I., Plácido de la Cruz, J.M. and Almaguer, S.P. (1989). Fisiografía y riesgo de erosión en la Reserva de la Biósfera "El Cielo". *Biotam, Univ. Autón. Tamaulipas, Méx.* 1(1): 28–37.

Chimal, A., González-Medrano, F., Díaz, I., Hernández, A., Noriega, R., Bravo, E., Pérez, J. and Vázquez, J. (1989). *Investigación sobre flora y fauna silvestres de la Reserva de la Biósfera "El Cielo", Tamaulipas*. Secretaría de Desarrollo Urbano y Ecología (SEDUE) and Universidad Autónoma Metropolitana-Xochimilco, Mexico, D.F. 106 pp. Unpublished technical report.

Contreras, A. (1991). *Conservación, producción y desarrollo rural: el caso de la Reserva de la Biósfera "El Cielo", Tamaulipas, México*. Thesis, División de Ciencias Sociales y Humanidades, Universidad Autónoma Metropolitana-Xochimilco, Mexico, D.F. 133 pp.

Ferrusquía-Villafranca, I. (1993). Geology of Mexico: a synopsis. In Ramamoorthy, T.P., Bye, R., Lot, A. and Fa, J.E. (eds), *Biological diversity of Mexico: origins and distribution*. Oxford University Press, New York. Pp. 3–107.

Heredia, G. (1989). Estudio de los hongos de la Reserva de la Biósfera El Cielo, Tamaulipas. Consideraciones sobre la distribución y ecología de algunas especies. *Acta Bot. Méx.* 7: 1–18.

Hernández-M., A. (1989). Importancia de la reserva "El Cielo" para los mamíferos de Tamaulipas. *Biotam, Univ. Autón. Tamaulipas, Méx.* 1: 13–20.

Hernández-Sandoval, L., González, C. and González-Medrano, F. (1991). Plantas útiles de Tamaulipas. *Anales Inst. Biol. Univ. Nac. Autón. México, Ser. Bot.* 62: 1–38.

Hernández-Xolocotzi, E., Crum, H.A., Fox Jr., W.B. and Sharp, A.J. (1951). A unique vegetation area in Tamaulipas. *Bull. Torrey Bot. Club* 78: 458–463.

Hooper, E.T. (1953). Notes on mammals of Tamaulipas, Mexico. *Occas. Papers Mus. Zool. Univ. Mich.* 544: 1–12.

Johnston, M.C., Nixon, K., Nesom, G.L. and Martínez, M. (1989). Listado de plantas vasculares conocidas de la Sierra de Guatemala, Gómez Farías, Tamaulipas, México. *Biotam, Univ. Autón. Tamaulipas, Méx.* 1(2): 21–33.

Lof, L.V. (1980). *The ferns of the Rancho del Cielo region.* M.S. thesis, Pan American University, Edinburg, Texas.

Lucas, G. and Synge, H. (1978). *The IUCN plant red data book.* IUCN, Morges, Switzerland. 540 pp.

MacNeish, R.S. (1964). Food-gathering and incipient agriculture stage in prehistoric Middle America. In Wauchope, R. (ed.), *Handbook of Middle American Indians*, Vol. 1. West, R.C. (ed.), *Natural environment and early cultures.* University of Texas Press, Austin. Pp. 413–426.

Marshall, N.T. (1989). Parlor palms. Increasing popularity threatens Central American species. *TRAFFIC USA* 9(3): 1–3.

Martin, P.S. (1958). A biogeography of reptiles and amphibians in the Gómez Farías region, Tamaulipas, México. *Misc. Public. Mus. Zool. Univ. Mich.* 101: 1–102.

Pepin-Lehalleur, M. (1986). *Formación y dinámica de un sistema agrario regional: la región del Mante, Tamaulipas.* El Colegio de México, Mexico, D.F. 24 pp. Unpublished report.

Puig, H. (1976). *Végétation de la Huasteca, Mexique.* Etudes mésoaméricaines Vol. V. Mission Archeologique et Ethnologique Française au Mexique. Mexico, D.F. 531 pp.

Puig, H. and Bracho, R. (1987). Climatología. In Puig, H. and Bracho, R. (eds), *El bosque mesófilo de montaña de Tamaulipas.* Instituto de Ecología, Public. 21, Mexico, D.F. Pp. 39–54.

Puig, H., Bracho, R. and Sosa, V.J. (1983). Composición florística y estructura del bosque mesófilo en Gómez Farías, Tamaulipas, México. *Biótica* 8: 339–359.

Puig, H., Bracho, R. and Sosa, V.J. (1987). Affinités phytogéographiques de la fôret tropicale humide de montagne de la réserve MAB "El Cielo" de Gómez-Farías, Tamaulipas, Mexique. *Compt. Rend. Séances Soc. Biogéogr.* 63: 115–140.

Rzedowski, J. (1978). *Vegetación de México.* Editorial Limusa, Mexico, D.F. 432 pp.

Rzedowski, J. (1993). Diversity and origins of the phanerogamic flora of Mexico. In Ramamoorthy, T.P., Bye, R., Lot, A. and Fa, J.E. (eds), *Biological diversity of Mexico: origins and distribution.* Oxford University Press, New York. Pp. 129–144.

SEDUE (1991). Listado de especies raras, amenazadas, en peligro de extinción, o sujetas a protección especial, y sus endemismos en la República Mexicana. Flora terrestre y acuática. *Diario Oficial* 17/05/91: 9–24.

Sosa, V.J. (1987). Generalidades de la región de Gómez Farías. In Puig, H. and Bracho, R. (eds), *El bosque mesófilo de montaña de Tamaulipas.* Instituto de Ecología, Public. 21, Mexico, D.F. Pp. 15–28.

Valenzuela, R. and Chacón-Jiménez, S. (1991). Los poliporáceos de México. III. Algunas especies de la Reserva de la Biósfera El Cielo, Tamaulipas. *Rev. Mex. Mic.* 7: 39–70.

Valiente-Banuet, A. (1984). *Análisis de la vegetación de la región de Gómez Farías, Tamaulipas.* Tesis de Licenciatura, Facultad de Ciencias, Universidad Nacional Autónoma de México, Mexico, D.F. 92 pp.

Vovides, A.P. (1981). Lista preliminar de plantas mexicanas raras o en peligro de extinción. *Biótica* 6: 219–228.

Wood, P. (1975a). A glorious botanical confusion. In Jackson, D.D., Wood, P. and the editors of Time-Life Books, *The Sierra Madre.* Time-Life Books, New York. Pp. 104–117.

Wood, P. (1975b). Where the birds are. In Jackson, D.D., Wood, P. and the editors of Time-Life Books, *The Sierra Madre.* Time-Life Books, New York. Pp. 84–103.

Authors

This Data Sheet was written by Vinicio J. Sosa, Arturo Hernández and Armando Contreras (Instituto de Ecología, Apdo. Postal 63, 91000 Xalapa, Veracruz, Mexico).

CUATRO CIENEGAS REGION
Mexico

Location: Intermontane basin in central Coahuila State, northern Mexico, between approximately latitudes 26°45'–27°05'N and longitudes 102°05'–102°20'W.

Area: c. 2000 km².

Altitude: From c. 740 m on basin floor to over 3000 m in Sierra de la Madera.

Vegetation: Grasslands with aquatic, semi-aquatic and gypsum-dune habitats in valley; desert scrub and chaparral on mountain slopes; oak-pine woodlands, and montane forests of pine, fir and Douglas fir.

Flora: Diverse flora of eastern Chihuahuan Desert, rich in endemics; 860 native species in 458 genera of 114 families, with 23 species endemic; threatened species.

Useful plants: Principally for grazing and forestry; also medicinals.

Other values: Tourism, aquatic resources, refuge for fauna – including relict populations and at least 33 endemics; threatened species.

Threats: Logging, mining, extraction of gypsum, grazing, agriculture, as well as contamination and depletion of water.

Conservation: None.

Geography

The region is within the Coahuilan subprovince of the Chihuahuan-Coahuilan Plateaux and Ranges morphotectonic province (Ferrusquía-Villafranca 1993). The Bolsón de Cuatro Ciénegas is a naturally closed-drainage intermontane basin measuring c. 40 km east-west and 25–30 km north-south (cf. Pinkava 1979). It is nearly bisected by the north-jutting Sierra de San Marcos (Map 21). North-centrally in the basin (at 26°59'N, 102°02'W), the town of Cuatro Ciénegas de Carranza prospers near the mouth of the Cañón River, with its fresh water flowing through Cañón del Agua, which is the northern portal between Sierra de la Madera and Sierra de Menchaca.

The basin is bounded on the east by Sierra San Vicente and Sierra de la Purísima, and on the west by Sierra de la Fragua. A railway (Ferrocarriles Nacionales de México) roughly parallels an east-west road across the basin from Puerto Salado to Puerto de Jora. Mexico Highway 30 extends across the municipality of Cuatro Ciénegas southward along the western flank of Sierra de San Marcos toward San Pedro by way of Puerto San Marcos. Minckley (1969) stated that the basin has abundant water, much of it subterranean.

There are seven major above-ground drainage systems, the largest being the San Marcos River (locally the Mesquites River). According to Rodríguez-González (1926), the valley had no outlet but drained through ponds into a large depression in the eastern lobe of the basin, giving the town of Cuatro Ciénegas its name ("Four Marshes"). However, the basin is now drained by a series of constructed canals which ultimately pass eastward through Puerto Salado to the Río Grande (= Río Bravo del Norte). The basin's rivers originate as cool or usually thermal springs (25°–38°C) emerging from travertine-lined tubes or from pits ("pozos", locally "posos")

(Minckley 1969). Churince River originates in a large poso and terminates in a large shallow mineralized lake, Laguna Grande. Here evaporation results in the production of nearly pure gypsum salts which dry along the shores and are blown into a very complex series of dunes of varied ages, particularly to the north and west.

There are also subterranean channels, notably along the bases of mountains. Localized foundering of channel roofs has resulted in hundreds of posos. This may be due to lowering of the water table and general sag of the basin floor (Minckley 1969). The posos vary in depth from less than 1 m to more than 10 m, and in diameter from but a few cm to more than 200 m by slumping of their walls; the largest generally are called "lagunas". Progressive foundering of subterranean channels in linear series of posos leads to open channels occupied by swift-flowing streams.

Usually a poso has inflow and outflow. If the outflow becomes plugged, the overflowing water produces travertine deposits, or even cone springs such as Poso Escobeda. If the inflow stops, eutrophication follows, resulting in extensive marshes (Minckley 1969). Large downflow lagunas salinize by evaporation and form miry lakes or playas such as Laguna Grande and Laguna Salada. Waterways are subject to great modification by salts, principally carbonates and sulfates (Minckley and Cole 1968; Wood 1975).

The Cañón River, however, is very different. Its fresh cool water forms well-developed pools and riffles, and supports a rich vegetation along its banks. It is the source of water for the town and irrigation. Although the central basin's rivers are also used for irrigation, some fallow fields south of town blown into dunes apparently testify to accumulation of salts from those sources.

Well-developed and arroyo-dissected alluvial slopes ("bajadas") rise above the basin floor. Towering above them are the massive limestone sierras, part of a system of mountains trending north-west to south-east that form a barrier, keeping moisture-laden winds from the Atlantic Ocean and Gulf of Mexico from passing on to the arid Central Plateau. The mean annual precipitation in the basin is less than 200 mm (Shreve 1944). Air temperatures range locally from below 0°C in winter to over 44°C in summer (Marsh 1984).

Vegetation

The vegetation of the Cuatro Ciénegas region may be divided into the following major groupings (Pinkava 1984):

1. Basin grassland
The saline valley floor supports an extensive grassland dominated by such salt-tolerant species as *Distichlis stricta*, *Monanthochloe littoralis* and *Sporobolus airoides* – largely replaced in less saline areas by *S. wrightii* and *S. spiciformis*. Portions of the grassland are cultivated or used for pasturage. Secondary succession on fallow or abandoned fields usually results in open stands of species of *Atriplex*, *Suaeda* and *Prosopis* (if dunes form) before returning to grassland.

2. Aquatic and semi-aquatic habitats
A series of vegetational changes may be witnessed indirectly by observing posos of varying ecological maturities. The slumping banks of the newly founded poso are soon populated by sedges (*Eleocharis*, *Carex*), additional grasses (*Phragmites*, *Spartina*, *Setaria*) and species of *Polygala*, *Eustoma* and *Flaveria*. The poso enlarges and matures and ultimately contains the aquatics *Nymphaea ampla*, *Utricularia obtusa* and *Chara*. The shores support *Fimbristylis thermalis*, *Fuirena simplex*, *Heliotropium curassavicum*, *Bacopa monnieri*, *Ludwigia octovalvis*, *Anemopsis californica*, *Ipomoea sagittata*, *Eupatorium betonicifolia*, and the trees and shrubs *Prosopis glandulosa* (mesquite), *Acacia greggii*, *Fraxinus berlandieriana* and *Salix nigra*.

3. Gypsum dunes
White gypsum salts blown from evaporating lakebeds form the dunes. Active dunes up to 6–9 m high encroach upon streams, posos, older dunes and the surrounding plains. Important in stabilizing the dunes are mesquite trees (often nearly completely buried in the sand) and *Acacia greggii*, *Yucca treculeana* and *Varilla mexicana*.

Upon stabilization, a greyish crust forms at the surface, apparently by recrystallization. Occupying the dunes are gypsophilous species, particularly some bizarre endemic Compositae – *Machaeranthera restiformis*, *M. gypsophila*, *Gaillardia gypsophila*, *Dyssodia gypsophila* and *Haploesthes robusta*. Also present are *Selinocarpus purpusianus*,

MAP 21. CUATRO CIENEGAS REGION (CPD SITE MA10), MEXICO (after Minckley 1969)

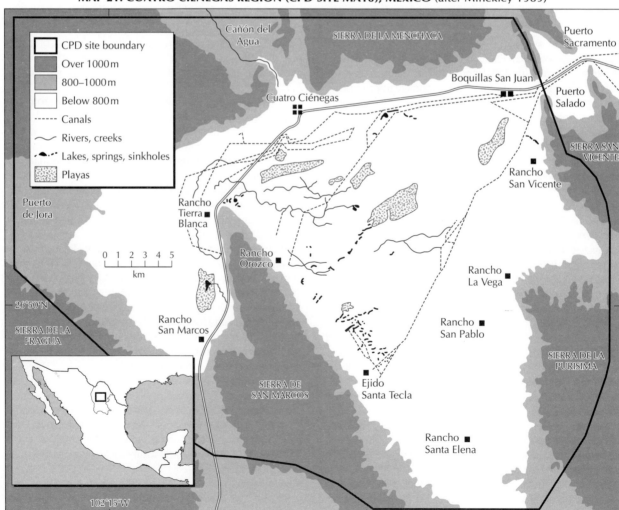

CPD Site MA10: Cuatro Ciénegas region, Mexico. Dunes formed from white gypsum salts blown from evaporating lakebeds. Photo: Patricia Almada-Villela.

Nerisyrenia incana, Petalonyx crenatus, Chamaesyce astyla, Fouquieria splendens and *Echinocereus enneacanthus*.

4. Transition zone

Encircling the basin and its lagunas is an interrupted band of shrubs and trees. These small islands or "mogotes" are comprised of *Condalia warnockii, Suaeda palmeri, Allenrolfea occidentalis, Atriplex canescens, Acacia greggii, A. neovernicosa* and *Prosopis glandulosa*.

5. Desert scrub

The bajadas and the lower mountain slopes and arroyos support a rich shrub flora. Characteristic are *Larrea tridentata* with varying combinations of co-dominants – *Agave lechuguilla, A. striata* subsp. *falcata, Hechtia scariosa, Opuntia bradtiana, Yucca rostrata, Cordia parviflora, Jatropha dioica, Euphorbia antisyphilitica, Parthenium incanum* and *Selaginella lepidophylla*. In certain areas *Flourensia cernua* and *Sericodes greggii* largely replace the *Larrea*.

6. Chaparral zone

Chaparral is widespread, but best developed on northerly and easterly exposures in protected arroyos and canyons and in areas near montane forests. Its often dense growth consists of oaks (*Quercus hypoxantha, Q. intricata, Q. invaginata, Q. pringlei*), heaths (*Arbutus texana,*

Arctostaphylos pungens), scattered pines (*Pinus remota*) and numerous shrubs and small trees.

On drier sites and on south-facing slopes occur *Quercus hypoxantha, Q. intricata, Flourensia retinophylla, Sophora secundiflora, Cercocarpus montanus, Nolina cespitifera* and species of *Salvia, Agave* and *Dasylirion*.

7. Pine-oak and oak woodlands

On higher, moister slopes in protected canyons and at cliff bases, chaparral grades into woodlands of oaks (*Quercus gravesii, Q. glaucoides, Q. pringlei*), pinyons (*Pinus johannis*) and junipers (*Juniperus erythrocarpa* var. *coahuilensis, J. flaccida*). Associates include *Garrya ovata, Cercocarpus* spp., *Rhamnus betulifolia, Prunus serotina, Arbutus texana, Vitis arizonica* and *Ptelea trifoliata*. The palm *Brahea berlandieri* borders pine-oak woodlands, chaparral and lower montane forests on mid-slopes of the sides of north-facing canyons.

8. Montane forest

The mesic montane forest is best developed in the upper north-facing canyons of the sierras. Forming the canopy are *Pseudotsuga menziesii, Abies durangensis* var. *coahuilensis* and *Pinus strobiformis*. The understorey includes primarily *Quercus gravesii, Arbutus texana, Acer grandidentatum* and *Cornus sericea*. The forest floor is moss-covered, with only scattered herbs and small shrubs.

On more exposed areas *Pinus arizonica* and *Cupressus arizonica* predominate. Below the montane forest on north-facing slopes are pine-oak and oak forests or chaparral. On exposed drier south-facing slopes below the crest and cliffs are scattered pines (*Pinus arizonica, P. remota*), oaks (*Quercus intricata, Q. bypoxantha, Q. greggii*) and shrubs (*Xerospiraea hartwegiana, Abelia coriacea, Arctostaphylos pungens, Garrya ovata, Nolina cespitifera* and species of *Agave, Opuntia* and *Dasylirion*).

Flora

The native vascular flora of the Bolsón de Cuatro Ciénegas and the surrounding sierras (except the unstudied Sierra de la Fragua) consists of 879 taxa distributed among 860 species in 458 genera of 114 families (Pinkava 1984). Plants normally cultivated but distant from habitation account for an additional 6 genera and species in 3 families.

The basin is known to support a flora and fauna rich in endemics (Marsh 1984). The very rich flora is one of the most varied in the Chihuahuan Desert; 49 taxa have been described with type localities in the region. Of these, 45 are considered accepted taxa, and 23 endemic (Pinkava 1984). An even greater number of taxa are only found in the region and immediately adjacent areas. Concentrations of endemic taxa include: (1) c. 12 endemics in the Laguna Grande-Laguna Churince complex, northward to the Mesquites River, eastward onto the north tip of Sierra de San Marcos; (2) c. 10 endemics in canyons and bajadas of canyons of the Sierra de la Madera's east-facing slope to the crest and partly down the west-facing slope; and (3) an additional 2 endemic taxa in south-facing bajadas and canyons just above Poso Anteojo. These areas need to be correlated with the endemic fauna and afforded protection, in coordination with the people living there.

Useful plants

Many plants of this flora are used in traditional medicine in the rural communities. The extraction of wood from the Sierra de la Madera is the most extensive exploitation, which little by little diminishes the sparse forest, and puts in danger the population of *Abies durangensis* var. *coahuilensis*. The endemic species are potential genetic resources whose uses and properties should be studied.

Social and environmental values

The Bolsón de Cuatro Ciénegas is unique in having such notable posos of thermal waters and white dunes of gypsum. There are only three such dune formations in North America. Since c. 1964 the striking scenery has attracted the attention of people who consider this oasis a place for recreation (Wood 1975).

The fauna and flora are special, including at least 56 endemics. Some species have been studied considerably; the c. 33 faunal endemics include fishes, turtles, lizards, crustaceans, snails and scorpions. Establishing a biological station in the region would help safeguard the biota, and would enable future generations to learn about the biota, which readily exhibits an interesting array of evolutionary processes such as

isolation, adaptation, sympatry, niche segregation, character displacement, speciation and endemism (Taylor and Minckley 1966; Marsh 1984).

Threats

The region has been occupied by agricultural people since the 16th century, but remained little developed by 1959. Contreras-Balderas (1984) has assessed the impacts of people on the environment of the Cuatro Ciénegas Basin. Its ecosystems and biota are now suffering accelerating anthropogenic damage because of the drainage of water, temperature imbalances and losses of habitat, population and species. The gypsum dunes are undergoing extraction and destruction, since 1979 on an industrial scale, affecting endemic species. Other notorious effects are irrational development of agriculture and pasturage, overgrazing by goats and horses and depositions of trash. The exploitation of Cactaceae is frequent for sale in the U.S.A., Europe and Japan. Increasing needs of the increasing human population threaten to escalate damage to local ecosystems and to endemic species, possibly causing extinctions.

Conservation

Mexico's Secretaría de Medio Ambiente, Recursos Naturales y Pesca (SEMARNAP) (formerly in SEDESOL), with funding from the World Bank, is studying the region to establish policies for protection and conservation, and planning to develop rationale use and management, including restoration and reforestation efforts (A. Rodríguez 1993, pers. comm.). Because of the rich desert biota, the large number of endemic taxa of plants and animals, the unusual and fragile habitats, as well as the intriguing scenery, there is urgent need to preserve representative portions of this region. This remarkable natural heritage is being degraded more and more, and much could be lost (Marsh 1984; Pinkava 1987). The Cuatro Ciénegas Basin is included in The (U.S.) Nature Conservancy's Parks in Peril Program to help establish a long-term base for preservation (TNC 1990).

References

Contreras-Balderas, S. (1984). Environmental impacts in Cuatro Ciénegas, Coahuila, México: a commentary. *J. Ariz.-Nev. Acad. Sci.* 19: 85–88.

Ferrusquía-Villafranca, I. (1993). Geology of Mexico: a synopsis. In Ramamoorthy, T.P., Bye, R., Lot, A. and Fa, J.E. (eds), *Biological diversity of Mexico: origins and distribution*. Oxford University Press, New York. Pp. 3–107.

Marsh, P.C. (ed.) (1984). Biota of Cuatro Ciénegas, Coahuila, México. Proceedings of a special symposium. *J. Ariz.-Nev. Acad. Sci.* 19(1): 1–90.

Minckley, W.L. (1969). *Environments of the Bolsón of Cuatro Ciénegas, Coahuila, México, with special reference to the aquatic biota*. University of Texas, El Paso. Science Series No. 2. 65 pp.

Minckley, W.L. and Cole, G.A. (1968). Preliminary limnologic information on waters of the Cuatro Cienegas Basin, Coahuila, Mexico. *Southwestern Nat.* 13: 421–431.

Pinkava, D.J. (1979). Vegetation and flora of the Bolsón of Cuatro Ciénegas region, Coahuila, México – I. *Bol. Soc. Bot. Méx.* 38: 35–73.

Pinkava, D.J. (1984). Vegetation and flora of the Bolsón of Cuatro Ciénegas region, Coahuila, México: IV. Summary, endemism and corrected catalogue. *J. Ariz.-Nev. Acad. Sci.* 19: 23–47.

Pinkava, D.J. (1987). An urgent need for preservation: Cuatro Cienegas. *Agave* 2: 7–9.

Rodríguez-González, J. (1926). *Geografía del estado de Coahuila.* Soc. Edición y Librería Franco-Americana, Mexico, D.F.

Shreve, F. (1944). Rainfall of northern Mexico. *Ecology* 25: 105–111.

Taylor, D.W. and Minckley, W.L. (1966). New world for biologists. *Pacific Discovery* 19(1): 18–22.

TNC (1990). *Parks in Peril: a conservation partnership for the Americas.* The Nature Conservancy (TNC), Arlington, Virginia. 24 pp.

Wood, P. (1975). A nature walk in Cuatro Ciénegas Basin. In Jackson, D.D., Wood, P. and the editors of Time-Life Books, *The Sierra Madre.* Time-Life Books, New York. Pp. 68–83.

Authors

This Data Sheet was written by Dr Donald J. Pinkava (Arizona State University, Department of Botany, Tempe, Arizona 85287-1601, U.S.A.) and José A. Villarreal-Quintanilla (Universidad Autónoma Agraria Antonio Narro, Departamento de Botánica, C.P. 25315, Buenavista, Saltillo, Coahuila, Mexico).

APACHIAN/MADREAN REGION OF SOUTH-WESTERN NORTH AMERICA
Mexico and U.S.A.

Location: Largely in north-west Mexico from Sinaloa border with Sonora and Chihuahua, centred on continental divide northward to south-east Arizona and south-west New Mexico in U.S.A.; about latitudes 26°–32°N and longitudes 111°–107°W.

Area: Approximately 180,000 km² – 600 km long and 300 km west to east.

Altitude: 500–3500 m; highest are Sierra Mohinora in south, several peaks c. 3000 m in north.

Vegetation: Nine physiognomic vegetation types: Madrean montane coniferous forest; oak-coniferous woodland; barrancan oak woodland; oak savanna; Madrean chaparral; short-grass prairie; tropical deciduous forest; subtropical thorn scrub; and subtropical desert fringe.

Flora: Northern portion of Sierra Madre Occidental phytogeographic province, where Apachian and Madrean floristic districts interdigitate. Approximately 4000 species of vascular plants; many endemics; a large number of species at northern limits, many at eastern or western limits.

Useful plants: At least 700–1000 wild plants useful; at least 250 estimated wild congeners of major crops; highest diversity of crop land-races of 18 pre-Columbian cultivated species north of neotropics. One endangered domesticate – *Panicum sonorum*; threatened medicinal plants.

Other values: Watershed protection; remote homeland of some of Mexico's least industrialized Amerindians; source of new industrial and medicinal crops; refuge for bats, Mexican wolf, thick-billed parrot, numerous threatened fishes and other vertebrates; ecotourism.

Threats: Pulping and lumber development (US$400–600 million funded) concomitant with North American Free Trade Agreement (NAFTA); species-selective over-harvesting of trees; erosion from agriculture, overgrazing and forest replacement with pastures of exotic grasses; extensive open-pit mining; increasing illicit drug traffic, associated violence and subjugation of native peoples to produce contraband.

Conservation: Much of region in U.S.A. variously protected – U.S. Forest Service for most montane habitats; also several National Monuments, private foundation for Animas Mountains. Less than 10% of area in Mexico protected – one National Park (40 km²), some mountains under SEMARNAP, several mountainous and other regions protected by private owners. Several areas designated for protection by State of Sonora. Informal conservation of less than 1% of region by Amerindian communities.

Geography

The region represents two floristic districts of the Madrean province centred on the Sierra Madre Occidental of northern Mexico (Map 22). Within or associated with this region are myriad individual ranges forming the Sierra Madre in north-western Mexico and more than two dozen major mountain ranges in south-eastern Arizona and far south-western New Mexico.

The Apachian district extends northward to the Mogollon Rim in south-western U.S.A. This district is represented by a northern archipelago of biological "sky-island" mountain ranges, including the Pinaleño, Galiuro, Santa Catalina, Santa Rita, Baboquivari, Patagonia, Huachuca, Chiricahua, Animas and smaller ranges. Also included is the "Deming Bridge" surrounding the Chiricahua and Animas mountains – it is the lowest place on the continental divide between Mexico and Canada. The adjacent mid-elevations and sky-island peaks in northern Mexico form the southern portion of the Apachian district, which merges with the northern reaches of the Madrean district. The Madrean district is characterized by the cordilleran flora between the Sierra Mohinora in the State of Chihuahua near the Sinaloa-Durango border and the northernmost edge of the Sierra Madre Occidental proper, c. 150 km south of the U.S. border.

Volcanic tuff and Laramide limestones dominate the surface geology. The region is drained to the west primarily by the Río Yaqui, Río Mayo and Río Fuerte watersheds, and to the east by the Río Bravo (Rio Grande) and Río Conchos. The northern outlier archipelago in south-eastern Arizona falls largely within the Gila River drainage. Precipitation ranges from 300 mm per year to more than 1200 mm, with summer monsoons predominating among the seasonal contributions. At the highest elevations most of the region experiences severe winter freezes, especially in the northern mountains. The lower elevations and even some areas at intermediate elevations toward the southern end of the region are essentially frost-free. Numerous frost-sensitive tropical plant species reach their northernmost limits in the region, especially in protected microenvironments. Near the northern end of the region many species reach their eastern or western limits on the Deming Bridge, the spill-over point on the continental divide (R.D. Worthington, pers. comm.).

Vegetation

The evergreen woodlands and savannas of this region evolved out of more generalized Madro-Tertiary vegetation before the end of the Pleistocene (Axelrod 1979). Despite the commonality of species in genera such as *Pinus, Juniperus* and *Quercus*, there are at least two distinct floristic assemblages of woodland species in the mid- to upper elevations (McLaughlin 1986). These allied but readily distinguishable floristic elements are the Apachian – in the northern horseshoe-shaped district rimming the northern Sierra Madre Occidental and extending as far north as the Mogollon Rim; and the Madrean – the more seminal flora characteristic of the Mexican mountains between the Sierra Mohinora and approaching the U.S. border (McLaughlin 1986, 1989).

Within these two districts, eight main physiognomic vegetation types can be found: montane evergreen forest, oak-coniferous (evergreen) woodland, oak savanna (the oaks mostly drought-deciduous), chaparral, short-grass prairie, tropical deciduous forest, subtropical thorn scrub and subtropical desert fringe (Marshall 1957; Rzedowski 1978; Brown 1982; Búrquez, Martínez-Yrizar and Felger, in press). A ninth type (which may be the counterpart of the oak savanna) is barrancan oak woodland, which forms a distinctive narrow belt on the western slopes of the Sierra Madre (Gentry 1942). Three of these types – montane evergreen forest, mixed evergreen woodland and chaparral – are strongly associated with the Madro-Tertiary flora

MAP 22. APACHIAN/MADREAN REGION OF SOUTH-WESTERN NORTH AMERICA (CPD SITE MA11)

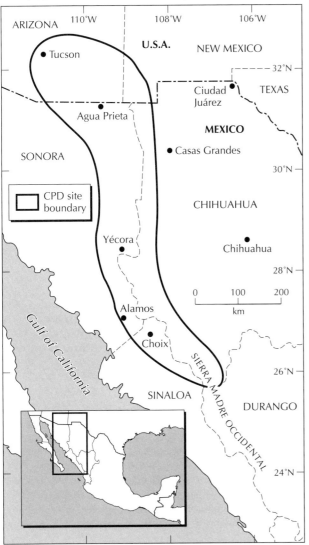

(Axelrod 1979). The Sonoran Desert reaches its north-eastern limit at the lower flanks of the western sky islands, and the Chihuahuan Desert reaches its north-western limit at the lower flanks of the eastern sky islands. The Pinaleño and Chiricahua mountains and a few north-facing peaks in south-western Chihuahua support spruce-fir forest at their highest elevations.

Flora

An estimated 4000 vascular plant species are found within this region (Felger and Wilson 1995). McLaughlin (1995) provides an excellent overview of the flora of the sky islands of the Apachian district. The Chiricahua Mountains (c. 1840 km²) support a flora of c. 1200 species (Reeves 1976; Bennett, Johnson and Kunzmann, in prep.); the Animas Mountains, c. one-fifth the size of the Chiricahuas, have a flora of c. 638 species (Wagner 1977). The Rincon and Huachuca mountains, with respectively 959 and 907 species, are rather rich whereas the Pinaleño Mountains, with 786 species, are comparatively depauperate (Bowers and McLaughlin 1987; McLaughlin and Bowers 1990; McLaughlin 1993, 1995).

The species-richness of the mountain floras increases farther to the south in Sonora and Chihuahua. The flora of the Sierra de los Ajos is estimated at 1000 species (Fishbein, Felger and Garza 1995), and White (1949) documented 1200 species and infraspecific taxa for the Sierra del Tigre and Río Bavispe region in north-eastern Sonora. The Río Mayo and surrounding areas in south-eastern Sonora and south-western Chihuahua include at least 2100 vascular plant species (Gentry 1942; Martin, in prep.), with many endemic species and a large number of "tropical" taxa at their northern limits. The rugged terrain and diverse habitats of the Parque Nacional de la Cascada de Basaseachic in south-western Chihuahua support c. 750 species in only c. 40 km² (Spellenberg, Lebgue and Corral-Diaz, in prep.). Nearby at Nabogame (18 km north-west of Yepachic) in an area of similar size but more arid and with less habitat diversity, Laferrière (1995) documented 601 species. The expansive central Sierra Tarahumara has c. 1900 species (Bye, unpublished).

The plant families with the largest numbers of species in the region are Asteraceae, Fabaceae, Poaceae and Euphorbiaceae. The Sierra Madre Occidental is a major centre of oak diversity – there are at least two dozen species of *Quercus* in the region.

The rugged Apachian/Madrean region is a zone of mass meeting and geographic termini of species and floras from the north and the south. At least 17 plant families reach their northern limits – at least within western North America: Begoniaceae, Bombacaceae, Bromeliaceae, Caricaceae, Clethraceae, Cochlospermaceae, Eriocaulaceae, Erythroxylaceae, Gesneriaceae, Magnoliaceae, Melastomataceae, Meliaceae, Myrtaceae, Olacaceae, Opiliaceae, Piperaceae and Zamiaceae. Seventy-two percent of the tree flora, 164 species of southern or tropical origin, reach their northern limits in the region. Examples of prominent genera of tropical trees at their northern limits in western North America include *Trophis, Chlorophora* and six species of *Ficus* (Moraceae), *Tabebuia* (Bignoniaceae), *Drypetes* (Euphorbiaceae), *Platymiscium* (Fabaceae),

Cinnamomum and *Persea* (Lauraceae) and *Aphananthe* (Ulmaceae) (Felger and Johnston 1995). Also at their northern limits are the pseudobulb-forming orchids, represented by ten genera, many of which are epiphytic.

The importance of the floristic contribution of the two major vegetation types of this region to the overall diversity of the Mexican flora has been described by Rzedowski (1991, 1993). Even though conifer-oak forest and tropical deciduous forest/thorn scrub account for respectively 21% and 17% of the Mexican territory, they contribute the largest number of vascular plant species to the flora. Of Mexico's estimated 22,000 vascular plant species, c. 7000 (24%) are in conifer-oak forests whereas c. 6000 species (20%) occur in tropical dry forests. These figures are higher than those for tropical humid forests and deserts of Mexico. The conifer and oak forest of Mexico, to which the Apachian/Madrean region generally belongs, has the highest endemism of all the major vegetation types in Mexico – 70% of the flora is endemic. If the conifer and oak forest is analysed in relation to "Mega-Mexico 3" of Rzedowski (1993), which encompasses the western U.S. borderlands (including those in our region) and also Central America into northern Nicaragua, 85% of the species are endemic. Hence, the flora of the Apachian/Madrean region is the north-western backbone of the two richest floras of Mexico – a country that ranks as one of the three top mega-diversity centres of the world.

Useful plants

The ethnobotanies of the Tarahumara, Guarijio, Mayo, Mountain Pima and Sonoran mestizos have been studied by Gentry (1942, 1963); Pennington (1963); Bye (1976); Laferrière (1991); Reina-Guerrero (1993); Yetman, Van Devender and López Estudillo (in press); and others. More than 300 food plants and 450 medicinal plants from this region have been documented ethnographically. These include large ethnofloral representations of Agavaceae, Cactaceae, Fabaceae and Solanaceae. Relatively unique phytochemical properties have been identified from analyses adjunct to these ethnobotanies, for instance the high papain content of *Jarilla chocola* (Tookey and Gentry 1969) and the high soluble-fibre content of *Hyptis*, *Plantago* and *Salvia* seed mucilages (Brand *et al.* 1990). Other utilitarian categories have generated considerable interest, such as fish poisons used by the Tarahumara (Pennington 1963). The "toloache" (*Datura lanosa*) of the western barrancas of Chihuahua has the highest content of hyoscine (scopolamine) of any Mexican species studied; this alkaloid is in great demand by the pharmaceutical industry (Bye, Mata and Pimentel-Vázquez 1991).

It has been estimated that among the regional floras in arid and semi-arid south-western North America, 18% of the species have been utilized by people for food and 20% for medicinal purposes (Felger and Nabhan 1978; Felger and Wilson 1995). About 10% of the edible species, or 1.8% of the flora, served as major food resources (Felger 1979). These estimates, based on compilation of known data, are verified by individual ethnobotanical studies (e.g. Gentry 1942, 1963; Bye 1976, 1985; Felger and Moser 1985; Laferrière 1991; Rea, in press). For example, the Tarahumara utilized at least 220 species of plants for food. Their pharmacopoeia

includes c. 300 plant species (Bye 1985), 47 of which are collected and sold in urban markets of northern Mexico (Bye 1986).

The region is the richest in wild congeners of domesticated crops of any New World area north of the Tropic of Cancer (Nabhan and Felger 1985; Nabhan 1991). The wild relatives of domesticated crop plants include more than 250 species (Felger and Wilson 1995). Eighteen crop species are well represented in the region, including domesticates of *Agave*, *Cucurbita*, *Lepidium*, *Hyptis*, *Panicum*, *Phaseolus*, *Prunus* and *Solanum*, and many of these are found exclusively in the region. The land-race diversity of the native crops is richer than in any other American region north of the tropics.

At Nabogame (near Sierra Mohinora) in Chihuahua, the northernmost population of "teosinte" is disjunct by several hundred km from the tropical range of these wild and weedy maize relatives, where they sometimes introgress with cultivated maize (Doebley and Nabhan 1989). Native Seed/SEARCH, a Tucson-based organization devoted to the preservation and dissemination of indigenous crop plants, has made the seed of the Nabogame teosinte available to plant breeders. The mosaic of wild montane vegetation and Amerindian fields has provided ideal settings for studying introgression between wild and domesticated *Capsicum*, *Cucurbita*, *Phaseolus* and *Zea*.

Nabhan (1990b) identified the geographic patterns of 18 wild *Phaseolus* species in the Sierra Madre Occidental – a richer assemblage than found anywhere else north of the Tropic of Cancer. The 13 species at the south-western edge of the region (southern Sonora and northern Sinaloa west of the divide) are far more than predicted by species richness/latitudinal distribution regressions. Several of these bean species are endemic to the sierras, yet the region's germplasm resources remain under-collected. Preliminary results from pollination, DNA and isozyme studies by R. Bye suggest reciprocal gene flow between wild *Phaseolus coccineus* subsp. *formosus* and the special domesticate "tekómari", the *P. coccineus* subsp. *coccineus* of the Tarahumara. Gene flow between the wild and cultivated forms is aided by Tarahumara agro-ecological practices (which include the management of nearby forests) and may be responsible for the maintenance of this productive scarlet runner bean, which is adapted to high mountain areas with short growing seasons.

Social and environmental values

Mt Graham, the summit of the massive Pinaleño Mountains, is a sacred site of religious and cultural importance to the San Carlos Apache (McCarthy 1991). Portions of the Apachian/Madrean region have been inhabited for centuries by some of the least industrialized indigenous peoples in North America. The Tarahumara are one of the largest groups of native peoples in Mexico; their detailed folk science of plants has been acclaimed since the studies of Lumholtz (1902). The Guarijio, Mountain Pima, Opata, Mayo and Apache inhabitants of the region also merit special interest for their knowledge and uses of botanical resources.

The Apachian-Madrean forest is the largest oxygen- and biomass-producing terrestrial ecosystem in south-western North America. Watersheds of the Sierra Madre Occidental provide water to the densely populated areas on the coastal plain, where much of Mexican commercial food production takes place. The northern Sinaloa – southern Sonora irrigated districts draw on the Yaqui, Mayo and Fuerte rivers and are major producers of wheat as well as winter vegetables for export. The Río Bravo (Grande) provides irrigation water for southern Texas and adjacent Tamaulipas, where citrus, melons, chilies and tomatoes are important crops. Protection of headwaters from logging and overgrazing can help to stabilize hydrological resources required by agriculture and urban populations.

This is a region of tremendous habitat diversity and species-richness. Over half the bird species north of Mexico in North America (including Greenland) occur in the Chiricahua Mountains (Kunzmann, Johnson and Bennett, in prep.). The sky islands of southern Arizona, New Mexico and northern Mexico provide forested places for recreation, environmental education and the spiritual solitude of urban inhabitants living in the nearby desert lowlands.

Economic assessment

Although only a few non-timber resources have been quantified in terms of their economic potential, there is potential for cottage industries based on the sustainable harvest of wild plants. For example, during a "dry" year roughly 20 tons of dry fruits of wild "chiltepines" (*Capsicum annuum* var. *aviculare*) are harvested in the State of Sonora; as much as 50 tons might be harvested during a "wet" year (De Witt 1991). The total export to U.S.A. is c. 6 tons, where the retail price in 1990 was US$72 per pound (Nabhan 1990a).

One of the best-known medicinal plants of this region is a lovage known as "chuchupate" (*Ligusticum porteri*, = *L. madrensis*). The Tarahumara value its aromatic roots for medicinal and ritual purposes. A local business is based upon the medicinal preparation "COPANGEL", two-fifths of which is the ground roots of this perennial herb. Clinical studies by the Mexican national health programme have determined that it is effective in the treatment of peptic ulcers (Mundo, Aizpuru and Lozoya, in press). The roots also are exported to the U.S.A., Japan and Germany. The species' popularity in medicinal herb markets has increased to the point that over-collection has diminished many local populations. This moisture- and shade-loving species also is declining as the forests are cleared and arroyo heads erode. Based upon the retail value of COPANGEL, one ha of chuchupate is worth c. US$75,000.

Acorns ("bellotas") of Emory oak (*Quercus emoryi*) are wild-harvested each summer in north-eastern Sonora and south-eastern Arizona. A favourite of Sonorans and many Arizonans, the seeds are eaten fresh. They are sold locally, and in Sonoran markets and Tucson they can be purchased for c. US$3.50 per kg. Unlike the acorns of many other oak species, they are palatable with no preparation due to their relatively low tannin content. These acorns have an extremely high glycemic index value (Brand *et al.* 1990). Various other Madrean oaks likewise have "sweet" acorns and represent a potentially significant resource. The current price of acorn meal can be as high as US$22 per kg in Korea, and the supply falls far short of demand.

Several species of columnar cacti in the lowland subtropical zones in Sonora and Chihuahua yield highly desirable fruits which can be eaten directly, dried or prepared as juice, condiments or wine. They are harvested for local consumption and occasionally reach marketplaces in nearby cities (such as Hermosillo, Sonora). The demand is high and vendors usually sell out very early when the fruit is available. Desirable species include mountain organpipe or "saguira" (*Stenocereus montanus*) and organpipe or "pitaya dulce" (*S. thurberi*). Organpipe jam sells retail for US$15 per kg; 10 kg of fresh fruits yield 1 kg of jam. In some regions of northern Sonora, the local Amerindian people annually harvest 45 kg of fruit per person. The harvest of fruits from wild populations of columnar cacti by native peoples seems to have virtually no effect on the populations, at least in a subsistence economy (Hastings and Turner 1965; Felger and Moser 1985).

In recent years ecotourism has provided a major economic resource for the region. This usually non-destructive industry is growing and expanding rapidly. The Chiricahua Mountains in Arizona and the Tarahumara region in south-western Chihuahua are especially popular, as is the Alamos region in south-eastern Sonora.

Threats

The remoteness of the rugged sierras in Mexico and their relatively small human populations and cultural conservatism have allowed the region to retain much of its ecosystems in a rather natural state until recently. However, in the Sierra Madre Occidental an ever-increasing human population is placing more demands on the environment. Whereas the New Mexico mountains and many areas in Arizona are well protected for the foreseeable future, certain Arizona mountains and many of the areas in northern Mexico are in a precarious situation. Funding of large-scale development by non-local national and multinational private and corporate sources poses threats to local cultures and environments.

Cultivation of opium poppies and marijuana, usually considered a legal problem, now poses major ecological, social and economic threats. The presence of armed drug-plant cultivators, buyers and distributors is a danger to the local inhabitants, tourists and scientists. Native peoples in the Sierra Madre Occidental have been forced by criminals from outside the mountain communities to cultivate and harvest drug crops (R. Gingrich, Forest Guardians, pers. comm.). Failure to cooperate with drug lords and their underlings has resulted in injury and death (Weisman 1994). Aerial spraying of herbicides by law enforcement agencies has caused extensive but undocumented and unstudied damage to the native vegetation and flora, and poses possible human health hazards. The spraying of illegal fields on Sierra de Alamos and elsewhere threatens unique relict populations of tropical species. Massive sprayings of the herbicide paraquat may be responsible for reduction in populations of *Rothschildia cincta*, the larvae of which feed on *Jatropha*. The cocoons of this large saturnid moth are much used by the Yaqui Amerindians of Sonora for ceremonial purposes (Peigler 1994).

Open-pit mining for copper, gold and other metals in many parts of this highly mineralized region poses a continuing major threat. Air pollution from smelters (such as at Cananea and Nacozari in north-eastern Sonora) is likewise threatening and has been linked to the serious decline of the Tarahumara frog (Hale *et al.* 1995). The ever-present demand for and shortage of fresh water in nearby desert communities puts a serious strain on the water resources of the region. Every river system has been dammed and all river-delta areas have been seriously damaged. Construction of an enormous dam on the Río Fuerte near Huites in Sinaloa portends disastrous environmental problems and will lead to massive loss of tropical deciduous forest and riparian habitats. Many of the large deep canyons and barrancas could be destroyed by dams, which would cause a tragic loss of biological diversity.

Soil erosion due to over-intensive agriculture and grazing is an accelerating problem. In subtropical southern parts of the region there is a long history of slash-and-burn agriculture for "rozas" (maize fields, e.g. the "milpas" of Gentry 1942). Expansion in area and decrease in regeneration cycles of the rozas along with other agricultural practises, especially wood-cutting and livestock-grazing, contribute to the ongoing problems of erosion.

The most serious cause of desertification (defined as the reduction in species richness or diversity) in desert and subtropical regions in north-western Mexico is replacement of the native vegetation with buffel grass (*Pennisetum ciliare*) for cattle-grazing. In Sonora, more than 5000 km² have been cleared and a government goal calls for as much as 60,000 additional km² to be cleared and planted with this species (Búrquez, in press).

Local environmental degradation has resulted from the cutting of hardwood legumes and other trees and shrubs for cooking fuel and to a lesser extent for home-heating. Extensive over-exploitation in Sonora of mesquite (*Prosopis velutina*, *P. glandulosa* var. *torreyana*) and desert ironwood (*Olneya tesota*) for charcoal, exported primarily to U.S.A., threatens lowlands. Removal of these trees is often an integral part of the conversion to buffel grass (Nabhan and Carr 1994). Many of the subtropical hardwood trees such as "amapa" (*Tabebuia impetiginosa*, *T. chrysantha*) yield highly prized lumber for roof beams ("vigas") and fine cabinet-making. The great "sabinos" or bald cypresses (*Taxodium mucronatum*) yield a durable wood of high quality which commands a high price. These and many other species are legally protected by the Mexican forestry department, but enforcement is difficult. The selective cutting of the tree croton (*Croton* cf. *niveus*) of south-eastern Sonora and nearby northern Sinaloa for tomato stakes and fence posts has made this once common plant scarce (Bye 1995; Steinmann and Felger, in prep.).

The most dire threats to the sierras, however, are the logging and pulping industries. Almost all of the coniferous forests of northern Mexico have been cut one to four times. Large-scale financing is supporting clear-cutting, harvesting and destruction of understorey trees and plants and pulping of diverse species including oaks and other hardwoods. Fortunately, a massive forest-harvesting project in the region promulgated by the World Bank (Seedhead News 1991) has been abandoned, at least for the time being. A large pulp mill at Anahuac, Chihuahua has been expanded and renovated with a bank loan of US$350 million.

Extensive road building in montane zones as well as recent political changes and increasing interest in the natural resources of the region continue. Accessibility heralds important environmental, economic and social changes in formerly isolated montane areas. This accessibility is allowing scientists to inventory the rich natural resources. We hope that not all of the change will be destructive to local peoples and the environment.

Conservation

Throughout the Apachian/Madrean region there are countless peaks and great canyons rich in biological diversity. All are ecologically significant. In diverse parts of the region *in situ* conservation by indigenous and rural peoples has been informal but relatively effective.

Most of the south-eastern Arizona sky-island mountains are administered by the U.S. Forest Service – principally in the Coronado National Forest – which is often faced with conflicting goals of conservation, providing access for recreation, and responsibilities to alot timber sales, grazing leases and leases for various developments. Potential development of mountain areas for housing and resorts is of recurrent concern to environmentalists and scientists. In the Pinaleño Mountains, the development of a large astronomy centre on Mt Graham has led to the mobilization of conservationists and traditional Native Americans, pitting these groups against various U.S. federal agencies and a consortium of institutions including the University of Arizona and the Vatican. Construction of the telescopes, support facilities and roads has led to destruction of old growth spruce-fir forest; further development is planned.

Most of the larger mountain ranges in Arizona and New Mexico enjoy protection by various government agencies and private entities. For example, the Gray Ranch in south-western New Mexico, which includes the Animas Mountains, was purchased by The Nature Conservancy in 1990; protection continues under the Animas Foundation to which this large property has been transferred. The Huachuca Mountains are protected by having portions incorporated into Ft. Huachuca Military Reservation as well as a National Monument, and Ramsey Canyon managed by The Nature Conservancy. The Chiricahua Mountains of south-eastern Arizona are well protected, with a substantial designated Wilderness Area, a Research Natural Area at Cave Creek (set aside under the aegis of the Arizona-Nevada Academy of Sciences), a National Monument and the Southwest Research Station of the American Museum of Natural History.

Mexico has excellent environmental laws, and environmental awareness among the general population has grown substantially in recent years. Unfortunately, conservation efforts suffer from acute under-funding and resultant neglect. Although several areas in north-western Mexico have a special conservation status, enforcement of environmental policy and protection of natural areas has been lax or non-existent. The Zona Protectora Ciudad de Hermosillo and Arroyo los Nogales Preserve have been all but forgotten and are essentially buried under urban sprawl (Búrquez, in press). Governmental policy encourages the clearing of land in concert with introduction of exotic plants such as buffel grass. Many of the most damaging exotics were introduced into the south-western U.S.A. by the U.S. Soil Conservation Service, and later taken to Mexico.

Mexico does not lack for talented and educated individuals and has a bulwark of governmental agencies that could become far more effective in the conservation of natural areas – with increased support. Many of the Mexican mountainous areas are under control of the federal government, either the Secretaría de Agricultura y Recursos Hidráulicos (SARH) which manages forest refuges, or the Secretaría de Medio Ambiente, Recursos Naturales y Pesca (SEMARNAP) (formerly SEDESOL's Subsecretaría de Ecología), which manages the

CPD Site MA11: Apachian/Madrean region of south-western North America. Foothills of Sierra Madre Occidental near Tepoca, Sonora, Mexico. Slopes with tropical deciduous forest. Although oaks and pines grow on hydrothermically altered soils nearby, the main mass of the sierra lies several km to the east. Photo: Antonio Cafiero.

national system of protected natural areas. However, SARH functions more in a supervisory capacity than in a direct management role. "Ejidos" (community lands), private enterprise and para-state companies have more influence on the day-to-day administration of the forests.

In Chihuahua, nine Unidades de Administración Forestal (UAFs) of SARH have issued permits to ejidos, small property owners and large corporate logging operations (Resendiz-Vázquez 1984). The UAFs, more recently renamed Unidades de Conservación y Desarrollo Forestal, have become locally controlled (by state and municipality). At least two units are developing conservation programmes (in addition to their forest management, exploitation and product development) in collaboration with the U.S. Forest Service and the Canadian Model Forest Program. In all cases, the units must comply with federal law. On lands with indigenous peoples, another government programme is in effect – for example the Programa de Desarrollo Forestal Chihuahua-Durango (INI 1993) for the Tarahumara; there are similar programmes among the Guarijio and Pima.

Various institutions such as the Centro Ecológico de Sonora in Hermosillo have been active in attempting to gain new or increase protection for nature preserves. There is an important undercurrent of conservation activity in governmental agencies and private organizations (e.g. Pronatura). Ultimately it will be through documentation of natural resources that protection will be achieved. Recent and increasing scientific interest in this region that, strangely,

has been largely ignored, is encouraging. However, even during this enlightened time, the protection of many areas is tenuous. There will be great losses if timely action is not taken.

Many unique areas in north-western Mexico with important biological, anthropological, historical and social values are worthy of protection; a few are mentioned. During the early 1990s a number of significant areas, some of substantial size, were designated by the Centro Ecológico de Sonora for protection and conservation management under the Sistema de Areas Naturales Protegidas del Estado de Sonora (SANPES). Some of them in the Apachian/Madrean region include Mesa el Campañero and Arroyo el Reparo near Yécora, Sierra de Alamos with the Upper Río Cuchujaqui area south-east of Alamos, Sierra Mazatán, Cañón la Cruz del Diablo, Sierra la Mariquita and Sierra San Luis.

Río Bavispe, Sierra de los Ajos and nearby Sierra de la Purica and Sierra Buenos Aires were granted protection in the 1930s; however, these areas were not managed as preserves and they have not escaped logging or cattle-ranching. The Sierra del Tigre and Sierra de los Ajos in northern Sonora are included in forestry preserves administered by SARH. Other mountain areas such as the Sierra San Luis are privately owned and well managed.

Tropical deciduous forest, one of the world's important biomes, is threatened globally. In Mexico a single area has obtained protection – the Chamela-Cuixmala Biosphere Reserve in Jalisco (CPD Site MA7, see Data Sheet). Areas for potential protection that include samples of the northernmost tropical deciduous forests on the continent include the Sierra de Alamos (rising above the colonial town of Alamos), portions of the Río Cuchujaqui including the upper region (Van Devender *et al.*, in press) and the lower elevations of the Arroyo el Reparo. The canyon of the Río Tepoca, c. 100 km west of Yécora (on Mexico Hwy 16), is likewise an important northern outlier of subtropical vegetation worthy of protection.

The spectacular deep and narrow Cañón la Cruz del Diablo just east of Guasabas is a unique scenic and biological region. Here flocks of military macaws fly over the northernmost *Brahea nitida* palms. The incomparable Barranca del Cobre and its associated canyons are to Mexico what the Grand Canyon is to U.S.A. The Río Conchos region is particularly deserving of greater protection. The incomparable 245-m high Basaseachic Waterfall and immediately surrounding area in south-western Chihuahua are in the Parque Nacional de la Cascada de Basaseachic (40 km²). Seriously under-funded and managed more to encourage tourism than for conservation, the vegetation has deteriorated (Sánchez-Vélez 1987). Funding of effective conservation which involves the indigenous people is essential. It is crucial to support local efforts to establish Biosphere Reserves in these and similar areas before it is too late.

References

Axelrod, D.I. (1979). Age and origin of Sonoran Desert vegetation. *Calif. Acad. Sciences Occasional Papers* 132: 1–74.

Bennett, P.S., Johnson, R.R. and Kunzmann, M.R. (in prep.). Annotated checklist of the vascular plants of southeastern Arizona Madrean Archipelago.

Bowers, J.E. and McLaughlin, S.P. (1987). Flora and vegetation of the Rincon Mountains, Pima County, Arizona. *Desert Plants* 8: 51–94.

Brand, J.C., Snow, B.J., Nabhan, G.P. and Truswell, A.S. (1990). Plasma glucose and insulin responses to traditional Pima Indian meals. *Amer. J. Clinical Nutrition* 5: 416–420.

Brown, D.E. (ed.) (1982). *Biotic communities of the American Southwest – United States and Mexico. Desert Plants* 4: 1–342.

Búrquez, A. (in press). Conservation and land use in Sonora. In Robichaux, R.H. (ed.), *Ecology and conservation of the Sonoran Desert flora: a tribute to the Desert Laboratory.* University of Arizona Press, Tucson.

Búrquez, A., Martínez-Yrizar, A. and Felger, R.S. (in press). Biodiversity at the southern desert edge in Sonora, Mexico. In Robichaux, R.H. (ed.), *Ecology and conservation of the Sonoran Desert flora: a tribute to the Desert Laboratory.* University of Arizona Press, Tucson.

Bye, R. (1976). *Ethnoecology of the Tarahumara of Chihuahua, Mexico.* Ph.D. dissertation. Department of Biology, Harvard University, Cambridge, Massachusetts. 344 pp.

Bye, R. (1985). Medicinal plants of the Tarahumara Indians of Chihuahua, Mexico. In Tyson, R.A. and Elerick, D.V. (eds), *Two mummies from Chihuahua: a multidisciplinary study.* San Diego Museum Paper No. 19: 77–104.

Bye, R. (1986). Medicinal plants of the Sierra Madre: comparative study of Tarahumara and Mexican market plants. *Economic Botany* 40: 103–124.

Bye, R. (1995). Ethnobotany of the Mexican dry tropical forest. In Bullock, S.H., Mooney, H.A. and Medina, E. (eds), *Seasonally dry tropical forests.* Cambridge University Press, Cambridge, U.K. Pp. 423–438.

Bye, R., Mata, R. and Pimentel-Vázquez, J.E. (1991). Botany, ethnobotany and chemistry of *Datura lanosa* (Solanaceae) in Mexico. *Anales Inst. Biol. Univ. Nac. Autónoma México, Serie Botánica* 61: 21–42.

De Witt, D. (1991). Yo soy un chiltepinero. *Chile Pepper* 5(3): 22–30.

Doebley, J. and Nabhan, G.P. (1989). Further evidence regarding gene flow between maize and teosinte. *Maize Genetics Cooperation Newsletter* No. 63: 107–108.

Felger, R.S. (1979). Ancient crops for the 21st century. In Ritchie, G. (ed.), *New agricultural crops.* AAAS Selected Symposium 38. Westview Press, Boulder, Colorado. Pp. 5–20.

Felger, R.S., Baker, M.A. and Wilson, M.F. (in prep.). *Medicinal plants of Arizona.*

Felger, R.S. and Johnston, M.B. (1995). The trees of the Madrean-Apachian region of southwestern North America. In DeBano, L.F., Ffolliott, P.F. and Hamre, R.H. (eds), *Biodiversity and management of the Madrean Archipelago: the sky islands of southwestern United States and northwestern Mexico.* USDA Forest Service, Rocky Mountain Forest and Range Experiment Station, Fort Collins, Colorado. Pp. 71–83.

Felger, R.S. and Moser, M.B. (1985). *People of the desert and sea: ethnobotany of the Seri Indians.* University of Arizona Press, Tucson. 435 pp.

Felger, R.S. and Nabhan, G.P. (1978). Agroecosystem diversity: a model from the Sonoran Desert. In Gonzalez, N.L. (ed.), *Social and technological management in dry lands.* AAAS Selected Symposium 10. Westview Press, Boulder, Colorado. Pp. 128–149.

Felger, R.S. and Wilson, M.F. (eds) (1995). Northern Sierra Madre Occidental and its Apachian outliers: a neglected center of biodiversity. In DeBano, L.F., Ffolliott, P.F. and Hamre, R.H. (eds), *Biodiversity and management of the Madrean Archipelago: the sky islands of southwestern United States and northwestern Mexico.* USDA Forest Service, Rocky Mountain Forest and Range Experiment Station, Fort Collins, Colorado. Pp. 36–59.

Fishbein, M., Felger, R.S. and Garza, F. (1995). Another jewel in the crown: a report on the flora of the Sierra de los Ajos, Sonora, Mexico. In DeBano, L.F., Ffolliott, P.F. and Hamre, R.H. (eds), *Biodiversity and management of the Madrean Archipelago: the sky islands of southwestern United States and northwestern Mexico*. USDA Forest Service, Rocky Mountain Forest and Range Experiment Station, Fort Collins, Colorado. Pp. 126–134.

Gentry, H.S. (1942). *Río Mayo plants*. Carnegie Institution of Washington Public. No. 527. Washington, D.C. 328 pp.

Gentry, H.S. (1963). The Warihio Indians of Sonora-Chihuahua: an ethnographic survey. *Bureau Amer. Ethnography Bull.* 186: 61–144.

Hale, S.F., Schwalbe, C.R., Jarchow, J.L., May, C., Lowe, C.H. and Johnson, T.B. (1995). Disappearance of the Tarahumara frog. In LaRoe, E.T., Farris, G.S., Puckett, C.E., Doran, P.D. and Mac, M.J. (eds), *Our living resources: a report to the nation on the distribution, abundance, and health of U.S. plants, animals, and ecosystems*. U.S. Department of the Interior, National Biological Service, Washington, D.C. Pp. 138–140.

Hastings, J.R. and Turner, R.M. (1965). *The changing mile*. University of Arizona Press, Tucson. 317 pp.

INI, Delegación Chihuahua (1993). *Pueblos indígenas y microdesarrollo en la Tarahumara – seminario permanente sobre indigenismo*. Instituto Nacional Indigenista (INI), Chihuahua, Chih. 109 pp., 5 annexes.

Kunzmann, M.R., Johnson, R.R. and Bennett, P.S. (in prep.). Annotated checklist of the birds of the Chiricahua Mountains.

Laferrière, J.E. (1991). *Optimal use of ethnobotanical resources by the Mountain Pima of Chihuahua, Mexico*. Ph.D. dissertation. University of Arizona, Tucson. 266 pp.

Laferrière, J.E. (1995). Vegetation and flora of the Mountain Pima village of Nabogame, Chihuahua, Mexico. *Phytologia* 77: 102–140.

Lumholtz, C. (1902). *Unknown Mexico*. Charles Scribner's Sons, New York. 2 vols. 530 pp. + 496 pp.

Marshall, J.T. (1957). *Birds of the pine-oak woodland in southern Arizona and adjacent Mexico*. Cooper Ornithological Society, Pacific Coast Avifauna 32. 125 pp.

Martin, P.S. (ed.) (in prep.). *Gentry's Río Mayo Flora*.

McCarthy, T. (2 August 1991). Apache tribe lives new vision in fight to save mountain. *National Catholic Reporter* Vol. 27, No. 36.

McLaughlin, S.P. (1986). Floristic analysis of the southwestern United States. *Great Basin Nat.* 46: 46–65.

McLaughlin, S.P. (1989). Natural floristic areas of the western United States. *Journal Biogeography* 16: 239–248.

McLaughlin, S.P. (1993). Additions to the flora of the Pinaleño Mountains, Arizona. *Journal Arizona Acad. Sciences* 27: 1–27.

McLaughlin, S.P. (1995). An overview of the flora of the sky islands, southeastern Arizona: diversity, affinities, and insularity. In DeBano, L.F., Ffolliott, P.F. and Hamre, R.H. (eds), *Biodiversity and management of the Madrean Archipelago: the sky islands of southwestern United States and northwestern Mexico*. USDA Forest Service, Rocky Mountain Forest and Range Experiment Station, Fort Collins, Colorado. Pp. 60–70.

McLaughlin, S.P. and Bowers, J.E. (1990). A floristic analysis and checklist for the northern Santa Rita Mountains, Pima Co., Arizona. *Southwestern Nat.* 35: 61–75.

Mundo, F., Aizpuru, V. and Lozoya, X. (in press). El uso de *Angelica archangelica* L. en el tratamiento de la enfermedad ulcerosa péptica. *Acta Médica Mexicana*.

Nabhan, G.P. (1990a). Conservationists and Forest Service join forces to save wild chiles. *Diversity* 6 (3–4): 47–48.

Nabhan, G.P. (1990b). Wild *Phaseolus* ecogeography in the Sierra Madre Occidental, Mexico. *Systematic and Ecogeographic Studies on Crop Genepools* 5. IBPGR, Rome. 35 pp.

Nabhan, G.P. (1991). Genetic resources of the U.S.-Mexican borderlands: wild relatives of crops, their uses and conservation. In Ganster, P. and Walter, H. (eds), *Environmental hazards and bioresource management in the United States-Mexico borderlands*. U.C.L.A. Latin American Center Publications, University of California, Los Angeles. Pp. 345–360.

Nabhan, G.P. and Carr, J.L. (eds) (1994). *Ironwood: an ecological and cultural keystone of the Sonoran Desert*. Occasional Papers in Conservation Biology 1. Conservation International, Washington, D.C. 92 pp.

Nabhan, G.P. and Felger, R.S. (1985). Wild desert relatives of crops: their direct use as food. In Wickens, G., Goodin, J.R. and Field, D.V. (eds), *Plants for arid lands*. George Allen and Unwin, London, U.K. Pp. 19–33.

Peigler, R.S. (1994). Non-sericultural uses of moth cocoons in diverse cultures. *Proc. Denver Mus. Nat. Hist. Ser. 3*, No. 5: 1–20.

Pennington, C.W. (1963). *The Tarahumar of Mexico*. University of Utah Press, Salt Lake City. 267 pp.

Rea, A.M. (in press). *At the desert's green edge: an ethnobotany of the Gila River Pima*. University of Arizona Press, Tucson.

Reeves, T. (1976). *Vegetation and flora of Chiricahua National Monument, Cochise County, Arizona*. M.S. thesis. Arizona State University, Tempe. 179 pp.

Reina-Guerrero, A.L. (1993). *Contribución a la introdución de nuevos cultivos en Sonora: las plantas medicinales de los Pimas Bajos del Municipio de Yécora, Sonora*. Thesis. Universidad de Sonora, Hermosillo. 256 pp.

Resendiz-Vázquez, P. (1984). *Resumen del inventario forestal de las U.A.F. del estado de Chihuahua*. Boletín Divulgativo No. 65. Instituto Nacional de Investigaciones Forestales, Mexico, D.F. 52 pp.

Rzedowski, J. (1978). *Vegetación de México*. Editorial Limusa, Mexico, D.F. 432 pp.

Rzedowski, J. (1991). Diversidad y orígenes de la flora fanerogámica de México. *Acta Botánica Mexicana* 14: 3–21.

Rzedowski, J. (1993). Diversity and origins of the phanerogamic flora of Mexico. In Ramamoorthy, T.P., Bye, R., Lot, A. and Fa, J.E. (eds), *Biological diversity in Mexico: origins and distribution*. Oxford University Press, New York. Pp. 129–144.

Sánchez-Vélez, A. (1987). *Conservación biológica en México*. Perspectivas. Colección Cuadernos Universitarios, Serie Agronomía No. 13, Universidad Autónoma Chapingo. Chapingo, Mexico. 136 pp.

Seedhead News (1991). Sierra Madre World Bank "development" or logging project? *The Seedhead News* Nos. 32 and 33: 1–11.

Spellenberg, R.W., Lebgue, T. and Corral-Diaz, R. (in prep.). Annotated checklist of the plants of the Parque Nacional de la Cascada de Basaseachic, southwest Chihuahua, Mexico.

Steinmann, V.W. and Felger, R.S. (in prep.). A synopsis of the Euphorbiaceae in Sonora, Mexico.

Tookey, H.L. and Gentry, H.S. (1969). Proteinase of *Jarilla chocola*, a relative of papaya. *Phytochemistry* 8: 989–991.

Van Devender, T.R., Sanders, A.C., Van Devender, R.K. and Meyer, S.A. (in press). Flora and vegetation of the Río Cuchuhaqui, a tropical deciduous forest near Alamos, Sonora, México. In Robichaux, R.H. (ed.), *The tropical deciduous forest of the Alamos, Sonora, region: ecology and conservation of a threatened ecosystem*. University of Arizona Press, Tucson.

Wagner, W.L. (1977). *Floristic affinities of Animas Mountain, southwestern New Mexico*. M.S. thesis. University of New Mexico, Albuquerque. 180 pp.

Weisman, A. (9 January 1994). The deadly harvest of the Sierra Madre. *Los Angeles Times Magazine*. Pp. 11–14, 33–34.

White, S.S. (1949). The vegetation and flora of the Rio Bavispe in northeastern Sonora, Mexico. *Lloydia* 11: 229–303.

Yetman, D.A., Van Devender, T.R. and López Estudillo, R.A. (in press). Monte Mojino: Mayos and trees in southern Sonora. In Robichaux, R.H. (ed.), *The tropical deciduous forest of the Alamos, Sonora, region: ecology and conservation of a threatened ecosystem*. University of Arizona Press, Tucson.

Authors

Dr Richard S. Felger (Drylands Institute, 2509 North Campbell Avenue #176, Tucson, Arizona 85719, U.S.A.), Dr Gary Paul Nabhan (Arizona-Sonora Desert Museum, 2021 North Kinney Road, Tucson, Arizona 85743, U.S.A.) and Dr Robert Bye [Universidad Nacional Autónoma de México (UNAM), Instituto de Biología, Jardín Botánico, Apartado Postal 70-614, 04510 Mexico, D.F., Mexico].

Acknowledgments

We thank Wallace Genetic Foundation, Pew Scholars on Conservation and Environment Program, and the Biodiversity Support Program of the World Wide Fund for Nature for their support. We also thank Michael F. Wilson for generous assistance.

CENTRAL REGION OF BAJA CALIFORNIA PENINSULA
Mexico

Location: Southern Baja California Norte and north-eastern Baja California Sur states, between latitudes 27°20'–28°55'N and longitudes 112°20'–114°10'W.

Area: c. 36,000 km².

Altitude: Mostly 0–1600 m, with highest peak 1985 m.

Vegetation: Xerophilous scrubland or brush: succulent-leaf and succulent-stem brushlands, halophilic brush, dune brush. Coastal lagoon communities.

Flora: Over 500 species of vascular plants; excluding lagoons, 496 species in El Vizcaíno Biosphere Reserve – 8% locally endemic, others endemic to peninsula. Transition zone within peninsula and between North American seasonal deserts and Mexican dryland tropics.

Useful plants: Fuelwood, food (e.g. cactus fruits), fodder, medicinals, ornamentals. Local traditional utility of species is mostly unknowable, as the original indigenous people are gone.

Other values: Refuge for wild fauna, including endemic and threatened taxa; high attraction for tourists.

Threats: Overgrazing by goats on slopes; expansion of agriculture on plains, causing depletion and contamination of groundwater; salt extraction, with saltwater inundation; road construction; gas, oil and mineral exploration; collection of ornamentals.

Conservation: c. 42% (15,000 km²) of the region is over half (c. 59%) of El Vizcaíno Biosphere Reserve.

Geography

In the Baja California Peninsula morphotectonic province (1300 km × 30–240 km), this mid-peninsular region includes portions of the Pacific Coastal and the Peninsular Ranges subprovinces (Ferrusquía-Villafranca 1993). Topography and climatic and edaphic conditions largely determine the distribution of the vegetation formations and plant communities.

Characteristic in the western and central portions of this region are lower elevations, nearly constant onshore wind, strong solar radiation with high day and low night temperatures, and annual precipitation usually less than 80 mm – predominantly in winter. The median annual temperature ranges between 18°–22°C. The cold marine current from California, U.S.A. influences the climate of this area, bringing cooling temperatures and moisture (regular fog).

Summer rains in the mid-peninsular region are often torrential, due to tropical depressions or cyclones ("chubascos"), which sometimes penetrate the Gulf of California. In contrast, winter rains are usually milder, resulting from the meeting of cold dry air masses with warm humid air masses (Salinas-Zavala, Coria and Díaz 1991). Sometimes storms from both north and south bypass this central region, which consequently can have irregular and patchy intensive droughts.

This region's central portion is within a closed watershed more extensive to its south-west, into which drain many watercourses from the bordering mountains. As well, several watercourses drain westward into the lagoon complex (Ojo de Liebre) near Guerrero Negro, and others flow southward into San Ignacio Lagoon. Few drainages directly reach the coast. Evaporation has caused high concentrations of several salts to deposit through much of the south-western portion of the region. The incessant transverse wind has contributed to

the formation of extensive, commonly parallel dunes. The presence of salt-tolerant vegetation in large low areas is considered evidence for shallow seas having occurred there during rather recent marine intrusions (Durham and Allison 1960; Ferrusquía-Villafranca 1993).

This region's eastern portion is more mountainous; it forms a crest (mostly not over 1600 m) that is nearer to the peninsula's coast on the Gulf of California, and it slopes more abruptly into the Sea of Cortes. These mountains are composed principally of extrusive volcanic and various sedimentary rocks (Ferrusquía-Villafranca 1993). This area's basic climate is fairly similar, but characterized by colder winters – occasionally with light snows that may linger on the peaks. The precipitation is very variable in amount and location, ranging from 0 mm (for 1–4 years) to a few hundred mm during the year, mostly in the summer.

Vegetation

The region is characterized by xerophilous scrubland, brush or thickets (Rzedowski 1978), and is almost entirely encompassed in two subdivisions of the Sonoran Desert, with a very small portion in the extreme north-east in a third subdivision (Shreve and Wiggins 1964; Wiggins 1980; Turner and Brown 1982). The vegetation occurs in four principal natural systems – three lowland deserts and highlands.

Vizcaíno Desert
The Vizcaíno Desert constitutes almost 70% of the mid-peninsular region, being represented in the west and centrally. This area is characterized by broad low-elevation plains made up mostly of Quaternary sedimentary formations from fine-

grained shallow marine and beach deposits, as well as older conglomerates.

Shreve called this biotic zone the Sarcophyllous Desert (Wiggins 1969, 1980), referring to the physiognomic dominance of species with succulent or thick leaves. *Agave* and *Ambrosia* (*Franseria*) are characteristic, and *Tillandsia recurvata* ("gallitos") and various lichens (e.g. *Rocella*) are abundant as epiphytes in the fog belt. Common taxa include *Agave shawii* subsp. *goldmaniana*, *A. sobria* subsp. *sobria*, *Yucca valida* ("datilillo"), *Fouquieria columnaris* (*Idria columnaris*), *Pachycormus discolor* (elephant tree or "copalquín"), *Stenocereus gummosus* ("pitaya agria"), *Ambrosia chenopodiifolia* (*Franseria chenopodiifolia*), *A. bryantii* (*F. bryantii*), *Atriplex magdalenae*, *A. polycarpa*, *Lycium* spp. and *Frankenia palmeri*.

Within this biotic zone is a brushy plant community on inland sand dunes, which are relatively unstable; the diversity of species is rather low. Among the characteristic species are *Chaenactis lacera*, *Nicolletia trifida* and *Dalea mollis*. These thickets provide refuge for the animals that inhabit this community and are sufficiently long-lasting for trophic relationships to have developed.

Gulf Coast Desert

The Gulf Coast Desert occupies c. 10% of the region in a narrow strip along the south-eastern side of the peninsula, from the coast (and some islands) to the crest of the mountains to the west. This area has irregular topography.

Shreve called this biotic zone the Sarcocaulescent Desert (Wiggins 1980); it is physiognomically dominated by species with succulent or similarly thick (pachycaul) trunks or stems, such as *Cercidium microphyllum* ("dipúa"), *Bursera hindsiana* ("copal"), *B. microphylla* ("torote"), *Jatropha cinerea* ("lomboy"), *J. cuneata*, *Pachycereus pringlei* ("cardón"), *Ferocactus* spp., *Opuntia cholla*, *O. molesta* and *O. bigelovii*. Also notable are *Fouquieria splendens* (ocotillo), *Lysiloma candida* and *Errazurizia megacarpa*.

San Felipe Desert

The north-eastern San Felipe Desert extends to about Bahía de las Animas, in the extreme north-east of the region. This desert is the driest and sunniest biotic zone in Baja California, being in the rain shadow of the northern highest mountains. Shreve called this the Microphyllous Desert (Wiggins 1980) because many conspicuous species have small leaves. The dominants are *Larrea tridentata* and *Ambrosia* (*Franseria*) spp.; frequent species include *Fouquieria splendens*, *Cercidium microphyllum*, *Olneya tesota* (ironwood or "palo fierro") and *Bursera microphylla*. *Opuntia cineracea* is a peninsular endemic found only in this desert.

Mountainous areas

Two principal mountainous areas above 1000 m (Map 23) constitute most of the remaining c. 20% of the mid-peninsular region, sharing floristic attributes with both the Vizcaíno and Gulf Coast deserts. The Sierra de La Libertad (also known as

CPD Site MA12: Central region of Baja California. The Vizcaíno Desert with an elephant tree or "copalquín" ***Pachycormus discolor*** **in the foreground.** Photo: Chris Mattison.

the Sierra de San Borja) is in the north and the Sierra de San Francisco in the south. The highest of Las Tres Vírgenes volcanoes reaches 1985 m; their lava field extends westward c. 60 km, and they may have erupted within historic times.

These highlands have mesas, hillsides with pronounced gradients, ravines, arroyos and canyons (Wiggins 1969, 1980). Xerophytic species and species more common to less arid climates occur. Characteristic taxa include *Fouquieria columnaris* (*Idria columnaris*), *Rhamnus crocea*, *Prunus ilicifolia*, *Prosopis glandulosa* var. *torreyana* (honey mesquite), *Yucca whipplei* ("lecheguilla"), *Xylococcus bicolor* (*Arctostaphylos bicolor*), *Ferocactus emoryi* var. *rectispinus*, *Croton ciliato-glanduliferum*, *Quercus oblongifolia*, *Sabal uresana*, *Cordia curassavica* and *Aralia scopulorum*.

Flora

Central Baja California has linkages with various nearby floristic regions (Wiggins 1960, 1980; Bowers, Delgadillo-M. and Sharp 1976), and many species reach their phytogeographic limit in approximately mid-peninsula (28°N). The mountains of southern California, U.S.A. have widely distributed genera shared with mountainous areas of Baja California, including *Rhus*, *Prunus*, *Penstemon*, *Rhamnus* and *Salvia* subgenus *Audibertia* (Nelson 1921). Affiliations with the arid southwestern U.S.A.'s Mojave Desert or Great Basin include the following genera well represented in drier regions:

CPD Site MA12: Central region of Baja California.
Boojum tree or "cirio" *Fouquieria columnaris* growing near the region's Gulf Coast. Photo: Barbara Tigar.

Astragalus, *Phacelia*, *Lotus*, *Cryptantha*, and to a lesser degree *Atriplex*, *Ephedra*, *Sphaeralcea*, *Chaenactis*, *Frankenia*, *Haplopappus*, *Chorizanthe* and *Eschscholzia* (Hastings, Turner and Warren 1972).

The Vizcaíno Desert and the other subdivisions of the Sonoran Desert share many species that are salt-tolerant, or are widely distributed – especially those that are subtropical, e.g. in *Bursera*, *Lycium*, *Stenocereus*, *Opuntia* and *Proboscidea*, as well as *Larrea tridentata* (Shreve and Wiggins 1964; Wiggins 1980).

The Baja California Peninsula is recognized as an area of moderate endemism (Rzedowski 1993), with approximately 10 genera and 20% of the species, and many of these peninsular endemics are present in the central region (Wiggins 1980). In Vizcaíno Desert Biosphere Reserve, excluding the flora of coastal lagoons, 496 species of vascular plants have been documented, 39 of which are endemic locally (León de la Luz, Cancino and Arriaga 1991). *Eriogonum* (Polygonaceae) is represented by 10 species, 8 of which are endemic. The families with the greatest species diversity are Compositae, Leguminosae, Gramineae, Cactaceae and Euphorbiaceae.

This central region of Baja California is thus a rich transitional zone (cf. Wiggins 1960), considering the lowland and highland habitats, the many species distributed on the peninsula mostly to the north or south and/or including species in northern North American and/or on the Mexican mainland, and the low proportion of local endemics (7.9%) but higher number of peninsular endemics.

Useful plants

The region was occupied by indigenous people, who are now extinct. There is only very limited indirect evidence to indicate how they utilized the natural resources. Nevertheless, ethnobotanical information on many species in this region is known from elsewhere (cf. Coyle and Roberts 1975), and a good many plants represent potential genetic reserves and aesthetic and commercial resources. For example, species such as *Fouquieria columnaris* (*Idria columnaris*) (boojum tree or "cirio") and *Pachycereus schottii* (*Lophocereus schottii*) forma *monstrosus* and many other succulents, pachycauls and dryland species are attractive for hobbyists, landscapers and others, and are in cultivation elsewhere.

Social and environmental values

In the mountainous areas ancient rock paintings are found, which are culturally significant and possibly among the earliest human documentations on the North American continent. The mountains are the principal renewable sources for water that supplies the towns and agricultural areas of the plains. Wild bighorn sheep (*Ovis canadensis*) inhabit the mountain slopes, and constitute a resource that when well managed for hunting, is important to obtain foreign currency for the state and federal governments.

Peninsular endemism in various groups of fauna is considerable (Ramamoorthy *et al.* 1993). Notably, the central region includes portions of two of the five mammalian primary areas on the peninsula, as determined by an index of faunistic change.

Baja California is one of five truly desert biotic provinces in Mexico, and still offers wilderness as well as rugged landscapes. The transverse dunes of the Vizcaíno Desert, resembling furrows in a mammoth ploughed field, occur in only a few places in the world.

Economic assessment

Tourism contributes to the economy of the region. Tourists (mainly from the U.S.A.) visit the lagoons around Guerrero Negro where grey whales (*Eschrichtius robustus*) come to reproduce. The lagoons also support hundreds of migratory birds from northern North America. The ancient rock paintings attract tourists to the sierras de Santa Martha and San Francisco.

Approximately 42,000 people live in central Baja California, mostly in rural areas. Santa Rosalía, San Ignacio and Guerrero Negro are the main towns, with a total population of c. 19,000. Only c. 8000 of the entire population are employed (BCS 1993). The main income-generating occupations in the general area are mining of salt and gypsum, harvesting of clams, and agriculture (Castellanos and Mendoza 1991).

The salt mines of Guerrero Negro are considered among the largest anywhere, with an annual production of 5.4 million tonnes. Gypsum mining on San Marcos Island (6 km south-east of Santa Rosalía) annually produces nearly 2.7 million tonnes. In the Vizcaíno plains, oil and gas reserves and deposits of magnesite, diatomite, chromite and asbestos have been found.

Agriculture is concentrated around the town of El Vizcaíno, which has c. 65 km² of irrigated land. Many of the field workers come from southern Mexico. However, too many demands are being placed on the limited groundwater aquifer, and it is estimated that fairly soon water problems will develop.

In the mountainous areas there are c. 24,000 goats, which cause significant damage to the native vegetation. At lower elevations, the cattle population is estimated to be c. 19,000, but they have no significant general impact on the native vegetation.

Threats

The ranchers do not control the grazing of their goats on the mountain slopes, which has led to serious plant damage and soil erosion. In some areas, overgrazing is so significant that only plants that are toxic to the goats can grow – e.g. *Astragalus francisquitensis* and *A. prorifer*.

On the western plains, roads have been built to bring in large equipment used for natural gas exploration by Petroleos Mexicanos. Reserves of natural gas have been found, but extraction is not profitable at present. On the western plains in Baja California Sur, six public grazing areas ("ejidos") encompassing 9500 km² have been set aside for 383 families. However, the pumped water supply is inadequate to support both agriculture and the livestock for these public grazing lands. Within a decade problems are expected to develop from brackish water intrusion into the groundwater.

The company Exportadora de Sal, S.A. manages the world's most productive salt works. Expansion of its activities has generated various impacts on the region's terrestrial and marine habitats. Terrestrial areas are inundated

MAP 23. CENTRAL REGION OF BAJA CALIFORNIA PENINSULA, MEXICO (CPD SITE MA12)

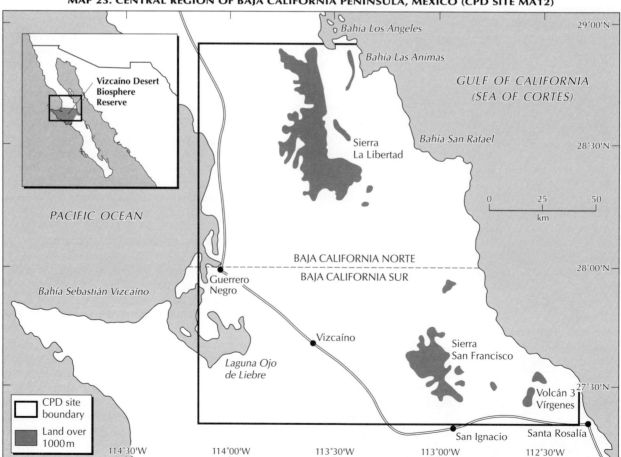

by saltwater, which for example affects the distribution of the peninsular pronghorn antelope (*Antilocapra americana* subsp. *peninsularis*), a deer-like mammal in danger of extinction.

Conservation

Approximately 42% (15,000 km²) of the central region of Baja California is protected in the Vizcaíno Desert Biosphere Reserve, which was declared in November 1988 and covers 25,500 km² (Ortega and Arriaga 1991) (see Map 23). Arid lands constitute nearly 40% of Mexico's territory, and c. 8% of these arid-land ecosystems are in conserved areas – c. 40% of the protected total is in this reserve. More research is needed on the region's biodiversity, including scientific studies to understand ecological interactions and adaptations of the species to this harsh environment.

Since goats are destroying extensive areas of native vegetation, special efforts are urgently needed to work with the ranchers to contain their animals and to convey the importance of maintaining naturally vegetated highlands. On the public-grazing lowlands, management plans for sustainable utilization of natural resources need to be provided. The personnel of the Centro de Investigaciones Biológicas de Baja California Sur in La Paz are familiar with these concerns, and although with limited support, they are attempting to address them.

References

BCS (1993). *Datos básicos 1992*. Gobierno Constitucional del Estado de Baja California Sur (BCS), La Paz, B.C.S., Mexico.

Bowers, F.D., Delgadillo-M., C. and Sharp, A.J. (1976). The mosses of Baja California. *J. Hattori Bot. Lab.* 40: 397–410.

Castellanos, A. and Mendoza, R. (1991). Aspectos socioeconómicos. In Ortega, R. and Arriaga, L. (eds), *La Reserva de la Biósfera El Vizcaíno en la Península de Baja California*. Publicación No. 4 del Centro de Investigaciones Biológicas de Baja California Sur (CIB-BCS), La Paz, B.C.S., Mexico. Pp. 33–52.

Coyle, J. and Roberts, N.C. (1975). *A field guide to the common and interesting plants of Baja California*. Natural History Publishing Co., La Jolla, Calif., U.S.A. 206 pp.

Durham, J.W. and Allison, E.C. (1960). The geologic history of Baja California and its marine fauna. In *Symposium: The biogeography of Baja California and adjacent seas. Syst. Zool.* 9: 47–91.

Ferrusquía-Villafranca, I. (1993). Geology of Mexico: a synopsis. In Ramamoorthy, T.P., Bye, R., Lot, A. and Fa, J.E. (eds), *Biological diversity of Mexico: origins and distribution*. Oxford University Press, New York. Pp. 3–107.

Hastings, J.R., Turner, R.M. and Warren, D.K. (1972). *An atlas of some plant distributions in the Sonoran Desert*. University of Arizona, Institute of Atmospheric Physics, Technical Reports on Meteorology and Climatology of Arid Regions No. 21. 255 pp.

León de la Luz, J.L., Cancino, J. and Arriaga, L. (1991). Asociaciones fisonómico-florísticas y flora. In Ortega, R. and Arriaga, L. (eds), *La Reserva de la Biósfera El Vizcaíno en la Península de Baja California*. Public. No. 4 del CIB-BCS, La Paz, B.C.S., Mexico. Pp. 123–154.

Nelson, E.W. (1921). *Lower California and its natural resources. Mem. Natl. Acad. Sci.* 16: 1–194.

Ortega, R. and Arriaga, L. (eds) (1991). *La Reserva de la Biósfera El Vizcaíno en la Península de Baja California*. Public. No. 4 del CIB-BCS, La Paz, B.C.S., Mexico. 317 pp.

Ramamoorthy, T.P., Bye, R., Lot, A. and Fa, J.E. (eds) (1993). *Biological diversity of Mexico: origins and distribution*. Oxford University Press, New York. 812 pp.

Rzedowski, J. (1978). *Vegetación de México*. Editorial Limusa, Mexico, D.F. 432 pp.

Rzedowski, J. (1993). Diversity and origins of the phanerogamic flora of Mexico. In Ramamoorthy, T.P., Bye, R., Lot, A. and Fa, J.E. (eds), *Biological diversity of Mexico: origins and distribution*. Oxford University Press, New York. Pp. 129–144.

Salinas-Zavala, C., Coria, R. and Díaz, E. (1991). Climatología y meteorología. In Ortega, R. and Arriaga, L. (eds), *La Reserva de la Biósfera El Vizcaíno en la Península de Baja California*. Public. No. 4 del CIB-BCS, La Paz, B.C.S., Mexico. Pp. 95–115.

Shreve, F. and Wiggins, I.L. (1964). *Vegetation and Flora of the Sonoran Desert*. Stanford University Press, Stanford, Calif., U.S.A. 2 vols., 1740 pp.

Turner, R.M. and Brown, D.E. (1982). Sonoran desert scrub. In Brown, D.E. (ed.), *Biotic communities of the American Southwest – United States and Mexico. Desert Plants* 4. Pp. 181–221.

Wiggins, I.L. (1960). The origins and relationships of the land flora. In *Symposium: The biogeography of Baja California and adjacent seas*. Part III, Terrestrial and fresh-water biotas. *Syst. Zool.* 9: 148–165.

Wiggins, I.L. (1969). Observations on the Vizcaíno Desert and its biota. *Proc. Calif. Acad. Sci.*, ser. 4, 36: 317–346.

Wiggins, I.L. (1980). *Flora of Baja California*. Stanford University Press, Stanford, Calif., U.S.A. 1025 pp.

Authors

This Data Sheet was written by José Luis León de la Luz (Centro de Investigaciones Biológicas de Baja California Sur, A.C., División de Biología Terrestre, Apdo. Postal 128, La Paz, Baja California Sur 23000, Mexico) and Olga Herrera-MacBryde (Smithsonian Institution, Department of Botany, NHB-166, Washington, DC 20560, U.S.A.).

PETEN REGION AND MAYA BIOSPHERE RESERVE
Guatemala

Location: In northern Guatemala, Department of Petén between about latitudes 16°–18°N and Maya Biosphere Reserve between latitudes 16°49–17°49'N, both between longitudes 89°08'–91°50'W.

Area: Petén Department c. 36,000 km², Maya Biosphere Reserve 15,000–16,000 km².

Altitude: c. 10–800 m.

Vegetation: Subtropical semi-deciduous moist forest, savanna, wetlands.

Flora: High diversity: over 3000 plant species in Maya Biosphere Reserve; distinct regional endemism; threatened species.

Useful plants: Timber species, fuelwood, fibres, fruits, medicinals; Maya Biosphere Reserve important for extraction of non-timber forest products: e.g. xate palm leaves, chicle, allspice.

Other values: Faunal refuge, including threatened species. Indigenous peoples, many major archaeological sites, tourist attractions.

Threats: Logging, colonization, agriculture, grazing, road building, oil exploration, fire, erosion, water pollution.

Conservation: National Parks and Monuments, Nature Reserves; Maya Biosphere Reserve includes 5 National Parks, 3 Biotopes and a multiple-use area – Laguna del Tigre is recognized under RAMSAR, and Tikal NP is a World Heritage Site.

Geography

The Department of Petén, which comprises the northern third of Guatemala, is one of the last remaining large wildland areas in Central America. Northern Petén is a plateau at an elevation of 200–400 m and forms the beginning of the Yucatán Peninsula of Mexico (Ferrusquía-Villafranca 1993). It is bounded to the south by a transverse chain of lakes extending eastward from near the Sierra del Lacandon (to 600 m elevation) and beyond Laguna Perdida to lakes Yaxhá and Sacnab. Central Petén, the low area between Lake Petén Itzá and the Subín River, encompasses only a small portion of the region. It is made up of Cretaceous limestone beds overlain with broad, deep clay. The area has many limestone hills and scattered sinkholes. Southern Petén for the most part is a broad and higher slightly undulating basin, except in the south-east where the Maya Mountains extend westward, descending from 1120 m in Belize – they are significant since the average elevation in the department is 200 m. The Petén is dominated by often thin soils of red clay ("tierra rosa") and black or dark brown clay ("rendzina"); humic gleys, grumusols and red-yellow podzols are interspersed throughout.

The Petén's main rivers – the San Pedro, Machaquilá (Santa Amelia) and La Pasión – flow westward through the department and empty into the Usumacinta River, which forms the western border with Mexico (see Lacandon Data Sheet, CPD Site MA1). The mean annual temperature is 26.5°C; the extremes are 12°C and 40°C. The mean annual precipitation ranges from 900–3500 mm, increasing from north-east to south. There is a pronounced dry season in January through April–May (Leyden 1984; Heinzman and Reining 1990; Schwartz 1990). Winds are stronger from February to June, and there are sporadic hurricanes.

The Maya Biosphere Reserve (15,000–16,000 km²) occupies the northern 40% of the Petén, encompassing nearly 10% of Guatemala's land area. It includes part of the municipalities La Libertad, San Andrés, San José, Flores and Melchor de Mencos. The landscape varies from gently undulating plains to karst topography with rounded to steep hills and narrow valleys. It is underlain by Early Tertiary limestone. Soils in the seasonally inundated lowlands are deep, poorly drained clay substrates, whereas in the uplands are shallow clay soils – neither are suitable for sustained low-input agricultural production. The precipitation averages 1200–1500 mm annually. The warmest period is April to September, with an average temperature of 32°C, and the coolest is November–January, with an average minimum of 20°C.

Vegetation

About 85% (30,000 km²) of the Petén was covered with semi-deciduous (seasonal) subtropical moist forest – the majority of the closed tropical broadleaved forest in Guatemala (cf. Nations and Komer 1984); less than 50% remains (Schwartz 1990). The northern Petén's vegetation (including the Maya BR) has much of the same flora as in the Yucatán Peninsula of Mexico and northern Belize. The Maya BR with the contiguous forests of Mexico and Belize (Map 24) is now Mesoamerica's last large lowland forest, c. 20,000 km². The canopy is 10–25 m high, being lower in seasonally flooded forest. Lundell (1937) defines three major tree associations within upland climax forest: (1) "ramonal" – groves of *Brosimum alicastrum*, especially found at sites of Maya ruins; (2) "caobal" – with *Swietenia macrophylla* (mahogany); and (3) "zapotal" – with *Manilkara zapota*, which is characteristic of dry rather than mesic upland forest. Epiphytes (e.g. orchids, aroids, bromeliads, cacti), ferns, bamboo and lianas are very abundant. In low-lying basins as at Laguna del Tigre, Laguna Perdida and lakes Petén Itzá and Yaxhá are swamps or marshes, which may be fringed by a "botanal" – with *Sabal* sp. Dense communities are found along rivers and the edges of lakes. Floating fern and sedge bogs occur in some lakes, and water-lilies in shallow open waters – the dominants are *Nymphaea ampla* and *Nymphoides humboldtianum*.

The central Petén has savanna with forested hills. Sinkholes are covered by herbaceous and subclimax forest (Lundell 1937). The savanna vegetation may have been created during the ancient Maya times of 3000–1700 BP (Leyden 1984), when they burned the forest for cultivation; continued burning prevents the forest from re-establishing. The savanna supports a diverse and complex herbaceous flora, most of which is fire resistant. The grasslands are surrounded by a barrier of scrub that acts as a buffer, protecting the mesophytic forest from fire.

MAP 24. PETEN REGION AND MAYA BIOSPHERE RESERVE, GUATEMALA (CPD SITE MA13) AND ADJACENT AREAS

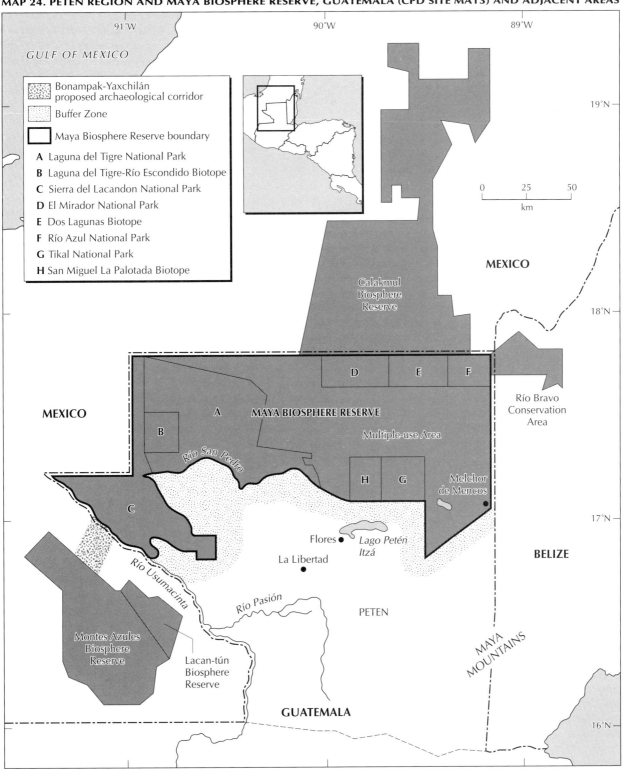

187

The southern Petén, which is the least explored botanically, has much of the same vegetation as the northern Petén but some species are less common, such as *Manilkara zapota*, perhaps replaced by *M. chicle*; *Swietenia macrophylla* has been logged extensively (Lundell 1937; Nations and Komer 1984). Species of Sapotaceae are most characteristic of this wetter area. The vegetation is allied more with southern Guatemala and southern Belize. Extensive aquatic and riparian vegetation occurs.

The tropical rain forests of Guatemala are of special value because their presence in northern Central America and southern Mexico is particularly complex. Guatemala has some rain forests no older than 10,000–11,000 years, and some in the Petén are considerably younger (1000 years old) due to former Mayan disturbance (Binford *et al.* 1987). Late Glacial vegetation consisted of marsh, savanna and juniper scrub. The Petén thus was not a Pleistocene refugium for mesophytic taxa (Leyden 1984), but the adjacent southern Izabal area of Guatemala and Belize may have been (Wendt 1993).

Flora

The Petén floristic associations are continuous with associations in the Mexican Lacandon region (CPD Site MA1). Since much of the Petén is not well known botanically, it is difficult to estimate the number of vascular plants or endemics. The floristic diversity of the Maya BR is considered exceptional, with over 3000 plant species (CONAP 1992). The overall region's flora is considered distinctive; e.g. many of the Petén's regionally endemic taxa are shared to varying extents with northern Belize and Mexico's majority of the Yucatán Peninsula to eastern Tabasco and the eastern highlands of Chiapas (Breedlove 1981; Wendt 1993). Most local Petén endemics have been found in areas that allow little human intervention, such as steep hills and swamps (Lundell 1937).

Information on the distribution of plants in the Petén, for example to determine which are the rare species, is still very limited and incomplete. Subtropical forests contain relatively low tree-species diversity and a higher number of individuals per species (Salafsky, Dugelby and Terborgh 1993). The high densities of species such as *Manilkara zapota*, *Chamaedorea* spp. and *Pimenta dioica* facilitate extractive industry.

Useful plants

The Petén region is rich in serviceable plants, for example thatching palms, construction materials, fuelwood, fibres – e.g. *Desmoncus* sp. ("bayal") and *Philodendron* sp. ("mimbre") for basketry and furniture, forest fruits, medicinal plants and species marketable from upland forests, such as *Manilkara zapota* ("chicozapote"), *Chamaedorea* spp. (mostly two understorey palms) and *Pimenta dioica*. A few studies have analysed the economic benefits of a conserving, sustainable use of Guatemala's tropical forests and renewable resources (Nations *et al.* 1988; Reining and Heinzman 1992; Salafsky, Dugelby and Terborgh 1993). An estimated 80% of the hardwoods in Guatemala occur in the Petén, such as *Swietenia macrophylla*, *Cedrela odorata*, *Calophyllum brasiliense* var. *rekoi*, *Pouteria* spp.,

Bursera simaruba, *Spondias* and *Ficus* (Leyden 1984). The Maya BR contains more than 300 species of useful trees (CONAP 1990).

A potentially important forest resource is *Brosimum alicastrum* ("ramón") – a common tree occasionally up to 30 m tall, which may have been nurtured by the Maya (Leyden 1984). The seeds were an important food source in pre-Columbian times, but present human consumption is quite low (Heinzman and Reining 1990); the fruits, foliage and bark are gathered as forage for mules and horses. These parts are rich in protein and other essential nutrients.

Economic assessment of non-timber forest products

Heinzman and Reining (1990) analysed some potentially sustainable rural extraction practices in the northern Petén. Collecting several products more or less sustainably for export represents a wage resource for over 6000 people who otherwise subsist mainly on slash-and-burn ("milpa") agriculture (see Lacandon Data Sheet, CPD Site MA1), but only 13% of the Petén soils are deep and well drained. The total economic return from these non-timber forest products is greater than if the forest were converted to pasture (Nations *et al.* 1988; Heinzman and Reining 1990).

In 1990 Guatemala passed a law (Decree 5-90) for a Maya Biosphere Reserve, designating 7500 km² of the BR for extractive industry based on non-timber forest products. During the past 30–100 years, three such major products have been harvested: (1) Millions of "xate" palm leaves (*Chamaedorea elegans*, *C. oblongata*) are exported through the year for greenery in floral arrangements. Xate produces US$4–6 million annually (Morell 1990). (2) Chicle, the latex of *Manilkara zapota* (a tree to more than 30 m tall), is extracted for the manufacture of chewing gum. The largest concentration of high-grade chicle is found in the Maya BR and sold primarily to the Japanese. Small quantities of latex from *Ficus lundellii*, *Bumelia mayana* and *Stemmadenia donnell-smithii* may be used as enhancing supplements. In 1990–1991 the high quality latex sold for US$3.75 per kg (Reining and Heinzman 1992). (3) Another important annual product harvested on a rather sustainable basis is allspice ("pimienta gorda", *Pimenta dioica*), a common tree rarely over 20 m tall. Currently Guatemala (and Mexico) supply almost 30% of the international market.

An important aspect of harvesting these renewable resources is that they promote conservation of the forest as well as providing income locally. A family can earn three times more as an average daily wage from the forest products than from clearing forest, planting maize and raising cattle. Heinzman and Reining (1990) recommend development of institutions to ensure sustainable use of these common resources, and to further diversify in non-timber forest products that are subject to sustainable exploitation.

Social and environmental values

The Petén has a rich fauna which is poorly known. The country's list so far includes 1453 vertebrate species (not including saltwater fishes); at least 333 bird species occur in the Petén. The Petén wetlands provide significant wintering grounds for many North American migratory bird species. The Laguna del Tigre complex of diverse wetlands represents one of the most extensive freshwater wetland areas in Central America, of which 484 km² are recognized under the world's RAMSAR convention.

The Petén is an important refuge for many species, such as howler monkey, ocelot, margay cat, jaguar, puma, Baird's tapir, harpy eagle, macaws, Moreletti's and American crocodiles, iguana, beaded lizard and boa constrictor. About 133 of the animal species are considered threatened; some species are listed in CITES appendices as at risk from international trade (URL 1984). No globally threatened bird species or Endemic Bird Areas are in this rain forest. Nevertheless, the area is of national importance for a number of species of birds of prey, including the near-threatened orange-breasted falcon (*Falco deiroleucus*).

Other economic benefits of conserving the biological diversity of the Petén include stabilization of hydrological functions, soil protection, tourism and the opportunities to create jobs. Tourism is a large and growing industry vital to the economy of Guatemala – in 1990 the country earned US$185 million. Tours in the Petén for nature and archaeological interest show that there is strong potential. In 1970 an all-weather road opened the central Petén to southern Guatemala. Tikal National Park draws 15% of the tourists who visit the country; it is estimated that 2500

archaeological sites occur in the Petén, perhaps half in the Maya BR. Several Maya peoples live in the Petén, e.g. the Lacandon, Itzá and Mopán.

The combination of the tropical forest environment and Maya history and culture is most appealing and should be conserved and promoted. A Mundo Maya (Maya World) or Ruta Maya (Maya Trail) is being organized, similar to the Inca Trail in South America. The Mundo Maya will connect the major archaeological sites and National Parks of Mexico, Guatemala, Belize, Honduras and El Salvador (Hagman 1989).

Threats

There are many threats to the vegetation and flora of the Petén, including colonization, road building, logging, fire, ranching cattle and pigs, oil exploration and over-collecting (D'Arcy 1977; Schwartz 1990; Stuart 1992). The population, which for decades was 15,000 persons or less in Petén Department (c. 36,000 km²), rose to 65,000 in 1973 and presently has reached over 300,000 people, increasing at the annual rate of 5.5% (compared to 2.9% for the rest of the nation) (Southgate and Basterrechea 1992). Colonists have included "ladinos" (mestizos) from elsewhere in Guatemala, refugees from El Salvador and Kekchí Amerindians from Alta Verapaz Department. The region is

CPD Site MA13: Petén region and Maya Biosphere Reserve, Guatemala. Rain-forest canopy and Maya temple ruins within Tikal National Park. Photo: Tony Morrison/South American Pictures.

becoming more accessible with construction of a major road linking the Petén with Guatemala City and Belize (*Prensa Libre* 1991).

Agricultural expansion and logging are the major reasons for forest loss. The rate of deforestation in the Petén is very high – in the last 10–15 years, 200–400 km² have been converted annually. Colonization is occurring in the western Petén and the Franja Transversal del Norte, which stretches along the southern Petén border. Two government agencies are granting parcels of land between 20 ha and 100 ha in the department for colonization. Milpa cultivation is destroying forest because the increasing human populations are not allowing adequate fallow periods for forest regeneration (cf. Lundell 1937). Severe erosion is evident in some places, e.g. the south-eastern Petén (Schwartz 1987, 1990).

Although logging permits within the Maya BR were revoked in March 1991, some residents are illegally cutting mahogany (*Swietenia macrophylla*) for construction or exporting the wood to Belize and Mexico for income; this valuable species is becoming scarce (Moser *et al.* 1975; Stuart 1992). *Chamaedorea* (xate) densities are being reduced (especially *C. elegans*) by over-collecting, and *Manilkara zapota* also probably has declined during the past 100 years. Inadequate transportation in the Petén still prevents large-scale cattle production, but there is some cattle-raising near Tikal and in the south-eastern and west-central portions of the department. Most of the cattle are shipped to Belize to be transported and sold in Mexico. Oil and mineral explorations have destroyed some forest with construction of roads and test sites; these threats are not intense, as the anticipated amounts of oil and mineral strikes have not been found.

Mexico and Guatemala have considered constructing a hydroelectric dam on the Usumacinta River. The proposed dam would have flooded more than 1000 km² of the Petén forest. The likely sediment load in the river would have required check dams farther upstream to keep the main dam from silting up (Nations and Komer 1984; Wilkerson 1985). The project was halted in 1992 by Mexico's president (see the Lacandon Data Sheet, CPD Site MA1).

Conservation

Between 1955 and 1988 Guatemala declared 52 conservation areas, although the majority did not meet international criteria for protected areas. Areas were reserved for scenic landscapes, pre-Columbian Maya ruins and particular plant or animal species (Biotopes) (Godoy 1988). For the last three decades, Guatemalan conservationists have been aware that much more of the country's biological diversity needed to be conserved and managed, and they recommended the creation of other reserves (Godoy 1987a, 1987b). A regional network of protected areas has been recommended for Petén Department – the Integrated System of Protected Areas of the Petén (SIAP). SIAP would comprise National Parks, Forest Reserves, Wildlife Refuges and other types of protected areas. These areas have been put in priority order for the instigation and development of conservation measures (Godoy and Castro 1990).

In 1989 Guatemala passed a Forestry Law (Decree No. 70-89) due to the increasing degradation of forests, and stated the importance of protecting and renovating forest resources.

A new agency was created to administer and manage forest resources, wild life and some National Parks – the Dirección General de Bosques y Vida Silvestre (DIGEBOS), incorporating the former Instituto Nacional Forestal (INAFOR). Also in 1989, a significant step was made towards increasing the number and effectiveness of conservation units with passage of a Protected Areas Law (Decree No. 4-89). An extensive national system of conservation units was created as the Guatemalan System of Protected Areas (SIGAP) and a new organization was established to manage them – the Consejo Nacional de Areas Protegidas (CONAP) (cf. Nations *et al.* 1988). This law established boundaries for 6 existing reserves and created 44 new Special Protection Areas (TNC 1989). About 44.4% of these areas are in the Petén (WCMC 1992) – c. 14,000 km² of the area north of 17°10'N and the Sierra del Lacandon in the western Petén.

In 1990, the Maya Biosphere Reserve was established (see Map 24). In three grouped areas, it has five National Parks and three Reserves (Biotopes): (1) Laguna del Tigre NP and Laguna del Tigre-Río Escondido Biotope (3508 km²), with Sierra del Lacandon NP nearby (1950 km²); (2) El Mirador NP, Dos Lagunas Biotope and Río Azul NP (1470 km²); and (3) Tikal NP (573 km²) and the adjacent San Miguel La Palotada (El Zotz) Biotope (435 km²). A multiple-use area (7500 km²) adjoins these protected areas for their protection and for the management of renewable forest resources, and next to the BR's southern boundary is a 10–15 km wide buffer zone (250 km²).

The Maya BR is cooperatively administered: the lead agency is CONAP; other key participants include the Centro de Estudios Conservacionistas (CECON) – the academic unit of the Universidad de San Carlos de Guatemala that is responsible for promoting field research and conservation of renewable resources; and the Instituto de Antropología e Historia (IDAEH) – the government institution responsible for administering the marvellous cultural heritage. The Maya BR is one of 11 areas given priority under the region's 1992 Convenio para la Conservación de la Biodiversidad y Protección de Areas Silvestres Prioritarias en América Central.

The Maya BR is among the key sites in The Nature Conservancy (TNC) Parks in Peril campaign, to build a conservation infrastructure and secure long-term funding to sustain local management of the protected areas and integrate them into local economies (Houseal 1990). The USAID Maya Resource Management project (MAYAREMA) is offering financial and technical assistance to CONAP to manage the resources of the BR more sustainably.

The Maya BR abuts two neighbouring protected areas: in Mexico, Calakmul Biosphere Reserve (8250 km²); in Belize, Río Bravo Conservation Area (610 km²). The U.S. Man and the Biosphere Program in 1992 approved a proposal to develop a regional approach for sustainable development and conservation of natural resources in the Maya tri-national region of Mexico, Guatemala and Belize. Paseo Pantera, a consortium of U.S. and Central American non-governmental and governmental organizations and institutions, helps regionally to enhance wildlands management and preserve natural diversity by means of a biological corridor from Guatemala through Panama, and Tikal NP is one of the five key areas in as many countries for its pilot work (Jukofsky 1992; Marynowski 1993).

Other international conservation organizations are supporting the Guatemalan efforts, including the Center for International Development and Environment/World Resources Institute (Nations *et al.* 1988), WWF – U.S. (1989; Cohn 1989) and Conservation International – which is aiding Guatemalan decision-makers and local communities by promoting low-impact tourism and developing markets for sustainably harvested products (CI 1991). The Nature Conservancy is preparing technical studies on 14 protection areas in the Petén to secure their status as National Parks and has established a Conservation Data Centre in cooperation with the Universidad de San Carlos (TNC 1989).

References

Binford, M.W., Brenner, M., Whitmore, T.J., Higuera-Gundy, A., Deevey, E.S. and Leyden, B.W. (1987). Ecosystems, paleoecology and human disturbance in subtropical and tropical America. *Quat. Sci. Rev.* 6: 115–128.

Breedlove, D.E. (1981). Introduction to the *Flora of Chiapas*. In Breedlove, D.E. (ed.), *Flora of Chiapas*. Part 1. California Academy of Sciences, San Francisco. 35 pp.

CI (1991). CI launches program in the Maya heartland. *Tropicus* 5(3): 3.

Cohn, J.P. (1989). A will to protect. *Américas* 41(2): 46–53.

CONAP (1990). *Reserva de la Biósfera Maya*. Consejo Nacional de Areas Protegidas (CONAP), Guatemala City. 8 pp.

CONAP (1992). *Plan maestro de la Reserva de la Biósfera Maya – RBM*. CONAP, Guatemala City. 25 pp.

D'Arcy, W.G. (1977). Endangered landscapes in Panama and Central America: the threat to plant species. In Prance, G.T. and Elias T.S. (eds), *Extinction is forever*. New York Botanical Garden, Bronx. Pp. 89–104.

Ferrusquía-Villafranca, I. (1993). Geology of Mexico: a synopsis. In Ramamoorthy, T.P., Bye, R., Lot, A. and Fa, J.E. (eds), *Biological diversity of Mexico: origins and distribution*. Oxford University Press, New York. Pp. 3–107.

Godoy, J.C. (1987a). Areas silvestres protegidas potenciales de El Petén, Guatemala. *Perspectiva* 8(1): 166–178.

Godoy, J.C. (1987b). *Memorias minitaller de áreas silvestres protegidas*. Centro de Estudios Conservacionistas (CECON), Centro Agronómico Tropical de Investigación y Enseñanza (CATIE) and WWF – U.S. Guatemala City. 21 pp.

Godoy, J.C. (1988). *Algunas consideraciones sobre áreas silvestres protegidas fronterizas*. CECON and CONAMA (Comisión Nacional del Medio Ambiente), Guatemala City. 3 pp.

Godoy, J.C. and Castro, F. (1990). *Plan estratégico del Sistema de Areas Protegidas de El Petén, Guatemala (SIAP)*. Proyecto de Conservación para el Desarrollo Sostenido en América Central. CATIE y el Unión Internacional para la Conservación de la Naturaleza y los Recursos Naturales (UICN). Turrialba, Costa Rica. 105 pp.

Hagman, H. (1989). Maya plan for tourists is, at heart, for ecology. *Washington Times* (9 October): E1, E5.

Heinzman, R.M. and Reining, C.C.S. (1990). *Sustained rural development: extractive forest reserves in the northern Petén of Guatemala*. Trop. Resources Inst. Working Paper No. 37, Yale School of Forestry and Environmental Studies, New Haven.

Houseal, B. (1990). Maya riches. *Nature Conservancy Mag.* 40(3): 16–21.

Jukofsky, D. (1992). Path of the Panther. *Wildlife Conservation* 95(5): 18–24.

Leyden, B.W. (1984). Guatemalan forest synthesis after Pleistocene aridity. *Proc. Natl. Acad. Sci.* 81: 4856–4859.

Lundell, C.L. (1937). *The vegetation of Petén*. Carnegie Institution of Washington Public. No. 478, Washington, D.C. 244 pp.

Marynowski, S. (1993). Paseo Pantera: the great American biotic interchange. *Wild Earth* (Special Issue): 71–74.

Morell, V. (1990). Bringing home a piece of the jungle. *Int. Wildlife* 20(5): 12–15.

Moser, D. and the editors of Time-Life Books (1975). *Central American jungles*. Time-Life Books, New York. 184 pp.

Nations, J.D. and Komer, D.I. (1984). *Conservation in Guatemala*. Center for Human Ecology, Austin, Texas. 170 pp.

Nations, J.D., Houseal, B., Ponciano, I., Billy, S., Godoy, J.C., Castro, F., Miller, G., Rose, D., Rosa, M.R. and Azurdia, C. (1988). *Biodiversity in Guatemala: biological diversity and tropical forest assessment*. Center for International Development and Environment/World Resources Institute, Washington, D.C. 185 pp.

Prensa Libre (13/9/1991). Carretera a Petén será una realidad.

Reining, C.C.S. and Heinzman, R.M. (1992). Nontimber forest products in the Petén, Guatemala: why extractive reserves are critical for both conservation and development. In Plotkin, M. and Famolare, L. (eds), *Sustainable harvest and marketing of rain forest products*. Island Press, Washington, D.C. Pp. 110–117.

Salafsky, N., Dugelby, B. and Terborgh, J. (1993). Can extractive reserves save the rain forest? An ecological and socioeconomic comparison of nontimber forest product extraction systems in Petén, Guatemala, and West Kalimantan, Indonesia. *Conserv. Biol.* 7(1): 39–52.

Schwartz, N.B. (1987). Colonization of northern Guatemala: the Petén. *J. Anthr. Res.* 43: 164–183.

Schwartz, N.B. (1990). *Forest society: a social history of Petén, Guatemala.* University of Pennsylvania Press, Philadelphia. 367 pp.

Southgate, D. and Basterrecha, M. (1992). Population growth, public policy and resource degradation: the case of Guatemala. *Ambio* 21: 460–464.

Stuart, G.E. (1992). Maya heartland under siege. *Natl. Geog.* 182(5): 95–106.

TNC (The Nature Conservancy) (1989). Guatemala enacts landmark conservation law. *Nature Conservancy Mag.* 39(3): 36.

URL (1984). *Perfil ambiental de la República de Guatemala*, Tomo II. Universidad Rafael Landívar (URL), Instituto de Ciencias Ambientales y Tecnología Agrícola (ICATA). URL/USAID-Guatemala-ROCAP Contract No. 596-0000-C-00-3060-00, Guatemala City. 249 pp.

WCMC (World Conservation Monitoring Centre) (1992). *Protected areas of the world. A review of national systems.* Vol. 4: *Nearctic and Neotropical.* IUCN, Gland, Switzerland and Cambridge, U.K. 459 pp.

Wendt, T. (1993). Composition, floristic affinities, and origins of the canopy tree flora of the Mexican Atlantic slope rain forests. In Ramamoorthy, T.P., Bye, R., Lot, A. and Fa, J.E. (eds), *Biological diversity of Mexico: origins and distribution.* Oxford University Press, London. Pp. 595–680.

Wilkerson, J.K. (1985). The Usumacinta River: troubles on a wild frontier. *Natl. Geog.* 168(4): 514–543.

WWF - U.S. (1989). Latin America and the Caribbean new and ongoing projects. In *WWF – U.S. 1989/90 Program.* Washington, D.C. Pp. 267–275.

Authors

This Data Sheet was written by Olga Herrera-MacBryde and Jane Villa-Lobos (Smithsonian Institution, Department of Botany, NHB-166, Washington, DC 20560, U.S.A.).

CENTRAL AMERICA: CPD SITE MA14

SIERRA DE LAS MINAS REGION AND BIOSPHERE RESERVE
Guatemala

Location: The Sierra de las Minas is in eastern Guatemala between about latitudes 15°07'–15°21'N and longitudes 89°18'–89°45'W.

Area: Region 4374 km², reserve 2363 km².

Altitude: 150–3015 m.

Vegetation: The region has cloud-forest associations, lower montane moist, wet and rain forests, premontane wet and rain forests, tropical and premontane dry forests, and thorn scrub.

Flora: Extremely diverse (over 2000 species recorded); high species endemism; several disjunct taxa; southern limit for several northern genera (e.g. *Acer*, *Taxus*).

Useful plants: Major timber reserves, especially conifers, and some remnants of lowland hardwood forests in north and south-east. 13 conifer species in region, which is a major centre for *Pinus*. Other useful species include tree ferns (for their adventitious roots), bamboo (for cottage industries) and many medicinal and food plants. Region potentially rich in phytogenetic resources.

Other values: Watershed protection – Sierra de las Minas is main source of water for local irrigation, light industry and household use. 62 permanent streams begin in upper elevations of BR. The scenery is important for development of tourism.

Threats: Expansion of agricultural frontier, logging, colonization, roads, over-hunting.

Conservation: Nearly 55% of region declared a Biosphere Reserve in 1990. A management plan approved in 1992 and active management being implemented, but facing heavy pressure from timber interests.

Geography

The Sierra de las Minas forms a mountain chain 15–30 km wide extending from Lake Izabal westward for 130 km (Map 25). It covers 4374 km² in Guatemala's departments of Izabal, Zacapa, El Progreso, Baja Verapaz and Alta Verapaz. The range is bordered to the south by the Motagua River and to the north by the Polochic River, both of which flow in valleys formed by geological faulting. Most of the 62 permanent streams arising in the Sierra de las Minas flow directly into either the Polochic or Motagua.

In the north, the sierra descends abruptly to the Polochic Valley. An unbroken ridge above 2100 m extends from near Chilascó through Cerro Raxón (3015 m) for 65 km east beyond Montaña El Imposible (2610 m). To the east, elevations drop gradually toward Lake Izabal and the Motagua River. The southern slopes are less steep and the forests more accessible. To the west, the sierra is bounded by the Salamá and San Jerónimo valleys, of igneous origin. To the north, Palaeozoic rock formations include schists and gneisses, which grade into Tertiary metamorphosed amphibolites and marbles to the south. There are belts of serpentine along the north-western margin and the southern side of the range (Weyl 1980).

The soils are generally shallow, often lateritic, 25–50 cm deep and consist of alluvial clays and loams. Sixty-five percent are on slopes highly susceptible to erosion. The Polochic Valley contains rich alluvial soils.

The Sierra de las Minas is heavily influenced by the north-east trade winds from the Caribbean Sea. Rainfall varies from over 4000 mm on mountain peaks and 2000 mm on Polochic-facing slopes to less than 500 mm in western Motagua Valley around Zacapa. In general rainfall is seasonal, especially on southern slopes, with marked decreases from January to April.

Temperatures vary considerably; information is lacking, especially for higher elevations. In the arid Motagua Valley mean temperatures average 24°C, ranging between 11.5°–41°C; at intermediate elevations temperatures may range between 5°–25°C; and at 1750 m and above light frosts are regularly experienced between December and March.

The higher slopes of the sierra are almost uninhabited, whereas the Motagua and Polochic valleys support ethnically distinct populations. To the south in the Motagua Valley the population is "ladino" (mestizo), with numerous small towns and villages, among them Río Hondo with c. 15,000 inhabitants. Subsistence crops are grown, such as maize, beans and cabbage, together with tomatoes, melons, tobacco and cucumber as cash crops. At mid-elevations coffee and cardamom farms may be found, together with low density cattle-farming and lumber extraction, particularly pines for electric posts. Hunting is a popular pastime.

The Polochic slopes are inhabited by the Q'eqchí to the north and the Poqomchí and Achí to the west. These lower slopes and valley floors are devoted to large farms with cattle, sugarcane and rice as major sources of income, and coffee and cardamom above c. 600 m. The steeply forested upper slopes are being increasingly invaded by Amerindian families displaced from lower elevations by the large farms. This is a significant threat to the Biosphere Reserve, which can only be mitigated by emphasis on sustainable resource use.

Vegetation

Two vegetation types of the sierra are especially significant in species diversity, endemic species and uniqueness to Mesoamerica: lower elevation thorn scrub and cloud forest.

The thorn scrub represents one of the driest areas of Central America, dominated by cactus and *Acacia* species, and is much threatened by light industry, cattle-ranching and irrigated crops. Only 200–300 km² of this formation remain; local communities have expressed interest in setting aside tracts for conservation. Most of the formation has already been disturbed to varying extents. The most abundant species include *Cephalocereus maxonii*, *Nyctocereus guatemalensis*, *Opuntia* spp., *Pereskiopsis*, *Acacia* spp. and *Guaiacum sanctum*.

The cloud forest, in the lower montane rain-forest zone (Holdridge system), covers c. 1300 km² – more than 65% is probably primary forest. This may represent the largest unbroken extent of cloud forest in Mesoamerica. The lower elevational limits are 1500 m on northern slopes, 1100 m on the south-eastern edge and 1900–1950 m on southern aspects of the range. Species composition varies depending on elevation. The highest elevations (2700 m and above) have forests with *Pinus ayacahuite*, *Abies guatemalensis*, *Quercus* spp., *Taxus globosa*, *Alnus*, etc. From c. 2000–2700 m broadleaved forests predominate, with important canopy trees being *Quercus* spp., Lauraceae (*Persea donnell-smithii*, *P. sessilis*, *P. schiedeana*), *Podocarpus oleifolius*, *Magnolia guatemalensis*, *Alfaroa costaricensis*, *Billia hippocastrum* and *Brunellia mexicana*.

Cloud-forest associations also occur at lower elevations in premontane rain forest and premontane wet forest down to 1000 m above Lake Izabal and the Polochic-facing slopes.

Most of these associations are as yet poorly known, but appear to have many interesting species and would repay further study. For instance the palm *Colpothrinax cookii*, extirpated at its type locality, was recently found here. There are several orchid species of mainly South American genera, such as *Paphinia* and *Kegeliella*.

The conifer forests, which include 13 species, are important lumber and germplasm reserves. Attempts are being made to incorporate these forests into sustainable harvesting systems, especially the mid-elevation lower montane moist forests dominated by *Pinus patula* subsp. *tecunumanii* and *P. oocarpa*. Associated species include *Quercus* spp., *Liquidambar*, *Acer skutchii* and *Tillandsia usneoides*. At lower elevations in premontane wet forests are scattered extensions of *Pinus caribea*, in association with *Curatella americana* and *Quercus* spp. especially where there are limestone outcrops. *Juniperus comitana* and *Cupressus lusitanica* occur as relicts in scattered stands at 1700–1900 m; and cloud forests contain *Abies guatemalensis*, *Taxus globosa*, *Podocarpus oleifolius* and *Pinus ayacahuite*. On the southern slopes at 1600–1900 m, a very interesting association includes several palm species.

To the south and west, lower montane wet forest occupies small extensions characterized by the presence of *Quercus* spp., *Pinus oocarpa*, *Alnus jorullensis*, *Prunus barbata* and *P. brachybotra*. On the dry southern slopes of the sierra up to c. 900 m, tracts of deciduous tropical dry forest grade into premontane dry forest. The species include *Quercus* spp., *Bursera simaruba*, occasional *Cedrela odorata*, *Ceiba aesculifolia*, *Leucaena guatemalensis*, *Cochlospermum vitifolium*, *Gliricidia sepium* and *Pseudobombax ellipticum*.

MAP 25. SIERRA DE LAS MINAS REGION AND BIOSPHERE RESERVE, GUATEMALA (CPD SITE MA14)

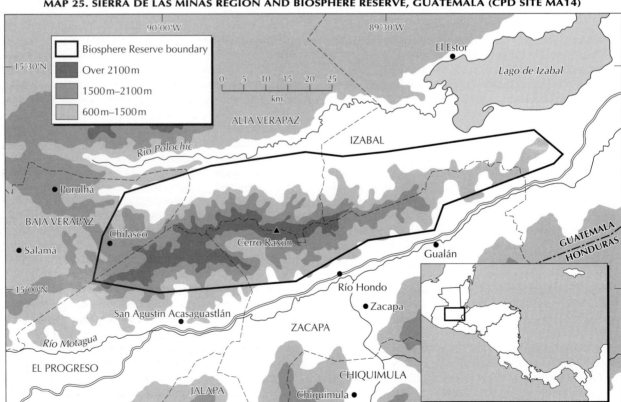

CPD Site MA14: Sierra de las Minas region and Biosphere Reserve, Guatemala, showing the threatened *Abies guatemalensis*. Photo: Shirley Keel.

In restricted areas of the sierra, marble outcrops support an especially interesting association that depends mainly on cloud-borne precipitation. Epiphyte-laden oak forests contain endemic agaves (e.g *Agave minarum*), *Beaucarnea guatemalensis* and epiphytic cacti. This association contrasts dramatically with surrounding forests of *Pinus oocarpa* and *P. patula* subsp. *tecunumanii* on dry soils.

Flora

The flora of the Sierra de las Minas is represented in the *Flora of Guatemala* (Standley, Steyermark and Williams 1946–1977). Steyermark collected extensively on the Motagua side of the sierra in Zacapa and El Progreso. Because of its general inaccessibility, the flora of the Polochic drainage is poorly known, and there are areas on both sides where there has been almost no exploration. The orchids are probably best known: 230 species in 25 genera (Ames and Correll 1952–1953; Correll 1965; Dix and Dix 1990); c. 30% of the species known from Guatemala occur in the Sierra de las Minas (Dix and Dix, in prep.). Among these are endemic species (e.g. *Epidendrum sobraloides*) and others known from Guatemala only in this region.

The Sierra de las Minas is surrounded by heavily deforested regions dedicated to coffee, cardamom and similar crops. As a result, this region represents a last refuge for many species (e.g. *Colpothrinax cookii*) described from the extensively explored and better known Alta Verapaz Department and Sierra de Chamá.

The Sierra de las Minas with close to 2000 recorded species is rich both in endemic species – at least 70 – and in species at the northern or southern limit of their ranges. North temperate elements extending to the sierra include *Acer skutchii*, *Liquidambar styraciflua* and *Taxus globosa*. Southern elements include *Podocarpus* species.

In addition to the species endemic to the sierra, 35% of the species in the *Flora of Guatemala* that were considered endemic to Guatemala can still be found in the sierra. Many of these species are gone from other Guatemalan localities.

Useful plants

In general, the resource potential for the region is high, but most species are under-utilized.

Timber extraction is a major source of income, especially from the southern slopes. The species most frequently exploited include *Pinus oocarpa*, *P. patula* subsp. *tecunumanii* and *P. caribea*, used especially for utility poles, railway sleepers and furniture. Most of the utility poles produced in Guatemala come from this region. There is a large potential for developing conifer plantations on deforested slopes in all of the sierra. The rich conifer diversity could lead to a sustainable seed-harvesting industry. In the Polochic drainage and the lower eastern slopes *Cedrela odorata*, *Dalbergia* (rosewood) and *Vochysia* spp. (San Juan) have been harvested in the past, but recent information on available resources is lacking.

The adventitious roots of several tree-fern species (*Dicksonia, Cyathea, Alsophila*) are harvested to produce pots or the fibre is used as substrate for growing ornamental plants. Bamboo is used in basket-making.

Medicinal plants abound. These include *Ocimum* spp., *Crescentia alata*, several Rubiaceae (e.g. *Borreria ocymoides, Randia armata, Hamelia patens*), *Dorstenia contrajerva, Neurolaena* spp. and many Solanaceae. Several species of Cucurbitaceae and Solanaceae represent potential germplasm resources of food plants, including local varieties of tomato.

With appropiate studies, there is development potential for producing and selling ornamental plants such as begonias, palms, orchids, bromeliads and peperomias, and medicinal plants, including species yielding contraceptives and anti-malarial drugs.

Social and environmental values

The Sierra de las Minas is important as a generator of orographic rainfall for the surrounding Polochic and Motagua valleys, giving rise to the 62 permanent streams. In the north, farmers depend on these rains for coffee, cardamom and rice production and for cattle-farming. To the south, this is the only rainfall reaching the north-eastern side of the arid Motagua Valley. As well as the basic needs of the local people, agricultural crops (e.g. melon, tobacco, grapes, citric fruits, tomatoes) depend on irrigation provided by the small rivers flowing from the sierra. Light industry, also dependent on a steady water supply, includes soft drinks, fertilizer and paper-recycling plants, and hydroelectricity is generated at the Río Hondo station. However, water flow has been reduced over 40% during the last 10 years and the water table has dropped, probably because of the loss of vegetation.

The fauna of the sierra is very diverse. The region harbours over 800 bird species. Significantly, the cloud forests support the largest remaining population of Guatemala's national bird, the quetzal (*Pharomacrus mocinne*). The Sierra de las Minas is in the Northern Central American Highlands Endemic Bird Area (EBA A14), which includes a number of mountain ranges in southern Mexico (e.g. Chiapas), Guatemala, El Salvador, Honduras and north-west Nicaragua. There are 21 species of birds confined to this EBA, and a large proportion occur in the Sierra de las Minas. The enigmatic and threatened horned guan (*Oreophasis derbianus*) was recently found in this mountain range (Howell and Webb 1992).

Other animals include white-tailed deer (*Odocoileus virginianus*), brocket deer (*Mazama americana*), peccaries (*Tayassu pecari*), jabalí (*Tayassu tajacu*), puma (*Felis concolor*), howler monkeys (*Alouatta*) and many small rodents and bats. The small mammals have not been carefully surveyed, but a detailed study began in 1992. Reptilian and amphibian diversity is high, with over 60% of the species known from Guatemala present in the sierra (Campbell 1982). Schuster (1992) described endemic insects (Passalidae) from the region and considers that the area may have high faunal diversity.

Scenic values are important and could form the basis for a sustainable tourist industry. The contrast in a drive of 3–4 hours between the Motagua Valley's arid thorn scrub dominated by cacti and acacias and the misty peaks of the epiphyte-laden cloud forests is impressive. Additional tourist attractions that could be sustainably managed include marketing of Amerindian textiles and handicrafts.

Three pre-Columbian archaeological sites within the BR are Río Zarquito – of the Classic period (Maya and Chorti), and Tinajas and Pueblo Viejo – of the post-Classic period (Maya and Q'eqchí). San Agustín Acasaguastlán represents an early Colonial mission.

Economic assessment

The water resources are the region's most valuable asset. The sierra provides water for irrigation in the dry Motagua Valley for growing melons, tobacco, tomatoes, grapes and other crops. The Río Hondo hydroelectric station produces 2% of Guatemala's electricity.

The main direct economic values are from the forest for lumber and fuelwood, which usually is the only energy source for domestic cooking. Some income is derived from bromeliad and tree-fern harvesting. Coffee and cardamom are important cultivated products.

Conserving the high-elevation forests is a first priority because both the water resources and the incipient tourist industry depend on them. Increased erosion, and decreased rainfall with 40% loss of water flow, have been documented already (Fundación Defensores de la Naturaleza 1992).

Threats

Deforestation with consequent erosion constantly threatens. Indiscriminate timber extraction has resulted in a removal rate beyond the system's capacity for sustainable management, with probable germplasm degradation. There also are insufficient reforestation projects.

On the Motagua Valley slopes, fire destroys large areas of forest every year as the agricultural frontier advances. Moreover, this devastation occurs as well on steep slopes that are not adequate for agriculture, and their natural regeneration is very slow.

Colonization by displaced Q'eqchís is a potential threat because of their non-sustainable cut, slash and burn monoculture and lack of soil conservation practices. Vegetable cultivation on a large scale is being promoted by development agencies, but is not necessarily a good idea because insufficient attention has been given to erosion control and soil degradation in these fields and to loss of water quality. Indiscriminate hunting for subsistence and recreation is a problem. Large-scale community education is needed, as well as ongoing studies on the effects of agricultural practices on soil structure.

Conservation

In October 1990, the Guatemalan legislature decreed an area of 2363 km² in the Sierra de las Minas as a Biosphere Reserve and appointed Fundación de los Defensores de la Naturaleza as its manager. The passage of this law was precedent setting (Fundación Defensores de la Naturaleza 1990). Not only is it the first time that a non-profit private organization (NGO) has been given the management of a major Guatemalan reserve, but there also is an innovative relationship with local communities, which share in overseeing the management process. UNESCO-MAB has recognized the reserve as a part of the world network of Biosphere Reserves.

Defensores de la Naturaleza was founded in 1983; its goals are preservation of the Sierra de las Minas Biosphere Reserve, environmental education of students and the general public, and education of Guatemalan leaders on the nation's conservation needs. Defensores works closely with other organizations through cooperative agreements.

The BR is zoned into a nuclear area (1057 km²), which contains 81% of the existing cloud forests; sustainable-use zone (346 km²); experimental forest recovery zone (42 km²); and buffer zone (920 km²).

The (U.S.) Nature Conservancy (TNC) has included the Sierra de las Minas in its Parks in Peril Program (TNC 1990) and contributed funds for land purchase by Defensores, and TNC also is actively involved in developing conservation strategy. It is creating information databases through the Centro de Datos para la Conservación (CDC) at the Centro de Estudios Conservacionistas (CECON), which is the academic unit of the Universidad de San Carlos de Guatemala that is responsible for promoting field research and conservation of renewable resources.

Other organizations involved in conservation and research in the sierra include the Universidad del Valle de Guatemala, with biodiversity studies on cloud-forest flora and fauna (such as epiphytes, insects and small mammals) and on sustainable use of tree fern and bamboo in the buffer zone. CARE, as a component of Proyecto PACA (Programa Ambiental para Centroamérica), is heavily involved in supporting Defensores in its environmental education and sustainable-use programmes.

Several local organizations have been formed. The oldest, FUNDEMABV (Fundación del Medio Ambiente para Baja Verapaz), has environmental education programmes in local schools, and others are developing programmes in Alta Verapaz, Zacapa and Izabal departments.

It appears that the BR is now established and effective programmes are underway. Hopes are high that good management, active local NGO participation, and positive relationships with indigenous peoples will allow this area to become a model for conservation and sustainable development in Guatemala.

References

Ames, O. and Correll, D.S. (1952-1953). *Orchids of Guatemala. Fieldiana, Botany* 26: 1–727.

Campbell, J.A. (1982). *The biogeography of the cloud forest herpetofauna of Middle America, with special references to Sierra de las Minas.* Ph.D. dissertation. University of Kansas, Lawrence. 322 pp.

Correll, D.S. (1965). Supplement to the orchids of Guatemala and British Honduras. *Fieldiana, Botany* 31(7): 177–221.

Dix, M.A. and Dix, M.W. (1990). La Sierra de las Minas: su diversidad orquideológica. In *Tercer Encuentro Latinoamericano de Orquideología.* Pp. 12–14.

Dix, M.A. and Dix, M.W. (in prep.). An annotated revised check list of the orchid flora of Guatemala.

Fundación Defensores de la Naturaleza (1990). *Estudio técnico para dar a Sierra de las Minas la categoría de reserva de la biósfera.* 44 pp.

Fundación Defensores de la Naturaleza (1992). *Reserva de la Biósfera Sierra de las Minas. Plan maestro 1992–1997.* 54 pp.

Howell, S.N.G. and Webb, S. (1992). New and noteworthy bird records from Guatemala and Honduras. *Bull. Brit. Orn. Club* 112: 42–49.

Schuster, J.C. (1992). Biotic areas and the distribution of passalid beetles (Coleoptera) in northern Central America: post-Pleistocene montane refuges. In *Biogeography of Mesoamerica.* Tulane Studies Zoology and Botany, Supplementary Publ. 1: 285–292.

Standley, P.C., Steyermark, J.A. and Williams, L.O. (1946-1977). *Flora of Guatemala. Fieldiana, Botany* 24.

TNC (1990). *Parks in Peril: a conservation partnership for the Americas.* The Nature Conservancy (TNC), Arlington, Virginia. 25 pp.

Weyl, R. (1980). *Geology of Central America.* Gebruder Borntraeger, Berlin. 373 pp.

Author

This Data Sheet was written by Dr Margaret A. Dix (Universidad del Valle de Guatemala, Departamento de Biología, Apdo. Postal 82, Guatemala City, Guatemala).

CENTRAL AMERICA: CPD SITE MA15

NORTH-EAST HONDURAS AND RIO PLATANO BIOSPHERE RESERVE
Honduras

Location: In Mosquitia region of north-east Honduras including nearby mountains, between latitudes 15°–16°N and longitudes 84°30'–85°30'W.

Area: c. 5250 km².

Altitude: 0–1500 m.

Vegetation: Mangrove and freshwater swamps and marshes; sedge prairie; pine savanna; gallery forest; tropical moist, subtropical moist and subtropical wet forests; elfin forest.

Flora: High diversity – probably over 2000 vascular plant species; threatened species.

Useful plants: Germplasm of timber species; medicinals.

Other values: Watershed protection, wetlands, faunal refuge, archaeological sites, indigenous cultures, dramatic scenery, tourist attractions.

Threats: Logging, colonization, cattle-grazing, road building, mining exploration.

Conservation: The Biosphere Reserve also is a World Heritage Site, Amerindian Reserve and Archaeological Park.

Geography

La Mosquitia (or the Miskito Coast) is essentially that part of north-eastern to eastern Honduras and eastern Nicaragua mainly occupied by the indigenous Miskito people, and usually corresponds to the area's lowland pine savanna. The Miskito pine savanna formation occurs on deeply weathered quartz sandy gravels of Pleistocene age in a strip averaging 45 km wide (to 180 km maximum) which extends from about 16°N, 85°30'W in Honduras for some 480 km southward to a few kilometres north of Bluefields, Nicaragua (12°N) (Parsons 1955; Daugherty 1989). The Honduran Mosquitia (16,630 km²) mainly encompasses most of the Department of Gracias a Dios and part of the Department of Colón (Clewell 1986).

The 5250 km² Río Plátano Biosphere Reserve is a 35 × 150 km area extending south-west (Map 26), which protects coastal Mosquitia plain and interior elevations including some of the Honduran disjunct north-easternmost highlands. The reserve is in three departments: over half in Gracias a Dios, the rest mostly in Colón, with less than 10% in Olancho (Glick 1980; Daugherty 1989). It includes the entire watershed of the 115 km-long Plátano River and portions of the watersheds of the lower Tinto, Paulaya, Wampú, Pao, Tuskruwás and lower Sikri (Sigre) rivers – which, along with a 5-km marine extension for the Caribbean Sea, basically form the BR's boundaries. The annual precipitation varies locally from perhaps less than 2850 mm to 3000–4000 mm, with a drier season around January–May; the average annual temperature is 26.6°C. In an average decade, the region is impacted by four intense tropical storms and two hurricanes.

About 75% of the Biosphere Reserve is mountainous, with many steep ridges; Pico Morrañanga reaches 1500 m and Punta de Piedra 1326 m. Remarkable geological formations are in the rugged upland region, such as the exposed granitic pinnacle El Viejo or Pico de Dama, which projects finger-like 150 m as the summit of Cerro Dama in the Cordillera Baltimore (Cruz 1986). Cataracts and cascading waterfalls are found, the highest (100–150 m) being the Cascada del Mirador in the headwaters of the Cuyamel River. In one cataract the Plátano River almost disappears among massive boulders in a gorge flanked by forested escarpments 100 m high (Cruz 1986).

The remaining 25% of the BR is an undulating to flat segment of the Caribbean coastal plain, which extends from a few to over 40 km inland and rises gradually from sea-level to almost 100 m in altitude, where the foothills begin abruptly. The Plátano River (which bisects the length of the reserve) meanders for 45 km through this area, forming oxbow lakes, backwater swamps and natural levees which are used for agricultural plots (Froehlich and Schwerin 1983; Glick and Betancourt 1983). Near the sea are freshwater and brackish lagoons and sandy beaches.

Vegetation

The Río Plátano Biosphere Reserve is the largest natural tract of forest remaining in the country. About 75% of the region is in the tropical moist-forest life zone, with 10–15% in the subtropical wet-forest life zone (cf. Houseal *et al.* 1985). Little is known as yet about the vegetation of the mountainous majority of the BR (Froehlich and Schwerin 1983). The limited knowledge of the reserve's plant species is reported in DIGERENARE and CATIE (1978), Froehlich and Schwerin (1983) and Glick and Betancourt (1983).

The most extensive mangrove ecosystems fringe the large coastal lagoons of Brus (brackish, 120 km²) and Ibans (freshwater, 63 km²). Although some mangroves have been cut, the area still retains much of the original formation, with *Rhizophora mangle* characteristic.

Inland from the beach is a broad coastal savanna, which in wetter locales consists of sedge prairie with abundant *Rhynchospora* spp., *Paspalum pulchellum, Tonina fluviatilis* and *Utricularia subulata*, and where drier has more grasses, *Fimbristylis paradoxa* and *Declieuxia fruticosa*. Thickets of the palm *Acoelorraphe wrightii* are common. In drier areas is savanna dominated by *Pinus caribaea* var. *hondurensis* (20–25 m tall), which farther inland becomes open woodland with an oak understorey (*Quercus oleoides*, to 12 m) and *Byrsonima crassifolia* (to 5 m) conspicuous, along with

several Melastomataceae, *Calliandra houstoniana* and the tree fern *Alsophila myosuroides* (Clewell 1986). The savanna is burned frequently to maintain pasturage for grazing and to keep game in the open for hunting.

Towards the large rivers are thickets dominated by *Miconia, Isertia, Psychotria* and *Helicteres*. Along the Plátano River and other alluvial rivers through the savanna, broadleaved gallery forest occurs in various successional stages, to 30–40 m high. Variously conspicuous taxa include *Albizia carbonaria, Calophyllum brasiliense* var. *rekoi*,

MAP 26. RIO PLATANO BIOSPHERE RESERVE, HONDURAS (IN CPD SITE MA15)

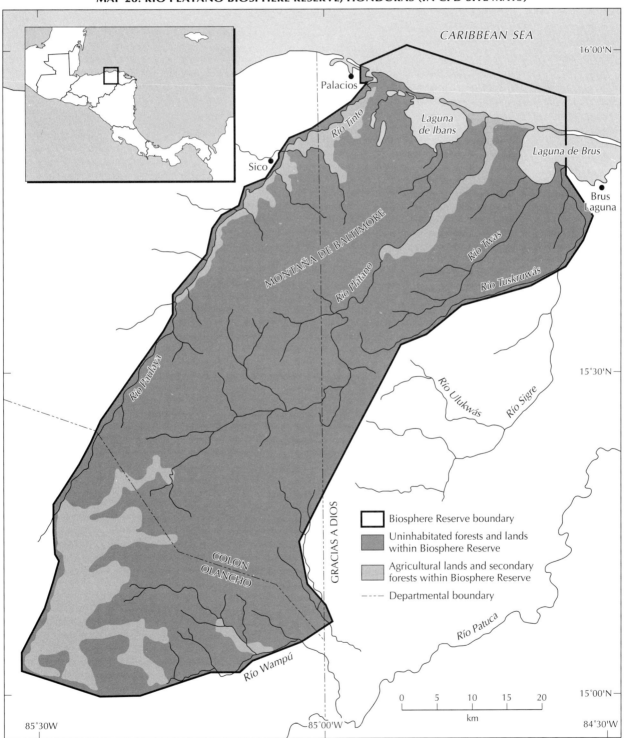

Cecropia, Ficus, Inga, Luehea seemannii, Lonchocarpus, Ochroma lagopus, Pachira aquatica and *Heliconia*. Small colluvial creeks are flanked by swamp forest with a dense canopy to 10 m dominated by Guttiferae (*Symphonia globulifera, Clusia* spp.) (Clewell 1986). On richer soils in moist forest that has been disturbed as a result of intermittent agriculture, the dominants are *Salix humboldtiana, Bambusa, Pithecellobium* and *Ceiba pentandra*.

The upland portion of the Plátano River watershed is covered by moist to wet forests which are poorly known. Common or characteristic within its lower elevations (among others) are *Apeiba membranacea, Bursera simaruba, Carapa guianensis, Casearia arborea, Cedrela odorata, Eugenia* sp., *Ficus insipida, Pourouma aspera, Pseudolmedia oxyphyllaria, Pterocarpus* sp., *Quararibea* sp., *Sloanea* spp., *Swietenia macrophylla* and *Vochysia hondurensis*. With increasing altitude, sampled sites included the following plentiful or notable species: at 250 m – *Garcinia intermedia, Pouteria* sp. and *Schizolobium parahybum*; at 450 m – *Ardisia tigrina, Pharus cornutus* (rare), *Smilax subpubescens* and *Ternstroemia tepezapote*; at 600 m – *Lobelia* sp., *Satyria warscewiczii* and *Welfia* sp.

Trunks and branches support a rich assortment of epiphytes which are more abundant on the trees at higher elevations. Some locales have very dense successional stages resulting from disturbance by storms. Elfin forests occur on exposed ridges where the prevailing trade winds from the Caribbean have strong effect – for example at 700 m with *Clusia salvinii, Magnolia sororum, Lacistema aggregatum* and *Psychotria elata*.

Flora

Honduras is estimated to have 6000 species of vascular plants (cf. Molina 1975; Nelson 1986). Forests remain over a third of the territory for a total of 40,000 km², approximately half broadleaved and half pine. No national Flora has been written, but many Honduran species are described in Nelson (1986), the *Flora of Guatemala* or other treatments (see Nelson 1989), and will be included in the *Flora de Nicaragua*.

Probably over 2000 vascular plant species occur in the BR. The Mosquitia may still be the least known region of Honduras, where species new to the country, phytogeographically distinctive and/or new to science are found whenever a research trip can be carried out (Nelson 1978; Proctor 1981, 1983; Froehlich and Schwerin 1983).

Useful plants

The reserve harbours populations of some important timber trees, such as *Calophyllum brasiliense* var. *rekoi, Carapa guianensis, Cedrela odorata, Swietenia macrophylla, Tabebuia rosea* and *Virola koschnyi*. The abundance of seemingly wild *Theobroma* ("cacao") near Las Crucitas del Río Aner suggests that it was cultivated there in ancient times. The region's diverse inhabitants use an array of native species for many essential purposes (cf. Froehlich and Schwerin 1983; Cruz 1991; Lagos-Witte 1992). For example, the pines and several palms are used for construction, and several of the timber species are made into dugout canoes – a major means of transportation in the region.

Social and environmental values

The different ecosystems of this large BR provide habitats for many species that are classified as globally rare or threatened. Mammals include Baird's tapir, manatee, jaguar, ocelot, margay, jaguarundi, cougar, southern river otter, collared and white-lipped peccaries, white-faced, mantled howler and spider monkeys, and giant anteater (Froehlich and Schwerin 1983). Over 375 bird species reside in or frequent the region (e.g. harpy eagle, scarlet and military macaws). Five restricted-range species occur in the vicinity of the Río Plátano; the pine savanna has received little ornithological attention despite the occurrence of a number of endemic subspecies. Almost 200 amphibian and reptile species occur in the BR (Glick and Betancourt 1983; Cruz 1991).

The region has long been a site of human occupation. It has many archaeological sites (e.g. Marañones, Lancetillal, Platanillales, Saguasón, Limeta), including mysterious petroglyphs carved into large boulders along the river edge. The Instituto Hondureño de Antropología e Historia (IHAH) is conducting studies in the BR. The village Las Crucitas del Río Aner in the reserve's south-east (on the Aner River a few km from the Wampú River) is established over one of the largest and most impressive archaeological sites (Lara-Pinto and Hasemann 1991). It is believed that the fabled ancient "Ciudad Blanca" awaits discovery within the reserve. Additional archaeological research in the region may confirm the surmise that its peoples were an important link between major pre-Columbian cultures in North America and South America.

The northern portion of the BR has c. 6000 inhabitants belonging to four cultural groups: the Miskito and Pesch (Paya) Amerindians, Garífunas (Afro-Caribbeans) and "ladinos" (mestizos). The Miskito are the dominant group in the reserve, with c. 4500 persons living in coastal settlements and two towns on the banks of the Tinto River. The few Pesch who inhabit the BR mostly live in a few foothill settlements between Las Marías and Baltiltuk. Several hundred Garífunas are established in the coastal town of Plaplaya. There also are several hundred ladinos established in eight small settlements along the Paulaya River. These groups have not caused serious impact on the forests of the reserve (Herlihy and Herlihy 1991).

Economic activity in the region is based primarily on agriculture, cattle and fisheries; agriculture is the basis of subsistence (Glick and Betancourt 1983; Houseal *et al.* 1985). Tourism is considered a key potential for the BR (Murphy 1991), although attractions such as the archaeological sites still remain largely undeveloped or with limited access. The 1980s war in neighbouring Nicaragua made it difficult to attract foreign tourists when the Biosphere Reserve was newly established.

Threats

The Río Plátano Biosphere Reserve has the largest natural forest remaining in Honduras, where current estimates indicate an annual deforestation of 645 km² (Daugherty 1989). There is significant pressure to use the reserve's natural resources, with particular interest in logging the lowland hardwoods. Migratory farmers and loggers enter the reserve, with resultant forest destruction. In c. 1982 stricter controls were enacted on mineral exploitation and mining by non-residents was eliminated (Glick and Betancourt 1983).

A significant impact to the forest along the eastern border would have been permanent relocation of 4000 Miskito from Nicaragua, which was under consideration by the Honduran Government refugee commission (Houseal *et al.* 1985). One of the most serious threats (c. 1982) was a proposed road to facilitate movement of military troops to the Honduras-Nicaragua border. The road would have crossed the BR's eastern boundary, which would have facilitated logging the region's hardwoods and opened the area to colonization. The plan was dropped as a consequence of international attention (FFPS 1983; Glick and Betancourt 1983; Nations and Komer 1983).

The most serious present problem for the BR's integrity is the cattle-frontier advancing from the south-west and into the reserve's south-western portion in the Wampú-Paulaya area. More than 6500 ladinos are established in 46 settlements and towns, roads reach beyond Dulce Nombre de Culmí into the area, and agricultural colonists continue to arrive (Herlihy and Herlihy 1991).

Conservation

In 1960 Honduras created the Ciudad Blanca Archaeological Reserve (c. 5250 km²) for the Plátano River region and in 1969 made it an Archaeological National Park. This recognition endorsed the importance of archaeological finds and legends of an ancient major Maya city in the area, although actual protection was minimal (Anon. 1979; WWF – U.S. 1988). The cultural significance led to other scientific evaluations, confirming that the Mosquitia region of Honduras and Nicaragua had the most virtually undisturbed forest of northern Central America. As a result, in 1980 the park was recognized as the Río Plátano Biosphere Reserve (the first in Central America) under UNESCO's Man and the Biosphere Programme (MAB), which provided stronger legal protection and a management planning process for the Plátano River watershed and adjacent slopes. In 1982 the BR was accepted on the World Heritage List (Houseal *et al.* 1985).

Under the Biosphere Reserve concept, not only the biotic and archaeological resources are protected but also the indigenous cultures. From the beginning the Miskito and Pesch have been involved with the Honduras Dirección General de Recursos Naturales Renovables (RENARE) in planning and management of the reserve, assisted by MAB, the Centro Agronómico Tropical de Investigación y Enseñanza (CATIE), the World Wildlife Fund – U.S. (WWF – U.S.) and the U.S. Peace Corps (Hartshorn 1983; Houseal *et al.* 1985). Management objectives include utilizing the BR as a model for studying human impact in the short and long term on tropical rain forests and identifying land-use practices that can be sustainable (Glick and Betancourt 1983). The BR has two basic zones to facilitate management: a core or natural zone (3180 km²) and a peripheral buffer zone (2070 km²), which has a cultural (lowland) portion.

WWF – U.S. has been involved in supporting projects for the Plátano River watershed for over a decade, including its establishment as a Biosphere Reserve. In 1987, the Honduran Ecological Association (AHE) organized an inter-institutional workshop (sponsored by WWF – U.S.) which produced a two-year detailed management plan for the Río Plátano Biosphere Reserve. The new plan called for the Honduran Corporation for Forest Development (COHDEFOR) to assume principal administrative duties. The plan outlines actions necessary to deal with serious threats to the BR's integrity and to make it the model of integrated resource development intended (WWF – U.S. 1988; Salaverri 1991).

In spite of some accomplishments and tremendous efforts by those involved with implementation of the management plan, degradation continues in the region (Salaverri 1991). To help alleviate the problem, the German Bank of Reconstruction and Development (KFW) in coordination with COHDEFOR undertook a feasibility study on the management plan for the Río Plátano BR; the 1991 results were accepted by the Honduran Government. Before receiving development aid for the reserve from Germany, Honduras is making revisions to the BR decree and relocating non-indigenous persons living in the Plátano River headwaters area. Logging is being curtailed, and the military is increasingly involved in protection of the environment (*El Heraldo* 1992). The Biosphere Reserve may be significantly expanded eastward, to the Patuca River (c. 84°18'W).

The probability is growing to expand protection for the tropical forest to threatened contiguous regions in eastern Olancho Department, by extending an ecological corridor south from the Wampú River beyond the middle Patuca River to the Coco River bordering Nicaragua and its Bosawas Biosphere Reserve. Paseo Pantera (Path of the Panther) (Marynowski 1993), a consortium of Wildlife Conservation International and the Caribbean Conservation Corporation, in collaboration with COHDEFOR and USAID, is engaged in trying to establish the corridor, which would link the Río Plátano BR, a proposed reserve (2300 km²) for the indigenous Tawahka Sumu (Herlihy and Leake 1990) and the adjacent (southward) proposed Patuca National Park (2200 km²) with the Bosawas BR (8000 km²) (Herlihy and Leake 1992). The Río Coco or Solidaridad Reserve is one of 11 areas given priority under the region's 1992 Convenio para la Conservación de la Biodiversidad y Protección de Areas Silvestres Prioritarias en América Central.

References

Anon. (1979). The Río Plátano Biosphere Reserve in Honduras. *Nature and Resources* 15(3): 24–26.

Clewell, A.F. (1986). Observations on the vegetation of the Mosquitia in Honduras. *Sida* 11: 258–270.

Cruz, G.A. (1986). *Areas silvestres de Honduras: guía de los parques nacionales, refugios de vida silvestre, reservas biológicas y monumentos naturales de Honduras*. Asociación Hondureña de Ecología, Tegucigalpa.

Cruz, G.A. (1991). La biodiversidad de la reserva. In Murphy, V. (ed.), *La Reserva de la Biósfera del Río Plátano: herencia de nuestro pasado*. Ventanas Tropicales, Tegucigalpa. Pp. 20–23.

Daugherty, H.E. (ed.) (1989). *Perfil ambiental de Honduras 1989*. SECPLAN, Tegucigalpa and USAID: DESFIL, Washington, D.C. 346 pp.

DIGERENARE and CATIE (1978). *La cuenca del río Plátano (Mosquitia, Honduras). Estudio preliminar de los recursos naturales y culturales de la cuenca y un plan para el desarrollo de una reserva de la biósfera en la región del río Plátano.* Dirección General de Recursos Naturales Renovables (DIGERENARE), Tegucigalpa, Honduras and Centro Agronómico Tropical de Investigación y Enseñanza (CATIE), Turrialba, Costa Rica. 133 pp.

El Heraldo (6/VII/92). [President Callejas announces ban on lumber activities.] Pg. 38.

FFPS (1983). Military road threatens Honduras virgin forest. *Oryx* 17: 110.

Froehlich, J.W. and Schwerin, K.H. (eds) (1983). *Conservation and indigenous human land use in the Río Plátano watershed, Northeast Honduras.* Research Paper Series No. 12, Latin American Institute, University of New Mexico, Albuquerque. 94 pp.

Glick, D. (1980). *Río Plátano Biosphere Reserve case study.* Integrative Studies Center, School of Natural Resources, University of Michigan, Ann Arbor. 120 pp.

Glick, D. and Betancourt, J. (1983). The Río Plátano Biosphere Reserve: unique resource, unique alternative. *Ambio* 12: 168–173.

Hartshorn, G.S. (1983). Wildlands conservation in Central America. In Sutton, S.L., Whitmore, T.C. and Chadwick, A.C. (eds), *Tropical rain forest: ecology and management.* Blackwell Scientific Publications, Oxford, U.K. Pp. 423–444.

Herlihy, P.H. and Herlihy, L.H. (1991). La herencia cultural de la Reserva de la Biósfera del Río Plátano: un área de confluencias étnicas en la Mosquitia. In Murphy, V. (ed.), *La Reserva de la Biósfera del Río Plátano.* Ventanas Tropicales, Tegucigalpa. Pp. 9–15.

Herlihy, P.H. and Leake, A.P. (1990). The Tawahka Sumu: a delicate balance in Mosquitia. *Cultural Survival Quarterly* 14(4): 13–16.

Herlihy, P.H. and Leake, A.P. (1992). *Situación actual del frente de colonización/deforestación en la región propuesta para el Parque Nacional Patuca.* Mosquitia Pawisa (MOPAWI), Tegucigalpa. 22 pp. + annexes.

Houseal, B., MacFarland, C., Archibold, G. and Chiari, A. (1985). Indigenous cultures and protected areas in Central America. *Cultural Survival Quarterly* 9(1): 10–20.

Lagos-Witte, S. (1992). Ethnobotanical contributions to the TRAMIL program in the Caribbean Basin: the case of Honduras. In Plotkin, M. and Famolare, L. (eds), *Sustainable harvest and marketing of rain forest products.* Island Press, Washington, D.C. Pp. 20–26.

Lara-Pinto, G. and Hasemann, G. (1991). Leyendas y arqueología: ¿cuántas ciudades blancas hay en la Mosquitia? In Murphy, V. (ed.), *La Reserva de la Biósfera del Río Plátano.* Ventanas Tropicales, Tegucigalpa. Pp. 16–19.

Marynowski, S. (1993). Paseo Pantera: the great American biotic interchange. *Wild Earth* (Special Issue): 71–74.

Molina, R.A. (1975). Enumeración de las plantas de Honduras. *Ceiba* 19(1): 1–118.

Murphy, V. (ed.) (1991). *La Reserva de la Biósfera del Río Plátano: herencia de nuestro pasado.* Ventanas Tropicales, Tegucigalpa. 26 pp.

Nations, J.D. and Komer, D.I. (1983). International action halts road through Honduras rainforest. *Ambio* 12: 124–125.

Nelson, C. (1978). Contribuciones a la flora de La Mosquitia, Honduras. *Ceiba* 22(1): 41–64.

Nelson, C. (1986). *Plantas comunes de Honduras.* 2 vols. Editorial Universidad Nacional Autónoma de Honduras, Tegucigalpa. 922 pp.

Nelson, C. (1989). Honduras. In Campbell, D.G. and Hammond, H.D. (eds), *Floristic inventory of tropical countries: the status of plant systematics, collections, and vegetation, plus recommendations for the future.* New York Botanical Garden, Bronx. Pp. 290–294.

Parsons, J.J. (1955). The Miskito pine savanna of Nicaragua and Honduras. *Ann. Assoc. Amer. Geogr.* 45: 36–63.

Proctor, G.R. (1981). Appendix 2. Mosquitia botanical collection list according to families. In Brunt, M.A. (ed.), *La Mosquitia, Honduras: resources and development potential.* Vol. 3. *Appendices.* Overseas Development Administration, Land Resources Development Centre, Project Report No. 110. Tolworth, Surbiton, U.K.

Proctor, G.R. (1983). New plant records from the Mosquitia region of Honduras. *Moscosoa* 2(1): 19–22.

Salaverri, J. (1991). La situación actual de la reserva. In Murphy, V. (ed.), *La Reserva de la Biósfera del Río Plátano.* Ventanas Tropicales, Tegucigalpa. Pp. 5–8.

WWF - U.S. (1988). Country plan-Honduras. In *1989/90 program.* Latin America and the Caribbean: new and ongoing projects. World Wildlife Fund – U.S. (WWF – U.S.), Washington, D.C. Pp. 291–312.

Author

This Data Sheet was written by Olga Herrera-MacBryde (Smithsonian Institution, Department of Botany, NHB-166, Washington, DC 20560, U.S.A.).

CENTRAL AMERICA: CPD SITE MA16

BRAULIO CARRILLO-LA SELVA REGION
Costa Rica

Location: On Caribbean slope primarily in Heredia Province of central Costa Rica, between about latitudes 10°00'–10°25'N and longitudes 83°50'–84°10'W.

Area: c. 500 km².

Altitude: c. 35–2906 m.

Vegetation: In 35 km and an altitudinal range of 2871 m, four life zones and two zonal transitions, from tropical wet forest through tropical premontane, lower montane and montane rain forests.

Flora: 4000–6000 vascular plant species in Braulio Carrillo National Park, 1900–2200 at La Selva Biological Station; threatened species.

Useful plants: Reserve for timber species, ornamentals, edible palmitos.

Other values: Refuge for fauna – including threatened species; germplasm reserve, watershed protection, extensive research base, scenery, tourism.

Threats: Ecological isolation; private land conversion and squatters; illegal logging, agriculture, grazing, hunting.

Conservation: A portion of Central Volcanic Cordillera Biosphere Reserve: Braulio Carrillo National Park and La Selva Biological Station. Nearby four private reserves.

Geography

Located in Heredia, San José, Cartago and Limón provinces of central Costa Rica, the c. 500 km² region includes La Selva Biological Station (15.6 km²) at the western edge of the Atlantic lowlands (55 km from the coast), and Braulio Carrillo National Park (465–475 km²). The station is c. 4 km south of Puerto Viejo de Sarapiquí – the limit of navigation from the Caribbean Sea on the Sarapiquí River, at the junction of the flat lowlands and the foothills, which ascend south-westward to Cacho Negro (2150 m) and Barva (2906 m) volcanoes.

The somewhat boot-shaped park has its largest portion in the highlands of the Central Volcanic Cordillera; 30 km to the south-west is the populous Central Valley (Meseta Central) and capital San José. The lower La Selva sector of the park, from 850 m downslope to the station in the north (at c. 135–35 m), is a narrow corridor (18 km × 4–6 km) mainly of primary rain forest lying between the Peje and Guácimo rivers. Virtually the entire watersheds of these rivers are within the park, and the upper watersheds of the Puerto Viejo and Sucio rivers are in the park highlands, which are strongly dissected, with deep valleys and many waterfalls. Barva Volcano is composed of several craters, now including the crystal-clear Barva Lake (70 m across) and Danta Lake (500 m across).

This region comprises what may be the most extensive altitudinal range (2871 m) of primary tropical forest protected in Central America (Pringle *et al.* 1984; Hartshorn and Peralta 1988; Pringle 1988). In the southern half of the station and farther south in the park's narrow La Selva sector, soils appear to be derived from andesites and basalt. At 2000 m some soils seem to be derived from ash, whereas at 2600 m near Barva Volcano there are considerable deposits of granular andesitic ash. With increasing elevation organic matter increases, and above 1400 m humus accumulates (Hartshorn and Peralta 1988; Pringle 1988). Landslides are common on steep slopes especially at the mid-elevations.

At La Selva station, annual rainfall averages c. 3962 mm (up to 150 mm have fallen in one day). The mean monthly air temperature is c. 25.8°C (varying annually by less than 3°C), with a diurnal range of 6°–12°C; the known daily maximum (in May) is 36.6°C, the known night-time minimum (in March) 16°C. At 1500–1800 m, the annual average precipitation could be 6000 mm; rainfall in the park may average 4500 m. Trade winds can be strong; a windstorm in June 1986 destroyed 1.5–2 km² of primary forests between 1100–1300 m. The wettest months are July and December, and somewhat less rain falls in February–April (the drier season) and perhaps September (the "veranillo"). During the drier season, intervals average c. 12 days between total daily rainfalls of more than 5 mm, although in 1983 there was an interval of 30 days without such precipitation.

Vegetation

The region is within four Holdridge life zones and two transitional areas; c. 75% of the park-station region is in the premontane and lower montane zones. Much more is known about the vegetation of the station than the connecting corridor or the more rugged, large upland portion of the park (CCT 1975; Boza and Mendoza 1981; Janzen 1983; Boza 1988; Hartshorn and Peralta 1988; McDade *et al.* 1994). Perhaps 75% remains in mature forest, and most of the rest has reverted to patches of secondary growth 6–25 or more years old.

1. The area of La Selva Biological Station is within the **tropical wet-forest life zone**. The station encompasses a variety of terrestrial and aquatic habitats – e.g. ridge and coastal plain forests, swamps, and riparian ecotones. In the mature forests *Pentaclethra macroloba* predominates, a canopy tree to 30–40 m. Other abundant canopy trees include *Apeiba membranacea*, *Brosimum lactescens*, *Goethalsia meiantha*, *Laetia procera* and *Pourouma bicolor* subsp. *scobina* (*P. aspera*). There may be over 30 palm species, which are strikingly abundant, such as the very common subcanopy palms *Welfia georgii*, *Socratea durissima* (*S. exorrhiza*) and *Iriartea deltoidea* (*I. gigantea*). In the understorey *Asterogyne martiana* and *Asplundia uncinata* are conspicuous. In swamp forest the characteristic, long-lived *Carapa guianensis* (*C. nicaraguensis*) (to 45 m tall and 2 m dbh) occurs in increased abundance, along with *Pterocarpus officinalis* (Hartshorn 1983; Clark 1988; Hartshorn and Peralta 1988; Pringle 1988; Lieberman *et al.* 1990).

2. **Tropical wet forest - cool transition** occurs at 250–600 m, commencing where *Pentaclethra macroloba* abruptly loses dominance. Other characteristic trees are *Euterpe macrospadix*, *Billia columbiana*, *Calophyllum brasiliense* var. *rekoi*, *Micropholis crotonoides* and *Vochysia ferruginea*. Subcanopy and understorey palms are less frequent.

3. At 600–800 m occurs **tropical premontane rain forest - perhumid transition**, where in addition to the three preceding species, *Alchornea latifolia* (which extends from 300–2300 m) and *Macrohasseltia macroterantha* become characteristic. The abundant lowland palm *Prestoea decurrens* reaches its limit in this belt.

4. The **tropical premontane rain-forest life zone** occurs between 800–1450 m. Prominent in the canopy are species of Lauraceae and Sapotaceae, together with *Hyeronima guatemalensis* and *Meliosma vernicosa*. A few lowland trees reach the upper limit of this zone, e.g. *Dendropanax arboreus*, *Dussia macroprophyllata* and *Pterocarpus rohrii* (*P. hayesii*). Palms continue to decline; *Prestoea longepetiolata* occurs at 1000–1400 m. Tree ferns increase markedly (e.g. species of *Cyathea*, *Cnemidaria*, *Dicksonia*).

5. The change to the distinctive **tropical lower montane rain-forest zone** (1450–2500 m) occurs gradually at the frost belt (1300–1600 m), above which brownish epiphytic mosses and the clambering understorey bamboo *Chusquea pohlii* are conspicuous. There are few palm species; *Prestoea allenii* and *Geonoma interrupta* are abundant. Typical trees include *Billia hippocastanum*, *Guatteria oliviformis*, *Hyeronima poasana*, *Ocotea austinii*, *Quercus tonduzii* and *Turpinia occidentalis*. In both of the montane zones occur *Alnus*, *Cornus*, *Magnolia*, *Prunus*, *Symplocos*, *Viburnum* and *Podocarpus*.

6. In the **tropical montane rain-forest zone** (2500–2906 m), characteristic trees are *Brunellia costaricensis*, *Didymopanax pittieri*, *Drimys winteri*, *Ilex vulcanicola*, *Quercus costaricensis* and *Weinmannia pinnata*. The canopy is about half as high (20–23 m) as at the station.

Flora

A checklist of the vascular plants of La Selva station is being maintained (Wilbur *et al.* 1994); in 1986, publication of the families began toward production of a complete Flora (see Hammel 1990), which may cover 1900–2200 species. The flora is rich in epiphytes and hemi-epiphytes, which make up 25% of the species, the majority being orchids (114 spp.), aroids and ferns. Over 460 species of trees are known from the station. A 1983 preliminary inventory of the park's La Selva sector (most of the former Zona Protectora) identified c. 200 tree species; 800 might occur in the entire park-station region, which is 40% of the tree species estimated for the entire country (Janzen 1983; Pringle 1988). Included is a recently discovered family of trees (Ticodendraceae). The phytogeographic affinities of the lowland flora are mostly with Panama and northern South America. Perhaps 10% of the lowland flora is regionally endemic.

La Selva sector forests may be essential habitat for at least 75 tree species in many families, together with some species new to science; 34 of those rare trees are named in Pringle *et al.* (1984). Documented rare herbaceous and understorey species occurring in this corridor (some known from the station, but otherwise unknown or seldom found in Costa Rica) include *Danaea carillensis*, *Justicia sarapiquensis*, *Philodendron rigidifolium*, *Potalia amara* and *Xylopia bocatorena*. At least 16 other species that are rare in Costa Rica occur in the corridor but are unknown from the station. During a just 10-day reconnaissance of the corridor in 1983 (Pringle *et al.* 1984; Pringle 1988), at least 28 plant species new to science and 12 new to Costa Rica were discovered, including the rare fern *Thelypteris valdepilosa* – previously known only from Colombia, and *Reedrollinsia* (known in southern Mexico). Some 4000–6000 plant species may occur in Braulio Carrillo National Park (Boza 1988; Clark 1990).

Useful plants

La Selva sector has species of economic including genetic-resource importance (e.g. the rare *Monstera deliciosa*; *Vanilla pauciflora*; two species of *Theobroma*). Some of the region's c. 56 reported palm species may be used for the vegetable palm hearts or are ornamental – e.g. *Geonoma epetiolata*, *Chamaedorea pumila* and *C. amabilis*. In the cool transition belt some valuable timber species are much more common than at the station – *Aspidosperma megalocarpon* (*A. cruentum*), *Calophyllum brasiliense* var. *rekoi*, *Dalbergia tucurrensis*, *Hyeronima oblonga*, *Lecythis ampla* and *Minquartia guianensis* (Hartshorn and Peralta 1988). *Alnus acuminata* (which reaches Mexico) is an important upland timber species not found farther north in Costa Rica (Janzen 1983).

Social and environmental values

The Braulio Carrillo National Park-La Selva region, together with other reserves that form the Central Volcanic Cordillera Biosphere Reserve, protects three sizeable watersheds of the Sarapiquí, Sucio and Chirripó rivers (MAB 1988; Pringle

1988). A panoramic major highway linking San José to Guápiles and the Caribbean port of Limón crosses through the park; it opened in 1987 (CCT 1975; Wallace 1992). The park (Boza 1988) is visited by an increasing number of tourists – over 500,000 in 1993 – including many international visitors.

The region forms a portion of two Endemic Bird Areas (EBAs A18 and A16). The relatively small highland region of Costa Rica and western Panama is home to an incredible 52 restricted-range bird species, many of which occur in the NP; c. 13 such species occur in the tropical wet forest of the Caribbean lowlands. Among the region's over 142 mammal species are 3 monkey species, 5 cat species and Baird's tapir (*Tapirus bairdii*) (Timm *et al.* 1989; Wilson 1990). The 87 known reptile species include 2 crocodilian and 56 snake species; there are 44 species of frogs (Guyer 1990). Over 4000 moth species and 500 butterfly species are estimated to be present.

La Selva Biological Station has become one of the best known neotropical sites, an investment in environmental research (summarized in McDade *et al.* 1994) that demonstrates the natural wealth in the region. At least 412 bird species have been reported from La Selva and nearby areas, including 75–80% of Costa Rica's land birds and 256 breeding species; c. 10% migrate from North America to winter within the complex, and c. 20% migrate altitudinally

CPD Site MA16: Braulio Carrillo-La Selva region, Costa Rica. Understorey of wet forest on Caribbean slope of Costa Rica at La Selva Biological Station. Photo: W.J. Kress.

within it – probably responding to the availability of fruits and flowers (nectar) (Pringle 1988; Stiles 1988). The complex is particularly important in this respect – it is one of few protected forest areas that caters for species such as the threatened bare-necked umbrellabird (*Cephalopterus glabricollis*), which winters in tropical lowlands and breeds in montane rain forests.

In 1980, La Selva Biological Station and the adjacent natural lands were recommended by the U.S. National Research Council as one of just four areas in which to concentrate tropical ecosystem studies. Between 1979 and 1986 the station gained 24-hour electricity from the national grid and convenient access by paving of the highway to Puerto Viejo, completion of an all-weather road to the east bank of the Puerto Viejo River and construction of a 100-m pedestrian suspension bridge to the station property. The 1953 Finca La Selva was on a remote frontier, only accessible by canoe, after a drive of very many hours on a tortuous road from San José; in 1993, the research station became just 1¼ hours away by paved highway. Also in the 1980s, the infrastructural facilities for research, teaching and environmental education were greatly improved.

Research includes (for example) systematics, evolutionary ecology, forest dynamics (e.g. Lieberman *et al.* 1990) and life-history studies of timber trees, and water, soil and nutrient processes in natural, disturbed and abandoned areas. Over 80 species of timber trees have been evaluated for usefulness in reforesting degraded soils; a dozen species have been chosen for use in farm forestry in the area, which makes the conservation of their germplasm particularly significant. Discovery of charcoal in soil cores from several areas of La Selva mature ("primary") forest initiated a project which is evaluating the influence of aboriginal peoples on the region's vegetation, from c. 3000 BP onward (Clark 1988, 1990; McDade *et al.* 1994).

Threats

The general region was transformed from connected forest to an ecological island between 1961 and 1977 (Gámez and Ugalde 1988). The area of La Selva station was largely surrounded by forest in 1965, but cattle pasturage and family farms now border the station except to the south (Greene 1988; Wallace 1992; Kohl 1993; Butterfield 1994). The station area is 56% primary forest, 18% high-graded and secondary forest and c. 25% early successional pastures, abandoned plantations and managed areas (Clark 1990). The 18-km long La Selva sector of the park (most of the former Zona Protectora), being a corridor only 4–6 km wide, may be too narrow to maintain some natural processes as it becomes even more isolated by forest conversion outside it (Greene 1988; Hartshorn and Peralta 1988; Pringle 1988; Stiles 1988). In 1983 the corridor was 73% primary forest, 10% mainly secondary forest, 16% patches of pasture and 1% crops (Pringle 1988). Since the majority of the Zona Protectora became part of the park in 1986, illegal logging and squatters have not been continuing problems, but poaching does occur (cf. Clark 1988). In c. 1981 the park highlands (320 km²) were 84% primary forest, 5% secondary forest and 11% ranches and farms in alluvial valleys (Boza and Mendoza 1981).

Conservation

Finca La Selva began in 1953 when Dr L.R. Holdridge purchased the Costa Rican homestead. The station (6.13 km²) was established in 1968 with its purchase by the Organization for Tropical Studies (OTS) (Stone 1988), and has since been expanded to 15.6 km². The highlands majority of the park (314–320 km²) was established in 1977–1978, having been prompted by the planned highway through the region (Boza and Mendoza 1981; Wallace 1992). La Selva Protection Zone (73.68 km² including the station) was declared in 1982 to create a natural corridor between the lowland station and upland park (Stone 1982). After extensive cooperation and fund-raising, the Peje sector (135 km² – the Zone, except for the station, plus a portion of the Central Volcanic Cordillera Forest Reserve) was incorporated into the park in 1986, and most private land within that 3–6 km wide Zone had been purchased by 1989 (Pringle 1988). In 1990 an additional narrow strip of 20 km² along the western border of the narrow La Selva sector was declared parkland; by October 1993, its purchase was c. 60% complete. Funds were raised by OTS, The Nature Conservancy (U.S.A.) and the World Wildlife Fund – U.S. working with Costa Rica's Fundación de Parques Nacionales, which maintains an integral function in assisting programmes of land purchase, protection and environmental education. Also, four private reserves totalling 14.22 km² are situated near the park-station complex (Butterfield 1994).

The government in 1987 through the National Parks Service began to integrate park management and community outreach in nine regional Conservation Areas (ACs), merging key protected areas and adjacent areas into units within a National System of Conservation Areas (SINAC). The Central Volcanic Cordillera AC (Map 27) combines the Braulio Carrillo-La Selva region with Volcán Poás National Park (56 km²), Volcán Irazú NP (23 km²) and the bordering Central Volcanic Cordillera Forest Reserve (910 km²) (Brandon and Umaña 1991). The entire complex (c. 1464 km²) has been recognized by UNESCO as the Central Volcanic Cordillera Biosphere Reserve (MAB 1988; Butterfield 1994). FUNDECOR, a regional non-governmental organization with a mandate to foster conservation and buffer-zone management of the BR, was established in 1989 with U.S. Agency for International Development financing, including an endowment.

The Organization for Tropical Studies facilitates education, research, and wise use of natural resources through a consortium of over 50 universities and institutions from U.S.A., Puerto Rico, Honduras and Costa Rica (Stone 1988; Clark 1990; Schnell 1993, pers. comm.). Efforts at La Selva Biological Station in research for ecodevelopment and in environmental education are expanding (Clark 1988, 1990; Kohl 1993; McDade *et al.* 1994). Investment in agency, institutional and human resources is increasing in Costa Rica and becoming well recognized as essential – so that protection and management of the park region may be lasting, with help from a carefully tailored institutional framework and its endowment, as pressures intensify with population growth (Clark 1988, 1990; Gámez and Ugalde 1988; Gómez 1988; Hartshorn and Peralta 1988; Stone 1988; Wallace 1992).

CPD Site MA16: Braulio Carrillo-La Selva region, Costa Rica. View, looking east towards Braulio Carrillo National Park, from road running north from San José. Photo: Sarah Fowler.

MAP 27. BRAULIO CARRILLO-LA SELVA REGION, COSTA RICA (CPD SITE MA16)

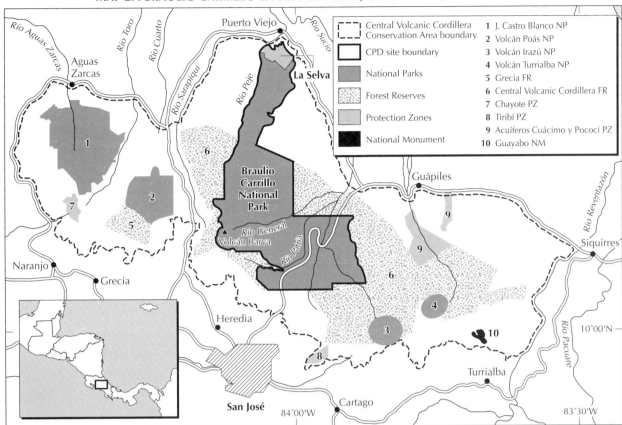

References

Boza, M.A. (1988). *Parques nacionales Costa Rica national parks*. Editorial Heliconia and Fundación Neotrópica, San José. 271 pp.

Boza, M.A. and Mendoza, R. (1981). *The national parks of Costa Rica*. Incafo, Madrid. 310 pp.

Brandon, K. and Umaña, A. (1991). Rooting for Costa Rica's megaparks. *Américas* 43(3): 20–31.

Butterfield, R.P. (1994). The regional context: land colonization and conservation in Sarapiquí. In McDade, L.A., Bawa, K.S., Hespenheide, H.A. and Hartshorn, G.S. (eds), *La Selva*. University of Chicago Press, Chicago. Pp. 299–306.

CCT (1975). *Estudio sobre bases ecológicas y legales para establecer medidas de protección y control a los bosques y aguas de la zona de influencia de la carretera San José-Guápiles*. Centro Científico Tropical (CCT), San José.

Clark, D.B. (1988). The search for solutions: research and education at the La Selva station and their relation to ecodevelopment. In Almeda, F. and Pringle, C.M. (eds), *Tropical rainforests: diversity and conservation*. California Academy of Sciences and American Association for Advancement of Science (AAAS) Pacific Division, San Francisco. Pp. 209–224.

Clark, D.B. (1990). La Selva Biological Station: a blueprint for stimulating tropical research. In Gentry, A.H. (ed.), *Four neotropical rainforests*. Yale University Press, New Haven. Pp. 9–27.

Gámez, R. and Ugalde, A. (1988). Costa Rica's national park system and the preservation of biological diversity: linking conservation with socio-economic development. In Almeda, F. and Pringle, C.M. (eds), *Tropical rainforests*. Calif. Acad. Sci. and AAAS Pacific Division, San Francisco. Pp. 131–142.

Gómez, L.D. (1988). The conservation of biological diversity: the case of Costa Rica in the year 2000. In Almeda, F. and Pringle, C.M. (eds), *Tropical rainforests*. Calif. Acad. Sci. and AAAS Pacific Division, San Francisco. Pp. 125–129.

Greene, H.W. (1988). Species richness in tropical predators. In Almeda, F. and Pringle, C.M. (eds), *Tropical rainforests*. Calif. Acad. Sci. and AAAS Pacific Division, San Francisco. Pp. 259–280.

Guyer, C. (1990). The herpetofauna of La Selva, Costa Rica. In Gentry, A.H. (ed.), *Four neotropical rainforests*. Yale University Press, New Haven. Pp. 371–385.

Hammel, B.E. (1990). The distribution of diversity among families, genera, and habit types in the La Selva flora. In Gentry, A.H. (ed.), *Four neotropical rainforests*. Yale University Press, New Haven. Pp. 75–84.

Hartshorn, G.S. (1983). Plants: introduction. Site descriptions: La Selva. In Janzen, D.H. (ed.), *Costa Rican natural history*. University of Chicago Press, Chicago. Pp. 136–140.

Hartshorn, G. and Peralta, R. (1988). Preliminary description of primary forests along the La Selva-Volcán Barva altitudinal transect, Costa Rica. In Almeda, F. and Pringle, C.M. (eds), *Tropical rainforests*. Calif. Acad. Sci. and AAAS Pacific Division, San Francisco. Pp. 281–295.

Janzen, D.H. (ed.) (1983). *Costa Rican natural history*. University of Chicago Press, Chicago. 816 pp.

Kohl, J. (1993). No reserve is an island. *Wildlife Conserv.* 96(5): 74–75, 82.

Lieberman, D., Hartshorn, G.S., Lieberman, M. and Peralta, R. (1990). Forest dynamics at La Selva Biological Station, Costa Rica, 1969–1985. In Gentry, A.H. (ed.), *Four neotropical rainforests*. Yale University Press, New Haven. Pp. 509–521.

MAB (Man and the Biosphere Programme) (1988). La Selva Biological Station designated as part of a Central Volcanic Cordillera Biosphere Reserve in Costa Rica. *MAB Bull.* 12(1): 6.

McDade, L.A., Bawa, K.S., Hespenheide, H.A. and Hartshorn, G.S. (eds) (1994). *La Selva: ecology and natural history of a neotropical rainforest*. University of Chicago Press, Chicago. 486 pp.

Pringle, C.M. (1988). History of conservation efforts and initial exploration of the lower extension of Parque Nacional Braulio Carrillo, Costa Rica. In Almeda, F. and Pringle, C.M. (eds), *Tropical rainforests*. Calif. Acad. Sci. and AAAS Pacific Division, San Francisco. Pp. 131–142.

Pringle, C.M., Chacón, I., Grayum, M.H., Greene, H.W., Hartshorn, G.S., Schatz, G.E., Stiles, F.G., Gómez, C. and Rodríguez, M. (1984). Natural history observations and ecological evaluation of the La Selva Protection Zone, Costa Rica. *Brenesia* 22: 189–206.

Stiles, F.G. (1988). Altitudinal movements of birds on the Caribbean slope of Costa Rica: implications for conservation. In Almeda, F. and Pringle, C.M. (eds), *Tropical rainforests*. Calif. Acad. Sci. and AAAS Pacific Division, San Francisco. Pp. 243–258.

Stone, D.E. (1982). *Conversion of Zona Protectora "La Selva" into an extension of Braulio Carrillo National Park, Costa Rica*. Proposal to WWF – U.S. (15 Oct.). Organization for Tropical Studies. 5 pp. + 3 appendices.

Stone, D.E. (1988). The Organization for Tropical Studies (OTS): a success story in graduate training and research. In Almeda, F. and Pringle, C.M. (eds), *Tropical rainforests*. Calif. Acad. Sci. and AAAS Pacific Division, San Francisco. Pp. 143–187.

Timm, R.M., Wilson, D.E., Clauson, B.L., LaVal, R.K. and Vaughan, C.S. (1989). *Mammals of the La Selva-Braulio Carrillo complex, Costa Rica*. North Amer. Fauna 75. 162 pp.

Wallace, D.R. (1992). *The quetzal and the macaw: the story of Costa Rica's national parks*. Sierra Club Books, San Francisco, Calif. 222 pp.

Wilbur, R.L. and collaborators (1994). Vascular plants: an interim checklist. In McDade, L.A., Bawa, K.S., Hespenheide, H.A. and Hartshorn, G.S. (eds), *La Selva*. University of Chicago Press, Chicago. Pp. 350–378.

Wilson, D.E. (1990). Mammals of La Selva, Costa Rica. In Gentry, A.H. (ed.), *Four neotropical rainforests*. Yale University Press, New Haven. Pp. 273–286.

Author

This Data Sheet was written by Olga Herrera-MacBryde (Smithsonian Institution, Department of Botany, NHB-166, Washington, DC 20560, U.S.A.).

LA AMISTAD BIOSPHERE RESERVE
Costa Rica and Panama

Location: In south-east Costa Rica and north-west Panama in the Talamanca range, including Pacific and Caribbean slopes and highest mountain in each country. Approximately within latitudes 8°44'–9°48'N and longitudes 82°16'–83°52'W.

Area: 6126 km² in Costa Rica in Biosphere Reserve, over 4000 km² in Panama planned for inclusion.

Altitude: 0–3819 m.

Vegetation: Ten life zones in altitudinal gradient from tropical humid forest to subalpine rain páramo. Includes exuberant oak forests and over 90% of Central American páramos.

Flora: Very rich – c. 10,000 vascular plant species; high endemism – c. 30%. Includes the conservation units of both countries with most diversity and endemism.

Useful plants: Species for timber, thatching, artisanal crafts; ornamentals; extracts for medicinal, ceremonial and dyeing purposes of indigenous inhabitants.

Other values: Several Amerindian peoples inhabit reserve; watershed protection; wilderness; refuge for many faunal species, including endangered species; genetic resources; ecotourism.

Threats: Conversion of forests to subsistence farms and pasturage; logging; human-set fires, growing of marijuana, pesticide runoff; mining prospects, planned transmontane highway and trans-isthmian oil pipeline.

Conservation: A World Heritage Site. In Costa Rica: 3 National Parks, 1 Protected Zone, 2 Biological Reserves, 1 Forest Reserve, 7 Amerindian Reserves, 1 Botanical Garden. In Panama: existing units planned for Biosphere Reserve core are 3 National Parks, 1 Forest Reserve, 1 Protection Forest, 1 Amerindian Reserve. Other areas being evaluated for addition.

Geography

La Amistad Biosphere Reserve (RBA) in south-eastern Costa Rica includes most of the Talamanca range and its slopes, in 15 conservation units. Six neighbouring protected areas in Panama's provinces of Chiriquí, Bocas del Toro and possibly Veraguas are planned for inclusion (MIRENEM *et al.* 1990) (Map 28).

The uplifting of the Talamanca range occurred 35–15 million years ago, mainly during the Oligocene and Miocene, with folding and intrusive events. The influence of Pleistocene glaciations is evident on the highest peaks (Cerro de la Muerte, Chirripó, Kamuk, Barú), which have for example glacial cirques, U-shaped valleys and moraines (Weyl 1955). The highest peak of each country is within the RBA: Cerro Chirripó (3819 m) in Costa Rica and Barú Volcano (3475 m) in Panama. Most soils within the reserve have loam or clay textures and are acid to very acid, with low permeability.

Climatic conditions within the RBA are very diverse (Herrera 1986), due to the region's large expanse, its geographic location which includes both the Pacific and Caribbean watersheds, the great altitudinal differences, and its irregular and abrupt topography, with slopes from 17° to over 40°. The predominant climate of the Pacific watershed is hot and humid, with a dry season from December to April. The Caribbean watershed is hot and wet throughout the year, with a not well-defined short dry season. At low and middle elevations mean annual temperatures may vary from 21°–26°C and mean annual precipitation from 2800–6840 mm. The climate in the central area of the RBA, which includes the high mountains, is predominantly cold and wet, with the mean annual precipitation 2800–5300 mm and mean annual temperatures 6°–23°C, including temperatures below freezing in the dry season.

Vegetation

Because of the extensive altitudinal and climatic ranges within the RBA, the majority of the Holdridge life zones of both countries are represented. The RBA lands are mainly above 2000 m in both watersheds. In Costa Rica eight of its 12 life zones are present: tropical humid and wet forests, premontane wet and rain forests, lower montane wet and rain forests, montane rain forest and subalpine rain páramo (Bolaños-M. and Watson-C. 1993). In Panama two other life zones are added: premontane humid and montane wet forests (Selles-A. 1992).

The tropical belt

The tropical belt, extending from sea-level to 600 m (–800 m), has forests with a canopy c. 30 m high. Emergent trees 45–55 m tall and 1–3 m in dbh with large buttresses are frequent. Representative tree species are *Dipteryx panamensis, Terminalia amazonia, Carapa nicaraguensis, Ceiba pentandra* and *Hyeronima laxiflora*. In the lower forest strata Rubiaceae, Leguminosae, Melastomataceae and Flacourtiaceae abound, together with many palm species. Common palm genera are *Socratea, Astrocaryum, Bactris, Geonoma* and *Asterogyne*. The understorey is relatively sparse, occupied by for example grasses, ferns and species of Marantaceae, Zingiberaceae and Araceae.

In Panama four special vegetation types in this belt are: mangroves, with large stilt-rooted *Rhizophora mangle*; homogeneous forests of *Prioria copaifera*; a *Campnosperma panamensis* association with trees 30 m tall near Bahía de Almirante; and a *Pterocarpus officinalis-Symphonia globulifera* association near Changuinola River (Selles-A. 1992; ANCON 1993).

The premontane belt

The premontane belt (600–800 m to 1300–1500 m) has the most biodiversity and physiognomic complexity in the RBA. These forests, with a canopy 30–40 m high, have large cylindric trunks with well-formed crowns, supporting many epiphytes. Representative species of the canopy and subcanopy include *Brosimum utile, Terminalia amazonia, Vochysia* spp., *Sacoglottis amazonica, Hirtella racemosa, Symphonia* sp. and *Mouriri* sp. Ferns and palms are abundant in the ground layer (Holdridge *et al.* 1971; MIRENEM *et al.* 1990; Selles-A. 1992).

In premontane wet forest, Las Tablas Protected Zone is one of the last remaining forests on acid intrusive and Quaternary materials over Pliocene terrain. The soils differ from others in the vicinity. This substrate and the area's distinctive Pacific-slope climate are responsible for a vegetation type with great richness and biodiversity (Gómez 1989).

MAP 28. LA AMISTAD BIOSPHERE RESERVE, COSTA RICA AND PANAMA (CPD SITE MA17)

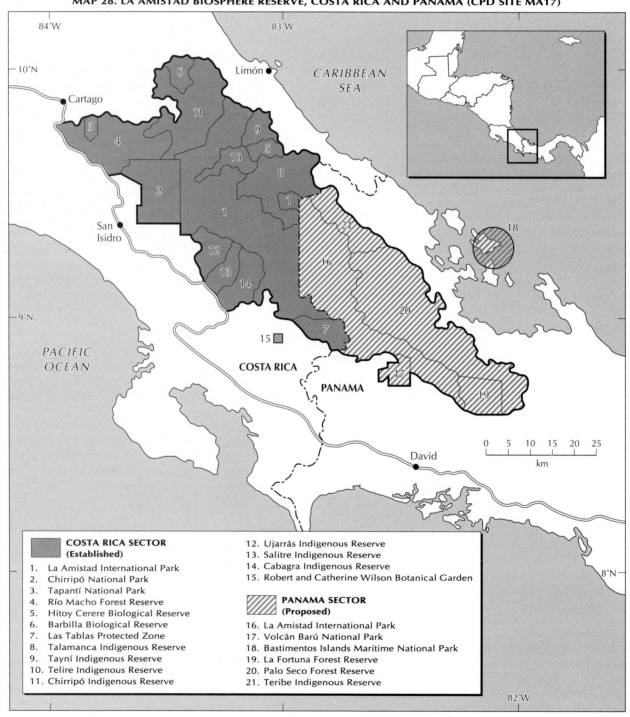

COSTA RICA SECTOR
(Established)

1. La Amistad International Park
2. Chirripó National Park
3. Tapantí National Park
4. Río Macho Forest Reserve
5. Hitoy Cerere Biological Reserve
6. Barbilla Biological Reserve
7. Las Tablas Protected Zone
8. Talamanca Indigenous Reserve
9. Tayní Indigenous Reserve
10. Telire Indigenous Reserve
11. Chirripó Indigenous Reserve
12. Ujarrás Indigenous Reserve
13. Salitre Indigenous Reserve
14. Cabagra Indigenous Reserve
15. Robert and Catherine Wilson Botanical Garden

PANAMA SECTOR
(Proposed)

16. La Amistad International Park
17. Volcán Barú National Park
18. Bastimentos Islands Maritime National Park
19. La Fortuna Forest Reserve
20. Palo Seco Forest Reserve
21. Teribe Indigenous Reserve

The lower montane belt

The lower montane belt starts at c. 1500 m on the Pacific slope and c. 1300 m (–1400 m) on the Caribbean slope. Covering the largest area within the reserve, it has a wide variety of species, especially in Lauraceae. Canopy species include *Cornus disciflora, Roupala complicata, Sapium* spp., *Hyeronima poasana, Magnolia poasana, Didymopanax pittieri* and *Podocarpus macrostachyus*. The oak *Quercus seemannii*, sometimes as tall as 40 m, is frequent, whereas *Q. oocarpa* and *Q. rapurahuensis* are common. *Quercus copeyensis* appears at the higher elevations (Jiménez-Marín and Chaverri-Polini 1991). Common epiphytes reported from Panama are *Anthurium carnosum, A. testaceum, Elleanthus glaucophyllus* and *Odontoglossum chiriquense*, as well as bryophytes and lichens (ANCON 1993).

The montane belt

The montane belt, from 2300 m (–2400 m) to 3200 m (–3300 m), includes montane wet forest and montane rain forest. The latter life zone covers large areas within the RBA and is relatively homogeneous floristically, with *Quercus* predominant. The two main oaks, *Q. copeyensis* and *Q. costaricensis*, show very high abundance and basal area (Jiménez *et al.* 1988; Orozco 1991; ANCON 1993; Koomen 1993). Some *Q. copeyensis* trees with large buttresses are 50 m in height and 1.5 m in dbh. Other common species in this belt are the conifers *Podocarpus macrostachyus* and *Prumnopitys standleyi*, along with *Magnolia poasana, Cleyera theaeoides*, several species of *Ilex* and *Weinmannia* and many species of Lauraceae (*Ocotea, Persea, Nectandra, Phoebe*) and Araliaceae.

Four types of forest communities in the montane belt have oak species associated with *Chusquea* bamboo. Within them, ten phytosociological groups have been described for the c. 190 plant species (Kappelle, Cleef and Chaverri 1989). In the understorey species of Rubiaceae, Melastomataceae and Myrsinaceae are common, as well as several species of *Chusquea* (Jiménez *et al.* 1988).

Subalpine rain páramo

Subalpine rain páramo, which occurs above 3100 m (–3400 m), is one of the life zones of least area within the RBA. Having many grass species and shrubs to 3 m tall, this vegetation characterizes the summits of Cerro de la Muerte, Chirripó (which has the largest páramo, c. 60 km²) and Kamuk in Costa Rica and Fábrega, Echandi and Barú in Panama. In general the shrubs are short and rounded, with robust strong branches and small sturdy leaves. The dwarf bamboo *Chusquea subtessellata* has the most local biomass (Weber 1959; Weston 1981).

Three distinct vegetational zones may be distinguished in the páramo: subpáramo (the subalpine dwarf forest, from 3100–3400 m), páramo proper and superpáramo. Floristically, c. 25 plant communities may be recognized – 10 zonal, c. 15 azonal. They are in very wet, very humid or rocky and dry habitats. Well-represented zonal species are *Chusquea subtessellata, Hypericum irazuense, H. costaricense, Festuca dolichophylla* and *Muhlenbergia flabellata*, whereas each of the species *Aciachne pulvinata, Azorella biloba, Isoetes storkii, Carex lehmanniana* and others predominate in separate azonal communities (Cleef and Chaverri-P., in prep.). Among the most commonly found families in the páramo, Gramineae, Asteraceae, Cyperaceae, Rosaceae and Ericaceae stand out for their large numbers of species.

Flora

High biodiversity and endemism are evident in the RBA in both Costa Rica and Panama (MIRENEM *et al.* 1990; Selles-A. 1992; ANCON 1993). The RBA is one of four regions with the highest endemism in Costa Rica – c. 90% of the country's known flora may be found there, with c. 30% being endemic. Some 10,000 vascular and 4000 non-vascular plant species are estimated to occur in the reserve. About 80% of Costa Rica's known ferns, 67% of the known orchids and almost all the known lichens are found in the RBA (Gómez 1989; MIRENEM *et al.* 1990).

The high biodiversity and endemism are the result of natural and disturbed vegetation occurring at elevations from sea-level to over 3800 m, in conjunction with the varied topography, array of soil types and range of different humidity regimes, which help create a myriad of habitats. Geological events related to formation of the Talamanca range, including long periods of isolation, contributed additionally to the development of a rich and diverse flora.

Several locations within the RBA have high endemism. Small populations of endemic ferns such as *Costaricia werkleana, Hyalotrichipteris* sp., *Polybotrya* sp., as well as tree ferns (*Nephelea, Cyathea, Trichipteris, Sphaeropteris*) grow within the Río Macho Forest Reserve (Gómez 1989). Endemism is also found in montane and páramo vegetation. Within the oak forests, *Quercus copeyensis, Q. costaricensis* and *Prumnopitys standleyi* as well as *Ilex tristis* and *Magnolia poasana* are endemic to the Talamanca range (Elizondo *et al.* 1989; MIRENEM *et al.* 1992). The epiphytes *Anthurium globosum* and *A. pittieri* var. *morii* are endemic in Panama (ANCON 1993). In the páramo, the abundant *Chusquea subtesselleta* is endemic, as well as rare plants such as *Wernera nubigena, Iltisia repens, Myrrhidendron donnell-smithii, Rumex costaricensis* and several species of *Westoniella* (MIRENEM *et al.* 1990; ANCON 1993; Cleef and Chaverri-P., in prep.).

Phytogeographically, in montane oak forests the 253 vascular plant genera collected (excluding bromeliads and orchids) show a stronger affinity with the tropics (75%) and neotropics (46%) than with temperate (17%) or cosmopolitan (8%) elements (Kappelle, Cleef and Chaverri 1992). A different pattern is evident for the 150 vascular plant genera collected from the páramo, which have more temperate affinity (53%) than tropical (36%) and neotropical (25%) or cosmopolitan (11%) elements. Almost 95% of these páramo genera also occur in the Andes (Cleef and Chaverri-P. 1992).

Useful plants

Many forest species within the RBA have a high commercial value as timber. Some species in the lowlands with potential for sustainable timber management are: *Carapa guianensis, Hyeronima alchorneoides, Aspidosperma megalocarpon, Terminalia amazonia, Virola* spp. and *Vochysia* spp.; at middle elevations: *Alnus acuminata* and *Cedrela tonduzii*; and in the high mountains: the oak species, which also have excellent qualities for charcoal, plus *Magnolia, Podocarpus* and several Lauraceae species. In the Panamanian highlands, *Magnolia sororum* is considered the most valuable tree species, producing an excellent timber.

Plants with medicinal value are used by indigenous and non-indigenous people. Medicine men ("awápas") use a large variety of plants to treat a wide range of ailments from anemia and ulcers to snake bites – e.g. *Dorstenia contrajerva*, *Petiveria alliacea*, *Psidium guajava*, *Quassia amara*, *Drimys granadensis*, *Senecio* spp., *Smilax* spp. and *Dioscorea* spp. Other plants for example in Palmae, Araceae, Moraceae and Bignoniaceae are used in handicrafts to construct baskets, hammocks, crates and bags; in wood carving to manufacture drums, bows and arrows and water containers; and in house construction. Many food plants are also found in the RBA, among them *Euterpe* sp. for its palm heart (Ocampo 1989; ANCON 1993).

Social and environmental values

The region is a portion of two Endemic Bird Areas (EBAs). The relatively small highlands of Costa Rica and western Panama (EBA A18) are home to an incredible 52 restricted-range bird species. La Amistad BR, in the Cordillera de Talamanca, protects over 10% of this highland zone. The RBA is one of the rare areas that has both montane and lowland forests in a continuous tract, thus allowing for the seasonal altitudinal migrations of many of the highland endemics. Some of the 13 restricted-range species of the Central American Caribbean Slope EBA (A16) also occur in the lower reaches of the cordillera.

The RBA offers adequate habitat for 70% of the wild animals found in Costa Rica. The Talamanca range in Costa Rica and Panama is also important for faunal endemism: 37 bird species and 75 subspecies, and 11 mammal species, are considered endemic (Elizondo *et al*. 1989). Endemism is also found in reptiles and amphibians, of which 75% of all the species known in Costa Rica are present in the RBA. The region has sufficient habitat for such endangered species as the West Indian manatee (*Trichechus manatus*), harpy eagle (*Harpia harpyja*), resplendent quetzal (*Pharomachrus mocinno costaricensis*) and harlequin frog (*Atelopus chiriquensis* – the only bufonid present in the 1800–2000 m cloud forests in Panama). Six cat species (including the jaguar), Baird's and Andean tapirs and the giant anteater (*Myrmecophaga tridactyla*) probably have ample habitat in the reserve. The insect fauna also is diverse; La Amistad is reputed to harbour the world's second most diverse butterfly fauna (MIRENEM *et al*. 1990; ANCON 1993).

Several large Amerindian groups inhabit the RBA. The Bribri and Cabecar peoples in Costa Rica, together numbering 12,000, represent almost two-thirds of Costa Rica's indigenous population. In Panama three of the country's five ethnic groups inhabit the reserve, including the Teribe and many Guaymí. About 165,000 indigenous inhabitants, c. 65% of the country's Amerindians, live and carry on activities within the reserve (MIRENEM *et al*. 1990; Castro-Chamberlain 1993).

Important watersheds with a high hydroelectric value occur in the RBA and supply water for drinking to the cities of Limón and San Isidro de El General (Costa Rica) and David (Panama), and for agriculture and animal husbandry in the surrounding regions. Ecotourism is important in the RBA, with potential socio-economic benefits for the surrounding communities (MIRENEM *et al*. 1990).

Economic assessment

Limitations in land use are due to high rainfall, abrupt slopes, acid soils and medium to low temperatures above 2000 m. Although much of this region is still covered by forests, especially in the Caribbean watershed, large areas have been cut for cattle-ranching and agriculture. Traditionally no special attention was given to using the land in an ecologically sound manner, and the soils have been degraded through erosion and fire. In recent years, however, long-term forestry has gained importance, including two projects by the Universidad Nacional and CATIE in Costa Rica for sustainable management of the montane oak forests.

The possibility is under study to produce secondary products sustainably from the forests. Many plants show economic potential, including ornamental, dyeing and medicinal species. Ornamental plants, including seeds, germinated seeds, seedlings or leaves in *Chamaedorea*, *Geonoma*, *Carludovica*, *Zamia*, *Anthurium* and *Syngonium*, are being marketed on a small scale and could be produced sustainably in the future, providing an economic gain to the inhabitants (Ocampo 1989). The infrastructure for ecotourism has placed a demand on palm leaves for thatching.

Threats

Although the RBA has experienced little development in comparison to most regions of both countries, several factors menace its stability and preservation. Permits for oil-mining exploration, which has provoked deforestation and pollution, have been issued since the 1980s. In Costa Rica, 23 additional permits for exploitation of coal and minerals have been presented (MIRENEM *et al*. 1990). Logging, growing marijuana (*Cannabis sativa* subsp. *indica*) and pesticide runoff from banana plantations also cause problems.

In the Amerindian Reserves in Costa Rica, pressure for the land and exploitation of natural resources by non-indigenous people is ever present. In Panama, the expansion of the agricultural frontier is also a major threat – an accelerated process of colonization and deforestation accompanied by soil and watershed degradation. Major areas being affected are Cerro Punta and the communities of Guadalupe, Las Nubes and the highlands of Boquete, in Chiriquí Province (ANCON 1993; Armien 1992). Hydroelectric projects are under evaluation in both countries. Two additional projects menace the biological integrity of the RBA: construction of a trans-Talamanca highway and a trans-isthmian oil pipeline.

Conservation

A resolution of the 1974 First Central American Meeting on Conservation of Natural and Cultural Resources was the establishment of an international park in the Talamanca range, including important natural vegetation in both countries. In 1979 the presidents of Costa Rica and Panama signed an agreement for cooperation in the Talamanca region, which involved binational efforts for preservation of the environment, management of natural resources, improvement of education and public health, and technical assistance in agriculture (MIRENEM *et al*. 1990). In 1992 La Amistad BR was one of 11 areas given priority under the region's Convenio para la Conservación de la Biodiversidad y Protección de Areas Silvestres Prioritarias en América Central. The development of binational projects and conservation are coordinated by the countries' Executive Secretariat of the Agreement for Border Cooperation, and the National Resources Bilateral Commission.

La Amistad International Park (PILA) was decreed in Costa Rica in March 1982, with a similar decree in Panama four years later. Also in 1982 UNESCO recognized PILA plus 14 adjacent protected areas in Costa Rica as a MAB Biosphere Reserve, and in 1983 as a World Heritage Site (Morales, Barborak and MacFarland 1983; MIRENEM *et al.* 1990; Wallace 1992). The Costa Rican sector of the reserve, which covers 6126 km² and 12% of the country's area, includes protection and sustainable management of 15 varied units: La Amistad International Park-Costa Rica (1939 km²), Chirripó National Park (501.5 km²), Tapantí NP (51 km²), Las Tablas Protected Zone (196 km²), Barbilla Biological Reserve (128 km²), Hitoy Cerere Biological Reserve (91.5 km²), Río Macho Forest Reserve (674 km²), Chirripó, Taymí, Telire, Talamanca, Ujarrás, Salitre and Cabagra Indigenous Reserves (2543 km²) and the Robert and Catherine Wilson Botanical Garden (1.4 km²), at Coto Brus.

In Panama, Volcán Barú National Park was established in 1976. Although the Biosphere Reserve has not yet been established in Panama, a conservation nucleus of six existing protected units has been suggested. The planned core for the Panamanian sector would add c. 4000 km² to the reserve: La Amistad International Park-Panama (2070 km²), Volcán Barú National Park (143 km²), La Fortuna Forest Reserve (200 km²), Palo Seco Protection Forest (1250 km²), Islas Bastimentos Maritime NP (132 km²) and Teribe Indigenous Reserve (c. 200 km²). The protection of new areas, such as the Cricamola watershed, is under discussion.

In 1990 Panama's Institute of Renewable Natural Resources (INRENARE) asked the Organization of American States (OAS/OEA) for cooperation in planning and development of the Panamanian sector of the PILA. This effort is being carried out by Panama's Ministry of National Planning and Economic Policy (MIDEPLAN) and INRENARE, with assistance from the OAS, Conservation International and the Asociación Nacional para la Conservación de la Naturaleza (ANCON). Activities include wildlands and regional planning, environmental education programmes for the surrounding communities (in conjunction with seven more conservation groups) and enforcement activities in four critical zones (ANCON 1993; Castro-Chamberlain 1993).

References

ANCON (1993). *La Amistad-Panamá*. Asociación Nacional para la Conservación de la Naturaleza (ANCON). 8 pp.

Armien, I.L. (1992). Determinación de los patrones de uso y tenencia de la tierra en las provincias de Chiriquí y Veraguas (distrito de Cañazas, La Palma y Santa Fe). In *Informe de diagnóstico: estrategia para la formulación e implementación de políticas para el ordenamiento ambiental de la región de La Amistad-Panamá*. Organización de Estados Americanos (OEA), San José, Costa Rica.

Bolaños-M., R.A. and Watson-C., V. (1993). *Mapa ecológico de Costa Rica, según el sistema de clasificación de zonas de vida del mundo de L.R. Holdridge*. Scale 1:200,000. Centro Científico Tropical, Instituto Costarricense de Electricidad, San José. 9 maps.

Castro-Chamberlain, J.J. (1993). *Proyecto de cooperación técnica binacional Reserva de la Biósfera La Amistad Costa Rica-Panamá*. OEA. 11 pp.

Cleef, A.M. and Chaverri-P., A. (1992). Phytogeography of the páramo flora of Cordillera de Talamanca, Costa Rica. In Balslev, H. and Luteyn, J.L. (eds), *Páramo: an Andean ecosystem under human influence*. Academic Press, London. Pp. 45–60.

Cleef, A.M. and Chaverri-P., A. (in prep.). An outline of the Chirripó and Buena Vista páramo vegetation, Cordillera de Talamanca, Costa Rica.

Elizondo, L.H., Jiménez, Q., Alfaro, R. and Chaves, R. (1989). *Contribución a la conservación de la biodiversidad de Costa Rica: 1. Areas de endemismo, 2. Vegetación natural*. Consultoría realizada para The Nature Conservancy y U.S. Fish and Wildlife Service. Fundación Neotrópica, San José, Costa Rica. 124 pp.

Gómez, L.D. (1989). *Unidades naturales; estado y uso actual de los ecosistemas y recursos naturales, beneficios potenciales y riegos naturales en la región de la Reserva de la Biósfera de Talamanca-Amistad*. Consultoría para la OEA, San José, Costa Rica. 13 pp.

Herrera, W. (1986). *Clima de Costa Rica*. Editorial Universidad Estatal a Distancia, San José, Costa Rica. 118 pp.

Holdridge, L.R., Grenke, W.C., Hatheway, W.H., Liang, T. and Tosi, J.A., Jr. (1971). *Forest environments in tropical life zones. A pilot study*. Pergamon Press, Oxford, U.K. 747 pp.

Jiménez, W., Chaverri, A., Miranda, R. and Rojas, I. (1988). Aproximaciones silviculturales al manejo de un robledal (*Quercus* spp.) en San Gerardo de Dota. *Turrialba* 38: 208–214.

Jiménez-Marín, W. and Chaverri-Polini, A. (1991). Consideraciones ecológicas y silviculturales acerca de los robles (*Quercus* spp.). *Ciencias Ambientales (Costa Rica)* 7: 49–63.

Kappelle, M., Cleef, A.M. and Chaverri, A. (1989). Phytosociology of montane *Chusquea-Quercus* forests, Cordillera de Talamanca, Costa Rica. *Brenesia* 32: 73–105.

Kappelle, M., Cleef, A.M. and Chaverri, A. (1992). Phytogeography of Talamanca montane *Quercus* forests, Costa Rica. *J. Biogeogr.* 19: 299–315.

Koomen, K. (1993). *Effect of overstory density and basal area on natural regeneration of* Quercus copeyensis *and* Quercus costaricensis *in a Costa Rica oak forest*. B.S. thesis, International Agricultural College Larenstein, Velp, The Netherlands. 66 pp.

MIRENEM (Ministerio de Recursos Naturales, Energía y Minas), MIDEPLAN (Ministerio de Planificación Nacional y Política Económica), CI and OEA (1990). *Estrategia para el desarrollo institucional de la Reserva de la Biósfera "La Amistad"*. Conservación Internacional (CI) and Organización de Estados Americanos (OEA), San José, Costa Rica. 174 pp.

MIRENEM, Museo Nacional de Costa Rica, and Instituto Nacional de Biodiversidad (1992). *Estudio nacional de biodiversidad. Costos, beneficios y necesidades de financiamiento de la conservación de la diversidad biológica en Costa Rica.* MIRENEM, San José. 164 pp.

Morales, R., Barborak, J.R. and MacFarland, C.G. (1983). *Planning and managing a multi-component, multi-category international biosphere: the case of the Amistad/ Talamanca Range/Bocas del Toro wildlands complex of Costa Rica and Panamá.* Centro Agronómico Tropical de Investigación y Enseñanza (CATIE), Turrialba, Costa Rica. 10 pp.

Ocampo, R. (1989). *Desarrollo de productos secundarios del bosque tropical: una alternativa económica.* Consultoría para la OEA, San José, Costa Rica. 14 pp.

Orozco, L. (1991). *Estudio ecológico y de estructura horizontal de seis comunidades boscosas de la Cordillera de Talamanca, Costa Rica.* CATIE, Turrialba, Costa Rica. 34 pp.

Selles-A., F.E. (1992). Recursos naturales. In *Informe de diagnóstico: estrategia para la formulación e implementación de políticas para el ordenamiento ambiental de la región de La Amistad-Panamá.* OEA, San José, Costa Rica. Pp. 10–45.

Wallace, D.R. (1992). *The quetzal and the macaw: the story of Costa Rica's national parks.* Sierra Club Books, San Francisco, Calif. 222 pp.

Weber, H. (1959). *Los páramos de Costa Rica y su concatenación fitogeográfica en los Andes suramericanos.* Instituto Geográfico de Costa Rica, San José. 67 pp.

Weston, A.S. (1981). *Páramos, ciénagas and subpáramo forest in the eastern part of the Cordillera de Talamanca.* Tropical Science Center, San José, Costa Rica. 15 pp.

Weyl, R. (1955). *Contribución a la geología de la Cordillera de Talamanca.* Instituto Geográfico de Costa Rica, San José. 77 pp.

Authors

This Data Sheet was written by Adelaida Chaverri and Bernal Herrera (Universidad Nacional, Escuela de Ciencias Ambientales, Programa ECOMA, Heredia, Costa Rica) and Olga Herrera-MacBryde (Smithsonian Institution, Department of Botany, NHB-166, Washington, DC 20560, U.S.A.).

OSA PENINSULA AND CORCOVADO NATIONAL PARK
Costa Rica

Location: In southern Costa Rica near south-western Panama, the park mainly between latitudes 8°27'–8°39'N and longitudes 83°25'–83°45'W.

Area: Park main sector 424 km², Bosque Esquinas sector 148 km²; peninsula c. 2330 km².

Altitude: 0–745 m.

Vegetation: Mostly tropical wet forest, also tropical premontane wet and rain forests; associations include marsh, mangrove and swamp forests, alluvial plains forest, cloud forest.

Flora: High diversity: 4000–5000 vascular plant species on peninsula, over 500 tree species in park. From recent fieldwork, 30 species and genus *Ruptiliocarpon* (Lepidobotryaceae) new to science. Phytogeographically unusual: species with various disjunct distributions.

Useful plants: Genetic reserve for species economically and/or ethnobotanically recognized (e.g. timbers, industrial chemicals, medicines, fruits), and for species with such potential.

Other values: Rich faunal diversity (two new insect genera recently discovered); threatened species; scientific benchmark; watershed protection; scenery, ecotourism.

Threats: Logging, access from an expanding network of roads, hunting, mining, squatters, tourism without adequate regulation.

Conservation: National Park, adjacent Forest and Amerindian Reserves; regional Conservation Area: integrated land-use planning for peninsula's general development.

Geography

The Osa Peninsula is a boot-like landmass of southern Costa Rica's Puntarenas Province, projecting eastward toward bordering Panama 40 km away. The main sector of the park covers nearly 418 (to c. 437) km² along the outer base of the peninsula (Map 29), 150 km south-east of Costa Rica's capital San José. The peninsula was once considered remote, and apart from the mainland due to initial mangrove forests, rough uplands and a freshwater swamp; it has become a directly engaged, developing frontier. The region's very wet natural systems evolved in partial isolation from the broadly drier tropical Pacific coast.

Within the park is the drainage of Corcovado Basin, a broad sediment-filled oceanic embayment between Punta Llorona and Punta Río Claro (near Sirena), which extends inland from the Pacific Ocean 2–10 km eastward. The basin's low plain covers c. 100 km² through which meander several rivers, and is rimmed except to the west by uplands (often below 500 m), which increase in altitude and irregular relief from an undulating plateau (below 200 m) in the north-west part of the park (north of Llorona), to 745 m in the south-east on the peninsula's highest cerros, Rincón and Mueller. The rugged uplands, produced by intensive tectonic activity and weathering (including frequent landslides), are dominated almost throughout by eroded narrow ridges and long steep slopes, with dense drainage networks (Tosi 1975; Herwitz 1981; Hartshorn 1983).

The Corcovado plain, which was formed in the Pliocene-Pleistocene, is composed of alluvial deposits and mollisol soils. The mountainous areas are composed of Late Cretaceous volcanic and sedimentary rocks; gold has deposited in many hillsides and streambeds; the soils are inceptisols and entisols. A virtually uninterrupted sandy beach extends for 20 km, with cliffs and pocket beaches at the northern and southern park headlands; there is a marine cave near the southern point. Up to ten tremors a day sometimes occur in the region, and crustal elevations have been observed. Seaward, the Cocos crustal tectonic plate descends beneath the Caribbean plate at the rate of a few cm per year (Boza and Mendoza 1981; Janzen 1983; Lew 1983; Gómez 1986).

There are few meteorological records from Corcovado National Park (Herwitz 1981; Coen 1983; Herrera 1986). The mean annual rainfall is estimated at 3000–3800 mm on the plain and 4000–5000 mm in the uplands, to 5500–6000 mm or more on the highest cerros (cf. Hartshorn 1983; Boza 1988). The c. 9-month rainy season's rainiest period is September–November; windstorms occur, rarely of tornado force (Boza and Mendoza 1981; Janzen 1983). The soils are completely saturated for c. 9–12 months of the year. The dry season is about (December–) January through March, with just a few rainy days per month, but the humidity stays high – although there is enough drying that in February–March set fires can be maintained (Janzen 1983; Jones 1990). The mean annual temperature may be 26°–27°C near the coast (varying daily by as much as 10°C) and 21°–23°C on the cerros, with March–May warmer and November–January cooler. The estimated daily sunshine averages 5 hours per day throughout the year, varying from 8 hours a day in February and 6 hours a day in January and March–May, to just 3–4 hours daily in June–November.

Vegetation

Corcovado National Park and adjacent areas conserve the most important tract of lowland tropical wet forest still extant on the Pacific side of Central America (Hartshorn 1983; WWF – U.S. 1988). The Corcovado plain in the Holdridge life-zone system is classified as tropical premontane wet forest – warm transition; the evergreen upland forest covering almost half the park is tropical wet forest, and the highest cerros probably harbour tropical premontane rain forest. During the region's c. 3-month dry season, the humidity stays unusually high and few species are deciduous. In certain areas (e.g. Cerro Rincón), epiphyte abundance may reach high levels similar to those for example at La Selva Biological Station on the Atlantic slope in central Costa Rica (in CPD Site MA16, see Braulio Carrillo-La Selva Region Data Sheet). The park's uplands can be considered the climatic association, whereas the plains comprise several edaphic and hydric associations (Hartshorn 1983). The rainy winds between May and August cause many tree falls.

Some 13 ecosystems have been discerned within the park (in the following account, proceeding essentially inland and upward) (Tosi 1975; Boza and Mendoza 1981; Vaughan 1981). With more research, 25–30 associations may become determinable (Hartshorn 1983; Gómez 1986),

including vegetations of the coastal strand, cliff and rock faces, landslips and waterfalls. Some of the main ecosystems and associations are:

1. **Mangrove swamps** (with the tallest trees to 30 m). Up to five tree species occur, including *Pelliciera rhizophorae* (cf. Jiménez 1984) and *Rhizophora racemosa*, as well as *Crinum erubescens* and *C. brevilobatum* (FNT 1992).

2. **Corcovado Lagoon**, which each rainy season is infused with up to 1 m of standing fresh water, and supports a floating central mat of ample herbaceous vegetation including *Eichhornia crassipes, Pistia stratiotes, Salvinia* sp. and *Utricularia* spp. (FNT 1992).

3. A **herbaceous marsh** c. 50 cm in height and 10 km² in expanse bordering the lagoon, with *Hymenachne* sp., *Panicum maximum, Ludwigia* sp., *Polygonum* sp. and *Aeschynomene* sp.

4. A **palm swamp (jolillo forest)** in a band around the marsh, dominated by virtually pure stands of *Raphia taedigera*, which has leaves nearly 15 m long (Devall and Kiester 1987).

MAP 29. OSA PENINSULA AND CORCOVADO NATIONAL PARK, COSTA RICA (CPD SITE MA18)

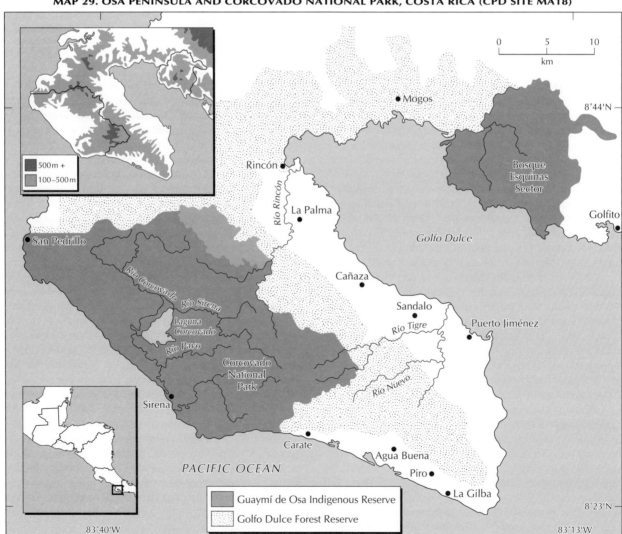

5. Varied **swamp forests**, for example peripheral to the palm swamp at first with *Andira inermis*, *Carapa guianensis*, *Crateva tapia* and *Luehea seemannii*, and beyond large well-buttressed canopy trees, stilt-rooted subcanopy trees and a fairly open palm understorey (*Crysophila guagara*, *Prestoea decurrens*), commonly including the trees *Carapa guianensis*, *Erythrina lanceolata*, *Grias fendleri*, *Mouriri* sp., *Pterocarpus officinalis* and *Virola koschnyi*. On particularly poorly drained alluvium and estuaries are lower forests, for example with *Mora oleifera* (which has the largest dicot seed – averaging c. 500 g), *Pterocarpus officinalis* (Janzen 1978), *Hasseltia* new sp. and *Pachira aquatica*.

6. **Gallery forest** on well-drained alluvial flats and terraces, with giant *Anacardium excelsum* (to 50 m tall and 3 m in dbh) and other large trees of *Caryocar costaricense*, *Hernandia didymantha*, *Pterygota excelsa*, *Terminalia oblonga* and *Ceiba pentandra* – an epiphyte-rich emergent of this species on the plain is possibly the largest tree in Central America, at over 70 m or even 80 m tall and over 3 m in diameter above buttresses 10 m high (Boza 1978, 1988).

7. **Plateau forest**: awe-inspiring with the very high density of large trees (a *Vantanea barbourii* was 65 m tall and 2 m in dbh). Other common trees include *Anaxagorea costaricensis*, *Aspidosperma spruceanum*, *Brosimum utile*, *Calophyllum longifolium*, *Carapa guianensis*, *Caryocar costaricense*, *Chrysochlamys* sp., *Couratari guianensis*, *Minquartia guianensis*, *Qualea paraense*, *Symphonia globulifera*, *Tetragastris panamensis*, *Trichilia* sp. and *Welfia georgii*.

8. **Mountain or uplands forest**: this extensive, 5-stratum ecosystem has a high diversity of species (c. 100–120 tree species per ha), very tall canopy trees (many over 50 m) sometimes unbranched up to 35 m or with spectacular buttresses, and abundant palms (*Iriartea deltoidea*, *Socratea exorrhiza*), lianas and vines. The absence of dominance is characteristic among the common tree species, which include *Ardisia cutteri*, *Aspidosperma spruceanum*, *Brosimum utile*, *Heisteria longipes*, *Poulsenia armata* and *Sorocea cufodontisii*. A site found north of the park (c. 5.5 km west of Rincón de Osa) had 22 species over 50 m tall, five over 60 m and an emergent of *Minquartia guianensis* that reached 73 m. On narrow ridges, *Brosimum utile* and *Scheelea rostrata* are characteristic; trees can still grow large, when well anchored upon both sides of a ridge (e.g. *Caryocar costaricense*, *Peltogyne purpurea*).

9. **Cloud forest**, with many epiphytes, tree ferns (perhaps *Cnemidaria choricarpa*, *Cyathea trichiata*), *Quercus* spp. (e.g. *Q. insignis*, *Q. rapurahuensis*), *Alfaroa guanacastensis*, *Oreomunnea pterocarpa* and *Ticodendron incognitum* (FNT 1992).

Flora

Few places on Earth have so much biotic diversity in such a small geographic area (Vaughan 1981), partly because of Costa Rica's role as a corridor for flora and fauna from north and south. So far over 1510 vascular plant species in 707 genera of 154 families have been documented (FNT 1992). There is an unusual mix of species on the peninsula, with their next closest population or related species far away (B.E. Hammel 1993, pers. comm.).

The Pacific lowland wet forests of Costa Rica have strong floristic affinities with the Chocó of north-western South America (CPD Site SA39, see Data Sheet) as well as the Venezuelan Amazon, and also have elements of the Guayanan flora that may have arrived before formation of the Panamanian isthmus (Gómez 1986). Primarily South American genera that reach their limit on the peninsula include *Huberodendron*, *Newtonia*, *Paramachaerium*, *Tachigali*, *Uribea*, *Williamodendron*, *Chaunochiton* and *Oenocarpus*. There also are distinctive floristic relationships: (1) with Costa Rica's mainland flora, where shared species are either at higher elevations (e.g. *Oreomunnea pterocarpa*, *Ticodendron incognitum*) or on the Atlantic slope (e.g. *Ruptiliocarpon*, which was recently discovered, in a family previously considered African); and as well with (2) Jamaica (*Ziziphus chloroxylon*), and with (3) several countries to the north, even as far as Mexico (e.g. *Recchia simplicifolia*).

The low-altitude cloud forests have unusual biogeographical characteristics and high endemism (FNT 1992). Perhaps 2–3% of the peninsular flora will be found to be endemic, which includes 25% endemism in Marantaceae, and the rare *Osa pulchra* (Rubiaceae). The Corcovado uplands forests undoubtedly have the greatest tree-species diversity in Central America – the park harbours at least 500 tree species (Hartshorn and Poveda 1983), a quarter of the tree species in all of Costa Rica (Boza 1988; Boza and Mendoza 1981).

Useful plants

The great breadth and depth of scientific knowledge accumulated about Costa Rica's natural wealth through decades of international and national efforts, including many significant studies within the park region (cf. Janzen 1983; Boza 1988), have made the park increasingly prominent for more complex and long-term biological and ecological research. The government in 1989 established a National Biodiversity Institute (INBio) to strongly accelerate both a complete national inventory (in 10 years) and investigation on the utility of the Costa Rican flora and fauna (Jones 1990; Brandon and Umaña 1991). The park region has been chosen as one of the first areas for an intensive inventory.

INBio has placed early emphasis on seeking pharmacological properties (Sittenfeld-Appel 1992). Locally a number of species provide medicinals, e.g. from seeds, bark or wood: *Quassia amara*, *Simaba cedron*, *Bursera simaruba*, *Hymenaea coubaril* and *Bauhinia manca*; from latex: *Brosimum utile* and *Ficus* spp.; and from the wounded trunk of *Copaifera camibar*, camíbar oil, which is extracted for trade.

Other commercial products include fibres for handicrafts, from the bark of *Apeiba tibourbou* and the aerial roots of climbing aroids and Cyclanthaceae. The region is a natural genetic reserve for example for *Persea americana* (avocado), *Byrsonima crassifolia* ("nance") and *Licania platypus* ("sonzapote"). The fruit of *Hymenaea coubaril* ("guapinol") may have major potential for sustainable extraction; the delicious pulp is edible and can be used in cakes and drinks, and the fruits are abundant (up to 45 kg per tree fall per day in February and March) and store well. Other non-timber products with economic potential are natural chewing gum (*Manilkara staminodella*), edible nuts (*Sterculia mexicana*), palm heart (*Iriartea*) and cosmetic oil (*Carapa guianensis*). *Osa pulchra* may have ornamental potential.

The Copenhagen Botanical Museum (K. Thomsen) in collaboration with Costa Rica's Fundación Neotrópica in 1992 initiated a study of the potential for non-timber forest products from the Osa Peninsula. The research is determining uses of the species as known locally and elsewhere, ecological quantification (both inventory and natural production), extractive potential and possible conflicts between extraction of the non-timber forest products and timber harvests.

Only 2000 km² of forests in Costa Rica might remain as of 1993 for the exploitation of timber – the 400–600 km² deforested yearly since 1977 is the fastest per capita conversion of any Latin American country. Timber in non-protected areas is likely to be gone before the year 2000 and Costa Rica may start to import lumber by 1995 (Tangley 1986; Omang 1987; Gámez and Ugalde 1988; Jones 1990). Tree replanting has already begun on the peninsula. The genetic stocks of many tree species in the park are increasingly valuable for critical reforestation or afforestation efforts (cf. Herwitz 1981).

Social and environmental values

The park protects various threatened plant and animal species, and is known to have about 124 species of mammals (over 50 bats); 375 species of birds (perhaps 5–8 endemics); 117 species of reptiles and amphibians (2 crocodilians, 4 sea-turtles); 66 freshwater fish species; and 70 species of marine crabs (Janzen 1983; Boza 1988; FNT 1992). Among the mammals are 4 monkey species (e.g. Central American squirrel monkey, white-faced capuchin), anteaters, sloths, southern river otter, crab-eating racoon, 5–6 cat species (e.g. ocelot, margay, jaguar), peccaries and Baird's tapir.

The birds include the country's largest population of scarlet macaws, the horned guan and possibly still the harpy eagle. The region is within the Southern Central American Pacific Slope Endemic Bird Area (EBA A17), which has 16 restricted-range species. At least 9 of them occur on the Osa Peninsula, of which the mangrove hummingbird (*Amazilia boucardi*) and yellow-billed cotinga (*Carpodectes antoniae*) are considered globally threatened and three others nearly threatened.

The Osa Peninsula harbours the most intact tropical lowland ecosystem for insects from central Panama to Mexico, with 6000–10,000 species – including certain bees, butterflies and beetles that need large areas to survive. Within the park are c. 6000 insect species, among them at least 220 species of butterflies (Boza and Mendoza 1981; Janzen 1983).

In the park are portions of the upper watersheds of the Rincón and Tigre rivers, which are important for the water supply of lowland communities. National and international tourism to the park is increasing, since it has become more accessible by a good road onto the peninsula to Puerto Jiménez, secondarily extending to Carate near the park's south-eastern boundary, and with the regular arrival of cruise ships. In the early days scientists were over 90% of the annual visitors (Cahn and Cahn 1979; cf. Wille-Trejos 1983); the growth of registered visitors increased from 324 in 1980 to 7863 in 1992, 80% of whom were from abroad.

Threats

Gold mining is a recurrent problem. There was a crisis a few years ago, with an influx of gold panners and some miners using more destructive modern methods of extraction from streambed alluvial deposits. The region had undergone a gold rush in the 1930s (Boza and Mendoza 1981) and in c. 1980–1985 the park was invaded by 1000–3000 people (mining squatters with their families) because of Costa Rica's troubled economy (especially the loss of local wages when banana prices dropped) and a rising price for gold (Tangley 1986; Wallace 1992). Dams, canals, even tunnels were made, and the mined tailings (soil sediment, some mercury) severely polluted aquatic ecosystems in the southern third of the park. Also, animals were sought by the influx of miners as well as other hunters and farmers and had already declined severely before 1975; they were virtually absent from the park's southern third – perhaps interrupting seed dispersal of some tree species (Tangley 1986). In 1986 the miners were relocated; restoration work and slow recovery are continuing (Tangley 1986; WWF – U.S. 1988; Jones 1990; Wallace 1992; cf. Boza and Mendoza 1981). Outside the park's boundaries modern mining goes on, for example just to the east.

Logging had occurred within the park area, and the sporadic removal of select trees is an ongoing risk. A major international operation was close to beginning in late 1975, when the park was established. The timber was estimated to have been worth more than the park's gold (Cahn and Cahn 1987). Beyond the park on the peninsula, deforestation first became significant between 1977 and 1983 (Gámez and Ugalde 1988). Already one-third of the forest is gone, and each year c. 10 km² more are cleared. The government's Institute of Agrarian Development (IDA) distributes land rights to those who will work the land, including tracts from a third (233 km²) of the neighbouring Golfo Dulce Forest Reserve, which regulates logging and tries to prevent the spread of slash-and-burn agriculture (Jones 1990).

Without sustainable development that is supported by the Costa Rican economy, the park will remain threatened: an oasis in a rising sea of societal pressures (Lewis 1982; Cahn and Cahn 1987; Gámez and Ugalde 1988). Meagre funding for the park and buffering Forest Reserve critically threaten both the capabilities to manage the planned maintenance and improvement of this rather new National Park (Vaughan 1981; MacFarland, Morales and Barborak 1984; Torres and Hurtado de Mendoza 1988) and to deal with intensifying societal stresses (Omang 1987; WWF – U.S. 1988; Jones 1990; Wallace 1992). Logging and mining roads have been built beyond the park; a side road approaches the park's mid-northern boundary and a road may be built to Drake Bay near the park's north-western boundary.

Conservation

The presence of yellow fever probably preserved the region from conversion to ranches and farms long ago (Carr and Carr

1983), and the absence of a road until recent years helped preservation. National and international organizations in 1962 began more intensively studying the flora, fauna and ecology of the Osa Peninsula. A wave of interest that arose in 1971 to establish a large protected area and the subsequent efforts were answered in October 1975, when the government established Corcovado National Park in response to the various escalating threats. In 1976, 100–300 families of settlers were relocated, and their cattle and pigs were removed from the Corcovado plain in 1978. Their (5–) 20–30 km² of pastures and croplands and 1–4 km² of under-cut forest (where the colonists had cut the undergrowth preparatory to felling and sowing for conversion to pasture), along with older degraded patches scattered through the plain, are recovering to diversified forest (cf. Herwitz 1981). The park was enlarged over 20% in 1980 to provide somewhat more natural boundaries and include part of the highlands of the peninsula (Wright 1976; Boza and Mendoza 1981; Hartshorn 1983), and the Bosque Esquinas sector which has a 2-km wide marine portion was recently established (see Map 29).

The government in 1987 began to integrate park management and community outreach for the country by means of nine regional Conservation Areas (ACs), which merged each key protected area and adjacent buffer zone into a developmental whole, within a National System of Conservation Areas (SINAC). The Osa Conservation Area (ACOSA) combines Corcovado NP (now 572 km²) and the Golfo Dulce Forest Reserve (592 km²) into unified work striving with local communities to define how to implement activities that are ecologically and economically sustainable (Brandon and Umaña 1991). An Amerindian Reserve (27.1 km²) for the Guaymí is adjacent to the park to the north. About where the peninsula meets the mainland to the west is Manglar Sierpe-Térraba Forest Reserve (227 km²). One-third of the peninsula is private land, which includes some ecotourism reserves; just 97 km² of IDA's available land is regarded as adequate for farming (only by working large, 20-ha parcels).

In 1988 an ecodevelopment programme (BOSCOSA) was initiated by Costa Rica's Fundación Neotrópica and the World Wildlife Fund – U.S. (WWF – U.S.) to improve matters on the peninsula, for example in the Golfo Dulce Forest Reserve where c. 5000 families live and in the Guaymí reserve where there are 24 indigenous families (and a few in-holdings). The BOSCOSA efforts include research, training, sustainable community forestry, natural forest management, land purchase, reforestation, agroforestry, cultivation of ornamentals, environmental education (including a Tropical Youth Centre), artisanry and ecotourism (Cabarle *et al.* 1992).

Work in conservation has received support from among others Conservation International, The Nature Conservancy (partly by means of its Parks in Peril campaign), WWF – U.S., rain-forest conservation groups in several countries, Catholic Relief Service, Organization of American States, and the Costa Rican, Danish, Dutch, Swedish and U.S. governments.

In 1990 Costa Rica advanced a cooperative National Strategy for Conservation and Sustainable Development (ECODES), which is dealing with such broad issues as land-capability assessment and integrated land-use processes, and developing new laws and regulations for the environment, forestry and SINAC. Recently a new law on biodiversity (including wild flora and fauna) was enacted. Funding sought from the Global Environment Facility (GEF) would strengthen the efforts such as BOSCOSA and INBio for the region. Such broad and integrated activities are critical for the survival and healthy functioning of adequately large natural systems of flora and fauna.

References

Boza, M.A. (1978). *Los parques nacionales de Costa Rica.* Ministerio de Agricultura y Ganadería, Servicio de Parques Nacionales, San José and Incafo, Madrid. 80 pp.

Boza, M.A. (1988). *Parques nacionales Costa Rica national parks.* Edit. Heliconia and Fundación Neotrópica, San José. Pp. 128–139.

Boza, M.A. and Mendoza, R. (1981). *The national parks of Costa Rica.* Incafo, Madrid. 310 pp.

Brandon, K. and Umaña, A. (1991). Rooting for Costa Rica's megaparks. *Américas* 43(3): 20–31.

Cabarle, B., Bauer, J., Palmer, P. and Symigton, M. (1992). *BOSCOSA, the program for forest management and conservation on the Osa Peninsula, Costa Rica.* Project Evaluation Report. USAID-Costa Rica. 92 pp.

Cahn, P. and Cahn, R. (1987). Coast of riches. *National Parks* 61(9–10): 18–20.

Cahn, R. and Cahn, P. (1979). Treasure of parks for a little country that really tries. *Smithsonian Mag.* 10(6): 64–73.

Carr, A. and Carr, D. (1983). A tiny country does things right. *National Wildlife* (Sept.): 19–26.

Coen, E. (1983). Climate. In Janzen, D.H. (ed.), *Costa Rican natural history.* University of Chicago Press, Chicago. Pp. 35–46.

Devall, M. and Kiester, R. (1987). Notes on *Raphia* at Corcovado. *Brenesia* 28: 89–96.

FNT (1992). *Evaluación ecológica rápida – Península de Osa.* Programa BOSCOSA, con la asistencia de WWF – U.S. Fundación Neotrópica (FNT), San José, Costa Rica. 136 pp. + annexes.

Gámez, R. and Ugalde, A. (1988). Costa Rica's national park system and the preservation of biological diversity: linking conservation with socio-economic development. In Almeda, F. and Pringle, C.M. (eds), *Tropical rainforests: diversity and conservation.* California Academy of Sciences and American Association for Advancement of Science Pacific Division, San Francisco. Pp. 131–142.

Gómez, L.D. (1986). *Vegetación de Costa Rica: apuntes para una biogeografía costarricense.* Edit. Universidad Estatal a Distancia, San José. 327 pp.

Hartshorn, G.S. (1983). Plants: introduction. In Janzen, D.H. (ed.), *Costa Rican natural history*. University of Chicago Press, Chicago. Pp. 118–157.

Hartshorn, G.S. and Poveda, L.J. (1983). Checklist of trees. In Janzen, D.H. (ed.), *Costa Rican natural history*. University of Chicago Press, Chicago. Pp. 158–183.

Herrera, W. (1986). *Clima de Costa Rica*. Edit. Universidad Estatal a Distancia, San José. 118 pp.

Herwitz, S.R. (1981). *Regeneration of selected tropical tree species in Corcovado National Park, Costa Rica*. Univ. Calif. Publ. Geogr. Vol. 24. 109 (+ 40) pp.

Janzen, D.H. (1978). Description of a *Pterocarpus officinalis* (Leguminosae) monoculture in Corcovado National Park, Costa Rica. *Brenesia* 14–15: 305–309.

Janzen, D.H. (ed.) (1983). *Costa Rican natural history*. University of Chicago Press, Chicago. 816 pp.

Jiménez, J.A. (1984). A hypothesis to explain the reduced distribution of the mangrove *Pelliciera rhizophorae* Tr. & Pl. *Biotropica* 16: 304–308.

Jones, L. (1990). Costa Rica: a vested promise in paradise. *Buzzworm: The Environmental Journal* 2(3): 30–39.

Lew, L.R. (1983). *The geology of the Osa Peninsula, Costa Rica*. M.S. thesis. Pennsylvania State University, University Park. 128 pp.

Lewis, B.C. (1982). *Land-use and conservation on the Osa Peninsula, Costa Rica*. Ph.D. dissertation. University of California, Berkeley.

MacFarland, C., Morales, R. and Barborak, J.R. (1984). Establishment, planning and implementation of a national wildlands system in Costa Rica. In McNeely, J.A. and Miller, K.R. (eds), *National parks, conservation, and development: the role of protected areas in sustaining society*. Smithsonian Institution Press, Washington, D.C. Pp. 592–599.

Omang, J. (1987). In the tropics, still rolling back the rain forest primeval. *Smithsonian* 17(12): 56–67.

Sittenfeld-Appel, A.M. (1992). Conservación de biodiversidad y prospección química: el caso del Instituto Nacional de Biodiversidad (INBio)/Biodiversity conservation and chemical prospecting: the case of the National Institute of Biodiversity (INBio). In *God, money and the rainforest. Proc. First Congr., April 20–22, 1992, San José, Costa Rica*. Center for Environmental Study, Grand Rapids, Michigan and University of Michigan Global Change Project, Ann Arbor. Pp. 67–76.

Tangley, L. (1986). Costa Rica – test case for the neotropics. *BioScience* 36: 296–300.

Torres, H. and Hurtado de Mendoza, L. (eds) (1988). *Parque Nacional Corcovado: plan general de manejo y desarrollo*. Fundación Parques Nac., San José. 407 pp.

Tosi, J.A., Jr. (1975). The Corcovado Basin on the Osa Peninsula. In Tosi, J.A., Jr., *Potential national parks, nature reserves, and wildlife sanctuary areas in Costa Rica: a survey of priorities*. Centro Científico Trop., San José. Separate pp. 12.

Vaughan, C.S. (1981). *Parque Nacional Corcovado: plan de manejo y desarrollo*. Universidad Nacional, Heredia, Costa Rica. 364 pp.

Wallace, D.R. (1992). *The quetzal and the macaw: the story of Costa Rica's national parks*. Sierra Club Books, San Francisco, Calif. 222 pp.

Wille-Trejos, A. (1983). *Corcovado: meditaciones de un biólogo*. Edit. Universidad Estatal a Distancia, San José. 230 pp.

Wright, R.M. (1976). Rain forest: Conservancy assists in the establishment of Costa Rica's Corcovado park. *Nature Conservancy News* 26(1): 17–21.

WWF – U.S. (1988). Miners moved from national park. *WWF News* No. 51 (Jan.–Feb.): 3.

Authors

This Data Sheet was written by Olga Herrera-MacBryde (Smithsonian Institution, Department of Botany, NHB-166, Washington, DC 20560, U.S.A.) and Tirso R. Maldonado, Valentín Jiménez and Karsten Thomsen (Fundación Neotrópica, Apdo. Postal 236-1002, San José, Costa Rica).

CERRO AZUL-CERRO JEFE REGION
Panama

Location: Central Panama north-east of Panama City, between latitudes 9°07'–9°17'N and longitudes 79°18'–79°27'W.

Area: 53 km².

Altitude: 300–1007 m.

Vegetation: Tropical premontane wet forest, tropical wet forest, tropical premontane rain forest.

Flora: 934 recorded species of ferns and flowering plants, very high endemism, disjunct taxa.

Useful plants: Timber trees, ornamentals, species of interest for chemical and pharmacological research.

Other values: Protection of tributaries in watersheds of Chagres River and Alajuela Lake for Panama Canal, and provision of potable water and hydroelectric energy for Panama City and Colón; local and international ecotourism.

Threats: Agro-industrial activities, settlement, fires.

Conservation: Within Chagres National Park, but some private in-holdings.

Geography

The Cerro Azul-Cerro Jefe region (Map 30) is in the Province of Panama c. 52 km north-east of the capital (Panama City). The region is located in the Cordillera de San Blas; Cerro Azul reaches 691 m, Cerro Jefe 1007 m. The topography is uneven, with ravines of varying depth.

The Cerro Azul-Cerro Jefe uplift is on the continental divide and the source of rivers flowing to both oceans – on the Pacific slope, including the Pacora, Tocumen and Juan Díaz rivers; on the Caribbean slope, including several rivers of the Chagres watershed which supplies major reservoirs. Gatún Lake (423 km²) was formed by damming the Chagres River in 1910 during construction of the Panama Canal and is an integral part of the watercourse for the transit of ships; Alajuela (Madden) Lake (57 km²) was formed in 1936.

Geologically, Cerro Azul and possibly Cerro Jefe are part of the Cerro Azul pluton. Comparison of magmatic rocks shows similarities between the plutons of cerros Azuero and Azul and the Pito (Darién) River (Destro 1986). This supports Recchi's hypothesis on the geological evolution of Panama, that the Azuero Peninsula and the area spreading out from Cerro Azul to the Pito River were aligned in pre-Tertiary and Palaeocene eras, and the pluton outcropped there, having arisen from a common magma. Later tectonic plate action moved the plutonic block northward that has become Cerro Azul and adjacent areas.

In nearby regions, there are various geologic faults due to past volcanic and tectonic activity. One of these faults is along the course of the Chagres River, interrupted upstream by the volcanic crater in Alajuela Lake.

According to the geologic map of Panama (IGN 1988), the bedrock of Cerro Azul is igneous-extrusive, including basalt, andesite, tuff and ignimbrite, whereas the bedrock of Cerro Jefe and its boundary areas is igneous-intrusive, including granodiorite, quartz-monzonite and diorites.

The soils are moderately to very stony latosols, and acid to very acid; they are non-arable (class VII), suitable only for forests and reserves. On the Cerro Jefe summit the considerably different concentrations of elements found in the soils have been sampled – especially iron, potassium and manganese (Valdespino 1988). There are acidic *Sphagnum* peat bogs, mainly on Cerro Jefe.

The forest is frequently foggy as a result of the climatic conditions that characterize this region. Winds from the north and north-east loaded with moisture from the Caribbean Sea prevail, resulting in an annual average rainfall of nearly 4000 mm (Valdespino 1988). The temperature during the year ranges between 17°C and 26°C.

Vegetation

This region includes three life zones in the Holdridge system: tropical premontane wet forest, tropical wet forest and tropical premontane rain forest (Tosi 1971).

In the **tropical premontane wet-forest zone** at c. 300–500 m, as a result of many years of human activities, generally the more or less fallow vegetation is mostly herbaceous. *Saccharum spontaneum*, an aggressive introduced herb to 3 m tall, has extended widely and partially displaced fodder pastures of the African grasses *Hyparrhenia rufa* and *Panicum maximum*, as well as native plants.

There are occurrences of shrubs if *Saccharum spontaneum* is absent, e.g. species of Dilleniaceae, Melastomataceae and Compositae, and sun-loving trees such as *Apeiba tibourbou*, *Xylopia aromatica*, *X. frutescens*, *Anacardium occidentale*, *Cecropia* sp., *Vismia* sp. and *Cordia alliodora*.

In disturbed older secondary vegetation occur trees such as *Enterolobium schomburgkii*, *Didymopanax morototonii*, *Spondias mombin*, *Pseudobombax septenatum* and *Calycophyllum candidissimum*. On degraded soils predominate *Roupala montana* and some species of *Clusia* and Melastomataceae.

At 600–800 m mature forest is found, interrupted by areas converted by the poultry-breeding industry, settlement and coffee cultivation. In this forest there are several arboreal strata and emergents 30 m or more tall, including *Calophyllum longifolium*, *Pouteria* sp., *Podocarpus* cf. *oleifolius* and the palms *Welfia georgii*, *Socratea durissima* (*S. exorrhiza*), *Euterpe precatoria* and *Wettinia augusta*.

The diversity of epiphytes is high, including for example bryophytes (liverworts and mosses), lichens, ferns, Bromeliaceae, Orchidaceae, Araceae, Cyclanthaceae and Ericaceae.

Toward the Caribbean slope mature forest is relatively better preserved, both **tropical wet forest and tropical premontane rain forest**, due partly to the rough topography and the abundant precipitation. Nonetheless, there is an area near Cerro Jefe known as Cerro Pelón where the vegetation is almost totally herbaceous, with *Rhynchospora cephalote* predominant, accompanied by species such as *Trachypogon plumosus*, *Andropogon bicornis*, *A. leucostachys* and *Scleria* sp. Occasionally the palm *Colpothrinax cookii* is found.

The forest on the summit of Cerro Jefe is influenced by frequent strong winds, having a vegetation of shrubby trees, generally 8–15 m tall with medium-sized to small leathery leaves; it shows a tendency towards sclerophylly (Gentry 1982). In this type of forest flourish *Ardisia* sp., *Alchornea* sp., *Myrsine* sp., *Clusia* spp. and some Sapotaceae, and endemics such as *Psychotria olgae*, *Licania jefensis* and *Vismia jefensis*. Epiphytic plants are abundant – Orchidaceae and Bromeliaceae are most representative (Torres 1989). At the leaf bases of *Vriesea* sp., *Guzmania* sp. and other bromeliads that retain water it is easy to find *Utricularia jamesoniana*, a carnivorous plant that feeds on organisms in the accumulated solutions.

Colpothrinax cookii (Read 1969) is prominent because of its dense populations; it is distributed up to 800 m. In primary forests with emergents it is infrequent or absent, and instead occur *Socratea durissima* (*S. exorrhiza*), *Wettinia augusta* and *Euterpe precatoria* – which tend to be shorter and stouter on hillsides and on the summit of Cerro Jefe. *Olyra standleyi* is sometimes concentrated in pure populations in open and disturbed areas.

Flora

Of the 1230 species endemic to Panama, 143 have been found on Cerro Jefe – including 45 local endemics. The angiosperm families with the highest number of endemic species are Rubiaceae (25), Araceae (13), Gesneriaceae (12), Ericaceae (8), Myrsinaceae (8), Compositae (7), Solanaceae (7) and Orchidaceae (5). Among the characteristic genera are *Psychotria* (16), *Anthurium* (13), *Columnea* (7) and *Ardisia* (4) (Carrasquilla 1987).

Lewis (1971) concluded that the Cerro Azul-Cerro Jefe region, like other relatively high regions in Panama, has been a site of refuge and evolution for many taxa that were geologically isolated from the North American range of mountains which reaches western Panama. The flora of western Panama is more allied with the flora to its north-west, because of the continental connection by the Middle Miocene that united Central America and North America. At that time the flora of present eastern Panama was still on groups of low volcanic islands, which included Cerro Jefe, and which were

CPD Site MA19: Cerro Azul-Cerro Jefe region, Panama. Summit of Cerro Jefe with the palm *Colpothrinax cookii* in the centre left of the photograph. Photo: Luis G. Carrasquilla.

populated by long-distance dispersal from nearby South America, as well as continental Panama. The Panamanian land-bridge between North America and South America became established c. 3.5-2.4 million years ago during the Late Pliocene (Graham 1972, 1985, 1993; Gentry 1982, 1985; Rich and Rich 1983).

Study of the flora of Cerro Azul-Cerro Jefe was initiated by P.H. Allen in the mid 1940s (Dwyer 1967); especially from 1965 onward, other foreign and Panamanian specialists have contributed much to the knowledge of the regional flora (cf. Martínez 1977–1978; Dwyer 1985; Hampshire 1989; Aranda 1991). Altogether, 836 species of flowering plants are recorded for Cerro Azul-Cerro Jefe. According to Carrasquilla (1987), on Cerro Jefe c. 486 species have been collected, 119 of which are epiphytes (cf. Torres 1989).

The pteridophytes on Cerro Azul-Cerro Jefe are frequent and quite diverse – 98 species have been identified. On Cerro Jefe most of the species are in Polypodiaceae (14), Hymenophyllaceae (10), Dryopteridaceae (6), Gleicheniaceae (5) and Cyatheaceae (4), and in the genera *Grammitis*, *Trichomanes* and *Elaphoglossum*. Tree ferns are distinctive components of the Cerro Jefe forests – *Trichipteris williamsii* is most abundant, then *Cyathea* sp. (Valdespino 1988).

Among the disjunct species on Cerro Azul-Cerro Jefe, *Hymenophyllum apiculatum* is also known from Venezuela (e.g. Guayana Highlands) and Colombia (Meta and Valle), so the population in Panama probably resulted from long-distance dispersal. *Licania affinis* also has been recorded in the Guiana area. A number of species seem to be disjuncts from the Guayana region and especially the Guayana

Lisianthus jefensis, a member of the Gentianaceae, on Cerro Jefe. Photo: Luis G. Carrasquilla.

Highlands, probably representing pre-Andean survivors of the flora of the pre-isthmian uplifted islands (Gentry 1985). This insular past also may be indicated by several species shared with the isolated Cerro Tacarcuna (1900 m) bordering Colombia: *Eleagnia nitidifolia*, *Conomorpha gentryi*, *Columnea mira* and *Vochysia jefensis*. *Grammitis randallii* is another disjunct fern (also on Cerro Tacarcuna), which is also rare in Jamaica (D.B. Lellinger, pers. comm.). *Colpothrinax cookii*, which in recent years has been found in San Blas and Chiriquí, occurs as well in certain areas of Guatemala (Gentry 1982).

Of the c. 486 flowering plant species documented on Cerro Jefe, 101 extend in distribution only to Costa Rica and Colombia. Slightly more of the species extend from just Costa Rica or Panama to South America than from Mexico to South America (Carrasquilla 1987). As an example of the greater southern affiliation, five species of *Miconia* are shared between Panama and South America, but only one species is shared with Central America.

The floristic affinities on Cerro Jefe partly reflect the relative likelihood of phytogeographic opportunities from the neighbouring regions. South America is much larger and more diverse than Central America, and the pre-isthmian islands were more or less to the west of South America – receiving the westward-prevailing air currents of the Intertropical Convergence Zone and oceanic current. Thus the greater dispersal of plant propagules was from east to west.

Useful plants

Because this region is within Chagres National Park, exploitation of the native plants is infrequent and local. There are timber trees such as *Calophyllum longifolium* ("María"), *Manilkara* sp. ("níspero") and *Podocarpus* cf. *oleifolius* ("pino de montaña"). Occasionally, the leaves and stalks of *Socratea durissima* (*S. exorrhiza*) and *Colpothrinax cookii* ("palma escoba") are used to make huts. Uses of the endemic plants are unknown. Some species, e.g. in *Rauvolfia*, *Cephaelis* and *Hamelia*, have been investigated for their chemical and pharmacological properties.

Social and environmental values

The region has great environmental value because its vegetation helps to maintain the climate. The cerros are also the sources of several rivers that flow into the Chagres River watershed, which mainly provides Alajuela Lake with water. Forty percent of this water is used for the Panama Canal, the generation of hydroelectric energy, and the water supplies to Panama City and Colón.

Because of the region's climatic and topographic conditions, farming and animal husbandry practices are naturally somewhat restricted. Telecommunication installations have been established on the summits of Cerro Azul and Cerro Jefe.

The Cerro Azul-Cerro Jefe region includes private lands, primarily owned by one family. Only 9 km² of the private lands are used for settlement and avicultural projects; the remainder is reserved for scientific purposes and ecotourism – such as "El Cantar" trail on Cerro Azul, which attracts national and foreign tourists, who are amazed by the biological diversity and as well by the priceless beautiful scenery of the surrounding area.

These cerros are at the western end of two Endemic Bird Areas (EBAs A19 and A20) primarily centred on Darién Province (CPD Site MA20). Perhaps ten restricted-range species are known from the vicinity of these two mountains. The lowlands of Cerro Azul and Cerro Jefe may be most important for the threatened speckled antshrike (*Xenornis setifrons*).

Threats

There are sometimes fires, both accidental and intentional, which usually result from traditional agricultural practices (D'Arcy and Correa-A. 1985). Agro-industrial activities also create disturbances in the Cerro Azul-Cerro Jefe region. Portions of the Cerro Jefe cloud forest have been cut for grazing (Hampshire 1989).

Erosional sites have resulted in the deforested and settled areas following the abundant rains, causing sediment deposition in several rivers that flow into Lake Alajuela. Residents are undertaking reforestation with *Pinus caribaea* and on a smaller scale *Araucaria* sp. to reduce the erosion. The Cerro Azul-Cerro Jefe forests, in addition to protecting some of the Chagres watershed, also contribute to diminishing the risk of floods from these cerros' rivers that flow southward into Panama Bay.

Conservation

For the protection of the Panama Canal and Lake Alajuela, creation of Chagres National Park (1290 km²) in 1984 was considered absolutely essential; it included Cerro Azul and Cerro Jefe. All the forests and lands were included that are profitable for water production and sustaining the Chagres

MAP 30. CERRO AZUL-CERRO JEFE REGION, PANAMA (CPD SITE MA19)

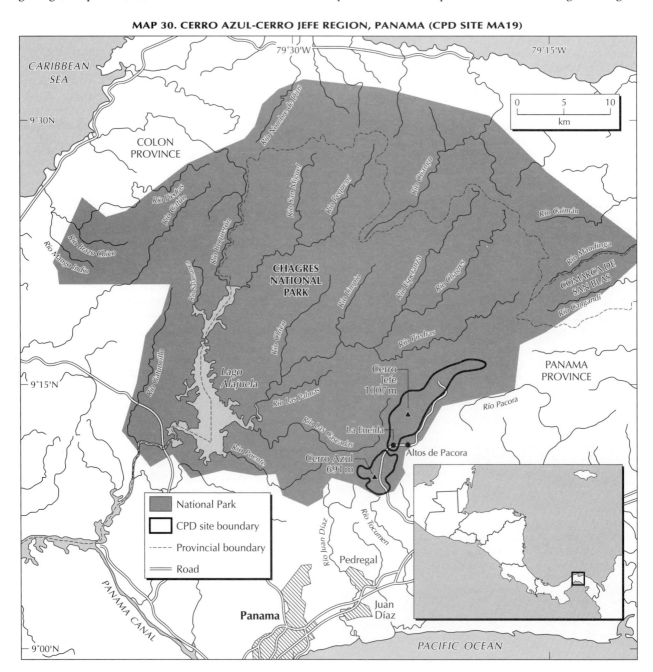

watershed, which is one of the most important rivers for the national economy.

Preservation of this region is essential because of its high level of plant endemism, and its general biodiversity, including a wealth of animals such as *Ramphastos*, bats (Gallardo and Jiménez 1990), *Alouatta*, *Saguinus* and *Bradypus*. A range of biological studies has been carried out and future projects are being considered, in this way providing better knowledge of the kinds, quantity and quality of the wild life that through millennia has survived and developed in this unusual region of Panama. A management plan for the park has been prepared (INRENARE 1987).

References

Aranda, J.E. (1991). *Estudio florístico de angiospermas terrestres en dos cuadrantes en Cerro Jefe, conjuntamente con una revisión del herbario y la Flora de Panamá para el mismo área*. Universidad de Panamá, Departamento de Botánica, Panama City. 302 pp.

Carrasquilla, L.G. (1987). *Características de la flora de angiospermas de Cerro Jefe, Provincia de Panamá*. Universidad de Panamá, Departamento de Botánica, Panama City. 28 pp.

D'Arcy, W.G. and Correa-A., M.D. (eds) (1985). *The botany and natural history of Panama: La botánica e historia natural de Panamá*. Monogr. Syst. Bot. No. 10, Missouri Botanical Garden, St. Louis. 455 pp.

Destro de, T. (1986). *El plutón de Cerro Azul. Boletín Informativo de la Facultad de Ingeniería Civil* No. 10, Universidad Tecnológica de Panamá, Panama City.

Dwyer, J.D. (1967). A new herbarium in the Canal Zone. *Taxon* 16: 158–159.

Dwyer, J.D. (1985). The history of plant collecting in Panama. In D'Arcy, W.G. and Correa-A., M.D. (eds), *The botany and natural history of Panama*. Monogr. Syst. Bot. Missouri Bot. Gard. 10. Pp. 179–183.

Gallardo, M. and Jiménez, Z. (1990). *Estudio de mamíferos del Chiroptera en un bosque secundario de cantera (sector de Alajuela), Parque Nacional Chagres*. Universidad de Panamá, Escuela de Biología, Panama City. 32 pp.

Gentry, A.H. (1982). Phytogeographic patterns as evidence for a Chocó refugium. In Prance, G.T. (ed.), *Biological diversification in the tropics*. Columbia University Press, New York. Pp. 112–136.

Gentry, A.H. (1985). Contrasting phytogeographic patterns of upland and lowland Panamanian plants. In D'Arcy, W.G. and Correa-A., M.D. (eds), *The botany and natural history of Panama*. Monogr. Syst. Bot. Missouri Bot. Gard. 10. Pp. 147–160.

Graham, A. (1972). Some aspects of Tertiary vegetational history about the Caribbean Basin. In *Memorias de Simposia, I Congreso Latinoamericano, V Mexicano de Botánica*, Mexico, D.F. Pp. 97–117.

Graham, A. (1985). Vegetational paleohistory studies in Panama and adjacent Central America. In D'Arcy, W.G. and Correa-A., M.D. (eds), *The botany and natural history of Panama*. Monogr. Syst. Bot. Missouri Bot. Gard. 10. Pp. 161–176.

Graham, A. (1993). Historical factors and biological diversity in Mexico. In Ramamoorthy, T.P., Bye, R., Lot, A. and Fa, J.E. (eds), *Biological diversity of Mexico: origins and distribution*. Oxford University Press, New York. Pp. 109–127.

Hampshire, R.J. (1989). Panama. In Campbell, D.G. and Hammond, H.D. (eds), *Floristic inventory of tropical countries: the status of plant systematics, collections, and vegetation, plus recommendations for the future*. New York Botanical Garden, Bronx. Pp. 309–312.

IGN (Instituto Geográfico Nacional "Tommy Guardia") (1988). *Atlas nacional de la República de Panamá*. Panama City, Panama.

INRENARE (1987). *Plan de manejo y desarrollo del Parque Nacional Chagres*. INRENARE, Departamento de Parques Nacionales y Vida Silvestre, Panama City.

Lewis, W.H. (1971). High floristic endemism in low cloud forests of Panama. *Biotropica* 3: 78–80.

Martínez, A. (1977-1978). *Estudio florístico en Cerro Jefe, Panamá*. Universidad de Panamá, Departamento de Botánica, Panama City. 69 pp.

Read, R.W. (1969). *Colpothrinax cookii* – a new species from Central America. *Principes* 13: 13–22.

Rich, P.V. and Rich, T.H. (1983). The Central American dispersal route: biotic history and paleogeography. In Janzen, D.H. (ed.), *Costa Rican natural history*. University of Chicago Press, Chicago. Pp. 12–34.

Torres, N. (1989). *Estudio sistemático de las especies de epífitas vasculares de Cerro Jefe, Provincia de Panamá*. Universidad de Panamá, Departamento de Botánica, Panama City. 93 pp.

Tosi, J.A., Jr. (1971). *Zonas de vida: una base ecológica para investigaciones silvícolas e inventariación forestal en la República de Panamá*. FAO, Rome.

Valdespino, I.A. (1988). *Estudio florístico de helechos en dos cuadrantes de Cerro Jefe, Provincia de Panamá*. Universidad de Panamá, Departamento de Botánica, Panama City. 130 pp.

Author

This Data Sheet was written by Prof. Luis G. Carrasquilla (Universidad de Panamá, Escuela de Biología, Departamento de Botánica, Estafeta Universitaria, Panama City, Panama).

CENTRAL AMERICA: CPD SITE MA20

DARIEN PROVINCE
AND DARIEN NATIONAL PARK
Panama

Location: Eastern Panama, the park in south-east Darién Province between latitudes 7°12'–8°31'N and longitudes 77°09'–78°25'W.

Area: Darién Province 16,671 km², Darién National Park 5790 km².

Altitude: 0–1875 m.

Vegetation: Tropical lowland dry, moist and wet forests, perhaps 500-year-old secondary rain forest; tropical premontane moist (warm transition), wet and pluvial forests; lower montane pluvial forest. Marshes and swamps, tall non-flooding forests, cloud and elfin forests.

Flora: Darién Province: 2440 species recorded; high endemism – e.g. Cerro Tacarcuna 23% endemism; threatened species.

Useful plants: Timber species; medicinals.

Other values: Wilderness; diverse fauna, including endemics and threatened species; watershed protection; Amerindian groups; intercontinental disease barrier; potential ecotourism.

Threats: Colonization, logging, agriculture, grazing, road building, mining.

Conservation: Darién National Park is a World Heritage Site and Biosphere Reserve, with a buffer zone and Punta Patiño Nature Reserve to east and Colombia's Los Katíos NP to west. Also, Canglón and Chepigana Forest Reserves; Comarca Emberá No. 1 (Cemaco District), Comarca Emberá No. 2 (Sambú District) and Kuna de Walá, Mortí y Nurrá Amerindian Reserve.

Geography

Located in the eastern portion of Panama which borders Colombia, Darién Province basically forms a long and flat lowland valley mostly fringed by mountains. Darién National Park is along 90% of the international border and extends from the Pacific coast to the continental divide, which is just 16 km from the Caribbean Sea.

The extensive central valley receives the drainages of the country's largest watershed (c. 13,371 km²), primarily from the south-east-flowing Chucunaque River and north-west-flowing Tuira River as well as the Balsas River, which join as the Tuira River to flow westward to the large Gulf of San Miguel on the Pacific Ocean (Map 31). Brackish to freshwater marshes and swamps extend well inland along the Tuira River's lower portion and the lower Chucunaque River. Twice a day strong Pacific tides (ranging 2–6 m) affect the Tuira and Chucunaque for many km inland.

The spine of Panama along the north-eastern border of the province is a fairly continuous mountain system, the Cordillera de San Blas and higher Serranía del Darién. The serranía emerges in the north with elevations of 300–600 m and in the south-east rises to 1875 m at Cerro Tacarcuna (8°10'N, 77°15'W), which is the highest mountain between the Andes and western Panama and only c. 32 km inland from Colombia's Caribbean coast at the Gulf of Urabá (which gives way to the vast lowland swamps and lagoons of the Atrato River).

The southern region of the province (Altos de Darién) is composed of a trident of somewhat discontinuous mountain systems extending northward. From east to west, they are: (1) Serranía de Pirre (with the flanking Serranía de Setetule), which extends into the central valley and contains the headwaters of the Tuira River; (2) Serranía de Jungurudó (with the Cordillera de Juradó to the south), extending northward to the Serranía de Bagre – which forms the western edge of the central valley; and (3) Serranía del Sapo, bordering the Pacific Ocean. The somewhat convergent southern ends of several of these mountain systems and the Altos de Quía help to demarcate the international border. The principal heights (c. 1150–1550 m) in the southern area include Cerro Pirre, Alturas de Nique, Cerro Piña and Cerro Sapo.

Along the Pacific Ocean, the south-western edge of the province is a somewhat undulating coastline with rocky shores and sandy beaches. Farther to the north-west, the coast is broken by the Gulf of San Miguel, where the wide Lower Tuira River meets the sea. North of the gulf, the border of the province is the south-eastern end of the Serranía de Majé at the Serranía de Cañazas.

During the Tertiary's Middle Eocene an insular volcanic arc emerged, and the regional lowlands emerged from tectonic activity probably in the Late Pliocene. During the Pleistocene the sea-level fluctuated from 100 m lower to 50 m higher than at present. The southern mountains and the Caribbean slope of the north-eastern mountains are largely of volcanic origin, whereas the inland slope of the Serranía del Darién is of sedimentary origin. The lowland clay soils are generally derived from Late Miocene shale with layers of dolomite and calcareous sandstone (Golley *et al.* 1975; Hartshorn 1981; INRENARE and ANCON 1988).

Rainfall is abundant (3000–4000 mm) on the mountains next to the Caribbean coast, still more abundant (4000–5000 mm) on the inland mountains, and diminishes (2800–1700 mm) in the central valley and gulf areas. There is a marked dry season (weaker near Colombia) with less than 100 mm monthly from January through March–April. The temperature generally varies between 16°–35°C, with an annual mean of c. 25°–27°C (OEA 1978; INRENARE and ANCON 1988).

Vegetation

The Darién is one of the most diverse and species-rich regions in Central America. It includes almost ten major vegetation types (see below), from the littoral strand and coastal dry forests to mangroves, brackish and freshwater swamps, and various lowland and premontane to lower montane rain-forest life zones, including cloud forests and elfin forests at rather low elevations because of the isthmian effect (see maps in Duke and Porter 1970; INRENARE and ANCON 1988). Five Holdridge life zones are represented in Darién National Park.

There is speculation whether the central Darién lowland forests are primary or much is secondary growth from a

MAP 31. DARIEN PROVINCE AND DARIEN NATIONAL PARK, PANAMA (CPD SITE MA20)

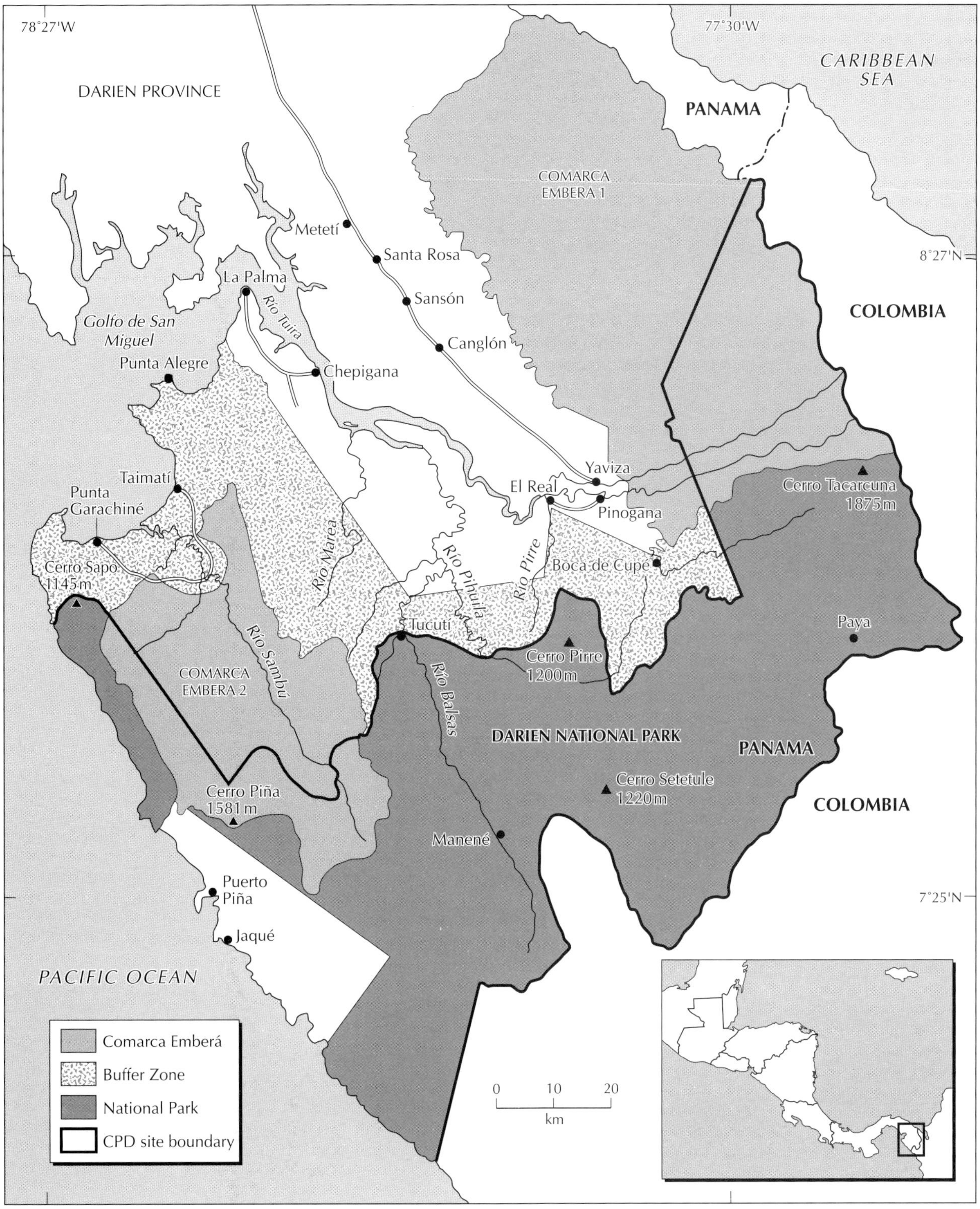

savanna created by the Amerindians and abandoned after the coming of the Spanish c. 500 years ago. Historical accounts, such as those of Balboa in 1513 and Wafer in 1681 are quoted in Sauer (1966) and Prebble (1968); also see Jaén-Suárez (1985). Some dominant trees may be secondary species that are not reproducing but have attained great size (Golley *et al.* 1975; INRENARE and ANCON 1988), although Hartshorn (1981) suggested that studies of forest dynamics and the regeneration potential of *Cavanillesia platanifolia* ("cuipo") did not support a secondary status.

Bordering the Gulf of San Miguel's southern embayments (from about Punta Garachiné eastward to Punta Alegre) is an isolated strip of **tropical dry forests** (thorn and seasonally strongly deciduous to semi-deciduous tropical dry forests). Characteristic trees include *Albizia caribaea*, *Bombacopsis quinata*, *Cochlospermum vitifolium*, *Prosopis juliflora* and *Sabal allenii*.

Mangrove forests occur along the Pacific and Atlantic coasts; their greatest representation is in the Gulf of San Miguel area, which has c. 46% of the mangrove forest of the entire country (G. Palacios 1992, pers. comm.). The major Pacific brackish water community is nearly pure stands of *Rhizophora brevistyla* to 41 m, which ranks them among the world's tallest mangroves. *Pelliciera rhizophorae* is an occasional associate. Minor brackish water communities develop in estuaries and rivers changing with decreasing salinity from *Avicennia germinans* to *Mora oleifera* and

then *Montrichardia arborescens*. Caribbean coastal islands, inlets and bays support narrow low bands of *Rhizophora mangle* and the less abundant and taller (to 5 m) *Laguncularia racemosa* (Golley *et al.* 1975; Hartshorn 1981).

In the Gulf of San Miguel area, saltmarsh vegetation gives way inland to freshwater **marshes and swamp forests**, often with palms (e.g. *Manicaria, Jessenia, Euterpe*) or stands of *Copaifera aromatica, C. panamensis, Pachira aquatica, Pterocarpus robrii* and various canopy dominants – *Prioria copaifera, Pterocarpus officinalis, Tabebuia rosea* and *Swartzia panamensis* (Porter 1973; Nations and Komer 1983). Along the Tuira, Chucunaque and Balsas rivers is broad evergreen **riparian forest** with many understorey palms and some pure stands of *Prioria copaifera* ("cativo"), which can reach 55 m tall (Hartshorn 1981); details on the cativo swamps are in Duke and Porter (1970).

The most widespread vegetation type in the Darién is seasonally **semi-deciduous tropical moist forest**, which occupies non-flooded lowlands of the Chucunaque-Tuira Basin and western areas; c. 10% of the park is in this life zone. The most dominant, emergent trees are especially the deciduous *Cavanillesia platanifolia* (to 40 m and 2 m in dbh) and *Ceiba pentandra*, as well as *Anacardium excelsum*. Frequent canopy trees include *Bombacopsis quinata, B. sessilis, Enterolobium cyclocarpum, Licania hypoleuca, Platypodium elegans, Pseudobombax septenatum, Sterculia*

CPD Site MA20: Darién Province and Darién National Park, Panama. *Cavanillesia platanifolia* grove (over 30 m high) in seasonal lowland forest during the dry season, with Cerro Pirre in the distance; central area of Darién National Park. Photo: Gwen Keller/ANCON.

apetala, Terminalia amazonia, Tetragastris panamensis and *Vitex cymosa*. In the subcanopy *Mouriri parviflora* is dominant and palms (e.g. *Sabal allenii*) are frequent. The dominant shrubs are *Faramea luteovirens, Mabea occidentalis* and *Piper pinoganense*. Lianas do very well in this formation (in the canopy and also the subcanopy), whereas epiphytes and ferns are relatively scarce (Golley *et al.* 1975; Hartshorn 1981).

In slightly higher (to c. 200–250 m) and/or wetter areas in the Darién occurs **premontane wet forest - warm transition**, and then to c. 500–600 m there is **tropical wet forest (seasonal yet evergreen tropical rain forest)**, which extends broadly into the northern Chocó of Colombia (in CPD Site SA39). About 60% of the park is in these two life zones. *Anacardium excelsum* is dominant in the canopy and the following trees are frequent: the above *Bombacopsis* spp., *Brosimum guianense, Ceiba pentandra, Cochlospermum williamsii, Dipteryx panamensis* and *Myroxylon balsamum*; the main subcanopy tree is *Oenocarpus panamanus*. The dominant understorey shrub is *Mabea occidentalis* and frequent shrubs include *Clidemia* spp., *Conostegia* spp. and *Miconia* spp. Epiphytes are abundant and canopy lianas common.

At higher elevations occurs **premontane pluvial forest** and on the highest peaks and ridges **lower montane pluvial forest**; these life zones cover c. 30% of the park. Cloud forest can commence at c. 750 m and elfin forest may be found on the highest exposed elevations. The premontane forest is low (30 m) and dense, with many subcanopy palms; lianas are common. The same *Brosimum* and *Dipteryx* species are frequent in the canopy and *Cephaelis elata* is a dominant shrub; ferns are most common in this zone. A distinctive montane oak forest (*Quercus humboldtii*) is present above 1400 m on Cerro Malí and 1500 m on nearby Cerro Tacarcuna. The dominant trees in cloud forest are *Oenocarpus panamanus*, in elfin forest *Clusia* spp. (Porter 1973; Golley *et al.* 1975; Gentry 1977, 1985).

Flora

Although the *Flora of Panama* has been finished, the coverage is very incomplete; a revised checklist being maintained by the Missouri Botanical Garden has 2440 species recorded for the Darién (ANCON 1993, pers. comm.). Since 1970, collecting in the wetter forests of Panama has revealed many new species (e.g. D'Arcy 1977; Gentry 1985, 1986). Limited collecting in the region of the park and the lack of synoptic studies preclude determining the total number of species or endemics (only 700 species have been recorded).

The diversity of the Pacific mangrove forest is pronounced, with c. 15 species compared to c. 6 on the Caribbean side. Duke (1975) provided a checklist for the half of collections identified for an ecological study of several Darién forest types; c. 400 species – including over 200 canopy and subcanopy trees – had been identified in tropical moist forest, and even riverine forest had c. 220 identified species – including 66 tree species. The samples suggest that these forests are as rich as forests in Amazonia.

Due to the mountainous conditions that reflect the Darién's past geological insularity and isolation, and as well its relatively recently formed lowlands, the Darién has its own characteristic flora. The flora has complex ancient and more recent affinities, which reflect the varying land's mixed history of location, isolation and connection (Gentry 1985). Upland elements show northern affinities with Central America (e.g. Costa Rica) that suggest floristic migration during cooler palaeohistoric periods. With Pleistocene climatic shifts, there may have been a refugium in this region. Lowland trees and lianas have Amazonian affinities, whereas the epiphytes and understorey shrubs have northern Andean relationships. The flora has particular affinities with the flora of the Chocó region of Colombia (in CPD Site SA39), as well as the ancient Guayana region (see CPD Site SA2).

Some studies have been conducted on and near Cerro Pirre and Cerro Tacarcuna, where many endemics occur (Lewis 1971; Gentry 1985). The limited collections (90% identified) from Cerro Tacarcuna show 23% endemism and c. 25% new species, including the new genus *Tacarcuna* (Euphorbiaceae) and such remarkable new species as *Freziera forerorum* (Theaceae), a tree which may have the most asymmetric leaf base among flowering plants (Gentry 1978). Other appropriate areas are presumed to have high angiosperm endemism, especially isolated cloud forests.

Useful plants

The Darién forests contain important reserves of timber, such as *Prioria copaifera* ("cativo"). This region has contributed 75% of the logs to the national market, with cativo comprising half of the total. In 1976, of the five species supplying 94% of all the logs Panama marketed (cativo, *Anacardium excelsum, Bombacopsis quinata, Vatairea* sp., *Hyeronima oblonga*), the first four were species of the Darién (Hartshorn 1981). A forest inventory of the province found adequate quantities of excellent timber trees and some that also can be used for plywood exports and paper and pulp products – e.g. *Anacardium excelsum, Capara guianensis, Cedrela odorata, Cordia alliodora, Dialium guianense, Myroxylon balsamum, Swietenia macrophylla, Tabebuia guayacan, T. rosea* and *Terminalia amazonia* (Lamb 1953; OEA 1978). About 35.5% of the province's land is in class VII, which is marginal for agriculture but suitable for forestry.

The indigenous peoples use many species, for many purposes (Duke 1968, 1970; Torres de Araúz 1985). Cooking oil extracted from the palm *Jessenia bataua* ("trupa") is particularly valued; the palm *Phytelephas seemannii* is used to make vegetable-ivory carvings.

Social and environmental values

The Darién includes three indigenous groups, Emberá and Wounaan (from the Colombian Chocó) and Kuna (Houseal *et al.* 1985), and early historic Negroes. Although just 400–500 Kuna are still in the Darién, 40,000–50,000 inhabit the nearby Comarca de Kuna Yala (San Blas). Within the park there were c. 2500 people in 1987: 250 Kuna, 1700 Emberá, few Wounaan, 400 Negroes and some colonists. The Amerindians have extensive knowledge of the forest, and despite long contact with outside societies have managed to keep some of their traditional forest cultures. The Kuna have been studied extensively (see D'Arcy and Correa-A. 1985); their language and behaviour are in part directed by their relationships with wild animals and plants and the symbolic and magical features they represent (Chapin 1991).

The lowland wet forest is of special importance because it may be a secondary forest almost 500 years old. If further research reveals that this forest is secondary, it may be possible to learn how to regenerate such tropical forest with greater success, more regenerative capacity and biodiversity.

The region has a rich fauna which is still poorly known. There are c. 770 vertebrate species in Darién Province, with considerable endemism; nearly all of them have been recorded in the park. Over 60% of Panama's mammal species occur in Darién Province, including six of the eight species of primates – three-banded douroucouli, Geoffroy's spider monkey, brown-headed spider monkey, crested bare-faced tamarin, mantled howler monkey and white-faced capuchin; and five or six cat species – puma, jaguar, ocelot, margay, jaguarundi and probably oncilla (*Felis tigrina*). Endemic reptile and amphibian species include two snakes, a lizard, a salamander and five frogs.

There are also many bird species in the region that are threatened elsewhere, such as crested guan, marbled wood-quail, harpy eagle, crested eagle, crested owl, brown violetear, green-crowned brilliant, great curassow, macaws and black-cheeked woodpecker. With its extensive lowland forests bounded by isolated mountains, the region is at the centre of two Endemic Bird Areas (EBAs): the Darién and Urabá Lowlands EBA (A19) below c. 900 m, and higher up the East Panama and Darién Highlands EBA (A20). Thirty restricted-range species occur in the two EBAs; some are confined to single mountain areas such as Cerro Tacarcuna.

The park protects one-third of the watersheds in Darién Province. Their conservation is highly important for its social and economic development. The region has potential for expanded ecotourism and for ecologically broad scientific research.

Threats

The eastern Darién and adjacent Colombia have the only gap (106 km) in the Pan American highway, which extends from Alaska to Tierra del Fuego. The highway penetrated the Darién in 1973, reaching Yaviza in 1984 – 24 km from the park. Secondary roads have opened some of the Darién interior, especially southward. The plans to complete the highway have been slowed more by limited funds than environmental concerns (Herlihy 1989). The almost inevitable opening of the Darién by a highway through the Tuira River Valley will bisect the park with a 22-km roadway and introduce numerous hazards and pressures on this fragile environment (D'Arcy 1977; Gentry 1977; INRENARE and ANCON 1988; Polsky 1992).

The Darién natural environment is already under continued and increasing pressure (Holz 1980; Herlihy 1989). On Panama's Azuero Peninsula west of Panama City, most forests have been cut down and burned; farmers are leaving and settling in other regions of the country – and the Darién is their first choice. In 1978, only 1.6% of Panama's population lived within Darién Province. The government has made the province a priority for major development of natural resources. Programmes have been started to make use of these resources as part of Panama's solution to alleviate immediate economic problems (OEA 1978; RARE 1983; INRENARE and ANCON 1988). Mining concessions have been granted in the park. Under the national colonization plan, settlers are steadily coming into the Darién and opening the frontier (Torres de Araúz 1970; D'Arcy 1977; Herlihy 1989). By 1986 the population had increased to c. 34,000, a third larger than in 1970, primarily by migration of colonists using access by road.

The areas predominantly under pressure from colonists and timber companies are wetlands and lowland forests (Mayo-Meléndez 1965; Duke 1968, 1975). The wetlands around the Tuira River were first settled, e.g. in the district of Chepigana. They are mostly cultivated with rice and maize (Duke 1968). Rice producers prefer creating new fields because of higher yields and lower weed control, rather than using the same fields year after year (Torres de Araúz 1970). New settlers have taught local inhabitants that clearing the land for pasture to raise cattle is as profitable if not more so than producing field crops. The environment of the region is undergoing enormous change; probably over 2000 km² are already farmlands and pasturage (Herlihy 1989), which is over 36% of the province's land suitable for such uses (i.e. classes II–VI – OEA 1978).

Darién lumber is an important factor in Panama's exports; it is mainly used for construction. The wetland areas are the prime logging areas (Mayo-Meléndez 1965). Mangroves also are used to make charcoal for cooking or to sell. The bark of *Rhizophora brevistyla* (red mangrove) is exported for use in the tannin industry. As the population of settlers increases, the mangroves are likely to be severely degraded, disrupting this fragile ecosystem – even though the mangrove forest is a breeding ground for white shrimp, the keystone of Panama's shrimp industry, which in 1984 resulted in exports of US$30.7 million.

From neighbouring Colombia potential threats could be realized with establishment of convenient access into the Darién. Most of the usable slopes on the Colombian border have been cleared (except in Los Katíos Natural National Park), but due to the natural barrier of mountains and swamps and no direct access, the Panamanian slopes have been relatively untouched. Disease is another problem that will confront Panama and northern America with completion of the Pan American highway. An example is the dreaded hoof-and-mouth disease ("aftosa") which affects cattle and pigs, and also over 50 plant diseases that may strike native species and crops.

Already Darién NP is facing encroachment by settlers. In the park's buffer zone there are c. 40 communities with more than 8000 people (INRENARE and ANCON 1988). Many inhabitants for example of the towns El Real de Santa María and Boca de Cupe have farms near the park boundaries. However, 78% of the park lands are in class VIII (48%), which is unsuitable for agriculture and forestry, or VII–VIII (30%), which is marginal for forestry (OEA 1978). The number of Emberá in the park has increased to 3000 (Polsky 1992).

The increasing population of colonists from western Panama disrupts cultural values of the indigenous peoples who have immigrated to or have been in the region (Herlihy 1989). The Chocó and the remaining Kuna have small agricultural plots predominantly along rivers and streams and do a limited amount of hunting. In the nearby territory Comarca de Kuna Yala (San Blas Province), the Kuna initiated a series of projects to secure their traditions and protect their forest from illegal encroachment. The projects include a wild-life reserve, a forest park and a botanical park (Nations and Komer 1983; Chapin 1984). Their endeavours have potential to be a model for protection of tropical rain forest by indigenous peoples, although the Kuna are being affected by the strong pressures of cultural change (cf. Chapin 1991).

Conservation

The heavy influx of settlers has prompted the Panamanian Government to recognize the need for more management and conservation. Darién National Park was established in 1980, partly having been preceded in 1972 by a Protection Forest. In 1973 Colombia established the contiguous Los Katíos Natural National Park (520 km²) for the southern end of the Serranía del Darién. The Panamanian region was accepted as a World Heritage Site in 1981 and a Biosphere Reserve in 1983. Definitive boundaries apparently still remain to be legalized by decree. Darién NP encompasses c. 5790 km², making it the largest Panamanian park and one of the largest protected areas in Central America, which conserves the largest tropical rain forest on the Pacific slope.

Other protected areas in Darién Province are: Kuna de Walá, Mortí y Nurrá Indigenous Reserve (in the headwaters area of the Chucunaque River, the boundaries being established); the Chocó homelands Comarca Emberá No. 1 (Distrito de Cemaco) (1826 km² in the Chucunaque-Tuira Basin) and Comarca Emberá No. 2 (Distrito de Sambú) (931 km² in the Sambú Basin); Canglón Forest Reserve (316 km²); Chepigana Forest Reserve (1460 km² in 1960, but the lands were reclassified and the smaller area is being redetermined); and Punta Patiño Nature Reserve (263 km²) (ANCON 1993, pers. comm.). Over half the province thus is under some management category involving environmental protection. The Darién Region is one of eleven areas given priority under the 1992 Convenio para la Conservación de la Biodiversidad y Protección de Areas Silvestres Prioritarias en América Central.

Standards have been established for entrance and use of the buffer zone (Map 31) and the Darién National Park, to develop in an orderly manner ecotourism with conservation, which will benefit both the park and the communities in and near the park. Darién NP has been given top priority for preservation of representative species of terrestrial and aquatic ecosystems (LaBastille 1978; INRENARE and ANCON 1988).

The park has an administrative office in El Real de Santa María and three ranger stations (Balsas, Cruce de Mono, Pirre) for protection, conservation and management. From 1989 INRENARE and ANCON have delimited and labelled 100 km of trails between the three ranger stations. The park rangers are provided with training and basic equipment for effective park protection and enforcement. There are monthly visits to the park's most critical areas to evaluate the effects of settlers, hunters, etc.

Environmental education is undertaken in every community in or neighbouring the park, including collaboration (by Panama and U.S.A.) in the Sistema Nacional para la Erradicación de la Malaria (SNEM) and the Comité para la Prevención de la Fiebre Aftosa (COPFA). Precautions have been taken (including regulations) to avoid and minimize impacts from the planned Pan American highway. The ecological dangers of opening roads and the highway without necessary controls are explained to the public.

In the buffer zone, ANCON (the Asociación Nacional para la Conservación de la Naturaleza) has established Peresénico Agroforestry Farm (near Pirre Ranger Station), a project of Finca Agroforestal. This provides opportunities for the neighbouring communities to benefit from a learning centre for agroforestry techniques and park information, and cheaper access tickets, etc., with a goal of reducing the pressure these communities exert on the protected zone. ANCON has an environmental education centre at El Real which works cooperatively with the park personnel and has research and development centres for ecotourism at Santa Cruz de Cana Mining Centre and Punta Patiño Nature Reserve.

Punta Patiño Nature Reserve, which is located in the park's buffer zone, has extensive mangrove forest and semi-deciduous dry forest. The reserve protects many mammal and bird species threatened with extinction (43 such species live in the area). ANCON is establishing an agroforestry programme in the area for habitat restoration using native plants and traditional crops. Notwithstanding the scientific character of this reserve, there is potential for developing some tourism without altering the natural ecosystems.

Conservation efforts in the Darién developed by INRENARE and ANCON are possible thanks to financial support of several international organizations such as The Nature Conservancy (partly through the Parks in Peril and the Last Great Places campaigns), U.S. Agency for International Development, Oro Verde, World Wildlife Fund – U.S. and World Wildlife Fund – U.K.

References

Chapin, M. (1984). Kuna Indians: initiate rain forest reserve in Panama. *Focus (WWF – U.S.)*: Pg. 6.

Chapin, M. (1991). Losing the way of the Great Father. *New Scientist* 131 (1781): 40–44.

D'Arcy, W.G. (1977). Endangered landscapes in Panama and Central America: the threat to plant species. In Prance, G.T. and Elias, T.S. (eds), *Extinction is forever*. New York Botanical Garden, Bronx. Pp. 89–104.

D'Arcy, W.G. and Correa-A., M.D. (eds) (1985). *The botany and natural history of Panama: La botánica e historia natural de Panamá*. Monogr. Syst. Bot. No. 10, Missouri Botanical Garden, St. Louis. 455 pp.

Duke, J.A. (1968). *Darien ethnobotanical dictionary*. Battelle Memorial Institute, Columbus, Ohio. 131 pp.

Duke, J.A. (1970). Ethnobotanical observations on the Choco Indians. *Econ. Bot.* 24: 344–366.

Duke, J.A. (1975). Plant species in the forest of Darien, Panama. In Golley, F.B., *et al.*, *Mineral cycling in a tropical moist forest ecosystem*. University of Georgia Press, Athens. Pp. 189–221.

Duke, J.A. and Porter, D.M. (1970). *Darien phytosociological dictionary*. Battelle Memorial Institute, Columbus, Ohio. 70 pp.

Gentry, A.H. (1977). Endangered plant species and habitats of Ecuador and Amazonian Peru. In Prance, G.T. and Elias, T.S. (eds), *Extinction is forever*. New York Botanical Garden, Bronx. Pp. 136–149.

Gentry, A.H. (1978). A new *Freziera* (Theaceae) from the Panama/Colombia border. *Ann. Missouri Bot. Gard.* 65: 773–774.

Gentry, A.H. (1985). Contrasting phytogeographic patterns of upland and lowland Panamanian plants. In D'Arcy, W.G. and Correa-A., M.D. (eds), *The botany and natural history of Panama: La botánica e historia natural de Panamá*. Monogr. Syst. Bot. 10, Missouri Botanical Garden, St. Louis. Pp. 147–160.

Gentry, A.H. (1986). Endemism in tropical versus temperate plant communities. In Soulé, M.E. (ed.), *Conservation biology: the science of scarcity and diversity*. Sinauer Associates, Sunderland, Massachusetts. Pp. 153–181.

Golley, F.B., McGinnis, J.T., Clements, R.G., Child, G.I. and Duever, M.J. (1975). *Mineral cycling in a tropical moist forest ecosystem*. University of Georgia Press, Athens. 248 pp.

Hartshorn, G.S. (1981). *Forests and forestry in Panama*. Institute of Current World Affairs, GSH-14. Hanover, New Hampshire. 17 pp.

Herlihy, P.H. (1989). Opening Panama's Darién gap. *J. Cultural Geogr.* 9(2): 41–59.

Holz, R.K. (1980). The Darién of Panama: the twilight of a unique environment. *Explorers J.* 58(4): 158–164.

Houseal, B.L., MacFarland, C., Archibold, G. and Chiari, A. (1985). Indigenous cultures and protected areas in Central America. *Cultural Survival Quarterly* 9(1): 10–20.

INRENARE and ANCON (1988). *[Plan de manejo y desarrollo integrado] Reserva de la Biósfera Darién*, basado en la labor de R.E. Weber. Instituto Nacional de Recursos Naturales Renovables (INRENARE) and Asociación Nacional para la Conservación de la Naturaleza (ANCON), Panama City. 176 pp.

Jaén-Suárez, O. (1985). Nuevos hombres y ganados y su impacto en el paisaje geográfico panameño entre 1500 y 1980. In D'Arcy, W.G. and Correa-A., M.D. (eds), *The botany and natural history of Panama*. Monogr. Syst. Bot. Missouri Bot. Gard. 10. Pp. 379–392.

LaBastille, A. (1978). *Facets of wildland conservation in Middle America*. Centro Agronómico Tropical de Investigación y Enseñanza (CATIE), Turrialba, Costa Rica. 37 pp.

Lamb, F.B. (1953). The forests of Darién, Panama. *Caribbean Forester* 14: 128–135.

Lewis, W.H. (1971). High floristic endemism in low cloud forests of Panama. *Biotropica* 3: 78–80.

Mayo-Meléndez, E. (1965). Algunas características ecológicas de los bosques inundables de Darién, Panamá, con miras a su posible utilización. *Turrialba* 15: 336–347.

Nations, J.D. and Komer, D.I. (1983). Central America's tropical rain forests: positive steps for survival. *Ambio* 12: 232–238.

OEA (1978). *Proyecto de desarrollo integrado de la región oriental de Panamá – Darién*. Secretaría General de la Organización de los Estados Americanos (OEA), Washington, D.C. 308 pp.

Polsky, C. (1992). Crossroads of the continents. *Nat. Conservancy* (M/A): 14–21.

Porter, D.M. (1973). The vegetation of Panama: a review. In Graham, A. (ed.), *Vegetation and vegetational history of northern Latin America*. Elsevier, New York. Pp. 167–201.

Prebble, J. (1968). *The Darien disaster*. Penguin Books, Middlesex, England, U.K. 384 pp.

RARE (1983). *Draft plan for the development of a private sector initiative in natural resource and environmental programs in the Republic of Panama*. RARE Report, Panama City. 55 pp.

Sauer, C.O. (1966). *The early Spanish Main*. University of California Press, Berkeley and Los Angeles. 306 pp.

Torres de Araúz, R. (1970). *Human ecology of route 17 (Sasardí-Mortí) region, Darién, Panama*. Battelle Memorial Institute, Columbus, Ohio. 200 pp.

Torres de Araúz, R. (1985). Etnobotánica Cuna. In D'Arcy, W.G. and Correa-A., M.D. (eds), *The botany and natural history of Panama*. Monogr. Syst. Bot. Missouri Bot. Gard. 10. Pp. 291–298.

Authors

This Data Sheet was written by Olga Herrera-MacBryde (Smithsonian Institution, Department of Botany, NHB-166, Washington, DC 20560, U.S.A.) and the Asociación Nacional para la Conservación de la Naturaleza (ANCON) (Apartado 1387, Panama City 1, Panama).

REGIONAL OVERVIEW: CARIBBEAN ISLANDS

Total land area: c. 240,000 km².

Population (1991): 35,000,000.

Number of islands: 115 + c. 3400 rocky islets and cays.

Altitude: -40 m (Lago de Enriquillo) to 3087 m (Pico Duarte), both in Hispaniola.

Natural vegetation: Lowland and montane tropical forest, evergreen thicket, savanna, cactus/thorn scrub, mangrove and riverine communities.

Number of vascular plants: c. 13,000 species.

Number of single island endemics: 6550 species.

Number of genera: c. 2500.

Number of endemic genera: 205.

Number of vascular plant families: 186 (excluding ferns).

Number of endemic families: 1 (Goetzeaceae).

Introduction

The Caribbean islands, with their spectacular land- and seascapes, and diverse flora and fauna have been, like many tropical islands, under European influence a little over 500 years. In that time they have been ruthlessly exploited for forest lumber, firewood, agricultural produce and minerals, mostly for the benefit of people who have never lived on them. They were stolen from the original inhabitants, fought over, bartered, bought, sold, colonized and eventually settled by people of African, Asiatic and European descent while the indigenous Amerindians were eliminated.

This history is similar to those of the Indian and Pacific Oceans, which have comparable islands derived from fragmented continents, ancient or active volcanoes, or living coral. The main distinguishing feature, outside the uniformity of coconut or sugar-cane plantations and contrasting with urban and resort developments, is the native flora which has evolved in each major tropical island cluster on independent lines.

All islands falling within the area of latitudes 10–27°N and longitudes 59–85°W are considered here, with Bermuda mentioned in context. The location, land-surface area and greatest altitude of the principal islands or island groups are given in Table 38. They are arranged from north to south in geographical units with some political adjustment for convenience of reference. Each entry includes a brief note on the number of islands or off-shore islets in the group.

The area of ocean encompassing the islands, excluding Bermuda, is nearly 5 million km². The total area of land is approximately 240,000 km², a ratio of sea:land of about 20:1. This ratio does not imply uniformity of dispersal nor size; on the contrary the locations, groupings and areas of the islands are very uneven. In all respects, whether considering physical or biological attributes, the key-word is diversity. The complexity makes it difficult to state reliably how many islands there are, but, allowing for the availability and accuracy of sources and some interpretation of definitions, it can be said that there are about 100 permanently inhabited islands and another 50 or so major vegetated cays (keys) and rocky islets with names that appear on maps at scales of 1:50,000 or smaller and in some gazetteers and atlases; in addition, several thousand low sandy or rocky cays, mostly in the northern half of the region, are recorded on larger-scale maps and Admiralty Charts. Distances between islands vary greatly, but none, except Bermuda, is more than 200 km from its nearest neighbour.

Geology

The oldest land areas of the modern Caribbean are at the extreme ends of the arc of islands in Cuba and Trinidad. Jurassic rocks have been identified in Cuba suggesting continental relationships and there are geological links between that island and Yucatán. Some biological distribution patterns support the hypothesis of a Cretaceous land-mass occupying the position of the Caribbean Sea (Weyl 1965) indicating direct links between northern South America and the Greater Antilles avoiding Central America (Stearn 1971).

More recently, renewed interest in plate tectonic research has led to some less speculative biogeographical conclusions. Jamaica was completely submerged between the mid-Eocene and early mid-Miocene and has at other times comprised several small islands. It is believed that, since its last emergence, Jamaica has had no connection with any other island nor a land-bridge to the continental mainland

TABLE 38. GEOGRAPHICAL DATA FOR THE CARIBBEAN ISLANDS

Island group	Area (km²)	Altitude (m)	Number of islands	References
BERMUDA				
Bermuda	55	79	6 + rocky islets	Britton 1918
BAHAMAS				
Bahamas	13,830 (10,010)*	65	35 + 700 cays + c. 2400 vegetated islets	Campbell 1978; Correll and Correll 1982
Turks and Caicos	430	47	5 + cays	Ray and Sprunt 1971
GREATER ANTILLES I				
Cuba	108,722	1974	1 + many large cays	Smith 1954; Borhidi and Muñiz 1980; Capote *et al.* 1989; Cuba 1989; Borhidi 1991
Isla de la Juventud	2126	310	1 + cays and mangrove islets	Jennings 1911, 1917; Britton 1916
Cayman Islands	264	43	3	Stoddart 1980; Sauer 1982; Brunt in Proctor 1984
GREATER ANTILLES II				
Jamaica	11,425 (10,830)*	2256	1 + c. 20 cays and rocky islets	Asprey and Robbins 1953; Thompson *et al.* 1986
Hispaniola:				
Haiti	27,749	2680	1 + 5 (incl. Navassa)	Ciferri 1936
Dominican Republic	48,441	3087	1 + 3 + rocky islets	Moscoso 1943
Puerto Rico	8897	1338	1 + 5 (incl. Mona)	Britton and Wilson 1923; Liogier and Martorell 1982; Liogier 1985–; Proctor 1989
Vieques	c.140	301	1 + c. 20 cays and rocks	Shafer 1914; Sorrié 1975
VIRGIN ISLANDS				
British Virgin Islands:				
Anegada	39	8.5	1 + Sombrero	Labastille and Richmond 1973; D'Arcy 1975
Tortola	62	542	1 + rocky islets	D'Arcy 1967; Howard and Kellogg 1987
Virgin Gorda	21	469	1 + rocky islets	Little *et al.* 1976
Jost Van Dyke	10	321	1 + rocky islets	Danforth 1935
U.S. Virgin Islands:				
St John	54	389		Woodbury and Weaver 1987
St Thomas	85	472	3 + c. 10 rocky islets and cays	
St Croix	215	355		Millspaugh 1902
LESSER ANTILLES – I LEEWARD ISLANDS				
Outer Leeward Islands:				
Anguilla	91	69	1 + cays	Box 1940; Harris 1965; Howard and Kellogg 1987
St Maarten	85	424	1 + islets	Boldingh 1909b; Stoffers 1956
St Barthélemy	21	281	1 + 6 islets	Questel 1941
Barbuda	160	45	1	Beard 1949
Antigua	279	402	1 + rocky limestone island	Wheeler 1916; Loveless 1960
Inner Leeward Islands:				
Saba	14	870	1	See St Maarten
St Eustatius/Statia	31	581	1	See St Maarten
St Kitts	168	1156	1	Beard 1949
Nevis	93	985	1	Beard 1949
Redonda	1.3	297	1 (unpopulated but goats present)	Tempany 1915; Box 1939a; Howard 1962a
Montserrat	98	914	1	Beard 1949
LESSER ANTILLES – II FRENCH ANTILLES				
Guadeloupe	1705	1465	6 + rocky islets	Duss 1897; Stehlé 1936; Beard 1949; Portecop 1979; Sastre and Portecop 1985
Martinique	1100	1463	1	See Guadeloupe
LESSER ANTILLES – III WINDWARD ISLANDS				
Dominica	751	1447	1 + 2 islets	Hodge 1954; Nicolson 1991
St Lucia	616	958	1 + 6 islets	Beard 1949
St Vincent	389	1234	1 + 4 islets	Beard 1949
Grenadines	90	308	16 + 100 or more islets and rocks	Howard 1952
Grenada	344	838	1 + rocky islets	Beard 1949; Groome 1970
LESSER ANTILLES – IV BARBADOS				
Barbados	430	336	1	Hardy 1932–1933; Gooding 1974
SOUTHERN OFFSHORE ISLANDS				
Southern Netherlands Antilles				
Aruba	175	188	1 + cays	Boldingh 1914; Stoffers 1956
Curaçao	425	375	1	See Aruba
Bonaire	265	240	2	See Aruba
Trinidad and Tobago				
Trinidad	4543	940	1 + 10 rocky islands and islets	Beard 1946; Richardson 1963
Tobago	295	576	1 + 2 large and a few small rocky islets	Beard 1944

Note: * refers to land area.

TOTAL LAND AREA: c. 240,000 km².　　　**TOTAL NUMBER OF ISLANDS:** 115 + cays and rocky islets.

(Buskirk 1985). Hispaniola (Haiti and the Dominican Republic) and Puerto Rico were continuous in the late Eocene and early Oligocene, but, since their separation, Hispaniola has been repeatedly divided and rejoined during the Pleistocene, at the level of the Cul de Sac – Enriquillo Valley, creating at times two islands of unequal size.

The Greater Antillean islands have all had a somewhat similar geological history but they differ from one another in the distribution, form and erosion patterns of the limestones deposited during several phases of submergence and uplift through the Tertiary period and Pleistocene. The richness of the flora of Cuba is explained, in part, by the wide range of soils derived from limestones, serpentine, dolomite, basalt, granite, diorite, gabbro, sandstone and slate (Borhidi 1991).

The Atlantic Plain of North America, including Florida and the Bahamas, was submerged until the Oligocene and has been much affected since by periodic changes in sea-level. Apart from sporadic unions between islands now separated by shallow seas, as within the Bahamas, Virgin Islands or Grenadines (as groups), the other islands have always been separate from one another. The Virgin Islands are peaks of a drowned mountain range created by volcanic activity in the Cretaceous.

The northern part of the Lesser Antilles comprises two more or less parallel arcs of islands. The irregular outer arc stretches from Anguilla to Antigua, the "Limestone Caribbees" (Harris 1965), and extends to include Grande-Terre of Guadeloupe, La Désirade and Marie Galante at the southernmost end. The complex geological history is represented by volcanic and sedimentary strata of Oligocene age, ranging to Pleistocene coral limestone and alluvium.

The inner arc of the Lesser Antilles is much longer, reaching from Saba in the north to Basse-Terre of Guadeloupe and on through the Windward Islands to Grenada. Formation of these exclusively volcanic islands began in the Eocene and still continues (Martin-Kaye 1963).

Trinidad belongs structurally to northern South America and a land connection with Venezuela existed almost into historical times. The general similarity with the coastal cordillera is seen in the mountainous ridges of metamorphosed sedimentary rocks of lower Cretaceous age with their igneous intrusions. At lower elevations the central plain of Trinidad combines erosion products from the nearby hills and exposed lacustrine sediments. The relationships of Tobago are less obvious but some connection with the Paria Peninsula, possibly even including faraway Barbados, should not be ruled out.

Climate

The latitude and western Atlantic position indicate warm moist conditions determined by easterly or north-easterly trade winds and the intertropical convergence. Rainfall and temperature data are given by most of the authors referred to in Table 38.

Rainfall depends on topography, ranging from less than 600 mm along leeward coasts in rain-shadow to over 5000 mm on windward slopes of mountains. Most inland areas have annual rainfall within the 1500–2000 mm per annum range. All climates in the region are seasonal with, at sea-level in most years, at least one dry month when rainfall is less than 100 mm. The main dry period is usually between January and April; there may be a second dry period in more southerly latitudes in July to September.

A serious deficiency in climatic data is the lack of rainfall records for higher altitudes. Moisture conditions are usually more uniform due to the formation of cloud around mountain peaks, but they are sometimes excessive. Sastre (1978, 1979a) and Sastre and Portecop (1985) have recorded maximum annual rainfall on the upper windward slopes, at altitudes of c. 1450 m, of the volcanic peaks of Guadeloupe and Martinique of the order of 8000–10,000 mm.

The temperature regime is equable as befits intertropical oceanic islands. At sea-level, temperature in February, which is often the coolest and driest month, rarely falls to 12°C. Rainy seasons are usually warmer, but maxima rarely reach 33°C. Overall averages at sea-level are mostly in the range 25–27°C.

With rare occurrences in Grenada and Tobago, and even rarer ones in Trinidad, hurricanes may from time to time devastate other islands (see Factors causing loss of biodiversity). The tracks pass mostly from south-east to north-west across the region.

Population

The original Amerindian inhabitants, Caribs and Arawaks, have long since ceased to influence or modify the land and vegetation of the Antilles. Fewer than 2000 Caribs live today in a reserve in Dominica and fewer than 3% of these are of pure race. In any case, the original Amerindians had very little influence on the environment in comparison with Europeans, who invaded from the time of Columbus, and the Africans and Asians brought by Europeans, originally as slaves or indentured plantation workers. The modification of the West Indian landscape has taken almost 500 years to reach its present condition.

The population of the Caribbean islands in 1991 was estimated to be 35 million (FAO 1990), projected to rise to about 40 million by the year 2000, and to nearly 60 million by 2025. It is difficult to make a reliable overall forecast because the rates of increase vary considerably from island to island depending on internal social and economic factors and international political links which affect emigration. Life expectancy ranges from 51 in Haiti to 74 in Puerto Rico.

In general, although all island-totals are increasing, the incremental rates and the proportions of populations economically active in agriculture have fallen over the past decade. These trends reflect increase in mechanization and efficiency on farms and in importations of foods replacing local produce, combined with migration to towns and coastal tourist resorts. Higher growth-rates are associated with larger rural populations and relatively smaller movements away from agricultural pursuits, as in Jamaica, Haiti and the Dominican Republic. The totals of people living in agricultural communities actually rose in Jamaica and Haiti (by 7.5%) during the decade, this trend matching several of the poorer Central American republics, where, however, numbers of those actually economically active on the land also fell.

There are 16 urban centres in the Caribbean with over 100,000 inhabitants: Bahamas – Grand Bahama, Nassau; Cuba – Habana, Camaguey, Santiago de Cuba, Guantanamo; Jamaica – Kingston, Montego Bay; Haiti – Port-au-Prince;

TABLE 39. CARIBBEAN POPULATION

Island or island group	Total (000s) 1980	1989	Incremental rates (% p.a.) 1980*	1989	Forecast year 2000 (000s)	Density: people/km² of arable land 1980	Economically active in agriculture, % of total population 1980	1989
Antigua-Barbuda	75	85	-	1.6	102	-	-	-
Bahamas	224	256	1.7	1.6	303	1488	3.6	2.7
Barbados	249	259	0.9	0.5	273	692	4.8	3.5
Bermuda	54	58	-	0.8	63	-	-	-
Cuba	9732	10,238	0.9	0.6	10,929	185	8.7	8.4
Dominica	73	80	1.6	1.1	90	416	-	-
Dominican Republic	5697	7019	2.8	2.6	9307	205	13.3	11.7
Grenada	92	101	1.8	1.1	113	700	-	-
Guadeloupe	327	339	1.1	0.4	354	435	6.1	4.7
Haiti	5413	6383	2.6	2.0	7936	425	32.0	28.8
Jamaica	2173	2478	2.1	1.6	2950	463	13.6	13.4
Martinique	326	330	0.9	0.1	335	592	5.5	3.9
Netherlands Antilles	238	274	2.2	1.7	333	162	-	-
Puerto Rico	3199	3650	1.6	1.6	4343	652	1.2	1.0
St Lucia	118	134	2.4	1.5	158	610	-	-
St Kitts-Nevis	52	56	-	0.9	62	-	-	-
St Vincent	99	109	2.8	1.1	123	626	-	-
Trinidad and Tobago	1095	1264	2.0	1.7	1522	702	3.7	3.0
Totals	**29,236**	**33,113**	-	-	**39,296**	-	-	-

Note:

* Incremental rates and density data for 1980 from Levine (1981); other figures from, or calculated from, *FAO Production Yearbook 1989* (1990).

Dominican Republic – Santo Domingo, Santiago de los Caballeros; Puerto Rico – San Juan; Guadeloupe – Pointe-à-Pitre; Martinique – Fort-de-France; Barbados – Bridgetown; Trinidad – Port-of-Spain.

Vegetation

The following descriptions of natural communities cover the range of vegetation types in the Caribbean. Further information on Caribbean vegetation may be found in the references accompanying Table 38. Topography, climate and edaphic features combine to determine the unique characteristics of the communities and the floristic diversity of each individual island. Determining those characteristics which are most in need of safeguarding requires analysis of each island; inference and extrapolation of data from other islands or similar mainland sites are rarely valid. Coastal and low-elevation communities have in general the greatest uniformity; the more inland and elevated communities have the highest diversity.

Coastal communities

The most common formation along the thousands of miles of coastline is the open pioneer vegetation of upper beaches and maritime rocks. Some differences depend on whether the substrate is of volcanic, coral-limestone or alluvial origin, or whether the rainfall is low or high. Most seashores have lower rainfall than the immediate hinterland, but, because of the continuous high humidity, woody vegetation is almost always evergreen. Floristic individuality is low among coastal pioneer communities and many species have an amphi-Atlantic or even pantropical distribution.

Cactus scrub, evergreen bushland and dry evergreen thicket occupy well drained, usually rocky, substrates. Succulents and low spiny shrub vegetation is characteristic of coastal areas with seasonally higher temperatures and rainfall under 700 mm, often falling in one short annual period. Unique floras of endemic cacti, other succulents and spiny shrubs exist along the leeward coasts of the larger islands. These places have mostly been spared from agricultural development on account of their infertility and lengthy droughts, but they may have been severely damaged by cutting for firewood and charcoal; resort development and gathering of cacti for horticultural use are contemporary risk factors.

Any forests which might have grown close to the sea have been long since removed, especially if they were near the first points of entry and settlement. Some undisturbed patches of marsh forest in the Negril area of Jamaica existed immediately inland from the strand in the early 1960s. They comprised mature examples of buttonwood (*Conocarpus erectus*) and *Tabebuia angustata*, both up to 25 m, accompanied by large woody climbers, such as *Dalbergia brownei*, and endemic epiphytes, such as *Hohenbergia negrilensis*. Standard Flora descriptions of buttonwood refer to a shrub or tree up to 7 m. This vegetation was eliminated very soon after a major road was built along the west coast of the island to enable easy access and tourist development.

Sand dunes are rare in the tropics (Morton 1957) because the beach sand scarcely becomes dry enough and the onshore winds strong enough to move it. Flattish particles of algal (*Halimeda*) or shell sand may offer broad surfaces to the wind but these are usually strongly cohesive in damp salty conditions. Dunes form in the Caribbean in Barbados (Gooding 1947), Aruba (Stoffers 1956) and Barbuda (Harris 1965). Given the rarity of these sites and their ecological interest, their vulnerability as natural accumulations of graded sand should be recognized.

Raised coral beaches, as for example along the north and south coasts of Jamaica, and in Cuba and southern Hispaniola, support a low suffruticose vegetation adapted to withstand wind, salt spray and high insolation. The rock pavements exemplify the harshest aspects of karst limestone, to which plants respond by becoming dwarfed and acquiring strongly rooted perennial stocks. Individuals of buttonwood here may flower and fruit at a height of 15 cm.

Mangrove, lagoon and riverine formations

In estuaries or sublittoral lagoons of low coastlines, mangroves are, or have been, extensive; many have been destroyed in recent years by reclamation and pollution. Lagoons and quiet coves with mangroves have suffered through competition for beach space, moorings and resort development. Mangroves are unhappy partners with human habitation on account of the unpleasant odour of the anaerobic muddy substrate and the mosquitoes which infest them, but, in relation to coastal resources, they have significant geomorphological and biological functions. They are dynamic communities showing diverse vegetational facies, as for example in the Cayman Islands (Brunt in Proctor 1984). Recent studies in Puerto Rico (Jiménez, Martinez, and Luis 1985), Guadeloupe (Renard 1976), Dominica (James 1980) and St Lucia (Portecop and Benito-Espinal 1986) emphasize their importance and the protection imperative. Mangrove swamps and sublittoral lagoons are fragile because of their sensitivity to any alteration of drainage systems. Early recognition of adverse change may save a valuable asset; an example may be cited of the Caroni Swamp in Trinidad where exploitation was recognized (Bacon 1970) and public awareness led eventually to designation as a National Park and legislated management (Thelen and Faizool 1979, 1980a).

Some mangroves have been reduced through the natural silting up of streams entering lagoons or the sea. Open barren saline areas (salinas) have resulted. These are frequent along the dipping south coast of Jamaica where their formation may have been accelerated through erosive cultivation methods on large sugar-cane plantations nearby. Adjacent to other coastlines are brackish or freshwater permanent swamps dominated by palms, tall grasses and sedges, as in Cuba (Cienaga Zapata), Jamaica (Black River Morass) and Trinidad (Nariva Swamp).

Because of the relatively small size of the islands compared with continental areas, rivers are short and have rather simple systems. Ponds and lakes are few but there are many streams and torrents especially in the higher wetter places; in limestone country the streams often flow underground. Some rivers have been dammed for water-supply and hydroelectric developments. Such developments frequently coincide with rugged terrain of scenic beauty and floristic richness which may be put at risk.

The four large rivers in the region are the Río Cauto in eastern Cuba, the Artibonite in Haiti, and the Río Yaque del Norte and Río Yaque del Sur in the Dominican Republic; they are between 150 and 200 km long. The Río Cauto is bordered in the lowland parts by swamps and gallery forests on rich black alluvial soils. The forests are characterized by the presence of *Roystonea* palms, a single canopy of partly deciduous trees to c. 20 m, and moisture-loving lianas. Removal of the gallery forest results in savanna with varying complements of tall grasses, sedges and palms of the genera *Copernicia* or *Sabal*. The original vegetation along the Río Yaque del Norte and Río Yaque del Sur has almost all been lost.

The shorter but floristically diverse Black River of Jamaica flows into a sunken landscape where it is associated with an extensive morass. The complex of communities here includes many facets of strictly aquatic vegetation both in flowing and still water. Herbaceous swamps comprise large grasses (*Phragmites*), sedges (*Cladium*, *Cyperus giganteus* and *Fuirena*) and other monocotyledons, such as *Typha*, *Alpinia*, *Sagittaria*, *Crinum* and *Thalia*. Marsh forest was once dominated by species of *Symphonia*, *Terminalia*, *Hibiscus elatus*, *Grias cauliflora* (the only West Indian representative of Lecythidaceae), and the endemic morass royal palm *Roystonea princeps*. Cutting of the trees followed by burning has produced a fire subclimax savanna with extensive areas of the sedge *Cladium jamaicense* and bull thatch palm *Sabal jamaicensis*. The riparian forest, which existed alongside the river further inland, comprised a number of rare and unusual trees and woody vines. It has now been virtually eliminated. The bignoniaceous *Tanaecium jaroba* and the endemic *Combretum robinsonii* are large lianas ecologically reminiscent of Amazonian riverine forests.

Forests and woodlands

The natural vegetation of the Antilles is essentially woody except where there is very low rainfall or the land becomes waterlogged during wet weather. The forest formation-types depend on temperature, related to altitude, and the amount and periodicity of rainfall. Different authors have classified the vegetation variously, frequently using non-comparable criteria to define the chosen categories. Some of the systems are physiognomic (Hodge 1941; Stehlé 1945–1946; Howard 1974, 1979; Cuba 1989) and others employ a more detailed analysis of floristic and phytosociological data (Beard 1944, 1946, 1949, 1955; Borhidi 1991). Where geological, geographical, climatic, edaphic, structural (life form) or taxonomic elements have been called upon in diverse arbitrary combinations, comparisons and logical ecological correlations are difficult. There is, as yet, no ideal classification of the vegetation based on plant data alone.

The most luxuriant forests are those at low elevations, up to c. 300 m altitude. Beard (1955) referred to these as seasonal forest, prefixed by the adjectives evergreen, semi-evergreen or deciduous, related to the behaviour of trees in places experiencing an increasing number of dry months. Lowland rain forest in its optimum evergreen non-seasonal expression does not occur in the Caribbean islands.

The 'climax' forests have different species of dominant trees with distinctive associated subordinate floras in each of the major island groups, related to high levels of intra-regional endemism (see Flora). The lowlands were easily accessible and ideal for pioneer development. Much, sometimes virtually all, of the original forest was replaced by plantations during the early days of colonization. As Beard (1949) pointed out "in practice to-day, natural forest and woodland is only found in areas which are too inaccessible or too unfavourable for cultivation".

On rugged terrain, such as the extensive karst limestone hills of the Greater Antilles, forests and woodland persist, although much has been cut for timber or fuelwood. Drier facies with rainfall under 1500 mm are particularly prone to this form of degradation (Kapos 1986). Generally, sites with easterly aspects backed by mountains have higher rainfall and carry more luxuriant forests (Dominica – Hodge 1954; Jamaica – Kelly 1986).

TABLE 40. AREAS OF NATURAL FOREST IN THE CARIBBEAN

Country	Area of forest (000s ha)		Total area (000s ha)	% of total land area (see Table 38)		
	Broadleaved 1980[1]	Conifer* 1980[1]	1980[1]	1920s[2]	1970s	1980[1]
Antigua and Barbuda	9	-	9	-	-	20.5
Bahamas (including Turks and Caicos)	40	283	323	?13	?28	30.9
Cuba	1255	200	1455	46	14	13.1
Dominica	41	-	41	-	-	54.6
Dominican Republic	444	185	629	77	23	13.0
Grenada	5	-	5	-	-	14.5
Guadeloupe	89	-	89	?12	?38	52.2
Haiti	36	12	48	60	7	1.7
Jamaica	67	-	67	30	?45	6.2
Martinique	[3.2]	-	[3.2]	-	?26	[2.9]
Montserrat	3	-	3	-	-	30.6
Puerto Rico	246	-	246	?20	?17	27.6
St Kitts-Nevis	5	-	5	-	-	19.2
St Lucia	8	-	8	-	-	13.0
St Vincent	12	-	12	-	-	30.8
Trinidad and Tobago	208	-	208	59	46	43.0
U.K. Virgin Islands	3	-	3	-	-	22.7
U.S. Virgin Islands	[3]	-	[3]	-	-	[8.5]
Totals	**2524**	**680**	**3204**			**13.4**

Notes:

[1] Data for 1980 principally from FAO document FO:MISC/88/7. That publication lacks figures for Martinique and the U.S. Virgin Islands for which estimates [ ] have been entered from other sources.

[2] Data for 1920s from Zon and Sparhawk (1923); for 1970s from Persson (1974). Entries marked with "?" show trends inconsistent with the usual expectation of shrinkage of natural forest; perhaps other measurement criteria were applied.

* Natural forests of pine exist only in Bahamas, Cuba and Hispaniola.

Various types of montane rain forest and lower montane rain forest have been described (Ciferri 1936; Beard 1942, 1949; Stehlé 1945; Asprey and Robbins 1953; Grubb and Tanner 1976; Tanner 1977, 1986). Although there have been clearances for crops such as coffee, vegetables, flowers and coniferous forestry, or in the past for quinine and tea, steeper slopes may still carry relatively less disturbed woodland than places at lower altitudes. Peaks often have a low vegetation of gnarled trees and shrubs, in some places referred to as elfin woodland, which is a response to strong wind, high insolation, minimal soil and low nutrient status (Howard 1968). Frost is a rare occurrence; water butts on the summit of Blue Mountain Peak (2256 m) in Jamaica may sometimes be found to have a thin film of ice on the surface in the early morning. The highest peak of the Caribbean islands is Pico Duarte (3087 m) in the west Cordillera Central of the Dominican Republic.

Accurate figures for the present forest cover for the Antillean islands are difficult to establish. Table 40 is compiled from data published by the FAO (1988) and other sources. Some sources do not distinguish between natural, managed or plantation forests and may not record recent felling or changes of composition through active management. Particular forestry policies may lead to reports of expanded areas of forest which might even comprise monocultures of exotic species.

Savannas

The savannas of Cuba have been described and discussed by Borhidi and Herrera (1977) and Borhidi (1991). They comprise a formation-type which is very well represented there. Concepts of what constitutes savanna range from semi-desert, on rocky soils experiencing rainfall of 300–600 mm per annum with 9–10 dry months, to artificial pastures in areas of much higher rainfall with shorter dry seasons. The latter exist mostly where original woodland has been removed and an herbaceous vegetation has become stabilized by repeated wood-cutting, fire and grazing. Estimates of how much of Cuba was savanna at the time of the first Spanish settlers vary from 5 to 30% or more of the land. None of these islands ever carried herds of grazing animals like the savannas of Africa, but the pre-Hispanic inhabitants may well have set fire to natural grassy areas alongside rivers and on marshes during dry weather.

A narrower concept of natural savanna was defined by Beard (1946, 1953) as edaphic. Seasonal, but not necessarily extremely dry, climatic conditions result in actual drought during periods of low rainfall and physiological drought due to impeded drainage and waterlogging during periods of high rainfall. Both situations are inimical to the growth of trees. These savannas are level and usually have sandy topsoils with impervious subsoil horizons. They occur in Cuba, Jamaica, Puerto Rico and Trinidad and are renowned for their local floristic diversity in contrast to derived savannas characterized by the presence of significant proportions of widespread adventive and exotic weeds.

In Cuba there are many endemic palms and their distribution is closely related to soil type. In areas of shallow infertile serpentine latosols many local endemics have evolved. Silica sands support short grass and sedge savanna with palms and sometimes with pines. Savanna vegetation with pines is reminiscent of physiographically comparable areas of open pine forests (pine barrens) in the Bahamas (Correll and Correll 1982) and Florida. The Bahamas, low oolite calcareous islands, have natural vegetation differing in floristic composition on the more fertile "blacklands" and the less fertile "whitelands". These Bahamian communities resemble more the glades and hammocks of Florida than the savannas of Cuba or Trinidad; strangely, the natural pinelands are confined to Grand Bahama, Abaco, New Providence and Andros, and disjunctly the Caicos Islands.

Siliceous soils are responsible for much individuality in the floras of Cuba and the Isla de la Juventud. Otherwise they are strictly localized in the Antilles. Two small sandy savannas exist in Jamaica and there is another at Bayamon in Puerto Rico, but the best known in the English-speaking islands are those of Trinidad (Beard 1946; Richardson 1963). The Aripo, Piarco, Mausica and Omeara Savannas occupy a series of low flattish dissected terraces derived from mainly siliceous erosion products of the Northern Range. In the last 30 years or so, the Mausica, Omeara and Piarco Savannas have been lost to housing, industrial, airport and water-supply developments. Only the Aripo Savannas remain and they have now been designated as areas protected for environmental conservation and scientific research (Thelen and Faizool 1980b). The sites of several endemic species will be preserved but some rare plants may have been lost on the other savannas.

The St Joseph Savanna, also in Trinidad, shares some species with the others but the habitat is a steep hillside. Thus, the factor of impeded drainage is replaced by the factor of shallow infertile soil; there are also more frequent fires. This vegetation is fairly typical *Byrsonima-Curatella* (Chaparro-Savanna Serrette) savanna, but there are local endemic and rare herbaceous species in the community. The Erin Savannas of the south of Trinidad have an intermediate ecological status between the "Aripo" and "St Joseph" types. All the savannas of Trinidad have their counterparts in Venezuela but the similarity of the floristic composition remains to be confirmed.

The colloquial understanding of the word "savanna" or "savannah" in the Caribbean is often in connection with a piece of flat land from which woody vegetation has been cleared; such places are frequently close to habitations and are used for recreation.

Flora

There are an estimated 13,000 native vascular plant species in the region. Table 41 includes flora statistics for 32 territories in descending order of the total number of vascular plant species recorded from them. The table provides counts of indigenous and naturalized species separately and, within these, phanerogams (seed plants) and pteridophytes (ferns and fern-allies) are also reported separately. The numbers of indigenous phanerogams and pteridophyta which are endemic are given in further separate columns (being already included in the totals of those categories), and the percentages are calculated accordingly. There is a small number of gymnosperms in most of the floras and these have been included numerically with the flowering plants.

The quality and availability of sources of the flora statistics in Table 41 is extremely variable but for most of the islands the information is relatively complete and up-to-date. Some

TABLE 41. FLORA STATISTICS OF INDIGENOUS AND NATURALISED VASCULAR PLANTS FOR SOME CARIBBEAN ISLANDS

| Island | Indigenous species | | | | | | Naturalized species | | Total vascular plants | Rare or threatened |
| | Phanerogams | | | Pteridophytes | | | Phanerogams | Pteridophytes | | |
	Total	Endemic	% endemic	Total	Endemic	% endemic				
Cuba	6015	3193	53	490	31	6	376	10	6891	960
Hispaniola	c.4685	c.1400	30	c.450	c.45	10	c.350	c.15	c.5500	276
Jamaica	2746	852	31	558	71	13	374	21	3699	428
Puerto Rico	2128	215	10	364	21	6	386	13	2891	515
Trinidad and Tobago	1982	215	11	277	21	8	c.300	9	2568	863
Guadeloupe	1400	19	1	272	4	2	283	4	1959	13 [a]
Martinique	1287	24	2	218	0		262	5	1772	13 [a]
Dominica	1027	10	1	200	2	1	120	5	1352	25
Bahamas	1068	117	11	43	1	2	203	2	1316	21
St Vincent	969	20	2	165	0		155	3	1292	
St Lucia	909	11	1	119	0		153	5	1186	19
Grenada	838	3	0.4	152	1	0.7	100	4	1094	
Antigua	808	0		37	0		70	1	916	
Montserrat	554	2	0.4	116	0		128	2	800	
St Kitts	533	0		126	1	1	69	3	731	
Barbados	542	3	0.6	30	0		151	5	728	
Cayman Islands	521	19	4	18	0		61	1	601	6
Saba	435	2	0.5	69	0				504	
Tortola	484 [b]	2 [b]	0.4						484 [b]	
Grenadines	467	0		6	0				473	
Bermuda	147	11	8	19	4	21	303	3	472	14
St Eustatius	427	2	0.4	37	0				464	
Aruba, Bonaire, Curaçao	460	25	5						460	
St Martin	431	1	0.2	15	0				446	
Mona	412	5	1	5	0				417	43
Virgin Gorda	397	1	0.3	6	0				403	
St Barthélemy	336	0		8	0				344	
Anguilla	321	1	0.3	0	0				321	
Nevis	182	0		78	1	1	15	1	276	
Barbuda	229	0		1	0				230	
Anegada	198	2	1						198	
Jost Van Dyke	73 [c]	1 [c]	1						73 [c]	

Notes:

[a] A combined figure is given for rare and threatened plants of Guadeloupe and Martinique.

[b] The figures for Tortola refer to dicotyledons only.

[c] The figures for Jost Van Dyke refer to trees only.

Many of the publications from which these data have been obtained may be found in references of Table 38.

of the data have been augmented and revised from more recent monographs or herbarium studies. Counts for the ferns of the Lesser Antilles have been taken from *The Flora of the Lesser Antilles*, volume 2 (Proctor in Howard 1977); for seed-plants, the same Flora (Howard 1974–1989) has been used, augmented and confirmed from other publications, independent local contributions and herbarium studies. Some islands such as Antigua, Grenada and St Lucia have never had an independent list published, while comprehensive works covering St Vincent and Trinidad and Tobago are long overdue for revision. References to published Floras are given in the bibliography to this overview.

Institutional bases for the study of Caribbean flora locally are patchy and most of them are limited in physical resources and capabilities. Fortunately, interest is growing and there is current activity in Cuba, Dominican Republic, Puerto Rico, St Lucia, Barbados, and Trinidad and Tobago. In recent years Floras, covering both flowering plants and ferns, have been published for Bahamas, Cayman Islands, Jamaica and the Lesser Antilles; a fern volume for Puerto Rico has recently appeared and a flowering plant Flora for Hispaniola is well advanced in production. A general Flora, to include all cryptogamic and phanerogamic groups, is proposed for the Greater Antilles under the direction of the New York Botanical Garden. Most of the current research on floristics in Cuba, Hispaniola, and Trinidad and Tobago is published in those islands but, for other territories, investigations and publication are institutionally based in North America and Europe.

Naturalized species

The distinction of species which are indigenous from those which have been introduced and have become naturalized is important from the point of view of conservation. The recognition of truly rare or threatened taxa is particularly necessary, especially if the threats might lead to global extinction. This is not to say that all plants restricted to small or clearly circumscribed areas are at risk, nor that every exotic introduction or widely dispersed species is thereby safe. One of the rarest orchids of Jamaica is not endemic there; it is known also from a few distant and scattered localities in South America. If possible, the status of each species in a territory should be assessed on its own merits because the potential of any introduction to become naturalized is unpredictable. In a complex of archipelagos extending over a latitudinal range of nearly 30°, local conditions favour the establishment of adventives differentially. Some introductions naturalize readily in some places and not in others. For example, Para rubber (*Hevea brasiliensis*) is totally at ease with conditions in Trinidad and reproduces there by means of vigorous seedlings, but this occurs only rarely in Jamaica. Introduced species known only in cultivation have been omitted from the flora statistics in this overview. Worldwide species, fully integrated in Caribbean floras, and not obviously or historically known to be of exotic derivation, are treated as native. The widespread seashore species and many synanthropic plants belong to this category.

The Caribbean species of mahogany (*Swietenia mahagoni*) is only truly native in southern Florida, Bahamas, Cayman Islands, Cuba, Jamaica and Hispaniola. It was much exploited in the early colonial era and is not now found in Jamaican forests (Howard and Proctor 1957). Well grown trees have become rare in the forests of Cuba where the species was once an important component of lowland karstic semi-deciduous woodlands; they are, however, not rare in Hispaniola. Because of its value, this species was introduced elsewhere and has now become naturalized. Watts (1978) has described the site-characteristics, structure and floristics of mahogany woodlots on coral limestone in Barbados. The accompanying ground cover that has developed is overwhelmingly of forest plants native to Barbados and adjacent islands and some native trees have also regenerated. These artificial but ecologically harmonious communities were regarded by the author as indicating the nature of the island's original forest cover, and were seen as refuge sites with a high potential conservation status.

Naturalized species have become fully integrated into some of the autochthonous island communities and in these situations exotic trees such as mango (*Mangifera indica*) or breadfruit (*Artocarpus altilis*) do not stand out as ecologically incompatible with the indigenous ones. Thus, floristically mixed communities have been created providing habitats for native birds, mammals and other animals, as well as indigenous epiphytes, parasites and climbers. Such artificial vegetation is regarded by many people as "natural".

Some species which do not seem to be fully integrated into the primeval vegetation were possibly introduced or moved into new areas, or even from island to island, in pre-Columbian times. The blue mahoe (*Hibiscus elatus*) could have been brought to Jamaica from Cuba by the Arawaks to be used for bark rope (Adams 1971). Hodge and Taylor (1957) suggested that the Island Caribs might have brought the larouman (or tirite) *Ischnosiphon arouma* from Trinidad or mainland South America to Dominica and Guadeloupe for use in basket-making.

One of the most interesting features of the collective statistics of the island floras is the observation that, whereas the naturalized flora of a small heavily exploited island like Barbados amounts, with 150 species, to c. 30% of the total, the absolute number of naturalized species increases much more slowly in relation to the size of the island than does the native flora. It then seems to reach a steady ceiling, in these examples, of not over 400 species. This means that Jamaica, Hispaniola and Cuba have 14%, 8% and 6% respectively of naturalized species – in inverse sequence to the sizes of these islands.

The total land area of the Lesser Antilles is c. 6500 km², and the total count of vascular plants for the archipelago is 2713. Of these, the number of naturalized species is 374, i.e. nearly 14%. There is no simple explanation for there being a standard limit on the number of naturalized species, and it is not suggested at this stage that there is a common pool of available exotic plants waiting for suitable niches to occur. Common elements in the patterns of human occupation of hitherto unexploited lands may result in correlated responses by vigorous and versatile plants and possibly there is only a limited number of such taxa which can adventively fill the gaps created by alteration or removal of native vegetation. A comprehensive list of all the species of these opportunistic floras occurring in a wide range of tropical situations would enable this question to be answered. A few remarks on the benefits or threats from naturalized species are made in the section later on Factors causing loss of biodiversity – Invasive exotics.

Endemism

The concept of endemism can be applied at three levels of dispersal, namely: (1) Continental – taxa extending to Florida, Central America or northern South America, (2) Antillean – taxa not extending, as far as known, beyond the Caribbean islands, and (3) Greater or Lesser Antillean groups, separately and exclusively. These divisions reflect the main events and trends of the flora history of the Neotropics.

Table 41 includes, for the large islands and many of the small ones, figures for the numbers of endemic species of phanerogams and pteridophytes. Each entry is accompanied by a calculated percentage of species endemism based on the indigenous total. Where the number of naturalized species is known, it seems to be more realistic to consider the native flora in its own right. The effect has been to present the endemism rate at a somewhat higher figure than that usually given (Borhidi 1991: 216).

Of the estimated 13,000 native vascular plant species in the region, about half are single island endemics. About half of these occur in Cuba. Some analyses of inter-island and multiple-island endemism have been made (Howard 1974; Borhidi 1991).

Endemic families
The Caribbean islands have 186 families of native seed plants. A clearly distinct endemic status for any one of them is difficult to maintain. Until recently Picrodendraceae (*Picrodendron*, 3 spp.) was upheld, but this taxon is now considered to belong to Euphorbiaceae. Goetzeaceae represents a group of 4 local genera but possibly with a Mexican element involved. This small group is considered by some authors to be intrinsically solanaceous (Mabberley 1987).

Goetzeaceae	
Coeloneurum	1 sp., Hispaniola
Espadaea	1 sp., Cuba
Goetzea	2 spp., Hispaniola, Puerto Rico
Henoonia	3 spp., Cuba

Endemic genera
There are about 2500 genera of seed plants in the Caribbean. These include 204 endemic flowering plants and one endemic gymnosperm (*Microcycas*). All the endemic genera are confined to the Greater Antilles (see Table 42). Of these, 118 are restricted to single islands. The endemic genera are classified in 53 families of which 19 families have one each, 14 have two and 20 have three or more of such genera (see Table 43). The number of species comprising the endemic genera is 766, about 12% of the total of endemic species and about 6% of all species found in the Caribbean islands. Thus, about 88% of the endemic species belong to non-endemic genera.

Comparison of the representation of endemic with non-endemic genera, and the distribution of the latter, leads to the conclusion that indigenous Antillean floras were recruited from pre-existing continental taxa. Northern migrations from Central America to the Greater Antilles took place largely independently of movements from South America to the more recent volcanic islands of the Lesser Antilles. Some of the most striking speciations have taken place locally in cosmopolitan genera like *Ipomoea*, *Solanum* and *Panicum*,

TABLE 42. GEOGRAPHICAL DISTRIBUTION OF SINGLE-ISLAND ENDEMIC GENERA

	Genera	Families	Species
Bahamas	0	-	-
Cuba	72	20	150
Hispaniola	37	21	58
Jamaica	7	5	13
Puerto Rico	2	2	2
Lesser Antilles	0	-	-
Total	**118**	-	**223**

TABLE 43. IMPORTANT FAMILIES WITH ENDEMIC GENERA

	Endemic genera	Species
Compositae	32	68
Rubiaceae	30	151
Euphorbiaceae	14	77
Leguminosae	13	39
Acanthaceae	7	14
Gramineae	7	11
Melastomataceae	6	87
Orchidaceae	6	11
Palmae	5	*14
Cactaceae	4	31
Myrtaceae	4	21

* *Coccothrinax* (Palmae) has 55 endemic Caribbean species but the genus is technically not endemic because there is one species in Mexico.

(Data largely from Borhidi 1991.)

and pantropical *Psychotria* (Adams 1990). Non-endemic genera, such as *Rondeletia* (Rubiaceae) and *Eugenia* (Myrtaceae), may be responsible for larger numbers of endemic species in the islands than are the endemic genera.

Endemic species
Comparison of Tables 38 and 41 reveals that all islands less than 2000 km² in area have fewer than 3% of endemics, while islands taken as groups, like Bermuda, Bahamas, Cayman Islands and Southern Netherlands Antilles, have slightly higher rates even if the combined area is quite small. Large single islands, over 8000 km², have much higher rates.

The number of single-island endemic species of vascular plants for the whole of the Caribbean is 6550, nearly half of them from Cuba (3224) and slightly fewer (3004) from Hispaniola, Jamaica and Puerto Rico taken together. The remaining single-island endemic species inhabit smaller islands, of which there are many and in which the endemic element in any one of them is negligible. Special habitats, such as exposed volcanic peaks and sulphur springs, where there are special floras, tend to have species which occur on more than one island. No accurate estimate has been made of the number of species endemic to the region as a whole.

Large islands have more endemic species than they have rare or threatened ones. This is presumably because the products of evolution have multiplied in more spacious niches and migrated relatively less often to neighbouring islands. Small islands tend to have fewer endemic species than they have rare or threatened ones. For the Lesser Antilles, nearly 70% of all endemism is spread to two or more islands in the archipelago (Table 44). Only the woody element, comprising distinctive climax forests, shows significant single-island endemism there; roughly

TABLE 44. VASCULAR PLANT ENDEMISM IN THE LESSER ANTILLES (ANGUILLA TO BARBADOS)

Number of islands or island groups to which species are endemic	Number of flowering plant species	Number of fern species	Total species
1	96	8	104
2	45	8	53
3	42	7	49
4	27	6	33
5	27	1	28
6	19	3	22
7	9	4	13
8	8	0	8
9	8	3	11
10	5	2	7
Totals	**286**	**42**	**328**

(Data from Fournet 1978; Howard 1974–1989; Nicolson 1991).

TABLE 45. FLOWERING PLANT GENERA WITH TWO OR MORE ENDEMIC SPECIES IN THE LESSER ANTILLES (ANGUILLA TO BARBADOS)

Genus (Family)	Total species	Endemic species
Stelis (Orchidaceae)	4	2
Lepanthes (Orchidaceae)	2	2
Geonoma (Palmae)	2	2
Asplundia (Cyclanthaceae)	3	2
Anthurium (Araceae)	8	3
Pitcairnia (Bromeliaceae)	5	4
Agave (Agavaceae)	6	6
Peperomia (Piperaceae)	18	4
Pilea (Urticaceae)	12	2
Siparuna (Monimiaceae)	3	3
Ocotea (Lauraceae)	15	7
Inga (Mimosaceae)	6	2
Galactia (Leguminosae)	7	6
Drypetes (Euphorbiaceae)	3	2
Sloanea (Elaeocarpaceae)	5	3
Freziera (Theaceae)	2	2
Clusia (Guttiferae)	5	3
Xylosma (Flacourtiaceae)	3	2
Begonia (Begoniaceae)	6	4
Calyptranthes (Myrtaceae)	5	2
Eugenia (Myrtaceae)	24	6
Myrcia (Myrtaceae)	8	2
Charianthus (Melastomataceae)	4	4
Clidemia (Melastomataceae)	5	3
Miconia (Melastomataceae)	18	11
Tibouchina (Melastomataceae)	5	4
Oreopanax (Araliaceae)	3	2
Schefflera (Araliaceae)	3	2
Cybianthus (Myrsinaceae)	4	4
Marsdenia (Asclepiadaceae)	3	2
Metastelma (Asclepiadaceae)	6	2
Besleria (Gesneriaceae)	6	5
Psychotria (Rubiaceae)	15	3
Rondeletia (Rubiaceae)	5	5
Rudgea (Rubiaceae)	3	2
Lobelia (Campanulaceae)	9	8
Eupatorium (Compositae)	17	9
Verbesina (Compositae)	5	3
Vernonia (Compositae)	4	2

(Data selected from Howard 1974–1989 in the order of appearance in Flora of Lesser Antilles.)

40% of arborescent endemic species are restricted to one island. Endemism in the Lesser Antilles has an overall rate of c. 12% and it is strongly associated with the forests of the wetter higher topography of the volcanic islands.

An independent count of Lesser Antillean endemic species (H. Synge 1991, pers. comm.) has a grand total of 327 species and 9 varieties. Genera with two or more endemic species are listed in Table 45. Ongoing exploration and taxonomic revision will produce revised statistics. A striking example of the consequences is in the recent discovery of the Barbados mastic *Mastichodendron sloaneanum* (Box and Philipson 1951) previously thought to be extinct. It has proved to be not different from the widespread West Indian mastic *Sideroxylon foetidissimum* (Howard 1989; Carrington 1991).

Useful plants

A summary of indigenous species with existing use and future development potential is presented in the various parts of Table 46. An asterisk denotes endemic status.

The natural ranges of Caribbean species, other than those known for certain to be endemic, are often obscure. Situations may change as taxonomic revisions are undertaken. For example, West Indian mahogany (*Swietenia mahagoni*) has only recently been accepted as specifically distinct from all of its continental mainland congeners and is, therefore, now considered to be an endemic species in the region.

A number of potential ornamental species, including shrubs (e.g. *Gesneria* spp., *Lisianthius* spp., *Portlandia* spp.), trees (e.g. *Charianthus fadyenii*, *Thespesia grandiflora*), cacti (e.g. *Rhodocactus cubensis*), epiphytes (e.g. many bromeliads and orchids), climbers (e.g. *Passiflora* spp., *Solandra* spp.) and ferns, all endemic, exist but are too numerous to be listed comprehensively here.

Naturally occurring food plants are extremely few. Whereas the native birds consume many, if not all, kinds of fruit with a fleshy pericarp and are the usual agents for the dispersal of the seeds, only a few of these species of trees and shrubs are used to any extent by people; some of these are listed in Table 46. Several berry-type fruits, acceptable to birds, are toxic to humans, for example the drupes of *Metopium brownii*, the burnwood of Jamaica.

Species which are used for both medicinal and culinary purposes have not been repeated in Table 46, parts 2 and 3.

Most medicinal plants in current use are widely dispersed, as Table 46, part 3, shows. Although admittedly not exhaustive, the list in Table 46 includes no species restricted to the Antilles. The notion is often voiced that native Caribbean flora might yield new medicinal plants and that forests should be conserved pending searches for them. This is fairly unrealistic because the use of "bush" for medicine is so general in the region that, at one time or another, everything is likely to have been tried. Empirical properties are well known and the higher priority is for these species to be tested systematically. Very few known medicinal plants are uncommon and many are ubiquitous weeds. If a new discovery were made in a rare local species, it would at once stand in danger of being over-exploited or possibly wiped out. Rare medicinal plants will only survive in cultivation. The people best fitted to do this are the local practitioners who already grow their own herbs, such as *Chenopodium ambrosioides* or *Aristolochia trilobata*, a tradition which should never be allowed to die out.

The same remarks as made above about the sustainability of medicinal plants also apply to craft materials which may only occur in the wild. In fact, many craft materials are obtained from common and readily available crops, such as coconut or maize, or from widespread introduced ornamentals

TABLE 46. SELECTED LIST OF ECONOMICALLY IMPORTANT PLANTS

1. Important timber trees

Alchornea latifolia	dovewood	Central America, Greater Antilles
Andira inermis	angelin	Tropical America, Tropical Africa
Beilschmiedia pendula	slugwood	Antilles
Brosimum alicastrum	breadnut	Tropical America
Brya ebenus	West Indian ebony	Cuba, Jamaica
Bucida buceras	black olive	Tropical America
Buchenavia capitata	wild olive	Tropical America
Byrsonima coriacea and spp.	hogberry	Tropical America
Caesalpinia violacea		Cuba
Calophyllum calaba	Santa Maria	Antilles
Calycogonium squamulosum		Puerto Rico
Calycophyllum candidissimum	dagame	Cuba
Carapa guianensis	crappo	Tropical America
Catalpa longissima	yokewood	Antilles
Cedrela odorata	cedar	Tropical America
Ceiba pentandra	silkcotton	Tropics
Chimarrhis cymosa	bois riviere	Antilles
Chrysophyllum spp. (4)		Tropical America
Citharexylum fruticosum	white fiddlewood	Tropical America
Clethra occidentalis	soapwood	Tropical America
Copaifera officinalis and spp.	balsam	Tropical America, Cuba?, Trinidad
Cordia alliodora	bois chypre	Tropical America
Cordia gerascanthus	baria, panchallon	Tropical America
Dacryodes excelsa	gommier	Antilles
Didymopanax morototoni	jereton	Tropical America
Dussia martinicensis	bois gamelle	Lesser Antilles
Eschweilera subglandulosa	guatacare	Trinidad, Guyana
Eugenia stahlii and spp.	rodwood	Antilles
Genipa americana	genip	Tropical America
Guaiacum officinale	lignum vitae	Tropical America
Guarea glabra and spp.	redwood, alligator wood	Antilles, northern South America
Guatteria caribaea	bois anglais	Lesser Antilles
Guazuma ulmifolia	bastard cedar	Tropical America
Guettarda valenzuelana and spp.		Antilles
Haenianthus salicifolius and spp.		Greater Antilles
Hernandia sonora	toporite	Tropical America
Hibiscus elatus	blue mahoe	Cuba, Jamaica
Homalium racemosum	cogwood	Tropical America
Hyeronima clusioides and spp.		Antilles, Trinidad
Hymenaea courbaril	West Indian locust	Tropical America
Inga laurina	pois doux	Tropical America
Inga vera		Jamaica, Hispaniola, Puerto Rico
Juglans insularis	nogel	Cuba
Juglans jamaicensis	walnut	Hispaniola, Puerto Rico
Lonchocarpus pentaphyllus	savonette jaune	Antilles
Lonchocarpus sericeus	savonette riviere	Tropical America
Lysiloma latisiliqua	sabicu	Bahamas, Cuba
Magnolia cubensis and spp.	mantequero	Northern Antilles
Manilkara bidentata and spp.	balata	Tropical America
Matayba apetala and spp.	pigeon wood	Tropical America
Meliosma herbertii		Tropical America
Micropholis guyanensis and spp.		Tropical America
Nectandra antillana and spp.	sweetwood	Antilles
Nectandra coriacea	timber sweetwood	Tropical America
Oxandra laurifolia	lancewood	Antilles
Ochroma pyramidale	balsa	Tropical America
Ocotea martinicensis and spp.	laurier	Antilles
Ormosia jamaicensis and spp.	red nickel	Jamaica, Antilles
Pachira spp.	wild chataigne	Trinidad
Petitia domingensis	fiddlewood	Northern Antilles
Phoebe montana (*Cinnamomum montanum*)		Greater Antilles
Piscidia piscipula	dogwood	Tropical America
Pithecellobium arboreum	wild tamarind	Tropical America
Pouteria multiflora and spp.	bullet	Tropical America
Pradosia spp. (3)		South America, Trinidad
Prunus occidentalis	almendro	Central America, Antilles

243

TABLE 46 ...continued

Sapindus saponaria	savonette	Tropical America
Sideroxylon spp. (25)	mastic	Antilles
Simarouba amara and spp.	bitter damson, bois blanc	Tropical America
Sloanea spp.	break-axe	Antilles
Spondias mombin	hog plum	Tropics
Stahlia monosperma		Dominican Republic, Puerto Rico
Swietenia mahagoni	West Indian mahogany	Northern Caribbean
Talauma dodecapetala	bois pin	Lesser Antilles
Tabebuia heterophylla and spp.	poui, roble	Antilles
Terminalia amazonia	yellow olivier	Northern South America, Trinidad
Terminalia latifolia and spp.	broadleaf	Antilles
Tetragastris balsamifera		Antilles
Thespesia grandiflora	montezuma	Puerto Rico
Thespesia populnea	seaside mahoe	Tropics
Trichilia moschata	muskwood	Jamaica, Guatemala
Vitex divaricata	black fiddlewood	Tropical America
Xylopia muricata	lancewood	Jamaica
Zanthoxylum spp.	satinwood	Antilles

2. Species used as food

a. Regularly used

Anacardium occidentale	cashew	Tropical America
Annona muricata	soursop	Tropical America
Annona reticulata	custard apple	Tropical America
Annona squamosa	sweetsop	Tropical America
Bixa orellana	annatto	Tropical America
Carica papaya	pawpaw	Tropical America
Chrysobalanus icaco	coco plum	Tropical America
Chrysophyllum cainito	star apple	Greater Antilles (widely cultivated elsewhere)
Colubrina arborescens	mauby	Tropical America
Colubrina elliptica	mauby	Tropical America
Malpighia emarginata	West Indian cherry	Antilles, northern South America
Mammea americana	mammey	Tropical America
Myrciaria floribunda		Tropical America
Pimenta dioica	pimento	Central America, West Indies
Pimenta racemosa	bay rum tree	South America, West Indies
Psidium guajava	guava	Tropical America
Spondias mombin	hog plum	? Tropical America
Spondias purpurea	Jamaica plum	Tropical America
Ximenia americana	tallow plum	Pantropical
Ziziphus rignonii		Antilles

b. Irregularly used

Acrocomia spinosa and spp.	maccafat	Tropical America
Annona glabra	pond apple	Tropical America, West Africa
Bomarea edulis	salsilla	Tropical America
Byrsonima coriacea	hogberry	Tropical America
Byrsonima crassifolia	savanna serette	Tropical America
Calathea allouia	topitambu	South America, West Indies
Crossopetalum rhacoma	poison cherry	Pan-Caribbean
Coccoloba uvifera	seagrape	Tropical America
Cucumis anguria	West Indian gherkin	Native to Africa
Eugenia ligustrina	rodwood	Tropical America
Ficus spp.		
Garcinia humilis	wild mammee	Antilles
Hymenaea courbaril	West Indian locust	Tropical America
Inga laurina	pois doux	Tropical America
Inga vera	pois doux	Greater Antilles (excluding Cuba)
Maclura tinctoria	fustic tree	Tropical America
Manilkara bidentata	balata	Tropical America
Melicoccus bijugatus	guinep	Tropical America
Miconia spp.	cotelette; fishleaf	Tropical America
Micropholis rugosa		
Opuntia spp.	tuna	Tropical America
Pereskia aculeata	West Indian gooseberry	Tropical America
Pilosocereus royeni	dildo pear	Antilles
Pouteria multiflora	bullet	Antilles
Rollinia mucosa	wild cashimar	Tropical America

TABLE 46 ...continued

3. Medicinal plants

Anacardium occidentale	cashew	Tropical America
Aristolochia trilobata	tref	Central America, West Indies
Bontia daphnoides	olive bush	South America, West Indies
Canella winterana	canella	Florida, Northern Antilles
Capraria biflora	thé-pays	Tropical America
Capsicum spp.	chilli pepper	Tropical America
Cassia alata	wild senna	Tropical America
Cecropia peltata	trumpet tree	Tropical America
Chione venosa	bois bandé	Antilles
Cordia curassavica	black sage	South America, West Indies
Eryngium foetidum	fitweed	Tropical America
Eupatorium odoratum	Christmas bush	Tropical America
Eupatorium triplinerve	japana	Tropical America
Fevillea cordifolia	antidote caccoon	Tropical America
Gossypium barbadense	cotton	Tropical South America
Gouania lupulina	chew stick	Tropical America
Guaiacum officinale	lignum vitae	Tropical America
Guaiacum sanctum	lignum vitae	Northern subtropical America
Hymenocallis spp.	spiderlily	Tropical America
Jatropha curcas	physicnut	Tropical America
Justicia pectoralis	garden balsam	Tropical America
Justicia secunda	St John bush	South America, West Indies
Microtea debilis		Tropical America
Mimosa pudica	sensitive plant	Tropical America
Myrica cerifera	waxwood	Central America, West Indies
Neurolaena lobata	zebapique	Tropical America
Parthenium hysterophorus		Tropical America
Petiveria alliacea	gully root	Tropical America
Phyllanthus amarus	seed-under-leaf	Tropical America
Picrasma excelsa	bitterwood	Greater Antilles, Venezuela
Pilocarpus racemosus		Antilles, mainland Caribbean coast
Piper spp.	jointers	Tropical America
Piscidia carthagenensis	dogwood	Tropical America
Pluchea carolinensis	wild tobacco	Tropical America
Porophyllum ruderale	shiny bush	Tropical America
Quassia amara	quassia	Tropical America
Richeria spp.	bois bande	Tropical America
Ryania speciosa	bois l'agli	Trinidad, Venezuela
Roupala montana	bois bande	South America, Trinidad
Ruellia tuberosa	minny root	Tropical America
Solanum americanum	gouma	Tropical America
Stachytarpheta jamaicensis	vervine	Tropical America
Tournefortia spp.	jigger bush	Tropical America

4. Craft materials

Abrus precatorius	crabs eyes	Tropics
Agave sobolifera and spp.	coratoe, sisal	Tropical America
Caesalpinia bonduc and spp.	nickel	Antilles
Carludovica palmata	jippi jappa	Tropical America
Coccothrinax jamaicensis and spp.	silver thatch	Antilles
Entada gigas and spp.	caccoon	Antilles
Hibiscus elatus	blue mahoe	Antilles
Hura crepitans	sandbox	Tropical America
Ischnosiphon arouma	larouman, tirite	South America, West Indies
Manicaria plukenetii and many other palms	timite	Trinidad

5. Genera of economic importance

The following genera include well known economic plants with species which are indigenous in the region. The value of the native Caribbean species as sources of genetical material has mostly still to be investigated.

Agave (Agavaceae)	Garcinia (Guttiferae)	Phaseolus (Leguminosae)
Canna (Cannaceae)	Glycine (Leguminosae)	Pimenta (Myrtaceae)
Cinnamomum (Lauraceae)	Gossypium (Malvaceae)	Podocarpus (Podocarpaceae)
Corchorus (Tiliaceae)	Ipomoea (Convolvulaceae)	Psidium (Myrtaceae)
Crotalaria (Leguminosae)	Juglans (Juglandaceae)	Solanum (Solanaceae)
Dioscorea (Dioscoreaceae)	Juniperus (Cupressaceae)	Vanilla (Orchidaceae)
Erythroxylum (Erythroxylaceae)	Magnolia (Magnoliaceae)	Vigna (Leguminosae)
Eugenia (Myrtaceae)	Persea (Lauraceae)	Xanthosoma (Araceae)

Note: * endemic species

or shade trees, such as flamboyant (*Delonix regia*) or red bead tree (*Adenanthera pavonina*). Other materials, used nowadays, include synthetic manufactured items like plastic sheeting and nylon cord. A healthy craft industry could be based on the added value of the skill of the crafts people rather than on any special merit of local raw materials.

The increased use of imported items could, on balance, result in the conservation of native species of plants. The depletion of natural populations of rare trees, such as *Ormosia jamaicensis* which is known to have been felled to obtain the beautiful seeds more easily, could be accelerated by the impaired reproductive potential caused by removing seeds and leaving the survivors to take their chance in natural competition. The seeds germinate readily in garden conditions and the tree grows vigorously in cultivation. Most carving and turning woods, like blue mahoe, *Hibiscus elatus* grow well enough in plantations to meet all demands.

In the past, dyestuff and rope have been obtained from the bark of mangroves and *Hibiscus elatus* and *H. tiliaceus* (*H. pernambucensis*) causing great damage to the trees. There are many precedents for the importation of synthetic imported materials for use in unique local crafts, for example the Kente cloth of Ghana.

Some native plants which are sources of insecticides have not yet, or have only sparingly, been brought into cultivation. They include *Dioclea mollicoma* and related species, *Piscidia* spp. and *Ryania speciosa*.

Factors causing loss of biodiversity

"Future developments in the West Indies will take place in environments which are already heavily modified, frequently degraded, and in which the natural resources are already depleted. There are no truly natural environments left in the Caribbean, only those which have survived 200–300 years of human impact" (Bacon 1985).

Natural environmental factors

Periodic earthquakes, volcanic eruptions and hurricanes are intrinsically associated with the physical instability of West Indian environments. Their occurrences have been reviewed by Tomblin (1981). Least predictable are the geological disturbances, but volcanic activity no longer affects the northern part of the region directly. Scars of the famous earthquake of 1692, which destroyed the town of Port Royal in Jamaica, are detectable where landslips still carry incompletely restored vegetation.

Volcanoes
Volcanic activity in the Lesser Antilles in this century was reviewed by Howard (1962b). The consequences of recent eruptions have been described for Guadeloupe (Howard, Portecop and Montaignac 1981; Sastre, Baudoin and Portecop 1983; Sastre 1985), Martinique (Sastre and Fiard 1986) and St Vincent (Rowley 1979). There are about 30 active or potentially active volcanoes in the Lesser Antilles but major events in this century have only taken place at the highest peaks of Guadeloupe, Martinique and St Vincent. The Antillean eruptions are characterized by the explosive emission of clouds of ash with boulders, and at times glowing avalanches ("nuées ardentes"), but not by lava flows. Falling stones and ash defoliate trees and bury shrubs and herbs; recovery occurs readily from persistent corms in plants like *Heliconia*, while some trees, such as breadfruit, regenerate easily from their roots. Glowing avalanches are more destructive; large volumes of gas, charged with fine ash, pebbles and boulders, flow downhill at speeds of up to 120 km/hr and temperatures within the cloud have been estimated to be of the order of 600–800°C. The force is great enough to topple walls and uproot trees, and the heat kills all living things. Following a major eruption, the vegetation returns to an appearance of normality after about 25 years (Rowley 1979).

Howard (1977) has suggested that some unusual species may have been eliminated by volcanic activity, but Sastre (1978) has shown that an endemic flora persists. If this is threatened by further volcanic activity, it is probably also at risk through tourism. Several named localities on the peaks of Guadeloupe have been recommended for protection (Sastre 1979b).

The vegetation close to permanent active fumaroles and sulphur springs, as in Montserrat, Dominica and St Lucia, is specialized and limited to a few species – *Clusia* and *Pitcairnia* are among the plants tolerant of sulphurous gases and hot, polluted water.

Hurricanes
There have been some disastrous hurricanes in recent years, events which have in a few hours destroyed more trees than decades of misuse of forest. These happenings are irregular but are much more frequent than geological upheavals. Jamaica is supposed to be at risk on average once every 9 years, Puerto Rico once every 10 years (Wadsworth and Englerth 1959), and Dominica once every 15 years (Neumann *et al.* 1978). Tobago has experienced only three in 200 years and Trinidad has no record of any event more severe than an occasional tropical storm.

The effects on plant life have been rarely recorded in as much detail as those of Hurricane David. This storm struck Dominica on 19 August 1979 and has been estimated to have been the most intense hurricane in the Caribbean during this century. Wind velocities averaged 92 km/hr, with peaks of 241 km/hr, over a minimum duration of 10.2 hr. On one-third of the island (246 km²), 5,100,000 trees were damaged (Lugo *et al.* 1983). The effects were seen to be greater on some species than others, but natural regeneration was rapid and this was attributed to the wetness of the localities where most damage was done. Groome (1970) studied some trees damaged by a hurricane in Grenada and found that some evergreen trees could not survive removal of all their leaves at one time, because the normal rate of replacement was too slow. It may well be that the frequency and intensity of hurricanes are environmental factors reflected in the growth characteristics of the trees and the ecology of climax forest in these islands.

The situation in Tobago seems to have been somewhat different. Hurricane Flora struck the Main Ridge Forest Reserve on 30 September 1963. This forest had been under protection for water supplies since 1765. Destruction was virtually complete: of the upper storey trees 75% were blown over and the remainder had their crowns completely removed or irreparably damaged. It was estimated that very few trees of the original canopy would reform a normal crown (Dardaine 1974). The ecological consequences were profound due to exposure of soil, drying out and erosion. The re-afforestation programme included planting of exotic species such as *Pinus caribaea, Tectona* (teak), *Swietenia* (mahogany), *Araucaria* and *Khaya* on a trial basis. It was estimated that replanting the whole of some 44.5 km² would take 18 years. Natural regeneration would surely have overtaken this programme, and it would be very revealing to have an up-to-date statement. It is possible that environmental harmony, especially of the quality needed for watershed protection, may be reached only through natural processes.

Heavy rainfall accompanies hurricanes and tropical storms, and may, especially in places where forest cover has been depleted, cause landslips on steep hillsides and result in flooding and further damage (Box 1939b). This type of environmental injury is little different from that brought about by road and reservoir construction and surface mining. Successions on the raw substrates often comprise tangled thickets of strong-stemmed gleicheniaceous ferns, colonization by woody species being slow.

Intensive studies of the regeneration of montane forest in Jamaica, following the destruction brought about by Hurricane Gilbert on 12–13 September 1988, are being carried out (Tanner 1989, pers. comm.). A collection of case studies documenting the damage to natural ecosystems and the initial recovery process from three recent hurricanes in the Caribbean has been published by Walker *et al.* (1991).

Human-made factors

Mining and quarrying
Threats to the landscape in the Caribbean arise mostly from mining activities. Whatever type of mining is carried out, vegetation is cleared and there is always some surface disturbance either from stripping operations or dumping of tailings.

An ancient activity was the collection of guano and phosphate from usually isolated rocky islands such as Alta Vela (Howard 1955, 1977), Navassa (Burne, Horsfield and Robinson 1974) and Redonda (Tempany 1915; Box 1939a; Howard 1962a). Goats, sheep and rats often came with the miners and persisted after the workings were abandoned, establishing an equilibrium with an altered vegetation.

Processing of bauxite to alumina produces large quantities of residual "red mud", containing excess sodium hydroxide with a pH in the range 12–13. It is totally hostile to any form of plant life. During the main years of alumina production in Jamaica from the early 1950s, this waste was pumped into mined-out pits and dammed valleys. Several years of leaching by rain and seepage led to conditions that plants could tolerate. In 1972, the oldest pond at the Kirkvine operation in Jamaica, to which no fresh residues had been added since 1957, had acquired an adventive vegetation of 166 species dominated by ferns, Compositae, grasses, and other mostly herbaceous plants. By 1975, successional changes showed primary colonizers replaced by species with different requirements and tolerances, but all were common, widespread, usually weedy species (Adams and Lawrence 1972; Adams and Ramjus 1975). Comparable mining operations, for example in Cuba and Hispaniola, produce landscapes from which natural vegetation has been removed permanently.

Riversides and beaches are often exploited destructively to obtain building sand and gravel and these removals may have secondary effects in the form of erosion, flooding, pollution and loss of visual amenity (Sastre and Portecop 1985). Extensive damage has occurred along the Aripo River

in Trinidad posing serious problems in savanna areas and downstream by silting.

Air and water pollution
A particular aspect of these universal hazards is, in the Caribbean, smothering of vegetation by bauxite or alumina dust especially where these products are loaded onto ships (Sidrak in Hudson, c. 1973). Large quantities of soluble carbohydrates result from sugar-cane extraction. The waste, called "dunder", is disposed of into drains and rivers, and causes eutrophication downstream. This is seasonal but, combined with rural runoff and urban sewage, coastal environments come under stress, particularly around the larger islands (Wade 1976; Schroeder and Thorhaug 1980; Rodriguez 1981; Provan, Wade and Mansingh 1987).

Fire
The natural vegetation of the West Indian islands has not evolved the fire-tolerant life forms that are found throughout much of tropical Africa or continental America, so fire is comparatively more destructive in the Caribbean. The few savanna species dependent on fire for their survival in competition with larger life forms are restricted in distribution and, generally, the woodlands and forests have experienced fire as a significant ecological factor only since human exploitation gained momentum within the past 500 years. Secondary savannas such as the *Maximiliana* (cocorite)/*Imperata* thickets and glades of the Central Plain of Trinidad are maintained and expanded by fire. They represent degraded forest on land of low natural fertility.

Fire is commonly used to clear land for agriculture and settlements, to "clean" undergrowth in forests, and to encourage new growth in savannas and bushland in the dry season for pasturage. Fires are also started in the belief that they will bring rain. Often such fires are allowed to spread unchecked into neighbouring areas. Efforts to protect forests or plant trees in grassy places are often thwarted by deliberate setting of fires, even in statutory Forest Reserves.

As most Caribbean islands, except Trinidad, lack fossil fuels and have very limited development of other sources of energy, the inhabitants are strongly dependent on fuelwood and charcoal.

Agricultural development
Soon after the Columbian discovery, a process of modification of the vegetation began which continues to this day. Plants suitable for human food are of negligible occurrence in the indigenous floras of any of the islands and those that occur are also found on mainland America and probably all originated there (see Table 46, part 2). Some staples such as cassava, maize and pulses had been brought from the mainland by Arawaks and Caribs at an earlier time. Citrus, ginger and plantains were among the first useful plants to be brought by Europeans from Asia. Sugar-cane was introduced into Santo Domingo in 1516. Crops to be grown for export, including indigo, cotton and tobacco, were established in Barbados by 1640. These crops replaced native woodlands, the timber being used for building and fuel for the sugar factories.

It was soon realized that clearing of forest had an adverse effect on water supplies, either by affecting the rainfall or by allowing less retention and greater loss through evaporation. Shifting cultivation on rainy mountain sides had serious effects on stream-flow and caused silting, comparable with damage done by hurricanes. Legislation to protect water supplies goes back to 1721 in Antigua and 1791 in St Vincent (Beard 1949).

Clearing of land was greatest at the time of the Napoleonic wars. After that, erosion and the depression of sugar as a commodity caused poor and degraded areas to be abandoned and to revert to pasture; some sugar estates began to wear out and were retired to pasture. Ultimately, various types of secondary woodland grew up. The history of land-use can be inferred through the presence in the woodlands of trees such as logwood, *Haematoxylum campechianum*, introduced from Central America in the early 18th century, and the stone walls of ancient field boundaries overgrown with woody vegetation. After the abolition of slavery, people scattered into the hills surrounding the plantations and settled there leading to further degradation of forests.

The philosophy that natural products of the land are divine bounty to be used directly or shared with whoever is in need is prevalent in the West Indies. Such belief tends to blur refinements of property and ownership, resulting in only occasional prosecutions and even rarer convictions for illicit felling of trees or environmental vandalism. Long-term culling for charcoal and uncommon forest products causes progressive damage. Tree fern trunks are removed and cut into blocks for the cultivation of ornamental orchids. Yam and bean poles are gathered from forests and these removals are selective of size-class, so that regenerative cycles are broken.

Coppice regrowth of trees cut for poles or firewood may develop multiple trunks producing an artificial shrub-like habit (Adams 1972). Some species of *Coccoloba* grow like this without being cut, as do the clustering *Bactris* palms, but more often the presence of coppice shoots is evidence of human disturbance. Other observations, such as paucity of field and ground layer species, openness of undergrowth and leaf-litter with saprophytes, may indicate the primary or mature status of forest (Adams 1977).

Tourism

The growth of tourism in many of the islands over the past 50 years has resulted in hotel developments along coastlines having white-sand beaches. This has often meant complete change to the landscape locally, involving the removal of natural vegetation and the planting of ornamental trees, shrubs and grass for lawns and golf courses. Mosquito control and marina developments have eliminated mangroves and littoral thickets in many places. New roads have often been constructed to give access to coastal areas which previously could only be reached on foot or by sea.

The movement of people from rural areas to towns and resorts, with the lure of employment opportunities in servicing the new tourism, has coincided with the decline of export-based plantation agriculture. Increased demand for fresh fruit and vegetables has often resulted in unacceptable levels of cultivation on unsuitable land (Rojas *et al.* 1988).

Introduced animals

The consequences of the introduction of grazing and browsing animals to Antigua, Barbuda, and Anguilla have been described by Harris (1965). The principle that man has priority in all matters relating to his own needs and comfort takes for granted that his attendant animals may also live off the land. Cattle are often allowed to roam free and damage caused by

them is usually impossible to redress. As far back as the buccaneers of the 17th century, cattle, goats and pigs had become feral on many islands. Predatory importations, as for example cats and mongoose, are destructive of other animals; several small mammals, mainly herbivores, birds and reptiles have become extinct because of them. A recent attempt to re-introduce the coney (*Geocapromys brownii*) into western parts of Jamaica have been unsuccessful (Wilkins 1991). Donkeys and deer (Harris 1965: 61) browse and alter the species relationships of secondary formations; introduced monkeys in Barbados, Grenada and Nevis live in balance with introduced citrus, guava and other fruit trees in artificial communities.

Other biological hazards

The worst environmental disaster to hit Bermuda in recent years has been the attack on the dominant endemic cedar trees (*Juniperus bermudiana*) by accidentally introduced scale insects. This has resulted in the loss of some 90% of the cedars. The infestation began in about 1943 and spread steadily. Although various attempts at biological control failed, some trees survived, apparently through resistance (Challinor and Wingate 1971).

Measures to control introduction and spread of pests and disease vectors are of paramount importance to agricultural authorities in the islands but sometimes protective practices may entail the destruction of indigenous species. Examples are the compulsory removal of malvaceous plants, such as *Thespesia*, on cotton-growing islands to reduce alternative hosts of insect pests. The destruction on sight of *Hippobroma longiflora* was at one time ordered in Haiti because it is so poisonous to stock. The systematic removal of the coastal manchineel tree *Hippomane mancinella* is often undertaken near resort developments because of the caustic sap, but this tree is not usually regarded as a hazard in Barbados nor on the French islands.

Invasive exotics

The successful naturalization of invasive exotic species in lowland areas, from which the original forest has long since been removed, may improve or diminish the environment in human terms. For example, the early introduction of guinea grass *Panicum maximum* was beneficial in providing cover and fodder. However, the recent invasion of Puerto Rico, Haiti, Cuba and Trinidad by the Asiatic grass *Dichanthium annulatum* (*Andropogon annulatus*) is unfortunate. This species competes successfully with guinea grass and is also one of the least palatable of pasture grasses (Goberdhan 1971). A greater disaster has been the introduction of the extremely aggressive shrub *Dichrostachys cinerea* from the savannas of Africa into Cuba and Marie Galante; the plant regenerates vigorously from the smallest fragments of roots. "A noxious tree that is a most aggressive invader" in the Bahamas is *Melaleuca quinquenervia* introduced from Australasia; *Schinus terebinthifolius* from Mexico is also invasive in those islands (Correll and Correll 1982).

Introductions from the Old World into the mountains of Hispaniola and Jamaica are numerous. Those which threaten native vegetation include mock orange *Pittosporum undulatum* from Australia in Jamaica (for example, see Data Sheet on the Blue and John Crow Mountains – CPD Site Cb10). Rose apple (*Syzygium jambos*) is a problem in Hispaniola and Jamaica, and already forms extensive secondary woodlands and shades out competitively most species of native shrubs and undershrubs; oddly, this species has become one of the most favoured support plants for a beautiful and uncommon local orchid *Comparettia falcata* in Puerto Rico (Rodriguez-Robles, Ackerman and Melendez 1990). Ginger lily *Hedychium gardnerianum* and its natural hybrids, and molasses grass *Melinis minutiflora*, the latter introduced into Jamaica about 1925 (Adams 1972), are strongly invasive and colonize disturbed areas rapidly. They appear to inhibit natural successions which might lead to the restoration of woody cover.

The local establishment of temperate herbaceous exotics in mountainous areas can also be exemplified by the situation in the Port Royal Mountains of Jamaica where in the mid-19th century many European and North American herbs were introduced accidentally with fodder imported to supplement feed for horses at a military hill station. None of these plants, which include *Plantago* spp., *Rubus* spp., *Rumex* spp., *Stellaria media*, *Taraxacum officinale* and several festucoid grasses, appear to pose much of a threat to indigenous flora in places which have already been developed, as footpaths and vegetable gardens, and they are not invasive.

Principles recently set out as suggestions for managing alien plants in high-altitude places (Dickson, Rodriguez and Machado 1988) are entirely worthy, but may be too exacting and expensive to implement effectively in the Caribbean region, even in a National Park system.

Conservation

Recognition and designation of protected areas

Encouragement to conserve natural fauna and vegetation on a formal basis in the Caribbean began largely with individuals and small groups with an amateur leaning towards natural history. Effective non-governmental organizations (NGOs) with concern for the preservation of native plants and animals are few. The Trinidad Field Naturalists' Club was founded in 1891 (the addition of "and Tobago" being made in 1975). The interests of the most of the founder members were predominantly zoological. Such groups have been formed in other islands from time to time, with variable levels of activity, continuity and support (e.g. Putney 1983).

Towards the end of the last century, it became clear that natural forests should come under some form of management (Hart 1891). Forest Departments began to be set up in the English-speaking islands some 25 years later, and eventually Forest Reserves were created. Although foresters were generally interested in only a few economic trees, the office of Chief Conservator of Forests was frequently combined with the function of Game Warden and this was a potentially useful link to promote respect for plant and animal wildlife equally.

The natural consequences of more coordinated concern were that some of the Forest Reserves came, in due course, to be looked at as protected areas. This tended to lead to the preservation of vegetation for environmental reasons, besides controlling the utilization of economic species and preventing the extinction of rare ones. Since international interests and their agencies have been giving encouragement to the establishment of National Parks and other protected areas, especially in tropical countries with significant amounts of natural vegetation still remaining, governments and NGOs have responded by submitting lists of designated and proposed locations. Unfortunately, in some cases, the selections encompass areas ranging from totally artificial to natural forests and the conservation ethic has been stretched to include units which are protected merely for their capability of producing commercial lumber, some even comprising a majority of exotic species. The Jamaican authorities list 126 Forest Reserves in Category VIII, 33 of them being less than 10 ha in area (see Table 47).

A special case of an area of relict forest comprising species, rare in one island but not globally threatened, was among the earliest to receive notice; this was Turner's Hall Wood in Barbados (Gooding 1944). The impact on attitudes of the realization that a flora may be under threat in a particular island, even if the species are common elsewhere, is significant for conservation generally. Many of the Caribbean islands do not and never have had a very distinctive flora, but what there is must still be worth protecting. Neither Turner's Hall Wood nor the Mason River Savanna in central Jamaica, which has several rare endemic species (Proctor 1970), have any association with Forest Reserves and their future management depends on other support.

The historical preoccupation with forest and present-day concern for tropical rain forest might divert attention from other vegetation of equally deserving botanical merit. For these other formations in the Caribbean to be brought to proper notice, it is necessary to take a broad view of the region as a whole to determine those areas which might have rare or limited physiographical characteristics. We are aware of: the special floras of serpentine soils in Cuba and Jamaica; the restricted existence of siliceous sand savannas outside Cuba; the excessively high rainfall locations of eastern Jamaica or highest ground of the Windward Islands; the general scarcity throughout the region of bodies of fresh water. There may be hitherto undiscovered discrete geographical units where accelerated local evolution has taken place. Some of them are not necessarily remnants of large disappearing formations but rather less obvious sites of unexpected interest, for example the Harris Savanna of Jamaica, or the Grand Etang crater lake of Grenada. Perhaps because they are not very distinctive or are not under obvious threat, they may be difficult to promote as protected areas.

Besides unique geological sites, non-conformity is a clue to the possibility of finding important biological communities and, by contrast, uniform situations are unlikely to furnish them. This means that, from the terrestrial flora viewpoint, inland areas may be of greater potential botanical interest than coastal mangroves, beach thickets and salinas.

In general, expectations of permanent or even moderately ongoing protection in these islands are weak. The principal reason for this is reluctance of local governments to make commitments to long-term funding. Whereas detailed surveys to enable the recognition and demarcation of National Parks or Wildlife Sanctuaries have been financed by international agencies (e.g. for Dominica, Honychurch 1978), follow-up from the same sources has been very limited or lacking. In view of this, it is difficult to know at any one time if designated areas are fully respected or really protected.

TABLE 47. SUMMARY OF TERRESTRIAL PROTECTED AREAS IN THE CARIBBEAN

Name of area	IUCN category	Number of terrestrial protected sites	Area protected (km²)
ANGUILLA			
No terrestrial protected areas have yet been established.			
ANTIGUA and BARBUDA			
National Parks	II	3	c. 41
Park Reserves	?	4	?
BAHAMAS			
Managed Nature Reserve	IV	1	18
National Parks	II	4	1216
BARBADOS			
No terrestrial protected areas have yet been established.			
BERMUDA			
Nature Reserves	?	12	c. 1
Preserves	IV	2	125
In addition there are many small parks and private reserves, mostly <0.1 km² each.			
BRITISH VIRGIN ISLANDS			
Bird Sanctuaries	IV	4	11
Forest Parks	II	5	1.5
CAYMAN ISLANDS			
Ecological Zone	I	1	17
Reserve	IV	1	3
CUBA			
Biosphere Reserves	IX	4	3335
Ecological Reserves	IV	3	76
Faunal Refuges	I, IV	9	1419
Integrated Management Areas	V, VIII	10	10,006
Managed Flora Reserves	IV	9	247
National Forests	?	3	>872
National Parks	II	7	1089
Nature Parks	V	4	1844
Nature Reserves	I	7	305
In addition there are 19 Touristic Nature Areas (mainly keys and lagoons) covering c. 1411 km² and 14 "Other Areas" covering 1188 km². A further 7 sites (covering 222 km² in total) are currently proposed for protection.			
DOMINICA			
Forest Reserves	VIII	2	92
National Parks	II	2	139
Protected Forest	VI	1	3
A proposed National Park covers c. 65 km².			
DOMINICAN REPUBLIC			
National Parks	II, V	12	5662
Natural Scientific Reserves	I, IV, V	7	676
In addition there is a Bird Sanctuary; 24 other sites are currently proposed for protection.			
GRENADA			
Forest Reserve	VIII	1	6
GUADELOUPE			
National Park	II	1	173
HAITI			
Natural National Parks	II, V	4	97

Name of area	IUCN category	Number of terrestrial protected sites	Area protected (km²)
JAMAICA			
Forest Reserves	VIII	126	>1000
Game Reserve	VIII	1	2
Natural Reserve	IV	1	0.5
MARTINIQUE			
Nature Reserve	I	1	5
Regional National Park	V	1	702
MONTSERRAT			
No terrestrial protected areas have yet been established – a National Park is proposed.			
NETHERLANDS ANTILLES (ARUBA, BONAIRE, CURAÇAO)			
National Parks	II	2	78
Ramsar Wetlands	R	5	19
PUERTO RICO			
Biosphere Reserves	IX	2	173
Commonwealth Forests	IV	14	243
National Forest	VIII	1	113
Nature Areas	IV	8	18
National Wildlife Reserves	IV	3	10
Nature Reserves	IV	7	134
Wildlife Refuges	IV	4	22
ST KITTS-NEVIS			
National Park	II	1	26
ST LUCIA			
Forest Reserves	VIII	15	c. 67
Nature Reserves	IV	2	5
Reserves	IV, VIII	2	1
ST VINCENT			
Forest Reserves	VIII	3	c. 1
Reserve	IV	1	44
TRINIDAD			
Scientific Reserve	I	1	18
Wildlife Sanctuaries	IV	8	2
Several National Parks, Scientific Reserves, Nature Reserves and Nature Conservation Reserves are proposed.			
TOBAGO			
Nature Reserve	I	1	7
Wildlife Sanctuaries	IV	2	c. 1
A National Park, a Scientific Reserve and a Nature Conservation Reserve are proposed.			
TURKS AND CAICOS ISLANDS			
10 National Parks (109 km²), 9 Nature Reserves (59 km²) and 1 Ramsar Site (2500 km²) comprise mainly marine and areas of importance for birds; some protected areas are complete island systems with littoral and sublittoral scrub communities.			
U.S. VIRGIN ISLANDS			
Biosphere Reserve	IX	1	61
National Park	II	1	53
National Wildlife Refuge	IV	1	1

Table 47 summarizes the designated protected areas of the Caribbean islands. A review of the protected areas system and conservation legislation in the Caribbean region is given in IUCN (1992).

Identification and population size of species at risk

Up to the end of the last century the only Flora in English was the general West Indian work of Grisebach (1859–1864). The initiation of Flora-writing projects for the larger islands (Jamaica in 1910; Puerto-Rico in 1923; Trinidad in 1928, etc.) produced independent awareness of whole floras, although some of these works are still incomplete or in great need of revision.

Deficiencies in the coverage of several of the Floras and the likelihood of a shortage of local personnel capable of making accurate identifications, lead to the possibility of errors of judgement in assessing claims for conservation status. One of the most difficult aims to fulfil is to estimate population size and area of dispersal of rare species. In countries where there are few people with appropriate specialized knowledge, decisions may be made on an *ad hoc* basis, rather than with good information and objective comparative analysis (Adams and Baksh 1981).

Centres of plant diversity and endemism

Cuba

Cb1. Coast from Juragoa to Casilda Peninsula; Trinidad Mountains; Sierra del Escambray

Area: c. 2700 km² in south-central Cuba. Altitude: 0–1156 m.

- ❖ Vegetation: Succulent and evergreen scrub thickets, including cacti; evergreen and semi-deciduous forests; seasonal and montane forests at higher elevations.
- ❖ Flora: >1200 vascular plant species, of which c. 40 endemic; 7 endemic genera.
- ❖ Threats: Proximity to towns and intensive cultivation pose threats. Native vegetation survives mostly in the mountains and on the coast, cays and less fertile land; otherwise native vegetation is fragmented.
- ❖ Conservation: Escambray Integrated Management Area (1870 km²) (IUCN Management Category: VIII); Topes de Collantes Natural Park (122 km²) (IUCN Management Category: V).

Cb2. Oriente

Area: 18,000 km², i.e. the Eastern Sub-Province of Borhidi (1991). Altitude: 0–1974 m.

- ❖ Vegetation: Seasonal evergreen forests, montane and submontane rain forests, pine forest, semi-evergreen scrub and semi-desert areas along the south coast.
- ❖ Flora: >3000 vascular plant species, of which more than 1500 strictly endemic; 24 known endemic genera. "This area is considered to be the cradle of the Cuban flora and, together with western Hispaniola (Haiti), the most prominent centre of speciation in the Antilles" (Borhidi 1991: 349). Referring to the Nipe-Baracoa Massif, which is within the sub-province: "The richest flora of the Caribbean, and one of the richest floras of the World" (Borhidi 1991: 351). Useful plants include pines, *Podocarpus*, palms and many plants of potential ornamental value.
- ❖ Threats: Removal of timber and fuelwood, mining and tourism. However, rugged topography, erosion and excessive drainage render most of the remaining important floristic areas unsuitable for agriculture.
- ❖ Conservation: In the south – Gran Parque Sierra Maestra (5270 km²), an Integrated Management Area (IUCN Management Category: VIII), comprising the National Parks of Desembarco del Granma (258 km²), Turquino (175 km²) and Gran Piedra (34 km²) (all of which are IUCN Management Category II) and the Biosphere Reserve of Baconao (846 km²) (IUCN Management Category: IX). In the north-east – Cuchillas del Toa Biosphere Reserve (1275 km²) (IUCN Management Category: IX).

Cb3. Pinar del Río

Area: 1150 km², situated in the western mountains. Altitude: 0–692 m.

- ❖ Vegetation: Coniferous forests, seasonal forests, forests and thickets on karstic limestone and serpentine, succulent and thorn scrub, swamp and oligotrophic lagoons and mangroves.

- ❖ Flora: Estimated 500 endemic vascular plant species and 16 endemic genera. Of particular interest is the endemic cycad *Microcycas calocoma*. Useful plants include pines and oaks.
- ❖ Threats: Removal of fuelwood, tourism. Forests on lower fertile soils degraded; thickets on rugged terrain and cliffs minimally at risk.
- ❖ Conservation: Mil Cumbres Integrated Management Area (166 km²) (IUCN Management Category: VIII), which includes **Cajálbana Tableland and Preluda Mountain region** (see Data Sheet); Sierra del Rosario Biosphere Reserve (100 km²) (IUCN Management Category: IX); Viñales National Park (containing Cuba's richest "mogotes" – limestone towers) (134 km²) (IUCN Management Category: II); Península de Guanahacabibes Biosphere Reserve (1015 km²) (IUCN Management Category: IX); Cabo Corrientes Natural Reserve (16 km²) (IUCN Management Category: I); Sur Isla de la Juventud Natural Park (800 km²) (IUCN Management Category: V); Penínsular de Zapata National Park and the Cienaga de Zapata Natural Reserve.

(Source: A. Leiva and R. Berazaín 1993, *in litt.*).

Cb3. (in part) Cajálbana Tableland and Preluda Mountain region

– see Data Sheet.

Dominica

Cb4. Morne Trois Pitons National Park

Area: 70 km², in south-central interior of Dominica. Altitude: c. 600–1383 m.

- ❖ Vegetation: Rain forest, montane forest, secondary palm brakes, elfin woodland. Other habitats provided by cold and hot (volcanic) lakes, sulphur springs, fumaroles.
- ❖ Flora: 500 vascular plant species. Plants provide craft materials; some potential ornamental species.
- ❖ Threats: Tourism and water supply development.
- ❖ Conservation: The area is a National Park which protects a vital watershed, but there are limited management resources.

Dominican Republic

Cb5. Cordillera Central

Area: >1530 km² already in designated protected areas in parts of Provinces La Vega, Santiago, San Juan de la Maguana, Azua and Valverde. Altitude: 1000–3087 m.

- ❖ Vegetation: Seasonal evergreen forest, submontane and montane rain forests with broadleaved hardwoods; *Prestoea montana* palm forest; *Pinus occidentalis* forest at higher elevations.
- ❖ Flora: Estimated 1500 vascular plant species, of which 25–30% probably endemic (T.A. Zanoni 1993, *in litt.*). Useful plants include timber trees, such as *Pinus occidentalis* (western pine), *Magnolia* spp. and Lauraceae.
- ❖ Threats: Logging, cattle grazing, tourism.

- Conservation: José Armando Bermúdez National Park (766 km²) and José del Carmen Ramírez National Park (738 km²) cover substantial parts of the upper Cordillera Central. Two Scientific Reserves (IUCN Management Category: IV) occur within the parks: Valle Nuevo (409 km²) and Ebano Verde Natural (23 km²).

Cb6. Los Haitises

Area: region covers 1315 km² in north-east Dominican Republic. Altitude: 0–380 m.
- Vegetation: Forests over limestone; however, only c. 10% of the native forests remain in and around the National Park.
- Flora: >500 vascular plant species, of which 138 are island endemics.
- Threats: Cattle grazing, agriculture, some logging, relocation of local population outside of National Park area.
- Conservation: Los Haitises National Park (IUCN Management Category II) currently covers 208 km², but there are plans to extend this to include an area along Samaná Bay and mangroves at the west end of bay. There are also plans to include much of the area and the Samaná Peninsula in a Biosphere Reserve.

(Source: T.A. Zanoni 1993, *in litt*. See also Zanoni *et al.* 1990).

Cb7. Sierra de Neiba

Situated in western Dominican Republic. Altitude: 1000–2000 m.
- Vegetation: Montane broadleaved forest, pine forest, cloud forest.
- Flora: 300–400 vascular plant species, of which 25–30% are endemic. Useful plants include timber trees (such as *Pinus occidentalis*) and tree ferns (*Cyathea* spp.).
- Threats: Clearance for agriculture, some grazing, removal of trees for timber and firewood.
- Conservation: Proposed National Park. Possibility of inclusion within a Biosphere Reserve in the future.

(Source: T.A. Zanoni 1993, *in litt*. See also Santana Ferreras 1993.)

Haiti

Cb8. Pic Macaya

Area: 55 km², in south-western Haiti in the Massif de la Hotte. Altitude: 900–2347 m.
- Vegetation: Wet broadleaved forest on limestone (900–c. 1250 m), complex mosaic of pine forest and cloud forest (c. 1250–2347 m).
- Flora: 665 vascular plant species (of which c. 30% are endemic to Hispaniola) and 165 bryophyte species so far recorded within National Park area (see below) (Judd and Skean 1987; Judd, Skean and McMullen 1990).
- Threats: Clearance for agriculture, some grazing, removal of trees for timber and firewood, and charcoal production.

- Conservation: Parc National Pic Macaya (20 km²) (IUCN Management Category: II); possible inclusion within a Biosphere Reserve (Sergile, Woods and Paryski 1992).

Cb9. Morne La Visite

Area: 20 km², in south-eastern Haiti in the Massif de la Selle. Altitude: 1600–2282 m.
- Vegetation: Pine forest and cloud forest.
- Flora: 337 vascular plant species (of which c. 34% are endemic to Hispaniola) and 95 bryophyte species so far recorded (Judd and Skean 1987).
- Threats: Clearance for agriculture, some grazing, removal of trees for timber and firewood, and charcoal production.
- Conservation: Parc National Morne La Visite (IUCN Management Category: II).

(Source: W.S. Judd 1993, *in litt*.).

Jamaica

Cb10. Blue and John Crow Mountains

– see Data Sheet.

Cb11. Cockpit Country

– see Data Sheet.

Trinidad

Cb12. Aripo Savannas Scientific Reserve

Area: 18 km², situated in the east-central lowland area of Caroni Plain. Altitude: 35–40 m.
- Vegetation: Marsh forest, palm-marsh and savanna.
- Flora: No available total but herbaceous flora includes at least 14 species of *Utricularia* (bladderworts), 5 species of Xyridaceae, and Eriocaulaceae, *Mayaca*, *Drosera* and ground orchids, not otherwise represented in the southern Caribbean. Useful plants include *Mauritia setigera* (Moriche palm); other palms for craft materials include *Euterpe precatoria* (manac), *Jessenia oligocarpa* (palm real) and *Manicaria plukenetii* (timite). Past forestry practices within the Long Stretch Forest Reserve, of which the savannas form a part, have removed most of the commercially valuable timber.
- Threats: Quarrying for sand and gravel takes place along the adjacent Aripo River and sometimes encroaches on the protected area. Fires occur regularly during the dry season in the grassy areas but their long-term significance is unknown; indigenous perennial herbaceous species are mostly naturally fire-adapted. Some hunting of indigenous mammals and birds takes place.
- Conservation: Scientific Reserve (IUCN Management Category: I). Access is restricted. Extremely poor soil quality and impeded drainage protect the area from agricultural exploitation.

References

Adams, C.D. (1971). *The blue mahoe and other bush.* Sangster's Bookstores, Kingston, Jamaica. 159 pp.

Adams, C.D. (1972). *Flowering plants of Jamaica.* University of the West Indies, Mona, Jamaica. 848 pp.

Adams, C.D. (1977). Terrestrial ecology. *Caroni River Basin study.* Water and Sewage Authority, Government of Trinidad and Tobago. 80 pp.; App. 89 pp.

Adams, C.D. (1990). Phytogeography of Jamaica. *Atti dei Convegni Lincei* 85: 681–693.

Adams, C.D. and Baksh, Y.S. (1981). What is an endangered plant? *Living World, Journal of the Trinidad and Tobago Field Naturalists' Club* 1981–1982: 9–14.

Adams, C.D. and Lawrence, W.A. (1972). The adventive vegetation of some red mud ponds. Report for Alcan Jamaica Limited. Kirkvine, Jamaica. 41 pp.

Adams, C.D. and Ramjus, H. (1975). A second report on the adventive vegetation of some red mud ponds. Report for Alcan Jamaica Limited. Kirkvine, Jamaica. 44 pp.

Alain, Hermano (1963). *Flora de Cuba* 5. Rio Piedras, Puerto Rico. 363 pp.

Asprey, G.F. and Loveless, A.R. (1958). The dry evergreen formations of Jamaica II. The raised coral beaches of the north coast. *Journal of Ecology* 46: 547–570.

Asprey, G.F. and Robbins, R.G. (1953). The vegetation of Jamaica. *Ecological Monographs* 23: 359–412.

Bacon, P.R. (1970). *The ecology of the Caroni Swamp, Trinidad.* Special Publication of the Central Statistical Office. Government Printer, Port of Spain, Trinidad. 68 pp.

Bacon, P.R. (1985). Environmental impact assessment. *Caribbean Conservation Association News* 4: 10–12.

Beard, J.S. (1942). Montane vegetation in the Antilles. *Caribbean Forester* 3: 61–74.

Beard, J.S. (1944). The natural vegetation of the island of Tobago, British West Indies. *Ecological Monographs* 14: 135–163.

Beard, J.S. (1946). The natural vegetation of Trinidad. *Oxford Forestry Memoirs* 20: 1–152.

Beard, J.S. (1949). Natural vegetation of the Windward and Leeward Islands. *Oxford Forestry Memoirs* 21: 1–191.

Beard, J.S. (1953). The savanna vegetation of northern tropical America. *Ecological Monographs* 23: 149–215.

Beard, J.S. (1955). The classification of tropical American vegetation types. *Ecology* 36: 89–100.

Boldingh, I. (1909a). A contribution to the knowledge of the flora of Anguilla (B.W.I.). *Recueil des Travaux Botaniques Néerlandais* 6: 1–36.

Boldingh, I. (1909b). The flora of St Eustatius, Saba and St Martin. *Flora of the Dutch West Indian Islands* 1: 1–321.

Boldingh, I. (1914). The flora of Curaçao, Aruba and Bonaire. *Flora of the Dutch West Indian Islands* 2: 1–197.

Borhidi, A. (1991). *Phytogeography and vegetation ecology of Cuba.* Akadémiai Kiadó, Budapest. 858 pp.

Borhidi, A. and Herrera, R.A. (1977). Génesis, características y clasificación de los ecosistemas de sabana en Cuba. *Ciencias Biológicas* 1: 115–130.

Borhidi, A. and Muñíz, O. (1980). Die Vegetationskarte von Kuba. *Acta Botanica Academiae Scientarum Hungaricae* 26: 25–53.

Borhidi, A. and Muñíz, O. (1983). *Catálogo de Plantas Cubanas Amenazadas o Extinguidas.* Academia de Ciencias de Cuba, Habana. 85 pp.

Box, H.E. (1938). A systematic enumeration of the flowering plants of Antigua. British Museum (Natural History), London. 439 pp. (Unpublished typescript.)

Box, H.E. (1939a). A note on the vegetation of Redonda, B.W.I. *Journal of Botany* 77: 311–313.

Box, H.E. (1939b). Observations on the landslides in St Lucia, B.W.I., in November 1938. *Empire Forestry Journal* 18: 119–121.

Box, H.E. (1940). Report upon a collection of plants from Anguilla, B.W.I. *Journal of Botany* 78: 14–16.

Box, H.E and Philipson, W.R. (1951). An undescribed species of *Mastichodendron* (Sapotaceae) from Barbados and Antigua. *Bulletin of the British Museum (Natural History), Botany* 1: 21–23.

Britton, N.L. (1916). The natural vegetation of the Isle of Pines, Cuba. *Journal of the New York Botanical Garden* 17: 64–71.

Britton, N.L. (1918). *Flora of Bermuda.* Scribners, New York. 585 pp.

Britton, N.L. and Wilson, P. (1923-1930). Botany of Porto Rico and the Virgin Islands. *Scientific Survey of Porto Rico and the Virgin Islands* 5–6. Academy of Sciences, New York.

Brunt, M.A. (1984). Environment and plant communities. In Proctor, G.R. Flora of the Cayman Islands. *Kew Bulletin, Additional Series* 11: 5–58.

Burne, R.V., Horsfield, W.T. and Robinson, E. (1974). The geology of Navassa Island. *Caribbean Journal of Science* 14: 109–114.

Buskirk, R.E. (1985). Zoogeographic patterns and tectonic history of Jamaica and the northern Caribbean. *Journal of Biogeography* 12: 445–461.

Campbell, D.G. (1978). *The Ephemeral Islands: a natural history of the Bahamas.* Macmillan, London. 151 pp.

Campbell, D.G. and Hammond, H.D. (eds) (1989). *Floristic inventory of tropical countries: the status of plant systematics, collections, and vegetation, plus recommendations for the future.* New York Botanical Garden, New York. 545 pp.

Capote, R.P., Berazaín, R. and Leiva, A. (1989). Cuba. In Campbell, D.G. and Hammond, H.D. (eds), *Floristic inventory of tropical countries: the status of plant systematics, collections, and vegetation, plus recommendations for the future.* New York Botanical Garden, New York. Pp. 316–335.

Carrington, C.M.S. (1991). New collections for the flora of Barbados II. *Journal of the Barbados Museum and Historical Society* 39: 60–71.

Challinor, D. and Wingate, D.B. (1971). The struggle for survival of the Bermuda cedar. *Biological Conservation* 3: 220–222.

Ciferri, R. (1936). Studio geobotanico dell'isola Hispaniola (Antille). *Atti Istituto Botanico 'Giovanni Briosi', ser. IV, 8:* 1–336.

Correll, D.S. and Correll, H.B. (1982). *Flora of the Bahama Archipelago (including the Turks and Caicos Islands).* J. Cramer, Vaduz, Liechtenstein. 1692 pp.

Cuba (1989). *Nuevo Atlas Nacional.*

Danforth, S.T. (1935). Supplementary account of the birds of the Virgin Islands including Culebra and adjacent islands pertaining to Puerto Rico, with notes on their food habits. *Journal of the Agricultural University of Puerto Rico* 19: 439–463.

D'Arcy, W.G. (1967). Annotated checklist of the dicotyledons of Tortola, Virgin Islands. *Rhodora* 69: 385–450.

D'Arcy, W.G. (1975). Anegada Island: vegetation and flora. *Atoll Research Bulletin* 188: 1–40.

Dardaine, S. (1974). *Annual Report of the Forestry Division for the year 1972.* Government Printery, Trinidad and Tobago. 92 pp.

Dickson, J.H., Rodriguez, J.C. and Machado, A. (1988). Invading plants at high altitudes on Tenerife especially in the Teide National Park. *Botanical Journal of the Linnean Society* 95: 155–179.

Duek, J.J. (1971). Lista de las especies cubanas de Lycopodiophyta, Psilophyta, Equisetophyta y Polypodiophyta (Pteridophyta). *Adansonia, sér. 2, 11:* 559–578, 717–731.

Duss, Rév. Père. (1897). Flore phanérogamique des Antilles françaises (Guadeloupe et Martinique). *Annales de l'Institut Colonial de Marseille* 3: 1–656.

FAO (1988). *An interim report on the state of forest resources in the developing countries.* FO: MISC/88/7. Forest Resources Division. FAO, Rome.

FAO (1990). *Production Yearbook 1989,* Vol. 43 [FAO Statistics Series No. 94]. FAO, Rome. 346 pp.

Fournet, J. (1978). *Flore illustrée des Phanérogames de Guadeloupe et de Martinique.* INRA, Paris. 1654 pp.

Garay, L.A. and Sweet, H.R. (1974). Orchidaceae. In Howard, R.A. (ed.), *Flora of the Lesser Antilles* 1: 1–235. Arnold Arboretum of Harvard University, Jamaica Plain, Massachusetts.

Gleason, H.A. and Cook, M.T. (1927). Plant ecology of Porto Rico. *Scientific Survey of Porto Rico and the Virgin Islands* 7: 1–173. Academy of Sciences, New York.

Goberdhan, L.C. (1971). *Andropogon annulatus* and *Ischaemum rugosum* – two new weeds of sugarcane in Trinidad. *PANS* 17: 178–179.

Gooding, E.G.B. (1944). Turner's Hall Wood, Barbados. *Caribbean Forester* 5: 153–170.

Gooding, E.G.B. (1947). Observations on the sand dunes of Barbados, British West Indies. *Journal of Ecology* 34: 111–125.

Gooding, E.G.B. (1974). *The plant communities of Barbados.* Ministry of Education, Barbados. 243 pp.

Gooding, E.G.B., Loveless, A.R. and Proctor, G.R. (1965). *Flora of Barbados.* HMSO, London. 486 pp.

Grisebach, A.H.R. (1859-1864). *Flora of the British West Indian Islands.* Lovell Reeve, London. 789 pp.

Groome, J.R. (1970). *A natural history of the island of Grenada, W.I.* Caribbean Printers, Arima, Trinidad. 115 pp.

Grubb, P.J. and Tanner, E.V.J. (1976). The montane forests and soils of Jamaica: a reassessment. *Journal of the Arnold Arboretum* 57: 313–368.

Hardy, F. (1932-1933). Some aspects of the flora of Barbados. *Agricultural Journal, Department of Science and Agriculture, Barbados* 1(1): 41–62; 1(2): 31–74; 1(3): 19–52; 1(4): 25–66; 2(1–2): 45–76.

Harris, D.R. (1965). Plants, animals, and man in the outer Leeward Islands, West Indies: an ecological study of Antigua, Barbuda, and Anguilla. *University of California Publications in Geography* 18: 1–164.

Hart, J.H. (1891). *Report on forest conservation.* Colonial Office, London.

Hodge, W.H. (1941). The natural resources of the Lesser Antilles. *Chronica Botanica* 6: 448–449.

Hodge, W.H. (1954). Flora of Dominica, B.W.I., Part I. *Lloydia* 17: 1–238.

Hodge, W.H. and Taylor, D. (1957). The ethnobotany of the island Caribs of Dominica. *Webbia* 12: 513–644.

Honychurch, P.N. (1978). *Morne Trois Pitons National Park – vegetation.* Dominica National Park Service, Roseau. 27 pp.

Howard, R.A. (1952). The vegetation of the Grenadines, Windward Islands, British West Indies. *Contributions from the Gray Herbarium of Harvard University* 174: 1–129.

Howard, R.A. (1955). The vegetation of Beata and Alta Vela Islands, Hispaniola. *Journal of the Arnold Arboretum* 36: 209–239.

Howard, R.A. (1962a). Botanical and other observations on Redonda, the West Indies. *Journal of the Arnold Arboretum* 43: 51–66.

Howard, R.A. (1962b). Volcanism and vegetation in the Lesser Antilles. *Journal of the Arnold Arboretum* 43: 279–314.

Howard, R.A. (1968). The ecology of an elfin forest in Puerto Rico I. Introduction and composition studies. *Journal of the Arnold Arboretum* 49: 381–418.

Howard, R.A. (1974). The vegetation of the Antilles. In Graham, A. (ed.), *Vegetation and vegetational history of northern Latin America.* Elsevier, Amsterdam. Pp. 1–38.

Howard, R.A. (1974-1989). *Flora of the Lesser Antilles* 1–6. Arnold Arboretum of Harvard University, Jamaica Plain, Massachusetts.

Howard, R.A. (1977). Conservation and the endangered species of plants in the Caribbean Islands. In Prance, G.T. and Elias, T.S. (eds), *Extinction is forever.* New York Botanical Garden, New York. Pp. 105–114.

Howard, R.A. (1979). Flora of the West Indies. In Larsen, K. and Holm-Nielsen, L.B. (eds), *Tropical botany.* Academic Press, London. Pp. 239–250.

Howard, R.A. and Kellogg, E.A. (1987). Contributions to a Flora of Anguilla and adjacent islets. *Journal of the Arnold Arboretum* 68: 105–131.

Howard, R.A., Portecop, J. and Montaignac, P. de (1981). The post-eruptive vegetation of La Soufrière, Guadeloupe, 1977–1979. *Journal of the Arnold Arboretum* 61: 749–764.

Howard, R.A. and Proctor, G.R. (1957). The vegetation on bauxite soils in Jamaica. *Journal of the Arnold Arboretum* 38: 1–41, 151–169.

Hudson, B. (ed.) (c. 1973). *Conservation in Jamaica: a symposium of the Jamaica Geographical Society.* J.G.S., Kingston, Jamaica. 48 pp.

IUCN (1992). *Protected areas of the world: a review of national systems. Volume 4: Nearctic and Neotropical.* IUCN, Gland, Switzerland and Cambridge, U.K. xxiv, 460 pp. (Prepared by the World Conservation Monitoring Centre.)

James, A. (1980). *Freshwater swamps and mangrove species in Dominica.* Forestry Division, Roseau, Dominica. 37 pp.

Jennings, O.E. (1911). Notes on the ferns of the Isle of Pines. *American Fern Journal* 1: 129–136.

Jennings, O.E. (1917). A contribution to the botany of the Isle of Pines, Cuba. *Annals of the Carnegie Museum* 11: 19–290.

Jiménez, J.A., Martinez, R. and Luis, E. (1985). Massive tree mortality in a Puerto Rican mangrove forest. *Caribbean Journal of Science* 21: 75–78.

Judd, W.S. and Skean, J.D., jr (1987). Floristic study of Morne La Visite and Pic Macaya National Parks, Haiti. *Bull. Florida Mus. Nat. Hist., Biol. Sci.* 32(1): 1–136.

Judd, W.S., Skean, J.D., jr and McMullen, C.K. (1990). The flora of Macaya Biosphere Reserve: additional taxa, taxonomic and nomenclatural changes. *Moscosoa* 6: 124–133.

Kapos, V. (1986). Dry limestone forests of Jamaica. In Thompson, D.A., Bretting, P.K. and Humphreys, M. (eds), *Forests of Jamaica: papers from the Caribbean Regional Seminar on Forests of Jamaica held in Kingston, Jamaica 1983.* The Jamaican Society of Scientists and Technologists, Kingston, Jamaica. Pp. 48–58, 144–148.

Kelly, D.L. (1986). Native forests on wet limestone in north-eastern Jamaica. In Thompson, D.A., Bretting, P.K. and Humphreys, M. (eds), *Forests of Jamaica: papers from the Caribbean Regional Seminar on Forests of Jamaica held in Kingston, Jamaica 1983.* The Jamaican Society of Scientists and Technologists, Kingston, Jamaica. Pp. 31–42, 133–139.

Kelly, D.L. (1988). The threatened flowering plants of Jamaica. *Biological Conservation* 46: 201–216.

Labastille, A. and Richmond, M. (1973). Birds and mammals of Anegada Island, British Virgin Islands. *Caribbean Journal of Science* 13: 91–109.

León, Hermano (1947). *Flora de Cuba* 1. Habana, Cuba. 441 pp.

León, Hermano and Alain, Hermano (1951-1956). *Flora de Cuba* 2–4. Habana, Cuba. 2: 456 pp. (1951); 3: 502 pp. (1953); 4: 556 pp. (1956).

Levine, B. (1981). Abundance and scarcity in the Caribbean. *Ambio* 10: 274–282.

Liogier, A. (1982-). *La Flora de la Española,* in progress. Universidad Central del Este, San Pedro de Macorís, Dominican Republic. 1: 317 pp. (1982); 2: 420 pp. (1983); 3: 431 pp. (1985); 4: 377 pp. (1986).

Liogier, A.H. (1985-). *Descriptive Flora of Puerto Rico and adjacent islands: Spermatophyta.* Editorial de la Universidad de Puerto Rico, Rio Piedras. 1: 352 pp. (1985); 2: 481 pp. (1988).

Liogier, A. and Martorell, L.F. (1982). *Flora of Puerto Rico and adjacent islands: a systematic synopsis.* Editorial de la Universidad de Puerto Rico. 342 pp.

Little, E.L., jr (1969). Trees of Jost Van Dyke (British Virgin Islands). *United States Department of Agriculture Forest Service Research Paper,* ITF-9: 1–12.

Little, E.L., jr and Wadsworth, F.H. (1964). Common trees of Puerto Rico and the Virgin Islands. *United States Department of Agriculture Forest Service Agricultural Handbook* No. 249: 548 pp.

Little, E.L., jr, Woodbury, R.O. and Wadsworth, F.H. (1974). Trees of Puerto Rico and the Virgin Islands, Second Volume. *United States Department of Agriculture Forest Service Agricultural Handbook* No. 449: 1024 pp.

Little, E.L., jr, Woodbury, R.O. and Wadsworth, F.H. (1976). Flora of Virgin Gorda (British Virgin Islands). *United States Department of Agriculture Forest Service Research Paper,* ITF-21: 1–36.

Loveless, A.R. (1960). The vegetation of Antigua, West Indies. *Journal of Ecology* 48: 495–527.

Loveless, A.R. and Asprey, G.F. (1957). The dry evergreen formations of Jamaica, I. The limestone hills of the south coast. *Journal of Ecology* 45: 799–822.

Lugo, A.E., Applefield, M., Pool, D.J., and McDonald, R.B. (1983). The impact of Hurricane David on the forests of Dominica. *Canadian Journal of Forest Research* 13: 201–211.

Lugo, A.E., Schmidt, R. and Brown, S. (1981). Tropical forests in the Caribbean. *Ambio* 10: 318–324.

Mabberley, D.J. (1987). *The plant-book.* Cambridge University Press. 706 pp.

Martin-Kaye, P.H.A. (1963). Accordant summit levels in the Lesser Antilles. *Caribbean Journal of Science* 3: 181–184.

Millspaugh, C.F. (1902). Flora of the island of St. Croix. *Field Museum, Botany* 1: 441–546.

Morton, J.K. (1957). Sand-dune formation on a tropical shore. *Journal of Ecology* 45: 495–497.

Moscoso, R.M. (1943). *Catalogus Florae Domingensis.* L. and S. Printing Co., New York. 732 pp.

Neumann, C.J., Cry, G.W., Caso, E.L. and Jarvian, B.R. (1978). *Tropical cyclones of the North Atlantic Ocean, 1871–1977.* National Climatic Center, Ashville, North Carolina.

Nicolson, D.H. (1991). Flora of Dominica, Part 2: Dicotyledoneae. *Smithsonian Contributions to Botany* 77: 1–274.

Persson, R. (1974). *Research notes No. 17.* Royal College of Forestry, Stockholm, Sweden.

Portecop, J. (1979). Phytogéographie, cartographie écologique et aménagement dans une île tropicale: le cas de la Martinique. *Documents de Cartographie Ecologique, Université de Grenoble* 21: 1–78.

Portecop, J. and Benito-Espinal, E. (1986). *The mangroves of St Lucia: a preliminary survey.* ECNAMP. 56 pp.

Proctor, G.R. (1970). Mason River field station. *Jamaica Journal* 4: 29–33.

Proctor, G.R. (1977). Pteridophytes. In Howard, R.A. (ed.), *Flora of the Lesser Antilles,* 2. Arnold Arboretum of Harvard University, Jamaica Plain, Massachusetts. 414 pp.

Proctor, G.R. (1982). More additions to the flora of Jamaica. *Journal of the Arnold Arboretum* 63: 199–315.

Proctor, G.R. (1984). Flora of the Cayman Islands. *Kew Bulletin, Additional Series* 11: 1–834.

Proctor, G.R. (1985). *Ferns of Jamaica.* British Museum (Natural History), London. 631 pp.

Proctor, G.R. (1986). Cockpit Country and its vegetation. In Thompson, D.A., Bretting, P.K. and Humphreys, M. (eds), *Forests of Jamaica: papers from the Caribbean Regional Seminar on Forests of Jamaica held in Kingston, Jamaica 1983.* The Jamaican Association of Scientists and Technologists, Kingston, Jamaica. Pp. 43–47, 140–143.

Proctor, G.R. (1986). Vegetation of the Black River Morass. In Thompson, D.A., Bretting, P.K. and Humphreys, M. (eds), *Forests of Jamaica: papers from the Caribbean Regional Seminar on Forests of Jamaica held in Kingston, Jamaica 1983.* The Jamaican Association of Scientists and Technologists, Kingston, Jamaica. Pp. 59–65, 149–150.

Proctor, G.R. (1989). Ferns of Puerto Rico and the Virgin Islands. *Memoirs of the New York Botanic Garden* 53: 1–389, App. 1–3.

Proctor, G.R. (1994). Phytogeography of the Cayman Islands. In Brunt, M.A. and Davies, J.R. (eds), *The Cayman Islands: Natural History and Biogeography.* Kluwer Academic Publishers, Dordrecht, Amsterdam. Pp. 237–244.

Proctor, G.R. (1996). Additions and corrections to "Flora of the Cayman Islands". *Kew Bulletin* 51: 483–507.

Provan, M., Wade, B. and Mansingh, A. (1987). Origin, nature and effects of oil pollution in Kingston Harbour, Jamaica. *Caribbean Journal of Science* 23: 105–113.

Putney, A.D. (1983). Basis for the selection of protected areas in the Lesser Antilles. In *Proceedings, Workshop on Biosphere Reserves.* St John, U.S. Virgin Islands. Pp. 1–11.

Questel, A. (1941). *The Flora of St Bartholomew (French West Indies) and its origin.* Imprimerie Catholique, Basse-Terre, Guadeloupe. 224 pp.

Ray, C. and Sprunt, A. (1971). *Parks and conservation in the Turks and Caicos Islands. A report on the ecology of the Turks and Caicos with particular emphasis upon the impact of development upon the natural environment.* Johns Hopkins University, Baltimore, Maryland. 45 pp.

Renard, Y. (1976). La Mangrove. *Parc Naturel de Guadeloupe, Office National des Forêts, Service Animation* 2: 1–32.

Richardson, W.D. (1963). Observations on the vegetation and ecology of the Aripo savannas, Trinidad. *Journal of Ecology* 51: 295–313.

Rodriguez, A. (1981). Marine and coastal environmental stress in the wider Caribbean region. *Ambio* 10: 283–294.

Rodriguez-Robles, J.A., Ackerman, J.D. and Melendez, E.J. (1990). Host distribution and hurricane damage to an orchid population at Toro Negro Forest, Puerto Rico, West Indies. *Caribbean Journal of Science* 26: 163–164.

Rojas, E., Wirshafter, R.M., Radke, J. and Hosier, R. (1988). Land conservation in small developing countries: computer assisted studies in Saint Lucia. *Ambio* 17: 282–288.

Rowley, K. (1979). Soufrière: a volcano in the Caribbean environment. *Trinidad Naturalist* 2(9): 19–28.

Santana Ferreras, B. (1993). Zonación de la vegetación en un transecto altitudino (La Descubierta – Hondo Valle) en Sierra de Neiba, República Dominicana. *Moscosoa* 7.

Sastre, C. (1978). Plantes menacées de Guadeloupe et de Martinique, 1. Espèces altitudinales. *Bulletin du Muséum National d'Histoire Naturelle*, sér. III (No. 519), *Ecologie générale* 42: 65–93.

Sastre, C. (1979a). Etudes biologiques sur la flore des formations altitudinales de Guadeloupe et de Martinique. *Panda*: 6–7.

Sastre, C. (1979b). Considérations phytogéographiques sur les sommets volcaniques antillais. *Comptes Rendues de la Société Biogéographique* 484: 127–135.

Sastre, C. (1985). Endemovicariance et spéciation: application à la systématique des *Lobelia* L. des petites Antilles. *Comptes Rendues de la Académie des Sciences, Paris,* t. 300, sér. III(4): 161–164.

Sastre, C., Baudoin, R. and Portecop, J. (1983). Evolution de la végétation de la Soufrière de Guadeloupe depuis les éruptions de 1976–77 par l'étude de la répartition d'espèces indicatrices. *Bulletin du Muséum National d'Histoire Naturelle*, sér. IV, *Section Adansonia* 5: 63–92.

Sastre, C. and Fiard, J.P. (1986). Evolution de la flore terrestre de la Montagne Pelée (Martinique) après les éruptions du XXe siècle. Mise en évidence de bio-indicateurs volcaniques. *Comptes Rendues de la Société Biogéographiques* 62: 19–42.

Sastre, C. and Portecop, J. (1985). *Plantes fabuleuses des Antilles.* Editions Caribéennes, Paris. 139 pp.

Sauer, J.D. (1982). Cayman Islands seashore vegetation: a study in comparative biogeography. *University of California Publications in Geography* 25: 1–161.

Schroeder, P.B. and Thorhaug, A. (1980). Trace metal cycling in tropical-subtropical estuaries dominated by the seagrass *Thalassia testudinum. American Journal of Botany* 67: 1075–1088.

Sergile, F.E., Woods, C.A. and Paryski, P.E. (1992). *Final report of the Macaya Biosphere Reserve project.* Florida Museum of Natural History, Gainesville, Florida. 130 pp.

Shafer, J.A. (1914). Botanical exploration on the island of Vieques, Porto Rico. *Journal of the New York Botanical Garden* 15: 103–105.

Smith, E.E. (1954). The forests of Cuba. *Maria Moors Cabot Foundation Publication* 2: 1–98.

Sorrié, B.A. (1975). Observations on the birds of Vieques Island, Puerto Rico. *Caribbean Journal of Science* 15: 89–103.

Stearn, W.T. (1971). A survey of the tropical genera *Oplonia* and *Psilanthele* (Acanthaceae). *Bulletin of the British Museum (Natural History), Botany* 4: 261–323.

Stehlé, H. (1936). *Flore de la Guadeloupe et Dépendences. 1. Essai d'Ecologie et de Géographie Botanique.* Basse-Terre, Guadeloupe. 284 pp.

Stehlé, H. (1945-1946). Forest types of the Caribbean islands. *Caribbean Forester* 6 (Supplement): 273–408.

Stehlé, H. and Stehlé, M. (1947). Liste complémentaire des arbres et arbustes des petites Antilles. *Caribbean Forester* 8: 91–111 (–123).

Stoddart, D.R. (1980). Vegetation of Little Cayman. *Atoll Research Bulletin* 241: 53–70.

Stoffers, A.L. (1956). The vegetation of the Netherlands Antilles: studies on the flora of Curaçao and other Caribbean islands. *Natuurwetenschapplijke Studiekring voor Suriname en de Nederlandse Antillen* 15: 1–142.

Sugden, A.M., Tanner, E.V.J. and Kapos, V. (1985). Regeneration following clearing in a Jamaican montane forest: results of a ten-year study. *Journal of Tropical Ecology* 1: 329–351.

Tanner, E.V.J. (1977). Four montane rain forests of Jamaica: a quantitative characterization of the floristics, the soils and the foliar mineral levels, and a discussion of the interrelations. *Journal of Ecology* 65: 883–918.

Tanner, E.V.J. (1986). Forests of the Blue Mountains and the Port Royal Mountains of Jamaica. In Thompson, D.A., Bretting, P.K. and Humphreys, M. (eds), *Forests of Jamaica: papers from the Caribbean Regional Seminar on Forests of Jamaica held in Kingston, Jamaica 1983.* The Jamaican Association of Scientists and Technologists, Kingston, Jamaica. Pp. 15–30, 127–132.

Tempany, H.A. (1915). Report on the island of Redonda. *West Indian Bulletin* 15: 22–26.

Thelen, K.D. and Faizool, S. (eds) (1979). *Management and development plan: Caroni Swamp National Park.* Forestry Division, Ministry of Agriculture, Lands and Fisheries, Port of Spain, Trinidad. 54 pp.

Thelen, K.D. and Faizool, S. (eds) (1980a). *Policy for the establishment and management of a national park system in Trinidad and Tobago.* Forestry Division, Ministry of Agriculture, Lands and Fisheries, Port of Spain, Trinidad. 26 pp.

Thelen, K.D. and Faizool, S. (eds) (1980b). *Management and development plan: Aripo Savannas Scientific Reserve.* Forestry Division, Ministry of Agriculture, Lands and Fisheries, Port of Spain, Trinidad. 43 pp.

Thompson, D.A., Bretting, P.K. and Humphreys, M. (eds) (1986). *Forests of Jamaica: papers from the Caribbean Regional Seminar on Forests of Jamaica held in Kingston, Jamaica 1983.* The Jamaican Society of Scientists and Technologists, Kingston, Jamaica. 162 pp.

Tomblin, J. (1981). Earthquakes, volcanoes and hurricanes: a review of natural hazards and vulnerability in the West Indies. *Ambio* 10: 340–345.

Wade, B.A. (1976). *The pollution ecology of Kingston Harbour.* University of the West Indies, Mona, Jamaica.

Wadsworth, F.H. and Englerth, G.H. (1959). Effects of the 1956 hurricane on forests in Puerto Rico. *Caribbean Forester* 20: 38–51.

Walker, L.R., Brokaw, N.V.L., Lodge, D.J. and Waide, R.B. (eds) (1991). Ecosystem, plant, and animal responses to hurricanes in the Caribbean. *Biotropica* 23(4a): 311–521.

Watts, D. (1978). Biogeographical variation in the mahogany (*Swietenia mahagoni* (L.) Jacq.) woodlots of Barbados, West Indies. *Journal of Biogeography* 5: 347–363.

Weyl, R. (1965). Die palaeogeographische Entwicklung des mittelamerikanisch-westindischen Raumes. *Geologische Rundschau* 54: 1213–1240.

Wheeler, L.R. (1916). The botany of Antigua. *Journal of Botany* 54: 41–52.

Wilkins, L. (1991). Notes on the Jamaican Hutia, *Geocapromys brownii* and a reintroduction of a captive bred population. *Jamaica Naturalist* 1(1): 10–14.

Williams, R.O. and Cheesman, E.E. (1928-). *Flora of Trinidad and Tobago* 1–3, in progress. Government Printer, Port of Spain, Trinidad.

Woodbury, R.O., Martorell, L.F. and García-Tuduri, J.C. (1977). The Flora of Mona and Monito Islands, Puerto Rico (West Indies). *Bulletin of the Agricultural Experiment Station, Rio Piedras, Puerto Rico* 252: 1–60.

Woodbury, R.O. and Weaver, P.L. (1987). The vegetation of St John and Hassel Island, U.S. Virgin Islands. *United States Department of the Interior, National Park Service, Southeast Regional Office. Research/Resources Management Report* SER-83: 1–103.

Zanoni, T.A. (1990). The flora and vegetation of Pico Duarte and Loma La Pelona, Dominican Republic – "the top of the Caribbean". *Mem. New York Bot. Gard.* 64: 279–289.

Zanoni, T.A., Mejia, M.M., Pimentel, J.D. and Garcia, R.G. (1990). La flora y la vegetación de Los Haitises, República Dominicana. *Moscosoa* 6: 46–98.

Zon, R. and Sparhawk, W.N. (1923). *Forest resources of the world.* McGraw Hill, New York.

Author

C. Dennis Adams, The Natural History Museum, Department of Botany, Cromwell Road, London SW7 5BD, U.K.

Acknowledgements

Grateful thanks are extended to Drs Angela Leiva and Rosalina Berazaín (Jardin Botanico Nacional de Cuba, Universidad de la Habana, Carrelera del Ricio Km 35, Calabazar, Habana, Cuba) for kindly providing information on important plant sites in Cuba; to Dr Walter S. Judd (Department of Botany, University of Miami, Department of Biology, PO Box 249118, Coral Gables, Florida 33124, U.S.A.) and T.A. Zanoni (New York Botanic Gardens, Bronx, New York 10458-5126, U.S.A.) for information on plant sites in Hispaniola; to Dr Joël Jérémie (Muséum National d'Histoire Naturelle, Laboratoire de Phanérogamie, 16 Rue Button, 75005 Paris, France) for information on the Lesser Antilles; and to Dr Richard A. Howard (Harvard University Herbaria, 22 Divinity Avenue, Cambridge, Massachusetts 02138, U.S.A.) for commenting on an earlier version of the text. The data on protected areas was kindly supplied by James Paine, Protected Areas Data Unit, World Conservation Monitoring Centre, 219 Huntingdon Road, Cambridge, CB3 0DL, U.K.

CAJALBANA TABLELAND AND PRELUDA MOUNTAIN REGION
Cuba

Location: Comprises two upland regions near the north coast of western Cuba: the Cajálbana Tableland and the Preluda Mountain region, located at latitude 22°40'N and longitude 83°16'W, in the province of Pinar del Rio, about 100 km from La Habana.

Area: c. 100 km².

Altitude: Cajálbana Tableland reaches 464 m above sea-level; Preluda Mountain is over 200 m above sea-level.

Vegetation: Pine forests (with *Pinus caribaea*), thorny xerophytic thickets, some gallery forests.

Flora: c. 353 vascular plant species, including c. 40 strictly endemic species and 4 endemic genera. The region includes the highest concentration of endemic taxa per unit area in Cuba.

Useful plants: Pines (e.g. *Pinus caribaea*) used for timber and as sources of resin.

Other values: Seed source of timber trees used in reforestation programmes, high landscape value, potential tourism, forestry education.

Threats: Fire, soil erosion, clearance for agriculture, timber exploitation, some tourist activities (e.g. camping).

Conservation: The region is traditionally managed as "Forestry Patrimony" and part is included in Mil Cumbres Integrated Management Area (166 km²) (IUCN Management Category: VIII).

Geography

The region comprises two upland areas: the Cajálbana Tableland and Preluda Mountain (Map 32). The Cajálbana Tableland is 15 km in length and 9 km wide, parallel to the north coast. The whole region is between 300 and 400 m above sea-level, the highest point reaching 464 m. The southern slopes of the tableland are abrupt. Preluda Mountain lies to the south-west of the Cajálbana Tableland, from which it is separated by the Tortuga River. The easternmost limit is the Puercas River. The mountain is 2 km long and 1 km wide.

Both areas contain old serpentine ultramafic igneous rock basement (peridotites), dating from the Late Cretaceous (100–65 million years BP). The Cajálbana Tableland has very deep red, lateritic or ferritic soils which are slightly acidic, contain high concentrations of nickel and oxides of iron and aluminium, and have a low Ca/Mg ratio. Preluda Mountain and the southern slopes of Cajálbana Tableland are rocky, with undifferentiated serpentine soils (magnesic black soils), which also contain high concentrations of nickel.

The annual rainfall is 1300–1500 mm, with a dry season from November to April and a rainy season from May to October. The mean annual temperature is 25°C (slightly lower at the highest points). The warmest month is August, with a mean temperature of 28°C, and the coldest is January (with a mean of 21.5°C).

Vegetation

Pine forests

Pine forests are the most extensive vegetation over deep lateritic soils on the Cajálbana Tableland (Map 33), covering nearly 70 km². The canopy is closed, the main emergent species being *Pinus caribaea*, which can reach 30 m. In moist areas, pines are mixed with broadleaved trees, such as *Calophyllum pinetorum* and *Clusia rosea*.

The understorey is very dense and rich in shrubs, most belonging to Rubiaceae, Euphorbiaceae, Myrtaceae and Melastomataceae. Palms of the genera *Copernicia* and *Coccothrinax* are particularly distinctive. The shrubs are mainly xerophytic (microphyllous) species, including *Jacquinia brunnescens*, *Vaccinium ramonii*, *Sauvallella immarginata*, *Neomazaea phialanthoides*, *Acuneanthus tinifolius* and *Malpighia horrida*. Mesophytic shrubs include *Purdiaea cubensis*, *Tetrazygia coriacea* and *Tabebuia lepidota*. The herbaceous layer is dominated by grasses of the genera *Aristida*, *Andropogon* and *Arthrostylidium*, herbaceous dicots and ferns, such as *Ayenia cajalbanensis*, *Mitracarpus glabrescens* and *Anemia cajalbanica*. The forests contain some lianas, such as *Lescaillea equisetiformis*, *Cynanchum* spp., *Aristolochia* spp., and the spiny fern *Odontosoria wrightiana*. Epiphytes include many species of *Tillandsia* spp.

Where clearances have been made in the past, the canopy of the forest is more open and the shrub layer is dominated by *Comocladia dentata* (a very poisonous member of the Anacardiaceae); grasses, lianas and epiphytes are poorly represented.

At the southern side of the Cajálbana Tableland, a dry type of pine forest occurs on rocky substrates. It includes pines, thorny shrubs and the succulent *Agave cajalbanensis*.

Thorny xerophytic thicket

Thorny xerophytic thicket contains the highest floristic diversity of all vegetation types in the region. It occurs on the rocky (magnesic black) soils of the southern slopes of Cajálbana Tableland and Preluda Mountain (Map 33). The thicket is a dense bushy formation, about 3 m high. It includes xeromorphic, sometimes thorny, shrubs and succulents, with some emergent trees and palms. Examples of the emergent species are: *Amyris lineata, Euphorbia cubensis, Pseudocarpidium ilicifolium* and *Cocccothrinax yuraguana*. The most abundant shrubs are: *Phyllanthus trigonocarpus, Erythroxylum minutifolium* and members of the Rubiaceae, Myrtaceae and Euphorbiaceae.

The herb layer is rich in species of *Aristida, Andropogon* and *Arthrostylidium*. Other interesting herbaceous plants include *Heptanthus ranunculoides, Lachnorhiza piloselloides* and the cycad *Zamia kickii*. Lianas include *Mesechites rosea* and *Jacquemontia* spp. Epiphytes are represented principally by species of *Tillandsia* and *Encyclia*.

Gallery forests

Gallery (riverine) forests are developed on alluvial deposits along rivers (especially the Tortuga River) and streams (Map 33). The tree layer has a dense canopy, with *Didymopanax morototoni, Bombacopsis emarginata,*

Syzygium jambos and *Dendropanax cuneifolius,* and the palms *Calyptrogyne dulcis* and *Copernicia glabrescens,* among others.

The shrub layer is dense and comprises mesophytic plants, such as *Rondeletia peduncularis* and *Calyptranthes enneantha.* The herb layer includes *Pinillosia berteri, Wedelia rugosa, Pinguicula albida* and grasses such as *Mniochloa strephioides* and *Arthostylidium capillifolium.* Species of *Adiantum, Blechnum, Polypodium* and *Thelypteris* are the most common ferns. Lianas include *Dioscorea wrightii, Vanilla dilloniana* and *Clematis dioica.* Orchids, such as *Encyclia* spp. and *Epidendrum* spp., are among the epiphytes.

Compared to the other formations, gallery forests have relatively low species endemism.

Flora

Borhidi (1991) considers the Cajálbana Mountains (Cajalbanense) to be a separate floristic district within the

MAP 32. CAJALBANA TABLELAND AND PRELUDA MOUNTAIN REGION, CUBA (CPD SITE Cb3, in part)

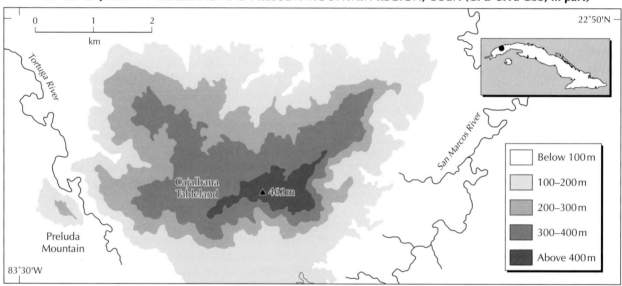

MAP 33. CAJALBANA TABLELAND AND PRELUDA MOUNTAIN REGION, CUBA (CPD SITE Cb3, in part), SHOWING VEGETATION ZONES

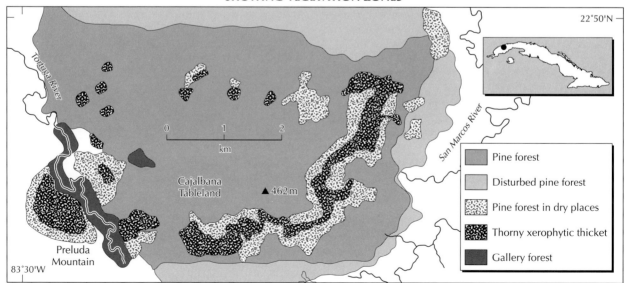

Pinar del Rio sector of the phytogeographic Sub-Province of Western Cuba (Occidento-Cubanicum) (see CPD Site Cb3 in Regional Overview).

The flora of Cajálbana Tableland and Preluda Mountain contains nearly 330 species of angiosperms, 3 species of gymnosperms and c. 20 species of ferns and fern allies. Of the 181 genera of angiosperms in Cuba, 75 are found in the district. Families with the most number of species are Rubiaceae (40), Compositae (27), Gramineae (24), Euphorbiaceae (18), Myrtaceae (15) and Leguminosae (15). There are 4 strictly endemic genera, namely: *Neomazaea*, *Sauvallella*, *Phyllacanthus* and *Lescaillea* (all of which are monotypic) and about 40 strictly endemic species, many of which are found on serpentine soils. Indeed, the Cajálbana Tableland and Preluda Mountain Region has the highest concentration of endemic taxa per unit area in the whole of Cuba.

Useful plants

The most important species are pines (especially *Pinus caribaea*), which are used as sources of timber, resin and other products. The region contains an important gene pool for species used in reforestation programmes in Cuba.

Local people use the leaves of the palm *Coccothinax yuraguana* (yuraguano) as roofing material for their houses.

Social and environmental values

Only a few people live in the upper part of the Cajálbana Tableland, where there is a Forestry School. The Preluda Mountain is uninhabited.

The region is the source of several rivers. The landscape is one of the most beautiful in western Cuba and has considerable potential for tourism.

Cuba and the Bahama Islands constitute one Endemic Bird Area (EBA) with a total of 24 restricted-range landbirds occurring of which 19 are endemic; 13 of these birds are confined to Cuba and the Isle of Pines only. A further nine species are endemic to Cuba, but occur on the island widely, and thus do not qualify as restricted-range. Habitat requirements vary but many species are reliant on forest (including pine forest) and woodland.

Threats

The most serious threat is fire, which can spread rapidly through the xeromorphic and resiniferous vegetation.

Agriculture is increasing around the perimeter of the area. Timber exploitation and, to some extent, tourism could increase soil erosion and open up areas to invasion by exotic plant species.

Table 48 shows the 18 strict endemic taxa that are considered to be rare or threatened according to the IUCN Red Data Book categories; several more taxa may also qualify for listing.

TABLE 48. RARE AND THREATENED ENDEMICS OF THE CAJALBANA TABLELAND AND PRELUDA MOUNTAIN REGION

Plant name	IUCN category
Acalypha nana (Euphorbiaceae)	I
Ayenia cajalbanensis (Sterculiaceae)	E
Cassia acunae (Leguminosae)	I
Calycogonium microphyllum (Melastomataceae)	I
Calyptranthes pozasiana (Myrtaceae)	E
Cynometra cubensis ssp. *ophiticola* (Leguminosae)	R
Euphorbia cubensis (Euphorbiaceae)	I
Ginoria thomasiana (Lythraceae)	I
Gochnatia intertexta (Compositae)	I
Heptanthus brevipes (Compositae)	I
Machaonia dumosa (Rubiaceae)	I
Maytenus lineata (Celastraceae)	I
Mimosa apleura (Leguminosae)	I
Nodocarpaea radicans (Rubiaceae)	E
Plinia dermatodes (Myrtaceae)	V
Rajania hermanii (Dioscoreaceae)	I
Sauvallella immarginata (Leguminosae)	I
Scolosanthus acunae (Rubiaceae)	I

Conservation

The region is traditionally managed as "Forestry Patrimony". The southern side of Cajálbana Tableland is included in the Mil Cumbres Integrated Management Area (of area 166 km²) (IUCN Management Category: VIII).

Fires are controlled by planting broadleaved plants as barriers and by creating firebreaks in the forests.

References

Alain, H. (1950). Notas sobre la vegetación de la loma de Cajálbana, Pinar del Rio. *Rev. Soc. Cub. Bot.* 7(1): 8–18.

Berazaín, R. (1987). Notas sobre la vegetación y flora de la Sierra de Cajálbana y Sierra Preluda (Pinar del Rio) Rev. *Jardín Bot. Nac.* 8(3): 39–68.

Borhidi, A. (1987). The main vegetation units of Cuba. *Acta Bot. Hung.* 33(3–4): 151–185.

Borhidi, A. (1991). *Phytogeography and vegetation ecology of Cuba*. Akadémiai Kiadó, Budapest. 858 pp.

Borhidi, A. and Muñiz, O. (1986). The phytogeographic survey of Cuba II. Floristic relations and phytogeographic subdivision. *Acta Bot. Hung.* 32 (1–4): 3–48.

Samek, V. (1973). Pinares de Cajálbana. Estudio Sinecológico. *Ser. Forestal, Academia de Ciencias* 13: 3–56.

Authors

This Data Sheet was prepared by Drs Angela Leiva and Rosalina Berazaín (Jardin Botanico Nacional de Cuba, Universidad de la Habana, Carrelera del Ricio Km 35, Calabazar, Habana, Cuba).

BLUE AND JOHN CROW MOUNTAINS
Jamaica

Location: Mountainous land in the eastern parishes of St Andrew, St Thomas, Portland and St Mary. The most biologically diverse areas are in the upper parts of the Blue Mountains and in the John Crow Mountains, but surrounding degraded areas of Crown Land have been included in the boundaries of the Blue and John Crow Mountains National Park, located between latitudes 17°57'–18°12'N and longitudes 76°49'–76°16'W.

Area: National Park: 782 km²; the area of greatest scientific interest is perhaps half to two-thirds of that.

Altitude: 380–2256 m (summit of Middle Peak, Blue Mountains). Blue Mountains: 1220–2256 m; John Crow Mountains: 380–1143 m.

Vegetation: Lower and upper montane rain forest, montane scrub, tall-grass montane savanna, cliff and landslide vegetation.

Flora: >600 species of flowering plants, of which 87 are strictly endemic to the National Park. About 275 Jamaican endemic species and 14 endemic varieties are found in the area.

Useful plants: Some plants used locally for medicinal purposes; potential ornamental species, including Gesneriaceae, Orchidaceae, Bromeliaceae, *Pilea* and *Peperomia* spp.

Other values: Major watershed for the capital city, Kingston, and for Port Antonio. Watershed for cash crops, including coffee, sugar cane and banana plantations.

Threats: Clearance of native forest for subsistence farming and for commercial crops, fire, invasion of exotic plant species, some collecting of epiphytes for local market.

Conservation: National Park established in 1991.

Geography

The Blue and John Crow Mountains are located at the eastern end of Jamaica. In the broad sense, the Blue Mountains include the Port Royal and Mount Telegraph Ranges (Kerr *et al.* 1992), and these have been included in the new Blue and John Crow Mountains National Park. These lands, which are all Government-owned, lie within the parishes of St Andrew, St Thomas, Portland and south-east St Mary, and are within about 20 km of the capital, Kingston.

The National Park is located between latitudes 17°57'–18°12'N and longitudes 76°49'–76°16'W. The National Park area is 782 km², but much of this has been altered from its natural state and is now used for forestry, coffee production or subsistence farming.

The Blue Mountains have a complex geology reflecting volcanic and marine influences. The rocks are of igneous and sedimentary origin and the soils are mainly siliceous. There are some knolls of limestone. In contrast, the John Crow Mountains comprise white limestone overlain by marine sandstones and shale.

The Grand Ridge of the Blue Mountains stretches for 16 km across the eastern part of Jamaica, ending in the east with the Rio Grande valley, and in the west with the Wag Water River trough. Much of the range is over 1800 m, the major peaks being: Blue Mountain Peak, comprising Middle Peak (2256 m), the highest point of Jamaica, and East Peak (2246 m); Sugar Loaf Peak (c. 2150 m); High Peak (2082 m); Mossman's Peak (2028 m); and Sir John

Peak (1927 m). Lesser peaks and ridges radiate from these and give way to slopes sometimes in excess of 70° and frequently over 50°. To the west of the Grand Ridge are the lower Port Royal Mountains and the Mount Telegraph Range. These include Mount Horeb (c. 1490 m) and Catherine's Peak (1539 m) in the vicinity of Hardwar Gap, and Mount Telegraph itself (1301 m) in a more northerly direction.

The John Crow Mountains run parallel to the eastern coast of the island, east of the Rio Grande valley. The range rises gently from the east to a maximum height of 1140 m, but ends abruptly along a steep escarpment to the west. The Rio Grande separates the John Crow Mountains from the Blue Mountains; the ranges join at Corn Puss Gap (640 m), the boundary of the parishes of Portland and St Thomas. Unlike the sharp peaks of the Blue Mountains, the summit of the John Crow Mountains is a slightly tilted plateau, with an unusual landscape of sinkholes and outcrops.

Average rainfall ranges between 1500 and 6000 mm per annum. The wettest parts of the park are the northern slopes of the Blue and John Crow Mountains, as a result of the prevailing moisture-laden winds (the trade winds) blowing from the Atlantic Ocean. On the northern slopes of the Blue Mountains, mist is present for about 70% of daylight hours for most of the year; the corresponding figure for the southern slopes is 30%. Average annual mean temperatures at 1500 m within the Blue Mountain forests are between 18.5° and 20.5°C (maximum: 24°C; minimum: 8.5°C).

Vegetation

Lower montane tropical forest

Wet limestone forest occurs in the John Crow Mountains at altitudes of less than 800–900 m. Dominant species are *Calophyllum calaba*, *Calyptronoma occidentalis*, *Drypetes alba*, *Heliconia caribaea* and *Cyathea grevilleana*.

Wet slope forest over sedimentary/igneous substrata occurs on (1) the steep western slopes of the John Crow Mountains and (2) on the northern slopes of the Blue Mountains below 1000 m. It is characterized by large trees (c. 26 m high, 70 cm dbh), including *Calophyllum calaba*, *Symphonia globulifera*, *Pouteria multiflora*, *Ficus* spp. and *Hernandia catalpifolia*, with smaller *Calyptronoma occidentalis* and species of Melastomataceae and Rubiaceae. Climbers are abundant.

Gully forest occurs in gullies in the Blue Mountains. The canopy height is 12–18 m. The most significant trees are *Laplacea haematoxylon*, *Solanum punctulatum* and *Turpinia occidentalis*. Melastomataceae and Rubiaceae are common smaller trees; *Cyathea pubescens* is frequent.

Upper montane tropical forest

Upper montane forest is the most extensive natural forest type of the Blue Mountains. Among the most abundant species are *Clethra occidentalis*, *Podocarpus urbanii*, *Alchornea latifolia* and *Cyrilla racemiflora*. In forests at very high altitudes, *Clethra alexandri* and *Ilex obcordata* are abundant. Mor Ridge Forest is an uncommon variant which includes the tree ferns *Blechnum underwoodianum* and *Cyathea gracilis*.

Upper montane wet limestone forest occurs on limestone rocks capping the western slopes of the John Crow Mountains. Tall trees are absent from this vegetation type.

Montane scrub

Montane limestone thicket covers much of the high plateau area of the John Crow Mountains and is dominated by *Clusia havetioides* and *C. portlandiana*. A variant of this type occurs on John Crow Peak in the Blue Mountains and on other limestone outcrops in that region.

Montane grassland

Tall-grass montane savanna is confined to High Peak in the Blue Mountains and is dominated by *Danthonia domingensis*.

Flora

The flora of the Blue and John Crow Mountains contains more than 600 species of flowering plants and includes

MAP 34. BLUE AND JOHN CROW MOUNTAINS, JAMAICA (CPD SITE Cb10)

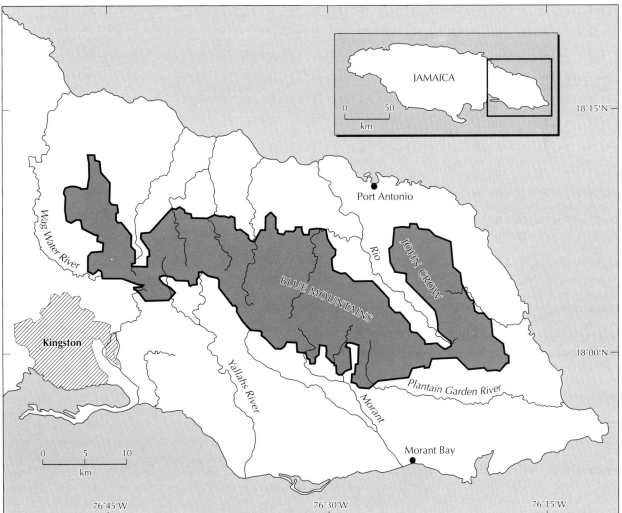

"Mor Ridge Forest", showing the characteristic cushions of the moss *Leucobryum giganteum*, the hanging moss *Phyllogonium fulgens* on the tree trunks and bromeliads.
Photo: Susan Iremonger.

Rondeletia subsessiliflora, an endemic treelet confined to the John Crow Mountains, and even there very scarce.
Photo: Susan Iremonger.

c. 275 vascular plants species and 14 varieties which are endemic to Jamaica (i.e. approximately 33% of the Jamaica's endemic vascular plant flora occurs in these mountains). About 87 vascular plant species are strictly endemic to the Blue and John Crow Mountains. Of these, 47 species are only known from the parish of Portland, 23 species only from St Andrew and 17 only from St Thomas.

Of Jamaica's seven endemic genera, representatives of at least one are found in the Blue and John Crow Mountains: *Odontocline laciniata*. Genera which are well represented by endemic species in the flora of the park are *Pilea* (12 spp.), *Lepanthes* (12 spp.), *Psychotria* (12 spp.) and *Eugenia* (11 spp.).

Due to differences in geology and topography, the Blue Mountains support a different flora to that of the John Crow Mountains. Thus, many of the endemic plants are confined to one or other of the two ranges. The National Park area, therefore, encompasses two centres of plant endemism.

Useful plants

Native forest plants have been used by local people as sources of timber, firewood, thatch, herbal medicine and food. Recently, the forests have also been used as a source of ornamental plants.

Timber trees include *Sideroxylon* spp., *Calophyllum calaba*, *Nectandra* spp., *Guarea glabra*, *Hibiscus elatus* and *Cedrela odorata*. *Eugenia* spp. (rodwood) are cut for firewood.

Hedyosmum spp. are used as a cure for headaches. *Passiflora* spp., *Rubus* spp. and *Vaccinium meridionale* provide edible fruits. Ornamental species include many orchids, bromeliads, *Pilea* and *Peperomia*.

Social and environmental values

The forests of the Blue Mountains protect the watershed of Jamaica's capital city, Kingston. Agricultural areas, including coffee plantations, on the lower slopes of both ranges depend upon the water supply assured by the forested slopes of the mountains.

Commercial forestry has occurred in these mountains in the past. Up to the 1970s, hardwoods such as *Hibiscus elatus* and *Cedrela odorata* were planted. In the 1970s and 1980s, *Pinus caribaea* was also planted on a very large scale. Commercial forest activities have been scaled down since Hurricane Gilbert destroyed many of the plantations in 1988.

A total of 36 restricted-range landbirds occur on Jamaica, of which 28 are endemic to the island (the highest total for any Caribbean island). Most of the birds occur in the mountain forests throughout the island, an exception being the recently split red-billed and black-billed streamertails, *Trochilus polytmus* and *T. scitulus*, which are confined to western/central and eastern Jamaica (including the John Crow Mountains) respectively.

Threats

The most immediate serious threat is deforestation as a result of large and small-scale cultivation, principally on the southern slopes of the Blue Mountains, but also in other areas. Deforestation causes massive soil erosion and the quality of the land deteriorates rapidly. Water courses leading from the mountains become heavily laden with sediment and water flows decrease and become more erratic. This results in water shortages alternating with floods at lower altitudes.

The deterioration in land quality renders it productive only for a few years, after which it is abandoned. The African grass *Melinis minutiflora* colonizes this land very successfully, to the exclusion of native species. Due to the susceptibility of this species to combustion, fires burn frequently on the mountainsides, eliminating any possibility of forest regeneration.

The gradual change of the floristic composition of the forests due to invasion by aggressive alien species is underway in some parts. Particularly troublesome in the Blue Mountains are *Pittosporum undulatum* and *Hedychium gardnerianum*.

A threat to some ornamental species is collecting for sale at local markets. Species of orchids and bromeliads are particularly at risk.

Some mineral prospecting licences extend within the park area and it is possible that there will be a conflict of interest if commercial deposits are discovered.

Conservation

Most of the natural forests of the Blue and John Crow mountains lie on Government-owned "Crown Lands". These lands have been the responsibility of the Government Forest Department, which has planted some of them with timber trees, leased parts to the Coffee Industries Board and left the more inaccessible parts in their natural state.

The entire area has now been declared a National Park under the jurisdiction of the Forest Department. Management plans are now being formulated. The plans include zoning for different land uses and creating a buffer zone where land uses will be controlled. The success of the park depends on the implementation of the management plan, along with the enforcement of regulations to prevent the removal of timber and other plants, and the inclusion of local people in decision-making and staffing arrangements.

References

Adams, C.D. (1972). *Flowering plants of Jamaica*. University of the West Indies, Mona, Jamaica. 848 pp.

Grossman, D., Iremonger, S.F. and Muchoney, D.M.M. (1992). *Jamaica: a rapid ecological assessment*. The Nature Conservancy, Virginia, U.S.A.

Kelly, D.L. (1986). Native forests on wet limestone in north-eastern Jamaica. In Thompson, D.A., Bretting, P.K. and M. Humphreys (eds), *Forests of Jamaica: papers from the Caribbean Regional Seminar on Forests of Jamaica held in Kingston, Jamaica 1983*. The Jamaican Society of Scientists and Technologists, Kingston, Jamaica. Pp. 31–42.

Kelly, D.L. (1988). The threatened flowering plants of Jamaica. *Biological Conservation* 46: 201–216.

Kerr, R., Lee, D., Walling, L., Green, G., Bellingham, P.J. and Iremonger, S.F. (1992). Management plan for the Blue Mountains/John Crow Mountains national park. Internal report, Forest Department, Government of Jamaica, Kingston, Jamaica.

Proctor, G.R. (1982). More additions to the flora of Jamaica. *Journal of the Arnold Arboretum* 63: 199–315.

Proctor, G.R. (1985). *Ferns of Jamaica*. British Museum (Natural History), London. 631 pp.

Tanner, E.V.J. (1986). Forests of the Blue Mountains and the Port Royal Mountains. In Thompson, D.A. Bretting, P.K. and Humphreys, M. (eds), *Forests of Jamaica: papers from the Caribbean Regional Seminar on Forests of Jamaica held in Kingston, Jamaica 1983*. The Jamaican Society of Scientists and Technologists, Kingston, Jamaica. Pp. 15–30.

Author

This Data Sheet was prepared by Dr Susan Iremonger (WCMC, 219 Huntingdon Road, Cambridge CB3 0DL, UK).

COCKPIT COUNTRY
Jamaica

Location: The karst limestone "cockpit" terrain of the parishes of Trelawny, St James, upper St Elizabeth, Manchester and Clarendon, together with parts of St Ann, between latitudes 18°06'–18°25'N and longitudes 77°27'–77°55'W.

Area: The total area of the Cockpit Country is 430 km², of which about 202.5 km² is still of high importance for biodiversity.

Altitude: 300–746 m.

Vegetation: Evergreen seasonal forest: mesic limestone forest and degraded mesic limestone forest. Limestone cliff and landslide vegetation. In valleys, human interference has resulted in areas of pasture and some agricultural crops.

Flora: The whole region is estimated to contain 1500 vascular plant species, of which 400 are endemic to Jamaica, including 100 species of angiosperms and one species of fern which are strictly endemic to the Cockpit Country.

Useful plants: Timber trees, medicinal plants, potential ornamental plants.

Other values: High landscape value.

Threats: Clearance for subsistence farming and for commercial crops, fire, road building, illegal timber cutting.

Conservation: Most of the area has Forest Reserve status.

Geography

The Cockpit Country is an expanse of land in central-western Jamaica which has an unusual "eggbox" topography of limestone karst hills and valleys. The area extends between approximately 18°06'–18°25'N latitude and 77°27'–77°55'W longitude, in the parishes of Trelawny, St James, upper St Elizabeth, Manchester, Clarendon and part of St Ann. To the east of the main block lies the "central inlier", composed of bedrock other than limestone, which supports different vegetation. This is surrounded on the north and south sides by extensions of the Cockpit Country limestone, which support vegetation similar to, and as diverse as, the Cockpits themselves.

The majority of the Cockpit Country is Government-owned Forest Reserve. The main areas are Cockpit (223.3 km²), Fyffe and Rankine (9.6 km²), Peru Mountain and Thicketts (2.5 km²), Chatsworth (3.8 km²) and Cooks Bottom (2 km²). These figures include unforested land: the total area of land worth conserving has yet to be delimited. Some privately-owned areas on the periphery are worth conserving.

Rainfall in the Cockpit Country area varies between c. 1900 and 3800 mm per annum. Little of this is retained on the steep dry slopes, but flows into underground aquifers beneath the more fertile valleys.

Vegetation

The forest vegetation may be described as mesic limestone forest, in the evergreen seasonal forest formation of Beard (1955). The forests are, however, poorly known. Kelly *et al.* (1988) recorded 75 tree species in a total area of 1000 m²; 27 of the 75 species were represented by one individual only. The canopy height was between 16 and 24 m, and was dominated (in patches) by species such as *Guapira fragrans*

(*Pisonia fragrans*). Rubiaceae and Myrtaceae were frequent in the tree and shrub layers, but the ground flora was dominated by ferns (22 of 30 ground herbs were pteridophytes). Orchidaceae and Bromeliaceae were frequent epiphytes and mosses were plentiful on rock outcrops and tree bases.

Limestone cliff vegetation supports many local endemics. There are some landslide formations as well as modified vegetation types, including pasture and cropland.

Flora

The native flora of the whole of the Cockpit Country includes an estimated 1500 vascular plant species (C.D. Adams, personal estimate 1992), about 42% of the Jamaica's native vascular flora. About 60% (500 species) of Jamaica's endemic vascular plant flora occurs in the Cockpit Country. In the more restricted area of high biodiversity, covering about 202.5 km² according to JCDT (1992), there are an estimated 800–900 vascular plant species (C.D. Adams, personal estimate 1992).

According to Proctor (1986), there are 106 species which, in Jamaica, are only found in the Cockpit Country. These include 100 species of angiosperms and one species of fern which are strictly endemic to the Cockpit Country and five other species which are not endemic to Jamaica. The strict endemics are best represented in the families Rubiaceae (11 spp.), Compositae (9 spp.), Gesneriaceae (8 spp.), Euphorbiaceae (7 spp.), Orchidaceae (7 spp., all in the genus *Lepanthes*) and Myrtaceae (6 spp.). Floristic studies indicate that each limestone knoll can support many different plants from the next, including plants which are endemic to just one knoll.

There have been few floristic studies in the forests of the Cockpit Country. A total area of 1000 m² of forest on the fringes of the Cockpit Country was found by Kelly *et al.* (1988)

to support 235 higher plant species (206 species of angiosperms and 29 of pteridophytes); 75 species of tree >2 m tall were recorded. The high diversity may be explained by the intermediate-rainfall character of the forests: species tolerant of very wet conditions (e.g. *Guzmania lingulata*) can co-exist with those adapted to dry conditions (e.g. *Hylocereus triangularis*). (The study area from which these figures were obtained has, unfortunately, been cut over and no forest remains there now.)

Useful plants

There are a number of ornamental endemic plant species with great horticultural potential. These include: *Portlandia coccinea*, *Lisianthius capitatus*, *Palicourea pulchra* and *Piper verrucosum*. Two strictly endemic trees are valuable for their timber: *Terminalia arbuscula* and *Manilkara excisa*. Other endemics, not confined to the Cockpit Country, are valuable timber species.

Two wild relatives of edible yam occur: *Rajania cyclophylla*, endemic to the Cockpit Country, and *R. cordata*, restricted in Jamaica to the Cockpits but also occurring in other West Indian islands.

Some plants are used locally for medicinal purposes, but there is no published information on these.

Social and environmental values

The population of the Cockpit Country is sparse and generally confined to places accessible by road. Of the estimated 4500 inhabitants, most are farmers (cultivation extending deep into the area). The rugged terrain and the paucity of surface water has helped to prevent forest clearances for timber and cultivation. Although much of the valley forests have been cleared, the steep slopes and hill summits have only been minimally disturbed.

A total of 36 restricted-range landbirds occur on Jamaica, of which 28 are endemic to the island (the highest total for

MAP 35. COCKPIT COUNTRY, JAMAICA (CPD SITE Cb11)

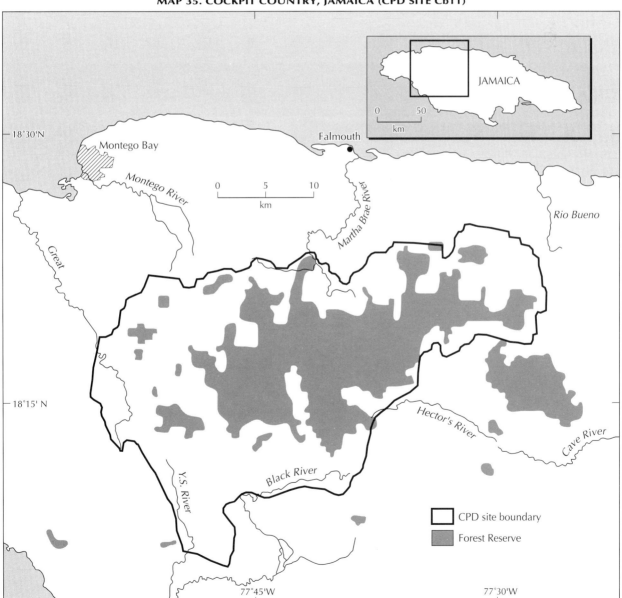

any Caribbean island). Most of the birds occur in the mountain forests throughout the island, an exception being the recently split red-billed and black-billed streamertails, *Trochilus polytmus* and *T. scitulus*, which are confined to western/central and eastern Jamaica (including the John Crow Mountains) respectively.

Threats

The greatest threats are clearance for agriculture, illegal cutting of timber and firewood collection. The rate of deforestation between 1981–1987 is estimated at 15.8% (2.8% per year) (Eyre 1989). Until recently, interior forests were very inaccessible. However, roads are being built into the area, which will inevitably lead to increased deforestation and selective cutting. The area is, therefore, severely threatened.

Conservation

The Cockpit Country has not yet been formally declared a protected area, although much of the area has Forest Reserve status – no plant may be removed without permission of the Forest Department.

Because of the amount of human activity in the area, as well as some of the area being in private ownership, designation of the area to allow for continuation of some human uses is possibly more appropriate than National Park status. Some form of protection is being recommended to the Government in JCDT (1992).

Conservation efforts on a more localized scale would be beneficial, as some of the knolls which are the type localities for a number of the endemics, are privately owned.

References

Adams, C.D. (1972). *Flowering plants of Jamaica*. University of the West Indies, Mona, Jamaica. 848 pp.

Beard, J.S. (1955). The classification of tropical American vegetation types. *Ecology* 36: 89–100.

Conrad Douglas and Associates (1992). *Plan for a system of protected natural areas for Jamaica*. Jamaica Conservation and Development Trust, Kingston, Jamaica.

Eyre, L.A. (1989). Slow death of a tropical rainforest: the Cockpit Country of Jamaica. *Environmental Quality and Ecosystem Stability* IV-A: 599–606.

JCDT (1992). *The plan for a system of protected areas in Jamaica (background documents prepared by Conrad Douglas and Associates)*. Jamaica Conservation and Development Trust, Kingston, Jamaica.

Kelly, D.L. (1988). The threatened flowering plants of Jamaica. *Biological Conservation* 46: 201–216.

Kelly, D.L., Tanner, E.V.J., Kapos, V., Dickinson, T.A., Goodfriend, G.A. and Fairbairn, P. (1988). Jamaican limestone forests: floristics, structure and environment of three examples along a rainfall gradient. *Journal of Tropical Ecology* 4: 121–156.

Proctor, G.R. (1982). More additions to the flora of Jamaica. *Journal of the Arnold Arboretum* 63: 199–315.

Proctor, G.R. (1985). *Ferns of Jamaica*. British Museum (Natural History), London. 631 pp.

Proctor, G.R. (1986). Cockpit Country and its vegetation. In Thompson, D.A., Bretting, P.K. and Humphreys, M. (eds), *Forests of Jamaica*. The Jamaican Association of Scientists and Technologists, Kingston, Jamaica. Pp. 43–47, 140–143.

Author

This Data Sheet was prepared by Dr Susan Iremonger (WCMC, 219 Huntingdon Road, Cambridge CB3 0DL, UK).

REGIONAL OVERVIEW: SOUTH AMERICA

Total land area: c.17,519,900–17,529,250 km².

Population (1995): 320,000,000.

Maximum altitude: 6959 m.

Vegetation: Essentially all life zones and physiognomic vegetational formations. Principal vegetation types are tropical evergreen and semi-evergreen moist forest, dry forest to woodland, open grassy savanna, desert and arid steppe, Mediterranean-climate communities, temperate evergreen forest, and several montane formations (e.g. páramo, puna).

Number of vascular plant species: Over 83,000.

Number of endemic vascular plant species: Over 50,000; regions with 5%–76% endemism.

Number of genera of flowering plants: 4200.

Introduction

The continent of South America has about one-eighth of the Earth's land surface, situated between latitudes 12°N–55°S and longitudes 80°–35°W; no other continent has a greater latitudinal span. Eighty percent of its land mass is within the tropical zone, yet it extends into the subantarctic. The extensive zones of temperate and cold climates in the vicinity of the Equator, in the Andes, are unique. The land area of about 17,519,900–17,529,250 km² is under the jurisdiction of 13 countries (Table 49); French Guiana is governed as an overseas department of France. The region's 1995 population of c. 320 million people is estimated to reach 452 million people in 2025. Three of the world's 21 megacities are in South America: São Paulo, Buenos Aires and Rio de Janeiro (WRI, UNEP and UNDP 1994).

TABLE 49. SOUTH AMERICAN COUNTRIES[1]

Country	Territory (km²)	Population (1995)
French Guiana[2]	90,976	127,500 (1992)
Surinam	156,000	460,000
Guyana	196,850	830,000
Venezuela	882,050	21,480,000
Brazil	8,456,510	161,380,000
Colombia	1,038,700	35,100,000
Ecuador	276,840	11,820,000
Peru	1,280,000	23,850,000
Bolivia	1,084,380	8,070,000
Paraguay	397,300	4,890,000
Uruguay	174,810	3,190,000
Argentina	2,736,690	34,260,000
Chile	748,800	14,240,000

Sources:
[1] WRI, UNEP and UNDP 1994
[2] 1994 almanac

Geological setting

Although the neotropics may be conveniently considered as a single phytogeographic unit, the region is geologically complex. The neotropics include not only the South American continental plate but the southern portion of the North American plate, as well as the independent Caribbean plate (Clapperton 1993). The complicated geological history of the region, for example as these plates intermittently separated and collided through the Cretaceous and the Tertiary, provides the milieu within which plant evolution has been superimposed.

South America has been an island continent during most of the period of angiosperm evolution, whereas Central America constitutes one of the two tropical parts of the Laurasian "world continent". Both South America and North America have been moving westward, roughly in tandem, since the breakup of Pangaea in the Mesozoic. In contrast, the Antillean plate with its flotsam of Antillean islands formed only during the Cenozoic and has moved in a retrograde eastern direction, at least with respect to its larger neighbours. Whereas South America and North America have been widely separated through most of their geological histories, there has been generally increasing contact between them through most of the Cenozoic, culminating in their coalescence with formation of the Isthmus of Panama c. 3.1 million years ago (Keigwin 1978). The date of this epochal event in neotropical geological history has been gradually estimated to be younger, with estimates of 5.7 million years ago giving way to as recently as 1.8 million years ago (Keller, Zenker and Stone 1989). In addition to their Pleistocene connection via the Isthmus of Panama, South America and North America apparently were more or less directly interconnected via the proto-Antilles for a short time near the end of the Cretaceous, prior to formation of the Caribbean plate (Buskirk 1992).

The outstanding geological feature of South America is the Andes, the longest mountain range in the world, which extends in a nearly straight line of over 7000 km from the north to the southern tip of the continent. The Andes have the highest mountain in the Western Hemisphere, the highest mountain in the world's tropics, and as measured from the centre of the Earth (rather than metres above sea-level), the highest mountain in the world. The most important break in the north-south sweep of the "cordillera" is the Huancabamba Depression in northern Peru, where the eastern chain of the cordillera is entirely ruptured (by the Marañón River) and even the western chain dips to 2145 m (at the Abra de Porculla). The existence of this massive mountain range has had profound effects on plant and animal evolution in South America, and consequently has profound effects on essential conservation priorities.

In essence, the Andes represent a classical plate tectonic upthrust of continental rock, as the leading edge of the westward-moving South American plate collides with the oceanic Pacific plates. The Southern Andes are the oldest, with significant uplift already present in early Cenozoic times, prior to the Oligocene. Most of the uplift of the Central Andes was in the Miocene or later, whereas most of the uplift of the northern portion of the cordillera has been Plio-Pleistocene (van der Hammen 1974). To the north the Andes become more geologically complex, breaking into three separate cordilleras on the Ecuador/Colombia border. Much of the north-western margin of South America, including Colombia's western and central cordilleras, appears to be amassed "suspect terrane" rather than an integral part of the South American continental plate (Juteau *et al.* 1977; McCourt, Aspden and Brook 1984).

Much of the rest of the South American continent consists of two great crystalline shields that represent the western portion of what was once Gondwanaland. The north-eastern portion of the continent constitutes the Guayana Shield, whereas much of Brazil south of Amazonia is underlain by the Brazilian Shield. These two major shields were formerly interconnected across what is today the Lower Amazon. They consist of a Precambrian igneous basement overlain by ancient much-eroded Precambrian sediments. The Guayana region has been the most heavily eroded, with basement elevations mostly below 500 m interrupted by massive flat-topped table mountains, the fabled "tepuis", typically rising to 2000 m or 2500 m. The peak of the highest of these, Cerro Neblina or Pico da Neblina on the Venezuela/Brazil border, reaches an altitude of 3015 m and is the highest point in South America outside the Andes. The tepuis and similar formations are highest and most extensive in southern Venezuela, becoming smaller and more isolated to the west and east where La Macarena near the base of the Andes in Colombia and the Inini-Camopi Range in French Guiana respectively represent their ultimate vestiges.

The quartzite and sandstone of the Guayana Shield erode into nutrient-poor sands, and much of the Guayana region is characterized by extreme impoverishment of soils. The rivers draining this region are largely very acidic black-water rivers, of which the Rio Negro is the most famous.

The Brazilian Shield is generally higher and less dissected, with much of central Brazil having an elevation of 800–1000 m. The Brazilian Shield is mostly drained by clear-water rivers such as the Tapajós and Xingú.

In contrast to these ancient shields, the Amazonian heartland of South America is low and geologically young. Prior to the Miocene most of Amazonia constituted a large inland sea opening to the Pacific. With uplift of the Central Andes, this sea became a giant lake that gradually filled with Andean sediments. When the Amazon River broke through the narrow connection between the Guayanan and Brazilian shields near Santarém, Brazil, Amazonia began to drain eastward into the Atlantic. Nevertheless, the region remains so flat that ocean-going ships can reach Iquitos, Peru, which is only 110 m above sea-level, yet 3000 km from the mouth of the Amazon and less than 800 km from the Pacific Ocean. Most of Amazonian Ecuador, Peru and Bolivia is below 200 m in elevation. The process of Amazonian sedimentation is continuing, as the sediment-laden white-water rivers course down from the Andes, continually changing their channels and depositing and redepositing their sediments along the way. About 26% of Peruvian Amazonia shows direct evidence of recent riverine reworking (Salo *et al.* 1986). With the lack of relief, it is not surprising that rather fine nuances of drainage, topography and depositional history are often major determinants of vegetation.

Like Amazonia, some other distinctive geological features of the South American continent are relatively low, flat and geologically young, such as the chaco/pantanal/pampa region to the south, the Venezuelan/Colombian Llanos to the north and the trans-Andean Chocó region of Colombia and Ecuador to the west. Large portions of these areas have been inundated during periods of high sea-level in the past, and large portions of all of these regions are seasonally inundated presently.

One aspect of the geological history of Latin America that has received much biogeographic attention is the series of Pleistocene climatic fluctuations and their effects on distribution and evolution of the present neotropical biota. It is clear from the palynological record that major changes in vegetation were associated with the cycles of Pleistocene glaciation (e.g. van der Hammen 1974), although to what extent lowland Amazonia was predominantly drier (e.g. Haffer 1969; van der Hammen 1974), colder (Colinvaux 1987; Liu and Colinvaux 1988) or both, and how this affected the Pleistocene distribution of tropical forest, remain hotly contested (Colinvaux 1987; Räsänen, Salo and Kalliola 1991). Although most of the corroborative geomorphological evidence for dry periods in the tropical lowlands during the Pleistocene is now otherwise interpreted (Irion 1989; Colinvaux 1987), some new data look promising. There are also several other theories that attempt to explain aspects of present biogeography on the basis of past geological events, including river-channel formation and migration (Capparella 1988; Salo *et al.* 1986; Salo and Räsänen 1989), hypothesized massive flooding in south-western Amazonia (Campbell and Frailey 1984), and the formation of a putative giant Pleistocene lake in Amazonia (Frailey *et al.* 1988).

Mesoamerica

For its size, Middle America is even more complex geologically than South America (see Middle America regional overview, p. 97). Nuclear Central America, an integral part of the North American continent, reaches south to central Nicaragua. The region from southern Nicaragua to the isthmus of Darién in Panama is geologically younger and presents recent volcanism, uplift and associated sedimentation.

Like South America, the northern neotropics have a mountainous spine that breaks into separate cordilleras in the north. In general the Middle American cordilleras are highest to the north in Mexico, and lowest in Panama to the south-east. In Mexico, the geological picture is complicated by a band of volcanoes that bisects the continent from east to west at the latitude of Mexico City. This "eje volcanico transversal" is associated with the Mexican megashear, along which the southern half of the country has gradually moved eastward with respect to the northern half.

In southern Central America, volcanism has been most intensive in Costa Rica, which has two sections of its Central Cordillera reaching above treeline. In northern Costa Rica and adjacent Nicaragua the volcanoes become gradually reduced in size and more isolated from each other to the north. Similarly in Panama the Central Cordillera is over 2000 m high to the west near the Costa Rican border but only about 500 m high in most of the eastern part of the country. In central Panama, the Panama Canal cuts through a continental divide of only 100 m elevation, and in the San Juan River/Lake Nicaragua area of Nicaragua the maximum elevation is even less. For montane organisms, these interruptions in the cordillera represent major biological discontinuities.

The Yucatán Peninsula area of Mexico, Guatemala and Belize represents a geologically anomalous portion of Middle America. It is a flat limestone formation more like the Greater Antilles or Peninsular Florida than the mountainous terrain and volcanic soil of most of Middle America. Limestone is otherwise relatively rare in the continental neotropics, in contrast to many other parts of the world, with small outcrops like those in the Madden Lake region of central Panama or the Coloso area of northern Colombia being associated with peculiar floras. These areas, like the Yucatán Peninsula, tend to show distinctly Antillean floristic affinities, paralleling the geological ones.

Caribbean

The Antillean islands constitute the third geologic unit of the neotropics (see Caribbean Islands regional overview, p. 233). The Antilles make up in geological complexity what they lack in size. The most striking geological anomaly is Hispaniola, which is a composite of what were three separate islands during much of the Cenozoic. In addition to being completely submerged during part of the mid-Cenozoic, the southern peninsula of Hispaniola was probably attached to Cuba instead of Hispaniola until the end of the Cenozoic. Jamaica too was completely submerged during much of the mid-Cenozoic, and has a different geological history from the rest of the Greater Antilles, with closer connections to Central America via the now-submerged Nicaraguan Rise. Possibly a collision of the western end of the Greater Antilles island arc with Mexico-Guatemala fragmented its western end to form Jamaica.

Also phytogeographically and conservationally important, some of the Antilles have extensive areas of distinctive substrates. In addition to large areas of limestone, most of the Greater Antilles (Cuba, Hispaniola, Puerto Rico) have significant areas of serpentine and other ultrabasic rocks formed from uplift of patches of oceanic crust during the north-eastward movement of the Caribbean plate. The Lesser Antilles are small and actively volcanic. Most of the other smaller islands are low limestone keys with little or no geological relief. These patterns are clearly reflected in the Antillean flora. The most striking concentrations of local endemism occur in areas of ultrabasic rocks or on unusual types of limestone on the larger islands. The Lesser Antilles, Bahamas and other smaller islands have only a depauperate subset of the generally most widespread Antillean taxa.

Vegetation

The neotropics include a broad array of vegetation types commensurate with their ecological diversity. Along the west coast of South America are both one of the wettest places in the world – Tutunendó in the Chocó region of Colombia, with 11,770 mm of annual precipitation, and the driest – no rain has been recorded in parts of the Atacama Desert of Chile. The largest tract of rain forest in the world is in the Amazon Basin, and Amazonia has received a perhaps disproportionate share of the world's conservation attention. While the forests of Upper Amazonia are the most diverse in the world for many kinds of organisms, including trees as well as butterflies, amphibians, reptiles, birds and mammals, other vegetation types have equal or greater concentrations of local endemism and are more acutely threatened. In particular, the plight of dry forests and of Andean montane forests are beginning to receive increased attention. Some isolated areas of lowland moist forest outside of Amazonia also have highly endemic floras and are currently much more threatened than Amazonia. In the following paragraphs are sketched the major neotropical vegetation types, followed by a conservation assessment of each.

At the very broadest level, the lowland vegetation types of South America and the rest of the neotropics may be summarized as:

1. **Tropical moist forest (evergreen or semi-evergreen rain forest)** in Amazonia, the coastal region of Brazil, the Chocó and the lower Magdalena Valley, and along the Atlantic coast of Central America to Mexico.

2. **Dry forest (intergrading into woodland)** along the Pacific side of Mexico and Central America, in northern Colombia and Venezuela, coastal Ecuador and adjacent Peru, the Velasco area (Chiquitania) of eastern Bolivia, a broad swath from north-west Argentina to north-east Brazil encompassing chaco, cerrado and caatinga, and with scattered smaller patches elsewhere.

3. **Open grassy savanna** in the pampas region of north-eastern Argentina and adjacent Uruguay and southernmost Brazil, the Llanos de Mojos and adjacent pantanal of Bolivia and Brazil, the Llanos of Colombia and Venezuela, and the Gran Sabana and Sipaliwini savanna in the Guayana region.

4. **Desert and arid steppe** in northern Mexico, the dry Sechura and Atacama regions along the west coast of South America between 5°S and 30°S, and in the monte and Patagonian steppes of the south-eastern part of the Southern Cone of South America.

5. The **Mediterranean-climate region** of central Chile.

6. The **temperate evergreen forests** of southern Chile with an adjacent fringe of Argentina.

More complex montane formations occur along the Andean Cordillera which stretches the length of the western periphery of South America, in the more interrupted Central American/Mexican cordilleran system, in the tepuis of the Guayana region and in the coastal cordillera of southern Brazil.

Moist and wet forests

In general, forests receiving more than 1600 mm (Gentry 1995) or 2000 mm (Holdridge 1967) of annual rainfall are evergreen or semi-evergreen and may be referred to as tropical moist forest. In the neotropics, lowland tropical moist forest is often further subdivided, following the Holdridge life-zone system, into moist forest (2000–4000 mm of precipitation annually), wet forest (4000–8000 mm) and pluvial forest (over 8000 mm). Nearly all of the Amazon Basin receives 2000 mm or more of annual rainfall and constitutes variants of the moist forest. There are also several major regions of lowland moist forest variously disjunct from the Amazonian core area. These include the region along the Atlantic coast of Central America (extending into Mexico), the lower Magdalena Valley of northern Colombia, the Chocó region along the Pacific coast of Colombia and northern Ecuador, and the coastal forests of Brazil.

Lowland moist forest is the most diverse neotropical vegetation type, structurally as well as taxonomically. In most lowland moist-forest and wet-forest regions around a quarter of the species are vines and lianas, a quarter to a half terrestrial herbs (including weeds), up to a quarter vascular epiphytes and only about a quarter trees (Gentry and Dodson 1987; Gentry 1990b). To the extent that smaller organisms such as herbs and epiphytes may demand different conservation strategies than large organisms like trees (or top predators), this habitat diversity assumes conservation importance.

Diversity patterns are also important for conservation planning. There is a strong correlation of plant community diversity with precipitation – wetter forests generally are more botanically diverse. For plants the most species-rich forests in the world are the aseasonal lowland moist and wet forests of Upper Amazonia and the Chocó region. For plants over 2.5 cm dbh in 0.1-ha samples, world record sites are in the pluvial-forest area of the Colombian Chocó (258–265 species); for plants over 10 cm dbh in 1-ha plots, the world record is near Iquitos, Peru (300 species out of 606 individual trees and lianas).

Concentrations of endemism do not necessarily follow those of diversity. Local endemism appears to be concentrated in cloud-forest regions along the base of the northern Andes and in adjacent southern Central America (cf. Vázquez-García 1995), and in the north-western sector of Amazonia where the substrate mosaic associated with sediments from the Guayana Shield is most complex (Gentry 1986a). Overall regional endemism in predominantly moist-forest areas is greatest in Amazonia, with an estimated 13,700 endemic species constituting 76% of the flora (Gentry 1992d). However many of these species are relatively widespread within Amazonia.

The much more restricted (and devastated, see below) Mata Atlântica forests of coastal Brazil have almost three-quarters as many endemic species (c. 9500) as Amazonia and similarly high endemism (73% of the flora) (Gentry 1992d). Moreover a larger proportion of the Mata Atlântica species probably are locally endemic.

On the other side of South America, the trans-Andean very wet to wet and moist forests of the Chocó and coastal Ecuador are also geographically isolated and highly endemic (cf. Terborgh and Winter 1982). Estimates of endemism in the Chocó phytogeographic region are c. 20% (Gentry 1982b). Probably about 1260 or 20% of western Ecuador's 6300 naturally occurring species also are endemic (Dodson and Gentry 1991). For the northern Andean region as a whole, including both the coastal lowlands of western Colombia and Ecuador and the adjacent uplands, Gentry (1992d) estimated over 8000 endemic species, constituting 56% of the flora. Moreover this is probably the floristically most poorly known part of the neotropics, perhaps of the world, surely with several thousand mostly endemic species awaiting discovery and description.

Dry forests

There are seven main areas of dry forest in the neotropics, and by some estimations this may be the most acutely threatened of all neotropical vegetations. The interior dry areas of South America are outstanding in their regional endemism, estimated at 73%. Two of the most extensive neotropical dry-forest areas represent manifestations of the standard interface between the subtropical high pressure desert areas and the moist equatorial tropics.

In Middle America, this area of strongly seasonal climate occurs mostly along the Pacific coast in a narrow but formerly continuous band from Mexico to the Guanacaste region of north-western Costa Rica. There are also outliers farther south in the Terraba Valley of Costa Rica, Azuero Peninsula of Panama, and even around Garachiné in the Darién (Panama), partially connecting the main Middle American dry forest with that of northern South America. These western Middle American dry forests are made up almost entirely of broadleaved deciduous species. In addition, the northern part of the Yucatán and large areas of the Antilles are covered by dry-forest variants. Most of the Caribbean dry forests are on limestone, and their woody species tend to be distinctively more sclerophyllous and smaller leaved than are the Pacific coast dry-forest plants. In the driest areas, both these types of dry forest tend to smaller stature and merge into various kinds of thorn-scrub matorral.

In South America, only the extreme northern parts of Colombia and Venezuela reach far enough from the Equator to enter the strongly seasonal subtropical zone. Floristically and physiognomically this northern dry area is very much like similarly dry areas of western Middle America. The strongly seasonal region of northern South America also includes the open savannas of the Llanos extending from the Orinoco River west and north to the base of the Eastern Cordillera of the Colombian Andes and the north slope of the Coast Range of Venezuela. Large areas of the low-lying, often poorly drained Llanos are seasonally inundated, especially in the Apure region.

The main area of tropical dry forest in South America is the chaco region, encompassing the western half of Paraguay and adjacent areas of Bolivia and Argentina, south of 17°S latitude. The "chaco" is physiognomically distinctive in being a dense scrubby vegetation of mostly small-leaved, spiny branched small trees interspersed with scattered large individuals of a few characteristic species of large trees. To the south, the chaco gives way to the desert scrub of the Argentine monte.

There is a distinctive but generally neglected area of dry forest at the interface between the chaco and Amazonia in Bolivia. The names Chiquitania and Velasco forest have been used locally in Bolivia to refer to this vegetation, which extends from the Tucuvaca Valley and Serranía de Chiquitos in easternmost Santa Cruz Department interruptedly westward to the base of the Andes and along much of the lower Andean slopes of the southern half of Bolivia. This region of closed-canopy dry forest is physiognomically similar to that of western Central America, with tall broadleaved completely deciduous (caducifolious) trees. Although it has been locally regarded as merely representing the transition between the chaco and Amazonia, it is a floristically and physiognomically distinctive unit that should be accorded equivalent conservation importance to the other major dry-forest vegetation types (Gentry 1994).

The chaco is adjoined to the north by two large and phytogeographically distinctive areas of dry forest, the cerrado and caatinga, which cover a small portion of easternmost Bolivia and most of the Brazilian Shield area of central and north-eastern Brazil. The typical vegetation of the "cerrado" region consists of wooded savanna with characteristically gnarled sclerophyllous-leaved trees with thick twisted branches and thick bark, widely enough separated to allow a ground cover of grass intermixed with a rich assortment of woody-rooted (xylopodial) subshrubs. The cerrado also includes areas where the trees form a nearly closed canopy ("cerradão"), and large open areas of grasses and subshrubs with no trees at all ("campo limpio" and "campo rupestre"). Although the cerrado is appropriately considered a kind of dry forest, some cerrado regions actually receive more rainfall than do adjacent forest regions; excess aluminium in the soil may be as important as the climate in determining its distribution.

The even drier forest of the caatinga of north-eastern Brazil extends from an appropriately subtropical 17°S latitude farther north to a surprisingly equatorial 3°S. Why this region should have such low rainfall remains poorly understood. Another climatic peculiarity is the irregularity of its rainfall, not only with low annual precipitation, but also with frequent years when the rains fail almost completely. The typical vegetation of the "caatinga" – relatively low, dense, small-leaved and completely deciduous in the dry season – is physiognomically similar to that of the chaco.

The final major South American dry-forest area is the coastal forest of north-western Peru and south-western Ecuador. Even more anomalous in its geographical setting than the caatinga, this dry-forest region is positioned almost on the Equator. The occurrence of dry forest so near the Equator is due to the offshore Humboldt Current. While similar cold-water currents occur along mid-latitude western coasts of other continents, the Humboldt Current is perhaps the strongest of these and is the only cold current reaching so near the Equator. The dry forest of coastal Peru and adjacent Ecuador is (or at least was, see below) physiognomically similar to that of western Central America, tall with a closed canopy of broadleaved completely deciduous trees.

There also are a number of scattered smaller patches of tropical dry forest and/or savanna in various inter-Andean valleys, around Tarapoto, Peru, the Trinidad region of Bolivia, Brazil's Roraima area, the Surinam/Brazil border region, on Marajó Island, and in the pantanal region of the upper Paraguay River.

Grasslands and deserts

Grasslands and deserts occupy smaller areas of the neotropics than they do in Africa or most higher latitude continents. The main grassland region of the neotropics is the pampas region between about 39°S and 28°S and encompassing most of Uruguay as well as adjacent eastern Argentina and southernmost Brazil. The other major grassland area is the llanos region of Colombia and Venezuela. Smaller predominantly grassland regions occur in north-eastern Bolivia (Llanos de Mojos) and the south-eastern Guayana region (Gran Sabana and Sipaliwini savanna). There are also areas with few or no trees and dominated by grasses in the cerrado and pantanal regions of Brazil, and scattered outliers associated with local edaphic peculiarities elsewhere.

None of the major grassland regions has many endemic species, in contrast to the campos rupestres of the Brazilian Shield and the Guayana area white-sand savannas, which have many endemics. This contrast is especially marked in southern Venezuela where some savanna patches have clay soils and a llanos-type flora of widespread species, whereas others have sandy soils and a flora of Amazonian affinities with many endemic species (Huber 1982).

The desert regions of Latin America are confined to northern Mexico, the monte (Morello 1958; Orians and Solbrig 1977) and Patagonian steppes of Argentina, and the narrow Pacific coastal strip of northern Chile and Peru. The 3500-km long South American coastal desert is one of the most arid in the world – most of it is largely devoid of vegetation. This region is saved from conservational obscurity, however, by the occurrence of island-like patches of mostly herbaceous vegetation in places where steep coastal slopes are regularly bathed in winter fog. Although these "lomas" formations are individually not very rich in species (mostly fewer than 100 spp.), they have a very high degree of endemism due to their insular nature. The overall lomas flora includes nearly 1000 species, mostly annuals or geophytes. Diversity and endemism in the lomas formations generally increase southward, where cacti and other succulents are also increasingly represented (Müller 1985; Rundel et al. 1991).

Montane vegetation

The main montane-forest area of the neotropics is associated with the Andes. A major but more interrupted montane-forest strip is associated with the mountainous backbone of Central America. Venezuela's Cordillera de la Costa phytogeographically is essentially an Andean extension, although geologically distinct from the Eastern Cordillera of the Colombian Andes. The tepui summits of the Guayana Highlands, though small in area, constitute a highly distinctive and phytogeographically fascinating montane environment. The Serra do Mar along Brazil's south-eastern coast is mostly low elevation but has a few peaks reaching above treeline with a depauperate páramo-like vegetation.

The Andes may be conveniently recognized in three segments: northern – Venezuela, Colombia and Ecuador; central – Peru and Bolivia; and southern – Chile and Argentina. In general the northern Andes are wetter, the central and southern regions drier. The main biogeographic discontinuity in the Andean forests is associated with the Huancabamba Depression in northern Peru, where the extensive system of dry inter-Andean valleys of the Marañón River and its tributaries entirely bisects the Eastern Cordillera and is associated with a topographically complex region having unusually high local endemism.

Treeline in the tropical Andes occurs around 3500 m, depending on latitude and local factors. Above treeline, the wet grass-dominated vegetation of the Venezuelan, Colombian and northern Ecuadorian Andes is termed "páramo"; this drier vegetation, occurring from Peru to Argentina and Chile, is the "puna". Colombian and Venezuelan páramos are characterized by *Espeletia* (Compositae) with its typical pachycaul-rosette growth form. The vegetation above treeline of most of Ecuador and northernmost Peru, locally called "jalca" in Peru, is ecologically as well as geographically intermediate; although generally called páramo in Ecuador, this region lacks the definitive *Espeletia* aspect of the typical northern páramos.

While individual high-Andean plant communities are not very rich in species, many different communities can occur in close proximity in broken montane terrain. Thus the several high-Andean sites for which Florulas are available (Cleef 1981; Smith 1988; Galeano 1990; Ruthsatz 1977) have between 500–800 species, approaching the size of some lowland tropical Florulas.

The moist Andean slopes generally show a distinctive floristic zonation, with woody plant diversity decreasing linearly with altitude from c. 1500 m to treeline. Below 1500 m Andean forests are generally similar both in floristic composition and diversity to equivalent samples of lowland forest. There are also structural changes at different elevations. For example hemi-epiphytic climbers show a strong peak in abundance between 1500–2400 m, epiphytes are usually more numerous in middle-elevation cloud forests, and the stem density of woody plants is usually greater at higher elevations (Gentry 1992a). While the northern Andes have cloud forest on both western and eastern slopes, increasing aridity south from the Equator limits cloud forest to an ever narrower band on the Pacific slope. South of 7°S latitude, forest on the western slopes of the Andes is restricted to isolated protected pockets, and the predominant slope vegetation becomes chaparral, thorn scrub and desert.

One of the most striking features of the Andes phytogeographically is the high level of floristic endemism. In part this is associated with the discontinuity of high-altitude vegetation types, which are strongly fragmented into habitat islands. In addition to microgeographic allopatric speciation related to habitat fragmentation, it seems likely that unusually dynamic speciation, perhaps associated with genetic drift in small founder populations, may be a prevalent evolutionary theme in Andean cloud forests (Gentry and Dodson 1987; Gentry 1989).

The combination of high local endemism (Gentry 1986a, 1993a; Luteyn 1989; Henderson, Churchill and Luteyn 1991) with major deforestation makes the Andes one of South America's conservationally most critical regions. As with the dry forests, the Andean forests have recently begun to receive greater conservation attention (Henderson, Churchill and Luteyn 1991; Young and Valencia 1992). Estimates of deforestation for the northern Andes as a whole are generally over 90%. Some areas are even more critical – perhaps less than 5% of Colombia's high-altitude montane forests remain (Hernández-C. 1990) and only c. 4% of the original forest persists on the western Andean slopes of Ecuador (Dodson and Gentry 1991). Most of the northern Peruvian Andes are similarly deforested (cf. Dillon 1994). Although relatively

extensive forests still remain on the Amazon-facing slopes of Peru and Bolivia, much of this area is being actively deforested, in large part to grow "coca" (*Erythroxylum coca*) and opium poppy (*Papaver somniferum*).

Flora

From a conservation perspective, the neotropical region merits very special attention. Just as South America is sometimes called the "bird continent", the neotropics might well be termed the "plant continent" in deference to their uniquely rich botanical diversity (Table 50). If current estimates are accurate, the neotropical region contains 90,000–100,000 plant species, twice to nearly three times as many as in either tropical Africa or tropical Australasia (cf. Prance 1994).

The last great places for plant collecting are in the northern half of South America (J. Wurdack 1995, pers. comm.), which is two to four times less documented by herbarium specimens than elsewhere in the tropics (cf. Campbell 1989). Some of the main relatively unexplored areas (according to Wurdack) are, in Brazil: Serra de Tumucumaque (Tumuc-Humac Mountains), along the border with Surinam and French Guiana; slopes, especially the eastern slopes, of Pico da Neblina; in north-western Mato Grosso State, along the Linea Telegráfica; in Venezuela: slopes and talus forests of the tepuis; páramos west of Piñango (north of Mérida); eastern slopes to Páramo de Tamá (State of Mérida, near border with Colombia); in Colombia: Páramo de Frontino (west of Medellín); Cuatrecasas' headwater localities of collection in western Colombia, particularly in the Department of Valle del Cauca (cf. Cuatrecasas 1958); upper elevations of the Serranía de La Macarena (Department of Meta); in Ecuador: Cordillera de Los Llanganates (which is east of Ambato) (cf. Kennerley and Bromley 1971); Cordillera de Cutucú (Province of Morona-Santiago); Cordillera del Cóndor, along the border with Peru; in Peru: elevations above 700 m of the Cerros Campanquiz, which are mostly in the Department of Amazonas; the eastern cordillera in the Department of Amazonas, Province of Chachapoyas (e.g. the Cerro de las Siete Lagunas east of Cerro Campanario); portions of the Cordillera de Vilcabamba (which is north-west of Cusco), including the northern Cutivireni region (Villa-l obos 1995); and in Bolivia: the easternmost Andes and granitic outliers in the Department of Santa Cruz.

TABLE 50. ESTIMATED FLORISTIC INVENTORY OF SOUTH AMERICA

Number of species of flowering plants[1]

Neotropics (excluding temperate South America)	90,000
South America (with 7000 in temperate South America)	81,000

Number of species of pteridophytes[2]

Neotropics	3500
South America	2000

Number of flowering plant genera[1] — 4200

Number of endemic neotropical families[1] — c. 25 (+ 8 with 1–2 spp. in Africa)

Sources:
[1] Gentry 1982a
[2] R. Moran, pers. comm.

TABLE 51. SPECIES DIVERSITY IN SOUTH AMERICA[1,3]

I. CARIBBEAN REGION (OF SOUTH AMERICA)			**VII. PACIFIC COAST**		
Number of angiosperm species	6000		**Chocó**		
Species endemism (%)	24		Number of species		8000
Endemic families	0		Species endemism (%)		20
			Coastal Ecuador		
			Number of species		6300
II. GUAYANA SHIELD[2]			Species endemism (%)		20
Guayanan Venezuela (3 states)			**Lomas and coastal desert**		
Number of species	9300		Number of species		1000
Species endemism (%)	24		Species endemism (%)		40
(40% endemic to Guayana Shield)			Endemic families, shared with central Chile		2
Pantepui (Venezuela over 1300–1500 m)			(Nolanaceae, Malesherbiaceae)		
Number of species	2300				
Species endemism (%)	65		**VIII. SOUTHERN CONE (CONOSUR)**		
Endemic families	1–3		**Chile**		
(Hymenophyllopsidaceae, and perhaps at			Number of species		5200
family level Tepuianthaceae and Saccifoliaceae)			Endemic families (below plus Calyceraceae)		5
Endemic genera	120		**Mediterranean Chile**		
			Number of species		1800–2400
			Species endemism		High
III. AMAZONIA sensu lato			Endemic families		2
Number of species	18,000		(Gomortegaceae, Malesherbiaceae)		
Species endemism (%)	76		**Altoandina (High Andes)**		
Endemic families	4		Number of species		?
(Dialypetalanthaceae, Rhabdodendraceae,			Species endemism		?
Duckeodendraceae, Thurniaceae)			Endemic families		0
			Monte		
IV. MATA ATLÂNTICA			Number of species		700
Number of angiosperm species	13,000		Species endemism (%)		5
Species endemism (%)	73		Endemic family, shared with Patagonia		1
Endemic families	0		(Halophytaceae)		
			Patagonia		
V. INTERIOR DRY AND MESIC FORESTS			Number of species		1200
Number of angiosperm species	9300		Species endemism (%)		30
Species endemism (%)	73		Endemic family, shared with monte		1
Endemic families	0		(Halophytaceae)		
			Temperate rain forest		
VI. (TROPICAL) ANDES			Number of species		450
Northern Andes (incl. Chocó)			Species endemism		?
Number of species	14,500		Endemic families (Myzodendraceae, Aextoxicaceae)		2
Species endemism (%)	56				
Central Andes (Peru-Bolivia)			Note:		
Number of species	13,000		[1] Data mostly from Gentry (1982a, 1992d)		
Species endemism (%)	54		[2] Data from Berry, Huber and Holst (1995)		
Endemic family (Columelliaceae)	1		[3] Vascular plants unless stated otherwise		

Floristic diversity is very asymmetrically distributed in South America (cf. Table 51). If the nine phytogeographic regions recognized by Gentry (1982a) for the neotropics are taken as a basis, Central America with Mexico (Mesoamerica) and Amazonia are the richest in species, with each of these two regions having about a quarter of the neotropical total. At the opposite extreme, the Antilles have an estimated 9% of the total neotropical flora and the Caribbean coastal region of Colombia and Venezuela has only 8%. The minuscule area of the Guayana Highlands (above 1500 m) accounts for only c. 2.5% of the neotropical flora, but has one of the highest rates of endemism (65%) in the region (Berry, Huber and Holst 1995). The three main tropical South American dry areas together include a relatively low 11% of the neotropical species total. Intermediate levels of regional plant species richness are found in the Northern Andean and Southern Andean regions and the Mata Atlântica area of Brazil, which each have between 16–18% of the tropical flora of the neotropical region.

Regional endemism is greatest in Amazonia including lowland Guayana (76%), but almost as great in coastal Brazil (73%) and the chaco-cerrado-caatinga dry areas (73%). In contrast, those two Andean subregions, Central America, and the Antilles have endemism levels of 54–60%, and the northern Colombia/Venezuela region only 24%.

Farther south in the Southern Cone of South America, the monte of Argentina is estimated to include 700 species with 5% endemism, and Patagonia 1200 species with 30% endemism. Chile as a whole has 5215 species (Marticorena and Quezada 1985; Marticorena 1990), with 1800–2400 in the Mediterranean-climate area of central Chile where endemism is high, perhaps greater than for any of the equivalent tropical regions.

The reasons for the unique floristic diversity of the neotropics as compared to Africa or tropical Australasia continue to be hotly debated. A popular theory is allopatric multiplication of species in habitat-island forest refugia during Pleistocene glacial advances (Haffer 1969; Prance 1973, 1982). Africa, which is higher and drier, would have had fewer refugia and more extinction. Tropical Asia was less affected, being buffered by the nearby ocean due to the island status of its components and by its proximity to a rain source from the Pacific (the world's largest ocean). Other theories, not necessarily mutually exclusive (cf. Terborgh and Winter 1982), focus on explosive speciation in the more extensive cloud-forest area of the neotropics (Gentry 1982a, 1989; Gentry and Dodson 1987); "Endlerian" speciation associated with habitat specialization in the uniquely complicated habitat mosaic of north-western and north-central Amazonia (Gentry 1986a, 1989; Gentry and Ortiz-S. 1993); speciation associated with riverine barriers to gene flow in the largest river system of the world (Capparella 1988; Ducke and Black 1953); or biogeographical phenomena associated with the Great American Interchange and stemming from the direct juxtaposition of Laurasian and Gondwanan elements via the Isthmus of Panama (Gentry 1982a; Marshall *et al.* 1979).

Social and environmental values, and economic importance

The indigenous groups (nations) of South America (Gray 1987) are varyingly diverse peoples who often partly depend directly on the natural environment for their biological and cultural well being or survival. Their approximate presence is shown in Table 52.

As the site of one of the Vavilovian centres of domestication, South America has played an important role in providing plants useful to people. The Andean centre of domestication rivals the Indo-Malayan and Mediterranean areas as the region that has produced the most important crop plants. Tobacco, potatoes, grain amaranths, quinoa, peanuts, lima beans, kidney beans, tomatoes and perhaps sweet potatoes and pineapples all derive from the Peruvian Andes and immediately adjacent regions (Anderson 1952). Based on land-race diversity, western Amazonia was the centre of domestication of a series of less well-known but increasingly important crops, including "pejibaye" or peach palm (*Bactris gasipaes*), "biriba" or "anona" (*Rollinia mucosa*), "abiu" or "caimito" (*Pouteria caimito*), "sapota" (*Quararibea cordata*), "araza" (*Eugenia stipitata*), "uvilla" (*Pourouma cecropiifolia*) and "cubiu" or "cocona" (*Solanum sessiliflorum*) (Clement 1989). Of the 86 major crops and their more than 100 species included in a summary of crop plant evolution (Simmonds 1976), 24 crops are neotropical in origin either wholly (19) or partly (5).

Also, a host of South American forest plants are used locally but have not reached world commerce. Amazonia is especially rich in wild fruits (e.g. Duke and Vásquez 1994). For example around Iquitos, Peru, 139 species of forest-harvested fruits are regularly consumed, 57 of them important enough to be sold in the local produce market (Vásquez and Gentry 1989). There are a multitude of other uses for neotropical plants. Gentry (1992b) notes that 38% of the Bignoniaceae species of north-western South America have specific ethnobotanical uses and suggests that this could be extrapolated to 10,000 species with uses in this part of the world alone.

Many studies have shown that the direct economic value of such products can be very high (e.g. Peters, Gentry and Mendelsohn 1989; Balick and Mendelsohn 1992). In a single

hectare of species-rich tropical forest near Iquitos, 454 of the 858 trees and lianas of dbh 10 cm or more have actual or potential uses (Gentry 1986c), with the hectare of forest potentially producing US$650 worth of fruit and US$50 worth of rubber per year. If the 93 m^3 of sellable timber worth US$1000 is included, the net present value of the hectare of forest is US$9000, far more than the net present value of managed plantations or cattle-ranching.

Additionally, the major role of forested areas in controlling erosion, recycling rainfall and as a carbon sink are now well known. As the territory with the largest tropical forest remaining in the world, South America plays a major role in providing such regional and planetary environmental services.

Loss, threats and conservation

Although the neotropical region has the most forest, it is also losing more forest each year than any other area of tropical forest (Myers 1982; Reid 1992). In western Ecuador only 4% of the original forest cover remains (Dodson and Gentry 1991). Much attention has focused on Brazil, which includes 48% of the South American area. Perhaps the most definitive satellite analysis of deforestation in Amazonia to date (Skole and Tucker 1993) indicates that as of 1988 only c. 10% of Brazilian Amazonia had been deforested, but if allowance is made for a 1-km edge effect, fully 20% of Brazilian Amazonia had been impacted. Deforestation in Rondônia alone has been c. 4000 km^2 per year, reaching almost 40,000 km^2 or 15% of the state by 1989 (Malingreau and Tucker 1988; Fearnside 1991). In coastal Brazil estimates of surviving forest range from 2% (IUCN and WWF 1982) to 12% (Brown and Brown 1992).

Burgeoning populations are the biggest factor in the ongoing losses, although political and economic instability in some areas, and short-sighted "development" programmes in other areas, also play significant roles. In most of the neotropics, unlike much of the Old World, commercial lumbering operations have played a relatively small role so far.

Conservational awareness throughout the region has increased dramatically in the past few years. Not only are increasing numbers of National Parks and similar conservation units being set aside, but there is also rapidly growing interest in the possibility of sustainable use of tropical forests as a conservation strategy.

Unfortunately many destructive and unsustainable uses of forest can masquerade behind the banner of sustainable use. Making this promising new concept fulfil its potential remains a major challenge. Similarly the growing appreciation of the potential value of biodiversity has been accompanied by too much political preoccupation and posturing about sovereignty over potential genetic resources.

Despite such problems, it is clear that the diversity of rainforest plant life is intrinsically valuable. South America, botanically the richest continent, is also the greatest repository of potentially useful plants. Conservation of South America's plant diversity is clearly a world conservational priority.

Centres of plant diversity and endemism

The following overview of South American species diversity is subdivided into eight large regions: Caribbean, Guayana Highlands, Amazonia, Mata Atlântica, Interior dry and mesic

TABLE 52. THE AMERINDIANS OF SOUTH AMERICA

Country	Population	Identity
French Guiana	4500	6 indigenous nations
Surinam	8000	c. 5 indigenous nations
Guyana	45,000	c. 6 indigenous nations
Venezuela	150,000	c. 30 language groups
Brazil	over 225,000	c. 225 indigenous nations
Colombia	300,000	over 60 indigenous nations
Ecuador	c. 3,000,000	c. 10 language groups
Peru	9,100,000	c. 62 indigenous nations
Bolivia	4,150,000	c. 32 indigenous nations
Paraguay	80,000	17 indigenous nations
Argentina	c. 350,000	16 indigenous nations
Chile	1,020,000	c. 4 indigenous nations

Source: Gray 1987

forests, Andes, Pacific Coast, and Southern Cone. The delimitation of these regions has been made mainly on geographical and floristic criteria. They do not represent phytogeographic regions in the strict sense, since in some cases it was necessary to join several geographical regions or subregions together, in order to allow a broad evaluation of the floristic resources of the South American continent.

The 46 sites that have been recognized as centres of plant diversity (see Map 36 and Table 53) of course do not represent all of the important centres in South America. Although some additional important sites are mentioned below, comprehensive coverage will require a great deal more effort.

I. CARIBBEAN REGION (OF SOUTH AMERICA)

The coastal area of northern Colombia and Venezuela is phytogeographically distinct from the rest of South America and more closely related floristically to Central America and the West Indies. The region includes the Cordillera de la Costa of Venezuela as well as the Llanos savanna grasslands, and the strongly seasonal lowland forests bordering the Caribbean coast. The latter includes a small area of dry thorn scrub in the most northern, driest part of the region on the Guajira Peninsula. The area north of the Orinoco River is open savanna interrupted by gallery forests along the major rivers, and is phytogeographically completely different from adjacent Amazonia. The region is delimited to the south by the Orinoco and its major Colombian tributary the Guaviare, and to the west the limit is the 500 m contour of the Andean Cordillera Oriental. The division between the Andes proper and the Coastal Cordillera is not the Colombia/Venezuela border, but near the Yaracuy Depression; thus the uplands of Táchira, Mérida and Trujillo states of Venezuela are included in the Andean region.

Around the northern margins of the Central Andean and Western Andean cordilleras in Colombia, delimitation of the Caribbean region from the Andes and from the western Colombian Chocó region is more complicated. West of the Sinu River and especially in the Atrato Delta region, the Caribbean coastal area is much less seasonal and phytogeographically might more appropriately be included in the Chocó region. The vegetation of much of the central Magdalena Valley of Colombia is also humid moist and wet forest and has phytogeographical affinities with both the Chocó and Amazonia. There is a significant endemic element in the moist part of the Magdalena, which also includes a large swampy zone with several distinctive floristic elements. However the upper Magdalena Valley is strongly seasonal dry forest and phytogeographically a part of the Caribbean coastal region. The wet-forest part of the Magdalena Valley and the adjacent Nechi area are referred to the Pacific Coast region, whereas the bulk of the Magdalena Valley up to 1000 m in elevation and south to Neiva at 3°N latitude is included in the Caribbean region. The narrower dry interior Cauca Valley can similarly be referred to the Caribbean region.

Flora

The Caribbean region of South America as a whole has perhaps 6000 angiosperm species and a relatively low regional endemism estimated at 24% (Gentry 1982a). This is by far the lowest level of regional endemism of the neotropical phytogeographic regions recognized by Gentry (1982a). The lowland forests of the Caribbean region are highly seasonal and not very diverse botanically. Similarly the Llanos have very little endemism, unlike their physiognomic counterpart the cerrado.

The Cordillera de la Costa, geologically not a part of the Andes, is geographically in essence an Andean outlier, but floristically may be more similar to Central America. There are also some disjunctions with Guayana and Amazonia (Steyermark 1982). This mountainous region has a more diversified flora and a significant complement of endemic species, estimated at 10%. Endemism and species richness are both concentrated in the wet montane areas. A hint of the floristic richness comes from the Florula of Cerro Avila above Caracas (Steyermark and Huber 1978), which includes 1892 species of vascular plants. This is the most species contained in any similar Florula, but the data are not strictly comparable, since the Avila Florula covers several different vegetation zones.

Most of the flora of the region is shared with that of Central America, and there seems little doubt that there must once have been more or direct connections between the seasonally dry forests of coastal Colombia and Venezuela and those of western Central America. There are also several small areas on limestone hills near the coast (e.g. Coloso, Colombia) which have a few taxa of distinctly Caribbean affinities, such as *Buxus*.

Species of economic importance

The indigenous peoples of the dry Caribbean coastal region eat the fruits of a variety of local species including *Bactris gasipaes, Chrysobalanus icaco, Geoffroea* and *Acrocomia*. In moister regions *Gustavia superba* is a familiar native species with an edible fruit. In more upland forests of the Sierra Nevada de Santa Marta, the wild-harvested fruits of *Metteniusa edulis* ("cantyi") and *Pouteria arguacoensium* ("manzana") are dietary staples (Cuadros 1990). The coastal region also has many tree species with useful wood, including both widespread taxa like *Cordia alliodora, Aspidosperma polyneuron, Tabebuia rosea*, and endemic species like *Tabebuia billbergii* – the national tree of Venezuela (Cuadros 1990; Gentry 1992a).

Threats

The coastal areas of northern Colombia and Venezuela have been intensively occupied, following the arrival of the first European colonizers in the early 16th century. They are now very densely populated, and have some of the largest cities of the region (Caracas, Maracaibo, Cartagena). The main impact on the flora and vegetation has been caused by shifting cultivation and large-scale deforestations for agricultural and urban purposes. In general the entire region is heavily affected, especially at lower elevations, but there are still some remnants of montane cloud forests protected in several National Parks and Natural Monuments.

The survival of representative and viable samples of arid vegetation in north-eastern Colombia and northern Venezuela is being heavily affected by large-scale conversion into agricultural lands, selective timber extraction – especially boles of certain arborescent cacti for furniture, and browsing

MAP 36. CENTRES OF PLANT DIVERSITY AND ENDEMISM: SOUTH AMERICA
The map shows the general locations of the CPD Data Sheet sites. See Table 53 for the key to the Data Sheet site numbers.

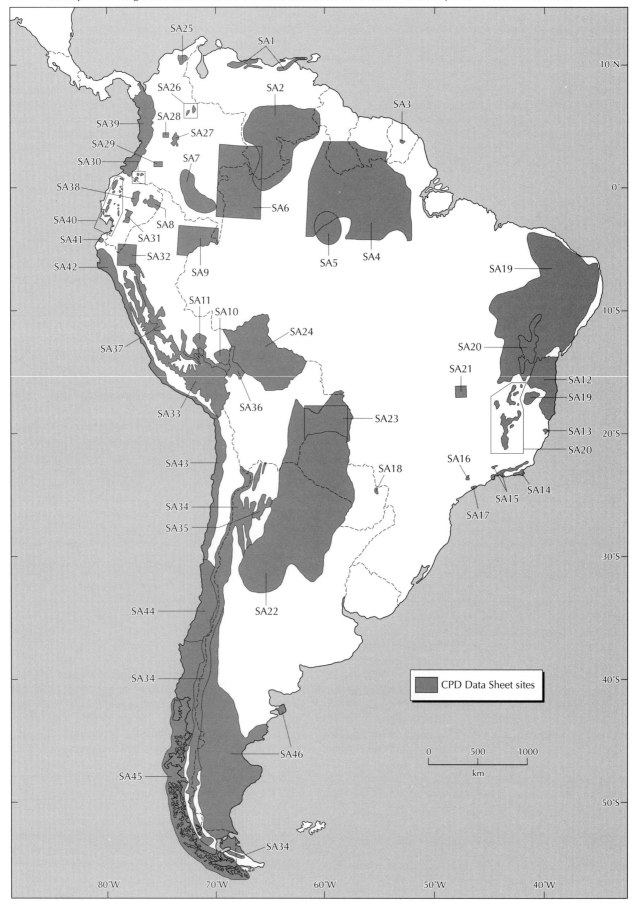

TABLE 53. SOUTH AMERICAN SITES IDENTIFIED AS CENTRES OF PLANT DIVERSITY AND ENDEMISM

The list of sites is arranged according to the sequence in the Regional Overview. All sites have been selected for Data Sheet treatment (see Map 36).

I. CARIBBEAN REGION (OF SOUTH AMERICA)
SA1.   **Coastal Cordillera** (Venezuela)

II. GUAYANA HIGHLANDS
SA2.   **Pantepui region** (Venezuela, Brazil, Guyana)

III. AMAZONIA
a. NORTH-EASTERN AMAZONIA–GUAYANA
SA3.   **Saül region** (French Guiana)
b. TRANSVERSE DRY BELT
SA4.   **Transverse Dry Belt** (Brazil, Guyana, Surinam)
c. CENTRAL AND GUAYANAN AMAZONIA
SA5.   **Manaus region** (Brazil)
SA6.   **Upper Rio Negro region** (Brazil, Colombia, Venezuela)
SA7.   **Chiribiquete-Araracuara-Cahuinarí region** (Colombia)
d. WESTERN AMAZONIA
SA8.   **Yasuní National Park and Waorani Ethnic Reserve** (Ecuador)
SA9.   **Iquitos region** (Peru, Colombia)
e. SOUTH-WESTERN AMAZONIA
SA10. **Tambopata region** (Peru)
f. PRE-ANDEAN AMAZONIA
SA11. **Lowlands of Manu National Park: Cocha Cashu Biological Station** (Peru)

IV. MATA ATLÂNTICA
a. NORTHERN REGION (Rio Grande do Norte to Bahia)
SA12. **Atlantic moist forest of southern Bahia** (Brazil)
b. CENTRAL REGION (Espírito Santo to São Paulo)
SA13. **Tabuleiro forests of northern Espírito Santo** (Brazil)
SA14. **Cabo Frio region** (Brazil)
SA15. **Mountain ranges of Rio de Janeiro** (Brazil)
c. SOUTHERN REGION (southern São Paulo to Rio Grande do Sul)
SA16. **Serra do Japi** (Brazil)
SA17. **Juréia-Itatins Ecological Station** (Brazil)
d. INTERIOR: PARANA BASIN
SA18. **Mbaracayú Reserve** (Paraguay)

V. INTERIOR DRY AND MESIC FORESTS
SA19. **Caatinga of north-eastern Brazil**
SA20. **Espinhaço Range region** (Brazil)
SA21. **Distrito Federal** (Brazil)
SA22. **Gran Chaco** (Argentina, Paraguay, Brazil, Bolivia)
SA23. **South-eastern Santa Cruz** (Bolivia)
SA24. **Llanos de Mojos region** (Bolivia)

VI. (TROPICAL) ANDES
a. PARAMO WITH ESPELETIINAE (e.g. Espeletia)
SA25. **Sierra Nevada de Santa Marta** (Colombia)
SA26. **Sierra Nevada del Cocuy-Guantiva** (Colombia)
SA27. **Páramo de Sumapaz region** (Colombia)
SA28. **Region of Los Nevados Natural National Park** (Colombia)
SA29. **Colombian Central Massif**
SA30. **Volcanoes of Nariñense Plateau** (Colombia, Ecuador)
b. PARAMO WITHOUT ESPELETIINAE TO PUNA
SA31. **Páramos and Andean forests of Sangay National Park** (Ecuador)
SA32. **Huancabamba region** (Peru and Ecuador)
SA33. **Peruvian puna**
SA34. **Altoandina** (Chile and Argentina) (brief Data Sheet)
c. TUCUMAN-BOLIVIAN REGION
SA35. **Anconquija region** (Argentina)
d. EASTERN SLOPE
SA36. **Madidi-Apolo region** (Bolivia)
SA37. **Eastern slopes of Peruvian Andes**
SA38. **Gran Sumaco and Upper Napo River region** (Ecuador)

VII. PACIFIC COAST
SA39. **Colombian Pacific Coast region (Chocó)**
SA40. **Ecuadorian Pacific Coast mesic forests**
SA41. **Cerros de Amotape National Park region** (Peru)
SA42. **Lomas formations** (Peru)
SA43. **Lomas formations of the Atacama Desert** (Chile)

VIII. SOUTHERN CONE
SA44. **Mediterranean region and La Campana National Park** (Chile)
SA45. **Temperate rain forest** (Chile and Argentina)
SA34. **Altoandina** (Chile and Argentina) (brief Data Sheet)
SA46. **Patagonia** (Argentina and Chile)

The list below is arranged alphabetically by country for purposes of cross-reference.

ARGENTINA
SA22. **Gran Chaco**
SA34. **Altoandina**
SA35. **Anconquija region**
SA45. **Temperate rain forest** (Data Sheet mainly on Chile)
SA46. **Patagonia**

BOLIVIA
SA22. **Gran Chaco**
SA23. **South-eastern Santa Cruz**
SA24. **Llanos de Mojos region**
SA36. **Madidi-Apolo region**

BRAZIL
SA2.   **Pantepui region** (Data Sheet mainly on Venezuela)
SA4.   **Transverse Dry Belt**
SA5.   **Manaus region**
SA6.   **Upper Rio Negro region**
SA12. **Atlantic moist forest of southern Bahia**
SA13. **Tabuleiro forests of northern Espírito Santo**
SA14. **Cabo Frio region**
SA15. **Mountain ranges of Rio de Janeiro**
SA16. **Serra do Japi**
SA17. **Juréia-Itatins Ecological Station**
SA19. **Caatinga of north-eastern Brazil**
SA20. **Espinhaço Range region**
SA21. **Distrito Federal**
SA22. **Gran Chaco**

CHILE
SA34. **Altoandina**
SA43. **Lomas formations of the Atacama Desert**
SA44. **Mediterranean region and La Campana National Park**
SA45. **Temperate rain forest**
SA46. **Patagonia**

COLOMBIA
SA6.   **Upper Rio Negro region**
SA7.   **Chiribiquete-Araracuara-Cahuinarí region**
SA9.   **Iquitos region**
SA25. **Sierra Nevada de Santa Marta**
SA26. **Sierra Nevada del Cocuy-Guantiva**
SA27. **Páramo de Sumapaz region**
SA28. **Region of Los Nevados Natural National Park**
SA29. **Colombian Central Massif**
SA30. **Volcanoes of Nariñense Plateau**
SA39. **Colombian Pacific Coast region (Chocó)**

ECUADOR
SA8.   **Yasuní National Park and Waorani Ethnic Reserve**
SA30. **Volcanoes of Nariñense Plateau**
SA31. **Páramos and Andean forests of Sangay National Park**
SA32. **Huancabamba region**
SA38. **Gran Sumaco and Upper Napo River region**
SA40. **Ecuadorian Pacific Coast mesic forests**

FRENCH GUIANA
SA3.   **Saül region**

GUYANA
SA2.   **Pantepui region** (Data Sheet mainly on Venezuela)
SA4.   **Transverse dry belt** (Data Sheet mainly on Brazil)

PARAGUAY
SA18. **Mbaracayú Reserve**
SA22. **Gran Chaco**

PERU
SA9.   **Iquitos region**
SA10. **Tambopata region**
SA11. **Lowlands of Manu National Park: Cocha Cashu Biological Station**
SA32. **Huancabamba region**
SA33. **Peruvian puna**
SA37. **Eastern slopes of Peruvian Andes**
SA41. **Cerros de Amotape National Park region**
SA42. **Lomas formations**

SURINAM
SA4.   **Transverse dry belt** (Data Sheet mainly on Brazil)

VENEZUELA
SA1.   **Coastal Cordillera**
SA2.   **Pantepui region**
SA6.   **Upper Rio Negro region**

by goats. Dry forests are threatened by logging and deforestation for pasture and agricultural lands.

Conservation

Most of this region has been devastated by human activity. The situation is especially critical in northern Colombia, where the largest patches of remnant natural vegetation include rather few km². The shining exception is the dry forest (not the moist forest which is mostly secondary) of Tayrona Natural National Park (Rangel-Ch. and Lowy-C. 1995) on the north flank of the Sierra Nevada de Santa Marta (CPD Site SA25); this is one of the best preserved dry forests in the neotropics and deserves more international conservation attention. (In contrast, the nearby Salamanca NNP protects mostly dead mangroves, killed as a result of construction of the coastal highway.)

In Venezuela six National Parks encompassing almost 5000 km² protect coastal and insular ecosystems along the Coastal Cordillera, and eight National Parks encompassing some 5130 km² protect more mountainous parts of the entire Coastal Cordillera between Paria and San Luis, although some of them have largely second-growth vegetation. Especially significant is Henri Pittier NP (established 1937 – one of the early National Parks in Latin America), which includes not only an area of rich cloud forest but also a transect of Caribbean-slope dry forest.

Centres of plant diversity and endemism

Venezuela

SA1. Coastal Cordillera

– see Data Sheet.

II. GUAYANA HIGHLANDS

The Guayana Highlands are defined here as a distinct floristic entity extending between (1300–) 1500–3000 m, comprising all upper table mountains ("tepuis") located within the area of the ancient Guayana Shield. There are approximately 50 different tepui summits ranging in elevation between 1800–2500 m, which are concentrated mostly in southern Venezuela, with a few outliers in Brazil (Pico da Neblina) and Guyana (Mounts Ayanganna and Wokumung). The Guayana Highlands thus occupy a fragmented area of approximately 20,000 km², which is embedded in the northern border region of Amazonia and is limited farther to the north by the Caribbean region. Others (e.g. Huber 1987, 1994) consider the Guayana Highlands instead as a province of a phytogeographically separate region, the Guayana region, calling this distinctive high-mountain area Pantepui.

Flora

Although the Guayana Highlands flora is not exceedingly rich (approximately 2300 species – Berry, Huber and Holst 1995), its level of endemism is high (c. 65%) and mainly restricted to the tepui summit vegetation. The mountain systems of Neblina, Chimantá, Duida-Marahuaca and Auyán-tepui have the highest endemism. Endemism is particularly concentrated in the families Theaceae, Rapateaceae, Ochnaceae, Compositae (tribe Mutisieae), Bromeliaceae, Xyridaceae and Eriocaulaceae; in addition there are one to three endemic families – the fern family Hymenophyllopsidaceae and the perhaps good families Tepuianthaceae and Saccifoliaceae. The high-tepui flora also contains a number of Andean taxa, especially amongst Ericaceae, Winteraceae and Podocarpaceae, with several locally endemic genera and species.

Another outstanding feature of the Guayana Highlands is its remarkable ecological diversity, which is most pronounced in the herbaceous and scrub formations. Many peculiar life forms are found in these ecosystems, such as the tubular herbs of *Brocchinia* (Bromeliaceae) and *Heliamphora* (Sarraceniaceae), the two-ranked (distichous) broadleaved herbs of *Stegolepis, Marahuacaea* and *Phelpsiella* (all Rapateaceae) and the strange stalked-rosette shrubs of *Bonnetia* (Theaceae) and *Chimantaea* (Compositae).

Species of economic importance

Many indigenous tribes of the surrounding lowlands consider the tepuis as mythical or sacred mountains with inaccessible or forbidden summits; therefore the plants of this region are unknown to the Amerindians rather than being used. The present economic importance of the tepui flora lies mainly in its genetic potential. Some species are beginning to be sought after by plant hunters for horticultural purposes, such as the species of *Heliamphora* and some particularly attractive orchids and bromeliads.

Threats

The main threat to the tepui flora and vegetation is increasing mass tourism. Many species are very fragile or brittle and do not withstand major physical impact (e.g. trampling or other mechanical damage). Furthermore, due to their low fire resistance, escaped or intentional fires originating from tourist camps represent one of the main threats to these unique vegetation types.

Conservation

At present all tepui summits of the Venezuelan Guayana Highlands are protected in six National Parks, 28 Natural Monuments and one Biosphere Reserve, which were declared during recent years. However, only two National Parks (Canaima and Duida-Marahuaca) have their management plans legally approved, and their implementation is limited by severe understaffing.

Centres of plant diversity and endemism

Venezuela, Brazil, Guyana

SA2. Pantepui region of Venezuela

– see Data Sheet.

III. AMAZONIA

The geological Amazon Basin is a large horseshoe-shaped lowland 3000 km long and 300–800 km wide (Leopoldo *et al.* 1987). It is bordered to the west and south-west by the Andes, and to the north and south respectively by the Precambrian crystalline shields of the Guianas (elevations to 1000–3000 m) and the Brazilian tableland (elevations mostly 600–1300 m) (Haffer 1987; Takeuchi 1961). Over the last 3000 km the river falls only 60 m (Leopoldo *et al.* 1987). The overall region's geology, topography, dynamics and climate are now known to be neither uniform nor simple, but the myth of a vast and homogenous stable forest is difficult to eradicate (Daly and Prance 1989; Prance 1987; Sioli 1984).

The basin is divided into three approximate regions based on differences in the rivers and the geological history of each region (Junk and Furch 1985; Putzer 1984; Sioli 1984):

1. Upper Amazonia (which drained westward until uplift of the Andes in the later Tertiary) extends as a broad funnel along the Andean foothills, narrowing eastward almost as far as Manaus and the confluence of the Negro and Amazonas/Solimões rivers. Much of its soil is sediment from the Andes (Anderson 1981; Daly and Prance 1989).

2. Much of Lower Amazonia is known as Central Amazonia (and the middle basin). Bound between the ancient crystalline shields, it is narrower (the funnel's spout) and extends from about Leticia in Colombia eastward beyond Manaus to about the mouth of the Xingú River. Central Amazonia is bordered to the north-west and north by white-sand areas, and to the east and south-east by a dry zone (Transverse Dry Belt) following the Trombetas River Basin and the Madeira/Tapajós interfluve, but has an undefined southern and south-western boundary.

3. The small easternmost region of Amazonia, from a bit east of the mouth of the Xingú River to the Atlantic Ocean, includes the river divided into several arms, with delta, estuaries and many islands including Marajó Island (50,000 km², the size of Switzerland). Hydrologically, Amazonia does not include the south-eastern basin (1 million km²) formed by the Araguaia-Tocantins rivers, which is often included in Amazonian estimates, but essentially drains independently into the Atlantic just south-east of the Amazon River (near Belém) (Leopoldo *et al.* 1987).

There has been considerable discussion as to the boundaries of Amazonia (UNDP 1992). While much of the Andean Cordillera drains into the Amazon River and is sometimes included in a broad concept of Amazonia, the upland Andes are phytogeographically distinct from lowland Amazonia. In contrast, the Guianas are not hydrologically a part of the Amazon Basin, since their rivers drain directly into the Atlantic Ocean, but the lowland Guianas phytogeographically can be considered part of the Amazon.

The broad limits of Amazonian vegetation with its tropical forests (the hylaea) extend well beyond the geological basin. The southern as well as north-western limits are imprecise – gradual, mosaic or abrupt transitions as species are replaced because of changes in climate and soils. To the north-east, the Guianas are included up to 900 m, where the Guayana Highlands are considered to begin (Daly and Prance 1989).

As defined here, Amazonia covers c. 7.6 million km², including lowland Guayana. This northern Amazonian limit is roughly contiguous with the south edge of the Llanos, approximately along the Vichada and Orinoco rivers. To the west Amazonia is delimited by the Andean escarpment. For phytogeographical analysis, the 500 m contour, above which many lowland tropical species do not ascend, may be used to divide the Amazonian and Andean regions. There are also narrow extensions of Amazonian forest along the base of the Andes, most strikingly in Bolivia where the sub-Andean tail of Amazonia reaches 17°S latitude. Amazonia is generally limited southward by the escarpment of the Brazilian Shield and the associated cerrado vegetation.

The landscape within the basin includes dissected sedimentary plains or plateaux and crystalline terrain as well as hilly landforms and wide alluvial terrains (Bigarella and Ferreira 1985). Amazonian soils are sedimentary deposits, to 300 m deep; in Lower Amazonia they have developed over 20 million years. Most of the soils are mixtures of kaolinite (high in aluminium), iron oxides and quartz; they are nutrient-poor owing to their origins and millennia of intense weathering. Calcium, magnesium and phosphorus are very scarce, with phosphorus a limiting factor for the entire ecosystem (Bigarella and Ferreira 1985; Chauvel, Lucas and Boulet 1987; Irmler 1977; Jordan 1985; Leopoldo *et al.* 1987). Nevertheless there is an extensive diversity of c. 22 soil units recognized in six classes. Latosols and podzols are frequent, with gleysols (clays and silts) nearer the rivers, and more sandy soils to the north-west, south and east (Brown 1987). Soils appear to correlate poorly with large-scale phytogeography in Amazonia, but can be critical in explaining local distributions of species (Daly and Prance 1989; Jordan 1985).

The tributary rivers of the Amazon differ in the physical and chemical properties of their waters; these variations strongly affect the vegetation. The three basic types are:

1. White-water rivers especially of Upper Amazonia, which erode and drain the Andes, carrying a heavy sediment load; they are neutral or slightly acidic. The Solimões River and Amazon River itself are examples. Their floodplains have relatively fertile clay soils (Leopoldo *et al.* 1987; Sioli 1984).

2. Clear-water rivers mainly of Lower Amazonia, draining the ancient crystalline shields. Already long since stripped of nutrients, they have a very low sediment load, and vary greatly from acidic to alkaline (e.g. Tapajós and Xingú rivers).

3. Black-water rivers (e.g. the Rio Negro), which drain the geologically old, nutrient-poor, low (over 200 m) sandy area especially to the north-west between the Andes and the Guayana Shield; their waters are very dark and acidic (especially with humic and fulvic acids) from incomplete decomposition of the organic forest litter, and also have very low sediment loads. Their floodplains tend to be sandy and poor in nutrients (Leopoldo *et al.* 1987; Sioli 1984).

Most tributaries of the Amazon are of clear or black water. Most white waters are from the south-western quarter of the basin (the Andean drainage that becomes the Solimões and Madeira rivers). The annual inundation, which affects at least 100,000 km² of forest, may last 2–10 months (rarely all year) and be up to 20 m in depth. In Lower Amazonia "terra firme" (i.e. uplands) and the boundaries of floodplains are generally distinct (Daly and Prance 1989; Goulding 1988).

Annual rainfall over most of Amazonia exceeds 2000 mm, with a single pronounced dry season. The annual precipitation in Central Amazonia is 2000–2500 mm. Rainfall generally increases and seasonality decreases to the west, with the area from Leticia to Iquitos and into Ecuador having little seasonality and rainfall in excess of 3000 mm. There is also a high-rainfall area centred in French Guiana, and there are local areas of relatively high rainfall scattered around the periphery of Amazonia.

Vegetation

Most of Amazonia is covered by various manifestations of tropical rain forest, but there are important phytogeographical differences between forests on different substrates. Seasonally inundated forests occur in an estimated 2% of Amazonia immediately adjacent to the major rivers. Forests inundated by black-water rivers ("igapó") have very different sets of species than those inundated by white-water rivers ("várzea, tahuampa"). Similarly forests on white-sand soils ("campinarana, wallaba, varillal") are floristically very different from those on clay soils, and both are distinct from alluvial-soil terra-firme forests. There are also a rich array of different successional stages of these various forest types. Finally, small patches of savanna or scrubland can occur on white sand or in small areas with less rainfall and stronger dry seasons. The vegetation types of Amazonia have been discussed in detail by a number of authors, including Prance (1979, 1987), Pires and Prance (1985), Daly and Prance (1989), Encarnación (1985) and Duivenwoorden et al. (1988).

Flora

The forests of Upper Amazonia are the most diverse in the world, with up to 300 species over 10 cm dbh per ha. Leguminosae is usually the dominant woody family; on the poor sandy soils of the Guayana Shield its subfamily Caesalpinioideae is especially prevalent. On the richest alluvial soils Moraceae may equal the Leguminosae in diversity. Other important families of trees include Lauraceae, Sapotaceae and Annonaceae. The dominant liana family is Bignoniaceae, followed in importance by Leguminosae. Epiphytes and forest-floor herbs are relatively poorly represented in most of Amazonia.

Endemism is often associated with substrate specificity, and concentrated in the habitat mosaic of north-central and north-west Amazonia. While there is also some endemism among lowland terra-firme taxa, a typical distributional pattern is for a species that occurs on clay soil to be widespread over a large portion of the region. There are also concentrations of endemic taxa in the relatively rich-soil area near the base of the Andes, the Guayana area and perhaps other locales. The reputed centres of endemism were instrumental in developing the hypothesis of Pleistocene refugia. Whereas it now appears that some putative centres of

endemism are artefacts of collection intensity (Nelson et al. 1990), the areas of endemism associated with habitat differentiation, especially around the periphery of the Guayana Highlands, clearly are real.

Species of economic importance

More than 2000 species of Amazonian plants have been reported to be useful to people (UNDP 1992; cf. Duke and Vásquez 1994; Bennett 1992). For example, at least 1250 species have been recorded to be used in Amazonian Peru (Vásquez 1989). Upper Amazonia is the centre of diversity of a number of important crop species (Clement 1989). In the Iquitos region of Peru c. 220 species of fruit are consumed (Vásquez and Gentry 1989 and unpublished). Floodplain and terra-firme forests near Iquitos can provide profits of US$400 per ha per year, mostly from sale of the native fruits (Peters, Gentry and Mendelsohn 1989), but near Manaus, Brazil one finds neither the high natural productivity of native fruits nor such market tastes. Extensive areas of intact terra-firme forest in Brazil's western State of Acre are utilized and occupied by rubber tappers. In eastern Acre they gather Brazil nuts (from the tree Bertholletia excelsa) in the rainy season when latex cannot be collected. For the country, Brazil nuts amounted to 22,000 tons and US$36 million of export value in 1983 (Müller and Calzavara 1986) and US$20 million in 1989 (Ryan 1991).

According to the Association of Exporters of the Manaus Free Zone, Brazil's State of Amazonas in 1984 exported overseas an FOB (free on board) value of US$26.6 million of regional products, mostly plant products, representing 50% of the total overseas exports from the state. These products are from Central Amazonia (see Lescure et al. 1992), with little from near Manaus itself. The most important extractive cash crop exported through Manaus fluctuates between "sôrva" (Couma spp.) and the Brazil nut ("castanha"), which respectively yielded US$6.2 million and US$3.8 million of annual exports. Aniba rosaeodora and A. duckei provide rosewood oil, an essential oil with US$1.4 million FOB value exported. The Madeira River region provides most of the Copaiba balsam, with US$263,000 FOB value exported.

For the Brazilian Amazon, extractive non-timber products effectively traded and recorded by government census amount to US$110 million per year (Vantomme 1990). This underestimates the real full value of such products and services – for example, a hut constructed from forest materials, fruits and fish consumed at home and probably many products sold in informal markets, all of which do not enter into the calculation.

In the Brazilian Amazon 260 species have some economic significance for timber and 50 are marketed in significant quantity, with c. 4000 tree species estimated to have some forestry potential (Sternadt, Ternadt and Camargos 1988; UNDP 1992). Extractive products provide far less than the US$1.2 billion annual proceeds from timber (Vantomme 1990). Though improperly managed, some of the timber is obtained in a non-destructive way, such as the wood for board and molding lumber floated from tidal forests near Breves in the State of Pará. In the Upper Amazon as well, logging of c. 6 species of fast-growing softwoods for peeler logs by floating them out of várzea floodplains is common and non-destructive. Fishes used for food that are dependent on forest fruits or forest insects are conserved in these areas.

Sustainable economic values

Prance (1990) suggested several sustainable economic activities for the Amazon. These include: "adaptation of indigenous agroforestry techniques to use mixed cultures where crops are grown; restoration with sustainable timber plantations of areas which have already been destroyed; the increase of extractive reserves and development of markets for ["green"] products that are extracted from the forest; the greater use of oligarchic forests which are abundant in some parts of Amazonia; and the development of agriculture only in the few places where the soils are suitable, such as in the flood plains of white-water rivers."

Extractivism and forest enrichment work well in certain situations, such as in the tidal forests near Belém where few crops could be grown and the large city provides markets for perishable palm hearts and fruits. As many as 40 people per km² can be found living off the tidal forests (A. Anderson 1989, pers. comm. to B. Nelson), where the most important economic species is the "açaí" palm (*Euterpe oleracea*). US$20 million per year is obtained from export of palm hearts of this species (Ryan 1991), which can be harvested in a sustainable manner, as the trees grow in clumps with juveniles around the base.

Often the major extractive activities have a questionable future. They are destined to be replaced by more efficient plantation or agricultural systems (Bunker 1985; Homma 1992), as is now happening with rubber (*Hevea brasiliensis*) planted south of Amazonia (Parfit 1989; Ryan 1991). Plantation technology has been developed for the Brazil nut (Müller and Calzavara 1986) but has not yet replaced native harvests. Selected graft stock in Brazil-nut plantations can produce consistently larger seeds, which have a much higher per-kg sale value than the average natural tree of some regions. Just 30,000 ha of such plantations could replace all of Brazil's present extractive production.

So far, very little agricultural activity is being developed on terra firme around Manaus. Most meat and garden vegetables can be transported economically from the south via Belém, because barged trucks return empty after delivering 80% of Brazil's consumer electronic items. This situation permitted Manaus to generate US$4–5 billion in goods and services each year until recently. Every 48 hours the city thus generates receipts equivalent to the total annual Brazil-nut exports. The extremely limited deforestation around such a large city is quite remarkable. An essential part of this system is the use of barge traffic rather than roads, which avoids the inevitable deforestation that penetrates from all trafficable roads in Amazonia.

Before 1991, 6% of all foreign tourism money spent in Brazil was spent in Manaus and the number of visitors grew by 5.8% per year. Two-thirds were attracted by the natural environment, and each spent US$760 during an average 5-day stay (*A Critica* 17/5/1992). The most spending is in the city itself, even by those tourists taken to jungle lodges a few hours away. Adventure cruises based on foreign vessels come fully outfitted and self-contained, but leave some receipts in small towns by purchasing handicrafts and hiring local entertainment and guides. Though tourism is a large business, actual ecotourism is an extremely small fraction, and employment in backwoods areas for potential deforesters constitutes a similarly small fraction of the ecotourism money.

Tourism receipts are not a significant factor in land-use choices made by those wielding axes in the far reaches of Amazonas State, nor by squatters, ranchers and timber harvesters along the hot dusty roads of southern Pará. As the population of the region increases, any green options for making a livelihood must produce greater receipts. Even if rubber tapping were to revive, it could only employ a small portion of the more than 15 million people in Brazil's designated "Legal Amazon".

Threats to extractive reserves have prompted proponents of extractivism to search for additional non-timber cash crops from intact forest. Over 80 promising species have been identified for research and promotion by the Instituto Nacional de Pesquisas da Amazônia (INPA) in Manaus (M. Allegretti 1991, pers. comm. to B. Nelson).

Exploitation of hydroelectric potential as well as mining of iron ore, bauxite, cassiterite and gold are furnishing the largest economic returns from the Amazon's natural resources. Access controlled by the Companhia Vale do Rio Doce (CVRD) to the large Carajás mining concession has kept that area an island of green within a patchwork of farms and pastures. When properly controlled and conducted by well-capitalized and regulated companies, hydroelectric activity and mining can be relatively benign, although synergistic effects, such as charcoal production for pig-iron smelting outside the Carajás concession, can be devastating. Deforestation for ranching and small farms far exceeds the loss of forest to hydroelectric and mining activities, which also generate much more wealth per hectare of deforested or flooded land. Unfortunately, such wealth is not well distributed, so tens of thousands of urban and rural poor turn to less efficient and environmentally damaging gold prospecting or itinerant agriculture.

Dry-season Landsat images indicate that in Maranhão, Tocantins, Pará east of Belém, Amazonas along várzea floodplain lakes, and parts of Roraima, most of the terra firme deforested for both pasture and small farms has been abandoned to regrowth (or is in a long fallow cycle). However in north-eastern Mato Grosso, in Rondônia and much of Acre, most deforested land is under intensive use, with satellite spectral signatures suggestive of dry pasture or bare soil.

In the Lower Amazon below Manaus to the mouth of the Xingú River, grazing of cattle takes place during low water on fertile floodplain savannas and deforested várzea. Terra-firme pastures are maintained as holding areas for grazing during the high-water months while the lowlands are flooded. This seasonal alternation means that the upland deforestation will penetrate only a limited distance from rivers, until (or unless) good roads are established. Large várzea/terra-firme landholdings often belong to owners located in Manaus, Parintins or Santarém where the beef is consumed. This extensive type of cattle-ranching is successful and has been established and maintained for generations, without infamous subsidies as are associated with non-sustainable terra-firme ranching in south-eastern Pará. Ranching also appears to be sustainable on the more fertile terra-firme soils of the Amazon Basin, such as in large parts of Acre and Rondônia to the south-west.

Threats

Extractivism and ecotourism are often suggested as ideally benign and sustainable activities for the Amazon, but their economic potential appears grossly insufficient. This suggests that Amazonia, far from being a demographic void, may already have exceeded an ideal population size compatible with maintaining most of its forest intact. Furthermore, masses of poverty-striken farmers from the populous north-east of Brazil and rural poor in the extreme south stand ready to swell the Amazonian population at the slightest evidence that one can make a marginal living.

After depleting timber in the Atlantic coastal forest, hundreds of lumber mills and mechanized timber harvesting operations are now using the closest available Amazonian hardwood stands, attracting labourers from the north-east for a US$112 monthly wage (Veríssimo *et al.* 1992). Eighty thousand jobs offered by the factories of the Manaus Free Zone at similarly low wage levels led in just 15 years to a large sprawl of shanty towns. And 350,000 gold and tin prospectors readily shifted into remote areas, ignoring laws, Amerindian rights and international boundaries (*Veja* 12/2/92).

Deforestation of the Brazilian portion of Amazonia in 1978, 1988, 1989, 1990 and 1991 has been measured precisely by Brazil's Instituto Nacional de Pesquisas Espaciais (National Space Research Institute) (INPE) (Fearnside, Tardin and Meira Filho 1990; INPE 1992; ESP 1992). The region includes 4.9 million km², of which 3.9 million km² were originally covered in natural forest according to INPE (or 3.6 million km² according to Nascimento and Homma 1984). The total area deforested was 426,000 km² as of August 1991 – 10.9% of the forests and 8.7% of the region. The rate was highest in 1987, although there are no reliable measurements for that year. Annual deforestation estimates were 21,500 km²/yr for the 11.6-year period 1978–1989, 18,800 km² for 1988–89, 13,800 km² for 1989–90 and 11,100 km² for 1990–91. The rates dropped after 1987 due to a number of possible factors, including a reduced threat of agrarian reform by 1988, heavy rainfall in 1989, drastic reduction in the money supply for investments in 1990 and widespread popular sentiment in Brazil and abroad opposed to deforestation, spurring more effective law enforcement. Yet over 10% of the c. 4 million km² of Amazonian forest has been cleared, and deforestation has continued at a rate of more than 10,000 km² per year (Fearnside 1991; Brown and Brown 1992).

Habitat degradation covers a larger area than deforestation *sensu stricto*, since the forests between closely spaced "fish-spine" colonization roads are intensively logged and hunted. Deforestation and habitat degradation are concentrated along southern and eastern flanks of the naturally forested areas (Skole, Chomentowski and Nobre 1992), where extensive road building, better soils, higher populations of nearby rural poor and proximity to a considerable domestic market for sawn lumber are the main causes. Large ranches, timber harvesting and small holdings of poor farmers are about equally important agents of deforestation/degradation in the Brazilian Amazon, which vary widely in relative importance from state to state.

There is a possibility as the eastern Amazon Basin continues to be deforested that the naturally dry zone could extend westward towards Manaus – with consequent massive forest mortality. Water vapour input to the Amazon Basin is from the Atlantic Ocean, carried westward by prevailing winds. Through evaporation and transpiration, terra-firme forests recycle 50–60% of the rain received back into the atmosphere, to fall again farther west (Salati and Marques 1984).

Although only four to eight trees are saleable in 1 ha of terra firme in eastern Pará, mechanized logging kills or damages 26% of preharvest standing trees (Uhl and Vieira 1989) and allows greater penetration of sunlight to the forest floor. Abundant dry slash fosters deep penetration of fires from nearby pastures (Uhl and Buschbacher 1985). No significant replanting of native terra-firme species occurs in the harvested heterogeneous forests, nor is it likely to occur until accessible primary forests are depleted. Extrapolation of recent logging rates indicates the depletion of commercial stocks for the entire State of Pará in 80 years (Uhl and Vieira 1989).

Road access along the southern and eastern rim of Amazonia almost always sets off an acceleration of deforestation for pastures and subsistence farming. The phenomenon has metastasized to more centrally located foci of road-based colonization in eastern and south-central Roraima, eastern Acre, along the Trans-Amazon highway at Sucunduri in south-eastern Amazonas, and between Itaituba and Altamira in central-western Pará.

Where there are few roads, poorer soils and great distance from domestic hardwood timber markets, as in Amazonas State, deforestation is largely limited to a 5-km deep rim on the terra-firme shores of dendritic lakes connected to the Amazon. These productive lakes have both abundant food fish and permanently dry shores. The 5-km limit is the distance subsistence farmers walk daily from their lakeside homes in search of primary forest to burn and obtain a better nutrient input for swidden manioc fields. Though much richer, floodplain soils do not permit a reliable manioc harvest (*Manihot esculenta*) due to occasional higher floods.

Attempts at ecologically benign development throughout Amazonia probably are close to futile if Brazil and the other Amazonian countries do not solve the problems of poverty, rapid population growth and overpopulation outside of this region.

Conservation

Some kind of protected status has been given to at least 1,812,510 km² or c. 23% of Amazonia (UNDP 1992). However, the total includes such areas as indigenous territories, Forest Reserves and Extractive Reserves, which often function more as sources of timber or other forest products than as conservation units for nature. Of that total, 245,560 km² or 3.2% of Amazonia constitutes National Parks and perhaps twice as much (562,230 km² or 7.4% of Amazonia) is included in various Ecological Reserves, Faunal Refuges and Biosphere Reserves with a primarily conservational purpose. Peres and Terborgh (1995) found 117 of all 459 conservation units in Amazonia or 25% to be strict nature reserves, accounting for 41% of the region's land that is under some form of institutional protection; their average size is 4765 km². The largest portion of Amazonia by far is in Brazil. About 143,390 km² or 3.6% of Brazilian Amazonia is totally protected (Brown and Brown 1992), including 97,000 km² in National Parks (UNDP 1992). While the total reserve area in Amazonia is impressive, nearly all the conservation units have been created in the last decade, and many exist only on paper. A map prepared by Conservation International (1991) from Workshop 90 (Manaus) pinpoints specific areas for conservational focus in Amazonia (cf. Rylands 1991).

Centres of plant diversity and endemism

a. NORTH-EASTERN AMAZONIA–GUAYANA

French Guiana

SA3. Saül region

– see Data Sheet.

b. TRANSVERSE DRY BELT

Brazil, Guyana, Surinam

SA4. Transverse Dry Belt of Brazil

– see Data Sheet.

c. CENTRAL AND GUAYANAN AMAZONIA

Brazil

SA5. Manaus region

– see Data Sheet.

Brazil, Colombia, Venezuela

SA6. Upper Rio Negro region

– see Data Sheet.

Colombia

SA7. Chiribiquete-Araracuara-Cahuinarí region

– see Data Sheet.

d. WESTERN AMAZONIA

Ecuador

SA8. Yasuní National Park and Waorani Ethnic Reserve

– see Data Sheet.

Peru, Colombia

SA9. Iquitos region

– see Data Sheet.

e. SOUTH-WESTERN AMAZONIA

Peru

SA10. Tambopata region

– see Data Sheet.

f. PRE-ANDEAN AMAZONIA

Peru

SA11. Lowlands of Manu National Park: Cocha Cashu Biological Station

– see Data Sheet.

IV. MATA ATLÂNTICA

The narrow Mata Atlântica formation originally extended from Brazil's north-eastern tip near Cabo de São Roque in Rio Grande do Norte State to Torres in Rio Grande do Sul, the country's southernmost state. Biologically the Mata Atlântica is often divided into northern, central and southern portions. The northern part extends from Rio Grande do Norte to Bahia, the central part from Espírito Santo to São Paulo and the southern part from southern São Paulo to north-eastern Rio Grande do Sul. In addition to the coastal forest, much of the drainage of the Paraná River is floristically a part of the Mata Atlântica, extending from about 20°S in southern Minas Gerais to 30°S in central Rio Grande do Sul, and also including most of Paraguay east of the Paraguay River and Argentina's north-easternmost province of Misiones. As thus defined the Mata Atlântica also includes the upland region of Paraná and adjacent states where *Araucaria* forest is the dominant vegetation.

Whereas most of the Mata Atlântica region has a relatively low annual rainfall, mostly below 2000 mm, the dry season is usually relatively weak. As a result annual precipitation as low as 1400 mm can support physiognomically and floristically typical tropical moist forest, as at Linhares in Espírito Santo (Peixoto and Gentry 1990). The tropical portion of the Mata Atlântica is also peculiar in its latitudinal extent, with species-rich forest crossing the Tropic of Capricorn and reaching at least 24°S latitude. Elsewhere in the world, tropical forests generally become less diverse poleward from 12° latitude (Gentry 1990b). South of São Paulo and in the Paraguay portion of the Mata Atlântica, species richness drops off abruptly, although the forests are still physiognomically tropical and endemism may remain high. For example, the most diverse Paraguayan forest at Mbaracayú (24°42'S) has only 85 species over 2.5 cm dbh in 0.1 ha (Keel, Gentry and Spinzi 1993). However, a local Florula (Barros *et al.* 1991) from Cardoso Island south of 25°S in extreme southern São Paulo State has a respectable 986 species. Rough montane topography and shallow soils, landslides due to high rainfall, along with natural forest dynamics and in the south the occurrence of frost (minimum -4°C) in 1–5 days each year, result in a mosaic of successional stages and consequently a highly variable species composition (Mantovani 1990).

Flora

The Mata Atlântica forests are very diverse botanically, with an estimated 13,000 angiosperm species, accounting for c. 16% of the neotropical flora, in a relatively small area (Gentry 1982a). At the level of local plant communities, the Mata Atlântica forests are also unusually diverse, presumably primarily because of the relatively weak dry season characterizing much of the region. Indeed 0.1-ha samples of the forests of Espírito Santo and Rio de Janeiro states, where precipitation is only 1400–1500 mm, have 160–200 species of plants over 2.5 cm dbh, as many as in all but the richest of Amazonian forests (Peixoto and Gentry 1990, and original data).

There are some conspicuous floristic differences between the Mata Atlântica and Amazonia. Most striking is the prevalence here of Myrtaceae, which generally replaces Leguminosae as the most prevalent woody family (Peixoto and Gentry 1990). Monimiaceae is also much represented when compared to Amazonia. On the other hand, other prevalent Amazonian families are poorly represented, such as Myristicaceae, Palmae, Moraceae, Meliaceae, Chrysobalanaceae and Lecythidaceae. Bignoniaceae is the dominant liana family, just as elsewhere in the lowland neotropics, but Hippocrateaceae is better represented than elsewhere, edging out Leguminosae as the second most prevalent liana family. On the other hand, Connaraceae and Dilleniaceae are generally more poorly represented among lianas than in Amazonia. Orchids are especially prevalent in many Mata Atlântica forests; for example Orchidaceae is by far the largest family on Cardoso Island, with 118 species, 12% of the entire flora (Barros *et al.* 1991).

Overall endemism is very high in the Mata Atlântica, with an estimated 9400 angiosperm species endemic (Gentry 1992d). This 73% regional endemism is second only to Amazonia among the neotropical phytogeographic regions (Gentry 1992d). As usual the endemism is concentrated in epiphytes and other herbaceous taxa. An estimated 67% of the Mata Atlântica canopy taxa are endemic to the region (53% of the trees *fide* Mori and Boom 1981), whereas an astounding 86% of the Mata Atlântica taxa of principally herbaceous and shrubby families are apparently endemic, a higher figure than for any other neotropical phytogeographic region (Gentry 1982a).

Local endemism appears to be equally high, but few actual data are available. Although many species are wide-ranging, there appears to be a disproportionate number of local endemics, especially in such specialized habitats as the coastal "restingas". Endemism may be concentrated in Rio de Janeiro State, although this could partially reflect collection artefact; for example, 36 species are endemic to its coastal restingas, with 11 endemic to the Cabo Frio region.

Species of economic importance

Perhaps the two most economically important plants of coastal Brazil are the legume timber trees *Caesalpinia echinata* (Brazil-wood) and *Dalbergia nigra* (Brazilian rosewood), the latter once considered Brazil's mos valuable wood. Both have been over-exploited and are now very rare. Also very important economically are several Bignoniaceae timber trees, including *Paratecoma peroba*, a locally endemic genus and species of the Doce River Valley, which was once the most important timber of Rio de Janeiro but is now nearly extinct (Record and Hess 1940; Gentry 1992b). A similar fate may be now befalling *Tabebuia heptaphylla* in eastern Paraguay, where it is the preferred timber species. Rhizomes from the tree fern *Dicksonia sellowiana* are used as substrate for the cultivation of ornamentals. Some widely consumed fruits like *Myrciaria cauliflora* are also native to the region. Nevertheless, it is curious that the economically valuable plants singled out in the CPD Data Sheets on the Mata Atlântica are mostly timber trees, whereas in other regions edible fruits and medicinal plants are often emphasized. This might be happenstance or reflect some kind of real cultural or biological difference.

Threats

The Atlantic forest region is the agricultural and industrial heart of Brazil, with c. 43% of the country's population. In north-eastern and south-eastern Brazil much of the conversion was completed with settlement by the end of the 19th century, although in the temperate southern area devastation has accelerated since the 1970s. The central Atlantic region (southern Bahia and northern Espírito Santo) mostly has been converted since c. 1965 with the opening of new access routes (Brown and Brown 1992).

From the viewpoint of plant conservation, the Mata Atlântica is one of the world's most critical regions; this area featured prominently as one of Myers' (1988) extinction hot spots. Of the original area of forest of c. 1.2 million km², well over 1 million km² had been deforested by 1990. Estimates of remaining forest range from 12% (Brown and Brown 1992) to 2–5% (IUCN and WWF 1982), most of which is highly fragmented.

Conservation

Almost 31,000 km² of the remaining forest, 2.5% of the original area, are variously protected in many National Parks and equivalent reserves, mostly in the southern part of the Mata Atlântica in the states São Paulo (11,540 km²) and Paraná (5160 km²) (Brown and Brown 1992). The Fundação SOS Mata Atlântica (SOS Atlantic Forest Foundation) was founded in 1986 to stimulate, assist and coordinate efforts to conserve the remaining important sites of the region (Collins 1990).

Centres of plant diversity and endemism

a. NORTHERN REGION (Rio Grande do Norte to Bahia)

Brazil

SA12. Atlantic moist forest of southern Bahia

– see Data Sheet.

b. CENTRAL REGION
(Espírito Santo to São Paulo)

Brazil

SA13. Tabuleiro forests of northern Espírito Santo

– see Data Sheet.

SA14. Cabo Frio region

– see Data Sheet.

SA15. Mountain ranges of Rio de Janeiro

– see Data Sheet.

c. SOUTHERN REGION (southern São Paulo to Rio Grande do Sul)

Brazil

SA16. Serra do Japi

– see Data Sheet.

SA17. Juréia-Itatins Ecological Station

– see Data Sheet.

d. INTERIOR: PARANA BASIN

Paraguay

SA18. Mbaracayú Reserve

– see Data Sheet.

V. INTERIOR DRY AND MESIC FORESTS

The dry-forest landscapes of the interior of the South American continent are traditionally divided into three major regions:

1. the caatinga of north-eastern Brazil, east of c. 45°W longitude;

2. the cerrado, covering the bulk of the Brazilian Shield, generally west of 45°W and/or south of 15°S; and

3. the chaco, west of the Paraguay River and south of c. 17°S.

At its southern margin in São Paulo State to the south of c. 20°S latitude, the "cerrado" vegetation becomes restricted to interrupted patches; west of the Paraná River, nearly continuous cerrado reaches the Tropic of Capricorn in the Sierra de Amambay of central-eastern Paraguay; and at its western margin well-developed cerrado occurs on Bolivia's Huanchaco Plateau with scattered patches on rocky outcrops farther south and west in Santa Cruz Department.

To these three main dry-forest units might appropriately be added a fourth: (4) the Bolivian Chiquitania, which constitutes what has been called the transition zone between the chaco and Amazonia, and covers most of the northern half of Santa Cruz Department as well as some adjacent areas (Gentry 1994).

Together the vegetations of the interior dry-forest general regions cover over 4 million km²:

❖ caatinga   800,000–1,000,000 km²
❖ cerrado   2 million km²
❖ chaco   1 million km²
❖ Chiquitania   100,000 km²
❖ Llanos de Mojos   100,000 km²
❖ pantanal   100,000 km²

Of these, the driest formations are the "caatinga" and the "chaco", mostly with below 800 mm of annual rainfall and a strong dry season of 5–8 months; both are characterized by irregular year-to-year fluctuations in precipitation (Gómes 1981; Ramella and Spichiger 1989). The cerrado, climatically intermediate between wetter tropical forest and the scrubbier chaco and caatinga formations, mostly has c. 800–1800 mm of annual rainfall with a strong dry season of 3–4 months (Eiten 1972; Gentry 1995). However, there is some overlap in precipitation, with parts of the chaco and caatinga having up to 1000 (–1300) mm of annual precipitation and the wettest part of the cerrado with up to 2000 mm (Spichiger et al. 1991; Eiten 1972), whereas parts of the adjacent Mata Atlântica with less than 1500 mm, but relatively evenly distributed precipitation, have well-developed semi-evergreen or even evergreen tropical forest.

Although there are few climatological records, the Chiquitania dry forest evidently occurs in areas with between 1000–1500 mm of rainfall and a strong dry season. Apparently its differentiation from the cerrado is based on soils rather than climate, as indicated by the presence of outliers of Chiquitania-type tall deciduous forest scattered through the Brazilian cerrado.

Flora

As a whole the South American interior dry forests are surprisingly diverse botanically. Together the caatinga-cerrado-chaco region accounts for an estimated 11% of the neotropical flora with perhaps c. 9300 angiosperm species.

Smaller areas of these dry forests can also be extremely diverse. A recent compilation by Filgueiras and Pereira (1990) of the plant species in Brazil's Distrito Federal included c. 2500 species of vascular plants, almost exactly the same number as in a roughly comparable Florula for three sites in the Iquitos area of Amazonian Peru (R. Vásquez, in prep.). The Serra do Cipó Florula (Giulietti et al. 1987), for a cerrado variant known as "campo rupestre", includes almost 1600 species. At a second campo-rupestre site in Bahia, Harley and Simmons (1986) found 663 vascular plant species. Emperaire's (1987) caatinga relevés may be taken as the equivalent of a local Florula in south-east Piaui of this vegetation type; again the flora is relatively diverse with 615 species recorded. On the other hand the more subtropical entire chaco has a relatively low floristic diversity, with only 1000–1200 species estimated to occur.

At yet smaller scales, these dry forests also can be strikingly diverse. The cerrado typically has 230–250 vascular plant species (occasionally as many as 300–350 species) in 0.1-ha samples (Eiten 1984), making this one of the most species-rich vegetations on Earth at this scale (Gentry and Dodson 1987), being exceeded only by some wet tropical forests. The caatinga and chaco, which have a much poorer ground cover of grasses and herbs, are much less diverse at the 0.1-ha scale. The Chiquitania dry forests are intermediate, being unusually diverse for dry-forest (87 species with dbh 2.5 cm or more in 0.1 ha), a distinction shared with the dry forests of western Mexico, but lacking the rich layer of herbs and subshrubs that makes the cerrado so species-rich.

These dry-forest regions are characterized by a number of floristic peculiarities, some shared and some not. In the cerrado (e.g. both campo-rupestre Florula sites as well as the Distrito Federal as a whole), Leguminosae, Compositae and Gramineae are the most speciose families, whereas only the Leguminosae is speciose in Amazonia. At all three of the sites Melastomataceae, Myrtaceae and Malpighiaceae are far more prevalent in the woody flora than in Amazonia and among the most speciose 6–12 families. Even more distinctive floristically are the monocot families Eriocaulaceae, Xyridaceae and Velloziaceae, which are among the most speciose families in the campo-rupestre sites and are hardly present at all beyond the Brazilian and Guayanan shields. In the caatinga, Leguminosae are even more prevalent than in cerrado or Amazonia, constituting a fifth of the entire south-east Piauí Florula (Emperaire 1987). Other families conspicuously more prevalent in caatinga than cerrado are the second and third most species-rich families – Euphorbiaceae and Bignoniaceae – as well as Cactaceae, Turneraceae, Boraginaceae and Combretaceae. The Leguminosae are also very prevalent in the chaco, but less speciose and less dominant. More striking is the conspicuous representation of Cactaceae, which is as speciose as Leguminosae in 0.1-ha samples of plants with dbh 2.5 cm or more in the chaco vegetation (Gentry 1993a). Other families generally better represented in chaco than in cerrado or caatinga include Capparidaceae, Nyctaginaceae and Zygophyllaceae. The Chiquitania dry forest may have some typical neotropical dry-forest familial compositions, with Leguminoae the most speciose woody family and Bignoniaceae the predominate family of lianas.

The interior dry areas of South America are outstanding in their high levels of endemism. Gentry's (1982a) estimate for regional endemism is 73%, as high as that of the Mata Atlântica and exceeded only by Amazonia among the major neotropical phytogeographic regions. Moreover there is remarkably little floristic overlap between the major dry areas at the specific level, although their floras share most of the same families and genera. Extrapolating from Emperaire (1987), the majority of the caatinga species is endemic to caatinga and almost two-thirds are restricted to caatinga plus cerrado. In a series of 0.1-ha samples of chaco vegetation, 59%, 77% and 85% of the completely identified species of 2.5 cm dbh or more are chaco endemics (Gentry 1995); these figures should prove higher for the still uncensused herbaceous taxa. Joly (1970) suggested that the campos rupestres of the cerrado region have the highest levels of endemism of any Brazilian vegetation type. In families like Velloziaceae, Eriocaulaceae and Xyridaceae,

60–70% of the world's species occur in the campos rupestres, mostly as local endemics (Giulietti and Pirani 1988). As an extreme case, 91.5% of the 59 species of Velloziaceae of the Serra do Cipó are locally endemic. Preliminary observations indicate that the virtually unstudied Chiquitania dry forests are also floristically distinctive, with a significant level of endemism (Gentry 1994, 1995).

In summary, all four of the main interior South American dry-forest regions have significant complements of endemic plant species and a concordant conservation importance. This contrasts with conservation priorities that might be drawn from other groups of organisms. Especially noteworthy is the high level of plant endemism in the caatinga (Sampaio 1995) as compared with very low levels of endemism (and presumably conservation importance) for better known taxa like vertebrates.

In contrast, the swampy areas associated with the dry-forest formations, including the pantanal of Mato Grosso and adjacent easternmost Bolivia and the llanos of Bolivian Beni, have floras consisting mostly of widespread species and very little plant endemism (Prance and Schaller 1982; Haase and Beck 1989), despite their importance for conservation from a zoological perspective on account of their high concentrations of large vertebrates. Prance and Schaller (1982) listed only four plant species endemic to the pantanal. The main conservation significance to these swampy regions from the botanical viewpoint relates to their interesting hydrological mosaics.

Species of economic importance

Major regional timber trees in the dry-forest areas include *Schinopsis* in the chaco, and *Tabebuia, Aspidosperma, Amburana, Swietenia* and *Cedrela* in the cerrado, Chiquitania and adjacent areas. In the cerrado and related vegetation types, the authors of the CPD Data Sheets have emphasized native fruits, as well as timber products. Some of these dry-area fruits are from well known edible-fruited genera like *Caryocar (C. brasiliensis), Byrsonima* and *Anacardium*. However, some of the fruit-producing genera are widespread but rarely reported to have edible fruit, here including *Salacia elliptica, Mouriri glazioviana* and several Myrtaceae, whose use may be relatively particular to dry-area species. Liqueurs prepared from these fruits are reported to be especially popular. In contrast, in the chaco region natural fibres are noted along with timber as economically important, rather than fruits. The main chaco fibre-producing plants are Bromeliaceae and *Trithrinax* palms. Easily the most interesting and region-specific economic use for native plants, which constitutes a major cottage industry, is the collection of everlasting plants for dried floral arrangements from the campo-rupestre region of the cerrado. Some 300 tons a year of some 40 species are collected, mostly Eriocaulaceae and Xyridaceae, including some rare endemics.

Threats

The conservation status of the South American dry-forest regions only can be described as critical. Tropical dry forest is now widely regarded as one of the most threatened – perhaps the most threatened – tropical vegetation type (Janzen 1988; Lerdau, Whitbeck and Holbrook 1991; Bullock,

Mooney and Medina 1995). This perception contrasts vividly with suggestions by conservationists a few years ago that an appropriate strategy for preserving the Amazonian rain forest might be to develop the cerrado instead!

The caatinga, where much of Brazil's population of rural poor is concentrated, has been over-exploited for centuries; now desertification is a serious threat. Huge areas of the cerrado have been devastated in the last decade by cattle-ranching and mechanized agriculture on a large scale. As of 1990, it was estimated that 56% of the original 2 million km² was in managed use and 37% under cultivation (Dias 1990). Similarly much of the chaco has been ravaged by over-grazing, and increasingly by mechanized agriculture.

There is no reserved area for the semi-deciduous dry forest here designated the Chiquitania formation, and the largest area of intact tall dry forest, the Tucuvaca Valley of Bolivia, is in the process of being subdivided for soybean farming (Gentry 1994). In Brazil these forests have been reduced to isolated stands, and this has been called the probably most endangered forest biome in Brazil (Ratter 1991).

Conservation

About 7% of the cerrado remained in pristine condition in 1990, including 1.5% in reserves (Dias 1990). Perhaps the only large pristine area of chaco left is in south-eastern Bolivia near the Paraguay border (Parker *et al.* 1993).

Fairly large areas of cerrado and chaco have been designated as reserves, but most are new, and only in part effective. The Brazilian cerrado has 132,994 km² or 6.6% of its area in National Parks or other conservation units falling in UNESCO categories I–IV (Dias 1990); Noel Kempf Mercado NP in Bolivia includes another 9000 km², a quarter in cerrado (T. Killeen, pers. comm.). In Paraguay 11,000 km² of chaco have been designated as National Parks and in Argentina an additional 950 km² as National Parks or Natural Reserves; in Bolivia, ironically the best preserved chaco has been totally unprotected.

Centres of plant diversity and endemism

Brazil

SA19. Caatinga of north-eastern Brazil

– see Data Sheet.

SA20. Espinhaço Range region

– see Data Sheet.

SA21. Distrito Federal

– see Data Sheet.

Argentina, Paraguay, Brazil, Bolivia

SA22. Gran Chaco

– see Data Sheet.

Bolivia

SA23. South-eastern Santa Cruz

– see Data Sheet.

SA24. Llanos de Mojos region

– see Data Sheet.

VI. (TROPICAL) ANDES

The Andes are by far the most significant mountain range in the world's tropics. They span 64° of latitude: from 11°N, where the highest point in Colombia overlooks the Caribbean Sea near the northernmost tip of South America, to the continent's southern tip near 53°S in Tierra del Fuego. The highest mountain in the Western Hemisphere is Aconcagua (6959 m), on the border between Chile and Argentina in the Altoandina region. The highest mountain in the world's tropics is Huascarán (6768 m) in Peru's Cordillera Blanca. Even the highest mountain in the world by some calculations is Andean, since the summit of Ecuador's Chimborazo Volcano (6267 m) is farther from the Earth's centre than is the summit of Mount Everest, due to Chimborazo's more equatorial location and the Earth's oblate form.

The Andes essentially represent the upthrust leading edge of the South American continent, formed as the South American continental plate collides with the adjacent Nazca plate. Parts of the western Andes of Colombia and Ecuador have a more complicated origin as suspect terrane. In general the Southern Andes are oldest, the Northern Andes youngest. The range generally broadens to the north, at the Nudo de Pasto splitting into three separate cordilleras in Colombia. The Sierra Nevada de Santa Marta is somewhat isolated from the rest of the Eastern Cordillera, but is here considered part of the Andean region. Although the Eastern Cordillera is rather continuous with Venezuela's Coastal Cordillera, that area has a different geological history and is here considered part of the Caribbean region, despite obvious Andean affinities.

Since vegetation is continuous from the Andean slopes into the adjacent lowlands, delimitation of the montane Andean region is necessarily somewhat subjective. The Andean region is defined here by the (900–) 1000 m contour line to the west (following Gentry 1993a; Dodson and Gentry 1991; Forero and Gentry 1989) and by the 500 m contour to the east (following Gentry 1982a, 1993a). The valleys of the Magdalena and Cauca rivers that separate the three Colombian cordilleras are included in the Caribbean region, approximately following the 1000 m contour. Even though the upper Marañón River and its tributaries nearly sever the Andes in northern Peru, the entire Huancabamba Depression is above 500 m in elevation and is included here in the Andes.

Vegetation

Andean vegetational zonation is largely defined by altitude, thus tending to form long narrow bands along the slopes of the cordillera. The most striking differentiation is at treeline, with the lower slopes of the Andes covered by forest and the upper slopes by herbaceous vegetation usually dominated by grasses. Treeline in the tropical Andes generally occurs around 3200–3500 m and is associated with an average annual temperature of c. 6°C (Holdridge 1967; van der Hammen 1974; Terborgh 1971; Rivas-Martínez and Tovar 1982). Tree-line tends to be slightly higher in the more subtropical Bolivian Andes, and gradually descends in the Altoandina region to near sea-level at the southern tip of the continent. On most of the dry Pacific-facing slopes of the central and southern Andes, trees are limited to tiny remnants in protected microsites and the differentiation between high-Andean and mid-elevation vegetation is much less distinct.

Precipitation in the high Andes generally decreases from north to south, with the high-Andean grass-dominated vegetations falling into three major types correlated with the precipitation differences. The wettest northern high-Andean formation, called "páramo", typically has the distinctive thick-stemmed (pachycaul) Compositae genus *Espeletia* as a co-dominant with grasses. The grass-dominated vegetation of the drier southern high Andes is called "puna". Páramo with *Espeletia* reaches south only to northernmost Ecuador; puna occurs from near the Huancabamba Depression (and Cajamarca) in north-eastern Peru southward to northern Argentina. The intermediate precipitation area, called "jalca" in north-eastern Peru and páramo in Ecuador, is floristically similar to the northern Andean páramos except for lack of the characteristically dominant *Espeletia*. There are also limited areas of a similar jalca-like wet puna along the upper forest margin of the eastern escarpment of the Peruvian Andes.

Peculiar growth forms characterize many puna and páramo plants. In addition to the pachycaul-rosette growth form of *Espeletia*, cushion plants are common, as are sessile rosette growth forms. The woody plants are mostly dense subshrubs with very small sclerophyllous leaves.

The forested tropical Andean slopes ("selva nublada" in Colombia, "ceja de la montaña" in Peru, "yungas" in Bolivia) form a continuous band along the eastern Andean slopes roughly between 500–3500 m. This forest band is usually subdivided into about three altitudinal zones: premontane or submontane forest extending up to (1000–) 1500 m, lower montane or tropical montane forest between c. 1500 m and c. (2300–) 2500 m, and upper montane forest between c. 2500 m and (3200–) 3500 m. Sometimes a fourth altitudinal zone is recognized, subpáramo or prepuna, consisting of the usually rather narrow shrub-dominated transition zone from forest to non-forest that generally occurs between 3000–3500 m (Cleef *et al.* 1984; Frahm and Gradstein 1991; van der Hammen 1974; Cuatrecasas 1958; Terborgh 1971). The correlations between these vegetational types and the altitude are not constant, and the zones generally occur at lower altitudes on narrower cordilleras and outlying ridges – the well-known Massenerhebung effect (Flenley 1995). The transition between premontane and montane forest around 1500 m elevation is usually associated with the dew-line or cloud-line and is characterized especially by an abruptly greater density of both vascular and non-vascular epiphytes (Terborgh 1971; Cleef *et al.* 1984; Frahm and Gradstein 1991). The transition from lower montane to upper montane forest around 2500 m tends to be associated with the 12°C isohyet of average annual temperature (Holdridge 1967; Rivas-Martínez and Tovar 1982). Above this altitude, soil becomes strongly peaty, vascular epiphytes are reduced, and bryophytes become much more prevalent, forming thick moss layers on the trees.

On the western slopes of the Andes similar cloud-forest formations occur south to central Ecuador, although their zonation is much less obvious in the super-saturated conditions of the Chocó region, where some cloud-forest features and elements extend down to sea-level (Gentry 1986b). In southern Ecuador and northernmost Peru, inland from the Humboldt Current and thus drier, the cloud-forest band becomes more restricted altitudinally. South of 5°S latitude in Peru, forest is restricted to small isolated patches in the most protected valleys. Farther south, the lower limit of these forest patches recedes upward. At the latitude of Lima (12°S), relict forest patches occur only at very high altitudes over 3000 m; farther south they are altogether lacking.

In addition to the prevalent cloud-forest vegetation, the Andes include important areas of dry shrub dominated vegetation in various inter-Andean valleys and on the Pacific-facing western slopes of Peru. The largest and most important of the inter-Andean dry areas is the Huancabamba Depression of northern Peru and southern Ecuador, where the main Andean cordilleras are extensively dissected by the Marañón River and its major tributaries. Similar but less extensive shrubby areas occur along the upper Apurimac and Huallaga rivers in Peru, in the Cochabamba area of Bolivia and on a smaller scale in Ecuador and Colombia.

Flora

In the context of South America, the Andes are unusual in their prevalence of Laurasian plant families and genera, many of which do not extend into the adjacent tropical lowlands (Gentry 1982a). These northern elements, mostly newly arrived from the north in Plio-Pleistocene times as the Isthmus of Panama closed, mixed complexly with tropical Gondwanan-derived and southern subtropical taxa like *Podocarpus, Drimys, Panopsis* and *Weinmannia* to give rise to today's Andean flora. Perhaps a quarter of the genera of the puna and páramo belong to the old austral-antarctic flora (van der Hammen and Cleef 1983). In the Southern Cone, the puna together with the Altoandina and Patagonia constitute the Andino-Patagonian Dominion (Cabrera and Willink 1973). The southern taxa are less prevalent in the Andean forests, which are largely a mixture of the Laurasian and tropical Gondwanan elements.

The floristic composition of the high-Andean páramo and puna vegetation is remarkably consistent at the familial level, with Compositae always the largest family, nearly always followed by Gramineae. The other most diverse families are Scrophulariaceae, Orchidaceae, Leguminosae, Melastomataceae, Solanaceae, Caryophyllaceae, Rosaceae, Cyperaceae, Cruciferae (and sometimes Ericaceae – e.g. in the Puracé, Colombia area), with over half of the species of a given high-Andean site belonging to these dozen most speciose families, nearly all of which tend to be of Laurasian origin and poorly represented in the tropical South American lowlands (Vareschi 1970; Cleef 1981; Smith 1988; Galeano 1990; Ruthsatz 1977). *Senecio* (Compositae) is consistently the most speciose high-Andean genus; *Solanum, Calceolaria* and *Valeriana* are also notably species-rich in the Andes (Gentry 1993a). Several of these genera are famous for their local endemism, presumably associated with the archipelago-like nature of the high-Andean vegetation.

A similar combination of predictable familial and generic composition combined with strong local endemism in certain groups characterizes the Andean cloud forests. The premontane vegetation zone below 1500 m is largely tropical in its familial and generic composition, contrasting with the middle and upper elevation forests which are floristically distinctive (Gentry 1992b).

The middle elevation forests are usually dominated by Lauraceae, Melastomataceae and Rubiaceae. They are also characterized by a small group of tree genera each belonging to a different family, including *Saurauia, Ilex, Alnus, Brunellia, Styloceras, Viburnum, Hedyosmum, Clethra, Weinmannia, Billia, Juglans, Prunus, Meliosma, Styrax, Symplocos* and *Gordonia*. Most of these genera belong to Laurasian families and most are absent or very poorly represented in the lowlands (Gentry 1992b). If complete species lists are compared, predominantly epiphytic taxa like Araceae (especially *Anthurium*), Ericaceae, *Peperomia* (Piperaceae), Orchidaceae (especially *Epidendrum, Maxillaria, Pleurothallis*) and ferns (especially *Polypodium*) are among the most diverse components of middle elevation Andean forests. Considering the shrubs and trees of upland Ecuador, Compositae, Melastomataceae, Ericaceae, Solanaceae and Rubiaceae are the largest families and *Miconia, Piper* and *Solanum* the largest genera (Ulloa-U. and Jørgensen 1993). These typically Andean taxa tend to be very poorly known taxonomically; some are very prone to local speciation (Gentry and Dodson 1987; Gentry 1992b).

The highest altitude Andean forests show similar patterns but with different taxa. Forests near treeline are dominated by Compositae, Ericaceae, Myrsinaceae, Melastomataceae and Rosaceae, and include genera like *Drimys, Cervantesia, Vallea, Hesperomeles, Polylepis, Escallonia* and *Myrica* that are poorly represented or absent in lower altitude forests.

It is conservationally relevant that although there is a general tendency for diversity to decrease with increasing altitude (Gentry 1988), individual páramo and puna Florulas can have 600–800 species and be almost as species-rich as many lowland tropical Florulas (Smith 1988; Galeano 1990; Ruthsatz 1977). Presumably this richness is due to the unusually close juxtaposition of different habitats in montane environments. Similarly the species lists for individual altitudinal transects in Colombia (by the EcoAndes project) include c. 1200 vascular plant species, again reflecting the telescoping of habitats. Young and León (see Data Sheet SA37) estimate that the 1500–3500 m strip of eastern Peruvian Andean forest includes 14% of Peru's flora in 5% of the country's area. Expanding the slope area to the 500 m contour, they estimate that about half of Peru's species might be included in the 20% of the country represented by the eastern Andean forests. Similarly, Balslev (1988) estimates that half of Ecuador's species occur in the 10% of the country represented by middle-elevation (900–3000 m) Andean forests.

Sparse data suggest that endemism generally increases in montane forests. For example Balslev (1988) found in his sample of recently monographed Ecuadorian plants that 40% of the high-altitude (over 3000 m) species and 39% of the middle-elevation (900–3000 m) species are endemic to Ecuador, compared to 16% of lowland species. Half of the lowland species are widespread, as compared to one-quarter of the mid-elevation taxa and only one-sixth of the high altitude taxa.

There are regional peculiarities between different areas of Andean vegetation. Thus it is imperative to apportion conservation efforts to include representation of many different, floristically distinctive vegetations in the Andean region. As examples, *Quercus* is dominant in many high-Andean forests in Colombia, but absent farther south. The western slope of the Colombian Andes is especially rich in Araceae, mostly still undescribed, and Ericaceae, of which Luteyn (1989) reported finding 41 species, 12 new to science, in a few days' collecting along a single 5-km stretch of road. Isolated massifs like Colombia's Sierra Nevada de Santa Marta tend to have unusually high rates of endemism (e.g. 11 endemic *Diplostephium* spp., Compositae) but relatively depauperate floras; Santa Marta even has endemic genera like *Cabreriella, Castanedia* and *Raouliopsis* (Compositae), but only half as many páramo genera as the Eastern Cordillera of the Colombian Andes (Cleef and Rangel 1984). The relatively young Eastern Cordillera has a much richer high-Andean flora and greater endemism than the older, but more actively volcanic Central Cordillera (Rangel, pers. comm.). Solanaceae are more dominant in relatively dry cloud forests in north-western Peru (Gentry 1992b). The dry western slopes of the Peruvian Andes and the dry inter-Andean valleys are dominated by shrubby and herbaceous Compositae.

To the extent that high-Andean endemism reflects broken local topography, such patterns can be very useful for conservation planning. For example, *Fuchsia* (Onagraceae) has its greatest diversity in the Northern Andes' Central Cordillera of Colombia and into Ecuador, but much higher endemism in the more dissected terrain of the Peruvian Andes (Berry 1982). *Telipogon* (Orchidaceae) has its greatest concentrations of locally endemic species in the more complicated topography of the Nudo de Loja and Nudo de Pasto than in intervening regions (Gentry and Dodson 1987). In a 200-km² area of the Huancabamba region in Peru's Cajamarca and Amazonas departments are concentrated 63 of the 181 neotropical species of *Calceolaria* (Scrophulariaceae), 21 of them with distributional areas less than 50 km across (Molau 1988). Complexly dissected regions like the Huancabamba area and the Nudo de Pasto are clearly of exceptional conservational significance.

Species of economic importance

The Peruvian and Bolivian Andes constitute one of the major Vavilovian centres of domestication. This Andean centre of domestication rivals the Indo-Malayan and Mediterranean regions as the area that has produced the world's most important crops. Thus it is hardly surprising that the Andean region has a large number of economically important plant species. Many of the high-Andean food plants are tubers. In addition to the potato (*Solanum tuberosum*) and several wild relatives, "ollucos" (*Ullucus tuberosus*) and "oca" (*Oxalis tuberosa*) are local staples with broader commercial potential. Similarly, the native grain amaranths (*Amaranthus*) have been singled out as more protein-rich than grain cereals. Edible fruits from the region include several species of *Rubus, Solanum* (especially *S. quitoense*), *Cyphomandra, Lycopersicon, Carica* and *Passiflora*. There are many medicinal plants in the Andes, with the famous antimalarial *Cinchona* one of the best known; several CPD Data Sheets especially emphasize how many high-altitude páramo and puna plants have medicinal uses.

Important timber trees from the Andes include *Cedrela, Juglans, Quercus* and *Podocarpus*. With so many epiphytes and spectacular small herbs, the Andean flora also has significant, as yet largely unexploited potential for horticulture. An example is the red-flowered ladyslipper orchid *Phragmipedium besseyi* which was discovered on the eastern slopes of the Peruvian Andes not too many years ago, causing a major horticultural stir, with the first plants selling for several thousand US$ each (C. Dodson, pers. comm.); it continues to be popular now that it is produced in cultivation. There also are numerous Andean species used in local handicrafts. A particularly interesting example is "barniz de Pasto", obtained from the stipule exudate of a species of *Elaeagia* (Rubiaceae) and used in a kind of traditional lacquer-work in the Pasto region of southern Colombia.

In the Sierra Nevada de Santa Marta Data Sheet (SA25), Rangel and Garzón cite an unpublished 1987 thesis by E. Carbonó that makes abundantly clear the economic importance of Andean plants for one group of Andean natives. The Kogui Amerindians of the Sierra Nevada consume 32 species and use 81 for medicine, 44 in construction, 12 for dyes and 36 mythologically.

Perhaps the most important economic use of Andean plants is to provide ecological services such as erosion control and water retention. While difficult to quantify, such ecological services are clearly of paramount value in steep montane terrain. Unfortunately this value tends to be more appreciated once these services have been lost. The current crisis in electricity-generating capacity throughout the Andean region, due to lack of sufficient water to fill the hydroelectric reservoirs and run the turbines, is now widely appreciated to be a direct result of deforestation and other destruction of natural Andean ecosystems.

Threats

The threats to biodiversity in the Andean region that are most often cited in the CPD Data Sheets are especially: clearing (legally and illicitly) of too steep land for agricultural cultivation, overgrazing and the wide-scale burning associated with cattle husbandry in páramo and puna, habitat loss from fuelwood collecting at the highest altitudes where trees are rare, and pollution from mining.

At the highest altitudes, cultivation of potatoes and other tubers to feed a burgeoning population may be the biggest threat. While even larger (and lower) areas of Andean forest were destroyed for coffee cultivation, currently little expansion of the coffee growing area seems to be taking place. However, in the cloud-forest areas increasing cultivation of coca in Bolivia and Peru, mostly to produce cocaine for the European and North American markets, and of marijuana (*Cannabis sativa*) and opium poppy in Colombia, also for export markets, have led to destruction of huge areas of forest. Ironically, attempts to control the drug crops have had no effect on the drug supply, but have had a devastating effect on natural ecosystems, both directly by indiscriminate spraying of chemicals from airplanes, and indirectly, since drug growers merely clear more forest for their preferred crops on steeper slopes in more inaccessible sites.

In all the Andean countries, failure to appreciate the constraints on agricultural development imposed by steep mountainous terrain has led to major losses of biodiversity through inappropriate colonization schemes by short-sighted government agencies, funded typically by international lending agencies.

Andean forests are among the most threatened of all tropical forest vegetations. Some entire forest types, like the *Podocarpus* forests that used to cover significant parts of the Andes, have already all but disappeared. In Colombia estimates suggest that less than 10% of the Andean forests remain intact (Henderson, Churchill and Luteyn 1991) – perhaps even less than 5% of the high-altitude upper montane forest. In Ecuador almost nothing is left of the natural forests of the central valley and only 4% of the forests on the western Andean slopes (Dodson and Gentry 1991). North-western Peru north of the Huancabamba Depression probably retains even less intact forest, with the last *Podocarpus* forests of the Jaén/San Ignacio region currently being cut. Although there are still areas of relatively intact forest on the eastern slopes of the Andes of Ecuador, Peru and Bolivia, all three countries have active road-building programmes and rampant deforestation in this region.

Conservation

The natural restriction or contraction of different vegetation types to relatively small areas is a general trend in the Andes, suggesting that appropriately planned conservation units on the Andean slopes would be likely to include more species than equivalent areas almost anywhere else.

A number of large and potentially significant conservation units exist in the Andean region, but only some of them are effective. Most of the National Parks, several of which are featured in CPD Data Sheets, include extensive areas of páramo or puna as well as Andean-slope cloud forest. A glaring omission from the park systems is the typically narrow strip of relatively wet fertile-soil forest that occurs in the 500–1000 m band. Most of the large upland parks either do not extend down to a low enough elevation to include this forest type or include it just on paper. Perhaps only the Sumaco region in Ecuador (CPD Site SA38) and Manu National Park in Peru (see CPD Site SA37) protect this critical zone, and even in Manu settlers are already making incursions. There are only a few places left in the Andes where the whole elevational sequence might be protected, perhaps including the Tambopata/Candamo region in southern Peru (see CPD Sites SA37 and SA10) and the Madidi region in Bolivia (CPD Site SA36). Clearly such areas need immediate evaluation and appropriate conservation, before colonization makes preservation unfeasible or impossible.

Perhaps a worse problem is that many Andean parks are "paper parks". The typical protection provided a National Park in the Andes is termed passive protection by Young and León (Data Sheet SA37) – inaccessibility rather than the park status provides a modicum of protection. They note that there are 20 park guards for the 20,000 km² conserved in the three relatively effective parks in the eastern Andes of Peru, an impossible total responsibility of 1000 km² per guard. They estimate that an additional 6000 km² of various other protected areas and reserves have no actual protection. To this total may be added Cutervo NP, which has been mostly clear-cut and had not a single guard when Gentry last visited it. Colombia's north-western Paramillo NNP has a single guard completely lacking in equipment or support and so without capability to even marginally effect the ongoing deforestation that has virtually eliminated natural vegetation from the great majority of the park. If National Parks are to be a useful mechanism for conservation of Andean biodiversity, major increases in their budgets and major changes in their administration will be necessary.

Protection of watershed forests in connection with hydroelectric generation is effective in some parts of the Andes, most notably of the Anchicaya watershed in Colombia by the Corporación Valle del Cauca (CVC); one of the advantages is that such protection is apparently politically more acceptable than is biodiversity protection *per se*. Private reserves like La Planada Nature Reserve (Stone 1985) in Colombia and Maquipucuna Reserve in Ecuador (Sarmiento 1995) seem to be more effective than most of the official parks, perhaps because the former tend to have relatively good financial support, but unfortunately the area of such reserves at present is minuscule. Several CPD Data Sheet authors emphasize the potential for ecotourism in the scenically spectacular Andes, although to date tourism has had little economic impact other than in the Cusco/Machu Picchu area of Peru. Clearly novel conservation strategies need to be implemented very rapidly in the Andean region; in many areas it is probably already too late.

Centres of plant diversity and endemism

a. PARAMO WITH ESPELETIINAE (e.g. Espeletia)

Colombia

SA25. Sierra Nevada de Santa Marta

– see Data Sheet.

SA26. Sierra Nevada del Cocuy-Guantiva

– see Data Sheet.

SA27. Páramo de Sumapaz region

– see Data Sheet.

SA28. Region of Los Nevados Natural National Park

– see Data Sheet.

SA29. Colombian Central Massif

– see Data Sheet.

Colombia, Ecuador

SA30. Volcanoes of Nariñense Plateau

– see Data Sheet.

b. PARAMO WITHOUT ESPELETIINAE TO PUNA

Ecuador

SA31. Páramos and Andean forests of Sangay National Park

– see Data Sheet.

Peru, Ecuador

SA32. Huancabamba region

– see Data Sheet.

Peru

SA33. Peruvian puna

– see Data Sheet.

Argentina, Chile

SA34. Altoandina

– see brief Data Sheet.

c. TUCUMAN-BOLIVIAN REGION

Argentina

SA35. Anconquija region

– see Data Sheet.

d. EASTERN SLOPE

Bolivia

SA36. Madidi-Apolo region

– see Data Sheet.

Peru

SA37. Eastern slopes of Peruvian Andes

– see Data Sheet.

Ecuador

SA38. Gran Sumaco and Upper Napo River region

– see Data Sheet.

VII. PACIFIC COAST

The Pacific coastal region of tropical South America, sometimes called the trans-Andean region, is a geographically coherent unit even though it is extremely heterogeneous phytogeographically due to its extreme variation in precipitation. It forms a mostly narrow strip along the western side of South America stretching from La Serena just north of 30°S latitude in northern Chile north across the Equator to include the drainage of the Atrato River in Colombia as well as the adjacent Urabá region and the climatically similar wettest part of the Magdalena Valley. Thus the "Pacific Coast" region includes the narrow strip of Caribbean coast that marks the northern limit of Chocó Department, and reaches a northern limit at Cabo Tiburón on the Panama border near 8°45'N.

To the west this region is delimited by the Pacific Ocean and the Panama border. To the east the region is less precisely delimited by the Andean escarpment. The separation of the coastal region from the Andean region is often placed near the 1000 m elevational contour (Gentry 1993a; Dodson and Gentry 1991; Forero and Gentry 1989).

In the geologically and geographically complex region of northern Colombia where the Andean Western and Central cordilleras reach their terminations, vegetation similar to that of the Chocó extends around the northern tips of the two cordilleras into the Magdalena Valley. The wet-forest Nechi area is clearly a part of the Chocó phytogeographically and is logically included in the Pacific coastal region of South America even though the area's rivers drain north into the Caribbean. The moist and wet forests of the central Magdalena Valley are phytogeographically more distinctive and might well be considered to constitute a separate region coordinate with Amazonia or the Mata Atlântica. However, for the purposes of this summary review, the part of the Magdalena Valley between 5°30'–7°30'N is considered under the rubric of the Pacific Coast region.

Ecologically this region runs the gamut from among the wettest to the driest extremes of the neotropics, and the world. In general the northern Chocó region is perhumid, whereas the Sechura and Atacama deserts of Peru and northern Chile are perarid. Coastal Ecuador is transitional, ranging from perhumid rain forest in the north to desert in the south-west on the Santa Elena Peninsula. One of the Earth's wettest places is at Tutunendó in the Colombian Chocó, which receives 11,770 mm of annual precipitation. No rainfall has ever been recorded in parts of northern Chile's Atacama Desert.

In Colombia there is a north-south gradient of lessening precipitation, with the Caribbean coastal region more strongly seasonal. In Ecuador, vegetational zonation depends largely on proximity to the offshore Humboldt Current; precipitation zones are generally parallel to the coast, with the westernmost extension of the country (between Cape San Mateo and the Santa Elena Peninsula) the driest and the Andean foothills the wettest. Due to the strength of the Humboldt Current, coastal Ecuador and the Galápagos Islands are the only places in the world where desert and near-desert conditions reach to the Equator. In Peru and northern Chile the precipitation gradient also decreases from north to south, with the tropical dry forest of the northernmost part of coastal Peru grading into progressively drier desert to the south.

Vegetation

The Chocó pluvial forest is one of the few places in the world with the over 8000 mm of annual rainfall that categorize true tropical rain forest under the widely used Holdridge system of vegetation classification. The perhumid, or pluvial, forest occupies most of the lowland coastal region of Colombia south of 7°N latitude, extending along the base of the Andes across the Ecuador border to a southern terminus near 0°30'N. The Chocó pluvial forest is physiognomically as well as floristically distinctive, being characterized by many features that are normally restricted to mid-elevation cloud forests, including extremely thick coverings of moss and other non-vascular epiphytes on branches and tree trunks and prevalence of hemi-epiphytic climbers rather than free-climbing lianas. There are unusually high densities of trees of small (2.5–10 cm dbh) and medium (10–30 cm dbh) sizes (Gentry 1993b), unusually low densities of large trees, and overall low biomass (Faber-Langendoen and Gentry 1991). Holdridge-system wet and moist forests, which are physiognomically (as well as floristically) distinct from pluvial forest, cover most of lowland north-western Colombia north of 7°N, and the north-western part of Ecuador, extending southward as a narrowing strip along the Andean foothills almost to the Peruvian border. Even in the Chocó itself there is a mosaic of wet and pluvial forests, with the low coastal cordilleras generally relatively drier and covered by wet forest. The wet-forest and moist-forest regions have the free-climbing lianas, large emergent trees and high biomass that characterize Amazonian and most other such neotropical forests.

Like Amazonia, the lowland Chocó includes a mosaic of different vegetation types. Several of the most distinctive are swampy, including the best developed and most species-rich mangroves of the neotropics, with at least 15 species of mangroves or mangrove-associated plants, most of which do not occur on the Atlantic coast (Gentry 1982b). In the lower Atrato River drainage of the northern Chocó, there are nearly pure stands of *Raphia* palm in frequently inundated areas, and in less swampy areas "cativales" with single-species dominance by the legume *Prioria copaifera*. There are also extensive herbaceous swamps. Scattered along the Pacific coast northward from Esmeraldas, Ecuador are single-species dominated freshwater and brackish-water swamp forests respectively dominated by *Campnosperma panamensis* and *Mora megistosperma*, in addition to the more continuous mangrove formations.

Not surprisingly considering the heavy rainfall, the soils of most of the Chocó region (see CPD Site SA39) are highly leached and nutrient-poor. Much of the region has typical red-clay lateritic soils but some areas, especially near the base of the Andes and in the floodplains of major rivers, have relatively young less-leached soils. Relatively rich soils are especially prevalent in the southern part of the Ecuadorian Chocó. A peculiar soil type of unusual botanical interest is the white clay that occurs in much of the Bajo Calima region of Colombia's Valle Department and also west of Lita in northern Ecuador. This clay is especially associated with such distinctive Chocó vegetational features as gigantic sclerophyllous leaves and unusually large fruits (Gentry 1986b, 1993b).

The relatively small, anomalously low-latitude, dry-forest area of south-western Ecuador (cf. Kessler 1992) and north-westernmost Peru is very different from the adjacent moist and wet forests in its almost complete loss of leaves during the dry season. The vegetation of this Tumbes region is distinctive among neotropical dry forests in the prevalence of large emergent Bombacaceae trees, which tend to have grotesque growth forms and give the forest a very distinctive aspect. Another peculiarity is the extreme density of herbaceous vines that cover trees and ground alike during the wet season.

The Pacific Coast desert region forms a continuous strip over 3500 km long from near the Ecuador/Peru border at 5°S latitude to La Serena, Chile near 30°S. In the coastal desert regions vegetation is extremely sparse, even entirely lacking over large areas which instead may be covered with drifting sands. Botanically the most interesting part of the coastal desert is the "lomas", a curious archipelago-like formation of mostly ephemeral and geophytic herbs which grow in the condensation zone where winter fogs are intercepted by steep hills adjacent to the coast (see CPD Sites SA42 and SA43). While the lomas have a lush seasonal growth of herbs for several months, for most of the year the plant life disappears entirely, leaving apparently bare desert.

Flora

Due to its isolation by the Andes from the rest of lowland South America, the trans-Andean Pacific coastal region is characterized by a high level of endemism, leading to a concomitant need for conservational focus. The flora of the humid Chocó has been very roughly estimated as likely to include 8000 vascular plant species, c. 20% of them strictly endemic (Gentry 1982b). Although there are no endemic families in the wet part of Pacific South America, several genera are endemic and some of them have undergone significant speciation, including *Trianaeopiper* (17 spp.) (Piperaceae) and *Cremosperma* (16 spp.) (Gesneriaceae). The flora of coastal Ecuador below 900 m elevation with the Tumbes area of Peru is estimated to consist of c. 6300 vascular plant species, with c. 20% or 1260 species strictly endemic to the region (Dodson and Gentry 1991). The dry-forest portion of coastal Ecuador contributes c. 1000 species to this total, most of them not shared with moist or wet forest, with a similar endemism rate of 19%. Several small genera are endemic to Ecuadorian dry forest, whereas the adjacent moist forest has no endemics.

The predominant families in Chocó are generally those well represented in other parts of the lowland neotropics. For example among the almost 4000 vascular plant species listed for Colombia's Chocó Department by Forero and Gentry (1989), the largest family is Orchidaceae, followed by Rubiaceae, Leguminosae, Melastomataceae, Piperaceae, Compositae, Gesneriaceae, Araceae, Palmae, Bromeliaceae, Solanaceae and Euphorbiaceae. Similarly for the Florula of the Río Palenque Science Center in the Ecuadorian Chocó band (Dodson and Gentry 1978), the largest family is Orchidaceae, followed by ferns, Leguminosae, Araceae, Piperaceae, Compositae, Moraceae, Solanaceae, Gesneriaceae, Gramineae, Rubiaceae, Euphorbiaceae, Melastomataceae and Bromeliaceae. Although somewhat less speciose, predominantly hemi-epiphytic families like Ericaceae and Guttiferae are conspicuously more prevalent in the lowland Chocó than in lowland forests of Amazonia or Central America.

When data for plants of dbh 2.5 cm or more in 0.1-ha samples are compared, Leguminosae, Rubiaceae, Palmae, Annonaceae, Melastomataceae, Sapotaceae and Guttiferae are the predominant families in Chocó pluvial forests (Gentry 1986b). Most noteworthy are the preponderance of Guttiferae, Melastomataceae, Myrtaceae and Bombacaceae as compared to similar samples in Amazonia and Central America. Other families, like Bignoniaceae and Meliaceae, are less prevalent in Chocó pluvial-forest samples than in most other moist lowland neotropical forests. Following this much-replicated sampling technique, the Chocó pluvial forests are the most species-rich in the entire world, although almost equalled by some Upper Amazonian samples.

Certain genera tend to be strikingly species-rich in the Chocó region, although most of them are also among the largest genera elsewhere in the lowland neotropics. The largest genus is *Piper*, both in Chocó Department and at Río Palenque. Other exceptionally speciose genera in Chocó Department are *Psychotria, Miconia, Anthurium, Thelypteris, Peperomia, Maxillaria, Cavendishia, Elaphoglossum, Solanum, Trichomanes, Heliconia, Epidendrum, Columnea, Clidemia* and *Guzmania* (Forero and Gentry 1989). At Río Palenque, *Piper* is followed by *Ficus, Solanum, Peperomia, Philodendron, Pleurothallis, Anthurium, Epidendrum, Maxillaria, Thelypteris* and *Heliconia* (Dodson and Gentry 1978; Gentry 1990a). Most of these species-rich genera are epiphytic, and most of the rest are understorey shrubs and herbs. It has been suggested that such taxa are prone to unusually rapid speciation leading to extreme local endemism, with the Chocó region's foothill cloud forests the epicentre of this phenomenon (Gentry 1986b, 1989; Gentry and Dodson 1987). In the extreme example of the Centinela ridge in central-western Ecuador, a 20-km² area of cloud forest may have supported 90 locally endemic species, about a tenth of the entire flora (Gentry 1986a; Gentry and Dodson 1987; Dodson and Gentry 1991).

The dry forest of coastal Ecuador and adjacent north-western Peru is floristically similar to continentally interior neotropical dry forests, with dominance of Leguminosae trees and Bignoniaceae lianas. Perhaps its most striking floristic feature is the unusual diversity of Bombacaceae (Gentry 1995). In the driest regions, as on Ecuador's Santa Elena Peninsula, Capparidaceae are especially prevalent. This overall region has unusually high endemism for dry forest, with 19% (c. 190) of its species endemic (Dodson and Gentry 1991).

The offshore Galápagos Islands (volume 2, CPD Data Sheet site PO7) constitute essentially a depauperate subset of this flora where only the bird-dispersed and a few wind-dispersed (pogonochore) species of the coastal dry forest arrived, and sometimes speciated. The entire Galápagos native flora includes only c. 541 species of vascular plants (Wiggins and Porter 1971; Porter 1983; Hamann 1995). Endemism is 41.4%, with most of the 224 endemic species being rather obscure herbs (Porter 1983); the dominant dry-forest trees are conspecific with those of the mainland.

Floristically, the coastal desert of Peru and northern Chile has little in common with other equatorial lowlands in South America and more in common with the high Andes. The predominant families are mostly herbaceous or shrubby, and include Compositae, Malvaceae, Cactaceae, Boraginaceae, Cruciferae, Aizoaceae, Portulacaceae, Solanaceae and Umbelliferae (Rundel *et al.* 1991). The coastal desert area is characterized by very strong local endemism in the lomas formation, with two speciose near-endemic families (Nolanaceae and Malesherbiaceae, both shared with Mediterranean Chile) and genera such as *Calceolaria, Palaua, Tiquilia* and *Nolana* showing significant lomas radiations (Rundel *et al.* 1991). Both diversity and endemism are greater in the lomas of the southern part of the coastal desert, with 62% of southern Peru's lomas flora endemic (Müller 1985). There is also a sharp floristic break near the Peru/Chile border with only 7% of the region's species occurring on both sides of this climatic barrier (Rundel *et al.* 1991). Although the individual lomas have relatively small floras usually of less than 100 species (with 230 in the richest, the Paposo area), there is such a strong degree of local endemism that the coastal desert contributes an estimated additional 1000 species to the Pacific coast total (Rundel *et al.* 1991).

Species of economic importance

The Chocó region has a very large complement of useful plants and presumably a very real potential for conservation via extractive reserves and other sustainable uses of forest products. Moreover, many of the useful species of this region appear to be locally endemic, and a very different set of genera and families are of economic importance in the coastal Pacific wet forests than in Amazonia or Central America. Included among the region's useful plants are many fruits such as the wildly popular local delicacy "borojo" (*Borojoa patinoi*). Romero-Castañeda (1985) lists 42 Chocó fruit species that are edible, including such otherwise rarely consumed genera as *Montrichardia, Aechmea, Crataeva, Leonia, Malvaviscus, Patinoa, Maripa* and *Pentagonia*. Several of the unusually large fruits that characterize Chocó species are consumed, including *Orbignya cuatrecasana* and *Compsoneura atopa*. Another peculiarity of the consumption is that various fruits are consumed as a source of starch rather than for their sweet sugary taste (Gentry 1992c). A surprising number of the edible species are recently scientifically described or have yet to be described (Gentry 1992c).

The region is rich in plants used for fibres, including many palms (Bernal 1992) and aroids (e.g. *Heteropsis, Philodendron* – Gentry 1992c); the fibres of *Carludovica palmata* (Cyclanthaceae) form the basis of Ecuador's "Panama hat" industry (Miller 1986). There are also plants with horticultural potential. A good example is *Anthurium andreanum*, among the world's most widely cultivated aroids, which is endemic to a small region of western Colombia and north-western Ecuador; its collection was an important source of income for the Quaiquer Amerindians until local populations were decimated. A major effort is currently underway to revitalize the harvest of "tagua" (*Phytelephas*), formerly much used for buttons and as an ivory substitute, as a conservation-motivated sustainable extractive industry (Calero-H. 1992; Bernal 1992; Ziffer 1992). Canning of palm hearts (from *Euterpe oleracea*) is a significant local industry with six factories along the Colombian coast (Bernal 1992). In coastal Peru, the manufacture of "algarrobina" from *Prosopis* fruits to flavour drinks forms an important local industry.

The Pacific coastal region's abundance of tree species includes many species that have been harvested for timber. Again a very different taxonomic complement is exploited locally than elsewhere in the neotropics. Among the most important timber trees are several of the species that form single-species dominated stands including *Prioria copaifera* ("cativo") in the north, which is rapidly disappearing; *Campnosperma panamensis* ("sajo"), which is the single most important central Chocó timber species, accounting for 16% of the regional timber production in 1982; and *Mora megistosperma* ("nato"), with 280 km² harvested by 1982 (CVC 1983). In the high-rainfall part of central Chocó, *Goupia* and various Humiriaceae are the major timber species. The coastal Ecuador wet forest has lost its timber industry along with its trees; the formerly most important timber tree of the region, the rather recently described endemic *Caryodaphnopsis theobromifolia*, has been reduced to a few trees in the 1-km² Río Palenque reserve. The extremely hard wood of *Tabebuia chrysantha* ("guayacán") is the most exploited timber of the coastal Ecuador dry forest; as it has disappeared, the handicraft industry formerly dependent on it has turned to painting inferior woods so as to create imitations of the prized species.

Threats

The conservation status of the Pacific coastal region of South America is generally grim. Both the Colombian Chocó and coastal Ecuador were singled out by Myers (1988) as two of the world's 15 evolutionary hot spots in need of special conservation focus. The lomas formations have also been heavily impacted by human use, including pollution from mining, overgrazing during the winter growing season and even construction of new towns, but fewer species may be at risk (Ferreyra 1977).

Overall, 96% of the forests of coastal Ecuador have been destroyed, in an area of high regional and local endemism; only 1% of the coastal dry forest remains. Isolated cloud-forested outlying ridges apparently have significant complements of extremely local endemics. When deforestation of one of them was completed, on the 600 m elevation Centinela ridge south of Santo Domingo de los Colorados, perhaps 90 species of plants went extinct (Gentry 1986a; Dodson and Gentry 1991), one of the most massive local extinctions from tropical deforestation to date. The concept of "Centinelan extinctions" has even entered the conservation literature (Wilson 1992).

While three-quarters of the Colombian Chocó forest remains intact, large areas in the north and south are completely deforested, as is the Bajo Calima area. If the suspected pattern of high local endemism prevails in floristically poorly known Chocó, similar extinctions may be anticipated there as well, if deforestation continues unabated. Also, the Magdalena Valley has significant endemism and has been mostly deforested.

As in many other parts of the tropics, deforestation is the major contributor to the loss of biodiversity in the Pacific coastal region of South America. Western Ecuador is a particularly poignant example of this change (Parker and Carr 1992). Even though the rate of population increase dropped from 3.2% per year a few years ago to 2.4% by 1992 (World Population Data Sheet), Ecuador's population continues to grow alarmingly. Aggressive road-building and agricultural settlement programmes have accelerated forest loss as has a sociopolitical climate where privately owned forested lands are considered "unused" and susceptible to expropriation, either officially or extraofficially via squatters (Dodson and

Gentry 1991). Unlike many tropical areas, the relatively rich soils of much of western Ecuador make agriculture sustainable as well as profitable, increasing the pressures on forest remnants.

The same processes that have already led to the devastation of western Ecuador's forests are currently in operation in western Colombia, where they are aggravated by environmental impact from large-scale commercial operations including oil-palm plantations (especially in Nariño), banana plantations (Urabá area), paper-pulp production (mostly in lowland Valle) and timbering (especially in northern Chocó). According to Myers (1988), 28% of the Colombian Chocó has already been deforested.

Over-exploitation of natural resources is also a problem. For example, mangroves are extensively harvested in the Buenaventura area for construction timber and tannins – virtually the entire 2420 km² of mangrove forest having been exploited at least once; 21,292 m³ per year of mangrove wood were being cut by 1982 (CVC 1983). Worse, in Ecuador mangroves are being replaced entirely by shrimp farming operations. The once extensive "cativales" of northern Chocó have been decimated by timbering operations as has much of the Atrato region's upland forest. By 1982, 23% of the Pacific coast forests of Colombia had been harvested for timber, including well over 5741 km² in the southern and central coastal region alone (CVC 1983). Most of the northern Chocó has been assigned to timber concessions.

Conservation

Several large conservation units have been designated in this region (see CPD Data Sheets), with a total of c. 11,000 km² officially protected in the Pacific coastal lowlands. About half of the total is in the wet northern region, including four lowland National Parks in Colombia plus the lowland part (to 100 m) of Paramillo NNP (which has 4600 km²) in Colombia's Córdoba and Antioquia departments and the Cotacachi-Cayapas/Awá reserve area in Ecuador. There is no conservation unit in the Magdalena Valley. Two large dry-forest National Parks total c. 1300 km² – Machalilla in Ecuador and Cerros de Amotape in northern Peru. There are two national Sanctuaries in the Peruvian Desert region, Lomas de Lachay and Paracas, totalling 360 km² – however 60% is marine. In addition most land in the Galápagos Islands is designated as a National Park. Thus the remaining natural vegetation of the Pacific coastal region, except the Magdalena Valley, might seem to be well protected.

However, attempts at direct conservation via National Parks have been notably ineffective in much of the Pacific Coast region of South America. For example the only legally protected dry forest in coastal Ecuador is Machalilla NP, but as much as 75% of its surface area remains effectively in private hands and the forest has been largely decimated since the park was established (in 1979) (contrary to IUCN 1982 – see Arriaga 1987; Parker and Carr 1992). Cerros de Amotape NP in Tumbes, Peru, still effectively protects some dry forest, but is seriously threatened by livestock browsing and fire. The large Cotacachi-Cayapas Ecological Reserve in northern Ecuador, which spans what was originally an intact ecological gradient from the high Andes to the coastal lowlands, is being seriously encroached along its boundaries (Parker and Carr 1992). Most of the lowland part of Paramillo NNP at the north tip of the Western Cordillera in Colombia has been clear-cut for cattle pasture, apparently since the park was established (in 1977). Most of Los Katíos NNP in the Urabá area of Chocó near the Panama border was deforested while the park was in the process of being established (in 1973). It is obvious that a much greater effort is needed if such reserves are to be effective.

A number of novel conservation strategies seem to have been more effective in conservation than the underfunded and understaffed National Parks. In Ecuador the best preserved dry forest is the Arenillas Reserve (200 km²) protected by the military as a buffer along the Peru border. In Colombia watershed protection has been exemplary by the Corporación Valle del Cauca (CVC) for its electricity-generating system on the Anchicaya River. Small private reserves like La Planada in Colombia and in Ecuador Río Palenque and the Universidad de Guayaquil's Jauneche are playing an important conservation role, which needs to be expanded.

Perhaps most significant of all, there is growing appreciation of the significance of conservation, with local sustainable-use initiatives, generally supported by NGOs (non-governmental organizations), seen in part as a mechanism for protecting forest populations of people and their cultures from being overrun by outsiders from adjacent montane areas. Recently established indigenous reserves that occupy a significant part of the still-forested Chocó will hopefully play a similar role. Even some of the large companies notorious for forest exploitation have begun to see the importance of forest conservation, to the extent of setting aside and protecting reserves on their own initiatives. Ecotourism, currently almost non-existent in this region, is perceived as a potential new source of income that could be compatible with conservation of biodiversity and local lifestyles as well. The connection between widespread deforestation and energy rationing due to lack of hydroelectric power from empty or silted-up reservoirs has become widely recognized.

Centres of plant diversity and endemism

Colombia

SA39. Colombian Pacific Coast region (Chocó)

– see Data Sheet.

Ecuador

SA40. Ecuadorian Pacific Coast mesic forests

– see Data Sheet.

Peru

SA41. Cerros de Amotape National Park region

– see Data Sheet.

SA42. Lomas formations

– see Data Sheet.

Chile

SA43. Lomas formations of the Atacama Desert

– see Data Sheet.

VIII. SOUTHERN CONE

The biogeographic Southern Cone includes most of extra-tropical South America from roughly south of 30°S latitude to Tierra del Fuego at 55°S. The Andes reach their highest altitude (Cerro Aconcagua, 6959 m) on the Argentina/Chile border, and the region also includes vast lowland areas (the pampas) and tablelands (Patagonia). In accord with the ample latitudinal and altitudinal ranges, the climate covers a whole spectrum of variants from generally dry and warm, dry and cold, and moist and cold. The annual precipitation varies from c. 100 mm (Patagonian steppe) to c. 5000 mm (Valdivian forest). The region has been treated phytogeographically with diverse criteria by different authors, but the subdivisions proposed by Cabrera and Willink (1973) are widely accepted.

Three phytogeographic units (provinces) are included in the north-eastern half of the Southern Cone: pampas, espinal and monte, all having strong floristic affinities with the chaco farther north, but clearly delimited on floristic-physiognomic criteria.

Mediterranean Chile but particularly Altoandina and Patagonia share numerous floristic and physiognomic characteristics; with the puna, the latter two regions constitute an independent Andino-Patagonian Dominion (Cabrera and Willink 1973). The Altoandina comprises the highest elevations of the southern Andes along the Chile/Argentina border, extending upwards from 4400 m in the north to 500 m in Tierra del Fuego. The vegetational limit is 5600 m in the north-east, at Famatina. The southernmost portion appears as islands interspersed among the subantarctic forest.

The temperate rain forest of Chile and Argentina comprises perhaps the most distinct floristic unit in South America, and the only one in which a considerably high number of elements alien to the neotropics can be found. It has been assigned conveniently to the Antarctic Region (Cabrera and Willink 1973; M.T.K. Arroyo 1994, pers. comm.).

Vegetation

Almost all of the main vegetation formations (except tropical forest) are represented in the Southern Cone: prairie, grass-steppe, shrub-steppe, scrub, desert, tundra, sclerophyllous forest, deciduous temperate forest and evergreen forest. A number of physiognomically well-characterized phytogeographic units are generally recognized.

The "pampas" cover c. 900,000 km² between latitudes 28°–39°S and longitudes 50°–65°W in the southernmost part of Brazil, the whole of Uruguay and the central-eastern part of Argentina. These vast plains are only interrupted by a few low mountain ranges; the highest mountains reach 1300 m, in Argentina's Sierra de la Ventana north of Bahía Blanca. The climate is mild with precipitation of 600–1200 mm more or less evenly distributed through the year. The soils are very rich. The dominant vegetation types are grassy prairie and grass-steppe in which numerous species of the Gramineae tribe Stipeae (*Stipa* and *Piptochaetium*) are particularly conspicuous. There is an almost absolute lack of native trees, except along main watercourses.

The "espinal" is a winter-deciduous dry-forest band that embraces the pampas on the west. Both this phytogeographic province and the pampas province are clearly defined on physiognomic terms due to the presence in the espinal of occasional trees, mainly species of *Prosopis*. There are palm communities dominated by *Syagrus yatay*. This region is also clearly separate from the neighbouring chaco due to the espinal's absence of large *Schinopsis* trees.

The "monte" occupies a great portion (c. 600,000 km²) of central-western Argentina between 28°–44°S, from sea-level in the south-east to 3400 m in the north-west (Hunziker 1952). The climate presents wide daily and yearly temperature fluctuations, with the precipitation below 200 mm per year – conditions of semi-aridity and aridity. The dominant vegetation is shrub-steppe in which several species of Zygophyllaceae (*Larrea, Bulnesia, Plectrocarpa*) are conspicuous. Succulents are abundant, mainly Cactaceae which become dominant in the north-west on the eastern slopes of the mountains. That area, the "prepuna", has sometimes been recognized as an independent phytogeographic unit (Cabrera 1976).

Mediterranean Chile extends between 29°–38°S from the lower western Andean slopes to the Pacific Ocean. The climate is dry in summer and cool and damp in winter. The Coastal Cordillera (with highest elevation c. 2200 m) parallels the Andes through the entire region. Various vegetation types are present: shrubby xerophytic communities, evergreen matorral, palm forest of *Jubaea chilensis*, bamboo-dominated (*Chusquea*) thickets, sclerophyllous forest, hygrophilous forest, relict deciduous *Nothofagus* forest at the northern limit of its distribution, and alpine-like steppe.

The Altoandina province has been divided into three districts: Quichua, Cuyano and Austral (Cabrera 1976). The climate is cold and dry (with more humidity southward), averaging annually e.g. c. 3°C in Mina Aguilar (Jujuy) and c. -2°C in Cristo Redentor (Mendoza); the scarce precipitation may come as snow, and winds are strong. The most important vegetation types are grass-steppe, chamaephyte-steppe (having low woody and herbaceous species with buds on aerial branches close to the ground), and shrub-steppe; there are also bogs, and semi-deserts with lichens.

Grass-steppe is composed of isolated plants, sometimes also including chamaephytes and hemi-cryptophytes in rosettes. Among the important elements are *Festuca, Poa, Stipa* and *Calamagrostis* (*Deyeuxia*) with dwarf dicots in mats (e.g. *Adesmia, Mulinum, Senecio, Azorella, Junellia, Verbena*) (Hunziker 1952; Ward and Dimitri 1966); there are also cacti in cushions (*Opuntia* subgenus *Tephrocactus*). Lichen semi-desert occurs on the most humid slopes and can reach 5900 m (Nevado de Chañi). Chamaephyte-steppe generally occupies high elevations with loose soil; dwarf plants are common, e.g. *Senecio* spp., and cushions of *Oxalis compacta, Valeriana* spp., *Pycnophyllum molle* and *Werneria* spp. In wet locales there are bogs of Cyperaceae, Juncaceae and Gramineae.

The Patagonian steppe is a vast semi-desert region from below 2000 m on the eastern slopes of the Andes to the Atlantic Ocean and northward to the monte. The vegetation is primarily xeric shrubland, large scrub areas of cushion-like and dwarf shrubs, and grassy steppe.

Temperate rain forest occupies the southernmost portion of the continent and Tierra del Fuego, along a narrow fringe east of the cordillera in Argentina and south of 40°S latitude and in Chile between the Andes and the Pacific Ocean.

Flora

The pampas, espinal and monte can be treated jointly with the chaco, which shares almost all the genera, as a single floristic unit. These four regions have been considered to constitute the Chaco Dominion (Cabrera and Willink 1973; Cabrera 1976). They are also related, although to a lesser degree, to the Patagonian and the Alto-Andean regions. No data are available on the number of species in this floristic region as a whole.

In the Argentinian portion of the pampas, the number of species of vascular plants probably reaches 1800 (Cabrera 1963–1970), among which there are very few ferns and fern allies, and only one genus of gymnosperms (*Ephedra*). The families with the largest number of species are Gramineae and Compositae, with other well-represented families being Amaranthaceae, Caryophyllaceae, Chenopodiaceae, Cyperaceae, Cruciferae, Leguminosae, Malvaceae, Solanaceae and Umbelliferae. Recent floristic surveys in the southern portion of the pampas have found c. 500–600 species in areas of 5000–10,000 km² (Villamil, unpublished). Endemism is low (5–15%).

In the monte *Bulnesia*, *Larrea* and *Plectocarpa* of the family Zygophyllaceae are dominant, accompanied by *Monttea aphylla*, *Bougainvillea spinosa*, *Cercidium australe*, etc. Xerophilous species of Bromeliaceae (*Deuterocohnia*, *Abromeitiella*, *Tillandsia*, *Puya*) and Cactaceae are also well represented. The number of endemic species is c. 30, some of which dominate certain areas. In spite of its great floristic uniformity, this region has phytogeographical interest due to its affinities with North American semi-deserts and as well the Chilean Mediterranean region (Morello 1958).

Mediterranean Chile includes c. 1800–2400 species, and its endemism may be one of the highest in South America.

Although there are no comprehensive data about the whole Alto-Andean region, species richness and diversity are surely high. Among the best represented families are Gramineae [with many species of *Calamagrostis (Deyeuxia)*, *Festuca*, *Poa*, *Stipa*, etc.], Leguminosae (*Astragalus*, *Adesmia*, etc.) and Compositae (*Senecio*, *Chuquiraga*, *Mutisia*, etc.). Most of the genera are of neotropical origin, but there is a good representation of holarctic elements, and in its southern sector the Altoandina has an increasing proportion of antarctic elements (Cabrera 1976). There are no endemic families, and various endemic genera – *Barneoudia*, *Pycnophyllum*, *Hexaptera*, *Nototriche* and *Werneria*.

The temperate rain forest includes numerous floristic elements of both the antarctic and the Andean floras. Several woody gymnosperm species constitute conspicuous elements of the local vegetation, among them the second longest-lived tree in the world (*Fitz-roya cupressoides*), which attains over 3600 years (Lara and Villalba 1993).

Species of economic importance

Extra-tropical South America has not been a centre of radiation of cultivated plants of comparable importance to the Peruvian area. Only recently are breeding programmes being carried on. Numerous species have been used traditionally as medicinal plants and others have attained relevance as ornamentals. The floristic richness of grasslands assures a most important genetic reservoir especially for forage species.

The pods of several species of *Prosopis* ("algarrobo") are important forage resources, and also are used as human food and to prepare fermented beverages, although their use is limited to local communities. "Madi" (*Madia sativa*) is an oilseed species from the southern Andean region. "Mango" (*Bromus mango*) is a cereal that had been considered extinct, but recently was rediscovered. *Fragaria chiloensis* is a wild ancestor of cultivated strawberries. The seeds of the endemic *Araucaria araucana* have been staple food for indigenous populations until present times. The endemic Chilean palm tree (*Jubaea chilensis*), with edible fruits, was threatened with extinction due to over-exploitation to obtain the sugar from its sap. Fruits of a variety of native species are edible or locally used to prepare confections, such as several species of *Cereus* and related genera, *Berberis* ("calafates"), *Zizyphus mistol* ("mistol") and *Geoffroea decorticans* ("chañar"). Other species are highly reputed for honey production, such as *Eucryphia cordifolia* ("ulmo"), which is endemic to the temperate rain forest. Numerous species are used for furniture-making (*Prosopis*), construction (*Araucaria*, *Nothofagus*, *Fitz-roya*), fuelwood or charcoal and as natural dyes. Textile fibres are obtained from various bromeliads (*Puya*, *Deuterocohnia*). Numerous species from the Alto-Andean region are used as forage, fuel or medicinals (Ruthsatz 1974).

Threats

Although awareness about the importance of conservation from ethical as well as social and economic perspectives is progressing, the negative impact of human activities is still considerable upon most of the ecosystems in the Southern Cone. Deforestation by farming and overgrazing, often followed by soil erosion and desertification, are the most important factors of ecosystem degradation and destruction of the original vegetation (IUCN 1993). Erosion by wind and water are the main problems in large portions of the pampas, espinal and monte regions. The Patagonian steppes are profoundly affected by overgrazing of sheep, whereas goats have a more noticeable impact on the mountainous areas and in the monte, where the best forage species are being replaced by less valuable plants. In the forested areas the most valuable timber species are selectively cut and replaced by farming, or planting of more rapidly growing exotic trees such as *Eucalyptus* spp. and *Pinus* spp.

In some regions such as the pampas and Mediterranean Chile, the impact of human activity has drastically transformed the original landscape, of which there remains only relicts. The once abundant megafauna (felids, canids, camelids, cervids) has virtually disappeared from these regions. In the espinal large areas have been or are being deforested for farming and fuelwood.

Conservation

Argentina was the first country in Latin America to create a National Park (Nahuel Huapi, in 1903), and has a centralized administration. There are a total of 190 protected areas which cover 4.35% of the national territory (Barzetti 1993). Many of the National Parks have been created for geopolitical rather than ecological or biogeographical reasons, and consequently some natural regions are proportionately over represented, such as the subantarctic forest, whereas others, such as the pampas, are not represented at all in the national system of protected areas. A series of large protected areas and Biosphere Reserves has been created recently; effective measures for their protection still have to be implemented.

In Uruguay 0.17% of the national territory is in eight protected areas under IMPARQUES administration (Barzetti 1993). In developing a country-wide national system of protected areas, priority in investigation and management plans has been given to 16 of 36 potentially important or protected areas. Efforts are concentrated in the south-eastern region (Atlantic plains), which is richest in biodiversity. The proposed national system would bring protection to 0.70% of the country. There is no integrated and definitive legislation for the management of protected areas (WCMC 1992).

In Chile the main conservational concern is related to over-exploitation of the native forests for timber, woodchips and fuel. The national system of protected areas is centralized, and 18.18% of the national territory is under some kind of conservation management, with 79 protected areas (Barzetti 1993). However the development of protective policies is too recent to have prevented the degradation inflicted upon a substantial portion of the Chilean forests.

In order to effectively accomplish conservation objectives, the systems of protected areas throughout the Southern Cone must be increased both in quality and quantity. Care should be taken to assign protective status to representative samples of all the ecosystems, including those with no obvious touristic or scenic appeal.

Centres of plant diversity and endemism

Chile

SA44. Mediterranean region and La Campana National Park

– see Data Sheet.

Argentina, Chile

SA34. Altoandina

– see brief Data Sheet.

SA45. Temperate rain forest of Chile

– see Data Sheet.

SA46. Patagonia

– see Data Sheet.

References

Anderson, A.B. (1981). White-sand vegetation of Brazilian Amazonia. *Biotropica* 13: 199–210.

Anderson, E. (1952, 1967). *Plants, man, and life*. University of California Press, Berkeley. 251 pp.

Arriaga, L. (1987). *Manejo de recursos costeros en el Ecuador. La pesca artesanal en el Ecuador*. ESPOL, Guayaquil. Pp. 3–10.

Balick, M.J. and Mendelsohn, R.O (1992). Assessing the economic value of traditional medicines from tropical rain forests. *Conserv. Biol.* 6: 128–130.

Balslev, H. (1988). Distribution patterns of Ecuadorean plant species. *Taxon* 37: 567–577.

Barros, F. de, Fiuza de Melo, M.M.R., Chiea, S.A.C., Kirizawa, M., Wanderley, M. das G.L. and Jung-Mendaçolli, S.L. (1991). *Flora fanerogâmica da Ilha do Cardoso*, Vol. 1: *Caracterização geral da vegetação e listagem das espécies ocorrentes*. Instituto de Botânica, São Paulo. 184 pp.

Barzetti, V. (ed.) (1993). *Parques y progreso. Areas protegidas y desarrollo económico en América Latina y el Caribe*. IV Congreso Mundial de Parques y Areas Protegidas, Caracas, Venezuela. IUCN and IDB, IUCN Publications Services Unit, Cambridge, U.K. 258 pp.

Bennett, B.C. (1992). Plants and people of the Amazonian rainforests: the role of ethnobotany in sustainable development. *BioScience* 42: 599–607.

Bernal, R.G. (1992). Colombian palm products. In Plotkin, M. and Famolare, L. (eds), *Sustainable harvest and marketing of rain forest products*. Island Press, Washington, D.C. Pp. 158–172.

Berry, P.E. (1982). The systematics and evolution of *Fuchsia* sect. *Fuchsia* (Onagraceae). *Ann. Missouri Bot. Garden* 69: 1–198.

Berry, P.E., Huber, O. and Holst, B.K. (1995). Floristic analysis and phytogeography. In Steyermark, J.A., Berry, P.E., Holst, B.K. and Yatskievych, K. (eds), *Flora of the Venezuelan Guayana*, Vol. 1: *Introduction*. Missouri Botanical Garden, St. Louis. Pp. 161–191.

Bigarella, J.J. and Ferreira, A.M.M. (1985). Amazonian geology and the Pleistocene and the Cenozoic environments and paleoclimates. In Prance, G.T. and Lovejoy, T.E. (eds), *Amazonia*. Key Environments Series. Pergamon Press, New York. Pp. 49–71.

Brown Jr., K.S. (1987). Soils. In Whitmore, T.C. and Prance, G.T. (eds), *Biogeography and Quaternary history in tropical America*. Clarendon Press, Oxford, U.K. Pp. 24–28.

Brown Jr., K.S. and Brown, G.G. (1992). Habitat alteration and species loss in Brazilian forests. In Whitmore, T.C. and Sayer, J.A. (eds), *Tropical deforestation and species extinction*. Chapman & Hall, London. Pp. 119–142.

Bullock, S.H., Mooney, H.A. and Medina, E. (eds) (1995). *Seasonally dry tropical forests*. Cambridge University Press, Cambridge, U.K. 521 pp.

Bunker, S.G. (1985). *Underdeveloping the Amazon*. University of Chicago Press, Chicago. 279 pp.

Buskirk, R.E. (1992). Zoogeographic and plate tectonic relationships of Jamaica to Mesoamerica. In Darwin, S. and Welden, A. (eds), *Biogeography of Mesoamerica*. Tulane Studies in Zoology and Botany, Suppl. Publ. 1. Pp. 9–16.

Cabrera, A.L. (1963-1970). *Flora de la provincia de Buenos Aires*, 6 vols. Instituto Nacional de Tecnología Agropecuaria (INTA), Buenos Aires.

Cabrera, A.L. (1976). *Regiones fitogeográficas argentinas*. In Parodi, L.R. (ed.), *Enciclopedia argentina de agricultura y jardinería*, 2nd edition. Vol.2(1). Editorial Acmé, Buenos Aires. Pp. 1–85.

Cabrera, A.L. and Willink, A. (1973). *Biogeografía de América Latina*. Organización de los Estados Americanos (OEA), Serie de Biología, Monogr. No. 13, Washington, D.C. 117 pp.

Calero-H., R. (1992). The Tagua Initiative in Ecuador: a community approach to tropical rain forest conservation and development. In Plotkin, M. and Famolare, L. (eds), *Sustainable harvest and marketing of rain forest products*. Island Press, Washington, D.C. Pp. 263–273.

Campbell, D.G. (1989). The importance of floristic inventory in the tropics. In Campbell, D.G. and Hammond, H.D. (eds), *Floristic inventory of tropical countries: the status of plant systematics, collections, and vegetation, plus recommendations for the future*. New York Botanical Garden, Bronx. Pp. 5–30.

Campbell Jr., K.E. and Frailey, D. (1984). Holocene flooding and species diversity in southwestern Amazonia. *Quaternary Research* 21: 369–375.

Capparella, A.P. (1988). Genetic variation in neotropical birds: implications for the speciation process. In Ouellet, H. (ed.), *Acta XIX Congr. Int. Ornith., Ottawa, Canada, 22-29 VI 1986*, Vol. 2. Natl. Mus. Nat. Sci., University of Ottawa Press, Ottawa. Pp. 1658–1664.

Chauvel, A., Lucas, Y. and Boulet, R. (1987). On the genesis of the soil mantle of the region of Manaus, Central Amazonia, Brazil. *Experientia* 43: 234–240.

Clapperton, C.M. (1993). *Quaternary geology and geomorphology of South America*. Elsevier, Amsterdam. 779 pp.

Cleef, A.M. (1981). *The vegetation of the páramos of the Colombian Cordillera Oriental*. Diss. Bot. 61. J. Cramer, Vaduz. 320 pp.

Cleef, A.M. and Rangel-Ch., J.O. (1984). La vegetación del páramo del noroeste de la Sierra Nevada de Santa Marta. In van der Hammen, T. and Ruiz-C., P.M. (eds), *La Sierra Nevada de Santa Marta (Colombia), transecto Buritaca – La Cumbre*. Studies on Tropical Andean Ecosystems Vol. 2. J. Cramer, Berlin. Pp. 24–72.

Cleef, A.M., Rangel-Ch., J.O., van der Hammen, T. and Jaramillo-M., R. (1984). La vegetación de las selvas del transecto Buritaca. In van der Hammen, T. and Ruiz-C., P.M. (eds), *La Sierra Nevada de Santa Marta (Colombia), transecto Buritaca – La Cumbre*. Studies on Tropical Andean Ecosystems Vol. 2. J. Cramer, Berlin. Pp. 267–406.

Clement, C.R. (1989). A center of crop genetic diversity in western Amazonia. *BioScience* 39: 624–631.

Colinvaux, P.A. (1987). Amazon diversity in light of the palaoecological record. *Quaternary Sci. Rev.* 6: 93–114.

Collins, M. (ed.) (1990). *The last rain forests: a world conservation atlas*. Oxford University Press, New York. 200 pp.

Conservation International (1991). *Workshop 90: biological priorities for conservation in Amazonia;* scale 1:5,000,000. Conservation International, Washington, D.C. Map.

Cuadros, H. (1990). Vegetación caribeña. In Jimeno, M. (ed.), *Caribe Colombia*. Financiera Eléctrica Nacional (FEN), Bogotá. Pp. 66–83.

Cuatrecasas, J. (1958). Aspectos de la vegetación natural de Colombia. *Rev. Acad. Col. Cienc. Exactas Fís. Nat.* 10: 221–268.

CVC (1983). *Plan de desarollo integral para la costa pacífica*. Corporación Valle del Cauca (CVC), Cali, Colombia.

Daly, D.C. and Prance, G.T. (1989). Brazilian Amazon. In Campbell, D.G. and Hammond, H.D. (eds), *Floristic inventory of tropical countries*. New York Botanical Garden, Bronx. Pp. 401–426.

Dias, B.F. de S. (1990). Conservação da natureza no cerrado brasileiro. In Novaes Pinto, M. (ed.), *Cerrado: caracterização, ocupação e perspectivas*. Editora Universidade de Brasília, Brasília. Pp. 583–640.

Dillon, M.O. (1994). Bosques húmedos del norte del Perú. *Arnaldoa* 2(1): 29–42.

Dodson, C.H. and Gentry, A.H. (1978). *Flora of the Río Palenque Science Center, Los Ríos, Ecuador. Selbyana* 4. 628 pp.

Dodson, C.H. and Gentry, A.H. (1991). Biological extinction in western Ecuador. *Ann. Missouri Bot. Garden* 78: 273–295.

Ducke, A. and Black, G.A. (1953). Phytogeographical notes on the Brazilian Amazon. *An. Acad. Brasil. Ciênc.* 25: 1–46.

Duivenvoorden, J.F., Lips, J.M., Palacios, P. and Saldarriaga, J. (1988). Levantamiento ecológico de parte de la cuenca del Medio Caquetá en la Amazonia colombiana. *Colomb. Amaz.* 3: 7–38.

Duke, J.A. and Vásquez, R. (1994). *Amazonian ethnobotanical dictionary.* CRC Press, Boca Raton, Florida. 215 pp.

Eiten, G. (1972). The cerrado vegetation of Brazil. *Bot. Review* 38: 201–341.

Eiten, G. (1984). Vegetation of Brasília. *Phytocoenologia* 12: 271–292.

Emperaire, L. (1987). *Vegetation et gestion des resources naturelles dans la caatinga du sud-est du Piaui (Bresil).* Ed. ORSTOM F7, TDM-52. 378 pp.

Encarnación, F. (1985). Introducción a la flora y vegetación de la Amazonia peruana: estado actual de los estudios, medio natural y ensayo de una clave de determinación de las formaciones vegetales en la llanura amazónica. *Candollea* 40: 237–252.

Estado de São Paulo (ESP), 28 March 1992.

Faber-Langendoen, D. and Gentry, A.H. (1991). The structure and diversity of rain forests at Bajo Calima, Chocó region, western Colombia. *Biotropica* 23: 2–11.

Fearnside, P.M. (1991). Rondônia: estradas que levam à devastação. *Ciência Hojea. Amazônia*: 114–122. Bloch Editores, Rio de Janeiro.

Fearnside, P.M., Tardin, A.T. and Meira Filho, L.G. (1990). *Deforestation rate in Brazilian Amazonia.* Instituto Nacional de Pesquisas da Amazônia (INPA)/Instituto Nacional de Pesquisas Espaciais (INPE). 8 pp. Mimeographed.

Ferreyra, R. (1977). Endangered species and plant communities in Andean and coastal Peru. In Prance, G.T. and Elias, T.S. (eds),*Extinction is forever: threatened and endangered species of plants in the Americas and their significance in ecosystems today and in the future.* New York Botanical Garden, Bronx. Pp. 150–157.

Filgueiras, T.S. and Pereira, B.A.S. (1990). Flora do Distrito Federal. In Novaes Pinto, M. (ed.), *Cerrado: caracterização, ocupação e perspectivas.* Editora Universidade de Brasília, Brasília. Pp. 331–388.

Flenley, J.R. (1995). Cloud forest, the Massenerhebung effect, and ultraviolet insolation. In Hamilton, L.S., Juvik, J.O. and Scatena, F.N. (eds), *Tropical montane cloud forests.* Ecol. Studies Vol. 110. Springer-Verlag, New York. Pp. 150–155.

Forero, E. and Gentry, A.H. (1989). *Lista anotada de las plantas del Departamento del Chocó, Colombia.* Biblioteca J.J. Triana No. 10. Instituto de Ciencias Naturales, Museo de Historia Natural, Universidad Nacional de Colombia, Bogotá. 142 pp.

Frahm, J.-P. and Gradstein, S.R. (1991). An altitudinal zonation of tropical rain forests using bryophytes. *J. Biogeogr.* 18: 669–678.

Frailey, C.D., Lavina, E.L., Rancy, A. and Souza Filho, J.P. (1988). A proposed Pleistocene/Holocene lake in the Amazon Basin and its significance to Amazonian geology and biogeography. *Acta Amazonica* 18: 119–143.

Galeano, W.H. (1990). *The Flora of Yanacocha, a tropical high-Andean forest in southern Peru.* M.S. thesis, University of Missouri, St. Louis. 270 pp.

Gentry, A.H. (1982a). Neotropical floristic diversity: phytogeographical connections between Central and South America, Pleistocene climatic fluctuations, or an accident of the Andean orogeny? *Ann. Missouri Bot. Garden* 69: 557–593.

Gentry, A.H. (1982b). Phytogeographic patterns as evidence for a Chocó refuge. In Prance, G.T. (ed.), *Biological diversification in the tropics.* Columbia University Press, New York. Pp. 112–136.

Gentry, A.H. (1986a). Endemism in tropical vs. temperate plant communities. In Soulé, M.E. (ed.), *Conservation biology: the science of scarcity and diversity.* Sinauer Associates, Sunderland, Massachusetts, U.S.A. Pp. 153–181.

Gentry, A.H. (1986b). Species richness and floristic composition of Chocó region plant communities. *Caldasia* 15: 71–91.

Gentry, A.H. (1986c). Sumario de patrones fitogeográficos neotropicales y sus implicaciones para el desarrollo de la Amazonia. *Rev. Acad. Col. Cienc. Exactas Fís. Nat.* 16: 101–116.

Gentry, A.H. (1988). Changes in plant community diversity and floristic composition on environmental and geographical gradients. *Ann. Missouri Bot. Garden* 75: 1–34.

Gentry, A.H. (1989). Speciation in tropical forests. In Holm-Nielsen, L.B., Nielsen, I.C. and Balslev, H. (eds), *Tropical forests: botanical dynamics, speciation and diversity.* Academic Press, London. Pp. 113–134.

Gentry, A.H. (1990a). Floristic similarities and differences between southern Central America and Upper and Central Amazonia. In Gentry, A.H. (ed.), *Four neotropical rainforests.* Yale University Press, New Haven. Pp. 141–157.

Gentry, A.H. (1990b). Tropical forests. In Kast, A. (ed.), *Biogeography and ecology of forest bird communities.* SPB Academic Publishing, The Hague, Netherlands. Pp. 35–43.

Gentry, A.H. (1992a). Bignoniaceae – Part II (tribe Tecomeae). Flora Neotropica Monogr. 25(2): 1–370.

Gentry, A.H. (1992b). Diversity and floristic composition of Andean forests. In Young, K.R. and Valencia, N. (eds), *Biogeografía, ecología y conservación del bosque montano en el Perú*. Univ. Nac. Mayor San Marcos, Mem. Museo Hist. Nat. (Lima) 21: 11–29.

Gentry, A.H. (1992c). New nontimber forest products from western South America. In Plotkin, M. and Famolare, L. (eds), *Sustainable harvest and marketing of rain forest products*. Island Press, Washington, D.C. Pp. 125–136.

Gentry, A.H. (1992d). Tropical forest biodiversity: distributional patterns and their conservational significance. *Oikos* 63: 19–28.

Gentry, A.H. (1993a). Overview of the Peruvian flora. In Brako, L. and Zarucchi, J.L., *Catalogue of the flowering plants and gymnosperms of Peru*. Monogr. Syst. Bot. Missouri Bot. Garden 45: xxix–xl.

Gentry, A.H. (1993b). Riqueza de especies y composición florística. In Leyva-F., P. (ed.), *Colombia Pacífico*, Vol. 1. Fondo Protección del Medio Ambiente José Celestino Mutis, Publicaciones Financiera Eléctrica Nacional (FEN), Santafé de Bogotá. Pp. 200–219.

Gentry, A.H. (1994). A new South American vegetation type: the conservational significance of the dry forest of Santa Cruz, Bolivia. *Conserv. Biol.* In press.

Gentry, A.H. (1995). Diversity and floristic composition of neotropical dry forests. In Bullock, S.H., Mooney, H.A. and Medina, E. (eds), *Seasonally dry tropical forests*. Cambridge University Press, Cambridge, U.K. Pp. 146–194.

Gentry, A.H. and Dodson, C.H. (1987). Diversity and biogeography of neotropical vascular epiphytes. *Ann. Missouri Bot. Garden* 74: 205–233.

Gentry, A.H. and Ortiz-S., R. (1993). Patrones de composición florística en la Amazonia peruana. In Kalliola, R., Puhakka, M. and Danjoy, W. (eds), *Amazonia peruana – vegetación húmeda tropical en el llano subandino*. Proyecto Amazonia Universidad de Turku (PAUT) and Oficina Nacional de Evaluación de Recursos Naturales (ONERN), Jyväskylä, Finland. Pp. 155–166.

Giulietti, A.M. and Pirani, J.R. (1988). Patterns of geographic distribution of some plant species from the Espinhaço Range, Minas Gerais and Bahia, Brazil. In Vanzolini, P.E. and Heyer, W.R. (eds), *Proceedings of a workshop on neotropical distribution patterns*. Academia Brasileira de Ciências, Rio de Janeiro. Pp. 39–69.

Giulietti, A.M., Menezes, N.L., Pirani, J.R., Meguro, M. and Wanderley, M.G.L. (1987). Flora da Serra do Cipó, Minas Gerais: caracterização e lista das espécies. *Bol. Botânica, Univ. São Paulo* 9: 1–151.

Gómes, M.A.F. (1981). *Padrões de caatinga nos Cariris Velhos, Paraíba*. M.S. thesis, Universidade Federal Rural de Pernambuco, Recife.

Goulding, M. (1988). Ecology and management of migratory food fishes of the Amazon Basin. In Almeda, F. and Pringle, C.M. (eds), *Tropical rainforests: diversity and conservation*. California Academy of Sciences and American Association for Advancement of Science Pacific Division, San Francisco. Pp. 71–85.

Gray, A. (1987). *The Amerindians of South America*. The Minority Rights Group Report No. 15. London, U.K. 27 pp.

Haase, R. and Beck, S. (1989). Structure and composition of savanna vegetation in northern Bolivia: a preliminary report. *Brittonia* 41: 80–100.

Haffer, J. (1969). Speciation in Amazonian forest birds. *Science* 165: 131–137.

Haffer, J. (1987). Quaternary history of tropical America. In Whitmore, T.C. and Prance, G.T. (eds), *Biogeography and Quaternary history in tropical America*. Clarendon Press, Oxford, U.K. Pp. 1–18.

Hamann, O. (1995). Galápagos Islands, Ecuador. In WWF and IUCN, *Centres of plant diversity: a guide and strategy for their conservation*, Vol. 2: *Asia, Australasia and the Pacific*. IUCN Publications Unit, Cambridge, U.K. Pp. 556–564.

Harley, R.M. and Simmons, N.A. (1986). *Florula of Mucugê, Chapada Diamantina, Bahia, Brazil*. Royal Botanic Gardens, Kew, Richmond, U.K. 227 pp.

Henderson, A., Churchill, S.P. and Luteyn, J.L. (1991). Neotropical plant diversity. *Nature* 351: 21–22.

Hernández-Camacho, J.I. (1990). Las selvas andinas de Colombia. In Carrizosa, J. and Hernández-C., J.I. (eds), *Selva y futuro*. INDERENA, Bogotá.

Holdridge, L.R. (1967). *Life zone ecology*. Tropical Science Center, San José, Costa Rica. 206 pp.

Homma, A.K.O. (1992). The dynamics of extraction in Amazonia: a historical perspective. *Advances Econ. Bot.* 9: 23–31.

Huber, O. (1982). Significance of savanna vegetation in the Amazon Territory of Venezuela. In Prance, G.T. (ed.), *Biological diversification in the tropics*. Columbia University Press, New York. Pp. 221–244.

Huber, O. (1987). Consideraciones sobre el concepto de Pantepui. *Pantepui* 1(2): 2–10.

Huber, O. (1994). Recent advances in the phytogeography of the Guayana region, South America. *Mém. Soc. Biogéogr. (3ème série)* 4: 53–63.

Hunziker, J.H. (1952). Las comunidades vegetales de la Cordillera de La Rioja. *Rev. Invest. Agric.* 6: 167–196.

INPE (Instituto Nacional de Pesquisas Espaciais) (1992). *Deforestation in Brazilian Amazonia*. São José dos Campos, São Paulo. 2 pp. Mimeographed.

Irion, G. (1989). Quaternary geological history of the Amazon lowlands. In Holm-Nielsen, L.B., Nielsen, I.C. and Balslev, H. (eds), *Tropical forests*. Academic Press, London. Pp. 23–34.

Irmler, U. (1977). Inundation-forest types in the vicinity of Manaus. In Schmithüsen, J. (ed.), *Biogeographica* Vol. 8, *Ecosystem research in South America*. W. Junk Publishers, The Hague. Pp. 17–29.

IUCN (1982). *IUCN directory of neotropical protected areas*. Tycooly International Publ. Ltd., Dublin, Ireland. 436 pp.

IUCN (1993). *Documento de Paraty. Brasil*. IUCN and IDB. 47 pp.

IUCN and WWF (World Wildlife Fund) (1982). *Tropical Forest Campaign fact sheet no. 14 – Brazil*. Washington, D.C. 2 pp.

Janzen, D.H. (1988). Tropical dry forests: the most endangered major tropical ecosystem. In Wilson, E.O. (ed.), *Biodiversity*. National Academy Press, Washington, D.C. Pp. 130–137.

Joly, A.B. (1970). *Conheça a vegetação brasileira*. EDUSP e Polígono, São Paulo.

Jordan, C.F. (1985). Soils of the Amazon rainforest. In Prance, G.T. and Lovejoy, T.E. (eds), *Amazonia*. Pergamon Press, New York. Pp. 83–94.

Junk, W.J. and Furch, K. (1985). The physical and chemical properties of Amazonian waters and their relationships with the biota. In Prance, G.T. and Lovejoy, T.E. (eds), *Amazonia*. Pergamon Press, New York. Pp. 3–17.

Juteau, T., Megard, M., Raharison, L. and Whitechurch, H. (1977). Les assemblages ophiolitiques de l'occident equatorien: nature petrographique et position structurale. *Bull. Soc. Geol. France, Ser. 7*, 1: 1127–1132.

Keel, S., Gentry, A.H. and Spinzi, L. (1993). Using vegetation analysis to select conservation sites in eastern Paraguay. *Conserv. Biol.* 7: 66–75.

Keigwin Jr., L.D. (1978). Pliocene closing of the Isthmus of Panama, based on biostratigraphical evidence from nearby Pacific Ocean and Caribbean Sea cores. *Geology* 6: 630–634.

Keller, G., Zenker, C.E. and Stone, S.M. (1989). Late Neogene history of the Pacific-Caribbean gateway. *J. South Amer. Earth Sci.* 2(1): 73–108.

Kennerley, J.B. and Bromley, R.J. (1971). *Geology and geomorphology of the Llanganati Mountains, Ecuador*. Instituto Ecuatoriano de Ciencias Naturales, Contrib. No. 73. Quito. 16 pp.

Kessler, M. (1992). The vegetation of south-west Ecuador. In Best, B.J. (ed.), *The threatened forests of south-west Ecuador*. Biosphere Publications, Leeds, U.K. Pp. 79–100.

Lara, A. and Villalba, R. (1993). A 3620-year temperature record from *Fitzroya cupressoides* tree rings in southern South America. *Science* 260: 1104–1106.

Leopoldo, P.R., Franken, W., Salati, E. and Ribeiro, M.N. (1987). Towards a water balance in the Central Amazonia region. *Experientia* 43: 222–233.

Lerdau, M., Whitbeck, J. and Holbrook, N.M. (1991). Tropical deciduous forest: death of a biome. *Trends in Ecology and Evolution* 6: 201–202.

Lescure, J.-P., Emperaire, L., Pinton, F. and Renault-Lescure, O. (1992). Nontimber forest products and extractive activities in the Middle Rio Negro region, Brazil. In Plotkin, M. and Famolare, L. (eds), *Sustainable harvest and marketing of rain forest products*. Island Press, Washington, D.C. Pp. 151–157.

Liu, K.-B. and Colinvaux, P.A. (1988). A 5200-year history of Amazon rain forest. *J. Biogeogr.* 15: 231–248.

Luteyn, J.L. (1989). Speciation and diversity of Ericaceae in neotropical montane vegetation. In Holm-Nielsen, L.B., Nielsen, I.C. and Balslev, H. (eds), *Tropical forests*. Academic Press, London. Pp. 297–310.

Malingreau, J.-P. and Tucker, C.J. (1988). Large-scale deforestation in the southeastern Amazon Basin of Brazil. *Ambio* 17: 49–55.

Mantovani, W. (1990). A dinámica das florescas na encosta atlántica. In *2. Simpósio de ecossistemas da costa sul e sudeste brasileira: estructura, função e manejo*. Publ. ACIESP No. 71-1. Anais Academia de Ciências do Estado de São Paulo. Pp. 304–313.

Marshall, L.G., Butler, R.F., Drake, R.E., Curtis, G.H. and Tedford, R.H. (1979). Calibration of the Great American Interchange. *Science* 204: 272–279.

Marticorena, C. (1990). Contribución a la estadística de la flora vascular de Chile. *Gayana, Bot.* 47: 85–113.

Marticorena, C. and Quezada, M. (1985). Catálogo de la flora vascular de Chile. *Gayana, Bot.* 42: 1–157.

McCourt, W.J., Aspden, J. and Brook, M. (1984). New geological and geochronological data from the Colombian Andes: continental growth by multiple accretion. *J. Geol. Soc. London* 141: 831–845.

Miller, T. (1986). *The Panama hat trail: a journey from South America*. William Morrow, New York. 271 pp.

Molau, U. (1988). Scrophulariaceae – Part I: Calceolarieae. *Flora Neotropica Monogr.* 47: 1–326.

Morello, J. (1958). La provincia fitogeográfica del Monte. *Opera Lilloana* 2: 1–155.

Mori, S.A. and Boom, B.M. (1981). *Final report to the World Wildlife Fund–U.S. on the botanical survey of the endangered moist forests of eastern Brazil.* New York Botanical Garden, Bronx. 109 pp.

Müller, C.H. and Calzavara, B.B.G. (1986). Castanha-do-Brasil: conhecimentos atuais. In Dantas, M. (ed.), *Proceedings*, Vol. IV, *Perennial crops. 1st Symposium on the Humid Tropics, 1984.* EMBRAPA/CPATU Documentos 36, Belém. Pp. 223–229.

Müller, G.K. (1985). Zur floristischen Analyse der peruanischen Loma-Vegetation. *Flora* 176: 153–165.

Myers, N. (1982). Depletion of tropical moist forests: a comparative review of rates and causes in the three main regions. *Acta Amazonica* 12: 745–758.

Myers, N. (1988). Threatened biotas: "hot-spots" in tropical forests. *Environmentalist* 8: 187–208.

Nascimento, C. and Homma, A.K.O. (1984). *Amazônia: meio ambiente e tecnologia agrícola.* EMBRAPA/CPATU, Belém. 282 pp.

Nelson, B.W., Ferreira, C.A.C., Silva, M.F. da and Kawasaki, M.L. (1990). Endemism centres, refugia and botanical collection density in Brazilian Amazonia. *Nature* 345: 714–716.

Orians, G.H. and Solbrig, O.T. (eds) (1977). *Convergent evolution in warm deserts: an examination of strategies and patterns in deserts of Argentina and the United States.* Dowden, Hutchinson & Ross, Stroudsburg, Pennsylvania, U.S.A. 333 pp.

Parfit, M. (1989). Whose hands will shape the future of the Amazon's green mansions? *Smithsonian* (Nov.): 58–75.

Parker III, T.A. and Carr, J.L. (eds) (1992). *Status of forest remnants in the Cordillera de la Costa and adjacent areas of southwestern Ecuador.* RAP Working Paper 2, Conservation International, Washington, D.C. 172 pp.

Parker III, T.A., Gentry, A.H., Foster, R.B., Emmons, L.H. and Remsen Jr., J.V. (1993). *The lowland dry forests of Santa Cruz, Bolivia: a global conservation priority.* RAP Working Paper 4, Conservation International, Washington, D.C. 104 pp.

Peixoto, A.L. and Gentry, A.H. (1990). Diversidade e composição florística da mata de tabuleiro na Reserva Florestal de Linhares (Espírito Santo, Brasil). *Rev. Brasil. Bot.* 13: 19–25.

Peres, C.A. and Terborgh, J.W. (1995). Amazonian nature reserves: an analysis of the defensibility status of existing conservation units and design criteria for the future. *Conserv. Biol.* 9: 34–46.

Peters, C.M., Gentry, A.H. and Mendelsohn, R.O. (1989). Valuation of an Amazonian rainforest. *Nature* 339: 655–656.

Pires, J.M. and Prance, G.T. (1985). The vegetation types of the Brazilian Amazon. In Prance, G.T. and Lovejoy, T.E. (eds), *Amazonia.* Pergamon Press, Oxford, U.K. Pp. 109–145.

Porter, D.M. (1983). Vascular plants of the Galápagos: origins and dispersal. In Bowman, R.I., Berson, M. and Leviton, A.E. (eds), *Patterns of evolution in Galápagos organisms.* American Association for Advancement of Science Pacific Division, San Francisco. Pp. 33–96.

Prance, G.T. (1973). Phytogeographic support for the theory of Pleistocene forest refuges in the Amazon Basin, based on evidence from distribution patterns in Caryocaraceae, Chrysobalanaceae, Dichapetalaceae, and Lecythidaceae. *Acta Amazonica* 3: 5–28.

Prance, G.T. (1979). Notes on the vegetation of Amazonia III. The terminology of Amazon forest types subject to inundation. *Brittonia* 31: 26–38.

Prance, G.T. (ed.) (1982). *Biological diversification in the tropics.* Columbia University Press, New York. 714 pp.

Prance, G.T. (1987). Vegetation. In Whitmore, T.C. and Prance, G.T. (eds), *Biogeography and Quaternary history in tropical America.* Clarendon Press, Oxford, U.K. Pp. 28–45.

Prance, G.T. (1990). Future of the Amazonian rainforest. *Futures* (Nov.): 891–903.

Prance, G.T. (1994). A comparison of the efficacy of higher taxa and species numbers in the assessment of biodiversity in the neotropics. *Phil. Trans. Royal Soc. London, Series B, Biol. Sci.* 345: 89–99.

Prance, G.T. and Schaller, G.B. (1982). Preliminary study of some vegetation types of the Pantanal, Mato Grosso, Brazil. *Brittonia* 34: 228–251.

Putzer, H. (1984). The geological evolution of the Amazon Basin and its mineral resources. In Sioli, H. (ed.), *The Amazon.* W. Junk Publishers, Dordrecht, The Netherlands. Pp. 15–46.

Ramella, L. and Spichiger, R. (1989). Interpretación preliminar del medio físico y de la vegetación del Chaco Boreal. Contribución al estudio de la flora y de la vegetación del Chaco. I. *Candollea* 44: 639–680.

Rangel-Ch., J.O. and Lowy-C., P.D. (1995). Parque Nacional Natural Tayrona. In Rangel-Ch., J.O. (ed.), *Colombia: diversidad biótica I.* INDERENA – Universidad Nacional de Colombia, Santafé de Bogotá. Pp. 233–238.

Räsänen, M.E., Salo, J.S. and Kalliola, R.J. (1991). Fluvial perturbance in the western Amazon Basin: regulation by long-term sub-Andean tectonics. *Science* 238: 1398–1401.

Ratter, J.A. (1991). *The conservation situation of the Brazilian cerrado vegetation*. Report for WWF (U.K.). Royal Botanic Garden, Edinburgh, Scotland, U.K. 19 pp.

Record, S.J. and Hess, R.W. (1940). American timbers of the family Bignoniaceae. *Trop. Woods* 63: 9–38.

Reid, W.V. (1992). How many species will there be? In Whitmore, T.C. and Sayer, J.A. (eds), *Tropical deforestation and species extinctions*. Chapman & Hall, London. Pp. 55–73.

Rivas-Martínez, S. and Tovar, O. (1982). Vegetatio Andinae, I. Datos sobre las comunidades vegetales altoandinas de los Andes Centrales del Perú. *Lazaroa* 4: 167–187.

Romero-Castañeda, R. (1985). *Frutas silvestres del Chocó*. Instituto Colombiano de Cultura Hispánica, Bogotá. 122 pp.

Rundel, P.W., Dillon, M.O., Palma, B., Mooney, H.A., Gulmon, S.L. and Ehleringer, J.R. (1991). The phytogeography and ecology of the coastal Atacama and Peruvian deserts. *Aliso* 13(1): 1–50.

Ruthsatz, B. (1974). Los arbustos de las estepas andinas del noroeste argentino y su uso actual. *Bol. Soc. Argent. Bot.* 16: 27–45.

Ruthsatz, B. (1977). *Pflanzengeschellschaften und ihre Lebensbedingungen in den Andinen Halfswusten Nordwest-Argentiniens*. Diss. Bot. 29. 168 pp.

Ryan, J.C. (1991). Goods from the woods. *World·Watch* 4(4): 19–26.

Rylands, A.B. (1991). *The status of conservation areas in the Brazilian Amazon*. World Wildlife Fund, Washington, D.C. 146 pp.

Salati, E. and Marques, J. (1984). Climatology of the Amazon region. In Sioli, H. (ed.), *The Amazon*. W. Junk Publishers, Dordrecht, The Netherlands. Pp. 85–126.

Salo, J.S. and Räsänen, M.E. (1989). Hierarchy of landscape patterns in western Amazon. In Holm-Nielsen, L.B., Nielsen, I.C. and Balslev, H. (eds), *Tropical forests*. Pp. 35–45.

Salo, J.S., Kalliola, R.J., Häkkinen, I., Mäkinen, Y., Niemelä, P., Puhakka, M. and Coley, P.D. (1986). River dynamics and the diversity of Amazon lowland forest. *Nature* 322: 254–258.

Sampaio, E.V.S.B. (1995). Overview of the Brazilian caatinga. In Bullock, S.H., Mooney, H.A. and Medina, E. (eds), *Seasonally dry tropical forests*. Cambridge University Press, Cambridge, U.K. Pp. 35–63.

Sarmiento, F.O. (1995). Human impacts on the cloud forests of the Upper Guayllabamba River Basin, Ecuador, and suggested management responses. In Hamilton, L.S., Juvik, J.O. and Scatena, F.N. (eds), *Tropical montane cloud forests*. Ecol. Studies Vol. 110. Springer-Verlag, New York. Pp. 284–295.

Simmonds, N.W. (ed.) (1976). *Evolution of crop plants*. Longman, London. 339 pp.

Sioli, H. (1984). The Amazon and its main affluents: hydrography, morphology of the river courses, and river types. In Sioli, H. (ed.), *The Amazon*. W. Junk Publishers, Dordrecht, The Netherlands. Pp. 127–165.

Skole, D.L. and Tucker, C.J. (1993). Tropical deforestation and habitat fragmentation in the Amazon: satellite data from 1978 to 1988. *Science* 260: 1905–1910.

Skole, D.L., Chomentowski, W.H. and Nobre, A.D. (1992). A remote sensing and GIS methodology for estimating the trace gas emissions from tropical deforestation: a case study from Amazonia. Paper presented at World Forest Watch Meeting, 27–30 May 1992, São José dos Campos, SP, Brazil. 19 pp. Mimeographed.

Smith, D.N. (1988). *Flora and vegetation of the Huascaran National Park, Ancash, Peru, with preliminary taxonomic studies for a manual of the flora*. Ph.D. thesis, Iowa State University, Ames. 281 pp.

Spichiger, R., Ramella, L., Palese, R. and Mereles, F. (1991). Proposición de leyenda para la cartografía de las formaciones vegetales del Chaco paraguayo. Contribución al estudio de la flora y de la vegetación del Chaco. III. *Candollea* 46: 541–564.

Sternadt, G.H., Ternadt, G.H. and Camargos, J. (1988). Novas perspectivas de utilização da cor da madeira amazonica e seu aproveitamento comercial. *Brasil Florestal* 65: 16–24.

Steyermark, J.A. (1982). Relationships of some Venezuelan forest refuges with lowland tropical floras. In Prance, G.T. (ed.), *Biological diversification in the tropics*. Columbia University Press, New York. Pp. 182–220.

Steyermark, J.A. and Huber, O. (1978). *Flora del Avila*. Sociedad Venezolana de Ciencias Naturales, Caracas. 971 pp.

Stone, R.D. (1985). Dateline: La Planada. *WWF Focus* 7(2): 3.

Takeuchi, M. (1961). The structure of the Amazonian vegetation II. Tropical rain forest. *J. Fac. Sci. Univ. Tokyo, Sect. III, Bot.* 8: 1–26.

Terborgh, J. (1971). Distribution on environmental gradients: theory and a preliminary interpretation of distributional patterns in the avifauna of the Cordillera Vilcabamba, Peru. *Ecology* 52: 23–40.

Terborgh, J. and Winter, B. (1982). Evolutionary circumstances of species with small ranges. In Prance, G.T. (ed.), *Biological diversification in the tropics*. Columbia University Press, New York. Pp. 587–600.

Uhl, C. and Buschbacher, R. (1985). A disturbing synergism between cattle ranch burning practices and selective tree harvesting in the eastern Amazon. *Biotropica* 17: 265–268.

Uhl, C. and Vieira, I.C.G. (1989). Ecological impacts of selective logging in the Brazilian Amazon: a case study from the Paragominas region of the state of Pará. *Biotropica* 21: 98–106.

Ulloa-U., C. and Jørgensen, P.M. (1993). *Arboles y arbustos de los Andes del Ecuador*. Rep. Bot. Inst., University of Aarhus No. 30. 264 pp.

UNDP (1992). *Amazonia without myths*. Commission on Development and Environment for Amazonia, UN Development Programme (UNDP), New York. 99 pp.

van der Hammen, T. (1974). The Pleistocene changes of vegetation in tropical South America. *J. Biogeogr.* 1: 3–26.

van der Hammen, T. and Cleef, A.M. (1983). Datos para la historia de la flora andina. *Rev. Chilena Hist. Nat.* 56: 97–107.

Vantomme, P. (1990). Forest extractivism in the Amazon: is it a sustainable and economically viable activity? In *Abstracts, Forest 90. First International Symposium on Environmental Studies on Tropical Rain Forests, October, 1990. Development Strategies for Amazonia*. Manaus. Pp. 39–40.

Vareschi, V. (1970). *Flora de los páramos de Venezuela*. Universidad de los Andes, Mérida, Venezuela. 429 pp.

Vásquez, R. (1989). *Plantas útiles de la Amazonía peruana*. Proyecto Flora del Perú, Iquitos. 195 pp.

Vásquez, R. and Gentry, A.H. (1989). Use and misuse of forest-harvested fruits in the Iquitos area. *Conserv. Biol.* 3: 1–12.

Vázquez-García, J.A. (1995). Cloud forest archipelagos: preservation of fragmented montane ecosystems in tropical America. In Hamilton, L.S., Juvik, J.O. and Scatena, F.N. (eds), *Tropical montane cloud forests*. Springer-Verlag, New York. Pp. 315–332.

Veja (12 February 1992). Pp. 34–41.

Veríssimo, A., Barreto, P., Mattos, M., Tarifa, R. and Uhl, C. (1992). Logging impacts and prospects for sustainable forest management in an old Amazonian frontier: the case of Paragominas. *Forest Ecology and Management* 55: 169–199.

Villa-Lobos, J. (1995). Peruvian group supports Ashaninka heritage and environment. *Biol. Conserv. Newsl.* No. 149: 1–2.

Ward, R.T. and Dimitri, M.J. (1966). Alpine tundra on Mt. Catedral in the southern Andes. *New Zealand J. Bot.* 4: 42–56.

WCMC (World Conservation Monitoring Centre) (1992). *Protected areas of the world. A review of national systems*, Vol. 4. *Nearctic and Neotropical*. IUCN, Gland, Switzerland and Cambridge, U.K. 459 pp.

Wiggins, I.L. and Porter, D.M. (1971). *Flora of the Galápagos Islands*. Stanford University Press, Stanford, Calif. 998 pp.

Wilson, E.O. (1992). *The diversity of life*. Harvard University Press, Cambridge, Massachusetts. 424 pp.

WRI, UNEP and UNDP (1994). *World resources 1994–95: a guide to the global environment*. World Resources Institute (WRI), United Nations Environment Programme (UNEP) and United Nations Development Programme (UNDP). Oxford University Press, New York. 403 pp.

Young, K. and Valencia, N. (eds) (1992). *Biogeografía, ecología y conservación del bosque montano en el Perú*. Memorias del Museo de Historia Natural Vol. 21, Universidad Nacional Mayor de San Marcos (UNMSM), Lima. 223 pp.

Ziffer, K. (1992). The Tagua Initiative: building the market for a rain forest product. In Plotkin, M. and Famolare, L. (eds), *Sustainable harvest and marketing of rain forest products*. Island Press, Washington, D.C. Pp. 274–279.

Authors

This Regional Overview was written by Dr Alwyn H. Gentry[1] (Missouri Botanical Garden, P.O. Box 299, St. Louis, MO 63166-0299, U.S.A.) with contributions by Olga Herrera-MacBryde (Smithsonian Institution, Department of Botany, NHB-166, Washington, DC 20560, U.S.A.), Dr Otto Huber (Research Associate, Apartado 80405, Caracas 1080-A, Venezuela), Dr Bruce W. Nelson [Instituto Nacional de Pesquisas da Amazônia (INPA), Caixa Postal 478, 69000 Manaus, AM, Brazil] and Dr Carlos B. Villamil (Universidad Nacional del Sur, Departamento de Biología, Perú 670, 8000 Bahía Blanca, Argentina).

[1] While this overview was in preparation, Dr Gentry died on 3 August 1993 in Ecuador in a plane crash, during a survey of an important coastal site of plant diversity.

COASTAL CORDILLERA
Venezuela

Location: Mountain system extending east-west along northern coast of Venezuela, about from Yaracuy Depression (latitude c. 10°20'N, longitude c. 68°30'W) eastward to tip of Paria Peninsula (latitude 10°45'N, longitude 61°50'W).

Area: c. 45,000 km².

Altitude: 0–2765 m.

Vegetation: Mangroves, coastal thorn scrub, hill savanna, deciduous forest, semi-deciduous lower montane forest, evergreen lower montane forest, evergreen montane cloud forest, upper montane elfin forest, upper montane scrub (subpáramo).

Flora: c. 5000 vascular plant species with numerous endemic taxa, especially in cloud forests.

Useful plants: Many medicinal and ornamental species.

Other values: Watershed protection, fauna refuge, tourist attraction.

Threats: High demographic pressure; colonization; shifting and industrial agriculture; road construction; set fires.

Conservation: 11 National Parks, 5 Natural Monuments, Fauna Refuges.

Geography

The Coastal Cordillera (Cordillera de la Costa) is the third largest mountain system in Venezuela, extending c. 720 km in an east-west direction along the northern (Caribbean) coast (Map 37). The easternmost extension lies in the Paria Peninsula opposite the island of Trinidad (and the system extends into Trinidad); the western end is in the State of Yaracuy where the depression of Yaracuy forms a natural separation from the south-westerly adjacent Andes mountains (and the Sierra de Aroa can also be considered coastal). The northern limit of the Coastal Cordillera is formed by the coastline of the Caribbean Sea, and to the south it is bordered by the extensive plains of the Llanos.

The Coastal Cordillera is articulated into two transversal sections, a larger western and a smaller eastern one. The western section in turn consists of two parallel east-west mountain chains: a northern Serranía del Litoral from the River Yaracuy, and a southerly adjacent and lower Serranía del Interior extending to the depression of Unare (c. 10°N, 65°W). Between them are the densely populated valleys of Caracas and Tuy, and the Valencia Lake Depression.

Several mountains north of Caracas in the coastal chain have received the most botanical attention: El Avila (2200 m), La Silla de Caracas (2650 m) and Pico Naiguatá (2765 m) – which is the highest peak of the entire Coastal Cordillera. Other botanically important mountains of the Serranía del Litoral are Pico Codazzi (c. 2200 m) near Colonia Tovar 60 km west of Caracas, and the mountains of Rancho Grande (the highest is Pico La Mesa, c. 2400 m) to the north of the city Maracay in Aragua State (120 km west of Caracas). The most important mountains of the Serranía del Interior are Guatopo (c. 1800 m) (30 km south-east of Caracas), and the Morros de San Juan (over 1700 m) (near the town San Juan in Guárico State). The eastern section comprises the large massif of Turimiquire (c. 2590 m) located to the east of the depression of Unare, followed farther east by the cerros Humo (c. 1350 m) and Patao on the Península de Paria, and continues in the island of Trinidad under the name Northern Range.

The geology of the Coastal Cordillera consists mainly of Tertiary schists and gneisses, underlain in some parts by granites; limestones are also present in several areas, especially in the Serranía del Interior, where the Morros de San Juan offer a characteristic landscape feature. The Coastal Cordillera was formed by strong tectonic movements during the Tertiary (60 million years ago) and is considerably older than the Andes (which uplifted 30 million years ago).

Soils are generally acidic and predominantly entisols and ultisols. Nutrient levels are generally low; in some soils under forest cover high concentrations of aluminium have been found (Zinck 1986).

The general climate of the Coastal Cordillera is strongly influenced by the north-eastern trade winds, especially during the dry season from December to April, whereas during the rainy season the Intertropical Convergence Zone reaches its northern limit in the area, accompanied by predominantly southern winds with high moisture. The local climate is very diverse, ranging from arid macrothermic conditions (average annual temperature over 24°C) near the coast to perhumid mesothermic conditions on the upper slopes (average annual temperature 20°–10°C). Above about 800 m on the windward and 1000 m on the leeward slopes, frequent mist occurs, extending usually 1000–1200 m upwards. Seasonality is strong in the lower regions, where the dry season is marked, but at higher elevations seasonality tends to become less pronounced.

Vegetation

A wide variety of vegetation types covers the entire Coastal Cordillera, showing a characteristic altitudinal zonation throughout the range. Predominant is the forest formation, which ranges from deciduous low forests to evergreen luxuriant and very tall upper montane forests, also known as cloud forests. Scrub is present in the arid coastal areas and on the uppermost peaks of higher mountains. Savannas and related herbaceous formations occur on lower slopes, mainly toward the interior valleys and plains.

The following altitudinal sequence of vegetation types represents the most common situation in both the central and eastern Coastal Cordillera, although variations are noticeable in various areas, due to local climatic, topographic and edaphic conditions.

This typical transect upwards from sea-level is in the central Coastal Cordillera (Schäfer 1952; Beebe and Crane 1947):

1. Mangrove forest
Where the coastline is not formed by steep rocky slopes emerging abruptly from the sea, small mangrove stands with *Rhizophora mangle, Laguncularia racemosa, Avicennia nitida* and *Conocarpus erecta* form low dense forests.

2. Coastal xerophytic thorn scrub
The lowermost slopes (c. 2–200 m above sea-level) are covered with a dense, 3–8 m high thorn scrub dominated by columnar cacti (*Stenocereus griseus*), spiny shrubs [essentially Mimosoideae such as *Prosopis juliflora, Acacia* spp., and Flacourtiaceae (*Ximenia americana*)], low sclerophyllous shrubs (*Capparis linearis, Jacquinia pungens, Plumeria* sp.) and some understorey herbs belonging mainly to Malvaceae, Euphorbiaceae and Erythroxylaceae.

3. Semi-deciduous lower montane forest
These usually two-storeyed, 15–25 m high forests occupy the slopes from c. 300–600 m. Dominant trees are *Anacardium excelsum, Ceiba pentandra, Bourreria cumanensis* and several legumes, such as *Erythrina* spp. (which are frequently planted as shade trees for cocoa plantations common in the area). The understorey is dense and lianas are frequent.

4. Evergreen montane forest
From 600–800 m to 900 m elevation, evergreen transition forests form a narrow belt between the lower montane semi-deciduous and upper montane cloud forests. Heavily buttressed trees to 60 m tall of the endemic *Gyranthera caribensis* form small stands, emerging above the general forest canopy dominated by *Trophis racemosa, Ficus macbridei, Tetragastris caracasana, Zanthoxylum ocumarense, Banara nitida,* etc. The understorey is dominated by shrubs of *Aphelandra micans, Besleria disgrega, Psychotria macrophylla* and

MAP 37. COASTAL CORDILLERA, VENEZUELA (CPD SITE SA1)

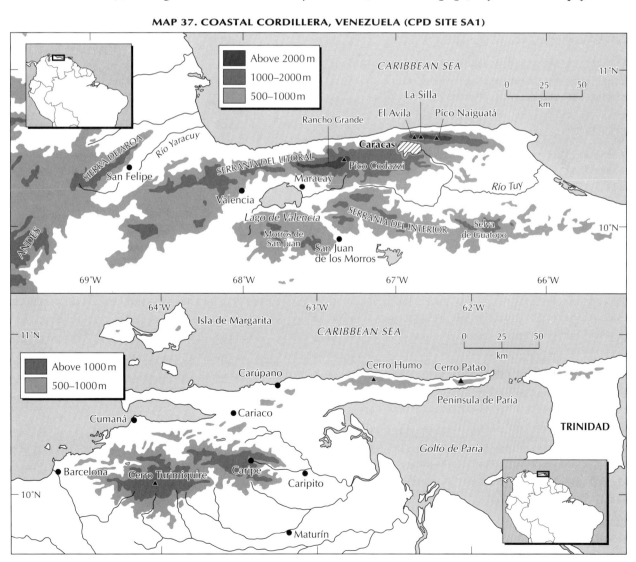

Hoffmannia apodantha, together with large colonies of giant herbs, e.g. Musaceae (*Heliconia bihai, H. revoluta*), Araceae (*Dieffenbachia maculata*), Marantaceae and ferns.

5. Evergreen montane cloud forest

These forests, extending from c. 1000–2200 m elevation, are by far the most species-rich plant communities in the entire Coastal Cordillera, representing also the most complex ecosystems with the most life forms and ecological niches (Vareschi 1986). They are irregularly structured in 2–3 layers, with a main canopy extending 15–20 m, overtopped by emergents to 40 m. Canopy trees include *Ecclinusa abbreviata, Sloanea* spp., *Neea* spp., *Eschweilera perumbonata, Graffenrieda latifolia* and *Ocotea* spp. Palms are very frequent in the canopy and in the understorey as well, either growing solitary (*Dictyocarium* sp., *Socratea* sp., *Geonoma* spp.) or in large clumps (*Hyospathe pittieri, Catoblastus praemorsus, Bactris* sp.). Also epiphytes are very abundant – mainly ferns, orchids, bromeliads, ericads and gesneriads. The shrub and herb layer is usually dense and dominated by many endemics, such as *Psychotria agostinii, P. costanensis, Palicourea fendleri, Schoenobiblos daphnoides, Geonoma spinescens* and *Anthurium bredemeyeri*. Terrestrial ferns, including tree ferns, are particularly abundant and diverse in these very moist, misty and cool forests.

6. Upper montane elfin forest and scrub (subpáramo)

Above c. 2200–2400 m, a low mossy forest grading into an open scrub vegetation covers the highest peaks of the Coastal Cordillera; its physiognomy resembles in some instances the subpáramo belt of the Andes, with which it even shares some significant species, such as *Libanothamnus neriifolius*. This primitive espeletioid composite dominates the open scrub, whereas in the elfin forests the most frequent tree species are *Clusia multiflora, Weinmannia* spp. and *Prumnopitys harmsiana*. In the upper montane scrub an irregular but often rather dense herbaceous layer is present, in which several species of Andean affinities grow, e.g. *Achyrocline flavida, Orthrosanthus chimboracensis* or *Juncus effusus*.

Flora

Phytogeographically, the flora of the Coastal Cordillera shows strong relationships to the Mesoamerican and Caribbean floristic regions, and to a lesser degree, to the northern Andean flora. Botanical collecting has been active in the Coastal Cordillera of northern Venezuela since Alexander von Humboldt and Aimé Bonpland visited in 1799–1800 to the mountains around Cumaná, Caracas and Valencia, with cumulative emphasis on the Avila and Silla mountains above Caracas, the Rancho Grande area near Maracay (Aragua State) and on Cerro Turimiquire. However, the cordillera as a whole is still far from being adequately explored botanically, especially in the middle and upper montane regions. The rich cloud forests on the Avila mountain and in Rancho Grande are only better known because of their easy access.

For the entire region only one modern Flora exists, covering not more than 5% of the entire range: the Avila, Silla and Naiguatá mountains above Caracas (Steyermark and Huber 1978). There is a 1984 checklist of species from Rancho Grande (Henri Pittier National Park) (Badillo, Rojas and Huber 1984).

According to Steyermark (1981, pers. comm.), probably no fewer than 5000 species of higher plants occur in the Coastal Cordillera; the *Flora del Avila* includes 1892 species of vascular plants. Undoubtedly, the various cloud forests of the Coastal Cordillera harbour the most species and endemics. In a 0.25-ha plot in the cloud forests of Rancho Grande, the 150 trees >10 cm dbh represented more than 60 different species; in a second plot only 1000 m from the former, a similar number of species was found, but more than 80% of the species were different from those in the first plot (Huber 1986). Even short visits reveal an overwhelming floristic diversity in the many different cloud-forest types existing along the entire 720 km of the Coastal Cordillera. Recent accounts of a few large families in the *Flora de Venezuela* (e.g. Pteridophyta, Piperaceae, Melastomataceae, Rubiaceae), confirm that a significantly large number of species belong to this ecoregion of the northern neotropics.

The flora of the Venezuelan Coastal Cordillera probably contains some 10% endemic taxa (Steyermark 1981, pers. comm.). Endemism, like floristic richness, is mainly concentrated in the montane cloud and rain forests, being practically absent from the dry flora of the lower montane belts. No attempts to more precisely quantify the degree of endemism of this ecoregion have been made, mainly because the floristic knowledge of the area is so incomplete.

Useful plants

Surely many species of the Coastal Cordillera were used by the indigenous population prior to European colonization; however, most of this knowledge has been lost. A limited number of herbs (native and introduced) are used in popular medicine, e.g. "cariaquito morado" (*Lantana trifolia*) (Rodríguez 1983). The red bark of the quinine tree (*Cinchona henleana*) was used as an antimalarial remedy. Many trees of the dry and semi-deciduous forests, such as certain Meliaceae (*Cedrela, Trichilia*), Anacardiaceae (*Spondias mombin*), Lauraceae and many Leguminosae were used over centuries for construction, furniture, etc. As a result of this long-time selective logging, many of the dry forests have been degraded to low forests or scrub, with much less species diversity.

The outstanding botanical richness of the vegetation of the Coastal Cordillera constitutes one of the most valuable biological and genetic resources of Venezuela. Since a consistent programme of pharmaceutical and biochemical screening of many promising plant taxa has not yet been undertaken, it is impossible to estimate the real genetic and medicinal potential of this resource.

Social and environmental values

Most of the steep mountain slopes are exposed to heavy soil erosion after elimination of the forest cover. The natural forests are not only an important protection of watersheds including many sources of drinking water, but also protect soils from further erosion.

Due to the high population density in the entire Coastal Range, demand for recreation areas and mountain tourism is very pronounced. A recreation and tourist industry is rapidly growing, benefiting greatly either directly or indirectly from the numerous natural landscapes and sceneries of the Coastal Cordillera already protected by National Parks and Natural Monuments.

Birds of restricted range in the Coastal Cordillera primarily inhabit the evergreen montane and cloud forests and elfin forest. The large western section of the range, the Cordillera de la Costa Central, forms an Endemic Bird Area (EBA B04) which has 18 restricted-range bird species (five confined to these mountains). The smaller eastern section forms the Cordillera de Caripe and Paria Peninsula EBA (B03), which is home to 14 restricted-range species, five found nowhere else. These five species are considered threatened, making this EBA one of the highest priorities for neotropical bird conservation.

Threats

The plants of the Coastal Cordillera in northern Venezuela have probably been altered most by the human impacts during the last five centuries, following European colonization and subsequent heavy occupation of the region. Although Amerindian groups had lived there for many more centuries, their impact on the vegetation is not known but probably was not as intense.

Deforestation for shifting cultivation was responsible for most of the destruction of the original forest on the lower and middle mountain slopes. Today more extensive deforestation is mainly caused by urban and industrial expansions, but these usually do not affect the upper mountain slopes. Deforestation is also caused by construction of more and more tourist and recreation resorts in the vicinity of the larger cities, also involving intensive road building into previously inaccessible areas. Another cause of deforestation is the establishment of large agricultural farms, mainly for fruits and vegetables.

A major threat to the remaining forests is frequent fires lit during the dry season in the already deforested and mostly savanna-covered, surrounding lower areas. Although fires usually do not penetrate into the cloud forests proper, in some places the adjacent montane evergreen forests are being severely reduced.

Conservation

For centuries, the lower montane forests of the Coastal Cordillera were used for cocoa plantations, whereas the cooler cloud forests were preferably used for coffee plantations. Today this is very restricted, and in many cases forests originally modified for plantations through selective logging, planting of shade trees and clearing of the understorey are slowly recovering their original structure, although not yet their original floristic composition.

Governmental protection of ecosystems in the Coastal Cordillera has acquired significant dimensions: there are 11 National Parks and five Natural Monuments, protecting large parts of the remaining natural vegetation (García 1989). Henri Pittiér National Park in Aragua State is the oldest park in Venezuela (since 1937). El Avila NP (declared in 1958) is one of the most extensive wilderness areas in the

region and serves as the most important recreation area for the capital Caracas.

National Parks protecting mainly mountain ecosystems in the Coastal Cordillera are: Yurubí (237 km²), Henri Pittier (Rancho Grande) (1078 km²), San Esteban (435 km²), El Avila (852 km²), Macarao (150 km²), Guatopo (1225 km²), El Guácharo (627 km²) and Península de Paria (375 km²). National Parks protecting mainly coastal areas of the Coastal Cordillera are: Morrocoy (321 km²), Laguna de Tacarigua (391 km²) and Mochima (949 km²) (Gabaldón 1992).

References

Badillo, V.M., Rojas, C.E.B. de and Huber, O. (1984). Lista preliminar de especies de antófitas del Parque Nacional 'Henri Pittier', Estado Aragua. *Ernstia* 26: 1–58.

Beebe, W. and Crane, J. (1947). Ecology of Rancho Grande, a subtropical cloud forest in northern Venezuela. *Zoologica* 32(5): 43–60.

Gabaldón, M. (1992). *Parques nacionales de Venezuela.* Parques Nacionales y Conservación Ambiental No. 1. Fundación Banco Consolidado, Caracas. 116 pp.

García, R. (1989). Los parques nacionales de Venezuela. *Encuentros* 6: 15–20.

Huber, O. (1986). Las selvas nubladas de Rancho Grande: observaciones sobre su fisionomía, estructura y fenología. In Huber, O. (ed.), *La selva nublada de Rancho Grande, Parque Nacional 'Henri Pittier'.* Fondo Editorial Acta Científica Venezolana, Caracas. Pp. 131–170.

Rodríguez-M., P. (1983). *Plantas de la medicina popular venezolana de venta en herbolarios.* Sociedad Venezolana de Ciencias Naturales, Caracas. 267 pp.

Schäfer, E. (1952). Ökologischer Querschnitt durch den 'Parque Nacional de Aragua'. *Journal für Ornithologie* 93(3/4): 313–352.

Steyermark, J.A. and Huber, O. (1978). *Flora del Avila.* Sociedad Venezolana de Ciencias Naturales, Caracas. 971 pp.

Vareschi, V. (1986). Cinco breves ensayos ecológicos acerca de la selva virgen de Rancho Grande. In Huber, O. (ed.), *La selva nublada de Rancho Grande, Parque Nacional 'Henri Pittier'.* Fondo Editorial Acta Científica Venezolana, Caracas. Pp. 171–187.

Zinck, A. (1986). Los suelos. In Huber, O. (ed.), *La selva nublada de Rancho Grande, Parque Nacional 'Henri Pittier'.* Fondo Editorial Acta Científica Venezolana, Caracas. Pp. 31–66.

Author

This Data Sheet was written by Dr Otto Huber (Research Associate, Apartado 80405, Caracas 1080-A, Venezuela).

GUAYANA HIGHLANDS: CPD SITE SA2
PANTEPUI REGION
of Venezuela

Location: Phytogeographic province of Guayana Highlands, mainly in southern Venezuela, and as well in north-western Guyana and northernmost Brazil, between latitudes 0°–7°N and longitudes 60°–67°W.

Area: Province covers 6000–7000 km².

Altitude: 1300–3015 m.

Vegetation: Upper montane evergreen tall forest, upper montane evergreen low forest ("tepui" forest), evergreen tepui scrub, tepui fields, open-rock pioneer vegetation.

Flora: Very high diversity (probably c. 3000 spp.), very high endemism of genera and species, with large number of disjunct taxa of tropical montane and extra-tropical floras.

Useful plants: Plant resources other than ornamentals yet to be utilized, but undoubtedly potentially rich in genetic resources.

Other values: Fauna refuge, endemic animals; watershed protection for drinking water and hydroelectric energy; tourist attraction.

Threats: Increasing tourist pressure, roads, airstrips; over-collecting; mining; fire.

Conservation: 6 National Parks, 28 Natural Monuments and 1 Biosphere Reserve cover this province in Venezuela.

Geography

The phytogeographic province of Pantepui forms part of the floristic region of Guayana (Huber 1994) and includes all upper slopes and summits of the characteristically flat-topped table mountains ("tepuis") of the Guayana Highlands. Pantepui, a term established by Mayr and Phelps (1967) and subsequently used with different meanings (Huber 1987), is mainly located in southern Venezuela, where there are more than 90% of the mountains of this type (Map 38), with some lower outliers in north-western Guyana and northernmost Brazil.

The Guayana Highlands are drained principally by the Orinoco River with its southern and eastern affluents, and to a minor extent by the Cuyuni and Mazaruni rivers in Guyana and the Branco and Negro rivers in northern Brazil. The regional geology consists of an igneous (mainly granitic) Precambrian base (the Guayana Shield), overlain by younger Precambrian meta-sedimentary sandstones and quartzites of variable thickness – 100 m to more than 3000 m. These predominantly horizontally bedded sandstone layers, belonging to the Roraima Formation, have been intruded repeatedly and in many places by more recent igneous and volcanic materials, resulting in a variety of mainly diabasic sills and dykes at intermediate and upper elevations (Schubert and Huber 1990).

The physiography of Pantepui typically shows an altitudinal sequence of more or less inclined piedmont slopes followed by vertical sandstone walls up to 1000 m high, surmounted by generally flat but broken summit plains, which in some places alternate with rounded hill-and-summit topography typical of diabasic intrusive areas. The mountains vary greatly in size and form, ranging from tower-like, needle-pointed isolated peaks (e.g. Cerro Autana, Wadakapiapue-tepui, Cerro Aratitiyope) with less than a hectare of summit surface, to huge mountain systems of up to 600 km² or more like those of Auyán-tepui, Chimantá massif, Jáua, Duida and Neblina. Extensive peats have developed on many summit areas. All the soils are highly acidic and extremely nutrient poor.

The climate reflects the equatorial position of Pantepui: a rather equilibrated annual thermic regime (with mean annual temperatures ranging between 20°–8°C according to altitude) contrasts with extremely pronounced variations in the daily regime (with variations of up to 20°C). The lowest recorded air temperatures are around 1°–2°C (on the summit of Mt Roraima, at 2740 m); frost has never been convincingly measured, though it may well occur occasionally on the highest peaks, above 2800 m. Annual rainfall varies between 2000 and 4000 mm, with a slight decrease during January to March, but no true dry season has been observed on any tepui summit. Strong winds (mainly north-east trade winds) during a greater part of the year, coupled with high irradiance, certainly cause high evapotranspiration, but probably compensated by permanently high humidity.

Vegetation

The vegetation of this extensive area varies greatly according to altitude and the fragmented nature of this discontinuous mountain region. On a recent vegetation map of Venezuela (Huber and Alarcón 1988), no fewer than 14 different vegetation types are shown (based on physiognomic, ecological and floristic criteria). Broadly, there are four main vegetation formations: arboreal (forests), shrub (scrub), herbaceous (meadows and savanna-like "campos") and pioneer (on rock outcrops and walls).

1. The **forest formation** includes a series of montane forest types, covering not only the upper slopes of the base of the tepuis, but also some portions of the summits, especially on diabasic hills, or along creeks or in depressions. Whereas the slope forests are amongst the tallest (up to 60 m) and densest reported for Venezuela (J.J. Wurdack, pers. comm.), the summit forests ("tepui" forests) are dense but seldom higher than 8–12 m and relatively simple in their structure and floristic composition. Typical and dominant trees of the summit forests belong to such genera as *Bonnetia*, *Schefflera* and *Stenopadus*, whereas in slope forests Lauraceae, Magnoliaceae, Elaeocarpaceae, Rubiaceae and Myrtaceae are among the most frequent families.

2. The **scrub formation** has the highest diversity, with a variety of both physiognomically and floristically distinct communities. Almost all the larger tepui summits have their own types of scrub, ranging from low (less than 2 m) and open, to tall (3–5 m) and very dense communities, with high floristic diversity. Especially remarkable are páramo-like scrub types on Chimantá formed by dense colonies of stem-rosette Compositae, and extremely scleromorphic and thick-stemmed *Bonnetia* populations on Neblina, Duida and Chimantá.

3. The **herbaceous formation** is also floristically and morphologically diverse. The dominance of various members of Rapateaceae (e.g. *Stegolepis*, *Kunhardtia*) in

MAP 38. PANTEPUI REGION, VENEZUELA (CPD SITE SA2), SHOWING PRINCIPAL MOUNTAIN SYSTEMS ("TEPUIS")

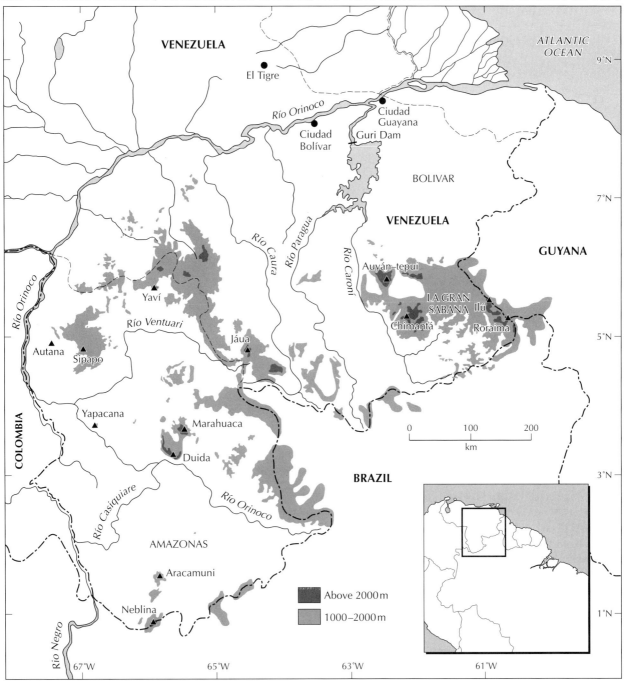

CPD Site SA2: Pantepui region, Venezuela. Acopán-tepui, near Chimantá-tepui. Photo: John F. Pruski.

these high mountain fields, together with Xyridaceae, Cyperaceae and Eriocaulaceae, is one of the outstanding features in almost all summit landscapes; grasses are usually not dominant and belong mainly to the subfamily Bambusoideae. These "campos", attaining heights of 0.5–2 m, are generally quite dense and best developed on deep organic substrates.

4. The **pioneer formation** colonizing extensive open-rock areas on summits, and on walls of the flanks of the tepuis, exhibits remarkable ecological and floristic diversity. The dominant life forms are terrestrial rosettes of Bromeliaceae (e.g. *Navia, Lindmania, Brocchinia*). There are also many different lichens, hepatics and mosses, which as colonizers on organic and inorganic substrates, occupy main niches.

Flora

A Flora of the entire region of Guayana in Venezuela is in an advanced stage of preparation; of the more than 9000 species, probably 3000 are found in Pantepui. According to Steyermark (1986), some 460 genera in 97 families are known to occur on the tepui summits and slopes. Numerically, the most important families are Orchidaceae, Bromeliaceae, Melastomataceae, Rubiaceae, Eriocaulaceae, Xyridaceae, Ericaceae and Compositae.

Apart from the species richness, the phytogeographic province also shows very pronounced floristic autochthony, resulting in a high degree of endemism and ecologically restricted populations of widely disjunct taxa. The Pantepui Province is probably one of the main centres of endemism of the neotropics; compared to its small overall surface, it probably contains the highest number of endemic taxa per unit area. Known endemics thus far include one fern family (Hymenophyllopsidaceae), one flowering plant family (Saccifoliaceae) and 39 summit genera, of which the most important are found in Theaceae, Rapateaceae, Ericaceae, Compositae, Bromeliaceae, Rubiaceae and Melastomataceae (Steyermark 1986).

Some parts of the Pantepui Province have a greater concentration of endemic taxa than others: Cerro de la Neblina in the extreme south is likely to hold the most endemic genera, followed by the Chimantá massif and Duida-Marahuaca. At least five well-defined centres of endemism can be distinguished in Pantepui (Map 38), ranging from the eastern tepui chain (Roraima-Ilú), to the central massifs of Jáua-Sarisariñama, passing through the north-western group of Yaví-Sipapo and ending in the southern Neblina-Aracamuni massif (Huber 1987).

Useful plants

Since no people lived in the upper regions of the Guayana Highlands, their plant resources have not been utilized. However, the many endemic taxa undoubtedly represent a considerable genetic resource, which might reveal interesting

314

biochemical properties when thorough analyses are realized. Furthermore, these complex high-mountain ecosystems contain an impressive number of plants with specialized adaptations to various environmental factors. As such, they represent important subjects for study of still unknown ecological and functional structures and processes.

Social and environmental values

The Pantepui vegetation is an important rainfall interceptor in the uppermost portions of the different river basins, which are increasingly utilized to generate hydroelectric power. It is necessary to maintain the hydrological equilibrium in the upper basins in order to assure a regular supply of water for power and drinking (e.g. Galán 1984).

The beautiful and impressive table mountains (tepuis) of the Guayana Highlands are a most attractive South American mountain landscape. Indeed, tourist interest in this region increases from year to year, due to easier access and wider publicity, both nationally and internationally.

The Pantepui region (Endemic Bird Area B02) is home to 36 bird species totally restricted to the vicinity of the tepui mountains. They are primarily montane species occurring in the humid forest on the piedmont slopes above c. 600 m; the swift *Cypseloides phelpsi* is confined to the vertical cliffs, and *Emberizoides duidae* is confined to the herbaceous vegetation of the summits (Cerro Duida). Most of the endemics are to be found on the Gran Sabana, although a number of species are confined to lone, isolated tepuis. Due to the limited habitat disturbance in this region, none of the endemics is considered threatened currently.

Threats

There are no serious threats as yet to the existence of the Pantepui flora and vegetation. However, mass tourism is just beginning, and already some mountains (such as Roraima and Auyán-tepui) show the impacts of uncontrolled visitation (e.g. accumulation of litter, destruction of vegetation for campsites, exploitation of crystals for souvenirs). Canaima, the largest National Park in Venezuela (30,000 km²), is affected by rapidly growing mass tourism, which is favoured by road construction, illegal airstrips and helicopter flights into previously inaccessible areas. There is a great danger of fire accidentally spreading from campsites, which could lead to the destruction of large sections of summit vegetation. Such vegetation seems to need extremely long periods for recuperation.

One south-western tepui, Cerro Yapacana, although of lower altitude and not actually within the Pantepui Province, is suffering heavily from uncontrolled mining activities at its base and on the slopes. Since 1978, when Yapacana National Park was declared, the gold mining has taken place illegally.

There is a risk of scientific over-collection in certain Pantepui ecosystems: some summits are only covered by a very sparse vegetation in which many species are extremely rare. Hobbyists and commercial collectors pose a threat to some of the more ornamental plants, such as orchids, bromeliads and various carnivorous plants (which remarkably include a few bromeliads).

Conservation

For 10–25 years, important parts (30–40%) of the Pantepui Province in Venezuela were included in five National Parks (Canaima, Jáua-Sarisariñama, Yapacana, Duida-Marahuaca, Neblina) totalling 52,200 km² and in one Natural Monument (Cerro Autana). From January 1991, the entire province was brought into such protection, and a Biosphere Reserve has been declared. However, more effective controls and enforcement of the regulations are needed, and biologists' evaluation of the impacts.

Regional development agencies sometimes act in an ambivalent manner, sponsoring tourism, mining, logging and other exploitive activities, but also strongly supporting river-basin protection for hydroelectric schemes. Fortunately in recent years national and international concern has grown vigorously for better and more thorough protection of this important region, leading to a powerful new legislative initiative to establish all tepuis over 800 m above sea-level as Natural Monuments. With their new status (6 NPs, 28 NMs, 1 BR) in Venezuela, the entire Pantepui Province may continue to exist as the outstanding and unique biological treasure of the Guayana region, and not become a "lost world", as it is often but quite inappropriately called.

References

Galán, C. (ed.) (1984). *La protección de la cuenca del Río Caroní.* C.V.G. Electrificación del Caroní, C.A. 51 pp.

Huber, O. (1987). Consideraciones sobre el concepto de Pantepui. *Pantepui* 2: 2–10.

Huber, O. (1994). Recent advances in the phytogeography of the Guayana region, South America. *Mém. Soc. Biogéogr. (3ème série)* 4: 53–63.

Huber, O. and Alarcón, C. (1988). *Mapa de vegetación de Venezuela.* 1:2,000,000. Ministerio del Ambiente y de los Recursos Naturales Renovables and Fundación BIOMA, Caracas.

Mayr, E. and Phelps Jr., W.H. (1967). The origin of the bird fauna of the South Venezuelan Highlands. *Bull. Amer. Mus. Nat. Hist.* 136: 269–328.

Schubert, C. and Huber, O. (1990, English ed.; 1989, Spanish ed.). *The Gran Sabana: panorama of a region.* Lagoven, S.A., Caracas. 107 pp.

Steyermark, J.A. (1986). Speciation and endemism in the flora of Venezuelan tepuis. In Vuilleumier, F. and Monasterio, M. (eds), *High altitude tropical biogeography.* Oxford University Press, New York. Pp. 317–373.

Author

This Data Sheet was written by Dr Otto Huber (Research Associate, Apartado 80405, Caracas 1080-A, Venezuela).

SAUL REGION
French Guiana

Location: Central French Guiana, surrounding village of Saül (latitude 3°37'N, longitude 53°12'W); proposed reserve within about latitudes 3°30'–3°45'N and longitudes 52°90'–53°30'W.

Area: Over 1340 km².

Altitude: c. 200–762 m.

Vegetation: Lowland moist forest, swamp forest, submontane forest, low summit forest, granitic-outcrop (inselberg) association.

Flora: 2000 species or more; representative of eastern Guayana Lowland Floristic Province, and most species-diverse region of French Guiana; some endemism on higher mountains, apparently also in lowlands.

Useful plants: Rich in medicinal and timber species; some extraction of rosewood oil (*Aniba rosaeodora*) and "balata" (*Chrysophyllum sanguinolentum*) in past, and palm hearts (*Euterpe oleracea*) for local consumption.

Other values: Watershed protection, genetic resources, ecotourism.

Threats: Road construction, agricultural-settlement schemes, charcoal production, gold mining; potential fuelwood cutting for electricity.

Conservation: Proposed reserve (600 km²) plus potential reserve (190 km²), near boundary of proposed National Park (1600–1800 km²).

Geography

The Guayana Lowland Floristic Province of north-eastern South America is bounded to the north and east by the Atlantic Ocean, to the south by the Amazon River and to the west and north-west by the Negro and Orinoco rivers. Many species of plants and animals occur only in all or part of this vast area (de Granville and Sanité 1992; Mori 1991). Characteristic of the eastern portion of this phytogeographic region is the Saül region, in the geographic centre of French Guiana (Map 39).

This is a zone of contact between granitic formations to the east and basic volcanic rocks of the Paramaca Series to the west (de Granville 1975). The areas on granite generally have rugged relief with steep slopes whereas the areas on volcanic rocks are characterized by gentle regular slopes, the highest capped by a lateritic crust (de Granville 1991). Although the soils are poor (as most tropical rain-forest soils), they are among the best in French Guiana because they are relatively deep with fairly good vertical drainage.

The region of Saül is at the headwaters of major tributaries of the Maroni, Mana and Approuague rivers. Most of the region is moderately dissected terrain between about 200 m and 400 m, ranging to 762 m on Monts Galbao. Also important are the tabletop mountain Mont Belvédère (760 m) and Pic Matecho (590 m), a granitic outcrop (inselberg) botanically explored only once. The proposed reserve around the village of Saül is more extensive north of the village (Map 39).

Climate is influenced by the relative position of the Intertropical Convergence Zone (ITCZ). During the dry season from August to November, the ITCZ lies north of French Guiana. From December to June, the period of heaviest rain, the ITCZ is directly over or south of French Guiana. The village of Saül receives an average of 2413 mm of rain yearly. In 1982, the average annual temperature was

27.1°C. Temperature is relatively constant throughout the year, with the daily fluctuations greater than the annual. The difference between the longest and shortest days of the year is 35 minutes (Mori and Prance 1987a).

Vegetation

The main vegetation of the region is lowland moist forest, for the most part dominated by trees of Burseraceae, Sapotaceae, Lecythidaceae, Mimosaceae, Caesalpiniaceae, Rubiaceae, Moraceae, Chrysobalanaceae, Meliaceae and Bombacaceae. The understorey is rich in Rubiaceae and Melastomataceae and sometimes dominated by *Astrocaryum* palms. Mori and Boom (1987) found 619 trees of 10 cm dbh or more, with a total basal area of 53.0 m², per ha. The tallest tree recorded in the region is a 56 m *Terminalia guyanensis* (Oldeman 1974). The tall stature and high basal area suggest that this forest has not undergone major disturbance for a long time.

Other vegetation types in the region, such as swamp forests along streams, submontane forests above 500 m, low forest on lateritic crust and an open granitic-outcrop association, have not been ecologically studied (cf. Lindeman and Mori 1989).

Flora

North-eastern South America is one of the last wilderness areas, and a high percentage of its plant species are only found there. A multinational project is producing a *Flora of the Guianas* which is slowly reducing our ignorance. The Guayana lowlands is an important source of genetic variability for many lowland South American tree families, and c. 25% of the species of some of these important neotropical families occur within the political boundaries of the Guianas: e.g. Chrysobalanaceae, Lecythidaceae, Meliaceae and Sapotaceae (Mori 1991).

The lowland Guayana flora is also important as a source of plant species that have migrated into the Amazon Basin. Mori and Prance (1987b) suggested that migration out of the Guayana lowlands has prevailed over immigration, and de Granville (1992) demonstrated that many Guayanan species have peri-Amazonian distributions. Several theories may help explain the present distribution of the Guayanan plants, including: the presence of Pleistocene refugia of forest during dry climatic periods (de Granville 1982), flooding of the Amazon Basin by various marine transgressions (de Granville 1992) and the presence of a large lake (Lago Amazonas) throughout much of the Amazon as recently as in the Late Pleistocene-Holocene (Frailey et al. 1988; Mori 1991).

The proposed reserve surrounding Saül village is representative of the eastern Guayana lowland flora (for the Guayana Highlands see CPD Site SA2, the Pantepui region). French Guiana has some 4000 species of vascular plants (de Granville 1990), of which at least 2000 are found in the proposed reserve area (Cremers et al. 1988). Sabatier (see de Granville 1990) calculated that c. 746 species of large trees (over 10 cm dbh) are in French Guiana, and as many as 531 of them may be found in the vicinity of Saül (Mori and Boom 1987). Of the species of Lecythidaceae known from the three Guianas, 54.7% would be protected if the proposed reserve were established (Mori 1991). According to de Granville (1990, 1991), 60% of the entire French Guianan flora may be represented in the Saül region. This relatively small area has a surprisingly high percentage of the entire eastern Guayana lowland flora.

Compared to other regions of French Guiana, Saül has the most endemics. French Guiana is considered to have more than 150 endemic species of vascular plants, although some may be found elsewhere when there has been adequate collecting throughout the lowland Guayana area. Because the Saül region is not separated phytogeographically, endemism within the proposed reserve is low: c. 50 of the endemics are in the Saül region and more than 20 have been found only within the proposed reserve (de Granville 1990). The outlying submontane forest floras on the summits of Monts Galbao (with some endemics) and Mont Belvédère show affinities with the Guayana Highlands and Andean floras.

Useful plants

These forests harbour many actual and potential medicinal plants (Grenand, Moretti and Jacquemin 1987), as well as species traditionally exploited as non-timber forest products such as palm hearts (*Euterpe oleracea*), rosewood oil (*Aniba rosaeodora*) and "balata" (*Chrysophyllum sanguinolentum*). The Guayana lowland flora also is rich in timber species (Gazel 1990).

Social and environmental values

Intact forests in the Saül region provide watershed protection for the headwaters of major tributaries of the Maroni, Mana and Approuague rivers, which are among the most important in French Guiana. This is one of the French Guianan regions most frequently visited for ecotourism.

The lowland moist forest that is characteristic of the region around Saül forms the primary vegetation for the Guayana Shield Endemic Bird Area (EBA B01), which is centred on the three Guianas and extends from eastern Venezuela into the State of Amapá in northern Brazil. Five restricted-range bird species occur in this large tract, while at least seven others with larger ranges are essentially confined to this area.

Economic assessment

The principal source of income has been small-scale gold mining. Attempts at other than subsistence agriculture have failed, however mostly because of the difficulty of transporting produce to market. Likewise timber production, although economically important in other parts of French Guiana, has

MAP 39. SAUL REGION, FRENCH GUIANA (CPD SITE SA3)

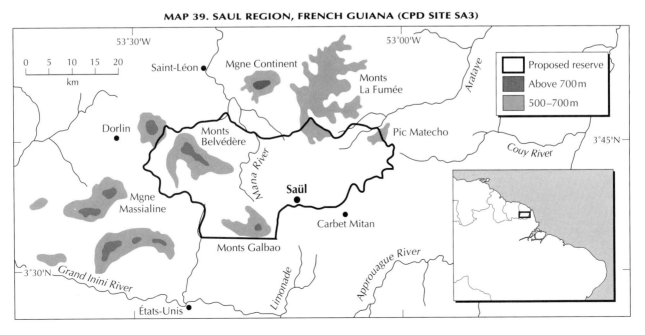

not been feasible because of the limited local market and high cost of transportation to coastal markets.

The greatest economic potential for the region is tourism. A large tour operator, based in Cayenne, offers trips on the Mana River which originate or end in Saül. Several individuals have developed moderately successful local tourist businesses and the village recently constructed a rustic hotel. Tourists, mostly from France, visit the region to see undisturbed rain forest and hike the nearly 100 km of trails around the village.

Threats

A road is under construction to link Cayenne to Saül. It even has been suggested that when the road is completed, Saül should be designated the capital. The road will bring radical changes to this pristine forest region. Current low-scale threats are slash-and-burn agriculture, cutting of trees for charcoal production, limited gold mining which has had minimal impact on the environment, and hunting especially to feed an ever-increasing number of tourists. Potentially, fuelwood may be cut to produce electricity.

Conservation

In 1993 the Department of French Guiana (which is an overseas department of France) was declaring 600 km² surrounding Saül village as an Arrêté de Protection du Biotope (protected area for natural habitat) (Map 39), and the Office National Forestière was considering establishment of a Nature Reserve of 190 km² in the same general region. Earlier a National Park of 1336 km² had been recommended, surrounding the village between 3°33'–3°49'N and 53°00'–53°27'W (de Granville 1975), with some attempts to establish that park since (de Granville 1990). These two declared and potential reserves are near/at the northern boundary of a proposed Southern National Park, which would include most of the southern half of French Guiana (1600–1800 km²).

References

Cremers, G., Feuillet, C., de Granville, J.-J., Hoff, M., Gracie, C.A. and Mori, S.A. (1988). Inventaire des phanérogames et des ptéridophytes de la région de Saül (Guyane Française). Communication de 'AUBLET', la Banque de Données botaniques de l'Herbier du Centre ORSTOM de Cayenne 11: 1–74.

Frailey, C.D., Lavina, E.L., Rancy, A. and Souza Filho, J.P. (1988). A proposed Pleistocene/Holocene lake in the Amazon Basin and its significance to Amazonian geology and biogeography. *Acta Amazonica* 18: 119–143.

Gazel, M. (1990). *Les bois.* La Documentation Guyanaise, Saga, Cayenne.

Granville, J.-J. de (1975). *Projets de réserves botaniques et forestières en Guyane.* Office de la Recherche Scientifique et Technique Outre-Mer (ORSTOM), Centre ORSTOM de Cayenne. 29 pp.

Granville, J.-J. de (1982). Rain forest and xeric flora refuges in French Guiana. In Prance, G.T. (ed.), *Biological diversification in the tropics.* Columbia University Press, New York. Pp. 159–181.

Granville, J.-J. de (1990). Priority conservation areas in French Guiana. Workshop 90, 10–20 Jan. 1990, Manaus, Brazil. Manuscript.

Granville, J.-J. de (1991). Remarks on the montane flora and vegetation types in the Guianas. *Willdenowia* 21: 201–213.

Granville, J.-J. de (1992). Un cas de distribution particulier: les espèces forestières péri-amazoniennes. *Compt. Rend. Séances Soc. Biogéogr.* 68: 1–33.

Granville, J.-J. de and Sanité, L.P. (1992). Guyana francesa: áreas protegidas y actividades humanas. In Amend, S. and Amend, T. (eds), *¿Espacio sin habitantes?* UICN/Editorial Nueva Sociedad. Pp. 265–287.

Grenand, P., Moretti, C. and Jacquemin, H. (1987). *Pharmacopées traditionnelles en Guyane: Créoles, Palikur, Wayãpi.* Édit. ORSTOM, Collection Mém. No. 108, Paris. 569 pp.

Lindeman, J.C. and Mori, S.A. (1989). The Guianas. In Campbell, D.G. and Hammond, H.D. (eds), *Floristic inventory of tropical countries: the status of plant systematics, collections, and vegetation, plus recommendations for the future.* New York Botanical Garden, Bronx. Pp. 375–390.

Mori, S.A. (1991). The Guayana Lowland Floristic Province. *Compt. Rend. Séances Soc. Biogéogr.* 67: 67–75.

Mori, S.A. and Boom, B.M. (1987). The forest. In Mori, S.A. and collaborators, The Lecythidaceae of a lowland neotropical forest: La Fumée Mountain, French Guiana. *Mem. New York Bot. Gard.* 44: 9–29.

Mori, S.A. and Prance, G.T. (1987a). Phenology. In Mori, S.A. and collaborators, The Lecythidaceae of a lowland neotropical forest. *Mem. New York Bot. Gard.* 44: 124–136.

Mori, S.A. and Prance, G.T. (1987b). Phytogeography. In Mori, S.A. and collaborators, The Lecythidaceae of a lowland neotropical forest. *Mem. New York Bot. Gard.* 44: 55–71.

Oldeman, R.A.A. (1974). *L'architecture de la Forêt Guyanaise.* Mém. ORSTOM No. 73, Paris. 204 pp.

Authors

This Data Sheet was written by Dr Scott A. Mori (New York Botanical Garden, Herbarium, Bronx, New York 10458-5126, U.S.A.) and Dr Jean-Jacques de Granville (Centre ORSTOM de Cayenne, 97323 Cayenne CEDEX France, French Guiana).

AMAZONIA: CPD SITE SA4

TRANSVERSE DRY BELT
of Brazil

Location: Mainly in northern Brazil, between latitudes 4°N–4°S and longitudes 62°–53°W, diagonally from Upper Branco River to Lower Xingú River.

Area: c. 500,000 km².

Altitude: Basically c. 5–60 m, with tabletop hills 60–200 m and a 400–600 m plateau.

Vegetation: Tropical lowland or "terra-firme" forests: dense mesophytic to mostly semi-open to open dry forests, savanna-forest; "várzea" and "igapó" inundation forests of white water and black or clear waters; various types of savanna ("campo"): terra-firme savanna, várzea savanna, "campina" (on hard substrate).

Flora: Major families of trees: Leguminosae, Sapotaceae, Burseraceae, Chrysobalanaceae, Lecythidaceae. High palm diversity in open forests. Dry corridor with some endemism.

Useful plants: Several Amerindian peoples' traditional utilization of many species; rich in genetic resources for timber, as well as cacao, "cupuaçú" (*Theobroma grandiflorum*) and other non-timber species.

Other values: Amerindian lands; river transport, watershed protection, hydroelectric energy, mineral resources, fisheries, tourism.

Threats: Mercury pollution from increasing number of gold miners; timber extraction; road building; corporate mining.

Conservation: Biological Reserve, 2 Ecological Stations, National Forest, Forest Experiment Station and Reserve; 4 Amerindian Reserves.

Geography

The Transverse Dry Belt is an approximate climatically defined region mostly south of the Guianas, between central and far eastern Amazonia (Nimer 1977; Pires and Prance 1977). This general region extends from north-west of Boa Vista on the Upper Branco River in Brazil's Roraima Territory south-eastward beyond the Lower Xingú River in Pará State, crossing southern Guyana and south-western Surinam (Map 40). Although with an annual precipitation as high as 1750–2100 mm, the region has a distinct Köppen Aw climate or in the Thornthwaite system primarily $B_1rA'a'$ climate edged by the greater potential evapotranspiration of $B_2rA'a'$ (SUDAM 1984), which is characterized by a seasonal (winter) drought (extending to as much as August–December) during which the rate of respiration exceeds photosynthesis. Drought coupled with deep sandy soils results in various types of savanna vegetation that are widely distributed in the region (Hoogmoed 1979), and characterize the Jari-Trombetas vegetation subprovince or phytogeographic region (Ducke and Black 1953, 1954; Rizzini 1963; Prance 1977, 1985; Whitmore and Prance 1987; IBGE 1990).

Except for flooded strips alongside rivers that were formed from Quaternary sediments, most of the region is Tertiary terrain. A remnant strip of Palaeozoic sand and siltstone occurs from the Jari River to the Rio Negro (Map 40), northward and parallel to the Amazon River over crystalline rocks of the ancient Uatumã group (DNPM 1981).

The topography of the Transverse Dry Belt is mainly lowlands, which can be classified into three basic landforms: wide alluvial terrain, low Tertiary plains and tabletop hills. The highest areas (400–600 m) correspond to the Maracanaquará Plateau (west of the middle Jari River), where the more ancient sediments occur.

Vegetation

The major vegetation type in the mosaic of the Transverse Dry Belt is semi-open forest, although dense mesophytic forests are found in some areas of higher elevation and as well on some ravine sides and valley bottoms. The Transverse Dry Belt additionally includes many open dry forests, large savannas and a number of small savanna enclaves. The most frequent vegetation types (Prance 1977, 1985; IBGE and IBDF 1988) are given below.

1. Lowland terra-firme (i.e. locally upland) forests

❖ *Dense lowland forest*
❖ *Semi-open lowland forest*
❖ *Lowland savanna-forest*
   Open forest or "mata seca", with palms or without palms. Amazonian caatinga or "campina" forest.

The locally upland forests are called "terra-firme" forests to distinguish them from the forests in flood areas. Also referred to as lowland tropical moist forests (when compared with premontane and montane forests), they are the most common forest type in Amazonia. However their physiognomy in the Transverse Dry Belt differs from other regions by being mostly semi-deciduous. Such forests are sometimes becoming recognized as transition forests (Whitmore and Prance 1987; Prance 1989).

The amount of basal area rather than deciduousness better differentiates these types of terra-firme forest (J.M. Pires 1994, pers. comm.). Considering trees with dbh 10 cm or more, their basal area per ha in dense forest is c. 40 m² to over 25 m², in semi-open forest 30–20 m² and in open forest less than 20 m². Amazonian "caatinga" is a special type of open forest occurring mainly on white sands in the Upper Rio Negro region (CPD Site SA6), with trees of small basal area.

The dense forest includes *Theobroma subincanum* and several species of *Manilkara* and *Protium*. In the predominant semi-open forests, the following species are characteristic: *Poupartia amazonica* (Anacardiaceae); *Pourouma minor* (Cecropiaceae); *Goupia glabra* (Celastraceae); *Bertholletia excelsa, Gustavia augusta* (Lecythidaceae); *Cedrela odorata* (Meliaceae); *Astrocaryum mumbaca, Bactris sphaerocarpa, Geonoma baculifera, Maximiliana maripa, Oenocarpus distichus, Orbignya (Attalea) spectabilis, Socratea exorrhiza* (Palmae); and *Cedrelinga cateniformis, Dinizia excelsa,*

Sclerolobium goeldianum (Leguminosae). The open forest includes *Couma guianensis* and *Hancornia speciosa.*

2. Inundation forests

❖ *Várzea swamp forest*
   Permanent or seasonal várzea forest.
❖ *Igapó swamp forest*
   Permanent or seasonal igapó forest.

As in the rest of Amazonia, the Transverse Dry Belt has its share of inundation forests. These can be separated into two main types based on the influence they receive from nearby water bodies: "várzea" forests occur with nutrient-rich white-water rivers and "igapó" forests with nutrient-poor black-water and nutrient-intermediate clear-water rivers. Each type can be subdivided into swamp forests that are permanent or seasonal.

MAP 40. CENTRAL-EASTERN AMAZONIA, SHOWING APPROXIMATE REGION OF TRANSVERSE DRY BELT (CPD SITE SA4)
(after DNPM 1981 and SUDAM 1984)

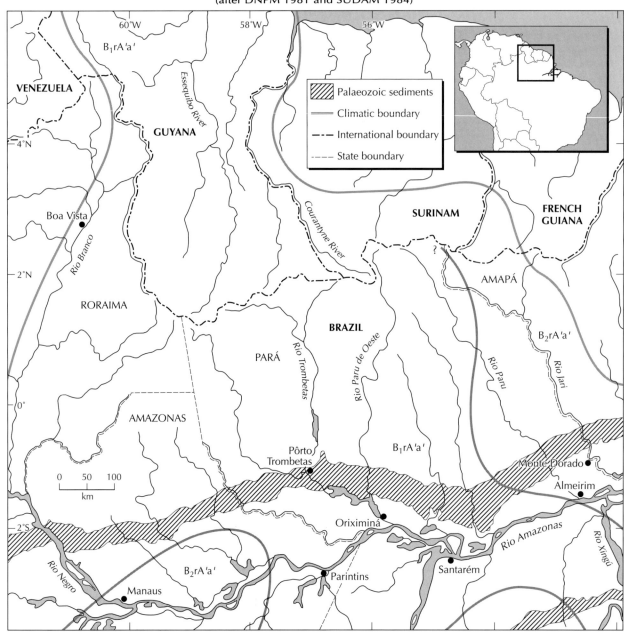

Several notable species in swamp forests are *Mauritia flexuosa* and *Virola surinamensis*; in várzea forest also occur *Euterpe oleracea* and *Hevea* spp., whereas igapó forest also has *Leopoldinia* spp.

The lacustrine-riverine history of the Amazon Basin suggests that igapó forests are much older than várzea forests, with the igapó forests existing since the Late Cretaceous, whilst the várzea forests developed only after the Andean orogeny caused the Atlantic drainage of the Amazonas graben to form the Amazon River (Kubitski 1989; Clapperton 1993).

3. Savannas or "campos" *sensu stricto*

❖ *Upland or terra-firme savanna*
❖ *Inundation or várzea savanna*
❖ *"Capinarana" or "campina"* (on hard lateritic soils or rock surfaces)

The savannas found in the Transverse Dry Belt region have an array of physiognomies (Ducke and Black 1953, 1954), but in essence were formed by similar processes involving imperfect soil drainage, impermeable subsoil layers and occurrence of sandy, gravelly or bleached soils on a deep water table.

The palaeohistory of Amazonia and the present distribution of ancient geological features of both Amazonia and Guayana suggest that the sands found in the present-day savannas ultimately originated from the Roraima sandstone, which underwent various erosion episodes by volcanism and weathering processes, with the resulting sediments being carried southward by extant tributaries of the Amazon River.

Typical savannas include *Byrsonima crassifolia* and various genera of Melastomataceae and Cyperaceae. While some of the savannas have been studied botanically, many have never been sampled (Daly and Prance 1989). The savannas of Roraima were studied by Takeuchi (1960), who reported finding vestiges of the use of fire to stimulate the growth of grasses. The savannas of Ariramba, near the Parú de Oeste (or Cuminá) River north of Oriximiná, were studied by Le Cointe (1922), Ducke and Black (1953, 1954) and Egler (1960). In the Jari area, a curious swampy savanna over a rock substrate was investigated by Pires-O'Brien (1992), and its flora compared with those of other savannas of Amazonia and Guayana.

During dry glacial periods of the Quaternary, the Transverse Dry Belt's deciduous forests and savannas may have served as corridors for exchange between the Llanos of Colombia and Venezuela and the savannas of Brazil's Central Plateau (Clapperton 1993; Daly and Prance 1989; Hoogmoed 1979). In the Trombetas area, moist forest probably persisted as a refugium (Prance 1982; Whitmore and Prance 1987).

Flora

The flora of the Transverse Dry Belt is best known along Highway BR-174, along the margins of the Amazon River, and around areas of greater development such as Pôrto Trombetas and Monte Dourado. Lists of the prominent species have been compiled for each of the vegetation types.

In the forests of the Transverse Dry Belt hundreds of tree species mix in various ways. Least rich are the dry semi-open forests and savanna-forests, which have from 120–140 different tree species in 1 ha. Many open forests have a large proportion of palm species. Most species-rich are the dense lowland forests, which have up to 240 different trees in 1 ha. These forests are found on terraces and in some ravines, and are characterized by fewer palms and more shade-tolerant trees.

The richest flora is probably over the geological strip of Palaeozoic sediments (Map 40). Along this strip are two centres of plant endemism summarized by Whitmore and Prance (1987) – in the Manaus region (CPD Site SA5) and in the Rio Trombetas Basin. The Jari Ecological Station, on the western side of the middle Jari River, also has high species richness and some vicariant taxa (Pires 1991; Pires-O'Brien 1992). A field survey from Trombetas to Jari along the Palaeozoic strip could determine whether the Trombetas area is a restricted centre of diversity or if such higher diversity occurs along the strip.

Useful plants

The region of the Transverse Dry Belt is the natural habitat of many plants producing spices and drugs that were exploited by the Portuguese from the second half of the 17th century to the first half of the 18th century. Among the economic products sought during that period were "wild clove" (*Dicypellium caryophyllatum*), an Amazon cinnamon (*Aniba canelilla*), vanilla (*Vanilla planifolia*), cassia (*Cassia* spp.), cocoa (*Theobroma cacao*) and "salsaparilha" (*Smilax* sp.). The exploration phase for spices and drugs was followed by an exploration phase for rubber. Raw rubber (*Hevea brasiliensis*), taken to Europe in 1744 by the French traveller and explorer Charles-Marie de La Condamine, soon came under study by various chemists. The world demand started in 1839 after invention of the process of vulcanization (von Hagen 1945).

The list of economic plants from the Amazon has increased substantially with the progress in chemical studies of plants suspected to have active compounds. For example, plants that yield the poison "curare" used on arrows by the Amerindians resulted in the discovery of three major groups of alkaloids as well as some glycosides with various pharmacological applications. Another example is the Lauraceae "pau-rosa" (*Aniba rosaeodora*), which has been heavily over-exploited for its fragrant oil linalol. Advances in phytochemical research on Amazonian plants continue to reveal species that have chemical compounds of scientific and economic interest (Gottlieb and Mors 1978). Interest in the rosewood oil from pau-rosa led to an extensive study of the essential oils of a large number of *Aniba* species, which resulted in new insights in plant evolutionary biology (Gottlieb 1985).

Since 1959 the Superintendência para o Desenvolvimento da Amazônia (SUDAM) has experimented with growing some useful species of trees south-east of Santarém, at the Curuá-Una Experiment Station. Good results have been found with *Vochysia maxima*, *Jacaranda copaia*, *Simaruba amara*, *Goupia glabra*, *Virola cuspidata*, *Anacardium giganteum*, *Bertholletia excelsa*, *Parkia multijuga*, *Didymopanax morototoni* and *Bagassa guianensis* (Pedrosso *et al.* 1982).

The Instituto Agronômico do Norte (IAN) in Belém, which is now called the Centro de Pesquisa Agroflorestal da Amazônia Oriental (acronym CPATU – since previously called the Centro de Pesquisa Agropecuária do Trópico Úmido), and which is linked with the Empresa Brasileira de Pesquisa Agropecuária (EMBRAPA), has a tradition of research in forest management at the Tapajós Forest Reserve south-west of Santarém. In 1975 permanent experimental plots were begun for commercial and diagnostic inventories on natural regeneration following exploitation. Among the most desirable trees are *Hymenaea courbaril, Piptadenia suaveolens, Didymopanax morototoni, Jacaranda copaia* and *Bixa arborea* (Carvalho *et al.* 1984).

Social and environmental values

This vast area is very poorly known ornithologically. It is difficult to judge the range of many of the bird species as they are often recorded from only a handful of widely spaced localities. The Upper Rio Branco Endemic Bird Area (EBA B56) is located in the northern part of this region. This EBA consists of the gallery forests of the Upper Branco River and associated rivers of northern Roraima in northernmost Brazil and extreme western Guyana. There are two restricted-range species confined to the EBA: Rio Branco antbird (*Cercomacra carbonaria*) and hoary-throated spinetail (*Synallaxis kollari*). Both are considered threatened due to their small ranges which render them vulnerable to habitat destruction. Farther south, the Central Amazonian Brazil EBA (B43) overlaps with this region.

The northern part of the Transverse Dry Belt region, mainly in areas recognized officially, is home for several indigenous groups – particularly Jacamim, Wai-Wai, Nhamundá-Mapuera, Tirió and Yanomami. The Yanomami live between the headwaters of the Orinoco River in Venezuela and some tributaries of the Negro and Branco, and are considered the last relatively isolated large indigenous group in the Amazon. Their total population is c. 17,000, of whom c. 10,000 live in Brazil (Albert 1992). Heavy migration of gold prospectors to areas close to indigenous territory is a threat to the survival of the Yanomami and other tribes, by bringing diseases to which the Amerindians are not resistant. In the region of the Trombetas River, there are some black communities formed over 100 years ago by runaway slaves known as "Quilombos".

In the Transverse Dry Belt the greater proportion of the Western population historically has been concentrated alongside the Amazon and other major rivers. In Amazonia as a whole the building of roads and development projects have broken this trend, and populations now concentrate near such areas. The Transverse Dry Belt region is cut by a section of the Manaus-Rio Branco road (BR-174). However, the Transverse Dry Belt still has fewer roads than many other regions in the Amazon, which may be a blessing since the most disturbed areas are alongside the major highways.

"Ribeirinhos" commonly dwell alongside the rivers, living predominantly from fishing and a diversified subsistence agriculture. There is a ribeirinho culture, formed by the influence of European ways on indigenous populations. "Caboclo" is another term used to characterize such local dwellers, although this word is also used for those inhabitants who represent mixtures involving Europeans and Amerindians. Turtles are important in the local diet, as the Trombetas region is a breeding centre for them. Small-scale agro-industry is also significant for ribeirinho or caboclo dwellers, producing sugarcane, "aguardente" (a type of rum), manioc flour, salted fish and extractive forest products.

Populations from areas under the influence of major roads and development projects have broken away from the traditional activities of subsistence agriculture and small-scale agro-industry into larger scale production of fewer goods. Paralleling the independent family-run farming properties, there has been an increase in large properties, saw mills, mines and commerce, resulting in an increased demand for labour and for land.

Economic assessment

The general region's macroeconomy is based mainly on mining activities, pulp production and extraction of timber. There also are many non-timber plant products that are commercially exploited (IBGE 1990) (see Table 54). Some of these native products already are cultivated for a larger economic return – such as "açaí", grown for production of heart-of-palm.

TABLE 54. MAJOR NON-TIMBER PLANT EXTRACTIVES FROM BRAZILIAN AMAZONIA

Product	Species	Amazon	Brazil
		(1000 US$)	
Açaí fruit	*Euterpe oleracea*	142,038	145,881
Heart-of-palm	*Euterpe oleracea*	138,744	142,060
Brazil nut	*Bertholletia excelsa*	35,534	36,241
Hard hevea rubber	*Hevea brasiliensis*	24,908	24,928
Liquid hevea rubber	*Hevea brasiliensis*	1513	1513
Sôrva gum	*Couma* spp.	1524	1524
Buriti fibre	*Mauritia flexuosa*	900	1155
Cumarú	*Dipteryx odorata*	332	333
Maçaranduba gum	*Manilkara* spp.	298	298
Balata gum	*Manilkara* spp.	19	19
Copaíba balm	*Copaifera* spp.	94	99

By leading to large-scale deforestation, the new social order is more damaging to the environment than the traditional practices. In the Transverse Dry Belt the trends are just beginning. More efforts must be put into promoting socio-economic activities that allow maintenance of natural vegetation cover, discouraging those which promote deforestation and poor development.

The region of the Jari River has c. 1100 km^2 of converted forests cultivated particularly for production of some pulpwood species. The land is owned by the Companhia Florestal Monte Dourado, which acquired it from D.K. Ludwig's Projeto Jari that began in 1968 (Rankin 1985). Although perennial-crop plantations have been considered as a most suitable form of development for Amazonia (Alvim 1981), ecological considerations such as the difficulties with monocultures and the need for applied research in semi-natural reserves must be included in sensible planning for enduring development (Hecht 1988; Daly and Prance 1989; cf. Pires 1991).

During the United Nations Conference on Environment and Development (UNCED) in June 1992 in Rio de Janeiro, the Brazilian authorities showed a disposition towards sustainable development of the Amazon region. A large project promoting economic-ecological zoning of the State of Pará is underway by the federal government (through SAE, the Secretariat of Strategic Studies) in conjunction with the State of Pará (through IDESP, the Institute for Economic and Social Development of Pará). The project aims to gain a better knowledge of Pará in order to plan its long-term, sustainable development.

Threats

A northern east-west highway known as Perimetral Norte (BR-210) was planned in 1973 by the Brazilian Government, but has not been fully constructed due to a lack of financial resources. The relative absence of roads at present makes the Transverse Dry Belt less threatened by deforestation than most other areas of Amazonia. The major threat is mercury pollution due to its increased use by prospectors to amalgamate gold particles (Martinelli et al. 1988). The areas most threatened by the pollution are the Upper Branco and the Lower Xingú, which have been subjected to heavy migration by gold and diamond prospectors. In the Alto Rio Branco area the migration has been facilitated by numerous airfields constructed by Brazil's Calha Norte project, aimed at patrolling the frontier. Corporate mining also could be a threat to the natural vegetation.

The area of Pôrto Trombetas has a large deposit of bauxite that is being exploited by the company Mineração Rio do Norte (MRN), which deforests 100–120 ha yearly; MRN makes efforts to regenerate forest on the stripped sites. The ore is transported on a 100-km railroad built for that purpose. An increasing migrant population is concentrated in a band south of the Amazon River, which has the highest demography. It is in this area where most forest-burning episodes have been reported.

Conservation

The Transverse Dry Belt general region has several conservation units (IBGE and IBDF 1988), including a Biological Reserve, two Ecological Stations and as well a National Forest (Table 55). There are other areas semi-protected, such as the Curuá-Una Experiment Station and Forest Reserve (712 km²), and at least four Amerindian Reserves: Jacamim, Wai-Wai, Nhamundá-Mapuera and Tumucumaque.

A centre of endemism has been proposed in the Trombetas River area (see Whitmore and Prance 1987). The region has many other areas that have been given priority for additional conservation efforts due to particular features (Prance 1990), such as the Palaeozoic strip.

TABLE 55. CONSERVATION AREAS IN THE TRANSVERSE DRY BELT REGION

	Year founded	Area (km²)
Maracá Ecological Station	1977	1013–3500
Rio Trombetas Biological Reserve	1979	38,500
Jari Ecological Station	1977	2271
Caxiuanã National Forest	-	2000–3150
Total		**43,784–47,421**

References

Albert, B. (1992). Terras indígenas, política ambiental e geopolítica militar no desenvolvimento da Amazônia: a propósito do caso Yanomami. In Léna, P. and Oliveira, A.E. de (eds), *Amazônia: a fronteira agrícola 20 anos depois*. Coleção Eduardo Galvão, Museu Paraense Emílio Goeldi, Belém. Pp. 37–57.

Alvim, P. de T. (1981). A perspective appraisal of perennial crops in the Amazon Basin. *Interciencia* 6: 139–145.

Carvalho, J.O.P. de, Silva, J.N.M. da, Lopes, J. do C.A. and Costa, H.B. da (1984). *Manejo de florestas naturais do trópico úmido com referência especial à Floresta Nacional do Tapajós no estado do Pará*. EMBRAPA-CPATU, Documentos No. 26. 14 pp.

Clapperton, C.M. (1993). *Quaternary geology and geomorphology of South America*. Elsevier, Amsterdam. 779 pp.

Daly, D.C. and Prance, G.T. (1989). Brazilian Amazon. In Campbell, D.G. and Hammond, H.D. (eds), *Floristic inventory of tropical countries: the status of plant systematics, collections, and vegetation, plus recommendations for the future*. New York Botanical Garden, Bronx. Pp. 401–426.

DNPM (1981). *Mapa geológico do Brasil*, scale 1:2,500,000. Coordinated by Schobbenhaus, C., Campos, D.A., Derze, G.R. and Asnius, H.E. Ministério das Minas e Energia, DNPM, Rio de Janeiro.

Ducke, A. and Black, G.A. (1953). Phytogeographical notes on the Brazilian Amazon. *An. Acad. Brasil. Ciênc.* 25: 1–46.

Ducke, A. and Black, G.A. (1954). *Notas sobre a fitogeografia da Amazônia brasileira*. Instituto Agronômico do Norte (IAN), Boletim Técnico 29. Belém. 62 pp.

Egler, W.A. (1960). Contribuiçoes ao conhecimento dos campos da Amazônia. I. Os campos do Ariramba. *Bol. Museu Paraense Emílio Goeldi, Nova Série Botânica* 4: 1–36.

Gottlieb, O.R. (1985). The chemical uses and chemical geography of Amazon plants. In Prance, G.T. and Lovejoy, T.E. (eds), *Amazonia*. Key Environments Series, Pergamon Press, Oxford, U.K. Pp. 218–238.

Gottlieb, O.R. and Mors, W.B. (1978). Fitoquímica amazonica: uma apreciação em perspectiva. *Interciencia* 3: 252–263.

Hecht, S.B. (1988). Jari at age 19: lesson's for Brazil's silvicultural plans at Carajás. *Interciencia* 13: 12–24.

Hoogmoed, M.S. (1979). The herpetofauna of the Guianan region. In Duellman, W.E. (ed.), *The South American herpetofauna: its origin, evolution, and dispersal*. Monogr. Mus. Nat. Hist. Univ. Kansas 7: 241–279.

IBGE (1990). *Anuário estatístico do Brasil 1989*. Instituto Brasileiro de Geografia e Estatística (IBGE), Rio de Janeiro.

IBGE and IBDF (1988). *Mapa de vegetação do Brasil*, scale 1:5,000,000. IBGE and Instituto Brasileiro de Desenvolvimento Florestal (IBDF), Rio de Janeiro.

Kubitski, K. (1989). The ecogeographical differentiation of Amazonian inundation forests. *Plant Syst. Evol.* 162: 285–304.

Le Cointe, P. (1922). *L'Amazonie brésilienne*, 2 vols. Editeur Augustin Challamel, Paris. 1025 pp.

Martinelli, L.A., Ferreira, J.R., Fosberg, B.R. and Victorio, R.L. (1988). Mercury contamination in the Amazon: a gold rush consequence. *Ambio* 17: 252–254.

Nimer, E. (1977). Clima. In IBGE, *Geografia do Brasil*, Vol. I. *Região Norte*. IBGE, Rio de Janeiro. Pp. 39–58.

Pedrosso, L.M., Lopes, C.A.C., Peres, A.S.G., Dourado, R.S.A. and Vasconcelos, P.C.S. (1982). Pesquisas silviculturais na região do trópico úmido brasileiro. *Superintendência do Desenvolvimento da Amazônia (SUDAM), Documenta* 4(1/2): 35–68.

Pires, J.M. and Prance, G.T. (1977). The Amazon forest: a natural heritage to be preserved. In Prance, G.T. and Elias, T.S. (eds), *Extinction is forever: threatened and endangered species of plants in the Americas and their significance in ecosystems today and in the future*. New York Botanical Garden, Bronx. Pp. 158–194.

Pires, M.J.P. (1991). *Phenology of selected tropical trees from Jari, Lower Amazon, Brazil*. Ph.D. thesis, University of London. 322 pp.

Pires-O'Brien, M.J. (1992). Report on a remote swampy rock savanna, at the mid-Jari River Basin, Lower Amazon. *Bot. J. Linnean Soc.* 108: 21–33.

Prance, G.T. (1977). The phytogeographic subdivisions of Amazonia and their influence on the selection of biological reserves. In Prance, G.T. and Elias, T.S. (eds), *Extinction is forever*. New York Botanical Garden, Bronx. Pp. 195–213.

Prance, G.T. (1982). Forest refuges: evidence from woody angiosperms. In Prance, G.T. (ed.), *Biological diversification in the tropics*. Columbia University Press, New York. Pp. 137–157.

Prance, G.T. (1985). The changing forests. In Prance, G.T. and Lovejoy, T.E. (eds), *Amazonia*. Pergamon Press, Oxford, U.K. Pp. 146–165.

Prance, G.T. (1989). American tropical forests. In Lieth, H. and Werger, M.J.A. (eds), *Tropical rain forest ecosystems*. Ecosystems of the World 14B. Elsevier Science Publishers, Amsterdam. Pp. 99–132.

Prance, G.T. (1990). Consensus for conservation. *Nature* 345: 384.

Rankin, J. McK. (1985). Forestry in the Brazilian Amazon. In Prance, G.T. and Lovejoy, T.E. (eds), *Amazonia*. Pergamon Press, Oxford, U.K. Pp. 369–392.

Rizzini, C.T. (1963). Nota prévia sôbre a divisão fitogeografica do Brasil. *Revista Brasil. Geogr.* 25(1): 1–64.

SUDAM (1984). *Projeto de hidrologia e climatologia da Amazônia brasileira*. Superintendência para o Desenvolvimento da Amazônia (SUDAM), Publicação 39. Belém.

Takeuchi, M. (1960). A estrutura da vegetação na Amazônia II. As savanas do norte da Amazônia. *Bol. Museu Paraense Emílio Goeldi, Nova Série Botânica* 7: 1–14.

von Hagen, V.W. (1945). *South America called them: explorations of the great naturalists*. Robert Hale, London. 311 pp.

Whitmore, T.C. and Prance, G.T. (eds) (1987). *Biogeography and Quaternary history in tropical America*. Oxford Science Publications, Oxford, U.K. 214 pp.

Author

This Data Sheet was written by Dra. Maria Joaquina Pires-O'Brien (Ministério da Educação e do Desporto, Faculdade de Ciências Agrárias do Pará, Departamento de Ciências Florestais, Caixa Postal 917, 66.077-530 Belém, PA, Brazil).

AMAZONIA: CPD SITE SA5

MANAUS REGION
Brazil

Location: In western Central Amazonia, a circle with 150-km radius around Manaus, between latitudes 1°30'–3°00'S and longitudes 58°15'–61°00'W.

Area: c. 70,700 km².

Altitude: 16–130 m.

Vegetation: Several types of rain forests on non-flooded and in variously inundated areas, forested to open shrubby formations on white sandy soils.

Flora: High diversity, especially of tree species; high endemism.

Useful plants: Genetic reserves, e.g. for timber trees; edible fruits, medicinal plants.

Other values: Fauna refuge, watershed protection, research stations, tourism.

Threats: Road building; colonization; fire; conversion to pasture or plantation; removal of sand and fuelwood.

Conservation: Ecological Research Stations, Forest Reserves, Environmental Protection Zone (APA); nearby National Park, Amerindian Reserve, Biological Reserve.

Geography

The Manaus region is in western Central Amazonia in Brazil's northern State of Amazonas, and includes the lower drainage area of the Rio Negro and its confluence with the Solimões River, which extends eastward as the Amazon River. Habitats have been selected within a circle with a radius of 150 km from the city of Manaus (3°S, 60°W), near the Rio Negro's mouth (Map 41) (G.T. Prance, E. Lleras and B.W. Nelson, pers. comm., Workshop 90, Manaus).

The mean annual temperature is 27.2°C, and the maximum is 37.8°C (Junk 1983; Leopoldo *et al.* 1987). The mean annual precipitation at Manaus is 2100 mm, with 73% received as heavy rains. About 26% of the water runs off, 20% becomes groundwater and the rest is evaporated and transpired by the vegetation, usually to recycle as rain in c. 5.5 days. The Lower Amazon Basin is somewhat drier from July to October (Daly and Prance 1989; Leopoldo *et al.* 1987).

Vegetation

Three main types of vegetation occur in the vicinity of Manaus: (1) rain forests on "terra firme" (non-flooding areas), as well as small terra-firme savannas on compacted clay soil; (2) forests of inundated areas, as well as grassland and floating mats; and (3) formations on white-sand soils (Prance 1987b).

1. Rain forests on terra firme (uplands)
This vegetation type covers c. 50% of Amazonia (Prance 1987b); the Manaus forests are typical of these Central Amazonian forests that have developed on the dominant clayey latosolic sediments of Tertiary plateaux and slopes. The soils are well structured and well drained but of poor fertility (Brickmann 1989). Some differences in structure and composition occur according to topography: tall terra-firme rain forest on clay soil, Amazonian caatinga forest (campinarana) and open Amazonian caatinga scrub (campina) on small patches of hydromorphic podzols, and 130 km north of Manaus, campinas and campinaranas on sandstone.

Two main terra-firme subunits are recognized: (i) **terra-firme lowland forest** on rather flat areas, which presents the characteristic heterogenous climax; and (ii) **terra-firme hill forest**, above 250 m (e.g. north of the Manaus region). Locally, in addition to forest on rather flat areas, there are distinctive forests on slopes (Brickmann 1989; Guillaumet 1987; Prance 1987b).

The climax forest has large biomass, a high closed canopy, high species diversity with much local variation in composition, large lianas (of somewhat limited frequency) and a relatively sparse herbaceous ground cover. Four layers may be detected: the canopy at 30–35 m with a few emergents to 44 m, a stratum of 12–15 m tall trees, a stratum of small trees and shrubs 7–12 m tall and a stratum of low shrubs and saplings to 7 m (Guillaumet 1987).

Guillaumet (1987) emphasized the difficulty in comparing the few site inventories made in the region. He lists the 12 tallest tree species (35–44 m), those of largest dbh (c. 20 species over 85–200 cm) and the density of trees of various sizes. About 20% of the larger trees have buttresses. Characteristically, palms are abundant and diverse in several strata. At one site, in each of seven families more than 10 species occurred, with the most in Lecythidaceae (18 spp.), Moraceae (15) and Sapotaceae (14). Typically, no species of tree is common; among 235 species of trees in the area, the most frequent species (individuals per ha) at the site were *Eschweilera coriacea* (only 26) and *Scleronema micranthum* (only 9) (Prance, Rodrigues and Silva 1976). Lianas especially of Leguminosae were frequent at one site. There are abundant epiphytes, especially ferns, Bromeliaceae and Orchidaceae. The forest is particularly rich in hemi-epiphytes and pseudo-epiphytes such as species of *Philodendron*, *Heteropsis* and

Anthurium (Araceae) and seven other families. Stranglers of *Ficus* spp. and Clusiaceae are abundant but not necessarily strangling their host trees. About 12 genera of semi-parasitic Loranthaceae are frequent in the canopy. Flowering plant saprophytes (Burmanniaceae, Gentianaceae, Triuridaceae) are also rich in species.

2. Inundated forests, grassland and floating vegetation

There are substantial Amazonian areas of both permanently and periodically (e.g. annually) flooded forests. Species diversity is less than in terra-firme forests, with less regional endemism (Prance 1987b). The permanent types are: (i) **permanent swamp forest of white water**; and (ii) **permanent "igapó" of black or clear water**. The underlying soil is eutrophic humic gley; in some areas of dystrophic humic gley, palm swamps occur. Periodically flooded forests include: (iii) **Seasonal "várzea"**, which is floodplain forest periodically flooded by fertile muddy-white waters, e.g. along the Solimões and Amazon floodplains. It is similar to the upland forest, with fewer species and more lianas. (iv) **Seasonal igapó**, which is floodplain forest periodically flooded by low-fertility black or clear water, e.g. along the Rio Negro. There are fewer, often different species than in seasonal várzea, and many of the trees have scleromorphic adaptations. Last, (v) **sporadic floodplain forest** occurs where there are quickly draining or flash floods at irregular times. This type is mainly in the upper portions of rivers and narrow streams and has greater species diversity, since many terra-firme species can survive such limited inundation.

The total flooded area of the Rio Negro and its tributaries probably exceeds 2000 km², and includes permanent and seasonal igapó, and sporadic floodplain forest (Goulding 1988). Annual fluctuation of the river-level is 9–12 m, with the inundation lasting 4–5 months along the upper and middle river and 7 months along the lower Rio Negro; the highest water is in June or July. Although data are limited, the annual variability appears large. Rarely forests remain flooded for 2–3 years, but this causes tree and shrub mortality; annual minimum and maximum fluctuations recorded at Manaus have been 5.5 m and 14 m (Goulding 1988; Irmler 1977). However, between November and February while the level of the Upper Rio Negro is falling with the dry season, the lower river is rising because it is dammed back by the rising Solimões-Amazon River (Goulding 1988), which results in somewhat intermediate vegetation from the influx of white water (Irmler 1977; Prance 1979).

Inundated forests have been described by Prance (1979, 1989; see also Junk 1983). Seasonal igapó has limited species diversity and generally lower biomass, since it is on poor sandy soils along riverbanks and the water lacks nutrients; the forest is usually not extensive and often is interspersed with open beaches. A few studies have been done on igapó vegetation near Manaus (e.g. Prance 1979, 1989; Keel and Prance 1979; Rodrigues 1961). The canopy is lower and the species tend to be different from those of várzea or terra firme. Only 54 species of trees and shrubs were found at one site. The most common trees are in Myrtaceae, such as *Eugenia inundata*; other common trees are *Alchornea castaneifolia*, *Copaifera martii*, *Piranhea trifoliata* and *Triplaris surinamensis* (Pires and Prance 1985; Prance 1989). Zonation of species occurs gradually along the moisture gradient away from the river from heavily to lightly flooded (Keel and Prance 1979).

Seasonal várzea is the most widespread type of inundated forest in Amazonia, often extending several km back from the riverbank (cf. Junk 1984, 1989). In Lower Amazonia it is associated with robust grass meadows (Pires and Prance 1985) and in the Manaus region with the Solimões River. Some characteristic trees of seasonal várzea around Manaus include *Astrocaryum jauari*, *Calycophyllum spruceanum*, *Carapa guianensis*, *Ceiba pentandra*, *Hevea brasiliensis*, *Hura crepitans* and *Macrolobium acaciifolium*. The herbaceous understorey is rich in individuals of *Heliconia* and *Costus*.

During the high-water season, vast **mats of vegetation** form on the extensive lakes and lake-like expanses where riverbanks often are completely inundated. Mats may cover several km² within a few months, yet with over 90% of the species reproducing and dying by the subsequent dry season. In white-water lakes and swamps that do not dry up, vegetational succession progresses from floating plants such as *Eichhornia crassipes* and *Salvinia auriculata* to a secondary community of grasses and sedges. *Montrichardia arborescens*, a tree-like Araceae to 6 m or more tall, then may colonize the mat, compacting it into a floating community from 20 cm above to partly 100 cm below the surface. Climbers such as *Ipomoea* spp. and even trees to 6–8 m (*Bombax, Cassia, Cecropia, Ficus*) may occur. Lago dos Patos (c. 100 km from Manaus) has a floating mat 200 m wide, 3 km long and more than 4 m thick. Usually, however, the mats sink at 1 m thickness or less (Junk 1983).

3. Formations on white-sand soils

Interspersed within lowland terra-firme rain forest of the Rio Negro region are white-sand "campinas", which are islands of open shrub to low-forest formations on podzols of leached quartz-sand soils or regosols. Campinas occur from the headwaters of the Rio Negro to Manaus near the Solimões-Amazon River (Goulding 1985), and in large or small areas scattered over much of Amazonia (Prance 1987b). Campinas often gradually grade (physiognomically and floristically) into the surrounding forest. These soils are not only extremely poor in nutrients, but too porous to retain water. The extreme ecological restrictions have resulted in a continuum of highly distinctive formations, generally characterized by scleromorphic leaves, gnarled trunks and unusual floristic composition (Macedo and Prance 1978). The formations may cause the black water, because they are unable to filter out humic acids from organic decomposition (Anderson 1981; Anderson, Prance and Albuquerque 1975; Goulding, Carvalho and Ferreira 1988).

There are no accurate estimates of the total area of Amazonia or the Rio Negro Basin occupied by these formations, but it is considerable – thousands of km². There are four phases to the vegetation: savanna; scrub; woodland with a patchy canopy 5–15 m high (with emergents to 20 m tall); and forest with a generally continuous canopy 20 m to almost 30 m high. The campina-forest areas around Manaus are lower (up to 15 m) and poorer in species than the immense relatively rich areas (especially in understorey and ground-cover species) found in the Upper Rio Negro region (CPD Site SA6) (Anderson 1981; Prance 1987b). Around Manaus the campina forest is dominated by tree species such as *Aldina heterophylla*, *Glycoxylon inophyllum* and *Humiria balsamifera*. The twisted, much-branched trees are loaded with many epiphytic orchids, bromeliads, Araceae, Gesneriaceae (e.g. *Codonanthe*) and pteridophytes. The many open areas of sand are covered by the blue-green alga *Stigonema tomentosum*. Lichens and mosses are abundant on branches and the soil surface. Sometimes the ground is covered with a spongy mat of *Cladonia*; *Sphagnum* is less frequent (Pires and Prance 1985; Prance 1987b, 1989).

Flora

Being at the confluence of the Negro and Amazon rivers, the Manaus region has extensive periodically flooded habitats (igapó and várzea), each with a mosaic of communities determined by the duration of flooding, water chemistry and sedimentation characteristics. The modern-age dry belt east of Manaus cutting north-west to south-east across the Amazon River (CPD Site SA4) may impede the eastward migration of species, causing decreased diversity near Belém (in the State of Pará) relative to Manaus and westward from Manaus. Finally, the first direct dated evidence for a dry Pleistocene Amazon, in Acre (Kronberg, Benchimol and Bird 1991) and at Carajás (Absy *et al.* 1989), as well as indirect evidence for relative aridity on the Guayanan and Brazilian shields (Veiga 1991), lend new weight to the Pleistocene refugium hypothesis (see Whitmore and Prance 1987) as an explanation for high α diversity, if Manaus was spared this dryness.

Each of the vegetation types is more or less floristically distinct, so there is high β diversity within this small circle. Total floristic diversity is unknown, but α diversity in terra-firme tree plots is high at c. 200–230 spp./ha (of trees with dbh 10 cm or more) and the total tree flora is well over 1000 species. The terra-firme list (represented by the Ducke Forest Reserve) is dominated by the speciose tree genera *Licania, Inga, Protium, Eschweilera, Swartzia, Aniba, Miconia, Ocotea, Casearia* and *Couepia*. Apparently there is high non-edaphic endemism in several families (e.g. Lecythidaceae, Chrysobalanaceae, Bignoniaceae, possibly Sapotaceae), and one endemic family – Duckeodendraceae.

The Manaus region is probably a centre of high diversity and possibly also local endemism. Species from western Amazonia, the Guianas and eastern Amazonia overlap, making this a priority area for conservation. There is a unique concentration of diversity in tree taxa, and families not speciose elsewhere: Chrysobalanaceae, Burseraceae, Lecythidaceae and Vochysiaceae.

The forests around Manaus and Belém are the most heavily collected sites in Amazonia (Daly and Prance 1989). Species lists have been compiled for some sites around Manaus and some detailed information has been gathered on composition and distribution (e.g. Prance, Rodrigues and Silva 1976; Anderson, Prance and Albuquerque 1975; Rodrigues 1961, 1967). The most important feature of the tropical rain forest on terra firme is its outstanding species diversity. Many studies have shown this richness throughout the region, with variation according to local rainfall. For example, Prance, Rodrigues and Silva (1976) in 1 ha of forest near Manaus recorded 179 tree species 15 cm or more in dbh; the 235 tree species in the general area were in 43 families. Fittkau and Klinge (1973) reported 502 species of trees and shrubs per ha in Manaus locations. Guillaumet (1987) provides summaries and analysis of this variable, complex ecosystem.

Because of physiological stress caused by the immersion, as well as the likelihood of species with adaptations to water dispersal being widely distributed, inundated forests have less species diversity than terra-firme forests and also less regional endemism (Prance 1987b). However, a significant number of species of trees and shrubs that occur in seasonal igapó are endemic to this habitat, and a few species are endemic to both seasonal igapó and várzea.

Plant diversity in campinas is reduced in comparison to contiguous rain forest. In the Lower Rio Negro region where the campinas are small and scattered, species richness is low in contrast to the extensive, older yet often isolated areas of the Upper Rio Negro region. The campinas are in effect ecological islands. Macedo and Prance (1978) showed that in the Manaus area, long-distance dispersal plays a very important role in distributing the species – 60% were dispersed by birds. Many species of plants are endemic to campinas (c. 55%) and occur in most of them; a few rare species are endemic to only one or two campinas (e.g. *Erythroxylum campinense*).

The Manaus region is the most controversial natural refugium of Amazonia. Some botanists regard it as an artefact of the large amount of collecting conducted there (Daly and Prance 1989). However, Prance (1987a) estimated that although the amount of endemism seemingly present could be greatly reduced by intensive collecting elsewhere, endemism would never drop to a level such that the Manaus region would no longer be considered an important centre.

Gentry (1990) compared a checklist (derived from Brazil's Programa Flora database) of 825 vascular plant species at the Adolpho Ducke Forest Reserve (100 km²) near Manaus (Prance 1990) with checklists from Central American wet and seasonal sites and a Peruvian Amazon seasonal site. The Central Amazonian flora represented by the Ducke reserve stands out as having a unique set of speciose families. The most speciose genera were almost exclusively trees, rather than herb, shrub or epiphyte genera which were most diversified at the three other sites. The peculiarity of the Manaus flora carries over into the family level: Manaus has the most unusual suite of families among the 20 most speciose families at each of the four sites. Chrysobalanaceae, Burseraceae, Lecythidaceae and Vochysiaceae are speciose at Manaus but not in the Peruvian Amazon or Central America, whereas Leguminosae, Moraceae, Lauraceae and Sapotaceae are speciose at Manaus and one or more of the other sites. The Manaus pattern of dominance by trees (in the listing of species) would probably hold as well for the less diversified flora of eastern Amazonia, but its checklists have been based solely on tree plots, so comparison of diversity by habit is not possible.

However, the comparison by simple presence or absence of families and genera for the Manaus flora with the flora of other sites is misleading since it is based on the single checklist (825 spp.) from the Ducke reserve. About 30% of the Instituto Nacional de Pesquisas da Amazônia (INPA) herbarium's 115,000 Brazilian Amazon specimens (as of 1989) are from the few degree-squares around Manaus (and a large percentage are not duplicated in other herbaria), and additionally there are more than 50,000 vouchers from the Minimum Critical Size of Ecosystems Project (MCSE) 70–90 km north of Manaus. Most of the families conspicuously absent from the Ducke reserve checklist (Gentry 1990) do occur regionally outside this reserve.

The monotypic Duckeodendraceae can be considered regionally endemic to Manaus – *Duckeodendron cestroides* penetrates at least as far as the lower Madeira River. Many endemic species are recorded for Manaus (Prance 1987a, 1990), but the very low density of scientific collecting for 1500 km west and south-west from Manaus does not allow a reliable demarcation of endemic limits in those directions (Nelson *et al.* 1990).

No fully vouchered inventory of plot diversity of trees has been published using a 10-cm dbh cutoff for 1 ha of terra-firme forest near Manaus. Using a 15-cm dbh cutoff, Prance, Rodrigues and Silva (1976) found 179 tree species in 1 ha. Milliken *et al.* (1992) in 1 ha c. 200 km north-west of Manaus found 201 species of trees ≥10 cm dbh. At the Ducke reserve, Alencar (1986) found 215 trees of ≥10 cm dbh or more in 1 ha and almost 400 different tree species in 2.5 ha. Gentry (in prep.), using a 2.5-cm dbh cutoff in a 0.1-ha plot at the Ducke reserve, found c. 233 tree, liana and acaulescent palm species among 365 individuals – about as high as the richest sites near Iquitos, Peru (CPD Site SA9).

These few data suggest that single hectares of well-drained terra-firme clay near Manaus have c. 200–230 tree and liana species 10 cm or more in dbh/ha, considerably fewer than the almost 300 species on 1 ha near Iquitos (Gentry 1988). On the other hand, Central Amazonia has very much higher plot diversity than Central America or eastern Amazonia. Fifty contiguous 1-ha plots on Barro Colorado Island (BCI) in Panama averaged just 93 species ≥10 cm dbh per ha (Foster and Hubbell 1990), and seven 1-ha tree plots in eastern Amazonia near Marabá and Carajás had 119–130 spp. per ha (Silva *et al.* 1988).

Of all the 0.1-ha inventories done by Gentry in neotropical forests, that at the Ducke reserve is the most above the predicted diversity based on an otherwise consistent relationship between increasing rainfall and diversity (cf. Gentry 1988). Again, this remarkably high diversity actually may be the norm for a great distance from Manaus – published plot-diversity data are lacking for terra firme between Manaus and Iquitos, but two fully vouchered discontinuous 1-ha plots of várzea floodplain forest at the lower Japurá River have been inventoried (Ayres 1986). For trees ≥10 cm dbh, they held 135 and 109 species separately and 176 species in total, similar to the diversity of terra-firme plots in eastern Amazonia.

For larger plot sizes the Manaus tree flora also ranks very high in diversity. On a family by family basis, Rankin-de-Mérona *et al.* (1992) so far have identified 698 tree species ≥10 cm dbh in 53 families for 70 fully vouchered non-contiguous hectares on terra-firme clay and sand soils at the MCSE Project; mean density is 636 individuals per ha. They estimate that the 70 ha eventually will yield 1000 tree species ≥10 cm dbh. This major study is the best indication of the total tree diversity of Manaus. Much lower numbers are known from other large inventories: 320 tree and liana species ≥10 cm dbh in 10 ha spread over a large area east and south of Belém (Salomão *et al.*, in press) and 306 species (trees and shrubs ≥2 cm dbh) for 50 ha of seasonal forest on Barro Colorado Island (Foster and Hubbell 1990). The entire known BCI tree and shrub flora is 409 species, which is probably exceeded by just the tree species with dbh 10 cm or more on 10 ha near Manaus.

Species × area data for the MCSE hectares along a 40-km strip are also indicative of very high diversity. Myristicaceae are evenly distributed – reaching 90% of their 70-ha species total after tallying just 7 ha, whereas Annonaceae require more than 36 ha to attain 90% of their species total in the 70 ha. Most remarkable is the linear ascending species × area relationship for Sapotaceae in the 35 ha analyzed so far. Most of the species in five major families (Burseraceae, Annonaceae, Apocynaceae, Melastomataceae, Myristicaceae) occur at an average density of less than one tree per ha. The most abundant species are often clumped and therefore absent from many of the hectares. The data are not yet available for many families, so they cannot yet be ranked by species diversity at the MCSE sites.

Several observations may explain the exceptionally high diversity of Central Amazonian tree plots. Part of this diversity near Manaus is due to its central location – a spillover effect providing species typical of the western Amazon, the Guianas and the eastern Amazon (Prance 1978, 1990), all added to the widespread Amazonian species. A heterogeneity of substrates near Manaus also contributes β diversity and then α diversity by smaller scale overlap. For example, the Guayana Shield lies only 140 km north of the city, apparently permitting some Guayanan elements to occur at least as far as the MCSE sites (Prance 1990). Off the southern edge of the Guayana Shield are patches of open xeromorphic flora on Palaeozoic sandstone, with some new records for Brazil of typically Venezuelan species (O. Huber, pers. comm.) as well as new species. On the opposite (southern) bank of the Rio Negro are two small clay-soil savannas which contribute typical widespread savanna elements such as *Physocalymma scaberrimum*, *Curatella americana* and *Curculigo scorzoneraefolia*, whereas podzol campinas and sandy igapó have provided niches for penetration of families and genera from the sandy lowlands and highlands of the Guayana sandstone floristic province, e.g. Humiriaceae, Theaceae, Rapateaceae, *Micrandra, Eperua, Pagamea, Glycoxylon* and *Neoythece* (Kubitzki 1989).

Useful plants

Cavalcante (1976, 1979) includes 171 species in his survey of edible fruits of the Amazon; 84 occur in the extensive terra-firme primary forests or along their small streams and 34 are in a preliminary checklist of vascular plants from a reserve near Manaus (Prance 1990). Nonetheless, very few native fruits (or medicinals) are sold in local Manaus markets, for cultural reasons and because native fruit species on the poor soils north of Manaus produce little edible biomass (M. van Roosmalen, pers. comm.). Three edible palms occur in high densities near Manaus, and could be exploited to a much greater degree: *Oenocarpus* (*Jessenia*) *bataua* (fruit, oil), *Oenocarpus bacaba* (fruit) and *Mauritia flexuosa* (fruit). Important species exploited by a limited number of extractivists near Manaus are *Minquartia guianensis* for durable posts; *Mezilaurus itauba* for boat construction; other timber species for local use; and hardwoods in general for fuelwood consumed by city bakeries.

Within 150 km of Manaus the only major extractive species exploited are *Aniba rosaeodora* and *A. duckei* for an essential oil. In total from the general region, rosewood oil valued at US$1.4 million FOB (free on board) was exported in 1984 through Manaus, according to the Association of Exporters of the Manaus Free Zone.

Economic assessment

Central Amazonia has no major concentrations of prime export hardwoods. Várzea-forest peeler-log species support three large plywood factories for export, but they are depleted near Manaus. The logs are extracted by floating them out of flood forest, causing little ecological damage. The State of Amazonas produces c. 20% of the Brazilian Amazon's annual 285,000 m³ of laminates and plywoods (Silva *et al.* 1991). These faster growing (softwood) species could be planted on floodplains, but presently there is no significant planting of native timber or plywood species anywhere in Amazonia.

In contrast with Acre, Iquitos and Belém, minor forest products are of little economic importance in Central Amazonia (Vantomme 1990). Nevertheless, extractive products and flooded-forest fish consumed locally and generally not tallied in statistics are important for the survival of hundreds of thousands of people. Floodplain forests are necessary for survival of the fruit-eating "tambaqui" (*Colossoma macropomum*), which leads the list in tonnage for food-fish species sold in Manaus markets. About 90% of the fish sold are taken from várzea lakes, forests and rivers (Bittencourt and Cox-Fernandes 1990); many fishes depend on food chains in the flooded forest for part of the year (Goulding 1980).

Green tourism attracts many foreign visitors, who leave a large share of the annual US$200 million (1989) of tourism receipts in Manaus. Jungle lodges and riverboat tours are growing in number to attend this market, though many are

TABLE 56. SELECTED CONSERVATION AREAS IN GENERAL MANAUS REGION
Numbers in bold refer to localities on Map 41

	Administrative unit	Area (km²)	Authority		Administrative unit	Area (km²)	Authority
1	Jaú National Park	22,720	IBAMA	8	Walter Egler Forest Reserve	8	INPA
2	Waimiri-Atroari Indigenous Reserve			9	Floresta Viva Native Animal		
3	Utumã Biological Reserve	5600	IBAMA		Rehabilitation Centre	12	Private
4	Anavilhanas Ecological Station	3500	IBAMA	10	Adolpho Ducke Forest Reserve	100	INPA
5	Tropical Silviculture Station	248	INPA	11	Jungle Warfare Training Centre	1150	CIGS/EB
6	Campina Forest Reserve	9	INPA	12	Maroaga Caverns APA		
7	CECAN (Centre for Breeding Native Animals)	141	IBAMA		(Environmental Protection Zone)	2562	IMA/AM

MAP 41. MANAUS REGION, BRAZIL (CPD SITE SA5), SHOWING SELECTED CONSERVATION AREAS
Numbers refer to localities mentioned in Table 56

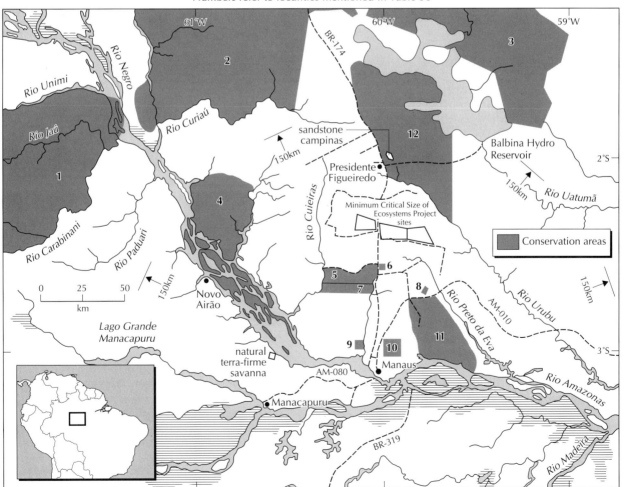

poorly organized and marketed. Although receipts in hard currency are grossly under-reported, thousands of people are directly employed in hotels, lodges, restaurants and transportation.

Threats

Daly and Prance (1989) provide a basic overview of the conversion of Amazonia, including construction of major roads and plans for hydroelectric, agricultural and agro-forestry development. Lovejoy (1985) gives an extensive review, and Hecht (1981) gives a detailed analysis especially of natural-resource uses and their effects on the Amazonian soils.

Road building stimulates subsistence colonization, hunting and invasion of reserves. Fragile dry sandstone flora and unique clear-water stream ecosystems on sandstone are threatened 130 km north of Manaus in Presidente Figueiredo by deforestation, erosion and sedimentation. Large-scale deforestation for subsidized pastures had been generally diminishing, but it is being renewed north of Manaus for establishment of oil-palm plantations. Várzea forests have had their economic species heavily exploited.

Urban expansion/invasion occurred into one INPA reserve near Manaus. Campina is the most threatened type of vegetation in the Manaus vicinity. Campinas have been used as a source of sand for road building and construction projects throughout Amazonia, resulting in complete loss of many areas. Many campinas have been burned, resulting in a considerable loss of habitat as well as the extirpation or extinction of many endemic species, and their replacement by secondary growth "capoeira". The degree, intensity and frequency of widespread burning are factors affecting regeneration potential. Recuperation may take many centuries. The long-term effects of these factors can be seen at a number of campina sites where radiocarbon-dated pottery shard remains and charcoal indicate former human occupation and cultivation practices 800–1100 years ago (Anderson 1981).

Conservation

The economy of the Manaus Free Zone has a conservation effect: it permits cheaper transport of foodstuffs from distant southern Brazil, and diverts capital from deforestation and agriculture. Brazilian institutions (e.g. INPA) with international collaboration are undertaking major ecological studies to understand the dynamics of the forests with a view to their long-term use (Schubart and Walker 1987). Just 1.48% of the State of Amazonas had been deforested as of August 1991 (INPE 1992), so most conservation areas are simply lines drawn through a continuous forest cover.

Four governmental research stations are established in the Manaus region and larger areas are being considered. Over 320 km² of INPA reserves include terra firme, campinarana and campina, and a 1150-km² military base with predominantly terra-firme forest is well protected (Table 56). The Anavilhanas Ecological Station has large protected tracts of terra firme and igapó. No várzea ecosystems near Manaus are protected, but westward on the Lower Japurá River, Mamirauá Ecological Station is on várzea.

References

Absy, M.L., van der Hammen, T., Soubies, F., Suguio, K., Martin, L., Fournier, M. and Turcq, B. (1989). Data on the history of vegetation and climate in Carajás, eastern Amazonia. In *International Symposium on Global Changes in South America During the Quaternary, São Paulo, 1989.* ABEQUA/INQUA, Special Publication No. 1, São Paulo. Pp. 129–131.

Alencar, J.C. (1986). *Análise de associação e estrutura de uma comunidade de floresta tropical úmida onde ocorre* Aniba rosaeodora *Ducke (Leguminosae).* Ph.D. thesis. INPA/Universidade do Amazonas, Manaus. 332 pp.

Anderson, A.B. (1981). White-sand vegetation of Brazilian Amazonia. *Biotropica* 13: 199–210.

Anderson, A.B., Prance, G.T. and Albuquerque, B.W.P. de (1975). Estudos sobre a vegetação das campinas Amazonicas III. A vegetação lenhosa da campina da Reserva Biológica INPA-SUFRAMA. *Acta Amazonica* 5: 225–246.

Ayres, J.M. (1986). *Uakaris and Amazonian flooded forest.* Ph.D. dissertation. Cambridge University, Sidney Sussex College, Cambridge, U.K. 337 pp.

Bittencourt, M.M. and Cox-Fernandes, C. (1990). Pesca comercial na Amazônia Central: uma atividade sustentada por peixes migradores. *Ciência Hoje* 11(64): 20–24.

Brickmann, W.L.F. (1989). System propulsion of an Amazonian lowland forest: an outline. *GeoJournal* 19: 369–380.

Cavalcante, P.B. (1976). *Frutos comestíveis da Amazônia,* Vols. 1 and 2. INPA, Manaus. 166 pp.

Cavalcante, P.B. (1979). *Frutos comestíveis da Amazônia,* Vol. 3. *Public. Avulsas* No. 33, *Museu Paraense Emílio Goeldi,* Belém. 61 pp.

Daly, D.C. and Prance, G.T. (1989). Brazilian Amazon. In Campbell, D.G. and Hammond, H.D. (eds), *Floristic inventory of tropical countries: the status of plant systematics, collections, and vegetation, plus recommendations for the future.* New York Botanical Garden, Bronx. Pp. 401–426.

Fittkau, E.J. and Klinge, H. (1973). On biomass and trophic structure of the Central Amazonia rain forest ecosystem. *Biotropica* 5: 2–14.

Foster, R.B. and Hubbell, S.P. (1990). The floristic composition of the Barro Colorado Island forest. In Gentry, A.H. (ed.), *Four neotropical rainforests.* Yale University Press, New Haven. Pp. 85–111.

Gentry, A.H. (1988). Changes in plant community diversity and floristic composition on geographical and environmental gradients. *Ann. Missouri Bot. Gard.* 75: 1–34.

Gentry, A.H. (1990). Floristic similarities and differences between southern Central America and Upper and Central Amazonia. In Gentry, A.H. (ed.), *Four neotropical rainforests*. Yale University Press, New Haven. Pp. 141–157.

Goulding, M. (1980). *The fishes and the forest: explorations in Amazonian natural history*. University of California Press, Berkeley. 280 pp.

Goulding, M. (1985). Forest fishes of the Amazon. In Prance, G.T. and Lovejoy, T.E. (eds), *Amazonia*. Key Environments Series. Pergamon Press, Oxford, U.K. Pp. 267–276.

Goulding, M. (1988). Ecology and management of migratory food fishes of the Amazon Basin. In Almeda, F. and Pringle, C.M. (eds), *Tropical rainforests: diversity and conservation*. California Academy of Sciences and American Association for Advancement of Science Pacific Division, San Francisco. Pp. 71–85.

Goulding, M., Carvalho, M.L. and Ferreira, E.G. (1988). *Rio Negro: rich life in poor water*. SPB Academic Publishing, The Hague, The Netherlands. 200 pp.

Guillaumet, J.L. (1987). Some structural and floristic aspects of the forest. *Experientia* 43: 241–250.

Hecht, S.B. (1981). Deforestation in the Amazon Basin: magnitude, dynamics and soil resource effects. *Studies in Third World Societies* 13: 61–108.

INPE (Instituto Nacional de Pesquisas Espaciais) (1992). *Deforestation in Brazilian Amazonia*. São José dos Campos, SP. 2 pp. Mimeographed.

Irmler, U. (1977). Inundation-forest types in the vicinity of Manaus. In Schmithüsen, J. (ed.), *Biogeographica* Vol. 8, *Ecosystem research in South America*. W. Junk Publishers, The Hague. Pp. 17–29.

Junk, W.J. (1983). Ecology of swamps on the Middle Amazon. In Gore, A.J.P. (ed.), *Mires: swamp, bog, fen and moor*. Ecosystems of the World 4B. Elsevier, Amsterdam. Pp. 269–294.

Junk, W.J. (1984). Ecology of the *várzea*, floodplain of Amazonian white-water rivers. In Sioli, H. (ed.), *The Amazon, limnology and landscape ecology of a mighty tropical river and its basin*. Monogr. Biol. Vol. 56. W. Junk Publishers, Dordrecht, The Netherlands. Pp. 215–243.

Junk, W.J. (1989). Flood tolerance and tree distribution in central Amazonian floodplains. In Holm-Nielsen, L.B., Nielsen, I.C. and Balslev, H. (eds), *Tropical forests: botanical dynamics, speciation and diversity*. Academic Press, London, U.K. Pp. 47–64.

Keel, S.H. and Prance, G.T. (1979). Studies of the vegetation of a white-sand black-water igapó (Rio Negro, Brazil). *Acta Amazonica* 9: 645–655.

Kronberg, B.I., Benchimol, R.E. and Bird, M.I. (1991). Geochemistry of Acre subbasin sediments: window on ice-age Amazonia. *Interciencia* 16: 138–141.

Kubitzki, K. (1989). Amazon lowland and Guayana highland. Historical and ecological aspects of their floristic development. *Rev. Acad. Colomb. Cienc. Exactas Fís. Nat.* 17: 271–276.

Leopoldo, P.R., Franken, W., Salati, E. and Ribeiro, M.N. (1987). Towards a water balance in the Central Amazonia region. *Experientia* 43: 222–233.

Lovejoy, T.E. (1985). Amazonia, people and today. In Prance, G.T. and Lovejoy, T.E. (eds), *Amazonia*. Pergamon Press, Oxford, U.K. Pp. 328–338.

Macedo, M. and Prance, G.T. (1978). Notes on the vegetation of Amazonia II. The dispersal of plants in Amazonian white-sand campinas: the campinas as functional islands. *Brittonia* 30: 203–215.

Milliken, W., Miller, R.P., Pollard, S.R. and Wandelli, E.V. (1992). *Ethnobotany of the Waimiri Atroari Indians of Brazil*. Royal Botanic Gardens, Kew, Richmond, U.K. 146 pp.

Nelson, B.W., Ferreira, C.A.C., Freitas, M.F. da and Kawasaki, M.L. (1990). Endemism centres, refugia and botanical collection density in Brazilian Amazonia. *Nature* 345: 714–716.

Pires, J.M. and Prance, G.T. (1985). The vegetation types of the Brazilian Amazon. In Prance, G.T. and Lovejoy, T.E. (eds), *Amazonia*. Pergamon Press, New York. Pp. 109–145.

Prance, G.T. (1978). The origin and evolution of the Amazon flora. *Interciencia* 3: 207–222.

Prance, G.T. (1979). Notes on the vegetation of Amazonia III. The terminology of Amazon forest types subject to inundation. *Brittonia* 31: 26–38.

Prance, G.T. (1987a). Biogeography of neotropical plants. In Whitmore, T.C. and Prance, G.T. (eds), *Biogeography and Quaternary history in tropical America*. Clarendon Press, Oxford, U.K. Pp. 46–65.

Prance, G.T. (1987b). Vegetation. In Whitmore, T.C. and Prance, G.T. (eds), *Biogeography and Quaternary history in tropical America*. Clarendon Press, Oxford, U.K. Pp. 28–45.

Prance, G.T. (1989). American tropical forests. In Lieth, H. and Werger, M.J.A. (eds), *Tropical rain forest ecosystems: biogeographical and ecological studies*. Ecosystems of the World 14B. Elsevier, New York. Pp. 99–132.

Prance, G.T. (1990). The floristic composition of the forests of Central Amazonian Brazil. In Gentry, A.H. (ed.), *Four neotropical rainforests*. Yale University Press, New Haven. Pp. 112–140.

Prance, G.T., Rodrigues, W.A. and Silva, M.F. da (1976). Inventário florestal de um hectare de mata de terra firma, Km 30 da Estrada Manaus-Itacoatiara. *Acta Amazonica* 6: 9–35.

Rankin-de-Mérona, J.M., Prance, G.T., Hutchings, R.W., Silva, M.F. da, Rodrigues, W.A. and Uehling, M.E. (1992). Preliminary results of a large-scale tree inventory of upland rain forest in the Central Amazon. *Acta Amazonica* 22: 493–534.

Rodrigues, W.A. (1961). Estudo preliminar de mata várzea alta de uma ilha do baixo Rio Negro de solo argiloso e úmido. *Publ. Bot. INPA* 10: 1–50.

Rodrigues, W.A. (1967). Inventário florestal pilôto ao longo da Estrada Manaus-Itacoatiara, estado do Amazonas: dados preliminares. In *Atas do Simp. Sobre a Biota Amazonica*, Vol. 7 *(Conservação da natureza e recursos naturais)*. Centro Nacional de Desenvolvimento Científico e Político (CNPq), Rio de Janeiro. Pp. 257–267.

Salomão, R.P. and collaborators (in press). *Bol. Mus. Paraense Emilio Goeldi, Série Botânica.*

Schubart, H.O.R. and Walker, I. (1987). The dynamics of the Amazonian terra-firme forest. *Experientia* 43: 221–222.

Silva, A.S.L. and ten collaborators (1988). *Estudo e preservação de recursos humanos e naturais da area do Projeto "Ferro Carajás".* Final report to CVRD, Subproject "Inventário Botânico".

Silva, D.A. da, Frazão, F.J., Rocha, J.S., Matos, J.L.M., Trugilho, P.F. and Iwakiri, S. (1991). A indústria de base florestal na Amazônia. In Val, A.L., Figlioulo, R. and Feldberg, E. (eds), *Bases científicas para estratégias de preservação e desenvolvimento da Amazônia: fatos e perspectivas.* INPA, Manaus. Pp. 239–249.

Vantomme, P. (1990). Forest extractivism in the Amazon: is it a sustainable and economically viable activity? In *Abstracts, Forest 90. First International Symposium on Environmental Studies on Tropical Rain Forests. Development Strategies for Amazonia, Manaus, October 1990.* Pp. 39–40.

Veiga, A.T.C. (1991). Paleoenvironmental and archeological significance of alluvial placers of the Brazilian Amazon. In *(Proceedings of the Symposium on Global Changes in South America During the Quaternary).* Boletim IG-USP, Publicação Especial No. 8. São Paulo. Pp. 213–222.

Whitmore, T.C. and Prance, G.T. (eds) (1987). *Biogeography and Quaternary history in tropical America.* Oxford Monogr. Biogeography No. 3. Clarendon Press, Oxford, U.K. 214 pp.

Authors

This Data Sheet was written by Dr Bruce W. Nelson [Instituto Nacional de Pesquisas da Amazônia (INPA), Caixa Postal 478, 69000 Manaus, AM, Brazil] and Olga Herrera-MacBryde (Smithsonian Institution, Department of Botany, NHB-166, Washington, DC 20560, U.S.A.).

AMAZONIA: CPD SITE SA6

UPPER RIO NEGRO REGION
Brazil, Colombia, Venezuela

Location: Western Colombia, south-western Venezuela and north-western Brazil, between latitudes 4°N–2°S and longitudes 70°–66°W.

Area: Over 250,000 km².

Altitude: Mostly from less than 100 m to 500 m, reaching 1000 m.

Vegetation: Amazon caatinga forest ("campinarana"), Amazon caatinga shrubland ("campina"), black-water flood forest ("igapó"), submontane rain forest.

Flora: Extremely high diversity (over 15,000 species); high generic and specific endemism.

Useful plants: Many species – e.g. *Hevea* rubber, Brazil nut, rosewood oil, *Caryocar*, several palms. Possible origin of some economically important families and genera – e.g. Sapotaceae, Lecythidaceae, Caryocaraceae, Vochysiaceae, *Hevea*, *Caryocar*.

Other values: Watershed protection; germplasm reserves; Amerindian lands, traditions and folklore; wilderness; ecotourism.

Threats: Gold mining; "coca" (*Erythroxylum*) plantations; selective logging.

Conservation: 5 National Parks, 2 National Nature Reserves, a Forest Reserve and several Amerindian Reserves established; additional conservation units needed to protect adequate habitat diversity.

Geography

Geomorphologically, the Upper Rio Negro region correlates fairly well with the Rio Branco-Rio Negro Depression, which is the western portion of the Northern Amazonian Depression (Bezerra *et al.* 1990). To the region's south is the Western Amazonian Depression, beginning near where Colombia's Caquetá River joins the Japurá River of Brazil. Generally, this region is a broad plain between the somewhat higher western Colombian plateaux (and mesas) and the Guayana Highlands farther east in Venezuela and Brazil. The north-eastern limit in Venezuela is roughly from Yapacana National Park and the Orinoco River to its junction with the Casiquiare Canal. Excluding the highlands of the Pico da Neblina, which are in the Pantepui region (CPD Site SA2) (Huber 1987), altitudes vary generally from less than 100 m to 500 m above sea-level.

The regional geology is an extension of the Pantepui, consisting of a highly eroded early Precambrian (Archean) igneous basement of the Guayana complex, overlain by more recent Precambrian marine sandstones belonging to the Roraima Formation (Bezerra *et al.* 1990). The soils are very acidic, nutrient poor and periodically waterlogged quartzic podzols, or dystrophic red-yellow laterites (SNLCS 1981; Jordan 1987).

Rivers are important features of the region. The Guaviare and Inírida flow to the Orinoco, whereas the Guainía, Negro, Vaupés/Uaupés and Caquetá-Japurá flow to the Amazon. Three types of rivers occur (Sioli 1984): predominating are tea-coloured or black-water rivers, which drain white-sand vegetation – the most important is the Rio Negro (Goulding, Carvalho and Ferreira 1988); clear-water rivers, draining mainly quartzitic highlands; and white-water eutrophic rivers (e.g. the Guaviare), draining sediment-rich areas.

The climate is humid tropical (Köppen's Afi) or superhumid tropical (Thornthwaite's ArA′a′), with a mean annual rainfall of 2500–3500 mm, 180–240 days of rain per year and no month with less than 100 mm of rain. The temperature is isothermic, with yearly means of 24°–25°C, and the mean relative humidity is 80–90% throughout the year (SUDAM 1984).

The Upper Rio Negro region corresponds to the core area of tropical extremely moist forest as defined by UNESCO (1981), which extends approximately from 4°N to 2°50′S and 63° to 74°10′W.

Vegetation

Throughout the Amazon Basin, the vegetation is strongly correlated with the geomorphology, soils and climate (Salgado and Brazão 1990). Although the Upper Rio Negro is noted for its mosaic of unique Amazon caatinga forest and Amazon caatinga shrubland on sandy soils (Anderson 1981), most of this low-lying area is an ecological transition between extensive submontane rain forest (tropical moist forest) and black-water flood forest (Daly and Prance 1989). Amazon caatinga forest and shrubland are predominant on the hydromorphic podzols, submontane rain forest is most commonly found on the red-yellow laterites and black-water flood forest is associated with dystrophic alluvial gley soils.

1. **Amazon caatinga forest ("campinarana")** is a type of caatinga formation unique to Amazonia. Fairly frequent north of the Amazon River, these forests extend to north-eastern Peru and also have been found as far south as the Serra do Cachimbo (6°–8°S, 57°–58°W) (Lleras and

Kirkbride 1978). Their core area, in the Upper Rio Negro region, is characterized by a tough-leaved, arborescent savanna vegetation adapted to very poor sandy soils that are periodically flooded (e.g. Klinge, Medina and Herrera 1977). Vegetation height correlates with the duration of annual flooding – shrubs and trees of c. 5 m characterize wetter sites, whereas trees up to 20 m tall occur on drier sites. The canopy is fairly open, with sufficient light reaching the ground so that a profusion of terrestrial "epiphytes" occurs. The topsoil is humic, with tufty patches formed by *Cladonia* lichens, *Trichomanes* filmy ferns, etc.

2. The Amazon caatinga forest phases into **Amazon caatinga shrubland** ("campina"), which is less high, more open and drier. The campinas are characterized by islands of vegetation less than 1 m² to several hundred m² surrounded by sandy open areas. The larger islands have one to several large trees. Species composition is similar throughout these shrublands.

3. **Black-water flood forest** ("igapó") borders all the watercourses and merges with the Amazon caatinga forest and shrubland. Based on data from around Manaus (CPD Site SA5), the black-water flood forests are very similar in species composition to these caatinga formations. The Amazon caatinga forests and shrublands may be relictual from black-water flood forests, having resulted from the change of river courses. Lleras and Kirkbride (1978) have found caatinga forest and shrubland atop bedrock as well, contributing to the soil development of the Serra do Cachimbo.

4. **Evergreen submontane rain forest** occurs dispersed through uplands ("terra firme"), to 1000 m above sea-level. The canopy rarely exceeds 30 m in height, with a few emergents such as *Manilkara huberi, Caryocar villosum* and *Hymenaea parviflora*. Other frequent species include *Carapa guianensis, Sacoglottis guianensis, Pouteria surinamensis, Ocotea roraimae* and several Vochysiaceae in *Qualea, Vochysia* and *Erisma*. Several genera of palms are associated with this type of forest – the *Jessenia/ Oenocarpus* complex comprises some of the largest populations. Occurring also are *Bertholletia excelsa* and several species of *Hevea*.

Flora

Although a precise estimate on the size of the flora of the Upper Rio Negro region is presently impossible, 50–70% of the species in the Amazon Basin probably are represented. Therefore, using G.T. Prance's estimate (pers. comm.) of 30,000 vascular plant species in the Amazon Basin, the Upper Rio Negro and adjacent superhumid forests may have 15,000 to 21,000 species – which is probably up to ten times more than occur in the highly diverse area just north of Manaus. (The latter area coincides with the southernmost intrusion of the Guayana Shield and has the same basic geomorphology, geology and age as the Upper Rio Negro.)

Floristic collections are inadequate to compile a Flora for the Upper Rio Negro region; in comparison, the collecting

intensity for the area around Manaus has been 10 to 20 times greater (Nelson *et al.* 1990). Nonetheless, in studying plant diversity throughout the Amazon Basin on the basis of existing collections, Lleras *et al.* (1992) recorded less than 5% difference between these two areas in the number of known species.

It is generally believed that most edaphic endemics in the region are associated with the white-sand vegetation. However, most research on this vegetation has shown a low species diversity when compared with the forests on richer soils (Pires 1957; Lleras and Kirkbride 1978). Moreover, many of the important taxa that originated in the region are components of the humid rain forests (Gentry 1982; Lleras *et al.* 1992). The importance of the Upper Rio Negro for plant diversity is not in the many endemics occurring on the white-sand vegetation, but that the region of which the Upper Rio Negro is the core, is the repository of the very old forest elements of the vegetation of the Guayana Shield. These high forests are much more diverse and with many more endemics than are in the white-sand vegetation, and they may be the centres of origin of a great part of the neotropical and palaeotropical lowland flora.

Lleras *et al.* (1992) proposed the Upper Rio Negro *sensu lato* (including most of the extremely moist forests of UNESCO 1981) as the centre of origin for many of its tropical families, including Caryocaraceae, Connaraceae, Sapotaceae, Meliaceae, Lecythidaceae, Dichapetalaceae and the tribe Henriquezieae of Rubiaceae. Gentry (1982) surmised that in Guayanan Pleistocene forest refugia the majority of families endemic to the neotropics probably originated, including as well the mainly neotropical Humiriaceae, Vochysiaceae and Bignoniaceae.

However, presumed refugia can be collection artefacts (Nelson *et al.* 1990). These extremely moist forests probably survived the climatic changes of the Pleistocene almost intact, which together with the great antiquity of the region as a whole, would argue further for excluding the concept of very recent Pleistocene forest refugia as the source of the biodiversity in the humid tropics. It is probable that the Upper Rio Negro and similar areas, and not such refugia, are the sites for the origin, evolution and long-term preservation of many neotropical taxa.

Henderson, Churchill and Luteyn (1991) probably are correct in proposing the Andes as the most diverse region of South America and probably of the world – the very many habitats and niches induced by the complex mosaic of soils, climate, geology, geomorphology and topography guarantee this. However, the Upper Rio Negro region, and indeed all of this extremely moist forest (UNESCO 1981) of which other important centres of diversity are covered in this volume, probably hold the highest floristic diversity on the planet for arborescent species.

This richness is due to three basic factors: (1) the region lies atop the continent's oldest geological formation, which is probably one of the world's oldest; (2) the region is more diverse in terms of biotic and abiotic factors than the rest of the Amazon Basin; and (3) it corresponds to an extensive transition zone where stress tends to cause diversification instead of intraspecific variation (Lleras, unpublished). Furthermore, Gentry (1986) has proposed a correlation between high rainfall and high diversity, and the Upper Rio Negro and its neighbouring regions are certainly the wettest portions of the Amazon; more significant than the high total rainfall is the lack of water stress throughout the year.

Useful plants

As one of the centres of origin of the neotropical lowland flora, the Upper Rio Negro region has many presently exploited as well as potentially useful species. A realistic assessment of economically important species is impossible due to the sparsity of data. However, important taxa found in the region include several species of *Hevea* (rubber), *Caryocar, Jessenia/Oenocarpus, Bertholletia excelsa* (Brazil nut) and *Aniba rosaeodora* (rosewood oil). *Paullinia cupana* var. *cupana*, a locally used wild form of "guaraná" with great potential for breeding purposes, is also found. Outside of the Upper Rio Negro core, the extremely moist forest is the centre of origin for *Theobroma* and *Pourouma*.

MAP 42. UPPER RIO NEGRO REGION *SENSU LATO* IN COLOMBIA, VENEZUELA AND BRAZIL (CPD SITE SA6), WITH EXISTING AND RECOMMENDED CONSERVATION UNITS

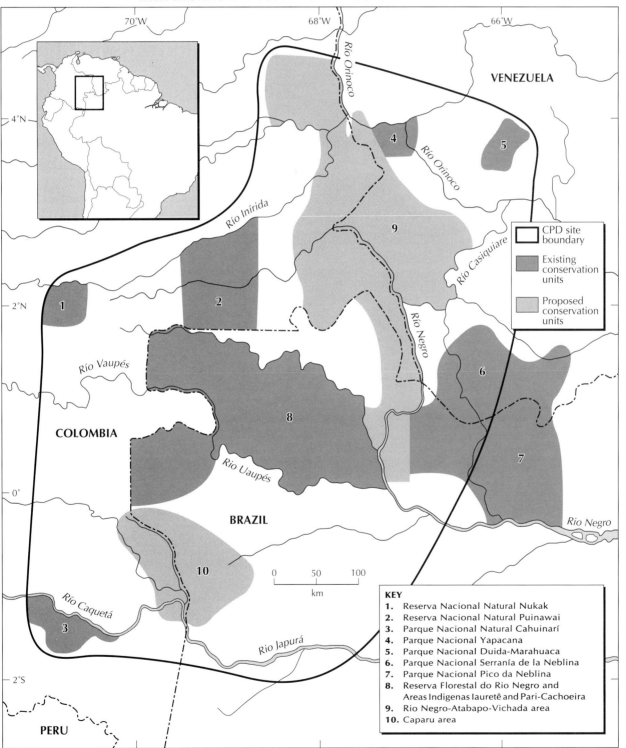

KEY
1. Reserva Nacional Natural Nukak
2. Reserva Nacional Natural Puinawai
3. Parque Nacional Natural Cahuinarí
4. Parque Nacional Yapacana
5. Parque Nacional Duida-Marahuaca
6. Parque Nacional Serranía de la Neblina
7. Parque Nacional Pico da Neblina
8. Reserva Florestal do Rio Negro and Areas Indigenas Iauretê and Pari-Cachoeira
9. Rio Negro-Atabapo-Vichada area
10. Caparu area

Social and environmental values

The Upper Rio Negro region straddles both the Orinoco and Negro river basins; 40% of the water in the Amazon River comes from the Rio Negro. Watershed conservation of this vital region should be a high priority.

The forests of the Upper Rio Negro regulate the hydrology of the Rio Negro Basin. Drastic changes in soil coverage in the Upper Rio Negro would affect much of the Amazon Basin. Variations in climate and rainfall in the Upper Rio Negro may cause flooding, drought or other climatic changes as far south as Manaus.

Many Amerindian communities in all three countries of this region have traditions and capabilities to manage the forest sustainably. These communities are presently the sole possessors of the knowledge and folklore that might permit humankind to exploit the floristic and faunistic resources of the region without destroying them (e.g. see Schultes and Raffauf 1990).

There have been few studies on the avifauna of the Upper Rio Negro and Orinoco white-sand forest (Endemic Bird Area B11), which supports 13 restricted-range species (two of them also occur in other areas). These birds are primarily confined to the humid tropical forest along the rivers, although some inhabit the adjacent Amazon caatinga forest and shrubland. Due to the relatively intact state of the vegetation in this region only the rare Orinoco softtail (*Thripophaga cherriei*) is considered threatened – but essentially because of a lack of knowledge rather than a perceived threat.

The unique mix of topography, waterways and vegetation make this region a valuable tourist asset, which could be exploited in conjunction with the Pantepui region.

Economic assessment

The geology of the region is not well known. Iron ore and manganese are believed to be present, but it is not known if they are commercially exploitable (Bezerra *et al.* 1990). Gold is found in localized pockets along the borders of Brazil, Colombia and Venezuela.

The soils are poor, so extensive agriculture is not appropriate (cf. Saldarriaga 1994). A preliminary map of soil usage (SNLCS 1990) recommended that areas of podzols with Amazon caatinga forest and shrubland should be set aside for conservation, whereas the areas of lateritic soils with high forest should be used for extractive activities and light agriculture with perennials.

The greatest value of the Upper Rio Negro region may be its wealth of species, including its role as a repository for many useful species, along with the indigenous knowledge of them. With world recognition of the importance of maintaining biodiversity, the Upper Rio Negro may become a major economic asset for Brazil, Colombia and Venezuela.

Threats

Presently, the Upper Rio Negro is not threatened, because access from all three countries is difficult. A recent survey of deforestation in Amazonia based on trace-gas emissions (Skole *et al.* 1992) showed that most of the Upper Rio Negro area was untouched – with deforestation between 1978 and 1988 having occurred only very locally along the Colombia-Brazil border, where an extensive Amerindian reserve coincides with gold mining (0°–1°N, 69°–70°W).

Threats may come from an increase in illegal gold mining, the cutting down of forests to establish plantations of "coca" (*Erythroxylum coca* var. *epadu*), and selective logging. Furthermore, it is very difficult to convince governments or the general public that it is important to preserve areas that we know so little about.

Conservation

The Upper Rio Negro region has one of the highest percentages of protected areas in the Amazon Basin north of the Equator (Map 42). These constitute the National Parks Serranía de la Neblina (13,600 km²) in Venezuela and in adjacent Brazil Pico da Neblina (22,000 km²), as well as the Rio Negro Forest Reserve (37,900 km²) and two extensive Amerindian Reserves. In Colombia two National Nature Reserves have been established, Puinawai (10,925 km²) and Nukak (8550 km²), and many Amerindian communities have strong participation in the administration of the region as a whole. In Venezuela there are two more National Parks, Yapacana (3200 km²) and Duida-Marahuaca (2100 km²).

Biologists meeting in Manaus in 1990 recommended a high priority for conservation of the large area from the border between Colombia and Venezuela (Rio Negro-Atabapo-Vichada, c. 4°N) extending south and joining with the existing conservation units in Brazil (CI 1991). At least one more conservation unit (Caparu) should be established south of the Equator, extending roughly from 0°20'N to 1°30'S and 69° to 71°30'W, following the Apaporis River to the Japurá Rive (CI 1991). Adjacent National Parks in Colombia and Brazil have been proposed for this area, which is fairly near Colombia's Cahuinarí Natural National Park (see CPD Site SA7).

The Upper Rio Negro region is part of a large area with high diversity, and meets with another – the Pantepui region (see CPD Site SA2); clear boundaries are not possible and many reserves coincide.

The most important priorities for conservation at present are extensive surveys and scientific collections of the whole area (Lleras *et al.* 1992). It is highly likely that less than 5% of the total flora is known, and it is extremely important to assess what is being conserved.

References

Anderson, A.B. (1981). White-sand vegetation of Brazilian Amazonia. *Biotropica* 13: 199–210.

Bezerra, P.E.L., da Cunha, B.C.C., Del'Arco, J.O., Drago, V.A. and de Montalvão, R.M.G. (1990). Geologia. In *Projeto Zoneamento das Potencialidades dos Recursos Naturais da Amazônia Legal*. IBGE/SUDAM (Instituto Brasileiro de Geografia e Estatística/Superintendência para o Desenvolvimento da Amazônia), Rio de Janeiro. Pp. 91–164.

CI (1991). Workshop 90. Biological priorities for conservation in Amazonia. Conservation International (CI), Washington, D.C. Map.

Daly, D.C. and Prance, G.T. (1989). Brazilian Amazon. In Campbell, D.G. and Hammond, H.D. (eds), *Floristic inventory of tropical countries: the status of plant systematics, collections, and vegetation, plus recommendations for the future.* New York Botanical Garden, Bronx. Pp. 401–426.

Gentry, A.H. (1982). Phytogeographic patterns as evidence for a Chocó refuge. In Prance, G.T. (ed.), *Biological diversification in the tropics.* Columbia University Press, New York. Pp. 112–136.

Gentry, A.H. (1986). An overview of neotropical phytogeographic patterns with an emphasis on Amazonia. In Dantas, M. (ed.), *Proceedings 1st Symposium on the Humid Tropics,* Vol. II, *Flora and forest.* EMBRAPA/CPATU (Empresa Brasileira de Pesquisa Agropecuária/Centro de Pesquisa Agropecuária do Trópico Umido), Documentos 36, Belém. Pp. 19–36.

Goulding, M., Carvalho, M.L. and Ferreira, E.G. (1988). *Rio Negro: rich life in poor water.* SPB Academic Publishing, The Hague, The Netherlands. 200 pp.

Henderson, A., Churchill, S.P. and Luteyn, J.L. (1991). Neotropical plant diversity. *Nature* 351(2 May): 21–22.

Huber, O. (1987). Consideraciones sobre el concepto de Pantepui. *Pantepui* 2: 2–10.

Jordan, C.F. (1987). Soils of the Amazon rainforest. In Whitmore, T.C. and Prance, G.T. (eds), *Biogeography and Quaternary history in tropical Latin America.* Oxford University Press, Oxford, U.K. Pp. 83–94.

Klinge, H., Medina, E. and Herrera, R. (1977). Studies on the ecology of Amazon Caatinga forest in southern Venezuela. I. General features. *Acta Ci. Venez.* 28: 270–276.

Lleras, E. and Kirkbride Jr., J.H. (1978). Alguns aspectos da vegetação da Serra do Cachimbo. *Acta Amazonica* 8: 51–65.

Lleras, E., Leite, A.M.C., Scariot, A.S. and de Sá Brandão, J.E. (1992). *Definição de áreas de alta diversidade vegetal e endemismos na Amazônia Brasileira.* Final Report to UN Food and Agriculture Organization (FAO), Brasília. 65 pp.

Nelson, B.W., Ferreira, C.A.C., da Silva, M.F. and Kawasaki, M.L. (1990). Endemism centres, refugia and botanical collection density in Brazilian Amazonia. *Nature* 345(6277): 714–716.

Pires, J.M. (1957). Noções sobre ecologia e fitogeografia da Amazônia. *Norte Agronômico* 3(3): 37–54.

Saldarriaga, J.G. (1994). *Recovery of the jungle on "Tierra Firme" in the upper Rio Negro region of Amazonia in Colombia and Venezuela.* Estudios en la Amazonia Colombiana, Vol. 5. Tropenbos-Colombia, Bogotá. 201 pp.

Salgado, L.M. and Brazão, J.E.M. (1990). Vegetação. In *Projeto Zoneamento das Potencialidades dos Recursos Naturais da Amazônia Legal.* IBGE/SUDAM, Rio de Janeiro. Pp. 189–211.

Schultes, R.E. and Raffauf, R.F. (1990). *The healing forest: medicinal and toxic plants of the Northwest Amazonia.* Dioscorides Press, Portland, Oregon, U.S.A. 484 pp.

Sioli, H. (1984). The Amazon and its main affluents: hydrology, morphology of the river courses, and river types. In Sioli, H. (ed.), *The Amazon: limnology and landscape ecology of a mighty tropical river and its basin.* Monogr. Biol. 56. Junk, Dordrecht, The Netherlands. Pp. 127–165.

Skole, D.L., Chomentowski, W.H., Nobre, A.D. and Tucker, C.J. (1992). A remote sensing and GIS methodology for estimating the trace gas emissions from tropical deforestation: a case study from Amazonia. Paper presented at the World Forest Watch Meeting, San José dos Campos, São Paulo, Brazil, 27–30 May 1992. 19 pp. Unpublished.

SNLCS (1981). *Mapa de solos do Brasil.* EMBRAPA/SNLCS (Serviço Nacional de Levantamento e Conservação de Solos). Rio de Janeiro.

SNLCS (1990). Delineamento macro-agroecológico do Brasil. *Globo Rural* No. 46, Anexo. Ed. Bloch, Rio de Janeiro.

SUDAM (1984). *Atlas climatológico da Amazônia Brasileira.* SUDAM Publ. No. 39, Belém. 125 pp.

UNESCO (1981). *Vegetation map of South America. Map and explanatory notes.* UNESCO, Natural Resources Research. UNESCO Press, Paris. 189 pp. + map.

Author

This Data Sheet was written by Dr Eduardo Lleras [Centro Nacional de Pesquisas de Recursos Genéticos e Biotecnologia (CENARGEN)/EMBRAPA, Caixa Postal 02-372, 70.849 Brasília, D.F., Brazil].

CHIRIBIQUETE-ARARACUARA-CAHUINARI REGION
Colombia

Location: In south-eastern Colombian Amazonia, between latitudes 1°40'N–1°50'S and longitudes 70°40'–73°30'W.

Area: 50,000 km².

Altitude: 150–700 m.

Vegetation: On floodplains to low uplands of Tertiary and Quaternary sediments: various moist tropical forest communities. On sandstone plateau outcrops: evergreen sclerophyllous scrub, open dwarf scrub, and graminoid and herb communities.

Flora: 12,000 vascular plants. Westernmost transition between floristic regions: many Guayanan Highlands taxa especially on sandstone outcrops, which also have many endemics; very species-rich dominantly Amazonian flora in lower areas. Many new species, and a new genus of Dipterocarpaceae, recently discovered.

Useful plants: *Hevea* spp. rubber; many forest products (fruits, palm hearts, woods, fibres, medicines, dyes) used by indigenous peoples, who also have many strains (selections) of cultivated plants developed for Amazonia.

Other values: Archaeological sites; Amerindian lands; vast wilderness; impressive scenery of table mountains and canyon; much faunal diversity.

Threats: No land-use and management plan for entire region; large expanse invites uncontrolled colonization; in Andean headwaters of Caquetá River, deforestation and industry might be contaminating water and sediments, affecting floodplain forests and fishery.

Conservation: 2 Natural National Parks; Amerindian Reserves.

Geography

The Chiribiquete-Araracuara-Cahuinarí region is in the north-western portion of the Amazon Basin. The largest of the three main rivers is the Caquetá, which originates in the Andes and is the region's only white-water river; together with the Apaporis, it drains into the Japurá River of Brazil. Farther north, the Vaupés River forms part of the catchment of Brazil's Rio Negro (CPD Site SA6).

The northern (Chiribiquete-Araracuara) portion of this region is dominated by north-south aligned outcrops of Palaeozoic sandstone formations in an extensive interrupted sandstone plateau at c. 400 m, rising locally to c. 700 m (confirmed by a recent Colombian-Spanish botanical expedition), with broad interior valleys which are partially filled with sandy colluvial sediments. The highest plateau outcrops or cerros (Cerro Azul, Cerro Quemado, Cerro Chiribiquete, Cerro Campana) appear as characteristic table mountains. Southward the overall plateau gradually descends: at Araracuara it has an elevation of c. 300 m, and beyond 1°40'S no sandstone outcrops are found.

The sandstone plateaux have a generally flat topography, dissected by numerous fissures of variable depth. Mostly along the plateau margins dissected portions occur with hilly to undulating topography. Dominant soils of the plateaux are moderately deep to shallow, greyish white and sandy in texture, poorly to somewhat excessively drained and with extremely poor nutrient chemical properties. The poorly drained soils might be called "groundwater podzols" or "wet white-sand" soils. At sites with hardrock outcrops, soil is virtually absent.

The southern and eastern (Araracuara-Cahuinarí) portion of the region consists mostly of slightly consolidated Tertiary sediments, which form a dissected sedimentary plain at c. 250 m (c. 60–80 m above the lowest riverwater levels). Most valleys are 30–60 m deep, with 20°–40° slopes. The drainage pattern is a tree-like branching of watercourses occurring in high density (3.1–4.7 km per km²). The soils of this hilly terrain are well drained, from loamy to clayey and chemically poor to very poor (Duivenvoorden and Lips 1993).

The alluvial plains of the larger rivers comprise less than 5% of the region (Duivenvoorden and Lips 1993), and have been built up by Pleistocene and Holocene deposits (van der Hammen *et al.* 1991a). They consist of flat to dissected terraces and floodplains c. 5–60 m above average riverwater levels. The soils of the Caquetá floodplain are relatively rich, whereas the soils of the floodplains of the black-water Amazonian rivers are chemically poor. Occasionally, particularly on flat sites of the higher fluvial terraces of the Caquetá River, small to extensive areas with podzolised soils are found, which are comparable to the groundwater podzols on top of the sandstone plateaux (Duivenvoorden and Lips 1993).

The tropical climate (Afi of Köppen 1936) has a mean daily temperature closely fluctuating around 26°C, to somewhat cooler at the more elevated sites in the Chiribiquete area. The annual precipitation is c. 3000 mm (the monthly mean exceeding 60 mm) and has a nearly unimodal distribution: most of the rainfall occurs in May–July, the least in December–February, which still averages well over 100 mm. The monthly average precipitation always exceeds the average potential evaporation. However, the sandy soils of the sandstone plateaux are likely to experience a water deficit almost every year for short intervals during the drier months (Duivenvoorden and Lips 1993, based on data from a meteorological station at Araracuara). In the Chiribiquete general area, rainfall might be somewhat lower due to the drier Llanos Orientales farther north, but on the high Chiribiquete cerros frequent low clouds and mist might increase the humidity and rainfall.

Vegetation

The evergreen moist tropical forests that cover most of the Tertiary sediments and Quaternary alluvial plains of the Chiribiquete area and Middle Caquetá River Basin represent one of the few remaining extensive, undisturbed rain-forest ecosystems in the Amazon Basin. Knowledge of the forests of the Middle Caquetá Basin has improved since 1986 from inventories due to the Tropenbos-Colombia Programme (Duivenvoorden and Lips 1993; Urrego 1990; Galeano 1991; Alvarez and Londoño, in prep.; and students' unpublished reports).

The forests on well-drained sites in the Middle Caquetá Basin generally are 25–30 m high; 0.1-ha samples have 65–85 trees (10 cm or more dbh) with a basal area of 3.0–3.5 m² and aerial biomass of 25–30 tonnes. On well-drained upland sites, at least two forest communities can be recognized; the more common trees are *Clathrotropis*

macrocarpa, Iryanthera tricornis, Micropholis guyanensis, Eschweilera rufifolia, Micranda spruceana, Pouteria sp. and *Swartzia schomburgkii*. At well-drained sites on the floodplains, at least six different forest communities occur; the most common species are *Cecropia membranacea, Annona hypoglauca, Iriartea deltoidea, Ficus* spp., *Astrocaryum jauari, Parkia multijuga, Oxandra mediocris, Mollia lepidota* and *Pouteria torta*. Other common trees include *Oenocarpus bataua, Euterpe precatoria, Eschweilera coriacea, Brosimum utile, Virola elongata* and *Hevea guianensis*. The understorey of the forest is particularly rich in species belonging mostly to Melastomataceae, Rubiaceae, Arecaceae, Annonaceae, etc.

The forests on poorly drained sites generally are lower, have less tree aerial biomass, higher densities of palms and (much) higher treelet densities. At least six different forest communities have been recognized; often the palm *Mauritia flexuosa* is a frequent tree. In the higher floodplain and low terraces of the Caquetá River, peat forests are found, dominated by among others *Clusia spathulifolia, Rhodognaphalopsis brevipes, Mauritiella aculeata* and *Euterpe catinga*. At the sites with podzolised soils on the high terraces of the Caquetá River, a sclerophyllous low forest and scrub vegetation is found (Amazonian caatinga according to Anderson 1981), which has a number of species in common with the wet white-sand vegetation of the sandstone plateaux.

Species diversity of the rain forests of the Middle Caquetá Basin is very high, both the α type (species richness

CPD Site SA7: Palm swamp forest with *Mauritia flexuosa* (Palmae) on alluvial sediments of the Caquetá River.
Photo: J.F. Duivenvoorden.

TABLE 57. NUMBER OF TREE SPECIES IN FORESTS ON WELL-DRAINED SOILS IN THREE MAJOR PHYSIOGRAPHIC AREAS OF THE MIDDLE CAQUETA RIVER BASIN (Duivenvoorden and Lips 1993)

Trees ≥10 cm dbh	Floodplain of Caquetá River	Floodplain of Amazon rivers	Uplands
Maximum	41	54	57
Average ± SE	28 ± 3 (n=13)	27 ± 4 (n=10)	38 ± 1 (n=39)

SE = standard error; n = number of 0.1-ha plots.

at small homogeneous sites) and the β type (overall number of species related to degree of habitat differentiation), particularly on the alluvial plains and the sandstone outcrops. Tree species richness of the forest on well-drained uplands (Table 57) is comparably high with some Peruvian rain forests (Gentry 1988, 1990).

On the sandstone outcropped plateaux, vegetation patterns are complex mosaics. Most of the more elevated plateaux in the Chiribiquete area are covered by open low tree and shrub communities with the following taxa common: *Graffenrieda fantastica, Clusia chiribiquetensis, C. schultesii, Tepuianthus savanensis, Pochota nitida, Ternstroemia campinicola, Miconia paradoxa, Hevea viridis* var. *toxicodendroides, Decagonocarpus cornutus* and *Ficus chiribiquetensis*. On the sandier soils *Vellozia phantasmagoria* forms an open low vegetation, whereas on flat sites with hardrock outcrops *Navia garcia-barrigae* is a dominant. At small sites of stagnant rainwater on shallow sandy soils, *Utricularia chiribiquetensis*, Burmanniaceae,

CPD Site SA7: *Cladonia* spp. growing on the soil of *Bonnetia martiana* (Theaceae) forest. Photo: J.F. Duivenvoorden.

Drosera sp. and *Abolboda macrostachya* form herb communities. *Utricularia neottioides* occurs alone in running water on hardrock (Fuertes and Estrada, in prep.). On the sandy colluvial deposits, a low forest of *Protium leptostachyum, Bonnetia martiana, Merania urceolata, Matayba macrolepis, Oenocarpus bataua* and *Pagamea coriacea* has been discerned.

At Araracuara, the top of the sandstone plateau is partially covered by low forest, generally with a high density of treelets (to 10 cm dbh); at least two floristic communities are found: a forest marked by *Dimorphandra cuprea, D. vernicosa* and *Ocotea esmeraldana* and a very distinct forest dominated by *Bonnetia martiana*. There is also a closed scrub vegetation 2–3 m high with *Bonnetia martiana* dominant. Common species in wet white-sand vegetation are *Ilex divaricata, Ternstroemia* spp., *Abolboda macrostachya, Macairea rufescens, Syngonanthus umbellatus, Lagenocarpus pendulus*, various species of *Clusia, Xyris* and *Utricularia*, and *Cladonia* lichens (Duivenvoorden and Cleef 1994). A considerable part of this poorly drained plateau is covered by a closed, medium-tall graminoid vegetation with *Axonopus schultesii* and *Schoenocephalium martianum* co-dominant.

On sites with very thin, colluvial sandy soils there is an open low herb community with among others *Xyris wurdackii* var. *caquetensis, X. savanensis, X. araracuare, Paspalum* sp., *Syngonanthus vaupesanus* and *Cladonia peltastica*. These graminoid and herb communities are clearly related to the wet white-sand savannas of Surinam (Heyligers 1963; van Donselaar 1965), and may represent the southernmost, lowest extension of the wet white-sand table-mountain vegetation of the Guayana floristic region. At sites on the Araracuara plateau with hardrock outcrops, an open vegetation with *Navia garcia-barrigae, Acanthella sprucei* and *Cladonia vareschii* occurs.

Flora

Much of the flora has still to be studied; the region may harbour 12,000 species of vascular plants. The region is rich not only in vascular plant species: a preliminary inventory of epiphyllous lichens at Araracuara yielded 135 species – over one-third of such lichens known in the world (Sipman 1990). The floristic research of the Tropenbos-Colombia Programme in the Middle Caquetá River area by 1992 had inventoried some 24 ha (well distributed over all the physiographic units); c. 3000 species of vascular plants in 129 families have been recorded. Among the quantitatively most important families are Melastomataceae, Rubiaceae, Annonaceae, Moraceae, Arecaceae, Fabaceae, Mimosaceae and Euphorbiaceae.

The Chiribiquete-Araracuara-Cahuinarí region is the westernmost transition between the floristic regions of Guayana and the Amazon Basin. The vegetation of the Tertiary sedimentary plain and the younger alluvial valley systems contains dominantly Amazonian lowland elements. The outcropped sandstone plateaux have a high percentage of elements of the Guayana region, and apparently represent its most south-western habitats for a number of taxa in *Xyris, Abolboda, Vellozia, Rapatea, Schoenocephalium, Brocchinia, Navia, Steyerbromelia, Duckeella, Decagonocarpus, Senefelderopsis, Gongylolepis, Retiniphyllum, Pagamea* and

Tepuianthus (the family Tepuianthaceae has been considered endemic to the Guayana floristic region).

The following 23 taxa presently are considered endemic to the sandstone plateaux of the Chiribiquete-Araracuara area: *Asplundia ponderosa, Clusia chiribiquetensis, C. schultesii, Combretum karijonorum, Euplassa saxicola, Ficus ajajuensis, F. chiribiquetensis, Graffenrieda fantastica, Hibiscus sebastianii, Justicia cuatrecasasii, Macairea schultesii, Navia schultesiana, Paullinia splendida, Paepalanthus moldenkeanus, Senefelderopsis chiribiquetensis, Solanum apaporum, Styrax rigidifolius, Tragia* new sp., *Raddiella molliculma, Utricularia chiribiquetensis, Vellozia phantasmagoria, Xyris araracuare* and *X. trachysperma* (Schultes 1944; Smith 1944; Fuertes 1992; Kral and Duivenvoorden 1993; Duivenvoorden and Cleef 1994).

The importance of the Chiribiquete-Araracuara-Cahuinarí region in botanical and phytogeographical research are illustrated by the recent discovery of 35 species new to science, including a new genus of the Dipterocarpaceae (only the second genus in the neotropics of this predominantly palaeotropical family), as well as the first records in Colombia for some 50 species; and the finding of several species in the alluvial plain of the Caquetá River that were hitherto known only from the Guayana Highlands (e.g. *Oxandra asbeckii, Ocotea neblinae*).

Useful plants

The indigenous peoples of the region extract products from the forest mostly for subsistence purposes. Nearly 90% of the trees (10 cm or more dbh) are used (Sánchez and Miraña 1991). Among the most important species and products are (1) fruits: *Unonopsis spectabilis, Couma catingae, Mauritia flexuosa, Pourouma tomentosa*; (2) palm hearts: *Euterpe precatoria, Oenocarpus bataua*; (3) woods: *Iryanthera juruensis, I. laevis, Ocotea acyphylla, Calophyllum longifolium*; (4) fibres: *Astrocaryum aculeatum, Eschweilera tessmannii, Cariniana decandra*; (5) medicines: *Protium nodulosum, Crepidospermum cuneifolium, Iryanthera ulei, Vismia macrophylla*; and (6) dyes: *Genipa americana, Duroia* spp.

Certain *Hevea* species have been commercially exploited and studied as genetic resources for rubber, particularly during the first three decades of the 1900s. The palm *Bactris gasipaes*, as well as *Pourouma cecropiifolia* and *Theobroma bicolor*, widely known as potentially valuable for commercial exploitation, are present in all the local home gardens. In the Amerindian peoples' shifting slash-and-burn agriculture, the most important species are *Manihot esculenta, Zea mays, Dioscorea trifida, Xanthosoma violaceum, Nicotianum tabacum, Erythroxylum coca* and *Ananas comosus*; often a large number of Amazonian selections can be distinguished, e.g. 18 kinds of hot peppers (*Capsicum chinense*) (Vélez 1991).

Social and environmental values

The several indigenous peoples in the region are very concerned to maintain their own Amazonian cultures, habits and ways of sustainable use of the forest. More than 90% of the area between Araracuara and the mouth of the Cahuinarí River has been declared Amerindian Reserves. Northward, surrounding Chiribiquete Natural National Park, no Amerindian Reserves exist. An intriguing aspect of the Chiribiquete area is impressive Amerindian wall paintings; their origin, exact age and meaning are not understood. Near Araracuara, archaeological research has detected one of the oldest dates (2700 BC) for the beginning of maize cultivation in the Amazon Basin (Mora *et al.* 1991). In the general region, several sites with anthropic soils occur.

The integral vastness of this rain forest provides important habitat for large Amazonian predators (e.g. jaguar, puma, harpy eagle). Without doubt attractive for tourists are the striking scenery of the cerros of Chiribiquete and the canyon of Araracuara, the petroglyphs on Precambrian rocks along the Caquetá River (Reichel de von Hildebrand 1975) and the beautiful, undisturbed wilderness of the basins of the Caquetá River and particularly the Cahuinarí River with its abundant animal diversity.

Economic assessment

At present, the population density in this region is very low (less than 0.2 person per km^2 along the Middle Caquetá River, still lower in the Chiribiquete area). The several indigenous groups are small. Few settlements of colonists exist. Tourism to the National Parks and other attractions is strenuous. It is extremely difficult to estimate the potential economic value of the plant resources, including the traditionally cultivated strains. Currently, hardly any forest products from the region reach the national market. Nonetheless, sustainable economic benefits from products certainly can be expected by careful development using forest species (e.g. medicines, fruits, dyes, fibres, woods) and plateau species (e.g. ornamentals) due to the high floristic diversity of the vegetation.

Threats

Fortunately, the region's remoteness and inaccessibility because of rapids in all of the major rivers (Vaupés, Apaporis, Caquetá) and the absence of roads have been safeguarding the Chiribiquete-Araracuara-Cahuinarí rain forest from almost all disturbance. However, the large expanse of the region makes it vulnerable to colonization, and the government has tried to encourage people to move into the Llanos Orientales and Amazon regions (Forero 1989). The potential exists to construct roads into the region, especially from San José del Guaviare (north of the Vaupés River). The easy entrance of colonists would lead to large-scale deforestation (pastoral development and timbering) (cf. Gentry 1989) and uncontrolled burning of the fragile dry sandstone vegetation.

The most immediate threat to the region comes from the Andean headwater areas, particularly of the Caquetá River. Increasing deforestation and the development of mining and industry are likely to be altering the flood regime of the Caquetá River (van der Hammen *et al.* 1991b) and causing contamination of the water and sediments, thus affecting the floodplain forests. Commercial fishing along the Caquetá River is the only natural resource use with products reaching the national market (Rodríguez 1991).

Conservation

Two Natural National Parks (totalling 18,550 km²) have been established in the region (Map 43): Chiribiquete NNP (12,800 km²) to the north-west – the largest park in the Colombian National Park system; and Cahuinarí NNP (5750 km²) in the south-east (Sánchez *et al.* 1990). The establishment of three Amerindian Reserves (comprising c. 3470 km²) in the Araracuara-Cahuinarí area also favours protection and conservation of the forest: for the Muinane,

Villa Azul (c. 598 km²); for the Andoke, Aduche (579 km²); and for the Wetoto, Monochoa (c. 2294 km²), and as well a small part of the sizeable Predio Putumayo (c. 57,992 km²) extends into the region. The environmental policy of the Colombian Government for the Amazon Basin tries to combine conservation with development, looking for improvement of living conditions for the human population within an ecologically viable framework (DNP 1991). Presently, however, there is no land-use and management plan for integrated conservation of the Chiribiquete-Araracuara-Cahuinarí region.

MAP 43. CHIRIBIQUETE-ARARACUARA-CAHUINARI REGION, COLOMBIA (CPD SITE SA7)

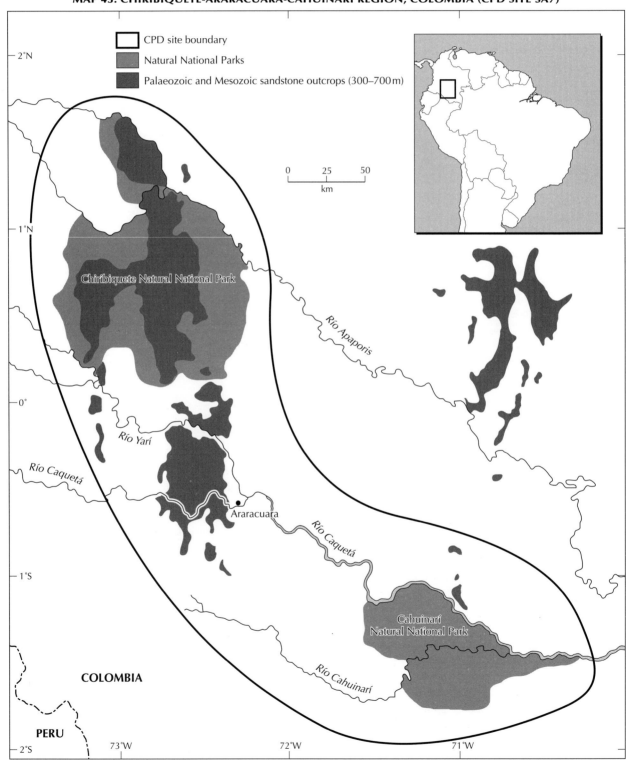

References

Anderson, A.B. (1981). White-sand vegetation of Brazilian Amazonia. *Biotropica* 13: 199–210.

DNP (1991). Política para el desarollo y la conservación de la Amazonia. Doc. Dirección Nacional de Planeación (DNP) 2545–UDT. *Colombia Amazónica* 5: 11–44.

Duivenvoorden, J.F. and Cleef, A.M. (1994). Amazonian savanna vegetation on the sandstone plateau near Araracuara, Colombia. *Phytocoenologia* 24: 197–232.

Duivenvoorden, J.F. and Lips, J.M. (1993). *Landscape ecology of the Middle Caquetá Basin. Explanatory notes to the maps.* Estudios en la Amazonia Colombiana, Vol. 3A. Tropenbos-Colombia, Bogotá. 301 pp.

Forero, E. (1989). Colombia. In Campbell, D.G. and Hammond, H.D. (eds), *Floristic inventory of tropical countries: the status of plant systematics, collections, and vegetation, plus recommendations for the future.* New York Botanical Garden, Bronx. Pp. 353–361.

Fuertes, J. (1992). Estudios botánicos en la Guayana colombiana, 1. Una nueva especie de *Hibiscus* sección *Furcaria* (Malvaceae). *Anales Jard. Bot. Madrid* 50: 65–72.

Galeano, G. (1991). *Las palmas de la región de Araracuara.* Estudios en la Amazonia Colombiana, Vol. 1. Tropenbos-Colombia, Bogotá. 181 pp.

Gentry, A.H. (1988). Tree species richness of Upper Amazonian forests. *Proc. Natl. Acad. Sci.* 85: 156–159.

Gentry, A.H. (1989). Northwest South America (Colombia, Ecuador and Peru). In Campbell, D.G. and Hammond, H.D. (eds), *Floristic inventory of tropical countries.* New York Botanical Garden, Bronx. Pp. 391–400.

Gentry, A.H. (1990). Floristic similarities and differences between southern Central America and Upper and Central Amazonia. In Gentry, A.H. (ed.), *Four neotropical rainforests.* Yale University Press, New Haven. Pp. 141–157.

Heyligers, P.C. (1963). Vegetation and soil of a white-sand savanna in Suriname. *Verh. Kon. Ned. Akad. Wetensch., Afd. Natuurk.*, Sect. 2, 54(3): 1–148 and in Hulster, I.A. and Lanjouw, J. (eds), *The vegetation of Suriname*, Vol. III. van Eedensfonds, Amsterdam.

Köppen, W. (1936). Das geographische System der Klimate. In Köppen, W. and Geiger, R., *Handbuch der Klimatologie*, Bd. I, Teil C. Berlin.

Kral, R. and Duivenvoorden, J. (1993). A new species of nematopoid *Xyris* (Xyridaceae) from the Araracuara area of Colombia. *Novon* 3: 55–57.

Mora, D., Herrera, L.F., Cavelier, I. and Rodríguez, C. (1991). *Cultivars, anthropic soils and stability. A preliminary report of archaeological research in Araracuara, Colombian Amazonia.* Univ. Pittsburgh Latin Amer. Archeol. Reports No. 2. 87 pp.

Reichel de von Hildebrand, E. (1975). Levantamiento de los petroglifos del río Caquetá entre La Pedrera y Araracuara. *Rev. Colombiana Antropol.* 19: 303–370.

Rodríguez, C.A. (1991). *Commercial fisheries in the Lower Caquetá River.* Estudios en la Amazonia Colombiana, Vol. 2. Tropenbos-Colombia, Bogotá. 152 pp.

Sánchez, H., Hernández, J.I., Rodríguez, J.V. and Castaño, C. (1990). *Nuevos parques nacionales, Colombia.* Instituto Nacional de los Recursos Naturales y Renovables y del Ambiente (INDERENA), Bogotá. 213 (+ 25) pp.

Sánchez, M. and Miraña, P. (1991). Utilización de la vegetación arbórea en el Medio Caquetá, A. El árbol dentro de las unidades de la tierra, un recurso para la comunidad Miraña. *Colombia Amazónica* 5: 151–160.

Schultes, R.E. (1944). Plantae colombianae, IX. *Caldasia* 3: 121–130.

Sipman, H. (1990). Colección preliminar de líquenes sobre hojas en Araracuara, Colombia. *Colombia Amazónica* 4: 59–65.

Smith, L.B. (1944). A new bromeliad (*Navia*) from Colombia. *Caldasia* 3: 131.

Urrego, L.E. (1990). Apuntes preliminares sobre la composición y estructura de los bosques inundables en el Medio Caquetá, Amazonas, Colombia. *Colombia Amazónica* 4: 23–30.

van der Hammen, T., Duivenvoorden, J.F., Lips, J.M., Urrego, L.E. and Espejo, N. (1991a). El Cuaternario Tardío del área del Medio Caquetá (Amazonia Colombiana). *Colombia Amazónica* 5: 63–90.

van der Hammen, T., Urrego, L.E., Espejo, N., Duivenvoorden, J.F. and Lips, J.M. (1991b). Fluctuaciones del nivel del agua del río y de la velocidad de sedimentación durante los últimos 13,000 años en el área del Medio Caquetá (Amazonia Colombiana). *Colombia Amazónica* 5: 91–130.

van Donselaar, J. (1965). An ecological and phytogeographic study of northern Surinam savannas. *Wentia* 14: 1–163.

Vélez, A.J. (1991). El ají (*Capsicum chinense* Jacq.), patrimonio cultural y fitogenético de las culturas amazónicas. *Colombia Amazónica* 5: 161–185.

Authors

This Data Sheet was written by M. Sánchez-S. [Corporación Colombiana para la Amazonia, Araracuara (COA), A.A. 34174, Bogotá, D.E., Colombia], L.E. Urrego, Dr J.G. Saldarriaga (Fundación Tropenbos-Colombia, A.A. 36062, Bogotá, D.E., Colombia), J. Fuertes, J. Estrada (Real Jardín Botánico, Consejo Superior de Investigaciones Científicas, Plaza de Murillo 2, 28014 Madrid, Spain) and Dr J.F. Duivenvoorden (University of Amsterdam, Hugo de Vries Laboratorium, Kruislaan 318, 1098 SM Amsterdam, The Netherlands).

AMAZONIA: CPD SITE SA8

YASUNI NATIONAL PARK AND WAORANI ETHNIC RESERVE
Ecuador

Location: Between Napo and Curaray rivers in Napo and Pastaza provinces, Amazonian Ecuador; south and east of town of Coca to near Nuevo Rocafuerte and Peru, between latitudes 0°26'–1°40'S and longitudes 77°30'–75°25'W.

Area: 15,920 km²: 9820 km² in park, 6100 km² in reserve.

Altitude: c. 200–350 m.

Vegetation: Tropical moist forest: tierra-firme, várzea, swamp and igapó forests.

Flora: c. 4000 species, high diversity; potentially high in regional endemics.

Useful plants: Medicinal, ornamental and timber species.

Other values: Large wilderness, diverse fauna including regional endemics and threatened species, potential genetic resources, Amerindian land, watershed protection, ecotourism.

Threats: Road construction and oil pipeline; illegal settlers and logging.

Conservation: National Park, Ethnic Reserve, Biosphere Reserve, forthcoming Research Station.

Geography

The eastern portion of Ecuador in the Amazon Basin (the "Oriente") comprises the lowlands that gradually slope downward from 600 m to less than 200 m at the eastern frontier with Peru (Balslev and Renner 1989; Tschopp 1953). The topography is low and undulating to slightly hilly terrain between broad swampy floodplains of the main rivers. Geologically the Oriente is part of the extensive area filled with Cretaceous-Tertiary sediments between the Andes and the Brazilian Shield (Tschopp 1953).

This region is drained by the Napo and Pastaza river systems, which diverge respectively toward the north-east to east and the south-east from the depression between the Andean uplifts of the Serranía del Napo (with Sumaco Volcano) and the Sierra de Cutucú (Tschopp 1953). The Napo is the major river, flowing eastward to join the Marañón River near Iquitos, Peru (CPD Site SA9) and form the Solimões River, which in turn flows eastward to Manaus, Brazil (CPD Site SA5).

Yasuní National Park covers 9820 km² south of the Napo River and north of the Curaray River in Napo and Pastaza provinces of central eastern Ecuador, extending eastward from c. 40 km east of the town of Coca (76°40'W) almost to Nuevo Rocafuerte near the border with Peru (Map 44). Much of the park's northern boundary is the Tiputini River and much of the southern boundary is the Curaray River. There is a roughly rectangular north-western extension of the park to the south bank of the Napo River at Añangu and westward to the Indillama River.

The adjacent Waorani Ethnic Reserve includes 6100 km². The eastern part of the reserve is largely encompassed to the north, east and south by the park. The reserve extends westward to c. 77°30'W, but is almost bisected by the Auca road which runs south from the town of Coca (or Puerto Francisco de Orellana), an oil centre and port. A broad swath of land 10 km wide on either side of the road is occupied by colonists, but more or less south of the road a corridor provides the Waorani Amerindians access between the eastern and western portions of their reserve.

Most of the park and reserve has low hills of red clay dystropept soil. Low humic gley soil probably occurs in the swampy or poorly drained areas in the eastern part of the park (Duellman 1978; Neill 1988b). There are no peat swamps or podzols (Balslev and Renner 1989).

Weather stations some distance west and east of the park suggest that the annual temperature averages 25°C (with extremes of 15° and 38°) and the annual rainfall is 2425–3145 mm, with a humidity of 88%. Although rarely rainless for more than c. 10 days, between August and February some months may be drier (Balslev et al. 1987; Blandin Landívar 1976; Duellman 1978). Flooding is not seasonal (Balslev and Renner 1989).

Vegetation

The entire region is within the tropical moist-forest life zone of the Holdridge system. The park and reserve are in the Solimões-Amazonas phytogeographic region (Nations 1988). Four main vegetation types have been recognized within the park and reserve, but the vegetation has not been mapped.

1. Probably more than 90% of the area is unflooded upland ("tierra-firme") forest, occurring on the low hills of red clay dystropept soil. The canopy is 25–30 m high, with emergents such as *Cedrelinga cateniformis* (to 45–50 m tall and 2–3 m dbh) and *Parkia* spp. Canopy trees include several Myristicaceae (*Otoba glycycarpa*, *Osteophloeum platyspermum*, *Virola* spp.). *Simaruba amara*, *Dussia tessmannii*, *Hymenaea oblongifolia* and several genera of Moraceae and Sapotaceae also occur. Trees with buttresses or stilt roots are frequent. The understorey on

344

hills may be quite open with small trees and shrubs, lianas may be abundant, and epiphytes are less diverse and abundant than in wetter forest nearer the Andes (Neill 1988b). The ground layer tends to be only weakly developed (Balslev *et al.* 1987).

2. Along the banks of the Napo River is a narrow strip (200–1000 m wide) of relatively fertile soil, enriched by sediments from the Andes when the river floods. This "**várzea**" **forest** is generally flooded only once every several years. The canopy layer is somewhat higher (35–40 m) than in the upland forest, with occasional emergents such as *Ceiba pentandra* and *Ficus* spp. to 50 m tall. Common canopy dominants include *Otoba parvifolia*, *Chimarrhis glabriflora*, *Celtis schippii* and *Guarea kunthiana*. The ivory-nut palm *Phytelephas macrocarpa* is a common small understorey tree.

3. The third type of vegetation is **swamp forest**, which occurs in extensive stands along the Napo River and the lower reaches of the Tiputini River, a main tributary of the Napo. Swamp forest is flooded for much of the year, but the ground is exposed during dry periods. Characteristic are nearly pure stands of the palm *Mauritia flexuosa*, as well as a few other swamp species such as *Virola surinamensis* and *Symphonia globulifera*.

4. The Yasuní River is a black-water river, which bears very little sediment because its headwaters are in the Amazon lowlands rather than the Andes. The waters are stained the colour of dark tea by tannic acids dissolved from riverside vegetation. Along the banks of this river and associated lagoons is "**igapó**" **forest**, which is almost totally floristically distinct from the upland and várzea forests. Common trees include *Macrolobium acaciifolium*, *Coussapoa trinervia* and the palm *Astrocaryum jauari*.

Flora

The Napo River region of Ecuador and Peru has been proposed as one of the primary Pleistocene forest refugia, characterized by a high degree of animal and plant endemism. The refugium extends from the foothills of the Andes eastwards to the "Trapecio Amazónico" of Colombia and Peru (Duellman 1978; Prance 1982) – the park and reserve are within a more finely delimited South Napo Pleistocene refugium.

The upper part of the Amazon Basin may have emerged from a mid-continental lake and become forested as recently as 1.8 million years ago, with the greatest uplift of the Andes. During climatic fluctuations of the Pleistocene and Holocene the forest may have become fragmented, and then rejoined north and south but been separated by drier areas to the east (e.g. Duellman 1978). For the Oriente as a whole, which however is just a part of the postulated Pleistocene refugium, Balslev and Renner (1989) estimated endemism at only 1%. The plant species composition of the park and reserve remain unknown, as does the extent of regional endemism represented within them.

The little collecting and study done so far at one area (Añangu) showed high but not exceptional richness for moist lowland forest (Balslev and Renner 1989). Some 394 species of trees over 10 cm dbh were found: 153–228 species per ha in unflooded forest and 146 in floodplain

MAP 44. YASUNÍ NATIONAL PARK AND WAORANI ETHNIC RESERVE, ECUADOR (CPD SITE SA8)

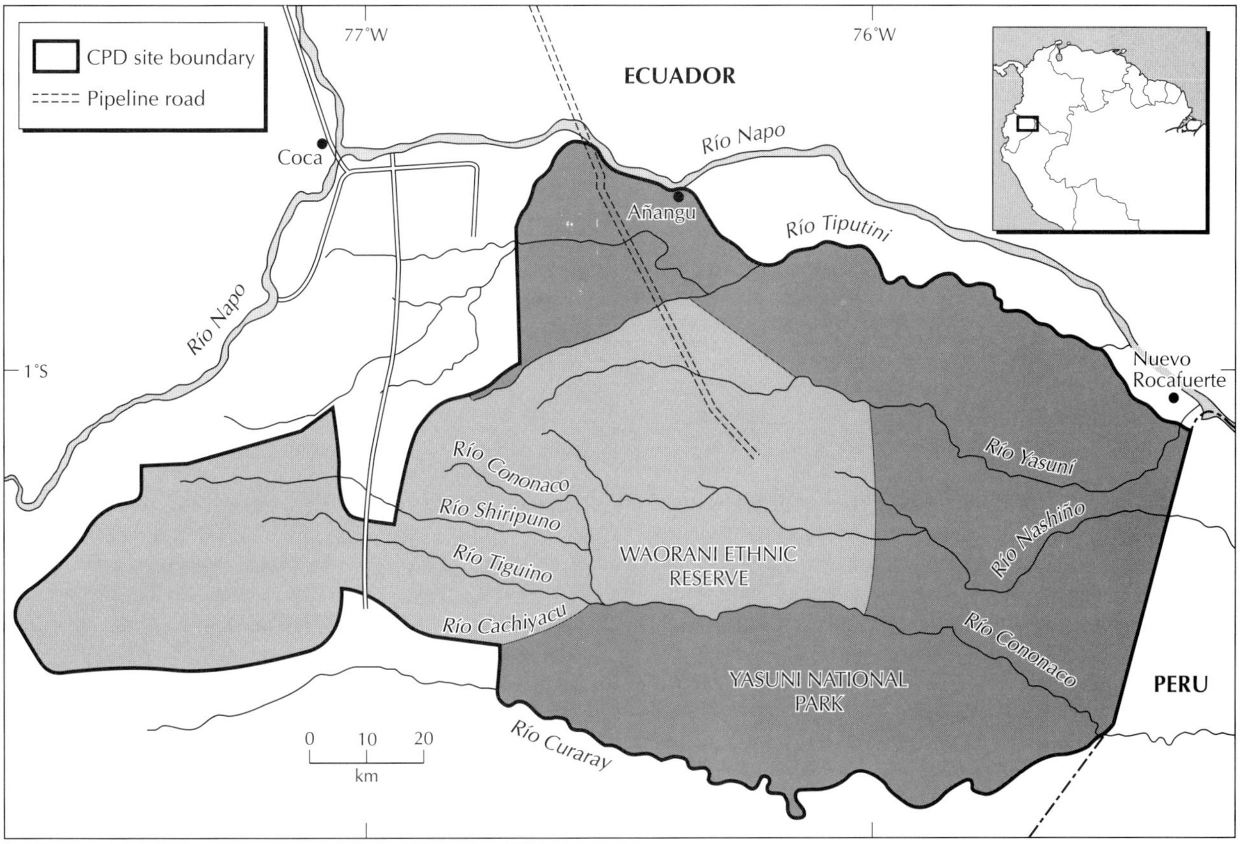

forest, with 19% shared. As usual in the neotropical lowlands, Moraceae and Leguminosae were most frequent. So little known is the region that in two weeks of field work near an exploratory oil well in the western part of the park, several new species of trees and at least two new orchid species were discovered, as well as over 15 records of orchids new for Ecuador (Neill 1988b). The flora of the region contains many species in common with the lowlands of the nearby Gran Sumaco and Upper Napo River region (see Gran Sumaco Data Sheet, CPD Site SA38), but the distribution of the flora in Amazonian Ecuador is highly heterogeneous – many species present in the wetter Gran Sumaco region do not occur in the Yasuní region, and vice versa.

In late 1992, botanists from the National Herbarium of Ecuador and the Missouri Botanical Garden initiated a large-scale floristic inventory along the oil-pipeline road which is being built through 120 km of primary forest in the Yasuní National Park and Waorani Ethnic Reserve. Specimens were collected from felled trees. This survey has continued for two years and will provide much more thorough knowledge of the flora.

Useful plants

Little that is definitive can be said until the flora is better known. *Hevea guianensis* is present – a less commercially desirable rubber tree than *H. brasiliensis*, but an important genetic resource. *Phytelephas macrocarpa* (the vegetable-ivory palm) also is found – a species that has received renewed international commercial interest. *Cedrelinga cateniformis*, which is prized for construction of dugout canoes, has potential as a commercial timber. This species might replace the dwindling Spanish cedar (*Cedrela odorata*) and bigleaf mahogany (*Swietenia macrophylla*), which are among the valuable timber trees selectively taken from accessible areas along the rivers.

Ethnobotanical studies in lowland Ecuador (Kvist and Holm-Nielsen 1987) include various medicinal and other uses of species by the Waorani (Davis and Yost 1983): e.g. *Bactris gasipaes* (blowgun wood), *Curarea tecunarum* (arrow poison and fungal diseases), *Minquartia guianensis* (fish poison), *Banisteriopsis muricata* (hallucinogen), *Renealmia* spp. and *Urera baccifera* (snake bites), *Piper augustum* and *P. conejoense* (toothbrush and decay preventive), *Sphaeropteris* sp. (local dental anesthetic), *Virola* spp. and *Iryanthera* spp. (fungal diseases).

Social and environmental values

The park and reserve provide extensive habitat for many animals, e.g. harpy eagles, macaws, jaguars, primates, freshwater dolphins and anacondas. The area's great expanse provides a rare chance to conserve unfragmented and undisturbed ecosystems and populations functioning naturally, including threatened species.

The two protected areas are embraced by the large Napo and Upper Amazon lowlands Endemic Bird Area (EBA B19), which extends from southernmost Colombia and eastern Ecuador eastward into northern Peru and westernmost Brazil. Ten species of birds are limited to this area, although they essentially represent the most restricted species of a (distributionally poorly known) suite of birds that are confined to the river islands, riverine forest and várzea forest of the Amazon Basin rivers. The birds in this EBA, just one of which is considered threatened, are seemingly confined to the tierra-firme or várzea forests.

The Waorani Ethnic Reserve protects tribal land of the Waorani ("Auca") Amerindians, some of whom have fiercely resisted all outside efforts to contact them (Nations 1988; Whitten 1981; Yost 1981). Oil exploration began in their area in the 1940s. The Waorani ethnobotany is notably different from that of neighbouring peoples, suggesting their past isolation. Several family groups of Waorani live in the eastern portion of Yasuní park.

Economic assessment

In 1992, a large-scale conservation programme known as SUBIR (Sustainable Use of Biological Resources) was initiated for the Yasuní region as well as two other protected areas in Ecuador. The goals of SUBIR are to promote conservation by increasing the capacity of Ecuadorian agencies to protect core areas, as well as encouraging non-destructive uses of natural resources by peoples living in buffer zones around the protected areas. Financed by the U.S. Agency for International Development, SUBIR is carried out by a consortium of organizations led by CARE International, with the collaboration of the Ecuadorian Ministerio de Agricultura y Ganadería (Ministry of Agriculture and Livestock) and local environmental and community organizations.

The SUBIR programme seeks to help develop viable economic alternatives that will enable people in the buffer zones to produce sufficient income without causing deforestation or other resource-destructive activities. These alternatives may include ecotourism, production of handicrafts and other goods from the forest, and improved agricultural techniques that obviate the need to continually clear more forested land. SUBIR is an experiment in its initial stages; its results will not be evident for several years.

Threats

The forests of Ecuador's Oriente are undergoing extensive deforestation as a result of oil exploration and production followed by colonization, which began with the most recent and successful phase of exploration in 1964–1969 (Schodt 1987). These activities led to the construction of a 420-km oil pipeline to transport petroleum from the Oriente oil fields over a 4300-m high pass in the Andes and down to the port of Esmeraldas on the Pacific coast. With the pipeline, the first roads were constructed into Ecuador's north-eastern Amazon and then south, in 1971 opening the region to colonization, e.g. through relocation of farmers from the over-crowded coastal and mountain regions of the country (Bromley 1973; Neill 1988a). Where a few thousand lived 20 years ago, now over 100,000 people live in Napo Province and are transforming large tracts of forest into agricultural fields and pastures (Uquillas 1984). African oil palm (*Elaeis guineensis*) is the favourite plantation crop. Selective logging occurs where the trees are accessible.

In the mid 1980s, new oil fields were located in the Pastaza and Napo river valleys, including a significant reserve of 150 million barrels of heavy crude petroleum beneath the Waorani Ethnic Reserve and Yasuní National Park. Ecuador's plan for extraction by building a road as well as a pipeline to the oil fields through the untouched Yasuní forest, instead of flying in materials, sparked intense controversy within Ecuador as well as internationally. The government's PetroEcuador awarded a concession for development of the oil reserves in petroleum block 16, which occupies 2000 km² within the Waorani reserve, to the U.S.-based company Maxus. In December 1992, construction of the road and pipeline began from the Napo River south into the centre of the Waorani reserve-Yasuní park territory, amid continuing opposition from Ecuadorian environmental organizations.

The environmental mitigation plan for the development project within the Yasuní-Waorani area includes some provisions for reducing negative impacts. Strict control of persons entering the road is planned to avoid settlers and logging, and wells are to be drilled in clusters to reduce deforestation. However, the feasibility of being able to prevent invasion of the park and reserve by colonists once the road is established is questionable. The events that often follow the building of a road in a protected area are sadly recorded nearby. Construction of oil-pipeline roads through the Cuyabeno Wildlife Reserve north of the Napo River led to colonization of the area by more than 1000 families.

Conservation

The discovery of oil in the Oriente has brought considerable prosperity to Ecuador (oil exports provide 70% of the country's income) (Schodt 1987), greatly increased opportunities for colonization in the region and emphasized the need to protect its diverse biological resources – which also are of economic significance. Two large reserves were created in 1979: Cuyabeno Wildlife Reserve (2547 km²) north of the Napo River and Yasuní National Park (originally 6797 km²) – the largest mainland park in Ecuador.

Due in part to the conflict between conservation of protected areas and development of oil fields, the size and shape of Yasuní park have been changed twice by governmental decrees. An Ethnic Reserve of 1600 km² for the Waorani was established in 1968, south-west of the original Yasuní park boundary (Whitten 1981). Until the 1960s, the Waorani were nomadic over c. 20,000 km² (nearly all of the territory between the Napo and Curaray rivers) and their hostile reactions to all outsiders had kept their land nearly undisturbed (e.g. Kvist and Holm-Nielsen 1987).

In 1990, the Waorani Ethnic Reserve was enlarged eastward to 6100 km², and a large part of Yasuní park was ceded to the Waorani reserve, including the major portion of the oil fields near the Yasuní River. As partial recompense for loss of the park lands, additional territory was added on the south-east of Yasuní NP. In 1992, the park was enlarged again to 9820 km². Together with the Waorani reserve, the officially protected Yasuní region now comprises almost 16,000 km².

An important step toward the legal protection of the region was the May 1989 declaration of the park and its buffer zone as a Biosphere Reserve, under UNESCO's Man and the Biosphere Programme (Coello Hinojosa and Nations 1989). The Biosphere Reserve now includes the Waorani Ethnic Reserve and the enlarged Yasuní NP. However, the future of the park and reserve remain uncertain. The oil reserves will last only 20 years. Over the long term it will be more productive to protect the genetic resources of the region and promote tourism that can generate steady income. Future generations of Ecuadorians especially could have the legacy of a great Amazonian park, the homeland of indigenous people, and an extraordinary representation of plant and animal species in one of the world's diverse large wilderness regions.

A preliminary master plan for the park has been prepared by the Departamento de Administración de Areas Naturales y Vida Silvestre of the Ministerio de Agricultura y Ganadería (Coello Hinojosa and Nations 1989). However, the changing boundaries of the park, the establishment of the Waorani Ethnic Reserve and the petroleum development necessitate thorough revision of the management plan for the region. Legal mechanisms for future management and environmental protection of the Waorani Ethnic Reserve, in particular, have not been clarified by the government. The conservation outlook for the Yasuní region is not yet bleak – the Maxus petroleum company which holds the development concession in the area has demonstrated a commitment to support conservation efforts.

Among other contributions, Maxus agreed to build a scientific research station within the park, which is to be managed by the Pontificia Universidad Católica del Ecuador, Quito. The first major research project of the station is studying the forest dynamics of a diverse 50-ha permanent plot established on the Tiputini River (Foster 1994).

References

Balslev, H., Luteyn, J., Øllgaard, B. and Holm-Nielsen, L.B. (1987). Composition and structure of adjacent unflooded and floodplain forest in Amazonian Ecuador. *Opera Botanica* 92: 37–57.

Balslev, H. and Renner, S.S. (1989). Diversity of east Ecuadorean lowland forests. In Holm-Nielsen, L.B., Nielsen, I. and Balslev, H. (eds), *Tropical forest: botanical dynamics, speciation and diversity*. Academic Press, London. Pp. 287–295.

Blandin Landívar, C. (1976). *El clima y sus características en el Ecuador*. Biblioteca Ecuador. XI Asamblea General del Instituto Panamericano de Geografía e Historia, Quito. 86 pp.

Bromley, R.J. (1973). Agricultural colonization in the upper Amazon Basin: the impact of oil discoveries. *Tijdschr. Econ. Soc. Geogr.* 63: 278–294.

Coello Hinojosa, F. and Nations, J.D. (1989). *Plan preliminar de manejo del Parque Nacional Yasuní 'Reserva de la Biósfera'*. Ministerio de Agricultura y Ganadería, Dirección Nacional Forestal, Departamento de Areas Naturales, Quito. 121 pp.

Davis, E.W. and Yost, J.A. (1983). The ethnobotany of the Waorani of Eastern Ecuador. *Bot. Mus. Leafl. Harvard Univ.* 29: 159–217.

Duellman, W.E. (1978). *The biology of an equatorial herpetofauna in Amazonian Ecuador.* University of Kansas Museum of Natural History, Miscellaneous Publication No. 65. 352 pp.

Foster, R.B. (1994). 50-hectare plot being established in Yasuní National Park, Ecuador. *Inside CTFS* (Fall 1994): 11.

Kvist, L.P. and Holm-Nielsen, L.B. (1987). Ethnobotanical aspects of lowland Ecuador. *Opera Botanica* 92: 83–107.

Nations, J.D. (1988). *Road construction and oil production in Ecuador's Yasuní National Park.* Center for Human Ecology, Antigua, Guatemala. 18 pp. + 3 maps. Unpublished.

Neill, D.A. (1988a). A field trip in Ecuador: oil wells, Indians and rainforests of the Upper Amazon. *Missouri Bot. Gard. Bull.* 76(3): 5–7.

Neill, D.A. (1988b). *The vegetation of the Amo 2 oil well site, Yasuní National Park, Napo Province, Ecuador.* 4 pp. Unpublished.

Prance, G.T. (1982). Forest refuges: evidence from woody angiosperms. In Prance, G.T. (ed.), *Biological diversification in the tropics.* Columbia University Press, New York. Pp. 137–158.

Schodt, D.W. (1987). *Ecuador: an Andean enigma.* Westview Press, Boulder, Colorado, U.S.A. 188 pp.

Tschopp, H.J. (1953). Oil explorations in the Oriente of Ecuador, 1938–1950. *Bull. Amer. Assoc. Petrol. Geol.* 37: 2303–2347.

Uquillas, J. (1984). Colonization and spontaneous settlement in the Ecuadorian Amazon. In Schmink, M. and Wood, C.H. (eds), *Frontier expansion in Amazonia.* University of Florida Press, Gainesville. Pp. 261–284.

Whitten Jr., N.E. (1981). Amazonia today at the base of the Andes: an ethnic interface in ecological, social and ideological perspectives. In Whitten Jr., N.E. (ed.), *Cultural transformations and ethnicity in modern Ecuador.* University of Illinois Press, Urbana. Pp. 121–161.

Yost, J.A. (1981). Twenty years of contact: the mechanisms of change in Wao ("Auca") culture. In Whitten Jr., N.E. (ed.), *Cultural transformations and ethnicity in modern Ecuador.* University of Illinois Press, Urbana. Pp. 677–704.

Authors

This Data Sheet was written by Olga Herrera-MacBryde (Smithsonian Institution, Department of Botany, NHB-166, Washington, DC 20560, U.S.A.) and Dr David A. Neill (Missouri Botanical Garden, P.O. Box 299, St. Louis, MO 63166, U.S.A. and Herbario Nacional del Ecuador, Casilla 17-12-867, Quito, Ecuador).

AMAZONIA: CPD SITE SA9

IQUITOS REGION
Peru and Colombia

Location: North-eastern Amazonian Peru and adjacent southernmost Colombia between latitudes 3°–5°S and longitudes 74°–70°W, from about Jenaro Herrera on Ucayali River east to Leticia.

Area: c. 80,000 km².

Altitude: c. 105–140 m.

Vegetation: Varied upland and wetland evergreen tropical moist forests and shrubby woodlands; swamps, marshes.

Flora: Over 2265 species recorded; exceptionally high diversity associated with complex mosaic of habitats; possibly world's highest diversity of tree species; many endemics.

Useful plants: Perhaps more native species used for food than anywhere else – 120 fruit-producing wild species especially important; timber trees for construction and fuelwood; fibres; extracts; medicinals.

Other values: High diversity of fauna; centre for research, including studies on valuation and sustainable utilization of Amazonian forests; ecotourism.

Threats: Deforestation due to population pressures.

Conservation: In Peru, Pacaya-Samiria National Reserve in part and Tamshiyacu-Tahuayo Communal Reserve; in Colombia, Amacayacu National Natural Park.

Geography

The Iquitos region of western Amazonia is mainly in Peru's Department of Loreto, including portions of the provinces Maynas, Loreto, Ramón Castilla and Requena and the eastern buffer zone of Pacaya-Samiria National Reserve. For conservation purposes, the southernmost extension of Colombia is considered part of the Iquitos region. Colombia is to the north of the region, Brazil to the south, and both countries are to the east (Map 45).

The major urban centre is the port city of Iquitos, which is located in the north-western part of the region on the west bank of the Upper Amazon River at 106 m in elevation. Iquitos is 600 km east of the base of the Andes and 3636 km west of the Atlantic Ocean (ONERN 1975). The Putumayo River marks the general area's northern boundary with Colombia and the Yavari/Javari River its southern boundary with Brazil. The principal rivers traversing the region are the Napo, Marañón and Ucayali, which become the Upper Amazon (to Solimões) River (Peñaherrera 1989).

The Iquitos region formerly constituted an east-west geosyncline, which in the Cambrian period formed a basin of water that connected to the Pacific Ocean. During the Miocene with the initial rise of the Central Andean Cordillera, drainage was redirected to the Atlantic Ocean, creating a giant lake which slowly filled with assorted sediments (Encarnación 1985). Thus formed the extensive lowland plane of degraded Tertiary and Quaternary alluvial deposits that cover much of the Central Amazon sedimentary basin.

As a result of these geological phenomena, there is very little altitudinal differentiation. The older deposits constitute the non-flooded "tierra firme" – terrain found in most of Amazonia, which has few if any essentially permanent lakes (Räsänen, Salo and Jungner 1991). Soils of more recent origin are restricted to active floodplains or várzeas (Meggers 1971).

"Várzea" is continually enriched by new deposits of Andean sediments, and has successional riverine features including a complex system of many lakes.

Due to different substrates, there is a complex habitat mosaic. Characteristic substrates include upland white sand; upland lateritic soil; non-inundated alluvial soil; and "tahuampa" – floodplains seasonally inundated by either suspension-rich white water or suspension-poor black water. This mosaic of habitats is more completely developed in the black-water river area of northern Loreto Department; it reaches its southern limit near Jenaro Herrera on the Ucayali River and extends eastward to Leticia (Map 45) (Gentry and Ortiz-S. 1993).

The climate of the region is relatively uniform (Encarnación 1985). The average annual temperatures are between 25°–27°C, with diel variations of less than 8°–10°C, and a minimum of 16°C occurring during June and July. Annual precipitation ranges from 2386 mm at Leticia to 2400–3400 mm. The Iquitos region has no well-marked dry season. The complete absence of seasonality of the rainfall in diverse warm habitats probably is a key factor responsible for considering this region and the Western Amazon in general as the world's highest in diversity of tree species and several other major groups of organisms (cf. Gentry 1988b; Gentry and Ortiz-S. 1993).

Vegetation

The distinguishing phytogeographic characteristic of the Amazon is its magnificent tropical rain forest, which A. von Humboldt called the hylaea. Detailed classification of the vegetation of the Peruvian Amazon is very incomplete. In the Holdridge life-zone system, it is considered primarily tropical moist forest ("bosque húmedo tropical"), and many subcategories have been recognized (ONERN 1976). Studies of the Iquitos region have resulted in various classifications (Tuomisto 1993). The major types found are upland forests and wetland forests, each of which also correlates with more shrubby vegetation (Encarnación 1985).

The Peruvian Amazonian region is unique as a tectonically active foreland occupied by often pristine tropical rain forest (Kalliola *et al.* 1991). The vegetation is highly dynamic, which implies that a large proportion of the forest is transitional. Even relatively stable sites can become destablized through abrupt changes in habitat conditions. Succession associated with this geologic dynamism plays an important role in the patterns of the Iquitos region, which has a very complex and intricate mosaic of habitats. Although the climax vegetation is generally tropical moist forest, on different substrates there are floristically different communities. While the species composition of the communities on the different substrates is very distinct, the familial composition of different communities is remarkably similar (Gentry 1988a).

Upland (tierra-firme) forests
(Ruokolainen and Tuomisto 1993)

1. **Vegetation on white-sand soils** has two main formations:

❖ "Varillal" forest. The rather slender trees are 3 cm or larger in dbh and 10–25 m tall; they typically have very sclerophyllous leaves. The understorey is relatively sparse, with few herbs except for terrestrial ferns, which are very common – especially *Trichomanes* spp., *Elaphoglossum* spp. and *Lindsaea divaricata*. The varillal can be either wet, with relatively poor drainage, much organic matter, roots covering the ground and few tree species – *Caraipa utilis* (Clusiaceae)

is often dominant; or dry, with good drainage, little organic matter and many more tree species.

❖ "Chamizal" vegetation is shorter, usually less than 3 m high, with dispersed emergent trees to nearly 8 m tall. It is dominated by shrubs, e.g. *Graffenrieda limbata* (Melastomataceae), and interspersed throughout with small palms, e.g. *Mauritia* and *Mauritiella*; there are few species of ferns.

2. **Vegetation on lateritic soils** is heterogeneous, and also less well known, although the most prevalent in the region. Edaphic conditions and variations in drainage appear influential. According to various studies (Gentry 1988a, 1988b), the lateritic soils have the highest diversity of plant species. Near Iquitos (4°S), while the white sand of a 1-ha sample at Mishana (near the Nanay River) had 83 trees 30 cm or more in dbh, the relatively richer lateritic soil at nearby Yanamono supported 110 such large trees (Gentry 1988b).

Wetland forests and woodlands, swamps and marshes

1. **Seasonally or sporadically inundated vegetation** is of three main types and several other successional types:

a) Flooded by regular annual cycles of rivers:

❖ Black-water tahuampa vegetation ("igapó") occurs in areas adjacent to the black-water or somewhat mixed-water rivers, creeks or lagoons (Gentry and Ortiz-S. 1993). The medium-sized trees and shrubs are much branched, with abundant adventitious or stilt roots.

MAP 45. IQUITOS REGION, PERU AND COLOMBIA (CPD SITE SA9)

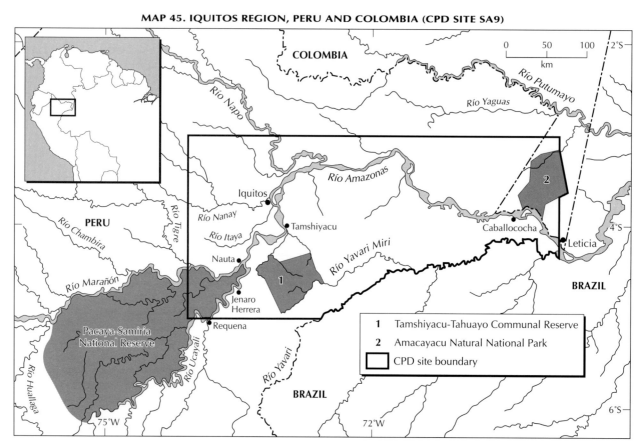

The species include *Campsiandra* sp., *Macrolobium acaciaefolium, Symmeria paniculata, Ficus* sp., *Alchornea castaneifolia* and *Myrciaria dubia*.

- ❖ White-water tahuampa vegetation ("várzea") occurs in areas adjacent to the white-water rivers, typically with shrubby vegetation and dispersed large trees, such as *Ficus insipida, Maquira coriacea* and *Hura crepitans*. The vegetation includes *Bactris* and *Desmoncus* palms and *Heliconia* spp., and the lower stratum has herbaceous annual species such as *Bambusa superba*.

- ❖ Successional "**vegetación de barrial**" occurs on muddy soil derived from recent alluvial accumulation. The species include *Paspalum* spp., *Echinochloa* sp., *Panicum* spp., *Ludwigia* spp., *Salix humboldtiana, Cyperus* spp. and *Tessaria integrifolia*. These areas are used to grow rice (*Oryza sativa*).

- ❖ "**Vegetación de playa**" occurs on recent sandy beaches adjacent to or between the mudflats ("barriales"). In natural succession *Gynerium sagittatum* is invasive. These areas are used to grow *Phaseolus* sp., *Vigna* sp. and *Arachis hypogaea* (peanut).

b) *Flooded irregularly by overflow of rivers or local torrential rains:*

- ❖ "**Bosque de restinga**" – forest flash-flooded by outlying to local rainfall at any time of year (Prance 1979). The vegetation is similar to the white-water tahuampa, with large trees such as *Ficus insipida, Maquira coriacea, Calycophyllum spruceanum* and *Sloanea* sp., and in the herbaceous stratum *Heliconia* sp.

2. **Permanently inundated vegetation (swamps and marshes)** is of three general kinds (Kalliola *et al.* 1991):

- ❖ **Palm swamp**, characterized particularly by *Mauritia flexuosa* ("aguaje") and with other common palms such as *Euterpe precatoria, Geonoma acaulis* and *Oenocarpus mapora*. Some dicotyledonous trees may be present, including *Ficus, Symphonia* and *Virola*.

- ❖ **Shrub swamp**, for example characterized by *Adenaria floribunda, Alchornea castaneifolia* and *Salix humboldtiana*, followed in succession by *Annona hypoglauca*, the palm *Astrocaryum jauari* and *Cecropia latiloba*.

- ❖ **Herbaceous wetland or marsh**. Rooted grasses such as *Paspalum repens* and *Echinochloa polystachya* form seasonally floating mats. Other taxa variously occurring in marshy communities include *Victoria*, Cyperaceae, more Gramineae, *Echinodorus, Ludwigia, Polygonum* and *Montrichardia*. Free-floating aquatics can be abundant at the edge of an open-water area – characteristic are *Hydrocotyle, Pistia, Eichhornea, Azolla* and *Salvinia*.

Flora

Preliminary data from several well-censused areas near Iquitos record 2265 species in ten sites, with 65% of the species known from only one of the sites (Gentry and Ortiz-S. 1993; Vásquez and Pipoly, in prep.). To the east in Colombia,

CPD Site SA9: Iquitos Region. *Meliosma vasquezii,* Sabiaceae, a locally endemic species. Photo: Alwyn Gentry.

CPD Site SA9: Iquitos Region. *Schlegelia cauliflora,* Bignoniaceae. Photo: Alwyn Gentry.

c. 1500 species have been recorded in Amacayacu Natural National Park (Rudas and Pipoly, in prep.); sometimes these are the same species as those found near Iquitos.

An outstanding aspect of Iquitos forests is their high α diversity (Gentry 1988a, 1988b). Two 1-ha plots on different soil types in the Iquitos region were determined to be very rich in species of trees: at Yanamono (on alluvial-terrace lateritic soil) there were 580 trees 10 cm or more in dbh representing 283 species – only 15% of the tree (and 17 liana) species were represented by more than two individuals; Mishana (on white sand) had 842 such trees of 275 species. Thus the Upper Amazon rain forest may be the world's richest in tree-species diversity (Gentry 1986, 1988b). The Iquitos region's forests are also among the richest known in all plant species 2.5 cm or more in dbh, with 0.1-ha samples averaging 218 species (Gentry and Ortiz-S. 1993).

Usually Leguminosae is the most diverse family in Iquitos forests, although in some of the richest-soil areas Moraceae may present almost as many species (Gentry 1988b). Other prevalent woody families are Lauraceae, Annonaceae, Rubiaceae, Myristicaceae, Sapotaceae and Meliaceae. Palms are well represented in most Iquitos forests and are especially prevalent on richer soils (Vásquez and Gentry 1989). The most important liana family is Bignoniaceae (as in most of the neotropics), and then Leguminosae, Hippocrateaceae, Menispermaceae, Sapindaceae and Malpighiaceae (Gentry 1991).

Locally endemic species include *Meliosma vasquezii* (Sabiaceae), *Caraipa utilis* (Clusiaceae) and *Aptandra caudata* (Olacaceae). Some endemics in this region known only on sandy soil are *Hirtella revillae* (Chrysobalanaceae), *Jacqueshubera loretensis* (Leguminosae), *Spathelia terminalioides* (Rutaceae) and *Ambelania occidentalis* (Apocynaceae). *Schlegelia cauliflora* (Bignoniaceae) was considered a local white-sand endemic until recently, when it was found at Araracuara in Colombia (in CPD Site SA7).

Useful plants

The usefulness of the Iquitos forests to the people who live in them is varied and vast – for food, remedies, construction, crafts, commerce, etc. Pinedo-Vásquez *et al.* (1990) studied 7.5 ha of the secondary communal forest adjacent to San Rafael (near Iquitos), finding that 60% of the 218 tree species (dbh 10 cm or more) and 66% (2511) of the individual trees were utilized for one or more purposes: for example, 28% of the species produce food; 10% remedies; 31% construction materials (including bark of *Duguetia lucida* and *Helicteres pentandra* for lashing to make houses); 5% crafts; and for commercial products – 24% timbers (only 26% of these trees were of the required 26 cm minimum dbh), 22% non-timber products and 7.6% exudates.

The forests around Iquitos include numerous edible species. Vásquez and Gentry (1989) recorded the uses of c. 220 fruits consumed by the local inhabitants, and determined that of 193 fruit species consumed regularly, 120 were exclusively wild-harvested. The taxonomic diversity of the fruits is outstanding: 39 families were represented and in 34 were species with wild-harvested fruits.

The family Palmae is preeminent, with 23 species producing wild-harvested fruits (Vásquez and Gentry 1989). The fruit of the palm *Mauritia flexuosa* is the most important. It provides a number of saleable products (Padoch 1988; Mejía 1992), including the raw ripe fruit ("aguaje"); "masa", a seedless mashed pulp; "curichi", a frozen drink; "chupetes" or popsicles and ice cream. Another important palm is *Jessenia bataua*, which produces an oil very similar to olive oil. The protein quality of its fruits is comparable to animal protein (Balick and Gershoff 1981); these fruits also are made into ice cream and sold in Iquitos (Kahn 1988). Young leaves of *Euterpe precatoria* (known as "palmito") are used to make salads, and its roots are used to treat some hepatic and renal problems (Mejía 1992).

Other notable fruit species found in the Iquitos region are *Myrciaria dubia* ("camu-camu") (Myrtaceae), which is exceptionally rich in vitamin C; and *Grias peruviana* ("sacha mangua") (Lecythidaceae), whose mesocarp is rich in vitamin A. Additional notable fruit-producing families in the region include Apocynaceae, Annonaceae, Moraceae and Sapotaceae.

Native traditions have been well preserved among the region's peoples, who generally continue to use many forest plants as medicinals. For example, *Martinella* is widely used as an eye medicine (Gentry and Cook 1984). A tea made from the leaves of *Spondias mombin* is used to treat diarrhoea, stomach ache, vaginal infections and dermatitis; a tonic made from its bark is claimed to be an effective contraceptive (Peters and Hammond 1990). Some medicinal plants are exported. For instance, much of the world's curare, which is used as a muscle relaxant in modern medicine, comes from the Peruvian Amazon. Also, there has been recent large-scale export of *Uncaria* bark from Amazonian Peru to treat prostate cancer (Gentry 1993). Other examples of species used medicinally are *Abuta grandifolia* – stems and/or roots to treat sterile women, post-menstrual haemorrhages and rheumatism; *Abuta solimoesensis*, bark to treat anaemia and rheumatism; *Alchornea castaneifolia*, bark to treat rheumatism, arthritis, muscle pain and colds; *Ficus insipida* var. *insipida*, latex as a vermifuge; and *Ficus trigona*, as an antidiarrhoetic (Duke and Vásquez 1994).

The Iquitos region has important timber species including *Caryocar*, *Hymenaea*, *Iryanthera*, *Virola*, *Endlichera*, *Eschweilera*, *Aniba*, *Ocotea* and *Swartzia*. Approximately 70% of the wood used to build houses in Iquitos is from *Caraipa utilis* (Vásquez 1991) and in Jenaro Herrera, from *Haploclathra cordata* (Vásquez 1993). Both of these species are endemic to the Iquitos region. Other species frequently used for construction include *Aspidosperma nitidum* (Apocynaceae); *Caraipa densifolia, C. tereticaulis* (Clusiaceae); *Duroia paraensis* (Rubiaceae); and *Eschweilera turbinata* (Lecythidaceae) (Soto and Vásquez 1989; cf. Pinedo-Vásquez, Zarin and Jipp 1992).

The climbing palm *Desmoncus* has been used locally as a fibre source for rattan products (Gentry 1986). Also, there is a thriving fibre industry based on the aerial roots of *Philodendron solimoesensis* (Araceae). Stems of various other monocotyledons including *Ischnosiphon* (Marantaceae) and *Heteropsis* (Araceae) are similarly used locally (Gentry 1992).

Social and environmental values

The Tamshiyacu-Tahuayo Communal Reserve, as well as similar areas not officially designated as reserves (Pinedo-Vásquez, Zarin and Jipp 1992), are a source of products from the fauna and flora for many rural inhabitants. The Iquitos region forests can offer important refuge for wild animals threatened by over-hunting. This is especially true in the areas north of the Napo; near the Brazilian border the hunting pressure has not been as severe.

The Iquitos region is embraced by the large Napo and Upper Amazon lowlands Endemic Bird Area (EBA B19), which extends from southernmost Colombia and eastern Ecuador eastward into northern Peru and westernmost Brazil. Ten species of birds are limited to this area, although they essentially represent the most restricted species of a (distributionally poorly known) suite of birds that are confined to the river islands, riverine forest and várzea forest of the Amazon Basin rivers. The birds in this EBA, just one of which is considered threatened, are seemingly confined to the tierra-firme or várzea forests.

Economic assessment

The city of Iquitos has a population of over 269,000, and is the capital of the Department of Loreto (now Amazonas region), which has a low population density – 1.4 inhabitants per km². Most residents of the many riverine villages of the region are largely agriculturalists and generalists who may also fish, collect forest products, hunt, create or prepare some products (e.g. handicrafts, alcoholic drinks, medicinals) and engage in some wage labour (Padoch 1992). Certain native fruit species are beginning to be grown as crops, including *Grias neuberthii, Inga minutula, Muntingia calabura, Passiflora quadrangularis, Pourouma cecropiifolia* and *Psidium guajava* (Vásquez and Gentry 1989).

Wild-harvested fruits are important products in Amazonian Peru. Including those consumed occasionally, 219 wild-harvested fruit species are presently used by the people of the Iquitos region (Vásquez and Gentry 1989 and unpublished). The annual value per ha of a population of *Myrciaria dubia* in flooded forest is estimated to be US$5700–7620; the value of the fruit produced by *Grias peruviana* is US$4242, whereas *Spondias mombin* produces US$378 worth of fruit (Peters, Gentry and Mendelsohn 1989).

Of the 32 tree species at San Rafael with non-monetary use-values of 2.0 or more (based on addition of major and minor categorical uses), 78% have markets in Iquitos. Over-exploitation can result: *Brosimum paraense* and *Heisteria pallida*, with use-values of 3.0 and dependable markets, have been so depleted that only two trees of each were found in the sampled 10% of the 50-year-old secondary forests. In 1984 the community of San Rafael established a communal forest reserve (8 km²), which has detailed community rules for sustainable use (Pinedo-Vásquez *et al.* 1990; Pinedo-Vásquez, Zarin and Jipp 1992).

Funds raised by small-scale ecological tourism were used to build Quebrada Sucusari, a research laboratory on the Napo River with 1012 km². It is operated by the Amazon Center for Environmental Education and Research (through the ACEER Foundation).

Threats

Habitat destruction by deforestation, especially along navigable rivers, is due to population pressure (Aramburú 1984). Some species are deleteriously exploited for their products; e.g. the palms *Mauritia flexuosa* and *Euterpe precatoria* are cut down (Padoch 1988; Pinedo-Vásquez, Zarin and Jipp 1992). Over-hunting for food (e.g. deer, peccary) and trade (e.g. live birds and primates) has reduced populations of important agents for dispersal in much of the region.

Conservation

The Tamshiyacu-Tahuayo Communal Reserve (3225 km²) is located in north-eastern Peru between the rivers Tamshiyacu, Tahuayo and Yavari Miri (or Mirim) (Map 45), approximately between 4°11'–4°56'S and 72°30'–73°17'W. It is predominantly non-flooded forest (tierra firme), and has 13 primate species; for two (*Saguinus mystax mystax* and *Cacajao calvus rubicundus*), this is the first protected area.

This reserve is managed in ways to benefit the local people; there are management programmes for the aguaje palm, ungulates, fisheries and swidden-fallow agroforestry (Bodmer *et al.* 1990). At least 44 village and inter-village forest and/or lake reserves which total 128 km² have been established by local communities to prevent misuse and develop sustainable extraction of their natural resources. Their efforts are aided by the grassroots organization FEDECANAL (Federación Departamental de Campesinos y Nativos de Loreto) (Pinedo-Vásquez, Zarin and Jipp 1992).

The Iquitos region includes only the eastern buffer zone of the Pacaya-Samiria National Reserve, which is an important preserve for wetland forests and the region's aquatic fauna, and the only officially designated large reserve in northern Amazonian Peru. Several national and international conservation organizations are conducting intensive research in the area in order to develop strategies to protect this reserve.

North of the Amazon River on the border in Colombia (Leticia municipality) is Amacayacu Natural National Park (1700 km²), between 3°02'–3°50'S and 69°54'–70°20'W (INDERENA 1990). Peru may extend the protection to form a large binational park, which could become one of the most important conservation units in Amazonia.

References

Aramburú, C.E. (1984). Expansion of the agrarian and demographic frontier in the Peruvian selva. In Schmink, M. and Wood, C.H. (eds), *Frontier expansion in Amazonia*. University of Florida Press, Gainesville. Pp. 153–179.

Balick, M.J. and Gershoff, S.N. (1981). Nutritional evaluation of the *Jessenia bataua* palm: source of high quality protein and oil from tropical America. *Econ. Bot.* 35: 261–271.

Bodmer, R., Penn, J., Fang, T.G. and Moya, L. (1990). Management programmes and protected areas: the case of the Reserva Comunal Tamshiyacu-Tahuayo, Peru. *Parks* 1: 22–25.

Duke, J.A. and Vásquez, R. (1994). *Amazonian ethnobotanical dictionary*. CRC Press, Boca Raton, Florida, U.S.A. 215 pp.

Encarnación, F. (1985). Introducción a la flora y vegetación de la Amazonia peruana: estado actual de los estudios, medio natural y ensayo de una clave de determinación de las formaciones vegetales en la llanura amazónica. *Candollea* 40: 237–252.

Gentry, A.H. (1986). Sumario de patrones fitogeográficos neotropicales y sus implicaciones para el desarrollo de la Amazonia. *Rev. Acad. Col. Cienc. Exactas Fís. Nat.* 16: 101–116.

Gentry, A.H. (1988a). Changes in plant community diversity and floristic composition on environmental and geographical gradients. *Ann. Missouri Bot. Gard.* 75: 1–34.

Gentry, A.H. (1988b). Tree species richness of Upper Amazonian forests. *Proc. Natl. Acad. Sci. USA* 85: 156–159.

Gentry, A.H. (1991). The distribution and evolution of climbing plants. In Putz, F.E. and Mooney, H.A. (eds), *Biology of vines*. Cambridge University Press, Cambridge, U.K. Pp. 3–52.

Gentry, A.H. (1992). New nontimber forest products from western South America. In Plotkin, M. and Famolare, L. (eds), *Sustainable harvest and marketing of rain forest products*. Island Press, Washington, D.C. Pp. 125–136.

Gentry, A.H. (1993). Tropical forest biodiversity and the potential for new medicinal plants. In Kinghorn, A.D. and Balandrin, M.F. (eds), *Human medicinal agents from plants*. American Chemical Society Symposium Series 534. Pp. 13–24.

Gentry, A.H. and Cook, K. (1984). *Martinella* (Bignoniaceae): a widely used eye medicine of South America. *J. Ethnopharmacology* 11: 337–343.

Gentry, A.H. and Ortiz-S., R. (1993). Patrones de composición florística en la Amazonia peruana. In Kalliola, R., Puhakka, M. and Danjoy, W. (eds), *Amazonia peruana— vegetación húmeda subtropical en el llano subandino*. Proyecto Amazonia Universidad de Turku (PAUT) and Oficina Nacional de Evaluación de Recursos Naturales (ONERN). Jyväskylä, Finland. Pp. 155–166.

INDERENA (1990). *Nuevos parques nacionales de Colombia*. Instituto Nacional de los Recursos Naturales Renovables y del Ambiente (INDERENA), Bogotá. 213 pp.

Kahn, F. (1988). Ecology of economically important palms in the Peruvian Amazon. *Advances Economic Botany* 6: 42–49.

Kalliola, R., Puhakka, M., Salo, J., Tuomisto, H. and Ruokolainen, K. (1991). The dynamics, distribution and classification of swamp vegetation in Peruvian Amazonia. *Ann. Bot. Fennici* 28: 225–239.

Meggers, B.J. (1971). *Amazonia: man and land in a counterfeit paradise*. Aldine and Atherton, Chicago. 182 pp.

Mejía, K. (1992). Las palmeras en los mercados de Iquitos. *Bull. Inst. Fr. Etudes Andines* 21: 755–769.

ONERN (1975). *Inventario, evaluación e integración de los recursos naturales de la zona de Iquitos, Nauta, Requena y Colonia Angamos*. Oficina Nacional de Evaluación de Recursos Naturales (ONERN), Lima. 269 pp.

ONERN (1976). *Mapa ecológico del Perú. Guía explicativa*. ONERN, Lima. 147 pp.

Padoch, C. (1988). Aguaje (*Mauritia flexuosa* L.f.) in the economy of Iquitos, Peru. *Advances Economic Botany* 6: 214–224.

Padoch, C. (1992). Marketing of non-timber forest products in Western Amazonia: general observations and research priorities. *Advances Economic Botany* 9: 43–50.

Peñaherrera, C. (1989). *Atlas del Perú*. Instituto Geográfico Nacional, Lima. 399 pp.

Peters, C.M. and Hammond, E.J. (1990). Fruits from the flooded forest of Peruvian Amazonia: yield estimates for natural populations of three promising species. *Advances Economic Botany* 8: 159–176.

Peters, C.M., Gentry, A.H. and Mendelsohn, R.O. (1989). Valuation of an Amazonian rainforest. *Nature* 339: 655–656.

Pinedo-Vásquez, M., Zarin, D. and Jipp, P. (1992). Community forest and lake reserves in the [Peruvian Amazon]: a local alternative for sustainable use of tropical forests. *Advances Economic Botany* 9: 79–86.

Pinedo-Vásquez, M., Zarin, D., Jipp, P. and Chota-Inuma, J. (1990). Use-values of tree species in a communal forest reserve in Northeast Peru. *Conserv. Biol.* 4: 405–416.

Prance, G.T. (1979). Notes on the vegetation of Amazonia III. The terminology of Amazonian forest types subject to inundation. *Brittonia* 31: 26–38.

Räsänen, M.E., Salo, J.S. and Jungner, H. (1991). Holocene floodplain lake sediments in the Amazon: ^{14}C dating and palaeoecological use. *Quaternary Science Reviews* 10: 363–372.

Rudas, A. and Pipoly, J. (in prep.). *Flórula del Parque Nacional Natural Amacayacu*. Missouri Botanical Garden, St. Louis, U.S.A.

Ruokolainen, K. and Tuomisto, H. (1993). La vegetación de terrenos no inundables (tierra firme) en la selva baja de la Amazonia peruana. In Kalliola, R., Puhakka, M. and Danjoy, W. (eds), *Amazonia peruana– vegetación húmeda subtropical en el llano subandino*. PAUT and ONERN. Jyväskylä, Finland. Pp. 139–153.

Soto, S.T. and Vásquez, R. (1989). *Maderas redondas de uso estructural: un material de construcción a revalorar en la selva peruana*. Consejo Nacional de Ciencia y Tecnología (CONCYTEC), Lima. 60 pp.

Tuomisto, H. (1993). Clasificación de vegetación en la selva baja peruana. In Kalliola, R., Puhakka, M. and Danjoy, W. (eds), *Amazonia peruana – vegetación húmeda subtropical en el llano subandino*. PAUT and ONERN. Jyväskylä, Finland. Pp. 103–112.

Vásquez, R. (1991). *Caraipa* (Guttiferae) del Perú. *Ann. Missouri Bot. Gard.* 78: 1002–1008.

Vásquez, R. (1993). Una nueva *Haploclathra* (Clusiaceae) de la Amazonia peruana. *Novon* 3: 499–501.

Vásquez, R. and Gentry, A.H. (1989). Use and misuse of forest-harvested fruits in the Iquitos area. *Conserv. Biol.* 3: 350–361.

Vásquez, R. and Gentry, A.H. (in prep.). Catálogo de los frutos comestibles de la Amazonia peruana. Universidad Nacional de la Amazonia Peruana, Iquitos, Peru and Missouri Botanical Garden, St. Louis, U.S.A.

Vásquez, R. and Pipoly, J. (in prep.). *Flórula de las reservas biológicas de Iquitos*. Universidad Nacional de la Amazonia Peruana, Iquitos, Peru and Missouri Botanical Garden, St. Louis, U.S.A.

Author

This Data Sheet was written by Rosa Ortiz-S. (Missouri Botanical Garden, P.O. Box 299, Saint Louis, MO 63166-0299, U.S.A.).

AMAZONIA: CPD SITE SA10

TAMBOPATA REGION
Peru

Location: South-eastern Peru, between about latitudes 12°30'–14°00'S and longitudes 68°50'–70°30'W.

Area: c. 15,000 km².

Altitude: 250–3000 m.

Vegetation: Subtropical montane to premontane wet and moist forests; below 500 m, tropical lowland moist forest, swamp forest, swamps, oxbow lakes and patches of open savanna.

Flora: 2500–3000 species; numerous endemic species.

Useful plants: Timber species, Brazil nut, *Hevea* rubber; many species provide fruits or other products used locally.

Other values: Entire river-drainage protected; Amerindian peoples and knowledge; potential germplasm resources; world's greatest lowland records of birds – 554 spp., and butterflies – 1217 spp.; ecotourism.

Threats: Subsistence agriculture and cattle-ranching, with a population explosion – mostly immigration from highlands; selective logging; gold mining pollutes Madre de Dios River and its major tributaries with mercury.

Conservation: 1 National Sanctuary (Pampas del Heath), 2 Reserved Areas (Tambopata Explorer's Inn Reserve and Tambopata-Candamo Reserved Zone).

Geography

The Tambopata region of the south-western Amazon Basin and nearby Andes comprises the lowlands of the area south of the Madre de Dios River and east of the town of Puerto Maldonado in Madre de Dios Department, and the montane headwaters of the Tambopata River in Puno Department (Map 46). The region could be expanded phytogeographically to include parts of adjacent Acre in western Brazil and Pando and La Paz departments in northern Bolivia.

The montane portion of this region (500–3000 m) belongs to the Madre de Dios physiographic province (Young 1992). Geologically, it is characterized by very steep slopes and Early Palaeozoic and Early Tertiary sedimentary bedrock (Peñaherrera 1989). The information on the vegetation, the flora and conservation of this upland portion of the region are covered in the Data Sheet for the Eastern Slopes of the Peruvian Andes (CPD Site SA37).

The lowland portion of the Tambopata region, below 500 m, is characterized by gentle relief and Quaternary sediments. Rainfall at Puerto Maldonado is near 2000 mm annually and the 5-month dry season has 3 months that average less than 60 mm (ONERN 1972). The median annual temperature is 12°–13°C, so the Tambopata region is considered subtropical in the Holdridge life-zone system (ONERN 1976). During the austral winter, strong cold fronts ("friajes") are common from May to July; temperatures can drop to 7°C, which is a potential limiting factor for many tropical plants.

Another geographical distinction of this region is the absence of many of the habitats that characterize northern Upper Amazonia. The Tambopata region does not have white-sand soils and the various seasonally inundated types of forest called "tahuampa" in Peru and "várzea" and "igapó" in Brazil. Instead, the habitat mosaic is related to riverine dynamics and associated successional changes. Riverine

meander loops may grow 25 m every year, implying that the entire meander belt is swept out and replaced by younger deposits every 500–1000 years (Terborgh 1990). There is evidence of these riverine processes in 26% of lowland Peruvian Amazonia (Salo *et al.* 1986).

In addition to the younger succession, there are more subtle vegetational differences, which are associated with relatively rich alluvial soils, older less fertile and leached clay soils, or presumably old and less fertile sandy-clay soils, each supporting distinctive floristic ensembles (Gentry 1988). Part of the habitat mosaic is also made up of a complex suite of sites with impeded drainage, ranging from oxbow lakes to swamp forests.

Open savannas of the Pampas del Heath occupy shallow basins along the Heath River on the Bolivian border. A variety of inceptisols constitute the soils, which are mostly acidic and with organic content ranging from 8.1% to 1.6% (Denevan 1980).

Vegetation

Above 500 m three upland subtropical forests occur: upper montane, lower montane and premontane (see Data Sheet for the Eastern Slopes of the Peruvian Andes, CPD Site SA37). The upper and lower montane forests occupy a narrow belt on steep slopes.

The dominant vegetation of the lowland portion of the Tambopata region is tropical moist forest. Since the area is at the south-western periphery of Amazonia, there is a strong dry season when many of the trees are deciduous. Nevertheless, the mature forest is physiognomically typical rain forest with a canopy 30 m high, frequent emergents (the most frequent being *Bertholletia excelsa*, the Brazil nut), and numerous palms, lianas and epiphytes.

CPD Site SA10: Tambopata region, Peru. Río Tambopata, in Tambopata-Candamo Reserved Zone.
Photo: WWF/André Bärtschi.

Natural successional processes associated with changes in river channels are prevalent in much of Amazonia (Salo *et al.* 1986; Kalliola, Puhakka and Danjoy 1993). Similar to other sites in southern Amazonian Peru, the lowland portion of the Tambopata region is especially notorious for the conspicuous role of bamboo (mostly *Guadua weberbaueri*), which covers large patches (e.g. Terborgh 1985). In addition, where succession is further complicated by cyclical post-flowering die-offs, there are various younger successional stages along the meander belts of larger rivers. Successional growth in these areas typically consists of riverside strips dominated by *Tessaria* and *Gynerium*, followed by dominance of *Cecropia membranacea*, and then forests dominated by *Ficus insipida* and/or *Cedrela odorata* (e.g. Foster, Arce-B. and Wachter 1986; Salo *et al.* 1986; Foster 1990).

While part of the conspicuous habitat mosaic that characterizes north-western Amazonia is absent in Madre de Dios, there remain a number of conspicuously distinct types of vegetation other than the successional areas. At Tambopata, there are at least four main types of mature forest, associated respectively with alluvial soil, clayey soil, sandy-clay soil and poorly drained swamps (Erwin 1985). One-ha plots of trees and lianas 10 cm or more in dbh document the striking floristic differences. The overlap in species between pairs of plots on similar substrates is over 30%, but on different substrates less than 10% (Gentry 1988). About 600 such species occur in each plot – which is impressive, but significantly lower than in similar samples from the Iquitos region (CPD Site SA9). The overall number of species at Tambopata is high due to the many distinctive habitats.

Another important type of vegetation in the lowlands is the open savanna that occurs near the border with Bolivia, east of Palma Real River. This savanna is surrounded by or mixed with three other vegetation types: palm swamp forest (with *Euterpe precatoria* and *Jessenia* sp.), open scrubland dominated by *Curatella americana* and seasonal forest with *Calophyllum* (Denevan 1980).

Flora

The Tambopata region is botanically almost unexplored. The upland portion may comprise c. 140 families of vascular plants (León, Young and Brako 1992); notable are Alzateaceae, Arecaceae, Lauraceae, Moraceae and Rubiaceae (see CPD Site SA37, on Eastern Slopes of the Peruvian Andes).

The flora of the lowland forests includes nearly 2000 species. Records from the Tambopata Explorer's Inn Reserve (5.5 km²) show almost 1400 species of vascular plants (Reynel and Gentry, in prep.). At this locality, the two largest families are Leguminosae (106 spp.) and Rubiaceae (101 spp.), as is typical in much of Amazonia. Bignoniaceae, the predominant liana family, is third with 58 species, followed closely by Moraceae and ferns and their allies with 55 and 54 species. While these are always among the most speciose families and groups in Amazonia, the preeminence of Bignoniaceae is greater than at any other site for which comparable data are available. Euphorbiaceae, Melastomataceae and Piperaceae each have 44 species in the Florula list, whereas Araceae, palms, Solanaceae, Gramineae, orchids, Annonaceae,

Lauraceae and Sapindaceae each have 28–33 species. Families with c. 20 species each include Meliaceae, Flacourtiaceae, Cyperaceae, Acanthaceae and Apocynaceae. Thus the familial composition of this Tambopata Reserve is typically Upper Amazonian (cf. Gentry 1990; Gentry and Ortiz-S. 1993).

At the generic level the flora of the Tambopata Explorer's Inn Reserve is also neotropically typical. The largest genera are *Piper* with 34 species and *Psychotria* and *Inga* with 32. These are the same genera that are most speciose on Barro Colorado Island, Panama and among the most speciose genera at all sites for which comparable data are available (Gentry 1990). Other especially speciose genera at Tambopata are *Miconia, Solanum, Arrabidaea, Paullinia, Ficus* and *Philodendron*. Of these, the 14 species of *Arrabidaea* are most distinctive, since it is usually the largest Bignoniaceae genus present but not among the most speciose genera at any other site known. While the data are very incomplete and not fully tabulated, c. 12% of the Tambopata Florula species may be endemic to the south-western part of Upper Amazonia (i.e. Madre de Dios in Peru, Acre in Brazil and parts of Pando and La Paz in Bolivia). Thus, even though its flora is typically Amazonian at the familial and generic levels, south-western Amazonia clearly warrants separate conservation focus.

Moreover, there are some striking floristic differences between Tambopata and Manu National Park (CPD Site SA11), which is also in Madre de Dios Department but farther north-west and nearer the base of the Andes. For example, *Bertholletia excelsa* is common at Tambopata but very rare at Manu; locally dominant species such as *Lueheopsis hoehnei* and *Tabebuia insignis* are unknown in Manu NP.

There are two additional floristically distinct habitats. The open savannas of the Pampas del Heath have a completely different flora. The majority of species of the Pampas del Heath are not represented in other areas of the Tambopata region and are otherwise unknown from Peru, although they are known from similar poorly drained habitats to the south. The foothill ridges along the Tavara River also have a flora that is quite different from that at the Tambopata Reserve (Conservation International Rapid Assessment Program team, unpublished data).

Useful plants

The upland portion of the Tambopata region was explored early by extractors of timber and quinine bark (*Cinchona* spp.); today only the former activity continues. Huge

MAP 46. TAMBOPATA REGION, PERU (CPD SITE SA10)

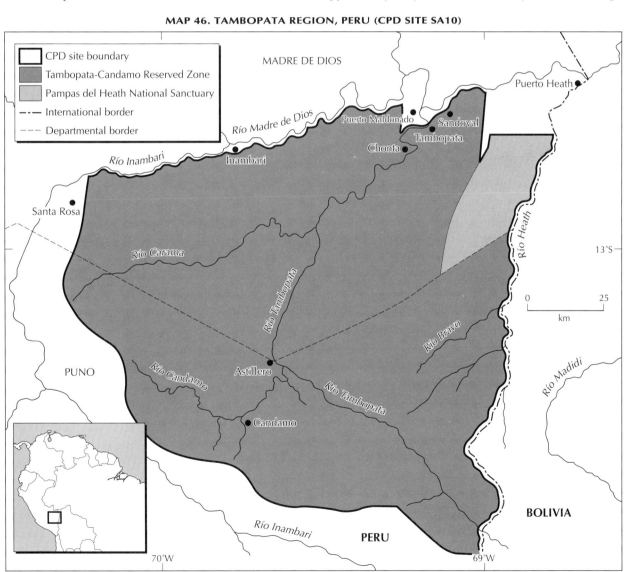

Podocarpus trees, perhaps the most valuable Andean timber, occur on the foothill ridges. A large number of species are commercialized for their timber, and saw-wood from at least 150 species mainly from the lowlands is sold in Puerto Maldonado. There are several wild conspecifics of cultivated fruit trees – including *Theobroma cacao, Bactris* aff. *gasipaes, Pourouma cecropiifolia, Inga edulis, Pouteria macrophylla* – mostly concentrated in alluvial-soil forests (Phillips *et al.* 1994). A wild pineapple (*Ananas* sp.) that might be the ancestor of the cultivated pineapple is common in the Pampas del Heath.

Harvesting of Brazil nuts constitutes a major local industry, the species occurring at low density (c. 1 tree per ha) in most of the lowland forest types. Rubber tapping was formerly important in the region and continues along the Brazilian border; the only rubber species in the reserve is probably *Hevea guianensis* (which produces inferior rubber to *H. brasiliensis*). Some local species provide useful fruits. Ethnobotanical studies indicate that some of the local "mestizos" know uses for up to 90% of the forest trees, mostly for subsistence construction, food and medicinal purposes (Phillips 1993; Phillips and Gentry 1993a, 1993b). The most useful wild species for local mestizo and indigenous peoples are palms, especially *Iriartea deltoidea* and *Euterpe precatoria*.

Social and environmental values

The human population in the lowlands of the Tambopata region consists of several indigenous groups (Arasaire, Ese-Eje, Huarayo), and an ethnically and culturally mixed group of recent migrants and longtime residents, mostly from the neighbouring highlands of Cuzco and Puno. The savanna in Pampas del Heath was inhabited by the Tiatinagua, a group considered to have gone extinct (Denevan 1980). Local settlers, especially the now largely acculturated Ese-Eje Amerindians, continue to harvest many products from the nearby forests, which are important to them not only economically but as part of their cultural heritage.

Living conditions in the region are at the medium poverty level. The agricultural products are maize, cassava, rice and plantains, which are sold in the local market or exported to Cuzco. All the major economic activities are extractive – gold mining and timber exploitation are the principal endeavours.

An important environmental value of the Tambopata-Candamo Reserved Zone is that it protects the entire drainage of a significant river system – one of the few places in the world where this has been established. Thus the now well-known values of natural forest in controlling flooding and erosion may be especially important here.

The South-east Peruvian lowlands Endemic Bird Area (EBA B30), which embraces the continuous lowlands from Manu south-eastward to the Tambopata region, is home to 15 restricted-range bird species, none of which occurs elsewhere. They are primarily confined to humid riverine and floodplain forests, with associated successional habitats also utilized. Two of these endemics are considered threatened, and the integrity of this reserve and the Manu region is essential for their long-term survival. The adjacent slopes of the Andes are home to a further 43 restricted-range species occurring in two EBAs.

Several tourist camps in operation along the Tambopata River and several on the Madre de Dios River near Puerto Maldonado call attention to the biotic richness of the region. The Tambopata Reserve is internationally famous as the site of the world's greatest lowland records of birds (554 spp., Parker 1991) and butterflies (1217 spp., Lamas, Robbins and Harvey 1991 and pers. comm.), and many of the tourists are drawn by this biological diversity. Puerto Maldonado proudly features signs proclaiming it the world's centre of biodiversity. The overall impact of tourism in the local economy is unknown.

Economic assessment

Due to the importance of Brazil-nut harvesting to the local economy, this is one of the parts of Amazonia where extractive reserves might be most economically feasible. Unfortunately, political unrest, low market prices and transportation problems (the harvest is flown out) have all but destroyed this industry. Only one of the several Brazil-nut processing companies that once operated in Puerto Maldonado is still functioning.

While a recent study of the value of fruits and other natural products produced by the forest at Tambopata was disappointing as compared to some estimated potential values of Amazonian minor forest products, it did show that there are many valuable forest resources in the region and that harvest values could be substantially increased with appropriate technology and market development (Phillips 1993). The region's many wild congeners of cultivated crops are also valuable as sources of germplasm, although their actual economic worth is difficult to quantify. This region is a part of one of the major Vavilov centres of plant domestication, suggesting that it has special significance.

Threats

Subsistence agriculture and cattle-ranching have expanded in the region in recent years with a population explosion, largely due to immigration from the adjacent highlands. It is especially unfortunate that the alluvial-soil forests along the Tambopata River have been the most impacted by subsistence agriculture, as they have the most potential as subsistence extractive reserves. Gold mining has been the main attraction for migrants, and increasingly pollutes the Madre de Dios River and its major tributaries just west of the Tambopata region with mercury used for extraction.

Several valuable timber species are locally threatened due to over-exploitation, including *Cedrela odorata, C. fissilis* and *Cedrelinga cateniformis*. There are few incentives for the local people to conserve valuable species, in part because the loggers receive only a small fraction of market prices and in part due to the difficulty of obtaining secure land tenure. Even the original Tambopata Reserve (5.5 km²) adjacent to the Explorer's Inn Tourist Camp has been subjected to poaching of wild animals and timber trees.

Conservation

The very interesting Pampas del Heath has been formally designated as a National Sanctuary encompassing 1021 km². The Tambopata-Candamo Reserved Zone (14,000 km²) was formally declared in 1990; its status has yet to be clarified by the Peruvian Government. Many settlers live within the boundaries of this zone but there is little or no effective control of their activities. There is hope that a significant portion of the region will be established as a National Park.

References

Denevan, W.M. (1980). The Rio Heath savannas of southeastern Peru. *Geoscience and Man* 21: 157–163.

Erwin, T.L. (1985). Tambopata Reserved Zone, Madre de Dios, Peru: history and description of the reserve. *Rev. Peruana Entomología* 27: 1–8.

Foster, R.B. (1990). Long-term change in the successional forest community of the Rio Manu floodplain. In Gentry, A.H. (ed.), *Four neotropical rainforests*. Yale University Press, New Haven. Pp. 565–572.

Foster, R.B., Arce-B., J. and Wachter, T.S. (1986). Dispersal and sequential plant communities in Amazonian Peru floodplain. In Estrada, A. and Fleming, T.H. (eds), *Frugivores and seed dispersal*. W. Junk, Dordrecht, The Netherlands. Pp. 357–370.

Gentry, A.H. (1988). Patterns of plant community diversity and floristic composition on environmental and geographical gradients. *Ann. Missouri Bot. Gard.* 75: 1–34.

Gentry, A.H. (1990). Floristic similarities and differences between southern Central America and Upper and Central Amazonia. In Gentry, A.H. (ed.), *Four neotropical rainforests*. Yale University Press, New Haven. Pp. 141–157.

Gentry, A.H. and Ortiz-S., R. (1993). Patrones de composición florística en la Amazonia peruana. In Kalliola, R., Puhakka, M. and Danjoy, W. (eds), *Amazonia peruana — vegetación húmeda tropical en el llano subandino*. PAUT and ONERN. Jyväskylä, Finland. Pp. 155–166.

Kalliola, R., Puhakka, M. and Danjoy, W. (eds) (1993). *Amazonia peruana – vegetación húmeda tropical en el llano subandino*. Proyecto Amazonia Universidad de Turku (PAUT) and Oficina Nacional de Evaluación de Recursos Naturales (ONERN). Jyväskylä, Finland. 265 pp.

Lamas, G., Robbins, R. and Harvey, D. (1991). A preliminary survey of the butterfly fauna of Pakitza, Parque Nacional Manu, Perú, with an estimate of its species richness. *Publ. Museo Hist. Nat., Ser. A. Zool.* 40: 1–19.

León, B., Young, K.R. and Brako, L. (1992). Análisis de la composición florística del bosque montano oriental del Perú. In Young, K.R. and Valencia, N. (eds), *Biogeografía, ecología y conservación del bosque montano en el Perú. Mem. Museo Hist. Nat.*, Universidad Nacional Mayor de San Marcos (UNMSM), Lima. Vol. 21. Pp. 141–154.

ONERN (1972). *Inventario, evaluación e integración de los recursos naturales de la zona de los ríos Inambari y Madre de Dios*. Oficina Nacional de Evaluación de Recursos Naturales (ONERN), Lima.

ONERN (1976). *Mapa ecológico del Perú. Guía explicativa*. ONERN, Lima. 147 pp.

Parker, T.A., III (1991). Birds of Alto Madidi. In Parker, T.A., III and Bailey, B. (eds), *A biological assessment of the Alto Madidi region and adjacent areas of Northwest Bolivia, May 18–June 15, 1990*. RAP (Rapid Assessment Program) Working Papers 1. Conservation International, Washington, D.C. Pp. 21–23.

Peñaherrera, C. (1989). *Atlas del Perú*. Instituto Geográfico Nacional, Lima. 399 pp.

Phillips, O. (1993). The potential for harvesting fruit in tropical rainforests: new data from Amazonian Peru. *Biodiversity and Conservation* 2: 18–38

Phillips, O. and Gentry, A.H. (1993a). The useful plants of Tambopata, Peru I: statistical hypothesis tests with a new quantitative technique. *Econ. Bot.* 47: 15–32.

Phillips, O. and Gentry, A.H. (1993b). The useful plants of Tambopata, Peru II: additional hypothesis testing in quantitative ethnobotany. *Econ. Bot.* 47: 33–43.

Phillips, O., Gentry, A.H., Reynel, C., Wilken, P. and Gálvez-D., C. (1994). Quantitative ethnobotany and Amazonian conservation. *Conserv. Biol.* 8: 225–248.

Salo, J., Kalliola, R., Häkkinen, I., Mäkinen, Y., Niemelä, P., Puhakka, M. and Coley, P.D. (1986). River dynamics and the diversity of Amazon lowland forest. *Nature* 322: 254–258.

Terborgh, J. (1985). Habitat selection in Amazonian birds. In Cody, M.L. (ed.), *Habitat selection in birds*. Academic Press, New York. Pp. 311–338.

Terborgh, J. (1990). An overview of research at Cocha Cashu Biological Station. In Gentry, A.H. (ed.), *Four neotropical rainforests*. Yale University Press, New Haven. Pp. 48–59.

Young, K.R. (1992). Biogeography of the montane forest zone of the eastern slopes of Peru. In Young, K.R. and Valencia, N. (eds), *Biogeografía, ecología y conservación del bosque montano en el Perú. Mem. Museo Hist. Nat.*, UNMSM, Lima. Vol. 21. Pp. 119–140.

Authors

This Data Sheet was written by Dr Alwyn H. Gentry (Missouri Botanical Garden, P.O. Box 299, St. Louis, MO 63166-0299, U.S.A.) and Dra. Blanca León (Universidad Nacional Mayor de San Marcos, Museo de Historia Natural, Avenida Arenales 1256, Apartado 14-0434, Lima-14, Peru).

AMAZONIA: CPD SITE SA11

LOWLANDS OF MANU NATIONAL PARK: COCHA CASHU BIOLOGICAL STATION
Peru

Location: Upper Amazon of southern Peru, between latitudes 11°16'–13°11'S and longitudes 71°10'–72°25'W.

Area: c. 7500 km²; Cocha Cashu Biological Station c. 10 km².

Altitude: c. 300–400 m.

Vegetation: Evergreen tropical rain forest.

Flora: c. 1900 species; numerous species locally endemic or restricted to south-western Amazonia.

Useful plants: Timber trees, edible fruits.

Other values: Wilderness; Amerindians; research base for large-scale and undisturbed natural processes; potential germplasm resources; ecotourism.

Threats: Potential road building, petroleum and mining exploration; inadequate budget for park protection.

Conservation: National Park, Biosphere Reserve.

Geography

Manu National Park, one of the largest and most significant National Parks in South America, extends from above 4000 m in the Andean Cordillera of Peru's Department of Cuzco to 300 m on the southern Amazonian floodplain in the Department of Madre de Dios (Map 47) (MacQuarrie 1992). The park includes the entire watershed of the Manu River and part of the catchment area of the Alto Madre de Dios River. Manu NP covers c.15,328 km², with roughly half in the pre-Andean subregion of Amazonia (below 500 m) where this site description is focused.

Most of the Manu River floodplain, including the Cocha Cashu field station, is at c. 400 m in elevation. The soil is among the most fertile in the tropics, judging from the density of animals and the growth rate of trees (Foster 1990a). Research in the park has been done mostly in the floodplain of the Manu River, especially at Cocha Cashu Biological Station (11°54'S, 71°22'W).

The Cocha Cashu study area is c. 10 km². In the area around the station the average annual temperature is 24°C and rainfall c. 2000 mm. Most of the rainfall occurs between November and May, while June through October normally receives less than 100 mm, though the year-to-year variation in the intensity and duration of the dry season is considerable (Terborgh 1990).

Vegetation

The Manu River floodplain is mostly covered by evergreen tropical forest, although a few species lose their leaves during the June–October dry season. One of the unique features of this almost pristine region is that the dynamic successional processes that characterize the vegetation of lowland Amazonia are preserved intact. While only a small fraction of the region is covered by obviously early successional vegetation, the entire floodplain may be eroded

and redeposited every 500–1000 years (Terborgh 1990). Thus most of the floodplain forest, including the physiognomically most cathedral-like forest around the Cocha Cashu field station, may represent long-term vegetational succession (Gentry and Terborgh 1990). The Manu River area is one of very few places where the habitat mosaic of beaches, oxbow lakes and various forest types that result from these riverine processes can be observed intact, and their effects on the maintenance of species richness evaluated (Terborgh 1983, 1990; Foster 1990b).

Palms are an especially characteristic element of Manu park's vegetation. Three of the five commonest tree species in the mature floodplain forest are palms, and the palms together constitute c. one-sixth of stems with 10 cm or more dbh in this forest (Gentry and Terborgh 1990). Lianas are also very prevalent in the mature forest at Cocha Cashu – 79 lianas 2.5 cm or more in diameter representing 43 species were found in a 0.1-ha sample (Gentry 1985). Another distinctive physiognomic aspect of the vegetation of the rich-soil floodplain forest is the prevalence of strangling figs (*Ficus*). Many more fully developed stranglers are represented in a 1-ha tree plot at Cocha Cashu than in similar plots elsewhere in Amazonia (Gentry and Terborgh 1990). Some are truly massive, simultaneously strangling several host trees, and the gaps between their root columns can be so large that there are trails passing through these boles.

Flora

The Upper Amazonian rain forest represented by Manu is one of the most species-rich forest types in the world, and one of the only remaining regions of the rich-soil forests that originally characterized much of the eastern Andean forelands (Gentry and Ortiz-S. 1993). The species diversity of this region is generally greater than in most of Central Amazonia, and only slightly less than in the completely aseasonal forest farther north around Iquitos (CPD Site SA9). Although many new plant species have been discovered in Manu NP, it is not known to what extent they reflect local endemism rather than inadequate collection elsewhere.

In many ways the rich-soil area of Manu park is floristically more similar to southern Central America than the geographically nearer poor-soil areas of Amazonia (Gentry 1985). Several species previously known only from Central America and northernmost South America have been found. Presumably they represent part of a band of distinctive rich-soil flora that once extended along the base of the Andean/Amazonian interface. There are also a number of unanticipated and unexplained disjunctions – e.g. *Clytostoma campanulatum* was known only from the São Paulo region of Brazil.

Manu NP as a whole has at least 3000 vascular plant species, with the lowlands below 500 m having a documented flora of 1856 species in 751 genera of 130 families (Foster

1990a). The documented vascular flora of the Cocha Cashu area on recent alluvial soils is 1370 species belonging to 637 genera of 119 families. Leguminosae is the most important family, with over 90 species found in the floodplain-forest flora and over 140 species for the entire Manu River flora. However, the legumes constitute only 8% of the species; a dozen other families have 30 or more species, which is a typical familial composition for lowland Amazonia: Moraceae, Rubiaceae, Orchidaceae, Acanthaceae, Sapindaceae, Bignoniaceae, Araceae, Myrtaceae, Piperaceae, Sapotaceae and Solanaceae. Among the trees predominate Leguminosae, Moraceae, Sapotaceae and Lauraceae (Foster 1990a).

The largest genus is *Ficus*, which has its neotropical record with 35 species in the floodplain forest near the Cocha Cashu station. *Inga*, *Piper* and *Pouteria* have over 20 species; the largest liana genus is *Paullinia* with 19 species. The large numbers of species in genera such as *Ficus*, *Inga*,

MAP 47. MANU BIOSPHERE RESERVE, PERU

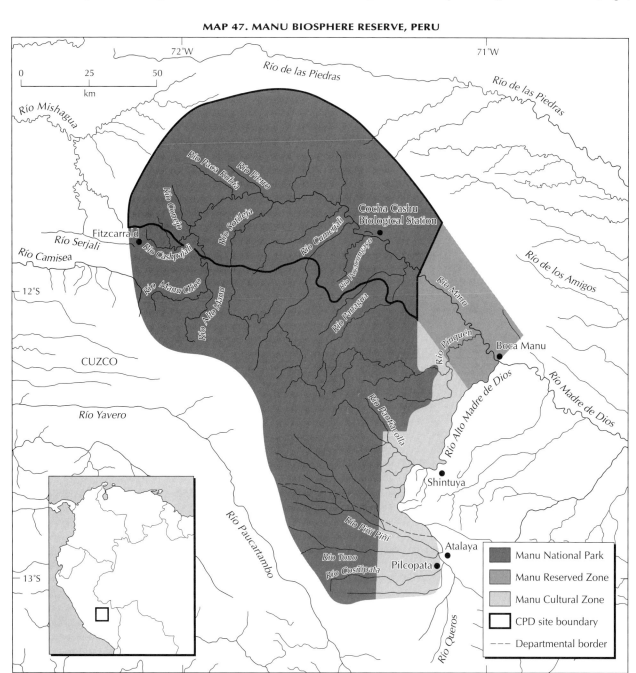

Piper, Pouteria and *Paullinia* are typical of neotropical forests on relatively rich soils. An interesting floristic peculiarity is the predominance of shrub species – the 375 shrub species known from the Manu floodplain constitute a quarter of the entire local flora.

Useful plants

The region includes nearly pure stands of mature *Cedrela odorata* (Spanish cedar) in successional forests on intermediate-age river terraces and has scattered large individuals of *Swietenia macrophylla* (bigleaf mahogany) in the floodplain forest. These have traditionally been perhaps the two most important Amazonian timber species, but they have been severely over-harvested, and today are usually encountered only rarely elsewhere. The now unique single-species stands of *Cedrela* are especially interesting, as they probably indicate that the well-known problem of *Hypsipyla* budworm attack on mahogany plantations is due more to their inappropriate placement on poor soils than to their density of planting.

A number of commercially important or potentially important fruit-tree species are very common in the Manu River floodplain forests (as well as other pre-Andean rich-soil forests). *Quararibea cordata* ("sapote") and *Theobroma cacao* (cacao) are among the ten commonest tree species at Cocha Cashu. Other frequently found tree species in genera with edible fruits include several palms (e.g. *Astrocaryum*), *Pouteria, Annona, Inga* and *Diospyros*. The forests of the rich-soil Upper Amazonian region represented by Manu NP probably include a disproportionate number of the general region's economically important plants, and they are exceptionally important to maintain germplasm for future programmes of genetic improvement (Gentry 1985; Phillips and Gentry 1993).

Social and environmental values

Between 300–500 Amerindians are believed to inhabit the Manu park region; they are isolated and have little contact with outsiders. The tribal groups represented are the Amahuaca, Machiguenga, Piro and Yaminahua (Baird 1984).

Cocha Cashu Biological Station was founded in 1969–1970, initially to facilitate investigation of the black caiman (*Melanosuchus niger*), South America's largest and commercially most valuable crocodilian. The station is one of the few research sites in the neotropics to offer an undisturbed ecosystem with a full complement of both top predators and their prey (Terborgh 1990). In the lowlands, the diversity of animal species includes 13 species of primates, giant armadillo, giant anteater, Amazon and giant otters, jaguar, ocelot, black and spectacled caimans and river turtles. Over 550 species of birds have been identified (Robinson and Terborgh 1990).

The Cocha Cashu field station enables scientists, students and managers to gain new insights on how an intact tropical forest ecosystem functions. More than 200 publications have been based on research at Cocha Cashu. Some processes can only be studied under near-pristine conditions in large intact forests, such as complex interactions involving successional dynamics over hundreds of years (Foster

1990b), seasonal changes in vertebrate foraging behaviours (Terborgh 1983) and the existence of keystone plant species that limit a forest's carrying capacity at critical times of the year (Terborgh 1986).

For many large vertebrate species, as well as such over-exploited plant species as in *Cedrela* and *Swietenia*, Manu NP provides one of the last conservation strongholds. Manu NP probably includes more species than any other conservation reserve in the world. Due to the vast area and altitudinal range within this important reserve, three Endemic Bird Areas (EBAs) are represented. The South-east Peruvian lowlands EBA (B30) is home to 15 restricted-range bird species, which inhabit the humid lowland forest up to c. 400 m; the Eastern Andean foothills of Peru EBA (B29) is centred between 600–2200 m, with 11 restricted-range species inhabiting humid forest; and the High Peruvian Andes EBA (B27), from 1800–4300 m, has 30 restricted-range birds occurring in the more semi-humid and arid vegetation. Three (primarily poorly known) threatened bird species occur at the lower altitudes in this region; no less than 12 such species inhabit the more degraded High Peruvian Andes EBA.

Economic assessment

The environment of the lowlands of Manu park owes its almost pristine quality to its location in the remote south-eastern corner of Peru. Cuzco, the nearest population centre, satisfies its need for food and wood from nearer sources. Larger markets (Lima and Arequipa) are 4–5 days away from the Manu region. Only the most valuable products, principally such prime hardwoods as *Cedrela, Swietenia* and *Cedrelinga*, can justify the high costs of transport.

The retention of large predators and high primate densities gives Manu NP a unique tourist potential. Ecotourism has recently been carried out around the park's fringes. Tourist camps exist within cultural and reserved zones adjacent to the park. Groom, Podolsky and Munn (1991) have considered this activity beneficial to the local economies; long-term effects on the indigenous populations and the wild animals are unknown.

Threats

According to Terborgh (1990), the short-term factors most likely to influence the development of south-eastern Peru are the possible discovery of mineral deposits and the price of petroleum. Various attempts to explore for oil or build a road through Manu NP have been blocked thus far. The greatest recent threat to the park's integrity has been a severe cutback in the number of park guards and its infrastructure.

Conservation

Manu National Park was established in 1973 covering 15,328 km², and in 1977 an area encompassing 18,812 km² was declared a Biosphere Reserve under UNESCO's Man and the Biosphere Programme. Manu NP is widely perceived as Peru's premier park. Public attention combined with significant financial support have made protection and management more effective for Manu NP than Peru's other conservation areas. Since 1970, the nearly constant presence of scientists at Cocha Cashu Biological Station has had a positive effect on the park's management. However, if more support for park guards and infrastructure is not forthcoming, the positive status may change radically. Moreover, the national economy will need to improve for long-term conservation to be secure.

References

Baird, V. (1984). Tropical treasure under threat – bid to s ave Peru's natural heritage. *Lima Times*, 5 October. Pp. 6–7.

Foster, R.B. (1990a). The floristic composition of the Rio Manu floodplain forest. In Gentry, A.H. (ed.), *Four neotropical rainforests*. Yale University Press, New Haven. Pp. 99–111.

Foster, R.B. (1990b). Long-term change in the successional forest community of the Rio Manu floodplain. In Gentry, A.H. (ed.), *Four neotropical rainforests*. Yale University Press, New Haven. Pp. 565–572.

Gentry, A.H. (1985). Algunos resultados preliminares de estudios botánicos en el Parque Nacional del Manu. In Ríos, M.A. (ed.), *Reporte Manu*. Centro de Datos para la Conservación, Universidad Agraria, La Molina, Peru. Pp. 2/1–2/22.

Gentry, A.H. and Ortiz-S., R. (1993). Patrones de composición florística en la Amazonia peruana. In Kalliola, R., Puhakka, M. and Danjoy, W. (eds), *Amazonia peruana – vegetación húmeda tropical en el llano subandino*. Proyecto Amazonia Universidad de Turku (PAUT) and Oficina Nacional de Evaluación de Recursos Naturales (ONERN). Jyväskylä, Finland. Pp. 155–166.

Gentry, A.H. and Terborgh, J. (1990). Composition and dynamics of the Cocha Cashu "mature" floodplain forest. In Gentry, A.H. (ed.), *Four neotropical rainforests*. Yale University Press, New Haven. Pp. 542–564.

Groom, M.J., Podolsky, R.D. and Munn, C.A. (1991). Tourism as a sustained use of wildlife: a case study of Madre de Dios, southeastern Peru. In Robinson, J.G. and Redford, K.H. (eds), *Neotropical wildlife use and conservation*. University of Chicago Press, Chicago and London. Pp. 393–412.

MacQuarrie, K. (1992). *El paraíso amazónico del Perú: Manu, parque nacional y reserva de la biosfera / Peru's Amazonian Eden: Manu National Park and Biosphere Reserve*. Francis O. Patthey e Hijos, Barcelona, Spain. 320 pp.

Phillips, O. and Gentry, A.H. (1993). The useful plants of Tambopata, Peru II: additional hypothesis testing in quantitative ethnobotany. *Econ. Bot.* 47: 33–43.

Robinson, S.K. and Terborgh, J. (1990). Bird communities of the Cocha Cashu Biological Station in Amazonian Peru. In Gentry, A.H. (ed.), *Four neotropical rainforests*. Yale University Press, New Haven. Pp. 199–216.

Terborgh, J. (1983). *Five new world primates: a study in comparative ecology*. Princeton University Press, Princeton, New Jersey. 260 pp.

Terborgh, J. (1986). Keystone plant resources in the tropical forest. In Soulé, M.E. (ed.), *Conservation biology: the science of scarcity and diversity*. Sinauer Associates, Sunderland, Massachusetts, U.S.A. Pp. 330–344.

Terborgh, J. (1990). An overview of research at Cocha Cashu Biological Station. In Gentry, A.H. (ed.), *Four neotropical rainforests*. Yale University Press, New Haven. Pp. 48–59.

Author

This Data Sheet was written by Dr Alwyn H. Gentry, Missouri Botanical Garden, P.O. Box 299, St. Louis, MO 63166-0299, U.S.A.).

ATLANTIC MOIST FOREST OF SOUTHERN BAHIA
South-eastern Brazil

Location: Southern Bahia State, extending c. 100–200 km inland from coast, between about latitudes 13°–18°S and longitudes 39°–41°30'W.

Area: Originally c. 70,500 km²: 33,500 km² wet forest and 37,000 km² mesophytic forest; fragmented extent remaining is unknown, probably 3–5%.

Altitude: 0–1000 m.

Vegetation: Sequentially inland: littoral "restinga" forest, moist (wet to mesophytic) forests, liana forest.

Flora: Related to rain forests of eastern Amazon but distinct, with very high percentage of endemics; extremely diverse – c. 440 tree species ≥5 cm dbh per ha. Because diversity so high and flora so poorly known, no useful estimate of number of species is possible.

Useful plants: Timber species, including now rare *Caesalpinia echinata* (Brazil-wood), *Dalbergia nigra* (Brazilian rosewood); epiphytes as breeding sites for cocoa pollinators; ornamentals.

Other values: Endemic and threatened fauna; rich in potential germplasm resources; possible ecological tourism.

Threats: Clearing for timber, cattle pastures, crops, plantations; gathering fuelwood, charcoal production. Brazil's Atlantic Coast forest is one of two most endangered forests on Earth – only 2–5% of original forest estimated worth saving. Certain forest types affected disproportionately, on soils good for economically important crops.

Conservation: Less than 300 km² or 0.1% of wet forest in some federal, state and private reserves; virtually none of mesophytic forest conserved.

Geography

The Atlantic Coast tropical forest originally extended along much of the Brazilian coast, from easternmost South America in the State of Rio Grande do Norte southward to Rio Grande do Sul, forming a narrow fringe between the ocean and the dry uplands of the planalto (Collins 1990). In southern Bahia, this forest occupied a zone c. 100–200 km wide, extending southward from south of Salvador to northern Espírito Santo State (Gouvêa, Silva and Hori 1976; Mori *et al.* 1983; Mori 1989). The remainder is mainly in seven separated tracts from c. 15°–17°30'S, between Ilhéus and Punta da Baleia (Collins 1990).

The forest becomes progressively drier inland, and is replaced by "caatinga" or "cerrado" of the planalto (see respectively CPD Sites SA19 and SA21). Within southern Bahia are areas with complex topography and very different soils. The result is a patchwork of diverse micro-habitats throughout the superficially uniform forest.

Vegetation

The Atlantic forest of southern Bahia consists of four types of forest (Gouvêa, Silva and Hori 1976; Mori *et al.* 1983; Mori 1989). Each type occupies a narrow strip to c. 50 km wide within the coastal forest zone (Map 48), and varies in composition depending upon elevation, soils and drainage. As the forest gradually becomes drier inland, it changes from (1) **littoral restinga forest** to (2) **southern Bahian wet forest** – characterized by over 1000 mm of rainfall annually and no distinct dry period, to (3) **southern Bahian mesophytic forest** – characterized by c. 1000 mm of rainfall annually and a distinct dry period, to (4) **liana forest** – seasonally dry, deciduous forest characterized by c. 800 mm of rainfall annually with clear wet and dry seasons.

The wet forest and mesophytic forest are collectively referred to as **moist forest**. This forest appears to be stratified, with lower, canopy and emergent layers; epiphytes (especially ferns, aroids and bromeliads) and lianas are common. The seven most important species of trees (based on relative frequency, density and dominance) in an area studied near Olivença (Una municipality) were an unidentified Myrtaceae, *Eriotheca macrophylla*, *Diploon cuspidatum*, *Rinorea bahiensis*, *Macrolobium latifolium*, *Eschweilera ovata* and an unidentified Lauraceae (Mori *et al.* 1983).

The original extent of the southern Bahian moist forest was c. 70,500 km², with about 33,500 km² wet forest ("mata higrofila") and 37,000 km² mesophytic forest ("mata mesofila"). In 1976 the remaining extent was estimated to total just 8300 km², comprising 5800 km² of wet forest and 2500 km² of mesophytic forest (Vinha, Soares Ramos and Hori 1976). The fragmented extent still persisting is not known, but is less than half of what existed in 1976 and probably closer to a quarter, or c. 3–5% of the original moist forest – patches totalling only c. 2000–3500 km². Pristine stands of forest probably no longer exist in the region.

Flora

Within the Atlantic Coast tropical forest separate centres of endemism have been recognized, which are considered to have resulted from forest fragmentation during dry periods of the Pleistocene when cerrado and caatinga expanded their distributions. Southern Bahia is considered part of one of these refugia (Mori *et al.* 1983; Mori 1989). Brazil's moist eastern forests also appear to have many evolutionarily primitive species; such bambusoid taxa suggest that Bahia may have been a source of bambusoid grasses that colonized Amazonia (Soderstrom and Calderón 1980).

A preliminary list of plants of the moist forest of southern Bahia was provided by Mori *et al.* (1983). The flora is related to the rain forests of the eastern Amazon but clearly distinct, with some shared species but a very high percentage of endemic species. Current research (Thomas and Carvalho, in prep.) indicates that the diversity of trees in this region is among the highest known, with c. 440 species 5 cm or more in dbh per ha. The most prominent family by far is Myrtaceae, comprising 20–25% of the species; other important and diverse families include Sapotaceae, Leguminosae and Euphorbiaceae.

About 53% of the tree species are thought to be endemic to Brazil's Atlantic Coast forest (Mori and Boom 1981; Mori 1989), with a large but undetermined percentage restricted to the moist forests of southern Bahia. Many apparently endemic species await scientific study of herbarium specimens and publication, especially in large, poorly understood families such as the Myrtaceae and Lauraceae. General scientific research is still revealing very large numbers of species new to science or known from only a few collections.

Striking endemics from the broad region include three genera of legumes – *Brodriguesia*, *Arapatiella*, *Harleyodendron*; a genus of composites – *Santosia*; and four genera of bambusoid grasses – *Atractantha*, *Anomochloa*, *Alvimia*, *Sucrea*. At the species level, the number of endemics is very high; some of the notable plants include at least two species of *Hornschuchia* (Annonaceae); *Couepia longipetiolata*, *Hirtella parviunguis*, *H. santosii*, *Licania santosii*, *Parinari alvimii* (Chrysobalanaceae); *Chamaechrista aspidiifolia*, *C. onusta*, *Zollernia magnifica*, *Z. modesta* (Leguminosae); *Ossaea capitata*, *O. marginata*, *Tibouchina bahiensis*, *T. morii*, *T. paulo-alvimii*, *T. stipulacea* (Melastomataceae); *Acanthosyris paulo-alvimii* (Santalaceae); *Aphelandra harleyi* (Acanthaceae); *Stifftia axillaris*, *Mikania belemii* (Compositae); and *Attalea funifera* (Palmae).

Useful plants

The Atlantic moist forest of southern Bahia is home to a number of useful species. The epiphytic flora may still provide breeding sites for midges that are essential for the pollination of cocoa (*Theobroma cacao*), which is the most important crop in the areas of moist forest in southern Bahia (Fish and Soria 1978; Mori 1989). Ornamental orchids and bromeliads from the region have enriched horticulture.

Some particularly valuable timber species are *Tabebuia* spp. ("ipê"), *Aspidosperma* spp. ("peroba"), *Cedrela odorata* ("cedro"), *Plathymenia foliolosa* ("vinhático"), *Astronium concinnum* ("aderno") and *Centrolobium* sp. ("putumuju"). Two species deserve special mention because of their over-exploitation and the present rarity of wild trees. *Caesalpinia echinata* (Brazil-wood, "pau-brasil") was one of Brazil's first economically important forest products, being used to extract a textile dye; the country was named after this species. As early as 1809, the species was considered threatened in some of coastal Brazil – e.g. the State of Alagôas (Fraga 1960). Its lumber is prized for making the bows of musical instruments. *Dalbergia nigra* (Brazilian rosewood, "jacarandá") may be the individually most valuable timber tree in South America, used for furniture-making, veneer and high quality musical instruments. Because of its rarity from habitat loss and over-use, it was placed in CITES Appendix I in March 1992 and its commercial export thus prohibited.

The endemic palm *Attalea funifera* ("piaçava" or "piassava") is an important component of the local economy, which is found in the sandy soils of the coastal restinga forest. Its long coarse fibres are harvested for thatch, brooms, mats and rope (Silva and Vinha 1982).

Social and environmental values

The extremely high biological diversity of tropical forests signifies an equally high genetic diversity; preserving such

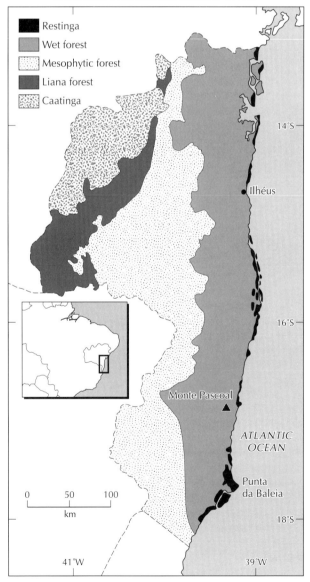

MAP 48. SOUTHERN BAHIA, BRAZIL, SHOWING THE DIFFERENT ORIGINAL FOREST TYPES

Restinga
Wet forest
Mesophytic forest
Liana forest
Caatinga

14°S
Ilhéus
16°S
Monte Pascoal
ATLANTIC OCEAN
Punta da Baleia
18°S

0   50   100
km

41°W   39°W

resource. In southern Bahia, where endemism and diversity are particularly high and the extent of remaining intact forest is so low, the forest preserves are especially precious.

The Una Biological Reserve was established to protect the endemic threatened golden-headed lion tamarin (*Leontopithecus chrysomelas*). The Atlantic Coast forests of southern Bahia are part of the South-east Brazilian lowland to foothills Endemic Bird Area (B52). More than 20 restricted-range bird species occur in this part of the forests; several are confined here, such as the fringe-backed fire-eye (*Pyriglena atra*), Stresemann's bristlefront (*Merulaxis stresemanni*) and Bahia tapaculo (*Scytalopus psychopompus*). Another 11 threatened birds are found with these forests; for many of them the only recent records are from the small protected areas in the region. Monte Pascoal National Park, Pau CVRD (Companhia Vale do Rio Doce) reserve (Fazenda Americana) and Pôrto Seguro CVRD reserve together harbour the threatened white-necked hawk (*Leucopternis lacernulata*), red-billed curassow (*Crax blumenbachii*), red-browed parrot (*Amazona rhodocorytha*), blue-throated parakeet (*Pyrrhura cruentata*), golden-tailed parrotlet (*Touit surda*), hook-billed hermit (*Ramphodon dohrnii*), black-hooded berryeater (*Carpornis melanocephalus*), banded cotinga (*Cotinga maculata*), cinnamon-vented piha (*Lipaugus lanioides*), white-winged cotinga (*Xipholena atropurpurea*) and Temminck's seedeater (*Sporophila falcirostris*).

Threats

European settlement in the region began almost 500 years ago, so the accessible fertile areas have long since been cleared for cash crops (primarily cocoa) and cattle pastures. About one-third of the Brazilians live in the eastern region, on just 6% of the country's land (Mori 1989) The remaining forested areas are predominantly on poorer soils or have been inaccessible, due to steep terrain or lack of roads. These forests on poor soils are suffering most from timber extraction and short-term subsistence agriculture and cattle-raising. Certain forests, however, particularly on good soils that can support economically important crops, have suffered disproportionately and are the most endangered.

Only 2–5% of the original forest throughout the Atlantic Coast forested region was estimated to remain in 1982 in a condition worth saving (IUCN and WWF 1982; Collins 1990). This fully agrees with the current assessment of deforestation in southern Bahia (cf. Mori 1989). Threats include conversion by timber exploitation, cash-crop agriculture (e.g. cocoa, sugarcane, rubber, oil palm, piassava palm – *Attalea funifera*, "guaraná" – *Paullinia cupana* var. *sorbilis*, cloves, black pepper), cattle pastures, subsistence agriculture, pulpwood plantations and housing – particularly along the coast, where tourism is an important industry.

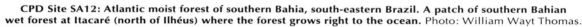

CPD Site SA12: Atlantic moist forest of southern Bahia, south-eastern Brazil. A patch of southern Bahian wet forest at Itacaré (north of Ilhéus) where the forest grows right to the ocean. Photo: William Wayt Thomas.

Monte Pascoal National Park was established in 1961; in 1980, 85 km² were returned to the Pataxó Amerindians. Within a decade, their largely forested land had been completely cleared, and much of it became abandoned pastures (Collins 1990).

Most of the forest reserves are not adequately staffed or patrolled, resulting in inadequate protection against wood cutters, hunters and squatters.

Conservation

The nature reserves in southern Bahia are quite limited in number and size (cf. Mori 1989), comprising less than 300 km² – less than 0.1% of the original 33,500 km² of wet forest (Mori *et al.* 1983). The situation for mesophytic forest is worse, as no significant reserves exist. On the fertile soils that support cocoa plantations there are few sizeable remnants of intact forest and no forest reserves. This forest type supports a unique assemblage of species but is poorly known.

Even for the wet forests, the reserves have not been created in a manner that ensures preservation of the maximum diversity of micro-habitats (Fundação SOS Mata Atlântica 1990; Mori 1989). The federally protected forest areas are Monte Pascoal NP (135 km²), Una Biological Reserve (60 km²) and Pau-brasil Ecological Station (11.45 km²). Two state-owned reserves exist, Reserva Estadual de Laracas (3 km²) and RE Wenceslau Guimarães (125 km² – but less than 20 km² remain in forest). Privately held reserves include the Pôrto Seguro Forest Reserve (60 km²).

Additional reserves must be determined carefully, in order to conserve as much as is still realistically possible of the complex array of forest types in this biologically rich and diverse region. Privately held areas of forest in need of more permanent preservation include those near Serra Grande (Uruçuca), Maruim (Una), Belmonte, Camacã, Trancoso (Pôrto Seguro), Cumuruxativa and Potiraguá (Fundação SOS Mata Atlântica 1990). More reserves are urgently needed.

References

Collins, M. (ed.) (1990). *The last rain forests: a world conservation atlas.* Oxford University Press, New York. 200 pp.

Fish, D. and Soria, S. (1978). Water-holding plants (phytotelmata) as larval habitats for Ceratopogonid pollinators of cacao in Bahia, Brazil. *Revis. Theobroma* 8: 133–146.

Fraga, M.V.G. (1960). A questão florestal ao tempo do Brasil-colônial. *Anuário Brasil. Econ. Florest.* 3(3): 7–96.

CPD Site SA12: Atlantic moist forest of southern Bahia, south-eastern Brazil. *Calliandra bella* (Mimosaceae), an endemic shrub of forest edges and clearings. Photo: William Wayt Thomas.

Fundação SOS Mata Atlântica (1990). *Workshop Mata Atlântica: problemas, diretrizes e estratégias de conservação.* SOS Mata Atlântica, São Paulo. 64 pp.

Gouvêa, J.B.S., Silva, L.A.M. and Hori, M. (1976). 1. Fitogeografia. In Comissão Executiva do Plano da Lavoura Cacaueira and Instituto Interamericano de Ciências Agrícolas–OEA, *Diagnóstico socioeconômico da região cacaueira, Recursos florestais.* Vol. 7. Ilhéus, Bahia. Pp. 1–7.

IUCN and WWF (World Wildlife Fund) (1982). *Tropical Forest Campaign fact sheet no. 14– Brazil.* Washington, D.C. 2 pp.

Mori, S.A. (1989). Eastern, extra-Amazonian Brazil. In Campbell, D.G. and Hammond, H.D. (eds), *Floristic inventory of tropical countries: the status of plant systematics, collections, and vegetation, plus recommendations for the future.* New York Botanical Garden, Bronx. Pp. 427–454.

Mori, S.A. and Boom, B.M. (1981). *Final report to the World Wildlife Fund-US on the botanical survey of the endangered moist forests of eastern Brazil.* New York Botanical Garden, Bronx. 109 pp.

Mori, S.A., Boom, B.M., Carvalho, A.M. de and Santos, T.S. dos (1983). Southern Bahian moist forests. *Bot. Review* 49: 155–232.

Silva, L.A.M. and Vinha, S.G. da (1982). A piaçaveira (*Attalea funifera* Mart.) e vegetação associada no município de Ilhéus, Bahia. *Bol. Técn., Centro de Pesquisas do Cacau* 101: 1–12.

Soderstrom, T.R. and Calderón, C.E. (1980). In search of primitive bamboos. *Natl. Geogr. Soc. Research Reports* 12: 647–654.

Vinha, S.G. da, Soares Ramos, T. de J. and Hori, M. (1976). 2. Inventário florestal. In Comissão Executiva do Plano da Lavoura Cacaueira and Instituto Interamericano de Ciências Agrícolas–OEA, *Diagnóstico socioeconômico da região cacaueira, Recursos florestais.* Vol. 7. Ilhéus, Bahia. Pp. 20–121.

Authors

This Data Sheet was written by Dr Wayt Thomas (New York Botanical Garden, Herbarium, Bronx, NY 10458-5126, U.S.A.), Dr André M. de Carvalho [Centro de Pesquisas do Cacau (CEPEC), Caixa Postal 7, 45660 Ilhéus, Bahia, Brazil] and Olga Herrera-MacBryde (Smithsonian Institution, Department of Botany, NHB-166, Washington, DC 20560, U.S.A.).

MATA ATLÂNTICA: CPD SITE SA13

TABULEIRO FORESTS
OF NORTHERN ESPIRITO SANTO
South-eastern Brazil

Location: Atlantic Coast forest north of Vitória and Linhares, between about latitudes 18°15'–19°20'S and longitudes 39°10'–40°25'W.

Area: c. 484 km² conserved, almost all within Linhares Forest Reserve and Sooretama Biological Reserve.

Altitude: 28–90 m.

Vegetation: Tall terra-firme rain forest, shorter mussununga forest, várzea forest, savannas.

Flora: High tree-species diversity – c. 637 species recorded; 372 species of shrubs, lianas and herbaceous plants recorded; many species and some genera endemic; disjuncts with Amazonia.

Useful plants: Valuable timber trees, also many species that produce resins, oils or medicinals.

Other values: Ecological islands of survival for fauna; potential genetic resources; important centre for environmental research, training and education.

Threats: Logging, poaching, access from highway through Sooretama reserve, escaped fires.

Conservation: Córrego do Veado Biological Reserve (24 km²), Linhares Forest Reserve (220 km²) and adjacent Sooretama BR (240 km²).

Geography

The tabuleiro forests of the State of Espírito Santo are an integral part of the Atlantic tropical rain forest (Monteiro and Kaz 1992). Their core area is in the north-eastern part of the state within the São Mateus, Barra Seca and Doce river basins (Map 49). The remaining forest is almost totally restricted to the north (Collins 1990), where the Linhares Forest Reserve and the Sooretama Biological Reserve are located, in the municipality Linhares. These two reserves comprise c. 460 km² of continuous forest. The Córrego do Veado BR located farther north in the Itaúnas River Basin (municipality Pinheiros) harbours 24 km² of tabuleiro forest.

The Tertiary sediments ("tabuleiros") of the Barreiras series which underlie this type of forest consist of low flat tablelands, varying between 28–90 m in elevation. The clayey sands are fairly deep, with somewhat indistinct horizons. They are poor to very poor in many nutrients, especially compared to the soils derived from decomposed crystalline rocks of the uplands in the western part of the state. The tabuleiro landscape is completed by shallow flat valleys dotted with marshes and lakes.

A hot humid climate prevails, differing from other regions of the Atlantic forest by its rather pronounced dry season (May to September), which makes the vegetation somewhat semi-deciduous. Annual rainfall is c. 1400 mm, with the rainy season October to March. The mean annual temperature is 22.8°C, with extremes (from the past 10 years) of 8.3°C (in August) and 37.1°C (in January). The mean annual relative humidity is c. 84% (Jesus 1987).

Vegetation

Four distinct vegetation types occur in the tabuleiro forests of northern Espírito Santo (Peixoto, Rosa and Joels 1995):

1. **Tall "terra-firme" rain forest**, which covers the greater part of the region, has the highest species diversity, especially for trees and lianas. There are three strata. The canopy towers c. 31 m above the forest floor, with emergents reaching 40 m. The c. 280 species of trees per ha with dbh 5 cm or more make this vegetation type one of the most diverse in South America (Peixoto 1992). The density of these trees is c. 1358 individuals per ha, and c. 15 trees per ha are 80 cm or more in dbh. Many species have latex or buttressed roots.

2. **"Mussununga" forest** is found on sandy ridges, and covers a smaller area than the tall terra-firme forest. The trees are usually smaller, less densely distributed, with lighter coloured trunks and stiffer leaves. This forest is less rich in species, but has a surprising number of endemics, both woody and herbaceous. The lower stratum is made up of canopy-tree seedlings and populations of species of Araceae, Marantaceae, Bromeliaceae and Gramineae (Bambusoideae).

3. **"Várzea" forest** is found in places where the water table is at the surface during most of the year. This is an open forest, with a discontinuous canopy c. 8 m above the forest floor, and is the poorest in species of all four vegetation types. *Tabebuia cassinoides* is clearly dominant, together with *Cecropia lyratiloba* and *Bactris setosa*. In the dense graminoid herbaceous layer, *Acrostichum danaefolium* and *Blechnum serrulatum* occur. Submersed or floating aquatic vegetation appears where water is more abundant. In the ecotone between várzea and the other vegetation types are populations of *Symphonia globulifera*, *Calophyllum brasiliense* and *Geonoma schottiana*.

4. **"Campos nativos"** are natural savanna enclaves within the tabuleiro forest in places where the sandy soil layers are

deepest. Generally this plant cover has two characteristic types: (a) discontinuous thickets which may reach heights of 4 m, interspersed with open spaces thinly covered by herbaceous vegetation; Cactaceae and Bromeliaceae are common; and (b) continuous, dense tall graminoid vegetation with scattered trees, mainly *Kielmeyera* spp.

Flora

Although the region's flora is poorly known, it is estimated that species diversity is high, especially among trees. Endemism is also high, and many widely distributed species have distinct biotypes in this region. An inventory of tree species registered c. 637 species (Jesus 1987); the five families with the highest numbers of species were Myrtaceae (96 spp.), Leguminosae (85), Sapotaceae (33), Lauraceae (30) and Rubiaceae (29). There were 372 species of herbaceous plants, shrubs and lianas recorded. The cryptogamic flora has not been surveyed.

An important characteristic of the flora of this region as compared to other areas of tropical wet forest is the richness of robust liana species. Some families have more species of lianas per ha in Linhares than any other neotropical area,

e.g. Hippocrateaceae, with eight spp. (Peixoto and Gentry 1990). There are a significant number of endemic herbaceous and arboreal species and genera. Some large tree species restricted to these forests are *Simira grazielae*, *Polygala pulcherrima* and *Plinia renatiana*. Endemic genera include the monotypic *Hydrogaster* (*H. trinervis*) (Tiliaceae) and *Grazielodendron* (*G. riodocensis*) (Leguminosae). There are many elements that are common to the floras of tabuleiro forest and Amazon forest, and also several vicariant taxa, which indicate former geological and climatic periods when the Atlantic and the Amazonian forests were linked (Peixoto 1992).

Useful plants

Economically valuable plants include timber trees, such as *Dalbergia nigra* (Brazilian rosewood) – which is most highly valued on the domestic and foreign markets, *Paratecoma peroba*, *Astronium graveolens*, *A. concinnum* and several *Tabebuia* spp. Resins and oils are extracted from some species, such as *Copaifera langsdorfii*, *Virola surinamensis* and *Protium macrophyllum*. Other species are used in popular medicine – e.g. *Tynanthus elegans*, *Herreria*

CPD Site SA13: Aerial view of tabuleiro forests of Linhares Forest Reserve, northern Espírito Santo, Brazil.
Photo: Ariane Luna Peixoto.

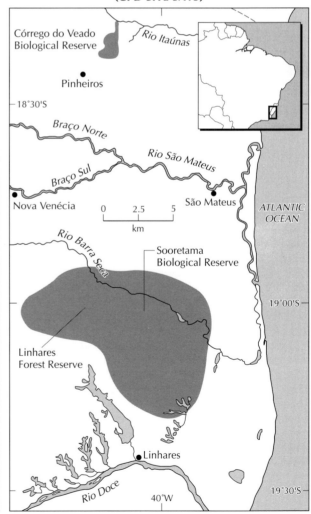

MAP 49. TABULEIRO FORESTS OF NORTHERN ESPIRITO SANTO STATE, SOUTH-EASTERN BRAZIL (CPD SITE SA13)

salsaparilha, Piper spp., *Tabebuia heptaphylla, Hymenaea courbaril, Geissospermum laeve* and *Cissampelos* spp.

Social and environmental values

The Linhares Forest Reserve and Sooretama Biological Reserve together represent the largest remnant of lowland forest in eastern Brazil. This tract contains 25% of the forested area (mostly mangroves) left in the State of Espírito Santo, and is one of the last remaining refuges for several threatened plant and animal species (Jesus 1987; Collar 1986).

About 281 birds are known to occur in the region. The Atlantic Coast forests of northern Espírito Santo are part of the South-east Brazilian lowland to foothills Endemic Bird Area (EBA B52). More than 20 restricted-range bird species from this EBA occur in this part of the forests. There are also a number of threatened species, many of which are also of restricted range. Important populations for several threatened species are in the Linhares CVRD reserve, Sooretama Biological Reserve and Rio Doce State Park (Minas Gerais), including white-necked hawk (*Leucopternis lacernulata*), red-billed curassow (*Crax blumenbachii*) – which has a third of its present population in these two reserves (Scott 1988), red-browed parrot (*Amazona rhodocorytha*), blue-throated parakeet (*Pyrrhura cruentata*), golden-tailed parrotlet (*Touit surda*), blue-bellied parrot (*Triclaria malachitacea*), hook-billed hermit (*Ramphodon dohrnii*), straited softtail (*Thripophaga macroura*), plumbeous antvireo (*Dysithamnus plumbeus*), black-hooded berryeater (*Carpornis melanocephalus*), banded cotinga (*Cotinga maculata*), cinnamon-vented piha (*Lipaugus lanioides*), white-winged cotinga (*Xipholena atropurpurea*) and buffy-fronted seedeater (*Sporophila frontalis*).

The Linhares and Sooretama reserves have an enormous potential for research on the structure, function and management of these ecosystems. Several factors make the Linhares Forest Reserve one of the most valued reserves in eastern Brazil for natural-area research, such as easy access, the availability of laboratories and lodging, extensive road and trail systems within the forest and an observation tower that overtops the canopy. This reserve is used as a training centre for administrators and staff of conservation units due to its administrative efficiency and the maintenance, enforcement and fire-control programmes. Both reserves are important regional and national centres for teaching students, developing projects in environmental education, and ecological tourism for children, university students and executives. The reserves are accessed by BR-101 highway, which links the states from Rio de Janeiro to Bahia.

Economic assessment

The tabuleiro forest is locally known as a production forest because of its potential and the intense timber exploitation that went on during the first half of the 20th century, as a result of the colonization of northern Espírito Santo State (Heinsdijk *et al.* 1965). Regional economic development received a considerable boost only after 1923, when a bridge was built over the Doce River, linking the city of Colatina to the areas northward. Most of the lumber and other forest products were taken out by this road, which depleted the forest resources of the region.

In 1943, the Instituto Brasileiro de Desenvolvimento Florestal (today IBAMA) created the Sooretama Biological Reserve in order to preserve the local flora and fauna from lumbermen and poachers (Pádua and Coimbra Filho 1979). In the early 1950s, the Companhia Vale do Rio Doce (CVRD) began to buy up tracts of land covered with natural forest to maintain a timber reserve for the production of railroad ties for the Vitória-Minas railway. However, the results of research carried out in the forest convinced the CVRD to maintain the natural forest, to use it as an environmental protection area, and to develop a series of experiments aimed at sustainable production – based on logging of only mature trees (Jesus 1987).

The Linhares Forest Reserve has become an economically viable enterprise for the CVRD, due to varied activities that contribute to sustained management of the area. Methods for collection and improvement of seeds, seedling production and silviculture have been developed for 160 native forest species. The CVRD greenhouse produces and commercializes c. 15,000 seedlings from 600 mostly native species. Generation of technological data from taxonomy to silviculture on species, harvest and sale of seeds from selected mother-trees, production and sale of seedlings, as well as other activities have made the reserve an economic asset. The development of programmes for the recovery of degraded areas as well as the other activities produce enough revenue to maintain this reserve and even supply funds for other activities.

Threats

Forests that grow on the dystrophic soils of the Barreiras series are extremely vulnerable, because rapid decay and recycling is the main source of nutrients for the plants. Therefore, the removal of the forest cover makes recovery of composition and structure extremely difficult due to the sandy, easily leached poor soils. The areas surrounding the Linhares and Sooretama reserves were deforested and converted to other uses, mostly sugarcane cultivation and low-density cattle-raising on large ranches, which has created the constant threat of fire during the dry season. The Sooretama BR also suffers from poaching pressures and logging along its entire perimeter, and especially along the highway BR-101, which cuts through the reserve. A restoration programme is urgently needed in the areas surrounding this reserve, which should include increased continuity with the Linhares FR, thus diminishing the ecological island effect and helping to ensure survival for many species.

Conservation

The Córrego do Veado Biological Reserve, the Linhares Forest Reserve and the Sooretama BR are permanently protected areas. The Linhares FR belongs to one of Brazil's largest mixed-economy companies, the Companhia Vale do Rio Doce, and the BRs are administered by a federal government agency – the Instituto Brasileiro do Meio Ambiente e dos Recursos Naturais Renováveis (IBAMA). According to Brazilian law, Biological Reserves are in preservation category A, which includes units for the most part unaltered by human activities, where the biota receives total protection and natural ecological and geological processes are not subject to interference. Scientific research may be done in these units, and public access is limited to controlled educational activities; recreation is not permitted.

The Linhares FR administration in the past few years has begun conservation programmes in several different parts of the state, besides the management and conservation of its own area. These programmes include collection and propagation of native species, especially those that are threatened or rare, restoration of degraded areas, and educational programmes.

References

Collar, N.J. (1986). The best-kept secret in Brazil. *World Birdwatch* 8(2): 14–15.

Collins, M. (ed.) (1990). *The last rain forests: a world conservation atlas.* Oxford University Press, New York. 200 pp.

Heinsdijk, D., Macedo, J.G. de, Andel, S. and Ascoly, R.B. (1965). A floresta do norte do Espírito Santo. *Bol. Rec. Nat. Renov., Ministério da Agricultura* 7: 1–69.

Jesus, R.M. de (1987). Mata Atlântica de Linhares: aspectos florestais. *Anais do Seminário – Desenvolvimento Econômico e Impacto Ambiental em Área do Trópico Úmido Brasileiro – A Experiência da CVRD, Linhares.* Companhia Vale do Rio Doce (CVRD), Rio de Janeiro. Pp. 35–71.

Monteiro, S. and Kaz, L. (eds) (1992). *Atlantic rain forest.* Edições Alumbramento, Rio de Janeiro. 180 pp.

Pádua, M.T.J. and Coimbra Filho, A.F. (1979). *Os parques nacionais do Brasil.* Instituto de Cooperação Iberoamericano. INCAFO, Madrid. 224 pp.

Peixoto, A.L. (1992). Vegetation of Atlantic forest. In Monteiro, S. and Kaz, L. (eds), *Atlantic rain forest.* Edições Alumbramento, Rio de Janeiro. Pp. 31–39.

Peixoto, A.L. and Gentry, A.H. (1990). Diversidade e composição florística da mata de tabuleiro na Reserva Florestal de Linhares (Espírito Santo, Brasil). *Rev. Brasil. Bot.* 13: 19–25.

Peixoto, A.L., Rosa, M.M. and Joels, L.C. (1995). Diagramas de perfil de cobertura de um trecho de floresta de tabuleiro na Reserva Florestal de Linhares (Espírito Santo, Brasil). *Acta Bot. Brasil.* 9(2): 1–17.

Scott, D.A. (1988). Preservação da natureza e pesquisa sobre a fauna pela CVRD. Observações e sugestões. *Espaço, Ambiente e Planejamento* 7: 1–52.

Authors

This Data Sheet was written by Dra. Ariane Luna Peixoto and Inês Machline Silva (Universidade Federal Rural do Rio de Janeiro, Caixa Postal 74582, 23851-970 Seropédica, Itaguaí, RJ, Brazil).

CABO FRIO REGION
South-eastern Brazil

Location: Coastal region east of city of Rio de Janeiro, between about latitudes 22°30'–23°00'S and longitudes 42°42'–41°52'W.

Area: c. 1500 km².

Altitude: Sea-level to c. 500 m.

Vegetation: Mangroves, coastal evergreen scrub to forest on sandy substrate, xeromorphic thickets on low hills, submontane rain forest.

Flora: 1500–2200 species of vascular plants estimated, 740 species recorded; high diversity, many disjuncts at their southern limit, high species endemism, relict species; threatened species.

Useful plants: *Caesalpinia echinata* (Brazil-wood), medicinals; intense local utilization of many plants by fishermen.

Other values: Dune stabilization, archaeological sites, tourism, important locale for global climate-change research.

Threats: Increasing land development – severely threatened by vacation homes, tourism, cattle-raising, agriculture.

Conservation: Environmental Protection Zones (APAs) and Ecological Reserves comprise c. 10% of region, but do not adequately conserve its biological diversity.

Geography

The Cabo Frio region is in the State of Rio de Janeiro, c. 120 km directly east of the city of Rio de Janeiro. This cape region covers c. 1500 km² from sea-level up to c. 500 m, with less than 10% above 100 m. The region is bordered on the east and south by the Atlantic Ocean, on the west by the local Serra do Mato Grosso, and on the north by the upper limits of the watershed of the Araruama Lagoon and the lower reaches of the Una and São João rivers.

Three distinct physiographic units are in this region (Map 50): (1) sandy coastal plains (beach ridges, dunes) and lowlands (tidal areas, lagoons, alluvial deposits); (2) low hills of the Búzios and Cabo Frio peninsulas and coastal islands; and (3) inland hills to c. 500 m.

The regional geology consists of a Precambrian crystalline basement of granitic-gneiss rocks, with local alkaline intrusions. The evolution of the coastal plain has been greatly affected by relative sea-level changes. The outer dune system was formed some 2000 years ago under colder and drier conditions (FEEMA 1988).

Inland and coastal climatic types prevail within the region (Barbiére 1984). From Cabo Frio Island northward to the Búzios Peninsula, the climate is greatly influenced by cold oceanic upwelling from the Falkland Current off the coast (Martin, Flexor and Valentin 1989), which causes reduced precipitation and moderated temperatures. The mean annual rainfall is c. 800 mm and the mean annual temperature 25°C, with minimum 12°C and maximum 36°C. The soil's water balance is in deficit throughout the year, in marked contrast to most of south-eastern coastal Brazil. However, relative humidity averages over 80% due to the moisture-laden winds from the ocean.

Vegetation

The vegetation varies according to its physiographic location and distance from the ocean (Ule 1901; Lacerda, Araújo and Maciel 1993).

The sea-land interface, which occupies a very small portion of the region, is covered by floristically poor low mangrove forests (*Avicennia, Rhizophora, Laguncularia*) and saltwater marshes (*Salicornia, Sesuvium, Triglochin*). Strand vegetation has many species common along tropical coasts.

Low areas between the beach ridges and dune slacks support characteristic marshy vegetation. Beach ridges are occupied by diverse vegetation types that vary from sparse open communities to dense evergreen forest. The remnant forest (15–20 m high) contains at least 110 tree species, according to a preliminary survey of c. 4 ha (Sá *et al.* 1992), and is dominated by Leguminosae (*Pterocarpus robrii, Pseudopiptadenia contorta, Albizia polycephala*) and Myrtaceae (*Eugenia, Myrciaria, Marlierea*). Most of the sandy coastal plain is covered by open scrub vegetation – probably caused by human activities; the Leguminosae and Myrtaceae are still prominent, together with Bromeliaceae and Euphorbiaceae.

The low xeromorphic forest on hillsides facing the ocean from Cabo Frio Island to the Búzios Peninsula is unique along this east coast. Columnar cacti give a characteristic appearance to the low thickets. Many endemic species occur. Present climatic conditions may have maintained this enclave as a remnant of vegetation that existed during the drier and colder glacial periods of the Pleistocene (Ab'Sáber 1974).

Farther from the ocean, the low mountains (to c. 500 m) support forests similar to the Atlantic Coast rain forest, but

MAP 50. CABO FRIO REGION, SOUTH-EASTERN BRAZIL (CPD SITE SA14)

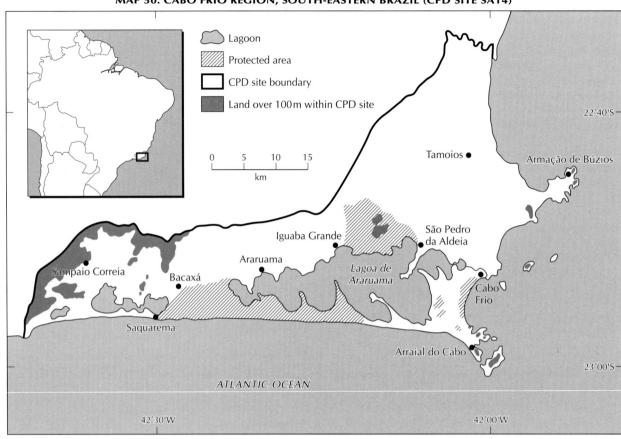

with fewer epiphytes. Little is known on the floristic composition of these forests, but many Atlantic forest and coastal plain species occur.

Flora

The flora of the region is poorly known, except on the sandy coastal plains ("restingas"). Of the 740 species listed for the region, c. 80% are from the coastal plain (Araújo and Henriques 1984). The total number of species will most certainly reach 1500–2200 when complete inventories are made.

Based on a comparison with other areas of restinga from the Rio de Janeiro coast, Cabo Frio is much richer in species, having 57% of the species represented in 12% of the total area. The Cabo Frio region also contains the most endemic coastal plain species: 26 of the 36 endemics listed for the restingas of Rio de Janeiro (Lacerda, Araújo and Maciel 1993). Eleven endemic species have been identified from the hillside thickets. Endemic trees of the region include *Chrysophyllum januariensis*, *Duguetia rhizantha*, *Erythroxylum glazioui*, *Marlierea schottii*, *Rollinia parviflora* and *Swartzia glazioviana*. Several disjunct species have their southern limit in the Cabo Frio region, e.g. *Bonnetia stricta*, *Connarus ovatifolius*, *Cathedra bahiensis* and *Stephanopodium blanchetianum*.

Useful plants

Individual trees of the now rare Brazil-wood (*Caesalpinia echinata*) are found in the Cabo Frio region. Once an important dyewood and still used for violin bows, this species was intimately connected with the history and colonization of Brazil (Cunha and Lima 1992). Local fishermen use many plants, not only in their work but also as a source of medicinals.

Social and environmental values

The restinga vegetation of the Cabo Frio region is very important for three restricted-range bird species (Stattersfield *et al.*, in prep.), which are threatened by the continuing destruction of this habitat. They are the golden-tailed parrotlet (*Touit surda*), restinga antwren (*Formicivora littoralis*) and black-backed tanager (*Tangara peruviana*). The antwren has a particularly small range, being endemic to the restinga of the State of Rio de Janeiro, so the Cabo Frio region comprises much of its range.

The dune vegetation is of great importance for maintaining environmental stability. The local climatic conditions promote formation of active dune systems, which in areas where the vegetation has been destroyed block roads and invade buildings.

The scenic beauty of the region – combining white sands, verdant vegetation and the azure blue waters of the ocean – has long been an attraction for tourism.

There are many archaeological sites in the region that contain important remains, both prehistoric (Kneip and Pallestrini 1984; Schmitz 1990) and historic (Cunha and Lima 1992). The region is also an important locale for the study of global climate change (Martin and Suguio 1989).

Threats

Cattle-raising and agriculture (e.g. growing sugarcane) are carried out in the Cabo Frio region. The ecosystems are seriously threatened by expanding land development and increasing pressure from tourism. Lack of regional planning strategy and environmental controls allows occupation of the sandy coastal plains by housing projects (for vacation homes) with inadequate infrastructure and by sanitary landfill operations. Salt-producing flats on the edge of the Araruama Lagoon, when abandoned, have been built up instead of being returned to the lagoon to improve the fishery. Remnants of the forest also are being cleared for summer homes for tourists. The lack of infrastructure and enforcement additionally threatens the few conservation units.

Conservation

There are two Ecological Reserves in the Cabo Frio region, the Reserva Ecológica de Jacarepiá (12.5 km²) and the Reserva Ecológica de Massambaba (13.7 km²), and two Environmental Protection Zones, the Area de Proteção Ambiental de Massambaba (111 km²) which includes the two Ecological Reserves and the APA de Sapiatiba (60 km²). The dunes are protected by decree from the Fundação Estadual de Engenharia do Meio Ambiente (FEEMA). However, these units are administered by the State of Rio de Janeiro with very limited funds. As a result, the boundaries have not been fenced, there are no rangers in the units and no local headquarters, etc.

Some hope for preservation of natural wealth of the Cabo Frio region lies in assistance from the locally active non-governmental organizations, whose members are attentive to infractions of the law and notify the appropriate State authorities.

References

Ab'Sáber, A.N. (1974). O domínio morfoclimático semi-árido das caatingas brasileiras. *Geomorfologia (São Paulo)* 43: 1–39.

Araújo, D.S. Dunn de and Henriques, R.P.B. (1984). Análise florística das restingas do estado do Rio de Janeiro. In Lacerda, L.D. de, Araújo, D.S. Dunn de, Cerqueira, R. and Turcq, B. (eds), *Restingas: origem, estrutura, processos*. Centro Educacional Universidade Federal Fluminense (CEUFF), Niterói. Pp. 159–193.

Barbiére, E.B. (1984). Cabo Frio e Iguaba Grande, dois microclimas distintos a um curto intervalo espacial. In Lacerda, L.D. de, Araújo, D.S. Dunn de, Cerqueira, R. and Turcq, B. (eds), *Restingas: origem, estrutura, processos*. CEUFF, Niterói. Pp. 3–13.

Cunha, M.W. da and Lima, H.C. de (1992). *Travels to the land of Brazilwood*. Ag. Bras. Cultura, Rio de Janeiro. 64 pp.

FEEMA (1988). *A importância da biota de Cabo Frio*. Fundação Estadual de Engenharia do Meio Ambiente (FEEMA), Rio de Janeiro. 50 pp.

Kneip, L.M. and Pallestrini, L. (1984). Restingas do estado do Rio de Janeiro (Niterói a Cabo Frio): 8 mil anos de ocupação humana. In Lacerda, L.D. de, Araújo, D.S. Dunn de, Cerqueira, R. and Turcq, B. (eds), *Restingas: origem, estrutura, processos*. CEUFF, Niterói. Pp. 139–146.

Lacerda, L.D. de, Araújo, D.S. Dunn de and Maciel, N.C. (1993). Dry coastal ecosystems of the tropical Brazilian coast. In van der Maarel, E. (ed.), *Dry coastal ecosystems: Africa, America, Asia and Oceania*. Ecosystems of the World 2B. Elsevier, Amsterdam. Pp. 477–493.

Martin, L. and Suguio, K. (1989). Excursion route along the Brazilian coast between Santos (State of São Paulo) and Campos (north of State of Rio de Janeiro). *International Symposium on Global Changes in South America during the Quaternary*. Special Publ. No. 2. Associação Brasileira de Estudos do Quaternário, São Paulo. Pp. 1–136.

Martin, L., Flexor, J.-M. and Valentin, J.L. (1989). The influence of the "El Niño" phenomenon on the enhancement or annihilation of Cabo Frio upwelling on the Brazilian coast of the State of Rio de Janeiro. *International Symposium on Global Changes in South America during the Quaternary*. Special Publ. No. 1. Associação Brasileira de Estudos do Quaternário, São Paulo. Pp. 225–227.

Sá, C.F.C., Araújo, D.S. Dunn de, Cavalcanti, M.J., Alves, T.F., Alvarez-Pereira, M.C., Lima, H.C. de and Fonseca, V.S. (1992). Estrutura da floresta de cordão arenoso da Reserva Ecológica Estadual de Jacarepiá, Saquarema, RJ. *Resumos do XLIII Congresso Nacional de Botânica da Sociedade Botânica do Brasil*, Aracajú, SE.

Schmitz, P.I. (1990). Caçadores e colectores antigos da região do cerrado. In Novaes Pinto, M. (ed.), *Cerrado: caracterização, ocupação e perspectivas*. Editora Universidade de Brasília, Brasília. Pp. 101–146.

Stattersfield, A.J., Crosby, M.J., Long, A.J. and Wege, D.C. (in prep.). *Global directory of Endemic Bird Areas*. BirdLife Conservation Series. BirdLife International, Cambridge, U.K.

Ule, E. (1901). Die Vegetation von Cabo Frio an der Küste von Brasilien. *Bot. Jahrb. Syst.* 28: 511–528.

Author

This Data Sheet was written by Dra. Dorothy Sue Dunn de Araújo (FEEMA, Serviço de Ecologia Aplicada, Estr. da Vista Chinesa 741, 20531-410 Rio de Janeiro, RJ, Brazil).

MOUNTAIN RANGES OF RIO DE JANEIRO
South-eastern Brazil

Location: In State of Rio de Janeiro, escarpments and valleys of Serra da Mantiqueira and Serra do Mar, between latitudes 21°35'–23°20'S and longitudes 44°55'–41°30'W.

Area: c. 7000 km².

Altitude: c. 60–2800 m, with mean of c. 800–900 m. Most outstanding are Pedra do Sino (Bell Rock) at 2263 m and Pico Agulhas Negras (Black Needles Peak) at 2787 m.

Vegetation: High-altitude fields, Atlantic Coast rain forests – including formations on higher slopes and lower mountains to edge of plains.

Flora: High diversity – 5000–6000 species; disjuncts with Amazon and Andes; high endemism; threatened species.

Useful plants: Trees for lumber and fuelwood; medicinals; ornamentals. Insufficient scientific and technological knowledge on most of the plants impairs sustainable uses.

Other values: Threatened fauna, watershed protection, erosion reduction, potential genetic resources, tourist attraction.

Threats: Increasing pressures of settlement; expansion of agriculture and cattle-raising; gathering fuelwood, charcoal production; logging; collecting ornamentals and palm hearts; road construction; excessive tourism; invasive exotics; fire.

Conservation: c. 3220 km² in c. 20 units: National and State Parks, Federal and State Ecological Stations and Biological Reserves, Environmental Protection Zones (APAs), other designated areas.

Geography

The State of Rio de Janeiro has much topographic diversity (Domingues 1976). The mountainous region is formed by the Serra do Mar and inland Serra da Mantiqueira, consisting of crystalline rocks of the Brazilian Shield.

The Serra do Mar often presents a steep slope, both abrupt and continuous, in crossing the state from west-south-west to east-north-east, from the border with the State of São Paulo to the municipality Campos (Map 51). The range emerges directly from the ocean; from this point northward, it distances itself from the coastline but remains parallel, separated by alluvial plains. In the north it finally is transformed into a series of peaks and isolated hills. The relief is very pronounced, with elevations from sea-level to c. 2300 m. The highest points are in the area of Teresópolis; outstanding are Pedra do Sino (2263 m), Pedra Acu (2230 m) and Pico Dedo de Deus (1695 m).

The Serra da Mantiqueira rises farther inland to the west, where the states of Rio de Janeiro and São Paulo meet the State of Minas Gerais, and continues from west-south-west to the north-east in Minas Gerais. In Rio de Janeiro this mountain range often is very dissected, having its more accentuated topography in the south, where the peaks of Itatiaia reach nearly 2800 m.

Between the escarpments of the Serra da Mantiqueira and Serra do Mar is the valley of the Paraíba do Sul River, which is much lower than the crests of the two ranges. Another important topographic feature is the lowlands along the coast ("baixadas"). Their width (to 25 km) varies greatly, as the proximity of the foothills of the Serra do Mar to the sea is variable.

The only major river is the Paraíba do Sul, which leaves the Serra do Mar *sensu lato* at the north-eastern end of Serra Grande and flows into the ocean. It drains part of the Serra da Mantiqueira and the Serra do Mar's north-western slope. The crest of the Serra do Mar is a watershed boundary that directs shorter south-eastern rivers to the ocean.

The climate of the state is very variable because of the elevation and orientation of the mountain ranges. The coastal lowlands, Serra do Mar, Paraíba do Sul Valley, and Serra da Mantiqueira are all oriented west-south-west to east-north-east and have dramatically different elevations. The frequent arrivals of cool fronts of polar origin and polar anticyclones also are important factors affecting the regional climate. These are all significant influences that result in an uneven distribution of rainfall in Rio de Janeiro.

Summer (December–March) is warmer and in general wetter than winter (June–September). In the lowlands the climate is predominantly hot and humid, with the mean annual temperature c. 24°–26°C and the mean annual rainfall usually a little above 1000 mm. In the Paraíba valley, the rainfall is c. 1500 mm and the temperature also between 24°–26°C. In the mountainous regions the rainfall is high (2000–2500 mm) with no dry months and the temperature normally c. 18°–19°C (Nimer 1989; Domingues 1976).

Vegetation

The evergreen vegetation of these mountain ranges shows a diversified physiognomy, shaped mainly by the topographic and climatic factors. There is considerable variation in height, stratification and floristic composition. The vegetation can be classified in three types which, although distinct, often show intermediate characteristics: low-mountain forest, high-mountain forest and high-altitude fields.

1. Low-mountain forest

The rounded hills and extremely carved hillsides between 60–800 m form the foothills above which the crests of the Serra do Mar and Serra da Mantiqueira rise. The foothills are covered by a dense forest with a continuous canopy which can reach 35 m. The canopy is characterized among others by *Cariniana estrellensis, Hyeronima alchorneoides, Virola oleifera, Jacaratia spinosa, Ocotea* spp., *Pseudopiptadenia inaequalis, Moldenhawera floribunda, Chrysophyllum imperiale, Eugenia* spp. and *Aspidosperma parvifolium*. In the understorey, the common taxa are *Miconia* spp., *Erythroxylum* spp., *Myrcia* spp., *Eugenia* spp., *Faramea* spp., *Psychotria* spp., *Geonoma* spp., *Euterpe edulis* and *Astrocaryum aculeatissimum*. In the valleys and ravines, some species common in the herbaceous strata are *Calathea aemula, Aphelandra squarrosa, Heliconia laneana, Anthurium* spp., *Philodendron* spp. and many ferns.

2. High-mountain forest

When the vegetation reaches c. 800 m, it becomes a shady humid forest with huge trees sustaining a wide variety of epiphytes and lianas. The most outstanding taxa composing the canopy include *Ocotea* spp., *Nectandra* spp., *Cinnamomum* spp., *Eugenia* spp., *Tibouchina* spp., *Solanum swartzianum, Vernonia arborea, Cabralea canjerana* and *Symplocos variabilis*. The understorey is composed of *Hedyosmum brasiliensis, Myrcia* spp., *Psychotria velloziana, Guatteria nigrescens, Euterpe edulis* in large numbers and populations of tree ferns (*Cyathea delgadii*).

Above 1400 m and extending to 1600–1800 m is a more open, upland forest formation with smaller trees. The soil is very shallow and there are huge uncovered rocks. The trees are represented by *Miconia* spp., *Rapanea* spp., *Lamanonia speciosa, Weinmannia* spp. and *Drimys brasiliensis*. The liana *Fuchsia regia* is well represented. Foliaceous and filamentous lichens appear in large quantities on the tree branches.

3. Open fields ("campos de altitude")

Above this upland open forest are open fields ("campos de altitude"), either directly on rock or on little islands of soil with a depth of 50 cm or less. In this landscape occur large populations of terrestrial bromeliads (*Vriesea* spp., *Pitcairnia* spp., *Tillandsia* spp.), Cyperaceae, Xyridaceae, Eriocaulaceae and Orchidaceae. In the shrub community, the very frequent taxa are Melastomataceae (*Tibouchina* spp., *Trembleya* spp.), Onagraceae (*Fuchsia* spp.), Compositae, Ericaceae and Gramineae (mainly *Chusquea pinnifolia*).

Flora

The flora of the mountain ranges of Rio de Janeiro has received considerable study since at least the beginning of the 19th century. Although a high number of taxa from the region were covered in *Flora Brasiliensis*, data were very incomplete. The first intensive floristic surveys were during the 1940s in the Serra dos Orgãos (Davis 1945; Veloso 1945; Rizzini 1953–1954) and in the Serra da Mantiqueira, particularly at Itatiaia (Brade 1956). These data, added to those recently obtained in other regions (JBRJ 1990, 1992; Guedes 1989; Martinelli 1989) are beginning to thoroughly demonstrate the great diversity and high level of endemism in the flora of Rio de Janeiro.

About 5000–6000 species of vascular plants occur, with 70–80% endemic to the tropical Atlantic coast. The most important families in numbers of species are Myrtaceae, Lauraceae, Leguminosae, Rubiaceae, Orchidaceae, Compositae, Euphorbiaceae and Sapotaceae. There is an urgent need for floristic inventories to build up precise knowledge about the distribution of the species.

Although data are still very incomplete, the State of Rio de Janeiro is considered one of the centres of endemism for

MAP 51. MOUNTAIN RANGES OF RIO DE JANEIRO, SOUTH-EASTERN BRAZIL (CPD SITE SA15)

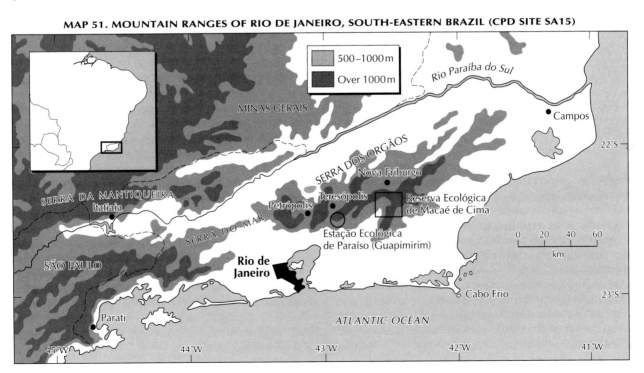

CPD Site SA15: Mountain ranges of Rio de Janeiro. Huge trees with bromeliads, epiphytes and lianas in the shady humid forest vegetation above 800 m. Photo: Tony Morrison/South American Pictures.

the Atlantic Coast rain forest (Mori, Boom and Prance 1981). These centres are being correlated with possible areas of forest refugia in arid periods during the Pleistocene (Mori 1989).

The high-altitude fields also possess many endemics, including palaeoendemics. Species found within this unique flora very often are restricted to only one mountain ridge. Classical examples include species of *Prepusa, Fuchsia, Tillandsia* and *Tibouchina*, and the well-known amaryllid *Worsleya rayneri* (Martinelli 1984) and bamboo *Glaziophyton mirabile*.

Useful plants

The forest resources of the Atlantic coast were heavily exploited during the initial colonization of Brazil. Plants that were useful for a wide variety of purposes such as timber, medicinals and food were intensely exploited as commercially important products. In south-eastern Brazil these valuable species included *Caesalpinia echinata* (Brazil-wood) and many other hardwoods – *Dalbergia nigra, Melanoxylum brauna, Mezilaurus navalium, Colubrina glandulosa* and species of *Ocotea, Nectandra, Aspidosperma* and *Tabebuia*.

Despite the over-exploitation of the last 500 years, the forest plants on the mountain ranges of Rio de Janeiro still have great potential. Recent preliminary studies have shown that c. 35% of the vascular plants in this region have some

economic value. This high percentage is even more significant when one considers that the areas covered with natural forest in Rio de Janeiro have been reduced to less than 15% of their former extent.

Social and environmental values

Forest preservation will protect watersheds in these mountains that provide water for many towns and cities (Domingues 1976). The forests occurring in the mountain ranges have an important role in preventing soil erosion, as they are located on lands where erosion susceptibility is moderate to severe, with slopes of 35° or higher ridges with slopes of 45°. These forests, while reducing damage to the roads and highways which are the economic link between Brazil's south-east and north-east, support the transportation of agricultural products of the mountains to the metropolitan region. Economic activities developed in the mountainous region of Rio de Janeiro are primarily directed toward production of legumes and horticultural products. There is also significant production of oranges, tangerines and bananas.

This is one of the most important regions of the world for biodiversity, holding a high number of endemic taxa, many of which are threatened because so little of their original habitats remains. Two very important Endemic Bird Areas (EBAs) are partially found here. The South-east

378

Brazilian lowland to foothills EBA (B52) stretches along the coast from Rio Grande do Sul to north-eastern Brazil. Many of the 60 bird species comprising this EBA are confined to south-eastern Brazil; 38 occur in Rio de Janeiro State, 17 of which are threatened. The South-east Brazilian mountains EBA (B53) is centred on the Serra do Mar and Serra da Mantiqueira; 18 of the 20 species comprising this EBA occur in the Rio de Janeiro ranges.

The scenic beauty of the landscapes and the richness and endemism of the fauna and flora support ecological tourism in the region. Considering that the city of Rio de Janeiro is one of the main tourist attractions of Brazil, the preservation of the forest in this region is basic to development of tourist activities.

Scientific knowledge of the coastal rain forest is still very limited. There is an urgent necessity for research on composition, structure, function and conservation for this ecosystem. The remaining forest is of fundamental importance for preservation of the genetic diversity of the species. A programme of conservation must be planned and initiated, including an inventory of species and their uses, biological studies on those species with the most potential and creation of genetic reserves. Development of studies on sustainable uses of the species is urgently needed.

Threats

The absence of a policy promoting settlement of populations in the countryside and stimulating agricultural production, particularly in the State of Rio de Janeiro, has caused a great impact in the areas near large urban centres. Immigrants come in large numbers (mostly from the country's north-east) in search of a new better life in the south-east. Nowadays, informed about the difficulties of living in the city of Rio de Janeiro, they opt for settlement in the adjacent mountainous region, where they generally become unskilled labourers.

The proximity of the city of Rio, as well as the mild climate and exuberant scenery, are key ingredients for land speculation. Construction of summer villas in forested areas belonging to non-working farms is often an additional problem. These construction projects normally have no plan for the land occupation, nor have they conducted any studies of environmental impact, and they may worsen the risk of landslides from seasonal strong showers. Since many of these dwellings are illegally built above the maximum elevation permitted (to ensure preservation of water springs), shortage of water during the dry season is a common problem.

The current habitat loss and exploitation of forests in Rio de Janeiro are destructive practices which are drastically reducing the size and the number of the state's remaining natural areas. Destruction of the forest around some high-altitude granitic outcrops has made their vegetation susceptible to fire, which is made worse by invasion of the African grass *Panicum maximum*.

Some cases of selective exploitation of species rate mention. The extraction of palm heart from *Euterpe edulis* ("palmito") is a source of money for poor people living on hillsides, along the roads and highways. Collecting of ornamentals (e.g. bromeliads, orchids, Araceae) is also alarming. Pteridophytes too are threatened, particularly species

of tree ferns of *Cyathea* (including *Trichipteris*) and *Dicksonia sellowiana* – the popular "xaxim", which is now listed as a species threatened with extinction. These taxa are collected in order to supply the florists established in mountain towns and countless orchid fanciers.

In addition there is the threat of excessive timber extraction. Areas near the towns of Nova Friburgo, Teresópolis and Resende provide 32% of the total extracted timber of the state, much of which is fuelwood to supply brick factories in the foothills as well as bakeries in metropolitan Rio de Janeiro (SECPLAN 1990–1991). The great volume of timber that used to occur along the Atlantic coast has been consumed almost totally as charcoal, ignoring the potential of each species for devising the rational use of these resources.

Another important factor in the inadequate use of natural resources is the lack of an effective policy protecting the areas officially under conservation. Unlike Brazilian hydroelectric plants, which were built by federal acquisition of the lands to be inundated, the many preserves – although legally created – are not adequately protected and do not effectively function because a large portion of their areas is privately owned.

Conservation

In 1990 the Government of the State of Rio de Janeiro decreed as preserved the remaining areas of Atlantic forest in the state. Thus Rio de Janeiro became the fourth state (1985, São Paulo and Paraná; 1990, Espírito Santo) to participate in preservation of the Serra do Mar *sensu lato*, promoting the conservation of a continuous forest (Collins 1990) with an area larger and under better protection than many others in the country. The demarcated area in Rio de Janeiro covers 6567 km², corresponding to 15.16% of the state's territory, with forest covering c. 5908 km². The parks, reserves, Ecological Stations and Environmental Protection Zones (Areas de Proteção Ambiental – APAs) under the responsibility of federal or state governments total c. 3220 km² in c. 20 units.

The existence of large protected areas grouping parks and reserves, such as Bocaina (1100 km²), Itatiaia (300 km²), Tinguá (260 km²) and Serra dos Orgãos (115 km²), does not necessarily signify effective conservation of natural areas, because little is known about their floristic and faunistic composition. This paucity of knowledge limits management strategies and environmental education, and in consequence there is poor supervision of the reserve borders. Effective conservation of the areas of coastal forest, as well as the other areas with ecological value, is probably linked to acquiring the land, without which the endeavours may be in vain.

References

Brade, A.C. (1956). A flora do Parque Nacional de Itatiaia. *Parque Nac. Itatiaia Bol.* 5: 1–85.

Collins, M. (ed.) (1990). *The last rain forests: a world conservation atlas.* Oxford University Press, New York. 200 pp.

Davis, D.E. (1945). The annual life cycle of plants, mosquitoes, birds and mammals in two Brazilian forests. *Ecol. Monogr.* 15(3): 243–295.

Domingues, A.J.P. (1976). Estudo do relevo, hidrografia, clima e vegetação das regiões programa do estado do Rio de Janeiro. *Bol. Geogr.* 248: 5–73.

Guedes, R.R. (1989). Composição florística e estrutura de um trecho de mata perturbada de baixada no município de Magé, Rio de Janeiro. *Arq. Jard. Bot. Rio de Janeiro* 29: 155–200.

JBRJ (1990). *Relatório anual do programa Mata Atlântica, Rio de Janeiro.* Jardim Botânico do Rio de Janeiro (JBRJ), Rio de Janeiro. 220 pp.

JBRJ (1992). *Relatório anual do programa Mata Atlântica, Rio de Janeiro.* JBRJ, Rio de Janeiro. 73 pp.

Martinelli, G. (1984). Nota sobre *Worsleya rayneri* (J.D. Hooker) Traub & Moldenke, espécie ameaçada de extinção. *Rodriguésia* 36(58): 65–72.

Martinelli, G. (1989). *Campos de altitude.* Editora Index, Rio de Janeiro. 160 pp.

Mori, S.A. (1989). Eastern, extra-Amazonian Brazil. In Campbell, D.G. and Hammond, H.D. (eds), *Floristic inventory of tropical countries: the status of plant systematics, collections, and vegetation, plus recommendations for the future.* New York Botanical Garden, Bronx. Pp. 427–454.

Mori, S.A., Boom, B.M. and Prance, G.T. (1981). Distribution patterns and conservation of eastern Brazilian coastal forest tree species. *Brittonia* 33: 233–245.

Nimer, E. (1989). *Climatologia do Brasil.* Instituto Brasileiro de Geografia e Estatística (IBGE), Departamento de Recursos Naturais e Estudos Ambientais, Rio de Janeiro. 422 pp.

Rizzini, C.T. (1953-1954). Flora organensis. Lista preliminar dos cormophyta da Serra dos Orgaos. *Arq. Jard. Bot. Rio de Janeiro* 13: 117–246.

SECPLAN (Secretaria de Planejamento e Controle) (1990-1991). *Anuário estatístico do estado do Rio de Janeiro (1990/91)*, 7/8. Governo do Estado do Rio de Janeiro, Rio de Janeiro.

Veloso, H.P. (1945). As comunidades e as estações botânicas de Teresópolis, estado do Rio de Janeiro. *Bol. Mus. Nac. Rio de Janeiro, Sér. Bot.* 3: 1–95.

Authors

This Data Sheet was written by Rejan R. Guedes-Bruni and Haroldo C. de Lima (Jardim Botânico do Rio de Janeiro, Seção de Botânica Sistemática, Rua Pacheco Leão 915, 22460 Rio de Janeiro, RJ, Brazil).

Acknowledgement

Map information supplied by Pedro Marcilio da Silva Leite, IBGE.

MATA ATLÂNTICA: CPD SITE SA16

SERRA DO JAPI
South-eastern Brazil

Location: Western part of Atlantic Upland between São Paulo and Campinas in eastern São Paulo State, about latitudes 23°12'–23°22'S and longitudes 46°57'–47°05'W.

Area: c. 354 km².

Altitude: c. 800–1300 m.

Vegetation: Semi-deciduous forest, semi-deciduous altitudinal forest, rocky outcrop vegetation.

Flora: Regionally high diversity; threatened species.

Useful plants: Reserve for timber trees; many species used for medicinals and handicrafts, some as ornamentals and for food.

Other values: Refuge for fauna; watershed protection for drinking water; potential genetic resources; scientific research; environmental education; tourist attraction.

Threats: Population growth, industrial development, air pollution, logging, fire, mining, increasing tourist pressure.

Conservation: c. 191 km² considered "historical patrimony" by government agency; municipal reserve Jundiaí; Environmental Protection Zone (APA); in Atlantic Forest Biosphere Reserve.

Geography

The Serra do Japi (Map 52) includes portions of the municipalities Jundiaí, Itupeva, Cabreúva, Pirapora do Bom Jesus and Cajamar. The region is drained mainly by the Tietê River, which flows to the north-west, and its tributaries – especially the Jundiaí River.

This mountain range is a quartzitic formation, originated from a Precambrian geosyncline (Ab'Sáber 1992). The soils are classified as red-yellow latosols (oxisol group). On the abrupt hillsides and flattened hilltops is shallow dystrophic soil, which is acidic and poor in nutrients, and outcrops of quartzitic rock or flint stones; at the base of the mountains is deep eutrophic soil, which is less acidic and rather rich in nutrients (Morellato 1992b; Rodrigues and Shepherd 1992).

Average annual precipitation in the region varies from 1907 mm at Cajamar to 1500 mm north-east and 1367 mm north-west of Jundiaí. The mean annual temperatures range between 19.2°C at the base of the range and 15.7°C on the hillsides (Pinto 1992).

There is a dry cool season (winter) from April to September, with mean monthly rainfall generally below 80 mm, and a wet warm season (summer) from October to March, with monthly rainfall above 100 mm. This bimodal rainfall is typical of the Atlantic Upland where semi-deciduous forest occurs. The rains concentrate mainly in the first months of summer (October–January), occasioned by the Atlantic Polar Front. The warmest month is January (mean temperatures 18.4°–22.2°C), the coolest month July (11.8°–15.3°C) (Pinto 1992).

Vegetation

The Serra do Japi is located between the two most expressive vegetation types of south-eastern Brazil (both within the Atlantic Domain): (1) the variously called semi-deciduous forest or upland forest (Leitão-Filho 1987), tropical seasonal forest (Longman and Jenik 1987) or subtropical moist forest (Holdridge 1947), which occurs at elevations of 600 m or higher; and (2) the Atlantic tropical moist forest or Atlantic forest *sensu stricto*, extending from lowlands along the Atlantic coast to the Serra do Mar (see CPD Site SA15) and Serra de Paranapiacaba (Joly, Leitão-Filho and Santos 1992).

The semi-deciduous forest presents different physiognomies, which are related to changes in soil, altitude and climate along its extensive distribution in the states of Paraná, São Paulo, Minas Gerais and Goiás, and is bounded by the Pantanal and Cerrado domains.

Along the base of the Serra do Japi are extensive areas of secondary forest disturbed by fire, logging, mining or agriculture. Successional species are common, such as *Trema micrantha*, *Acacia polyphylla*, *Cecropia pachystachya*, *C. glazioui*, *Piptadenia gonoacantha*, *Solanum swartzianum*, *Aegiphila sellowiana* and *Celtis iguanae*. The Serra do Japi topography and soil are inadequate for agriculture, which has preserved the upland to the present with the major area of continuous relatively undisturbed semi-deciduous forest in south-eastern Brazil. Three vegetation types are recognized (Rodrigues *et al.* 1989; Leitão-Filho 1992):

1. **Semi-deciduous forest (upland forest)** dominates the landscape, mostly between 700–900 m. The forest is markedly seasonal, with the most leaf fall occurring during the dry season and c. 40% of the tree species deciduous (Morellato *et al.* 1989; Morellato and Leitão-Filho 1992). The canopy is composed of trees 20–25 m tall, mostly species of Myrtaceae, Lauraceae, Meliaceae, Caesalpiniaceae, Mimosaceae, Euphorbiaceae and Fabaceae. Some families such as Anacardiaceae and Myrsinaceae are represented by few species but many individuals. The typical emergent trees include *Cariniana estrellensis* and *Cedrela fissilis*. The lowest tree stratum is

discontinuous, dominated by Rubiaceae, Myrtaceae and Meliaceae. The shrub and herbaceous strata are dense and characterized by shade-tolerant species of Piperaceae, Araceae and Violaceae.

2. **Semi-deciduous altitudinal forest** occurs generally up to 1000 m. It has a closed canopy 8–10 m high with some emergent trees up to 15 m tall, such as *Vernonia diffusa*, *Aspidosperma olivaceum* and *Prunus sellowii*. There is no conspicuous lower tree stratum. The most common tree families are Anacardiaceae, Clethraceae, Compositae, Cunoniaceae, Euphorbiaceae, Fabaceae *sensu stricto*, Myrtaceae and Vochysiaceae. The shrub and herbaceous strata have scattered plants.

    The flowering peak of the canopy trees occurs in the altitudinal forest (on the hilltops) at the onset of first rains (September–October), and then in the semi-deciduous (upland) forest in November (Morellato *et al.* 1989).

3. **Rocky outcrop vegetation.** Outcrops of rock occur sporadically in the Serra do Japi. Their vegetation is typical of semi-arid zones and stone fields, dominated by Cactaceae, Bromeliaceae, Cyperaceae, Eriocaulaceae and Piperaceae, and some shrubs of Celastraceae, Compositae, Ericaceae, Melastomataceae and Myrtaceae. This vegetation is probably a relict of the vegetation occurring in the region during semi-arid periods of the Quaternary (Ab'Sáber 1992).

Flora

The flora of the entire Serra do Japi region is poorly known. Floristic studies have investigated only tree species on the western side of the range (Rodrigues *et al.* 1989; Rodrigues and Shepherd 1992). According to Leitão-Filho (1992), 303 tree species belonging to 176 genera of 63 families are known, so far, in the Serra do Japi. They represent 86% of the families, 70.9% of the genera and 45.6% of the tree species known from the entire semi-deciduous forest. The diversity of the range's flora must be even higher since the data are restricted to the trees and the sites studied, which demonstrates the importance for conservation of the region. A great number of lianas, shrubs and herbaceous species occur, whereas epiphytes are less diverse and abundant than in the Atlantic forest *sensu stricto*.

The number of tree species characteristic of the Atlantic forest that occur at Serra do Japi is low. The Atlantic moist forest has the greatest species diversity and more endemics. Nonetheless, there is a decreasing gradient of tree-species diversity from the semi-deciduous forests of Serra do Japi to the interior of São Paulo, revealing that this region is more diverse than other semi-deciduous forests.

Useful plants

The vegetation of the Serra do Japi has traditionally offered various products important to local populations: fine hardwoods from timber trees (species of *Cedrela*, *Aspidosperma*, *Balfourodendron*, *Cariniana*, *Machaerium*); other wood (*Metrodorea*, *Callisthene*, *Guarea*, *Miconia*); fruit trees (species of *Eugenia*, *Psidium*, *Campomanesia*, *Diclidanthera*, *Rheedia*); ornamentals (e.g. *Tabebuia chrysotricha*, *Tibouchina*

sellowiana, *Lithrea moleoides*); and other species for handicrafts. Medicinal plants are particularly important to rural populations from the region of Serra do Japi, and a great number of medicinals are found in its rich flora.

Social and environmental values

Because of its importance, the natural history of the Serra do Japi is the subject of a comprehensive book (Morellato 1992a) which describes the environment, flora and fauna, and the status of conservation for this mountain range. The region has high animal diversity, and is a refuge for several threatened animals, e.g. the titi monkey (*Callicebus personatus*) ("sauá"), and the black hawk-eagle (*Spizaetus tyrannus*) and king vulture (*Sarcoramphus papa*) (Morellato 1992a). The mountain range is a watershed for drinking water, an important reserve for *in situ* genetic conservation and has been used for environmental education.

The Serra do Japi region has high landscape diversity and scenic quality and could provide the basis for development of a planned tourist industry, which could benefit the regional economy.

Economic assessment

Ecological tourism may be one (perhaps the main) important economic asset of the region. The Serra do Japi has excellent potential for tourism, being close to population centres (40 km from São Paulo, 50 km from Campinas), easily accessible by two important state roads and with a beautiful landscape. Forest management, with the sustained production of seeds, seedlings, timber, ornamentals and medicinal plants, could bring other economic assets.

Threats

During the 1970s and 1980s, the main significant impacts on Serra do Japi were mining, logging, deforestation and fire; in the late 1980s, tourist pressure increased.

Most of the region's population resides in the municipality Jundiaí. During the 1990s it is projected that the Jundiaí area will experience the most rapid population growth and industrial development in the State of São Paulo. The major threats to the plant resources will be more urbanization, industrialization, air pollution, deforestation and tourist pressure. Nevertheless, a regional development plan for the Jundiaí area could reduce forest degradation and might prevent total forest destruction.

Conservation

Efforts to preserve the Serra do Japi, the largest and best preserved area of semi-deciduous forest within the State of São Paulo, were initiated in the 1970s; in 1983 c. 191 km² were declared an area of "historical patrimony" by the Conselho de Defesa do Patrimônio Histórico, Artístico, Arquitetônico e Turístico (CONDEPHAAT) (see Map 52 – recommended protection zone 3). This state government agency regulates development activities at Serra do Japi such as land use, tourism, mining and utilization of any natural resource, but does not exclude the land owners. By 1984 part of the Serra do Japi and the urban area of Jundiaí and Cabreúva municipalities (see area within protection zones 1–3) were considered an Environmental Protection Zone (Area de Proteção Ambiental – APA). A small municipal reserve (Jundiaí) was established at the end of 1992. In 1993 the Serra do Japi was included in an Atlantic Forest Biosphere Reserve under the UNESCO-MAB Programme.

Unfortunately these acts of legislation are not effective enough to ensure the region's conservation (Joly 1992). An official plan is urgently needed to protect the Serra do Japi. All the land above 800 m within protection zone 3 (see Map 52) needs to be decreed a park or reserve; recommended protection zone 2 needs to be declared a transition area of restricted activities including ecotourism, education, forest management and sustained-yield timber production; and recommended protection zone 1 should be used as a buffer zone for controlled activities, such as tourism, fruit growing and sustainable agriculture. Inclusion of the Serra do Japi in the Atlantic Forest Biosphere Reserve may help to bring about such changes.

References

Ab'Sáber, A.N. (1992). A Serra do Japi, sua origem geomorfológica e a teoria dos refúgios. In Morellato, L.P.C. (ed.), *História natural da Serra do Japi: ecologia e preservação de uma área florestal no sudeste do Brasil.* Editora da Unicamp/Fapesp [Fundação de Amparo a Pesquisa do Estado de São Paulo], Campinas. Pp. 12–23.

Holdridge, L.R. (1947). Determination of world plant formations from simple climatic data. *Science* 105: 367–368.

Joly, C.A. (1992). A preservação da Serra do Japi. In Morellato, L.P.C. (ed.), *História natural da Serra do Japi: ecologia e preservação de uma área florestal no sudeste do Brasil.* Editora da Unicamp/Fapesp, Campinas. Pp. 310–321.

Joly, C.A., Leitão-Filho, H.F. and Santos, S.M. (1992). O patrimônio florístico. In Camara, I.G. (ed.), *Mata Atlântica.* Fundação SOS Mata Atlântica, Editora Index, São Paulo. Pp. 96–128.

MAP 52. SERRA DO JAPI, SOUTH-EASTERN BRAZIL (CPD SITE SA16)

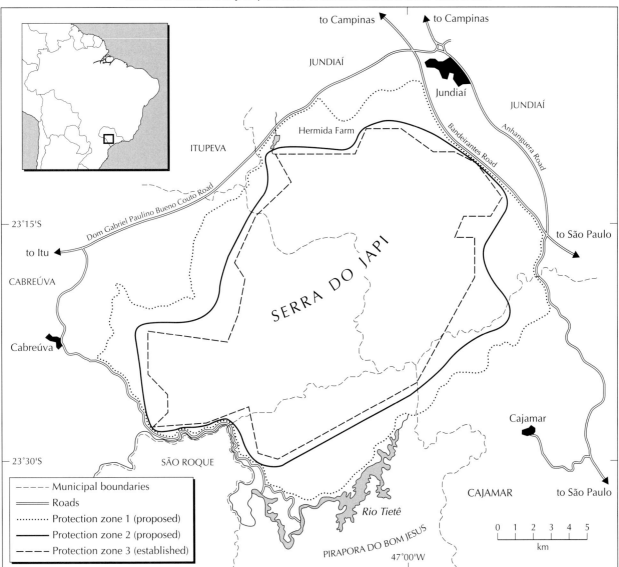

Leitão-Filho, H.F. (1987). Considerações sobre a composição florística das matas brasileiras. *Bol. Inst. Pesquisas Florestais* 12: 21–32.

Leitão-Filho, H.F. (1992). A flora arbórea da Serra do Japi. In Morellato, L.P.C. (ed.), *História natural da Serra do Japi: ecologia e preservação de uma área florestal no sudeste do Brasil.* Editora da Unicamp/Fapesp, Campinas. Pp. 40–62.

Longman, K.A. and Jenik, J. (1987). *Tropical forest and its environment.* Longman Singapore Publishers, Singapore. 347 pp.

Morellato, L.P.C. (ed.) (1992a). *História natural da Serra do Japi: ecologia e preservação de uma área florestal no sudeste do Brasil.* Editora da Unicamp/Fapesp, Campinas. 321 pp.

Morellato, L.P.C. (1992b). Nutrient cycling in two southeastern Brazilian forests. I. Litterfall and litter standing crop. *Journal Trop. Ecol.* 8: 202–215.

Morellato, L.P.C. and Leitão-Filho, H.F. (1992). Padrões de frutificação e dispersão na Serra do Japi. In Morellato, L.P.C. (ed.), *História natural da Serra do Japi: ecologia e preservação de uma área florestal no sudeste do Brasil.* Editora da Unicamp/Fapesp, Campinas. Pp. 112–140.

Morellato, L.P.C., Rodrigues, R.R., Leitão-Filho, H.F. and Joly, C.A. (1989). Estudo fenológico comparativo de espécies arbóreas de floresta de altitude e floresta mesófila semidecídua na Serra do Japi, Jundiaí, SP. *Revista Brasil. Botânica* 12: 85–98.

Pinto, H.S. (1992). O clima da Serra do Japi. In Morellato, L.P.C. (ed.), *História natural da Serra do Japi: ecologia e preservação de uma área florestal no sudeste do Brasil.* Editora da Unicamp/Fapesp, Campinas. Pp. 30–38.

Rodrigues, R.R., Morellato, L.P.C., Joly, C.A. and Leitão-Filho, H.F. (1989). Estudo florístico e fitossociológico em um gradiente altitudinal de mata estacional mesófila semidecídua na Serra do Japi, Jundiaí, SP. *Revista Brasil. Botânica* 12: 71–84.

Rodrigues, R.R. and Shepherd, G.J. (1992). Análise da variação estrutural e fisionômica da vegetação e das características edáficas, num gradiente altitudinal, na Serra do Japi. In Morellato, L.P.C. (ed.), *História natural da Serra do Japi: ecologia e preservação de uma área florestal no sudeste do Brasil.* Editora da Unicamp/Fapesp, Campinas. Pp. 64–96.

Authors

This Data Sheet was written by Dr Hermógenes F. Leitão Filho [Departamento de Botânica – IB, Universidade Estadual de Campinas (UNICAMP), Caixa Postal 6109, 13081 Campinas, SP, Brazil] and Dra. L. Patrícia C. Morellato [CNPq/Departamento de Botânica, Instituto de Biociências de Rio Claro, Universidade Estadual Paulista (UNESP), Caixa Postal 199, 13506-900 Rio Claro, SP, Brazil].

Acknowledgement

Map information supplied by L.R. Jordão.

JURÉIA-ITATINS ECOLOGICAL STATION
South-eastern Brazil

Location: In southern coastal area of São Paulo State, between latitudes 24°17'–24°40'S and longitudes 47°00'–47°30'W.

Area: 792 km².

Altitude: 0–800 m.

Vegetation: Open "campo" on mountaintops, moist forests on mountainsides, "restinga" vegetation and mangrove swamps near ocean.

Flora: High diversity and high endemism expected, at least 500–600 species; some disjuncts from Amazonia; threatened species.

Useful plants: Rich in medicinals; species for fibres, timbers (especially *Tabebuia cassinoides*); edible species (especially *Euterpe edulis*); ornamentals.

Other values: Ecological sanctuary for threatened fauna; research station; potential genetic resources; ecotourism; historical sites.

Threats: Palm-heart gathering, logging, hunting.

Conservation: Assured with establishment of Juréia-Itatins Ecological Station.

Geography

The Juréia-Itatins Ecological Station (Estação Ecológica Juréia-Itatins) (EEJI) occupies portions of the municipalities Peruíbe, Itariri, Miracatu and Iguape in the valley of the Ribeira de Iguape River, 210 km south-west of the city São Paulo. The EEJI has roughly the shape of an inverted triangle 90 km wide and 45 km from north to south (Map 53). It is crossed by the Una do Prelado River, a black-water river that winds north-eastward more or less parallel to the Atlantic coast for 80 km, isolating the Serra da Juréia. To the north-east the region is delimited by the Paranapuã massif, a buttress of the Serra dos Itatins.

The Serra da Juréia (which means prominent point in Tupi-Guaraní) is an inselberg connected to the rest of the mainland by a plain of alluvial sands, which formed during a post-glacial submersion by the sea. This is known as the Cananéia Transgression, which took place some 5100 years ago when the sea was c. 3 m above its present level (Por and Imperatriz-Fonseca 1984). For long periods the Una do Prelado River and its principal tributary the Cacunduva River were for much of their length turned into saltwater gulfs, which cut off the Serra da Juréia from the mainland. Evidence of this insular past includes the occurrence of "sambaquis" – deposits of shells and other debris left by prehistoric indigenous populations c. 30 km inland from the estuary of the Una. The isolation brought about by the Cananéia Transgression and the fact that the forests on the flanks and top of this massif are even now separated from the body of the Atlantic forest by the alluvial plains, may be interpreted as conditions suitable for speciation.

The Serra da Juréia is a Precambrian horst. The Juréia massif covers an area of 58 km² and reaches elevations of 400–800 m. It is divided by a depression occupied by the

Verde River, whose clear water and steep southern course are fed by the waterfalls of Juréia. The eastern face of the massif falls abruptly to the ocean, forming a steeply sloping coastline. The Una do Prelado River rises in the region of Banhado Grande south-west of the Serra da Juréia and is formed almost exclusively by rainwater; it is nutrient poor though rich in humic substances, with a pH of 3.7 (Por 1986; Por *et al.* 1984). The Una runs over a low plain (up to 4 m above sea-level) and is influenced by the tides over almost its entire course. In times of drought, seawater penetrates as far as 30 km into the estuary. The salinity of the waters of the Una and Cacunduva rivers is regulated entirely by rainfall and tides. Mangrove swamps extend as far as c. 5 km inland from the estuary.

The watercourses that arise on the Serra da Juréia contain clear nutrient-poor water with a pH of 5–6.5. There are springs in the natural campo, but most of the streams are intermittent and come from rainfall. During periods of intense rain, small waterfalls may turn into large cataracts. The hydrological wealth of the region results from two principal factors – lithology and climate.

The Ribeira lowland (Baixada do Ribeira) is characterized by average annual temperatures of c. 21°–22°C and the higher elevations with c. 17°–18°C. Relative humidity is above 80% and the average annual rainfall is c. 2200 mm (Camargo, Pinto and Troppmair 1972). High rainfall is due to the influence of two air masses: the Tropical Atlantic Mass originates in the South Atlantic and is active throughout the year, directly influencing the distribution and quantity of rainfall; the Polar Atlantic Mass arises in Patagonia and is more limited in its effect but still of great importance, causing abrupt drops in temperature that may cause frost at the higher elevations (Tôha 1985).

On the tops of the mountains the climate has favoured the formation of soils, which have an average depth of 2–3 m. On the slopes, which may be as steep as 35°, the soils are shallower and in the absence of vegetation are soon eroded. In the lowlands between the mountains lie the alluvial soils that result from recent sedimentation. The nature of the mountain soils hinders the infiltration of water to deeper layers; thus the water table does not lie deep, and water surfaces as a large number of springs and watercourses.

Vegetation

The vegetation of the region varies according to altitude and soil. There are five main types: open campo, forest, scrub, herbaceous restinga vegetation, and mangrove (Eiten 1970).

The **open "campo" vegetation**, which covers the tops of the mountains at elevations over 300 m, is grass-herb-subshrub fields with scattered low to medium-tall xeromorphic shrubs.

The forest formation includes a series of **evergreen tropical forest** types covering the slopes and base of the mountains (Coutinho 1962). The three most important types are: (1) a **moist tall forest** on the seaward slope of the Serra do Mar; (2) **littoral tall forest**, a low-altitude forest on alluvial and lacustrine clays (and occasionally sands) of the inner part

of the littoral plain; and (3) **restinga forest**, a low to medium-tall forest, with a composition very different from the slopes of the Serra do Mar.

The other formations that occur in the "restinga", a littoral sandy plain c. 40 km in extent (Silva and Leitão-Filho 1982), are: (1) **restinga closed scrub**, a low to tall and very dense evergreen scrub forest; (2) **restinga open scrub ("nhundu")**, composed of evergreen shrubs and low twisted trees; and (3) **herbaceous beach vegetation**, closed to sparse evergreen grass-herb fields covering the low dunes and beaches (Hueck 1955).

The **mangrove vegetation** is an evergreen low forest, or low to tall and closed scrub, of *Rhizophora mangle, Avicennia schaueriana*, *Laguncularia racemosa* and *Hibiscus pernambucensis*, usually with distinct boundaries between it and the restinga forest.

Flora

The flora of Brazil's Atlantic Coast forest is incompletely known, with new species and even new genera being discovered. Random samples of the moist forest indicate significant endemism, which has been estimated as 53.5%

MAP 53. JURÉIA-ITATINS ECOLOGICAL STATION, SOUTH-EASTERN BRAZIL (CPD SITE SA17)

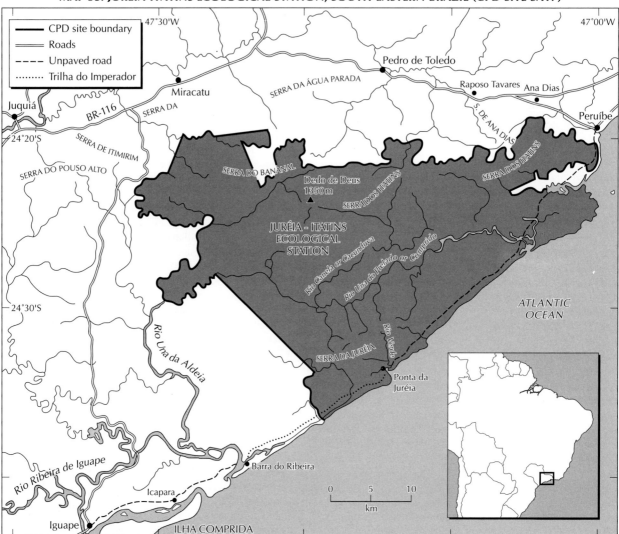

among tree species (Mori *et al.* 1983) and 37.5% in non-arborescent families – or 74.4% including Bromeliaceae. Endemism of palms is a rather high 64%, with 49 species endemic to the Atlantic coastal forests; 11 are considered threatened. Endemism for bamboo genera has been estimated at 40.9%; the region has been a refuge for various primitive species of bambusoid grasses and some woody groups (Soderstrom and Calderón 1974). The isolation of the Atlantic forest from Amazonia took place in the late Tertiary, which explains why this region has many plant species in common with the Amazon, but also many endemics.

A Flora of the Serra da Juréia is in preparation. Preliminary results on the slope forest indicate the presence of 76 families of dicotyledons, 21 of monocotyledons and 14 of ferns and fern allies; c. 500–600 species are estimated for the region. The canopy varies in height from 20–30 m, with emergents reaching 35 m. The families most numerous in tree species are Myrtaceae (18 spp.), Leguminosae (16), Melastomataceae (10) and Annonaceae (9). The shrub layer varies in height from 2 m to 5 m, and is distinguished by many species of Rubiaceae (15), Acanthaceae (9), Piperaceae (5), Rutaceae (5) and Myrtaceae (5). The tree and shrub layers also contain a wide diversity of lianas, creepers and epiphytes. Among the epiphytes, most of the species are in Orchidaceae, Bromeliaceae, Araceae and Gesneriaceae or ferns, whereas most lianas and vines are in Compositae, Malpighiaceae and Bignoniaceae.

Results of a three-year inventory programme confirm the wealth of the flora and a number of facts point to the importance of its conservation: the occurrence of rare species (known from few specimens), such as *Passiflora watsoniana* and *Piper bowei*; the occurrence of species with disjunct distributions, such as *Loreya spruceana* (Melastomataceae), known from the Amazon region; a large number of saprophytes of the families Orchidaceae, Burmanniaceae, Gentianaceae, and Triuridaceae (so uncommon in herbarium collections); and the discovery of new species – *Anthurium jureianum* (Araceae) and others not yet described in *Sinningia* (Gesneriaceae), *Merostachys* (Poaceae – Bambusoideae), *Sorocea* (Moraceae) and *Barbacenia* (Velloziaceae).

Useful plants

The region includes important reserves of timber, such as *Ocotea* spp., *Nectandra* spp. ("canelas, niúva, sassafrás"), *Macrosamanea pedicellaris* ("timboúva"), *Hymenaea courbaril* ("jatobá"), *Cariniana estrellensis* ("jequitibá"), *Cabralea canjerana* ("canjerana"), *Tabebuia cassinoides* ("caxeta"); fibres, such as *Eriotheca pentaphylla* ("ouvira"), *Bactris setosa* ("tucum"), *Hibiscus pernambucensis*; and edible plants, such as *Euterpe edulis* (palm heart), *Psidium cattleianum* ("araçá"), *Rheedia gardneriana* ("bacupari"). These resources were traditionally used by local people, but this usage is now prohibited by environmental legislation.

Most of the inhabitants in the region ("caiçaras") have a rich pharmacopoeia based mostly on medicinal plants. Species used include *Psidium guajava* ("goiaba"), *P. cattleianum* and *Solanum inaequale* ("quina-branca") for diarrhoea; *Renealmia petasites* ("capitiu") for malaria; *Rheedia pachiptera* ("bálsamo") and *Struthanthus* spp. ("enxertinho") for external wounds; *Aphelandra ornata* ("erva-de-lagarto") for snake bite; and *Casearia sylvestris* ("erva-de-macaco") and *Bactris* sp. ("tucum-

branco") for bruises (Born, Diniz and Rossi 1989). Studies begun during the last four years will certainly reveal more species with medicinal properties.

Social and environmental values

Despite its status as an Ecological Station, Juréia is inhabited by 365 families of caiçaras, the descendants of an old Amerindian-European mixture, who are grouped in 22 villages or live in widely isolated huts along the beaches and rivers (Por and Imperatriz-Fonseca 1984). There is just one small fishing community, at the mouth of the Una do Prelado River. The Government of São Paulo State has a plan for management of the station, to make possible the maintenance of the communities with minimal disturbance to the environment, and intends to eradicate almost all banana plantations and insure forest regeneration.

Geographical isolation has helped to allow persistence of sites of historical interest, such as the Caminho do Imperador (Imperial Highway), which was opened in 1545 at the behest of the Governor General of the province to link the townships of Cananéia and Iguape with São Vicente. The road is still used by local people, mostly for religious pilgrimages.

The region has been used for ecological tourism on account of its scenic beauty and the presence of several threatened species of animals and plants. Routine tourism occurs at only a few sites such as Cachoeira do Paraíso, yet brings a considerable number of people to the region each year.

The Serra da Juréia is particularly useful for research on speciation because of its natural isolation from the rest of the Atlantic forest.

Economic assessment

In the long run, the greatest economic value of the Juréia-Itatins Ecological Station may be the preservation of a good sample of Atlantic forest biodiversity, allowing for studies and sustainable exploitation of indigenous species in the future. Other uses include the academic training of scientists and environmental education through ecological tourism.

Threats

The forest has undergone relatively few disturbances, mainly from agriculture (banana plantations) developed in the lowlands more intensively in the past, before establishment of the Ecological Station. Despite its isolation, hunting and logging do occur. Nowadays the area is threatened mostly by palm-heart gatherers.

Conservation

The natural conservation of the region can be explained by its isolation. The mountains of the Juréia-Itatins massif make access to c. 40 km of restinga impossible by road. The area has thus been preserved over the years.

On 8 April 1958, State Decree No. 31,650 created Serra dos Itatins Reserva Florestal Estadual (State Forest Reserve).

On 28 January 1963, SD No. 41,538 turned the area of the Serra dos Itatins over to the Guaraní Amerindians. In 1979 the Secretaria Especial do Meio Ambiente (an agency of the federal government) was granted an area in the Juréia massif, where a Biological Reserve was created. In 1980 the president signed a decree whereby 236 km² of Juréia were disappropriated and could not be used for construction of two nuclear power plants (Iguape I and II); in 1985 the idea of constructing the power plants and the area were abandoned by the federal government. On 20 January 1986, the state government of São Paulo created the Juréia-Itatins Ecological Station by SD No. 24,646. On 28 April 1987, the government approved Law No. 5649, whereby the EEJI actual area of 792 km² was defined (indigenous and other lands were excluded).

The region is now an official ecological sanctuary. Juréia provides protection for example for timber species and palmito palm, and dozens of rare and threatened mammal and bird species, including spider monkey, jaguar, otter, capybara and tinamou.

References

Born, G.C.C., Diniz, P.S.N.B. and Rossi, L. (1989). *Levantamento etnofarmacológico e etnobotânico nas comunidades da Cachoeira do Guilherme e parte do Rio Comprido (sítio Ribeirão Branco – sítio Morrote de Fora) da Estação Ecológica de Juréia-Itatins, Iguape, São Paulo.* Secretaria do Meio Ambiente do Estado de São Paulo, São Paulo. 94 pp.

Camargo, J.C.G., Pinto, S.A.F. and Troppmair, H. (1972). Estudo fitogeográfico e ecológico da bacia hidrográfica paulista do Rio da Ribeira. *Biogeografia, Inst. Geogr., Univ. São Paulo* 5: 1–30.

Capobianco, J.P. (1988). Juréia. *Horizonte Geográfico* 1(2): 30–39.

Coutinho, L.M. (1962). Contribuição ao conhecimento da mata pluvial tropical. *Bol. Fac. Filos. Ciênc. Letr., Univ. São Paulo* 257, *Botânica* 18: 1–219.

Eiten, G. (1970). A vegetação do estado de São Paulo. *Bol. Inst. Bot.* 7: 1–147.

Hueck, K. (1955). *Plantas e formação organogênica das dunas do litoral paulista.* Parte 1. Instituto de Botânica, Secretaria da Agricultura do Estado de São Paulo. 130 pp.

Mori, S.A., Boom, B.M., Carvalho, A.M. de and Santos, T.S. dos (1983). Southern Bahian moist forests. *Bot. Rev.* 49: 155–232.

Por, F.D. (1986). Stream type diversity in the Atlantic lowland of the Juréia area (subtropical Brazil). *Hydrobiologia* 131: 39–45.

Por, F.D. and Imperatriz-Fonseca, V.L. (1984). The Juréia Ecological Reserve, São Paulo, Brazil – facts and plans. *Environm. Conserv.* 11: 67–70.

Por, F.D., Shimizu, G.Y., Almeida-Por, M.S., Tôha, F.A.L. and Rocha Oliveira, I. (1984). The blackwater river estuary of Rio Una do Prelado (São Paulo, Brazil): preliminary hydrobiological data. *Rev. Hydrobiol. Trop.* 17: 245–258.

Silva, A.F. da and Leitão Filho, H.F. (1982). Composição florística e estrutura de um trecho da Mata Atlântica de encosta no município de Ubatuba (São Paulo, Brasil). *Revista Brasil. Bot.* 5: 43–52.

Soderstrom, T.R. and Calderón, C.E. (1974). Primitive forest grasses and evolution of the Bambusoideae. *Biotropica* 6: 141–153.

Tôha, F.A.L. (1985). *Ecologia de zooplâncton do estuário do Rio Una do Prelado (São Paulo, Brasil).* Ph.D. thesis, Universidade de São Paulo, São Paulo.

Authors

This Data Sheet was written by Dra. Maria Candida H. Mamede, Dra. Inês Cordeiro and Lucia Rossi (Secretaria de Estado do Meio Ambiente, Instituto de Botânica, Caixa Postal 4005, 01061-970 São Paulo, SP, Brazil).

MATA ATLÂNTICA: CPD SITE SA18

MBARACAYU RESERVE
Paraguay

Location: Mostly south-west of Cordillera de Mbaracayú in north-central Canindeyú Department, eastern Paraguay; extending approximately 30 km north-south and 20 km east-west between latitudes 23°59'–24°16'S and longitudes 55°20'–55°33'W.

Area: c. 580–600 km².

Altitude: c. 140–450 m.

Vegetation: Subtropical moist non-flooded, riverine and flooded forests; warm-temperate transitional forest; "cerrado" woodland; grassland; wetlands.

Flora: Highest diversity in eastern Paraguay; many tropical taxa at southern limit of their ranges; regionally endemic species; threatened species.

Useful plants: Some 250 species: e.g. timbers; industrial substances; *Ilex paraguariensis* ("yerba mate") tea; wild fruits and relatives of commercial fruits; extensive local pharmacopoeia (e.g. *Tabebuia impetiginosa* – "pau d'arco"); ornamentals.

Other values: Large area for fauna of moist-forest habitats, including threatened species; potential genetic resources; watershed protection for upper Jejuí River of Paraguay River system.

Threats: Deforestation in bordering areas, illegal logging, clandestine cultivation of marijuana (*Cannabis sativa*), over-hunting.

Conservation: Governmentally recognized private reserve; overall management plan, potential buffer areas.

Geography

Paraguay is entirely within the upper watershed that drains southward to become La Plata River. The Paraguay River begins north of the country in the pantanal wetlands and bisects Paraguay into distinct natural regions: to the west are vast alluvial plains of the drier chaco (264,925 km²) (CPD Site SA22); to the east is more humid, geologically varied terrain (159,827 km²), commonly called Paraguay Oriental (Eastern Paraguay).

Eastern Paraguay is bounded by the Apa River to the north, the Amambay-Mbaracayú mountain ranges to the north-east and the Paraná River to the east and south, and is subdivided into two fairly distinctive areas by a series of low (300–860 m) rolling hills and mountain chains which form an S-shaped watershed from the Sierra de San Joaquín and Cordillera de Ybytyruzú southward to the Cordillera de San Rafael. East of the watershed is the Paraná Plateau; to the west opens a broad undulating plain which gradually descends south-westward to become the immense Neembucú wetlands. The Mbaracayú Reserve is mostly south-west of the Cordillera de Mbaracayú, in north-central Canindeyú Department (Map 54).

Eastern Paraguay is largely underlain by Triassic basaltic lava which has given rise to red-brown lateritic or very dark colluvial and hydromorphic soils, and Precarboniferous sandstone where red-yellow sandy podzols prevail. The region's annual rainfall averages 1600–1800 mm. The rainy season extends from October to March, and July and August are driest. The mean annual temperature is 20°–25°C, with extreme highs near 40°C and lows c. 0°C although frost is infrequent and limited to early mornings. There is distinct biseasonality, with hot humid summers and cool less humid winters (IIED, STP and USAID 1985).

Vegetation

Eastern Paraguay was once covered with extensive forests (Zardini 1993). The remnants are some of the continent's last larger subtropical moist forests. Semi-deciduous, they harbour several endemic subtropical genera, and some tropical and cerrado species in the southern extreme of their ranges. These forests are also phytogeographically intriguing because of the intermingled taxa with tropical and temperate affinities. An inventory of the flora and fauna of Mbaracayú was conducted in 1989–1990 using the procedure of Rapid Ecological Assessment (TNC 1992) to identify the natural communities and assess their natural status; 19 plant communities were recognized (CDC-Paraguay 1991).

The reserve harbours several types of forest. Non-flooded climax forest covers c. 87% of the area and reaches 15–30 m in height. Bignoniaceae, Leguminosae and Meliaceae are dominant. There are two tree strata and some canopy trees are deciduous. Emergents include *Tabebuia heptaphylla*, *Cedrela fissilis*, *Balfourodendron riedelianum*, *Piptadenia rigida* and *Peltophorum dubium*; understorey trees include *Sorocea bonplandii*, *Guarea kunthiana* and *Pilocarpus pennatifolius*. Epiphytes occur, and lianas are notably important – their density of 92 individuals per 0.1 ha is one of the highest reported for the neotropics (Keel, Gentry and Spinzi 1993).

The climax forest grades into riverine or flooded forests in wetter areas. *Bambusa guadua* (= *Guadua angustifolia*) can be important in these wetter habitats, forming dense and often nearly pure stands. Riverbanks are often dominated by *Luehea divaricata*, *Inga marginata* and *Esenbeckia grandiflora*. In areas of waterlogged soil, the canopy is lower (20 m high) with more evergreen species, commonly including *Rapanea* sp., more Myrtaceae (e.g. *Myrciaria baporeti*, *Myrcia bombycina*, *Eugenia uniflora*) and palms (e.g. *Geonoma schottiana*, *Syagrus romanzoffiana*).

On extremely sandy or poor soil, species characteristic of cerrado vegetation are found in dense abundance. They are short (not over 5 m high); some show tortuous growth habits and some have corky bark. Diagnostic are *Butia yatay*, *Allagoptera* sp., *Anadenanthera peregrina* and *Gochnatia polymorpha*. Sometimes (in "campo sucio") only subshrubs and herbs are dominant.

Grasslands are dominated by graminoids such as *Andropogon* and *Axonopus*. In wetlands (marshes, lagoons), sedges (e.g. *Eleocharis*) and low herbs are dominant and the plants may form dense mats.

Flora

Located on a phytogeographic ecotone and possibly having served as a refuge for subtropical species during past climatic fluctuations, the reserve protects many regional endemics. At the generic level some of the endemics are monotypic: *Hennecartia* (Monimiaceae), *Bastardiopsis* (Malvaceae), *Holocalyx* (Leguminosae), *Balfourodendron* (Rutaceae) and *Pseudananas* (Bromeliaceae). There are many endemic species

in genera that are widely distributed and highly specialized (e.g. *Citronella gongonha*, *Arrabidaea mutabilis*, *Macfadyena mollis*) as well as relictual (*Diatenopteryx sorbifolia*, *Melicoccus lepidoctela*, *Patagonula americana*).

The floristic affinities of the region with tropical areas to the north are clearly indicated by the presence of such tropical woody species as *Annona amambayensis*, *Aspidosperma polyneuron*, *Cariniana estrellensis* and *Geonoma schottiana*, and the predominance of major tropical families such as Myrtaceae, Bignoniaceae, Leguminosae, Moraceae and Meliaceae (Keel, Gentry and Spinzi 1993). Temperate affiliations can be seen among the diverse herbaceous taxa such as in Ranunculaceae, Scrophulariaceae, Polygonaceae and the predominantly woody genera *Prunus* and *Hexachlamys*.

This area is thought to have the highest floristic diversity in eastern Paraguay. Though the species diversity of the Mbaracayú region is relatively low presumably due to the subtropical climate and rather uniform topography, the regionally endemic species and especially genera, the mixture of tropical and temperate species, and as well the high density of lianas, make the forests of these wildlands significant for the conservation of plant diversity.

Useful plants

The Mbaracayú region contains rich timber resources and precious woods. Species marketed internationally include *Tabebuia heptaphylla* ("lapacho rosado, tajy"), *T. impetiginosa*, *Cedrela fissilis* ("cedro, ygary"), *Balfourodendron riedelianum* ("guatambu, yvyra ñeti"), *Myrocarpus frondosus* ("incienso, yvyra pajé") and *Aspidosperma polyneuron* ("peroba"). *Balfourodendron* makes up 9.3% of the commercial volume of timber in the region. These resources are in critical need of protection from excessive logging (Tortorelli 1966; López *et al.* 1987; Peace Corps Paraguay, undated; Zardini 1993).

Some forests also have extensive stands of *Copaifera langsdorfii* ("kupay"), whose resin can be transformed into diesel fuel (López *et al.* 1987). Other forests harbour wild populations of *Ilex paraguariensis* ("yerba mate" or Paraguayan tea), which is of considerable economic importance in the Southern Cone (FMB 1989). Fruit crops such as the pineapple (*Ananas comosus*) and custard apple (*Annona* spp.) have wild relatives (*Pseudananas ananasoides* and *Annona amambayensis*), providing other examples of the need for *in situ* conservation of the germplasm diversity in the Mbaracayú Reserve.

The indigenous Aché and Guaraní peoples make extensive use of the flora and have an abundant pharmacopoeia that merits further study for compounds with medicinal value. The powdered bark of *Tabebuia impetiginosa* ("pau d'arco", "taheebo" or "lapacho" tea), which is said to improve the immune system (e.g. Tierra 1988), is sold as a medicinal in the U.S.A. and Europe. A preliminary species list of the economic plants in the Mbaracayú region has been prepared (CDC-Paraguay 1991; Keel, Gentry and Spinzi 1993).

Social and environmental values

At present there are two ethnic groups settled near the Mbaracayú Reserve, the Aché and the Guaraní (FMB 1989).

MAP 54. MBARACAYU RESERVE, PARAGUAY (CPD SITE SA18)

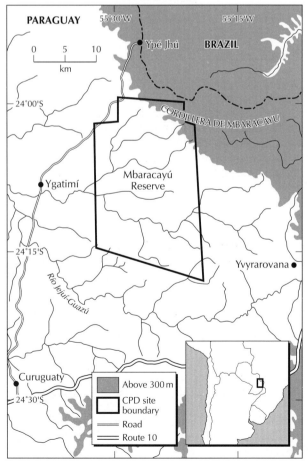

The traditionally nomadic Aché lived in the eastern forests as hunters and gatherers until 1975–1979. They are slowly adapting to the life of small farmers, although they still need the remnant forests' fauna and fruits for an adequate diet (Homer 1992). The Guaraní survive on subsistence farming. They go to the reserve to collect yerba-mate leaves and less often to hunt. The Fundación Moisés Bertoni and The Nature Conservancy (TNC) have developed a land management programme for the indigenous peoples to continue their traditional uses of the Mbaracayú forests (FMB and TNC 1991).

These forests protect the Jejuí watershed. The Mbaracayú Reserve is the source for various tributaries of the Jejuí-mí River, which later as the Jejuí-Gauzú River joins the Paraguay River. Designed to protect one of the few remaining subtropical moist forests in Eastern Paraguay, the Mbaracayú Reserve has an important role as well in securing diverse habitats for many species of the regional fauna, including threatened species and some of the prominent mammals and birds of Paraguay such as jaguar, ocelot, margay, bush dog, tapir, giant armadillo, king vulture, macaw, bare-throated bellbird, bare-faced curassow and burrito (CDC-Paraguay 1991).

The general region has important areas of evergreen humid forest ("Atlantic" forest) and remnant *Araucaria* forest which support a number of birds that have restricted ranges – including the canebrake groundcreeper (*Clibanornis dendrocolaptes*) and creamy-bellied gnatcatcher (*Polioptila lactea*), or that are threatened. The region is important for the following five threatened birds: black-fronted piping-guan (*Pipile jacutinga*), vinaceous amazon (*Amazona vinacea*), helmeted woodpecker (*Dryocopus galeatus*), São Paulo tyrannulet (*Phylloscartes paulistus*) and russet-winged spadebill (*Platyrinchus leucoryphus*).

Threats

In 1945 as much as 43% or 68,000 km² of Eastern Paraguay was still forested, but half that forest is now gone, as the land has been converted to other uses (Keel, Gentry and Spinzi 1993; Zardini 1993). The Jejuí Basin including the Mbaracayú area was used to obtain natural products such as animal hides, timber and particularly yerba mate until 1970 when the construction of roads began, which encouraged new settlements. Typically their purpose is to establish cattle ranches, or cultivate cash crops that often cause soil infertility (FMB and TNC 1991; FMB 1989). Deforestation, at a rate surpassing 1000–1500 km² yearly (Zardini 1993), in some places has reached the reserve boundary. Measures to prevent the reserve turning into an ecological island are urgently needed.

Except for the cordillera in the north-east near the border with Brazil, there are no effective natural barriers that protect the reserve, which also has several roads. Unless boundaries are well marked and an effective patrol system is put in place, theft of timber and fuelwood and poaching will continue to degrade the reserve.

CPD Site SA18: Mbaracayú Reserve, Paraguay. Lagunita, a wetland community within the reserve.
Photo: Shirley Keel.

Conservation

In 1988, 0.13% of Eastern Paraguay was under the national protected area system. The Conservation Data Center (CDC) of Paraguay with support of The (U.S.) Nature Conservancy identified 23 potential areas for conservation in six ecoregions, with the Mbaracayú Wildlands figuring as a priority site (CDC-Paraguay 1990). Acting on the CDC-Paraguay recommendations, Paraguay's Ministerio de Agricultura y Ganadería (Ministry of Agriculture and Livestock) decreed five new National Parks, one new Wildlife Refuge and greatly expanded protected areas in Eastern Paraguay. As a result, lands in the national protected area system have increased to c. 1.2% (2000 km²) of Eastern Paraguay.

The Fundación Moisés Bertoni is developing a system of private reserves in Paraguay. In 1991 the Conservancy and Fundación agreed to purchase the Mbaracayú land (with funding from private sources and USAID) from the International Finance Corporation (for US$2 million). This World Bank subsidiary was acting under its new conservation policy, and controlled the land because a forestry company in 1979 defaulted on a loan (FMB and TNC 1991; Homer 1992). The property has been made a patrimony of the Fundación Mbaracayú, which holds title and is in charge of managing the reserve. The Mbaracayú Reserve has been recognized by the Paraguayan Government, and has received national media attention and nearby community support.

As a priority site selected for the USAID-supported Parks in Peril Program, the Fundación Bertoni working with TNC has been able to begin implementing the management programme. The 1989–1990 inventory (CDC-Paraguay 1991) established the baseline information used for managing the reserve. Land easement or the full purchase of intact forested areas as buffer zones will add effectiveness to the system of protection. Detailed ecological studies in several forest remnants in the adjacent eastern Alto Paraná Department provide some insights into forest dynamics and life histories of many species in the Mbaracayú Reserve (Stutz de Ortega 1986, 1987). The reserve's future as a sanctuary in eastern Paraguay may be bright.

References

CDC-Paraguay (1990). *Areas prioritarias para la conservación en la Región Oriental del Paraguay.* Centro de Datos para la Conservación (CDC), Asunción. 99 pp.

CDC-Paraguay (1991). *Estudios biológicos en el área del Proyecto Mbaracayú, Canindeyú, República del Paraguay.* CDC, Asunción. Unpublished report. 115 pp.

FMB (1989). *Análisis socioeconómico y cultural de las poblaciones asentadas en el área de influencia del Proyecto Mbaracayú.* Fundación Moisés Bertoni para la Conservación de la Naturaleza (FMB), Asunción. 51 + xlv pp.

FMB and TNC (1991). *Mbaracayú Nature Reserve management program. A Park in Peril work plan.* FMB, Asunción and The Nature Conservancy (TNC), Arlington, Virginia, U.S.A. Unpublished.

Homer, S. (1992). The last hunt: on the trail with Paraguay's forest people. *Nature Conservancy* (Nov.–Dec.): 24–29.

IIED, STP and USAID; Raidán, G. (1985). *Perfil ambiental del Paraguay.* International Institute for Environment and Development (IIED), Secretaría Técnica de Planificación (STP) and U.S. Agency for International Development (USAID). Cromos, Asunción. 162 pp.

Keel, S., Gentry, A.H. and Spinzi, L. (1993). Using vegetation analysis to facilitate the selection of conservation sites in eastern Paraguay. *Conservation Biology* 7: 66–75.

López, J.A., Little Jr., E.L., Ritz, G.F., Rombold, J.S. and Hahn, W. (1987). *Arboles comunes del Paraguay.* Cuerpo de Paz, Washington, D.C. 425 pp.

Peace Corps Paraguay (n.d.). *Forestry training manual.* Paraguay.

Stutz de Ortega, L.C. (1986). Etudes floristiques de divers stades secondaires des formations forestières du Haut Parana (Paraguay oriental). Floraison, fructification et dispersion des espèces forestières. *Candollea* 41: 121–144.

Stutz de Ortega, L.C. (1987). Etudes floristiques de divers stades secondaires des formations forestières du Haut Parana (Paraguay oriental). Structure, composition floristique et régéneration naturelle: comparaison entre la forêt primaire et la forêt sélectivement exploitée. *Candollea* 42: 205–262.

Tierra, M. (1988). *Planetary herbology.* Lotus Press, Santa Fe, New Mexico, U.S.A. 485 pp.

TNC (1992). *Evaluación ecológica rápida. Un manual para usuarios de América Latina y el Caribe.* TNC, Arlington, Virginia, U.S.A.

Tortorelli, L.A. (1966). *Formaciones forestales y maderas del Paraguay.* Universidad de Asunción, Facultad de Agronomía y Veterinaria, Asunción.

Zardini, E.M. (1993). Paraguay's floristic inventory. *Natl. Geogr. Res. Explor.* 9(1): 128–131.

Authors

This Data Sheet was written by Dr Shirley Keel (The Nature Conservancy, Latin America Science Program, 1815 North Lynn St., Arlington, VA 22209, U.S.A.) and Olga Herrera-MacBryde, Smithsonian Institution, Department of Botany, NHB-166, Washington, DC 20560, U.S.A.).

CAATINGA OF NORTH-EASTERN BRAZIL
Brazil

Location: Between about latitudes 3°–17°S and longitudes 45°–35°W, covering part or all of eight states: Piauí, Ceará, Rio Grande do Norte, Paraíba, Pernambuco, Alagôas, Sergipe and Bahia.

Area: c. 1,000,000 km².

Altitude: 0–1000 m.

Vegetation: Mosaic of many different types, divided into xerophytic "caatinga" proper and remnants of older floras replaced with gradual climatic drying of region. Other types include gallery forest, cerrado, humid montane forest ("brejo de altitude"), grassland.

Flora: Varied species, adapted to many local climatic conditions: xerophytes, including cacti and bromeliads, to elements of tropical rain forests and cerrado, also many grasses and legumes; specific and generic endemism.

Useful plants: Many grass and legume forages that may be of importance in other arid and semi-arid regions; several local fruits; many very important palms; various timber species.

Other values: Many species of animals losing habitats at alarming rates, some species on verge of extinction (e.g. Spix's macaw); watershed and soil protection; important archaeological sites.

Threats: Overgrazing, timber and fuelwood extraction, plant collecting, desertification, irrigation, modern agricultural expansion, salinization.

Conservation: c. 3168 km² (c. 0.31% of region) in 16 main conservation units of different types.

Geography

The "caatinga", the north-eastern floristic province of Brazil (Andrade-Lima 1981), constitutes one of three arid nuclei in South America, together with the Guajira Peninsula along the Caribbean coast of Colombia and Venezuela, and the dry belt that extends from Argentina through Chile and Peru into Ecuador (Ab'Sáber 1980). Each of these three regions represents a different geological province and is subject to a different climatic regime, and each has originated very distinct xeromorphic vegetation.

The geology of the caatinga is in essence not much different from that of the Upper Rio Negro region (CPD Site SA6). Both originated from very old Precambrian rocks, severely degraded during the Tertiary, and overlain by more recent marine sandstones and other sediments. As in the Upper Rio Negro, there are remnant crystalline outcrops, including monolithic mesas and isolated mountain ranges (Ab'Sáber 1977; Fernandes and Bezerra 1990).

Climatically the caatinga region can be considered one of the more complex areas in the world, being at the crossroads or convergence of several highly unstable winds coming from north, south, west and east (Nimer 1969; Andrade-Lima 1977). The most significant winds are the northern currents, with an intertropical convergence centre at c. 5°N, which influence rainfall as far as 9°–10°S. The climate is largely Köppen BShw (hot, semi-arid, with summer rains) or BShw' (hot, semi-arid, with summer and autumn rains). The rainfall is influenced by the equatorial Atlantic, southern Atlantic and equatorial continental masses, and varies between 200–800 mm (rarely 1000 mm) annually, with a 3–5 month rainy season and 7–9 dry months (roughly May–November). Rainfall is highly irregular, leading to

catastrophic droughts and floods. The temperature is isothermic, with means varying between 25°–29°C.

Vegetation

The caatinga as delimited in Map 55 does not correspond to a single type of vegetation, but is a broad mosaic of types. Towards the coast, the caatinga is replaced by remnants of the Atlantic Coast forest (Mata Atlântica); inland, the caatinga merges with no clear limits into the "cerrado" (see CPD Site SA21). Interspersed with the caatinga are low mountains with uplands that are much more humid, containing elements ("brejos de altitude") of the Atlantic and Amazonian forests, with trees 30–35 m tall.

General characteristics of the caatinga elements include total loss of leaves during the dry season, small and firm (xeric) leaves, intense branching of the trees from the base (giving them a shrubby appearance) and the presence of succulent and crassulaceous species (Romariz 1974). Most authors recognize two main types of caatinga: **dry caatinga** ("sertão") located in the interior and more **humid caatinga** ("agreste") toward the coast. Eiten (1983) divided the caatinga into the following eight categories:

1. **Caatinga forest**, or low (8–10 m) xerophytic deciduous tropical broadleaved forest, with closed canopies, and the trees having a ground coverage over 60%. This robust formation occurs where there is sufficient rain and the soils are deep enough.

2. **Arborescent caatinga**, with the shrubby subcanopy not closed, and tree coverage 10–60%.

3. **Arborescent-shrubby closed caatinga**, or low xerophytic deciduous open tropical broadleaved forest with closed scrub, where the tree coverage is 10–60%. This is the most common form of undisturbed caatinga, sometimes called "carrasco".

4. **Arborescent-shrubby open caatinga**, with the total ground coverage of trees, shrubs, cacti, bromeliads, etc. between 10–60%.

5. **Shrubby closed caatinga**, or xerophytic deciduous or semi-deciduous closed tropical broadleaved scrub; the thoroughly deciduous scrub is more common.

6. **Shrubby open caatinga**, or xerophytic open tropical scrub, which can be composed of deciduous broadleaved species, cacti and bromeliads, or mixtures of the same. Coverage varies between 10–60%. Common throughout the caatinga on very shallow soils or rocky outcrops.

7. **Caatinga savanna** or xerophytic short-graminose tropical savanna with deciduous broadleaved scrub; this formation is usually called "seridó".

8. **Rocky caatinga savanna** or xerophytic sparse tropical scrub, in which both scrub and graminose elements have ground coverage of less than 10%. This formation occurs on pavements and outcrops of massive rock, with the plants interspersed in cracks and hollows.

MAP 55. CAATINGA OF NORTH-EASTERN BRAZIL (CPD SITE SA19)

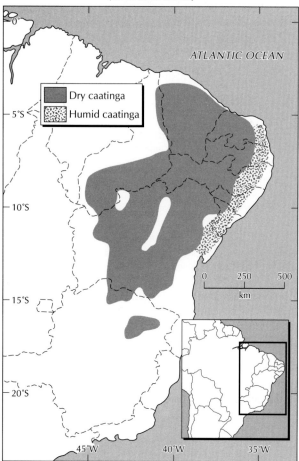

Flora

The flora of the caatinga region is poorly known. Humid sierras have elements (brejos de altitude) of both the Amazonian and the Atlantic Coast forest floras. Fernandes and Bezerra (1990) postulate that before the breakup of Gondwanaland in the Cretaceous (c. 90–95 million years ago), the north-east of Brazil together with the adjacent part of Africa (e.g. Nigeria) constituted a bridge between those two floras, which were probably not distinct. Certainly, fossil evidence indicates that a great part of the north-east was covered by tropical rain forests (Bigarella, Andrade-Lima and Riehs 1979; Lima 1979; Fernandes and Bezerra 1990). Pockets of cerrado also occur, suggesting that in the past this vegetation type occupied a significantly larger area (Fernandes and Bezerra 1990). Moreover, at some time in the past the dry caatingas formed part of a much larger dry belt, as evidenced by: (1) the occurrence of the same species in the chaco/pantanal complex, for example *Apuleia leiocarpa*, *Geoffroea superba* and *Peltophorum dubium*, (2) the presence of vicariant species such as *Copernicia prunifera/C. alba* and *Ziziphus joazeiro/Z. mistol*, and (3) many other species affinities in genera such as *Astronium*, *Aspidosperma*, *Schinopsis*, *Capparis*, *Mimosa* and *Jatropha* (Rizzini 1963; Fernandes and Bezerra 1990).

In the caatinga itself, the most representative tree species include *Mimosa tenuiflora*, *M. caesalpinifolia*, *Caesalpinia microphylla*, *C. bracteosa*, *C. ferrea*, *C. macrophylla*, *Auxemma oncocalyx*, *A. glazioviana*, *Patagonula bahiana*, *Fraunhofera multiflora*, *Amburana cearensis*, *Schinopsis brasiliensis*, *Cavanillesia arborea*, *Cereus jamacaru* and *Astronium urundeuva*.

The shrubby-scrub elements include *Caesalpinia bracteosa*, *C. pyramidalis*, *Capparis ico*, *C. jacobinensis*, *Croton compressus*, *C. sonderianus*, *Lantana camara*, *Jatropha gossypifolia*, *J. mollissima*, *Piptadenia aculeata*, *Mimosa malacocentra*, *M. tenuiflora*, *Cordia leucocephala*, *Combretum leprosum*, *Luetzelburgia auriculata*, *Stylosanthes humilis*, *S. scabra*, *S. gracilis*, *S. guianensis*, *Aspidosperma pyrifolium*, *Hyptis suaveolens*, *Borreria verticillata*, *Centrosema brasilianum*, *Ipomoea asarifolia*, *Aeschynomene monteroi*, *A. filosa*, *A. americana*, *Zornia latifolia*, *Z. reticulata*, *Sida glomerata*, *S. galheirensis*, *Senna uniflora*, *Waltheria ferruginea*, *Melochia americana*, *Melocactus bahiensis*, *Discocactus placentiformis* and *Opuntia inamoena*. The grasses include *Paspalum*, *Aristida*, *Cenchrus* and *Setaria*.

Other discrete vegetation types associated with the caatinga are distinctive palm stands. The most common are "carnaúba" stands ("carnaubais"), where the principal component is *Copernicia prunifera*, and "babassu" stands ("babaçuais"), formed mainly by *Orbignya phalerata*. Other important stands include "tucúm" palms (*Astrocaryum aculeatissimum*) and "macaúba" palms (*Acrocomia aculeata*).

The dry forests associated with the caatinga are between 500–600 m above sea-level and differ in general vegetation although caatinga elements are present. The most characteristic species include *Triplaris gardneriana*, *Tabebuia serratifolia*, *Tallisia esculenta*, *Spondias lutea*, *Ceiba glaziovii*, *Astronium fraxinifolium*, *Guazuma ulmifolia*, *Anadenanthera macrocarpa*, *Pithecellobium polycephalum* and *Melanoxylum brauna*.

Among the most common species in the humid sierras, which are between 600–1000 m above sea-level, are *Cedrela odorata, Didymopanax morototoni, Lecythis pisonii, Pithecellobium polycephalum, Manilkara rufula, M. subtriflora, Hymenaea courbaril, H. martiana, Inga fagifolia, Gallesia gorasema, Lonchocarpus obtusus, L. sericeus, Cordia tricotoma, Machaerium amplum, Symphonia globulifera, Pterocarpus violaceus, Zollernia ilicifolia, Dalbergia variabilis, Jacaratia dodecaphylla* and *Pilocarpus jaborandii.*

Evidence of the caatinga as a separate floristic province is given by its endemics (Fernandes and Bezerra 1990; Andrade-Lima 1981). Endemic genera include *Moldenhawera, Cranocarpus* (Leguminosae); *Fraunhofera* (Celastraceae); *Apterokarpos* (Anacardiaceae); *Auxemma* (Boraginaceae); and *Neoglaziovia* (Bromeliaceae). Among the endemic species are *Cavanillesia arborea, Bursera leptophloeos, Ceiba glaziovii, Aeschynomene monteroi, Mimosa caesalpinifolia, M. tenuiflora, Pilosocereus gounellei, P. squamosus* (= *Facheiroa squamosa*), *Cereus jamacaru, Patagonula bahiana* and *Calliandra depauperata.*

Useful plants

The caatinga has many useful species. The legumes and grasses are of great importance as sources of forage for arid lands, including *Stylosanthes, Zornia, Paspalum, Macroptilium, Galactia, Vigna, Centrosema, Aeschynomene, Chamaecrista, Desmanthus, Panicum* and *Digitaria*. Many species are valuable sources of food, with many used locally as fruit, e.g. *Tallisia esculenta, Spondias lutea* ("umbú"), *Spondias tuberosa, Lecythis pisonii, Manilkara rufa* and *Hancornia speciosa* ("mangabá").

Important caatinga timber species include *Anadenanthera macrocarpa, Ziziphus joazeiro, Amburana cearensis, Astronium fraxinifolium, A. urundeuva, Tabebuia impetiginosa, T. caraiba* and *Schinopsis brasiliensis*, as well as those found in patches of other types of vegetation, e.g. *Cedrela odorata, Dalbergia variabilis, Didymopanax morototoni* and *Pithecellobium polycephalum*. Among the medicinals are *Pilocarpus jaborandii*, which as well as *Amburana cearensis* is officially listed as threatened.

The palms are of special importance in that they constitute the backbone of the local domestic economy in many parts of the North-east. Rural populations rely heavily on extraction from babassu (babaçú) (*Orbignya phalerata*) (May 1992), carnaúba (*Copernicia prunifera*), tucúm (*Astrocaryum aculeatissimum*) and to a lesser extent macaúba (*Acrocomia aculeata*), and many species of *Syagrus, Scheelea* and *Attalea*.

Social and environmental values

The region contains a complete sequence of human occupation extending from c. 12,000 years ago, including the most abundant examples and diversified traditions of rock art from South America (Schmitz 1987). North-east of Brasília in the State of Piauí, in Serra da Capivara National Park (1000 km²) is an archaeological site claimed to be the most ancient in the Americas, purportedly some 50,000 years old, but its dating has not yet been settled (Meltzer, Adovasio and Dillehay 1994).

The north-east of Brazil has the largest population concentration in the country, as well as the poorest people. The so-called triangle of drought, which includes practically all the vegetation described, supports over 60 million inhabitants. The social situation is extremely complex, with a large majority of the rural population depending upon extractivism for a large part of their livelihoods. For example, over half of the population of Maranhão and Piauí states depend on babassu for c. 50% of their income (cf. May 1992). In other areas of the North-east, the same can be said for other species.

Decline of caraiba woodland (dominated by *Tabebuia caraiba*) along the São Francisco River in Pernambuco and Bahia between Juazeiro (or perhaps Remanso) and Abaré may have contributed to the endangerment of the nearly extinct Spix's macaw (*Cyanopsitta spixii*) (Juniper and Yamashita 1991).

Economic assessment

The local flora is the source of livelihood for many people. Some of the extractive species, including most of the palms and fruits, have a fair to high potential under cultivation. The babassu, macaúba and tucúm palms are all potential commercial sources of high quality lauric oil, with macaúba also having a high percentage of oleic oil. Among the fruits, "umbú" (*Spondias lutea*) and "mangabá" (*Hancornia speciosa*) probably are the most promising.

Large parts of the region are being put under irrigation, especially in the valley of the São Francisco River. Overall this is very promising, as the soils for the most part are fertile, and without the severe aluminium limitations of the cerrado and Amazon Basin. However, with fairly saline water and salt pans near the water table, salinization of the soil is a major threat. The region is beginning to undergo an economic boom, based primarily on irrigated agriculture; if the trend continues, it may become an important breadbasket, not only for Brazil but the world. The São Francisco Valley is exporting fruit, including grapes, papayas and melons, and will probably end up competing in world markets of these and other products. Paradoxically, this new North-east will probably co-exist with the losses described below. At the same time that some areas develop their economic potential, others will continue to be degraded, in terms of the environment and their carrying capacity for human populations.

Threats

Overgrazing by cattle and goats and excessive harvesting of fruits are seriously affecting the population structure of most of the more important species. Indiscriminate timbering for industry as well as fuelwood and charcoal is decimating the original vegetation. The region is rapidly approaching the situation of Africa in the Sahara and Sahel, where chronic drought and misuse of the environment are threatening a major catastrophe. Desertification is a serious threat, as is gradual salinization of the soils due to irrigation. Furthermore, most if not all of the over 16 conservation units in the North-east are under severe pressure from the local human populations.

Conservation

As shown in Table 58, with c. 0.31% of its area (c. 3168 km²) in official principal conservation units, the North-east falls below Brazil's national average which is c. 4%. Since this region has the highest population pressure, and the population is most dependent on the natural vegetation, conservation efforts fall short of what is necessary (Andrade-Lima 1977).

TABLE 58. PRINCIPAL CONSERVATION UNITS IN NORTH-EAST OF BRAZIL

Conservation unit	State	Area (ha)
Biological Reserve		
Guaribas	Paraíba	4321
Serra Negra	Pernambuco	1100
Saltinho	Pernambuco	548
Pedra Talhada	Pernambuco/Alagôas	4469
Santa Isabel	Sergipe	2766
Ecological Station		
Seridó	Rio Grande do Norte	1116
Mamanguape	Paraíba	2670
Foz São Francisco/		
Praia do Peba	Alagôas	5322
Itabaiana	Sergipe	1100
Uruçuí-Una	Piauí	135,000
National Park		
Sete Cidades	Piauí	6221
Ubajara	Ceará	563
Serra da Capivara	Piauí	97,993
National Forest		
Araripe	Ceará	38,262
Environmental Protection Zone (APA)		
Jericoacoara	Ceará	6800
Piacabuçu	Alagôas	8600
Total		**316,851**

References

Ab'Sáber, A.N. (1977). *Potencialidades paisagisticas brasileiras.* Universidade de São Paulo, São Paulo.

Ab'Sáber, A.N. (1980). O dominio morfoclimático semi-árido das caatingas brasileiras. *Craton and Intracraton* 6. Universidade Estadual Paulista "Julio de Mesquita Filho", São José de Rio Preto.

Andrade-Lima, D. de (1977). Preservation of the flora of northeastern Brazil. In Prance, G.T. and Elias, T.S. (eds), *Extinction is forever: threatened and endangered species of plants in the Americas and their significance in ecosystems today and in the future.* New York Botanical Garden, Bronx. Pp. 234–239.

Andrade-Lima, D. de (1981). The caatingas dominium. *Rev. Bras. Bot.* 4: 149–163.

Bigarella, J.J., Andrade-Lima, D. de and Riehs, P.J. (1975) [1979]. Considerações a respeito das mudanças páleoambientais na distribuição de algumas espécies vegetais e animais no Brasil. *An. Acad. Bras. Ciênc.* 47 (supl.): 411–464.

Eiten, G. (1983). *Classificação da vegetação do Brasil.* CNPq/ Coordenação Editorial, Brasília.

Fernandes, A. and Bezerra, P. (1990). *Estudo fitogeográfico do Brasil.* Stylus Comunicações, Fortaleza, Ceará.

Juniper, A.T. and Yamashita, C. (1991). The habitat and status of Spix's macaw *Cyanopsitta spixii. Bird Conserv. Internat.* 1: 1–9.

Lima, M.R. (1979). *Paleontologia da formação Santana (Cretáceo do nordeste do Brasil). Estágio atual do conhecimento.* Universidade de São Paulo, São Paulo.

May, P.H. (1992). Babassu palm product markets. In Plotkin, M. and Famolare, L. (eds), *Sustainable harvest and marketing of rain forest products.* Island Press, Washington, D.C. Pp. 143–150.

Meltzer, D.J., Adovasio, J.M. and Dillehay, T.D. (1994). On a Pleistocene human occupation at Pedra Furada, Brazil. *Antiquity* 68: 695–714.

Nimer, E. (1969). *Clima – circulação atmosférica. Paisagens do Brasil.* Instituto Brasileiro de Geografia e Estatística (IBGE), Série D, No. 2. Rio de Janeiro.

Rizzini, C.T. (1963). Nota prévia sôbre a divisão fitogeográfica do Brasil. *Rev. Bras. Geogr.* 25(1): 1–64.

Romariz, D.A. (1974). *Aspectos da vegetação do Brasil.* Instituto Brasileiro de Desenvolvimento Florestal (IBDF), Rio de Janeiro.

Schmitz, P.I. (1987). Prehistoric hunters and gatherers of Brazil. *Journal World Prehist.* 1: 53–126.

Author

This Data Sheet was written by Dr Eduardo Lleras [Centro Nacional de Pesquisas de Recursos Genéticos e Biotecnologia (CENARGEN)/EMBRAPA, Caixa Postal 02-372, 70.849 Brasília, D.F., Brazil].

INTERIOR DRY AND MESIC FORESTS: CPD SITE SA20

ESPINHAÇO RANGE REGION
Eastern Brazil

Location: Phytogeographic province of Espinhaço mountains in Bahia and Minas Gerais states, between latitudes 10°–20°35'S and longitudes 40°10'–44°30'W.

Area: 6000–7000 km².

Altitude: Especially 1000–2107 m.

Vegetation: Mosaic of formations with extensive ecotonal areas. Largely "campos rupestres": open-rock pioneer vegetation and rock-dwelling plants; oligotrophic marshes; gallery forests; (savanna-like) "cerrado"; montane (cloud) forests; semi-deciduous to deciduous forests.

Flora: Very high diversity, probably more than 4000 species of vascular plants. Campos rupestres with very high generic and specific endemism, and large number of disjunct taxa of coastal "restingas", other mountains of central Brazil (mainly Goiás) or Guayana Highlands.

Useful plants: Ornamentals, especially everlasting plants and Velloziaceae, and other tepaloid monocotyledons – e.g. orchids, bromeliads; medicinals; potentially, other genetic resources.

Other values: Watershed protection for drinking water, hydroelectric energy; tourist attraction.

Threats: Fire; cattle-grazing; mining; timber extraction for charcoal, building, fences, etc.; over-collecting of horticultural species; increasing tourist pressure; inappropriate hydroelectric schemes and road construction.

Conservation: Serra do Cipó National Park (Minas Gerais), Chapada Diamantina NP decreed (Bahia). An Ecological Station, State Parks, Environmental Protection Zones (APAs).

Geography

The Cadeia do Espinhaço (Espinhaço Range) forms the watershed between the Atlantic Ocean and the basin of the São Francisco River in eastern Brazil (Map 56). From 50 km to 100 km wide, it is formed of numerous low mountains or "serras" (900–1500 m high, peaking at over 2000 m), interrupted by river valleys that can be extensive and deep. The region naturally falls into a number of mountainous areas which represent local centres of biodiversity. The southern limit (at 20°35'S) is in the central State of Minas Gerais south-east of Belo Horizonte in the Serra do Ouro Branco; six other important centres in this state are Serra da Piedade, Serra do Caraça, Serra do Cipó, Diamantina Plateau, Serra do Cabral and Serra do Grão-Mogol. In the State of Bahia 1100 km to the north (inland from Salvador), the Chapada Diamantina is an extensive elevated area which includes a number of important centres, such as in the west the Serra do Rio de Contas, Pico das Almas and Pico de Itabira and in the east the Serra do Sincorá. Farther north the Chapada dies away (at c. 10°S) in a number of isolated and scattered massifs, such as Morro do Chapéu (Serra do Tombador) and Serra da Jacobina.

The Espinhaço Range was formed intermittently in segments from the Palaeozoic onwards (King 1956). The highest peaks (1800–2107 m) are remnants of older surfaces that have been worn down. The oldest known denudation-surface dates from the Early Cretaceous; at its base are large areas of much older rocks. The structural blocks of the Espinhaço Range date to the Precambrian

(Abreu 1984). The large-scale folding along a north-west/south-east axis underwent a long process of erosion and was remodelled by tectonic movement at the beginning of the Tertiary. Relief is highly accentuated, with broad deep valleys, especially in schistose and filite formations, whereas the highest ridges and crests occur in quartzite and sandstone formations (Moreira 1965). Soils are in general shallow and sandy, highly acidic and extremely nutrient poor.

Between the two main areas of the Espinhaço Range (one in Bahia, the other in Minas Gerais) lies lowland with deeper soils, c. 300 km broad from north to south and mostly over 500 m above sea-level, although dissected by large river systems such as the Contas River, with the Pardo and Jequitinhonha rivers farther south. This lowland is a barrier to migration of the northern (Bahian) Espinhaço flora to the south and vice versa, and probably acted as such in the past when it could have been drier or covered by forest (Harley 1988).

The climate is mesothermic (Cwb of Köppen 1931), with mild summers, a rainy season in the summer and average annual temperatures between 17.4°–19.8°C; the average in the hottest month is below 22°C. The rainy season lasts 7–8 months, while the dry period may last 3–4 months and coincides with winter. Annual rainfall in the southern part of the region is c. 1500 mm. Northwards the dry season increases in duration, until at least in the lowlands it may occupy the major part of the year. Although a strongly seasonal climate prevails, moist clouds provide dew and some rain during the dry season, allowing some soils to remain wet throughout the year.

MAP 56. ESPINHAÇO RANGE REGION, BRAZIL (CPD SITE SA20), SHOWING LAND ABOVE 1000 m IN THE ESPINHAÇO RANGE, IN THE STATES OF MINAS GERAIS AND BAHIA

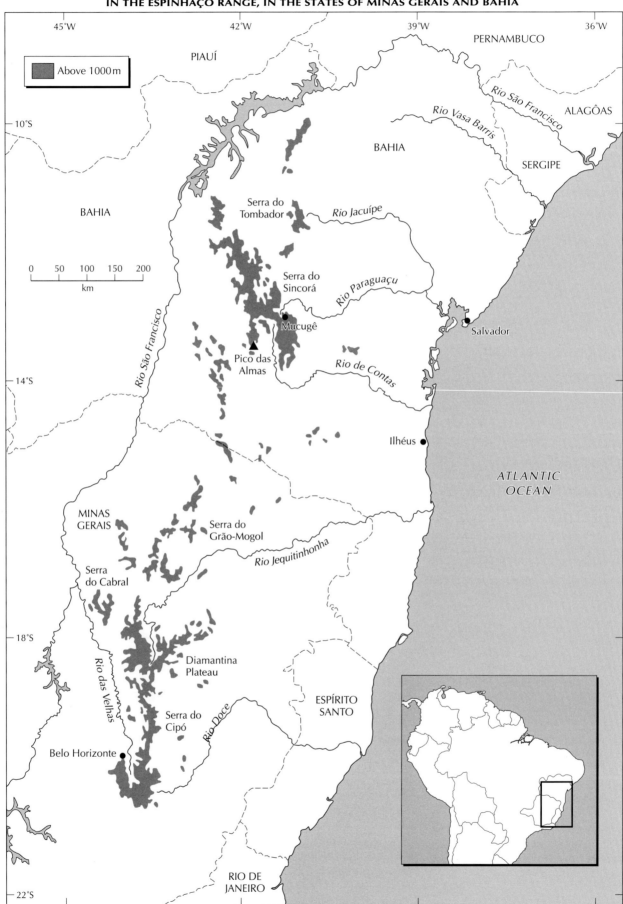

Vegetation

Although the Espinhaço Range is entirely within the tropics, its landscape is far removed from most of tropical Brazil. The physiognomy varies greatly according to the locally prevailing physiography, with the largest vegetation formation being "campos rupestres" (rocky fields) (in the terminology of Magalhães 1966).

Various types of forest occur in the Cadeia do Espinhaço. Gallery forests are alongside streams, with a tree stratum 7–15 m high. The widespread dominant trees are *Tapirira guianensis, Xylopia emarginata, Richeria grandis, Copaifera langsdorffii, Protium almecega, P. brasiliense, Vismia guianensis, Tibouchina candolleana, Hyeronima alchorneoides, Croton urucurana, Guapira opposita, Humiria balsamifera, Guatteria sellowiana, Cabralea canjerana* and *Vochysia acuminata*. At higher altitudes with increased humidity often occur such species as *Drimys brasiliensis, Podocarpus* spp., *Weinmannia* spp. and *Lamanonia* spp. At the margins and in the lower strata appear small trees (*Trembleya parviflora*, many Myrtaceae and Rubiaceae) and palms (*Geonoma* spp.), together with tree ferns (*Cyathea, Trichipteris*). The gallery forests may extend upwards and link with usually isolated forest clumps ("capões"), which occur in open grasslands at higher altitudes. The passage from gallery forests to capões presents a gradual modification in the floristic composition and the amount of vegetation that is deciduous. At higher altitudes, forests are strongly influenced by atmospheric humidity, clouds, etc. and may be rich in epiphytes, including orchids, bromeliads, *Anthurium, Peperomia*, ferns and bryophytes.

Where local geological conditions permit, for example on calcareous soils, deciduous or semi-deciduous forests with a very different species composition may occur. On the Bambuí limestone formation scattered through the region dry deciduous forest appears, including *Cedrela fissilis, Apeiba tibourbou, Luehea divaricata, Guazuma ulmifolia, Molopanthera paniculata, Phyllanthus acuminatus, Cnidoscolus urens, Hymenaea courbaril, Chorisia* sp., *Allophyllus sericeus, Aloysia virgata, Trema micrantha, Trichilia* spp. and the tall columnar cactus *Cereus calcirupicola*.

On slopes between 860–1000 m occur areas of "campo cerrado", in which the predominance of arboreal-shrubby forms declines with increase in altitude and greater development of the soil profile, and gradually gives way to "campo sujo" and campo rupestre. These cerrado patches have a peculiar physiognomy, marked by sparse tortuous treelets and shrubs often with thick bark, mainly *Vochysia thyrsoidea, Qualea cordata, Q. parviflora, Byrsonima verbascifolia, Neea theifera, Caryocar brasiliense, Kielmeyera coriacea, Didymopanax vinosus, Hymenaea stigonocarpa, Dalbergia miscolobium, Campomanesia adamantium* and *C. pubescens*. These species are scattered through a continuous ground cover of a variety of herbaceous species, particularly the grasses *Aristida, Paspalum stellatum* and *Axonopus brasiliensis*.

In the northern (Bahia) part of the Espinhaço Range, a form of campo cerrado occurs at these altitudes that is locally known as "campos gerais", in which the tree stratum is usually absent and the ground stratum rich in stemless palms and with fewer grasses. The "campo rupestre", which forms the bulk of the vegetation on the higher slopes (1000–2000 m) or even from 700–900 m in some areas, is primarily composed of a more or less continuous herbaceous stratum and sclerophyllous evergreen small shrubs and subshrubs often with imbricate leaves, showing much morphological convergence in various families (Menezes and Giulietti 1986; A. Giulietti *et al.* 1987). However, rather than being a homogeneous vegetation type, these campos are an assemblage of communities forming a rich mosaic, under control of local topography, the nature of the substrate, and the microclimate.

There are patches of pure sand intermixed with patches of pebbles or gravel. Here and there stand rocky outcrops of different sizes that provide shade or shelter from prevailing winds and hold precious moisture in crevices, although the outcrops may increase heat by reflection. The interaction of all these factors is instrumental in the formation of a huge variety of micro-habitats. The bare rock surfaces and crevices are colonised by many lichens and other epilithic plants adapted to exploit dew and mist.

The adaptations include specialized roots and pseudobulbs in orchids, water tanks and specialized hairs in bromeliads and pseudo-trunks covered with the remains of leaf-sheaths between which are numerous adventitious roots in Velloziaceae. Other adaptive strategies for survival in this harsh landscape can be seen in species of many different families. These include waxy leaf surfaces which reflect heat, thick layers of hairs which protect against radiation and reduce transpiration, tightly furled rosettes of leaves and chemical secretions which protect against fire, and many anatomical and physiological modifications that give greater water-storage capacity.

Certain families, genera and species are very typical of the sandy or rocky substrates that form much of the campos rupestres, often occurring in large and sometimes showy populations and contributing to the general physiognomy of the vegetation. These taxa include Velloziaceae (*Barbacenia, Vellozia, Pleurostima*); Eriocaulaceae (*Leiothrix, Paepalanthus, Syngonanthus*); Xyridaceae (*Xyris*); Compositae (*Lychnophora, Proteopsis, Wunderlichia*, many other genera); Melastomataceae (e.g. *Cambessedesia, Microlicia, Marcetia*); Ericaceae (*Gaylussacia, Leucothoe*); Labiatae (*Hyptis, Eriope*); Leguminosae (*Chamaecrista, Calliandra, Mimosa, Camptosema*); Rubiaceae (*Declieuxia*); Lythraceae (*Cuphea, Diplusodon*); Malpighiaceae (*Camarea, Byrsonima*); Verbenaceae (*Lippia, Stachytarpheta*); Myrtaceae (*Campomanesia, Myrcia*); Euphorbiaceae (*Croton, Phyllanthus*); Orchidaceae (*Laelia, Cleistes, Oncidium*); Bromeliaceae (*Dyckia, Vriesea*, tank species of *Tillandsia*); Cyperaceae (*Bulbostylis, Lagenocarpus*); Gramineae (*Aristida, Panicum, Paspalum, Axonopus* and many other genera, including the bambusoid *Aulonemia effusa* and *Chusquea* spp.); and certain ferns and their allies (*Anemia, Doryopteris, Huperzia*).

On plateaux with poor drainage acid bogs may form, having Cyperaceae (*Rhynchospora*); Rapateaceae (*Cephalostemon*); Eriocaulaceae (*Syngonanthus, Leiothrix*); Xyridaceae (*Xyris, Abolboda*); Orchidaceae (*Habenaria, Stenorrhynchus*); Lentibulariaceae (*Utricularia, Genlisea*); and Droseraceae (*Drosera*).

The Espinhaço Range is crossed by a large number of streams supporting a rich and varied aquatic vegetation, with species of *Eriocaulon, Utricularia, Ludwigia, Eleocharis, Mayaca, Nymphoides, Lycopodium, Lycopodiella, Apinagia* and *Laurenbergia tetrandra*.

The striking differences between the Espinhaço vegetation and the surrounding lowlands are clearly determined by the differences in geological and topographic conditions, as well as the upland presence of clouds and dew, which are absent in the lowlands during the dry season. In Bahia, on descending from the serras, the luxuriant campos usually give way to a band of cerrado or campos gerais, soon replaced at lower altitudes by xerophilous deciduous forest known as "caatinga" (CPD Site SA19), which is the dominant vegetation formation of the semi-arid regions of north-eastern Brazil. In Minas Gerais, the upland formations are surrounded by lower regions with deeper latosols occupied by cerrado (see CPD Site SA21), the dominant formation on the Brazilian Central Plateau.

These factors are important in explaining the composition, affinities and divergences of these adjacent floras, but the patterns also reflect past history, such as former migration routes and the Pleistocene climatic changes in eastern Brazil.

Flora

The first naturalists who visited eastern and central Brazil in the 19th century were greatly impressed by the wealth of the flora of the Espinhaço Range. Taxonomic and floristic researchers are now providing detailed evidence of the great floristic richness. Teams of the Universidade de São Paulo and the Royal Botanic Gardens, Kew, U.K. are elaborating inventories of the vascular flora of selected sites within the Espinhaço, to produce local Florulas that will provide better understanding of the distribution and dynamics of the species. Inventories have been published for Mucugê, Bahia (Harley and Simmons 1986) and the Serra do Cipó, Minas Gerais (A. Giulietti *et al.* 1987). Other areas being studied are the Catolés and Pico das Almas massifs in Bahia, and the Serra do Grão-Mogol, Diamantina Plateau and Serra do Ambrósio in Minas Gerais.

The flora of the Espinhaço Range probably has more than 4000 species. About 200 km² of the Serra do Cipó show the extraordinary number of 1590 species (A. Giulietti *et al.* 1987), in 103 families of dicotyledons (402 genera), 24 families of monocotyledons (118 genera), 1 family of gymnosperms (Podocarpaceae), 10 of ferns and 11 of bryophytes. The best represented families among dicotyledons are Compositae (169 spp.), Melastomataceae (90), Rubiaceae (47), Fabaceae *s.s.* (47), Myrtaceae (45), Malpighiaceae (42), Caesalpiniaceae (34), Euphorbiaceae (31), Verbenaceae (26) and Mimosaceae (26), whereas in monocotyledons Gramineae are the most speciose family (130 spp.), followed by Eriocaulaceae (84), Orchidaceae (80), Velloziaceae (58), Xyridaceae (47), Bromeliaceae (36) and Cyperaceae (32).

Elsewhere in the Espinhaço zone the same families are dominant, with variations in relative species richness, as observed for Mucugê (Harley and Simmons 1986). Provisional figures for the Pico das Almas are also broadly the same, though the total of species for this smaller area is c. 1200. Compositae is again represented by the most species, with Melastomataceae, Rubiaceae, Leguminosae and Myrtaceae also well represented among dicotyledons; a greater number of Lentibulariaceae is notable. Among monocotyledons Cyperaceae predominate, with relatively fewer Orchidaceae, Gramineae, Eriocaulaceae, Xyridaceae and Velloziaceae.

A number of phytogeographical elements can be found within the flora of the Espinhaço Range. The majority of forest species show a wide distribution through South America, e.g. *Tapirira guianensis, Protium heptaphyllum, Cabralea canjerana, Richeria grandis, Hyeronima alchorneoides* and *Simarouba amara* (Giulietti and Pirani 1988). Some genera have a greater number of species in the Andes and reach south-eastern Brazil (e.g. *Hedyosmum, Clethra, Weinmannia, Podocarpus*), whereas others show a greater concentration in eastern Brazil although occurring also in northern South America and Central America (e.g. *Eugenia, Euplassa*). The gallery forests, capões and cloud forests may represent relict communities from a period when forest covered much more of eastern Brazil and probably they acted – and still act – as migration routes for tree species. The epiphytic flora of the more humid altitudinal forest shows a clear link with the Atlantic Coast forest (Mata Atlântica) (e.g. with *Oncidium crispum, Peperomia* spp., *Billbergia amoena, B. vittata*).

The flora of the campos rupestres is largely composed of an endemic element, both in genera and species. They may show floristic links to other parts of South America, where vicariant taxa occur; in some cases such patterns can be found in individual species, which have a disjunct distribution between the campo rupestre and other regions. Probably the greater portion of the campo-rupestre flora has common elements with the cerrado flora, which in many areas surrounds it. This cannot be said of the caatinga flora, which is often adjacent to campo rupestre in Bahia but has a very different floristic composition, except for widespread genera that occur in a wide range of ecosystems.

Among genera in which a majority of the species are characteristic of either or both cerrado and campo rupestre are *Eremanthus, Vanillosmopsis, Qualea, Campomanesia, Camarea, Peixotoa, Hyptis, Ctenium, Aristida, Mimosa, Mandevilla, Macrosiphonia, Diplusodon, Kielmeyera, Trimezia, Jacaranda, Gomphrena, Declieuxia* and the nearly stemless *Syagrus* spp.

Another element in the campo-rupestre flora contains those species or groups of species that show a disjunct distribution between the montane campos rupestres and the coastal sands ("restingas"). Both habitats share certain edaphic and climatic factors – a quartzitic substrate in a very open habitat with high insolation and frequent periods of high atmospheric humidity. Species in common include the leafless *Phyllanthus angustissimus* and *P. klotzschianus, Mandevilla moricandiana, Marcetia taxifolia, Mimosa lewisii, Lagenocarpus rigidus, Vellozia dasypus, Leiothrix rufula, Paepalanthus ramosus* and most species of *Syngonanthus* section *Thysanocephalus*. Certain genera show a marked preference for these two ecosystems, including *Gaylussacia, Leucothoe, Allagoptera, Bonnetia* and *Moldenhawera*. A genus very typical of the campo rupestre is *Eriope*, although *E. blanchetii* is endemic to the restingas of Bahia and Sergipe.

The "tepui" region of the Guayana Highlands (CPD Site SA2) shares many similarities with the campo rupestre; both are formed in part from Precambrian shields that occur north and south of the Amazon. Both regions contain open areas of rock outcrop surrounded by low-lying vegetation that forms a barrier to migration and both experience high atmospheric humidity. Owing to the great distance that separates these regions, few species are restricted to both. Nonetheless, certain families that have their main centre of diversity in the tepuis have outliers or secondary centres in the campo rupestre of the Espinhaço Range. Examples include *Cottendorfia*, with *C. florida* in the campos rupestres of Bahia; *Bonnetia*, with *B. stricta* in the campo rupestre of Bahia and also occurring in the restinga. In the Rapateaceae, which is centred in the Guayana Highlands (16 genera, c. 100 spp.), *Cephalostemon riedelianus* only occurs in the campo rupestre of Minas Gerais. Families such as Xyridaceae and Eriocaulaceae have two main centres of diversity – in the campos rupestres of Brazil and the Guayana Highlands. In Xyridaceae, the major generic diversity occurs in the latter region with three small endemic genera, although the number of species in the largest and most widespread genus *Xyris* is fewer there than in the campos rupestres of Brazil. In Eriocaulaceae, the greatest morphological diversity also can be found in the Guayana Highlands, with up to two or perhaps three endemic genera, although the number of species, notably in *Leiothrix* but also *Paepalanthus* and *Syngonanthus*, is much higher in the campos rupestres. *Marcetia taxifolia* shows disjunction (reflecting that found in *Bonnetia*), with a distribution not only in the campos rupestres but also in the restingas of Brazil and the tepuis. For *Marcetia taxifolia*, however, the isolated occurrence is in the Guayanas, with the main centre of diversity of the genus in the Espinhaço Range.

Elements with an Andean or a southern Brazilian connection occur sporadically within the campo-rupestre flora. In *Paepalanthus* subgenus *Platycaulon*, with 46 species, 28 are restricted to the campos rupestres of Minas Gerais, whereas 12 are endemic to the mountains of Colombia. *Vellozia* section *Radia* has centres of diversity in the Brazilian campos rupestres and in some mountains of Venezuela and Colombia. Similar links can be found in Labiatae: *Hyptis irwinii* and *H. stachydifolia* have their closest relations in the Andes, and *Eriope macrostachya*, which occurs in campo rupestre along forest margins, also occurs (as *Eriope macrostachya* var. *platanthera*) in the Venezuelan Andes.

Endemism can be extremely high in the groups with a primary or important centre of diversity in the campos rupestres of the Espinhaço Range. The very characteristic Velloziaceae has c. 250 species in South America and Central America and only 29 in Africa. More than 70% of the species (173 spp.) are concentrated in Minas Gerais, essentially on the Diamantina Plateau and Serra do Cipó (Mello-Silva 1989). *Barbacenia* has c. 65 species mostly in the southern sector of the Espinhaço (Minas Gerais), with only three in Bahia. *Burlemarxia*, described by Menezes and Semir (1991) with three (two new) species, is endemic to the Diamantina Plateau. The data presented by A. Giulietti *et al.* (1987) show that among the 59 species of Velloziaceae found on the Serra do Cipó, 46.5% are restricted to that range, and 91.5% are endemic to the Espinhaço; 27 have been described in the last 23 years. In the Grão-Mogol area, 19 species of Velloziaceae include nine endemics and three new taxa (Mello-Silva 1989).

In Eriocaulaceae, *Paepalanthus* subgenus *Xeractis* (27 spp.) and *Leiothrix* subgenus *Leiothrix* (13 spp.) are wholly restricted to the Espinhaço Range in Minas Gerais. For the Serra do Cipó, A. Giulietti *et al.* (1987) refer to 84 species in this family – 68.7% being endemic to the Espinhaço and 32.5% restricted to the Serra do Cipó. In the Serra do Ambrósio (near Rio Vermelho) occur three endemic and related species of *Syngonanthus*, with high economic value, which are only now being described. In Xyridaceae, Wanderley (1992) pointed out that of the 152 species of *Xyris* found in Brazil,

46 are endemic to the Espinhaço Range and 14 exclusive to the Serra do Cipó.

Many other groups of flowering plants show numbers of endemic taxa in the Espinhaço Range that are as high. Striking examples are *Pseudotrimezia* (Iridaceae) and *Cipocereus* (Cactaceae), genera endemic to the southern (Minas Gerais) sector of the Espinhaço. The monotypic genera *Raylea* (Sterculiaceae), *Morithamnus* and *Bishopiella* (Compositae) are confined to the Bahian sector.

The species composing these and many other genera in the campos rupestres of the Espinhaço are generally found as disjunct populations, often restricted to single small serras – isolated geographically and in gene exchange by natural barriers. This situation is the outcome of a long phase of climatic fluctuation during the Quaternary, when populations expanded and contracted as conditions improved or deteriorated, with resultant irregular gene exchange between neighbouring populations and the evolution of new species, often with very limited present distributions.

Useful plants

The Espinhaço Range has a long history of occupation and exploitation of minerals. The plant resources have barely been utilized for food or medicinals on a large scale. The large number of taxa in the region undoubtedly represents a vast genetic resource. Plants used locally for food mainly include many fruits, e.g. "goiabinha" and "araçá" (*Psidium* spp.), "guabiroba" (*Campomanesia* spp.), "puçá" or "mandapuçá" (*Mouriri glazioviana*), "jatobá" (*Hymenaea* spp.), "pequi" (*Caryocar brasiliense*), "mangabá" (*Hancornia speciosa*), "murici" (*Byrsonima* spp.), "cajuzinho-do-campo" (*Anacardium* spp.) and nuts of a number of palm species (e.g. "licuri" – *Syagrus coronata*, "catolé" – *Attalea* spp., "macaúba" – *Acrocomia aculeata*).

Local people prepare traditional remedies from many species, for instance: dried leaves of arnica (*Lychnophora pinaster*) are used to prepare a tea with hepatic properties; fresh leaves of *L. ericoides* in alcohol are said to heal wounds; shoots and leaves of "chapé" (*Hyptis lutescens*) are believed to cure flu; leaves of "paneira" (*Norantea adamantium*) provide a tea for kidney problems; roots of "quina-de-vaca" (*Remijea ferruginea*) are used for stomach diseases; "sambaibinha" (*Davilla rugosa*) is for baths; roots and leaflets of "caroba" and "carobinha" (*Jacaranda* spp.) are put into "pinga" (an alcoholic liquor) to cleanse the blood; and "catuaba" (*Anemopaegma arvense*) is said to be an effective tonic and aphrodisiac.

Other products obtained in the region include gums from jatobá (*Hymenaea* spp.), resins from "pau-d'óleo" or "copaíba" (*Copaifera langsdorffii*) and tannins from murici (*Byrsonima* spp.). The resinous stems of the larger species of *Vellozia* ("canela-de-ema") were used as fuel for locomotives on the Curvelo-Diamantina railway and these days are stacked to sell as an effective fire lighter for wood stoves and to make torches. The densely woolly leaves of some *Lychnophora* species provide combustibles for home fires and stuffing for pillows and cushions. The powdered roots of *Palicourea marcgravii* are a very effective rat poison.

A fairly recent development has been expansion of the trade in everlasting plants ("sempre-vivas") – collection and trade in dried inflorescences of some 40 wild species, several of them rare endemics, that are in demand particularly for export to U.S.A., Japan and Europe (N. Giulietti *et al.* 1987). About 300 metric tonnes dry weight are being obtained yearly. The species most prized for their beauty are the Eriocaulaceae *Syngonanthus elegans* ("sempre-viva pé-de-ouro"), *S. venustus* ("brejeira"), *S. xeranthemoides* ("jazida"), *S. magnificus* ("sempre-viva gigante"), *S. suberosus* ("margarida"), *S. brasiliana* ("brasiliana"), *Paepalanthus macrocephalus* ("botão-branco"), *Leiothrix flavescens* ("botão-bolinha"); the grasses *Aristida riparia* ("rabo-de-raposa"), *Aulonemia effusa* ("andrequicé"), *Axonopus brasiliensis* ("pingo-de-neve"), *Gynerium sagittatum* ("cana-brava"); the Xyridaceae *Xyris nigricans* ("coroinha"), *X. cipoensis* ("abacaxi-dourado"), *X. coutensis* ("cacau"); and the sedges *Rhynchospora globosa* ("espeta-nariz"), *R. speciosa* ("capim-estrela").

Many showy species are uprooted by horticulturists and tourists, especially orchids (e.g. *Laelia, Cattleya, Oncidium, Encyclia advena* – "sumaré"), bromeliads (e.g. *Dyckia, Orthophytum, Vriesea*), cacti (e.g. *Uebelmannia gummifera, Melocactus, Micranthocereus auri-azureus*) and "lilies" (mainly *Hippeastrum*).

Many other species in a wide range of families have potential as ornamentals. Many endemic species have never been investigated for possible uses – this a rich field for future research.

Social and environmental values

Up to 1890 the greatest part of the population of the south-east was concentrated in Minas Gerais. More than a million gold and diamond prospectors lived in the region, principally distributed from Ouro Preto to Diamantina. In Bahia on the Chapada Diamantina, the population was large for the same reasons, with the mines principally from Rio de Contas to Catolés in the west and Mucugê to Lençóis in the east, with the road at that time extending northward to Jacobina. This period of mineral extraction, with indiscriminate exploitation of the natural resources, coincided with expansion of small towns. With the decline of this industry at the end of the last century, the towns fell on hard times. In recent years as means of transportation and communication have improved, many of these towns have become attractive centres of tourism, due to their picturesque surroundings and old colonial-style houses, as well as the rich diversity of the local fauna and flora. Ouro Preto and Diamantina in Minas Gerais and Lençóis in Bahia are examples where the main income is from tourism. In spite of the decline of gold and diamond mining, this activity still occurs on a small scale in the whole length of the Espinhaço Range; for example in Diamantina, diamonds still remain as the principal mining resource of the municipality. In the last few decades the extraction of quartz crystals for the electronics industry has developed in a number of areas, such as Serra do Cabral in Minas Gerais and Barra da Estiva and near Catolés in Bahia.

Various rivers have their source in the mountains of the Espinhaço Range and are of vital importance for the economic development of north-eastern and south-eastern Brazil. Among those flowing to the Atlantic are the Itapicuru, Jacuípe, Paraguaçu and Contas rivers in Bahia and Jequitinhonha, Mucuri and Doce rivers in Minas Gerais, and those draining west that are tributaries on the east bank of the São Francisco River: Salitre, Jacaré, Verde and Paramirim rivers in Bahia and Verde Grande, Pacuí and Velhas in Minas Gerais. Protection of their mountain sources is essential to maintain water quality and avoid excessive erosion and silting.

The Espinhaço Range is not a region of high soil fertility, being mostly suitable for subsistence agriculture with such crops as beans and cassava, and coffee production in some forested areas, principally the Chapada Diamantina. Cotton is cultivated in many areas, particularly the Rio das Velhas Basin to the north of Belo Horizonte. Various areas of campo rupestre have experienced the introduction (in the last two decades) of exotics such as *Eucalyptus* and *Pinus*. These species have not proved profitable – which was completely predictable, given the soil characteristics.

Throughout the region there is extensive cattle-ranching, using natural pastures which are annually burnt to stimulate young growth of herbs and grasses. In Minas Gerais a total of c. 80,000 km² , principally with cerrados, are now used for this purpose, producing both milk and meat; in Bahia the occurrence is on a smaller scale. Upland pastures that retain forage during the dry season are invaded by grazing cattle from the lowland areas.

Over the last 15 or so years a large number of local communities in the Espinhaço Range have become involved in the collection and commercialization of sempre-vivas, especially in the neighbourhoods of the Serra do Cipó, Diamantina and the Serra do Cabral in Minas Gerais and from Mucugê and Piatã in Bahia. This activity is seasonal, almost entirely restricted to January to August when the species of greatest commercial value can be found in flower. The flowering stems are collected, dried, sorted into bundles and sold for decoration in homes and floral displays. In the Diamantina district, this industry has become the second greatest source of income (after diamonds) from natural resources in the municipality. The entire city of Mucugê turns its attention during July to the collection and drying of the sempre-viva de Mucugê, with the drying carried out in the streets. This species became scientifically recognized only recently, as *Syngonanthus mucugensis*. On upper slopes of the Chapada do Couto (Minas Gerais), temporary settlements sometimes of over 900 inhabitants occur during April–May solely to collect and dry three species: *Syngonanthus elegans, S.* cf. *bisulcatus* ("sempre-viva chapadeira") and *Xyris coutensis* – the latter two are endemic to the Chapada do Couto. In the Serra do Ambrósio there are also many endemic species, four of which are commercially important: *Uebelmannia gummifera*, a globose cactus that has been much exported to Europe, and *Syngonanthus brasiliana, S. magnificus* and *S. suberosus*. The latter three species, which are also new to science, have the greatest wholesale value among all those used and are found only in that limited area. Species of sempre-vivas unknown to science have been discovered in local storage.

This region is the main part of the Bahia-Minas Gerais tablelands Endemic Bird Area (EBA B50), which has five restricted-range bird species confined to it: the hooded visorbearer (*Augastes lumachellus*), hyacinth visorbearer (*A. scutatus*), Cipó canastero (*Asthenes luizae*), grey-backed tachuri (*Polystictus superciliaris*) and pale-throated pampa-finch (*Embernagra longicauda*). The Cipó canastero was discovered as recently as 1990 and is endemic to a small area in the Serra do Cipó. It is the only one of these species currently listed as threatened, but the others are listed as near-threatened. Habitats are decreasing throughout this EBA.

Economic assessment

The Cadeia do Espinhaço has contributed its natural resources and services to the national economy in various ways. At present, current values on plant products have either not been calculated or may be difficult to extricate from the current literature. (No attempt is made here to produce figures for mineral extraction and other general forms of activity.)

For the sempre-vivas industry, which is almost exclusive to the Espinhaço Range, some figures have been supplied (N. Giulietti *et al.* 1987). Cases showing the situation in 1984 can be given: *Syngonanthus elegans* – 40,000 kg/year; *S. venustus* plus *S. dealbatus* ("brejeira") – 20,000 kg/yr; *S. xeranthemoides* – 12,000 kg/yr; *Paepalanthus macrocephalus* – 10,000 kg/yr; *S. magnificus* and *S. suberosus* – each 4000 kg/yr; *Aristida jubata* ("barba-de-bode") – 40,000 kg/yr; *A. riparia* – 20,000 kg/yr; *Aulonemia effusa* and *Axonopus brasiliensis* – each 12,000 kg/yr; *Xyris nigricans* – 5000 kg/yr; *X. cipoensis* – 1000 kg/yr; *Rhynchospora globosa* – 20,000 kg/yr; *R. speciosa* – 5000 kg/yr. Of 40 species with high economic value, those in *Syngonanthus* command the highest wholesale prices. Between 1974–1986, an average of 645,000 kg of sempre-vivas were exported annually. However, whereas 1,007,449 kg were exported in 1978, there were only 320,003 kg in 1986. This might be explained by a fall in product value and/or a reduction in the populations. In the same period the price per kg varied from just US$4.54–4.97, showing that an instability of prices cannot be the reason for the dramatic fall in trade.

The sempre-vivas industry may be an important factor in ensuring the survival of many local communities in the Espinhaço zone and reducing emigration to the towns. Usually whole families are involved in harvesting from the wild. To support the populace, the temporary settlements include markets and shops to sell food and other necessities to aid those working in the field. New strategies must be developed to rationalize the conservation of threatened species and the needs of the local communities. One solution may be cultivation of suitable species, such as *Syngonanthus elegans* in the Diamantina district where it is native. This species has already proved very productive in experimental cultivation, but most species of Eriocaulaceae appear to have very special growing requirements. Cultivation of suitable species, together with a policy for the sustainable collection of the wild plants, could go far in ensuring the survival of their native populations while having a beneficial effect on the local economies.

Threats

Familiar mining activities, once carried out on a large scale in the Espinhaço Range, still persist localized to some areas, with development of new industries in others (manganese, etc.). The impact of major soil disturbances in the search for gold and diamonds is notorious, as long-abandoned mining areas remain bare and devoid of their former vegetation for decades and in some cases even longer. Soil erosion is greatly accentuated by construction of badly planned highways; frequently soil has to be brought from other areas for their construction, encouraging invasive weeds.

Today the main activity throughout the Espinhaço Range is cattle-raising, which is favoured by the existence of natural grassland. The destruction of these areas is accentuated by the trampling of cattle and annual burning carried out by local farmers to stimulate regrowth of the herbaceous vegetation. This has a selective effect on the species diversity of the grasslands, because although many species have adaptations to resist fire, its frequent and regular application eventually favours few species of animals and plants.

Due to the low fertility and shallowness of the soils, the impact of agricultural activities is greatly restricted in the mountains of the Espinhaço Range, compared to other areas such as the cerrado. Where the soils are deeper and more fertile, however, subsistence agriculture can be practised, mainly with sugarcane, beans, rice and maize. In a few places in the Chapada Diamantina on more extensive areas of level ground where water is available, cash crops also may be raised on a small scale. Particularly where deeper soils support forest, coffee and other crops have been planted after removal of the natural vegetation.

Forest and cerrado formations in the Espinhaço Range have been intensively exploited for fuelwood for domestic use and especially in production of charcoal. This is an important material in the iron and steel industries of Minas Gerais. Even in the campo rupestre, which is much less rich in woody plants, the collection of the resinous branches of various species of *Vellozia* (e.g. *V. sincorana* – "candombá", in the Serra do Sincorá) has caused a drastic reduction in the size of natural populations of these species, which probably take decades to reach maturity.

The indiscriminate collection of whole plants for horticultural use, particularly orchids, bromeliads, ferns and cacti, undoubtedly threatens the survival of a number of species. Those species that are the rarest, with small populations endemic to restricted areas, are most sought after for that very reason, as they will fetch the highest prices in the horticultural market. An example is the small orchid *Constantia cipoensis*, endemic to the Serra do Cipó and occurring as an epiphyte only on an undescribed species of *Vellozia*. As this *Vellozia* (also endemic to the mountain) is extremely tall, collectors not only strip the orchids from its branches, but also destroy it in the process.

Finally, the constant collection of sempre-vivas, which completely removes the inflorescences, is reducing the size of natural populations. For example, whereas the volume of *Syngonanthus magnificus* exported in 1984 was 4000 kg, in 1986 it fell to 1500 kg. This spectacular species, which fetched the highest prices, has a very restricted range, being found only in the Serra do Ambrósio, and is now greatly diminished in the wild.

Conservation

The Espinhaço Range has two National Parks. Chapada Diamantina National Park (Bahia), created in 1985, comprises 1520 km² already delimited; it is still in the initial phase of organization. Serra do Cipó NP (338 km²) (Santana do Riacho and Jaboticatubas, Minas Gerais) was created in 1984, and is delimited and in operation; it is surrounded by a buffer zone c. 30 km wide. However, the park's infrastructure is not yet sufficient for effective protection.

Other variously conserved areas include Gruta dos Brehões APA (Environmental Protection Zone), 119 km² in Bahia; and in Minas Gerais, Cavernas do Peruaçu APA, 1500 km² (Januária and Itacarambi); Cachoeira Andorinhas APA, 187 km²; Rio Doce State Park, 360 km²; Itacolomi State Park, 70 km² (Mariana and Ouro Preto); and Tripuí Ecological Station, 7 km² (Ouro Preto).

Considering the high endemism of the flora and fauna, these variously designated areas form only part of what needs to be conserved in the region. Areas in Bahia, such as Pico das Almas (1850 m) and Serra de Barbado (2107 m) – the highest point in the Espinhaço Range, and areas in Minas Gerais, such as in the vicinity of Diamantina in the central part of the range and Grão-Mogol in the north, are a few of the centres of extremely high diversity that are unprotected. Such natural areas must be included in a more coordinated plan for the conservation of the biodiversity in the region. Such a strategy needs a much greater knowledge of the diversity and distribution of the fauna and flora and a better understanding of the dynamics of the campo-rupestre ecosystem.

References

Abreu, A.A. (1984). O Planalto de Diamantina: um setor da Serra do Espinhaço em Minas Gerais. Orientação – Instituto de Geografia Univ. São Paulo 5: 75–79.

Giulietti, A.M. and Pirani, J.R. (1988). Patterns of geographic distribution of some plant species from the Espinhaço Range, Minas Gerais and Bahia, Brazil. In Vanzolini, P.E. and Heyer, W.R. (eds), *Proceedings of a workshop on neotropical distribution patterns.* Academia Brasileira de Ciências, Rio de Janeiro. Pp. 39–69.

Giulietti, A.M., Menezes, N.L., Pirani, J.R., Meguro, M. and Wanderley, M.G.L. (1987). Flora da Serra do Cipó, Minas Gerais: caracterização e lista das espécies. *Bol. Botânica, Univ. São Paulo* 9: 1–151.

Giulietti, N., Giulietti, A.M., Pirani, J.R. and Menezes, N.L. (1987). Estudos em sempre-vivas: importância econômica do extrativismo em Minas Gerais, Brasil. *Acta Bot. Bras.* 1(2) (supl.): 179–193.

Harley, R.M. (1988). Evolution and distribution of *Eriope* (Labiatae), and its relatives, in Brazil. In Vanzolini, P.E. and Heyer, W.R. (eds), *Proceedings of a workshop on neotropical distribution patterns.* Academia Brasileira de Ciências, Rio de Janeiro. Pp. 71–120.

Harley, R.M. and Simmons, N.A. (1986). *Florula of Mucugê. Chapada Diamantina – Bahia, Brazil.* Royal Botanic Gardens, Kew. 227 pp.

King, L.C. (1956). A geomorfologia do Brasil Oriental. *Rev. Brasil. Geogr.* 18: 147–265.

Köppen, W. (1931). *Climatología.* Fondo de Cultura Económica, Buenos Aires.

Magalhães, G.M. (1966). Sobre os cerrados de Minas Gerais. *An. Acad. Brasil. Ciênc.* 38 (supl.): 59–70.

Mello-Silva, R. (1989). *Velloziaceae de Grão-Mogol, Minas Gerais, Brasil.* M.S. thesis. Universidade de São Paulo, São Paulo.

Menezes, N.L. and Giulietti, A.M. (1986). Campos rupestres. Paraíso botânico na Serra do Cipó. *Ciência Hoje* 4(26): 38–44.

Menezes, N.L. and Semir, J. (1991). *Burlemarxia,* a new genus of Velloziaceae. *Taxon* 40: 413–426.

Moreira, A.N. (1965). Relevo. In Instituto Brasileiro de Geografia e Estatística (IBGE), *Geografia do Brasil,* vol. 5. *Grande Região Leste.* IBGE, Rio de Janeiro. Pp. 5–54.

Wanderley, M.G.L. (1992). *Estudos taxonômicos no gênero* Xyris L. (Xyridaceae) da Serra do Cipó, Minas Gerais, Brasil. Ph.D. thesis. Universidade de São Paulo, São Paulo.

Authors

This Data Sheet was written by Dra. Ana Maria Giulietti and Dr José R. Pirani (Universidade de São Paulo, Departamento de Botânica, Caixa Postal 11461, 05499 São Paulo, SP, Brazil) and Dr Raymond M. Harley (Royal Botanic Gardens, Kew, Richmond, Surrey TW9 3AB, U.K.).

INTERIOR DRY AND MESIC FORESTS: CPD SITE SA21

DISTRITO FEDERAL
Brazil

Location: Brazil's Federal District is a rectangular area in eastern central part of State of Goiás bordering State of Minas Gerais, at latitudes 15°31'–16°03'S and longitudes 47°02'–48°15'W.

Area: 5814 km².

Altitude: 750–1336 m.

Vegetation: Mosaic of many vegetation types, mostly in (1) forest habitats: gallery forests, mesophytic forests and woodland form of cerrado (cerradão); and (2) savanna habitats: gradient of decreasing tree density under generality of cerrado – cerrado denso, cerrado *sensu stricto*, campo cerrado, campo sujo de cerrado, campo limpo de cerrado; plus (3) localized formations such as seasonal and permanent marshes, veredas, campos de murunduns.

Flora: Very high diversity – at least 3000 species of vascular plants expected; few endemic and rare species.

Useful plants: Native fruits (trees and shrubs), medicinals, forage (grasses and legumes), ornamentals (e.g. for dried flower arrangements), genetic resources (especially trees).

Other values: Watershed protection for drinking water, hydroelectric energy and irrigation projects; high faunal diversity, including rare and threatened species.

Threats: Increasing population pressures, set fires, erosion, pollution, planned dam construction.

Conservation: 1 National Park, 3 Areas of Outstanding Ecological Interest (ARIEs), 5 Ecological Reserves, 3 Ecological Stations, 1 Biological Experiment Station, 5 Environmental Protection Zones (APAs).

Geography

The Distrito Federal (DF) is in the core area of the cerrado region. Brasília, the capital city of Brazil, is in the DF, which also encompasses seven satellite cities and additional settlements and rural communities. The DF is situated in the Brazilian Uplands on the Pratinha Plateau, east of the Planalto de Mato Grosso and Middle Massif of Goiás and west of the São Francisco Basin.

The area is an integral part of the central sector of the geological Tocantins province, and is composed of rocks of the Canastra Formation and Paranoá Group of Middle to Late Precambrian age, with a lateritic detritus cover of Tertiary-Quaternary age and recent Quaternary alluvial deposits (Barros 1990). The most extensive soil class is latosols, which occupy 54% of the region. The deep, well-drained latosols are restricted to essentially level terrain (slopes of less than 8%). They are associated with the cerrado *sensu stricto* (Haridasan 1990), and are the most extensively cultivated. Other soil types in the DF are dystrophic cambisols, podzols, hydromorphic soils, and very small areas of limestone-derived soils and limestone outcrops.

Hypsometric data indicate that the DF is 57% uplands above 1000 m (Novaes Pinto 1990a), which are a divider for the Paraná-La Plata, Tocantins-Amazon and São Francisco drainages. Because of the predominance of metamorphic rocks, aquifers are mainly fissural, in which each year about 1.2 billion m³ of water percolate (Barros 1990). These rocks act as impervious water tanks with special characteristics. The lentic environment is represented by small natural lakes (Lagoa Bonita, Lagoa Joaquim Medeiros, Lagoa Carás, Lagoa QL3 Norte), as well as three reservoirs (Descoberto, Santa Maria, Paranoá). The DF has four hydrographic basins:

Descoberto, São Bartolomeu, Maranhão and Preto. The first and second drain southward into La Plata Basin, the third northward into the Amazon Basin and the last eastward into the São Francisco Basin.

The climate in the cerrado region is tropical (Köppen's Aw). Mean annual precipitation generally varies from 1100 to 1600 mm, with c. 90% occurring in the warmer wet season (October–April); there is a marked dry season from May to September. The annual temperature averages from 18°–20°C, with July being 16°–18°C and September and October 20°–22°C.

Vegetation

The DF has a mosaic of different vegetation types that can be put into two groups: forest and savanna. There are gallery forests, mesophytic forests, semi-deciduous forests, and "cerradão" (cerrado woodlands). Savanna ("cerrado") includes a gradient of decreasing density of trees: cerrado denso, cerrado *sensu stricto*, campo cerrado, campo sujo de cerrado and campo limpo de cerrado (Eiten 1972, 1990). The "campo limpo de cerrado" has no trees, being dominated by herbs (primarily grasses), with subshrubs and shrubs. Accessory vegetation types include seasonal marshes (wet campos), permanent marshes, "veredas" (swamps), campos de murunduns and a few rocky or litholic fields (Eiten 1990).

The vegetation is overwhelmingly dominated by cerrado *sensu stricto* (cerrado típico), which is floristically heterogeneous (Felfili and Silva 1993). Gallery forests occur along rivers and semi-deciduous forests on limestone-derived soils and limestone outcrops. Mesophytic forest and cerradão were once fairly common, but are now the rarest types of vegetation in the DF – because their soil is richer and more productive, they have been cleared and transformed into agricultural fields and pastures. A few tracts are in protected areas; those outside the conservation system face a bleak future.

The gallery forests are of two distinct types: swampy and dry. The swamp forest remains waterlogged all year; the dry forest is never waterlogged. Some key species characterize swampy gallery forests: *Talauma ovata, Euterpe edulis, Geonoma schottiana* and *Mauritia vinifera* – a palm which is also characteristic of veredas. The "veredas" (or "brejos") are ecologically very important because they are the headwaters of many watercourses. As they have surface water all year, they are oases critical to the survival of many species. Animals gather at veredas to drink during the driest months.

"Campos de murunduns" consist of areas with rounded earth mounds (murunduns) covered by woody cerrado plants, and between the mounds predominately grasses, sedges and *Xyris* spp. Mound height varies from 0.05–2 m and the mounds are generally semi-elliptical. Their origin appears to be related to drainage patterns and differential erosion (Araújo Neto *et al.* 1986).

The rocky and litholic fields resemble campo-rupestre areas. They are dominated by *Vellozia* spp., *Lychnophora ericoides, Paepalanthus* spp., *Xyris* spp. and *Dyckia* spp. These habitats occur in small patches amongst the general cerrado vegetation and are generally overlooked. Their flora is very different from the surrounding areas and their limited soil is rocky and sandy.

Flora

The DF flora is only partially known. There is a checklist with 2500 native species of vascular plants (Filgueiras and Pereira 1990) but several areas have never been adequately surveyed; at least 3000 species are expected to occur. A project for an illustrated Flora is under way (Cavalcanti and Proença, pers. comm.). Extensive field collecting was carried out in 1992 and projected to continue in an effort to sample the remaining native vegetation.

Families with the largest number of species are Leguminosae, Gramineae (305 spp. – Filgueiras 1991), Compositae, Orchidaceae (233 spp. – Bianchetti *et al.* 1991), Rubiaceae, Myrtaceae and Melastomataceae. Genera especially diverse in species are *Paspalum* and *Panicum* (Gramineae), *Habenaria* (Orchidaceae), *Vernonia* (Compositae), *Chamaecrista* and *Mimosa* (Leguminosae), *Miconia* (Melastomataceae) and *Hyptis* (Labiatae).

Several endemic, rare or threatened species or populations are known. A single gymnosperm *Podocarpus brasiliensis* occurs in the DF, where it is considered threatened; only two small populations of this relict species have been found. Other examples of species rare in the DF are *Wunderlichia mirabilis, Weinmannia organensis, Manihot anomala, Apoclada cannavieira* (a single population), *Arthropogon filifolius, Hymenolobium heringerianum, Lupinus insignis* and *Drimys winteri*. Among the endemics are *Calea heringeri, Panicum subtiramulosum, Hymenolobium heringerianum*

CPD Site SA21: Distrito Federal, Brazil. Cerrado woodland within Brasília National Park.
Photo: Tony Morrison/South American Pictures.

and *Cedrela odorata* var. *xerogeiton*. *Lychnophora ericoides* is threatened, even though three sizeable populations have been located, because it is over-exploited by those who collect and commercialize medicinal plants. The herbaceous bambusoid grasses *Aulonemia aristulata* and *Pharus lappulaceus* are common through their overall ranges, but only a single population of each has been found in the DF.

Useful plants

Many useful plants occur in the DF. Most have been used traditionally, yet new uses are constantly being sought (Almeida, Silva and Ribeiro 1987). These plants fall into various categories: timber, charcoal, fibre, cork, food, forage, oil, honey, medicinal, tannin and ornamental.

A number of tree species are commercialized for their wood. Especially sought are *Hymenaea* spp., *Blepharocalyx salicifolius* (*B. suaveolens*), *Ocotea* spp., *Pterodon pubescens*, *Copaifera langsdorffii* and *Calophyllum brasiliense*. Most of the cerrado trees can produce good quality charcoal, which has become a quick source of income to farmers who own tracts of undisturbed cerrado. The collection and commercialization of wild plants for dried flower arrangements employs hundreds of people during the year.

Fruits of many tree species are harvested and sold in street fairs or local markets. Examples are "pequi" (*Caryocar brasiliense*), "mangabá" (*Hancornia speciosa*), "araticum" (*Annona crassiflora*) and "cagaita" (*Eugenia dysenterica*). Liqueurs and spirits of native fruit species can be found in any local market. Especially famous are the liqueurs made from pequi, "murici" (*Byrsonima* spp.) and "genipapo" (*Genipa americana*).

Medicinal plants are an important aspect of Brazilian culture, especially in rural areas. The number of medicinal plants is considered high in the cerrado in general (Siqueira 1981) and in the DF (Barros 1982). These plants (e.g. *Lychnophora ericoides, Centrosema bracteosum, Pterodon pubescens, Anemopaegma arvense*) are routinely collected for private use as well as sold both locally and elsewhere in the country.

Social and environmental values

The Distrito Federal represents only 0.07% of Brazilian territory, but its political, social, cultural and economic importance is enormous. It is a melting pot for Brazilian culture, attracting people from all parts of the country. Besides permanent residents, Brasília also has a large fluctuating population of politicians, lobbyists and executives, who have temporary homes in the city and pay premium prices for top quality housing. Conservation issues in the DF are especially significant because what happens is likely to be regarded as a model for the entire cerrado region and perhaps the whole country. There is considerable pressure to create an industrial district that will generate numerous jobs, which will bring still more people.

The diversity of fauna recorded from the DF is quite high. In the cerrado region of Brazil there are c. 300 species of mammals; in the DF, notable species include 30 bats, four monkeys, several anteaters (including the giant anteater) and five cats. Over 400 species of birds are known from the DF,

c. 200 of which are associated with the cerrado biome; c. 14 species are regionally endemic or rare. The threatened endemic bird known as the Brasília tapaculo (*Scytalopus novacapitalis*) survives locally in swampy gallery forest and dense streamside vegetation in the states of Goiás, the Distrito Federal and Minas Gerais. The DF has over 50 species of Lepidoptera (Rocha *et al.* 1990).

The area's diverse biological wealth has attracted the attention of a great number of scientists, to the extent that the DF is much better known than any other area in the cerrado region (Novaes Pinto 1990b; Dias 1990). The project Flora do Distrito Federal to prepare an illustrated Flora is led by local scientists who work in close cooperation with international botanical institutions.

The DF has a tradition of developing cooperative research programmes among different institutions, which has proven very effective in disciplines such as ecology. The following have a history of conducting joint research projects: Universidade de Brasília, EMBRAPA (Empresa Brasileira de Pesquisa Agropecuária), Reserva Ecológica do IBGE, Jardim Botânico de Brasília and SEMATEC (Secretaria do Meio Ambiente, Ciência e Tecnologia). Non-governmental organizations such as FUNATURA (Fundação Pró-Natureza) are also very active in promoting research and related activities.

Economic assessment

The DF has considerable economic activity. Commerce and services predominate, but agriculture also has an important role, especially for vegetables, fruits (oranges – *Citrus*) and recently soybeans (*Glycine max*) and dairy products. Although the cerrado soils are poor in some essential minerals (e.g. nitrogen, phosphorus, calcium, magnesium), they may become quite productive through technologies such as liming, fertilizing and the general use of pesticides.

Cattle production is carried out using native pasturage in the wet season; 135 grass species are of forage value (Filgueiras and Wechsler 1992; Filgueiras 1992). During the dry season, planted pastures supply the bulk of the forage (mostly *Andropogon gayanus, Brachiaria* spp., *Hyparrhenia rufa, Panicum maximum*). Several legume species (commonly *Stylosanthes* spp. and *Leucaena* spp.) are also planted in the pastures to increase their nutritional value.

Despite increasing environmental awareness, charcoal is still being produced in the DF and is very profitable. It has become a cash crop – yet no investment is needed and there is no risk. The farmer normally sells the raw material as a type of standing crop, i.e. the cerrado in its natural state; the woody plants are extracted and made into the charcoal, especially for use in the steel industry.

The flourishing business in wild plants for flower arrangements is a source of income for many families. In 1984 to the south-east in the region of Diamantina, Minas Gerais (within CPD Site SA20), 257 tonnes of dried plants were collected (Burman 1991). A sizeable portion of these plants is destined for international markets, especially in U.S.A. and Japan. As these resources are becoming scarce within the DF, collectors are slowly moving to other areas where the plants are still abundant.

The fruits of some native species are readily harvested and sell at a good price wherever they occur in reasonable quantities, e.g. *Caryocar brasiliense, Annona crassiflora,*

Hancornia speciosa and *Mauritia vinifera* ("buriti"). Medicinal plants are gathered in the DF in large quantities and sold locally or exported to other areas (Goiás, Minas Gerais). The trade involves both untreated plants and liquid extracts from them ("garrafadas"). These medicinals can be purchased in any marketplace in Central Brazil; in the DF they are readily available in all open markets and in speciality shops.

Quarrying is restricted to extraction of limestone, gravel and sand. The limestone is converted into cement, which is the principal building material. Cement production is one of the major industrial activities in the DF. The gravel and sand are used in road construction and repair as well as building homes. There is increasing demand for gravel and sand in the area because Brasília and the satellite cities are growing rapidly. Thus major income for many families in operating their businesses depends on these public-land resources.

Threats

The biggest threat to the biodiversity of the Distrito Federal is the steady increase of the human population. Brasília was projected to accommodate 500,000 people towards the year 2000. However in 1990, 30 years after its foundation, the city's population was over 1 million. The estimated DF population is now 2 million, which encompasses the satellite cities Ceilândia, Taguatinga, Guará I, Guará II, Núcleo Bandeirante, Sobradinho and Planaltina, as well as several settlements and many rural communities. New settlements appear every year through the DF – some legal, some illegal. Migration has become common because Brasília attracts people from all parts of the country looking for jobs, health facilities, schools and housing.

The shortage of water is already a problem and demand is growing with the population. In 1992 there were projects to build a new dam to store water for human consumption and generating electricity. The São Bartolomeu River has been selected not because of the quality of its water – it is simply the largest river available in the region. Plans exist to bring water from outside the DF in the near future, because very serious water shortages are projected. Drainage of wetland areas (especially veredas) for agriculture additionally threatens the water supply as well as this specialized vegetation. Water pollution and contamination (mostly from pesticides) already are major concerns. Farms in the Descoberto River Basin supply most of the vegetables consumed in the DF, but also contribute significantly to the pollution in the area.

The price of land in the DF is extremely high for Brazil. There is a chronic housing shortage mainly for those with low and middle incomes and tremendous pressure for new housing developments. Closed condominiums ("condomínios fechados") are appearing as a viable alternative for middle-class families; they are in rural and semi-rural areas on public and private lands. The condominiums are primarily situated in Áreas de Proteção Ambiental (APAs), making their establishment a serious threat to these legally conserved zones. As the condominiums were not included in the original plans for the DF, a battle was under way in 1992 to legalize them.

The classification as an APA can be misleading; it is a weak type of conservation unit. Entire neighbourhoods, large settlements, agriculture and some industries are found within such areas. The APA do Lago Paranoá can hardly be considered

even semi-rural; it is heavily populated and includes the two most prestigious residential sections of Brasília, with large estates and country homes. The APA do Rio São Bartolomeu has become almost totally colonized by condominiums, small farms and new settlements. Consequently, the APAs at best should be considered as semi-rural or rural multiple-use zones rather than nature conservation units.

Additionally, various conservation units are becoming progressively isolated as they are surrounded by farms and settlements. The human influences on these areas are notable. Some of the problems are an increase in the number of human-provoked fires, illegal hunting and fishing, the presence of feral animals (dogs and cats), and weeds. Feral animals are a serious problem in Brasília National Park, where dog packs have been reported to prey on large mammals such as tapir, deer and capybara. Habitat alteration is a serious problem caused by weeds such as *Melinis minutiflora, Panicum maximum, Andropogon gayanus, Hyparrhenia rufa* and *Brachiaria decumbens*. These African grasses are commonly cultivated in the pastures but because of their aggressiveness, they invade conserved areas and compete with native species, sometimes eliminating them (Filgueiras 1989).

Fire is a serious threat, particularly to the conservation units. Human-provoked fires occur every year and some are very damaging to the animals and plants alike. A long-term cooperative project involving national and international scientists is assessing the effects of different fire regimes on the cerrado environment. The results of their research will help to establish management policies for conservation units throughout the region where fire has been a regular natural occurrence.

Conservation

The main conservation units in the DF are indicated in Table 59 (cf. Map 57) (SEMATEC and GDF 1992; Dias 1990). About 3.2% of the DF is in totally protected areas (which is higher than the c. 2.5% of the country as a whole). However, the entire range of the DF vegetation types is not conserved. For example, calcareous outcrops are not protected – they are aggressively exploited for their limestone. Also litholic fields and campos de murundus are not adequately protected. Representative portions of these habitats should be legally and effectively preserved – while good areas still remain.

Three contiguous areas (Estação Ecológica do Jardim Botânico, Reserva Ecológica do IBGE, Estação Ecológica da Universidade de Brasília) within the APA Gama-Cabeça de Veado are the site of several ecological research projects, from floristic surveys to the effects of fire on plants, animals, soil and water. The Reserva Ecológica do IBGE is perhaps the best studied of all the DF conservation units (more than 30 theses and dissertations have been carried out), and it has the best data on flora (1700 species), fauna, soil, and fire practices. A management plan for this reserve was nearing completion in 1993.

Brasília National Park is the only conservation unit that has had a (preliminary) management plan, which is being revised by a team of experts. The APA de Cafuringa (which was created in 1988 but presently exists without protection) is particularly important ecologically because it links the park and the Chapada da Contagem, and thus plays a key role in preventing the park's isolation.

TABLE 59. PRINCIPAL CONSERVATION UNITS IN BRAZIL'S DISTRITO FEDERAL

Conservation unit	Area (ha)
Estação Ecológica de Águas Emendadas	10,547
Estação Ecológica do Jardim Botânico de Brasília (Estação Florestal Cabeça de Veado)	4000
Estação Ecológica da Universidade de Brasília [Fazenda Experimental Água Limpa; Área de Relevante Interesse Ecológico (ARIE) Capetinga-Taquara]	2100
Estação Experimental de Biologia da Universidade de Brasília	c. 50
Reserva Ecológica do Guará	147
Reserva Ecológica do Instituto Brasileiro de Geografia e Estatística (IBGE) (Reserva Ecológica do Roncador)	(1263 or) c. 1360
Reserva Ecológica do Gama	136
Reserva Ecológica do Serrinha	c. 200
Reserva Ecológica do Cerradão	c. 50
ARIE dos Córregos Taguatinga-Cortado	210
ARIE Santuário de Vida Silvestre do Riacho Fundo	c. 400
ARIE do Paranoá Sul	144
Parque Nacional de Brasília	28,000
Subtotal = c. 473 km² or	**c. 47,344**
Área de Proteção Ambiental (APA) do Rio Descoberto	39,100
APA de Cafuringa	(30,000 or) 39,000
APA do Lago Paranoá	16,000
APA das Bacias do Gama e Cabeça de Veado	25,000
APA do Rio São Bartolomeu	84,100
Total = c. 2425 km² or	**c. 242,490**

As a result of a symposium on alternatives for developing the cerrado region carried out by FUNATURA (Dias 1992), it was recommended to the federal government that the APA de Cafuringa, Brasília National Park, Estação Ecológica de Águas Emendadas, APA do Rio São Bartolomeu and APA Gama-Cabeça de Veado (totalling c. 1866 km²) be designated as a Biosphere Reserve, due to their biological significance as well as their economic, social, political and cultural values.

The government is making an effort to create new protected areas and responding to environmental groups and the local scientific community, which has a history of acting politically together with community leaders on important conservation matters. At least 8% of the Distrito Federal needs to be well preserved through legal protection and management of the best natural areas. The present goals are to bring new natural areas into conservation (e.g. the limestone outcrops, the litholic campos, the campos de murunduns) and to procure funds to assist the nature conservation units so that they may become a full reality rather than mainly existing legally. Financial support is urgently needed to provide the conservation units with the basic infrastructure to secure effective protection of biodiversity. This includes the purchase of some land and of equipment for fire control, construction of buildings and fences and hiring of forest guards and administrative personnel.

The urban population of Brasília is becoming increasingly aware of conservation issues. Environmental education is becoming very important in Brasília and several institutions are beginning to develop their own programmes to educate school children, teenagers and adults. Once available, these programmes become very popular and much sought after especially by the school system. Ecology-related topics have a special appeal that intrigues and fascinates both younger and older generations. Special attention should be paid to the populations living around the conservation units. If well motivated and

MAP 57. DISTRITO FEDERAL, BRAZIL (CPD SITE SA21) (after SEMATEC and GDF 1992)

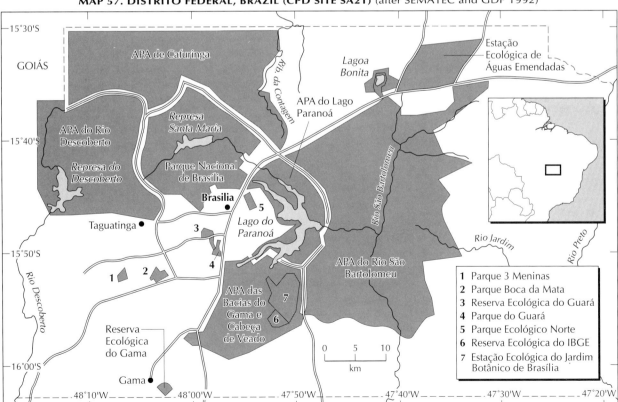

educated, they can have key roles in preserving the biodiversity of the areas they live near.

References

Almeida, S.P., Silva, J.A. da and Ribeiro, J.F. (1987). *Aproveitamento alimentar de espécies nativas dos cerrados: araticum, baru, cagaita e jatobá.* Documentos 26, EMBRAPA-CPAC (Centro de Pesquisa Agropecuária dos Cerrados), Planaltina, D.F., Brazil.

Araújo Neto, M.D., Furley, P.A., Haridasan, M. and Johnson, C.E. (1986). The murunduns of the cerrado region of Central Brazil. *Journal Trop. Ecol.* 2: 17–35.

Barros, J.G.C. (1990). Caracterização geológica e hidrogeológica do Distrito Federal. In Novaes Pinto, M. (ed.), *Cerrado: caracterização, ocupação e perspectivas.* Editora Universidade de Brasília, Brasília. Pp. 257–275.

Barros, M.A.G. (1982). Flora medicinal do Distrito Federal. *Brasil Florestal* 12: 35–45.

Bianchetti, L., Batista, J.A.N., Salles, A.H., Maury, C.M.R.F. and Andrade, F.A.T. (1991). Contribuição ao conhecimento da família Orchidaceae no Distrito Federal – novas citações. In Rizzo, J.A. (ed.), *Resumos dos trabalhos do Congresso Nacional de Botânica.* Universidade Federal de Goiás, Goiânia. Pg. 384.

Burman, A.G. (1991). Saving Brazil's savannas. *New Scientist* 1758: 30–34.

Dias, B.F.S. (1990). Conservação da natureza no cerrado brasileiro. In Novaes Pinto, M. (ed.), *Cerrado.* Editora Universidade de Brasília, Brasília. Pp. 582–640.

Dias, B.F.S. (ed.) (1992). *Alternativas de desenvolvimento dos cerrados: manejo e conservação dos recursos naturais renováveis.* FUNATURA (Fundação Pró-Natureza); IBAMA (Instituto Brasileiro do Meio Ambiente e dos Recursos Naturais Renováveis), Brasília.

Eiten, G. (1972). The cerrado vegetation of Brazil. *Bot. Review* 38: 201–341.

Eiten, G. (1990). Vegetação do cerrado. In Novaes Pinto, M. (ed.), *Cerrado.* Editora Universidade de Brasília, Brasília. Pp. 9–65.

Felfili, J.M. and Silva Jr., M.C. da (1993). A comparative study of cerrado (*sensu stricto*) vegetation in Central Brazil. *Journal Trop. Ecol.* 9: 277–289.

Filgueiras, T.S. (1989). Africanas no Brasil: gramíneas introduzidas da África. *Cadernos de Geociências* 2: 41–46.

Filgueiras, T.S. (1991). A floristic analysis of the Gramineae of Brazil's Distrito Federal and a list of the species occurring in the area. *Edinburgh Journal Bot.* 48: 73–80.

Filgueiras, T.S. (1992). Gramíneas forrageiras nativas no Distrito Federal. *Pesquisa Agropecuária Brasileira* 27: 1103–1111.

Filgueiras, T.S. and Pereira, B.A.S. (1990). Flora do Distrito Federal. In Novaes Pinto, M. (ed.), *Cerrado.* Editora Universidade de Brasília, Brasília. Pp. 331–388.

Filgueiras, T.S. and Wechsler, F.S. (1992). Pastagens nativas. In Dias, B.F.S. (ed.), *Alternativas de desenvolvimento dos cerrados: manejo e conservação dos recursos naturais renováveis.* FUNATURA; IBAMA, Brasília. Pp. 47–49.

Haridasan, M. (1990). Solos do Distrito Federal. In Novaes Pinto, M. (ed.), *Cerrado.* Editora Universidade de Brasília, Brasília. Pp. 309–330.

Novaes Pinto, M. (1990a). Caracterização geomorfológica do Distrito Federal. In Novaes Pinto, M. (ed.), *Cerrado.* Editora Universidade de Brasília, Brasília. Pp. 277–308.

Novaes Pinto, M. (ed.) (1990b). *Cerrado: caracterização, ocupação e perspectivas.* Editora Universidade de Brasília, Brasília. 657 pp.

Rocha, I.R.D., Cavalcanti, R.B., Marinho Filho, J.S. and Kitayama, K. (1990). Fauna do Distrito Federal. In Novaes Pinto, M. (ed.), *Cerrado.* Editora Universidade de Brasília, Brasília. Pp. 389–411.

SEMATEC and GDF (1992). *Mapa ambiental do Distrito Federal 92.* Scale 1:150,000. Secretaria do Meio Ambiente, Ciência e Tecnologia (SEMATEC), IEMA ICT and Governo do Distrito Federal (GDF), Brasília.

Siqueira, J.C. de (1981). *Utilização popular das plantas dos cerrados.* Ed. Loyola, São Paulo.

Author

This Data Sheet was written by Dr Tarciso S. Filgueiras [Instituto Brasileiro de Geografia e Estatística (IBGE), Caixa Postal 08-770, 70200-200 Brasília, D.F., Brazil].

GRAN CHACO
Argentina, Paraguay, Brazil, Bolivia

Location: Vast plain in parts of northern Argentina, north-western Paraguay and south-eastern Bolivia with a small area in south-western Brazil; between about latitudes 17°00'–33°20'S and longitudes 57°10'–67°00'W.

Area: 1,010,000 km².

Altitude: Nearly flat plain at 100–500 m, with occasional mountains – e.g. Cerro León in Paraguay to c. 720 m, Sierras de Córdoba in Argentina to c. 2795 m.

Vegetation: Predominantly xerophytic deciduous forests; also semi-evergreen riverine forests, palm woodlands, savannas, halophytic shrubby steppes, grassy steppes, wetlands.

Flora: c. 1000–1200 species, with endemic genera and species and high species diversity in certain genera; many taxa with disjunct distributions.

Useful plants: For lumber, charcoal, fibres, industrial and medicinal chemicals; ornamentals.

Other values: Watershed protection, soil conservation, genetic resources, knowledge of indigenous inhabitants, impressive landscape, habitats for diverse fauna.

Threats: Timber exploitation, agricultural development, oil and gas exploration, road construction and consequent development, frequent fires, erosion.

Conservation: c. 1% of region (12,720 km²) in 6 National Parks and 1 National Nature Reserve in Argentina and Paraguay; several recommended protected areas.

Geography

The Gran Chaco constitutes the phytogeographic Chaquenian province, and occupies 500,000 km² in northern Argentina, 350,000 km² in north-western Paraguay, a few km² around Porto Murtinho in south-western Brazil (Mato Grosso do Sul State) and 160,000 km² in south-eastern Bolivia (in Tarija, Chuquisaca and Santa Cruz departments) (cf. Prado 1993). The landscape is almost flat, sloping gradually eastward (0.04% gradient). Geologically, the Chaco plain is a tectonic depression filled with 3000 m of sediments from the Palaeozoic, Mesozoic and Tertiary covered by fine unconsolidated Quaternary deposits (FAO and UNESCO 1971; FAO and UNEP 1985). Several interruptions of older rocks are elevated approximately 400 m to 2400 m above the plain, including the Pampean mountains in Argentina, Cerro León in Paraguay and the Chiquitos in Bolivia.

Soils are usually neutral or slightly alkaline, with a high base level of saturation (90–100%). In the west the soils may be acidic and are more open to sandy and with good drainage; in the east they are mainly clayish and with poor drainage. Because of the semi-arid climate, primary minerals and soluble salts are abundant and result in areas of saline soils (FAO and UNESCO 1971).

The topography is influenced by water and wind. Eighty percent of the region is in La Plata River watershed; its major tributaries crossing the Gran Chaco are the Pilcomayo, Bermejo (Teuco) and Juramento-Salado rivers. Significant confined watersheds are Sali-Dulce River and Salinas Grandes in Argentina, Salinas de San Miguel and San José in Bolivia and Timané (Lagerenza) River in Paraguay. In dry areas of the Gran Chaco, the watercourses are ephemeral, often changing locally from year to year; in wet areas, rivers persist. Winds blowing across the plain sometimes form undulating sand-dunes, which are best represented on the frontier between Paraguay and Bolivia.

Intense, long-lasting sheet lightning occurs in the Gran Chaco. These powerful discharges are considered to cause many of the large fires in the region (Schwerdtfeger 1976). The summers are hot and humid, the winters mild but with possible frost and drier – with a strong dry season to the west. Precipitation generally decreases westward, with the least rainfall just east of the sierras on the western plains (Prado 1993). During the irregular rainy season from October to April, inundations cover vast areas of the land – up to 15% and for several months (FAO and UNESCO 1971). Based on climatic conditions, three broad zones of the Gran Chaco can be recognized (Vargas 1988):

1. Humid-subhumid zone (Eastern Chaco), annually receiving from 1250 mm of rainfall in the east to 700 mm towards the west. Median temperatures vary from 23°C in the north to 19.5°C in the south, with extremes of 43°C and -2.5°C.

2. Semi-arid to arid zone (Central and Western Chaco), with 650–350 mm of rainfall from its east to west. Median temperatures are 28°C in the north to 12°C in the south. The isotherm of absolute maximum above 47°C passes through the Argentine provinces of Salta, Formosa and Chaco and surrounds Santiago del Estero Province, which has had lows to -7.2°C (Prado 1993).

3. Montane or highland zone (Sierran Chaco), with an average of 500–900 mm of rainfall and some sites receiving up to 1100 mm or more. Temperatures of these areas are very variable; the annual mean is 17°C or less.

MAP 58. GRAN CHACO (CPD SITE SA22) (after FAO and UNEP 1985)

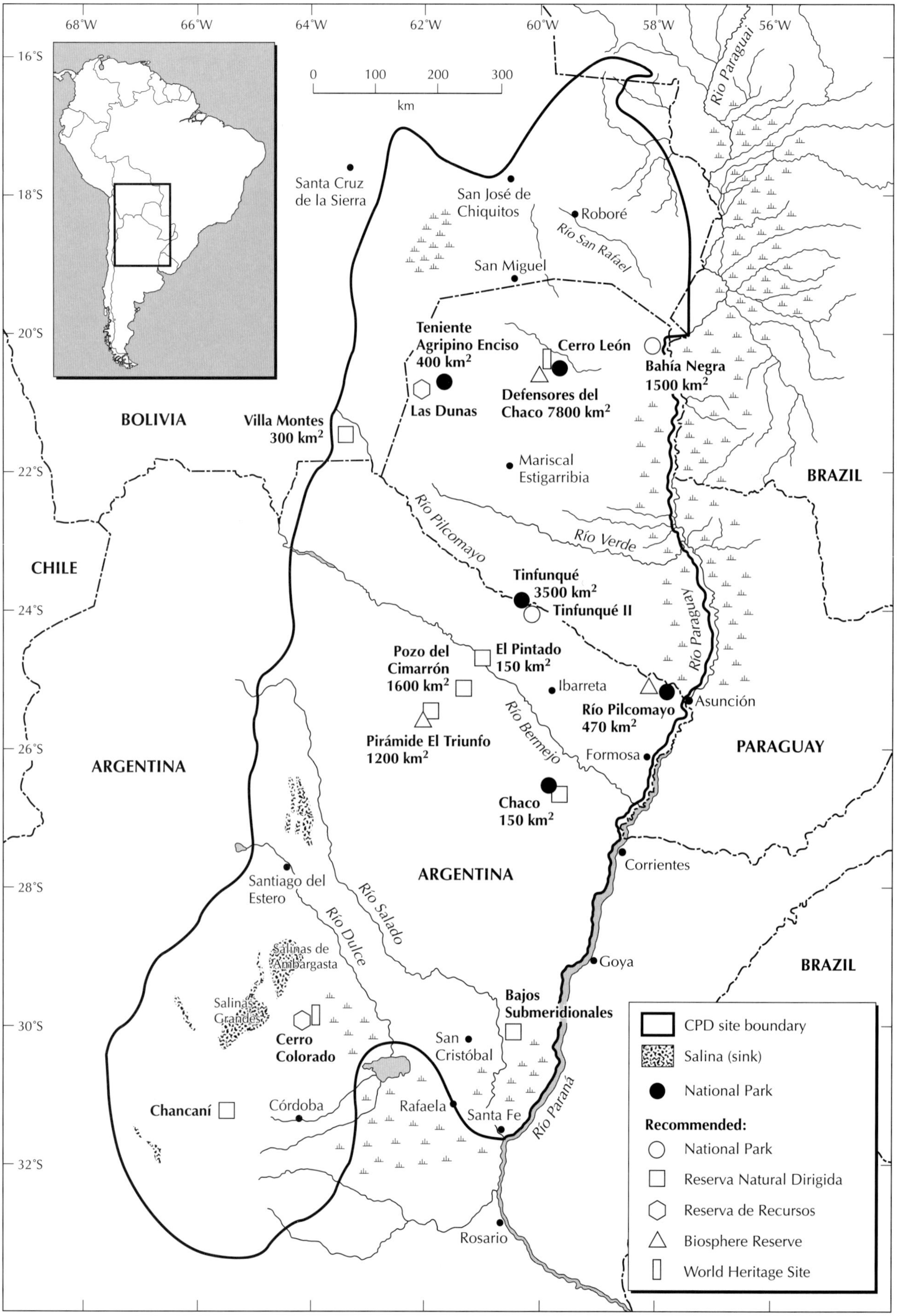

16°S

Santa Cruz de la Sierra

San José de Chiquitos

Roboré

San Miguel

Río San Rafael

BOLIVIA

18°S

Teniente Agripino Enciso 400 km²

Cerro León

Las Dunas

Defensores del Chaco 7800 km²

Bahía Negra 1500 km²

20°S

Villa Montes 300 km²

Mariscal Estigarribia

Río Pilcomayo

BRAZIL

22°S

CHILE

Río Verde

Tinfunqué 3500 km²

Tinfunqué II

Río Paraguay

24°S

Pozo del Cimarrón 1600 km²

El Pintado 150 km²

Ibarreta

Río Bermejo

Asunción

Río Pilcomayo 470 km²

Pirámide El Triunfo 1200 km²

Formosa

PARAGUAY

26°S

ARGENTINA

Chaco 150 km²

ARGENTINA

Corrientes

28°S

Santiago del Estero

Río Salado

Río Dulce

Salinas de Ambargasta

Goya

BRAZIL

Salinas Grandes

Bajos Submeridionales

30°S

Cerro Colorado

San Cristóbal

Chancaní

Córdoba

Rafaela

Santa Fe

Río Paraná

	CPD site boundary
	Salina (sink)
●	National Park

Recommended:

○	National Park
□	Reserva Natural Dirigida
⬡	Reserva de Recursos
△	Biosphere Reserve
▯	World Heritage Site

32°S

Rosario

0 100 200 300
km

68°W 66°W 64°W 62°W 60°W 58°W 56°W

Vegetation

The predominating vegetation is xerophytic deciduous forest with species in 3–4 strata: canopy and subcanopy trees, shrubby understorey and herbaceous grasses, with cacti and bromeliads. Localized edaphic and climatic conditions determine the vegetation in some parts of the region. Other types of vegetation include non-flooding and annually flooded riverine forests, wetlands, palm woodlands, savannas, grasslands, halophytic shrubby steppes, and cactus stands (Prado 1993; Herzog 1923; Spichiger *et al.* 1991; Ramella and Spichiger 1989; Cabrera 1976). The vegetation types are described below, based on the three climatic zones.

1. Humid-subhumid Chaco

On higher ground with well-developed soils, the mature forest community is transitional from wetter forests of southern Brazil. Notable species are *Tabebuia impetiginosa*, *Patagonula americana*, *Gleditsia amorphoides* and several Myrtaceae (*Myrcianthes pungens*, *M. cisplatensis*, *Eugenia uniflora*), as well as *Ruprechtia laxiflora*, *Pisonia zapallo*, *Scutia buxifolia* and in the north also *Astronium balansae*, *Diplokeleba floribunda*, *Ceiba speciosa* and *Pithecellobium scalare*. Lianas and epiphytes are frequent (Prado 1993).

On heavy-textured, sodium-ion-rich soils usually waterlogged during the rainy season, the stabilized forest community is dominated by *Schinopsis balansae*. Lianas are scarce, but epiphytes (especially *Tillandsia* spp.) are frequent. Other common canopy trees are *Aspidosperma quebracho-blanco* and *Caesalpinia paraguariensis*; the second storey includes *Prosopis nigra*, *Acacia praecox*, *A. aroma*, *Geoffroea decorticans*, *Ziziphus mistol* and *Sideroxylon obtusifolium* (Prado 1993). The common understorey species are *Schinus* sp., *Castella coccinea*, *Opuntia retrorsa* and *Cereus* sp. The common herbaceous species are *Dyckia ferox*, *Aechmea distachantha* and grasses. At the eastern beginning of the subhumid zone appear *Schinopsis lorentzii* and *S. heterophylla*.

Along rivers, diverse forest vegetation develops, for example woodlands of *Tessaria integrifolia* and *Salix humboldtiana*. There are subtropical semi-evergreen gallery forests which are flooded annually, and taller (to 30 m) riverine forests on non-flooding higher ground; both types are rich in lianas and vascular epiphytes (Spichiger *et al.* 1991; Prado 1993).

On alkaline soils that are seasonally flooded, palm savannas of *Copernicia alba* occur, with dispersed individuals of *Prosopis alba*, *P. algarobilla*, *Celtis tala*, etc. The herbaceous species are primarily grasses – *Spartina*, *Sporobolus* and *Paspalum*.

Several kinds of bunch grasses grow in both flooded and dry areas of the Gran Chaco. Important species are *Sorghastrum agrostoides* and *Paspalum intermedium*. On saline soils that are periodically flooded, "esparto" zones are dominated by savanna of *Elionurus muticus*, generally with other grasses including *Bothriochloa*, *Chloris* and *Schizachyrium*.

2. Semi-arid to arid Chaco

On mature soils, stabilized forest communities are dominated by *Schinopsis quebracho-colorado* and/or *Aspidosperma quebracho-blanco*. Fewer epiphytes occur in these types of forests. Other common tree species include *Ziziphus mistol* and *Prosopis* spp. The understorey may include *Caesalpinia paraguariensis*, *Cercidium australis*, *Jodina rhombifolia*, *Ruprechtia triflora* and *Castella coccinea*. Herbaceous species include *Bromelia serra*, *B. hieronymi* and *Deinacanthon urbanianum*.

Other types of vegetation in this zone include: on limey-clayish soils, forests of *Bulnesia sarmientoi* and *Tabebuia nodosa*; along riverbanks, forests dominated by *Enterolobium contortisiliquum*, *Acacia caven* or *Calycophyllum multiflorum*; in sandy areas, grassy savannas with *Schinopsis heterophylla*, as well as *Jacaranda mimosifolia* and *Schinopsis quebracho-colorado*; on somewhat saline soils, the giant cactus *Stetsonia coryne* in association with shrubs of *Bulnesia*, *Maytenus* and *Suaeda*; and on very saline soils, shrubby steppes of *Heterostachys ritteriana*, *Allenrolfea patagonica*, *Atriplex* and *Prosopis*.

3. Highland Chaco

In these Argentinian uplands, three zones of vegetation are recognized. In dry sunny areas up to 1800 m is a forest mainly of *Schinopsis haenkeana* ("horco-quebracho"); on shady cooler slopes is a forest of *Lithrea ternifolia* with *Fagara coco* and other species; and above 1800 m on shallow soils is grassland dominated by *Festuca hieronymi* and diverse species of *Stipa*. About 190 species are found in this high-elevation zone, and these areas are rich in endemics.

In the Paraguayan uplands (cerros León and Cabrera), three zones of vegetation are recognized as well. On slopes with sufficient water there is deciduous forest dominated by *Anadenanthera colubrina*, with *Pterogyne nitens*, *Amburana cearensis* and *Aspidosperma pyriformis*; on hilltops with shallow soils is shrubby cerrado-like savanna with *Pseudobombax campestre*; and on higher elevation tablelands with shallow soils occurs species-rich grass savanna of *Chloris*, *Digitaria* and *Stipa* along with *Tabebuia aurea*.

Flora

The floristic inventory of the Gran Chaco is incomplete (Zellweger *et al.* 1990), but the region is estimated to have 1000–1200 species. The most important families and genera are Leguminosae (*Prosopis*, *Acacia*, *Caesalpinia*, *Cercidium*, *Geoffroea*); Anacardiaceae (*Schinopsis*, *Lithrea*); Apocynaceae (*Aspidosperma*); Palmae (*Copernicia*, *Trithrinax*); Zygophyllaceae (*Bulnesia*); Rhamnaceae (*Ziziphus*); Santalaceae (*Jodina*, *Acanthosyris*); Cactaceae (*Opuntia*, *Cereus*, *Stetsonia*, *Pereskia*, *Quiabentia*); Bromeliaceae (*Dyckia*, *Bromelia*, *Puya*); and Gramineae (e.g. *Elionurus*, *Paspalum*, *Chloris*, *Trichloris*) (Cabido, Acosta and Díaz 1990).

The region has floristic affinities with the neighbouring phytogeographic provinces – espinal, monte, prepuna, caatinga and campos. Certain associations, such as those in savannas or sandy or upland areas, share floristic elements with the caatinga and cerrado (cf. Prado 1993) (see CPD Sites SA19 and SA21).

Among the rare and/or endemic plants are *Stetsonia*, *Lophocarpinia*, *Mimozyganthus*, *Stenodrepanum*, *Trachypteris pinnata*, *Berberis hieronymi*, *Arenaria achalensis*,

Cnicothamnus azafran, *Soliva triniifolia* and *Jatropha matacensis* (Prado 1993; Cabido, Acosta and Díaz 1990).

Useful plants

Large areas of the Gran Chaco have been intensively utilized. Many species that are commercially exploited may be at risk; appropriate multiple-use management and conservation efforts are needed. Chemical compounds that are useful for medicines, dyes and other industrial purposes are produced by many of the region's native species.

The genus *Schinopsis* ("quebracho"), represented by six species (out of a total of nine), is extensively exploited for timber and as an international source of tannin. Twenty of the 44 species of *Prosopis* ("algarrobo") are native to the region, and are used in furniture-making, as forage, and for industrial and medicinal purposes. The wood of *Bulnesia sarmientoi* ("palo santo"), from the Central Chaco, is the source of oil of guaiac, a fragrance used in soaps. Several species of Bromeliaceae are used by indigenous people as a source of fibre (not in the market economy). *Trithrinax campestris* leaves are processed industrially for fibre. Trunks of *Copernicia alba* ("caranday") are utilized for telephone and electric poles and in construction, and carnuba wax is extracted from the palm's leaves. Epiphytic Bromeliaceae and Orchidaceae are important in the horticultural trade, including the export market. Many species of bromeliads and orchids in the Gran Chaco may be scientifically undescribed. Promising agroforestry species include *Caesalpinia paraguariensis*, *Prosopis alba* var. *panta* and *P. chilensis* which are used as fodder; wild species of *Ananas* (pineapple) and *Oryza* (rice) are among the germplasm resources.

Social and environmental values

Population density decreases westward and northward. In Argentina, indigenous Mataco inhabit Salta and Tobas in the provinces of Formosa and Chaco. In Paraguay, 13 ethnic groups inhabit the Central Chaco, including several which until recently had little involvement with Western civilization (Williams 1982; FAO and UNEP 1985).

The highland areas are the source of water for many of the streams that supply the arid areas of the Gran Chaco. Adequate management of these watersheds is important to sustain the people, plants and animals of the more arid areas and for hydroelectric power.

The diversity of fauna is high, and subsistence and sport hunting are popular. The word "chaco" comes from the Quechua language and means hunting land. As recently as 1975, a new species of peccary was recognized, the taguá or giant peccary. Other native mammals include giant anteater, ocelot, Geoffroy's cat, pampas cat, jaguarundi, puma and jaguar. Further research on the ecology of the Gran Chaco and adaptations of species to the environmental stresses are needed. Some plants have evolved special adaptations to cope with the rigorous environmental conditions and other factors. The northern areas (in Bolivia and Paraguay) generally require greater scientific research.

Tourism is growing in the region, including ecotourism. Around Cerro Colorado in Córdoba, Argentina, 35,000 ancient rock paintings have been discovered.

Threats

The ecosystems of the Gran Chaco are relatively fragile and susceptible to erosion by wind and water. In some areas erosion has caused severe damage, for example north-west of Tarija in Bolivia.

Logging combined with livestock grazing is reducing natural diversity, decreasing populations and the genetic potential of species. For example, over a period of 60 years, the exploitation of *Schinopsis balansae* in Argentina decimated 104,000 km² (Bunstorf 1971). The massive felling of trees together with grazing create shrubland, which has much thorny *Prosopis ruscifolia* in humid areas and *Acacia furcatispina* in arid areas. This process can lead to desertification. People who depend on the forests for subsistence are losing a source of food (e.g. fruit, meat, honey), shelter and other amenities (Morello *et al.* 1978). Land clearing and over-hunting have become problems along the paved Trans-Chaco highway in Paraguay (Taber 1989).

Road construction for oil and gas exploration in Paraguay and Bolivia and for selective logging in Argentina affects the fauna and vegetation. Stones used in road construction are quarried from the scarce lower hills on the Chaco plain, causing further habitat loss.

Conservation

Some environmental education is carried out in the region, and just over 1% of the Gran Chaco is represented in protected areas (see Map 58). In Argentina, where there was the greatest portion of the Chaco, c. 720 km² are in three protected areas: in the humid zone, Chaco National Park (c. 150 km²), with *Schinopsis balansae*; in the subhumid zone, Río Pilcomayo NP (c. 470 km² near Asunción), with *Copernicia alba*; and in the subhumid and semi-arid zones, Formosa National Nature Reserve (100 km² of south-western Formosa Province).

In Paraguay, Chaco ecosystems are better represented in protected areas, with four National Parks covering c. 12,000 km² (DPNVS 1993): in the humid zone, Ypoá NP (1000 km²), and Tinfunqué NP (c. 2800 km²) with *Copernicia alba*; in the subhumid and semi-arid zones, Defensores del Chaco NP (7800 km²), which includes Cerro León (c. 700 km²) and has *Schinopsis balansae* and *Aspidosperma quebracho-blanco*; and in the semi-arid zone, Teniente Agripino Enciso NP (400 km²), with *Aspidosperma quebracho-blanco* and *Chorisia insignis*. Bolivia has no protected areas within its portion of the Chaco (WCMC 1992).

New protected areas for the Gran Chaco have been recommended in all three countries (FAO and UNEP 1985) (Map 58). In northern Argentina, three recommended National Nature Reserves would encompass 2950–3150 km²: Pirámide El Triunfo, Pozo del Cimarrón and El Pintado. In Paraguay there are recommendations (DPNVS 1993) for five National Parks totalling 14,600 km², two Ecological Reserves (3500 km²) and a Scientific Reserve (1500 km²), and in the west of Nueva Asunción Department, a sand-dune area (Las Dunas) also has been recommended for conservation. In Bolivia, 300 km² have been recommended as a Nature Reserve in the Villamontes-Sachapera area of Tarija Department, which would protect highland and semi-arid zones. In Santa Cruz Department, several conservation units have been recommended (see Data Sheet for CPD Site SA23).

The ecosystems of the sand dunes on the frontier of Bolivia and Paraguay warrant more attention. Several sites within the Gran Chaco merit attention that have not yet been recommended for protected status. Forests of *Prosopis* (five species and hybrids) in floodplain areas of the Pilcomayo River are at risk because of the recent popularity of algarrobo wood to manufacture furniture for use locally and export. In the arid zone in north-western Córdoba Province (Argentina), a Nature Reserve of 20,000 km² would protect a very high diversity of *Prosopis* species. Pampa de Achala, an extensive high plateau in Córdoba Province, has tremendous plant diversity and endemics. Cerros San Miguel (839 m high, c. 10 km²), Cabrera (720 m high, c. 7 km²) and Caimán in Bolivia and north-western Paraguay deserve consideration for their plant endemism and status as ancient refugia and for the maintenance of watersheds.

References

Bunstorf, J. (1971). Tanningewinnung und Landerschliessung in argentinischen Gran Chaco. *Geogr. Zeitschr.* 59: 117–204.

Cabido, M., Acosta, A. and Díaz, S. (1990). The vascular flora and vegetation of granitic outcrops in the upper Córdoba mountains, Argentina. *Phytocoenologia* 19: 267–281.

Cabrera, A.L. (1976). *Regiones fitogeográficas argentinas.* In Parodi, L.R. (ed.), *Enciclopedia argentina de agricultura y jardinería,* 2nd edition. Vol. 2(1). Editorial Acmé, Buenos Aires. Pp. 1–85.

DPNVS (1993). *SINASIP. Plan estratégico del sistema nacional de áreas silvestres protegidas.* Dirección de Parques Nacionales y Vida Silvestre (DPNVS) and Fundación Moisés Bertoni para la Conservación de la Naturaleza, Asunción, Paraguay.

FAO and UNEP (1985). *Un sistema de áreas silvestres protegidas para el Gran Chaco.* Proyecto FAO y PNUMA FP 6105-85 – 01, Documento Técnico No. 1. FAO, Santiago, Chile. 159 pp.

FAO and UNESCO (1971). *Soil map of the world 1:5,000,000.* Vol. IV, *South America.* UNESCO, Paris. 193 pp.

Herzog, T. (1923). Die Pflanzenwelt der bolivischen Anden und ihres östlichen Vorlandes. In Engler, A. and Drude, O. (eds), *Die Vegetation der Erde.* Vol. 15. Engelmann, Leipzig. Pp. 84–105.

Morello, J. *et al.* (1978). *Estudio de factibilidad para la creación de una reserva de ecosistemas en la Provincia del Chaco argentino.* SISAGRO, Buenos Aires.

Prado, D.E. (1993). What is the Gran Chaco vegetation in South America? I. A review. Contribution to the study of flora and vegetation of the Chaco. V. *Candollea* 48: 145–172.

Ramella, L. and Spichiger, R. (1989). Interpretación preliminar del medio físico y de la vegetación del Chaco Boreal. Contribución al estudio de la flora y de la vegetación del Chaco. I. *Candollea* 44: 639–680.

Schwerdtfeger, W. (ed.) (1976). *World survey of climatology.* Vol. 12: *Climates of Central and South America.* Elsevier Scientific Publishing, Amsterdam, The Netherlands. 532 pp.

Spichiger, R., Ramella, L., Palese, R. and Mereles, F. (1991). Proposición de leyenda para la cartografía de las formaciones vegetales del Chaco paraguayo. Contribución al estudio de la flora y de la vegetación del Chaco. III. *Candollea* 46: 541–564.

Taber, A.B. (1989). Pig from green hell. *Animal Kingdom* 92(4): 20–27.

Vargas, G. (1988). Chaco sudamericano: regiones naturales. In Alessandría, E., Bernardón, A., Díaz, R., Elisetch, M. and Virasoro, J. (eds), *X Reunión Grupo Campos y Chaco: memoria.* FAO, UNESCO-MAB, INTA (Instituto Nacional de Tecnología Agropecuaria) and UNC (Universidad Nacional de Córdoba). Córdoba, Argentina. Pp. 16–20.

WCMC (1992). *Protected areas of the world. A review of national systems.* Vol. 4. *Nearctic and Neotropical.* IUCN, Gland, Switzerland and Cambridge, U.K. 459 pp.

Williams, J.H. (1982). Paraguay's unchanging Chaco. *Américas* 34(4): 14–19.

Zellweger, C., Palese, R., Perret, P., Ramella, L. and Spichiger, R. (1990). Concept and use of an integrated database system for the Chaco. Application to a preliminary checklist. Contribution to the study of the flora and vegetation of the Chaco. II. *Candollea* 45: 681–690.

Authors

This Data Sheet was written by Dra. Francisca M. Galera (Universidad Nacional de Córdoba, Facultad de Ciencias Exactas, Físicas y Naturales, Centro de Ecología y Recursos Naturales Renovables, Avda. Vélez Sarsfield 299, 5000 Córdoba, Argentina) and Dr Lorenzo Ramella (Conservatoire et Jardin botaniques de la Ville de Genève, Case postale 60, CH-1292 Chambésy-Geneva, Switzerland).

SOUTH-EASTERN SANTA CRUZ
Bolivia

Location: Centrally in South America, in south-eastern Bolivia in Department of Santa Cruz, between about latitudes 17°–20°S and longitudes 62°–58°W in provinces of Velasco, Angel Sandoval, Chiquitos, Cordillera and Germán Busch.

Area: c. 70,000 km².

Altitude: 350–1290 m.

Vegetation: Transitions between phytogeographic provinces of Cerrado of central Brazil, Gran Pantanal and Gran Chaco. Mosaic of dry forests, savannas (cerrado and campo rupestre), savanna wetlands (valley-side campo and seasonally inundated savanna) and thorn scrub.

Flora: 2000–2500 species estimated. Overall biological diversity is high; each vegetation type has a distinct flora with moderate diversity. Endemism is probably a significant component of campo-rupestre savannas, which are disjunct from similar formations of central and eastern Brazil.

Useful plants: For lumber, particularly decay-resistant woods; palm telephone poles; forage from numerous grass species. Amerindian uses largely unstudied.

Other values: Watershed protection for Paraguay River and Gran Pantanal; pre-Columbian petroglyphs; Amerindian lands; birds and other fauna; tourism.

Threats: Construction of natural gas pipeline and highway; unsustainable logging; poorly located mechanized soybean cultivation and cattle-ranching; mining probable.

Conservation: National Park, Historical Park; proposals for 3 Biological Reserves, and Forest Reserves.

Geography

South-eastern Santa Cruz is near the centre of the South American continent. There are several distinct landforms. From the region's north-west to the mid-eastern Serranía de Sunsas extends the ancient Brazilian Shield, where hilly terrain and highly weathered soils have formed from a complex of various granitic and metamorphic rocks (Oblitas and Brockmann 1978). Next southward, just to the north of the Serranía de Chiquitos and Serranía de Santiago, lies a narrow plain with a complex of soil types from sandy to loamy clays. Its western portion drains into the San Julián River, a tributary of the Amazon, whereas its eastern valley is drained by the Tucavaca River emptying into the Otuquis wetlands and so the Paraguay River. The southern portion of the region is a rolling plain with extremely sandy soils of the Gran Chaco, bordered by the seasonally inundated alluvial plain of the Upper Paraguay River and eastward by the Gran Pantanal.

In the centre of the region, several small mountain chains with almost east-west orientation provide striking contrast to the surrounding plains. The Chiquitos and Santiago ranges together extend nearly 300 km and are composed of Ordovician, Devonian and Cretaceous sandstones. Mostly they are flat-topped or table mountains having a nearly vertical southern escarpment and a steep (30°–40°) northern slope. The Serranía de Santiago is the highest topographical feature in eastern Bolivia, with maximum elevation of 1290 m on El Portón.

The climate is typical of a tropical savanna region. The mean annual temperature is 25°C (in Roboré) and c. 1000 mm of precipitation occur yearly (Navarro 1992). A pronounced dry season coincides with the southern

winter. There are seasonal cold fronts ("surazos"), in June and July rarely associated with light frosts, which do not cause leaf drop in the evergreen tree species.

Vegetation

South-eastern Santa Cruz is situated in a transition zone between the Cerrado, the Gran Pantanal and the Gran Chaco regions of central South America. The vegetation is a complex mosaic of dry forests, savannas, savanna wetlands and thorn scrub correlated with local landforms (Map 59). The Serranía de Sunsas is the least studied of the areas; satellite images indicate it is covered by savanna ("cerrado") and dry forest. These rolling hills have a complicated geology, and forest is probably found on richer soils which have developed from the more easily weathered metamorphic formations. In other regions of Santa Cruz, the cerrado vegetation is found on hilltops and on the more acidic soils which have developed from granitic rocks (Killeen, Louman and Grimwood 1990).

The lower slopes of the Chiquitos and Santiago mountains are covered by forest vegetation similar in structure (and probably composition) to that on the Brazilian Shield (Navarro 1992). In the adjacent Tucavaca Valley, a distinctive closed-canopy dry forest is on the alluvial soils derived from the shield (Gentry 1993, 1994). In certain isolated locales of these areas, there is comparatively humid and evergreen forest, even with tree ferns (probably *Nephelea cuspidata*). The upper vegetation on the serranías varies. The western ridges (near San José de Chiquitos) are lower (to 600–800 m), more heavily eroded (not flat-topped) and covered

416

to their crests with low forest or cerrado. In contrast, the eastern flat-topped mesetas (to 900–1157 m) of the Serranía de Santiago support open savanna reminiscent of the campo rupestre of eastern Brazil (e.g. see CPD Site SA20).

Just 50 km south of the Serranía de Chiquitos, the Gran Chaco vegetation reaches its northern (non-Andean) limit: the shift from the dry forest to thorn scrub is relatively abrupt (Navarro 1992). Soil type, as well as climate, affect the transition. South of the Serranía de Chiquitos the soils are much more sandy and the vegetation rapidly changes in structure and composition to the xerophytic deciduous forest (thorn scrub) characteristic of the Gran Chaco (CPD Site SA22).

Eastward, the Tucavaca River drains into the Bañados de Otuquis which are contiguous to the east with the Gran Pantanal wetlands. In this Bolivian portion one finds a rich mosaic of vegetation types according to different levels of inundation and/or the duration of seasonal dryness. Characteristic for much of this wetland complex is the palm *Copernicia alba*. Palm savannas ("palmares") also occur north-west of San José de Chiquitos where meandering watercourses flow into Lake Concepción.

Flora

Little is known about the flora of this region. For bordering regions only two incomplete checklists and several descriptions of vegetation have been published (Cárdenas 1951; Prance and Schaller 1982; Killeen and Nee 1991; Killeen, Louman and Grimwood 1990; Spichiger *et al.* 1991; Ramella and Spichiger 1989). Botanical collecting expeditions, such as those organized by A. Orbigny (1834–1847), T. Herzog (1907), M. Cárdenas (1950), D. Daly (1983) and Conservation International's Rapid Assessment Program (1991) (see Parker *et al.* 1993) have resulted in many interesting specimens including recently discovered probably undescribed species, especially from the campo-rupestre savannas of the Serranía de Santiago.

Those open savannas are dominated by Gramineae; important also are herbaceous or subshrub species in Cyperaceae, Leguminosae, Compositae, Polygalaceae, Melastomataceae and Rubiaceae. Although not abundant, species in Velloziaceae, Xyridaceae and Eriocaulaceae occur. The grass flora is better investigated and indicates strong similarities with campo rupestre, having some of its characteristic species: *Axonopus brasiliensis*, *Paspalum gardnerianum*, *P. pectinatum*, *P. polyphyllum* and *Sporobolus sprengelii* (Killeen 1990).

The floras of the other vegetation types also are poorly known, but are assumed to be similar to the better studied floras in Bolivia and Brazil (Cerrado, dry forest, Gran Pantanal) and Paraguay and Argentina (Gran Chaco) (see CPD Sites SA21 and SA22). Yet Gentry (1993, 1994) found the dry forest in the Tucavaca Valley to be one of the most diverse in the neotropics (97 species over 2.5 cm dbh in 0.1 ha). Several taxa believed to be only in southern Brazil were relatively abundant; endemics also were important.

Due to the lack of even a preliminary checklist for south-eastern Santa Cruz, the amount of endemism is unknown. The only published endemics are *Frailea chiquitana* (Cactaceae) and *Axonopus herzogii* (Gramineae), but likely new species have been found in *Norantea*

(Marcgraviaceae), *Palicourea* (Rubiaceae) and *Andropogon* and *Anthaenantiopsis* (Gramineae).

Useful plants

Much of the region is sparsely populated by indigenous peoples (Chiquitano to the north, Ayoreo near the Serranía de Santiago, and Guaraní to the south-west), who have a rich heritage of using native plants. This usage is largely undocumented and their Westernization is proceeding rapidly. Much of this knowledge will be lost with the adult generation unless ethnobotanical studies are undertaken.

Copernicia alba palms are used to make telephone poles. Plants of the region in the national economy are timber and forage resources. Timber harvest had been selective and non-intensive (e.g. for the decay-resistant woods of *Astronium* spp., *Schinopsis* spp. and *Tabebuia* spp.), while grazing and browsing of natural vegetation by cattle have been extensive.

Social and environmental values

Indigenous peoples still are surviving with their traditional ways. The general region has broad potential attractiveness, offering the large Lake Concepción to its north, the scenic beauty of different serranías along with the regional highpoint,

CPD Site SA23: South-eastern Santa Cruz. *Chorisia* spp. ("toborochi") occur in dry and xerophytic forests. Photo: Tony Morrison/South American Pictures.

and the diversity of plants and animals, with cultural interests of pre-Columbian petroglyphs and recently restored Jesuit missions. Bird watchers are quite drawn to the mosaic Bañados de Otuquis and adjacent Gran Pantanal wetlands. An inexpensive railroad makes the region accessible to tourists from neighbouring regions. The vegetation and wetlands also provide catchment protection for the Upper Paraguay River and the Gran Pantanal.

This region covers a vast area within which is the southern end of the East Bolivian lowlands Endemic Bird Area (EBA B36). This is one of the least ornithologically known areas in Bolivia. The only restricted-range species of the EBA that definitely occurs in this region is the buff-bellied hermit (*Phaethornis subochraceus*).

Economic assessment

The antiquated railroad traversing the region roughly parallels the serranías (Map 59). This transportation system makes the region accessible to many Bolivian nationals and others who depend on low-cost methods of access. An imminent modern highway to Brazil will stimulate overall development.

Shifting agriculture by indigenous groups continues to be limited to restricted areas. Sustainable development would be similar to traditional land-use patterns that have occurred over the 250 years since settlement by Jesuit missionaries. Although cattle-ranching is suitable for many of the forest soils in South-eastern Santa Cruz, rotational grazing needs to be encouraged to improve the productivity of native species. Native vegetation could be supplemented by limited planting of cultivated forage grasses. Moreover, given the last decade's more intensive logging practices, modern forestry methods need to be employed to ensure the long-term productivity of the hardwoods.

Threats

In Santa Cruz Department (excluding the Gran Chaco vegetation), 17% of the forested area had been converted to cultivation and cattle-ranching by 1991: 13,759 km²; the yearly deforestation is 329 km² (based on 1985 and 1990 satellite images). Almost all forest is gone near the capital Santa Cruz de la Sierra in the Andean piedmont, but less than 1% had been deforested in South-eastern Santa Cruz (in the 1990 images).

The principal threat to the biological diversity of the region is expansion of the booming mechanized agricultural economy from central Santa Cruz Department. The local government and international agencies are promoting cultivation of soybeans (*Glycine soja*) on the central plain of Santa Cruz, where fertile alluvial soils can be used over the long term. Unfortunately, indiscriminate general expansion of mechanized agriculture is resulting in environmental degradation due to inappropriate land use in adjacent areas. The perception that the alluvial soils of the Tucavaca Valley are fertile has brought that area to the attention of investors. However, most of its soils are unsuited for mechanized agriculture because they are either too sandy or have subsoil clay hardpans. Nevertheless, speculators are purchasing land or obtaining governmental land at low cost through the Bolivian land-reform system.

Agricultural development in Santa Cruz also occurs from intensive cattle-raising operations using cultivated grasses and legumes. These operations can be viable over the long term with use of the appropriate soil types, and proper management techniques (Killeen 1991). Operations that attempt soybean cultivation might change eventually to cattle production, but probably after substantial damage has been done to the soil resources.

Two separate, major development projects are driving agricultural expansion eastward from the central plain into this region: construction of a modern highway to link Bolivia and Brazil, and construction of a natural gas pipeline between the Bolivian gas fields and the industrialized states of Brazil, scheduled for 1993–1994. The pipeline project will require an extensive road network for its construction, operation and maintenance. Unfortunately, the highway has been proposed to be approximately 40 km north of the railroad on the opposite, northern side of the Serranía de Santiago where it would extend through the Tucavaca Valley; conversely, the pipeline system would be approximately 40 km south of the railroad. If financing for these two separate projects is approved without broadened environmental consideration, significant damage will occur needlessly.

Important mining initiatives probably will take place by 1998 scattered throughout the Bolivian portion of the Brazilian Shield (C. Brockmann, pers. comm.). Semi-precious stones are widely found; north-west of the Serranía de Sunsas modern bulk gold mining is in operation; and world-class iron-ore deposits have been found near Santo Corazón and 25 km south of Puerto Suárez (at Mutún). Claims have been filed for rights to the subsoil minerals in the geologically complex Serranía de Sunsas, which has ultramafic rock formations: platinum, copper, manganese, nickel and other minerals are likely to be discovered in commercially suitable deposits.

Conservation

Separate environmental impact studies of the highway and the pipeline projects were made, without considering the synergistic effects of the two works, or the construction of them in the same region at the same time. Combining the projects and placing them within the existing developed corridor surrounding the railroad right-of-way would appreciably decrease their negative environmental effects.

Bolivia is making genuine efforts to conserve a substantial portion of its biological resources and recently formed a Sistema Nacional de Areas Protegidas (SNAP). Recent actions provide confidence that protected areas will be established in South-eastern Santa Cruz. Three basic areas (updated below) had been recommended for designation as conservation units by Navarro (1992) and the Gran Chaco is one of nine priority areas in Bolivia identified by the Global Environment Facility (GEF) for conservation. Although all the recommendations of Navarro (1992) would bring some 25% of Santa Cruz Department into SNAP (the National System of Protected Areas), land uses would differ – within many of his basic areas there are even developed localities: e.g. places settled with cattlemen or farmers, or the railroad corridor extending across the recommended Serranía de Santiago-Otuquis National Park.

Environmental organizations are working with scientists and government authorities to determine exact boundaries and additional areas for conservation, and the appropriate protective status for each area. Agricultural and timber interests work toward limiting the size of proposed reserves. Santa Cruz Department has the double distinction of leading the country in deforestation, but as well in the creation of functioning National Parks and Wildlife Reserves.

In the north-east (see Map 59 area 1), a San Matías Biological Reserve would incorporate an altitudinal transect extending from cerrado savanna and dry forest to inundated marsh and forest of the Gran Pantanal. In Brazil farther east are the Taiamã Ecological Station (112 km²) and the Mato Grosso Pantanal National Park (1350 km²) (IBGE and IBDF 1988). In south-east Santa Cruz (Map 59 areas 2 and 3), Santiago de Chiquitos and Otuquis-Pantanal Biological Reserves would contain portions of the Serranía de Sunsas, the Tucavaca Valley, the Serranía de Santiago and the Bañados de Otuquis. In the south-west (Map 59 area 4), the recently established Kaa-Iya National Park

includes the northern part of the Gran Chaco, the Serranía de Chiquitos and the Bañados de Izozog. Near this park and San José de Chiquitos is Santa Cruz la Vieja Historic National Park (c. 170 km²) (Map 59 area 5), which was the region's only protected area (although unmanaged), with dry-forest vegetation and archaeological ruins from the 1600s.

Bolivia contains more intact dry forest (sometimes called semi-deciduous forest) than any other Latin American country (Parker *et al.* 1993). However, no protected area in Bolivia encompassed a substantial portion, and insufficient dry forest has been planned for protection. Even though a substantial portion has become private holdings, much of the c. 40,000 km² of dry-forest formation in eastern Santa Cruz Department is still intact, and about half of this poorly studied formation is in South-eastern Santa Cruz. Some 2000–3000 km² of dry forest have been recommended for protection, mainly in the San Matías Biological Reserve. Most of the land between the three recommended protected areas and the new National Park has been recommended for incorporation into a series of

MAP 59. SOUTH-EASTERN SANTA CRUZ, BOLIVIA (CPD SITE SA23)

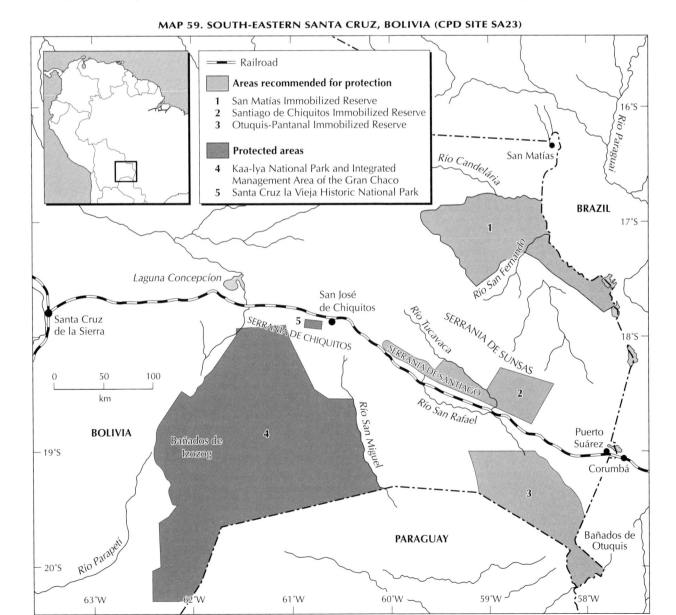

Forest Reserves (Navarro 1992). However, no governmental action has been taken and these forests are being exploited without a sustainable management plan.

References

Cárdenas, M. (1951). Un viaje botánico por la Provincia Chiquitos del oriente boliviano. *Revista Agric. (Cochabamba)* 7(6): 3–17.

Gentry, A.H. (1993). Dry forest vegetation and phytogeography in the Tucavaca Valley. In Parker III, T.A., Gentry, A.H., Foster, R.B., Emmons, L.H. and Remsen Jr., J.V., *The lowland dry forests of Santa Cruz, Bolivia*. RAP Working Pap. 4, CI, Washington, D.C. Pp. 40–42.

Gentry, A.H. (1994). A new South American vegetation type: the conservational significance of the dry forest of Santa Cruz, Bolivia. Manuscript.

IBGE and IBDF (1988). *Mapa de Vegetação do Brasil*. Scale 1:5,000,000. Fundação Instituto Brasileiro de Geografia e Estatística (IBGE) and Instituto Brasileiro de Desenvolvimento Florestal (IBDF), Rio de Janeiro.

Killeen, T.J. (1990). The grasses of Chiquitanía, Santa Cruz, Bolivia. *Ann. Missouri Bot. Gard.* 77: 125–201.

Killeen, T.J. (1991). Range management and land-use practices in Chiquitanía, Santa Cruz, Bolivia. *Rangelands* 13: 73–77.

Killeen, T.J., Louman, B.T. and Grimwood, T. (1990). La ecología paisajística de la región de Concepción y Lomerio en la Provincia Ñuflo de Chávez, Santa Cruz, Bolivia. *Ecología en Bolivia* 16: 1–46.

Killeen, T.J. and Nee, M. (1991). Catálogo de las plantas sabaneras de Concepción, Depto. Santa Cruz, Bolivia. *Ecología en Bolivia* 17: 53–71.

Navarro-S., G. (1992). *Estudio de parques nacionales y otras áreas protegidas (borrador final)*. Proyecto de Protección de Los Recursos Naturales en el Departmento de Santa Cruz (Componente Proyecto Tierras Bajas). Corp. Regional de Desarrollo de Santa Cruz (CORDECRUZ)-KFW-Consorcio IP/SCD/KWC, Santa Cruz. 96 pp.

Oblitas-G., J. and Brockmann-H., C.E. (1978). *Mapa geológico de Bolivia*. Servicio Geológico de Bolivia (GEOBOL) and Yacimientos Petrolíferos Fiscales de Bolivia (YPFB), La Paz.

Parker III, T.A., Gentry, A.H., Foster, R.B., Emmons, L.H. and Remsen Jr., J.V. (1993). *The lowland dry forests of Santa Cruz, Bolivia: a global conservation priority*. RAP (Rapid Assessment Program) Working Papers 4, Conservation International (CI), Washington, D.C. 104 pp.

Prance, G.T. and Schaller, G.B. (1982). Preliminary study on some vegetation types of the Pantanal, Mato Grosso, Brazil. *Brittonia* 34: 224–251.

Ramella, L. and Spichiger, R. (1989). Interpretación preliminar del medio físico y de la vegetación del Chaco Boreal. Contribución al estudio de la flora y de la vegetación del Chaco. I. *Candollea* 44: 640–680.

Spichiger, R., Ramella, L., Palese, R. and Mereles, F. (1991). Proposición de leyenda para la cartografía de las formaciones vegetales del Chaco paraguayo. Contribución al estudio de la flora y de la vegetación del Chaco. III. *Candollea* 46: 542–564.

Author

This Data Sheet was written by Dr Timothy J. Killeen (Universidad Mayor de San Andrés, Instituto de Ecología, Herbario Nacional de Bolivia, Casilla 20127, La Paz, Bolivia and Missouri Botanical Garden, P.O. Box 299, St. Louis, Missouri 63166-0299, U.S.A.).

INTERIOR DRY AND MESIC FORESTS: CPD SITE SA24

LLANOS DE MOJOS REGION
Bolivia

Location: North-eastern Bolivia, primarily in Beni Department, extending into Iturralde Province of La Paz Department and a little of Pando and Cochabamba departments, between latitudes 11°–16°S and longitudes 63°–69°W.

Area: c. 270,000 km²: savannas 150,000 km², forests 120,000 km².

Altitude: 130–235 m.

Vegetation: Mosaic of about eight different basic types of savannas and forests, many seasonally flooded; wetlands (marshes, swamps, lagoons).

Flora: c. 5000 species, 1500 in savannas; species highly adapted to changing hydrological, edaphic and climatic conditions; transitional zone with floristic elements of cerrado from the west, southern limit for Amazon forest and northern limit for Gran Chaco.

Useful plants: Forage grasses and legumes, timber species, *Hevea* rubber, Brazil nuts (*Bertholettia*), fruit trees, local cultivated strains of peanut (*Arachis*) and cassava (*Manihot*).

Other values: Potential genetic resources; abundance of many species of animals, including threatened species; several Amerindian peoples; ecotourism.

Threats: Overgrazing, fire, altered drainage, logging, road construction.

Conservation: Biosphere Reserve and Biological Station; Amerindian territories; Forestry Reserves; Regional (rangeland) Park, private reserves.

Geography

The Llanos de Mojos (Moxos) region of northern Bolivia is north-east of the Andes in the lowlands, mostly east of the Beni River, and west of the Serra dos Pacaás Novos and Chapada dos Parecis of Brazil (Rondônia and Mato Grosso). This region is the southernmost extension of the Amazon Basin, and the third largest complex of savannas in South America. The sediments of the Llanos are mostly of alluvial origin from the late Pleistocene and Quaternary (Campbell, Frailey and Arellano-L. 1985); beneath is a Tertiary marine molasse. The region belongs to a pericratonic basin underlain by the ancient Brazilian Shield, with various faults and diaclasations (Hanagarth and Sarmiento 1990). The plain has little local relief, but elevational change of a few meters may determine an area's water regime and biota.

Three main river systems drain the Llanos de Mojos, uniting northward to form the Madeira River of Brazil, which is the major south-western tributary of the Amazon. From Andean watercourses in western Bolivia derives the Beni River, while the Mamoré flows centrally from watercourses along the rest of these Andes; the Guaporé (or Iténez) drains from the low east bordering Brazil. The decline of these rivers on the plain is extremely slight, e.g. the Mamoré falls only 170 m during its course of c. 1500 km. Numerous meanders and abandoned channels exist.

The region of the Llanos de Mojos is transitional between the equatorial zone and the tropical summer-rain zone; it experiences high rainfall during the summer months, and dry periods of 2–3 months between June and August. The mean annual precipitation decreases eastward with greater distance from the Andes, e.g. from over 2000 mm in Rurrenabaque to 1300 mm in Magdalena. Nonetheless, any given locality has large fluctuations; e.g. at Estancia Espíritu (with 18 years of data), the annual precipitation varies from 1322 mm to 2454 mm. Very heavy rainfall for a few hours, which can reach more than 200 mm, is also characteristic in the Llanos (Hanagarth and Sarmiento 1990).

Various savannas and forests are inundated seasonally from overflow of several large Andean rivers, as well as accumulation of local rainwater. Due to outcroppings of the Brazilian Shield along the Madeira, Beni and Guaporé rivers, the waters slow; occasionally secondary tributaries even flow backward during the height of the rainy season (Beck 1983)! During the rainy season the rivers carry substantial loads of sediments and organic substrates. The floods cover large areas of the Llanos, and may remain for 5–7 months in some areas. Every 6–12 years, 80–90% of the region is inundated. This greater cyclical flooding may be an effect of the Pacific Ocean's El Niño current, but there has been no study correlating the phenomena.

The mean annual temperature is c. 26°C. Prevailing winds come from the north to north-east, but between May and September contrasting southern cold fronts ("surazos") frequently reach the Llanos de Mojos. The temperature can drop 10°C or more in a few hours and the cooler misty weather can last a day to about a week. The minimum temperature reported is 6°C in the central Llanos at Santa Ana de Yacuma.

Vegetation

The Llanos de Mojos region is an expansive mosaic of different vegetation types. Humid savannas predominate, interrupted and bordered by a variety of forest communities

(Ribera 1992; Foster 1989). Preliminary studies in several areas have revealed a high diversity of distinct savannas and forests, each type varying in species adapted to a particular habitat complex usually with a radically changing water regime – waterlogged to varyingly inundated or draining, and seasonal drought (Haase and Beck 1989; Beck 1984).

The savannas of the western to north-western portion of the Llanos de Mojos ("sabanas oligotrofas") (see Map 60) consist of grasslands dominated by bunchgrasses and termite mounds with scrub vegetation, due to the poor, generally acid soils. Characteristic species are *Leptocoryphium lanatum* and *Trachypogon plumosus* (Gramineae). One vegetation type ("sartenejal") within this association is characterized by a hummock-and-gully relief (varying by a few dm), with *Mesosetum penicillatum* (Gramineae), *Bulbostylis juncoides* (Cyperaceae) and shrubs such as *Macairea scabra* (Melastomataceae). Better draining areas have numerous large termite mounds with several species of shrubby Melastomataceae and small trees, e.g. *Xylopia aromatica* (Annonaceae). Limited areas of palm swamp occur, with *Mauritia flexuosa* and *Mauritiella aculeata*.

The central portion of the Llanos de Mojos, between the Beni and Mamoré rivers, generally is flooded much more than the north-western and northern portions. These floodplains ("sabanas inundables") have a high diversity of aquatic and marsh vegetation with good forage grasses, such as *Luziola*, *Hymenachne* and *Leersia*. Permanently somewhat wet areas consist of swamps dominated by sedges and *Thalia geniculata* (Marantaceae). Several enormous waterlogged swamps ("humedales") marked by *Cyperus giganteus* occur, e.g. surrounding Rogaguado Lake. In the drier "semialturas" large areas with open stands of the palm *Copernicia alba* are found. East of the Mamoré River research is lacking, but open palm forests (or "sabanas de palmares") appear to predominate and there apparently are waterlogged swamps like the humedales of the Magdalena area with a variety of different species (Ribera 1992; cf. IBGE and IBDF 1988).

The savannas of the northern Llanos de Mojos ("sabanas del cerrado") between the Beni and Iténez/Guaporé or Mamoré rivers are less heavily inundated and have affinities to the cerrado vegetation of central Brazil (see CPD Site SA21). These grasslands have a lateritic hardpan and typical cerrado genera: *Kielmeyera* and *Caraipa* (Guttiferae), *Byrsonima* (Malpighiaceae) and *Qualea* (Vochysiaceae).

Based on broad geographic and ecological conditions, four main types of forest may be distinguished in the region,

MAP 60. VEGETATION TYPES IN THE LLANOS DE MOJOS REGION, BOLIVIA (CPD SITE SA24) (after Ribera 1992)

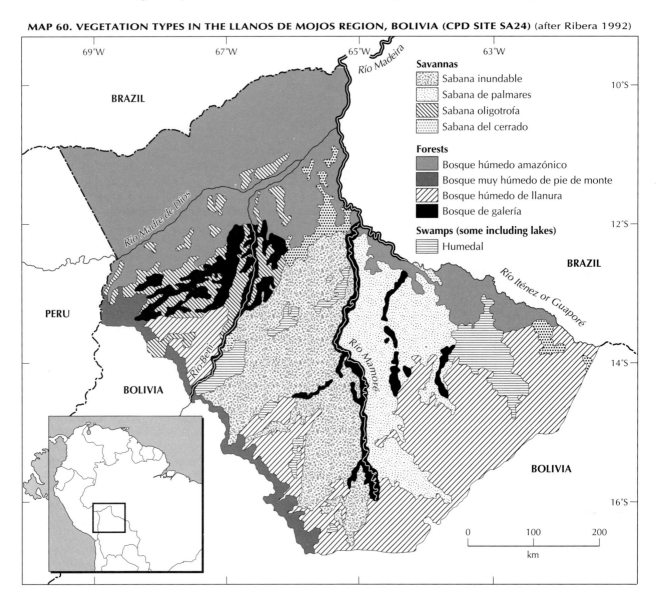

which meet or intermingle with the savannas (Map 60). In the far north (mainly in Pando Department), extending southward across the Madre de Dios and lower Beni rivers and narrowly along the Iténez/Guaporé, is the southernmost portion of the Amazon forest ("bosque húmedo amazónico"), with typical species including *Hevea brasiliensis* ("siringa") and *Bertholletia excelsa* ("castaña"). Along the Andean foothills occurs more humid rain forest ("bosque muy húmedo de pie de monte") (e.g. see Madidi-Apolo Data Sheet, CPD Site SA36), which contrasts strikingly with the oligotrophic savannas (and other forest types) of the Madidi River to Ixiamas area west of the Beni River (cf. Parker and Bailey 1991).

On relatively recently formed dikes along active rivers and their abandoned meanders occurs gallery forest ("bosque de galería"). A more diverse forest considered to be older gallery forest is found in large to small strips on long-abandoned river courses, and forest islands covering a few to several thousand ha are found throughout on higher ground. Perhaps some forest strips became segmented into islands by fires (mostly human-set). Scattered through the Llanos on small patches of sodic soils are thorny forest islands with *Machaerium hirtum* (Leguminosae).

The southern portion of the region is bordered with a seasonal, but mostly evergreen forest ("bosque húmedo de llanura"). In the midwestern Mamoré Basin 50 km east of the Andean foothills, this forest is being studied in the Beni Biosphere Reserve at the Beni Biological Station (Forsyth 1989; Campos-Dudley 1992). Six to ten forest communities of relatively limited species diversity have been distinguished, from 5–15 m to 30–35 m in height, well-draining to varyingly inundated, and some even with deciduous trees predominant.

Flora

Only a few studies have been made on the flora of the Llanos de Mojos. There may be 1500 species in the savannas of the whole region, and 5000 species mostly in the forests in the Chimane Ecosystem area of the Beni Biosphere Reserve with adjacent units to the south (Foster 1989). Phytosociological research near the Yacuma River (a major midwestern branch of the Mamoré River) documented c. 500 species (Beck 1983) and some 150 km north-west across the Beni River c. 600 species were found, with less than 20% in common (Moraes and Beck 1992). These areas yielded 6–8 new species: in *Boelckea* – a new genus of Scrophulariaceae (Rossow 1992), and *Bellucia* (Melastomataceae), *Casimirella* (Icacinaceae), *Lantana* (Verbenaceae), *Peltodon* (Labiatae), *Wolffiella* (Lemnaceae) and *Andropogon* (Gramineae). The most abundant families were Gramineae, Cyperaceae and Leguminosae; Xyridaceae, Eriocaulaceae, Lentibulariaceae and Melastomataceae were also important.

The number of endemic species is unknown, but probably not high. The rain-forest flora is essentially Amazonian, where there is a continuing yield of intriguing discoveries, e.g. the shrub *Styloceras brokawii* (Buxaceae) (Gentry and Foster 1981). In areas with less inundation, there are elements of the Gran Chaco flora (CPD Site SA22). The cerrado species are more common in the drier (anthropogenic) savannas at forest margins and on the lateritic soils in the north. Most of the savanna species seem to be widespread, occurring in similar wetland complexes in central Brazil and northern South America.

Useful plants

Fruits of some trees and shrubs are locally consumed, particularly species of *Rheedia* (Guttiferae), *Salacia* (Hippocrateaceae), *Myrcia* (Myrtaceae) and various palms, which are also used for extraction of oils, palm hearts, fibres (e.g. "jatata", *Geonoma diversa* – Rioja 1992) and construction. Wild relatives of pineapple (*Ananas comosus*) are found and numerous cultivars of peanut (*Arachis hypogaea*) are potential genetic resources. Historical works describe a wide range of economically important species and cultivars (Eder c. 1772); some are still grown, such as sweet and bitter cassava (*Manihot*), red pepper (*Capsicum*), squash (*Cucurbita*), gourd (*Lagenaria*) and cotton (*Gossypium*).

The high diversity of grasses, sedges and legumes is a potential source for selection and breeding of improved forage varieties for tropical to subtropical climates, particularly in the grasses *Paspalum*, *Panicum* and *Schizachyrium*, as well as the legumes *Centrosema* and *Aeschynomene*. Some valuable timber species such as "tajibo" (*Tabebuia* spp., Bignoniaceae) and "palo María" (*Calophyllum brasiliense*, Guttiferae) are still relatively abundant; others have become scarce. The Llanos could be an important genetic provenance of the species that have reached their distributional limits here.

Social and environmental values

The region's transitional mosaic of different ecosystems with the large amount of edge habitats provides refuges for multitudes of wild animals, including 13 of Bolivia's 18 threatened tropical species. Among the richness of birds (c. 500 species, Parker 1989) are an incredible diversity of waders (over 18 species) and large populations of regionally distributed birds (e.g. the rhea, *Rhea americana*), as well as several species from the Northern Hemisphere known to over-winter or migrate through the Llanos. A preliminary bird survey of the savannas near Ixiamas documented 135 species, and this avifauna had little in common with the birds of the savannas 150 km to the south-east near Yacuma (Parker and Bailey 1991).

The main portion of the East Bolivian lowlands Endemic Bird Area (EBA B36) is in the Llanos de Mojos region. This EBA comprises just three bird species, blue-throated macaw (*Ara glaucogularis*), buff-bellied hermit (*Phaethornis subochraceus*) and unicoloured thrush (*Turdus haplochrous*). The macaw and thrush are listed as threatened. The macaw was known only from trade until a small breeding population was found in 1992, and there is just a handful of sightings of the thrush from four localities – in 1992 from just inside the Beni Biosphere Reserve.

The many mammals living in the region include eight monkey species, five cat species, maned wolf (*Chrysocyon brachyurus*), giant armadillo (*Priodontes maximus*), 40 bat species, South American tapir (*Tapirus terrestris*) and "boutu" or pink river-dolphin (*Inia geoffrensis boliviensis*). New species of fishes, amphibians, reptiles and butterflies are being described (Hanagarth, pers. comm.). Due to the abundance and visibility of many wild animals, including several crocodilians and the anaconda (*Eunectes murinus*), the Llanos de Mojos could be very interesting for ecotourists.

Economic assessment

The population density is c. 1.2 persons per km² in Beni Department. The indigenous population is not large and mostly in forests or near the edge of savannas. The most numerous ethnic peoples are Arawak (particularly the Baure, Trinitario and Ignaciano or Mojo), with some Tacana peoples, as well as Movima, Itonama and Chimane (cf. Añez 1992). Forest islands and gallery forests could be used for the cultivation of vanilla (*Vanilla planifolia* and others), for which Beni Department was once renowned. According to the 1992 national census (*Presencia* 1992), just 30% of the 250,000 residents of the Beni reside in the countryside and the actually rural population had decreased.

Nevertheless, the region is a major centre for beef production: 1–2 million cattle graze the native forage species. The impact of this activity seems to be patchy, as extensive areas are not populated by the cattle. Cultivated grasses facilitate the cattle-management operations and provide green forage during critical months of the dry season. Good possibilities exist to develop sustainable management for harvest of wild animals such as rhea, capybara (*Hydrochaeris hydrochaeris*), "lagarto" or yacaré (*Caiman crocodilus yacare*) and "caimán negro" or black caiman (*Melanosuchus niger*). Cattle and horses can be near these wild animals without substantially affecting their behaviour or reproductive biology.

Threats

Some areas are influenced by road construction, which makes them accessible for colonization (Solomon 1989; Redford and Stearman 1989; Campos-Dudley 1992), and also may change the natural water flow and ecology of more distant plant communities. Large areas of forested land are being cleared for conversion to pasture. Exploitation for timber and fuelwood is constantly increasing, and may pose major threats to forest islands and gallery forests. The most valuable timber species, *Swietenia macrophylla* ("mara", bigleaf mahogany) (Forsyth 1989) and *Cedrela odorata* ("cedro colorado", tropical red-cedar) (Meliaceae), have been over-exploited and are becoming rare wherever accessible.

Major threats to the flora also result from overgrazing and abuse of fire as a management tool by the cattlemen. The overgrazing tends to occur near settlements. The fires sometimes get out of control and enter areas that should not be burned. Fire can eventually cause the replacement of forest by savanna. Nonetheless, plant diversity may be higher in moderately grazed and/or burned savannas. Currently, there is almost no pressure to convert natural grasslands into artificial pastures.

Conservation

There are virtually no managed protected areas for the Llanos de Mojos rich natural diversity. The ranches Elsner with Espíritu and San Rafael and the Estancia El Dorado, together some 2700 km² of wildlife reserves, are privately managed; their long-term status is not assured. The Rogaguado Lake area has been proposed as a National Reserve for marsh deer (*Blastocerus dichotomus*), which might need over 200 km² (Montes de Oca 1989). A Decreto Supremo in 1961 declared all lakes in Beni (and Pando) departments to be National Reserves (Suárez-Morales 1986), but management plans have not been elaborated or implemented.

Beni Biosphere Reserve (c. 14°30'–14°45'S and 66°00'–66°45'W), recognized in 1986, is the only functioning Biological Reserve because of the Beni Biological Station, established in 1982 (CI 1988). The 1350 km² reserve conserves 90% forest vegetation, between the Maniqui (Nuevo) and Curiraba rivers. Adjacent areas of c. 16,190 km² are committed to compatible uses (Amerindian territories; permanent production forests; Yacuma Regional Park, which is mostly private rangelands), coordinated in the Chimane Ecosystem Programme (Campos-Dudley 1992; Forsyth 1989; Redford and Stearman 1989; Walsh 1987). It is critical to determine and preserve representative savanna habitats west and north-east near this reserve, and in the other areas of the Llanos de Mojos. Across the Guaporé River to the east in Brazil (upriver from the Baures River), the Guaporé Biological Reserve near the Branco River includes 6000 km², half cerrado (IBGE and IBDF 1988).

References

Añez, J. (1992). The Chimane experience in selling jatata. In Plotkin, M. and Famolare, L. (eds), *Sustainable harvest and marketing of rain forest products.* Island Press, Washington, D.C. Pp. 197–198.

Beck, S.G. (1983). Vegetationsökologische Grundlagen der Viehwirtschaft in den Überschwemmungs-Savannen des Río Yacuma (Departamento Beni, Bolivien). *Dissertationes Bot.* 80: 1–186.

Beck, S.G. (1984). Comunidades vegetales de las sabanas inundadizas en el noreste de Bolivia. *Phytocoenologia* 12: 321–350.

Campbell Jr., K.E., Frailey, D. and Arellano-L., J. (1985). The geology of the Río Beni: further evidence for Holocene flooding in Amazonia. *Contr. Sci. Nat. Hist. Mus. Los Angeles Co.* 364: 1–18.

Campos-Dudley, L. (1992). Beni: surviving the crosswinds of conservation. *Américas* 44(3): 6–15.

CI (Conservation International) (1988). Faces of the Beni: a visit to the biosphere reserve. *Tropicus* 5(winter): 4–5.

Eder, F.J. (c. 1772/1985). *Breve descripción de las reducciones de Mojos.* Traducción y edición de J.M. Barnadas, Cochabamba. 424 pp.

Forsyth, A. (1989). The Beni: impressions from the field. *Orion* (summer): 30–39.

Foster, R.B. (1989). Vegetation of the Beni. *Orion* (summer): 39.

Gentry, A.H. and Foster, R.B. (1981). A new *Styloceras* (Buxaceae) from Peru: a phytogeographic missing link. *Ann. Missouri Bot. Gard.* 68: 122–124.

Haase, R. and Beck, S.G. (1989). Structure and composition of savanna vegetation in northern Bolivia: a preliminary report. *Brittonia* 41: 80–100.

Hanagarth, W. and Sarmiento, J. (1990). Reporte preliminar sobre la geoecología de la sabana de Espíritu y sus alrededores (Llanos de Mojos, Departamento del Beni, Bolivia). *Ecología en Bolivia* 16: 47–75.

IBGE and IBDF (1988). *Mapa de Vegetação do Brasil.* Scale 1:5,000,000. Fundação Instituto Brasileiro de Geografia e Estatística (IBGE) and Instituto Brasileiro de Desenvolvimento Florestal (IBDF), Rio de Janeiro.

Montes de Oca, I. (1989). *Geografía y recursos naturales de Bolivia*, 2nd edition. Edit. Educacional, La Paz. 574 pp.

Moraes, M. and Beck, S.G. (1992). Diversidad florística de Bolivia. In Marconi, M. (ed.), *Conservación de la diversidad biológica en Bolivia.* CDC-Bolivia, La Paz. Pp. 73–111.

Parker III, T.A. (1989). Beni avifauna. *Orion* (summer): 35.

Parker III, T.A. and Bailey, B. (eds) (1991). *A biological assessment of the Alto Madidi Region and adjacent areas of northwest Bolivia, May 18-June 15, 1990.* RAP (Rapid Assessment Program) Working Papers 1. Conservation International, Washington, D.C. 108 pp.

Presencia (6 August 1992). Cuántos y quiénes somos.

Redford, K.H. and Stearman, A.M. (1989). Local peoples and the Beni Biosphere Reserve, Bolivia. *Vida Silvestre Neotropical* 2(1): 49–56.

Ribera, M.O. (1992). Regiones ecológicas. In Marconi, M. (ed.), *Conservación de la diversidad biológica en Bolivia.* CDC-Bolivia, La Paz. Pp. 9–71.

Rioja, G. (1992). The Jatata Project: the pilot experience of Chimane empowerment. In Plotkin, M. and Famolare, L. (eds), *Sustainable harvest and marketing of rain forest products.* Island Press, Washington, D.C. Pp. 192–196.

Rossow, R. (1992). *Boelckea*, nuevo género de Scrophulariaceae de Bolivia. *Parodiana* 7: 15–24.

Solomon, J.C. (1989). Bolivia. In Campbell, D.G. and Hammond, H.D. (eds), *Floristic inventory of tropical countries: the status of plant systematics, collections, and vegetation, plus recommendations for the future.* New York Botanical Garden, Bronx. Pp. 455–463.

Suárez-Morales, O. (1986). *Parques nacionales y afines de Bolivia.* Academia Nacional de Ciencias de Bolivia, La Paz. 134 pp.

Walsh, J. (1987). Bolivia swaps debt for conservation. *Science* 237: 596–597.

Authors

This Data Sheet was written by Dr Stephan G. Beck and Mónica Moraes-R. (Universidad Mayor de San Andrés, Instituto de Ecología, Herbario Nacional de Bolivia, Correo Central - Casilla 10077, La Paz, Bolivia).

SIERRA NEVADA DE SANTA MARTA
Colombia

Location: North coast near Caribbean Sea, between latitudes 10°10'–11°20'N and longitudes 72°30'–74°15'W.

Area: 12,232 km².

Altitude: 0–5776 m.

Vegetation: Along 500–4300 m transect: premontane forest in lower tropical area; montane to dwarf forests in middle area; fields and thickets in páramo.

Flora: c. 1800 recorded species of vascular plants; high endemism.

Useful plants: Diverse medicinals, foods, and for ceremonies, dyes, construction, fuelwood.

Other values: Endemic fauna, watershed protection, Amerindian lands, archaeological sites, glacial landscapes, tourism.

Threats: Cattle-ranching, erosion, logging, accelerated land-clearing, illegal crops, guerrilla settlements, pollution; insufficient measures to preserve areas with high biodiversity.

Conservation: Tayrona Natural National Park (150 km²), Sierra Nevada de Santa Marta NNP (3830 km²) – partially overlapping two Amerindian Reserves: Resguardo Indígena Kogi-Malayo (Arsario) (c. 3618 km²), RI Arhuaco (1959 km²).

Geography

The Sierra Nevada de Santa Marta is on Colombia's Caribbean coast in the departments Magdalena, La Guajira and Cesar, between the cities Barranquilla and Maracaibo (Map 61). Diverse thermic regimes (or life zones) exist on the massif in a gradient ranging from low and warm tropical or equatorial areas through temperate regimes up to cold to frigid areas of the páramo, culminating in perpetual snow and ice (Fundación Pro-Sierra Nevada de Santa Marta 1991; van der Hammen and Ruiz-C. 1984).

The isolated massif was formed between the Early Miocene and Late Pleistocene. Very different rocky outcroppings are presented: granite batholiths, diorite and quartzomonzonite of the Mesozoic and Tertiary; rhyolitic and ignimbritic volcanic rocks, along with a varied sequence of sediments – limestones, sandstones and siltstones (Bartels 1984).

Along the altitudinal transect of this mountain system, different soils occur as do varied temperatures and precipitation (Sevink 1984; van der Hammen 1984a, 1984b). The thermal gradient changes 0.56°C per 100 m in elevation, from somewhat over 27°C at sea-level to 0°C at c. 4850 m. Between 500 m and 1300–1500 m, the soils are classified as dystropepts, with pH over 5; the median annual temperature (stable soil temperature) is 19.5°C and the precipitation reaches over 4000 mm (at 1400 m elevation). From 1500–2500 m, the soils belong to the humitropept group, with pH below 5; the temperature averages 14.5°C, the precipitation 3500 mm. From 2500–3300 m, the soils are tropaquepts, also with pH below 5; the temperature averages 9°C, the precipitation 2200 mm. From 2700–2900 m is the zone of greatest condensation. From 3300–4100 m, cryaquept and placaquept soils are found, with pH over 5; the temperature averages 7°C, the precipitation 1300 mm.

The average monthly temperature varies through the year by less than 2.5°C. In the lower part of the transect area there is a water deficit during three or four months (December–March).

Vegetation

The massif intercepts trade winds from the north-east, creating a differential distribution of vegetation belts and ecological segregation. In general, the seaward northern slope has conditions of greater moisture than the inland southern slope.

Along a transect from 4300 m to 500 m on the massif, the vegetation ranges through fields and thickets of the páramos, dwarf to montane forests in the middle elevations, and premontane forests in the lower equatorial region.

The syntaxonomic arrangement of the vegetation on the northern slope is as follows (Rangel-Ch. 1991; Cleef and Rangel-Ch. 1984; Cleef et al. 1984):

1. "Páramo" (4500-3300 m)
Bunchgrass fields, thickets (cf. Sturm and Rangel-Ch. 1985).

1.1. Zonal vegetation
Class Espeletio-Calamagrostietea. Order Calamagrostietalia effusae. Alliance 1: Drabo cheiranthoidis-Calamagrostion effusae (= Luzulo racemosae-Calamagrostion effusae). Association Drabo cheranthoidis-Calamogrostietum effusae, with other important species such as *Carex sanctae-marthae* (Cyperaceae) and *Diplostephium anactinotum* and *Oligandra chrysocoma* (Asteraceae). Alliance 2: Hyperico-Calamagrostion effusae. Two Associations: Perissocoelo-Calamagrostietum effusae and Spirantho-Pernettyetum prostratae, with species such as *Perissocoeleum purdiei* (Apiaceae) and *Gnaphalium graveolens* (Asteraceae).

On the southern flank (the dry zone), "pajonales" and thickets are described by the two Associations Stevio lucidae-

426

Calamagrostietum effusae and Valeriano karstenii-Libanothamnetum glossophylli.

1.2. Azonal vegetation
Aquatic and marshy communities. Order Oritrophio-Wernerietalia. Alliance 3: Wernerion crassae-pygmaeae. Association Oritrophio-Wernerietum pygmaeae.

2. Andean region (3300-2500 m)
The middle zone of the gradient, with low (5–8 m) forests, and montane forests (15–20 m high) that usually have three storeys.
Alliance 4: Myriantho ternifoliae-Weinmannion pinnatae. Two Associations: Chaetolepido santamartensis-Myrcianthetum ternifoliae, with the Subassociation Libanothamnetosum glossophylli at the treeline, and Clusio multiflorae-Weinmannietum pinnatae.

3. Sub-Andean region (2500-1150 m)
Alliance 5: Gustavio speciosae-Tovomition weddellianae. Two Associations: Cavendishio callistae-Tovomitetum weddellianae (2500–1600 m), with the two Subassociations Stylogynetosum and Graffenriedetosum santamartensis, and Calatolo costaricensis-Dictyocaryetum schultzei (1600–1150 m).

4. Tropical or equatorial region (1100-500 m)
The lower zone, with premontane forests to 20–35 m high.
Alliance 6: Zygio longifoliae-Virolion sebiferae. Two Associations: Dictyocaryo scultzei-Zygietum longifoliae (1150–800 m) and Poulsenio armatae-Perseetum americanae (800–400 m).

5. Below 500 m
Tropical mesic forest, tropical deciduous dry forest, and succulent and thorny matorral occur.

Flora

Although no exhaustive inventory of the flora exists for the entire massif, preliminary figures have reported 1800 vascular plant species (Rangel-Ch., Cleef and Salamanca-V. 1996).

MAP 61. SIERRA NEVADA DE SANTA MARTA, COLOMBIA (CPD SITE SA25)

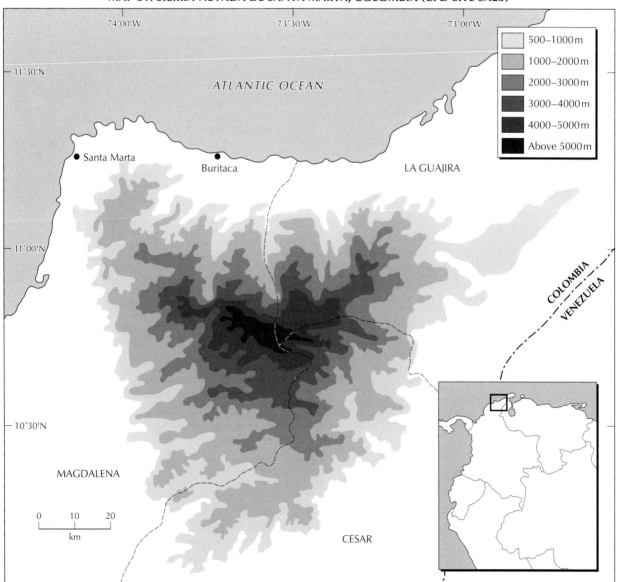

Table 60 presents the families and genera with the greatest numbers of species along the transect and the dominant families in each of the four life zones (Cuatrecasas 1958).

In the zone of the páramo exist species whose areas of distribution show endemism, such as in the genera *Berberis* (Berberidaceae); *Draba* (Brassicaceae); *Symplocos* (Symplocaceae); *Perissocoeleum, Micropleura, Cotopaxia* (Apiaceae); *Satureja* (Lamiaceae); *Aragoa* (Scrophulariaceae); *Senecio, Oligandra, Diplostephium, Hinterhubera* (Asteraceae); and *Carex* (Cyperaceae). Moreover, there are also endemic genera, such as *Cabreriella, Castanedia* and *Raouliopsis* (Asteraceae). In the Andean and sub-Andean regions, endemics are represented particularly by Melastomataceae, Arecaceae, Aquifoliaceae and Myrsinaceae.

TABLE 60. ESTIMATES OF FLORISTIC DIVERSITY ON THE SIERRA NEVADA DE SANTA MARTA (MAINLY NORTHERN SLOPE) BETWEEN 500–4500 m

1. Ten families with largest numbers of species and genera

Family	Number of species	Family	Number of genera
Asteraceae	156	Asteraceae	70
Orchidaceae	87	Leguminosae	30
Leguminosae	68	Poaceae	29
Polypodiaceae *s.l.*	67	Orchidaceae	28
Melastomataceae	57	Rubiaceae	25
Poaceae	55	Polypodiaceae *s.l.*	18
Rubiaceae	53	Melastomataceae	15
Solanaceae	51	Acanthaceae	11
Piperaceae	41	Solanaceae	11
Bromeliaceae	31	Ericaceae	10

2. Ten genera with the most species

Genus	Number of species	Family
Solanum	28	Solanaceae
Miconia	27	Melastomataceae
Peperomia	22	Piperaceae
Pleurothallis	21	Orchidaceae
Piper	18	Piperaceae
Tillandsia	16	Bromeliaceae
Ficus	16	Moraceae
Polypodium	15	Polypodiaceae
Diplostephium	12	Asteraceae
Asplenium	12	Polypodiaceae *s.l.*

3. Families with the most species by life zone

Tropical (550–1100 m) Family	Number of species	Sub-Andean (1150–2500 m) Family	Number of species
Leguminosae	13	Orchidaceae	51
Piperaceae	7	Asteraceae	45
Asteraceae	6	Rubiaceae	39
Melastomataceae	6	Leguminosae	35
Rubiaceae	5	Polypodiaceae *s.l.*	33
Acanthaceae	4	Melastomataceae	31
Bignoniaceae	4	Solanaceae	31
Poaceae	4	Poaceae	28
Bromeliaceae	3	Piperaceae	24
Solanaceae	3	Bromeliaceae	10

Andean (2500–3300 m) Family	Number of species	Páramo (3300–4500 m) Family	Number of species
Asteraceae	75	Asteraceae	24
Orchidaceae	29	Poaceae	9
Polypodiaceae *s.l.*	24	Polypodiaceae *s.l.*	7
Ericaceae	19	Rosaceae	6
Solanaceae	16	Cyperaceae	5
Bromeliaceae	15	Brassicaceae	4
Scrophulariaceae	14	Melastomataceae	4
Poaceae	13	Orchidaceae	3
Melastomataceae	12	Valerianaceae	3
Rosaceae	12	Gentianaceae	2

Useful plants

Indigenous communities, especially the Kogi, use some natural resources in the following ways (Carbonó 1987):

❖ **Medicinal:** 81 species, among which the following are significant: *Oreopanax floribundus* (Araliaceae); *Blepharodon mucronatum* (Asclepiadaceae); *Achyrocline satureioides, Hypochoeris sessiliflora, Stevia lucida, Verbesina* sp. (Asteraceae); *Protium neglectum* (Burseraceae); *Clusia* sp. (Clusiaceae); *Weinmannia pinnata* (Cunoniaceae); *Pernettya prostrata, Vaccinium meridionale* (Ericaceae); *Satureja caerulescens* (Lamiaceae); *Peperomia rotundifolia, Piper arboreum* (Piperaceae); *Muehlenbeckia tamnifolia* (Polygonaceae); and *Acaena cylindrostachya, Lachemilla* sp. (Rosaceae).

❖ **Alimentary:** 32 species, including *Annona muricata*, "guanábana" (Annonaceae); *Ananas comosus*, "piña" (Bromeliaceae); *Carica microcarpa, C. papaya*, papaya (Caricaceae); *Hirtella trianda* (Chrysobalancaceae); *Ipomoea batatas*, "batata" (Convolvulaceae); *Cucurbita moschata*, "ahuyama" (Cucurbitaceae); *Dioscorea alata*, "ñame" (Dioscoreaceae); *Manihot esculenta*, "yuca" (Euphorbiaceae); *Inga* spp. – "guama", *Vigna sinensis* (Leguminosae); *Bunchosia* spp., "ciruela" (Malpighiaceae); *Pourouma aspera* (Moraceae); *Psidium guajava* (Myrtaceae); *Passiflora mollissima*, "maracuya", *P. vitifolia* (Passifloraceae); *Pouteria arguacoensium*, "manzana" (Sapotaceae); and *Capsicum* spp., "ají" (Solanaceae).

❖ **For construction:** 44 species, including *Oreopanax diguensis*; *Cecropia peltata* (Cecropiaceae); *Clethra repanda* (Clethraceae); *Weinmannia pinnata*; *Doliocarpus dentatus* (Dilleniaceae); *Aniba* sp. (Lauraceae); *Miconia dodecandra* (Melastomataceae); *Rapanea dependens, R. ferruginea* (Myrsinaceae); *Psidium caudatum; Roupala suaveolens* (Symplocaceae); and *Trema micrantha* (Ulmaceae).

❖ **Fuelwood:** 6 species, among which are *Oreopanax floribundus; Pithecellobium latifolium* (Leguminosae); and *Cupania americana* (Sapindaceae).

❖ **Dyes:** 12 species, including *Bidens triplinervia* (Asteraceae); *Berberis* sp. (Berberidaceae); *Protium neglectum; Coriaria thymifolia* (Coriariaceae); and *Weinmannia pinnata*.

❖ **For ceremonials:** 36 species, including *Oreopanax diguensis, O. floribundus; Bursera tomentosa* (Burseraceae); *Lagenaria siceraria* (Cucurbitaceae); *Erythroxylum novogranatense* (Erythroxylaceae); *Crotalaria micans, C. pilosa, Lupinus carrikerri* (Leguminosae); *Monochaetum venosum* (Melastomataceae); *Psychotria horizontalis* (Rubiaceae); *Cardiospermum microcarpum* (Sapindaceae); and *Anemia ferruginea* (Schizaeaceae).

Some plant types in the massif are relicts for Colombia, e.g. *Dictyocaryum schultzei* (ivory-nut palm). Other palms have economic potential, such as *Scheelea magdalenica* ("palma curúa"), *Astrocaryum malybo* (straw palm), *Copernicia tectorum* ("palma zarare"), *Aiphanes aculeata* ("cohune" palm), and in Cyclanthaceae *Carludovica palmata* ("palma giraca"). The feasibility of utilizing other plant resources, such as *Vanilla planifolia* (vanilla) and *Furcraea* spp. ("pita"), is under study (Herrera de Turbay 1984).

Social and environmental values

The massif is characterized by the cultural heterogenity of its inhabitants. In higher elevations dwell the indigenous Kogi or Kággaba, Ijka or Arhuaco, and Sánha or Wiwa (Arsario or Malayo), who together make up 10% of the region's c. 200,000 inhabitants. The Kogi (from the northern slope) are the most traditional and least in contact with modern civilization. The lower part of the massif is inhabited by colonists from different regions of the country – Antioquia, Santander, Tolima and Guajira (Fundación Pro-Sierra de Santa Marta 1991).

The zones of the páramo provide indispensible vegetative resources for survival of the indigenous cultures. Of the 187 plants utilized by the Kogi, 51 grow in the páramo – 27% of that total. Of the 77 plants utilized medicinally, 43 grow in the páramo – 56%. Of the 12 plants used for dyes or construction, 8 are from the páramo – 66% of those types of resources (Carbonó 1987).

Extensive areas of the Sierra Nevada de Santa Marta served as home to pre-Columbian cultures, peoples who managed the environment in relative accord with nature. The preservation and in certain cases reconstruction of their cities (e.g. Pocigüeica, Bonda, Taironaca) is a contribution to the history of Colombia's culture, and as well tourist activities could be derived from them and related to implementation of centres of ecological tourism.

The massif constitutes a natural barrier, which blocks the circulation of the trade winds from the north-east and has a decisive effect on maintenance of the hydric regime of a large portion of the northern region of Colombia. The landscape of the glaciers – which are on the highest peaks of Colombia – is of great cultural value. There are various high-altitude lakes where important rivers originate, such as the Sevilla, Frío, Tucurinca and Buritaca. These constitute important sources for the provision of water to cities such as Santa Marta and Fundación.

The diversity of the upper zones is high. There are many endemic animals, such as the butterfly genera *Reliquia* and *Paramo*; the enigmatic frog genus *Geobatrachus* – 76% of the 21 known species of amphibians and reptiles from the páramo and cloud-forest regions are endemic; c. 70 subspecies and species of birds are endemic – including the Santa Marta conure (*Pyrrhura viridicata*), black-backed thornbill (*Ramphomicron dorsale*), Santa Marta whitestart (*Myioborus flavivertex*); and several mammals – e.g. a white-fronted capuchin monkey (*Cebus albifrons malitiosus*), rodents in *Sciurus*, *Oryzomys*, *Thomasomys*, *Proechimys* and *Diplomys*, a páramo brocket deer (*Mazama americana carrikeri*) (INDERENA 1984; Fundación Pro-Sierra de Santa Marta 1991).

The Santa Marta mountains are a discrete Endemic Bird Area (EBA B08); 15 bird species are restricted to the EBA and a further seven species shared with other EBAs. The large number of subspecies also confined to this region emphasizes the probable process of recent colonization and differentiation from species in the main Andean chain. Most of the endemics occupy wide altitudinal bands; just two species are confined to the tropical zone forest and two to the páramo and treeline scrub. One species in the EBA is considered threatened, but with recent reports on the poor state of the forests on the sierra, reassessment for the restricted-range species may be necessary.

Economic assessment

The most important economic value is related to the utilization of generated hydrological resources, which can only last if adequate areas covered by vegetation will persist. Also, timber trees are extracted.

Legal and illegal cultivation take place. Principal among the former is coffee, on the part of the massif between 700–1300 m. The indigenous peoples have cattle in the páramos, along with crops of potato, garlic and "arracacha" (*Arracacia xanthorrhiza*). In the mid-part of the massif, beans and maize are planted, whereas in the lower parts, sugarcane, avocado, yuca and cotton are grown. The illegal cultivation relates to the commerce of marijuana (*Cannabis sativa* subsp. *indica*) in the 1970s and more recently to "coca" (*Erythroxylum coca*), a crop linked to the cultural ancestors of the Amerindians (Fundación Pro-Sierra Nevada de Santa Marta 1991).

Threats

In the páramo zone, intensive cattle-ranching by indigenous peoples and colonists has affected extensive areas, in which all original vegetation has disappeared – severe problems of erosion with loss of soil are occurring (Rangel-Ch. 1989). In the mid-parts, the areas of the Andean life zones, cultivation and cattle-ranching have affected the native forests, which also suffer from timber exploitation. The planting of illegal crops such as marijuana and coca affect the middle and lower zones of the massif. Government control campaigns on these crops use highly toxic poisons, which also disturb the biotic systems. In the lower part of the massif the colonists have notably changed the landscape – there are problems of deforestation, unprotected soil, erosion and decrease in the volume of rivers.

The amount of land in the massif covered by forests has been reduced drastically during the last 50 years. In certain cases, the reduction is considered close to 50% – some have estimated that the deterioration is even greater, reaching 70–80% of the original area (Fundación Pro-Sierra Nevada de Santa Marta 1991; Herrera de Turbay 1985).

Conservation

In the area of the massif the Instituto Nacional de los Recursos Naturales Renovables y del Medio Ambiente (INDERENA) promoted the establishment of two National Parks: Sierra Nevada de Santa Marta Natural NP (1964, enlarged 1977) and Tayrona NNP (1969). There also are two indigenous reserves, which overlap with much of the Santa Marta NNP: Resguardo Indígena Arhuaco (1974, enlarged 1983) and RI Kogi-Malayo (Arsario) (1980, modified 1990). In 1989 INDERENA developed a park management plan, but it has not been fully implemented due to needs for financial and human resources – in 1993 there were six park rangers and a park director.

Fundación Pro-Sierra Nevada de Santa Marta (a non-governmental organization) assists INDERENA in the management of Alto de Mira and Filo Cartagena ranger districts through the Parks in Peril Program of The Nature Conservancy (U.S.), which has as a principal objective conservation of biological diversity, by financial and technical support to ensure adequate on-the-ground protection. This programme provides for creation and improvement of the existing infrastructure, a permanent conservation presence, inventory and monitoring of hunting activities and extraction of natural resources, involvement of local communities in the decision-making process, environmental education activities and work opportunities for local communities.

References

Bartels, G. (1984). Los pisos morfoclimáticos de la Sierra Nevada de Santa Marta (Colombia). In van der Hammen, T. and Ruiz-C., P.M. (eds), *La Sierra Nevada de Santa Marta (Colombia): transecto Buritaca– La Cumbre*. Studies on Tropical Andean Ecosystems Vol. 2. J. Cramer, Berlin. Pp. 99–129.

Carbonó, E. (1987). *Estudio etnobotánico entre los Kogui de la Sierra Nevada de Santa Marta*. M.S. thesis in Systematics. Instituto de Ciencias Naturales, Museo de Historia Natural, Universidad Nacional de Colombia, Bogotá. 154 pp.

Cleef, A.M. and Rangel-Ch., J.O. (1984). La vegetación del páramo del noroeste de la Sierra Nevada de Santa Marta. In van der Hammen, T. and Ruiz-C., P.M. (eds), *La Sierra Nevada de Santa Marta (Colombia): transecto Buritaca– La Cumbre*. Studies on Tropical Andean Ecosystems Vol. 2. J. Cramer, Berlin. Pp. 203–266.

Cleef, A.M., Rangel-Ch., J.O., van der Hammen, T. and Jaramillo-M., R. (1984). La vegetación de las selvas del transecto Buritaca. In van der Hammen, T. and Ruiz-C., P.M. (eds), *La Sierra Nevada de Santa Marta (Colombia): transecto Buritaca – La Cumbre*. Studies on Tropical Andean Ecosystems Vol. 2. J. Cramer, Berlin. Pp. 267–406.

Cuatrecasas, J. (1958). Aspectos de la vegetación natural de Colombia. *Rev. Acad. Col. Cienc. Exactas Fís. Nat.* 10(40): 221–268.

Fundación Pro-Sierra Nevada de Santa Marta (1991). *Historia y geografía: Sierra Nevada de Santa Marta*. Fondo Editorial Pro-Sierra Nevada de Santa Marta, Bogotá. 48 pp.

Herrera de Turbay, L.F. (1984). La actividad agrícola en la Sierra Nevada de Santa Marta (Colombia): perspectiva histórica. In van der Hammen, T. and Ruiz-C., P.M. (eds), *La Sierra Nevada de Santa Marta (Colombia): transecto Buritaca – La Cumbre*. Studies on Tropical Andean Ecosystems Vol. 2. J. Cramer, Berlin. Pp. 501–530.

Herrera de Turbay, L.F. (1985). *Agricultura aborigen y cambios de vegetación en la Sierra Nevada de Santa Marta*. Fundación de Investigaciones Arqueológicas Nacionales, Bogotá. 260 pp.

INDERENA (1984). *Colombia parques nacionales*. Instituto Nacional de los Recursos Naturales Renovables y del Medio Ambiente (INDERENA), Bogotá. 263 pp.

Rangel-Ch., J.O. (1989). Características bioecológicas y problemática de manejo de la región paramuna de Colombia. Memorias del seminario sobre "Los Páramos de Colombia". Suelos ecuatoriales. *Revista Sociedad Colombiana Ciencia del Suelo* 19(1): 11–19.

Rangel-Ch., J.O. (1991). *Vegetación y ambiente en tres gradientes montañosos de Colombia*. Ph.D. thesis. University of Amsterdam, Amsterdam. 349 pp.

Rangel-Ch., J.O., Cleef, A.M. and Salamanca-V., S. (1996). The equatorial interandean and subandean forests of the Parque Los Nevados transect, Cordillera Central, Colombia. In van der Hammen, T. and dos Santos, A. (eds), *La Cordillera Central Colombiana: transecto Parque Los Nevados*. Studies on Tropical Andean Ecosystems/Estudios de Ecosistemas Tropandinos Vol. 4/5. J. Cramer, Berlin. In press.

Sevink, J. (1984). An altitudinal sequence of soils in the Sierra Nevada de Santa Marta. In van der Hammen, T. and Ruiz-C., P.M. (eds), *La Sierra Nevada de Santa Marta (Colombia): transecto Buritaca – La Cumbre*. Studies on Tropical Andean Ecosystems Vol. 2. J. Cramer, Berlin. Pp. 131–138.

Sturm, H. and Rangel-Ch., J.O. (1985). *Ecología de los páramos andinos: una visión preliminar integrada*. Biblioteca J.J. Triana No. 9. Instituto de Ciencias Naturales, Museo de Historia Natural, Universidad Nacional de Colombia, Bogotá. 292 pp.

van der Hammen, T. (1984a). Datos eco-climatológicos del transecto Buritaca y alrededores (Sierra Nevada de Santa Marta). In van der Hammen, T. and Ruiz-C., P.M. (eds), *La Sierra Nevada de Santa Marta (Colombia): transecto Buritaca – La Cumbre*. Studies on Tropical Andean Ecosystems Vol. 2. J. Cramer, Berlin. Pp. 45–66.

van der Hammen, T. (1984b). Temperaturas de suelo en el transecto Buritaca – La Cumbre. In van der Hammen, T. and Ruiz-C., P.M. (eds), *La Sierra Nevada de Santa Marta (Colombia): transecto Buritaca – La Cumbre*. Studies on Tropical Andean Ecosystems Vol. 2. J. Cramer, Berlin. Pp. 67–74.

van der Hammen, T. and Ruiz-C., P.M. (eds) (1984). *La Sierra Nevada de Santa Marta (Colombia): transecto Buritaca – La Cumbre*. Studies on Tropical Andean Ecosystems/Estudios de Ecosistemas Tropandinos Vol. 2. J. Cramer, Berlin. 603 pp.

Authors

This Data Sheet was written by Dr J. Orlando Rangel-Ch. and Aída Garzón-C. (Universidad Nacional de Colombia, Instituto de Ciencias Naturales, Museo de Historia Natural, Apartado Aéreo 7495, Santafé de Bogotá, Colombia).

(Tropical) Andes: CPD Site SA26

SIERRA NEVADA DEL COCUY-GUANTIVA
Colombia

Location: In Eastern Cordillera, Cocuy region c. 280 km north-east and Guantiva region c. 230 km north to north-east of Bogotá. Cocuy region about latitudes 6°10'–6°45'N and longitudes 72°00'–72°25'W; Guantiva region about latitudes 6°03'–6°20'N and longitudes 72°40'–73°00'W.

Area: Cocuy region 3060 km², with 33 km north-south line of 22 snow capped peaks; Guantiva region 1200 km².

Altitude: Cocuy region c. 500–5493 m; Guantiva region from c. 2200 m on western slope to 4270 m.

Vegetation: Number of vegetation types and diversity highest of Eastern Cordillera: (1) on lower eastern slope (per)humid Andean rain forests, on lower western slope broad diversity of dry xerophytic scrub and low forest types of Chicamocha Valley; (2) on upper eastern slope *Espeletia* stem-rosette – bamboo páramo, on upper western slope *Espeletia* stem-rosette – bunchgrass páramo; (3) extensive superpáramo best developed of Eastern Cordillera.

Flora: Highest diversity – Cocuy páramo with over 220 genera of vascular plants – and highest proportion of endemism in páramos (including superpáramos) of Eastern Cordillera; endemic species also prominent in uppermost forests; threatened species.

Useful plants: Medicinals – e.g. widespread in regional use are "lítamo real" (*Draba litamo*, *D. cocuyensis*), "granizo" (*Hedyosmum* spp.), "árnica" (*Senecio* spp.); spices; and for construction.

Other values: Endemic animals, threatened species; watershed protection and water resources for villages and settlements; indigenous people; spectacular mountain scenery.

Threats: Too much sheep-grazing, occasionally combined with burning and hunting; cutting in uppermost forests, especially on western slope; upslope shift to c. 4000 m of low-income agriculture; ecotourism, when not adequately managed.

Conservation: El Cocuy Natural National Park (3060 km²), with 216 km² also an Amerindian Reserve.

Geography

The Cocuy-Guantiva region comprises large areas of Andean forests and páramos in mountain ranges on both sides of the Chicamocha River Valley in Boyacá Department. The Sierra Nevada del Cocuy is east of this deep valley and the Páramo de Guantiva west of the valley (Map 62). Included approximately are the areas between Chita/Sácama and Güicán-Las Mercedes/Bócota, and between Belén and Onzaga.

El Cocuy Natural National Park is on the boundary of Boyacá, Casanare and Arauca departments; on the north-west, Santander Department borders the region. The Sierra Nevada del Cocuy (or Güicán or Chita), between c. 6°20'–6°35'N latitudes, is the highest mountain range of the Colombian Eastern Cordillera. It is the only range in the cordillera carrying a snowcap, which extends over some 33 km following the main divide (Notestein and King 1932). Among the snow peaks are Ritacuva Blanco (5330 m), Picacho (5030 m), Puntiagudo (5200 m), El Castillo (5100 m) and "picos sin nombre" (c. 5000 m). Glacial valleys contain lakes, among which stand out for beautiful scenery e.g. La Plaza, Laguna Grande de la Sierra, Laguna Grande de Los Verdes and the Lagunillas Valley lakes (La Pintada, La Cuadrada, La Parada, La Atravesada). In the northern section of the Cocuy range appears a parallel ridge extending north, separated by the upper Ratoncito Valley, with spectacular glacial lakes and cushion bogs. Also, the impressive El Castillo peak is separated to the east from the Cocuy main ridge. This branch, which forms the boundary between Boyacá and Arauca departments, continues eastward and some 18 km east of Ritacuva Blanco constitutes the isolated snowcapped Sirará summit (5200 m).

Among the rivers in the northern sector are the Tunebo, Ratoncito, Rudiván (or Cubugón), Orozco and Derrumbada. On the eastern slope the main rivers are the Cusay, Tame, Purare and Mortiñal/San Lope; on the south to south-eastern side is the upper Casanare River with the tributaries Quebrada Los Osos, Q. Maicillo and Q. El Playón. The main rivers on the western side are tributaries of the Nevado River, which flows into the Chicamocha River: Quebrada Rechiniga and the rivers Cardenillo, Corralitos and Lagunillas.

The Cocuy range consists mainly of folded sedimentary Cretaceous (Albian-Aptian) quartzitic and sandstone rocks, shales and occasional limestone inclusions (van der Hammen *et al.* 1981; INDERENA 1984). The western dip slope of Cocuy has quartzite. In the westernmost section appears coal of the Guaduas (Maestrichtien) and Socha (Palaeocene) formations, while north to north-west of the main range are Palaeozoic (Mid-Devonian to Permian) sedimentary rocks. On the eastern slope, siltstones and calcareous layers near Patio Bolos contain numerous plant fossils. On the eastern slope between 5000–1000 m sedimentary rock is present, consisting of conglomerates and clayey-sandy layers related to Tertiary (Oligocene to Pliocene) river deposition. At lower elevations appear alluvial terraces and cones.

Quaternary deposits as a consequence of glacial action are common in the highest parts of the Cocuy range (Kraus and van der Hammen 1960; González, van der Hammen and Flint 1965; van der Hammen *et al.* 1981). In the southern

431

section of the range there are at least five (possibly six) glacial drift bodies. Drifts numbered 2–5 are of Last Glacial age. Drift 6 represents the historic Neoglacial and its terminal moraines mark the grass-páramo – superpáramo border, e.g. at páramo Cóncavo. Probably in the period between 45,000–25,000 BP the Cocuy glaciers expanded the most, under a relatively wet climate – down to elevations of at least 2700 m, where they were in contact with forest. At that time the narrow páramo belt contained abundant *Polylepis* forests. Today the Cocuy icecap and glaciers are still shrinking [see van der Hammen *et al.* (1981) Figs. 9–11 and the 1:80,000 scale glacio-morphological map].

The climate is very wet on the eastern slope of the Cocuy range, with yearly precipitation up to 4000–5000 mm at around 1000 m; the western slope is dry, with a yearly average (over 10 years) of c. 901 mm at El Cocuy (2749 m) and 959 mm (over 5 years) at Chita (3005 m). Permanently high atmospheric humidity and generally just one dry period (December to February) are present (van der Hammen *et al.* 1981). Annual average temperatures in El Cocuy NNP are in a wide spectrum, between 23.6°C and -3°C, with an average of 0°C at c. 4800 m.

The Páramo de Guantiva region (roughly 40 km × 30 km) according to local inhabitants largely encloses the land on the eastern border between Soatá, Susacón and Sátivasur, to the

MAP 62. SIERRA NEVADA DEL COCUY-GUANTIVA (CPD SITE SA26)

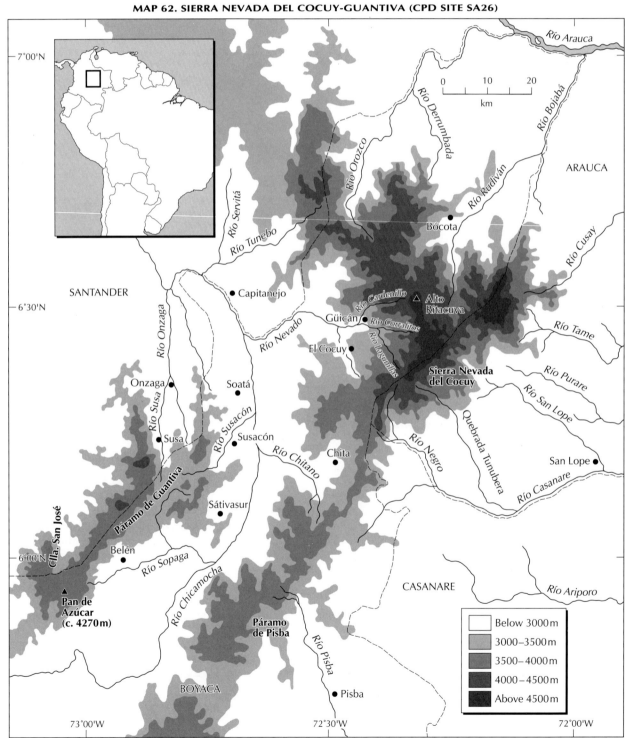

432

mountains west of Belén at the south-western edge, to the Cuchilla de San José and Onzaga at the northern border. The Páramo de Guantiva itself is some 5 km west to south-west of Susacón and reaches c. 3900 m. The mountain range north-west of Belén has a number of glacial lakes (e.g. Grande, El Alcohol, Redonda, Negra Cazadero) in deep U-shaped valleys with moraines. This section contains the highest summit, Pan de Azúcar (c. 4270 m). The mountain range runs in a north to north-east direction and is split into the south-north valleys of the Susa, Chaguasca and Chicamocha rivers.

With respect to the geology of the Belén cordillera area, Cleef (1981) referred to reddish sandstones, siltstones and conglomerates of the Girón Formation. The summit of Pan de Azúcar consists of Palaeozoic (Devonian to Permian) gneiss and granites. West and north-west of this peak are intrusive and extrusive rocks of Triassic to Jurassic age, and to the north is a narrow zone of Cretaceous sandstone.

Vegetation

Sierra Nevada del Cocuy

The vegetation of the Sierra Nevada del Cocuy includes montane vegetation – mainly forest types, and tropical alpine páramo. The treeline is c. 3500–3700 m on the dry western slope, but close to 3000 m on the wet eastern slope (Cleef 1981; van der Hammen and Cleef 1986). Probably there is a strong relationship with local climate. The more upslope treeline on the western slope may be related to the ascending warm dry air masses from the deep Chicamocha Valley; the much lower treeline of the eastern slope is related to the superwet environmental conditions prevailing there. The great height of the Sierra Nevada del Cocuy apparently largely prevents exchange from the dry and wet climates to their opposite sides.

Montane forest zone
On the Cocuy western slope the following observations have been made (van der Hammen *et al.* 1981; Cleef, unpublished):

❖ In the lower northern part near Capitanejo is dry xerophytic vegetation. From the Chicamocha Valley floor to 2800 m, most natural forest has been cleared, except for small stands on very steep slopes. Between c. 2800–3500 m *Weinmannia fagaroides* seems dominant in the patches of Andean forest (c. 15 m high), associated with species of *Ilex*, *Styrax* and *Prunus*, as well as *Clusia*, *Clethra*, *Myrsine*, *Rhamnus*, *Viburnum*, *Vallea*, *Psychotria*, *Xylosma*, *Cestrum*, *Oreopanax* and Lauraceae and Melastomataceae. Low (6–8 m) high-Andean forest of *Hesperomeles lanuginosa* occurs to c. 3750 m, interchanged with *Polylepis quadrijuga* forest to over 4000 m.

❖ In addition Cleef (unpublished) noted above the town of El Cocuy (2760 m): *Alnus acuminata* common (3080 m) along streams; residual forests of *Gynoxys*, *Oreopanax* and *Vallea stipularis* (c. 3250–3350 m); solitary *Polylepis quadrijuga* from 3550 m upslope. At 3800 m near Alto de la Cueva, are frequent *Espeletiopsis colombiana* patches (from 3600 m) and a shrubby species of *Lupinus*. Locally in the Andean forest belt (e.g. c. 2800 m at El Claval), there are small *Ludwigia peruviana-Carex acutata* swamps.

The Andean rain-forest zonation on the Cocuy eastern slope according to van der Hammen *et al.* (1981) is as follows:

1. From 2100–2550 m, very wet sub-Andean rain forest with *Weinmannia* cf. *pinnata* as the dominant tree, associated with cyatheaceous tree ferns and species of *Cecropia* and *Heliocarpus* (to 2200 m) and *Alchornea* and *Acalypha* (to 2400 m). In addition species of *Brunellia*, *Stylogyne*, *Myrsine*, *Eugenia*, *Hedyosmum*, *Ocotea*, *Ternstroemia*, *Guarea*, *Billia*, *Piper*, *Saurauia*, *Freziera*, *Psychotria* and *Sapium* are recorded from this forest (to 40 m high) by T. van der Hammen and R. Jaramillo-Mejía (in prep.) and van der Hammen *et al.* (1981).

2. From c. 2550–3100 (–3300) m, *Weinmannia rollottii* dominates the Andean rain forest (20–35 m high), associated with *W.* cf. *pinnata* and Cyatheaceae (up to 2950 m) – replaced at higher altitudes by *Blechnum* sp. The entire-leaved *Geonoma weberbaueri* and *Chusquea* bamboos are very conspicuous. Additional associates are species of *Miconia*, *Clusia*, *Clethra*, *Brunellia*, *Myrsine*, *Drimys*, *Ternstroemia*, *Freziera*, *Geissanthus*, *Ocotea*, *Hedyosmum*, *Piper*, *Monnina* and *Oreopanax*. In the uppermost wet forest, *Escallonia myrtilloides* is common.

Páramo vegetation
The zonal páramo contains a number of vegetation types, grouped into subpáramo, grass páramo and superpáramo; an asterisk (*) means the community has not been reported from elsewhere.

1. Subpáramo communities include (i) on the western slope, *Myrica parvifolia* shrubs on lowermost páramo Cóncavo (3700 m); *Pentacalia vaccinioides* shrub-páramo with *Espeletiopsis jimenez-quesadae*; between c. 3750–3900 m, dense *Arcytophyllum nitidum* dwarf shrubs with *Masdevallia coriacea* and *Scaphosepalum* sp.; and (ii) on the eastern slope, *Hypericum magniflorum* (3000–3350 m) and *H. lycopodioides* (3250–3700 m) shrubs with *Pentacalia cacaosensis*; and *Chusquea-Ageratina tinifolia* shrubs.

2. Grass-páramo communities include *Acaena cylindristachya-Plantago sericea* subsp. *argyrophylla* herb fields (c. 4000– 4100 m); only on the eastern slope, *Espeletia annemariana/E. cleefii-Chusquea tessellata* bamboo páramo (c. 3400– 3900 m); upper páramo of *Espeletiopsis colombiana/Espeletia cleefii – Calamagrostis effusa* [3900–4250 m according to Sturm and Rangel (1985), locally to 4500 m]; *Stipa hans-meyeri* patches on western slope, c. 4350 m on páramo Cóncavo – this vegetation type had only been reported from Sabana de los Leones, Chirripó páramo, Costa Rica and from Ecuador; on the western slope (4300–4400 m), *Diplostephium rhomboidale-Pentacalia vaccinioides* shrubs on terminal moraines.

3. Superpáramo communities include low shrubs of *Loricaria complanata* (c. 4200–4400 m) and of *Niphogeton josei* (4270–4340 m); on the eastern slope at c. 4250–4300 m, *Espeletiopsis colombiana/Espeletia cleefii* lower superpáramo stands with *Geranium sibbaldioides*; *Agrostis breviculmis-Acaulimalva* spp., lower superpáramo herb

field; on eastern slope at c. 4350 m, *Racomitrium crispulum-Valeriana plantaginea* lower superpáramo vegetation; low scrub of *Pentacalia guicanensis* in small patches from 4200 m upslope; *Senecio niveo-aureus* communities (4250 m up to snowcap); *Luzula racemosa-Pernettya prostrata* superpáramo moraine vegetation (from c. 4250 m up to icecap).

Noteworthy among many azonal páramo vegetation types are:

4. Aquatic and wetland communities

❖ Glacial-lake bottom vegetation dominated by respectively *Isoetes karstenii, I. glacialis* and *I. socia,* among others.
❖ Marsh vegetation with *Carex peucophila* and *Werneria crassa* subsp. *orientalis* (c. 4000–4425 m); occasionally peaty facies with the rare monotypic *Floscaldasia hypsophila* (shared only with Los Nevados NNP of the Central Cordillera) and/or with dense stands of stem-rosettes of *Espeletia lopezii* (Cleef 1981; Sturm and Rangel 1985).
❖ Cushion bogs of the juncaceous *Distichia muscoides* (altitudinally the highest cushion bog), *Plantago rigida,* or the cyperaceous *Oreobolus cleefii* (of limited occurrence). On the western slope, reported only from páramo Cóncavo (3550–3770m), *Puya aristiguietae-Sphagnum* bog. In the eastern-slope bamboo páramo, *Chusquea tessellata-Sphagnum* bogs with *Espeletia lopezii.*

5. Dwarf forests

❖ About 2700–3700 m, *Escallonia myrtilloides* dwarf forest; from 3450–4250 m, *Gynoxys albivestita* dwarf forest with an unknown *Echeveria* sp. (Crassulaceae); at 4000 m and upslope, *Polylepis quadrijuga* dwarf forest; from 4100–4300 m, gnarled dwarfed tree patches of *Pentacalia flos-fragrans.*
❖ Only on western side between c. 3700–4050 m, apparently caused by frequent grazing and trampling, *Aciachne aicularis* cushion meadows as subseral vegetation. On the western slope up to 3800 m, *Hesperomeles lanuginosa* dwarf forest; from 3800–4400 m, *Diplostephium rhomboidale* dwarf forest patches. Mainly on the eastern slope, *Sphagnum-Diplostephium revolutum* dwarf forest.

Páramo de Guantiva

The Andean forest vegetation in the Guantiva Páramo region also has been studied by van der Hammen and Jaramillo-Mejía (in prep.). Grabandt (1980) provided a summary, noting that oak (*Quercus*) forests are directly in contact with the páramo in the Onzaga area (van der Hammen and González 1965; van der Hammen 1962).

The páramo belt includes subpáramo thickets, dwarf shrubs and grass páramo; on the western and north-western Magdalena Valley slopes, *Chusquea tessellata* bamboo páramo; and on drier eastern slopes, *Calamagrostis* bunchgrass páramo. The summit area of Pan de Azúcar at c. 4270 m just reaches into the lowermost superpáramo belt.

1. In the subpáramo, among other types appear shrub-páramo with *Espeletia muiska,* dwarf shrubs of *Arcytophyllum nitidum* and *Sporobolus lasiophyllus.*

2. From the grass páramo have been reported different types of *Chusquea tessellata* bamboo páramo, *Espeletiopsis muiska/Espeletia boyacensis-Calamagrostis effusa* bunchgrass páramo with *Oreobolus* sp. and *Castratella piloselloides, Acaena cylindristachya-Plantago sericea* herb field and *Espeletiopsis guacharaco/Espeletia azucarina-Calamagrostis effusa* bunchgrass upper páramo.

3. The lower superpáramo at Pan de Azúcar consists of fragments of *Loricaria complanata* shrubs and scree vegetation of endemic *Poa* and *Halenia* sp., species of *Arenaria* and *Cerastium, Montia meridensis, Rhacocarpus purpurascens* and *Racomitrium crispulum,* giving evidence of the upper condensation zone, and small valley floors covered by *Senecio canescens.* On the opposite Morro Verde probably appear stands of *Lupinus alopecuroides* (Guillermo Merchán, pers. comm.).

Extrazonal and azonal vegetation types

Important are isolated patches of *Polylepis quadrijuga* dwarf forest in the páramo, which are common north-west of Belén. In the same area are a number of crystalline lakes and *Plantago rigida* cushion bogs on valley floors. Peaty valley floors near the treeline contain extensive *Sphagnum* peat-bog swamps with endemic species such as *Espeletia nemenkenii* or *Espeletia arbelaeziana.*

Flora

Sierra Nevada del Cocuy

In the páramo flora of the Sierra Nevada del Cocuy have been reported (thus far) c. 220 vascular plant genera (cf. Cleef 1983). In the superpáramo have been recorded 110 vascular plant species (in c. 55 genera), c. 20 of which are endemic, including the genus *Floscaldasia* (Compositae) – whereas in the superpáramo of Sumapaz (southward) are three endemic species and in Almorzadero (to the north-west) just two endemics. About two-thirds of the superpáramo species are shared with the Sumapaz superpáramo (see CPD Site SA27). The first sample collections of the region were made in 1939 in the upper Lagunillas Valley.

Endemic species of the Cocuy mountain range (cf. Cuatrecasas and Cleef 1978; Al-Shehbaz 1989; Rangel-Ch. and Santana-C. 1989) include *Aragoa hammenii, Draba cocuyensis, D. hammenii, D. litamo, Diplostephium rhomboidale, Hypericum lycopodioides, H. papillosum, Niphogeton josei, Salvia nubigena, Oritrophium cocuyense, Paepalanthus lodiculoides* var. *floccosus, Pentacalia guicanensis, P. cleefii, Senecio cocuyanus, S. adglacialis, S. pasqui-andinus, S. tergolanatus, S. virido-albus* (also at Guantiva), *Espeletia cleefii, Espeletiopsis jimenez-quesadae, Puya cleefii* and *Acaulimalva* species.

From the headwaters of the Casanare River on the southern slope of the Cocuy have been reported endemics such as *Aragoa dugandii, Castratella rosea, Pentacalia cacaosensis* and *Espeletia curialensis.*

The Sierra Nevada del Cocuy is the southernmost outpost known (so far) of e.g. *Carex peucophila, Ilex tamana* and *Libanothamnus tamanus.* Highly interesting was the 1977 find of the aquatic liverwort *Herbertus oblongifolius* predominant in a small lake at 4060 m in the headwaters of Quebrada El Amarillal. This rare species was known only from the Itatiaia massif near Rio de Janeiro, Brazil.

Páramo de Guantiva

Endemic species of the Guantiva Páramo include *Hypericum sabiniforme*, *Niphogeton fruticosa*, *Espeletia arbelaezii*, *E. azucarina*, *E. brachyaxiantha*, *E. discoidea*, *E. nemenkenii* and *Espeletiopsis muiska*. Also different oak (*Quercus*) species have been described from the Onzaga forests.

Useful plants

A number of plants are gathered – e.g. for medicinals, spices, construction – and frequently sold by inhabitants of the region, as well as by indigenous tribes (e.g. Tunebo) on the lower Llanos side of the eastern slope of the Cocuy range. "Lítamo real" (*Draba litamo*) is the medicinal plant most known from the region; also well known are "chichoria" (*Hypochoeris sessiliflora*), "verdolaga" (*Peperomia* sp.), "granizo" (*Hedyosmum* spp.) and "árnica" (*Senecio* spp.). "Palo colorado" (*Polylepis quadrijuga*) is not only used for fuelwood and fences but also construction of walls or dwellings, together with *Espeletia lopezii* stem-rosettes (trunks).

Social and environmental values

El Cocuy NNP harbours a great diversity of large mammals, such as four primate species, all the cats known from Colombia, the tapir *Tapirus terrestris* and two species of deer. There also are many species of birds in the region (Borrero 1955; Olivares 1973; INDERENA 1984).

The Sierra Nevada del Cocuy region was originally inhabited by indigenous groups of Lache (now extinct) and Tunebo. People living at these high altitudes survive in marginal conditions, because of low income and the harsh environment. Also the Tunebo Amerindians who cross the cordillera are very poor and in poor health. Development of the Cocuy NNP area must be in accordance with the needs, and enhance possibilities and the quality of life, of the local peasants. Presently the conditions in the Cocuy NNP area are unsafe and do not allow studies towards this goal. Peasants living in the Guantiva region are also poor, but probably in better conditions than those living in the Cocuy.

Economic assessment

The Sierra Nevada del Cocuy apparently has the largest sheep population of all the páramos of the Colombian Eastern Cordillera, where cattle-ranging usually prevails. Herds of sheep are mostly grazed on the dry western slope of the Cocuy, and their influence on the páramo vegetation is considerable.

The same region seems also to be the only place in the Eastern Cordillera where potato agriculture with some onions and other Andean tubers is possible up to 4000 m. The potato production for the outside market (Cúcuta near Venezuela) is important.

However, most important for the region may be the permanent water supply from the high snowcapped Cocuy mountains, not only for the eastern slope Orinoco drainage, but especially for the dry and almost deforested lands on the western slope towards the deep Chicamocha Valley. Coal and salt (near Salinas, headwaters of the Casanare River) are of minor importance.

The Guantiva Páramo is also an important regional watershed area, which is used as well for extensive ranging of cattle, its wood supply for construction, fences and fuel, for the gathering of medicinal and spice plants, and for hunting.

Threats

The relatively dense human population on the western side has had a very strong impact on the residual montane forests, whereas the eastern slope rain forests have so far remained largely intact. The shift upslope to c. 4000 m of low-income agriculture has converted habitats. Ecotourism, when not adequately managed, can be destructive.

Major threats are uncontrolled activities such as burning, grazing, hunting and especially tree cutting in the uppermost forest belt and forest patches, since the trees grow very slowly because of prevailing low temperatures.

Espeletia nemenkenii is a highly endangered species endemic in the Belén páramos; its habitat of treeline *Sphagnum-Blechnum* bogs has almost been drained and disappeared. The same situation may apply for the endangered *Espeletia arbelaeziana*.

Conservation

El Cocuy Natural National Park (3060 km²) was established in 1977; in 1974/1979, 216 km² were set aside in the Tunebos Indigenous Reserve (INDERENA 1984). The NNP could use a conservation and management plan, taking into special account the needs and priorities of the local poor inhabitants.

The Guantiva Páramo region does not so far have conservation status, through which a conservation and management plan could establish wise-use practices and conservation of representative areas of páramo and Andean forest. It is strongly recommended that the páramo-forest border be preserved along the western slope of the Guantiva region down to at least 2200–2500 m, in order to allow for the exchange of biotic elements.

References

Al-Shehbaz, I.A. (1989). New or noteworthy *Draba* (Brassicaceae) from South America. *J. Arnold Arboretum* 70: 427–437.

Borrero, J.I. (1955). Avifauna de la región de Soatá, departamento de Boyacá, Colombia. *Caldasia* 7: 52–86.

Cleef, A.M. (1978). Characteristics of neotropical páramo vegetation and its subantarctic relations between the southern temperate zone and tropical mountains. *Erdwissenschaftliche Forschung (Wiesbaden)* 11: 365–390.

Cleef, A.M. (1981). *The vegetation of the páramos of the Colombian Cordillera Oriental.* Dissert. Bot. 61. J. Cramer, Vaduz. 321 pp.

Cleef, A.M. (1983). Fitogeografía y composición de la flora vascular de los páramos de la Cordillera Oriental colombiana (Estudio comparativo con otras altas montañas del trópico). *Rev. Acad. Colomb. Cienc. Exactas Fís. Nat.* 15: 23–29.

Cuatrecasas, J. and Cleef, A.M. (1978). Una nueva Crucifera de la Sierra Nevada del Cocuy (Colombia). *Caldasia* 12: 145–158.

González, E., van der Hammen, T. and Flint, R.F. (1965). Late Quaternary glacial and vegetational sequence in Valle de Lagunillas, Sierra Nevada del Cocuy, Colombia. *Leidse Geologische Mededelingen* 32: 157–182.

Grabandt, R.A.J. (1980). Pollen rain in relation to arboreal vegetation in the Cordilleria Oriental. *Rev. Palaeobot. Palynol.* 29: 65–147.

INDERENA (1984). *Colombia parques nacionales.* Instituto Nacional de los Recursos Naturales Renovables y del Medio Ambiente (INDERENA), Bogotá. 263 pp.

Kraus, E. and van der Hammen, T. (1960). *Las expediciones de glaciología del A.G.I. a las Sierras Nevadas de Santa Marta y El Cocuy.* Comité Nac. del Año Geofísico, Instituto Geográfico Agustín Codazzi, Colombia. 9 pp.

Notestein, F.B. and King, R.E. (1932). The Sierra Nevada del Cocuy. *Geogr. Review* 22: 423–430.

Olivares, A. (1973). Aves de la Sierra Nevada del Cocuy, Colombia. *Rev. Acad. Colomb. Cienc. Exactas Fís. Nat.* 14: 39–48.

Rangel-Ch., J.O. and Santana-C., E. (1989). Estudios en *Draba* (Cruciferae) de Colombia. I. Cuatro especies nuevas de la Cordillera Oriental. *Rev. Acad. Colomb. Cienc. Exactas Fís. Nat.* 17: 347–355.

Sturm, H. and Rangel-Ch., J.O. (1985). *Ecología de los páramos andinos: una visión preliminar integrada.* Biblioteca J.J. Triana No. 9. Instituto de Ciencias Naturales, Museo de Historia Natural, Universidad Nacional de Colombia, Bogotá. 292 pp.

van der Hammen, T. (1962). Palinología de la región de Laguna de los Bobos. Historia de su clima, vegetación y agricultura durante los ultimos 5000 años. *Rev. Acad. Colomb. Cienc. Exactas Fís. Nat.* 11: 359–361.

van der Hammen, T. and Cleef, A.M. (1986). Development of the high Andean páramo flora and vegetation. In Vuilleumier, F. and Monasterio, M. (eds), *High altitude tropical biogeography.* Oxford University Press, New York. Pp. 153–201.

van der Hammen, T. and González, E. (1965). A Late-Glacial and Holocene pollen diagram from Ciénaga del Visitador (Dept. Boyacá, Colombia). *Leidse Geologische Mededelingen* 32: 193–201.

van der Hammen, T., Barelds, J., de Jong, H. and de Veer, A.A. (1981). Glacial sequence and environmental history in the Sierra Nevada del Cocuy (Colombia). *Palaeogeography, Palaeoclimatogy, Palaeoecology* 32: 247–340.

Author

This Data Sheet was written by Dr Antoine M. Cleef (University of Amsterdam, Hugo de Vries-Laboratorium, Kruislaan 318, 1098 SM Amsterdam, The Netherlands).

(TROPICAL) ANDES: CPD SITE SA27

PARAMO DE SUMAPAZ REGION
Colombia

Location: Eastern Cordillera, from c. 40–120 km south of Bogotá, between latitudes c. 4°50'–3°20'N and longitudes c. 74°40'–74°00'W.

Area: c. 15,000–16,000 km²; páramo c. 4000 km².

Altitude: About 300 to 4250 (–4300) m.

Vegetation: On lower western slope a wide variety of dry forests, on mid-slope lower and upper montane humid to wet forests, on top of watershed bordering wet bamboo and humid bunchgrass páramos; on eastern slope, lower montane rain forest grades into Amazonian rain forest.

Flora: High diversity – over 200 vascular plant genera, with substantial species endemism.

Useful plants: Many species, for construction, as medicinals or in indigenous agriculture.

Other values: Habitat for many vertebrate taxa, including endemics; threatened species; watershed protection and water resources for Bogotá and densely populated towns and villages bordering western and northern slopes; archaeological sites.

Threats: Road construction in western and northern areas much damages fragile uppermost forest, allows for extensive and locally intensive cattle-grazing. Extractions of timber, fuelwood and water increasing; mining becoming more important. On western slope, colonization relatively dense below and above treeline.

Conservation: Central-northern part of Sumapaz páramo a Natural National Park (1540 km²) since 1977.

Geography

The Páramo de Sumapaz region comprises c. 15,000–16,000 km² south of Bogotá (Map 63). The large ecological island of páramo comprises the upper part of much of the Sumapaz massif, which is formed primarily by the south-north main ridge of the Eastern Cordillera of the Andes. The top of the watershed is between c. 4100–3800 m in elevation, and extends southward with the high points Boca Grande, Chisacá, Medio Naranja, Alto Caicedo, Andabobos, La Rabona, Torquita or Fraile, Cáqueza and Alto San Mateo to Alto de Las Oseras (3830 m), where the departments Cundinamarca, Huila and Meta meet (Guhl 1964). The calcareous summit of Cerro El Nevado de Sumapaz rises to c. 4250 m (or nearly 4300 m) some 20 km east of the main ridge.

A number of tributary rivers drain the Sumapaz massif in two major directions: (1) the Chisacá, Blanco (Oeste), Sumapaz and Cabrera flow variously westward to the Magdalena River in the adjacent lowlands at c. 300 m elevation; and (2) all in the Orinoco River drainage, the Blanco (Este) flows north-eastward to the Meta River, and the Nevado and Guape eastward to the Ariari River and the Duda to the Guayabero River, which together form the Guaviare River.

Access from the north to the highest part is only by road from Usme via Chisacá and Nazareth to Andabobos and San Juan. A number of smaller roads are present mainly in the northern part. The eastern flanks are mostly still under mature forest cover, whereas on the western flank a road penetrates along the parallel valley of the Sumapaz River. At the southern edge of the Páramo de Sumapaz is the lowest part of the Eastern Cordillera (at 1874 m); here the Neiva-Uribe road crosses the cordillera at Las Cruces.

Geologically the main part of the Sumapaz massif consists of sedimentary rocks, mostly Cretaceous sandstone of the Villeta and Guadalupe groups, Guaduas Formation (Maestrichtien). Dark calcareous rocks of Palaeozoic origin that are rich in fossil corals (e.g. *Chaetetes* sp.) and molluscs give rise to a sharp parallel ridge east of the main watershed – El Nevado de Sumapaz is part of that system.

Quaternary glaciations have largely defined the landscape in the Páramo de Sumapaz region (Guhl 1964; Melief 1985; Helmens 1990). In the highest parts, cirque lakes and roches moutonnées (fleecy rocks) are common. Moraines are found along wide U-shaped valleys. On valley bottoms in the area west and north of El Nevado de Sumapaz (and also elsewhere) is a sequence of glacial lakes dammed by terminal moraines.

The median annual temperature in Sumapaz Natural National Park varies from 19°C at 1500 m to 2°C at 4300 m (INDERENA 1984). The average annual precipitation also varies a great deal in the Sumapaz region. The incidence of fog is very high, reducing irradiance and plant transpiration; the fog is an important factor for páramo and montane cloud forests. Detailed weather records are only available from páramo and western-slope stations. In the páramo above 3000 m on the drier Magdalena slope is a bimodal annual rain pattern (January and February are driest) with up to 1500 mm of precipitation, which shifts in the easternmost highlands to a unimodal rain pattern with up to 3000 mm or more. Cerro El Nevado de Sumapaz has a snow mantle during October and November. At c. 1700–2000 m occurs the highest annual montane precipitation. On Amazonian foothills, annual rainfall exceeding 4000 mm has been registered (Bates 1948).

Vegetation

The natural vegetation of the Páramo de Sumapaz region consists basically of warm lowland forests, montane rain forests, and páramo. The altitudinal vegetation sequence and composition are quite different on the relatively dry Magdalena slope and the extremely wet Amazon slope. Rather few vegetation studies had been conducted in the Páramo de Sumapaz (e.g. Sturm and Rangel-Ch. 1985; Ernst and Seljée 1988). Most of the present vegetation data are based on Cleef (1981) and in preparation, for the páramos and the uppermost forests; an EcoAndes transect was carried out mainly in the montane forest zone (van der Hammen and dos Santos 1996).

Zonal vegetation

On the drier Magdalena slope six distinct vegetation types have been distinguished: (1) Dry leguminous forest with spiny understorey species on the lowermost foothills in the Magdalena Valley, at 470 m to c. 750 m; (2) Dry meliaceous-sapindaceous forest (c. 750–1150 m); (3) A dry sub-Andean *Quercus humboldtii* forest belt, between c. 1150–1400 m; (4) Mixed sub-Andean or lower montane rain forest (c. 1400–2550 m), and also a fringe of *Quercus humboldtii* forest at c. 1900 m; (5) A wet *Gordonia-Clusia* Andean forest belt (c. 2500–3300 m) with an understorey of *Neurolepis aperta* bamboo; (6) High-Andean rain forest (c. 3300–3500 m). Considerable stretches of this uppermost forest belt consist of a humid type of dwarf forest of *Escallonia myrtilloides-Polylepis quadrijuga* which grades into a low forest of species of *Weinmannia, Clusia* and *Brunellia*.

The wet Amazon slope includes the following forest types: (1) Wet lowland rain forest, covering adjacent lowlands and the easternmost foothills (c. 500–1000 m); (2) Wet sub-Andean or lower montane rain forest (c. 1000–2250 m) with Moraceae, Sapotaceae, Palmae, Bombacaceae, Euphorbiaceae, Meliaceae and Rubiaceae, and in the uppermost part Guttiferae, more Euphorbiaceae and Cunoniaceae; (3.1) Wet Andean or upper montane rain forest (2250–3200 m) consisting of species of *Clusia* and *Weinmannia* and *Drimys granadensis* with tall bamboos (*Chusquea* and *Neurolepis*); (3.2) Uppermost dwarf forest (3200–3500 m) of *Diplostephium fosbergii, Myrsine dependens, Argeratina tinifolia* and species of *Miconia* and *Gynoxys* and tall *Neurolepis aristata* bamboo.

Transitions to treeline shrub-páramo and dwarf shrub-páramo develop where conditions allow (Fosberg 1944). *Pentacalia vernicosa* shrubs and open low scrub of *Diplostephium rupestre* are limited to the Sumapaz páramo and Central Cordillera volcanoes. *Gaultheria ramosissima-Aragoa perez-arbelaeziana* dwarf forest has only been located and studied at 3780 m on the western slope of the Sumapaz massif. *Gynoxys* cf. *subhirsuta* dwarf forest, apparently also limited to the Sumapaz region, is common in rock shelters, where also are characteristic high-altitude patches of *Erythrophyllopsis andina* and *Senecio niveo-aureus* – a remarkable reddish globular moss carpet with whitish *Senecio* ground rosettes.

The páramo constitutes the uppermost tropical alpine vegetation belt. Basically three types of zonal grass páramo are present: bunchgrass páramo on the drier westernmost side of the Páramo de Sumapaz, transitional bunchgrass-bamboo páramo covering most of the central area, and bamboo páramo on the easternmost fringe.

Probably the most extant zonal páramo vegetation type on the Sumapaz massif is *Chusquea tessellata* bamboo páramo with *Eryngium humile* and tall *Rhynchospora ruiziana*. Also usual for the Sumapaz páramo and connecting páramos surrounding the Bogotá high plain is a peaty type of *Calamagrostis effusa* bunchgrass páramo with *Oreobolus* cf. *goeppingeri*.

The superpáramo is well represented (100 ha) on El Nevado de Sumapaz, where under wet and foggy high-altitude conditions a special type of vegetation thrives that is different from elsewhere in the Colombian Andes (Cleef 1981). This superpáramo is fully developed from c. 4100 m upward. Some characteristic species include *Loricaria complanata* and *Diplostephium rupestre* (low scrub); *Senecio summus, S. niveo-aureus, Valeriana plantaginea, Draba cuatrecasana* and *Lachemilla nivalis* (ground rosettes); *Azorella multifida* (large cushions); the mosses *Rhacomytrium crispulum, Rhacocarpus purpurascens* and *Breutelia* spp.; and in the summit area, the reddish liverwort *Herbertus subdentatus* (Cleef 1978, 1981).

Extrazonal and azonal vegetation

Cushion communities of *Isoetes andicola* are limited to the páramo lakes of La Primavera and Del Medio of the central-east Sumapaz massif. In Sumapaz swamps and bogs, which are mainly on glacial valley floors, very characteristic are e.g. *Pentacalia reissiana* shrubs, *Carex pichinchensis* sedge swamp, a *Geranium confertum-Calamagrostis ligulata* mire type with extensive mats of *Draba sericea*, and a rare calciphytic *Calamagrostis ligulata* mire with predominance of the aquatic mosses *Drepanocladus aduncus* and *Calliergonella cuspidata. Plantago rigida* cushion bogs are common and have considerable extension on flat valley floors, frequently overgrown by species of the mosses *Breutelia* and *Campylopus*, and by an open dwarf forest of *Diplostephium revolutum* which apparently represents the end of succession (Cleef 1978, 1981). *Senecio summus* crevice communities (also present in the Central Cordillera) and *Azorella multifida* cushion vegetation are patchy and well developed above 4000 m in the summit area of El Nevado de Sumapaz. Cushions of the juncaceous *Distichia muscoides* have only been found in one locale.

Flora

The Páramo de Sumapaz constitutes the second biogeographic centre of the Eastern Cordillera, after the Sierra Nevada del Cocuy (CPD Site SA26). The vascular flora has over 200 genera and stands out by having a substantial number of endemic species, especially in the páramo zone. Endemics to the Sumapaz ecological island include *Hypericum prostratum* (Guttiferae) and *Pernettya hirta* (Ericaceae). Some of the endemics have also been recorded from páramos surrounding the Bogotá high plain. Endemic species have been reported as well in the uppermost dwarf forests and include *Diplostephium fosbergii, Habracanthus cleefii* and *Miconia cleefii*. A complete record is not available yet for the forest zone, which has been poorly explored.

The endemic species occurring in the páramos include *Draba sericea* and *D. cuatrecasana* (Brassicaceae) and *Aragoa perez-arbelaeziana* and *A. corrugatifolia* (Scrophulariaceae). A number of endemic *Espeletia* species occur, among which

**CPD Site SA27: Páramo de Sumapaz region.
A number of *Espeletia* species occur in the páramo;
this is *Espeletia grandiflora*.**
Photo: Tony Morrison/South American Pictures.

are *E. summapacis, E. tapirophila, E. miradorensis* and *E. cabrerensis*; other endemic Compositae are *Laestadia pinifolia, Pentacalia reissiana* and *P. nitida*. In the superpáramo 3 of c. 75 species (in almost 55 genera) are endemic, and about two-thirds of the species are shared with the Cocuy superpáramo farther to the north.

Among the bryophytes, some species were described from the Sumapaz massif but also might be found elsewhere, e.g. *Dendrocryphaea latifolia, Riccardia metaensis, Blindia gradsteinii, Gradsteinia andicola* (Oschyra 1990), *Plagiochila cleefii* and *Daltonia fenestrellata*.

Useful plants

A number of useful species occur in the region. Most of them are used as material in construction for houses, bridges and fences – particularly *Polylepis quadrijuga, Aragoa perez-arbelaeziana, Escallonia myrtilloides* and species of *Weinmannia, Clusia* and *Quercus*. There is apparently some commercial logging on the western slope, where forest is being converted to cattle pastures. Forests are particularly important in providing fuelwood. Both the páramo and the Andean forests contain many species for local medicinal use. A thorough ethnobotanical survey is lacking, as well as a systematic screening of species for commercial drugs.

Social and environmental values

The large number of glacial lakes and bogs constitutes an important drinking-water resource for Bogotá (Guhl 1968). The core area of El Nevado de Sumapaz contains important resources of limestone and some coal.

The northern part of the Páramo de Sumapaz up to c. 3500 m is inhabited by settlers, who live in small dwellings. People of the Pasca area (western slope) traditionally use the páramo for grazing cattle and hunting, and have small farms in the headwaters of the rivers draining the wet eastern slopes of the Páramo de Sumapaz. There are almost no prospects here for sustainable agriculture, except for a few crops such as onions, "papas criollas" (local strains of potatoes) and some other Andean tubers. Pre-Columbian archaeological sites have been studied in the páramos above Pasca (INDERENA 1984).

There are several regionally endemic animals, such as mammals and lizards. The Páramo de Sumapaz is also an important migration route for exchange of fauna between the western and eastern slopes. An unusual example from early in 1973 is a big anteater observed at c. 3800 m elevation between Andabobos and La Rabona crossing a bamboo páramo from west to east.

Threats

The chief threats to the Sumapaz region are logging activities, mainly on the western slope, and the subsequent change from forest into pasture. Degradation of such areas is commonly observed. Wood gathering for fuel is another significant threat to the uppermost forests, which only may recover at a very slow rate. Ranging cattle and burning are increasing throughout the region, which constitute also serious threats to wild vertebrates. Hunting takes place without discrimination; there is no real management. Mining is becoming more important in the region.

The water resources of lakes in the core area of El Nevado de Sumapaz will be used in the near future to supply Bogotá. The necessary infrastructure (e.g. access roads) will affect the wild glaciated scenery of Sumapaz, and attract more human disturbance. Consequently, the wild mammals and birds will be impacted, if the region is not carefully and adequately managed.

Conservation

The central-northern part of the Sumapaz páramo has been a National Park since 1977. The park (1540 km²) extends from about the forest line on the western slope across the divide and onto the wet eastern slope (to 1500 m), including its wet Andean rain forests (INDERENA 1984; Castaño-Uribe 1989).

It is strongly recommended that the core area of El Nevado de Sumapaz and the adjacent watershed highlands and valley systems be established as a sanctuary, which needs special protection because most of the endemism of the páramo species is concentrated here. The wet superpáramo vegetation (c. 4000–4250 m) covering the calcareous summit of El Nevado de Sumapaz is apparently unique in the northern Andes of Colombia and Venezuela, and wet páramo vegetation overlying calcareous rock is rare in the northern Andes.

It also is strongly recommended that the mainly pristine wet forests of the eastern Sumapaz slope be protected in order to maintain complete valley systems under forest cover, which would allow for the exchange of fauna and flora elements. This might guarantee the preservation of functional communities of the threatened spectacled bear (*Tremarctos ornatus*), oncilla (*Felis tigrina*) and mountain tapir (*Tapirus pinchaque*) among others. The lakes in the core area are important for migrating water birds from both hemispheres. Significant in this respect as well is conservation of the last dry and wet forest remnants of the western slope of the Sumapaz massif. Particularly important forest communities here are high-Andean *Aragoa perez-arbelaeziana* dwarf forest, *Escallonia-Polylepis quadrijuga* forest, *Weinmannia fagaroides-Hesperomeles lanuginosa* forest with *Polylepis quadrijuga*, and *Brunellia* sp.-*Gordonia-Clusia* forest with *Neurolepis aperta* bamboo

MAP 63. PARAMO DE SUMAPAZ REGION, COLOMBIA (CPD SITE SA27)

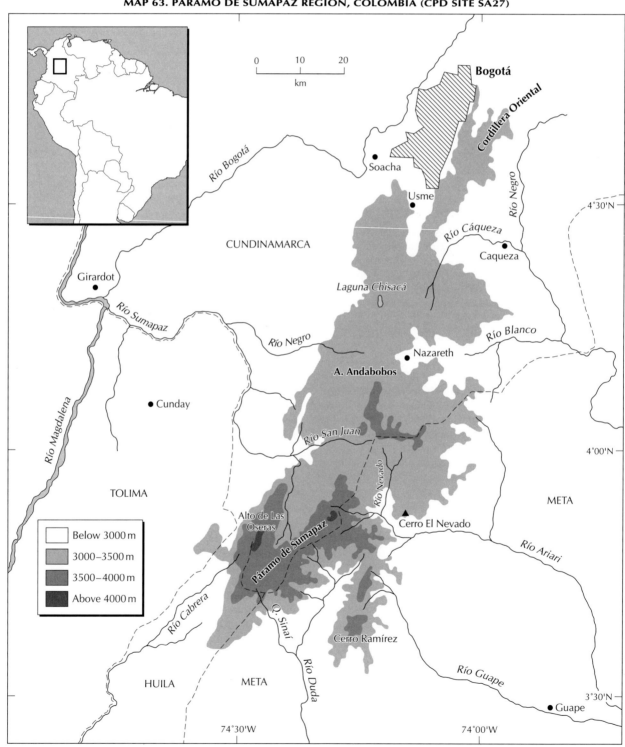

understorey, and downslope the last patches of the sub-Andean forest (including oak forest).

References

Bates, M. (1948). Climate and vegetation in the Villavicencio region of eastern Colombia. *Geogr. Rev.* 38: 555–574.

Castaño-Uribe, C. (ed.) (1989). *A guide to the National Natural Parks System of Colombia.* INDERENA, Bogotá. 198 pp.

Cleef, A.M. (1978). Characteristics of neotropical páramo vegetation and its subantarctic relations between the southern temperate zone and tropical mountains. *Erdwissenschaftliche Forschung (Wiesbaden)* 11: 365–390.

Cleef, A.M. (1981). *The vegetation of the páramos of the Colombian Cordillera Oriental.* University of Amsterdam, The Netherlands. 321 pp.

Ernst, M.T. and Seljée, O.J. (1988). *Vegetation mapping in Páramo Boca Grande (Cundinamarca, Colombia).* Internal report 253. Hugo de Vries Laboratory, University of Amsterdam, The Netherlands. 70 pp.

Fosberg, F.R. (1944). El Páramo de Sumapaz, Colombia. *J. New York Bot. Garden* 45: 226–234.

Guhl, E. (1964). Aspectos geográficos y humanos de la región de Sumapaz en la Cordillera Oriental de Colombia. *Rev. Acad. Colomb. Cienc. Exactas Fís. Nat.* 12: 153–161.

Guhl, E. (1968). Los páramos circundantes de la Sabana de Bogotá. Su ecología y su importancia para el régimen hidrológico de la misma. In Troll, C. (ed.), *Geo-ecology of the mountainous regions of the tropical Americas.* Coll. Geogr. 9, Bonn. Pp. 195–212.

Helmens, K.F. (1990). *Neogene-Quaternary geology of the high plain of Bogotá, Eastern Cordillera, Colombia.* Diss. Bot. 163. J. Cramer, Berlin. 202 pp.

INDERENA (1984). *Colombia parques nacionales.* Instituto Nacional de los Recursos Naturales Renovables y del Medio Ambiente (INDERENA), Bogotá. 263 pp.

Melief, A.B.M. (1985). *Late Quaternary paleoecology of the Parque Nacional Natural Los Nevados (Cordillera Central) and Sumapaz (Cordillera Oriental) areas, Colombia.* Ph.D. thesis, University of Amsterdam, The Netherlands. 162 pp.

Oschyra, O. (1990). *Gradsteinia andicola,* a remarkable aquatic moss from South America. *Tropical Bryology* 3: 19–28.

Sturm, H. and Rangel-Ch., J.O. (1985). *Ecología de los páramos andinos: una visión preliminar integrada.* Biblioteca J.J. Triana No. 9. Instituto de Ciencias Naturales, Museo de Historia Natural, Universidad Nacional de Colombia, Bogotá. 292 pp.

van der Hammen, T. and dos Santos, A.G. (eds) (1996). *La Cordillera Central Colombiana: transecto Parque Los Nevados.* Studies on Tropical Andean Ecosystems/ Estudios de Ecosistemas Tropandinos Vol. 4/5. J. Cramer, Berlin. In press.

Author

This Data Sheet was written by Dr Antoine M. Cleef (University of Amsterdam, Hugo de Vries-Laboratorium, Kruislaan 318, 1098 SM Amsterdam, The Netherlands).

REGION OF LOS NEVADOS NATURAL NATIONAL PARK
Colombia

Location: West of Bogotá in central Andean Colombia (4°10'–5°12'N), a transect including both slopes of Central Cordillera between Magdalena and Cauca rivers, between latitudes 4°42'–4°47'N and longitudes 74°49'–75°54'W.

Area: Region c. 12,200 km²; National Park 583 km², three other parks total c. 115 km².

Altitude: Transect on eastern slope 300–4600 m, on western slope 4600–1000 m.

Vegetation: In highlands, bunchgrass fields and "frailejones"; in mid-altitude zone, oak forests of *Quercus humboldtii* and forests with *Weinmannia rollottii*; in medium-low zone, forests of Lauraceae and *Hedyosmum racemosum*; in lower zone, dry forests with *Ochroma*, *Cecropia* and Leguminosae.

Flora: Species of vascular plants recorded c. 1250, lichens 300, bryophytes 200, macroscopic fungi 180; highlands with high endemism.

Useful plants: Species used for medicinals, food, crafts, construction, timber, replanting of degraded areas.

Other values: Threatened animal species; permanent snow and icecaps for mountain sports; thermal springs of Santa Rosa and Nevado del Ruiz; páramo lakes – important reservoirs for water and sport fishing.

Threats: Volcanic eruptions, burning of páramo, cattle-grazing, logging for charcoal production and fuelwood, erosion, park in-holdings.

Conservation: Los Nevados Natural National Park; Ucumarí Regional Park; Cañón del Quindío and another protected area (in Línea sector) administered by Corporación Regional de Quindío.

Geography

The Ruiz-Tolima massif has formed from the Pliocene to Holocene over an undulating high plateau slightly inclined toward the east and dipping more to the west, which was derived from a surface of polygenic flattening of the pre-Cretaceous Mesozoic and Oligocene.

There are eight principal volcanoes aligned approximately south-north along 350 km between latitudes 4°30'–5°15'N and longitudes 75°20'–75°30'W: Cerro Machín, Nevado del Tolima, Nevado del Quindío, Páramo de Santa Rosa, Nevado de Santa Isabel, Nevado del Cisne, Nevado del Ruiz and Cerro Bravo. Los Nevados Natural National Park (extending altitudinally from 2600–5400 m) is within latitudes 4°50'–5°00'N and longitudes 75°15'–75°45'W; it includes part of the municipality Villamaría in the Department Caldas, Santa Rosa de Cabal and Pereira in Department Risaralda, Salento in Department Quindío, and Ibagué, Líbano, Villahermosa and Casabianca in Department Tolima (Map 64) (INDERENA 1984).

The heritage of the glaciers affects the mountains above 3000 m. Glaciers presently cover a total of 35–40 km² on the summits of El Ruiz (5400 m), Santa Isabel (5100 m) and El Tolima (5200 m). Characteristics such as U-shaped valleys and moraines are remnants of the glacial activity in the Pleistocene covering 860 km², down to 2700 m on the eastern slope and 3200 m on the western slope.

The geomorphology of the transect of Los Nevados NNP basically includes relief, sculpture and lithology. The western slope is steeper and shorter than the eastern slope, especially above 2200 m. In the higher elevations above 3800–3500 m on the slopes of the volcanoes, the sculpturing is produced from lava flats covered by ashes which dried out from surface drainage. Between 3800–2300 m, the sculpture occurs as steps and sierra with strong dissection especially along the western slope. From 2300–1200 m there are hills more or less strongly dissected; along the western slope is a series of small north-south ranges separated by wide valleys. Below 1300 m are recent alluvial formations (Pérez-P. and van der Hammen 1983).

On the higher parts of the massif, above 3500 m on the eastern slope and 2200 m on the western slope, volcanic extrusive rocks are dominant, basically andesitic lava. The remainder of the eastern slope is formed by plutonic intrusive rocks of the Bosque, Santa Isabel and Ibagué batholiths, interrupted by outcrops of metamorphic rocks (amphibolites, schist, gneissic quartz-feldspar). The Magdalena River Valley is filled by Tertiary continental formations, covered in large part by cones and recent alluvial formations. The western slope has metamorphic rocks (schist, graphitic quartz-mica, amphibolite, amphibolitic schist) and metasedimentary rocks. The volcanism that developed from the mid-Miocene in the region is characterized by thick deposits of andesitic lava over c. 1500 km²; a continuous coating of volcanic ash sometimes 10 m thick is found in some sectors (Thouret 1989). The soil has developed from volcanic ashes.

Along the altitudinal transect in this mountain system are eight main soil units having nineteen main groups of soils, and significant variations of temperature and rainfall (Thouret 1983; Thouret and Faivre 1989). Below 700 m, the soils are eutrophic (eutropept, eutrustox, haplustalf), alluvial (fluvaquent, ustifluvent), andosolic (andic eutropept) or hydromorphic (tropaquent, tropaquept, fluvaquent); the average annual temperature varies between 24°–28°C and the minimum annual precipitation is 1000–1500 mm. From 1100–2300 m, the soils are humitropept, dystropept, haplustalf, eutropept or ustropept; the average temperature and precipitation vary between 14°–20°C and 2000–2300 mm. From 2300–3800 m, the soils are cryandept, dystrandept, andaquept, humitropept or aquic dystropept; the temperature and precipitation vary between 6°–14°C and 2000–3000 mm. From 3800–4500 m, the soils are andosoles (andepts) – andosolic (vitrandept) or vitric andosoles (cryandept); the temperature and precipitation vary between 0°–6°C and 1500–2000 mm. On the western slope the thermal gradient of the air changes 0.54°C with every 100 m rise, on the eastern slope by 0.51°C.

The relation between soil and air temperature along the gradient allows for the establishment of four zones, with "thermal ruptures" at 1280 m, 1980 m and 3700 m. Areas with thermal homogeneity have higher precipitation; in the two zones with less precipitation, there are strong temperature contrasts: below 1000 m and above 3500 m (Thouret 1983). The most rain falls in April–May and October–November, and the driest periods are in January–February and July–August. Because of humid winds from the Pacific Ocean, the western slope receives more moisture than the eastern slope.

Vegetation

Pioneer communities are established on pebbly soil in open spaces of the superpáramo, dominated by *Draba* spp. (Brassicaceae) and *Pentacalia gelida*, *Senecio canescens* and *Erigeron chionophyllus* (Asteraceae). In the páramo region are bunchgrass fields (*Calamagrostis effusa*, *C. recta*), bunchgrass-frailejones and frailejones-shrubland characterized by the "frailejón" *Espeletia hartwegiana* subsp. *centroandina*.

In the high fringe of the Andean montane region are dwarf forests (3–8 m high) of *Hesperomeles lanuginosa*. From this forest border downward to the base of the gradient, different compositions and areas of distribution of vegetation types occur, with well-developed mesic forests of two (to three) storeys – the canopy 20–35 m high, the lower arboreal stratum 10–15 m high. Eastern-slope forests are dominated by *Quercus humboldtii*, which is replaced at the lower boundary by

forests of Lauraceae (*Ocotea*, *Nectandra*). On the western slope, the forests of *Hesperomeles* are replaced by forests of *Weinmannia rollottii*, and in the mid-zone, by forests of *Hedyosmum racemosum*. In the lower zone, sparse dry forests occur dominated by species of *Cecropia* (Cecropiaceae), *Mauria* (Anacardiaceae) and *Casearia* (Flacourtiaceae) (Rangel-Ch. 1991).

The syntaxonomic arrangement for the transect vegetation of Los Nevados NNP is as follows (Sturm and Rangel-Ch. 1985; Cleef, Rangel-Ch. and Salamanca-V. 1983, 1996; Salamanca-V., Cleef and Rangel-Ch. 1996):

Mostly eastern and western slopes

1. **General páramo region (4700 m to 3800–3750 m)**

1.1. High fringe or superpáramo (4700–4100 m)
Three Associations: Senecio canescentis-Cerastietum floccosi. Lupino alopecuroidis-Agrostietum araucanae. Diplostephio eriophori-Loricarietum colombianae.

1.2. Middle area or páramo proper (4100–3750 m)
Five Associations: Calandrinio acaulis-Calamagrostietum rectae. Calamagrostietum effusae-rectae. Espeletio hartwegianae-Calamagrostietum effusae. Festuco dolichophyllae-Calamagrostietum effusae. Only on western slope: an association of *Aciachne pulvinata* and *Escallonia myrtilloides*.

Eastern slope, Santa Isabel Volcano to Venadillo in Tolima

2. **Andean region (3700-2500 m)**

2.1. High-Andean fringe (3700–3300 m)
Alliance 1: Diplostephio floribundi-bicoloris-Hesperomelion lanuginosae. Two Associations: Gynoxyo baccharoidis-Diplostephietum floribundi (3700–3600 m). Chusqueo scandentis-Hedyosmetum bonplandiani (3500–3300 m).

MAP 64. REGION OF LOS NEVADOS NATURAL NATIONAL PARK, COLOMBIA (CPD SITE SA28), SHOWING TRANSECT

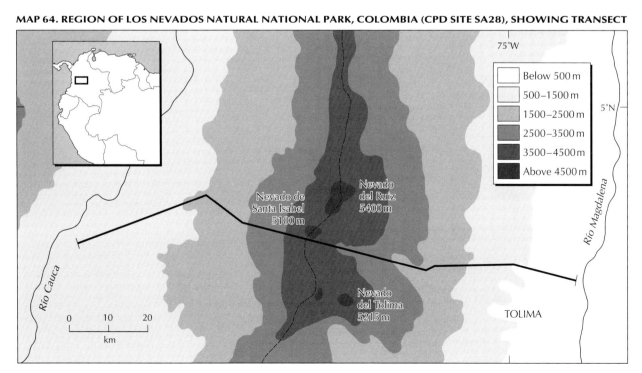

2.2. Mid-belt or typical Andean zone (3200–2500 m)
Alliance 2: Monotropo uniflorae-Quercion humboldtii. Two Associations: Weinmannio magnifoliae-Quercetum humboldtii (3110–2940 m). Clusio minoris-Quercetum humboldtii (2700–2500 m).

3. Sub-Andean region (2300-1200 m)

3.1. Mid-belt or typical sub-Andean zone (2300–1500 m)
Alliance 3: Hedyosmo racemosi-Nectandrion caucanae. Two Associations: Nectandretum acutifoliae-caucanae (2300–1900 m). Chrysochlamydo dependentis-Nectandretum globosae (1700–1500 m).

3.2. High tropical fringe
Association Protio macrophyllae-Rheedietum madruñae (1300–1200 m).

4. Tropical or equatorial region (900-300 m)
Order Erythroxylo citrifolii-Coccolobietalia obovatae. Alliance 4: Carludovico palmatae-Acalyphion villosae. Association with two Subassociations: Ardisio foetidae-Cupanietum latifoliae-Ochrometosum (840 m), Ardisio foetidae-Cupanietum latifoliae-Maurietosum (710 m). Also the Association Mayno suaveolentis-Casearietum corymbosae (500–300 m).

Western slope, Santa Rosa Volcano to Puerto Caldas in Risaralda

2. Andean region (3700-2900 m)

2.1. High-Andean fringe (3700–3500 m)
Alliance 5: Neurolepido aristatae-Oreopanacion nitidi. Three Associations, the first (3700–3650 m) with two Subassociations: Gynoxyo baccharoidis-Hesperomeletum lanuginosae-Pentacalietosum (3700 m), Gynoxyo baccharoidis-Hesperomeletum lanuginosae-Myrsinetosum perreticulatae (3650 m). Monochaeto lindeniani-Weinmannietum mariquitae (3500 m). Gordonio speciosae-Weinmannietum pubescentis (3300 m).

2.2. Typical Andean zone (3100–2900 m)
Alliance 6: Chusqueo scandentis-Weinmannion rollottii. Two Associations: Chusqueo scandentis-Brunellietum goudotii (3100 m). Tovomito guianensis-Clusietum multiflorae (2900 m).

3. Sub-Andean region (2700-1300 m)
Alliance 7: Palicoureo angustifoliae-Hedyosmion racemosi. Two Associations: Brunellio occidentalis-Moretum insignis (2700–2000 m). Ocoteo discoloris-Huertetum glandulosae (2000–1300 m), Subassociation Ocoteo discoloris-Huertetum glandulosae-Ladenbergietosum.

4. Tropical or equatorial region (1000 m)
Association Amyro pinnatae-Crotonetum glabelli.

Flora

In the transect and nearby páramo localities, the inventory recorded 1250 species of vascular plants, 200 of bryophytes,

300 of lichens and 180 of macroscopic fungi (Rangel-Ch. *et al.* 1996; Wolf 1989; Boekhout and Pulido-L. 1989). In the high terrain, diversity and concentration of species is greater than in the middle and lower zones. Roughly grouping families with similar herbaceous-shrubby growth characteristics (such as Polypodiaceae *s.l.*, Araceae, Poaceae, Asteraceae, Piperaceae, Orchidaceae), and families with arborescent (woody) growth characteristics (such as Rubiaceae, Leguminosae, Melastomataceae, Lauraceae, Moraceae), the groups are balanced in number of species. Table 61 gives the characteristic dominant features at the family and generic levels (Rangel-Ch. 1996).

Species with endemic distribution include: In the superpáramo, *Draba pachythyrsa*, *D. pennell-hazenii* (Brassicaceae); and *Senecio isabelis*, *Pentacalia gelida*,

TABLE 61. FLORISTIC DIVERSITY IN LOS NEVADOS PARK TRANSECT, BETWEEN 550–4500 m

1. Ten families with largest numbers of species and genera

Family	Number of species	Family	Number of genera
Asteraceae	161	Asteraceae	50
Polypodiaceae *s.l.*	90	Rubiaceae	26
Rubiaceae	76	Leguminosae	23
Leguminosae	55	Poaceae	23
Solanaceae	53	Orchidaceae	18
Melastomataceae	52	Polypodiaceae *s.l.*	15
Poaceae	47	Moraceae	11
Piperaceae	45	Solanaceae	10
Orchidaceae	41	Apocynaceae	9
Araceae	40	Araceae	9

2. Ten genera with the most species (morphospecies = *)

Genus	Number of species	Family
Miconia	50	Melastomataceae
Polypodium	26 *	Polypodiaceae
Piper	25	Piperaceae
Solanum	21	Solanaceae
Anthurium	20	Araceae
Inga	17 *	Leguminosae
Passiflora	17	Passifloraceae
Peperomia	17	Piperaceae
Psychotria	17	Rubiaceae
Oreopanax	14	Araliaceae

3. Families with the most species by life zone

Tropical (550–1100 m) Family	Number of species	Sub-Andean (1150–2500 m) Family	Number of species
Leguminosae	17	Leguminosae	30
Rubiaceae	10	Polypodiaceae *s.l.*	28
Sapindaceae	6	Rubiaceae	28
Asteraceae	5	Araceae	23
Piperaceae	5	Solanaceae	22
Solanaceae	5	Moraceae	20
Menispermaceae	4	Piperaceae	20
Polygonaceae	4	Lauraceae	19
Rutaceae	4	Melastomataceae	15
Apocynaceae	3	Orchidaceae	15

Andean (2500–3300 m) Family	Number of species	Páramo (over 3300 m) Family	Number of species
Asteraceae	45	Asteraceae	109
Polypodiaceae *s.l.*	37	Poaceae	31
Rubiaceae	33	Polypodiaceae *s.l.*	26
Melastomataceae	29	Scrophulariaceae	22
Orchidaceae	19	Brassicaceae	17
Piperaceae	19	Apiaceae	15
Araceae	17	Ericaceae	11
Solanaceae	16	Caryophyllaceae	10
Lauraceae	13	Cyperaceae	9
Cunoniaceae	12	Melastomataceae	9

Diplostephium eriophorum (Asteraceae). In the páramo and the high-Andean areas, *Berberis diazii* (Berberidaceae); *Oreopanax ruizianus* (Araliaceae); *Siphocampylus tolimanus* (Campanulaceae); *Peperomia pennellii* (Piperaceae); *Aphelandra trianae* (Acanthaceae); *Gunnera magnifica* (Gunneraceae); *Lupinus ruizensis* (Leguminosae); *Tibouchina andreana* (Melastomataceae); *Acaena ovalifolia* (Rosaceae); *Bartsia* cf. *pedicularioides*, *Pedicularis incurva* (Scrophulariaceae); *Gentianella dasyantha*, *Halenia tolimae* (Gentianaceae); *Diplostephium rupestre*, *D. violaceum*; *Guzmania vamvolxemii* (Bromeliaceae); and *Altensteinia rostrata*, *Platystele schmidtchenii* (Orchidaceae). In the Andean region, *Passiflora quindiensis* (Passifloraceae); *Myrrhidendron glaucescens* (Apiaceae); *Aegiphila pennellii* (Verbenaceae); and *Valeriana quindiensis* (Valerianaceae) (Rangel-Ch., Cleef and Salamanca-V. 1996).

Useful plants

❖ In folk medicine: *Aspidosperma polyneuron*, *Mandevilla trianae* (Apocynaceae); *Ilex* sp. (Aquifoliaceae); *Oreopanax capitatus*, *O. floribundus*, *O. ruizianus* (Araliaceae); *Achyrocline satureioides*, *Espeletia hartwegiana*, *Gnaphalium americanum*, *G. elegans*, *G. graveolens*, *G. pellitum*, *Hypochoeris sessiliflora*, *Senecio formosus* (Asteraceae); *Berberis glauca* (Berberidaceae); *Cordia alliodora*, *C. bogotensis*, *C. cylindrostachya*, *C. riparia*, *C. spinescens*, *Tournefortia fuliginosa*, *T. scabrida* (Boraginaceae); *Draba pennell-hazenii* (Brassicaceae); *Siphocampylus giganteus* (Campanulaceae); *Viburnum pichinchense*, *V. triphyllum* (Caprifoliaceae); *Cecropia arachnoidea* (Cecropiaceae); *Hedyosmum bonplandianum* (Chloranthaceae); *Hirtella americana* (Chrysobalanaceae); *Clusia alata*, *C. columnaris*, *C. multiflora*, *Rheedia madruño* (Clusiaceae); *Pernettya prostrata* (Ericaceae); *Croton glabellus*, *C. leptostachyus* (Euphorbiaceae); *Mayna suaveolens* (Flacourtiaceae); *Besleria* sp. (Gesneriaceae); *Oryctanthus botryostachys*, *Phoradendron piperoides* (Loranthaceae); *Lycopodium clavatum* (Lycopodiaceae); *Miconia ligustrina* (Melastomataceae); *Chlorophora tinctoria*, *Ficus glabrata*, *F. radula*, *Olmedia aspera* (Moraceae); *Ardisia sapida* (Myrsinaceae); *Peperomia glabella* (Piperaceae); *Monnina phytolaccaefolia* (Polygalaceae); *Asplenium occidentale*, *A. monanthes*, *Blechnum occidentale*, *Polypodium glaucophyllum*, *P. lanceolatum*, *P. percussum* (Polypodiaceae *s.l.*); *Ranunculus geranioides* (Ranunculaceae); *Lachemilla pectinata*, *L. fulvescens*, *Rubus bogotensis*, *R. glabratus* (Rosaceae); *Cinchona pubescens* (Rubiaceae); *Amyris pinnata* (Rutaceae); and *Viola stipularis* (Violaceae).

❖ Food: *Passiflora vitifolia*; and *Annona* sp. (Annonaceae).

❖ For home construction: *Aspidosperma polyneuron*.

❖ Dyes: *Viola stipularis*; and *Senecio formosus*.

❖ Forest species: *Cordia alliodora*; *Tovomita guianensis* (Clusiaceae); *Hyeronima colombiana* (Euphorbiaceae); *Quercus humboldtii* (Fagaceae); *Ficus glabrata*, *F. velutina*; *Podocarpus oleifolius* (Podocarpaceae); *Cinchona pubescens*, *Rondeletia pubescens* (Rubiaceae); *Pouteria caimito* (Sapotaceae); and *Trema micrantha* (Ulmaceae).

The references consulted were García-B. (1974–1975); Correa-Q. and Bernal (1989–1991); Murillo-P. (1983); and Mozo-M. (1968).

Social and environmental values

At present there is no knowledge of indigenous settlements. The eastern slope of the Central Cordillera was inhabited by the Panches, the western slope by the Quimbayas (INDERENA 1984).

In the area of the transect there are diverse environments and landscapes that merit conservation for their sociocultural values. In the high páramo, glaciers and snow cover the mountain crests; numerous lakes constitute an important reservoir for water that supports the agriculture of major parts of the region. Lake Otún and surrounding marshes are the source of the Otún River, whose waters are utilized by the Pereira aqueduct and surrounding municipalities.

Threatened fauna in the region include mountain tapir (*Tapirus pinchaque*) and spectacled bear (*Tremarctos ornatus*). Among the endemic animals are rufous-fronted parakeet (*Bolborhynchus ferrugineifrons*) and a bearded helmetcrest (*Oxypogon guerinii strubelii*) and the frog *Osornophryne percrassa*.

The Central Andean páramo Endemic Bird Area (EBA B60) and Subtropical inter-Andean Colombia EBA (B12) embrace the entire Central Andean cordillera from the páramo zone down to the subtropical (mid-Andean) forest. Twenty-seven restricted-range bird species occur in these two EBAs, 14 of which are entirely confined to the region. Due to increasing disturbance of the specialized vegetation, 15 bird species are considered threatened. In the drier low-lying zones of the adjacent Cauca and Magdalena valleys, this area overlaps with the Dry inter-Andean valleys EBA (B13), where four restricted-range birds occur.

Economic assessment

Generally, the only zone free of direct use of the soil is the superpáramo, between 4200–4800 m. In the páramo (3500–4200 m), extensive cattle-raising is the principal activity. In the Andean montane region (2500–3500 m), most of the land is utilized for potato crops. Between 1800–2500 m, cattle-raising and agriculture (especially maize) are the main production; between 1100–1800 m, the most extensive cultivation in the sub-Andean zone takes place – mostly coffee trees. Between 500–1100 m, sugarcane is cultivated; below 500 m, mechanized methods are applied in cultivation of rice, cotton and sorghum (Pérez-P. 1983).

Near the glaciers there are tourist shelters and winter sports are encouraged. The lake zone, especially Lake Otún, is used for sport fishing of introduced rainbow trout (*Salmo gardneri*). In the mid-part of the western slope, in Santa Rosa de Cabal and near Nevado del Ruiz, natural thermal spas are used commercially.

Two resources could be exploited: sulphur deposits in the Santa Rosa Volcano zone, provided appropriate technologies are applied that do not destroy the landscape, and the generation of electric energy from thermal water sources. But preferentially, tourism is to be encouraged, principally in the zones where geomorphology offers a wide variety of environments in relatively short distances. The highest ice and snow of El Ruiz are readily accessible and well known to Colombians (INDERENA 1984).

Threats

In high terrain, the principal threat is volcanic eruptions. The palaeoecological history of the zone shows several strong eruptions, and three of the eight volcanoes are still active (Salamanca-V. 1991; Salomons 1986; Melief 1985). Burning of grass to provide more palatable grazing for cattle diminishes the diversity of the vegetation, making it uniform and indirectly producing soil erosion (Verweij and Budde 1992). The eroded sediments reach the lakes, causing siltation. Trampling by cattle on fragile natural systems, such as the peat bogs and marshes of the páramo, produces profound changes in the physical medium and the biotic composition. Currently the livestock is cattle – introduction of sheep and goats must be discouraged. Pressures exerted by gathering fuelwood, grazing cattle and growing potatoes in the zone between the Andean montane region and the páramo have produced rapid transformation of the landscape and altered the hydrologic cycle (cf. Verweij and Beukema 1992).

Changes in precipitation can be associated with the accelerated transformation of natural ecologic systems, especially in the high terrain. In the Andean and sub-Andean zones, the exhaustion of forest resources has diminished the flow in rivers and ravines that provide water to municipal aqueducts.

In the mid-zone the felling of trees has reached alarming proportions. On the eastern slope, the oak groves (*Quercus humboldtii*) are very depleted, whereas on the western slopes, where soils have better hydromorphic conditions, conservation of the forest is acceptable, but in the last 50 years this zone has suffered from logging and charcoal production. The middle and lower zones (the coffee-growing belt) need the greatest attention, since there are very few places with original flora and fauna. In the lowest region, conditions are similar to the coffee-growing belt. In these areas there are larger human settlements and the pressure on the natural resources is almost unbearable; the use of fuelwood for domestic and in some cases semi-industrial applications (bakery ovens, artisan kilns) threatens the very existence of the small forest remnants.

In Los Nevados Natural National Park (which was established in 1973, with some protection from 1959), ownership of the land is not always clear, and there is private property within the park that creates obstacles to any programme of recovery.

Conservation

In the transect region and surroundings, there are the following protected areas: Los Nevados Natural National Park, Ucumarí Regional Park (42.4 km²) under administration of Corporación Autónoma Regional de Risaralda (CARDER) and two areas under protection of Corporación Regional de Quindío (CRQ): Cañón del Quindío (c. 38 km²) contiguous to the NNP, and in Línea sector, 35 km². The areas surrounding Lake Otún, with peat bogs and páramo marshes, have been incorporated into the system of supervision and control of CARDER, so preservation is guaranteed. In Marsella (Risaralda),

a botanical garden has generated local interest to replant critical areas in order to manage water for human use.

The creation and impulse generated by ecological groups, especially in Pereira, Manizales and Santa Rosa de Cabal, have favourably influenced environmental education and conscience-raising within the communities. It is an urgent necessity to delimit Forest Reserves (buy land) in the sub-Andean region between 1200–2500 m; although this zone has greater human presence, it presents the largest diversity in the vegetation.

References

Boekhout, T. and Pulido-L., M.M. (1989). The occurrence of macrofungi and their habitats in vegetations along the Parque Los Nevados transect. In van der Hammen, T., Díaz-P., S. and Alvarez, V.J. (eds), *La Cordillera Central Colombiana: transecto Parque Los Nevados (segunda parte)*. Studies on Tropical Andean Ecosystems/Estudios de Ecosistemas Tropandinos Vol. 3. J. Cramer, Berlin. Pp. 507–515.

Cleef, A.M., Rangel-Ch., J.O. and Salamanca-V., S. (1983). Reconocimiento de la vegetación de la parte alta del transecto Parque Los Nevados. In van der Hammen, T., Pérez-P., A. and Pinto-E., P. (eds), *La Cordillera Central Colombiana: transecto Parque Los Nevados (introducción y datos iniciales)*. Studies on Tropical Andean Ecosystems/Estudios de Ecosistemas Tropandinos Vol. 1. J. Cramer, Vaduz. Pp. 150–173.

Cleef, A.M., Rangel-Ch., J.O. and Salamanca-V., S. (1996). The Andean forests of the Parque Los Nevados transect, Cordillera Central, Colombia. In van der Hammen, T. and dos Santos, A. (eds), *La Cordillera Central Colombiana: transecto Parque Los Nevados*. Studies on Tropical Andean Ecosystems/Estudios de Ecosistemas Tropandinos Vol. 4/5. J. Cramer, Berlin. In press.

Correa-Q., J.E. and Bernal, H.Y. (1989-1991). *Especies vegetales promisorias de los países del Convenio Andrés Bello*. Vols. I, V and VI. Junta del Acuerdo de Cartagena (JUNAC). Ministerio de Educación y Ciencia, Spain and Secretaría Ejecutiva del Convenio Andrés Bello (SECAB), Bogotá. 541 pp., 569 pp., 507 pp.

García-B., H. (1974-1975). *Flora medicinal de Colombia*, Vols. 1–3. Instituto de Ciencias Naturales, Universidad Nacional de Colombia, Bogotá. 561 pp., 538 pp., 495 pp.

INDERENA (1984). *Colombia parques nacionales*. Instituto Nacional de los Recursos Naturales Renovables y del Medio Ambiente (INDERENA), Bogotá. 263 pp.

Melief, A.B.M. (1985). *Late Quaternary paleoecology of the Parque Nacional Natural Los Nevados (Cordillera Central) and Sumapaz (Cordillera Oriental) areas, Colombia*. Thesis. University of Amsterdam, Amsterdam. 162 pp.

Mozo-M., T. (1968). *Catálogos de especies forestales colombianas. Nombres vernáculos y científicos*, 2nd edition. Mimeographed.

Murillo-P., M.T. (1983). *Usos de los helechos en Suramérica con especial referencia a Colombia.* Biblioteca J.J. Triana No. 5. Instituto de Ciencias Naturales, Museo de Historia Natural, Universidad Nacional de Colombia, Bogotá. 156 pp.

Pérez-P., A. (1983). Algunos aspectos del clima. In van der Hammen, T., Pérez-P., A. and Pinto-E., P. (eds), *La Cordillera Central Colombiana: transecto Parque Los Nevados (introducción y datos iniciales).* Studies on Tropical Andean Ecosystems Vol. 1. J. Cramer, Vaduz. Pp. 38–47.

Pérez-P., A. and van der Hammen, T. (1983). Unidades eco-geográficas y ecosistemas en el Parque Natural Los Nevados: una síntesis inicial. In van der Hammen, T., Pérez-P., A. and Pinto-E., P. (eds), *La Cordillera Central Colombiana: transecto Parque Los Nevados (introducción y datos iniciales).* Studies on Tropical Andean Ecosystems Vol. 1. J. Cramer, Vaduz. Pp. 277–300.

Rangel-Ch., J.O. (1991). *Vegetación y ambiente en tres gradientes montañosos de Colombia.* Ph.D. thesis. University of Amsterdam, Amsterdam. 349 pp.

Rangel-Ch., J.O. (1996). Diversidad, frecuencia de las familias, géneros y especies de plantas superiores en el transecto del Parque Los Nevados. In van der Hammen, T. and dos Santos, A. (eds), *La Cordillera Central Colombiana: transecto Parque Los Nevados.* Studies on Tropical Andean Ecosystems Vol. 4/5. J. Cramer, Berlin. In press.

Rangel-Ch., J.O., Cleef, A.M. and Salamanca-V., S. (1996). The equatorial interandean and subandean forests of the Parque Los Nevados transect, Cordillera Central, Colombia. In van der Hammen, T. and dos Santos, A. (eds), *La Cordillera Central Colombiana: transecto Parque Los Nevados.* Studies on Tropical Andean Ecosystems Vol. 4/5. J. Cramer, Berlin. In press.

Rangel-Ch., J.O., Idrobo, J., Cleef, A.M. and van der Hammen, T. (1996). Lista del material herborizado en el transecto Parque Los Nevados. In van der Hammen, T. and dos Santos, A. (eds), *La Cordillera Central Colombiana: transecto Parque Los Nevados.* Studies on Tropical Andean Ecosystems Vol. 4/5. J. Cramer, Berlin. In press.

Salamanca-V., S. (1991). *The vegetation of the páramo and its dynamics in the volcanic massif Ruiz-Tolima (Cordillera Central, Colombia).* Ph.D. thesis. University of Amsterdam, Amsterdam. 122 pp.

Salamanca-V., S., Cleef, A.M. and Rangel-Ch., J.O. (1996). The páramo vegetation of the Parque Nacional Los Nevados, Cordillera Central, Colombia. In van der Hammen, T. and dos Santos, A. (eds), *La Cordillera Central Colombiana: transecto Parque Los Nevados.* Studies on Tropical Andean Ecosystems Vol. 4/5. J. Cramer, Berlin. In press.

Salomons, J.B. (1986). *Paleoecology of volcanic soils in the Colombian Central Cordillera (Parque Nacional Natural de Los Nevados). Dissert. Bot.* 95. 212 pp.

Sturm, H. and Rangel-Ch., J.O. (1985). *Ecología de los páramos andinos: una visión preliminar integrada.* Biblioteca J.J. Triana No. 9. Instituto de Ciencias Naturales, Museo de Historia Natural, Universidad Nacional de Colombia, Bogotá. 292 pp.

Thouret, J.-C. (1983). La temperatura de los suelos: temperatura estabilizada en profundidad y correlaciones térmicas y pluviométricas. In van der Hammen, T., Pérez-P., A. and Pinto-E., P. (eds), *La Cordillera Central Colombiana: transecto Parque Los Nevados (introducción y datos iniciales).* Studies on Tropical Andean Ecosystems Vol. 1. J. Cramer, Vaduz. Pp. 142–149.

Thouret, J.-C. (1989). Geomorfología y crono-estratigrafía del macizo volcánico Ruiz-Tolima (Cordillera Central Colombiana). In van der Hammen, T., Díaz-P., S. and Alvarez, V.J. (eds), *La Cordillera Central Colombiana: transecto Parque Los Nevados (segunda parte).* Studies on Tropical Andean Ecosystems Vol. 3. J. Cramer, Berlin. Pp. 257–277.

Thouret, J.-C. and Faivre, P. (1989). Suelos de la Cordillera Central, transecto Parque Los Nevados. In van der Hammen, T., Díaz-P., S. and Alvarez, V.J. (eds), *La Cordillera Central Colombiana: transecto Parque Los Nevados (segunda parte).* Studies on Tropical Andean Ecosystems Vol. 3. J. Cramer, Berlin. Pp. 293–441.

Verweij, P.A. and Beukema, H. (1992). Aspects of human influence on upper-Andean forest line vegetation. In Balslev, H. and Luteyn, J.L. (eds), *Páramo: an Andean ecosystem under human influence.* Academic Press, London. Pp. 171–175.

Verweij, P.A. and Budde, P.E. (1992). Burning and grazing gradients in páramo vegetation: initial ordination analyses. In Balslev, H. and Luteyn, J.L. (eds), *Páramo: an Andean ecosystem under human influence.* Academic Press, London. Pp. 179–195.

Wolf, J.H.D. (1989). Comunidades epífitas en un transecto altitudinal en la Cordillera Central, Colombia: datos iniciales sobre la cantidad de especies de briófitos y líquenes. In van der Hammen, T., Díaz-P., S. and Alvarez, V.J. (eds), *La Cordillera Central Colombiana: transecto Parque Los Nevados (segunda parte).* Studies on Tropical Andean Ecosystems Vol. 3. J. Cramer, Berlin. Pp. 455–459.

Authors

This Data Sheet was written by Dr J. Orlando Rangel-Ch. and Aída Garzón-C. (Universidad Nacional de Colombia, Instituto de Ciencias Naturales, Museo de Historia Natural, Apartado Aéreo 7495, Santafé de Bogotá, Colombia).

COLOMBIAN CENTRAL MASSIF
Colombia

Location: South-western Colombia east of Popayán: transect from Puracé Volcano eastward to Magdalena River Valley, from municipality Puracé (Department Cauca) to Tesalia (Department Huila), between about latitudes 2°21'–2°29'N and longitudes 76°23'–75°44'W.

Area: 1500–2000 km².

Altitude: Transect mainly c. 4500–1000 m.

Vegetation: In upper reaches, páramo – "pajonales" of *Calamagrostis* spp., "frailejones" of *Espeletia hartwegiana*. In high-Andean forests, *Weinmannia brachystachya, Miconia puracensis*; in mid-Andean forests, *Quercus humboldtii*. Lower fringe in tropical zone, with open forests of *Guarea guidonia* and semi-arid areas with columnar cacti.

Flora: c. 1200 vascular plant species; 270 recorded bryophytes and lichens.

Useful plants: For folk medicine, fuelwood, lumber, germplasm for reforestation programmes.

Other values: Threatened fauna; watershed protection; scenic landscapes; thermal springs of San Juan and Pilimbalá; archaeological sites; Amerindian homelands.

Threats: Mineral wastes from mining sulphur; burning of bunchgrasses; cattle-ranching; draining páramo marshes and peat bogs; felling trees; erosion; in mid-sector, accelerated land clearing.

Conservation: Puracé Natural National Park – a portion of Cinturón Andino Biosphere Reserve; Merenberg Nature Reserve; Amerindian Reserves.

Geography

The transect of La Plata Valley (Map 65) is between the crater of the volcano Puracé and the Serranía de las Minas, which is north of the Upper Magdalena River Valley – where the Central and Eastern cordilleras diverge northward from the Colombian Central Massif (Gran Macizo Colombiano). The transect includes part of Puracé Natural National Park, which extends from c. 2600–5000 m with the Serranía de Los Coconucos and covers c. 830 km² between latitudes 1°50'–2°24'N and longitudes 76°07'–76°37'W in the departments Huila and Cauca. Diverse life zones are present in this mountainous gradient; the high Andes and the páramo are biologically better known.

The regional geology presents a sequence: rocks from the Triassic-Jurassic, intrusive Jurassic rocks, marine sedimentary rocks of the Cretaceous, Tertiary molasse deposits and Quaternary deposits, which are related to orogenic volcanism or post-orogenic events in the rise of the Central and the Eastern cordilleras (Kroonenberg and Diederix 1985). Among the orogenic events, ignimbritic deposits characterize the landscape of the high plains, which are associated with lava flows of small volcanoes such as Merenberg and El Pensil. West of Puracé Volcano (4780 m) in the upper elevations of the massif is the Popayán formation of lavas, ashes, conglomerates, tufa and fluvial deposits.

Along the altitudinal transect of these mountain systems are different types of soil, as well as a wide variety of annual average temperatures and precipitation (Botero 1985; Rangel-Ch. and Espejo-B. 1989). From 500–1100 m in the tropical or equatorial region, in the alluvial plain of the Magdalena River and the foothills of Tesalia and Paicol, the prevailing soils are haplustalfs; the annual average temperature is 30°C, the precipitation varies between 1702–2108 mm. From 1150–2500 m in the higher tropical fringe and the sub-Andean region on recent colluvium, dystropept and haplustalf soils prevail; the temperature averages 18°C, the precipitation 1974 mm. From 2500–3300 m in the Andean montane region, in landscapes of ignimbritic high plains of old colluvium and ash mantles, dystropepts and dystrandepts prevail; the temperature is 15.2°C, the precipitation 1975 mm. From 3300–4400 m in the páramo region are moraines, deposits of pyroclasts, mud-flows and colluvium, with troposaprists, tropofluvents, hydrandepts and dystrandepts prevailing; the average temperature varies from 8°C to 5°C, the precipitation averages 2263 mm.

Although most often the rainiest month is May and the driest is January, and the mean monthly temperature may be lowest in about July to August and highest in about February or March, there is extreme diversity and even contrast from one part of the transect to another. Data on the seasonal variations with altitude through the year are presented in Rangel-Ch. and Espejo-B. (1989).

Vegetation

In the páramo different types of vegetation occur, principally bunchgrass fields ("pajonales") particularly on the slopes dominated by *Calamagrostis* species, "frailejones" of *Espeletia hartwegiana* subsp. *centroandina* (Asteraceae) on sites with well-watered soils, and peat bogs of *Distichia muscoides*. In the mid-elevations, the high-Andean forests (with trees to 10 m high) are dominated by *Weinmannia* and *Miconia*, whereas farther down slope are higher forests (over 25 m) of two strata in which *Quercus humboldtii* (Fagaceae) prevails. In adjacent zones following the gradient are forests of Lauraceae. In the tropical fringe prevail forests of deciduous foliage with *Guarea guidonia* (Meliaceae) and Leguminosae, and in semi-arid sites columnar cacti.

The syntaxonomic arrangement for part of the transect vegetation is as follows (Duque-N. and Rangel-Ch. 1989; Rangel-Ch. and Lozano-C. 1989; Rangel-Ch. and Franco-R. 1985):

1.   Páramo region (4600-3500 m), with superpáramo fringe (4350-3800 m)

1.1. Azonal vegetation
Alliance Oritrophio-Distichion muscoidis (4380-3600 m). Community groups of two physiognomic types: meadows and open thickets, characteristically with *Oritrophium peruvianum* (Asteraceae), *Valeriana microphylla* (Valerianaceae) and *Distichia muscoides* (Juncaceae). Three Associations: Lachemillo pectinatae-Loricarietum colombianae with *Lasiocephalus otophorus* (Asteraceae), *Bartsia stricta* (Scrophulariaceae) and *Carex bonplandii* (Cyperaceae). Lupino alopecuroidis-Valerianetum microphyllae with *Poa pauciflora* (Poaceae) and *Lachemilla nivalis* (Rosaceae). Agrostio boyacensis-Distichietum muscoidis bogs with *Sphagnum magellanicum* and *Pernettya prostrata* (Ericaceae).

1.2. Zonal vegetation
Order Calamagrostietalia effusae. Alliance Calamagrostio-Espeletion hartwegianae (3800–3200 m). Community groups with the three physiognomic types bunchgrass field, frailejón and semi-open thicket. Four Associations: Blechno loxensis-Espeletietum hartwegianae with additional associated species *Baccharis tricuneata* (Asteraceae), *Hypericum laricifolium* (Clusiaceae) and *Calamagrostis macrophylla* (Poaceae). Blechno loxensis-Diplostephietum floribundi with *Sibthorpia repens* (Scrophulariaceae) and many non-vascular epiphytes (*Frullania, Usnea*). Calamagrostietum effusae-macrophyllae with *Oreobolus venezuelensis* (Cyperaceae), *Ranunculus nubigenus* (Ranunculaceae) and *Bromus catharticus* (Poaceae). Chusquetum tessellatae ("chuscales") with the páramo bamboo (*Chusquea tessellata*) and other species such as *Calamagrostis effusa*, *Niphogeton ternata* (Apiaceae) and *Geranium sibbaldioides* (Geraniaceae).

2.   Andean montane region (3400-2400 m)
High-Andean forest. Association Weinmannio brachystachyae-Miconietum cuneifoliae (3400–3300 m). Three communities: *Weinmannia mariquitae* (Cunoniaceae) and *Miconia cuneifolia* (Melastomataceae) (3320 m); *Myrica pubescens* (Myricaceae), *Weinmannia subvelutina* and *Drimys granadensis* (Winteraceae) (3050 m); *Brunellia macrophylla* (Brunelliaceae), *Weinmannia pubescens*, *Clethra* aff.

MAP 65. NORTH-EAST OF COLOMBIAN CENTRAL MASSIF (CPD SITE SA29)

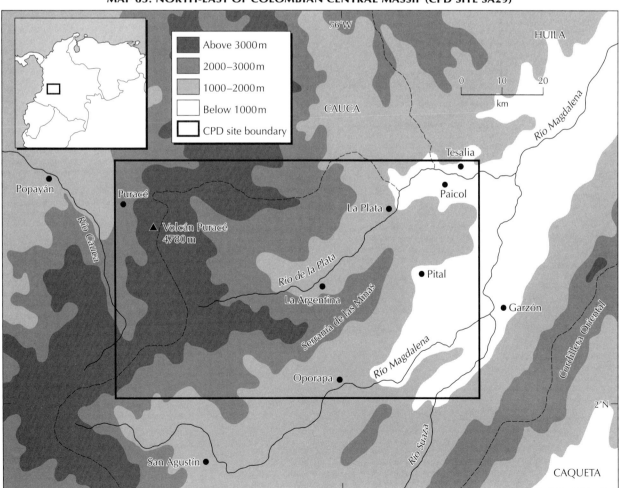

revoluta (Clethraceae) and *Hedyosmum bonplandianum* (Chloranthaceae) (2980 m).

3. Mid-zone

Mid-zone forest. Alliance Monotropo-Quercion humboldtii (2600–1800 m). Association Hedyosmo-Quercetum humboldtii (2450–2200 m). Community of *Hedyosmum huilense*, *Clethra fagifolia* and *Billia columbiana* (Hippocastanaceae) (2450 m).

There are enclaves of páramo vegetation at low altitudes, in marshy or boggy sites originating from the damming of streambeds by volcanic flows (Kroonenberg and Diederix 1985). A typical example is at 2380 m with the peat bog of La Candelaria, close to the Merenberg Nature Reserve, where in a short stretch are different stages of the sequential vegetation process – from marshy sites with a few water holes where *Guzmania gracilior*, *Blechnum columbiense* and *Oreobolus venezuelensis* prevail, to frailejones physiognomically and floristically similar to those of the páramo zone at 3400 m, with *Espeletia hartwegiana* and *Hypericum lancioides*. In the ecotone with forest vegetation, *Hedyosmum huilense* and Lauraceae prevail, and thickets are established of *Diplostephium floribundum*, *Weinmannia* sp. and *Miconia floribunda* (Rangel-Ch. and Lozano-C. 1986).

4. Tropical region

In the nearby tropical region, especially along the eastern slopes facing the Magdalena Valley, different types of vegetation are established, with prevalence at 1000 m of *Guarea guidonia* and *Perebea* sp. Three other communities along the gradient are: (i) *Spondias mombin*, *Hirtella americana* and *Mouriri myrtilloides* at 1000–900 m in moist locales of the dry valley of the Magdalena River; (ii) *Pithecellobium dulce*, *Xylosma velutinum* and *Croton argyrophyllus* at 780 m; and (iii) *Stenocereus* (*Lemaireocereus*) *griseus* and *Randia aculeata* at 530 m.

Flora

The eastern slope of the Puracé Volcano and the Magdalena Valley are better known and these data refer mainly to them. The western slope facing the city Popayán is less known and the environmental impact in the surrounding area has been greater. The estimate of 1200 species of seed plants and ferns in the transect between 1000–4450 m corresponds to 431 genera of 167 families. Table 62 presents data on the dominant families and genera and those families' distributions in the four life zones of the gradient (Rangel-Ch. 1991; Lozano-C. and Rangel-Ch. 1989).

On the higher elevations of the massif, there is endemism or restricted distribution for *Gunnera caucana* (Gunneraceae); *Miconia puracensis* (Melastomataceae); *Aphelandra grangeri* (Acanthaceae); and *Aequatorium latibracteolatum*, *Ageratina paezensis*, *Lasiocephalus puracensis*, *Pentacalia arbutifolia* (Asteraceae). Between 3100–2000 m, endemics include *Peperomia aguabonitensis*, *Piper subflavum* var. *longipedunculatum* (Piperaceae); *Ficus guntheri* (Moraceae); *Saurauia pulchra* (Actinidiaceae); *Begonia killipiana*, *B. tiliaefolia* (Begoniaceae); *Cavendishia divaricata*, *C. vinacea* (Ericaceae); *Salvia rufula* var. *paezorum* (Lamiaceae); *Aphelandra huilensis*; and species

TABLE 62. ESTIMATES OF FLORAL DIVERSITY ON PURACE VOLCANO AND IN SURROUNDING LOCALITIES

1. Ten families with largest numbers of species and genera

Family	Number of species	Family	Number of genera
Orchidaceae	76	Asteraceae	30
Asteraceae	75	Orchidaceae	23
Ericaceae	57	Ericaceae	15
Polypodiaceae *s.l.*	52	Rubiaceae	15
Melastomataceae	40	Poaceae	14
Piperaceae	40	Solanaceae	11
Rubiaceae	38	Euphorbiaceae	8
Solanaceae	38	Cyperaceae	7
Bromeliaceae	22	Gesneriaceae	7
Poaceae	21	Melastomataceae	7

2. Eleven genera with the most species

Genus	Number of species	Family
Peperomia	22	Piperaceae
Solanum	22	Solanaceae
Miconia	20	Melastomataceae
Polypodium	15	Polypodiaceae
Piper	14	Piperaceae
Anthurium	12	Araceae
Pleurothallis	11	Orchidaceae
Guzmania	10	Bromeliaceae
Pilea	10	Urticaceae
Bomarea	9	Alstroemeriaceae
Centropogon	9	Campanulaceae

3. Families with the most species by life zone

Tropical (550–1100 m) Family	Number of species	Sub-Andean (1150–2500 m) Family	Number of species
Myrtaceae	7	Orchidaceae	43
Orchidaceae	6	Polypodiaceae *s.l.*	33
Piperaceae	6	Solanaceae	32
Rubiaceae	5	Piperaceae	28
Asteraceae	4	Asteraceae	26
Euphorbiaceae	4	Rubiaceae	22
Flacourtiaceae	4	Ericaceae	19
Araceae	3	Melastomataceae	16
Bromeliaceae	3	Araceae	14
Sapindaceae	3	Bromeliaceae	14

Andean (2500–3000 m) Family	Number of species	Páramo (over 3300 m) Family	Number of species
Asteraceae	46	Asteraceae	17
Ericaceae	45	Poaceae	10
Orchidaceae	34	Ericaceae	5
Polypodiaceae *s.l.*	30	Rosaceae	5
Melastomataceae	28	Apiaceae	4
Piperaceae	28	Cyperaceae	3
Solanaceae	25	Clusiaceae	2
Rubiaceae	21	Leguminosae	2
Gesneriaceae	16	Melastomataceae	2
Cunoniaceae	12	Scrophulariaceae	2

of *Meliosma* (Sabiaceae), *Alfaroa* (Juglandaceae), *Myrcia* (Myrtaceae) and *Solanum* (Solanaceae).

Useful plants

In the high zones or páramo, bunches of *Calamagrostis macrophylla* and *C. effusa* are used to roof huts and for fodder. Species used to spice meats include *Hypochoeris sessiliflora* (Asteraceae) and *Azorella* spp. (Apiaceae). In the treeline open woods, the woody species are used for fuelwood. In the mid-Andean montane region, reforestation programmes use the native *Quercus humboldtii*, *Billia columbiana* ("cariseco"), *Alnus acuminata* ("aliso" – Betulaceae) and *Beilschmiedia* cf. *sulcata* (Lauraceae). Living hedges are made with *Sapium cuatrecasasii* ("lechero" – Euphorbiaceae), *Cedrela montana* (Meliaceae), *Guazuma ulmifolia* (Ulmaceae), *Spondias mombin* (Anacardiaceae) and *Styrax leptactinosus* (Styracaceae).

For folk medicinals, species include *Oreopanax floribundus* (Araliaceae); *Baccharis genistelloides, B. nitida, Barnadesia spinosa, Bidens pilosa, Gnaphalium antennarioides, Mikania antioquensis* (Asteraceae); *Befaria aestuans, Gaultheria anastomosans* (Ericaceae); *Gentiana corymbosa, G. sedifolia, Gentianella dasyantha, G. diffusa* (Gentianaceae); *Dendrophthora clavata, Gaiadendron punctatum, Oryctanthus botryostachys* (Loranthaceae); *Oxalis latoides* (Oxalidaceae); *Passiflora cumbalensis* (Passifloraceae); *Peperomia* cf. *macrostachya* (Piperaceae); *Lachemilla galeioides, L. hispidula, L. nivalis, L. pectinata, Rubus glabratus, R. guyanensis* (Rosaceae); *Cinchona officinalis, Ladenbergia macrocarpa, Nertera granadensis* (*N. depressa*), *Relbunium hypocarpium* (Rubiaceae); *Solanum lepidotum, S. lycioides, S. nigrum, S. quitoense* (Solanaceae); and *Viola stipularis* (Violaceae) (García-B. 1974–1975).

Social and environmental values

In localities of La Argentina and Serranía de las Minas are archaeological remains of houses and tombs of pre-Columbian settlements. The earliest known complex societies of the northern Andean highlands are in the Alto Magdalena region, which was densely populated by indigenous communities well before the arrival of the Spanish. The early cultures like those of San Agustín and Saladoblanco are of great archaeological and historical value, showing year-round settlements in the Upper Magdalena region for 2500 years (Drennan 1985a; Herrera, Drennan and Uribe 1989).

On the eastern slopes in the region of La Plata Valley, the inhabitants were Yalcones, Moscopanes and Quinchinas. The western slopes were inhabited by Paeces, Guambianos, Puracés, Coconucos and Popayanes. Most of these groups are extinct. A community of the Paece now lives in the neighbourhood of Puracé Volcano and Guambianos dwell in localities near Silvia (INDERENA 1984). Their homelands must be protected.

The páramo region close to Puracé Volcano is an important landscape resource. In the mid-elevation Andean region, the oak groves and other vegetation with *Quercus humboldtii* are the last remnants of the original vegetation; the associated fauna, especially birds and small mammals, find food there. Reforestation programmes with native species obtain necessary germplasm in these sites. The peat-bog zone of La Candelaria is a biological heritage that must be protected and studied in detail. There are over 50 lakes in the park. In localities of the transect several rivers flow – some originating on the mountaintops of the central massif. Conservation of these watersheds is vital to the surrounding municipalities. Several sites have thermal waters; their careful exploitation could bring in resources for conservation.

Threatened fauna in the region include mountain tapir (*Tapirus pinchaque*), rabbit deer (*Pudu mephistophiles*), spectacled bear (*Tremarctos ornatus*), Andean condor (*Vultur gryphus*) and black-and-chestnut eagle (*Oroaetus isidori*). Regional endemics include a bicolored antpitta (*Grallaria rufocinerea romeroana*). The Central Andean páramo Endemic Bird Area (EBA B60) and Subtropical inter-Andean Colombia EBA (B12) embrace the entire Central Andean cordillera from the páramo down to the subtropical (mid-Andean) forest. Twenty-seven restricted-range bird species occur in these two EBAs, 14 of which are entirely confined to the region. Due to increasing disturbance of the specialized vegetation, 15 bird species are considered threatened. In the drier low-lying zones of the adjacent Cauca and Magdalena valleys, this area overlaps with the Dry inter-Andean valleys EBA (B13), where four restricted-range birds occur.

Economic assessment

In the páramo region close to the Puracé Volcano, sulphur is extracted by mining. Indigenous people make use of the bunchgrasses for extensive cattle-ranching, and on drier slopes, cultivate potatoes and barley. In the mid-part of the transect in the Andean montane region, a clay is utilized in arts and crafts. In this same region, coffee plantations exist, sometimes combined with grazing of cattle. In forested areas there is timbering; the forests of mid-elevations and those of the altitudinal ecotone between dense and open vegetation could be managed silviculturally, so that wood for the local populations would be a permanent resource. In the last few years an illegal crop (*Papaver somniferum*, opium poppy) has become a problem. Tourism is another resource – to visit the thermal spas of San Juan and Pilimbalá, the scenic páramo landscape and the areas of pre-Columbian settlement.

Threats

All areas suffer pressure from one or more factors of human origin. In the páramo, the mining causes access roads to be constructed without concern for biology or conservation, uncontrolled release of gases and solid residues into the atmosphere and disposal of untreated wastes into creeks and gullies. Although on the slopes that face Popayán there are indications of use as long as 2500 years ago (Drennan 1985b), the present cultivation of potatoes has transformed large areas.

Also responsible for modifying the landscape is disorderly construction of roads to the Puracé crater, and extensive cattle-ranching – which entails periodic burning of bunchgrass fields and frailejones. This produces soil erosion. The drying-out of páramo marshes and peat bogs constitutes irreparable damage, since it is in the highlands that the rivers originate. In the mid-part of the region, deforestation and excessive felling of trees have produced extensive cleared sectors, where soil erosion is worrisome.

Conservation

Cinturón Andino Biosphere Reserve (declared in 1979) encompasses 8550 km² from Nevado del Huila Natural National Park (1580 km²) southward as far as Cueva de Los

Guácharos NNP (90 km²) (Olaya-Amaya c. 1984), thus including Puracé NNP (830 km²), which was first protected in 1961 and established in 1968 – it should be extended to protect Paletará Valley (in Cauca), where areas of great biodiversity exist. A Forest Reserve should be strongly supported to supply seeds and seedlings necessary for reforestation of sites particularly in the mid-zone. The private Merenberg Nature Reserve (c. 3 km²), which began in 1932, conserves remnants of oak forest from 2300–2800 m; while protecting the reserve, in 1975 Mrs Mechthild Buch was murdered. Near the oak groves at 2300–2400 m, a buffer zone should be created to include peat bogs with enclaves of páramo vegetation and a representative portion of the zonal forest. In the vicinity of the Serranía de las Minas localities with very humid atmosphere exist, due to orographic precipitation; oak groves occur together with species of *Alfaroa* and *Clusia* in a unique situation (Rangel-Ch. and Lozano-C. 1989).

References

Botero, P.J. (1985). Soilscapes: preliminary study/Paisajes-suelos: estudio preliminar. In Drennan, R.D. (ed.), *Regional archaeology in the Valle de la Plata, Colombia*. Museum of Anthropology, University of Michigan, Ann Arbor. Tech. Reports 16. Pp. 41–79.

Drennan, R.D. (1985a). Archeological survey and excavation/Excavación y reconocimiento arqueológico. In Drennan, R.D. (ed.), *Regional archaeology in the Valle de la Plata, Colombia*. Museum of Anthropology, University of Michigan, Ann Arbor. Tech. Reports 16. Pp. 117–180.

Drennan, R.D. (ed.) (1985b). *Regional archaeology in the Valle de la Plata, Colombia: a preliminary report on the 1984 season of the Proyecto Arqueológico Valle de la Plata/Arqueología regional en el Valle de la Plata, Colombia: informe preliminar sobre la temporada de 1984 del Proyecto Arqueológico Valle de la Plata*. Museum of Anthropology, University of Michigan, Ann Arbor. Technical Reports No. 16. 195 pp.

Duque-N., A. and Rangel-Ch., J.O. (1989). Phytosociological analysis of the páramo vegetation of Puracé Natural Park/Análisis fitosociológico de la vegetación paramuna del Parque Natural Puracé. In Herrera, L.F., Drennan, R.D. and Uribe, C.A. (eds), *Prehispanic chiefdoms in the Valle de la Plata, Volume 1*. University of Pittsburgh Memoirs in Latin American Archaeology 2. Pp. 69–93.

García-B., H. (1974-1975). *Flora medicinal de Colombia*, Vols. 1–3. Instituto de Ciencias Naturales, Universidad Nacional de Colombia, Bogotá. 561 pp., 538 pp., 495 pp.

Herrera, L.F., Drennan, R.D. and Uribe, C.A. (eds) (1989). *Prehispanic chiefdoms in the Valle de la Plata, Volume 1: the environmental context of human habitation/Cacicazgos prehispánicos del Valle de la Plata, Tomo I: el contexto medioambiental de la ocupación humana*. University of Pittsburgh Memoirs in Latin American Archaeology No. 2. 238 pp.

INDERENA (1984). *Colombia parques nacionales*. Instituto Nacional de los Recursos Naturales Renovables y del Medio Ambiente (INDERENA), Bogotá. 263 pp.

Kroonenberg, S.B. and Diederix, H. (1985). Geology/Geología. In Drennan, R.D. (ed.), *Regional archaeology in the Valle de la Plata, Colombia*. Museum of Anthropology, University of Michigan, Ann Arbor. Tech. Reports 16. Pp. 23–40.

Lozano-C., G. and Rangel-Ch., J.O. (1989). Floral inventory of the Valle de la Plata: vegetation profile from La Plata to the Puracé Volcano/Inventario florístico del Valle de la Plata: perfil de vegetación entre el municipio de La Plata (Huila) y el Volcán Puracé (Cauca). In Herrera, L.F., Drennan, R.D. and Uribe, C.A. (eds), *Prehispanic chiefdoms in the Valle de la Plata, Volume 1*. University of Pittsburgh Memoirs in Latin American Archaeology 2. Pp. 39–68.

Olaya-Amaya, A. (c. 1984). Parques nacionales en El Huila. Nevado del Huila, Puracé y Cueva de Los Guácharos: Reserva de Biósfera Agrupada del Cinturón Andino. *Huila. Rev. Acad. Huilense Hist.* 3: 15–25.

Rangel-Ch., J.O. (1991). *Vegetación y ambiente en tres gradientes montañosos de Colombia*. Ph.D. thesis. University of Amsterdam, Amsterdam. 349 pp.

Rangel-Ch., J.O. and Espejo-B., N.E. (1989). Climate/Clima. In Herrera, L.F., Drennan, R.D. and Uribe, C.A. (eds), *Prehispanic chiefdoms in the Valle de la Plata, Volume 1*. University of Pittsburgh Memoirs in Latin American Archaeology 2. Pp. 14–38.

Rangel-Ch., J.O. and Franco-R., P. (1985). Modern flora: plant communities on the Paicol-Puracé transect (Cordillera Central)/Flora actual: comunidades vegetales en el transecto Paicol-Puracé (Cordillera Central). In Drennan, R.D. (ed.), *Regional archaeology in the Valle de la Plata, Colombia*. Museum of Anthropology, University of Michigan, Ann Arbor. Tech. Reports 16. Pp. 81–108.

Rangel-Ch., J.O. and Lozano-C., G. (1986). Un perfil de vegetación entre La Plata (Huila) y el Volcán del Puracé. *Caldasia* 14 (68–70): 503–547.

Rangel-Ch., J.O. and Lozano-C., G. (1989). The forest vegetation of the Valle de la Plata/La vegetación selvática y boscosa del Valle de la Plata. In Herrera, L.F., Drennan, R.D. and Uribe, C.A. (eds), *Prehispanic chiefdoms in the Valle de la Plata, Volume 1*. University of Pittsburgh Memoirs in Latin American Archaeology 2. Pp. 95–118.

Authors

This Data Sheet was written by Dr J. Orlando Rangel-Ch. and Aída Garzón-C. (Universidad Nacional de Colombia, Instituto de Ciencias Naturales, Museo de Historia Natural, Apartado Aéreo 7495, Santafé de Bogotá, Colombia).

VOLCANOES OF NARIÑENSE PLATEAU
Colombia and Ecuador

Location: Four main volcanoes mostly in Nariño Department of south-western Colombia, with bordering northern Ecuador:
1. Galeras: eastern slope – municipality Pasto, western slope – municipalities La Florida, Sandoná, Consacá, Yacuanquér; latitude 1°13'N, longitude 77°22'W, altitude 4276 m.
2. Azufral: municipality Túquerres; latitude 1°05'N, longitude 77°41'W, altitude 4070 m.
3. Cumbal: municipality Cumbal; latitude 0°59'N, longitude 77°53'W, altitude 4850 m.
4. Chiles: municipality Cumbal, Police Department Chiles and Ecuador's Province Carchi; latitude 0°49'N, longitude 77°56'W, altitude 4761 m.

Area: c. 500 km².

Altitude: Gradients studied c. 3000–4500 m.

Vegetation: Forest with *Polylepis, Miconia, Diplostephium*; thickets with *Loricaria thujoides*; "frailejones" with *Espeletia hartwegiana, E. pycnophylla*; bunchgrass fields with *Calamagrostis effusa*.

Flora: c. 450 species of vascular plants, 80 species of bryophytes.

Useful plants: Many species used for medicinals; also for food, crafts, poison.

Other values: Fauna refuge, including threatened species; watershed protection; Amerindian lands; extraction of sulphur; scenery, tourism.

Threats: Volcanic eruptions (especially Galeras); potato farming; extensive cattle ranching; gathering fuelwood; reforestation with exotic species (*Pinus*).

Conservation: Galeras Fauna and Flora Sanctuary (176 km²).

Geography

Central-eastern Nariño is bounded by two mountain ranges with south-west – north-east axes (Map 66), and has major elevations as the four volcanoes Chiles (4761 m), Cumbal (4850 m), Azufral (4070 m) and Galeras (4276 m) (Cuellar-R. and Ramírez-L. 1986). The high Túquerres-Ipiales plateau of the inter-Andean depression is an inclined terrace starting with the mountainsides of Cumbal and Azufral and forms a series of low hills eastward down to the deep Guáitara River Valley (Luna-Z. and Carlhoum 1986). The area of these volcanoes is west and south-west of the city Pasto in Colombia's municipalities El Encano, Pasto, La Florida, Sandoná, Ancúya, Consacá, Yacuanquér, Tángua, Guaitarilla, Túquerres, Guachucal, Cumbal and Carlosama, with Chiles Volcano also in Ecuador's Province Carchi.

The landscape of the Nariño high plateau has resulted from a sequence of uplifts and sinkings during the Tertiary, processes of erosion and sedimentation due to orogenic movements, and volcanic activity of the Pleistocene continued during the Quaternary (cf. van der Hammen and Cleef 1986). The most important geomorphological factor relates to volcanic activity; the inter-Andean depression is filled with material from old eruptions. Quaternary (also volcanic) material is composed of andesitic tufa, agglomerite and gravel (pumice stone and sandy tufa). The tufic layers are of great extent and varied thickness (Grosse 1935). Toward the southern end of the region in the foothills of Azufral and Cumbal there are also pyroclastic deposits. The soils have originated from the volcanism of the Tertiary and the Quaternary.

Differences in soil types and annual mean temperature and precipitation are found along the gradients of these mountain systems (Luna-Z. and Carlhoum 1986). From 2000–3000 m, the soils are classified as andic humitropepts and typic dystrandepts; the mean temperature varies between 15°–10°C, the precipitation between 600–2000 mm. From 3000–3500 m, the soils belong to the umbric vitrandept and dystric cryandept groups; the mean temperature varies between 10°–6°C, the precipitation between 1000–2000 mm. From 3500–4200 m, the soils belong to the humitropept, dystropept and cryumbret groups; the mean temperature varies between 6°–0°C, and the thermal gradient changes 0.6°C per 100 m in elevation.

Vegetation

The high-Andean fringe and the páramo region are considered for these four volcanoes; they are best conserved along the gradients studied (Sturm and Rangel-Ch. 1985; Erazo-N. *et al.* 1991; Rangel-Ch. 1995).

Galeras Volcano

1. High-Andean region (3100–3500 m)
On rocky sites grow woods of *Weinmannia* cf. *microphylla* with arching trunks, the treetops covered with epiphytic bryophytes. In the lower stratum is abundant *Greigia* aff. *exserta* (Bromeliaceae), with *Rhynchospora aristata* and *Diplostephium glandulosum*.

There are also thickets of *Diplostephium floribundum*, *Miconia salicifolia*, *Pentacalia* sp. and *Solanum bogotense*. In the lower stratum, the dominants are *Coriaria thymifolia*, *Vaccinium* sp. and *Siphocampylus giganteus*.

2. Páramo region (3600-4400 m)

2.1. Páramo proper
"Pajonal-frailepajonal" (bunchgrass – frailejón-bunchgrass) of *Calamagrostis effusa* and *Espeletia hartwegiana*, with a shrubby stratum of *Puya hamata*, *Blechnum loxense* and *Diplostephium glandulosum*. In the herb stratum are *Loricaria thujoides*, *Lupinus colombiensis* (Leguminosae) and *Vaccinium floribundum*. In the ground stratum are *Arcytophyllum muticum*, *Hypochoeris setosus*, *Geranium rhomboidale* (Geraniaceae) and bryophytes in *Breutellia*, *Riccardia* and *Polytrichum*.

Thickets of *Loricaria thujoides* and *Arcytophyllum capitatum* prevail in páramo above 3800 m, with *Vaccinium floribundum*, *Hesperomeles heterophylla*, *Brachyotum strigosum* and *Gynoxys sancti-antoni*. In the herbaceous stratum are *Calamagrostis effusa*, *Halenia* sp. (Gentianaceae), *Rhynchospora macrochaeta* and *Hypericum lancioides* (Clusiaceae). In the ground stratum are *Gunnera magellanica*, *Geranium sibbaldioides* and *Nertera granadensis* (Rubiaceae).

2.2. Superpáramo (over 4200 m)
On the slopes, vegetation is discontinuous and dominated by mats of *Werneria humilis*, *Azorella pedunculata*, *Arcytophyllum muticum* and the more widely distributed *Hypochoeris sessiliflora* with *H. setosus*. At 4350 m, 30% of the terrain is covered with species such as *W. humilis*, *Hypochoeris* sp. (robust, with yellow flowers), *Agrostis* cf. *araucana* and the bryophytes *Racomitrium crispulum* and *Campylopus* sp.

Azufral Volcano

1. High-Andean region (3200-3600 m)
On mountaintops at 3200–3300 m are sparse patches of arborescent shrubs (to 6 m tall). The dominants are *Saurauia bullosa* (Actinidiaceae); *Oligactis coriacea*, *Barnadesia spinosa*, *Gynoxys sancti-antoni* (Asteraceae); *Hesperomeles glabrata*; *Viburnum pichinchense*; and *Miconia* spp. In the lower stratum are *Geissanthus serrulatus* (Myrsinaceae), *Piper lacunosum* (Piperaceae), *Cestrum* sp. (Solanaceae), *Monnina arborescens* (Polygalaceae) and species of *Anthurium*, *Peperomia* (Piperaceae) and *Polypodium* (Polypodiaceae).

On the hillsides (3540 m), shrub-like forests are dominated by *Escallonia myrtilloides* (Saxifragaceae) and *Weinmannia microphylla*. In the lower stratum are species of *Diplostephium* and *Gynoxys*. In the ground stratum is *Lachemilla orbiculata* together with bryophytes such as *Pleurozium schreberi* and *Campylopus* spp.

2. Páramo region (3600-4200 m)
The dominant vegetation consists of tussocks of *Calamagrostis effusa*; other communities include: (1) Tussocks of *Calamagrostis effusa* and *Cortaderia sericantha*, distributed

MAP 66. VOLCANOES OF NARIÑENSE PLATEAU (CPD SITE SA30)

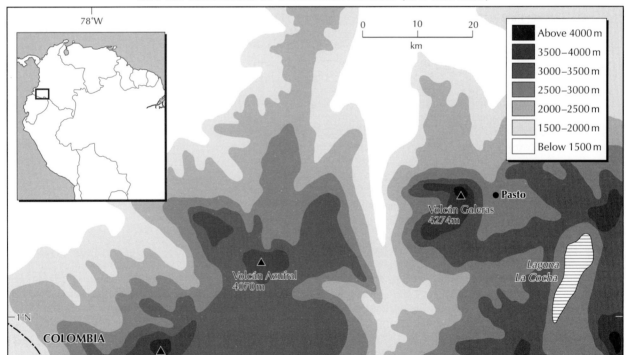

over drier areas with *Blechnum columbiense* and *Jamesonia robusta* (Polypodiaceae *s.l.*); (2) Bunchgrass-shrubland of *Calamagrostis effusa* and *Loricaria* cf. *complanata* together with a shrub-like stratum of *Pentacalia vaccinioides*, *Diplostephium* cf. *schultzei* and *Hypericum strictum*. In the herbaceous stratum are *Pernettya prostrata* and *Rhynchospora macrochaeta*. In the ground stratum are *Werneria humilis*, *Nertera granadensis* and *Oreobolus obtusangulus*.

In fields and peat bogs in marshy sites predominate *Werneria humilis*, *Loricaria thujoides* and *Luzula racemosa* (Juncaceae), associated with *Oritrophium limnophyllum*, *Eryngium humile* and the moss *Rhacocarpus purpurascens*. Above 4000 m, peat-bog vegetation is dominant. Occasionally there are patches of *Oritrophium limnophyllum*, mosses such as *Rhacocarpus purpurascens* and *Racomitrium crispulum*, and *Disterigma empetrifolium*, *Lachemilla hispidula* and *Loricaria* spp.

Cumbal Volcano

1. High-Andean region (3450–3600 m)
Shrubland is dominated by *Diplostephium schultzei* and *D. floribundum*, together with *Hypericum laricifolium*, *Blechnum loxense* and *Miconia salicifolia*.

Forests of *Polylepis* cf. *incana* grow in protected sites with *Diplostephium floribundum*, *Macleania rupestris*, *Myrsine dependens* (Myrsinaceae) and *Miconia salicifolia*. In the lower stratum are *Rhynchospora aristata*, *Carex bonplandii* and other *Miconia* spp.

Shrub-like forest is dominated by *Miconia salicifolia*, *Solanum bogotense* and species of Asteraceae.

2. Páramo region (3600–4100 m)
Pajonal-frailepajonal of *Calamagrostis effusa* and *Espeletia pycnophylla* with a shrub stratum of *Blechnum loxense*, *Pentacalia andicola* and *Castilleja fissifolia*. In the herbaceous stratum predominate *Agrostis tolucensis*, *Baccharis genestilloides* and *Rhynchospora macrochaeta*. In the ground stratum are *Satureja nubigena* (Lamiaceae), *Geranium sibbaldioides* and *Hypochoeris sessiliflora*.

Thickets are generally dominated by *Pentacalia vernicosa*. Thickets dominated by *Loricaria thujoides* continue above 3950 m.

Fields and peat-bog vegetation have *Werneria humilis*, *Cotula minuta*, *Eryngium humile*, *Juncus effusus* (Juncaceae) and *Sphagnum* sp. In marshy sites, *Espeletia pycnophylla* also occurs; where the water diminishes, the vegetation is dominated by *J. effusus*, *Rhynchospora macrochaeta*, *Loricaria thujoides* and *Pleurozium schreberi*.

In the superpáramo, patches of vegetation with *Werneria* spp., *Draba* and *Azorella* predominate (Rangel-Ch. 1995).

Chiles Volcano

2. Páramo region

2.1. Páramo proper (3800–4200 m)
Some thickets have *Loricaria thujoides* and *Pentacalia vernicosa*, together with *Calamagrostis* cf. *bogotensis*, *Lupinus* sp. and *Diplostephium eriophorum*. Other thickets have *Loricaria colombiana* and *Agrostis foliata*, together with *Pentacalia tephrosoides*, *Distichia muscoides* (Juncaceae) and *Gentiana sedifolia* (Gentianaceae). At 4100–4200 m,

sparse thickets have *Valeriana microphylla* (Valerianaceae), *Diplostephium eriophorum*, *Ranunculus guzmanii* and *Pentacalia* sp. In the ground stratum are *Disterigma empetrifolium*, *Hypochoeris sessiliflora* and *Breutellia tomentosa* and various other mosses.

On very moist sites in gullies is frailepajonal scrub of *Espeletia pycnophylla* and *Arcytophyllum capitatum*, with *Festuca* cf. *sublimis* and *Pernettya prostrata*. In the ground stratum are *Ourisia chamaedrifolia*, *Disterigma empetrifolium*, *Azorella aretioides* and *Plantago rigida* (Plantaginaceae).

Peat-bog vegetation covers considerable surface from the middle páramo to superpáramo, with *Plantago rigida*, *Oritrophium limnophyllum*, *Gentiana sedifolia*, *Distichia muscoides*, *Werneria humilis* and *Lachemilla hispidula*.

Vegetation forming mats on firm (not marshy) ground in the zone between mid-páramo and superpáramo is characterized by *Azorella pedunculata*, *A. diapensioides*, *Werneria humilis*, *Werneria* sp., *Hypochoeris sessiliflora* and *Plantago rigida*.

2.2. Superpáramo
Above 4250 m, rocky superpáramo and bare terrain are frequent; up to 4400 m, vegetation in mats prevails. At 4300 m occur patches of vegetation of *Draba hallii* together with *Agrostis araucana*, *Colobanthus quitensis*, *Pernettya prostrata* and *Lycopodium* sp. Above 4300 m the vegetation is discontinuous and 40% of the terrain unprotected; at 4450 m occur *D. hallii*, *Senecio* cf. *glacialis*, *Luzula racemosa*, *Lachemilla hispidula* and species of bryophytes.

Flora

This geographic region does not yet have a detailed floristic inventory. It is estimated to have 450 vascular plant species and 80 species of bryophytes (mosses and liverworts) (Erazo-N. *et al.* 1991). The topography and the spatial continuity from sub-Andean and Andean life zones up to the páramo with the upper limit in permanent snow (on Cumbal) give special characteristics to the volcanoes, resulting in diversification. These environments experience volcanic eruptions, which may have a stimulating effect on speciation or a controlling effect on local species – which may disappear.

There is small-scale landscape diversity on these volcanoes that yields close to 25 types of vegetation. Furthermore, these areas are the most northerly for some species of the Austral-Antarctic geographic region (van der Hammen and Cleef 1986). Outstanding characteristics of the Galeras and Chiles volcanoes are: (1) the diversity of the plants growing in mats; (2) extensive and vigorous thicket communities with prevalence of *Loricaria* spp. (Asteraceae); and (3) presence of elements unusual for páramo vegetation, i.e. species that in other geographic areas of Colombia belong to the montane Andean region – such as *Fuchsia* (Onagraceae), *Siphocampylus* (Campanulaceae) and *Weinmannia* (Cunoniaceae).

Most of the typically páramo species of Colombia are found on the Nariño volcanoes (Rangel-Ch. 1995). Comparison of the 11 families with seven or more genera in the biogeographic region of páramo with those taxa known on the páramos of the Nariño volcanoes shows at least the following representation (Sturm and Rangel-Ch. 1985; Luteyn,

Cleef and Rangel-Ch. 1992; cf. van der Hammen and Cleef 1986):

❖ Asteraceae: 84 páramo genera; on the Nariño páramos, 21 genera (*Hypochoeris, Chaptalia, Munnozia, Lourteigia, Mikania, Gynoxys, Lasiocephalus, Pentacalia, Senecio, Werneria, Calea, Espeletia, Vasquezia, Verbesina, Gnaphalium, Loricaria, Cotula, Baccharis, Conyza, Diplostephium, Oritrophium*).
❖ Poaceae: 5 of 53 genera (*Cortaderia, Bromus, Agrostis, Calamagrostis, Festuca*).
❖ Orchidaceae: 3 of 22 genera (*Altensteinia, Elleanthus, Epidendrum*).
❖ Apiaceae: 5 of 18 genera (*Azorella, Hydrocotyle, Eryngium, Niphogeton, Oreomyrrhis*).
❖ Ericaceae: 6 of 15 genera (*Befaria, Disterigma, Macleania, Gaultheria, Pernettya, Vaccinium*).
❖ Scrophulariaceae: 4 of 15 genera (*Calceolaria, Ourisia, Bartsia, Castilleja*).
❖ Brassicaceae: 2 of 14 genera (*Cardamine, Draba*).
❖ Melastomataceae: 2 of 11 genera (*Brachyotum, Miconia*).
❖ Caryophyllaceae: 4 of 10 genera (*Drymaria, Colobanthus, Cerastium, Arenaria*).
❖ Cyperaceae: 3 of 10 genera (*Oreobolus, Rhynchospora, Carex*).
❖ Rosaceae: 4 of 7 genera (*Hesperomeles, Polylepis, Rubus, Acaena*).

The following species are restricted in distribution or endemic to this region (Rangel-Ch., in prep.): *Ranunculus guzmanii* (Ranunculaceae); *Begonia pastoensis* (Begoniaceae); *Draba pycnophylla* (Brassicaceae); *Cavendishia oligantha, Disterigma dumontii* (Ericaceae); *Brunellia bullata* (Brunelliaceae); *Gunnera tajumbina* (Gunneraceae); *Ottoa oenanthoides* (Apiaceae); *Lepechinia vulcanicola, Salvia sagittata, Satureja jamesonii, S. tonella* (Lamiaceae); *Aphelandra mutisii* (Acanthaceae); *Arcytophyllum filiforme* (Rubiaceae); *Gynoxys sancti-antoni, Espeletia pycnophylla* (Asteraceae); *Anthurium carchiense* (Araceae); *Pitcairnia bakeri, Puya gigas, P. vestita, Guzmania wittmackii, Tillandsia pectinata* (Bromeliaceae); and *Epidendrum cernuum, E. scolptum* (Orchidaceae).

Useful plants

The use of plants in folk medicine, crafts, etc. is related to indigenous traditions (García-B. 1974–1975; Correa-Q. and Bernal 1989–1991; Kathleen 1978):

❖ For medicinals: *Hydrocotyle humboldtii* var. *pubescens, H. lehmannii, Niphogeton ternata* (Apiaceae); *Baccharis tricuneata, Espeletia pycnophylla, Hypochoeris sessiliflora, Senecio formosus* (Asteraceae); *Berberis rigidifolia* (Berberidaceae); *Tournefortia fuliginosa* (Boraginaceae); *Cardamine bonariensis* (Brassicaceae); *Siphocampylus giganteus* (Campanulaceae); *Viburnum pichinchense, V. triphyllum* (Caprifoliaceae); *Chenopodium ambrosioides, C. quinoa* (Chenopodiaceae); *Coriaria thymifolia* (Coriariaceae); *Befaria aestuans, Macleania rupestris, Pernettya prostrata* (Ericaceae); *Gunnera pilosa* (Gunneraceae); *Salvia tortuosa* (Lamiaceae); *Myrteola oxycoccoides*

(Myrtaceae); *Passiflora cumbalensis* (Passifloraceae); *Plantago linearis* (Plantaginaceae); *Axonopus micay, Paspalum plicatulum* (Poaceae); *Ranunculus nubigenus* (Ranunculaceae); *Lachemilla fulvescens, L. moritziana, L. nivalis* (Rosaceae); *Brugmansia sanguinea* (Solanaceae); *Viola arguta, V. scandens* (Violaceae); and *Drimys granadensis* (Winteraceae).
❖ For food: *Chenopodium quinoa* and *Passiflora mollissima*.
❖ For crafts: *Elaegia pastoensis* (Rubiaceae).
❖ Ornamental: *Senecio formosus*.
❖ For poison: *Pernettya prostrata*.

Social and environmental values

The Pasto, indigenous people who have inhabited the region since before the Spanish, occupy the high and middle valley of the Guáitara River down to Ancúya. Their territory extended along the high plateau to the Colombia-Ecuador border, limited to the east and west by the summits of the cordilleras. The Quillacinga had land north of the Pasto along the eastern side of the Guáitara, in the valley of the Juanambú River and along the foothills to the middle and high parts of the Mayo River (Kathleen 1978; Cerón 1987). The Kwaiker presently also live in the region (Cerón 1987).

The major importance of this volcanic region is its biological and cultural legacy. The great diversity of the vegetation communities, the sites as zones that represent important biogeographic boundaries, the decisive role of the high zones in the ecological behaviour and the economy of the human populations in the lower zones, are reasons enough for the preservation of these highlands.

Although the diversity of the fauna is less than in other páramo regions of Colombia, it is important to highlight the presence of spectacled bear (*Tremarctos ornatus*), páramo deer (*Odocoileus virginianus* cf. *goudotii*) and diverse species of amphibians in *Eleutherodactylus, Centrolenella* and *Phrynopus* (Sánchez-P. *et al.* 1990).

The Central Andean páramo Endemic Bird Area (EBA B60) includes the highest peaks of the Central Andes in Colombia and Ecuador. Ten bird species of restricted range occur in this EBA (nine are confined to it); six of them probably occur on the Nariño volcanoes. All these restricted-range species are confined to páramo grassland and *Polylepis* scrub, and the elfin forest–páramo ecotone. Due to the extensive destruction or disturbance of these habitats, five of the species are considered threatened.

Along the high-Andean fringe and lower páramo (3000–3400 m), potatoes and to a lesser extent broad beans (*Vicia faba*) are cultivated and extensive cattle-ranching is practised. The woody vegetation of the ecotone between the montane Andean and páramo regions has been much depleted for domestic fuelwood. In Cumbal Volcano there are reserves of sulphur estimated at three million tons; 20% of Colombia's production comes from Cumbal (IGAC 1967). The ice from Cumbal's glacier is extracted under rudimentary conditions by nearby farmers for the manufacture of ice cream and cool drinks.

The most relevant environmental values are the scenic landscapes and their floristic diversity. Galeras Volcano is readily accessible, with roads that lead close to the crater; it is a natural laboratory for understanding the ecologic gradation in mountainous Colombia, and should be used more frequently for environmental education. Other environmental values include watercourses that originate in the high elevations and supply water to the aqueducts of the communities in the foothills; ecotourism; and in a similar fashion, controlled use of the lakes for water sports. An alternative use in the region can be attained by means of equipment for electronic data transmission (radio, television).

Threats

Natural threats are related to volcanic eruptions, particularly from Galeras. The extension of the agricultural frontier into páramo (cattle-ranching and farming potatoes) has greatly modified some areas of Azufral Volcano and to a lesser extent the other volcanic areas. Exploitation of the lower páramo fringes (e.g. for fuelwood) has resulted in plantations of exotic pine (*Pinus*), but in areas that for climatic reasons do not make this cultivation economically viable. In several localities, lakes and marshes have been drained to make agricultural land.

Conservation

The Galeras Fauna and Flora Sanctuary (176 km²) was established in 1985 and is under administration of the Instituto Nacional de los Recursos Naturales Renovables y del Medio Ambiente (INDERENA). Protected areas are needed in the remaining locales. In Túquerres, there is interest and concern by the community to preserve the Azufral volcanic area. On Cumbal and Chiles, protected zones should be established for the areas that have not already been greatly transformed. In Ecuador, the highlands above 4500 m are legally considered to be national public property.

References

Cerón, B. (1987). *Kwaiker. Introducción a la Colombia Amerindia.* Instituto Colombiano de Antropología, Bogotá. Pp. 203–216.

Correa-Q., J.E. and Bernal, H.Y. (1989–1991). *Especies vegetales promisorias de los países del Convenio Andrés Bello,* Vols. 2, 3, 6. Junta del Acuerdo de Cartagena (JUNAC). Ministerio de Educación y Ciencia, Spain and Secretaría Ejecutiva del Convenio Andrés Bello (SECAB), Bogotá. 462 pp., 485 pp., 507 pp.

Cuellar-R., J. and Ramírez-L., C. (1986). *Descripción de los volcanes colombianos.* Postgrado en geofísica. Publicación interna, Universidad Nacional de Colombia, Bogotá. Pp. 2–12.

Erazo-N., G., de la Cruz-G., A., Delgado-M., A. and Montenegro-P., L. (1991). *Caracterización de la vegetación paramuna de los volcanes Azufral y Galeras.* Trabajo de grado presentado como requisito parcial para optar el título de Especialista en Ecología. Escuela de Postgrado, Universidad de Nariño, Pasto. 242 pp.

García-B., H. (1974–1975). *Flora medicinal de Colombia,* Vols. 1–3. Instituto de Ciencias Naturales, Universidad Nacional de Colombia, Bogotá. 561 pp., 538 pp., 495 pp.

Grosse, E. (1935). Acerca de la geología del sur de Colombia. Informe rendido al Ministerio de Minas, sobre un viaje al departamento de Nariño. *Compilación de Estudios Geológicos Oficiales en Colombia (Bogotá),* Vol. 3. Pp. 139–231.

IGAC (1967). *Atlas de Colombia.* Instituto Geográfico Agustín Codazzi (IGAC), Bogotá. 204 pp.

Kathleen, R. (1978). Las tribus de la antigua jurisdicción de Pasto en el siglo XVI. *Rev. Col. Antropol. (Bogotá)* 21: 11–14.

Luna-Z., C. and Carlhoum, F. (1986). Suelos derivados de cenizas volcánicas del departamento de Nariño. IGAC, Dirección Agrológica 9(2): 1–131.

Luteyn, J.L., Cleef, A.M. and Rangel-Ch., J.O. (1992). Plant diversity in páramo: towards a checklist of páramo plants and generic flora. In Balslev, H. and Luteyn, J.L. (eds), *Páramo: an Andean ecosystem under human influence.* Academic Press, London. Pp. 71–84.

Rangel-Ch., J.O. (ed.) (1995). *Colombia: diversidad biótica I.* INDERENA – Universidad Nacional de Colombia, Santafé de Bogotá. 442 pp.

Rangel-Ch., J.O. (in prep.). Vegetación y flora en los volcanes del sur de Colombia. *Caldasia.*

Sánchez-P., H., Hernández-C., J.I., Rodríguez-M., J.V. and Castaño-U., C. (1990). *Nuevos parques nacionales Colombia.* Instituto Nacional de los Recursos Naturales Renovables y del Medio Ambiente (INDERENA), Bogotá. 213 pp.

Sturm, H. and Rangel-Ch., J.O. (1985). *Ecología de los páramos andinos: una visión preliminar integrada.* Biblioteca J.J. Triana No. 9. Instituto de Ciencias Naturales, Museo de Historia Natural, Universidad Nacional de Colombia, Bogotá. 292 pp.

van der Hammen, T. and Cleef, A.M. (1986). Development of the high Andean páramo flora and vegetation. In Vuilleumier, F. and Monasterio, M. (eds), *High altitude tropical biogeography.* Oxford University Press, New York. Pp. 153–201.

Authors

This Data Sheet was written by Dr J. Orlando Rangel-Ch. and Aída Garzón-C. (Universidad Nacional de Colombia, Instituto de Ciencias Naturales, Museo de Historia Natural, Apartado Aéreo 7495, Santafé de Bogotá, Colombia).

PARAMOS AND ANDEAN FORESTS OF SANGAY NATIONAL PARK
Ecuador

Location: In central Ecuador beginning 160 km south-east of the capital of Quito, between about latitudes 1°25'–2°40'S and longitudes 79°–78°W.

Area: 5177 km² in park.

Altitude: 1000–5319 m.

Vegetation: Wide range of vegetation formations in various páramos, wetlands and Andean to sub-Andean forests.

Flora: High diversity and high endemism expected, with at least 3000 species in park.

Useful plants: In nearby communities many plants used for fuelwood, construction, fibre, forage, food and medicinals.

Other values: Protection of watersheds and fauna, including threatened species; wilderness; potential genetic resources; scenery; ecotourism; education.

Threats: Road construction, colonization, overgrazing, logging, poaching, set fires, mining, tourism.

Conservation: National Park and UNESCO-MAB World Natural Heritage Site.

Geography

Sangay National Park (Map 67), which occupies parts of the provinces Tungurahua, Chimborazo, Cañar and Morona-Santiago, includes páramos and Andean and sub-Andean forests of central Ecuador's Eastern Cordillera (Cordillera Oriental or Real) with an eastern outlier – Sangay Volcano, and in the south extends westward as the Nudo del Azuay across the continental divide to the Pacific slope (SFRNR n.d.). Altitudes range from 5319 m atop El Altar (which also has Amerindian names of Collanes and Cápac-urcu) to c. 1000 m on the eastern boundary toward the Amazonian lowlands.

The region has spectacular scenic beauty, which is dominated by three snow capped volcanoes, two of which are active: Tungurahua (5016 m) and Sangay (5230 m); the third is El Altar (5319 m), which has an impressively eroded and glaciated caldera collapsed to the west (Meyer 1907) and is extinct. In the southern part of the park are highlands with older mountains or "cerros" (Wolf 1892), some of which usually have some snow; they include Achipungo (4630 m), Sorochi (4730 m) and Ayapungo (4699 m). From southern Ecuador southward, active Quaternary volcanism is absent for 1600 km until latitude 17°S in southern Peru (Clapperton 1993). Sangay (called Tungur by the Shuar or Jívaro) is one of the world's most active volcanoes, with fumaroles observed continuously and glowing tephra and hot rocks ejected often (cf. Sauer 1965, 1971); the most recent violent eruptions of Tungurahua occurred in 1916–1925.

Around the volcanoes are glacial valleys (Clapperton 1993; Sauer 1965, 1971), which mostly trend eastward toward the Amazonian lowlands. Cirques, moraines and U-shaped valleys with extensive meadows are common highland features (Armstrong and Macey 1979). Most of the region's rivers eventually drain into the Amazon Basin.

Among the many rivers are the Llushín Grande which flows into the Pastaza River, the Palora with its tributary the Sangay which collect drainage from El Altar and Sangay, the Upano with its major tributary the Abanico, and in the south the Juval which flows into the partially bordering Paute River. The hydrography is complemented by over 100 glacial lakes in the páramo zone: among the prominent lakes are Pintada, Verde Cocha, Atillo (or Colaycocha), Magtayán, Cubillín, Culebrillas and the Aucococha Lakes. At the unusual elevation of 1800 m, in sub-Andean forest to the east of Sangay are the Sardinayacu Lakes.

The eastern part of the park is formed of Mesozoic and Tertiary volcanic and sedimentary rocks. The highlands are formed by pre-Cretaceous metamorphic and plutonic rocks strongly compressed vertically. The three strato-volcanoes originated from Tertiary and intensive Quaternary volcanism, emitting andesites and pyroclasts (Banco Central 1982; Schuerholz et al. 1980/1982). Explosions, hot lava flows and strong fumarolic activity are characteristic, and the surrounding land has frequently been devastated by mud flows and ash eruptions. The soils are black Andean rocky lithosols, with latosols in the foothills toward the Amazon (Armstrong and Macey 1979).

The climate is variable, with significant local differences. There is no well-defined wet or dry season particularly on the Amazonian side, although for example in the Culebrillas River area north-west of Sangay Volcano, May–August is wetter and October–January drier (Downer 1995). The annual rainfall in the east is as much as 5000 mm, whereas on the western slopes it is no more than 600 mm. Annual mean temperature also varies dramatically, from 24°C on the lower slopes to 0°C at 4750–4800 m and well below 0°C on the highest mountain peaks, which have glaciers (Armstrong and Macey 1979; SFRNR 1991, 1992; Acosta-Solís 1984; Clapperton 1993).

Vegetation

The extensive expanse of the region provides for a diversity of vegetation that is extremely rich in ecological variation – at least nine life zones from subalpine to premontane have been designated using the Holdridge system (Cañadas-Cruz 1983). Three major vegetation types are recognized: páramos, which may include forested patches; Andean forests, including the Ceja Andina; and sub-Andean forest. The upper montane forest is generally called cloud forest because these forests are nearly constantly covered by fog or misty precipitation.

Just below the snow-line at c. 4800 m, lichens and bryophytes are characteristic. Below them is the "páramo" (Acosta-Solís 1984), treeless vegetation dominated by bunchgrasses (*Calamagrostis*, *Festuca*, *Stipa*) and characterized by cushion plants (*Azorella pedunculata*, *Plantago rigida*, *Werneria humilis*), which occur with xerophytic shrubs such

MAP 67. SANGAY NATIONAL PARK, ECUADOR (CPD SITE SA31)

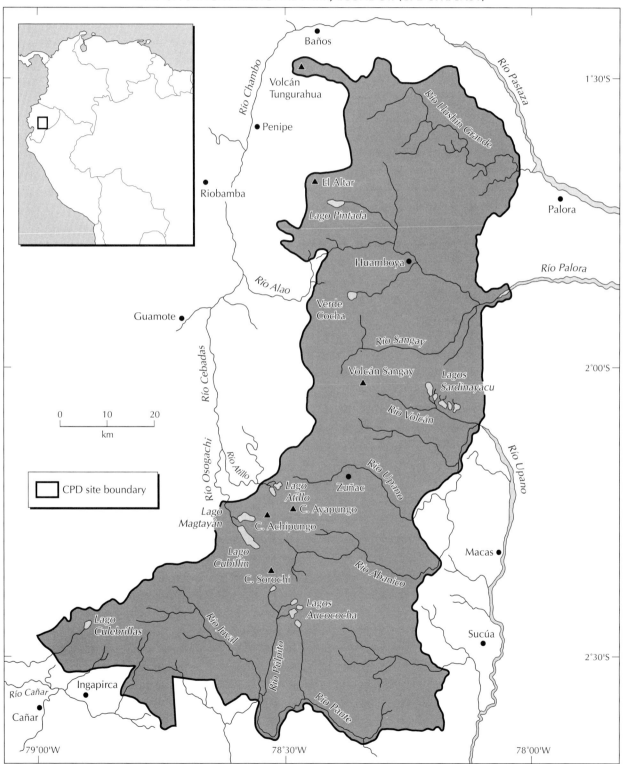

as *Arcytophyllum aristatum*, *Baccharis genistelloides*, *Calceolaria gossypina*, *Chuquiraga jussieui*, *Disterigma codonanthum*, *Gaultheria sclerophylla*, *Hypericum sprucei*, *Loricaria complanata*, *Myrteola nummularia*, *Pernettya prostrata*, *Vaccinium floribundum* and *Valeriana microphylla*; ferns such as *Blechnum loxense*, *Elaphoglossum cardiophyllum* and *Jamesonia escamanae*; and many herbs such as *Carex microglochin*, *Equisetum bogotense*, *Gentiana sedifolia*, *Gentianella splendens*, *Geranium cucullatum*, *Lupinus pubescens*, *Rostkovia magellanica*, *Oreobolus ecuadorensis*, *Pinguicula caliptrata*, *Ranunculus gusmanii*, *Valeriana bracteata* and *Viola glandulifera*. In undisturbed páramo may be found the bamboo *Neurolepis aristata*.

Sometimes patchy remnants of Andean forest occur as ecological islands in the páramo, dominated by small slow-growing *Polylepis reticulata*, the region's only tree that can grow up to 4100 m (e.g. in Alao Valley). **Stunted forest** 10 m high **known as the Ceja Andina** arises below the páramo, occurring from 3800–3200 m. Common species are *Brachyotum lindenii*, *Buddleja incana*, *Coriaria ruscifolia*, *Diplostephium glandulosum*, *Escallonia myrtilloides*, *Freziera microphylla*, *Gaiadendron punctatum*, *Gynoxys buxifolia*, *Hesperomeles lanuginosa*, *Miconia salicifolia*, *Monnina crassifolia*, *Muehlenbeckia vulcanica*, *Pentacalia vaccinoides*, *Ribes lehmannii*, *Saracha quitensis* and *Tristerix longebracteatus*.

The **Andean forest (cloud forest)** occurs with local variations from 3200–2800 m. These dense forests 15–20 m in height are adapted to steep slopes and high humidity. Characteristic woody species are *Alnus acuminata*, *Clethra obovata*, *Miconia latifolia*, *Myrcianthes hallii*, *Myrica pubescens*, *Myrsine andina*, *Oreopanax sprucei*, *Passiflora cumbalensis*, *Piper lanceaefolium*, *Saurauia aequatoriensis*, *Vallea stipularis* and *Weinmannia mariquitae*. There are also *Sphaeropteris atahuallpa* tree ferns. The abundance of epiphytes is remarkable, including bromeliads such as *Tillandsia tetrantha*, orchids and ferns. Landslides (natural or human-caused) are colonized especially by the high-Andean bamboo *Chusquea scandens*.

Sub-Andean forest occurs below 2800 m and near 1000 m overlaps with Amazonian floristic elements. This vegetation type is probably the most species-rich, but is also less known. The vegetation is luxurious, with trees 25 m or more tall which may include several endemic species, and also with valuable timber species (e.g. *Cedrella*, *Cordia*). Characteristic woody species are *Saurauia aequatoriensis*, *Cedrella montana*, *Clusia* spp., *Croton lechleri*, *Geonoma* spp., *Hedyosmum sprucei*, *Nectandra* spp., *Ocotea* spp., *Podocarpus oleifolius*, *Tibouchina lepidota* and *Vismia tomentosa*.

Much of the eastern area bordering the park has been cleared in recent years in colonization projects (Rudel and Horowitz 1993). Common species of secondary forests are *Cecropia* spp., *Pollalesta* spp. and *Tessaria integrifolia*. There is a strip of colonized clearings in the forest along the

CPD Site SA31: Western slope of Sangay Volcano viewed from the Culebrillas area of Sangay National Park, Ecuador.
Photo: Craig C. Downer.

Guamote-Macas road/trail (Upper Upano River) (Armstrong and Macey 1979). The original forests of the slopes to the west have long been replaced by grasslands for cattle.

Flora

The flora of Sangay National Park is poorly known, but at least 3000 species are expected to occur. Studies of the Andean forests of Ecuador above 2400 m recognize 93 families, 292 genera and 1566 species of trees and shrubs (Ulloa-Ulloa and Jørgensen 1993). Most of these genera are represented in the Sangay region. From a checklist of the high-Andean flora of Ecuador which contains 4430 species of seed plants (Jørgensen and Ulloa-Ulloa 1994), only 400 species were documented as occurring in the NP (with its pre-1992 boundaries). Collections have been made especially in the region's localities of Tungurahua, Alao across to Huamboya, Sangay, and Osogachi (Acosta-Solís 1968; Jørgensen and Jaramillo 1989; Schuerholz *et al.* 1980/1982). A thorough inventory of the park's flora is needed.

The southern area of the park, particularly at the Nudo del Azuay, is transitional between the northern and southern zones of the Ecuadorian Andes, which have been recognized as physiographically and floristically different (e.g. Wolf 1892; Acosta-Solís 1984). This area's floristic composition may be particularly diverse and interesting.

The eastern slopes of the tropical Andes have a high percentage of the neotropical flora, and their endemism appears to be high (Gentry 1989). Species considered endemic to the park or with its periphery as well include *Calceolaria martinezii*, *Centropogon trachyanthus*, *Miconia caseariata*, *Oritrophium ollgaardii*, *Polylepis microphylla* and *Saurauia aequatoriensis*. A beautiful high-Andean species collected only a few times is *Mutisia rimbachii*.

Useful plants

Although studies have not been made directly within the region, the area is used by Amerindian groups of the surroundings. Several timber resources and various plants are certainly used for fuelwood, construction, clothing, food, medicinals and rituals (cf. Cordero 1911; Kvist and Holm-Nielsen 1987). The great diversity of undamaged vegetation also is an enormous genetic reserve, and surely a source for wild relatives of crops and potentially valuable medicines.

Social and environmental values

Sangay National Park harbours a very diverse fauna, which is also not well documented. A major scientific investigation is in great need, as many species await study or discovery. The

CPD Site SA31: Andean forests and páramo of Sangay National Park, Ecuador. Upper Purshi forest to páramo (3500–4200 m) in Upano River watershed. Photo: Craig C. Downer.

mammals are as varied as the Andean fox (*Pseudalopex culpaeus*), mountain paca (*Agouti taczanowskii*), a guinea pig (*Cavia* sp.), the eastern South American tapir (*Tapirus terrestris*) and red howler monkey (*Alouatta seniculus*). The region is one of the last large refuges for many threatened species of mammals and birds, as well as the plants. The 28 species of large mammals include several classed in the IUCN *Red Data Book* as threatened. Among them are the severely endangered mountain tapir (*Tapirus pinchaque*), which has one of its few large populations in the park – a collection and list have been made of plants that it eats and in some cases helps to disperse or that are notable in its habitat (Downer 1995 and pers. comm.); and northern pudú (*Pudu mephistophiles*), spectacled bear (*Tremarctos ornatus*) and jaguar (*Panthera onca*).

Some 500 species of birds are likely, including several of world interest such as the Andean condor (*Vultur gryphus*), Andean cock-of-the-rock (*Rupicola peruviana*), torrent duck (*Merganetta armata*) and giant hummingbird (*Patagona gigas*) – the world's largest, with a wing-span of 30 cm (Armstrong and Macey 1979; SFRNR 1991). Sangay NP embraces two Endemic Bird Areas (EBAs). The Central Andean páramo EBA (B60) is home to ten bird species of restricted range, all confined to páramo and the páramo–Andean forest ecotone; five are considered threatened. At lower altitudes, this park is home to some of the 15 restricted-range species of the Eastern Andes of Ecuador and northern Peru EBA (B18), which primarily inhabit the sub-Andean forests; three of them are thought to be threatened.

The abundant vegetation is a very important protector for many watersheds, acting as a huge sponge which captures the rain and mist that is an important source of water in lower regions for direct human use, irrigation and development work. Proper management is essential to maintain this resource.

The region has archaeological importance of unknown extent – evidence of pre-Spanish settlement has been found in the Palora Valley. Near the park to the south-west (east of Cañar) are the Inca ruins of Ingapirca. Due to the region's extreme inaccessibility and difficult terrain and climate, most of the area is uninhabited and human influences are mainly on the periphery. Lowland parts of the park are used by Canelos Quichua Amerindians in the north-east (cf. Whitten 1985) and Shuar in the south-east (cf. Rudel and Horowitz 1993). In the park highlands Atillo settlement has some 400 people, mostly Andean Quichua, and along the park's western edge there are several long-established agricultural villages and large farms raising cattle, which use the páramos and also extract wood (FN 1992b; Cifuentes *et al.* 1989; Armstrong and Macey 1979; Downer 1995).

The landscape, geology, fauna and flora of Sangay NP are strong attractions but have not been appropriately developed even though the region attracts national and international tourists, mountain climbers and scientists (Macey *et al.* 1976; Schuerholz *et al.* 1980/1982; Anhalzer 1989; Martínez 1933; Snailham 1978).

Economic assessment

The economic values that this region has for the country have not been formally evaluated, but at present are underestimated. Local inhabitants have not regarded the resources of this legally protected area in a sustainable, economically significant manner. The value of the region's ecological services – such as protection of valley soils – is surely very high, taking into account the numerous natural water reservoirs and rivers, and agricultural activities in the lowlands adjacent to the rivers.

The Shuar use the land mostly for subsistence agricultural purposes, whereas the colonizers use their lands with more intensity. Close to the north-western part of the region are two tea-producing plantations. In the adjoining inter-Andean zone, usual vegetables are cultivated (maize, potatoes, etc.) and cattle ranches are operated (Macey *et al.* 1976). The páramos are partially used for the raising of cattle.

As a world-class protected area with many biologic resources and impressive scenery, good management within the constraints of sustainable development can transform this park into an important source of income both for nearby locations and more broadly. Nonetheless the income from tourism will be less significant than the region's overall ecological and biological rewards.

Threats

The Andean ecosystem has been greatly affected by excessive human activities. A large proportion of the Andean forests has been lost, especially as a result of deforestation caused by gathering fuelwood and by expansion of the agricultural frontier. Damaging factors include fires and grazing cattle which locally alter the vegetation or compact the soils.

The predominant threat to the region is the forthcoming opening of the Guamote-Atillo-Macas road (which is slowly being built but on schedule). The road will extend across the middle of the park and divide it into two sectors, transforming access with an easy route for logging and colonization – spontaneous or perhaps supported by official agencies (Downer 1991). Many natural areas are likely to be rapidly devastated (FN 1992a; cf. Rudel and Horowitz 1993).

Legal artisan mining occurs in the region and perhaps is harmful particularly because it is without regulation and careful technology. Projects for prospecting and exploiting auriferous and non-metallic mined resources are expected to develop, and mineral concessions are a potential near-term threat. Although the park is protected legally, the government in other protected natural areas has considered other resources more important and permitted some exploration and exploitation. This situation could get worse because of conflicting laws and very different political support concerning biological resources versus mined or other economic resources.

The impact of nearby communities does not seem major, but could become so. Reduction of surrounding natural habitats is forcing nearby communities to hunt more within the park. The provinces of Chimborazo and in recent years Morona-Santiago have been the main sources of colonists invading the NP. The impact of tourism has not been studied, but does not seem to be a major problem. Increasing tourism could represent a large near-term difficulty for administration of the park (Cifuentes *et al.* 1989).

Conservation

The greater Sangay region is one of the few large, highly diversified Andean ecosystems remaining in Ecuador; although disturbed in places (cf. Armstrong and Macey 1979), it has the best remaining natural páramos and Andean forests. A National Reserve was decreed in 1975, and Sangay National Park was established in 1979 with an area of c. 2720 km² (SFRNR 1991); it was expanded in 1992 to 5177 km² (by extending south and westward along the Nudo del Azuay to Cañar) (SFRNR n.d.). The region was recognized in 1983 as a UNESCO-MAB World Natural Heritage Site. The park's administrative centre is in Riobamba, with six peripheral guardianships. The NP is included in the Parks in Peril campaign (TNC 1990).

Sangay has an old management plan (Schuerholz *et al.* 1980/1982; SFRNR 1991), which requires revision and improvement. Education, ecotourism and research to show the value of the park should be encouraged. Facilities for visitors only in designated areas, and guidance by trained personnel – which may include local people – are required, as well as complete protection of the unaltered vegetation.

Ecuador's system of protected areas consists of 16 reserves (SFRNR 1991), and altogether 32 are expected to be established. In the Ministerio de Agricultura y Ganadería the Instituto Ecuatoriano Forestal y de Areas Naturales y Vida Silvestre (INEFAN) (formerly the Subsecretaría Forestal y de Recursos Naturales Renovables), which is responsible for the administration of these protected areas, works with many problems such as limited infrastructure and few professional personnel, uneven political support and lack of funds. Yet the efforts have achieved some success in conservation of remarkable natural resources, sometimes aided by the conditions of difficult access, as is presently the situation for Sangay NP (Cifuentes *et al.* 1989).

Throughout society there is a growing concern for the conservation of biodiversity, which is expected to exert more pressure on the government to adopt appropriate environmental policies. UNESCO's disturbing designation of Sangay National Park as a World Heritage Site – In Danger, should help to protect this extraordinary region.

References

Acosta-Solís, M. (1968). *Divisiones fitogeográficas y formaciones geobotánicas del Ecuador.* Casa de la Cultura Ecuatoriana, Quito. 307 pp.

Acosta-Solís, M. (1984). *Los páramos andinos del Ecuador.* Publicaciones Científicas MAS, Quito. 222 pp.

Anhalzer, J. (1989). *Sangay, parque nacional.* Azuca, Quito. 54 pp.

Armstrong, G.D. and Macey, A. (1979). Proposals for a Sangay National Park in Ecuador. *Biol. Conserv.* 16: 43–61.

Banco Central del Ecuador (1982). *Atlas del Ecuador.* Les éditions J.A., Paris.

Cañadas-Cruz, L. (1983). *El mapa bioclimático y ecológico del Ecuador.* Ministerio de Agricultura y Ganadería, Programa Nacional de Regionalización (MAG-PRONAREG), Quito. 210 pp.

Cifuentes, M., Ponce, A., Albán, F., Mena, P., Mosquera, G., Rodríguez, J., Silva, D., Suárez, L. and Torres, J. (1989). *Estrategia nacional de áreas protegidas II fase, contexto nacional.* Fundación Natura and Ministerio de Agricultura y Ganadería (MAG), Quito. Pp. 62–64.

Clapperton, C.M. (1993). *Quaternary geology and geomorphology of South America.* Elsevier, Amsterdam. 779 pp.

Cordero, L. (1911). *Enumeración botánica de las principales plantas, así útiles como nocivas, indígenas o aclimatadas que se dan en las provincias del Azuay y de Cañar de la República del Ecuador.* Retrato del Autor, Cuenca. 305 pp.

Downer, C.C. (1991). Andean highway violates international treaty. *E-Sheet* 2(52) (15/05/91): 1.

Downer, C.C. (1995). The gentle botanist. *Wildlife Conserv.* 98(4): 30–35.

FN (1992a). *Acciones de desarrollo y áreas naturales protegidas en el Ecuador: 2. Parque Nacional Sangay.* Fundación Natura (FN), Quito. 11 pp.

FN (1992b). Parque Nacional Sangay. In *Parques nacionales y otras áreas naturales protegidas del Ecuador.* FN and MAG, Quito. Pp. 20–25.

Gentry, A.H. (1989). Northwest South America (Colombia, Ecuador and Peru). In Campbell, D.G. and Hammond, H.D. (eds), *Floristic inventory of tropical countries: the status of plant systematics, collections, and vegetation, plus recommendations for the future.* New York Botanical Garden, Bronx. Pp. 391–400.

Jørgensen, P.M. and Jaramillo, J. (eds) (1989). *Informe final del proyecto "Estudios botánicos sobre la taxonomía del bosque montano".* Pontificia Universidad Católica del Ecuador, Quito. 205 pp. Unpublished report.

Jørgensen, P.M. and Ulloa-Ulloa, C. (1994). *Seed plants of the high Andes of Ecuador – a checklist.* AAU Reports 34: 1–443.

Kvist, L.P. and Holm-Nielsen, L.B. (1987). Ethnobotanical aspects of lowland Ecuador. *Opera Botanica* 92: 83–107.

Macey, A., Armstrong, G.D., Gallo, N. and Hall, M.L. (1976). *Sangay, un estudio de las alternativas de manejo.* MAG, Dirección General de Desarrollo Forestal, Departamento de Parques Nacionales y Vida Silvestre, Quito. 98 pp.

Martínez, N.G. (1933). *Exploraciones en los Andes ecuatorianos. El Tungurahua.* Publ. del Observatorio Astronómico y Meteorológico, Sección Geofísica, Quito. 118 pp.

Meyer, H. (1907). *In den Hoch-Anden von Ecuador*. Reimer, Berlin. 552 pp./(1938-1939 and 1993). *En los Altos Andes del Ecuador* (Guerrero, J., transl.). *Anales Univ. Central (Quito)* 60: 779–915, 61: 183–394, 1363–1569 and Ediciones Abya-Yala, Quito. 750 pp.

Rudel, T.K. and Horowitz, B. (1993). *Tropical deforestation: small farmers and land clearing in the Ecuadorian Amazon*. Columbia University Press, New York. 234 pp.

Sauer, W. (1965). *Geología del Ecuador*. Ministerio de Educación, Quito. 384 pp./(1971). *Geologie von Ecuador*. Gebruder Bortraeger, Berlin. 316 pp.

Schuerholz, G., Paucar, A., Huber, R. and Soria, J. (1980/1982). *Plan de manejo del Parque Nacional "Sangay"/Parque Nacional "Sangay"*. MAG, Dirección General de Desarrollo Forestal, Departamento de Administración de Parques Nacionales y Vida Silvestre, Quito. 190 pp.

SFRNR (1991). Parque Nacional Sangay. In *Sistema Nacional de Areas Protegidas y la vida silvestre del Ecuador*. MAG, Subsecretaría Forestal y de Recursos Naturales Renovables (SFRNR) and U.S. Fish and Wildlife Service, Quito. Pp. 15–17.

SFRNR (1992). *Parque Nacional Sangay*. MAG, SFRNR, Unidad de Educación Ambiental, Quito. Educational pamphlet.

SFRNR (n.d.). *Parque Nacional Sangay*. [Scale 1:200,000.] MAG, SFRNR, Quito. Draft map (enlarged NP).

Snailham, R. (1978). *Sangay survived*. Hutchinson, U.K.

TNC (1990). *Parks in Peril: a conservation partnership for the Americas*. The Nature Conservancy (TNC), Arlington, Virginia, U.S.A. 24 pp.

Ulloa-Ulloa, C. and Jørgensen, P.M. (1993). *Arboles y arbustos de los Andes del Ecuador*. AAU Reports 30: 1–263.

Whitten Jr., N.E. (1985). *Sicuanga Runa*. University of Illinois Press, Urbana and Chicago. 315 pp.

Wolf, T. (1892). *Geografía y geología del Ecuador*. Brockhaus, Leipzig. 671 pp./(1933). *Geography and geology of Ecuador* (Flanagan, J.W., transl.). Grand and Toy, Toronto. 684 pp.

Authors

This Data Sheet was written by Patricio Mena (Fundación Ecociencia, Casilla 17-12-257, Quito, Ecuador), Dra. Carmen Ulloa Ulloa (Missouri Botanical Garden, P.O. Box 299, St. Louis, MO 63166-0299, U.S.A.) and Olga Herrera-MacBryde (Smithsonian Institution, Department of Botany, NHB-166, Washington, DC 20560, U.S.A.).

(TROPICAL) ANDES: CPD SITE SA32

HUANCABAMBA REGION
Peru and Ecuador

Location: Andean slopes and valleys of northern Peru and southern Ecuador between latitudes 6°–4°30'S and longitudes 78°–79°40'W.

Area: 29,000 km².

Altitude: c. 1000–4000 m.

Vegetation: Dry forest at lower elevations, montane cloud forest, páramo-like above 3200–3500 m.

Flora: At least 2000–2500 vascular plant species; northern or southern limit for many species; many endemic species.

Useful plants: Valuable timber species; fuelwood; medicinals, ornamentals.

Other values: Watershed protection; distributional limit for numerous species; habitats for many endemic fauna.

Threats: Logging, expansion of agricultural frontier.

Conservation: 1 National Sanctuary (295 km²).

Geography

The Huancabamba region (Map 68) is a mountainous area of moderate but complex relief of northern Peru and southernmost Ecuador in the general area where the Central Andes and the Northern Andes diverge. The general area is tectonically multifarious, with distinctive terranes and two transverse mega-shears (Feininger 1987; Clapperton 1993). The overall region is part of an Early Palaeozoic and Precambrian schist belt (Atherton, Pitcher and Warden 1983) that forms a complicated system of isolated low ranges and basins referred to as the Huancabamba Depression. Here the older Central Andes (Miocene to Pliocene) and younger Northern Andes (Late Pliocene to Pleistocene) fragment into ranges usually less than 3500 m high separated by valleys mostly between 1000–2000 m above sea-level; the continental divide dips to 2145 m at the Abra de Porculla. Huancabamba also refers to a Peruvian province in the Department of Piura that lies at the headwaters of two Pacific-basin rivers, the Chira and the Piura. On the eastern side the Huancabamba Deflection is notable, where there are major changes in the overall geology of the Andes (Sillitoe 1974; Feininger 1987) and the Marañón River turns eastward into the Amazon Basin.

Because of the complexity and the high number of species known only from the Huancabamba region, Duellman (1979) and Berry (1982) consider it separately from other Andean regions. The geological history and topography of the region have resulted in one of the most important barriers and filters affecting biotic migration in the Andes. For example, the Huancabamba Depression "probably accounts for a proportionately larger increase in the number of [bird] species than any other barrier along the Andes" (Vuilleumier 1969). Many northern or southern terminations of distributions are seen among species of rodents (Reig 1986), birds (Vuilleumier 1969, 1977; Vuilleumier and Simberloff 1980; Parker et al. 1985), reptiles and amphibians (Duellman 1979; Cannatella 1982; Cadle 1991; Duellman and Wild 1993), butterflies

(Descimon 1986) and plants (Simpson 1975; Molau 1978, 1988; Baumann 1988).

The region also has served as an east-west corridor between the Amazon and Pacific basins (Gentry 1982; Cadle 1991; Patterson, Pacheco and Ashley 1992). In addition, isolation of the region's mountain ranges and inter-Andean valleys has promoted speciation – Panero (1992) includes a discussion of how speciation may have occurred in this and similar regions.

The region receives climatic and biogeographic influences from the Amazon Basin, the highlands of southern Ecuador and northern Peru, Pacific Ocean drainages and inter-Andean dry forests. Seasonal rains occur between December and March, and are exacerbated by the oceanic El Niño flow, which is associated with significantly more rainfall. The mean annual precipitation in non-El Niño years is 800–2500 mm. In areas above 2000 m elevation, high humidity is common due to the persistence of clouds during many months each year.

Vegetation

There is great biological heterogeneity because the region is a major biogeographical crossroads, and each vegetation type is of limited extent. Dry forests and scrub occur at lower elevations and in the valleys. Areas above c. 1700 m are predominantly montane cloud forests. "Páramo" grasslands are found above the treeline, which is at 3200–3500 m.

Dry forest extends from the valleys to c. 1700 m, and is dominated by a scattered to dense cover of legume trees of *Prosopis*, *Acacia*, *Piptadenia* and *Parkinsonia*. Other characteristic genera include *Capparis* (Capparaceae), *Cordia* (Boraginaceae), *Eriotheca* (Bombacaceae), *Loxopterygium* (Anacardiaceae) and *Muntingia* (Tiliaceae). The herbaceous layer is dominated by species of Asteraceae, Fabaceae, Solanaceae and Poaceae.

Montane cloud forest is the major vegetation type and typically extends from 1700–3300 m. Clouds from the Amazon produce high humidity on the slopes of the Andes, creating a dense, epiphyte-laden forest, often 15–25 m high. Characteristic trees include *Weinmannia* (Cunoniaceae); *Cassia* (Fabaceae); *Ocotea, Persea* (Lauraceae); *Miconia* (Melastomataceae); *Cedrela, Guarea, Ruagea, Schmardaea* (Meliaceae); *Myrsine* (Myrsinaceae); *Cinchona* (Rubiaceae); *Prunus, Hesperomeles* (Rosaceae); *Zanthoxylum* (Rutaceae); *Ficus, Morus* (Moraceae); *Cupania* (Sapindaceae); and *Solanum* (Solanaceae). The epiphytes are mostly orchids and bromeliads.

Andean highland vegetation, often called "jalca" or páramo in this region (cf. CPD Site SA33), is a shrubby grassland occurring above the montane cloud forest, around 3200 m and higher. Species of Poaceae and Asteraceae dominate except in bogs, where species of Cyperaceae predominate.

Flora

Based on the diversity of comparable regions in Ecuador and Peru, at least 2000–2500 species of vascular plants occur in this region. The evidence suggests that species-level endemism is quite high – the Huancabamba region has been identified as an Andean centre of endemism, despite the fact that the region is still insufficiently inventoried (Berry 1982). In addition, many Northern Andean species reach their southern distributional limit and many Central Andean species reach their northern limit. Data on floristic composition can be extracted or extrapolated from Weberbauer (1945), Valencia (1990), Cano and Valencia (1992), Gentry (1992), Jørgensen and Ulloa-Ulloa (1992) and Kessler (1992).

Dry forests have tree species such as *Caesalpinia corymbosa, Capparis angulata, C. mollis* and *Eriotheca discolor*, in addition to cereoid columnar cacti (Weberbauer 1945).

Important woody plant families in the humid montane forests include Araliaceae, Asteraceae, Clusiaceae, Cunoniaceae, Ericaceae, Euphorbiaceae, Lauraceae, Melastomataceae, Moraceae, Myrsinaceae, Myrtaceae, Rubiaceae and Solanaceae. Important genera of trees and shrubs include *Ilex, Oreopanax, Schefflera, Baccharis, Diplostephium, Gynoxys, Berberis, Brunellia, Centropogon, Siphocampylos, Hedyosmum, Clethra, Clusia, Weinmannia, Vallea, Gaultheria, Vaccinium, Escallonia, Ocotea, Persea, Desfontainia, Gaiadendron, Brachyotum, Miconia, Myrica, Myrcianthes, Myrsine, Piper, Podocarpus, Prumnopitys, Monnina, Oreocallis, Hesperomeles, Palicourea, Cestrum, Solanum* and *Symplocos*. Important genera of climbing and epiphytic plants include *Bomarea, Anthurium, Mikania, Munnozia, Tillandsia, Coriaria, Fuchsia, Elleanthus, Epidendrum, Masdevallia, Odontoglossum, Oncidium, Pleurothallis, Telipogon, Oxalis, Passiflora, Peperomia, Muehlenbeckia* and *Calceolaria*.

In the montane forest, other significant taxa include the rare Andean cedar *Cedrela lilloi*, two endemic species of *Zanthoxylum*, and Bignoniaceae endemics: *Jacaranda sprucei* (the highest-altitude species of the genus) and an undescribed *Tabebuia* species.

In the ecotone between the montane cloud forest and high-elevation zone, Asteraceae and Ericaceae become especially evident. Páramo species include grasses (*Calamagrostis, Festuca*); sedges (*Carex, Oreobolus, Scirpus*); composites (*Baccharis, Bidens, Diplostephium, Hieracium, Senecio, Werneria*); ericads (*Befaria, Disterigma, Gaultheria*); and scrophs (*Bartsia, Calceolaria*).

MAP 68. HUANCABAMBA REGION, NORTHERN PERU AND SOUTHERN ECUADOR (CPD SITE SA32)

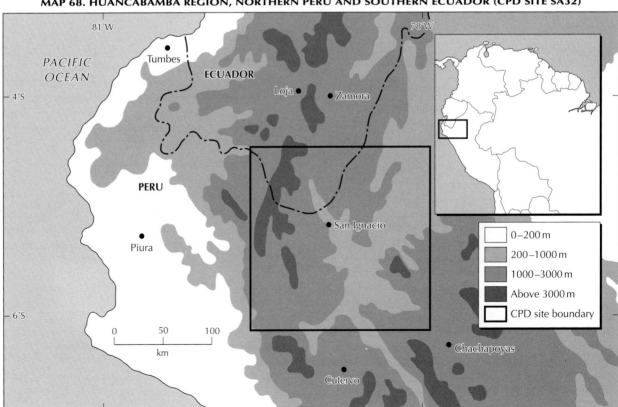

CPD Site SA32: Huancabamba region, Peru and Ecuador. Lower montane rain forest in Department of Amazonas, northern Peru. Photo: Kenneth R. Young.

Useful plants

Prance (1989) recognized "*Podocarpus* forest" in this region, because its mesic montane forests have important large populations of the podocarps *Podocarpus oleifolius* and *Prumnopitys montana*. These timber species are eagerly logged. They grow as large trees with numerous other tree species (e.g. Zevallos-Pollito 1988). *Podocarpus sprucei* is found in seasonally drier montane forests. Other timber species of high market value are also found in the region, including the rare "cedar" *Cedrela lilloi* and *Schmardaea microphylla*, a relative of mahogany (*Swietenia*) (Pennington *et al.* 1990).

Traditionally, the region has been a source of fuelwood and medicinal plants for local people. Many species of orchids and bromeliads have potential use as ornamentals.

Social and environmental values

The most conspicuous value of the natural ecosystems is probably for water production. All the agricultural development projects of northern Peru and southern Ecuador depend on irrigation of the dry inter-Andean valleys and the arid coastal plain. Only by keeping the upper watersheds protected can these water sources be assured, for municipal drinking water and huge irrigation projects.

Extracted timber tends to leave few economic benefits for local inhabitants, because the forests are not managed sustainably and reforestation is generally not practised. The region's flora includes useful plants and wild relatives of crop species, many of which are not found elsewhere. In addition, the natural ecosystems provide necessary habitats for many animal species, including many endemic species.

Economic assessment

Approximately 400,000 people live in the Peruvian portion of the region and perhaps another 150,000 are in the Ecuadorian portion. All indications are that this region will become increasingly attractive for both government-sponsored and spontaneous colonization.

Following road building and concomitant timber extraction, land at low to middle elevations is put into production of agricultural products such as coffee, fruits, maize and rice, often with irrigation. Land at high elevations is used as a source of timber and fuelwood or is converted into cattle rangeland. Development planning for this region appears to have neglected or underestimated the economic value of intact forests as watershed protection for the water supply.

Threats

The dry forests and semi-arid vegetation types are unique and unprotected. Probably all of them have been significantly altered by human activities.

The humid montane forests are the most threatened of this type of ecosystem in Peru; those of southern Ecuador appear to be already severely degraded. These forests are naturally isolated (Dillon 1994), and have been additionally fragmented by clearing for agriculture and rangeland. Perhaps 75% of the original humid forest has been removed and replaced by agricultural systems or scrub. The anthropogenic fragmentation is an ongoing process, but some fairly large tracts of forest are still continuous, especially in Peru near the border with Ecuador. In the past few years these forests have become increasingly threatened by logging for the valuable timber species.

Expansion of the agricultural frontier in both southern Ecuador and northern Peru continues to put great pressure on the remaining natural areas. A result is loss of habitat for animal and plant species. Gold mining also takes place in the region, and there are border disputes between the two countries.

Conservation

The Huancabamba region needs an expanded and improved park and nature reserve system. One nature reserve exists: the Tabaconas-Namballe National Sanctuary (295 km²) to the west of the town San Ignacio (Map 68), which was established in 1988 by the Peruvian Government. This potentially valuable conservation unit is unprotected and unstudied. In nearby regions, parks and reserves include: in Peru, Cutervo National Park near the town of that name (cf. Dillon 1994) and Noroeste Peruano Biosphere Reserve (see Data Sheet SA41); and in Ecuador, Podocarpus National Park between Loja and Zamora.

Existing laws concerning logging contracts, sale of timber products and reforestation require strong enforcement. National and international attention, plus funding, are needed to ensure that these laws and regulations are implemented. These forests are often used by local people, so conservation requires programmes that involve resource extraction and restoration projects. Local programmes of reforestation and watershed protection that include native tree species and management of the natural forests are needed. Vargas (1938) and Veillón (1962) have promoted the development of silvicultural practices for podocarps, but little has been done in Peru or Ecuador.

Also important are transnational biodiversity assessments and conservation strategies. For example, establishment of binational reserves or parks in this region could lead to an improvement in relations between the governments of Peru and Ecuador.

References

Atherton, M.P., Pitcher, W.S. and Warden, V. (1983). The Mesozoic marginal basin of central Peru. *Nature* 305: 303–306.

Baumann, F. (1988). Geographische Verbreitung und Ökologie Sudamerikanisher Hochgebirgspflanzen. *Physische Geographie (Zurich)* 28: 1–206.

Berry, P.E. (1982). The systematics and evolution of *Fuchsia* sect. *Fuchsia* (Onagraceae). *Ann. Missouri Bot. Gard.* 69: 1–198.

Cadle, J.E. (1991). Systematics of lizards of the genus *Stenocercus* (Iguania: Tropiduridae) from northern Peru: new species and comments on relationships and distribution patterns. *Proc. Acad. Nat. Sci. Philadelphia* 143: 1–96.

Cannatella, D.C. (1982). Leaf-frogs of the *Phyllomedusa perinesos* group (Anura: Hylidae). *Copeia* 1982: 501–513.

Cano, A. and Valencia, N. (1992). Composición florística de los bosques nublados secos de la vertiente occidental de los Andes peruanos. In Young, K.R. and Valencia, N. (eds), *Biogeografía, ecología y conservación del bosque montano en el Perú*. Memorias del Museo de Historia Natural, Vol. 21. Universidad Nacional Mayor de San Marcos (UNMSM), Lima. Pp. 171–180.

Clapperton, C.M. (1993). *Quaternary geology and geomorphology of South America*. Elsevier, Amsterdam. 779 pp.

Descimon, H. (1986). Origins of lepidopteran faunas in the high tropical Andes. In Vuilleumier, F. and Monasterio, M. (eds), *High altitude tropical biogeography*. Oxford University Press, New York. Pp. 500–532.

Dillon, M.O. (1994). Bosques húmedos del norte del Perú. *Arnaldoa* 2(1): 29–42.

Duellman, W.E. (1979). The herpetofauna of the Andes: patterns of distribution, origin, differentiation, and present communities. In Duellman, W.E. (ed.), *The South American herpetofauna: its origin, evolution, and dispersal*. Monogr. Mus. Nat. Hist. Univ. Kansas 7: 371–459.

Duellman, W.E. and Wild, E.R. (1993). Anuran amphibians from the Cordillera de Huancabamba, northern Peru: systematics, ecology, and biogeography. *Occas. Papers Mus. Nat. Hist. Univ. Kansas* 157: 1–53.

Feininger, T. (1987). Allochthonous terranes in the Andes of Ecuador and northwestern Peru. *Canadian J. Earth Sci.* 24: 266–278.

Gentry, A.H. (1982). Phytogeographic patterns as evidence for a Chocó refuge. In Prance, G.T. (ed.), *Biological diversification in the tropics*. Columbia University Press, New York. Pp. 112–136.

Gentry, A.H. (1992). Diversity and floristic composition of Andean forests of Peru and adjacent countries: implications for their conservation. In Young, K.R. and Valencia, N. (eds), *Biogeografía, ecología y conservación del bosque montano en el Perú*. Memorias Museo de Historia Natural, Vol. 21. UNMSM, Lima. Pp. 11–29.

Jørgensen, P.M. and Ulloa-Ulloa, C. (1992). Especies encontradas sobre los 2400 m.s.n.m. en las provincias de Loja y Zamora-Chinchipe. *Parque Nacional Podocarpus: Bol. Informativo sobre Biología, Conservación y Vida Silvestre* 3: 127–186.

Kessler, M. (1992). The vegetation of south-west Ecuador. In Best, B.J. (ed.), *The threatened forests of south-west Ecuador*. Biosphere Publications, Leeds, U.K. Pp. 79–100.

Molau, U. (1978). The genus *Calceolaria* in NW South America. *Bot. Notiser* 131: 219–227.

Molau, U. (1988). Scrophulariaceae – Part I: Calceolarieae. Flora Neotropica Monogr. 47. New York Botanical Garden, Bronx. 326 pp.

Panero, J.L. (1992). Systematics of *Pappobolus* (Asteraceae-Heliantheae). *Systematic Bot. Monogr.* 36: 1–195.

Parker III, T.A., Schulenberg, T.S., Graves, G.R. and Braun, M.J. (1985). The avifauna of the Huancabamba region, northern Peru. *Ornithol. Monogr.* 36: 169–197.

Patterson, B.D., Pacheco, V. and Ashley, M.V. (1992). On the origins of the western slope region of endemism: systematics of fig-eating bats, genus *Artibeus*. In Young, K.R. and Valencia, N. (eds), *Biogeografía, ecología y conservación del bosque montano en el Perú*. Mem. Museo Hist. Nat., Vol. 21. UNMSM, Lima. Pp. 189–205.

Pennington, T., Timana, M., Díaz, C. and Reynel, C. (1990). Un raro "cedro" redescubierto en el Perú. *Boletín de Lima* 67: 41–46.

Prance, G.T. (1989). American tropical forests. In Lieth, H. and Werger, M.J.A. (eds), *Tropical rain forest ecosystems*. Ecosystems of the World 14B. Elsevier, Amsterdam. Pp. 99–132.

Reig, O.A. (1986). Diversity patterns and differentiation of high Andean rodents. In Vuilleumier, F. and Monasterio, M. (eds), *High altitude tropical biogeography*. Oxford University Press, New York. Pp. 404–439.

Sillitoe, R.H. (1974). Tectonic segmentation of the Andes: implications for magmatism and metallogeny. *Nature* 250: 542–545.

Simpson, B.B. (1975). Pleistocene changes in the flora of the high tropical Andes. *Paleobiology* 1: 273–294.

Valencia, N. (1990). *Ecology of the forests on the western slopes of the Peruvian Andes*. Ph.D. thesis. University of Aberdeen, Aberdeen, Scotland, U.K.

Vargas, C.C. (1938). El *Podocarpus glomeratus* Don (intimpa), y la silvicultura nacional. *Acad. Cien. Exactas Físicas Nat. Lima (Perú)* 1: 27–32.

Veillón, J.P. (1962). *Coníferas autóctonas de Venezuela, los Podocarpus*. Talleres Gráficos Universitarios, Mérida, Venezuela.

Vuilleumier, F. (1969). Pleistocene speciation in birds living in the high Andes. *Nature* 223: 1179–1180.

Vuilleumier, F. (1977). Barrières écogéographiques permettant la spéciation des oiseaux des hautes Andes. In Descimon, H. (ed.), *Biogéographie et evolution en Amérique tropicale*. Ecole Normale Supérieure, Publ. Lab. Zool. 9, Paris. Pp. 29–51.

Vuilleumier, F. and Simberloff, S. (1980). Ecology versus history as determinants of patchy and insular distributions in high Andean birds. *Evolutionary Biol.* 12: 235–379.

Weberbauer, A. (1945). *El mundo vegetal de los Andes peruanos*. Ministerio de Agricultura, Dirección de Agricultura, Estación Experimental Agrícola de La Molina, Lima. 776 pp.

Zevallos-Pollito, P.A. (1988). *Estudio dendrológico de las podocarpáceas y otras especies forestales de Jaén y San Ignacio*. Consejo Nacional de Ciencia y Tecnología, Lima.

Authors

This Data Sheet was written by Dr Kenneth R. Young (University of Maryland Baltimore County, Department of Geography, Baltimore, MD 21228, U.S.A.) and Carlos Reynel (Universidad Nacional Agraria, Departamento de Biología, Secc. Botánica, Apartado 456, La Molina, Lima, Peru).

(TROPICAL) ANDES: CPD SITE SA33

PERUVIAN PUNA
Peru

Location: Andean highlands of Western and Eastern cordilleras, between latitudes 7°–18°S.

Area: 230,000 km².

Altitude: 3300–5000 m.

Vegetation: Tropical-alpine vegetation: grasslands, woodland patches, scrub, wetlands.

Flora: 1000–1500 species; several endemic genera, numerous endemic species.

Useful plants: Many traditional crops, medicinals, over 70 fodder species, fruits, spices, fuels.

Other values: Scenery, archaeological sites, habitats of endemic animal species, threatened fauna, minerals.

Threats: Overgrazing, burning, erosion, gathering fuelwood, pollution from mining.

Conservation: Huascarán National Park (3400 km²) in Huascarán Biosphere Reserve, parts (980 km²) of Río Abiseo NP and Manu NP; several reserves, although none strictly protected.

Geography

Tropical alpine-like vegetation is found in the Andes above the elevational limit of closed-canopy continuous forest and below the permanent snow-line. In Peru and southward this relatively dry altitudinal zone is known as "puna", although the wetter phases in northern Peru also are commonly distinguished as "jalca" (Weberbauer 1945) and near the border with Ecuador as "páramo" (Brack 1986). Because of ambiguity as to floristic boundaries and evaluation of human impacts on the Central Andean flora, this Data Sheet takes an inclusive view of Peruvian puna, including all habitats and vegetation types above 3300 m (see Map 69). The puna thus includes most of the departments Huancavelica, Ayacucho, Apurímac and Puno; large sections of Cajamarca, La Libertad, Ancash, Huánuco, Pasco, Junín, Lima, Ica, Arequipa and Cusco; and smaller portions of Lambayeque, Moquegua and Tacna. Map 69 and most of our discussion exclude the small area of páramo near the border with Ecuador in the northernmost portions of the departments of Piura and Cajamarca (cf. Luteyn 1992).

Puna is found in three geographic subregions in Peru. The main subregion is on the Western Cordillera of the Andes from c. 8°–18°S. This includes part of the Puno plateau and Lake Titicaca; the rest of both are in Bolivia. From 8° to 31°S (c. 2800 km), there is only one pass below 4000 m (Duellman 1979). Topographic relief of the Peruvian puna is moderate compared to the steep escarpments on the eastern and western slopes of the Andes. Most of the puna has a rolling topography, with a wide variety of substrate types and drainage classes. Much of the region is underlain by bedrock of Tertiary and Quaternary volcanics (Peñaherrera 1989). Active volcanoes are present in the south. The highest elevations are rocky, glacially modified horns, some with icecaps; the highest peak is Nevado Huascarán (6745 m).

The Eastern Cordillera is on average lower and narrower than the Western Cordillera and consequently has a smaller area of puna, which extends from c. 7°–15°S. Topography is relatively moderate between 3300–3800 m. Higher elevations are steep rocky cirques. Only in the south are peaks high enough to have icecaps; the highest is Nevado Salcantaya (6271 m). Bedrock is of metamorphized Precambrian and Palaeozoic sedimentary rock. Several deep valleys cut through the puna of the Eastern Cordillera, creating biogeographical barriers.

Northern Peru constitutes the third subregion. The puna (jalca) is quite fragmented among the high elevations of several relatively low mountain ranges. Bedrock is often of Cretaceous metamorphic and sedimentary rocks.

In general, there is great heterogeneity in puna substrates depending on the bedrock underneath. There are also great differences in drainage, with many resulting types of soils – from histosols in boggy sites to well-drained mollisols. Some areas, particularly in northern Peru, have deposits as much as 1 m deep of glacial loess. Glacial moraines and associated landforms are found in almost all of the puna (cf. Clapperton 1993). In cold and arid climatic cycles of the Pleistocene, much of the highlands above 4000 m at times was covered by ice and some glacial tongues descended below 3000 m.

The current climatic regimes apparently date back only several thousand years. The last major deglaciation began 12,000 years ago, and temperatures were apparently warmer than at present until c. 5000 years ago (Hansen, Wright and Bradbury 1984). Median annual temperatures range from c. 8°C at 3300 m to 0°C at snow-line. Temperature conditions are mostly constant during the year, but show considerable diurnal variation (Sarmiento 1986). Temperatures may reach below -2°C at night and between 10°–20°C during daylight. Near Lake Titicaca, the temperature and humidity extremes are moderated by the large lake.

The annual precipitation is widely variable, from c. 150 mm to over 1500 mm. Troll (1968) used the north-to-south decline in precipitation in the Central Andes and strong west-to-east increase in cloud cover and annual rainfall to recognize three types of puna in Peru. We designate them as (1) **wet puna**, in two major locations: along the Eastern

Cordillera, where it is influenced by moist air uplifted from the Amazon Basin, and in northern Peru, where air masses come from both the Pacific Ocean and the Amazon Basin; (2) **moist puna**, found in most of Peru, particularly on the Western Cordillera; and (3) **dry puna**, to the south of 15°S on the Western Cordillera. The wet puna receives 800 mm to more than 1500 mm of annual precipitation; the moist puna receives 400–800 mm, and the dry puna less than 400 mm.

The drier phases of puna are very seasonal, with most (70–80%) of the precipitation falling during 4–5 months, usually between December and April. Even the wet to moist puna (jalca) of northern Peru has a pronounced dry season from May to September. Only the wet puna areas of the Eastern Cordillera that are directly exposed to cloud banks on

the upper edges of the Amazon Basin are almost aseasonal, with just a short dry season in July or August.

Vegetation

The three types of puna correspond to at least 25 different Holdridge life zones (Tosi 1960; ONERN 1976). Relative air humidity is often less than 30%, except when fog is present. Puna plants (cf. Smith and Young 1987; Cabrera 1968) must have adaptations to resist or avoid desiccation. Plant growth also is cold-limited. Often temperatures drop to near or below freezing at night, particularly if there are no clouds. On cool days, species can continue to photosynthesize if

MAP 69. PERUVIAN PUNA (CPD SITE SA33)

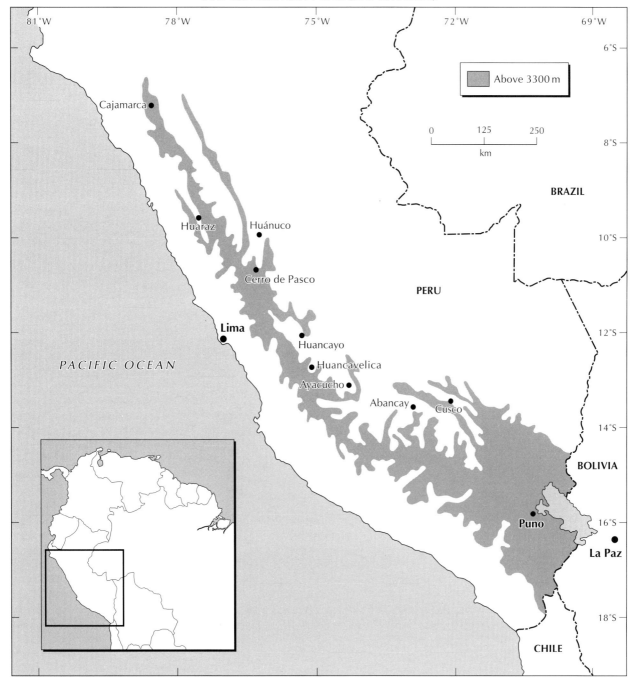

their leaves are in protected microclimates, for example by being near the ground or protected by the plant's pubescence.

From 3300 m to c. 4200 m the predominant vegetation types of puna are grasslands. Bunchgrasses of *Calamagrostis*, *Festuca* and *Stipa* form tussocks that can grow to occupy 0.25–0.75 m² in basal area and reach 1 m in height (if not burned or grazed). In the dry puna, *Festuca orthophylla* predominates. In wet puna on the Eastern Cordillera, several species of *Cortaderia* can be important. Spaces around the tussocks are often filled by a number of herbs, including non-tussock-forming grasses and sedges, prostrate or low-growing forbs, lichens, mosses, and ferns and their allies such as *Jamesonia* and *Lycopodiella*.

From 3300–3800 m it is common to find patches of Andean forest growing in sites protected from frost and fire. Genera frequently present as trees or shrubs include *Baccharis*, *Berberis*, *Brachyotum*, *Chuquiraga*, *Clethra*, *Escallonia*, *Gynoxys*, *Miconia*, *Myrsine* and *Weinmannia*. Sometimes *Alnus acuminata* can be found. From 3600 m to more than 4400 m, especially in the moist puna of the Western Cordillera, are forest patches dominated by *Polylepis* spp. Trees and shrubs of *Gynoxys*, *Miconia*, *Myrsine* and *Ribes* are also usually present. In the south, scrub dominated by *Buddleja* spp. may be present (Tovar-Serpa 1973).

Puna grasslands found in the proximity of these woodlands may be invaded by woody species, if they are not regularly grazed or burned. The frequency of natural fire regimes in Peruvian puna ecosystems is unknown. Generally people burn the puna grasslands every one to five years to renew forage for their livestock. Occasionally shrubs can be locally common in puna grasslands despite frequent burning; the shrubs are species of the microphyllous genera *Baccharis*, *Diplostephium*, *Hypericum*, *Loricaria* and *Parastrephia*.

Areas with poor drainage are usually dominated by sedges and rushes. Below 4000 m these genera include *Carex*, *Juncus*, *Oreobolus* and *Scirpus*. Above 4000 m, vegetation in the wet areas is dominated by floating and submerged cushion plants. *Distichia muscoides* and *Plantago rigida* are often conspicuous, forming large cushions. Other genera include *Gentiana*, *Hypsela*, *Isoetes*, *Lilaeopsis*, *Ourisia*, *Oxychloe* and *Scirpus*. Aquatic plants are common in ponds and streams, including *Crassula venezuelensis*, *Myriophyllum quitense*, *Potamogeton* spp., *Ranunculus* spp. and *Zannichellia andina* (León 1993).

Grasses begin to lose their dominance on well-drained substrates above c. 4200 m. The altitudinal zone of 4200–4800 m is what Weberbauer (1945) apparently considered typical puna. The predominant life forms are prostrate, cushion and rosette herbs of genera such as *Azorella*, *Baccharis*, *Daucus*, *Draba*, *Echinopsis* (*Trichocereus*), *Gentiana*, *Geranium*, *Lupinus*, *Nototriche*, *Plettkea*, *Valeriana* and *Werneria*. Elevations above 4800 m have vegetation types with these same taxa as dispersed individual plants. Often bare ground predominates, especially on scree slopes made unstable by needle-ice.

The dry puna is characterized by sparse vegetation and typically dominated by shrubs. Bunchgrasses and cushion

CPD Site SA33: Puna grassland in the Department of Cusco, southern Peru.
Photo: Kenneth R. Young.

plants are also present. Plant genera include *Aciachne, Adesmia, Margyricarpus, Parastrephia* and *Tetraglochin.*

Flora

Approximately 1000–1500 species comprise the vascular flora of Peru's puna. Many species are endemic to one or more of the puna's subregions, and many genera have their centres of diversity here, e.g. *Culcitium, Perezia* and *Polylepis*. Endemic genera include *Alpaminia* and *Weberbauera* (Brassicaceae) and *Mniodes* (Asteraceae) (Rivas-Martínez and Tovar-Serpa 1983).

About 10% of the flora is represented by pteridophytes. Conspicuous or speciose genera include *Elaphoglossum, Jamesonia, Huperzia* and *Lycopodium*. In wetlands and lakes, *Isoetes* spp. often are present.

Monocotyledons account for perhaps 30–40% of the flora. Grasses are very conspicuous, especially genera such as *Agrostis, Calamagrostis, Festuca, Paspalum* and *Stipa,* and in more humid areas *Chusquea* and *Cortaderia*. Also conspicuous are the sedges, such as *Carex, Oreobolus* and *Scirpus*; and the bromeliad *Puya*. Present, although usually inconspicuous, are orchids of the genera *Aa, Odontoglossum, Pleurothallis, Pterichis* and *Trichoceros*. Several species of *Bomarea* and *Sisyrinchium* are usually present. *Excremis coarctata* is common on the Eastern Cordillera (Young and León 1990), whereas several *Orthrosanthus* species are found on the Western Cordillera (Sagástegui-Alva 1988).

The dicotyledons are represented by more than 175 genera. Particularly diverse or conspicuous families include Asteraceae (*Baccharis, Bidens, Culcitium, Chaptalia, Diplostephium, Gnaphalium, Hieracium, Liabum, Oritrophium, Senecio, Stevia, Werneria*); Brassicaceae (*Cardamine, Draba, Weberbauera*); Campanulaceae (*Lysipomia*); Caryophyllaceae (*Arenaria, Cerastium, Pycnophyllum*); Rosaceae (*Acaena, Alchemilla, Potentilla*); Rubiaceae (*Arcytophyllum, Galium*); and Scrophulariaceae (*Agalinis, Bartsia, Calceolaria*).

Many of the herbs have adaptations to the cold: prostrate and rosette life forms, considerable pubescence and/or small leaves. Often they have surfaces that discourage grazing animals, such as prickles, spines or thorns on *Chuquiraga, Opuntia* and *Puya*. Plants that resprout after being grazed are also preadapted to survive repeated fires (Young and León 1991), which presumably explains the success of bunchgrasses in the Peruvian puna.

Useful plants

Native grasslands in Peru are used mainly for the grazing of cattle, horses, sheep and South American camelids. More than 70 native fodder species are recognized in the puna; among the more important are *Bromus* spp., *Calamagrostis* spp., *Festuca* spp., *Muhlenbergia* spp., *Nasella* spp., *Poa* spp. and *Stipa* spp. (Ruiz-Canales and Tapia-Núñez 1987; Becker, Terrones-H. and Tapia 1989). Also consumed by

CPD Site SA33: A flowering specimen of *Miconia rotundifolia* in Río Abiseo National Park, northern Peru.
Photo: Kenneth R. Young.

the livestock are *Alchemilla* spp., *Hypochoeris* spp., *Luzula* spp. and *Nototriche* spp., and some aquatics such as *Elodea potamogeton* and *Myriophyllum quitense*.

The lower elevations of puna are often used for growing native and introduced crops. Native Andean crops include tubers, represented by several kinds of potatoes (*Solanum acaule, S. andigenum, S. curtilobum, S. juzepczukii, S. tuberosum*), "ollucos" (*Ullucus tuberosus*), "oca" (*Oxalis tuberosa*); and pseudo-cereals (*Amaranthus caudatus, Chenopodium quinoa, C. pallidicaule*). Other Andean species are planted in mixed fields with tubers and cereals. These species, such as *Lupinus* spp., *Phaseolus* spp. and *Vicia* spp., may represent a potential source for new commercial crops.

Besides traditional Andean crops, many native species are harvested for food as fruits, e.g. *Ribes brachybotris* and *Salpichroa hirsuta*, or as part of a dietary complement, such as the algae *Nostoc* spp. Some of the puna plants are used as spices (*Hypochoeris, Tagetes*). Medicinal plants from the puna are highly appreciated in Peru. Common medicinals include the pteridophytes *Huperzia* and *Jamesonia*, and dicotyledons such as *Baccharis, Perezia, Tagetes* (Asteraceae); *Draba, Lepidium* (Brassicaceae); *Gentianella* (Gentianaceae); and *Lepechinia, Minthostachys, Salvia, Satureja* (Lamiaceae).

Most of the highland residents use fuelwood for cooking and heating. Several species of shrubs are commonly exploited, including *Baccharis tricuneata* and *Parastrephia lepidophylla* (Reynel 1988). Species of *Azorella* are used for fuel in Arequipa's puna (Hodge 1948). In the central and northern part of the puna region, fuel sources include the cushion plant *Distichia muscoides* and the small trees *Escallonia* spp. and *Kageneckia lanceolata*. In the south, boats and houses are constructed using *Scirpus californicus* subsp. *tatora*.

Social and environmental values

The long experience of the Andean people in selecting and maintaining crops is the basis for their rich agricultural heritage. In addition, the puna is a genetic storehouse due to the presence of wild relatives of present-day and potential subsistence and commercial crops (Altieri, Anderson and Merrick 1987). However, because of social and economic conditions, many rural communities in or near the puna are being abandoned in favour of migration to the cities.

Land-use practises date from pre-Inca times (e.g. Schjellerup 1992). During the Inca empire terracing and irrigation were widely used below 4000 m, thus dramatically changing the landscape (Seibert 1983). Most of the puna supports intensive grazing. The highest elevations are mainly grazed by llamas and alpacas, whereas at lower elevations these animals share pastures with sheep, and to a lesser extent, cattle.

Tourism is limited to archaeological sites, with the partial exception of Huascarán National Park, which receives many hikers and alpinists.

Thirty-five bird species of restricted range occur in the woodlands of *Polylepis, Weinmannia, Gynoxys* and *Escallonia*

CPD Site SA33: Cattle in puna grassland. Photo: Kenneth R. Young.

and the puna vegetation throughout the Peruvian Andes. Two Endemic Bird Areas (EBAs) have been defined. Many of the species restricted to the patchy high-altitude woodlands and puna in the High Peruvian Andes EBA (B27) have very specific ecological requirements; due to the widespread degradation of this zone, no less than 15 of them are considered threatened. The Junín puna EBA (B28) covers two discrete areas of central Peru in a region that was ice-free during the last glaciation; five bird species are confined to the area around Lago (Chinchaycocha) de Junín (about 11°S, 76°30'W).

Economic assessment

The puna of Peru is used by 4–6 million people, at least indirectly. Today, four large cities with a long history are located in the puna: Cajamarca, Cerro de Pasco, Cusco and Puno (Map 69). Puna is one of the critical resource zones for the rural people of the Central Andes (cf. Winterhalder and Thomas 1978). Scenery and archaeological or cultural attractions confer a high potential value for tourism. However, development of tourism in the region should be accompanied by improvement of living standards for the involved rural communities. Some native animals of the puna, such as the vicuña (*Lama vicugna*), represent a possibly sustainable resource.

Introduced crop species include barley (*Hordeum vulgare*), oats (*Avena sativa*), rye (*Secale cereale*), wheat (*Triticum*) and broad beans (*Vicia faba*). Introduced fodder species include *Dactylis glomerata*, *Festuca pratensis*, *Lolium perenne*, *Medicago sativa*, *Phalaris tuberosa* and *Trifolium pratense* (Ruiz-Canales and Tapia-Núñez 1987). Several introduced plants have become naturalized in lower elevations of puna. These include *Brassica* spp., *Lolium perenne*, *Medicago* spp. and *Trifolium* spp. *Pennisetum clandestinum* and *Taraxacum officinale* are among the worst weedy plants in the country.

Threats

The puna is one of the most heavily altered natural regions in Peru. The long history of human settlement goes back some 10,000 years. Recent social and economic conditions in the country have damaged preservation of genetic material of native Andean crops. Degradation of habitats at present is mostly related to overgrazing and contamination from mining.

Overgrazing increases soil erosion, especially in areas where drought has impacted in recent decades, as in the Department of Puno. Overgrazing has been related to cultural factors that cause over-stocking and to the lack of alternative sources for forage (e.g. Le Baron *et al.* 1979). Grassland burning associated with this grazing affects populations of native non-fodder plants, including for example the few existing stands of the giant bromeliad *Puya raimondii* (Cerrate de Ferreyra 1979).

Pollution by mining activities can be seen most clearly in water bodies and poorly drained areas. Aquatic and semi-aquatic vegetation is affected, thus threatening populations of geographically restricted plants, such as species of *Isoetes* (Young and León 1993).

The scarcity of fuelwood causes high pressure on the native vegetation: trees of *Polylepis* spp. and *Buddleja* spp. are cut for this purpose. Reforestation programmes have relied heavily on introduced species, mainly eucalypts (*Eucalyptus*), which has probably also adversely affected the local fauna and flora.

Conservation

Puna is officially protected in three National Parks. The vegetation within Huascarán National Park (3400 km²) consists exclusively of puna (Smith 1988); the NP is the core of Huascarán Biosphere Reserve (3992 km²), which is recognized by UNESCO-MAB. There are 184 km² classified as puna in Río Abiseo NP and 796 km² in Manu NP. Other protected areas with puna include five National Reserves (Calipuy, Junín, Pampa Galeras, Salinas – Aguada Blanca, Titicaca), three National Sanctuaries (Huayllay, Calipuy, Ampay) and three Historical Sanctuaries (Chacramarca, Pampas de Ayacucho, part of Machu Picchu).

These designated areas total approximately 9500 km² of puna and represent c. 4% of the general area of puna, or just 0.7% of the area of Peru. However, actual protection of puna in these parks and reserves is sadly deficient; there is a lack of infrastructure and personnel. In addition, there has been no attempt to preserve pristine habitat types or restore them.

Outside of the reserved areas, there need to be many more programmes that seek to alleviate the causes of over-grazing and soil erosion. For example, environmental impact assessment and mitigation should be carried out for mining operations located in puna. Reforestation programmes should incorporate more native species and be directed toward appropriate fuel species.

References

Altieri, M.A., Anderson, M.K. and Merrick, L.C. (1987). Peasant agriculture and the conservation of crop and wild plant resources. *Conserv. Biol.* 1: 49–58.

Becker, B., Terrones-H., F.M. and Tapia, M.E. (1989). *Los pastizales y producción forrajera en la Sierra de Cajamarca*. Proyecto Piloto de Ecosistemas Andinos, Cajamarca, Peru. 247 pp.

Brack, A. (1986). Ecología de un país complejo. In *Gran geografía del Perú: naturaleza y hombre*, Vol. II. Manfer-Juan Mejía Baca, Spain. Pp. 175–319.

Cabrera, A.L. (1968). Ecología vegetal de la puna. In Troll, C. (ed.), *Geo-ecology of the mountainous regions of the tropical Andes*. Colloquium Geographicum 9, Bonn. Pp. 91–116.

Cerrate de Ferreyra, E. (1979). *Vegetación del Valle de Chiquian, Provincia Bolognesi, Departamento de Ancash*. Editorial Los Pinos, Lima. 65 pp.

Clapperton, C.M. (1993). *Quaternary geology and geomorphology of South America*. Elsevier, Amsterdam. 779 pp.

Duellman, W.E. (1979). The herpetofauna of the Andes: patterns of distribution, origin, differentiation, and present communities. In Duellman, W.E. (ed.), *The South American herpetofauna: its origin, evolution, and dispersal*. Monogr. Mus. Nat. Hist. Univ. Kansas 7: 371–459.

Hansen, B.C.S., Wright, H.E. and Bradbury, J.P. (1984). Pollen studies in the Junín area, central Peruvian Andes. *Bull. Geol. Soc. Amer.* 95: 1454–1465.

Hodge, W. (1948). Yareta – fuel umbellifer of the Andean puna. *Economic Botany* 2: 113–118.

Le Baron, A., Bond, L.K., Aitken-S., P. and Michaelsen, L. (1979). An explanation of the Bolivian highlands grazing-erosion syndrome. *Journal Range Managem.* 32: 201–208.

León, B. (1993). Catálogo anotado de las fanerógamas acuáticas del Perú. In Kahn, F., León, B. and Young, K.R. (eds), *Las plantas vasculares en las aguas continentales del Perú*. Instituto Francés de Estudios Andinos (IFEA) – ORSTOM, Lima. Pp. 11–128.

Luteyn, J.L. (1992). Páramos: why study them? In Balslev, H. and Luteyn, J.L. (eds), *Páramo: an Andean ecosystem under human influence*. Academic Press, London. Pp. 1–14.

ONERN (1976). *Mapa ecológico del Perú. Guía explicativa*. Oficina Nacional de Evaluación de Recursos Naturales (ONERN), Lima. 117 pp.

Peñaherrera, C. (1989). *Atlas del Perú*. Instituto Geográfico Nacional, Lima. 399 pp.

Reynel, C. (1988). *Plantas para leña en el sur-occidente de Puno*. Proyecto Arbolandino, Puno, Peru. 165 pp.

Rivas-Martínez, S. and Tovar-Serpa, O. (1983). Síntesis biogeográfica de los Andes. *Collectanea Botánica (Barcelona)* 14: 515–521.

Ruiz-Canales, C. and Tapia-Núñez, M.E. (1987). *Producción y manejo de forrajes en los Andes del Perú*. Proyecto de Investigaciones de Sistemas Agropecuarios Andinos, Lima. 304 pp.

Sagástegui-Alva, A. (1988). *Vegetación y flora de la provincia de Contumazá*. CONCYTEC, Trujillo, Peru. 76 pp.

Sarmiento, G. (1986). Ecological features of climate in high tropical mountains. In Vuilleumier, F. and Monasterio, M. (eds), *High altitude tropical biogeography*. Oxford University Press, New York. Pp. 11–45.

Schjellerup, I. (1992). Pre-Columbian field systems and vegetation in the jalca of northeastern Peru. In Balslev, H. and Luteyn, J.L. (eds), *Páramo: an Andean ecosystem under human influence*. Academic Press, London. Pp. 137–150.

Seibert, P. (1983). Human impact on landscape and vegetation in the central High Andes. In Holzner, W., Werger, M.J.A. and Ikusima, I. (eds), *Man's impact on vegetation*. W. Junk Publishers, The Hague, The Netherlands. Pp. 261–276.

Smith, A.P. and Young, T.P. (1987). Tropical alpine plant ecology. *Ann. Rev. Ecol. Syst.* 18: 137–158.

Smith, D.N. (1988). *Flora and vegetation of the Huascarán National Park, Ancash, Peru, with preliminary taxonomic studies for a manual of the flora*. Ph.D. dissertation. Iowa State University, Ames.

Tosi Jr., J. (1960). *Zonas de vida natural en el Perú*. Organización de Estados Americanos, Boletín Técnico No. 5. Lima, Peru. 271 pp.

Tovar-Serpa, O. (1973). Comunidades vegetales de la reserva nacional de vicuñas de Pampa Galeras, Ayacucho, Peru. *Publ. Museo Hist. Natural "Javier Prado", Serie B* 27: 1–32.

Troll, C. (1968). The cordilleras of the tropical Americas. In Troll, C. (ed.), *Geo-ecology of the mountainous regions of the tropical Andes*. Colloquium Geographicum 9, Bonn. Pp. 15–55.

Weberbauer, A. (1945). *El mundo vegetal de los Andes peruanos*. Ministerio de Agricultura, Dirección de Agricultura, Estación Experimental Agrícola de La Molina, Lima. 776 pp.

Winterhalder, B.P. and Thomas, R.B. (1978). Geoecology of southern highland Peru: a human adaptive perspective. *Inst. Arctic Alpine Res. Occas. Pap.* 27: 1–91.

Young, K.R. and León, B. (1990). Catálogo de las plantas de la zona alta del Parque Nacional Río Abiseo, Perú. *Publ. Museo Hist. Natural (Lima), Ser. B* 34: 1–37.

Young, K.R. and León, B. (1991). Diversity, ecology and distribution of high-elevation pteridophytes within Río Abiseo National Park, north-central Peru. *Fern Gazette* 14: 25–39.

Young, K.R. and León, B. (1993). Distribución geográfica y conservación de las plantas acuáticas vasculares del Perú. In Kahn, F., León, B. and Young, K.R. (eds), *Las plantas vasculares en las aguas continentales del Perú*. IFEA-ORSTOM, Lima. Pp. 153–173.

Authors

This Data Sheet was written by Dr Kenneth R. Young (University of Maryland Baltimore County, Department of Geography, Baltimore, MD 21228, U.S.A.), Dra. Blanca León and Asunción Cano (Museo de Historia Natural, Apartado 14-0434, Lima-14, Peru) and Olga Herrera-MacBryde (Smithsonian Institution, Department of Botany, NHB-166, Washington, DC 20560, U.S.A.).

(Tropical) Andes and Southern Cone: CPD Site SA34

ALTOANDINA
Argentina, Chile

Location: Southern part of the high Andes, between latitudes c. 25° and 55°S and longitudes c. 66° and 74°W.
Area: No information.
Altitude: Lower limit 500 m in south to 4400 m in north; upper limit of lichens 5900 m.
Vegetation: Grass-steppe, chamaephyte-steppe, shrub-steppe; locally: lichen semi-desert, bog.
Flora: Includes neotropical, holarctic and antarctic floristic elements.
Useful plants: Forage, fuel and medicinal species.
Other values: No information.
Threats: No information.
Conservation: No information.

Geography

The Altoandina comprises the upper part of the southern Andes, from latitude c. 25°S to the tip of the continent in Tierra del Fuego (55°S) (see Map 36, p. 278). This southern part of the Andes includes the highest mountain in the Western Hemisphere (Aconcagua – 6959 m). The lower limit of the Altoandina vegetation descends from c. 4400 m in the north to c. 500 m in the extreme south; in the south it occurs as ecological islands surrounded by forest. The climate is cold and dry, though more humid southward. The scarce precipitation sometimes falls as snow. Winds are strong.

Vegetation

The most important vegetation types are grass-steppe, chamaephyte-steppe and shrub-steppe. Apart from grasses, the grass-steppe sometimes includes mat-forming species, such as *Adesmia*, *Azorella*, *Junellia*, *Mulinum*, *Senecio* and *Verbena* (Hunziker 1952; Ward and Dimitri 1966). Chamaephyte-steppe is generally found on loose soil at high elevations; common dwarf plants include *Senecio* spp. and cushion-forming *Oxalis compacta*, *Pycnophyllum molle*, *Valeriana* spp. and *Werneria* spp. Lichen semi-desert occurs on the most humid slopes, up to 5900 m. Bogs with Cyperaceae, Juncaceae and Gramineae are found in wet places.

Flora

The puna, Altoandina and Patagonia constitute the Andino-Patagonian Floristic Dominion (Cabrera 1976; Cabrera and Willink 1973). The flora includes neotropical, holarctic and (especially to the south) antarctic elements (Cabrera 1976). Well-represented families include Gramineae (*Calamagrostis, Festuca, Poa, Stipa*), Leguminosae (*Adesmia, Astragalus*) and Compositae (*Chuquiraga, Mutisia, Senecio*). There are no endemic families; endemic genera include *Barneoudia, Hexaptera, Nototriche, Pycnophyllum* and *Werneria*.

Useful plants

Numerous species are used as forage, fuel or medicinal (Ruthsatz 1974).

Social and environmental values, Threats, Conservation

No information.

References

Cabrera, A.L. (1976). *Regiones fitogeográficas argentinas*. In Parodi, L.R. (ed.), *Enciclopedia argentina de agricultura y jardinería*, 2nd edition. Vol. 2(1). Editorial Acmé, Buenos Aires. Pp. 1–85.

Cabrera, A.L. and Willink, A. (1973). *Biogeografía de América Latina*. Organización de los Estados Americanos (OEA), Serie de Biología, Monogr. No. 13, Washington, D.C. 117 pp.

Hunziker, J.H. (1952). Las comunidades vegetales de la Cordillera de La Rioja. *Rev. Invest. Agric.* 6: 167–196.

Ruthsatz, B. (1974). Los arbustos de las estepas andinas del noroeste argentino y su uso actual. *Bol. Soc. Argent. Bot.* 16: 27–45.

Ward, R.T. and Dimitri, M.J. (1966). Alpine tundra on Mt. Catedral in the southern Andes. *New Zealand J. Bot.* 4: 42–56.

Note: Information here is summarized from the South America overview.

ANCONQUIJA REGION
North-western Argentina

Location: Western Tucumán Province, with southern Salta and eastern Catamarca provinces, between latitudes 25°48'–27°38'S and longitudes 65°28'–66°50'W.

Area: c. 6000 km².

Altitude: c. 400–5550 m.

Vegetation: Six biogeographic provinces; vegetation types from Amazonian winter-dry rain forest through temperate cloud forest, Andean "páramo" grassland, high-Andean vegetation, and in western rain shadow, dry "prepuna" (spiny shrubland with tree cacti) and "monte" (semi-arid shrubland).

Flora: High diversity – close to 2000 species; high endemism, ancient species and habitats, southern extreme of ranges; neotropical and subantarctic elements intermingled with Andean flora; threatened species.

Useful plants: Potential genetic resources, e.g. for food crops; timbers, forages, medicinal plants, ornamentals.

Other values: Watershed protection, endemic and threatened animal species, archaeological sites, recreation, tourism.

Threats: Logging; clearing for agriculture and silviculture; fires; grazing and trampling; erosion; invasion of exotics; commercial collecting of medicinal and aromatic plants, fuelwood and ornamentals; collecting mosses, lichens and humus for gardening; road building; dams; unorganized tourism; potential mining.

Conservation: Soil and forest conservation laws, but not well enforced; several designated protected areas, but most inadequately or not enforced; proposed National Park for half of region (c. 3000 km²).

Geography

The Sierra de Anconquija and Cumbres Calchaquíes are isolated mountain ranges of the Pampean geological formation which extend 300 km along the western edge of Tucumán Province, including areas of Catamarca and a small portion of Salta (triprovincial area – see Map 70). The Precambrian rocks of these mountains were elevated long before the Andes and harbour an older flora. Glaciation acted upon the upper reaches to make breath-taking cirques, crests and peaks, and gently undulating high plateaux above 4000 m. The Sierra de Anconquija has peaks reaching 5550 m and at 3000 m joins to the north with the Cumbres Calchaquíes, which reach 4650 m and include a remarkable plateau at 4250 m. There are 40 or more glacial lakes.

The climate is very diverse. The eastern foothills are subtropical, with abundant summer rains and a winter dry season; annual precipitation is 1000–3000 mm. The eastern temperate mountain slopes (1500–2500 m to 3700 m) have summer mists for at least six months; in winter they occasionally receive thick snow. At higher elevations are typical mountain variability and intensity of sun, rain and snow. On the western slopes, rain shadows cause an arid climate and some dunes, with precipitation as low as 100 mm.

The Anconquija's eastern slopes rise from 500 m to 5550 m in less than 20 km (a world record) – there are spectacular slopes, gorges and waterfalls. High-Andean snowfields, lakes and bogs together with the vegetation ensure adequate regulation of water. Erosion is a permanent risk when the vegetation is degraded. These mountain ranges, together with the connected spurs of the sierras Medina, Nogalito and Santa Ana, constitute a geographical unit of c. 6000 km², nearly half of which is proposed for protection

as a National Park. Detailed descriptions are in Domínguez, Halloy and Terán (1978); Halloy, Grau and González (1982); Halloy (1985b); and Halloy, González and Grau (1994). The Anconquija group and its flora are closely associated with the Sierra de Ambato to the south and Sierra de Metán to the north.

Vegetation

This broadly defined Anconquija region includes six biogeographical provinces in three different dominions (Cabrera 1976). Subtropical forests ("lower yungas") cover the eastern base of the mountains, forming the southernmost exclave of Amazonian elements, which is isolated from Amazonia by drier vegetation. Temperate cloud forests ("upper yungas") occur above 1500–2500 m, then "páramo" grasslands. The high-Andean vegetation occurs between 3700–5100 m, and still higher is the aeolian zone with lichens. On the western side, the high-Andean vegetation gives way to thorny shrubs and candelabra cacti ("prepuna") and lower down to resinous shrubs and dune communities ("monte"). Each biogeographical province includes varied types of vegetation, described below from east to west.

In the Amazonian Dominion are the yungas and páramo provinces:

1. The "lower yungas" (Meyer 1963) has two forest types:

i) Lowland subtropical microphyllous forest on the foothills (300–500 m), with 1000 mm of rain annually. This formation was dominated by *Tipuana tipu* and

Enterolobium contortisiliquum; it has been almost completely eliminated to grow sugarcane (*Saccharum officinarum*) and for housing and industry. Although within the Anconquija region, the proposed park cannot include this formation due to lack of continuity. A portion must be preserved separately and possibly can be connected by corridors.

ii) **Subtropical summer-wet forests on the lower slopes** (500–1500 m), dominated by evergreens on southern exposures (*Phoebe porphyria, Ilex argentina*, many Myrtaceae – including the rare *Myrcianthes callicoma*), and deciduous trees on northern slopes (*Parapiptadenia excelsa, Jacaranda mimosifolia*) (e.g. Roldán 1991). Rain can be as much as 3000 mm annually. Trees support a profusion of epiphytes (e.g. orchids, bromeliads, ferns including Hymenophyllaceae) and lianas (of more species than in other Amazonian forests – Meyer 1963). There may be 2–3 strata of trees, an understorey of ferns and shrubs including *Psychotria carthagenensis, Miconia ioneura, Pteris deflexa* and *Dryopteris parallelogramma*, and a herbaceous layer that includes mosses.

2. **The "upper yungas" or temperate cloud forests** (1500–2500 m) (Hueck 1954; Vides 1985) are of three main types:

i) **Alder forest** (*Alnus acuminata*), which is the most extensive (Bell 1987).

ii) **Conifer forest** (*Podocarpus parlatorei*).

iii) **"Queñoa" forest** (*Polylepis australis*) – these contorted trees with red exfoliating bark and many epiphytes are an eerie sight in the cloud-shrouded mountains. The upper yungas have fewer tree species, but include such near endemics and rarities as *Crinodendron tucumanum* and *Escallonia schreiteri*. Epiphytes are still abundant, with mosses and ferns dominant. The undergrowth is quite diverse, having up to 40–50 species per 100 m².

3. **The "páramo"** (c. 2000–3700 m) is still in the Amazonian Dominion although with substantial differences (Halloy 1985a, 1985c), and includes three main formations:

i) **Páramo grassland** is most frequent, dominated by tuft grasses to 1 m high on sandy-loamy slopes. A rich undergrowth of plants covers the ground and prevents erosion.

ii) **Mesophytic shrubland** to 2 m high on northern exposures, extending as low as 1500 m, and on southern slopes interdigitating with the alder forests. Typical are *Baccharis* spp., *Ophryosporus charrua* and *Eupatorium* spp.

iii) **Microphyllous shrubland** c. 1 m high occurs above 2500 m on rocky slopes. The shrubs are intermingled with grasses, herbs and cushion plants.

Along mountain streams, marshes and rocky outcrops are galleries of *Cortaderia* grasses with a great variety of lichens, mosses, ferns, *Begonia, Satureja, Urtica*, etc.

4. In the Andino-Patagonian Dominion is the **high-Andean province**, from 3700–4600 m and sparsely to 5100 m (Halloy 1985a). There are again three main types of vegetation:

i) **Spiny grassland** is the main formation on sloping ground with sandy soils. The dominant *Festuca orthophylla* grasses often form circular or wave patterns that may be thousands of years old. These grasslands are highly susceptible to fires.

ii) **The cryptofruticetum** is dominated by woody and herbaceous plants growing flat on the ground, and rosette genera and circular-growing grasses occur. The high diversity of plants reaches 20–27 species per 1 m². This formation is well developed on the high plateau of the Cumbres Calchaquíes; it is highly susceptible to trampling.

iii) **Bogs ("ciénagas")**. Around springs and brooks, dense growths of cushion-forming Juncaceae, carpet-forming Cyperaceae and showy flowering plants form peat bogs and meadows, which retain and control water. Most of the bogs are many thousands of years old.

CPD Site SA35: Anconquija region, Argentina. Tussock grassland and shrubland at 3770 m in the Cumbres Calchaquíes. Photo: Stephan Halloy.

Also present are microphyllous shrub communities, cushion plants (as *Azorella compacta*) reaching ages of 3000 years or more, rocky outcrops, and lakes – which have five endemic *Isoetes* spp. (Halloy 1979a).

5. **Aeolian zone** (to 5500 m); this worldwide biome was named by Swan (1963). Above 4600 m plants become very scarce, with some vascular plants reaching 5100 m. Extremely rare local endemics are *Nototriche rohmederi*, *N. tucumana* and *Geranium* new sp.

The Chaco Dominion (Cabrera 1976) is represented in this region by the prepuna and monte biogeographical provinces:

6. **"Prepuna"** (Morello 1958). This restricted semi-arid area on the western slopes is characterized by spiny shrubs, columnar cacti (e.g. *Trichocereus pasacana*) and cliff-dwelling bromeliads. The cushion-forming bromeliad *Abromeitiella* is almost endemic. In spite of its aridity, the measured diversity of herbs and shrubs is up to 41 species per 100 m² – which can be higher than in eastern lowland forests.

7. **"Monte"** (Morello 1958). On the foothills to the west, evergreen and non-leafy shrubs dominate. They are related to much greater extensions of this province farther south, as well as to the chaco (CPD Site SA22). Annuals flower briefly and sometimes spectacularly during the rainy summer.

Flora

Many botanists have collected in the region, resulting in a well-known flora (large collections are at the Fundación Miguel Lillo Herbarium, Tucumán), but published information is dispersed. No compendium of the flora has been published, and there is only one 1977 volume in the *Flora Ilustrada de la Provincia de Tucumán*. Halloy (1985b) estimated that 1700 species of vascular plants were known within the proposed Anconquija park. Nearly 4% are known to be endemic, but knowledge by life zones suggests that endemism may reach closer to 20%.

Most of the region's diversity relates to the mountains' great age with little modification (a refugium). Repeated isolations and contacts through time have fostered great geographical variety and speciation. Since the two mountain ranges are isolated by a pass at 3000 m, vicariant speciation has occurred in high-Andean groups (e.g. *Nototriche*, *Isoetes* and lizards such as *Liolaemus*). The species-richness for vascular plants amounts to 0.57 species per km² (which is for example six times that in the Galápagos Islands and twelve times that in Sri Lanka). Above 4000 m in the Cumbres

CPD Site SA35: Anconquija region, Argentina. *Acaulimalva nubigena*, a familiar dwarf mallow of the high slopes in the Anconquija region (3680 m, Cumbres Calchaquíes). Photo: Stephan Halloy.

Cachalquíes, an area of c. 150 km², more than 200 vascular plant species occur, with at least 20% endemism. The diversity of 1.3 species per km² is particularly impressive considering its proximity to the altitudinal limit of vascular plants.

The Anconquija region is the southernmost limit for many Amazonian forest elements (e.g. *Phoebe porphyria*, *Tabebuia*, *Cedrela*, *Trichomanes*). Holarctic elements are frequent in the temperate forest and higher up (e.g. *Alnus acuminata*, *Geranium*, *Alchemilla*), many with several endemic species. Typical endemics of the Andes are strongly represented (e.g. *Polylepis*, *Werneria*, *Barneoudia*, *Nototriche*), and many groups have several diversified species. Many subantarctic elements are also present (e.g. *Asplenium tucumanense*, *Podocarpus parlatorei*, *Fuchsia boliviensis*, *Blechnum magellanicum*).

As a very preliminary underestimate, Halloy (1985b) listed possibly five extinct, 100 endangered and 150 rare plant species within the proposed Anconquija National Park.

Useful plants

At least 96 species of direct potential use occur in the region, including medicinal plants, ornamentals, foods, valuable woods and forages. In addition, there are crop relatives which can help increase crop yield and tolerance in such genera as *Fragaria*, *Solanum*, *Carica*, *Juglans* and many others. Valuable woods include *Tabebuia avellanedae*, *Juglans australis*, *Fagara coco* and *Cedrela lilloi*. The exploitation of these timber species has impoverished large tracts of forest. Medicinal plants are numerous, e.g. *Satureja parvifolia*, *Lycopodium saururus*, *Senecio graveolens* and *Werneria paposa*. In addition to local use, many of these plants are now collected on a commercial scale and sold in large cities.

The Biodiversity programme at Invermay Agricultural Centre in New Zealand, under a bilateral agreement with Argentina's Instituto Nacional de Tecnología Agropecuaria (INTA), is screening a wide range of plants of the Anconquija region for potential uses of all types – medicinals and aromatics, fruits, nuts, tubers, forages, timbers, ornamentals, etc. (Halloy 1992). Plants collected in the region are being grown, multiplied and their germplasm then evaluated for agronomic potential and product value. Learning to cultivate the promising species may help to save them from unsustainable exploitation.

Social and environmental values

Tucumán is the most densely populated Argentinian province. The Anconquija region (itself only sparsely populated) has had a lot to do with this, for centuries providing bountiful supplies of natural resources – good soils, wildlife, timber and water – to the surrounding lowlands. Abundant Inca and pre-Inca archaeological remains of major interest show the region's importance well before European colonization (Halloy, González and Grau 1994). Little derived from the ancient cultures continues.

The region is sparsely populated by local people, who either are sedentary or migrate seasonally with their herds of sheep, goats, cattle and occasionally llamas for supplies of fresh forage and fuel (Molinillo 1988). They have a strong attachment to the land and a good knowledge of local medicinal plants. They also grow small amounts of food plants for their own consumption. Much of the local knowledge is being lost through imposition of Western style education and values, migration to the lowlands to work for an income, and compulsory military service. Larger indigenous populations occupy lower valleys, one of which (Amaicha) still functions to a certain degree as an indigenous commune.

The Anconquija region is a major regulator of water, capturing rain for an area with a population of well over 1 million (Halloy 1984a, 1985b). Continuing deforestation and overgrazing already have resulted in large-scale erosion, repeated flooding during summer torrential rains and rapid filling of expensive dams with silt (Halloy, Grau and González 1982; Halloy, González and Grau 1994).

For the majority of the European population below and around Anconquija, the region only exists as a distant range of mountains emerging on ever-more-rare clear days from the sweltering polluted lowland atmosphere. Nonetheless a large number of people make use of the region for recreation, either driving through to small villages in the valleys, where they find some tourist facilities (e.g. Tafí del Valle, Amaicha) or enjoying tramping, fishing, hunting or mountaineering. Tourism is a strong earner for the three provinces surrounding the Anconquija region, and careful management and promotion could substantially enhance its attractiveness. Testimony to the appeal of the region is the name's use for important streets and businesses (but commonly distorted to Aconquija).

The region is within two Endemic Bird Areas (EBAs). The humid evergreen forests and alder forests between 800–2500 m are in the Boliviano-Tucumán Yungas EBA (B57) and the higher areas in the Argentine Prepuna and Puna zone EBA (B37). Nine species of restricted range are in the Yungas EBA, most of which occur in the Anconquija region. The alder forests are especially important for the near-threatened Tucumán parrot (*Amazona tucumana*) and threatened rufous-throated dipper (*Cinclus schulzi*). There are habitats of both within the proposed Anconquija National Park.

The Anconquija region is the only area within the Argentine Prepuna and Puna zone EBA that has all five of the restricted-range species in this EBA, including the threatened Tucumán mountain-finch or "chivi andino" (*Poospiza baeri*), which prefers dense growth along steep-sided ravines bordering streams in the puna. Only six ravines are known to have populations; some are in a reserve at El Infiernillo, which is inside the proposed NP. Grasslands adjacent to this species' habitat are susceptible to fire, which could reach its habitat.

Economic assessment

The economic assets of the region include: (1) genetic resources of crop relatives; (2) genetic diversity of potential new drugs, crops, timbers, forages, ornamentals, etc.; (3) source and regulator of most of the region's water and a strong influence on its soils; (4) source of ecological balance (especially, providing biological control agents); (5) relatively unmodified baseline communities for scientific research, e.g. to determine better management strategies for the whole region; (6) source of incomparable beauty and therefore tourism and recreation; (7) area for expansive open-air activities including education, sports, trekking and mountain climbing; and (8) source for physical and mental health.

The 1985 estimated total one-time cost of establishing the park, including expropriations and infrastructure, was US$2 million, with annual operating costs c. US$100,000 (Halloy 1985b).

Comprehensive economic assessment of ecological services and natural resources is in its infancy, but an attempt was made for the Anconquija (Halloy 1985b; Halloy, González and Grau 1994). The Anconquija region could yield tens of millions of U.S. dollars annually, and its proper management would result in saved expenses of tens to perhaps hundreds of millions (mainly on flood control and repair measures, reduced losses of soil and decreased inputs in pest control and fertilizers). Among the considerations used in the assessment were:

1. The proposed protected area of c. 3000 km^2 is a major controlling factor for the hydrology of over 100,000 km^2, which are mostly under intensive cultivation, with over 15% badly eroded.

2. The population living in the proposed protected area is less than a hundred. Their potential influence and importance is mostly unknown to them. The challenge is to maintain their traditions while ensuring that their aspirations for a Western life-style do not jeopardize the natural systems which support them.

3. The areas which the proposed park directly influences include over 1 million people, and the indirect influence of hydrological regulation affects as many as 20 million people. Refugees from flooding in these areas number several thousand almost every year, and sometimes hundreds of thousands.

4. Infrastructure damage in 1981 due to flash flooding because of degraded watersheds was US$13 million for Tucumán alone. Such damage is repeated almost every year, and can amount to billions of dollars (e.g. Paraná in 1983). This does not consider losses of topsoil, crops and human lives, nor diseases. For example, the loss of topsoil in eastern Tucumán from 1972 to 1981 was estimated to cost US$360 million (land reclamation $180 million, organic matter $100 million, nutrients $80 million) (J. Madozo, pers. comm. 1992). Large dams in the area can cost US$60 million (El Cadillal dam) to US$200 million (estimated for the projected Potrero del Clavillo dam). Conserving watershed vegetation would ensure a longer return on the investment.

5. Production was estimated to be tens of millions of dollars annually, considering (i) direct sustained extraction (mainly high-value items for propagation, in small volume and with low impact – e.g. seeds and cuttings); (ii) biological control in adjacent areas; (iii) tourism; and (iv) water use. This did not take into account possible benefits from royalties, e.g. on developing new crops. These benefits would have the bonus of better social distribution.

6. No attempt was made to assess the economic value of beauty, peace, pure air and the conservation of species for their own sakes, to which this park would contribute substantially (i.e. aesthetic and other so-called existence values). A comparison suggested was the cost of having a museum, or the price paid for an artistic masterpiece – each could be tens to hundreds of millions of dollars.

Threats

Logging of valuable woods and deforestation for agriculture and plantations of *Pinus* have accelerated (under a national credit policy favouring exotics). Nearly 10% of the forests of Tucumán Province were cut down from 1973 to 1982 (Grau 1984). Since an Anconquija park was proposed in 1913, at least 60% of the yungas is probably gone, especially the lower yungas (Vervoorst 1979, 1982). Of c. 100 km^2 of yungas forest within the proposed park area, 25 km^2 have been authorized for logging and 1.3 km^2 for complete deforestation (Halloy 1985b), which is going on. Other areas are logged without explicit authorization (Vides 1984; Halloy 1984b). Road building (Vides 1984), colonization and uncontrolled tourism favour the acceleration.

Overgrazing and trampling by cattle, sheep and goats are widespread in forest understorey and especially grassland and high-Andean vegetation. There is little fencing and no control, and the livestock densities are too high for adequate regeneration. Burning is a widespread problem, mainly within páramo and high-Andean vegetation; in 1980 alone c. 250 km^2 were burned.

Exotic plant species (e.g. *Ligustrum*, *Cynoglossum*, *Asparagus*, *Crataegus*) are aggressively invading parts of the region, especially the higher forest and lower páramo zones. Red deer and Axis deer introduced to the region in 1976 are great potential threats (Grau *et al.* 1982); there current status needs assessment.

Erosion, flooding, silting, mud flows and drought are direct consequences of the loss of vegetation. In the monte, grazing, trampling and fire have seriously affected the waterflow, with disastrous consequences to villages and dams (Halloy, Grau and González 1982; Halloy 1984a; Halloy, González and Grau 1994).

There is strong collecting pressure on some plant species (Cabrera 1977; Sota 1977) for: fuelwood, especially shrubs and cushion plants like *Azorella compacta* – one of the slowest growing species, which attains more than 3000 years of age; medicinals, especially *Werneria paposa*, *Lycopodium saururus* and *Senecio graveolens*, which are slow-growing and already rare; ornamentals (bulbs of Liliaceae and Iridaceae, *Chuquiraga*, ferns, orchids, bromeliads, cacti); and construction (*Trichocereus pasacana*, *T. terscheckii*) for roofing, doors, fine furniture and souvenirs. Tuft grasses of several species also are used for roofing, but this may be waning with the advent of corrugated metal roofs. Epiphytic mosses, lichens, etc. and forest humus are collected as substrate for plant growers and sold in Tucumán as well as exported to Buenos Aires (Cabrera 1977).

Many rare species are at risk, compounded by slow growth of those at high altitudes. Some are extremely susceptible to one big fire or concentration of cattle, logging, etc. – e.g. *Barneoudia balliana*, *Geranium* new sp., *Botrychium australe* and *Myrcianthes callicoma*.

Conservation

Before European colonization the region was used by Amerindians, who had considerable conservation knowledge. Provincial and federal laws requiring soil conservation and forest protection exist, but have not been adequately enforced. Miguel Lillo and Julio López Mañán first advanced a National Park in 1913, which would have been much larger than what now remains to conserve (Grau *et al.* 1982). Occasional new proposals for protection have been made, mostly for smaller areas. Five perhaps were

accepted as Provincial Parks, but only two actually exist (Map 70) – Santa Ana (185 km²) to the extreme south and La Florida (29 km²) to the centre-east of Anconquija. Also to the east, a 140-km² area of Lillo's originally planned park is owned by the Universidad Nacional de Tucumán and managed as San Javier Biological Park, and two small scientific reserves are owned by Tucumán Province – Aguas Chiquitas (7.4 km²) and Los Sosa (8.9 km²).

Through the university, San Javier park has a motivated corps, but most of these reserved areas are little more than mapped boundary lines due to lack of infrastructure and

MAP 70. ANCONQUIJA REGION, ARGENTINA (CPD SITE SA35)

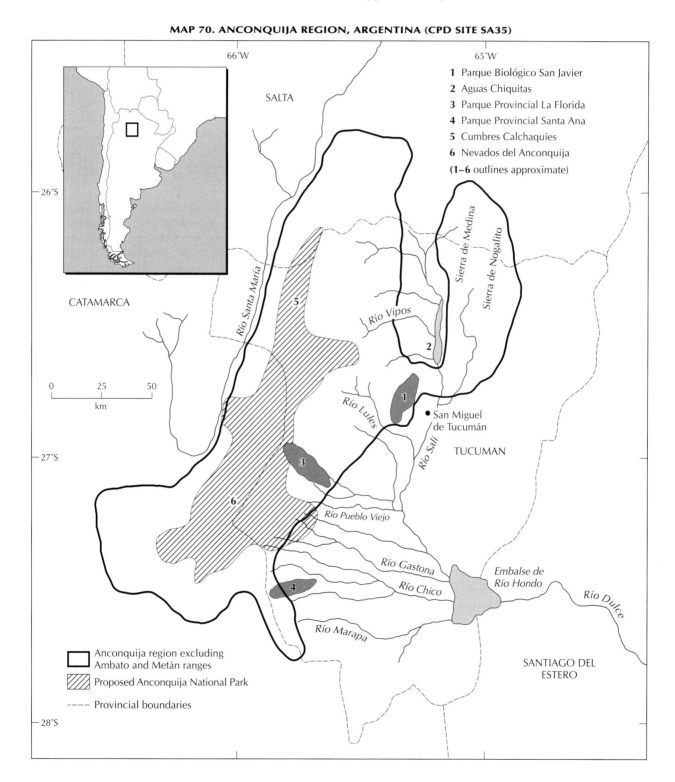

1 Parque Biológico San Javier
2 Aguas Chiquitas
3 Parque Provincial La Florida
4 Parque Provincial Santa Ana
5 Cumbres Calchaquíes
6 Nevados del Anconquija
(1–6 outlines approximate)

Anconquija region excluding Ambato and Metán ranges
Proposed Anconquija National Park
Provincial boundaries

funding. The Biodiversity programme at the Invermay Agricultural Centre in New Zealand has over 300 accessions of living plants from north-west Argentina, most of which originated from or occur in the Anconquija region. This collection will help to preserve some rare Anconquija species at least *ex situ* (Halloy 1992).

The proposed Anconquija National Park (Domínguez, Halloy and Terán 1978; Halloy 1979b, 1985b; Halloy, Grau and González 1982; Halloy, González and Grau 1994) includes c. 3000 km² of the best remaining habitats in the region, all above 3000 m except for a south-eastern area with the Cochuna River that descends to 1150 m. To facilitate efficient protection for the park, there are also suggestions for zoning and a management plan (Halloy 1979c). The diverse Anconquija region would be adequately conserved if the area was well protected and managed, together with the establishment of corridors for gene flow, and other projects including controlled tourism, research and education.

The Anconquija Park Project has been approved by the National House of Representatives (Trámite Parlamentario No. 79, 1984); in 1993 it was presented to the National House of Senators (Halloy, González and Grau 1994). However, it has been resisted by the Province of Tucumán, in which c. 60% of the park would be located. The provinces of Catamarca (which has over a third of the proposed area) and Salta also have to approve for the park to be created in its entirety.

References

Bell, D.A. (1987). *Distribución del bosque de aliso (*Alnus acuminata *H.B.K.) en la provincia de Tucumán, Argentina*. Seminario, Fac. Ciencias Nat., Universidad Nacional de Tucumán (UNT), Tucumán. 107 pp.

Cabrera, A.L. (1976). *Regiones fitogeográficas argentinas*. In Parodi, L.R. (ed.), *Enciclopedia argentina de agricultura y jardinería*, 2nd edition. Vol. 2(1). Editorial Acmé, Buenos Aires. Pp. 1–85.

Cabrera, A.L. (1977). Threatened and endangered species in Argentina. In Prance, G.T. and Elias, T.S. (eds), *Extinction is forever: threatened and endangered species of plants in the Americas and their significance in ecosystems today and in the future*. New York Botanical Garden, Bronx. Pp. 245–247.

Domínguez, E., Halloy, S. and Terán, E. (1978). *Informe sobre el estado actual del proyectado Parque Nacional del Anconquija*. Fundación Miguel Lillo, UNT Fac. Ciencias Nat. and Asociación Tucumana de Andinismo, Tucumán. 17 pp. Mimeographed.

Grau, A. (1984). *Informe preliminar – Subcomisión de Estudio: vegetación y fauna*. Senado, Provincia de Tucumán, Tucumán. 22 pp. Unpublished.

Grau, A., Halloy, S., Domínguez, E., González, J.A. and Vides, R. (1982). *Ciervos introducidos – estudio de su impacto ambiental en el noroeste argentino*. Fundación Miguel Lillo and UNT Fac. Ciencias Nat., Tucumán. 16 pp. Unpublished.

Halloy, S. (1979a). Dos nuevos *Isoëtes* (Lycopsida) de Alta Montaña, con datos ecológicos de las lagunas Muerta y Escondida. *Lilloa* 35: 65–95.

Halloy, S. (1979b). *Informe No. 2 sobre el proyectado Parque Nacional del Anconquija*. Fundación Miguel Lillo, Tucumán. 3 pp. Unpublished.

Halloy, S. (1979c). *Sugerencias para un plan de manejo y/o reglamentaciones para el proyectado Parque Nacional del Anconquija*. Servicio Nacional de Parques Nacionales, Tucumán. 5 pp. + map. Unpublished.

Halloy, S. (1984a). La importancia del régimen hidrológico y la estabilidad de ecosistemas de altura para la regulación del agua en los Valles Calchaquíes. *IV jornadas culturales del Valle Calchaquí. Tema: "el agua y la vida"*. Tucumán. Pp. 143–154.

Halloy, S. (1984b). *Solicitud de acción para proteger los bosques de Tucumán y Catamarca*. UNT, Tucumán. 8 pp. Unpublished.

Halloy, S. (1985a). *Climatología y edafología de Alta Montaña en relación con la composición y adaptación de las comunidades bióticas (con especial referencia a las Cumbres Calchaquíes, Tucumán)*. Ph.D. dissertation. Univ. Microfilm Internat. Publ. No. 85-02967, Ann Arbor, Michigan, U.S.A. 839 pp.

Halloy, S. (1985b). Exposición sobre el estado del proyecto del Parque Nacional Anconquija. UNT, Tucumán. 30 pp. Unpublished.

Halloy, S. (1985c). Reencuentro de *Azorella biloba* (Schlecht.) Wedd. en Tucumán. *Lilloa* 36: 267–269.

Halloy, S. (1987). Anconquija – Argentina. *Biol. Conserv. Newsletter (Washington)* 48: 1–2.

Halloy, S. (1992). El programa de biodiversidad de Nueva Zelandia. Investigación y conservación de recursos genéticos sudamericanos. *Yungas* 2(3): 4–6.

Halloy, S., González, J.A. and Grau, A. (1994). *Proyecto de creación del Parque Nacional Aconquija (Tucumán – Argentina). Informe No. 4*. Fundación Miguel Lillo, Serie Conserv. Naturaleza No. 9. Tucumán. 55 pp.

Halloy, S., Grau, A. and González, J.A. (1982). *Proyecto del Parque Nacional Anconquija. Informe No. 3*. Fundación Miguel Lillo, UNT Fac. Ciencias Nat. and Asociación Tucumana de Andinismo, Tucumán. 14 pp. Mimeographed.

Hueck, K. (1954). Der Anden-Erlenwald (die *Alnus jorullensis* Assoziation) in der Provinz Tucuman. *Angew. Pflanzensoz.* 1: 512–572.

Meyer, T. (1963). *Estudios sobre la selva tucumana*. Opera Lilloana 10. 144 pp.

Molinillo, M.F. (1988). *Aportes a la ecología antropológica de las Cumbres Calchaquíes de la provincia de Tucumán – uso de los recursos naturales en el Valle de Lara.* Seminario, UNT Fac. Ciencias Nat., Tucumán. 181 pp. Unpublished.

Morello, J.H. (1958). *La provincia fitogeográfica del Monte.* Opera Lilloana 2. 154 pp.

Roldán, A.I. (1991). *Estructura y principios de demografía de la selva montana en el Parque Biológico San Javier (Tucumán – Argentina).* Seminario, UNT Fac. Ciencias Nat., Tucumán. 88 pp.

Sota, E.R. de la (1977). The problems of threatened and endangered plant species and plant communities in Argentina. In Prance, G.T. and Elias, T.S. (eds), *Extinction is forever.* New York Botanical Garden, Bronx. Pp. 240–243.

Swan, L. (1963). The aeolian zone. *Science* 140: 77–78.

Vervoorst, F. (1979). La vegetación del noroeste argentino y su degradación. Fundación Miguel Lillo, Serie Conserv. Naturaleza 1: 5–9 + map.

Vervoorst, F. (1982). Noroeste. In Vervoorst, F. (ed.), *Conservación de la vegetación natural en la República Argentina.* Fundación Miguel Lillo, Serie Conserv. Naturaleza 2: 9–24.

Vides, R. (1984). Roban el tesoro de la laguna. *Huakaicha Puy (Revista del Grupo de Conservación)* No. 1: 5–7.

Vides, R. (1985). *Estudio de las taxocenosis de aves del bosque de aliso de Tucumán.* Seminario, UNT Fac. Ciencias Nat., Tucumán. 252 pp. Unpublished.

Author

This Data Sheet was written by Dr Stephan Halloy (Invermay Agricultural Centre, Crop and Food Research, Private Bag 50034, Mosgiel, New Zealand).

(TROPICAL) ANDES: CPD SITE SA36

MADIDI-APOLO REGION
Bolivia

Location: Eastern front range of Andes and adjacent alluvial plain of northern Bolivia; La Paz Department between latitudes 13°20'–14°00'S and longitudes 68°10'–69°10'W in Iturralde and Franz Tamayo provinces.

Area: 30,000 km².

Altitude: 250–2000 m.

Vegetation: Humid forest with different communities on montane slopes, piedmont and river margins; cloud forest on crests of higher south-western ranges; dry forest and savanna in intermontane valleys; humid savanna and marshes on alluvial plain to north-east.

Flora: Highest documented plant diversity in Bolivia – probably more than 5000 species of vascular plants; humid forest similar to montane forests of southern Peru and/or central Amazon; degree of endemism unknown, possibly with taxa known only from Bolivian yungas to south-west.

Useful plants: For quinine; mahogany and other timbers; palm oils and thatch.

Other values: Watershed protection for Beni River; indigenous peoples; attractive for ecotourism due to accessibility by river into pristine areas; many birds and other fauna, some threatened.

Threats: Forest exploitation based on few high-value timber species, leading to extensive road building and colonization; preliminary oil exploration in Andean front ranges may lead to more extensive development.

Conservation: Madidi National Park established in December 1995.

Geography

The Alto Madidi region is in the sub-Andean belt, consisting of a series of parallel ridges to 800–2000 m and valleys at 300–500 m, with a generally north-westerly orientation (Map 71). The ridges are anticlines composed of Ordovician, Devonian, Carboniferous and Cretaceous sandstones and mudstones; the valleys are synclines with Tertiary sediments, conglomerates and rocks (Oblitas and Brockmann 1978). East of the Andes lie Quaternary sediments of the Andean piedmont (at 200–300 m) and the extensive Beni-Chaco plain (at 130–200 m).

The most conspicuous geomorphological feature in the region is the Madidi-Quiquebey syncline, between the Serranía del Tutumo (also known as the Serranía del Tigre) and the serranías of Chepite and Eslabón. The southern end of this broad valley is drained by the Tuichi River, which joins the Beni River upstream from Rurrenabaque. The Alto Madidi River flows north-westerly to the termination of the Serranía del Tutumo, where it makes a broad arc to flow eastward across the alluvial plain of the Beni, eventually uniting with the Beni River.

Soils in the region vary depending upon geomorphology. Steep-sided mountain ridges have shallow soils with numerous sandstone outcrops; these soils are susceptible to erosion and vary from strongly acidic to neutral. Lower hills with Tertiary substratum have deeper soils and are only moderately acidic. Valleys, abandoned terraces and the alluvial plains are characterized by deep soils that vary from heavy clays to sandy loams and are strongly to slightly acidic (R. Lara, pers. comm.).

The climate is humid to very humid with the mean precipitation estimated at over 2000 mm per year. Prevailing winds are from the north, causing north-eastern slopes to have the most rainfall; there is a marked rain shadow to the south-west near Apolo. The mean annual temperature is estimated to be 26°C. There is a dry season coinciding with the austral winter, but the southerly cold fronts characteristic of other parts of lowland Bolivia have little impact in the Madidi-Apolo region.

Vegetation

The vegetation in the Madidi region has been described by Foster (1991); the following account includes additional observations along the Tuichi River and adjacent areas. The plant communities are largely correlated with altitude and topography.

The highest western mountain ranges have extensive cloud forest on north-eastern faces; fern brakes and meadows can be observed on some ridgetops. The steep-sided mountains have slippery clay soils, and with the high rainfall there are numerous landslides; their vegetation is a patchwork of communities in different successional stages.

Montane humid forest occurs on the ridge slopes and intergrades into a lowland forest with Amazonian affinities; similar forest types are found on abandoned river terraces in the piedmont. The floristic diversity of this lowland forest is considerable, with 204 species of 2.5 cm or more dbh per 0.1 ha (Foster and Gentry 1991).

Floodplain forest and plant communities in various successional stages occur along the rivers; presumably large areas of swamp forest associated with river meanders exist downstream along the Madidi and Heath rivers.

The alluvial plains north-east of the Andean foothills support vegetation that is similar in structure and physiography to the widespread and extensive savannas of the Beni lowlands. These grasslands are a complex mosaic of plant communities resulting from the interaction of edaphic conditions (including duration and degree of inundation) and fire. The forests occur as patches and along rivers, adding to the diversity of habitat types (see Data Sheet for Llanos de Mojos, CPD Site SA24). Savanna complexes are found near the town of Ixiamas and along the Heath River on the Peruvian border. In addition, the region has a variety of riverine and non-riverine lakes and permanent marshes which support aquatic vegetation.

A "dry ridgetop forest" community occurs below the cloud level; although less diverse, there is little overlap in floristic composition between this forest type and the more widespread premontane forest communities. Dry forest and well-drained savanna occur on the downwind side of the south-western ridges near the town of Apolo. This area appears similar to much of the La Paz Yungas where cloud forest, humid forest, dry forest and montane savanna exist within relatively short distances of one another.

Flora

Preliminary data indicate that the Madidi-Apolo region is diverse and interesting, probably with more than 5000 species of vascular plants. *Podocarpus* and *Prumnopitys* (Podocarpaceae) have been found at elevations as low as 1400 m; also common are typical cloud-forest elements such as *Cyathea* (Cyatheaceae); *Clusia* (Guttiferae); *Schefflera, Dendropanax* (Araliaceae); *Hedyosmum* (Chloranthaceae); and *Clethra* (Clethraceae), as well as numerous species of Melastomataceae and Rubiaceae. Epiphytic mosses, orchids and ferns are particularly abundant.

Montane forests at 400–700 m are highly diverse in species of Sapotaceae and Lauraceae; a species of *Ampelocera* (Ulmaceae) is common, as are *Poulsenia, Clarisia* and *Pseudolmedia* (Moraceae). Also well represented are Leguminosae (particularly *Inga*), Meliaceae, Myrsinaceae, Rubiaceae, Melastomataceae and Guttiferae. As at the higher elevations, epiphytes are abundant and diverse – particularly orchids and ferns.

The premontane forest has numerous interesting plants not previously known to occur in Bolivia (Foster, Gentry and Beck 1991). These include *Wettinia* and *Wendlandiella* (Palmae), *Anthodiscus* (Caryocaraceae), *Pterygota* (Sterculiaceae) and *Huberodendron* (Bombacaceae). The predominant family is Moraceae; also well represented are Annonaceae, Araceae, Bignoniaceae, Euphorbiaceae, Leguminosae, Melastomataceae, Myristicaceae, Palmae and Rubiaceae.

On the dry ridgetops two generic novelties for Bolivia were discovered – *Lecointea* (Leguminosae) and *Caryodendron* (Euphorbiaceae). One of the most common species is an undescribed taxon in the Malvales; lack of good flowering material has impeded its determination as either a *Reevesia* (Sterculiaceae) or an undescribed genus in Malvaceae.

Montane forests near Apolo apparently are similar in composition to "yungas" forests in the adjacent Province of Larecaja, where dozens of endemic species have been described (Rusby 1893–1896, 1907). Fortunately, unlike that region, much of the vegetation near Apolo remains relatively undisturbed. Montane savanna near Apolo seems to have some affinity to the "cerrado" of Brazil, as evidenced by *Dilodendron bipinnatum* (Sapindaceae), a species previously thought to be restricted to the Brazilian Shield region.

Humid savannas near Ixiamas have species new for Bolivia (as yet unidentified) in the Burmanniaceae, Eriocaulaceae and Xyridaceae, and *Schizaea incurvata* – a fern known in northern South America, with a single report from Peru. Intriguing as well is the apparent lack of floristic similarity of the Ixiamas pampas to other nearby savannas in the Beni (S. Beck, pers. comm.) or to the Heath pampas in Peru (A. Gentry, pers. comm.).

Useful plants

The Amerindian residents of the region have much knowledge about the native plants and their uses. The Alto Madidi is sparsely populated, and utilization of plant products has been largely restricted to local needs, such as construction timber (many tree species), palm thatch, fibres from tree bark (Annonaceae, Lecythidaceae, Tiliaceae), and medicinal plants. Many trees with edible fruits are used as a means to increase hunting success.

Economically, palms are very diverse, and many have oil-rich fruits or seeds with potential as non-timber forest products (Moraes 1993). Until the 1960s the region was a major centre of quinine-bark production (*Cinchona officinalis*). Currently there is a great deal of logging activity centred on *Swietenia macrophylla* (Meliaceae), with secondary interest in *Cedrela odorata* (Meliaceae) and *Amburana cearensis* (Leguminosae). Many other valuable timber species occur in the region but have not been exploited commercially.

Social and environmental values

The Alto Madidi is home to two groups of indigenous peoples. The Tacana reside near Ixiamas and Tumupasa in the piedmont of the Serranía del Tutumo. Lowland Quechua are found in several small settlements along the Tuichi River and near Apolo. Both peoples have a long history of commerce and interaction dating at least from Spanish settlement of Apolo in the 17th century. These groups are subsistence agriculturists, who depend extensively on the forest to provide a substantial part of their diet (R.M. Ruiz, pers. comm.). Although no detailed ethnobotanical studies have been made, anecdotal information indicates that the Tacana practise silviculture in the forests surrounding their villages to promote the growth of desirable fruit-bearing trees.

The Tuichi watershed is probably important for several endemic species of fishes restricted to the Beni River watershed (J. Sarmiento, pers. comm.). The Tuichi River is the only major tributary free of heavy-metal pollution. Much of the Beni watershed is polluted by effluent from La Paz (Boopi River-La Paz River) or by mercury pollution from gold-mining activities along the Tipuani-Mapiri River.

The fauna of Alto Madidi is comparable to the flora for diversity and interest (cf. Pearman 1993), and is similar in composition to adjacent areas of Peru. Bird diversity in the less than 100,000 km² of south-western Amazonia that includes this region is thought to be one of the highest on the continent, with c. 10% endemism (Parker 1991). Large mammals are unusually abundant, particularly tapirs and spider monkeys (Emmons 1991). Several endangered species are known to occur in the region, among them spectacled bears (*Tremarctos ornatus*) and short-eared dogs (*Atelocynus microtis*).

Tourism dates from at least the early 20th century when the Mulford biological expedition passed through Rurrenabaque and Ixiamas on its way to Manaus (Rusby 1922). Currently there are two ecotourism companies in Rurrenabaque that specialize in river trips up the Tuichi River. Salt licks along the Tuichi have been identified by C. Munn (Wildlife Conservation International), and the potential for showing macaws and other impressive animals is likely to increase ecotourism.

Threats

As in so much of Latin America, the Madidi-Apolo region is undergoing profound changes in land use; here the impetus is timber exploitation and colonization. These two factors will have a particularly synergistic and potentially destructive effect in northern La Paz. Comparison of satellite images from 1985 and 1990 revealed that 3697 km² of lowland and montane forests had been cleared in La Paz Department (11.7% of the total forested area), with an annual rate of deforestation of 91.3 km² (CUMAT 1992).

Bolivia is one of the world's largest producers of mahogany (*Swietenia macrophylla*), but over-exploitation and deforestation in Santa Cruz and the Beni have forced searchers to more remote regions of the country. Mahogany tends to be found in relatively dense stands (1–3 trees per ha) that are widely dispersed over a large geographic area. Timber companies typically build extensive road networks, log the mahogany populations and then abandon the forest (Synnott and Cassells 1991; cf. Rice, Gullison and Howard 1995).

The creation of roads in Bolivia is inevitably followed by colonization. This is particularly true in the Department of La Paz, where soil depletion and expanding populations in the Yungas and Alto Beni provide a constant stream of migrants to the newly opened lands of the frontier. Currently, the southern (i.e. Rurrenabaque-Ixiamas) portion of the La Paz-Pando highway is heavily settled, and it is inevitable that colonization will expand as new roads are made. Unlike the indigenous peoples of Iturralde Province, the colonists are market-oriented agriculturists with strong commercial contacts in the city of La Paz. They usually clear and cultivate five to ten times as much land per year as do the subsistence indigenous farmers.

Multinational oil corporations have been awarded long-term concessions along the Madre de Dios River and in the front ranges of the Andes. It is yet unclear whether there are sufficient petroleum reserves to justify extraction. Due to the region's remoteness, it is unlikely that any potential extraction will occur prior to the year 2000. Pipeline construction would necessitate extensive road building; the oil companies have expressed an interest in working with Bolivian agencies to minimize deforestation.

Conservation

Madidi was recognized in 1990 as the most diverse humid forest ecosystem in Bolivia by Conservation International's Rapid Assessment Program (Parker and Bailey 1991). Since then the Global Environment Facility (GEF) (World Bank) has designated the Alto Madidi as one of nine priority conservation areas in Bolivia.

In 1992, the Bolivian Government contracted a team of biologists and anthropologists to recommend boundaries for a National Park. They recommended establishment of an 18,000 km² reserve (CDC 1992). The proposed Madidi National Park, which has strong support from the indigenous people in the area, would encompass most of the region's vegetation types. The proposed park is adjacent to Peruvian protected areas (Santuario Nacional Pampas del Heath, Zona Reservada Tambopata Candamo – see CPD Site SA10), as well as other designated protected areas in Bolivia (Pilón Lajas NP, Manuripi-Heath NP). The Madidi park would unite these reserves into a very large preserved region.

Opposition to the park has been expressed by colonist unions and the forest industry. Lumber companies have short-term concessions within the proposed park boundaries and are demanding that these pre-existing concessions be honoured. The actual boundaries for the park, as well as permitted activities within the area, would be determined by Presidential decree. It is approaching four years since the formal proposal was made by the Environmental Secretariat of the Bolivian Government to create a National Park in the region. Definitive action was taken in December 1995 to grant this important area formal status as a National Park.

MAP 71. MADIDI-APOLO REGION, BOLIVIA (CPD SITE SA36), SHOWING MADIDI NATIONAL PARK (AS PROPOSED)

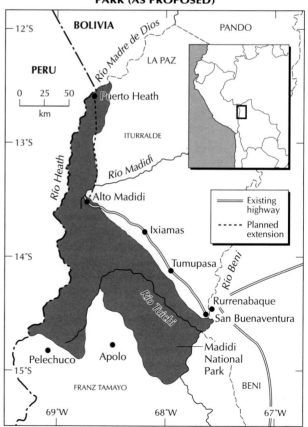

The recent legal formalization of the status and boundaries of the NP included creation of a participatory commission where the local inhabitants are directly involved in the park's management. This approach should provide a forum where land-use conflicts can be equitably dealt with. Although all of Iturralde Province was declared a Forest Reserve and is legally off-limits to agriculture, colonization is proceeding at a very rapid rate. The areas surrounding the park will continue to face substantial pressure from timber companies and colonists. At present, there is no integrated management programme being implemented in the region or developed. Thus, the long-term prospects for conservation of biological diversity in the whole region remain in doubt.

References

CDC (1992). *Propuesta del Parque Nacional Madidi.* Report by the Centro de Datos para la Conservación de Bolivia (CDC), Instituto de Ecología, Museo Nacional de Historia Natural, La Paz.

CUMAT (1992). *Desbosque de la Amazonía boliviana.* Centro de Investigaciones de la Capacidad de Uso Mayor de la Tierra (CUMAT), La Paz.

Emmons, L.H. (1991). Mammals of Alto Madidi. In Parker III, T.A. and Bailey, B. (eds), *A biological assessment of the Alto Madidi region and adjacent areas of Northwest Bolivia, May 18–June 15, 1990.* RAP (Rapid Assessment Program) Working Papers 1, Conservation International (CI), Washington, D.C. Pp. 23–25.

Foster, R.B. (1991). Plant communities of Alto Madidi, Bajo Tuichi, and the foothill ridges. In Parker III, T.A. and Bailey, B. (eds), *A biological assessment of the Alto Madidi region and adjacent areas of Northwest Bolivia, May 18–June 15, 1990.* RAP Working Paper 1, CI, Washington, D.C. Pp. 15–19.

Foster, R.B. and Gentry, A.H. (1991). Plant diversity. In Parker III, T.A. and Bailey, B. (eds), *A biological assessment of the Alto Madidi region and adjacent areas of Northwest Bolivia, May 18–June 15, 1990.* RAP Working Paper 1, CI, Washington, D.C. Pp. 20–21.

Foster, R.B., Gentry, A.H. and Beck, S. (1991). Plant list: Alto Madidi, Bajo Tuichi, and the foothill ridges. In Parker III, T.A. and Bailey, B. (eds), *A biological assessment of the Alto Madidi region and adjacent areas of Northwest Bolivia, May 18–June 15, 1990.* RAP Working Paper 1, CI, Washington, D.C. Pp. 75–92.

Moraes-R., M. (1993). *Diversity and uses of Bolivian palms. I. Southern area of Iturralde Province, Dept. La Paz.* International Foundation for Science D/1585-1. 8 pp.

Oblitas-G., J. and Brockmann-H., C.E. (1978). *Mapa geológico de Bolivia.* Servicio Geológico de Bolivia (GEOBOL) and Yacimientos Petrolíferos Fiscales de Bolivia (YPFB), La Paz.

Parker III, T.A. (1991). Birds of Alto Madidi. In Parker III, T.A. and Bailey, B. (eds), *A biological assessment of the Alto Madidi region and adjacent areas of Northwest Bolivia, May 18–June 15, 1990.* RAP Working Paper 1, CI, Washington, D.C. Pp. 21–23.

Parker III, T.A. and Bailey, B. (eds) (1991). *A biological assessment of the Alto Madidi region and adjacent areas of Northwest Bolivia, May 18–June 15, 1990.* RAP Working Paper 1, Conservation International, Washington, D.C. 108 pp.

Pearman, M. (1993). The avifauna of the Río Machariapo dry forest, northern La Paz Department, Bolivia: a preliminary investigation. *Bird Conserv. Internat.* 3: 105–117.

Rice, R.E., Gullison, T. and Howard, A.F. (1995). Ecology, economics and the unsustainable harvest of tropical timbers: the case of mahogany in the Chimanes Forest, Bolivia. In MacBryde, O.H. (ed.), *Measuring and Monitoring Forest Biological Diversity: The International Network of Biodiversity Plots. International Symposium, May 23–25, 1995.* Smithsonian Institution/Man and the Biosphere Biodiversity Program, Washington, D.C. P. 92.

Rusby, H.H. (1893-1896). On the collections of Mr. Miguel Bang in Bolivia, I–III. *Mem. Torrey Bot. Club* 3: 1–67, 4: 203–274, 6: 1–130.

Rusby, H.H. (1907). An enumeration of the plants collected by Miguel Bang, IV. *Bull. New York Bot. Gard.* 4: 309–470.

Rusby, H.H. (1922). Report of work on the Mulford Biological exploration of 1921–1922. *Mem. New York Bot. Gard.* 29: 101–112.

Synnott, T.J. and Cassells, D.S. (1991). *Evaluation report on Project PD 34/88 rev.1 (F), Conservation, management, utilization and sustained use of the Chimanes region, Department of Beni, Bolivia.* International Tropical Timber Council, Tokyo.

Author

This Data Sheet was written by Dr Timothy J. Killeen (Universidad Mayor de San Andrés, Herbario Nacional de Bolivia, Casilla 20127, La Paz, Bolivia and Missouri Botanical Garden. P.O. Box 299, St. Louis, MO 63166-0299, U.S.A., or Casilla 8854, La Paz, Bolivia).

EASTERN SLOPES OF PERUVIAN ANDES
Peru

Location: Amazonian side of Peru's eastern Andean cordillera with associated sub-Andean mountain belt and foothills, between latitudes 4.5°–14.5°S and longitudes 78°–69°W.

Area: c. 250,000 km².

Altitude: 400–3500 m.

Vegetation: Tropical and subtropical lowland, premontane, and montane forests – each with dry, moist, wet and pluvial formations.

Flora: 7000–10,000 vascular plant species; many endemic species, few endemic genera; threatened species.

Useful plants: Broad spectrum of many plant resources, but few used in sustainable manner.

Other values: Critical region for watershed and fauna protection; Amerindian lands.

Threats: Poorly designed highways and development projects; spontaneous colonization; logging; agriculture in "coca" (*Erythroxylum coca*) for cocaine trade.

Conservation: 3 extant National Parks, 6 other reserves.

Geography

The eastern slopes of the Peruvian Andes form a complex, rather natural geographical and biogeographical region (Map 72). Similar topography and vegetation formations are present on the Andean slopes facing the Amazon Basin from Venezuela to northern Bolivia, from c. 400 m elevation in the lowlands to often more than 3500 m. In Peru at c. 5°S, a natural barrier often called the Huancabamba Depression (see CPD Site SA32) is formed between the Northern Andes and Central Andes by the Marañón River (Vuilleumier 1977; Duellman 1979). This is our northern limit of the Peruvian eastern-slope region. The southern limit is arbitrarily the boundary between Peru and Bolivia.

Although the region can be delimited with these altitudinal, topographical and political boundaries (Map 72), the eastern slopes are not homogeneous. They have great vegetational and floristic diversity which is directly related to the region's geological and climatic diversity.

There are two major geological subdivisions. From 6°–11°S is a largely Precambrian metamorphized sedimentary rock mass with some areas of Tertiary-Quaternary volcanism, some Palaeozoic intrusives, and some sedimentary formations of Late Triassic-Jurassic, Tertiary and Quaternary ages (Peñaherrera 1989). The other subdivision extends from 11°–15°S and consists mostly of Early Palaeozoic sedimentary rocks, often locally metamorphized, with significant Permian-Triassic intrusives, and Tertiary sedimentary and metamorphic rocks.

General edaphic conditions in the two subdivisions might be different, because of the preponderance of different rock types. Surely greater edaphic diversity will be found in the transition zone (around 11°S) between the two geological areas, where large intrusive blocks are present amid a complex mix of sedimentary and metamorphic rocks. This expectation is supported by observations on the relationship between great edaphic variation and the resulting high floristic diversity in Yanachaga-Chemillén National Park (10.5°S, 75.5°W) (R.B. Foster, pers. comm.).

The Peruvian eastern slopes can be subdivided preliminarily into six physiographic provinces (Young 1992):

1. The Chachapoyan province, from 4.5°–7°S, consists of a series of north-west-trending mountain ranges.

2. The Western Huallaga province, from 7°–9.5°S, is located in the upper watershed of the Huallaga River and contains two parallel mountain chains – the Eastern Cordillera of the Andes and farther east the Cordillera Azul.

3. The Upper Pachitea River province, from 9°–11°S, includes two mountain ranges originating from the Cerro de Pasco area, and the headwater rivers draining into the Pachitea River.

4. The Tambo River province, from 11°–13.5°S, includes the headwaters of the Perené and Ene rivers.

5. The Urubamba River province, from 11°–14°S, is in the drainage basin of the Cusco highlands.

6. The Madre de Dios province, from 12°–14.5°S, includes the watershed of the Madre de Dios River.

The climate of the eastern slopes has been little studied; long-term records of precipitation and temperature do not exist. Based on extrapolated isolines of median temperatures and rainfall, local climatic regimes appear to vary from 9°–25°C and 500–7000 mm per year. Circulation is out of the Amazon Basin from east to west, and the most rain occurs from November to April, when the Intertropical Convergence Zone is south of the Equator. In the southernmost portion of the region, climatic patterns also are affected by cyclone systems originating in southern South America during its winter (ONERN-AID 1986).

Vegetation

More than 25 life zones in the Holdridge system have been designated for this region, using extrapolations of the limited climatic data, and separated by latitude (tropical to the north of 12°S and subtropical to the south), humidity province (dry, moist, wet, pluvial) and elevation (Tosi 1960; ONERN 1976). Drier life zones are in rain shadows in deep north-south trending valleys. Perhumid zones occur where the cloud belt forms: 1500–2500 m in the south, often 2500–3000 m in the north, except on lower peaks or mountain ranges.

An informal version of this life-zone system can be helpful to name the vegetation formations: **Lowland tropical or subtropical forest**, found from 200–500 m; a **premontane forest** zone, occurring roughly from 500–1500 m; **lower montane forest** formations, located from 1500–2500 m; and **upper montane forest**, found from approximately 2500 m to local treeline – which is at 3000–4000 m, with the highest elevations in the south and in areas not repeatedly burned by fires originating in adjacent high-elevation grasslands. The physiognomy of a particular forest and its species composition depend greatly on the local precipitation. Trees in cloud-belt forests tend to be somewhat shorter and covered with epiphytes.

Each of the six physiographic provinces is unique in the relative importance of these vegetation formations (Young 1992). The three northern provinces include elevations from 400–4000 m, whereas the southern provinces range from 400–5000 m and can be 30% covered by "puna" grasslands in their highest elevations (see CPD Site SA33).

The Chachapoyan province originally was covered mostly by moist premontane and lower montane forests, and includes some dry forest on its western and northern boundaries with the Marañón River. The Western Huallaga River province is about half covered by moist to wet montane forests, with the lower elevations originally dominated by dry to moist premontane and lowland forests. The Upper Pachitea River province is extremely humid, with moist, wet and pluvial forests occurring at all elevations. The Tambo River province once had extensive areas of dry to moist premontane and montane forests, but the remaining forests are located mostly at lower elevations and in more humid areas; one sizeable lowland area, the Gran Pajonal, is a savanna maintained by human-induced burning (Scott 1978). The Urubamba River province was originally more than 50% lowland forest, and the rest is rugged terrain with dry to pluvial premontane and montane forests. The Madre de Dios province is about equally lowland forest and premontane to montane forests.

CPD Site SA37: Eastern slopes of Peruvian Andes, showing montane wet forest on steep slopes in Río Abiseo National Park, northern Peru. Photo: Kenneth R. Young.

Flora

Certain broad generalizations can be made, but there is little information with which to propose floristic provinces for the region. There are no endemic plant families and few endemic genera: e.g. *Guraniopsis* (Cucurbitaceae); *Pterocladon* (Melastomataceae); *Anodiscus* (Gesneriaceae); *Cylindrosolenium* (Acanthaceae); and *Neokoehleria* and *Sauroglossum* (Orchidaceae). However there are numerous endemic species, undoubtedly due to topographical and ecological barriers. A sizeable number of the species are restricted to one of the six physiographic provinces.

A very conservative evaluation of vascular plant species richness in the moist to pluvial life zones of Peru's eastern slopes estimated 1700–2000 species at 1500–2500 m and 1000–1200 species at 2500–3500 m (Young 1991; León, Young and Brako 1992). The combined elevational zone (1500–3500 m) is c. 5% of Peru's surface area and may have 2400–2800 species, or c. 14% of the 18,000–20,000 species estimated for all of Peru (Gentry 1980). By assuming similar species richness for the altitudinal belt of premontane forest at 500–1500 m, adding another 1000 species present only in the Amazon bottomlands, plus perhaps 300 species restricted to the dry-forest life zones, 7000–10,000 vascular plant species probably comprise the eastern-slope flora of Peru. Thus the eastern slopes, which represent c. 20% of Peru's area, support half of its plant species.

From 400–1500 m, diverse families include pteridophytes such as Dryopteridaceae and Pteridaceae; monocotyledons such as Araceae, Arecaceae and Orchidaceae; and dicotyledons such as Acanthaceae, Annonaceae, Asteraceae, Bignoniaceae, Euphorbiaceae, Fabaceae, Lauraceae, Menispermaceae, Moraceae, Rubiaceae and Solanaceae. Speciose genera include *Adiantum, Polypodium, Calathea, Philodendron, Ficus, Inga, Miconia, Mikania, Paullinia, Piper, Psychotria, Solanum* and *Trichilia.*

From 1500–2500 m, diverse families include pteridophytes such as Cyatheaceae, Dryopteridaceae, Hymenophyllaceae, Polypodiaceae and Pteridaceae; monocotyledons such as Araceae, Bromeliaceae, Orchidaceae and Poaceae; and dicotyledons such as Asteraceae, Campanulaceae, Fabaceae, Lauraceae, Melastomataceae, Moraceae, Piperaceae, Rubiaceae, Solanaceae and Urticaceae. The most speciose genera are *Hymenophyllum, Tillandsia, Epidendrum, Maxillaria, Pleurothallis, Calceolaria, Miconia, Peperomia, Piper* and *Solanum.*

Above 2500 m, diverse plant families include pteridophytes such as Dryopteridaceae, Hymenophyllaceae and Pteridaceae; monocotyledons such as Bromeliaceae, Orchidaceae and Poaceae; and dicotyledons such as Asteraceae, Ericaceae, Melastomataceae and Scrophulariaceae. The most speciose genera are *Huperzia, Hymenophyllum, Bomarea, Tillandsia, Epidendrum, Pleurothallis, Calceolaria* and *Weinmannia.*

CPD Site SA37: Eastern slopes of Peruvian Andes, showing interior of montane wet forest at 3400 m in Manu National Park, southern Peru. Photo: Kenneth R. Young.

CPD Site SA37: Eastern slopes of Peruvian Andes, showing epiphytic bromeliads in Machu Picchu Sanctuary, southern Peru. Photo: Kenneth R. Young.

Useful plants

There is little sustainable use of the native plant resources in the region, except among traditional tribal peoples. Several groups of tribal peoples are present, mostly using land below 800 m. From north to south, important linguistic groups are the Aguaruna, Nantipa, Chayahuita, Lowland Quechua, Cashibo, Amuesha, Machinguenga, Huachipairi, Maschos and Arasaire. Numerous useful species and wild relatives of economic plants are present, probably including many actual and potential medicinals.

In general, timber trees (Lauraceae, Meliaceae, Podocarpaceae) and ornamental plants (Acanthaceae, Bromeliaceae, Gesneriaceae, Orchidaceae) are removed wherever accessible. The upper elevations in particular harbour many ornamental plants, e.g. ferns, bromeliads and orchids, which are transported to Lima and exported.

Social and environmental values

Because of the steep slopes and need to protect upper watersheds, essentially all of the region above c. 2000 m should be classified as Protection Forest. Tribal lands and hunting grounds are only protected in a natural state in Manu National Park (see CPD Site SA11 and Gentry 1990) and to a lesser extent near Yanachaga-Chemillén National Park. Tourism could be much more developed for this region, but most benefits from such activity usually have not remained with local peoples.

No less than 13 bird species of restricted range in two Endemic Bird Areas (EBAs) of this region are considered threatened; they are primarily in the fragile high-altitude habitats near treeline. From treeline to 2200 m in the High Peruvian Andes EBA (B27), 30 restricted-range bird species occupy the treeline forest patches or montane forests. In humid forest from 2200–600 m on the eastern slopes of the Peruvian Andes, in the Eastern Andean foothills of Peru EBA (B29), 11 restricted-range birds occur.

Economic assessment

Approximately 6% of Peru's population lives in the region – 1.5 million people. Below c. 2000 m, and especially in the dry and moist life zones, are important agricultural areas for raising cattle and growing maize, rice and tree crops such as coffee and oranges. The most important agricultural area is in the Department of San Martín. The region's major agricultural products are cassava, cattle, coca leaves, coffee, maize, plantains (*Musa* × *paradisiaca*) and rice (Young 1992).

Human activities in the region are extractive or destructive, and few have left lasting gains. At lower elevations in parts of the region there is great demand for land to use for agricultural purposes both illegal (extensive coca plantations) and legal (coffee, fruits, maize, rice). The areas most transformed are the lower elevations of the Western Huallaga River and Urubamba River physiographic provinces; the middle elevations of the Chachapoyan province; and the higher elevations of the Tambo River province (much of which has long been deforested and used in highland agronomic and animal-husbandry systems).

Threats

The right of Peruvians, especially local people, to develop this region does not negate the need for careful planning and mitigation of environmental impacts. Highway construction has been a major cause or accomplice of destruction of the natural vegetation and fauna on the eastern slopes (Young 1992). The region's only safe areas are those that are inaccessible. However, if not accessible from the existing highways, much of the eastern slopes is within reach of potential highways, which are already in proposals prepared particularly for the international donor agencies.

There is considerable spontaneous migration into the region (INADE-APODESA 1991; CNP and CIPA 1984; Aramburu, Bedoya Garland and Recharte-B. 1982). Most of the colonists are from highland communities where the possibilities for land acquisition and social services are even more limited than they face on the frontier. Amazonas and San Martín departments receive many people each year from the highlands of Cajamarca. The Selva Central sector, especially in Pasco Department, and the lowlands of Ayacucho have received hundreds from the highlands of Junín and Ayacucho who are fleeing the violence associated with guerilla groups. Madre de Dios Department has numerous colonists from the highlands of Cusco, plus c. 25,000 placer-gold miners. Very few colonists bring agricultural techniques appropriate to the ecological conditions of their new homes. Instead, they often try to recreate the deforested environments they were familiar with in the highlands (ONERN 1985). Natural vegetation remaining in the drier areas (dry and moist life zones) is particularly at risk and unlikely to be spared during the next several decades.

Large development projects in the 1960s and 1970s in the Huallaga Valley and Tingo María-Pucallpa area left a poorly designed and maintained highway system, deforestation and social conditions conducive to turning the area into what is now the world's leading supplier of the illegal coca leaves, which are processed into cocaine (Morales 1989). Development projects in the 1980s in the Selva Central (departments of Pasco and Huánuco) have since been virtually abandoned and almost certainly will have the same fate.

The kind of coca most used for the narcotics trade is grown chiefly from 500–1500 m, and occupies 2500–3000 km² of Peru's eastern slopes (Dourojeanni 1989). Control efforts tend to disperse the coca-deforestation front into more isolated valleys and onto steeper slopes. The Perené River area (departments of Ayacucho and Huánuco) also is being affected, and there is little doubt (unless the demand for cocaine should drop) that the Quillabamba (Cusco) and Madre de Dios areas in the south will be dramatically changed over the next several years.

Conservation

Three National Parks, Río Abiseo NP (8°S), Yanachaga-Chemillén NP (10.5°S) and Manu NP (11.5°–13°S), include about 18,300 km² or 7% of the eastern slopes, including land in three of the six physiographic provinces (Western Huallaga, Upper Pachitea, Madre de Dios). However, protection is mostly passive, achieved by the parks' inaccessibility; only c. 20 park guards cover this vast area

(1000 km² per guard!). Also, many important habitats and endemic species are not within these parks.

Other protected areas total c. 6000 km² of the region. They include two designated Sanctuaries (Machu Picchu, Ampay), two Reserved Zones (Apurímac, Tambopata-Candamo), one Protection Forest (San Matías – San Carlos) and one National Forest (Biabo). However, the areas are too small (Ampay, 36 km²; Machu Picchu, 326 km²) or receive no actual protection (Apurímac, Tambopata-Candamo, San Matías – San Carlos). Tingo María National Park (180 km²) probably has been destroyed; it was situated at 10°S, which

is in the midst of the coca-field area of the Upper Huallaga Valley.

There is little hope for conservation programmes in sectors where the government does not exert control, such as areas with coca-cocaine production and those occupied by guerrilla groups. In other areas, regional development interests, e.g. in building highways, mining, exploiting the timber and then speculating on land values, are currently much stronger than local and international conservation groups. This assessment is based on the recent history of poorly planned development and ineffective conservation

MAP 72. EASTERN SLOPES OF THE PERUVIAN ANDES (CPD SITE SA37)

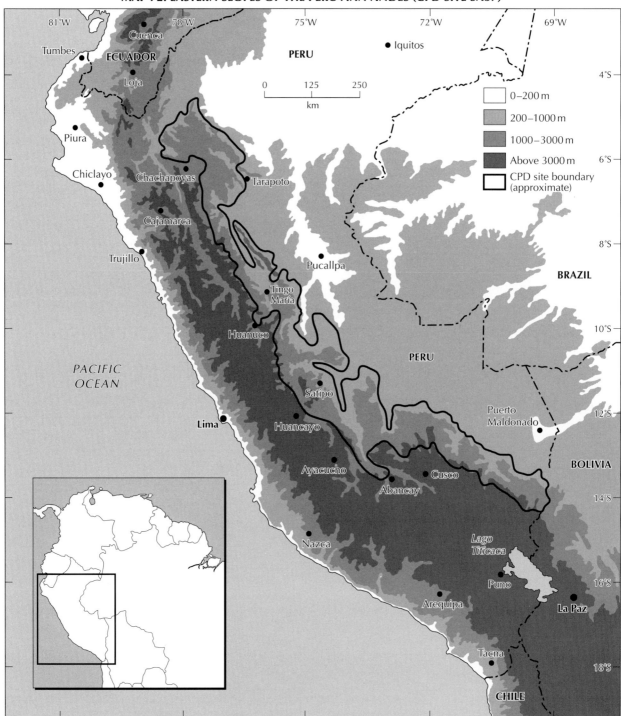

efforts in the region. There are, however, large tracts of wilderness on the eastern slopes that potentially could still be protected.

References

Aramburu, C.E.E., Bedoya Garland, E. and Recharte-B., J. (1982). *Colonización en la Amazonía.* Ediciones CIPA, Lima. 161 pp.

CNP and CIPA (1984). *Población y colonización en la alta Amazonía peruana.* Consejo Nacional de Población (CNP) y el Centro de Investigación y Promoción Amazónica (CIPA), Lima. 281 pp.

Dourojeanni, M.J. (1989). Impactos ambientales del cultivo de la coca y la producción de cocaína en la Amazonía peruana. In León, F.R. and Castro de la Mata, R. (eds), *Pasta básica de cocaína: un estudio multidisciplinario.* CEDRO, Lima. Pp. 281–289.

Duellman, W.E. (1979). The herpetofauna of the Andes: patterns of distribution, origin, differentiation, and present communities. In Duellman, W.E. (ed.), *The South American herpetofauna: its origin, evolution, and dispersal.* Monogr. Mus. Nat. Hist. Univ. Kansas No. 7. Pp. 371–459.

Gentry, A.H. (1980). The Flora of Peru. Conspectus and index to families. *Fieldiana Bot., New Series* 5: 1–11.

Gentry, A.H. (ed.) (1990). *Four neotropical rainforests.* Yale University Press, New Haven. 627 pp.

INADE-APODESA (1991). *Sistema de información geográfica de la selva alta del Perú: resultados primera etapa.* Instituto Nacional de Desarrollo (INADE), Proyecto: Apoyo a la Política de Desarrollo Selva Alta (APODESA), Lima.

León, B., Young, K.R. and Brako, L. (1992). Análisis de la composición florística del bosque montano oriental. In Young, K.R. and Valencia, N. (eds), *Biogeografía, ecología y conservación del bosque montano en el Perú.* Memorias del Museo de Historia Natural Vol. 21. Universidad Nacional Mayor de San Marcos (UNMSM), Lima. Pp. 141–154.

Morales, E. (1989). *Cocaine: white gold rush in Peru.* University of Arizona Press, Tucson.

ONERN (1976). *Mapa ecológico del Perú. Guía explicativa.* Oficina Nacional de Evaluación de Recursos Naturales (ONERN), Lima. 117 pp.

ONERN (1985). *Los recursos naturales del Perú.* ONERN, Lima.

ONERN-AID (1986). *Perfil ambiental del Perú.* Lima, Peru.

Peñaherrera, C. (1989). *Atlas del Perú.* Instituto Geográfico Nacional, Lima. 399 pp.

Scott, G. (1978). Grassland development in the Gran Pajonal of eastern Peru. *Hawaii Monographs Geography* 1: 1–187.

Tosi Jr., J. (1960). *Zonas de vida natural en el Perú.* Organización de Estados Americanos, Boletín Técnico No. 5. Lima, Peru. 271 pp.

Vuilleumier, F. (1977). Barrières écogéographiques permettant la spéciation des oiseaux des hautes Andes. In Descimon, H. (ed.), *Biogéographie et evolution en Amérique tropicale.* Ecole Normale Supérieure, Publ. Lab. Zool. No. 9, Paris. Pp. 29–51.

Young, K.R. (1991). Floristic diversity on the eastern slopes of the Peruvian Andes. *Candollea* 46: 125–143.

Young, K.R. (1992). Biogeography of the montane forest zone of the eastern slopes of Peru. In Young, K.R. and Valencia, N. (eds), *Biogeografía, ecología y conservación del bosque montano en el Perú.* Mem. Museo Hist. Nat. Vol. 21. UNMSM, Lima. Pp. 119–140.

Authors

This Data Sheet was written by Dr Kenneth R. Young (University of Maryland Baltimore County, Department of Geography, Baltimore, MD 21228, U.S.A.) and Dra. Blanca León (Universidad Nacional Mayor de San Marcos, Museo de Historia Natural, Apartado 14-0434, Lima-14, Peru).

GRAN SUMACO AND UPPER NAPO RIVER REGION
Ecuador

Location: Over 100 km south-east of Quito – eastern flanks of Andes surrounding Pan de Azúcar and Sumaco volcanoes and adjacent Amazonian lowlands including Upper Napo River Valley, between latitudes 0°00'–1°10'S and longitudes 77°55'–77°10'W.

Area: c. 9000 km² (proposed Biosphere Reserve and buffer zones).

Altitude: 300–3732 m.

Vegetation: Range of vegetation types along altitudinal gradient, from lowland tropical wet forest to undisturbed wet páramo.

Flora: Very high diversity – c. 6000 species; high endemism.

Useful plants: Many species used for medicinal and artisanal purposes by indigenous inhabitants; fruit trees; wild relatives of important crops.

Other values: Watershed protection for Upper Napo River Basin; habitat for many vertebrate taxa; genetic resources.

Threats: Road construction and colonization, conversion of forest to subsistence farms and cattle pastures; possible mining development.

Conservation: Sumaco – Napo Galeras National Park (2052 km²), Jatun Sacha Biological Station (15 km²).

Geography

The Gran Sumaco region, including the Upper Napo River Valley, comprises the outlying eastern slopes of the Ecuadorian Andes and adjacent piedmont at the extreme western edge of the Amazon Basin. Physiographically, the region is dominated by Sumaco Volcano (which means beautiful mountain in Quechua), an isolated cone-shaped volcano c. 50 km east of the main eastern range of the Andes. The summit of Sumaco (3732 m) is visible almost everywhere in the region; Cerro Pan de Azúcar is to its north. Both may be extinct. "Gran Sumaco" has been used in recent years by conservation planners (DESFIL 1990; AHT 1993) and refers to the entire region. The geographic concept of Gran Sumaco is here extended 20 km farther south to include the Upper Napo River Valley. This slightly expanded delimitation of the region is justified from a floristic standpoint as well as for conservation purposes.

The Gran Sumaco region of Napo Province is not easily defined, but the following boundaries may by considered (Map 73): to the north, the Coca River; to the east, a north-south line at the longitude of the lowland town of Loreto (c. 77°15'W); to the south, an east-west line c. 10 km south of the Napo River (c. 1°10'S); to·the west, the Tena-Baeza-El Chaco road. These are the approximate limits of the proposed Gran Sumaco Biosphere Reserve.

The cone of Sumaco Volcano rises at the northern end of a relatively level plateau at 1000–1200 m. The top of the plateau is a thick cap of distinctive volcanic basalt. Beneath are the Napo and the Hollín formations, which are Cretaceous deposits respectively of limestone and sandstone. The Napo limestone is exposed at the bottom of deep gorges cut through the plateau by numerous rivers, including the Hollín, Huamaní, Pucuno and Suno which arise on the slopes of Sumaco – these are tributaries of the Upper Napo River. Thirty km south of Sumaco, separated by the Hollín and Pucuno river valleys, is the isolated limestone massif of the Cordillera Galeras (north-east of the town of Tena).

Average annual precipitation in the region ranges from c. 3500 mm at Loreto to over 6000 mm on the slopes of Sumaco. There is no true dry season, but rainfall generally lessens during July–August and December–January (exhibiting the bimodal rainfall distribution typical of the equatorial zone). Even during these relatively dry spells, monthly precipitation rarely falls below 120 mm. Although most plants probably do not experience significant moisture stress, the drier months are marked by distinct peaks in flowering and fruiting of the canopy tree species.

Vegetation

The region is covered by tropical rain forest that varies in physiognomy and floristic composition with changes in altitude, geological substrate and precipitation. The vegetation map of the region (DESFIL 1990) distinguishes four principal Holdridge life zones; four corresponding vegetation types are recognized (Neill and Palacios 1990).

Tropical wet forest
Tropical wet forest is lowland forest, mostly below 600 m, with annual precipitation exceeding 4000 mm. In the Amazon Basin, this vegetation type is confined to the piedmont region, a narrow belt (50–100 km wide) at the base of the Andes. Farther east on the Amazonian plains, the climate is somewhat drier and tropical moist forest is dominant.

The canopy of the wet forest is composed of trees 35–40 m tall such as *Cedrelinga cateniformis*, *Parkia multijuga*, *Erisma uncinatum* and *Phragmotheca ecuadoriensis*. Myristicaceae are very abundant in the canopy, especially *Otoba glycycarpa* and *Virola* spp. On fertile alluvial soil near river margins, *Ceiba pentandra* is the most conspicuous emergent, and *Chimarrhis glabriflora*, *Guarea kunthiana* and *Celtis schippii* are common canopy trees. In the subcanopy (15–25 m high), the stilt-root palm *Iriartea deltoidea* is very common and a good zone indicator.

Premontane rain forest

Premontane rain forest covers a large area between 1000–1600 m elevation. The forest canopy is somewhat lower (30–40 m) than in tropical wet forest (Valencia *et al.*, in prep.). The largest trees are strangler figs (hemi-epiphytic *Ficus*). *Meriania hexamera*, a canopy tree with spectacular lilac-coloured blooms in December–February, is a good zone indicator. Numerous species of Lauraceae are present, many of which are new, undescribed taxa. On flat areas *Dacryodes cupularis* is the most abundant canopy tree. Common subcanopy trees include *Metteniusa tessmanniana*, *Ocotea javitensis* and the palms *Wettinia maynensis* and *Catoblastus praemorsus*. On ridges above 1300 m, the very tall palm *Dictyocaryum lamarckianum* is common, occasionally in nearly pure stands. In areas of natural disturbance are pure stands of bamboo, *Bambusa* sp. In the forest understorey are many shrubby species of Rubiaceae, Melastomataceae and Piperaceae. Epiphytes are abundant, especially Orchidaceae, Araceae and ferns.

Lower montane rain forest

Lower montane rain forest (true cloud forest) occurs from 1600–2800 m. Epiphytes are even more abundant than in the premontane rain forest. In the lower portion of this zone, trees are large (to 35 m tall and 1 m diameter); the forest canopy is progressively lower at higher elevations. Predominant are species of *Ficus*, *Clusia*, *Weinmannia* and Lauraceae. Rather surprisingly, very large individuals of *Cedrela odorata* occur at elevations up to 1800 m. In the subcanopy, *Morus insignis* is very common as well as *Croton* and *Alchornea*. Shrubby Melastomataceae are dominant in the understorey. On windswept ridges and poorly drained sites, the tall forest is replaced by dwarf dense scrub dominated by *Clusia*, *Weinmannia* and *Hedyosmum*. On steep slopes are large pure patches of dwarf bamboo (*Chusquea*).

Montane rain forest, and páramo

Montane forest occurs above 2800 m on the upper slopes of Sumaco Volcano and Cerro Pan de Azúcar. Forest cover on Sumaco extends to c. 3300 m; above, treeless "páramo" vegetation extends to the summit at 3732 m. No direct information is available for the upper forest zone of Sumaco. Aerial reconnaissance and knowledge of similar areas on the eastern slopes of the Andes indicate that near the upper limits is a very dense elfin forest with trees 5–10 m tall, dominated by species of *Clusia*. The Sumaco páramo has never been disturbed by grazing, fire or other human activities, which is probably unique among Ecuadorian páramos. A botanical expedition reached the summit of Sumaco by helicopter in 1979 (Løjtnant and Molau 1982). The páramo is dominated by the shrubby fern *Blechnum loxense* and the bunchgrass *Cortaderia nitida*.

Flora

The flora of the Gran Sumaco and Upper Napo River region, from the tropical zone to the páramo, probably exceeds 6000 species. The most thorough floristic inventories have been carried out in the lowland tropical wet forest, particularly since 1985 on the south bank of the Napo River at Jatun Sacha Biological Station, by botanists from Missouri Botanical Garden and the Herbario Nacional del Ecuador. More than 1800 species of vascular plants have been collected and identified from the 15 km² reserve (Neill, Cerón and Palacios, in prep.) and hundreds more species are yet to be identified. Based on these collections, since 1987 more than 25 species of trees new to science have been described. Some of them, e.g. *Pleurothyrium insigne*, are relatively widely distributed in western Amazonian Brazil, Peru and Ecuador. Others, such as *Rollinia helosioides*, are locally endemic to the Upper Napo River Valley and have not been found outside of the immediate vicinity of the Jatun Sacha reserve.

The flora of the higher elevation forests in the Gran Sumaco region has been much less well studied. Floristic inventories in the premontane rain forest, along the Hollín-Loreto road at 800–1400 m elevation, were initiated during 1988–1989 by Missouri Botanical Garden in conjunction with the Herbario Nacional of the Museo Ecuatoriano de Ciencias Naturales. Forests above 1400 m have been sampled very little. The Galeras massif has not been explored botanically at all, but is expected to contain many locally endemic species adapted to the exposed limestone substrate. Many new species of trees, aroids and orchids are being described from the recent collections in the premontane zone of Gran Sumaco. Species endemism at middle and upper elevations is undoubtedly quite high, but cannot be estimated reliably without more complete floristic data.

The forests of the region are highly diverse on a local as well as a regional scale. Several 1-ha permanent study plots have been established in lowland forest at the Jatun Sacha Biological Station (Neill *et al.* 1993), with all trees 10 cm or more in dbh marked, measured and identified within each plot. Up to 250 tree species occur in a single hectare. At higher elevations, tree diversity diminishes; a similar 1-ha plot at 1200 m on the slopes of Sumaco Volcano included c. 150 tree species (Valencia *et al.*, in prep.).

Useful plants

The Gran Sumaco and Upper Napo River region contains a wealth of plant species used traditionally by the native Quichua inhabitants for medicine, food, construction, crafts and clothing. A recent survey of Quichua ethnopharmacology (Marles, Neill and Farnsworth 1988) documented 120 plant species used medicinally in the region. Some of these are already marketed locally and internationally, such as *Croton lechleri*, a common second-growth tree which produces a dark red latex used traditionally to speed healing of wounds, and for other medicinal purposes. A U.S. pharmaceutical firm is carrying out clinical trials with an extract of *Croton* used to treat infant respiratory diseases. In anticipation of commercial demand, a *Croton* silvicultural programme is being initiated with indigenous communities in the region.

The Quichua also use numerous native species of food plants, particularly fruit trees. Western Amazonia in general has been cited as an important centre for the domestication of crop plants (Clements 1990). Domesticated and semi-domesticated land-races of fruit-bearing trees such as *Bactris gasipaes*, *Rollinia mucosa* and *Gustavia macarenensis*, and *Chrysophyllum venezuelanense* and *Pouteria caimito*, are commonly grown in Quichua house-gardens. These selected land-races are important genetic resources, and some of the edible fruit trees merit consideration for cultivation in other tropical regions.

A botanical garden and plant conservation centre is being established at Jatun Sacha Biological Station. Wild species of economically useful plants native to Amazonian Ecuador, as well as local native land-races of crop plants, are being brought into cultivation for purposes of research, agronomic improvement, education and conservation of germplasm.

Social and environmental values

Sumaco Volcano is the source area for many tributary rivers of the Upper Napo River and the forest cover is vital to the stability of the regional hydrological cycle. The volcano figures prominently in many Amerindian legends. The Sumaco region has been a traditional hunting-and-gathering area for Quijos Quichua communities of the Upper Napo River Basin (cf. Whitten 1976).

Preservation of a swath of intact forest from the lowlands to the upper limit of arborescent vegetation may be important for the future of fruit-eating birds (such as trogons, cotingas, cracids) that migrate up and down the slopes and, in turn, play a vital role in seed dispersal for many of the tree species.

The region around Sumaco Volcano falls within the Eastern Andes of Ecuador and northern Peru Endemic Bird Area (EBA B18), which is centred between 800–2000 m. Fifteen bird species of restricted range occur in this EBA, most inhabiting premontane and lower montane rain forest on the Andean slopes. Due to the widespread, increasing forest destruction throughout the Eastern Andes, three of these species are considered threatened.

Economic assessment

The mid-elevation forests of the Gran Sumaco region have been recently colonized by small farmers along the new Hollín-Loreto road, which transverses the plateau and gorges along the southern flanks of Sumaco Volcano. However, prospects for truly sustainable agriculture in the premontane and montane zones (above c. 1000 m) appear limited. Few crops perform well in the extremely wet climate and waterlogged soils. The main initial commercial crop is "naranjilla" (*Solanum quitoense*), which can be grown for about three years and then the site must be left fallow for a number of years before replanting. Ecotourism may have limited potential, given the difficulty of access to many areas and the rainy climate.

An alternative that is being developed on an experimental basis is sustained-yield timber production. A preliminary assessment of possibilities for sustainable timber production in the region was made by Palacios and Simione (1990). Cultural Survival and FOIN (the Federation of Indigenous Organizations of the Napo, which is for local Quichua-speaking communities) have begun a programme for forest management and sustained-yield timber harvest in three indigenous communities along the Hollín-Loreto road. Included are training in dendrology, silviculture and ecologically sound extraction methods for members of the local community and a technical team of Quichua foresters. As of October 1992, this programme is in training and inventory phases; timber extraction has not begun.

Economic assessment of non-timber forest products suggests that the establishment of "extractive reserves" within which fruits, medicinal barks, resins and the like are collected from the forest could be a viable economic alternative for inhabitants of the region. The forest would remain intact in such reserves. A recent study assessed the annual yield, local market value and extraction costs of such non-timber products in the three 1-ha permanent forest plots at Jatun Sacha Biological Station (Grimes *et al.* 1994). The net annual value of products that could be extracted from each hectare ranged from US$60 to US$156. These values compare very favourably on a per-hectare basis with alternative land uses such as timber extraction and cattle-ranching, particularly because extraction of non-timber products is sustainable on a long-term basis whereas the other uses are not. The most valuable product cited was neither edible fruit nor medicinal bark, but rather the resin of *Protium* spp., which is used by the Quichua to make a traditional ceramic varnish.

Threats

The Sumaco region was isolated and sparsely populated until late 1987, when the Hollín-Loreto road was completed. (Construction accelerated following a March 1987 earthquake that destroyed long sections of the Baeza-Lago Agrio road, the main route from Quito to the oil fields in Napo Province.) Following completion of the Hollín-Loreto road, several thousand colonists settled along it and cut the forest to establish cattle pastures and cultivate cash crops, principally *Solanum quitoense*. Many settlers were Quichua Amerindians from the Archidona area who had used the Sumaco region only as hunting grounds.

Most original forest is gone in a swath up to 1 km wide on both sides of the Hollín-Loreto road. Several subsidiary roads are being built, and deforestation is proceeding along them. The land-tenure situation exacerbates deforestation. Many of the indigenous communities and mestizo colonists do not yet have legal title to the land they occupy, so there is little incentive to undertake conservation measures.

Conservation

Preservation of tropical wet forest in Upper Amazonia is a high conservation priority. The vegetation and floristic composition of this high-rainfall forest are notably different. The tropical wet-forest life zone, a rather narrow belt below 600 m at the eastern base of the Andes (the piedmont region), was the major vegetation type in Amazonian

Ecuador that lacked sufficient legal protection – until 1994. The Sumaco region is the only remaining large area of undisturbed forest within this life zone in eastern Ecuador. Cayambe-Coca Ecological Reserve and Sangay National Park (CPD Site SA31), both also located on the eastern Andean slopes, do not include significant areas below 1000 m. Cuyabeno Wildlife Reserve and Yasuní National Park (CPD Site SA8), both farther east on the Amazonian plain, do not include terrain above 250 m and are covered by tropical moist forest, which has lower precipitation and somewhat lower species diversity at the community level.

An important feature in the Gran Sumaco region is the altitudinal transect of undisturbed forests and páramo from the lowlands (at 400 m) to the summit of Sumaco (3732 m). An undisturbed transect of this sort does not exist elsewhere in the eastern equatorial Andes – a comparative transect is in southern Peru at 16°S latitude, in Manu National Park (see CPD Sites SA11 and SA37).

Much of the Gran Sumaco region is classified officially as Protection Forest (which is owned by the government). Although deforestation and agricultural activity are prohibited, the regulations are not enforced. A small (15 km²) private

MAP 73. GRAN SUMACO AND UPPER NAPO RIVER REGION, ECUADOR (CPD SITE SA38)

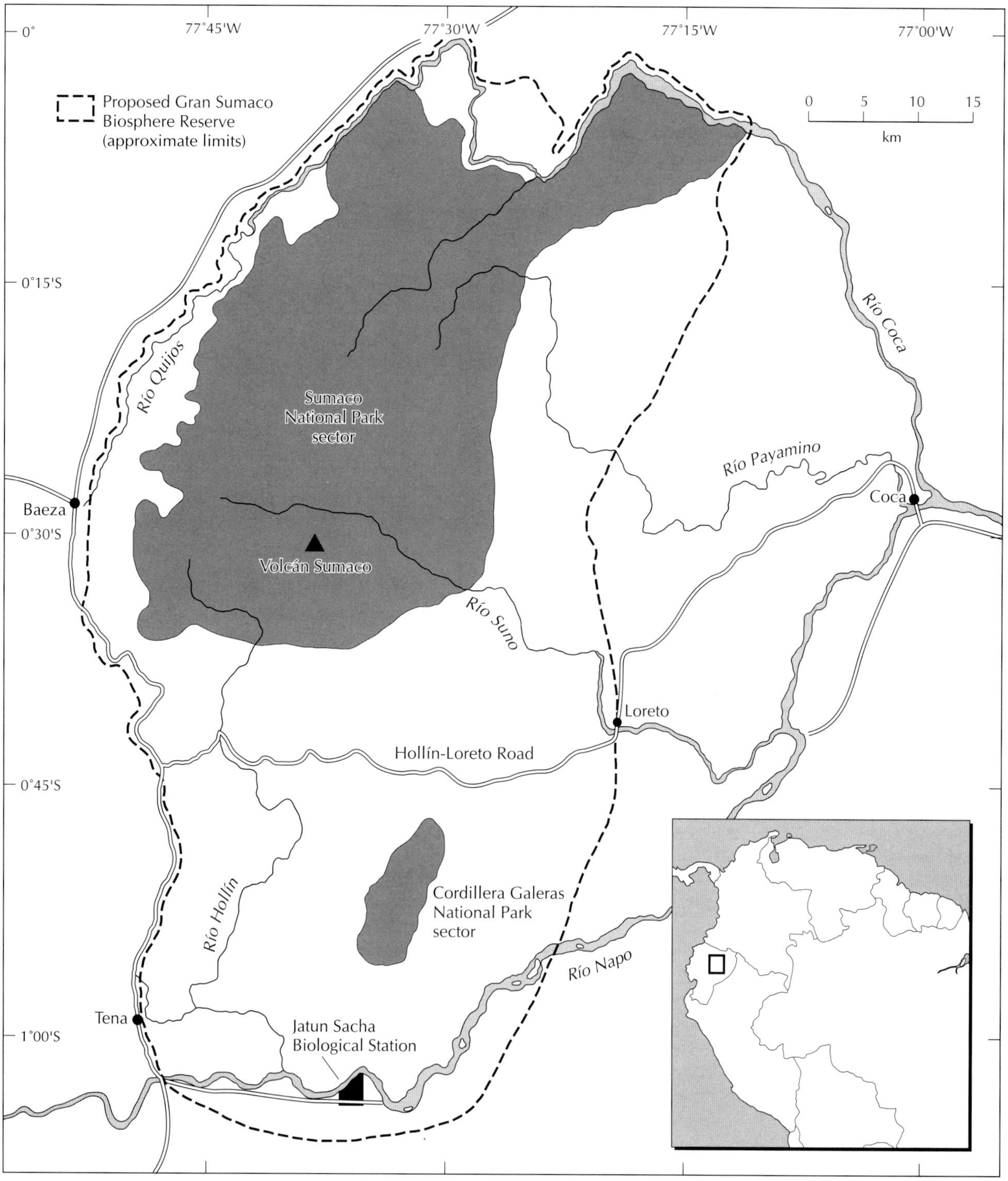

reserve protecting tropical wet forest is the Jatun Sacha Biological Station at the southern limit of the Gran Sumaco region. Jatun Sacha is currently in process of expansion through purchase of adjacent properties, with funding by donations from Children's Rainforest organizations in Europe and North America. This reserve may be able to protect 30 km² of forest.

A major conservation project for the Gran Sumaco region has been initiated, with funding provided by the German Government through its financial assistance bureau KFW. Earlier, the U.S. Agency for International Development, in conjunction with the Dirección Nacional Forestal (DINAF) of the Ecuadorian Ministerio de Agricultura y Ganadería, had commissioned two preliminary studies of the region: an overview by Fundación Natura and DINAF (1989) and the diagnostic management plan by DESFIL (1990). The KFW, in conjunction with the Instituto Ecuatoriano Forestal y de Areas Naturales y Vida Silvestre (INEFAN), commissioned a third study which included a detailed plan for conservation and sustainable development of the Gran Sumaco region (AHT 1993). Concurrently, in a separate agreement with INEFAN, the Australian Rainforest Information Centre and Izu Mangallpa Urcu (an Ecuadorian foundation formed by local Quichua families) initiated a management plan for the conservation of the Galeras area.

The Gran Sumaco conservation project proposed establishment of two National Parks as core protected areas (Map 73); in 1994, Sumaco – Napo Galeras National Park (2052 km²) was established. The Sumaco sector of the park (1906 km²) protects a continuous altitudinal transect from lowland forest to páramo on the eastern slopes of Sumaco. The Napo Galeras sector (147 km²) protects the unique flora on the limestone Galeras massif. Furthermore, the entire Gran Sumaco region (9000 km²) is expected to be declared as a Biosphere Reserve by UNESCO-MAB. The inhabited areas will be managed as buffer zones where sustainable agriculture, forestry and other activities will be promoted that help to protect the core areas.

References

AHT (Agrar- und Hydrotechnik Gmbh.) (1993). *Proyecto protección de la selva tropical Gran Sumaco: estudio de factibilidad.* Quito. 251 pp. + appendices.

Clements, C.R. (1990). A center of crop genetic diversity in western Amazonia. *BioScience* 39: 624–631.

DESFIL (1990). *Manejo de la zona del Gran Sumaco, Provincia del Napo, Ecuador. Development Strategies for Fragile Lands* (DESFIL), Washington, D.C. 104 pp.

Fundación Natura and DINAF (Dirección Nacional Forestal) (1989). *Proyecto Sumaco – Informe para la definición del área a protegerse y la selección de alternativas de manejo del Sumaco.* Fundación Natura, Quito.

Grimes, A., Loomis, S., Jahnige, P., Burnham, M., Onthank, K., Alarcón, R., Palacios, W.A., Cerón, C.E., Neill, D.A., Balick, M., Bennett, B. and Mendelsohn, R. (1994). Valuing the rain forest: the economic value of nontimber forest products in Ecuador. *Ambio* 23: 405–410.

Løjtnant, B. and Molau, U. (1982). Analysis of a virgin páramo plant community of Volcán Sumaco, Ecuador. *Nordic J. Bot.* 2: 567–574.

Marles, R.J., Neill, D.A. and Farnsworth, N.R. (1988). A contribution to the ethnopharmacology of the lowland Quichua people of Amazonian Ecuador. *Rev. Acad. Colombiana Cienc. Exactas Fís. Nat.* 63: 111–120.

Neill, D.A. and Palacios, W.A. (1990). Características naturales. In DESFIL, *Manejo de la zona del Gran Sumaco, Provincia del Napo, Ecuador.* DESFIL, Washington, D.C. Pp. 11–25.

Neill, D.A., Cerón, C.E. and Palacios, W.A. (in prep.). Flora de la Estación Biológica Jatun Sacha: lista preliminar.

Neill, D.A., Palacios, W.A., Cerón, C.E. and Mejía, L. (1993). Composition and structure of tropical wet forest in Amazonian Ecuador: diversity and edaphic differentiation. [*Abstracts: Association for Tropical Biology Annual Meeting, June 1–4, 1993.*] San Juan, Puerto Rico. Pp. 117–118.

Palacios, W.A. and Simione, R. (1990). Técnicas adecuadas y sostenidas para el manejo natural de los bosques. In DESFIL, *Manejo de la zona del Gran Sumaco, Provincia del Napo, Ecuador.* DESFIL, Washington, D.C. Pp. 62–72.

Valencia, R., Balslev, H., Palacios, W.A., Neill, D.A., Josse, C., Tirado, M. and Skov, F. (in prep.). Diversity and family composition of trees in different regions of Ecuador: a sample of 18 one-hectare plots. In Dallmeier, F. and Comiskey, J. (eds), *Measuring and Monitoring Forest Biological Diversity: The International Network of Biodiversity Plots. Proceedings of the International Symposium, May 23–25, 1995, Washington, D.C.* Smithsonian Institution Press, Washington, D.C.

Whitten Jr., N.E. (1976). *Sacha Runa: ethnicity and adaptation of Ecuadorian Jungle Quichua.* University of Illinois Press, Urbana and Chicago. 348 pp.

Authors

This Data Sheet was written by Dr David A. Neill (Herbario Nacional, Casilla 17-12-867, Quito, Ecuador and Missouri Botanical Garden, P.O. Box 299, St. Louis, MO 63166-0299, U.S.A.) and Walter A. Palacios (Herbario Nacional, Casilla 17-12-867, Quito, Ecuador).

PACIFIC COAST: CPD SITE SA39

COLOMBIAN PACIFIC COAST REGION (CHOCO)
Colombia

Location: Eastern Panama to northern Ecuador, between about latitudes 8°45'–1°15'N and longitudes 79°–76°15'W.

Area: c. 130,000 km².

Altitude: 0–1000 m.

Vegetation: Mangroves; aquatic and marshy communities along rivers and in lentic environments; on riverbanks and beaches, communities on sandy terrain left by rivers; tropical wet and pluvial forests on firm (upland) terrain.

Flora: High diversity – 8000–9000 species of vascular plants estimated, 4638 recorded; high endemism – 20%; many threatened species.

Useful plants: For folk medicine, food, fuelwood, lumber, fibres, dyes, tannin, rubber, gums.

Other values: Recorded so far are 577 bird species, 52 snake spp., 45 lizard and allied spp. and 127 amphibian spp.; high endemism; threatened species; Amerindian lands; gold, platinum, other minerals; tourism.

Threats: Indiscriminate felling of trees; raising cattle; permanent agriculture; pollution from mining; road building; erosion; destruction of mangroves (shrimp-farming, charcoal, tannin).

Conservation: Los Katíos Natural National Park, Utría NNP, low parts of Farallones de Cali NNP and Munchique NNP, Gorgona Island NNP, Sanquianga NNP; Amerindian Reserves.

Geography

The Colombian Pacific Coast region (Chocó *sensu lato*) is the stretch of land mainly between the Pacific Ocean and Cordillera Occidental of the Andes, from west of the mouth of the Atrato River near Panama to the Mataje River bordering north-western Ecuador (Rangel-Ch. and Aguilar 1993). The Colombian departments represented are primarily Chocó, Valle del Cauca, Cauca and Nariño.

The region is composed of lowlands, with elevations rarely exceeding 600 m, and highlands, which are usually above 600 m. There are three major landforms: (1) plains of recent alluvium; (2) low hills formed by relatively recent stream dissection of Tertiary and Pleistocene sediments; and (3) complex mountainous areas of Mesozoic rocks (West 1957).

Five subregional zones may be distinguished (Map 74) (Rangel-Ch. *et al.* 1994):

1. Northern coast, with areas of the Serranía del Darién and Urabá;

2. Seaboard generally up to 500 m elevation, along the Pacific coast;

3. Central strip, which includes northern wet forest, central pluvial forest and the San Juan River area;

4. So-called highlands of El Carmen del Atrato, and the San José del Palmar area; and

5. Localities of Pacific slope from 500–1000 m in the departments of Valle del Cauca, Cauca and Nariño.

The region has a complex of vast river basins, the most important being the Atrato and San Juan, and in the south the (San Juan de) Micay, Patía and Mira. Alluvial plains extend along the valleys of the Atrato and the San Juan rivers, which are separated by a narrow watershed divide (c. 100 m in elevation) called the Isthmus of San Pablo, and empty respectively into the Caribbean Sea (Gulf of Urabá) and the Pacific Ocean. Since these two rivers parallel the Cordillera Occidental and receive numerous Andean subsidiaries, they are swollen torrents for much of the year. The waters of many of the rivers to the Pacific carve deep gorges through the mountain flanks, forming numerous spectacular rapids and waterfalls. The San Juan River discharges the most water to the Pacific of all the South American rivers (Barnes 1993), and the Atrato is the second largest river in South America in terms of the volume of water.

Swamps and shallow lakes ("ciénagas") cover the greater part of the lower half of the Atrato Basin. In contrast, along much of the San Juan River, swamps are absent or confined. Bordering the Atrato and San Juan rivers are wide belts of hills c. 100–300 m high (West 1957; Forero 1982). The widest hilly tract is to the west between the floodplains and the Serranía de Baudó which edges the Pacific. The hills present several landforms; terraces often occur, and are sites for settlement and agriculture (West 1957; Whitten 1974).

Mountainous areas include: in the east, the lower slopes of the Andean Cordillera Occidental, and large somewhat isolated massifs such as Cerro Torrá (2700 m) in the San José del Palmar area of the San Juan Basin; to the west, Serranía del Darién (highest at Cerro Tacarcuna, 1875 m, on the border with Panama) and Serranía de Baudó (highest at Alto del Buey, approximately 1400 m) (Forero and Gentry 1989; Sánchez-P. *et al.* 1990; Sota 1972).

MAP 74. COLOMBIAN PACIFIC COAST REGION (CHOCO) (CPD SITE SA39), SHOWING SUBREGIONAL ZONES

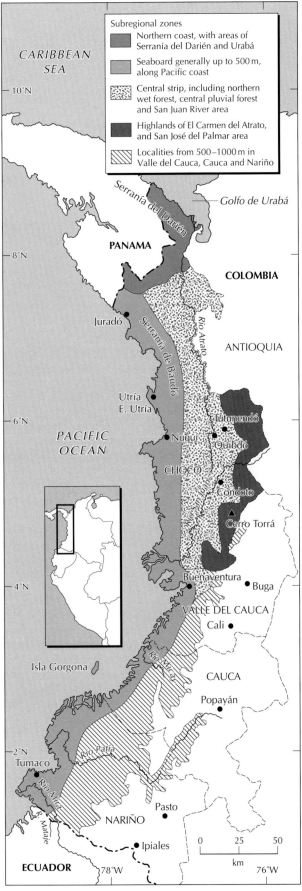

The typical landscape of the Pacific platform is hills and mountains. Erosion processes are enhanced in sectors with more of an altitudinal gradient and particularly where drying is more intense due to the instability of the geologic materials, and on occasion due to scarcity of vegetation. The alluvial soils, which occupy extended areas along both banks of large rivers and smaller tributaries and marine flats, contain easily alterable minerals (feldspars and ferromagnesiums) in the sands and amorphous materials in the clays. Relatively high levels of nutrients are present in such areas, reflected in a greater capacity of cation exchange, high values of base saturation and low content of exchangeable aluminium; these soils are generally of open-clayey and clayey textures (Castiblanco 1990). In contrast, an ecological distinction is a tendency to highly leached nutrient-poor soils; the white-clay soil that occurs in some areas is one of the poorest in such essential elements as phosphorus as well as trace elements like boron and zinc (Faber-Langendoen and Gentry 1991).

The Chocó is probably the wettest sizeable region on Earth, with various locales receiving average yearly precipitation of 4000 mm to over 9000 mm; Tutunendó, central Chocó, has averaged 11,770 mm (Sota 1972). West (1957) assembled records of maximum precipitation for the region – Quibdó (capital of Chocó Department), with an average annual precipitation of 8558 mm, in 1939 reached 15,058 mm. The Chocó is one of few places in the neotropics where tropical pluvial forest occurs (tropical rain forest *sensu* Holdridge) (Gentry 1982).

There is substantial precipitation in the north (where it is locally more varied), the greatest discharge of rain centrally, and the least in the south. At Riosucio rainfall averages 3700 mm, in the mid-region at Istmina 7560 mm and in the southern zone at La Florida 2250 mm. Dry seasons occur in the north toward Panama and the Caribbean Sea (about December–March), and in the south toward the Ecuadorian border. The precipitation is bimodal-tetraseasonal, with a greater concentration in the first half of the year. The rain tends to occur in the afternoon and at night. The climate is classified as $B_2rA'a'$ (Thornthwaite system) with little deficiency of megathermal water (Rangel-Ch. and Aguilar 1993; Sota 1972).

Sunshine averages 110 hours per month through the year. The annual mean temperature is 23.6°C, the average maximum is 29.9°C and minimum 18.6°C; the monthly means rarely exceed 28°C (West 1957). At Quibdó, the maximum temperature is 41°C, the minimum 15°C (Forero 1982). Periods of high relative humidity (September–June) average 89%.

Vegetation

The Chocó phytogeographic region (Gentry 1982) may be broadly considered to include the border area of eastern Panama (cf. Darién Data Sheet, CPD Site MA20), and the coastal lowlands of western Colombia and north-western Ecuador mainly covered by wet forest and pluvial forest (cf. Ecuadorian Mesic Forests Data Sheet, CPD Site SA40). Based on the altitudinal gradient, the variation of water content in the soil and the influence of the sea, several vegetation types are recognized (Rangel-Ch. and Lowy 1993; West

1957; Cuatrecasas 1958; Zuluaga-R. 1987; Von Prahl, Guhl and Grogl 1979; Aguirre-C. and Rangel-Ch. 1990; Acosta-Solís 1970):

1. **Mangroves** ("manglares"): arboreal or shrubby associations that are amphibious in character, with evergreen foliage. They occupy large expanses along the coastal Pacific seaboard. The vegetation corresponds to the syntaxonomic Order Rhizophoretalia. These mangroves are the most diverse in species in the Western Hemisphere (Gentry 1982). The species that are characteristic and diagnostic of the vegetational syntaxa belong to *Rhizophora* (Rhizophoraceae), *Pelliciera* (Pellicieraceae), *Avicennia* (Verbenaceae) and *Laguncularia* and *Conocarpus* (Combretaceae).

   Communities with the following dominants are present (Von Prahl, Guhl and Grogl 1979): (i) species of *Rhizophora*, which develop on the seaboard front and penetrate; (ii) species of *Avicennia*, established landward next to *Rhizophora* on more consolidated terrain; (iii) *Laguncularia racemosa*, which forms a strip next to the woods of *Avicennia*; and (iv) transitional communities – "natales" of *Mora megistosperma* (Leguminosae) and "cativales" of *Prioria copaifera* (Leguminosae).

2. **Aquatic and marshy communities**; in the sequential process, their point of origin is rivers and lentic environments. The following communities are present: (i) *Eichhornia crassipes* (Pontederiaceae) and *Pistia stratiotes* (Araceae); (ii) *Marathrum haenkeanum* (Podostemaceae) and *Dicranopygium crinitum* (Cyclanthaceae); and (iii) *Polygonum acuminatum* (Polygonaceae).

3. **Riverbank and beach communities** established on sandy terrain left by rivers: (i) "pajonales" with *Panicum* spp.; and (ii) canebrakes of *Gynerium sagittatum* in tall dense associations on riverbanks.

4. **Firm-terrain rain-forest vegetation in the lowlands**: in the northern zone, the vegetation is in the Alliance Brosimion utilis, which includes communities dominated by (i) *Cavanillesia platanifolia* (Bombacaceae); (ii) *Anacardium excelsum* (Anacardiaceae) and *Castilla elastica* (Moraceae); (iii) *Anacardium excelsum* and *Pseudolmedia laevigata* (Moraceae); and (iv) *Chrysophyllum* sp. (Sapotaceae) and *Brosimum guianense* (Moraceae). In the southern zone, the forest is pluri-stratified, with two arboreal strata, and climbing plants and epiphytes of vigourous growth.

   The lowland pluvial forests are among the world's most dense and species-rich. Samples of 0.1 ha halfway between Quibdó and Tutunendó and at Bajo Calima averaged 510 plants 2.5 cm or more in dbh of 262 species, including 88 trees (and lianas) 10 cm or more in dbh (Gentry 1986). These wettest forests tend to have more but smaller trees (and lower biomass) than do equivalent forests elsewhere in the neotropics. They are relatively poor in free-climbing lianas, which are mostly replaced by woody hemi-epiphytes belonging to families like Guttiferae, Marcgraviaceae and Melastomataceae (Gentry 1986), all of which have world centres of diversity in this region.

5. **Highlands belt**; communities with the following species: (i) *Inga* sp. (Leguminosae), *Billia columbiana* (Hippocastanaceae), *Brosimum* sp.; (ii) *Sorocea* sp. (Moraceae), *Jacaranda hesperia* (Bignoniaceae), *Pourouma chocoana* (Urticaceae); (iii) *Guatteria ferruginea* (Annonaceae), *Cecropia* sp. (Cecropiaceae), *Inga* spp.; and (iv) *Elaegia utilis* (Rubiaceae), *Brunellia* sp. (Brunelliaceae), *Pourouma* cf. *aspera*, *Inga* spp.

Flora

The flora of the Chocó is poorly known mainly because of difficult accessibility and the very wet climate, which create arduous conditions for exploration and collection of specimens (Lellinger and Sota 1972). Although there were some collections in the 19th century, the first thorough botanical survey was in 1971 for pteridophytes (Sota 1972; Lellinger 1975). Current data suggest that the Chocó is floristically rich, with 8000–9000 vascular plant species (Forero 1982, 1985; Gentry 1982, 1993; Forero and Gentry 1989).

In the entire region, so far 4638 species of vascular plants have been recorded, belonging to 201 families and 1376 genera (cf. Murillo and Lozano 1989). The families with greatest generic diversity are Orchidaceae, Leguminosae, Asteraceae and Rubiaceae (Table 63) (Rangel-Ch. and Lowy 1993; Forero and Gentry 1989). Palm species and Bombacaceae (especially species of *Quararibea* and its segregate *Phragmotheca*) are notably prevalent.

TABLE 63. FAMILIES AND GENERA WITH HIGHEST DIVERSITY IN COLOMBIAN PACIFIC (CHOCO) REGION

Family	Number of genera
Orchidaceae	106
Leguminosae	71
Asteraceae	69
Rubiaceae	67
Polypodiaceae *sensu lato*	45
Gramineae	39
Bignoniaceae	32
Palmae	32
Euphorbiaceae	31
Melastomataceae	31

Genus	Number of species
Piper	102
Psychotria	78
Miconia	63
Anthurium	57
Thelypteris	56
Peperomia	52
Solanum	46
Clusia	42
Maxillaria	40
Cavendishia	38

According to the subdivisions established for the Pacific region (Rangel-Ch. *et al.* 1994), the central subregion (zone 3) has the greatest concentration of plant species: 1440 species in 639 genera of 131 families; next is the coastal lowlands subregion (zone 2), with 1389 species in 733 genera of 165 families. Within zone 3, the greatest diversity of species is in the central pluvial forest at the centre of Chocó Department, in the Atrato River Valley in the vicinity of Quibdó, which receives 7200–7600 mm of rain annually.

A unique feature of the Chocó pluvial forests is unusually large leaves (Gentry 1986). The largest, which may represent the largest simple leaves of any woody dicotyledon, are individual masses of tissue 1 m long and over 50 cm wide of *Psittacanthus gigas* (Loranthaceae). The Chocó has a dozen family or genus world records for leaf size, especially at Bajo Calima where both large size and extreme sclerophylly characterize the leaves of many unrelated species. Some of these large-leaved taxa range to eastern Panama or the Ecuadorian Chocó, but most seem to be narrow endemics in the unusual phosphorus-lacking white clay of the Bajo Calima area. Species of Bajo Calima with (putatively) the largest leaves in their families include *Iryanthera megistophylla* (Myristicaceae), *Pleurothyrium* new sp. (Lauraceae), *Licania gentryi* (Chrysobalanaceae), *Psittacanthus gigas*, *Ilex* new sp. (with leaves 15–25 cm × c. 10 cm) (Aquifoliaceae), *Protium amplum* (Burseraceae), *Guarea cartaguenya* (Meliaceae), *Schlegelia dressleri* (Bignoniaceae) and *Pentagonia grandiflora* (Rubiaceae).

Chocó forests are poor in wind-dispersal and rich in bird- and mammal-dispersal of fruits and seeds; well over 90% of the woody species are dispersed by these animals (Gentry 1986). Presumably the relatively large seeds (which dispersal by vertebrates made possible) have a strong selective advantage in the highly leached, nutrient-poor soils of the lowland Chocó. Several families and genera have larger seeds in the Chocó than elsewhere. *Sacoglottis ovicarpa* is the largest-fruited Humiriaceae. *Orbignya cuatrecasana* is exceeded in fruit size among palms only by coconut (*Cocos*) and double coconut (*Lodoicea*). Several endemic species of Myristicaceae have the largest fruits in their genera (*Compsoneura*, *Iryanthera*).

The region's flora has strong affinities with the flora of Central America (particularly Panama), which gradually decrease southward (Forero 1982; Gentry 1982; Lellinger and Sota 1978). In general the tree species belong to Amazonian genera, and sometimes the same species occur. The Amazonian affinities within the Chocó also decrease southward (Gentry 1982, 1989).

An interesting peculiarity of the Chocó lowlands is that families usually restricted to the Andean uplands occur here near sea-level (Gentry 1986). Typical montane taxa with lowland representatives are *Podocarpus*, *Talauma*, *Hedyosmum*, *Meliosma*, *Brunellia*, *Panopsis* and *Ilex*. The presence of these generally montane taxa is an indication of the cloud-forest-like nature of the wet Chocó forests.

Collections in the central Chocó from Valle del Cauca Department are rich in endemic species (Gentry 1982). The southern Chocó probably has the entire continent's highest levels of biological endemism (Gentry 1993, pers. comm.). About 20% of the Chocó species are regionally endemic; also, high very local endemism within Chocó areas may be characteristic (Gentry 1989, 1992). This high endemism is the result of active speciation among epiphytes (e.g. Araceae, Orchidaceae, Bromeliaceae), palmettos (e.g. *Heliconia*, *Renealmia*, *Costus*) and shrubs (e.g. Melastomataceae, *Psychotria*). At the generic level the Chocó also is a significant centre of endemism – genera endemic or strongly centred here include *Otoba* (Myristicaceae), *Trianaeopiper* (Piperaceae), *Schlegelia* (Scrophulariaceae) and *Cremosperma* (Gesneriaceae). The high endemism is strong evidence suggesting there was a Pleistocene Chocó refugium (Forero 1982; Gentry 1982).

Useful plants

Some of the unusually large fruits and seeds of Chocó species are also edible for humans. The large mammal-dispersed fruit of "borojó" (*Borojoa patinoi*) has resulted in the recent advent through Colombia of a delicious "refresco". Extractive reserves based on extra-large-fruited species like borojó, that are adapted to flourish on ultra-poor soils under high rainfall, might be a better developmental option than typical practices.

The vegetational resources of this extensive region are used for many purposes, such as food, folk medicine, lumber, fibre, dyes and tannin, and rubber and gums. The following is just a sketch:

❖ **For food**: *Gustavia nana* ("membrillo"), *G. superba* ("pacó") (Lecythidaceae); *Compsoneura atopa* (Myristicaceae); *Euterpe oleracea* (syn. *E. cuatrecasana*) ("naidí"), *Jessenia bataua* (Palmae); and *Borojoa patinoi* (borojó) (Rubiaceae) (Romero-C. 1985).

❖ **For medicinals**: *Amaranthus hybridus* (Amaranthaceae); *Schefflera sphaerocoma* (Araliaceae); *Carica papaya* (Caricaceae); *Dichorisandra angustifolia* (Commelinaceae); *Xiphidium caeruleum* (Haemodoraceae); *Bauhinia glabra* (Leguminosae); *Calathea lutea*, *Stromanthe jacquinii* (Marantaceae); *Triolena spicata* (Melastomataceae); *Mendoncia retusa* (Mendonciaceae); *Piper sternii* (Piperaceae); *Andropogon bicornis* (Poaceae); *Cephaelis* spp., *Hamelia patens*, *Psychotria nervosa*, *Warscewiczia coccinea* (Rubiaceae); *Gloeospermum sphaerocarpum* (Violaceae); and *Costus pulverulentus* (Zingiberaceae) (Forero-Pinto 1980).

❖ **For lumber**: *Anacardium excelsum* ("espavé"), *Campnosperma panamensis* ("sajo"), *Tapirira guianensis* (Anacardiaceae); *Aspidosperma cruentum* (Apocynaceae); *Tabebuia rosea* (Bignoniaceae); *Symphonia globulifera* (Guttiferae); *Sacoglottis procera* (Humiriaceae); *Aniba guianensis* ("chachajo"), *Nectandra* spp. ("jigua") (Lauraceae); *Cariniana pyriformis* ("chibugá") (Lecythidaceae); *Prioria copaifera* ("cativo"), *Schizolobium parahybum* (Leguminosae); *Carapa guianensis* ("tangare") (Meliaceae); *Virola* spp. ("cuángare") (Myristicaceae); *Wettinia quinaria* (Palmae); *Guadua angustifolia* (= *Bambusa guadua*) (Poaceae); and *Genipa americana* (Rubiaceae) (Acero 1977; West 1957).

❖ **In industry or crafts**: *Couma macrocarpa* ("popa") (Apocynaceae); *Castilla elastica* (Moraceae); *Ammandra decasperma* ("antá"), *Astrocaryum standleyanum* ("güérregue"), *Attalea* (= *Ynesa*) *colenda*, *Jessenia polycarpa* ("milpesos"), *Manicaria saccifera* ("jícara"), *Orbignya cuatrecasana* ("táparo") (Palmae); and *Rhizophora* spp. ("mangles rojos") (Rhizophoraceae).

Social and environmental values

The inhabitants of the Pacific coast represent 2.7% of Colombia's population of 33 million. The population generally is sparse, and is 90% blacks (Whitten 1974) with 5% mestizos and 5% indigenous peoples. Most of the inhabitants depend for their livelihoods on subsistence agriculture, hunting, fishing, forestry and mining (Barnes 1993; West 1957). The most important urban areas are Buenaventura, Tumaco and Quibdó (Castiblanco 1990).

The vast Pacific region is rich in mineral and natural resources, which have attracted outsiders since the Spanish conquest. This region is the country's main producer of platinum and second largest producer of gold, which have been exploited especially since the 19th century. There are considerable deposits of bauxite, manganese, tin, zinc, nickel, tungsten, copper and chromium, as well as possible petroleum reserves (Barnes 1993). The forest resources have been over-exploited; the Amerindian groups (Emberá, Waunana, Eperara), blacks and mestizos have not reaped benefit from these exploitations. The great biodiversity is probably what contributes most to the richness of the region.

Birds recorded total 577 species in 353 genera of 67 families; Tyrannidae is the most diverse family with 28 genera and 60 species, and the most speciose genus is *Tangara* (16 tanager spp.) (Roda and Styles 1993). The Chocó region is one of the richest areas of avian endemism in the neotropics; there are two Endemic Bird Areas (EBAs). No fewer than 62 species of restricted range occur within the Chocó and Pacific slope of the Andes EBA (B14), which embraces the lowlands and Andes south of the Serranía de Baudó. A further 14 such species occur in the lowlands of eastern Panama and northern Colombia (north of the Serranía de Baudó) in the Darién and Urabá lowlands EBA (A19), which embraces the entire Atrato drainage; six are restricted to the lowland and foothill forests bordering the serranía. At least 28 of the restricted-range species are confined to lowlands below 800–900 m. Generally widespread forest destruction has caused 18 species to be threatened; three of the endemics are considered threatened but primarily because of poor observer coverage: Chocó tinamou (*Crypturellus kerriae*), speckled antshrike (*Xenornis setifrons*) and Baudó oropendola (*Psarocolius cassini*) – none of them has been recorded in Colombia since the 1940s.

There have been 97 species of reptiles recorded: the suborder Serpentes has 52 – Colubridae is the most diverse family with 25 genera and 35 species, and the most speciose genus is *Micrurus* (7 spp.); the suborder Sauria has 45 species representing 17 genera in six families – Iguanidae is the most diverse family with six genera and 26 species, and the most speciose genus is *Anolis* (20 spp.) (Sánchez and Castaño 1994). There are 127 recorded species of amphibians in 26 genera of seven families; Leptodactylidae is the most diverse family with three genera and 45 species, and the most speciose genus is *Eleutherodactylus* (40 spp.) (Roa and Ruiz 1993).

Threats

The principal threats are indiscriminate felling of trees; expanding permanent agriculture with plantations especially of African oil palm (*Elaeis guineensis*); raising cattle; dredging of rivers; shrimp-farming that deforests the mangroves; excessive direct exploitation of mangroves and palm groves; and road building (cf. Budowski 1990). The current rate of deforestation is 600 km² annually.

The rain forests of the Chocó have been a source of lumber since the beginning of the 20th century; West (1957) found that hardwoods scattered through the region and more concentrated softer woods of swamp forests had already been logged out from most accessible areas; see

also Lellinger and Sota (1972). Deforestation has intensified since 1976. Gentry (1989) reported that most slopes in the Darién Gap on the border with Panama have been cleared. In the Urabá area, only small patches of forest are left after expansion of the production of banana and star-fruit (*Averrhoa carambola*). North of Buenaventura in the Bajo Calima area, commercial paper-pulp operations have clear-cut extensive areas, and considerable land has been deforested in the Tumaco area of Nariño Department. Installation of commercial shrimp farms particularly in the south has caused destruction of mangroves (Barnes 1993). By 1984, over 15,000 km² of coastal forest had been destroyed for domestic use for wood and paper or to make way for agro-industrial production of the African oil palm.

Uncontrolled gold mining has produced massive sedimentation in rivers and contaminated them with mercury. Motorized equipment is eroding riverbanks, causing the riverbeds to drop, and threatening fish stocks and the ability to travel and transport goods between communities (Barnes 1993).

As part of Colombia's development strategy, the Chocó region is coming under increasing pressure (WWF 1993; Barnes 1993). The overall region is managed by regional autonomous corporations: CORPURABA, CODECHOCO, CVC (for the Dept. Valle del Cauca), CRC (for Cauca) and CORPONARIÑO. Forestry research is assisted by the Corporación Nacional de Investigaciones Forestales (CONIF) (Budowski 1990). Development is underway, with new roads promoting an influx of settlers, conversion of primary forests into pasturage and agricultural fields, mining, and new exploitation by timber companies.

Although not being advanced presently, the government in 1984 approved a plan to construct an interoceanic canal (similar to the Panama Canal) by dredging the Atrato and Truandó rivers (Palacios-M. 1985; Ramírez 1967). With this project would come major changes, such as hydroelectric plants and intensification of road building and land conversion (Convers-P. 1985).

Conservation

The four Natural National Parks Los Katíos (720 km²), Utría (543 km², terrestrial and marine), Isla de Gorgona (including Gorgonilla Islet) (492 km², 16 km² terrestrial) and Sanquianga (800 km²) protect 2013 km² of the region's terrain, which also includes the 200–1000 m portion of Farallones de Cali NNP (totalling 1500 km²) and 500–1000 m portion of Munchique NNP (440 km²) (INDERENA 1984; Sánchez-P. *et al.* 1990). The relatively new (1987) Utría Natural National Park (Chávez 1988) is receiving fundamental help through the Parks in Peril Program of The Nature Conservancy (U.S.) and Fundación Natura Colombia working with INDERENA (the Instituto Nacional de los Recursos Naturales Renovables y del Medio Ambiente) and the local communities (FNC 1993).

In a 1985 meeting in Bogotá (organized by WWF-U.S.), national and international experts assessed conservation priorities for the Chocó region, evaluating established reserves and recommending new ones. The recommendations and priorities were formulated to be used in the region's future development and management (García-Kirkbride 1986a, 1986b; Hernández-Camacho 1984; Gentry 1989). The government in 1993 initiated a 6-year Proyecto Biopacífico to fuse a strategy for conservation and sustainable management of the region's biodiversity (INDERENA 1993).

Additional areas need to be delineated and their protection promoted (cf. INDERENA 1993), e.g. (1) in the northernmost zone of the Chocó beyond the banks of the Atrato River; (2) an area along the Atrato riverbanks that includes the plant communities "manglares, natales, sajales"; (3) between the Atrato, Baudó and San Juan rivers; (4) in Valle del Cauca Department between the Calima River and Buenaventura (cf. Faber-Langendoen and Gentry 1991); and (5) Isla de Gallo in Tumaco Cove. The central Chocó, except where accessible from the Atrato and San Juan rivers or the expanding road system around Quibdó, currently remains covered by essentially intact forest. Nariño Department needs to receive special consideration to preserve representative areas of its biotic conditions.

References

Acero, L. (1977). *Estudio dendrológico de algunas especies del Departamento del Chocó.* Tesis de grado. Universidad Francisco José de Caldas, Bogotá.

Acosta-Solís, M. (1970). La selva del noroccidente ecuatoriano. *Rev. Acad. Col. Cienc. Exactas Fís. Nat.* 13: 499–533.

Aguirre-C., J. and Rangel-Ch., J.O. (eds) (1990). *Biota y ecosistemas de Gorgona.* Fondo FEN-Colombia, Bogotá. 303 pp.

Barnes, J. (1993). Driving roads through land rights. The Colombian Plan Pacífico. *The Ecologist* 23(4): 135–140.

Budowski, G. (1990). Desarrollo sostenible: el caso de la Provincia Fitogeográfica del Chocó. In INDERENA, *ECOBIOS, Colombia 88. El desarrollo sostenible: estrategias, políticas y acciones.* Septiembre 20–23, 1988. Memorias del Simposio Internacional, Bogotá, Colombia. Ministerio de Agricultura, INDERENA, Bogotá. Pp. 152–168.

Castiblanco, A. (1990). Ocupación de la región del Pacífico. *Colombia, Sus Gentes y Regiones, Rev. Instituto Geográfico Agustín Codazzi* 18: 98–113.

Chávez, M.E. (1988). Utría: the bay of beauty becomes a national park. *The Nature Conservancy (TNC) Internatl. News* (Spr.): 1, 10.

Convers-P., R. (1985). El canal interoceánico a nivel Atrato-Truandó y las hidroeléctricas del Atrato pueden cambiar la suerte del país. *Bol. Soc. Geogr. Col.* 37(119–120): 23–34.

Cuatrecasas, J. (1958). Aspectos de la vegetación natural de Colombia. *Rev. Acad. Col. Cienc. Exactas Fís. Nat.* 10: 221–268.

Faber-Langendoen, D. and Gentry, A.H. (1991). The structure and diversity of rain forests at Bajo Calima, Chocó region, western Colombia. *Biotropica* 23: 2–11.

FNC (1993). *Fundación Natura Colombia: annual report 1992.* Fundación Natura Colombia (FNC), Bogotá. 20 pp.

Forero, E. (1982). La flora y la vegetación del Chocó y sus relaciones fitogeográficas. *Colombia Geográfica, Rev. Inst. Geogr. Agustín Codazzi* 10: 77–90.

Forero, E. (1985). Estado actual del conocimiento de la vegetación y la flora del Chocó. In D'Arcy, W.G. and Correa-A., M.D. (eds), *The botany and natural history of Panama/La botánica e historia natural de Panamá.* Monogr. Syst. Bot. 10, Missouri Botanical Garden, St. Louis. Pp. 185–191.

Forero, E. and Gentry, A.H. (1989). *Lista anotada de las plantas del Departamento del Chocó, Colombia.* Biblioteca J.J. Triana No. 10. Instituto de Ciencias Naturales, Museo de Historia Natural, Universidad Nacional de Colombia, Bogotá. 142 pp.

Forero-Pinto, L.E. (1980). Etnobotánica de las comunidades indígenas Cuna y Waunana (Chocó, Colombia). *Cespedesia* 9: 115–301.

García-Kirkbride, M.C. (1986a). *Biological evaluation of the Chocó Biogeographic Region in Colombia.* WWF-U.S., Washington, D.C. 61 pp.

García-Kirkbride, M.C. (1986b). Conserving the Chocó. *Threatened Plants Newsletter* 17: 4–7.

Gentry, A.H. (1982). Phytogeographic patterns as evidence for a Chocó refuge. In Prance, G.T. (ed.), *Biological diversification in the tropics.* Columbia University Press, New York. Pp. 112–136.

Gentry, A.H. (1986). Species richness and floristic composition of Chocó region plant communities. *Caldasia* 15: 71–91.

Gentry, A.H. (1989). Northwest South America (Colombia, Ecuador and Peru). In Campbell, D.G. and Hammond, H.D. (eds), *Floristic inventory of tropical countries: the status of plant systematics, collections, and vegetation, plus recommendations for the future.* New York Botanical Garden, Bronx. Pp. 391–400.

Gentry, A.H. (1992). Tropical forest biodiversity: distributional patterns and their conservational significance. *Oikos* 63: 19–28.

Gentry, A.H. (1993). Riqueza de especies y composición florística. In Leyva-F., P. (ed.), *Colombia Pacífico,* Vol. 1. Fondo Protección del Medio Ambiente José Celestino Mutis, Publicaciones Financiera Eléctrica Nacional (FEN), Santafé de Bogotá. Pp. 200–219.

Hernández-Camacho, J.I. (1984). *Vistazo general sobre la protección de la naturaleza en Colombia.* INDERENA, Bogotá. 75 pp.

INDERENA (1984). *Colombia parques nacionales.* Instituto Nacional de los Recursos Naturales Renovables y del Medio Ambiente (INDERENA), Bogotá. 263 pp.

INDERENA (1993). *Conservación de la biodiversidad del Chocó Biogeográfico. Proyecto Biopacífico. Plan operativo.* INDERENA, Bogotá. 135 pp.

Lellinger, D.B. (1975). A phytogeographic analysis of Chocó pteridophytes. *Fern Gaz.* 11: 105–114.

Lellinger, D.B. and Sota, E.R. de la (1972). Collecting ferns in the Chocó, Colombia. *Amer. Fern J.* 62: 1–8.

Lellinger, D.B. and Sota, E.R. de la (1978). The phytogeography of the pteridophytes of the Departamento del Chocó, Colombia. *Natl. Geogr. Soc. Res. Rep.*, 1969 projects: 381–387.

Murillo, M.T. and Lozano, G. (1989). Hacia una Flórula del Parque Nacional Natural Islas de Gorgona y Gorgonilla (Cauca, Colombia). *Rev. Acad. Col. Cienc. Exactas Fís. Nat.* 17: 277–403.

Palacios-M., D. (1985). El canal Atrato-Truandó ordenado por la Ley 53 de 1984. *Bol. Soc. Geogr. Col.* 37(119–120): 15–19.

Ramírez, J.E. (1967). Proyecto del canal interoceánico Atrato-Truandó (Colombia) a nivel del mar. *Bol. Soc. Geogr. Col.* 25(95–96): 110–119.

Rangel-Ch., J.O. and Aguilar, M. (1993). Clima de la Costa Pacífica. In Rangel-Ch., J.O. (ed.), *Informe Proyecto Estudio de la Biodiversidad de Colombia.* Convenio INDERENA – Universidad Nacional de Colombia, Bogotá. Internal document.

Rangel-Ch., J.O. and Lowy, P. (1993). Tipos de vegetación y rasgos fitogeográficos en la Región Pacífica de Colombia. In Leyva-F., P. (ed.), *Colombia Pacífico*, Vol. 1. Fondo Protección del Medio Ambiente José Celestino Mutis, Publicaciones FEN, Santafé de Bogotá. Pp. 182–198.

Rangel-Ch., J.O., Castaño, O., Stiles, F.G., Ruiz, P., Lowy, P., Garzón-C., A., Sánchez, H. and Aguilar, M. (1994). Diagnóstico inicial de la biodiversidad en Colombia. *Resúmenes del IV Congreso Colombiano de Ecología.* Sociedad Colombiana de Ecología. Melgar, Colombia. Pp. 37–38.

Roa, S. and Ruiz, P. (1993). Anfibios. In Rangel-Ch., J.O. (ed.), *Informe Proyecto Estudio de la Biodiversidad de Colombia.* Convenio INDERENA – Universidad Nacional de Colombia, Bogotá. Internal document.

Roda, J. and Stiles, G. (1993). Aves. In Rangel-Ch., J.O. (ed.), *Informe Proyecto Estudio de la Biodiversidad de Colombia.* Convenio INDERENA – Universidad Nacional de Colombia, Bogotá. Internal document.

Romero-Castañeda, R. (1985). *Frutas silvestres del Chocó.* Instituto Colombiano de Cultura Hispánica, Bogotá. 122 pp.

Sánchez, H. and Castaño, O. (1994). La biodiversidad de los reptiles en Colombia. In Rangel-Ch., J.O. (ed.), *Informe Proyecto Estudio de la Biodiversidad de Colombia.* Convenio INDERENA – Universidad Nacional de Colombia, Bogotá. Internal document.

Sánchez-P., H., Hernández-C., J.I., Rodríguez-M., J.V. and Castaño-U., C. (1990). *Nuevos parques nacionales Colombia.* INDERENA, Bogotá. 213 pp.

Sota, E.R. de la (1972). Las pteridófitas y el epifitismo en el Departamento del Chocó (Colombia). *Anales Soc. Cient. Argentina* 194: 245–278.

Von Prahl, H., Guhl, F. and Grogl, M. (1979). *Gorgona.* Grupo Futura Editorial, Bogotá. 279 pp.

West, R.C. (1957). *The Pacific lowlands of Colombia: a Negroid area of the American tropics.* Louisiana State Univ. Studies, Soc. Sci. Ser. No. 8. Louisiana State University Press, Baton Rouge. 278 pp.

Whitten Jr., N.E. (1974). *Black frontiersmen: a South American case.* John Wiley and Sons, New York. 221 pp.

WWF (1993). Colombia's biologically rich Chocó forest faces increasing threats. *Focus* 15(2): 1, 6.

Zuluaga-R., S. (1987). Observaciones fitoecológicas en el Darién colombiano. *Pérez-Arbelaezia* 1: 85–145.

Authors

This Data Sheet was written by Olga Herrera-MacBryde (Smithsonian Institution, Department of Botany, NHB-166, Washington, DC 20560, U.S.A.), Orlando Rangel-Ch., Mauricio Aguilar-P., Hernán Sánchez-C., Petter Lowy-C., David Cuartas-Ch. and A. Garzón-C. (Universidad Nacional de Colombia, Instituto de Ciencias Naturales, Museo de Historia Natural, Apartado Aéreo 7495, Santafé de Bogotá, Colombia).

ECUADORIAN PACIFIC COAST MESIC FORESTS
Ecuador

Location: North-western Ecuador's Pacific coastal range, coastal plain and Andean flanks up to 900 m, within latitudes 1°25'N–2°00'(–3°30')S and longitudes 80°00'–78°40'W.

Area: 3700 km² of forest remaining from 52,000 km².

Altitude: 0–900 m.

Vegetation: Tropical lowland moist, wet and pluvial forests.

Flora: c. 5300 species – high diversity particularly in epiphytes; high endemism – c. 1060 species; threatened species.

Useful plants: Timber species, ivory-nut palm, ornamentals.

Other values: Homelands of Awá and Chachi indigenous peoples; potential genetic resources.

Threats: Logging, agriculture, grazing, colonization.

Conservation: Awá Ethnic Forest Reserve (1300 km²); Cotacachi-Capayas Ecological Reserve (2040 km²); ENDESA company reserves (c. 3 km²); Río Palenque Science Center (1 km²); Jauneche Reserve (1.30 km²).

Geography

The Pacific coastal region of Ecuador is mainly a relatively broad plain (30–70 km wide) west of the Andes. This region is drained by two major river systems: the Esmeraldas River in the north, and the Guayas River system in the central portion, which drains southward. West of the coastal plain and parallel to the Andes is the Cordillera de la Costa, with peaks over 800 m.

Climatically, the coastal region is characterized by a steep gradient of annual precipitation. The wettest area, near the Colombian border in the extreme north, receives more than 8000 mm of rainfall annually. Due to the influence of the marine Humboldt Current off the coast of South America, the coastal Ecuadorian climate is progressively drier to the south and west. In the south-west, the Santa Elena Peninsula receives less than 100 mm of rain annually and is classified as tropical desert (Cañadas-C. 1983).

This site description includes the relatively moist to pluvial northern and central parts of coastal Ecuador with annual rainfall of 2000 mm or more (zonal maps are in Dodson and Gentry 1991). This mesic region (Map 75) is in the coastal range, coastal plain and western slopes of the Andes to 900 m elevation. Pluvial forest, the wettest zone in the Holdridge life-zone system, covered a relatively small area (8000 km²) in extreme northern Ecuador along the base of the Andes. Wet forest occupied 12,000 km² in a zone broad in the north that tapers in southern Ecuador to a band less than 1 km wide at the base of the Andes, and also occurs in small isolated patches on higher parts of the coastal range. South and west of the wet-forest life zone, moist forest occupied a large area (32,000 km²) of the plain and coastal range. To the south and west of the moist-forest life zone is dry forest, with annual rainfall less than 2000 mm and a strong dry season of more than four months. (The Ecuadorian coastal dry forest is not included in this Data Sheet – this formation extends southward into Peru's Tumbes Province and is described in the Data Sheet on Cerros de Amotape National Park region, CPD Site SA41.)

Vegetation

The once continuous lowland wetter forests of coastal Ecuador have been fragmented and reduced to a few remnants (Dodson and Gentry 1991). The remaining patches of intact forest are quite distinct from one another in structure and species composition. Certainly the original forests varied considerably, depending upon local soils and climatic conditions. Very little quantitative information is available on the vegetation of this region. Botanists from the Herbario Nacional del Ecuador and Missouri Botanical Garden established a series of permanent 1-ha study plots in 1993 in forests of the Cotacachi-Cayapas Ecological Reserve, which can provide data for comparison with similar plots established through the neotropics (Gentry 1988; Campbell 1989).

The vegetation (cf. Neill 1992) on a forest remnant of 2.00 km² maintained as a reserve by the timber company ENDESA (Enchapes Decorativos) on its Pitzara River lands (0°20'N, 79°20'W) just south of the Guayllabamba River probably is rather typical of the forest that once covered this region. The canopy is c. 40 m high and fairly continuous, with few gaps. The absence of gaps and the relative abundance of large trees (dbh 70 cm or more) are striking features in comparison with Amazonian Ecuador's forest, where gaps appear to be more frequent and there are fewer large trees.

Among the large and common canopy tree species in primary forest are *Brosimum utile*, *Guarea kunthiana*, *Carapa guianensis*, *Dacryodes* sp., *Pouteria capacifolia*, *Virola dixonii* and *Huberodendron patinoi*. Occasional emergent trees over 60 m tall are present, e.g. the strangler fig *Ficus dugandii*. The subcanopy is dominated by two species of very abundant palms, the stilt-rooted *Iriartea deltoidea* and *Wettinia quinaria*; also very common are several species of *Matisia* (Bombacaceae). The understorey is dense and composed of many species of Rubiaceae and small palms, mostly *Geonoma* species. Covering the lower trunks of most of the forest's trees are diverse and very abundant epiphytic Araceae, Cyclanthaceae and ferns.

Flora

Although there are several partial floristic studies for western Ecuador, floristically the region remains quite poorly known. An illustrated guide to common trees of Esmeraldas Province was prepared (Little and Dixon 1969) as part of a forestry development project in the 1960s supported by the United Nations Food and Agriculture Organization (FAO). For the Río Palenque Science Center's Forest Reserve (1 km²) in the tropical wet forest of the Guayas River Basin, Dodson and Gentry (1978) wrote an illustrated Florula. Farther south, for the Jauneche Reserve (1.30 km²) in the tropical moist forest, Dodson, Gentry and Valverde (1986) published a comparable Florula. Considering the 283 genera found at one or both sites, the generic similarity between the latter two sites is fairly high at 0.59, using the Sørensen coefficient (Dillon 1994).

Botanists from the Pontificia Universidad Católica del Ecuador, Quito and the University of Aarhus, Denmark have made floristic studies in a Forest Reserve (0.80 km²) of the ENDESA timber company in western Pichincha Province (Jørgensen and Ulloa 1989). The Herbario Nacional del Ecuador and Missouri Botanical Garden are carrying out intensive floristic inventories in Awá Ethnic Forest Reserve and Cotacachi-Cayapas Ecological Reserve.

The Río Palenque Science Center (0°30'S, 79°20'W), between Quevedo and Santo Domingo de los Colorados, is the most thoroughly studied site in coastal Ecuador. At Río Palenque is the only remaining primary forest on the fertile soils of the Upper Guayas Basin – a fragment of just 1 km². Over 1100 species in 123 families have been recorded at Río Palenque; 20% of the species are local endemics, and 23 are known only from this reserve (Dodson and Gentry 1978).

Non-tree species – including epiphytes, lianas, shrubs and terrestrial herbs – are a very significant portion of the overall diversity of the plants in tropical forests. In the forests of coastal Ecuador, epiphytes are a particularly species-rich element of the floras. At Río Palenque, a 0.1-ha sample had 365 vascular plant species, the world's most species-rich site studied with the method: the species were 35% epiphytes (including hemi-epiphytes); 31% trees (including all juveniles), 24% with dbh 2.5–10 cm, 9% with dbh 10 cm or more; 14% terrestrial herbs (including palmettos); 11% shrubs; and 10% climbers (excluding the hemi-epiphytes) (Gentry and Dodson 1987).

In contrast to the very high diversity of epiphytes in these forests, the diversity of trees and shrubs is relatively low. Gentry (1992) carried out a series of sample transects in western Ecuadorian forests, recording plants with dbh 2.5 cm or more in 0.1-ha plots. The three moist-forest sites had a mean of 96 species and the four wet-forest sites a mean of 125 species. Neotropical lowland moist/wet forest averages 152 species in equivalent samples (Gentry 1988).

Another method of estimating α diversity in tropical forests involves enumerating all trees with dbh 10 cm or more in 1-ha sample plots (Campbell 1989). In Upper Amazonia and in the Colombian Chocó region north of coastal Ecuador (CPD Site SA39), 250–300 species of trees may occur in such plots (Gentry 1988; Faber-Langendoen and Gentry 1991). In western Ecuador, preliminary information indicates that the number of tree species probably exceeds 150 per ha (Palacios and Neill, in prep.). With respect to the diversity of tree species, the wet forests of western Ecuador may be more similar to those of Central America than Upper Amazonia.

The extant overall floristic diversity of coastal Ecuador will not be known unless more thorough inventories are carried out in all its vegetation types (cf. Dillon 1994). Dodson and Gentry (1991) estimated that 6300 vascular plant species occur in coastal Ecuador including the lower western slopes of the Andes up to 900 m, with 20% endemic (1260 species). About 1000 of the species occur only in dry forests, so the estimated total for moist, wet and pluvial forests is 5300 species, with 20% endemic – 1060 species. Although the α diversity of trees in this region is low compared to the diversity in Upper Amazonia, the broad range of habitats does support a very diverse regional flora with a high degree of endemism.

Gentry (1992) provided a concise phytogeographic overview of coastal Ecuador based on information that became available in the preceding several years: "Coastal Ecuador is of conservation significance for its high plant endemism ... What has not been appreciated previously is that the unusual high endemism of western Ecuador moist and wet forests is associated with relatively species-poor forests ... The pattern that results is a new and interesting one that focuses on western Ecuador (i.e., south of the town of Esmeraldas) as a unique and distinctive floristic region for wet and moist forest as well as for dry-forest vegetation, rather than as the tail-end of the [Colombian] Chocó flora as I had previously interpreted it ... This forest is characterized by low species diversity of trees (but high diversity of epiphytes), high endemism, a predominance of hemi-epiphytic climbers (also characteristic of the Chocó), unusually low levels of such characteristic taxa as Bignoniaceae and Leguminosae, and high levels of Araceae, Piperaceae, Moraceae and Cucurbitaceae".

Useful plants

North-western Ecuador and especially the lowland forests of Esmeraldas Province have provided almost all timber for the country for the last several decades. One of the most important trees, a prime timber for construction in Quito and other Ecuadorian cities, has been the locally endemic *Humiriastrum procerum* (Humiriaceae), which occurs only in wet forests north of the Guayllabamba River. The supply of this hardwood has been severely depleted by logging and the natural regeneration of this slow-growing species in logged-over forests appears to be very poor. Other trees important to the timber industry include soft-wooded species used for plywood, such as *Brosimum utile*, *Dacryodes* sp. and *Virola dixonii*.

One of the most characteristic plants of coastal Ecuador is the endemic ivory-nut palm or "tagua" *Palandra aequatorialis* (*Phytelephas aequatorialis*). This small tree (to 15 m) is common in the understorey of the natural forests throughout the region. The mature seed's endosperm is like ivory – white and hard. In the 1920s–1930s before the invention of synthetic plastics, vast quantities of ivory nuts were exported from Ecuador to the U.S.A. and Europe where they were used by the garment industry to manufacture buttons. Recently, with the worldwide CITES ban on elephant ivory, there has been renewed commercial interest in this vegetable ivory. Conservation International is carrying out the Tagua Initiative project to promote sustainable harvest of the ivory nuts in rural communities in Esmeraldas Province as a means of protecting the forests in which this species occurs (Calero-H. 1992; Ziffer 1992).

The flora of coastal Ecuador includes many plant species with potential as ornamentals, especially the diverse epiphytic

MAP 75. ECUADORIAN PACIFIC COAST MESIC FORESTS (CPD SITE SA40)

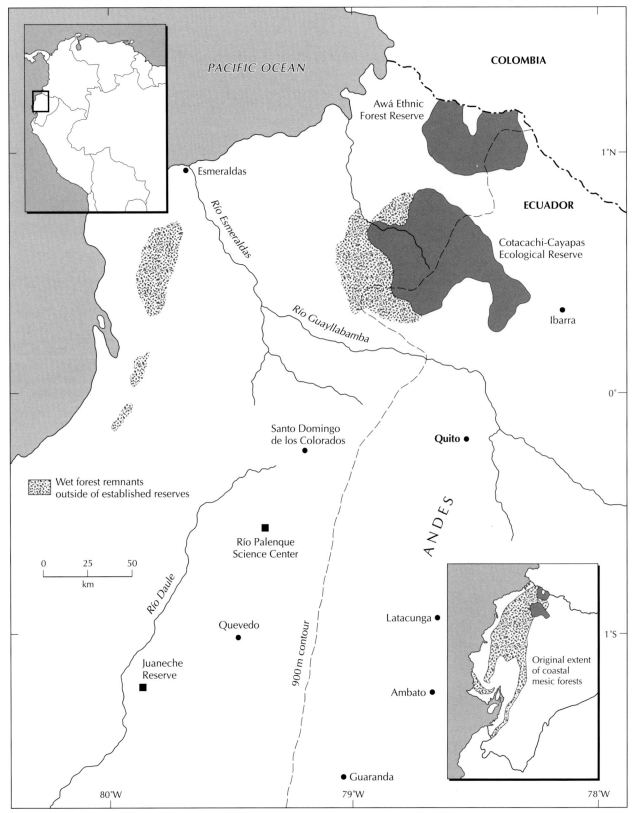

taxa. *Anthurium andraeanum*, one of the world's most widely cultivated house-plants, is native to the north-western corner of Ecuador as well as the southern Colombian Chocó. Numerous endemic species of Gesneriaceae in coastal Ecuador have potential as ornamentals, but relatively few have been brought into cultivation.

Social and environmental values

The wet forests of north-western Ecuador are home to two indigenous ethnic groups, which practise traditional subsistence agriculture and hunting and gathering in relatively large tracts of intact forest. The Awá (Coaquier) live on both sides of the Colombia-Ecuador border. In Ecuador, c. 2200 Awá inhabit the Awá Ethnic Forest Reserve (1300 km²) (Map 75). They clear only small patches for cultivation of plantains (*Musa × paradisiaca*) and other subsistence crops; their ethnobotany is being studied.

Further south, in the area around the Cayapas River, live the Chachi (Cayapa). About 3000 Chachis occupy 300 km² of lowland wet forest just west of the Cotacachi-Cayapas Ecological Reserve.

Both the Awá and the Chachi require large tracts of undisturbed forest to continue their traditional life-styles. These indigenous peoples should have an increasingly important role in conservation of these remaining moist and wet forests.

In the north of this region (southward to northern Guayas), bird species of restricted range from the Chocó and Pacific slope of the Andes Endemic Bird Area (EBA B14) are present in the wet and moist forests; c. 37 of 62 restricted-range species occur. Farther south in the moist and dry forests, species from the Tumbesian western Ecuador and Peru EBA (B20) predominate, with up to 55 restricted-range species present. Currently 30 species from these EBAs are considered threatened.

Threats

As early as the 1600s, the forests of western Ecuador were logged to provide wood for the ship-building industry that had started on the coast of Ecuador. Later cacao plantations replaced much of the forests.

Acosta-Solís (1947) mentions that original forest was only to be found in remote locations. No known reports suggest to what extent those forests once existed, but it is likely that they covered large portions of the country. He also suggested that only five regions were worth logging and each of them contained some wet forest. Plantations of fast-growing species to be used for paper pulp were recommended to replace the cleared native forest.

Until 1960, transportation into the wet forests was mainly by canoe using the extensive system of rivers throughout the forest. Since then roads have been constructed and improved, and under Ecuador's colonization law, people were allowed to clear forested areas and plant crops. The most prevalent cash crops are plantations of oil palms and bananas. By 1970, virtually all of the moist forest and much of the wet forest had been lost. Myers (1988) pointed out that due to this region's species diversity and high endemism, it is of considerable scientific interest. With the

increased and rapid destruction of habitats, he ranked the Ecuadorian coastal wet forest as one of the world's areas most in need of protection.

Logging and colonization continue in the region, mostly in remote areas of Esmeraldas Province in the north. The largest continuous tract of forest remaining in coastal Ecuador outside of established reserves is the lands (300 km²) of the Chachi Amerindians along the Cayapas River. Ecuadorian timber companies are attempting to make agreements with the Chachis for the sale of timber rights on these lands. Conservation efforts such as the programme SUBIR (Sustainable Use of Biological Resources) are attempting to collaborate with the Chachi to develop long-term, sustainable management plans for their forests.

Conservation

The largest conservation unit in the region is the Cotacachi-Cayapas Ecological Reserve (2040 km²), established in 1970 and accorded permanent status in 1979. Cotacachi-Cayapas preserves the greatest diversity of ecological regions in the country, ranging from high-Andean páramos and cloud forests on the western slopes of the Andes down to coastal forests at 200 m elevation (SFRNR 1991). About 40% of the reserve (c. 800 km²) is below 900 m in the lowland forest zone (Map 75).

In 1992, the large-scale conservation programme SUBIR was initiated in the region around the Cotacachi-Cayapas Ecological Reserve, as well as for two other protected areas in Ecuador. SUBIR's goals are to promote conservation by increasing the capacity of Ecuadorian agencies to protect the core natural areas, as well as by non-destructive use of the natural resources by the people living in buffer zones around the protected areas. (For more on the SUBIR programme, see the Data Sheet on Yasuní National Park and Waorani Ethnic Reserve, CPD Site SA8.)

The Awá Ethnic Forest Reserve (1300 km²) is another important conservation unit in western Ecuador; c. 800 km² are below 900 m (Map 75). This reserve was established as part of a binational agreement of Colombia and Ecuador; additional territory has been set aside for the Awá on the Colombian side of the border. Both governments are participating with Awá indigenous organizations in carrying out conservation as well as rural development programmes in this region. Most of the Awá reserve is accessible only by footpaths.

The ENDESA timber company has 50 km² of land in western Ecuador. Most was purchased from colonists who had cut the timber and grown crops or established pastures; the company has reforested the land for industrial timber plantations to produce plywood. ENDESA's holdings include c. 3 km² of primary forests, which it has promised to maintain as reserves. The ENDESA forests are virtually the only remaining undisturbed forests in lowland Pichincha Province, north of the city of Santo Domingo de los Colorados.

South of the Equator in coastal Ecuador there are very few patches of intact forest remaining, and they are very small. The forest of 1 km² at the privately owned Río Palenque Science Center is all that remains of primary wet forest in the Guayas River Basin. Since the 1960s the forest surrounding the field station has been converted into plantations of oil palms and bananas. Similarly, the forest of 1.30 km² at the

Jauneche Reserve, which is owned by the Universidad de Guayaquil, is virtually all that remains of primary moist forest in the Guayas Basin.

Some larger fragments of wet and moist forests still exist in as yet inaccessible areas of the coastal range, west of the Guayas Basin and south of the city of Esmeraldas (Map 75). However, these remnants are rapidly disappearing as colonists cut the forests higher and higher on the slopes. There is as yet no formal protection for forest remnants in the coastal range.

References

Acosta-Solís, M. (1947). Commercial possibilities of the forests of Ecuador – mainly Esmeraldas Province. *Tropical Woods* 89: 1–47.

Calero-H., R. (1992). The Tagua Initiative in Ecuador: a community approach to tropical rain forest conservation and development. In Plotkin, M. and Famolare, L. (eds), *Sustainable harvest and marketing of rain forest products.* Island Press, Washington, D.C. Pp. 263–273.

Campbell, D.G. (1989). Quantitative inventory of tropical forests. In Campbell, D.G. and Hammond, H.D. (eds), *Floristic inventory of tropical countries: the status of plant systematics, collections, and vegetation, plus recommendations for the future.* New York Botanical Garden, Bronx. Pp. 524–533.

Cañadas-C., L. (1983). *El mapa bioclimático y ecológico del Ecuador.* Ministerio de Agricultura y Ganadería, Programa Nacional de Regionalización (MAG-PRONAREG), Quito. 210 pp.

Dillon, M.O. (1994). Bosques húmedos del norte del Perú. *Arnaldoa* 2(1): 29–42.

Dodson, C.H. and Gentry, A.H. (1978). *Flora of the Río Palenque Science Center, Los Ríos, Ecuador. Selbyana* 4: 1–628.

Dodson, C.H. and Gentry, A.H. (1991). Biological extinction in western Ecuador. *Ann. Missouri Bot. Gard.* 78: 273–295.

Dodson, C.H., Gentry, A.H. and Valverde, F. de M. (1986). *La Flora de Jauneche, Los Ríos, Ecuador. Selbyana* 8: 1–512.

Faber-Langendoen, D. and Gentry, A.H. (1991). The structure and diversity of rain forests at Bajo Calima, Chocó region, western Colombia. *Biotropica* 23: 2–11.

Gentry, A.H. (1988). Changes in plant community diversity and floristic composition on geographical and environmental gradients. *Ann. Missouri Bot. Gard.* 75: 1–34.

Gentry, A.H. (1992). Phytogeographic overview. In Parker III, T.A. and Carr, J.L. (eds), *Status of forest remnants in the Cordillera de la Costa and adjacent areas of southwestern Ecuador.* RAP Working Paper 2, Conservation International, Washington, D.C. Pp. 56–58.

Gentry, A.H. and Dodson, C.H. (1987). Contribution of nontrees to species richness of a tropical rain forest. *Biotropica* 19: 149–156.

Jørgensen, P.M. and Ulloa-U., C. (1989). *Estudios botánicos en la "Reserva ENDESA" Pichincha, Ecuador.* Botanical Institute, University of Aarhus, Denmark. AAU Reports 22. 216 pp.

Little Jr., E.L. and Dixon, R.G. (1969). *Arboles comunes de la Provincia de Esmeraldas, Ecuador.* Estudio de preinversión para el desarrollo forestal del noroccidente, Ecuador; Informe final, Tomo IV. FAO/SF: 76/ECU 13. United Nations Development Program, Food and Agriculture Organization (FAO), Rome. 536 pp.

Myers, N. (1988). Threatened biotas: hot-spots in tropical forests. *The Environmentalist* 8: 187–208.

Neill, D.A. (1992). *ENDESA/BOTROSA [Enchapes Decorativos, S.A./Bosques Tropicales, S.A.] reforestation project, Ecuador: study of project's impact on regional plant diversity.* Herbario Nacional del Ecuador, Quito. 13 pp. Unpublished report.

SFRNR (1991). Reserva Ecológica Cotacachi-Cayapas. In *Sistema Nacional de Areas Protegidas y la vida silvestre del Ecuador.* MAG, Subsecretaría Forestal y de Recursos Naturales Renovables (SFRNR) and U.S. Fish and Wildlife Service, Quito. Pp. 32–34.

Ziffer, K. (1992). The Tagua Initiative: building the market for a rain forest product. In Plotkin, M. and Famolare, L. (eds), *Sustainable harvest and marketing of rain forest products.* Island Press, Washington, D.C. Pp. 274–279.

Author

This Data Sheet was written by Dr David A. Neill (Missouri Botanical Garden, P.O. Box 299, St. Louis, MO 63166-0299, U.S.A. and Herbario Nacional del Ecuador, Casilla 17-12-867, Quito, Ecuador).

CERROS DE AMOTAPE NATIONAL PARK REGION
North-western Peru

Location: In Grau region bordering Ecuador within Tumbes and Piura departments, between about latitudes 3°30'–4°30'S and longitudes 81°–80°W.

Area: Region c. 2314 km², park 913 km².

Altitude: 100–1618 m.

Vegetation: Dry equatorial forest with five life zones: very dry and premontane dry tropical forests, lowland and premontane tropical thorn forests, premontane-to-low tropical desert matorral.

Flora: Recorded so far are 510 species of vascular plants in National Forest, and 404 spp. including c. 80 forest spp. in National Park; high diversity and endemism in herbaceous stratum; threatened species.

Useful plants: Forest resources include timber, fuelwood, charcoal, materials for crafts, forage, medicinal and ornamental purposes.

Other values: Refuge for fauna, including many threatened spp. – especially birds; watershed protection; potential germplasm resources; archaeological and fossil sites, interesting landscapes, research, education, tourism.

Threats: Access by several roads, timber and fuelwood extraction, poaching, overgrazing, subsistence agriculture, set fires, soil erosion, desertification, potential dam construction.

Conservation: Tumbes National Forest (751 km²), Cerros de Amotape National Park (913 km²) and El Angolo Game Preserve (650 km²) together form Noroeste Peruano Biosphere Reserve (2314 km²).

Geography

Cerros de Amotape National Park is south of the Gulf of Guayaquil in extreme north-western Peru (3°46'–4°20'S, 80°50'–80°20'W), in the Grau region of Tumbes Department (Tumbes and Contralmirante Villar provinces) and Piura Department (Sullana Province). The NP is the core of Noroeste Peruano (Peruvian North-west) Biosphere Reserve, which also includes the bracketing Tumbes National Forest and El Angolo Game Preserve (WWF-FPCN 1989) (Map 76). The BR is between latitudes 3°28'–4°27'S and longitudes 80°58'–80°08'W, encompassing the largest conserved remnant of dry equatorial forest (cf. Kessler 1992) – which mainly extends over a physiographic region composed of an extensive span of desert coast, and the lower slopes of the western flank of the Cordillera de los Andes. Here a transverse tectonic mega-shear, the Amotape Cross, separates the Northern Andes and the Central Andes.

The Pacific coast has a flat or lightly undulated topography and inland contains three predominant Cenozoic marine terraces ("tablazos") now elevated 30–300 m and much dissected – Máncora, Talara and Lobitos, which successively formed steep slopes at the sea. There is a narrow littoral fringe between the Chira River and southern portion of Tumbes Department. All across the region from Máncora to Matapalo, large expanses are covered by rolling hills. North of Zorritos along the coast is an extremely arid alluvial plain that extends north-eastward to the border with Ecuador (WWF-FPCN 1989; Clapperton 1993).

The Andean Cordillera has its western limit where the topography changes from undulated to flat along spurs of the cordillera (e.g. Cordillera Larga) and the Cerros de Amotape (Best 1992). The Amotapes, extending from northern Piura Department into eastern Tumbes Department, form a low cordillera parallel to the coast. The topography is rugged, with deep canyons, gullies and rough uneven terrain, which reaches 1534 m in the park at Cerro La Concha (cf. WWF-FPCN 1989).

The south-eastern portion of the NP is above 1400 m, especially the small massif shared with El Angolo Game Preserve, with the cerros El Viento, Los Antiguos, El Padre and Carrizal (1618 m), Machete, El Perro, El Barco and Cocinas. The NP's eastern flank is below 800 m, where the Cuzco and Cazaderos gorges ("quebradas") are located. The northern portion of the NP descends to 200–100 m and the bordering Tumbes River, which originates eastward (as the Puyango River) in the Ecuadorian Andes (Best 1992).

The Tumbes River is the only permanent water source; in the Quebrada Cuzco water lasts almost year-round, and the extensive system of seasonal creeks includes Bocapán, Casitas, Seca and Fernández. The soils, which formed from Tertiary and Quaternary sediments, drain poorly. There is a strong correlation between topography and soil type and formation. In low flat areas, the soils are deep and of medium texture; often they contain high salt concentrations. On terraces and rolling terrain, the soils are shallower and less saline. The soils belong to the vertisol and yermosol groups; mainly vertisols sustain the forest biomass.

The climate is transitional between the desert climate of more southern coastal Peru (cf. CPD Site SA42) and the subhumid tropical climate of Ecuador (cf. CPD Site SA40). Given the NP's proximity to the Equator, a warm and humid climate might be expected, with high precipitation. However the climate is modified by several factors – the cold marine Humboldt Current and the Andean Cordillera impose an environment that is predominantly subarid, with clouds and fine precipitation (Best 1992). Fossils found in La Brea tarsands (seeps) imply there was a more moist Quaternary climate (Lemon and Churcher 1961).

Seasonal rains occur during December–March; the average annual precipitation within the NP is 900 mm. In some years there is little precipitation, whereas in years influenced by El Niño events it is heavy. Great breccia fans

of outwash gravel from the Amotape Cordillera, and the badlands topography of tablazos and Tertiary strata, suggest that in the past infrequent very heavy rainstorms have been significant (Clapperton 1993). The average annual temperature is 24°C and relative humidity 80% (ONERN 1992); during the rainy season the temperature can reach 35°C.

MAP 76. CERROS DE AMOTAPE NATIONAL PARK REGION, PERU (CPD SITE SA41)

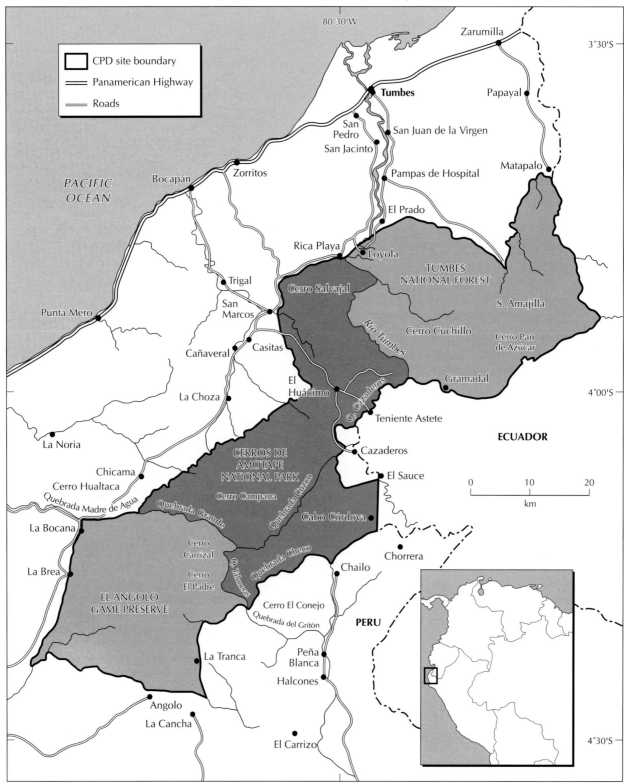

Vegetation

Depending upon the season, the physiognomy of the vegetation varies. During the rainy (summer) season, the woody vegetation is covered with leaves, giving the appearance of evergreen forest. In undulated terrain *Loxopterygium huasango* is a typical tree species. Flatlands and areas of rough terrain are covered with trees such as *Prosopis pallida* and *Capparis scabrida*. The low herbaceous stratum, which has the most floristic diversity, consists of annual species primarily of Poaceae, Asteraceae, Leguminosae, Amaranthaceae, Acanthaceae and Cucurbitaceae. During the long dry season, the shrub-like and arboreal vegetation lose their leaves and the grassy stratum becomes dry. Then the region assumes the appearance of a very poor floristic landscape.

Cerros de Amotape National Park includes five Holdridge life zones: transitional premontane-tropical to tropical desert matorral (18 km², 2% of the NP); premontane tropical thorn forest (35 km², 4%); tropical thorn forest (188 km², 21%); premontane tropical dry forest (397 km², 43%); and tropical very dry forest (275 km², 30%) (CDC-UNALM 1992a; WWF-FPCN 1989; ONERN 1976; cf. Kessler 1992).

Transitional premontane-tropical to tropical desert matorral (150–200 m), which occurs in a small area in the park's north-west, is characterized by shrubs and spiny (semi) deciduous trees and columnar cacti; the lower stratum is primarily grasses. Some of the principal species are *Capparis avicenniifolia* (syn. *C. ovalifolia*), *C. scabrida* (syn. *C. angulata*), *Prosopis juliflora*, *Acacia macracantha*, *Loxopterygium huasango* and *Cordia lutea*, as well as *Cercidium praecox* subsp. *praecox*, *Parkinsonia aculeata*, *Piptadenia flava* and the columnar cacti *Armatocereus cartwrightianus* and *Neoraimondia arequipensis* (syn. *Cereus macrostibas*). Due to deforestation and possibly the nature of the soil, large areas have become covered by a shrubby stratum dominated by the weedy *Ipomoea carnea*.

Tropical thorn forest occurs between 200–1000 m mainly in the western lowlands. The vegetation is sparse to open with shrubs, cacti, and trees averaging 8 m tall (with emergents to 10–12 m). The predominant trees include *Prosopis pallida*, *P. juliflora*, *Caesalpinia paipai*, *Bursera graveolens* and *Capparis mollis*; in more moist areas predominate *Eriotheca discolor* (= *Bombax discolor*), *E. ruizii*, *Tabebuia chrysantha*, *Ziziphus piurensis* and *Opuntia pubescens* (syn. *O. pestifer*). Also notable are *Cochlospermum vitifolium*, *Pithecellobium excelsum*, *Terminalia valverdeae* and *Armatocereus cartwrightianus*. The vegetation is degraded by human exploitation such as ranching.

Tropical very dry forest (300–800 m) is in a rather low band embracing the northern half of the Amotape range and a section along the Tumbes River near the northern limit of the NP – more of this life zone occurs in the adjacent Tumbes National Forest (Guerra 1957; Saavedra and Green 1987). The vegetation includes many shrubs (e.g. *Encelia canescens*, *Grabowskia boerhaaviaefolia*) and cacti, with deciduous trees such as *Eriotheca discolor*, *E. ruizii*, *Tabebuia chrysantha* and *Bursera graveolens* as well as *Alseis peruviana*, *Centrolobium ochroxylum*, *Cochlospermum vitifolium*, *Myroxylon peruiferum* and *Triplaris cumingiana* (syn. *T. guayaquilensis*) and many epiphytes, including *Tillandsia usneoides*.

Premontane tropical dry forest (400–1534 m) is the most widespread life zone in the NP. Common trees up to 20 m tall and 0.4–1 m dbh include *Alseis peruviana*, *Centrolobium ochroxylum*, *Eriotheca ruizii* and *E. discolor*; *Tillandsia usneoides* is conspicuous.

Premontane tropical thorn forest (600–1000 m) is in a limited band in the southern section of the NP and more prevalent in the adjacent El Angolo Game Preserve. Trees to 7–9 m tall include *Loxopterygium huasango*, *Bursera graveolens*, *Caesalpinia paipai*, *Acacia macracantha*, *Eriotheca discolor* and *Cochlospermum vitifolium* as well as *Eriotheca ruizii* and *Prosopis pallida*, along with several *Tillandsia* spp. (including *T. usneoides*).

Flora

The entire Biosphere Reserve is in the biogeographic province of dry equatorial forest, which covers 23,342 km² in far north-western Peru (1.8% of the national territory) (WWF-FPCN 1989). Quantitative data on the floristic diversity of the north-western Peruvian dry forests were obtained by reviewing published and unpublished lists and inventories, including compilations of Macbride (1936–1961), Weberbauer (1945) and Ferreyra (1983). Valuable information was obtained from botanical collections of D. Simpson (Field Museum of Natural History, Chicago) and R. Lao (La Molina Herbarium, Lima) and field evaluations of the Centro de Datos para la Conservación – Perú (CDC), Universidad Nacional Agraria La Molina (UNALM).

Cerros de Amotape NP was found to harbour 75 families, 273 genera, 404 species; Tumbes National Forest had 90 families, 349 genera, 510 species; and El Angolo Game Preserve 60 families, 151 genera, 179 species. The dominant families in the NP, according to decreasing diversity of genera and/or species, are Leguminosae (41 genera, 82 spp.); Poaceae (29 genera, 52 spp.); Asteraceae (25 genera, 27 spp.); Solanaceae (13 genera, 33 spp.); Malvaceae (10 genera, 26 spp.); Cactaceae (c. 11 genera, 13 spp.); and other families such as Amaranthaceae, Acanthaceae, Cucurbitaceae, Nyctaginaceae and Scrophulariaceae (CDC-UNALM 1992a, 1992b).

Other than the data for El Angolo Game Preserve, which were based on rather recent research (Ríos 1989), the compilations may be inaccurate due to a lack of recent floristic studies. Effort under the Flora of Peru Project is being carried out in the region, with collaboration of institutions such as Missouri Botanical Garden, Field Museum and the herbaria of San Marcos, La Molina and Trujillo. Analysis of the *Catalogue of the Flowering Plants and Gymnosperms of Peru* (Brako and Zarucchi 1993) may provide further data. There is considered to be high diversity and endemism in the region's herbaceous stratum.

Flora in critical danger are *Loxopterygium huasango*, *Tabebuia billbergii* subsp. *ampla*, *T. chrysantha* and *Ziziphus piurensis*. Populations of these dry-forest species have been reduced by selective logging, even though they occur in protected areas and legal mechanisms can regulate their utilization. Other threatened forest species include *Prosopis pallida*, *Capparis scabrida*, *Bursera graveolens*, *Eriotheca ruizii*, *Alseis peruviana* and *Centrolobium ochroxylum*. Species of restricted distribution include *Tecoma weberbaueriana* and *Macranthisiphon longiflorus* (CDC-UNALM 1991, 1992a; cf. Ferreyra 1977).

Useful plants

Cerros de Amotape is rich in potential germplasm resources. A comprehensive list of useful plants of the dry forest of north-western Peru was compiled by CDC-UNALM (1992a). Eighty forest species have been recorded in the NP, many of which are exploited. Those of greatest commercial value are *Tabebuia* spp. ("guayacán" or "madero negro"), *Loxopterygium huasango* ("hualtaco"), *Bursera graveolens* ("palo santo"), *Ziziphus piurensis* ("ébano"), *Geoffroea striata* ("almendro"), *Eriotheca ruizii* ("pasallo") and *Prosopis pallida* ("algarrobo"). These species are almost continuously extracted, for multiple uses – including young trees which appeared by natural regeneration after the strong 1982–1983 El Niño event (Aguilar 1990).

Species used for parquet flooring are *Ziziphus piurensis*, *Loxopterygium huasango*, *Ocotea piurensis*, *Alseis peruviana* ("palo de vaca" or "oreja de león") and *Tabebuia* spp.; for crafts: *Centrolobium ochroxylum*, *Leucaena trichodes*, *Sapindus saponaria* ("checo negro"), *Cordia lutea* and *Capparis scabrida*.

Species for charcoal include *Pithecellobium multiflorum* ("angolo"), *Myroxylon peruiferum*, *Prosopis* spp., *Bursera graveolens*, *Eriotheca ruizii* and *Cochlospermum vitifolium*.

There are many forage species, including *Prosopis pallida*, *Capparis scabrida*, *C. avicenniifolia*, *Cercidium praecox*, *Leucaena trichodes*, *Acacia macracantha* and *Cordia lutea*.

Plants used as medicinals include *Acacia macracantha*, *Bursera graveolens*, *Capparis mollis*, *C. avicenniifolia*, *Celtis* sp., *Cordia lutea*, *Ficus jacobii* ("mata palo"), *Ficus* sp., *Gallesia integrifolia* and *Loxopterygium huasango*. Ornamentals include *Tabebuia chrysantha*, *Cochlospermum vitifolium* and *Pithecellobium excelsum*.

Social and environmental values

Cerros de Amotape National Park protects a large remnant of Pacific dry forest (Gentry 1995), including its valuable fauna – with some species reaching their range limit in the vicinity. The park hosts a unique fauna, including species found in tropical forests, arid lands and the Andes. The NP protects c. 100 vertebrate species, such as the endangered mantled howler monkey (*Alouatta palliata*) and southern river otter [*Lontra longicaudis* (syn. *Lutra annectens*)]. Some of the vulnerable species are northern anteater (*Tamandua mexicana*), jaguar (*Panthera onca*), gato montés (*Oncifelis colocolo*) and ocelot (*Leopardus pardalis*). Well-known vulnerable birds include Andean condor (*Vultur gryphus*) and king vulture (*Sarcoramphus papa*) (CDC-UNALM 1992a; WWF-FPCN 1989).

The fauna of Tumbes National Forest and El Angolo Game Preserve are similarly important, sharing with the NP various threatened species. The BR might be habitat for the white-winged guan (*Penelope albipennis*), which had been thought extinct and is critically endangered (Macedo 1979; Best 1992; Collar *et al.* 1992). Recent inventories for the NF record c. 230 vertebrate species: 36 mammal spp.; 175 bird spp. – with 11 more or less threatened species such as the endangered grey-backed hawk (*Leucopternis occidentalis*) and the henna-hooded foliage-gleaner (*Hylocryptus erythrocephalus*), which has a stronghold here; six reptile spp.; and four amphibian spp. Inventories for the game preserve so far have recorded 24 species of mammals, 125

of birds, 12 of reptiles and seven of amphibians (CDC-UNALM 1992a, 1994; Collar *et al.* 1992).

Potential use of wild fauna in some sections of the region is fairly high, because of sport hunting as a tourist attraction. The Cordillera de los Amotapes and Tumbes NF have special relevance in potential management and use of some wild fauna in this fashion (e.g. collared peccary, *Tayassu tajacu* and white-tailed deer, *Odocoileus virginianus*). These activities offer the possibility of integrating as economically productive the marginal areas being used for agriculture and extensive cattle-ranching, which are developing with scarce or non-existent management. Areas adjacent to the Cerros de Amotape NP and forests that are open for public use are primarily used as pasturage for raising cattle and goats for both meat and milk.

The NP is important for archaeological remains, which have been discovered in Cazaderos, Guineal, Platanal and Modroño valleys. Oral tradition indicates that ruins of undetermined (perhaps Tumpis) origin exist on Cerro El Barco, which is also particularly rich in biodiversity. There are no indigenous people keeping ancestral life-styles. About thirty mestizos inhabited the park in 1987 near Teniente Astete village. Some settlements (Rica Playa, San Marcos, Carrizalillo, Cazaderos) are located on borders of the NP, which are critical locations at risk of encroachment because the people raise cattle and practise agriculture (WWF-FPCN 1989).

Local and regional fairs, some involving the Peru-Ecuador border area, are held every year to promote commerce, culture, tourism and recreation. South of Tumbes on the Salina Plains is where the Spaniards led by Francisco Pizarro made their landing in 1532. The NP can provide for recreation, education, ecotourism and scientific research; its zone of regional influence covers c. 11,670 km² (WWF-FPCN 1989).

Threats

The dry forest is subject to many threats, especially seasonal overgrazing, soil erosion and timber extraction. The predominant vegetation types such as thorn forest and very dry forest are destroyed due to overgrazing (by cattle and goats) and heavy exploitation of forest resources for domestic and industrial uses. The southern part of the park is increasingly deforested by wood extraction for use as fuelwood and construction materials. Legal mechanisms exist to regulate forest resources in the general region (Tumbes, Piura, Lambayeque), and the logging of commercial timber reached its peak in the past, but illegal extraction of timber for commercial use remains significant (Castillo 1973; DGFF 1981; Gonzáles *et al.* 1981; Jara and Otivo 1989; WWF-FPCN 1989; CDC-UNALM 1992a).

Near the Tumbes River in the park's northern area of Ucumares, there is a government plan under study to construct a dam (WWF-FPCN 1988, 1989). The NP is also at risk of some desertification. The southern border, where there is primarily thorny vegetation, is threatened by the encroaching 12,000 km² Sechura Desert.

As a result of deforestation in the dry forests of the north-western region of Peru, scrub vegetation has grown considerably in various areas during the past few years and now totals over c. 86 km². The shrub-like weed *Ipomoea carnea* ("borrachera") has invaded Cazaderos (in the NP's eastern area) and other areas adjacent to the NP such as Papayal, Ciénego Norte, San Marcos and Rica Playa. This weed is poisonous to cattle and goats, which graze in the areas.

Poaching of the rich fauna, including species endemic or restricted in Peru and Ecuador, is a problem that has contributed to several species being officially listed as at risk. They include mantled howler monkey (*Alouatta palliata*), southern river otter (*Lontra longicaudis*), American crocodile (*Crocodylus acutus*) and even the white-winged guan (*Penelope albipennis*) (Pulido 1991). Groups such as birds (especially parakeets) and increasingly reptiles, which supply the growing export market, need to have measures such as CITES implemented to manage their utilization.

The legal status of the BR locally is not well defined. There is a lack of coordination among the region's public and private entities that might affect the NP, and there is no master plan for the park. The NP's 1989–1990 operative plan (WWF and FPCN 1989) identified four critical areas needing attention: the park's upper portion with the Tumbes River – habitat for several endangered species; Teniente Astete and Cazaderos – subject to agriculture and cattle-grazing; El Huásimo, Cazaderos and the southern part – subject to invasion of *Ipomoea carnea*; and the southern region (El Chaylo) – subject to timber extraction.

Conservation

Cerros de Amotape National Park was established (by Supreme Decree No. 0800-75-AG, 22 July 1975) to protect one of the last remnants of Pacific dry forest, and species of flora and fauna threatened with extinction at national or regional levels (Brack-E., Ríos and Reyes 1975). The NP protects 3.91% of the biogeographic province of dry equatorial forest of Peru. Most similar formations in neighbouring western Ecuador have been converted to agriculture and ranching (see CPD South America Overview, and Best 1992; Collar *et al*. 1992; Kessler 1992). The NP is part of Peru's Sistema Nacional de Unidades de Conservación (SINUC) and strictly protected legally, which prohibits harmful land-use activities. The NP constitutes the core of the Noroeste Peruano (Peruvian North-west) Biosphere Reserve, which includes two adjoining areas: Tumbes National Forest (751 km²), established 8 July 1957; and El Angolo Game Preserve (650 km²), established 1 July 1975. The three conservation units (2314 km²) are in Peru's Sistema Nacional de Areas Naturales Protegidas por el Estado (SINANPE). Within the NF, the eastern area El Caucho – Campo Verde (c. 126 km²) on Cordillera Larga is recommended for increased protection (CDC-UNALM 1992a; cf. Wiedenfeld, Schulenberg and Robbins 1985; Collar *et al*. 1992), and an area of 27.5 km² between the borders of the NP and Ecuador is recommended for acquisition (WWF and FPCN 1989).

The Noroeste Peruano BR, which was recognized by UNESCO on 1 March 1977, has been given high conservation priority by the World Wildlife Fund (WWF-U.S.), which has been the principal funding source for the BR's conservation activities. With the Fundación Peruana para la Conservación de la Naturaleza (FPCN) and administrative and technical support of Peru's Dirección General Forestal y de Fauna (DGFF), WWF developed an operational plan for Cerros de Amotape NP (WWF and FPCN 1989). During the past

several years the support has resulted in collaboration of local communities with park personnel to guard the NP; construction of control posts; trials on methods to eradicate borrachera (*Ipomoea carnea*); preliminary attempts at goat management; implementation of "carob" (algarrobo) forest management; and establishment of apiaries and tree and medicinal plant nurseries.

Currently, the NP has three technical administrators and nine park guards. According to the proposed directive plan for SINUC (CDC-UNALM 1991), the NP should receive the highest conservation priority – due to its high biological diversity, medium degree of threat, and reasonable management of the region.

References

Aguilar, P. (1990). Sinópsis sobre los eventos del fenómeno "El Niño" en el Perú. *Boletín de Lima* 70: 69–84.

Best, B.J. (ed.) (1992). *The threatened forests of south-west Ecuador: the final report of the Ecuadorian Dry Forest Project 1991*. Biosphere Publications, Leeds, U.K. 240 pp.

Brack-E., A., Ríos, M.A. and Reyes, F. (1975). *Evaluación y bases para el establecimiento de un coto de caza y un parque nacional en la Cordillera de los Amotapes (Piura – Tumbes)*. Dirección General Forestal y de Caza (DGFC), Lima. 52 pp.

Brako, L. and Zarucchi, J.L. (1993). *Catalogue of the flowering plants and gymnosperms of Peru/Catálogo de las angiospermas y gimnospermas del Perú*. Monogr. Syst. Bot. Missouri Bot. Gard. 45. 1286 pp.

Castillo, M. (1973). *Diagnóstico forestal de los departamentos de Tumbes y Piura*. Ministerio de Agricultura, DGFC, Dirección de Recursos Forestales, Lima. 38 pp.

CDC-UNALM (1991). *Plan director del Sistema Nacional de Unidades de Conservación (SINUC), una aproximación desde la diversidad biológica (Propuesta del CDC-UNALM)*. Centro de Datos para la Conservación (CDC) – Universidad Nacional Agraria La Molina (UNALM), La Molina, Lima, Peru. 153 pp.

CDC-UNALM (1992a). *Estado de conservación de la diversidad natural de la región noroeste del Perú*. CDC-UNALM, La Molina. 211 pp.

CDC-UNALM (1992b). Lista de la flora en áreas manejadas del Bosque Seco del Noroeste. In *Estado de conservación de la diversidad natural de la región noroeste del Perú*. CDC-UNALM, La Molina. Pp. 181–198.

CDC-UNALM (1994). *Manual de las especies de fauna silvestre del Coto de Caza El Angolo (Sullana-Piura)*. CDC-UNALM, La Molina. Unpublished.

Clapperton, C.M. (1993). *Quaternary geology and geomorphology of South America*. Elsevier, Amsterdam. 779 pp.

Collar, N.J., Gonzaga, L.P., Krabbe, N., Madroño Nieto, A., Naranjo, L.G., Parker III, T.A. and Wege, D.C. (1992). *Threatened birds of the Americas: the ICPB/IUCN Red Data Book*, 3rd Edition, Part 2. Smithsonian Institution Press, Washington, D.C. and International Council for Bird Preservation (ICBP), Cambridge, U.K. 1150 pp.

DGFF (1981). *Inventario forestal del Bosque Seco del Norte: Tumbes– Piura– Lambayeque*. Ministerio de Agricultura, Dirección General Forestal y de Fauna (DGFF), Lima. 71 pp.

Ferreyra, R. (1977). Endangered species and plant communities in Andean and coastal Peru. In Prance, G.T. and Elias, T.S. (eds), *Extinction is forever: threatened and endangered species of plants in the Americas and their significance in ecosystems today and in the future*. New York Botanical Garden, Bronx. Pp. 150–157.

Ferreyra, R. (1983). Los tipos de vegetación de la costa peruana. *Anales Jardín Botánico Madrid* 40: 241–256.

Gentry, A.H. (1995). Diversity and floristic composition of neotropical dry forests. In Bullock, S.H., Mooney, H.A. and Medina, E. (eds), *Seasonally dry tropical forests*. Cambridge University Press, Cambridge, U.K. Pp. 146–194.

Gonzáles, M., Ríos, M.A., Ponce, C., Chung, A., Sabogal, C. and Vásquez-R., P. (1981). *Plan maestro de manejo forestal para el noroeste del Perú*. DGFF and UNALM, Lima. 218 pp.

Guerra, W. (1957). *Estudio forestal del Bosque Nacional de Tumbes*. Servicio Forestal y de Caza, Lima. 188 pp.

Jara, F. and Otivo, J. (1989). *Potencial forestal de la región Grau*. Centro de Investigación y Promoción del Campesinado (CIPCA), DGFF, Programa Nacional de Acción Forestal de Piura. 113 pp.

Kessler, M. (1992). The vegetation of south-west Ecuador. In Best, B.J. (ed.), *The threatened forests of south-west Ecuador*. Biosphere Publications, Leeds, U.K. Pp. 79–100.

Lemon, R.R. and Churcher, C.S. (1961). Pleistocene geology and paleontology of the Talara region, northwest Peru. *Amer. J. Sci.* 259: 410–429.

Macbride, J.F. (1936-1961). *Flora of Peru. Publ. Field Mus. Nat. Hist., Bot. Ser.* 13.

Macedo, H. de (1979). Redescubrimiento de la pava aliblanca (*Penelope albipennis* Taczanowski 1877). *Boletín de Lima* 1: 5–11.

ONERN (1976). *Mapa ecológico del Perú. Guía explicativa*. Oficina Nacional de Evaluación de Recursos Naturales (ONERN), Lima. 147 pp.

ONERN (1992). *Evaluación de los recursos naturales del Departamento de Tumbes*. ONERN, Lima. Unpublished.

Pulido, V. (1991). *El Libro Rojo de la fauna silvestre del Perú*. Instituto Nacional de Investigación Agropecuaria y Agroindustrial (INIAA), Lima. 219 pp.

Ríos, J. (1989). *Análisis del habitat del Coto de Caza El Angolo*. M.S. thesis in Forest Management. Escuela de Post-Grado, UNALM, Lima. 266 pp.

Saavedra, C. and Green, K. (1987). El Bosque Nacional de Tumbes – Perú. Aplicación de imágenes satélite en la evaluación de habitat para primates. *Boletín de Lima* 51: 81–87.

Simpson, D.R. (1990). *Donald Simpson's Peruvian collections. Departamento de Tumbes (Perú)*. Field Museum of Natural History, Chicago. 8 pp.

Weberbauer, A. (1945). *El mundo vegetal de los Andes peruanos*. Ministerio de Agricultura, Dirección de Agricultura, Estación Experimental Agrícola de La Molina, Lima. 776 pp.

Wiedenfeld, D.A., Schulenberg, T.S. and Robbins, M.B. (1985). Birds of a tropical deciduous forest in extreme northwestern Peru. In Buckley, P.A., Foster, M.S., Morton, E.S., Ridgely, R.S. and Buckley, F.G. (eds), *Neotropical ornithology*. Ornith. Monogr. No. 36. American Ornithologists' Union, Washington, D.C. Pp. 305–315.

WWF (World Wildlife Fund) and FPCN (Fundación Peruana para la Conservación de la Naturaleza) (1988). *Tropical Andes program: protecting a global center of biological diversity*. World Wildlife Fund-U.S., Washington, D.C. Pg. 11.

WWF and FPCN (1989). *Plan operativo del Parque Nacional Cerros de Amotape 1989-1990*. WWF, FPCN and DGFF, Lima. 154 pp.

Authors

This Data Sheet was written by Centro de Datos para la Conservación – Perú (CDC-Perú) (Universidad Nacional Agraria La Molina, Departamento de Manejo Forestal, Apartado 456, La Molina, Lima, Peru), Fundación Peruana para la Conservación de la Naturaleza (FPCN) (Apartado 18-1393, Lima, Peru) and Olga Herrera-MacBryde (Smithsonian Institution, Department of Botany, NHB-166, Washington, DC 20560, U.S.A.).

LOMAS FORMATIONS
Peru

Location: Series of over 40 isolated localities along coastal Peru, between latitudes 6°52'–18°24'S.

Area: Less than 2000 km² of lomas vegetation dispersed in 144,000 km² of desert.

Altitude: 0–1000 m.

Vegetation: Community composition highly variable: mixtures of annual, short-lived perennial and woody scrub vegetation.

Flora: c. 600 spp. of lichens, ferns, gymnosperms and flowering plants; high endemism (over 40%); threatened species.

Useful plants: Unique ecophysiological characteristics make the plants (e.g. *Solanum*, *Lycopersicon*) valuable as potential genetic resources.

Other values: Communities highly endemic; support for mammal, bird and insect communities; ecotourism.

Threats: Human pressure through urbanization, mining, livestock grazing, fuelwood collecting.

Conservation: Reserva Nacional de Lachay (51 km²); parts of Reserva Nacional de Paracas (1170 km²), Santuario Nacional de Mejía (7 km²).

Geography

The greater part of the west coast of South America (5°00'–29°55'S) is occupied by the Peruvian and Atacama deserts, which form a continuous belt for more than 3500 km along the western escarpment of the Andes from northern Peru to northern Chile. The phytogeography and ecology of these deserts have been reviewed in detail (Rauh 1985; Rundel *et al.* 1991), and a floristic inventory of the entire area has been completed (Dillon, in prep. for publication). While desert is continuous from Peru to Chile, the topography, climate and vegetation of each of the deserts is distinct (Rundel *et al.* 1991; Duncan and Dillon 1991). The Peruvian Desert is presented here; for the Atacama Desert, see the separate Data Sheet (CPD Site SA43).

The Peruvian Desert is a narrow band of hyper-arid habitat confined to the western coast and extends over 2000 km (5°–18°S), covering c. 140,000 km² or 11% of Peru (Library of Congress 1979). Three climatic anomalies are largely responsible for development of the hyper-arid conditions along the west coast of South America (Trewartha 1961; Johnson 1976).

There is remarkable homogeneity of temperature along the entire latitudinal extent of the desert. This pattern of temperature stability results from the influence of cool sea-surface temperatures associated with the northward flow of the coastal Humboldt (Peruvian) Current.

Since the extreme conditions exist for such an extended latitudinal distance (c. 3500 km), there are relatively abrupt climatic transitions both to the north and the south. As a result, a steppe climate is poorly developed along those margins.

Brief periods of heavy rainfall and relatively high temperatures occasionally affect parts of the desert, bringing wet tropical conditions. These periods are associated with rare yet recurrent proximity to the continent of the oceanic current El Niño (Dillon 1985a, 1985b; Dillon and Rundel 1989; Quinn, Neal and Antúnez de Mayolo 1987; Caviedes 1975).

Important also is the influence of strong atmospheric subsidence associated with a positionally stable, subtropical anticyclone. The result of the several factors is a mild, uniform coastal climate with regular formation of thick stratus cloud banks below 1000 m during the winter months (Prohaska 1973). Where coastal topography is low and flat, this stratus layer dissipates inward over broad areas with little biological effect, but where isolated mountains or steep coastal slopes intercept the clouds, a fog zone develops with a stratus layer concentrated against the hillsides.

These fogs, termed "garúas" in Peru and "camanchacas" in Chile, are the key to the extent and diversity of vegetation throughout the deserts of the western coast. The moisture allows the development of fog-zone plant communities termed "lomas" (small hills) between sea-level and 1000 m. Other authors have referred to these plant formations as the fertile belt (Johnston 1929), fog oases (Ellenberg 1959) or meadows on the desert (Goodspeed 1961; Goodspeed and Stork 1955). In Peru, there exist over 40 discrete localities in the desert that support vegetation (Map 77), including offshore islands such as Las Viejas, San Gallán and San Lorenzo. The actual area covered by vegetation, even during periods of optimal weather and maximum development, is probably less than 2000 km².

Vegetation

Lomas formations occur as discrete communities or islands of vegetation separated by hyper-arid territory devoid of plant life. Since growth is dependent upon available moisture, climatic patterns determine plant distributions. Topography and substrate combine to influence patterns of moisture availability and the area of suitable habitat. Ecological

MAP 77. LOMAS FORMATIONS, PERU (CPD SITE SA42)
AND LOMAS FORMATIONS OF THE ATACAMA DESERT, NORTHERN CHILE (CPD SITE SA43)

requirements and tolerances of individual species ultimately determine community composition; species endemism exceeds 40% (Rundel *et al.* 1991). This account provides brief descriptions of the larger lomas formations, with data on species diversity where available.

Northern coast

The geomorphology of north-western Peru consists of a broad coastal plain extending from the Ecuadorian border to near Casma (9°28'S). Some of this region has been geologically uplifted (Robinson 1964), but only a few areas are above 1000 m (cf. CPD Site SA41). Although the coastal fog zone only extends from 0–500 (–800) m in northern Peru (Rauh 1958), the low topography provides a broad arid zone, much of it without vegetation. Low shrubs dominate the sparse vegetation below 100 m, and cacti increase in importance with higher elevation. An interior desert at intermediate elevations is absent from northern Peru, unlike central and southern Peru.

North of 6°S, the prevailing current turns west to the open ocean; Clüsener and Breckle (1987) report small coastal stands of *Avicennia germinans* at 5°30'S near the mouth of the Piura River, and *Rhizophora mangle* (*R. harrisonii*) at 3°44'S near Bocapán. The vegetation of the Tumbes region (near Ecuador) has been described by Ferreyra (1957).

Much of the northern coastal plain (5°–7°S) is covered by the semi-arid Sechura Desert, with extensive areas of flat sandy plains and active dunes (Silbitol and Ericksen 1953; Kinzl 1958; Broggi 1961). Large expanses support little or no vegetation; sometimes *Parkinsonia aculeata* and the narrow endemic *Alternanthera peruviana* are common. A brief description of vegetation in the area of Cerro Illescas along the coast near Bayóvar (5°50'S) has been given by Ferreyra (1979) and Huey (1979). Flat sandy terrain is relatively bare, whereas low hummocky dunes are stabilized by *Capparis scabrida* and *C. avicenniifolia*. Dunes along the immediate coast are stabilized by *Distichlis spicata* and *Cryptocarpus pyriformis* (Weberbauer 1911).

River valleys flowing through the southern Sechura Desert are now largely irrigated agricultural land but once supported riparian communities with thickets of *Acacia macracantha, Salix humboldtiana, Schinus molle, Sapindus saponaria, Muntingia calabura* and a variety of woody and semi-woody shrubs (Weberbauer 1911; Ferreyra 1983).

Inland from the Sechura Desert along the coastal plain of northern Peru, semi-desert vegetation is well developed on slopes between the ranges of the Andes. Weberbauer (1911) and Ferreyra (1979, 1983) described this vegetation, which has dwarf trees of *Eriotheca discolor, Bursera graveolens* and *Acacia huarango*, a diversity of columnar cacti, and terrestrial bromeliads. The phytogeographical significance of these woodlands has been discussed by several botanists (Weberbauer 1936; Ferreyra 1960; Simpson 1975; Dillon and Cadle 1991).

Cerro Reque (6°52'S, 80 spp.), a small mountain near Chiclayo, marks the most northern extension of typical lomas vegetation (Llatas, unpublished data), including the narrow endemic *Sisymbrium llatasii*. North-west of Trujillo, Cerro Campana (7°58'S, 185 spp.) and Cerro Cabezón (7°54'S, 120 spp.) are the northernmost extension of well-developed lomas vegetation. On the lower sandy slopes below 200 m occurs a *Tillandsia* zone with a mixture of

cacti (*Espostoa melanostele, Haageocereus decumbens, Melocactus trujilloensis*), small trees and prostrate shrubs. Above 400 m, moisture conditions are more favourable and a diverse community of shrubs and herbaceous perennials occurs. These include several species not usually found within coastal desert environments which are present in large numbers, such as *Tillandsia multiflora, Puya ferruginea, Hypericum uliginosum, Pelexia matucanensis, Peperomia dolabriformis* and *Valeriana pinnatifida*. Endemic species from this region include *Pitcairnia lopezii* – the only member of this large genus to occur in the lomas formations, and *Senecio truxillensis, Apodanthera ferreyrana, Matelea alicae* and *Solanum mochiquense* (Sagástegui, Mostacero and López 1988).

To the south continues a series of small discrete mountains within the coastal plain, each with vegetation between 200–600 m, including Cerro Prieto (7°59'S, 30 spp.); Cerro Cabras (8°03'S, 48 spp.); Cerro Chiputur (8°10'S, 95 spp.); Cerro Negro (8°18'S, 26 spp.); and Lomas de Virú (8°19'S). These extensively studied formations (Sagástegui, Mostacero and López 1988) are a mixture of shrubs and herbaceous species similar to the composition on Cerro Campana. At Cerro Chimbote (9°04'S, 10 spp.) and Casma (9°28'S, 25 spp.), near where the coastal plain becomes narrow, is less lomas vegetation than the well-developed formation on Lomas de Lupín (10°33'S, 57 spp.). The lower-elevation lomas zone is dominated by cryptogams (Ferreyra 1953) and no endemics are recorded.

Central coast

Perhaps the most famous lomas formation in Peru is the Lomas de Lachay (11°21'S, 100 spp.), c. 60 km north of Lima; this site is a National Reserve. This formation was described glowingly and in detail by Ruiz in the late 18th century (see Jaramillo-Arango 1952). Ferreyra (1953) recognized two distinctive lomas zones here. The lower zone (100–300 m) is dominated by a cryptogamic community of *Nostoc commune* and a diverse array of foliose and fruticose lichens. Herbaceous and semi-woody vascular plants are less common. The perennial vegetation becomes much richer above 300 m, where distinctive communities occur on hillsides or rocky slopes in canyons. Dry rocky slopes support stands of terrestrial bromeliads (*Tillandsia latifolia, Puya ferruginea*) and lichens. Moist micro-habitats on these slopes support a very different flora with many herbaceous perennials. Canyons and valleys in the lomas zone are characterized by relatively dense stands of small trees, notably *Caesalpinia spinosa, Capparis prisca, Senna birostris* and *Carica candicans*. A remarkable feature of these small woods is the dense accumulation of epiphytes of all types – mosses, lichens, ferns, *Peperomia hillii, Calceolaria pinnata* and *Begonia geranifolia* are common on branches and trunks of the low trees.

Directly to the south, the Lomas de Chancay and Iguanil (11°24'S, 77 spp.) comprise a rich fertile zone between 160–400 m (Ferreyra 1953). True succulents are less common and a diverse group of herbaceous species and shrubby perennials dominate. The Lomas de Pasamayo (11°38'S, 20 spp.) occupies the upper limit of a large bluff affronting the ocean at nearly 500 m. This small herbaceous community includes *Loasa urens, Solanum multifidum, Palaua moschata, Erigeron leptorhizon, Acmella alba, Nolana gayana, N. humifusa, Verbena litoralis* and *Cryptantha limensis.*

The city of Lima is situated where the Rimac and Chillón rivers merge; from Lima southward, frequency and amplitude of precipitation decrease significantly and the lomas communities have lower overall diversity. South of Lima, the Lomas de Amancaes (12°01'S, 50 spp.), Atocongo (12°08'S, 80 spp.) and Lomas de Lurín (12°17'S, 30 spp.) support varied communities (Ocrospoma-Jara 1990; Cuya and Sánchez 1991).

Three of the guano islands offshore from the central Peruvian coast (San Lorenzo, San Gallán, Las Viejas) are high enough to support sparse lomas communities. Johnston (1931) recorded only 19 species of vascular plants on these islands, including one endemic – *Nolana insularis* (Johnston 1936; Ferreyra 1961b). Between Lima and Cañete (13°05'S), the coastal plain disappears and coastal geomorphology is dominated by the foothills of the Cordillera Occidental of the Andes. Ridges of these foothills separate numerous river valleys that drain the Andes in this region. Geological evidence suggests past land subsidence, with broad valley floors restricted to a few small coastal areas (Robinson 1964). A broad coastal plain extends from around Pisco to just north of Atiquipa.

Steep coastal ridges with a series of marine terraces are present from the Paracas Peninsula near Pisco to south of Ica (Craig and Psuty 1968). Inland from these ridges the

CPD Site SA42: Lomas formations, Peru.
Lomas de Lachay (shown), perhaps the most famous
lomas formation in Peru, is a National Reserve.
Photo: Tony Morrison/South American Pictures.

coastal plain lies largely below 300 m, though individual terraces occur up to 700 m (Robinson 1964). The sparse vegetation of the Lomas de Amara (13°42'S) and adjacent communities near Pisco and Ica have been described by Craig and Psuty (1968) and several government reports (ONERN 1971b, 1971c).

Above the lomas communities, *Neoraimondia arequipensis* forms a characteristic belt of columnar cacti of low density. As at Nazca (14°50'S), extensive stands of *Acacia macracantha, Prosopis chilensis* and other phreatophytes (with very long roots to reach the water table) are present along the channels of the Pisco and Ica rivers. Scattered stands of cultivated date palms (*Phoenix dactylifera*) suggest a dependable supply of groundwater. Marshy soils are present below the fog zone along the coast, with saline areas dominated by *Distichlis spicata* and freshwater sites supporting stands of *Scirpus.* Weakly developed dunes adjacent to the coast are frequently stabilized by *Distichlis* and *Sesuvium portulacastrum*, a succulent halophyte. *Distichlis* is a particularly effective stabilizer since its roots extend several meters deep (Craig and Psuty 1968).

Lomas vegetation is not well developed inland around Nazca, but extensive riparian forests of *Prosopis chilensis* and *Acacia macracantha* are present in river channels from near the coast up to 2000 m (ONERN 1971a). Other important riparian species include *Salix humboldtiana* and *Arundo donax*, as well as a variety of low shrubs and herbaceous plants. About 12 km south of Nazca, a small population of *Bulnesia retama* represents an unusual disjunction from populations in central Argentina (Weberbauer 1939).

Near the port of Lomas, the Lomas de Jahuay (15°22'S, 63 spp.) are found on windy flats at 300–900 m, and include several southern Peruvian endemics such as *Nolana plicata, Modiolastrum sandemanii, Coursetia weberbaueri, Malesherbia angustisecta, Nolana tomentella* and *Ambrosia dentata.*

The Lomas de Atiquipa (15°48'S, 120 spp.) and Chala (15°53'S, 20 spp.) in the northern part of the Department of Arequipa form a continuous formation broken only by a broad dry river channel. These were studied early and intensively by Peruvian botanists because of their rich diversity, and include *Arcytophyllum thymifolia, Calceolaria ajugoides, Galvezia fruticosa, Encelia canescens, Neoporteria islayensis (Islaya paucispina), Senna brongniartii, Heterosperma ferreyrii, Nolana inflata, Croton alnifolius, Heliotropium pilosum, Senecio smithianus* and the narrow endemic *Astragalus neobarnebyanus* (Gómez-Sosa 1986). This marks the northernmost station for the predominantly Chilean *Argylia radiata*. Inland to the north-east of Chala, the Lomas de Cháparra (Taimara) (15°50'S, 36 spp.) contain relict stands of small trees to 5 m tall occupying ravines at upper elevations, including *Maytenus octogona, Caesalpinia spinosa* and the endangered endemic *Myrcianthes ferreyrae*, which has been reduced to a few hundred individuals at a handful of local sites.

Southern coast

North of Camaná, the maritime slopes of an arm of the Andes extend near the coast and break up the coastal plain. The Lomas de Atico (16°14'S, 67 spp.) and Ocoña (16°30'S,

14 spp.) contain sparsely distributed shrubs such as *Coursetia weberbaueri* and herbaceous species including *Nolana spathulata*, *N. pallida*, *N. mariarosae* and *Sesuvium portulacastrum*. South of Atico a sizeable population of *Neoraimondia aticensis* occurs near the ocean, with *Heliotropium krauseanum*, *Calandrinia paniculata*, *Cleistocactus sextonianus* (*Loxanthocereus aticensis*) and the endemics *Eremocharis ferreyrae*, *Domeykoa amplexicaulis*, *Helogyne hutchinsonii*, *Nolana aticoana*, *Mathewsia peruviana* and *Hoffmanseggia arequipensis*.

Lomas south of the city of Camaná (16°35'S, 83 spp.) are well developed where the loose sandy soils are bathed by fog, and form a primarily herbaceous community from 20–800 m. The most common species is *Eragrostis peruviana*, which in 1983 because of a strong El Niño formed large pure communities. Other common herbaceous species include *Tiquilia paronychioides*, *Pasithea coerulea*, *Atriplex rotundifolia*, *Geranium limae*, *Cenchrus humilis*, *Astragalus triflorus*, *Loasa urens* and *Cristaria multifida*, as well as *Cleistocactus sextonianus* (*Loxanthocereus camanaensis*) and the narrow endemics *Palaua camanensis*, *P. trisepala* and *Nolana cerrateana*.

The Lomas de Mollendo (16°55'S, 122 spp.) and Islay, 20 km north of Mejía, support a large and diverse flora (Péfaur 1982). The community extends from 150 m to c. 1100 m in elevation within a narrow quebrada. On the slopes near the ocean frequent species include *Frankenia chilensis*, *Spergularia congestifolia*, *Verbena* (*Glandularia*) *clavata* and the narrow endemic *Viguiera weberbaueri*. At higher elevations are numerous woody species (*Carica candicans*, *Calliandra prostrata*, *Gaya pilosa*, *Lycium stenophyllum*, *Nolana pilosa*, *Monnina weberbaueri*) and annuals and perennials [*Alstroemeria paupercula*, *Lupinus mollendoensis*, *Caesalpinia* (*Hoffmanseggia*) *miranda*, *Polyachyrus annuus* and *Ophryosporus hoppii*]. Vegetation ceases near 1100 m – well above the fog zone. To the east, toward the Andean Cordillera, stretches the hyper-arid Pampa del Sacramento, an area of extensive sand dunes virtually devoid of vegetation (Barclay 1917; Finkel 1959). Similar formations are also found immediately to the north (Gay 1962).

Extensive lomas reappear at Mejía (17°07'S, 62 spp.) where lomas vegetation occurs sporadically on low sandy hills up to 600 m and more extensively on the higher hills farther inland to 1000 m. Along the coast, perennial halophytic endemics such as *Nolana adansonii*, *N. thinophila* and *Tiquilia conspicua* are common along the beaches. On slopes between 200–600 m, a herbaceous community includes *Palaua velutina*, *Portulaca pilosissima* and *Weberbauerella brongniartioides*. At elevations above 600 m occasional cacti are encountered, including *Neoraimondia arequipensis* and the night-blooming *Haageocereus decumbens* (*H. australis*). From 800–1000 m, nearly continuous grasses and perennial herbs occupy the slopes, including several rare species such as *Centaurium lomae*, *Microcala quadrangularis* and *Sisyrinchium micrantha*. Woody species include the small tree *Caesalpinia spinosa*, a montane species common to lomas formations of central Peru. Some 20 km to the interior north, the Lomas de Cachendo (17°00'S, 20 spp.) receive sufficient fog to develop a small community including *Palaua dissecta*, *Eragrostis peruviana*, *Oxalis megalorhiza* and occasionally *Neoraimondia arequipensis*.

The pattern of dominance by herbaceous species is broken in the coastal Lomas de Mostazal near the port of Ilo

(17°45'S, 53 spp.). Rich herbaceous communities similar to those described above frequently cover the entire ground surface, yet in addition a shrub community dominated by *Croton alnifolius* and *Grindelia glutinosa* occurs. During 1983 there were vast numbers of *Palaua weberbaueri*, *P. dissecta*, *Urocarpidium peruvianum*, *Nolana pallidula* and *N. spathulata*. East of these formations, large colonies of *Tillandsia purpurea* cover the sand dunes.

North-east of Tacna, slightly rolling hillsides (200–672 m) near Sama Grande and Sama Morro (17°48'–17°50'S, 100 spp.) support a flora first inventoried by Ferreyra (1961a). Rich herbaceous communities develop in response to sufficient fog or rare aperiodic rains. During the strong El Niño event of 1982–1983, dense mixed communities of annuals and herbaceous perennials developed, including *Nolana spathulata*, *N. arenicola*, *N. gracillima*, *Portulaca pilosissima*, *Calandrinia paniculata*, *Cristaria multifida*, *Eragrostis weberbaueri*, *Argylia radiata*, *Leptoglossis darcyana*, *Allionia incarnata*, *Tiquilia litoralis*, *Palaua dissecta*, *P. pusilla*, *Caesalpinia* (*Hoffmanseggia*) *prostrata*, *Mirabilis elegans* and *Monnina weberbaueri*. Woody species included *Ephedra americana*, *Encelia canescens* and the suffrutescent *Chenopodium petiolare*, *Nolana confinis*, *N. lycioides* and *Senna brongniartii*. The endemic cactus group *Islaya* of *Neoporteria sensu lato* has its southernmost species in this region – *Neoporteria krainziana* (*I. krainziana*).

Flora

Current estimates (Dillon, unpublished) for the total number of species represented within the Peruvian lomas formations are 72 families, 284 genera, 557 species; total numbers within the Atacama formations are 80 families, 225 genera, 550 species. Of the total of nearly 1000 species, only 68 species (c. 7%) are distributed within both the Peruvian and Chilean formations. The families containing the greatest diversity throughout the Peruvian and Atacama deserts include Asteraceae (57 genera, 147 spp.); Cactaceae (c. 14 genera, c. 92 spp.); Nolanaceae (2 genera, 79 spp.); Leguminosae (26 genera, 71 spp.); Solanaceae (19 genera, 64 spp.); Malvaceae (9 genera, 56 spp.); Apiaceae (13 genera, 51 spp.); and Boraginaceae (7 genera, 51 spp.).

The lomas species can be grouped into five broad categories (Dillon 1989): (1) wide-ranging pantropical or weedy species found within and outside the lomas (Müller 1988); (2) amphitropic disjuncts, primarily from arid and semi-arid regions of North America; (3) lomas disjuncts from adjacent Andean source populations; (4) lomas endemics (Peruvian and Chilean) with no extra-lomas populations, but widely distributed within the lomas; and (5) narrow lomas endemics found in only one or a few lomas formations.

While the arid environment is continuous between Peru and Chile, there seems to be a significant climatic barrier to dispersal between 18° and 22°S latitude in extreme northern Chile. The lack of topographic barriers to provide fog in the region between Arica and Antofagasta produces conditions with drought too severe for most plants to survive. Less than 7% of the combined Peruvian and Chilean lomas flora occur on both sides of this zone (Duncan and Dillon 1991). The four most important groups of coastal Peruvian cacti (*Haageocereus*, *Neoraimondia*, the *Islaya* group of

Neoporteria, the *Loxanthocereus* group of *Cleistocactus*) do not occur south of Arica. Conversely, three of the most important cactus groups of northern Chile (*Eulychnia*, *Copiapoa*, *Neoporteria sensu stricto*) fail to cross into Peru (Rauh 1958). Terrestrial *Tillandsia* species, which are a dominant aspect of the coastal vegetation of southern Peru, are of very limited distribution in the Atacama; the southernmost limit of the terrestrials is between Iquique and Tocopilla where pure stands of *Tillandsia landbeckii* occur.

Nolana occurs from central Chile to northern Peru and the Galápagos Islands, with peaks of species diversity near Paposo in Chile and Mollendo in Peru, but only *Nolana lycioides* is distributed both in Chile and Peru. A number of genera containing coastal perennials have crossed this floristic barrier in the zone south of Arica, including *Alternanthera*, *Ophryosporus*, *Tetragonia*, *Oxalis*, *Calandrinia*, *Senna*, *Palaua*, *Tiquilia*, *Heliotropium*, *Caesalpinia*, *Tephrocactus sensu lato* and *Urocarpidium*.

Despite the barrier, a number of species have become successfully established over the entire length of the Peruvian and Atacama deserts, such as *Apium* (*Ciclospermum*) *laciniatum*, *Encelia canescens*, *Chenopodium petiolare*, *Mirabilis prostrata* and *Loasa urens*. *Alstroemeria paupercula* occurs from Caldera in Chile to Chala in south-central Peru. *Pasithea coerulea* and *Fortunatia biflora* occur from central Chile to southern Peru.

Although many of the Peruvian and Atacama desert species have obvious floristic affinities with the Andean Cordillera, the level of endemism in the isolated coastal vegetation is extremely high for some families, e.g. Bromeliaceae, Cactaceae, Malvaceae, Aizoaceae, Portulacaceae, Solanaceae and Poaceae. Two endemic families of Andean flora, the Nolanaceae and Malesherbiaceae, are largely Atacama/Peruvian desert groups (Solbrig 1976). Most of the important desert families, however, are widespread: Apiaceae, Asteraceae, Boraginaceae, Cruciferae and Leguminosae.

The most endemic genera are to be found in southern Peru between 15°–18°S, and in northern Atacama formations, specifically between 24°14'–26°21'S. These include the largely Peruvian genera *Mathewsia* and *Dictyophragmus* (Cruciferae) and *Weberbaueriella* (Leguminosae), and the largely Chilean genera *Copiapoa* and *Eulychnia* (Cactaceae), *Dinemandra* (Malpighiaceae), *Domeykoa* and *Gymnophyton* (Apiaceae) and *Gypothamnium* and *Oxyphyllum* (Asteraceae). A considerable number of endemics occur within a wide range of genera, including species of *Ambrosia*, *Argylia*, *Astragalus*, *Nolana*, *Calceolaria*, *Palaua*, *Cristaria*, *Tiquilia*, *Dinemandra* and *Eremocharis*. Müller (1985) calculated 42% overall endemism within the Peruvian lomas formations, with 62% endemism for southern Peru and 22% for central Peru.

In contrast, in northern Peru between 8°S and 12°S occurs the greatest number of species with coastal disjunct populations and principally Andean distributions. In this region, various families with typically more mesic requirements are represented – e.g. Orchidaceae, Passifloraceae, Piperaceae and Begoniaceae.

A few North American species with amphitropic disjunct distributions are present in the lomas formations. For the most part these are not the usual Mediterranean-climate species commonly disjunct from California, U.S.A. to central Chile (Constance 1963; Raven 1963, 1972). Rather, the species have their origins in the Sonora or Mojave deserts and their distributions are often restricted to the lomas formations. Also several primarily North American genera have endemic species within the coastal deserts, including *Ambrosia dentata*, *Encelia canescens*, *Viguiera weberbaueri* and several species of *Tiquilia*. Other primarily North American taxa represented within the Peruvian and/or Atacama deserts include *Nama dichotoma*, *Phyla nodiflora*, *Linaria canadensis*, *Microcala quadrangularis*, *Triodanis perfoliata*, *Cressa truxillensis*, *Malacothrix clevelandii*, *Bahia ambrosioides*, *Amblyopappus pusillus* and *Perityle emoryi*.

A few taxa are derived from other arid regions of the world. For example, the widespread succulents *Carpobrotus chilensis* and *Mesembryanthemum crystallinum* likely have origins in South Africa.

Useful plants

Archaeologists have evidence that indigenous peoples have used the lomas formations as a temporal resource for over 5000 years (Moseley, Feldman and Ortloff 1981); in their original state, the lomas provided essential sites for foraging and the periodic cultivation of crops. No sustainable agriculture is currently conducted within the lomas formations.

As opportunities for genetic engineering expand, many taxa within the lomas formations will be sources of germplasm for agriculture and horticulture. For example, the economically important genera *Solanum* and *Lycopersicon* contain several lomas endemics with potentially useful ecophysiological characteristics (Hanson, pers. comm.).

Social and environmental values

The intrinsic value of the lomas formations is the unique nature of their composition and their high levels of endemism. The fauna of the lomas has not been studied in detail, but is potentially as rich in endemics as the flora. For these reasons, the lomas formations provide opportunities for scientific research and ecotourism. The Reserva Nacional de Lachay is one example of the successful management of a lomas formation, while providing recreational opportunities. If the impact on delicate communities is controlled through supervision, other lomas formations can be preserved for the future while being enjoyed by the public – especially the striking display resulting from El Niño events.

Twelve bird species of restricted range occur in the South Peruvian and north Chilean Pacific slope Endemic Bird Area (EBA B32), which extends from Peru's Lima Department southward to northern Chile. The birds restricted to this coastal strip and the Pacific slope foothills inhabit most of the vegetation associations, with at least three species (*Geositta crassirostris*, *Anairetes reguloides*, *Sicalis raimondii*) known to use the ephemeral lomas vegetation. Three species restricted to this EBA are presently considered threatened, but primarily due to their reliance on riparian vegetation rather than the lomas.

Economic assessment

The lomas formations are not generally considered of great economic potential because of the unpredictable and strongly fluctuating environmental conditions within the coastal desert. Nevertheless, investigations of the water potential from wind-blown fog have shown that considerable moisture is available (Oka 1986). However, continuing agriculture or silviculture in a lomas formation would be extremely damaging. Rather, their value lies in the potential for germplasm sources, ecotourism and continued scientific investigations.

Threats

The lomas formations are periodically used for grazing livestock, always to the detriment of the natural vegetational communities. Local inhabitants utilize many of the lomas formations for forage in favourable years, and they often gather the woody species for fuelwood (Ferreyra 1977). Destruction of natural communities by overgrazing is quite common (Dillon and Rundel 1991; Dillon 1991). During the exceptional El Niño of 1982–1983, farmers moved large numbers of cattle, sheep and goats from the drought-stricken sierra to various lomas formations in southern Peru. In addition to large-scale destruction of the perennials, these events foster the introduction of more weedy taxa common to higher elevation Andean environments. Both maize and wheat were cultivated in the Lomas de Tacna; the effects of plowing on the soil's seed banks have not been studied (Ohga 1982, 1986).

Urbanization and population growth in coastal cities have placed many lomas localities in peril. In the immediate vicinity of Lima, a number of lomas formations have been severely disturbed or eliminated (e.g. Amancaes, Chorrillos, Cajamarquilla, Cerro Agustino, Manzano), due to the population expansion of the last 60 years (Cuya and Sánchez 1991). Mining has had an impact on both the terrestrial and marine communities (Echegary et al. 1990). Other threats are introduction of orchard species, construction of irrigation canals and mining for the materials used in housing and road construction.

Conservation

Only a few areas within the desert coast are under the management and protection of the Peruvian Government (Díaz-Cartagena 1988). The Reserva Nacional de Lachay (Department Lima, 51 km²), which was established in 1977, is the only lomas with more than 100 species of vascular plants currently protected. The Reserva Nacional de Paracas (1170 km²) and Santuario Nacional de Mejía (7 km²), both primarily aquatic environments, include some adjacent lomas elements. In the *Plan Director del Sistema Nacional de Unidades de Conservación* (i.e. SINUC), several other coastal sites have been proposed for conservation, including Península Illescas, Department Piura (925 km²); Cerro Campana, Department La Libertad (42 km², 185 spp.); Albuferas de Playa Chica y Paraíso, Department Lima (3 km²); San Fernando, Department Ica (145 km²); and Lomas de Atiquipa y Taimara, Department Arequipa (47 km², 120 spp.) (CDC 1991).

A specimen-oriented computerized database of over 7000 records (LOMAFLOR, Dillon unpublished) has been constructed and the analysis is underway to determine patterns of diversity and endemism within the lomas formations. These data will provide recognition of the areas with maximum biodiversity to establish conservation priorities. The total area of immediately threatened habitat cannot be excessively large (since it will be less than 2000 km²). Prospects for successful conservation efforts in the lomas formations are excellent if the recommended biodiversity priorities are followed (Keel 1987; CDC 1991).

References

Barclay, W.S. (1917). Sand dunes in the Peruvian desert. *Geogr. J.* (London) 49: 53–56.

Broggi, J.A. (1961). Las ciclópeas dunas compuestas de la costa peruana, su origen y significación climática. *Bol. Soc. Geol. Perú* 36: 61–66.

Caviedes, C.L. (1975). Secas and El Niño: two simultaneous climatic hazards in South America. *Proc. Assoc. Amer. Geogr.* 7: 44–49.

CDC (1991). *Plan Director del Sistema Nacional de Unidades de Conservación: una aproximación desde la diversidad biológica.* Centro de Datos para la Conservación (CDC)-Universidad Nacional Agraria La Molina (UNALM), Lima.

Clüsener, M. and Breckle, S.W. (1987). Reasons for the limitation of mangrove along the west coast of northern Peru. *Vegetatio* 68: 173–177.

Constance, L. (1963). Amphitropical relationships in the herbaceous flora of the Pacific coast of North and South America: a symposium. *Quart. Rev. Biol.* 38: 109–116.

Craig, A.K. and Psuty, N.P. (1968). *The Paracas papers.* Vol. 1, No. 2 *Reconnaisance report.* Florida Atlantic University, Dept. Geography Occasional Publication No. 1, Boca Raton. 196 pp.

Cuya, O. and Sánchez, S. (1991). Flor de Amancaes: lomas que deben conservarse. *Bol. Lima* 76: 59–64.

Díaz-Cartagena, A. (1988). El Sistema Nacional de Unidades de Conservación en el Perú. *Bol. Lima* 59: 13–22.

Dillon, M.O. (1985a). The botanical response of the Andean desert lomas formations to the 1982–83 El Niño event. Abstr. *Amer. J. Bot.* 72: 950.

Dillon, M.O. (1985b). The silver lining of a very dark cloud, botanical studies in coastal Peru during the 1982–83 El Niño event. *Field Mus. Nat. Hist. Bull.* 56: 6–10.

Dillon, M.O. (1989). Origins and diversity of the lomas formations in the Atacama and Peruvian deserts of western South America. Abstr. *Amer. J. Bot.* 76: 212.

Dillon, M.O. (1991). A new species of *Tillandsia* (Bromeliaceae) from the Atacama Desert of northern Chile. *Brittonia* 43: 11–16.

Dillon, M.O. (in prep.). Flora of the lomas formations of Chile and Peru. *Fieldiana: Botany.*

Dillon, M.O. and Cadle, J.E. (1991). Biological inventory of bosque monteseco (Cajamarca, Peru) from a diversity and biogeographic perspective. Abstr. *Amer. J. Bot.* 78: 180.

Dillon, M.O. and Rundel, P.W. (1989). The botanical response of the Atacama and Peruvian desert floras to the 1982–83 El Niño event. In Glynn, P.W. (ed.), *Global ecological consequences of the 1982–83 El Niño-Southern Oscillation.* Elsevier Oceanography Series, Amsterdam. Pp. 487–504.

Duncan, T. and Dillon, M.O. (1991). Numerical analysis of the floristic relationships of the lomas of Peru and Chile. Abstr. *Amer. J. Bot.* 78: 183.

Echegary, M., Hinojosa, I., Ormeño, D., Zambrano, W. and Taype, L. (1990). Contaminación por cobre en el litoral sur del Perú. *Bol. Lima* 72: 23–27.

Ellenberg, H. (1959). Über den Wasserhaushalt tropischer Nebeloasen in der Küstenwüste Perus. *Ber. Geobot. Forsch. Inst.* Rübel 1958: 47–74.

Ferreyra, R. (1953). Comunidades de vegetales de algunas lomas costaneras del Perú. *Estac. Exp. Agrícola "La Molina", Bol.* 53: 1–88.

Ferreyra, R. (1957). Contribución al conocimiento de la flora costanera del norte peruano (Departamento de Tumbes). *Bol. Soc. Argent. Bot.* 6: 194–206.

Ferreyra, R. (1960). Algunos aspectos fitogeográficos del Perú. *Rev. Inst. Geogr. Univ. Nac. Mayor San Marcos (Lima)* 6: 41–88.

Ferreyra, R. (1961a). Las lomas costaneras del extremo sur del Perú. *Bol. Soc. Argent. Bot.* 9: 87–120.

Ferreyra, R. (1961b). Revisión de las especies peruanas del género *Nolana. Mem. Mus. Hist. Nat. "Javier Prado"* 12: 1–53.

Ferreyra, R. (1977). Endangered species and plant communities in Andean and coastal Peru. In Prance, G.T. and Elias, T.S. (eds), *Extinction is forever.* New York Botanical Garden, Bronx. Pp. 150–157.

Ferreyra, R. (1979). El algarrobal y manglar de la costa norte del Perú. *Bol. Lima* 1: 12–18.

Ferreyra, R. (1983). Los tipos de vegetación de la costa peruana. *Anales Jardín Bot. Madrid* 40: 241–256.

Finkel, H.J. (1959). The barchans of southern Peru. *J. Geol.* 67: 614–647.

Gay, P. (1962). Origen, distribución y movimiento de las arenas eólicas en el área de Yauca a Palpa. *Bol. Soc. Geol. Perú* 37: 37–58.

Gómez-Sosa, E. (1986). *Astragalus neobarnebyanus* (Leguminosae): a new species from Peru. *Brittonia* 38: 427–429.

Goodspeed, T.H. (1961). *Plant hunters in the Andes.* University of California Press, Berkeley. 378 pp.

Goodspeed, T.H. and Stork, H.E. (1955). The University of California Botanical Garden expeditions to the Andes (1935–1952): with observations on the phytogeography of Peru. *Univ. Calif. Publ. Bot.* 28(3): 79–142.

Huey, R.B. (1979). Parapatry and niche complementarity of Peruvian desert geckos (*Phyllodactylus*): the ambiguous role of competition. *Oecologia* 38: 249–259.

Jaramillo-Arango, J. (ed.) (1952). *Relación histórica del viage, que hizo a los reynos del Perú y Chile el botánico Dn. Hipólito Ruiz en el año de 1777 hasta el de 1788, en cuya época regresó a Madrid,* 2nd edition. Real Acad. Ci. Madrid, Madrid. 526 + 245 pp.

Johnson, A.M. (1976). The climate of Peru, Bolivia and Ecuador. In Schwerdtfeger, W. (ed.), *Climates of Central and South America.* World Survey of Climatology Vol. 12. Elsevier, Amsterdam. Pp. 147–218.

Johnston, I.M. (1929). Papers on the flora of northern Chile. *Contr. Gray Herb.* 85: 1–172.

Johnston, I.M. (1931). The vascular flora of the Guano Islands of Peru. *Contr. Gray Herb.* 95: 26–35.

Johnston, I.M. (1936). A study of the Nolanaceae. *Contr. Gray Herb.* 112: 1–83.

Keel, S. (1987). The ephemeral lomas of Peru. *The Nature Conservancy Mag.* 37(5): 16–20.

Kinzl, H. (1958). Die Dünen in der Küstenlandschaft von Peru. *Mitt. Geogr. Ges. Wien* 100: 5–17.

Library of Congress, Science and Technology Division (1979). Draft, *Environmental report on Peru.* U.S. Department of State, AID/DS/ST Contract No. SA/TOA 1–77, with U.S. Man and the Biosphere Secretariat. Washington, D.C. 110 pp. + 5. Appendices A and 1B.

Moseley, M.E., Feldman, R.A. and Ortloff, C.R. (1981). Living with crisis: human perception of process and time. In Nitecki, M. (ed.), *Biotic crises in ecological and evolutionary time.* Academic Press, New York. Pp. 231–267.

Müller, G.K. (1985). Zur floristischen Analyse der peruanischen Loma-Vegetation. *Flora* 176: 153–165.

Müller, G.K. (1988). Anthropogene Veränderungen der Loma-Vegetation Perus. *Flora* 180: 37–40.

Ocrospoma-Jara, M. (1990). Líquenes de las Lomas de Lurín, Lima. *Bol. Lima* 72: 28–29.

Ohga, N. (1982). Buried seed population in soil in the lomas vegetation. In Ono, M. (ed.), *A preliminary report of taxonomic and ecological studies on the lomas vegetation in the Pacific coast of Peru*. Makino Herbarium, Tokyo Metropol. University, Tokyo. Pp. 53–80.

Ohga, N. (1986). Dynamics of the buried seed population in soil, and the mechanisms of maintenance of the herbaceous lomas vegetation in the coastal desert of central Peru. In Ono, M. (ed.), *Taxonomic and ecological studies on the lomas vegetation in the Pacific coast of Peru*. Makino Herbarium, Tokyo Metropol. University, Tokyo. Pp. 53–78.

Oka, S. (1986). On trial measurements of the moisture in fog on Loma Ancon – in relation to an investigation into the conditions required for development of lomas communities. In Ono, M. (ed.), *Taxonomic and ecological studies on the lomas vegetation in the Pacific coast of Peru*. Makino Herbarium, Tokyo Metropol. University, Tokyo. Pp. 41–51.

ONERN (1971a). *Inventario, evaluación y uso racional de los recursos naturales de la costa. Cuenca del río Grande (Nazca)*. Oficina Nacional de Evaluación de Recursos Naturales (ONERN), Lima.

ONERN (1971b). *Inventario, evaluación y uso racional de los recursos naturales de la costa. Cuenca del río Ica*. ONERN, Lima.

ONERN (1971c). *Inventario, evaluación y uso racional de los recursos naturales de la costa. Cuenca del río Pisco*. ONERN, Lima.

Péfaur, J.E. (1982). Dynamics of plant communities in the lomas of southern Peru. *Vegetatio* 49: 163–171.

Prohaska, F. (1973). New evidence on the climatic controls along the Peruvian coast. In Amiran, D.H.K. and Wilson, A.W. (eds), *Coastal deserts, their natural and human environments*. University of Arizona Press, Tucson. Pp. 91–107.

Quinn, W.H., Neal, V.T. and Antúnez de Mayolo, S.E. (1987). El Niño occurrences over the past four and a half centuries. *J. Geophys. Res.* 92: 14,449–14,461.

Rauh, W. (1958). *Beitrag zur Kenntnis der peruanischen Kakteenvegetation. Sitzungsber. Heidelberger Akad. Wiss., Math.-Naturwiss. Kl.* 1958(1). 542 pp.

Rauh, W. (1985). The Peruvian-Chilean deserts. In Evenari, M., Noy-Meir, I. and Goodall, D.W. (eds), *Hot deserts and arid shrublands*, Vol. A. Ecosystems of the World, Vol. 12A. Elsevier, Amsterdam. Pp. 239–267.

Raven, P.H. (1963). Amphitropical relationships in the floras of North and South America. *Quart. Rev. Biol.* 38: 151–177.

Raven, P.H. (1972). Plant species disjunctions: a summary. *Ann. Missouri Bot. Gard.* 59: 234–246.

Robinson, D.A. (1964). *Peru in four dimensions*. American Studies Press, Lima. 424 pp.

Rundel, P.W., Dillon, M.O., Palma, B., Mooney, H.A., Gulmon, S.L. and Ehleringer, J.R. (1991). The phytogeography and ecology of the coastal Atacama and Peruvian deserts. *Aliso* 13(1): 1–50.

Sagástegui, A., Mostacero, J. and López, S. (1988). Fitoecología del Cerro Campana. *Bol. Soc. Bot. La Libertad* 14: 1–47.

Silbitol, R.H. and Ericksen, G.E. (1953). Some desert features of northwest central Peru. *Bol. Soc. Geol. Perú* 26: 225–246.

Simpson, B.B. (1975). Pleistocene changes in the flora of the high tropical Andes. *Paleobiology* 1: 273–294.

Solbrig, O.T. (1976). The origin and floristic affinities of the South American temperate desert and semidesert regions. In Goodall, D.W. (ed.), *Evolution of desert biota*. University of Texas Press, Austin. Pp. 7–50.

Trewartha, G.T. (1961). *The earth's problem climates*. University of Wisconsin Press, Madison. 371 pp.

Weberbauer, A. (1911). *Die Pflanzenwelt der peruanischen Anden*. Vegetation der Erde 12. Englemann, Leipzig. 355 pp.

Weberbauer, A. (1936). Phytogeography of the Peruvian Andes. *Field Mus. Nat. Hist., Bot. Ser.* 13: 13–81.

Weberbauer, A. (1939). La influencia de cambios climáticos y geológicos sobre la flora de la costa peruana. *Acad. Nac. Cienc. Exactas Fí. Nat.* 2: 201–209.

Weberbauer, A. (1945). *El mundo vegetal de los Andes peruanos*. Ministerio de Agricultura, Dirección de Agricultura, Estación Experimental Agrícola de La Molina, Lima. 776 pp.

Author

This Data Sheet was written by Dr Michael O. Dillon (Field Museum of Natural History, Center for Evolutionary and Environmental Biology, Department of Botany, Chicago, IL 60605-2496, U.S.A.).

LOMAS FORMATIONS OF THE ATACAMA DESERT
Northern Chile

Location: Series of almost 50 isolated localities in Atacama Desert along coastal Chile, between latitudes 18°24'–29°54'S.

Area: Less than 5000 km² of vegetation formations dispersed in western fringe of 291,000 km² of arid lands.

Altitude: 0–1100 m.

Vegetation: Community compositions are highly variable: mixtures of annual, short-lived perennial and woody scrub vegetation.

Flora: c. 550 spp. of ferns, gymnosperms and flowering plants; very high endemism (over 60%); many threatened species.

Useful plants: Unusual ecophysiological characteristics make the plants valuable as potential genetic resources.

Other values: Communities highly endemic; support for mammal, bird and insect communities; tourism.

Threats: Urbanization, mining, pollution, road construction, livestock grazing – especially numerous goats, fuelwood gathering, commercial plant collecting, erosion.

Conservation: La Chimba National Reserve (c. 30 km²); Pan de Azúcar National Park (c. 438 km²).

Geography

The Atacama Desert forms a continuous strip for nearly 1600 km along the narrow coast of the northern third of Chile in its Regions I–IV: Tarapacá, Antofagasta, Atacama, northern Coquimbo, from near Arica (18°24'S) southward to near La Serena (29°55'S). Adjacent to the Pacific are uplifted marine terraces or the coastal range arising directly from the ocean; farther eastward is a large alluvial plain, which is bordered by the Andes. Although arid conditions extend in the Andes up to c. 3500–4000 m, the Atacama Desert is usually deemed to end at their base or near 1500 m on the drier slopes (Börgel 1973). The faulted coastal mountains (mostly 500–1000 m high) are composed of Cretaceous sediments (limestone and sandstone) over more ancient masses of crystalline rocks. Whereas desert conditions are continuous along the coast of Peru and Chile, the topography, climate and vegetation of each desert are distinct (Harrington 1961; Rundel *et al.* 1991; Duncan and Dillon 1991; Rauh 1985). For some generalities and comparison with the coastal Peruvian Desert, see the Data Sheet on Lomas Formations – Peru, CPD Site SA42.

The conditions for development of hyper-arid habitats in this region are similar to those in coastal Peru, including the influence of cool sea-surface temperatures associated with the northward-flowing oceanic Humboldt Current, and strong atmospheric subsidence associated with a positionally stable, subtropical anticyclone. The result is a mild, uniform coastal climate with regular thick stratus cloud banks well below 1000 m during the winter months. The northern seaport Iquique has average monthly temperatures between 14.5°C (September) and 21°C (March); the recorded absolute extremes have been 8°C and 31.3°C. The Tropic of Capricorn crosses the region slightly north of the city of Antofagasta.

Where the coastal topography is low and flat, the winter stratus layer dissipates inward over broad areas with little biological effect, but where isolated mountains or steep coastal slopes intercept the clouds, a fog zone develops with a stratus layer concentrated against the hillsides. These fogs ("camanchacas") are the key to the extent and diversity of vegetation throughout the desert. The moisture allows the development of fog-zone plant communities termed "lomas" (small hills) between sea-level and 1100 m. These plant formations also have been called the fertile belt, fog oases or meadows on the desert. In Chile almost 50 discrete localities support such vegetation (see combined Map 77 on p. 520 with the Data Sheet Lomas Formations – Peru, CPD Site SA42). The Chilean area actually covered by this vegetation probably is less than 5000 km², even during periods of optimal weather and maximum development.

Much of western South America is influenced by brief periods of heavy rainfall and relatively high temperature associated with rare but recurrent El Niño events (Dillon and Rundel 1989; Quinn, Neal and Antúnez de Mayolo 1987; McGlone, Kershaw and Markgraf 1992). However, heavy storms are virtually unknown along the north coast of Chile, whereas northern Peruvian coastal cities occasionally receive torrential rains. The maximum precipitations for a 24-hour period recorded in Chile at Arica, Iquique and Antofagasta are 10, 13 and 28 mm respectively (Miller 1976). Johnston (1929) reported 17 mm at Antofagasta during a single day in 1925, a strong El Niño year. However, El Niño years can fail to produce increased precipitation in northern Chile. The El Niño event of 1983 caused heavy rains from northern to southern Peru, and a relatively high 7.3 mm in Iquique, but no notable rains farther south (Romero and Garrido 1985; Rutllant 1985). Nonetheless, that year's fogs appear to have been unusually dense, resulting in excellent flowering in the Chilean lomas formations (Prenafeta 1984).

Vegetation

Lomas formations occur as discrete communities or islands of vegetation separated by hyper-arid territory devoid of plant life. Since growth is dependent upon available moisture, the climatic patterns determine plant distributions. Topography and substrate combine to influence the patterns of moisture availability and areas of suitable habitat. The vegetation is largely restricted to the coastal escarpment and lower portions of numerous gorges ("quebradas") that traverse it. Ecological requirements and tolerances of individual species ultimately determine community composition; species endemism exceeds 60% (Rundel *et al.* 1991). Broad overviews of the vegetation have been provided by Schmithüsen (1956) and Rauh (1985). The following account briefly describes the larger lomas formations, including the data available on species diversity.

Northern Atacama Desert

The coastal zone from just north of Arica (near Peru) to the seaport Antofagasta is essentially barren, having extremely low species diversity and density. The values of mean annual precipitation reported for northern Chile are the lowest of any long-term records in the world, with Arica and Iquique averaging 0.6 mm and 2.1 mm respectively; in several consecutive years there may be no rainfall. Along the escarpment above 500 m grow scattered individuals of cacti – the columnar *Eulychnia iquiquensis*, and *Copiapoa* species. Near Iquique the vegetation is only weakly developed; some 20 species have been recorded. An exception 10 km east of Iquique at 990–1100 m is a large community of *Tillandsia landbeckii* on pure sand. The communities near Tocopilla with over 60 species recorded and Cobija with 15 species have only a few endemics, including the rare *Malesherbia tocopillana*, *Mathewsia collina* and *Nolana tocopillensis* (Jaffuel 1936).

The slopes and coastal plain around Antofagasta are virtually free of any vegetation except scattered individuals of *Eulychnia iquiquensis* and *Copiapoa* species. North of the city within Quebrada La Chimba vegetation tends to occur along the margins of the dry river canyon where runoff concentrates. Johnston (1929) collected around the city in the spring of 1925 following heavy rains and found 34 species; herbarium records and recent fieldwork indicate that nearly 60 species occur (Dillon, unpublished).

Cerro Moreno, a prominent rocky headland north-west of Antofagasta, has a small fog-dependent plant community on its south-western slope at a relatively high elevation (Follmann 1967). The vascular plant species recorded number just 28, less than half as many as at nearby La Chimba. The lower margin of the fog zone on Cerro Moreno is near 600 m and marked by scattered populations of *Copiapoa*. Although *Eulychnia iquiquensis* is dominant, its vigour is poor. The shrub stratum is most notably absent within the zone – only a few shrubby species occur, such as *Heliotropium pycnophyllum*, *Ephedra breana* and *Lycium deserti*. Annual and perennial herbs include *Cynanchum viride*, *Viola polypoda* and *Argythamnia canescens*. Two essentially non-desert species disjunct from central Chile have small populations at the site: *Acaena trifida* and *Colliguaja odorifera*. Vegetation along the coast below this fog zone is virtually absent, with only a few scattered individuals of *Nolana peruviana* and *Tetragonia angustifolia*.

Southern Atacama Desert

The richest development of the fog-zone vegetation in northern Chile is 50 km north of Taltal near the mining village Paposo (25°03'S), with no fewer than 230 associated species of vascular plants. The central portion of this fog zone is dominated by *Euphorbia lactiflua* and *Eulychnia iquiquensis* (Rundel and Mahu 1976). Other important shrubby species are *Echinopsis coquimbana* (*Trichocereus coquimbanus*), *Oxalis gigantea*, *Lycium stenophyllum*, *Proustia cuneifolia*, *Croton chilensis*, *Balbisia peduncularis* and *Tillandsia geissei*. Annuals include *Viola litoralis*, *V. polypoda*, *Cruckshanksia pumila*, *Alstroemeria graminea*, *Malesherbia humilis* and *Chaetanthera glabrata*. Above and below the central fog zone the coverage of species drops sharply and growth forms of the dominants change. Whereas drought-deciduous tall shrubs dominate within the zone, the importance of cacti and low semi-woody subshrubs increases away from the centre.

Dense mounds of the bromeliads *Deuterocohnia chrysantha* and *Puya boliviensis* are locally common along the coastal flats and mark the lower margin of the fog zone at c. 300 m. The coastal plain below the fog supports broad stands of *Copiapoa cinerea* var. *haseltoniana* and scattered shrubs, in communities similar to those at Pan de Azúcar to the south. Above the uppermost limits of regular fog at c. 800 m, *Copiapoa* stands again become dominant along with *Polyachyrus cinereus*, *Oxalis caesia*, *Nolana stenophylla*, *N. villosa*, *N. sedifolia* and *N. peruviana*. Some shrub species such as *Gypothamnium pinifolium* and *Chuquiraga ulicina* occur below and above this fog zone (Rundel and Mahu 1976; Johnston 1929). Large areas of coastal terraces from north of Paposo to Pan de Azúcar are covered by annuals, predominantly *Nolana aplocaryoides*.

Pan de Azúcar National Park (c. 26°04–26°09'S, 29 km north of Chañaral) has been described in some detail; the mean annual temperature is 16.4°C. Two types of plant communities are present (Rundel *et al.* 1980). One is composed of low succulents, with virtually mono-specific stands of *Copiapoa cinerea* var. *cinerea* (form "*columna-alba*") or *C. cinerea* var. *haseltoniana* (Mooney, Weisser and Gulmon 1977; Gulmon *et al.* 1979; Ehleringer *et al.* 1980; Anderson *et al.* 1990). Small mostly subterranean cacti such as *Neoporteria* sp. (= *Neochilenia malleolata*, *Thelocephala krausii*) are also present (Weisser 1967a, 1967b). A second type of community consists of open stands of less succulent shrubs on "bajadas" and in washes. The dominants are *Nolana mollis*, *Heliotropium linearifolium*, *Gypothamnium pinifolium*, *Oxyphyllum ulicinium*, *Euphorbia lactiflua*, *Tetragonia maritima* and *Eremocharis fruticosa*.

The inland region from Chañaral south to the Copiapó River marks the northernmost portion of El Norte Chico (Muñoz-Schick 1985). This region has open communities of scattered low shrubs of *Skytanthus acutus*, *Encelia canescens*, *Frankenia chilensis* and *Nolana* (*Alona*) *rostrata*. Annuals and short-lived perennials in favourable habitats include *Perityle emoryi*, *Oenothera coquimbensis*, *Adesmia latistipula*, *Astragalus coquimbensis*, *Cruckshanksia verticillata*, *Fagonia chilensis* and *Tetragonia angustifolia*. Numerous geophytes (with bulbs or corms) are also conspicuous in the region when sufficient moisture becomes available.

Immediately to the south of Chañaral several species have their southernmost distribution, including *Gypothamnium pinifolium*, *Dinemandra ericoides*, *Nolana aplocaryoides* and *Tiquilia litoralis*. The coastal strand and dune formations of the littoral belt from Chañaral southward to Caldera and the Copiapó River have many of the same species that occur in the interior, and additional characteristic species such as *Nolana divaricata, Heliotropium stenophyllum, H. linearifolium, H. pycnophyllum, Oxalis gigantea, Ophryosporus triangularis, Ephedra andina, Euphorbia lactiflua, Bahia ambrosioides, Senna cumingii* var. *coquimbensis, Tetragonia maritima, Echinopsis coquimbana* and *Eulychnia acida* (Kohler 1970; Reiche 1911; Opazo and Reiche 1909).

Quebrada El León (26°59'S) in the coastal area 15 km north of Caldera is a canyon with extraordinary rocky formations, running water and pools. The rich and diverse vegetation includes *Euphorbia lactiflua, Opuntia sphaerica* (*O. berteri auct.*), *Eulychnia breviflora, Copiapoa* spp. and the perhaps endemic *Neoporteria* sp. (= *Neochilenia transitensis*) (Anderson *et al.* 1990).

The broad lower valley of the Copiapó River funnels sufficient moist maritime air masses nearly 50 km inland to support scattered populations of coastal lichen species on the rocky hills around the city. However between Copiapó and Vallenar, vegetation is virtually or totally absent over extensive areas. Thin veneers of sand cover rocky gentle slopes over much of this region. The climate as well as the effects of human activities have been important in producing this arid landscape. The only significant species along many km of the Pan American highway are perennials, such as *Argylia radiata, Bulnesia chilensis, Encelia canescens, Frankenia chilensis, Caesalpinia angulata, Polyachyrus fuscus, P. poeppigii, Nolana rostrata* and *N. pterocarpa* (Kohler 1970).

Quebrada Carrizal Bajo (28°08'S) is in the coastal area between Totoral and Huasco. The coastal formations in this region have vegetation corresponding to the Desierto Costero Huasco (Gajardo 1987). Due to occasional precipitation, this vegetation has more permanence and continuity than the vegetation occurring farther north. The endemics include *Copiapoa echinoides, C. malletiana, Neoporteria carrizalensis* and *N. villosa* var. *laniceps* (Anderson *et al.* 1990). At times this region is dominated by geophytes, which can cover expanses of the desert spectacularly. Near Huasco, dune formations support a well-developed cover of halophytic species, notably *Sarcocornia fruticosa, Carpobrotus chilensis* and *Distichlis spicata* (Kohler 1970).

The geographical southern limit of the Atacama Desert is subject to a variety of views, with most opinions favouring either just north of La Serena (29°54'S) or 280 km farther north at the Copiapó River (27°20'S). La Serena receives erratic precipitation averaging less than 130 mm per year, compared with only 29 mm at Copiapó and almost 0 mm through much of northern Chile. On a floristic basis, La

CPD Site SA43: Lomas formations of the Atacama Desert. Cerro Perales, near Taltal, southern Atacama Desert, Chile.
Photo: Michael O. Dillon.

Serena is clearly the most appropriate periphery. Near La Serena north of the Elqui River Valley, semi-arid coastal scrub vegetation is replaced by desert succulent communities with floristic affinities to the regions northward (Rundel *et al.* 1991).

Desert vegetation of interior valleys between Copiapó and La Serena is sparse but regularly present, compared to the more arid regions farther north. Beginning c. 40 km south of Vallenar, the vegetation coverage increases markedly and continues to north of La Serena. Succulent communities with mound-forming species of *Eulychnia* and *Opuntia* (*Tephrocactus*) are dominant, and the shrubby *Heliotropium stenophyllum, Balsamocarpon brevifolium* and *Bulnesia chilensis* are common. In wash habitats important shrubs include *Cordia decandra, Adesmia argentea, A. microphylla* and *Ephedra*.

North of La Serena along the coast, precipitation is sufficient to support relatively high vegetation coverage in a transitional community with representatives from the coastal matorral of central Chile (Rundel 1981). Frequently encountered species include *Echinopsis coquimbana, Oxalis gigantea, Lobelia polyphylla, Myrcianthes coquimbensis, Puya chilensis, Lithrea caustica, Heliotropium stenophyllum, Nolana coelestis, N. crassulifolia* and *N. sedifolia*. Coastal dunes are dominated by *Nolana divaricata* and *Tetragonia maritima* (Kohler 1970), and associated species include *Nolana* (*Alona*) *carnosa, Ephedra breana* and *Skytanthus acutus*. Numerous ephemeral geophytes are manifest when rare rains fall, including *Alstroemeria* spp., *Tecophilaea* spp., *Leucocoryne* spp., *Rhodophiala* spp. and *Leontochir ovallei* (Muñoz-Schick 1985).

Flora

Recent catalogues on the vascular flora of Chile have been compiled (Arroyo, Marticorena and Muñoz 1990; Marticorena and Quezada 1985; Gajardo 1987), and a floristic inventory of the c. 7000 km² of lomas formations in Chile and Peru has been completed (Dillon, in preparation for publication). The northern Atacama (El Norte Grande) is a centre of pronounced endemism – a large number of species range only slightly to its north or south. The flora of this region was the focus of R.A. Philippi's first collecting efforts in Chile in 1853–1854, culminating in the *Florula Atacamensis* (1860). This is the type locality for nearly 200 of his newly described species. Reiche (1907, 1911, 1934–1937) contributed a series of papers with descriptions and notes on the flora of northern Chile. Johnston (1929) and Werdermann (1931) described the flora of the northern area from field studies in 1925 (a strong El Niño year). A photographically illustrated Flora includes about the southern

CPD Site SA43: Lomas formations of the Atacama Desert. Population of *Copiapoa cinerea* in Parque Nacional Pan de Azúcar, southern Atacama Desert, Chile. Photo: Michael O. Dillon.

quarter of the Atacama, in El Norte Chico (south of Chañaral to north of Petorca-La Ligua) (Muñoz-Schick 1985).

The estimated total numbers of taxa of vascular plants represented within the Atacama Desert's lomas formations are 80 families, 225 genera, 550 species (Dillon, in prep.). The families with the most species diversity are Asteraceae (33 genera, 65 spp.); Nolanaceae (2 genera, 37 spp.); Leguminosae (13 genera, 35 spp.); Cactaceae (7 genera, c. 30–40 spp. – cf. Hoffmann-J. 1989, Anderson *et al.* 1990); Boraginaceae (5 genera, 26 spp.); Solanaceae (9 genera, 19 spp.); and Apiaceae (8 genera, 17 spp.). Only 68 of the species (c. 7%) reach the lomas in Peru north of 18°S latitude.

The lomas formations from Arica to Antofagasta have little floristic affinity to the Peruvian formations immediately to the north. Of the 117 species of vascular plants recorded from this region by Johnston (1929), only five species (4%) were reaching their southern limit. In contrast, 89 species (76%) represented northern extensions of the richer flora south of Antofagasta. The remaining 23 species (20%) were endemic to the region, most notably species of *Nolana*.

The area around Arica itself clearly is related floristically to the Peruvian lomas, but many genera and species drop out at or immediately to the south of Arica. For example, the four most important groups of coastal Peruvian cacti either do not cross into Chile (the *Loxanthocereus* group of

Cleistocactus, Neoraimondia) or do not occur south of Arica (the *Islaya* group of *Neoporteria, Haageocereus* – except *H. fascicularis* occurs well above the lomas inland in Tarapacá); and similarly three of the most important cactus groups of northern Chile either do not cross into Peru (*Copiapoa*) or hardly do so (*Eulychnia, Neoporteria sensu stricto*) (Rauh 1958; Hoffmann-J. 1989). Terrestrial *Tillandsia* species are a dominant aspect of the coastal vegetation of southern Peru, but of very limited distribution in the Atacama Desert. The southernmost limit of these bromeliads is between Iquique and Tocopilla, where pure stands of *T. landbeckii* occur.

Important shrubby species primarily confined to the Atacama Desert include *Berberis litoralis, Anisomeria littoralis, Atriplex taltalensis, Adesmia viscidissima, Croton chilensis, Balbisia peduncularis, Nicotiana solanifolia, Teucrium nudicaule, Monttea chilensis, Stevia byssopifolia, Senecio almeidae, Gutierrezia taltalensis* and *Haplopappus deserticola*. The region is a centre of diversity for the Nolanaceae (Johnston 1936), with several species typically occurring sympatrically. Frequently encountered nolanas include *Nolana mollis, N. salsoloides, N. sedifolia, N. peruviana, N. leptophylla, N. villosa* and *N. aplocaryoides*.

A number of genera with coastal perennials cross the floristic barrier south of Arica, including *Alternanthera, Ophryosporus, Tetragonia, Oxalis, Calandrinia, Senna,*

CPD Site SA43: Lomas formations of the Atacama Desert. *Nolana rupicola* (Solanaceae) in Quebrada de la Cachina, south of Taltal, southern Atacama Desert, Chile. Photo: Michael O. Dillon.

Palaua, Tiquilia, Heliotropium, Caesalpinia, Opuntia (*Tephrocactus sensu lato*) and *Urocarpidium. Nolana* is distributed from northern Peru (as well as the Galápagos Islands) to central Chile, with peaks of species diversity near Mollendo in Peru and Paposo in Chile – only *N. lycioides* is distributed in both countries.

Despite the barrier, a number of species also are successfully established over the entire length of the Peruvian and Atacama deserts, such as *Apium* (*Ciclospermum*) *laciniatum, Encelia canescens, Chenopodium petiolare, Mirabilis prostrata* and *Loasa urens. Alstroemeria paupercula* occurs from Chala in south-central Peru to Caldera in Chile. *Pasithea coerulea* and *Fortunatia biflora* occur from southern Peru to central Chile. Although Solbrig (1976) drew a strong significance to the presence of floristic elements of the Chaco region of Argentina in the Atacama Desert, few of the species mentioned actually enter the coastal desert region – most of them are found in central Chile or on the higher slopes of the Andes.

Useful plants

This highly endemic flora is of inestimable importance. There are some traditional uses of species by the local inhabitants (Aronson 1990; Bittmann 1988), e.g. food from *Oxalis* spp., medicinals from *Salvia tubiflora* and *Ephedra* spp. Some plants of the Atacama Desert have developed unique ecophysiological traits through their evolutionary histories that will be of interest particularly in the genetic engineering of crops for agriculture and horticulture.

Social and environmental values

Global warming is difficult to document, yet important. Floristic distributional data combined with long-term precipitation records suggest that there has been a steady loss of biodiversity in northern Chile over the last 100 years (Rundel and Dillon, unpublished). While the trends will be difficult to reverse, these communities provide unusual opportunities to study the ongoing process of global climate change.

The intrinsic value of the Atacama Desert's plant and animal communities lies in the unique nature of their composition, the high levels of endemism and some species' remarkable adaptations for survival in some of the planet's most demanding conditions. Even rolling lichens are found (e.g. *Rocella cervicornis*) (Follmann 1966). These islands of habitat are essential for very many of Chile's threatened species (e.g. Benoit-C. 1989), and the lomas species have often been little studied or are even undiscovered. Recent scientific interest has led to more detailed research on the dynamics of this flora.

Some areas nearby have archaeological importance. The beauty and rarity of the lomas formations provide opportunities for tourism combined with scientific studies. If the impact on the delicate communities is controlled through supervision, lomas formations can be enjoyed by the public and preserved for the coming generations. Environmental education, on the importance, the rarity and the unusual characteristics of these natural resources, is desperately needed.

For example, Quebrada El León needs some recuperation from overuse and could become a lasting and informative oasis as a nature reserve for residents of Caldera and Copiapó. Punta de Teatinos (29°48'S), just north of La Serena, has interesting flora (e.g. *Neoporteria jussieui* var. *dimorpha*) and fauna – an estuary attracts many freshwater and marine birds. Both the land and water ecosystems require adequate protection, and this area could be a focus of nature recreation for residents of La Serena and Coquimbo.

Economic assessment

The primary utility of the Atacama Desert has been in mineral extraction and harvesting marine resources. The lomas formations are not generally considered of great economic potential due to the harsh, highly variable and unpredictable environmental conditions in the coastal desert.

Investigations of the potential water from wind-blown fog have shown that considerable moisture is available (Muñoz 1967). The generation of water by condensation from the available fog sources recently has stimulated local agriculture, which supplements costly produce brought in from several thousand km to the south.

Nonetheless, the poor soils and scarce water reserves in the northern Atacama Desert will not allow large-scale agricultural production immediately along the coast. Persistent agriculture or silviculture in lomas formations would be extremely damaging.

Threats

Since many sites have become accessible by road only recently (e.g. within the past c. 12 years), the Atacama's specialized ecosystems remained well preserved until recent times. Road construction in association with mining operations is increasing human occupation in the region. With the rise in copper prices during the 1980s, reactivation of mining activities utilizing large quantities of sulphuric acid has had an essentially undocumented impact on terrestrial and marine life (e.g. Anderson *et al.* 1990).

The mining, overgrazing by goats, extraction of fuelwood and commercial gathering of plants and animals have all combined to impact upon most regions in northern Chile. Efforts are needed to curtail hunting of rare species, such as some cacti and bulbs (Benoit-C. 1989; Anderson *et al.* 1990), guanacos, chinchillas and various birds including Andean condors, and in the regulation of the exploitation of marine species, e.g. shellfish, fish and algae.

Conservation

La Chimba National Reserve (c. 30 km²) was recently established; it is c. 15 km north of Antofagasta and regularly visited. Creation of a National Park in the exceptional area around Paposo (2000 km², with c. 345 species) would be a major step for protecting a substantial portion of the region's diversity and endemics (Reyes and Zizka 1989;

Hoffmann-J. 1991; Anderson *et al.* 1990). One large nationally protected area exists within the extreme desert region – Pan de Azúcar National Park (established in 1986), which covers c. 438 km². It has been recommended (Anderson *et al.* 1990) that this park be expanded northward to include Quebrada Esmeralda (25°50'S) and Quebrada de Las Lozas (25°41'S), which would protect areas very rich in cacti, and 15 km southward.

The recent discovery of rare and highly endemic species (e.g. *Copiapoa* spp., *Griselinia carlomunozii* and *Tillandsia tragophoba* – Anderson *et al.* 1990; Dillon and Muñoz-Schick 1993; Dillon 1991) shows the need for additional conservation measures. Since Atacama plants have narrowly confined geographic distributions, there needs to be a series of National Reserves or Parks to help ensure the future of these unique communities (Benoit-C. 1989; Anderson *et al.* 1990). Many of the cacti are propagated in nurseries at least abroad, which diminishes pressures to collect them from the wild, and export of cacti is regulated under CITES.

The estimated 550 species of vascular plants, most of which are largely endemic to the Atacama Desert, represent over 10% of the estimated 5100 continental species in Chile (Marticorena 1990). Conserving the flora of the Atacama Desert should be an immediate priority for local and national authorities and is internationally worthwhile.

References

Anderson, E.F., Bonilla-F., M., Hoffmann-J., A.E. and Taylor, N.P. (1990). *Succulent plant conservation studies and training in Chile.* World Wildlife Fund-U.S., Washington, D.C. 136 pp.

Aronson, J. (1990). Desert plants of use and charm from northern Chile. *Desert Plants* 10(2): 65–74, 79–86.

Arroyo, M.T. Kalin, Marticorena, C. and Muñoz, M. (1990). A checklist of the native annual flora of continental Chile. *Gayana, Bot.* 47: 119–135.

Benoit-C., I.L. (ed.) (1989). *Red Book on Chilean terrestrial flora (Part One).* Corporación Nacional Forestal (CONAF), Santiago. 151 pp.

Bittmann, B. (1988). Recursos y supervivencia en el Desierto de Atacama. In Masuda, S. (ed.), *Recursos naturales Andinos.* University of Tokyo, Tokyo. Pp. 153–208.

Börgel, R. (1973). The coastal desert of Chile. In Amiran, D.H.K. and Wilson, A.W. (eds), *Coastal deserts: their natural and human environments.* University of Arizona Press, Tucson. Pp. 111–114.

Dillon, M.O. (1991). A new species of *Tillandsia* (Bromeliaceae) from the Atacama Desert of northern Chile. *Brittonia* 43: 11–16.

Dillon, M.O. (in prep.). Flora of the lomas formations of Chile and Peru. *Fieldiana: Botany.*

Dillon, M.O. and Muñoz-Schick, M. (1993). A revision of the dioecious genus *Griselinia* (Griseliniaceae), including a new species from the coastal Atacama Desert of northern Chile. *Brittonia* 45: 261–274.

Dillon, M.O. and Rundel, P.W. (1989). The botanical response of the Atacama and Peruvian desert floras to the 1982–83 El Niño event. In Glynn, P.W. (ed.), *Global ecological consequences of the 1982–83 El Niño-Southern Oscillation.* Elsevier Oceanography Series, Amsterdam. Pp. 487–504.

Duncan, T. and Dillon, M.O. (1991). Numerical analysis of the floristic relationships of the lomas of Peru and Chile. Abstr. *Amer. J. Bot.* 78: 183.

Ehleringer, J.R., Mooney, H.A., Gulmon, S.L. and Rundel, P.W. (1980). Orientation and its consequences for *Copiapoa* (Cactaceae) in the Atacama Desert. *Oecologia* 46: 63–67.

Follmann, G. (1966). Chilenishe Wanderflechten. *Ber. Deutsch. Bot. Ges.* 79: 452–462.

Follmann, G. (1967). Die Flectenflora der nordchilenischen Nebeloasen Cerro Moreno. *Nova Hedwigia* 14: 215–281.

Gajardo, R. (1987). *La vegetation naturelle du Chile: proposition d'un systeme de classification et representation de la distribution géographique.* Ph.D. thesis. Université de Droit, d'Economie et des Sciences d'Aix-Marseille. 282 pp.

Gulmon, S.L., Rundel, P.W., Ehleringer, J.R. and Mooney, H.A. (1979). Spatial relations and competition in a Chilean desert cactus. *Oecologia* 44: 40–43.

Harrington, H.J. (1961). Geology of parts of Antofagasta and Atacama provinces, northern Chile. *Bull. Amer. Assoc. Petroleum Geol.* 45: 168–197.

Hoffmann-J., A.E. (1989). *Cactáceas en la flora silvestre de Chile.* Ediciones Fundación Claudio Gay, Santiago. 272 pp.

Hoffmann-J., A.E. (1991). Paposo: aguas cristalinas. Eco-Ambiente, *El Mercurio (Santiago)* 31/3/91.

Jaffuel, P.F. (1936). Excursiones botánicas a los alrededores de Tocopilla. *Rev. Chilena Hist. Nat.* 40: 265–274.

Johnston, I.M. (1929). Papers on the flora of northern Chile. *Contr. Gray Herb.* 85: 1–172.

Johnston, I.M. (1936). A study of the Nolanaceae. *Contr. Gray Herb.* 112: 1–83.

Kohler, A. (1970). Geobotanische Untersuchungen an Küstendünen Chiles zwischen 27 und 42 Grad Südl. Breite. *Bot. Jahrb. Syst.* 90: 55–200.

Marticorena, C. (1990). Contribución a la estadística de la flora vascular de Chile. *Gayana, Bot.* 47: 85–113.

Marticorena, C. and Quezada, M. (1985). Catálogo de la flora vascular de Chile. *Gayana, Bot.* 42: 1–157.

McGlone, M.S., Kershaw, A.P. and Markgraf, V. (1992). El Niño/Southern Oscillation climatic variability in Australasian and South American paleoenvironmental records. In Diaz, H.F. and Markgraf, V. (eds), *El Niño: historical and paleoclimatic aspects of the Southern Oscillation.* Cambridge University Press, Cambridge, U.K. Pp. 435–462.

Miller, A. (1976). The climate of Chile. In Schwerdtfeger, W. (ed.), *Climates of Central and South America.* World Survey of Climatology Vol. 12. Elsevier, Amsterdam. Pp. 113–145.

Mooney, H.A., Weisser, P.J. and Gulmon, S.L. (1977). Environmental adaptations of the Atacama Desert cactus *Copiapoa haseltoniana. Flora (Jena)* 166: 117–124.

Muñoz, H.R. (1967). *Captación de agua en la provincia de Antofagasta.* Rev. Universidad del Norte, Antofagasta. Unpublished.

Muñoz-Schick, M. (1985). *Flores del Norte Chico.* Dirección de Bibliotecas, Archivos y Museos, La Serena, Chile. 95 pp.

Opazo, A. and Reiche, C. (1909). Descripción y resultados de un viaje de estudio de Caldera a Paposo en busca de plantas que contengan caucho. *Anal. Agron. Santiago* 4: 189–237.

Philippi, R.A. (1860). *Die Reise durch die Wüste Atacama auf Befehl der chilenischen Regierung im Sommer 1853–1854.* Verlag Anton, Halle. 254 pp.

Prenafeta, S. (1984). El paraíso se mudó de casa. *Creces* 5: 12–13.

Quinn, I.H., Neal, V.T. and Antúnez de Mayolo, S.E. (1987). El Niño occurrences over the past four and a half centuries. *J. Geophys. Res.* 92: 14,449–14,461.

Rauh, W. (1958). *Beitrag zur Kenntnis der peruanischen Kakteenvegetation. Sitzungsber. Heidelberger Akad. Wiss., Math.-Naturwiss. Kl.* 1958(1). 542 pp.

Rauh, W. (1985). The Peruvian-Chilean deserts. In Evenari, M., Noy-Meir, I. and Goodall, D.W. (eds), *Hot deserts and arid shrublands,* Vol. A. Ecosystems of the World, Vol. 12A. Elsevier, Amsterdam. Pp. 239–267.

Reiche, K.F. (1907). *Grundzüge der Pflanzenverbreitung in Chile.* In Engler, A. and Drude, O. (eds), Die Vegetation der Erde, 8. Engelmann, Leipzig. 394 pp.

Reiche, K.F. (1911). Ein Frühlingsausflug in das Küstengebiet der Atacama (Chile). *Bot. Jahrb. Syst.* 45: 340–353.

Reiche, K.F. (1934-1937) [1938]. *Geografia botánica de Chile,* Vol. 1: 3–423(424), Vol. 2: 3–149(151). Imprenta Universitaria, Santiago.

Reyes, J.P. and Zizka, G. (1989). Nebelwüstenvegetation bei Paposo in Nord-Chile – Standort für einen neuen Nationalpark. *Der Palmengarten* 1/89: 52–63.

Romero, H. and Garrido, A.M. (1985). Influencias genéticas del fenómeno El Niño sobre los patrones climáticos de Chile. *Invest. Pesq. (Chile)* 32: 19–35.

Rundel, P.W. (1981). The matorral zone of central Chile. In di Castri, F., Goodall, D.W. and Specht, R.L. (eds), *Mediterranean-type shrublands.* Ecosystems of the World Vol. 11. Elsevier, Amsterdam. Pp. 175–201.

Rundel, P.W. and Mahu, M. (1976). Community structure and diversity in a coastal fog desert in northern Chile. *Flora (Jena)* 165: 493–505.

Rundel, P.W., Dillon, M.O., Palma, B., Mooney, H.A., Gulmon, S.L. and Ehleringer, J.R. (1991). The phytogeography and ecology of the coastal Atacama and Peruvian deserts. *Aliso* 13(1): 1–50.

Rundel, P.W., Ehleringer, J.R., Gulmon, S.L. and Mooney, H.A. (1980). Patterns of drought response in leaf-succulent shrubs of the coastal Atacama Desert in northern Chile. *Oecologia* 46: 196–200.

Rutllant, J. (1985). Algunos aspectos de la influencia climática, a nivel mundial y regional, del fenómeno El Niño. *Invest. Pesq. (Chile)* 32: 9–17.

Schmithüsen, J. (1956). Die räumliche Ordnung der chilenischen Vegetation. *Bonner Geogr. Abhandl.* 17: 1–86.

Solbrig, O.T. (1976). The origin and floristic affinities of the South American temperate desert and semidesert regions. In Goodall, D.W. (ed.), *Evolution of desert biota.* University of Texas Press, Austin. Pp. 7–50.

Weisser, P.J. (1967a). Chilenische Erdkakteen und ihre Umweltfaktoren. *Umschau Wiss. Tech.* 24: 808–809.

Weisser, P.J. (1967b). Zur Kenntnis der Erdkakteen in Chile. *Ber. Deutsch. Bot. Ges.* 80: 331–338.

Werdermann, E. (1931). Die Pflanzenwelt Nord- und Mittelchiles. *Vegetationsbilder* 21. Reihe 6/7: 31–42.

Authors

This Data Sheet was written by Dr Michael O. Dillon (Field Museum of Natural History, Center for Evolutionary and Environmental Biology, Department of Botany, Chicago, IL 60605-2496, U.S.A.) and Adriana E. Hoffmann-J. (Fundación Claudio Gay, Alvaro Casanova 613 – Peñalolén, Santiago, Chile).

MEDITERRANEAN REGION AND LA CAMPANA NATIONAL PARK
Central Chile

Location: Region between latitudes c. 29°–38°S; La Campana National Park and Vizcachas mountains north-west to west of Santiago in Cordillera de la Costa: latitudes c. 32°45'–33°40'S, longitudes c. 70°45'–71°20'W, in portions of Valparaíso and Metropolitan regions between Aconcagua and Mapocho rivers.

Area: Potential reserve of c. 5000 km² includes Cerro El Roble, Ocoa palm forest, La Campana NP (80 km²) and Coast Range southward to Talagante.

Altitude: c. 300–2222 m.

Vegetation: Alpine-like vegetation, *Nothofagus* forest, hygrophilous forest, sclerophyllous forest, matorral, xerophytic spiny shrubland, bamboo thicket, palm forest.

Flora: 1800–2400 vascular plant species; high diversity; 545 species recorded so far in La Campana NP, with c. 37 Chilean endemic genera represented; northern limit of *Nothofagus* in South America.

Useful plants: Timber species for construction, furniture-making, fuelwood, charcoal; species for edible fruits, medicinals, dyes, ornamentals.

Other values: Watershed protection, germplasm reserves, research, education, tourism.

Threats: Agriculture, grazing, gathering fuelwood, charcoal-making, excessive logging, erosion, urbanization, road construction, mining, pollution, unnatural fires, invasive exotics, reforestation with exotics; lack of funding for protected areas.

Conservation: c. 956 km² in 2 National Parks, 5 National Reserves – including separate NP and NR as La Campana-Peñuelas Biosphere Reserve, 1 Nature Sanctuary. Potential reserve (5000 km²) would encompass La Campana NP; 11 areas recommended for strong conservation.

Geography

The world's five areas of Mediterranean climate basically occur on the western coasts of continents north and south of the Equator between the latitudes 27°–40°. The climate characteristically has warm dry summers with high solar radiation and evaporation, and cold but benign damp winters with low solar radiation and evaporation. The Mediterranean region in central Chile (Rundel 1981) is a strip c. 100 km wide along the continental margin, comprising three physiographic areas: Cordillera de la Costa, Valle Central and Andean Cordillera up to c. 1800 m (CONAF 1982). The soils are poorly developed, coarse and shallow.

The coast along the Pacific is characterized by either a narrow plain or a series of marine terraces extending several hundred meters upward, which were formed during the Quaternary. Also sand dunes are widespread along the immediate coast. About 40 km inland the Cordillera de la Costa (Coast Range) begins abruptly. It was formed from Precambrian and Cretaceous crystalline rocks thrust up during the Tertiary. One of the highest elevations is Cerro El Roble (2222 m), north-west of Santiago on the eastern edge of the mountain range. North of Santiago the area is mountainous, with several spurs and transverse valleys from the Andes descending westward.

The Valle Central (Central Valley) is a longitudinal basin filled by sediments from the surrounding mountains, especially the Andes. The c. 20-km wide valley extends uninterruptedly from just north of Santiago southward for 1100 km to Puerto Montt, with typical elevations of 400–700 m. Its landform structure is very similar to the Central Valley of California,

U.S.A. except for the drainage. In Chile major streams arising in the Andes have cut perpendicularly across the Valle Central and broken through the Coast Range to reach the ocean, whereas in California the major rivers flow parallel to the valley.

The Andean Cordillera is the consequence of a complex series of tectonic uplifts from the Late Cretaceous accompanied by volcanic activity in the Oligocene and Miocene, followed by widespread erosion in the Late Tertiary. The modern form of the Andes is the result of major Pleistocene uplift which caused extensive folding of Mesozoic sediments. In central Chile the Andes reaches nearly 7000 m and consists of metamorphosed sedimentary rock underlain by granitic batholithic formations. Between 33°–37°S, about 40 volcanoes are known to have been active in the past 1 million years, and at least 20 during the last 12,000 years – with nine erupting between 1900–1980 (Fuentes and Espinosa 1986).

The defining climate of mild rainy winters (June–August) and summer droughts in the Mediterranean region of Chile varies greatly from year to year and within a season (Miller *et al.* 1977). The characteristic temperature patterns are defined primarily by topography. Masses of cold air from the sea penetrate beneath warmer air along the coast, creating a stable temperature-inversion stratum 400–700 m thick. Temperatures below the inversion layer are similar in the coastal and Andean ranges. In the interior the average annual maximum fluctuates between 20°–25°C, and the minimum frequently is below freezing. Near the coast the cool oceanic Humboldt Current moderates the average annual maximum temperature to 17°–19°C, and the minimum is rarely below 0°C.

The region's annual precipitation fluctuates between 100–800 mm, increasing from the coast toward the Andes and southward. Most precipitation occurs during the colder months as a few rainfalls. Convectional thunderstorms and lightning are rare. La Campana annual averages of precipitation vary between 400–600 mm, with temperatures of 9°–19°C (CONAF 1982). The annual average precipitation in the Santiago area is c. 350 mm, with c. 70% relative humidity.

Vegetation

The area between Combarbalá (31°11'S) and Linares (35°51'S) includes La Campana National Park and is characteristic for the Mediterranean-type climate and shrub vegetation. Most of the significant ecological communities of Mediterranean Chile are represented in the park, which includes eight main vegetation formations (Rundel and Weisser 1975; Rundel 1981; Villaseñor-C. 1979, 1980; Villaseñor-C. and Serey-E. 1980–1981).

1. **Alpine-like steppe vegetation** is found above 1600 m on the park's highest two mountains, on rocky slopes adjacent to *Nothofagus* stands. Dominants include tussock grasses such as *Stipa* and *Festuca*, *Carex setifolia* var. *berteroana* and low shrubs of *Chuquiraga oppositifolia*, *Mulinum spinosum*, *Valenzuelia* (*Guindilia*) *trinervis*, *Schinus montanus* and *Colliguaja integerrima*.

2. *Nothofagus* forest is the most unusual formation and represents the northern limit for the genus in South America. *Nothofagus obliqua* var. *macrocarpa* ("roble") occurs in scattered relictual stands 6–10 m high on south-facing slopes from c. 1000–1600 m on La Campanita (1510 m), La Campana (1890 m) and El Roble (2222 m) mountains. The thoroughly dominant roble is associated with the woody species *Lomatia hirsuta*, *Schinus montanus*, *Aristotelia chilensis*, *Azara petiolaris*, *Berberis actinacantha* and *Ribes punctatum* and the vines *Vicia magnifolia*, *Mutisia ilicifolia* and *Tropaeolum azureum*.

   At lower elevations, the *Nothofagus* forest is replaced by hygrophilous forest where water is plentiful most of the year (e.g. along watercourses in canyons and on foggy slopes), and under moderately mesic conditions by either matorral or bamboo thicket.

3. **Hygrophilous forest** occurs on the western flank of the Coast Range in several associations. The formation is best developed above Olmué on Cerro La Campana. These closed-canopy communities have mixed, broadleaved evergreens (2–) 10–20 m tall. The dominants in three distinctive associations are: (i) *Crinodendron patagua*, *Drimys winteri*, *Aristotelia chilensis* and *Rhaphithamnus spinosus*; (ii) *Persea lingue* (which is also at its northern limit) and *Luma* (*Myrceugenella*) *chequen*; and (iii) *Dasyphyllum excelsum*, *Beilschmiedia miersii*, *Myrceugenia obtusa* and *Sophora macrocarpa*. High humidity from fog between 600–700 m on La Campana provides conditions for luxuriant growth of cryptogamic epiphytes and bromeliads, e.g. *Tillandsia usneoides*. Lichens have high species diversity in the fog belt.

CPD Site SA44: Mediterranean region and La Campana National Park, central Chile. Evergreen sclerophyllous forest in Quebrada El Tigre (near Zapallar), in the Valparaíso (V) Region of the Cordillera de la Costa.
Photo: Adriana Hoffmann.

4. **Sclerophyllous forest** is less moist than hygrophilous forest, and predominates on south-facing low to moderate slopes between 400–1000 m. The dominants include *Cryptocarya alba*, *Peumus boldus*, *Beilschmiedia miersii* and *Azara petiolaris*; there are many lianas (*Proustia pyrifolia*, *Lardizabala biternata*, *Cissus striata*) and vines (*Bomarea salsilla*). Extensive stands are now rare in central Chile, yet good examples occur along the lower slopes of La Campana.

   Where sclerophyllous forest is disturbed, an open secondary vegetation is dominated by *Acacia caven* with *Sophora macrocarpa*, *Trevoa trinervis* and *Podanthus mitiqui*. A climax savanna of *Acacia caven* with *Prosopis chilensis* occurs on low arid plains and valley bottoms (Rundel 1981).

5. **Matorral** is variable in composition, and characterized by evergreen shrubs 1–3 m in height surrounded by a seasonal cover of herbaceous perennials (bulbs, ferns and vines) as well as some annuals (Rundel 1981). The matorral has many physiognomic and ecological similarities (at community to species levels) with the less diverse chaparral of southern California (Mooney 1977). Dominants include *Lithrea caustica*, *Quillaja saponaria*, *Kageneckia oblonga*, *Escallonia pulverulenta* and *Maytenus boaria*. Although most matorral communities of central Chile have been severely degraded by

CPD Site SA44: Mediterranean region and La Campana National Park, central Chile. In Ocoa area of La Campana National Park, with sclerophyllous matorral and *Puya berteroniana* in the foreground.
Photo: Adriana Hoffmann.

wood-cutting, charcoal-making and overgrazing, some natural stands still exist near cerros La Campana and El Roble.

6. **Xerophytic spiny shrubs and succulents** are found on north-facing slopes in the matorral areas. Species composition varies depending upon slope and altitude. Between 1000–1600 m occurs *Puya coerulea* var. *violacea*, a terrestrial bromeliad which is locally endemic to Chile's Mediterranean region, accompanied by *Flourensia thurifera*, *Baccharis linearis*, *Eryngium paniculatum* and *Satureja gilliesii*. Between 300–1000 m, the 2–4 m tall arborescent cactus *Echinopsis* (*Trichocereus*) *chiloensis* is common, along with *Puya berteroniana* and *P. chilensis*, and in spring an abundant herbaceous flora.

7. Extensive areas of **bamboo thicket** dominated by *Chusquea cumingii* form nearly impenetrable stands sometimes up to 3 m high in most community types in La Campana area.

8. **Palm forest** in the Ocoa Valley north of La Campana has the largest of the few remaining stands of *Jubaea chilensis*. The thousands of palms are one of the outstanding features of La Campana area. Individual trees may reach 1.5–2 m in diameter and 20–25 m tall, making it the largest

palm in the world in trunk volume (Rundel 1981). Once covering extensive areas in valleys of the Cordillera de la Costa, this species was heavily exploited and is now restricted to a belt along the Coast Range of central Chile from the southern Coquimbo (IV) Region southward to the Maule (VII) Region. It is found from near sea-level to above 1000 m, but absent from the Andean Cordillera.

Flora

The flora of Mediterranean Chile has three main phytogeographic relationships in addition to its endemics: taxa with affinities to the tropics, the south (Antarctic), or the Andes – an element that evolved with the forming of these mountains late in the Tertiary and Pleistocene (Raven 1973; Solbrig *et al.* 1977).

The relictual *Nothofagus* stands relate to cool-temperate Valdivian forests of southern Chile (Cassasa 1985). The flora (e.g. *Lomatia*, *Nothofagus*) also is related to the flora of southern Australia. Many of the Andean elements are related to North American migrants that reached South America with the joining of the two continents in the Late Pleistocene. The flora of central Chile has little taxonomic affiliation with the Mediterranean flora of California; most of the over 130 species now shared are weeds of European origin which were introduced by people in recent historical periods (Mooney 1977).

No complete Flora of Mediterranean Chile has been undertaken. Hoffmann-J. (1979) estimated that 1800–2000 vascular plant species may occur, yet between about 31°–31°15'S in an area of 104,000 km², Arroyo *et al.* (1994) estimated that at least 2395 species occurred. In this area there are 70 species of trees in 47 genera, and more tree species in the Coast Range than the Andes.

There are more woody species in a given area in central Chile than in the Mediterranean region of California, U.S.A., whereas total species richness is slightly lower in Chile than in equivalent California (Arroyo *et al.* 1994). Tree-species richness is highest between 34°–37°S latitudes, and is higher than in rain forest of equivalent latitudinal range (Arroyo 1994, pers. comm.).

Based on earlier studies near Santiago, about 10% of the genera and 95% of the species are Chilean endemics (Rundel 1981). Arroyo *et al.* (1994) found endemic tree species to be more frequent in the Coast Range – e.g. *Gomortega keule*, *Pitavia punctata*, *Nothofagus alessandrii*, *Beilschmiedia* spp. and *Jubaea chilensis*.

A preliminary partial list for La Campana National Park has 545 species: 16 pteridophytes; one gymnosperm; and 528 angiosperms, including 433 dicotyledons and 95 monocotyledons (R. Villaseñor-C. 1993, pers. comm.). The park well represents the floristic confluence of the four phytogeographic elements of the region (Villaseñor-C. and Serey-E. 1980). In the potential reserve at c. 33°S latitude there are a large number of endemics, including five annual and 18 perennial herbaceous genera, 11 genera of shrubs and three species of trees (Arroyo 1994, pers. comm.).

Species with lignotubers (e.g. *Cryptocarya alba*, *Lithrea caustica*, *Kageneckia oblonga*, *Quillaja saponaria*, *Colliguaja odorifera*, *Satureja gilliesii*, *Trevoa trinervis*) provide for their resprouting with basal underground buds (Montenegro, Avila and Schatte 1983), which perhaps evolved as an adaptation to volcanically induced fires (Fuentes and Espinosa 1986) and/or drought, frost and herbivory (Mooney 1977).

Mediterranean Chile has many threatened taxa (Benoit-C. 1989). *Tecophilaea cyanocrocus* is believed to be extinct in the wild due to collection of its bulbs. Endangered species include *Adiantum gertrudis*, *Avellanita bustillosii* and *Beilschmiedia berteroana*. Vulnerable taxa include *Beilschmiedia miersii*, *Crinodendron patagua*, *Cryptocarya alba*, *Dasyphyllum excelsum*, *Jubaea chilensis*, *Krameria cistoidea*, *Lomatia hirsuta*, *Nothofagus obliqua* var. *macrocarpa*, *Persea meyeniana*, *Porlieria chilensis*, *Prosopis* spp., *Puya coquimbensis* and *P. venusta*. Rare taxa include *Adesmia balsamica*, *A. resinosa*, *Myrceugenia correifolia*, *M. exsucca* and *M. rufa*. Many of these taxa occur in the potential reserve.

Useful plants

In Mediterranean Chile and the potential reserve there are many species of economic importance (Rundel and Weisser 1975; Muñoz-Pizarro 1971; Benoit-C. 1989). Timber trees used in construction and furniture manufacture include *Nothofagus obliqua* var. *macrocarpa*, *Sophora macrocarpa* and *Persea lingue*. Trees used for fuelwood and charcoal include *Lithrea caustica*, *Quillaja saponaria*, *Cryptocarya alba* and *Maytenus boaria*. Numerous species are used medicinally, e.g. *Peumus boldus*, *Quillaja saponaria*, *Trevoa trinervis*, *Maytenus boaria* and *Drimys winteri*. Species with edible fruits include *Jubaea chilensis* and *Beilschmiedia berteroana*; species with sap or extracts used for dye, soap or food include *Lomatia hirsuta*, *Quillaja saponaria*, *Jubaea chilensis* and *Kageneckia oblonga*. Ornamentals include cacti (e.g. *Neoporteria curvispina*), bromeliads (e.g. *Puya* spp. – some of which are also used for fibre) and many genera of bulbs (Hoffmann-J. 1989).

Social and environmental values

Central Chile is rich in natural resources which have been exploited over the centuries. La Campana region contains active mines for copper, quartz, feldspar, pyrites and molybdenum (Rundel and Weisser 1975). Long before the arrival of the Spanish in the 1530s, indigenous groups in this region (the Diaguita and the Picunche) practised agriculture near brooks and rivers in protected valleys. Principal crops included a grain (*Bromus mango*) and an oilseed (*Madia chilensis*). In addition to slash-and-burn winter farming, they constructed canals and irrigation ditches to water some crops during the dry summer (Aschmann and Bahre 1977).

Charles Darwin visited La Campana in August 1834 and in *The Voyage of the Beagle* described the beautiful scenery seen from atop this peak in the coastal mountain range. Today Andean condors are still observed flying over the summits. The area also harbours the world's largest hummingbird (*Patagona gigas gigas*) and pumas – which are gone from most of Chile. La Campana NP is important for its ecological value and as a recreational, educational and historical centre for much of the population. Over 9.5 million people (80% of Chile's population) live in this central region (Fuentes 1990) which includes Santiago and Valparaíso, both c. 40 km from the park.

Threats

Human activity has affected the Mediterranean ecosystems of central Chile for over 10,000 years. Use of fire and the hunting of large mammals, especially guanacos, have affected the vegetation. During colonial times land was cleared for agriculture and raising cattle. As villages and towns were established, large areas of forest and shrubs were burned, riverbeds modified, and the land altered for various purposes (Fuentes 1990; Rundel 1981; Aschmann and Bahre 1977; Cunill 1971).

The landscape of the Mediterranean region is greatly altered, particularly the Cordillera de la Costa, the Valle Central and the Andean precordillera. Hillsides are devoid of original vegetation, soil is eroded and a considerable percentage of the native flora is basically gone. Large native animals are now almost non-existent in the region, with the exception of some birds. The Valle Central is used for intensive agriculture that makes efficient use of water and soil.

Native vegetation has been affected directly by mining, logging, road construction, garbage dumps, recent urbanization, and pollution of the air, water and soil. Native forest has been replaced with exotic trees and the introduction of exotic weeds and fauna; e.g. the European rabbit (*Oryctolagus cuniculus*) has drastically changed the composition of some native communities.

The sclerophyllous vegetation of the Coast Range between 800–1000 m has been changed due to wood-cutting, charcoal-making and grazing of goats and sheep (Mooney *et al.* 1972; Fuentes and Hajek 1979). On the cordillera slopes, dense forests of broadleaved evergreen and deciduous species that occurred long ago have been transformed into degraded brush, which provides fuelwood and charcoal. Also leaves of *Peumus boldus* and other medicinal and/or aromatic species, including the dead leaves primarily of *Lithrea caustica* and bark of *Quillaja saponaria* (for saponin), are collected on these slopes. One important species of this region, *Jubaea chilensis*, was approaching extinction during the 17th and 18th centuries due to extensive felling of the trees especially for their sap, which supplied sugar for the entire country (Fuentes 1990).

The region (34°–37°S) with highest tree-species richness, between Rancagua and Concepción, is an area of intensive agriculture, and the region with the lowest proportion of land protected (0.1–22%) (Armesto *et al.* 1992).

Conservation

Chile has an extensive system of National Parks and protected areas administered by the Corporación Nacional Forestal (CONAF), but the Mediterranean region is one of the least represented in the Sistema Nacional de Areas Silvestres Protegidas del Estado (National System of Protected Wild Areas) (SNASPE). Eight principal areas are protected within Mediterranean Chile: (1) in Coquimbo (IV) Region near Combarbalá: Las Chinchillas National Reserve (42 km²); (2) in Valparaíso (V) Region: La Campana National Park (80 km²), Lago Peñuelas NR (91 km²), Río Blanco NR (102 km²); (3) in Metropolitan Region: Yerba Loca Nature Sanctuary (116 km²), Río Clarillo NR (102 km²); and (4) in O'Higgins (VI) Region: Las Palmas de Cocalán NP (37 km²) – which includes an important population of *Jubaea chilensis*, Río de los Cipreses

NR (386 km²). These areas in IUCN management categories II or IV (WCMC 1992) total c. 956 km², but they do not adequately conserve all the vegetation types of the Mediterranean region of Chile (cf. Balduzzi *et al.* 1981; Donoso-Z. 1982).

La Campana National Park was established in 1967. Protection was urgent because of deterioration as a result of population pressures and mining. La Campana area served for over a decade as a primary research site of the International Biological Programme (IBP), specifically for study of convergent evolution in the Chilean and Californian Mediterranean ecosystems (e.g. Mooney 1977). La Campana-Peñuelas Biosphere Reserve (171 km²), which was declared in 1984, links the non-contiguous NP and NR to the park's south-west.

In early 1993 CONAF held a symposium with c. 100 specialists to recommend priority sites for conservation of the biodiversity in Chile. The recommendation was accepted to incorporate into La Campana NP: Cerro El Roble, the Vizcachas range (reaching 2045 m), Cuesta La Dormida (1000–1500 m) and Alto de Chicauma (1990 m).

MAP 78. PORTION OF MEDITERRANEAN REGION, CENTRAL CHILE, SHOWING GENERAL AREA FOR A POTENTIAL RESERVE THAT WOULD INCLUDE CERRO EL ROBLE, LA CAMPANA NP AND PART OF COAST RANGE

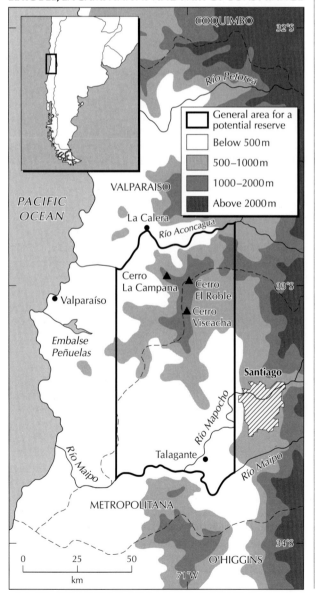

In addition, the following 11 areas in the Mediterranean region were recommended for strong protection: (1) in Coquimbo and Valparaíso (IV–V) regions near Pichidangui: Cerro Santa Inés – Los Molles (c. 40 km²); (2) in Valparaíso (V) Region: Alicahue in precordillera Petorca, Quebrada El Tigre in Zapallar, Cuesta El Melón, Bosque de Quintero, Bosque de Mantagua, Yali Marsh (c. 5.5 km²) and Laguna del Rey – important for nesting of birds on the Santiago coast; (3) in Metropolitan Region: Laguna de Aculeo; (4) in Metropolitan and O'Higgins (VI) regions: Altos (Macizo) de Cantillana (reaching 2318 m); and (5) in Maule (VII) Region in precordillera Molina: Radal – Siete Tazas Protected Area (77 km²) – from which a c. 52 km² NR will be declared (CODEFF 1994, pers. comm.), and in precordillera Talca: Altos de Vilches PA (169 km²) (cf. Ormazábal-Pagliotti 1985, 1988, 1989; CONAF 1982).

The potential reserve area (Map 78) including La Campana NP and approaching Lago Peñuelas NR (which is south-east of Valparaíso) has a mosaic of vegetation types, which have undergone various degrees of disturbance. The area could be managed as an enlarged UNESCO Man and the Biosphere (MAB) Biosphere Reserve with buffer zones for ecosystem restoration, as well as strictly protected zones. In order to ensure conservation, a realistic management plan and financial support must be provided. Since no other National Park occurs within a reasonable distance from the country's most populated cities (Santiago and Valparaíso), the recreational and educational values of this area also are significant.

References

Armesto, J.J., Smith-Ramírez, C., León, P. and Arroyo, M.T. Kalin (1992). Biodiversidad y conservación del bosque templado de Chile. *Ambiente y Desarrollo (Chile)* 8: 19–24.

Arroyo, M.T. Kalin, Cavieres, L., Marticorena, C. and Muñoz-Schick, M. (1994). Convergence in the mediterranean floras in central Chile and California: insights from comparative biogeography. In Arroyo, M.T. Kalin, Zedler, P. and Fox, M. (eds), *Ecology and biogeography of Mediterranean ecosystems in Chile, California and Australia*. Springer-Verlag, New York. Pp. 43–88.

Aschmann, H. and Bahre, C.J. (1977). Man's impact on the wild landscape. In Mooney, H.A. (ed.), *Convergent evolution in Chile and California. Mediterranean climate ecosystems*. Dowden, Hutchinson & Ross, Stroudsburg, Pennsylvania, U.S.A. Pp. 73–84.

Balduzzi, A., Serey-E., I., Tomaselli, R. and Villaseñor-C., R. (1981). New phytosociological observations on the Mediterranean type of climax vegetation of central Chile. *Atti Ist. Bot. Univ. Lab. Critt. Pavia*, serie 6, 14: 93–112.

Benoit-C., I.L. (ed.) (1989). *Red Book on Chilean terrestrial flora (Part One)*. Corporación Nacional Forestal (CONAF), Santiago. 151 pp.

Cassasa, I. (1985). *Estudio demográfico y florístico de los bosques de Nothofagus obliqua en Chile central*. Thesis, Universidad de Chile, Santiago.

CONAF (1982). *Chile. Sus parques nacionales y otras areas naturales.* CONAF, Santiago. 224 pp.

Cunill, P.G. (1971). Factores en la destrucción del paisaje chileno: recolección, caza y tala coloniales. *Informaciones Geográficas.* Universidad de Chile, Santiago. Número Especial: 235–264.

Donoso-Z., C. (1982). Reseña ecológica de los bosques mediterráneos de Chile. *Bosque (Valdivia)* 4(2): 117–146.

Fuentes, E.R. (1990). Landscape change in Mediterranean-type habitats of Chile: patterns and processes. In Zonneveld, I. and Forman, R.T. (eds), *Changing landscapes: an ecological perspective.* Springer-Verlag, New York. Pp. 165–190.

Fuentes, E.R. and Espinosa, G. (1986). Resilience of central Chile shrublands: a vulcanism-related hypothesis. *Interciencia* 11(4): 164–165.

Fuentes, E.R. and Hajek, E.R. (1979). Patterns of landscape modification in relation to agricultural practice in Central Chile. *Environm. Conserv.* 6: 265–271.

Hoffmann-J., A.E. (1979). *Flora silvestre de Chile: zona central.* Editorial Fundación Claudio Gay, Santiago. 254 pp.

Hoffmann-J., A.E. (1989). Chilean geophyte monocotyledons: taxonomic synopsis and conservation status. In Benoit-C., I.L. (ed.), *Red Book on Chilean terrestrial flora (Part One).* CONAF, Santiago. Pp. 141–151.

Miller, P.C., Bradbury, D.E., Hajek, E.R., LaMarche, V. and Thrower, N.J.W. (1977). Past and present environment. In Mooney, H.A. (ed.), *Convergent evolution in Chile and California. Mediterranean climate ecosystems.* Dowden, Hutchinson & Ross, Stroudsburg, Pennsylvania, U.S.A. Pp. 27–72.

Montenegro, G., Avila, G. and Schatte, P. (1983). Presence and development of lignotubers in shrubs of the Chilean matorral. *Canad. J. Bot.* 61: 1804–1808.

Mooney, H.A. (ed.) (1977). *Convergent evolution in Chile and California. Mediterranean climate ecosystems.* US/IBP Synthesis Series Vol. 5. Dowden, Hutchinson & Ross, Stroudsburg, Pennsylvania, U.S.A. 224 pp.

Mooney, H.A., Dunn, E.L., Shropshire, F. and Song Jr., L. (1972). Land-use history of California and Chile as related to the structure of the sclerophyll scrub vegetations. *Madroño* 21: 305–319.

Muñoz-Pizarro, C. (1971). *Chile: plantas en extinción.* Editorial Universitaria, Santiago. 247 pp.

Ormazábal-Pagliotti, C.S. (1985). *Análisis de las prioridades y posibilidades de mejorar en el corto plazo la cobertura ecológica y geográfico-administrativa del Sistema Nacional de Areas Silvestres Protegidas del Estado.* CONAF, Santiago. 63 pp.

Ormazábal-Pagliotti, C.S. (1988). Un moderno concepto de protección: sitios de interés botánico. *Chile Forestal* 11/88: 16–18.

Ormazábal-Pagliotti, C.S. (1989). Threatened plant sites and vegetation types in Chile. A proposal. In Benoit-C., I.L. (ed.), *Red Book on Chilean terrestrial flora (Part One).* CONAF, Santiago. Pp. 97–104.

Raven, P.H. (1973). The evolution of Mediterranean floras. In di Castri, F. and Mooney, H.A. (eds), *Mediterranean-type ecosystems. Origin and structure.* Ecological Studies 7. Springer-Verlag, New York. Pp. 213–224.

Rundel, P.W. (1981). The matorral zone of central Chile. In di Castri, F., Goodall, D.W. and Specht, R.L. (eds), *Mediterranean-type shrublands.* Ecosystems of the World Vol. 11. Elsevier, Amsterdam. Pp. 175–201.

Rundel, P.W. and Weisser, P.J. (1975). La Campana, a new national park in central Chile. *Biol. Conserv.* 8: 35–46.

Solbrig, O.T., Cody, M.L., Fuentes, E.R., Glanz, W.E., Hunt, J.H. and Moldenke, A.R. (1977). The origin of the biota. In Mooney, H.A. (ed.), *Convergent evolution in Chile and California. Mediterranean climate ecosystems.* Dowden, Hutchinson & Ross, Stroudsburg, Pennsylvania, U.S.A. Pp. 13–26.

Villaseñor-C., R. (1979). Estudio florístico de las formaciones vegetacionales del Cerro La Campana (Parque Nacional La Campana). *Arch. Biol. Med. Exp.* 12: 643.

Villaseñor-C., R. (1980). Unidades fisionómicas y florísticas del Parque Nacional La Campana. *Anales Mus. Hist. Nat. Valparaíso* 13: 65–70.

Villaseñor-C., R. and Serey-E., I. (1980). Importancia biogeográfica del Parque Nacional La Campana, en Chile central. *Arch. Biol. Med. Exp.* 13: 118.

Villaseñor-C., R. and Serey-E., I. (1980-1981). Estudio fitosociológico de la vegetación del Cerro La Campana (Parque Nacional La Campana), en Chile central. *Atti Ist. Bot. Univ. Lab. Critt. Pavia,* serie 6, 14: 69–91.

WCMC (World Conservation Monitoring Centre) (1992). *Protected areas of the world: a review of national systems.* Vol. 4: *Nearctic and Neotropical.* IUCN, Gland, Switzerland and Cambridge, U.K. 459 pp.

Authors

This Data Sheet was written by Adriana E. Hoffmann-J. (Fundación Claudio Gay, Alvaro Casanova 613 – Peñalolén, Santiago, Chile) and Olga Herrera-MacBryde (Smithsonian Institution, Department of Botany, NHB-166, Washington, DC 20560, U.S.A.).

TEMPERATE RAIN FOREST
of Chile

Location: Narrow band mainly along western Andean slopes from Bío-Bío and Arauco provinces in south-central Chile southward to Chiloé Province and into Tierra del Fuego Province, with a small portion in south-western Argentina; from latitude 38°30'S to 55°30'S around Cape Horn Islands (longitude c. 67°W).

Area: Over 70,000 km² but less than 110,000 km².

Altitude: c. 0–2000 m.

Vegetation: Evergreen and deciduous forests, including Valdivian, North Patagonian and Magellanic cool-temperate rain forests.

Flora: c. 450 vascular plant species; high endemism, including quasi-endemic family Aextoxicaceae and 17 monotypic regionally endemic genera; ecologically major threatened species.

Useful plants: Important tree species used in construction, others for woodchips (much exported) and fuelwood.

Other values: Habitat for threatened fauna; watershed protection.

Threats: Legal and illegal logging, clearing, afforestation, fire, agriculture, grazing.

Conservation: 25 National Parks, 14 National Reserves, 1 National Monument.

Geography

The western margin of southern South America has cool-temperate rain forest on a narrow landmass extending over c. 17.5° of latitude. The rain-forest zone constitutes a remarkable ecological island of wet forest, which is isolated by c. 1500–2000 km from other wet closed-canopy forests on the South American continent (Arroyo *et al.* 1996).

The rain forest occurs predominantly in Chile along the Andean and Coastal cordilleras, with small extensions across the Andes into neighbouring Argentina. The Andean Cordillera reaches c. 2000 m in height whereas the Coastal Cordillera rarely exceeds 1000 m. The Andean Cordillera is heavily glaciated, steep volcanic terrain which is highly prone to avalanches, with many of the glacial surfaces now covered by andesitic deposits (Veblen and Ashton 1978; Veblen, Schlegel and Oltremari 1983). In the far south the dominant physiographic features are deep fjords, the coastal archipelago, the North and South Patagonian icefields, and tidewater glaciers (Alaback 1991). The Coastal Cordillera of central Chile consists predominantly of Late Palaeozoic and Mesozoic metamorphic and plutonic rocks, intruded by Mesozoic plutons and dikes (Irwin 1989). Unlike the Andes, there is no evidence of glaciation (Villagrán 1990).

Conditions for rain forest in southern South America are provided by the Andean Cordillera intercepting strong westerly winds along the Pacific coast during winter and summer months and the cool northward-flowing oceanic Humboldt Current maintaining humid and foggy conditions in coastal locations. Throughout south-western South America, as a result of the low land/ocean ratio of southern South America and strong oceanic influences, the climate is highly equable, there being no latitudinal increase in annual temperature range (Arroyo *et al.* 1996). Delimited climatically according to criteria developed by Alaback (1991), the rain-forest region

has its northern limit close to treeline in the Andes at c. 38°30'S and close to the coast at 38°54'S, with its southern limit around the Cape Horn Islands off the mainland at 55°30'S. The rain forest extends from sea-level to the tree limit except in the Central Valley north of 41°S (Map 79), where conditions are too warm for the general forest to be considered cool-temperate rain forest.

The eastern limit at c. 39°S is on the western side of the Andes close to the treeline (1500 m) and then extends across the Andes into Argentina to 42°–43°S. Southward, the eastern limit again lies on the Chilean side of the Andes initially below the tree limit, to extend again to the treeline (c. 600 m) on the western margins of the North and South Patagonian icefields and the mountains of Tierra del Fuego. The region so delimited has a January isotherm of less than 16°C and annual precipitation from 1400–4900 mm (Veblen, Schlegel and Oltremari 1983), with more than 9–10% received over the summer months (Alaback 1991; Arroyo *et al.* 1996).

Based on floristic criteria (e.g. Oberdorfer 1960), the northern limit of the rain-forest zone is at 37°30'S and the south-eastern limit in south-eastern Tierra del Fuego. Relict stands of rain forest occur intermittently on steep south-facing slopes along the Chilean coast in the Mediterranean-type climate region (CPD Site SA44). Disjunct stands may be found as far north as 30°S, where they persist on fog-bound summits (Pérez and Villagrán 1985).

The rain forest, whether defined climatically or floristically, gradually intergrades to the north into Mediterranean-climate forests, and to the east to drier areas of *Austrocedrus chilensis* and deciduous *Nothofagus* forests. The ecotone between rain forest and Mediterranean climate in south-central Chile is broad and uneven, with many rain-forest elements extending far north of the climatic rain-forest boundary (Arroyo *et al.* 1996). Along the south-western margin of the continent where precipitation may exceed 4000 mm, rain forest gives way to Magellanic moorland containing patches with rainforest elements (Pisano 1977). Higher elevations on the island of Chiloé (Ruthsatz and Villagrán 1991) and the Coastal Cordillera farther north (Ramírez 1968) may also bear Magellanic moorland.

Geographic Information System (GIS) and other estimates suggest that cool-temperate rain forest originally covered from 70,000 km² to less than 110,000 km² (Kellogg 1993) – the larger GIS-derived figure is an over-estimate since alpine and moorland areas were not excluded. The actual area of native forest is estimated at 76,000 km² by INFOR (1991), which probably includes all the rain forests and Mediterranean forests with *Nothofagus* (but probably not sclerophyllous Mediterranean formations).

Vegetation

In this account the climatic definition of rain forest is followed; the rain-forest region as so defined includes both evergreen (Valdivian, North Patagonian, and Magellanic or Sub-Antarctic) and deciduous rain-forest types. Kellogg (1993) has suggested further division into seasonal and non-seasonal types, with the latter having perhumid and boreal sections (Map 79). On floristic criteria, only the predominantly evergreen Valdivian, North Patagonian and Magellanic rain-forest areas are classified as rain forest.

Much early vegetation study in the southern South American rain forest emphasized the phytosociological approach, with many plant associations described (Oberdorfer 1960). More recent work has emphasized physiognomic characteristics in combination with key dominants (Veblen and Schlegel 1982; Veblen, Schlegel and Oltremari 1983). The cool-temperate rain forest is most appropriately considered a mosaic of floristic types responding locally to key environmental gradients and soil conditions, reflected in gradual shifts in dominance and species richness. An outstanding feature is structural complexity, expressed in an unusually high representation for temperate forest of woody vines (lianas), climbing shrubs and hemi-parasites (Arroyo *et al.* 1996). In the Valdivian rainforest zone, the key environmental variables are steep altitudinal temperature gradients in the Andean and Coastal cordilleras, east-west trending valleys with contrasting microclimates, and in the Andes heavy avalanching and volcanism (Veblen and Schlegel 1982; Veblen, Schlegel and Oltremari 1983).

In the Coastal Cordillera and the Andean range (western flank), succession with increasing altitude is from high-diversity, evergreen closed-canopy rain forest to mixed angiosperm-conifer rain forest and then evergreen conifer forest. In the Andes at high elevations, the conifer forest is eventually succeeded by deciduous forest, which occurs to the treeline (c. 900–1200 m).

Low- to mid-elevation closed-canopy rain forest contains many species of vines, vining shrubs and woody epiphytes, e.g. *Asteranthera ovata*, *Mitraria coccinea*, *Sarmienta repens* (Gesneriaceae); *Campsidium valdivicum* (Bignoniaceae); *Hydrangea integerrima* (Hydrangeaceae); *Luzuriaga* spp. (Philesiaceae); and *Griselinia racemosa* (Cornaceae); hemi-parasites; and epiphytic ferns, especially Hymenophyllaceae. On well-drained sites the dominant angiosperm trees are *Aextoxicon punctatum*, *Laurelia sempervirens*, *Nothofagus obliqua*, *Eucryphia cordifolia* and *Laureliopsis philippiana*, with Proteaceae (*Gevuina avellana*, *Embothrium coccineum*,

Lomatia spp.), Myrtaceae (such as *Amomyrtus luma*, *Myrceugenia* spp., *Luma* spp.), *Dasyphyllum diacanthoides* (Asteraceae) and *Rhaphithamnus spinosus* (Verbenaceae) well represented throughout. *Nothofagus* is lacking in low-elevation forests on Chiloé Island (Donoso, Escobar and Urrutia 1985). Important tree species in structurally simpler, well-drained mid-elevation rain forest include *Nothofagus dombeyi*, *Laureliopsis philippiana*, *Weinmannia trichosperma*, *Caldcluvia paniculata*, *Saxegothaea conspicua*,

MAP 79. ORIGINAL DISTRIBUTION OF COOL-TEMPERATE RAIN FOREST IN SOUTHERN SOUTH AMERICA
(after Kellogg 1993)

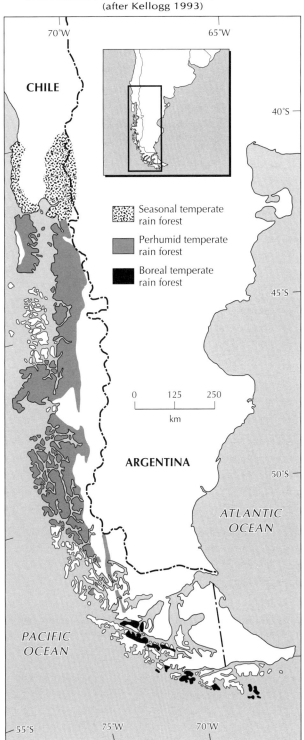

Nothofagus alpina, Drimys winteri and *Podocarpus nubigena*. The understorey is dark, with few herbaceous species (e.g. *Nertera grandensis*) and low shrub cover (Arroyo *et al.* 1996). *Chusquea* spp. are conspicuous in forest gaps and along the edges.

On poorly drained low- and mid-elevation sites *Tepualia stipularis* and *Pilgerodendron uviferum* become prominent, forming extensive inundated forests (Martínez 1981; Veblen and Schlegel 1982). In the Central Valley other inundated areas ("ñadis" and "hualves") occurring over an iron hardpan and drying in the summer months under higher temperatures are typically dominated by species of Myrtaceae and *Drimys winteri* (Winteraceae) (Ramírez *et al.* 1991).

Gymnosperm-dominated forests in the Valdivian rain-forest zone principally contain *Fitz-roya cupressoides* (Cupressaceae), which occurs discontinuously on the Coastal Cordillera from 39°50'S to 43°30'S and in the Andes from 40°S to 43°30'S. *Araucaria araucana* (Araucariaceae) forests appear only at the northern end of the rain-forest zone in the Andean Cordillera, whereas *Pilgerodendron* sometimes becomes dominant on swampy sites. Important variants of *Fitz-roya* forest include *Fitz-roya–Nothofagus betuloides* forest on upland slopes with winter snow; *Fitz-roya–Nothofagus nitida* forest on gentle slopes with over 4500 mm annual precipitation; *Fitz-roya–Pilgerodendron* forest on saturated soil on level marine terraces; marginal mixed forests at lower elevations in the Coastal Cordillera with such species as *Weinmannia trichosperma*, *Drimys winteri* and *Nothofagus nitida*; and

upland forest in the Coastal Cordillera with strong dominance of *Fitz-roya* (Donoso *et al.* 1993).

In the Andes, the evergreen angiosperm and mixed angiosperm-gymnosperm forests are eventually succeeded by predominantly deciduous forests comprised of *Nothofagus pumilio* and *N. antarctica*. A dense understorey of *Chusquea* spp. and typical shrub species (e.g. *Drimys andina*) usually prevails to treeline.

In the northern portion of the evergreen rain-forest zone, *Nothofagus betuloides*, more characteristic of the Magellanic rain forest, may also occur at treeline. East of the Andes below the deciduous forest belt, the rain forest is floristically impoverished; important species include *Nothofagus dombeyi*, *Saxegothaea conspicua*, *Laureliopsis philippiana* and *Podocarpus nubigena* (Dimitri and Correa 1966–1967). The *Podocarpus* intergrades into a band of *N. dombeyi* dominated forest subtended by drier *Austrocedrus chilensis* forest, which eventually intergrades with Patagonian steppe (see CPD Site SA46). Small areas of *Fitz-roya* and *Pilgerodendron* forest continue to appear on the eastern side of the Andes.

North Patagonian (43°20'S–47°30'S) and Magellanic (47°30'S–55°30'S) rain forests represent successively floristically impoverished and structurally simpler versions of the Valdivian rain forest. The North Patagonian rain forest occurs principally in heavily glaciated fjords and on numerous coastal islands on very thin glacial soils and exposed rock surfaces. The tree species typical of low-elevation Valdivian rain forest such as *Eucryphia cordifolia*, *Aextoxicon punctatum*, *Nothofagus*

CPD Site SA45: Temperate rain forest of Chile. Valdivian cool-temperate rain forest along the coast, in the Los Lagos (X) Region of Chile. Photo: Adriana Hoffmann.

CPD Site SA45: Evergreen cool-temperate rain forest ("tipo chilote"); dominant species are *Nothofagus nitida*, *N. betuloides*, *Weinmannia trichosperma* and *Fitz-roya cupressoides*. Photo: Adriana Hoffmann.

alpina, Dasyphyllum diacanthoides and *Laurelia sempervirens* are reduced in importance or absent, leaving such dominants as *Nothofagus dombeyi*, *Weinmannia trichosperma* and *Saxegothaea*, with abundant *Maytenus magellanica*, *Podocarpus nubigena* and *Drimys winteri*. Epiphytic ferns and frondose lichens are especially abundant, as are *Chusquea* spp. in the understorey. The Magellanic rain forest develops on thin podzolized soils of fluvioglacial origin. The northern *Nothofagus dombeyi* is replaced by the vicariant *N. betuloides*; *Pseudopanax laetevirens*, *Maytenus magellanica* and *Podocarpus nubiqena* become very abundant and *Chusquea* disappears. A plentiful herbaceous and shrubby understorey appears under the shorter and more open crowns (Arroyo *et al.* 1996).

On poorly drained coastal sites throughout the North Patagonian and Magellanic rain-forest zones, swamp forests with *Tepualia* and *Pilgerodendron* continue to occur. In spite of its high-latitude position, six species of vines/climbing shrubs may be found in Magellanic rain forest, constituting some of the world's highest latitudinal records for climbing plants: *Campsidium valdivicum*, *Griselinia ruscifolia*, *Lebetanthus myrsinites*, *Luzuriaga polyphylla*, *Philesia magellanica* and *Mitraria coccinea*.

Flora

Southern South American cool-temperate rain forest has a rich flora with a predominance of angiosperms and diverse life forms. Factors that have influenced its present taxonomic composition include: (1) a long-standing Gondwanan connection, which involved South America, Antarctica and other southern continents throughout the Cretaceous and Early to Mid Tertiary; (2) direct vegetational continuity between southern South America and tropical latitudes of South America in the Tertiary; (3) the present-day high climatic equability; and (4) relatively recent differentiation of a Mediterranean-type climate to the north of the rain-forest region. These features together have resulted in a rain-forest flora with unusually high generic richness and a high degree of relictual regional endemism.

Arroyo *et al.* (1996) estimated that the true rain-forest flora – those taxa occurring in or under the rain-forest canopy or within its immediate influence – comprises some 205 genera in 96 families, which includes 82 genera of woody plants in 50 families. At least one-third of the woody genera are of southern Gondwanan origin, with their nearest extant or fossil relatives in the general region of Australia, New Zealand, New Caledonia and New Guinea – such as *Aristotelia* (Elaeocarpaceae); *Caldcluvia* (Cunoniaceae); *Discaria* (Rhamnaceae); *Eucryphia* (Eucryphiaceae); *Gevuina, Lomatia* (Proteaceae); *Nothofagus* (Fagaceae); and *Pseudopanax* (Araliaceae). One-quarter of the woody genera have neotropical affinities, with their nearest congeners in the northern Andes or in south-eastern Brazil/north-eastern Argentina-Uruguay – such as *Antidaphne* (Eremolepidaceae), *Azara* (Flacourtiaceae), *Crinodendron* (Elaeocarpaceae), *Dasyphyllum* (Asteraceae), *Myrceugenia* (Myrtaceae), *Rhaphithamnus* (Verbenaceae) and *Tristerix* (Loranthaceae).

Most outstandingly, a third of the woody genera and one family (Aextoxicaceae) are endemic to the southern temperate area of Chile and Argentina (including Mediterranean-climate scrublands). Important monotypic regionally endemic genera

occurring in the rain forest include *Fitz-roya*, *Pilgerodendron* (Cupressaceae); *Saxegothaea* (Podocarpaceae); *Laureliopsis* (Monimiaceae); *Lebetanthus* (Epacridaceae); *Embothrium* (Proteaceae); *Tepualia* (Myrtaceae); *Desmaria, Notanthera* (Loranthaceae); *Aextoxicon* (Aextoxicaceae); *Latua* (Solanaceae); *Asteranthera, Sarmienta, Mitraria* (Gesneriaceae); *Campsidium* (Bignoniaceae); and *Lapageria, Philesia* (Philesiaceae). Close to 80% of the regionally endemic genera are monotypic (Arroyo *et al.* 1996). Such monotypic genera, by in large, represent ancient relictual taxa that were more widely distributed over southern South America in the Tertiary and have survived in the cool and equable coastal climatic conditions present today. The high concentration of relict endemic genera in this rain forest is analogous to the many palaeoendemic genera on some islands (Axelrod, Arroyo and Raven 1991).

Only two woody endemic genera (*Fitz-roya* and *Pilgerodendron*) are totally restricted to the rain-forest habitat. The majority of the regional endemics also occur farther north in the sclerophyllous forest in the Mediterranean-climate region, which together with the rain forest forms the more natural biogeographic province.

Although the flora of the general rain-forest area is fairly well known (Dimitri 1972), knowledge has only been recent on the total species richness for the rain-forest habitat (species in or under the forest canopy or within its immediate influence). South of 40°S, the vascular plant flora in Chile is estimated to total c. 1300 species (Arroyo, unpublished). A conservative estimate suggests that close to 450 vascular plant species may be found in the cool-temperate rain forest (Arroyo *et al.* 1996).

TABLE 64. FLORISTIC COMPOSITION OF CHILEAN COOL-TEMPERATE RAIN FOREST

Group	Families	Genera	Species
Angiosperms	81	175	369
Dicotyledons	68	133	275
Monocotyledons	13	42	94
Gymnosperms	3	7	8
Pteridophytes	12	23	66
Totals	**96**	**205**	**443**

Probably close to one-third of all the species of vascular plants in southern Chile thus depend to some extent on the rain-forest habitat. The woody rain-forest flora comprises c. 160 species; 44 are mostly evergreen trees, 81 shrubs and 35 vines. Herbaceous species number close to 300, of which 31 are epiphytes – mainly hymenophyllaceous ferns. Nonetheless, many herbaceous species are only occasional in the rain forest. Estimates of species richness are unavailable for the lichen and moss floras which are poorly known, but the region is very rich in lichens (Galloway 1992).

Strong reductions in species richness are seen in comparing the succession of rain-forest types along the latitudinal gradient (Arroyo *et al.* 1996). Maximum tree-species richness is at 40°–41°S, where degradation of the forest is most intense. Around 45°S, an area under imminent threat due to recent construction of penetration routes, the rain forest has some 20 species of trees and a rich vine flora. The largest woody family is Myrtaceae (15 spp.); the largest tree genus is *Nothofagus* (Fagaceae) (7 spp.); followed by *Lomatia* (Proteaceae) (3 spp.); and then *Podocarpus* (Podocarpaceae), *Amomyrtus* and *Myrceugenia* (Myrtaceae) and *Maytenus* (Celastraceae) (2 spp. each).

Total species richness in the Chilean cool-temperate rain forest far exceeds that reported in the much smaller area of such rain forest in Tasmania (Jarman, Kantvilas and Brown 1991). However since the sizes are very distinct for the regions compared, it cannot as yet be ascertained whether South American cool-temperate rain forest is intrinsically richer. Greater richness, however, is evident in comparison with the cool-temperate rain forest of the Pacific Northwest of the U.S.A., where there are far fewer species present in an area of about twice the size (Alaback, pers. comm.).

A very large proportion of the woody species occurring in southern South American rain forest, and many of the herbaceous species, are endemic to southern South America. Relatively few species in the rain forest are entirely restricted to the rain-forest habitat, as defined climatically. The low rain-forest endemism reflects the very gradual ecotone between the Mediterranean sclerophyllous forest and the rain forest, with many species being in both vegetation types. The following trees may be considered restricted to the rain forest: *Nothofagus nitida*, *Fitz-roya*, *Crinodendron hookerianum* and *Pilgerodendron*, and *Podocarpus nubigena* is almost restricted to it. Among shrubs, *Baccharis elaeoides* (Asteraceae) and *Latua pubiflora* are endemic to the rain-forest zone, although they also are found in non-forest habitats. Other virtually endemic rain-forest species occurring in the rain forest and/or adjacent vegetation types in the rain-forest zone include *Blechnum corralense* (Blechnaceae); *Megalastrum spectabile* var. *philippianum* (Dryopteridaceae); *Gleichenia litoralis* (Gleicheniaceae); *Hymenophyllum quetrihuense*, *H. umbratile* (Hymenophyllaceae); *Anemone hepaticifolia* (Ranunculaceae); *Viola corralensis* (Violaceae); *Loasa acerifolia* (Loasaceae); *Valdivia gayana* (Saxifragaceae); *Misodendrum macrolepis* (Misodendraceae); *Ercilla syncarpellata* (Apocynaceae); *Cynanchum lancifolium* (Asclepiadaceae); *Lobelia bridgesii* (Campanulaceae); *Uncinia negeri* (Cyperaceae); *Dioscorea cissophylla* (Dioscoreaceae); and *Arachnitis quetrihuensis* (Corsiaceae). Although *Gaultheria antarctica* (Ericaceae) and *Tapeinia pumila* (Iridaceae) are endemic to the general region and occur in rain forest, they are more typical of alpine or moorland habitats.

To these species can be added many more that occur only marginally outside the rain-forest area, such as *Philesia magellanica*, *Hymenophyllum nahuelhuapense*, *Griselinia racemosa*, *Lebetanthus myrsinites* and *Gunnera lobata*, and a number of species virtually restricted to the rain forest that appear again on the Juan Fernández Islands – *Hymenophyllum cuneatum*, *Trichomanes exsectum*, *Serpyllopsis caespitosa* (Hymenophyllaceae); and *Peperomia nummularioides* (Piperaceae). The above-mentioned endemics have their greatest concentration in the Coastal Cordillera around latitudes 40°–42°S, which on palynological grounds is considered to have provided an important refugium for rain-forest species in the Pleistocene (Villagrán 1990). This area undoubtedly should be an important focus for conservation efforts.

There are no records of plant species extinct within historic times. The most threatened species are *Araucaria araucana* and *Fitz-roya cupressoides*. They are the largest trees in Chile and the most valued for their wood. *Araucaria araucana* (Chilean population), *Fitz-roya* and *Pilgerodendron*

are in Appendix I of the Convention on International Trade in Endangered Species of Wild Fauna and Flora (CITES), and *Fitz-roya* is listed as threatened (i.e. IUCN Vulnerable) under the U.S. Endangered Species Act (which additionally controls import of its wood into U.S.A.).

Other species in the rain forest are considered to be variously threatened: among the endangered are *Pitavia punctata*, *Berberidopsis corallina* and *Valdivia gayana*; those vulnerable include *Austrocedrus chilensis*, *Persea lingue*, *Laurelia philippiana* and *Nothofagus alpina*; and those rare include *Citronella mucronata*, *Corynabutilon ochsenii*, *Eucryphia glutinosa*, *Hebe salicifolia*, *Lobelia bridgesii*, *Maytenus chubutensis*, *Myrceugenia colchaquensis*, *M. leptospermoides*, *M. pinifolia*, *Orites myrtoidea*, *Prumnopitys andina*, *Ribes integrifolium*, *Satureja multiflora*, *Schinus marchandii* and *Scutellaria valdiviana*.

Useful plants

Fitz-roya cupressoides ("alerce") is used for construction of various kinds. Because the wood is unaffected by heat and humidity, it has been the most valued Chilean native timber. The wood is used for musical instruments, shingles and light carpentry; the inner bark for ship caulking; and the resin is burned as incense (Bernath 1937). The wood of *Araucaria araucana* is used in construction of ceilings, floorings, furniture and pulp. *Nothofagus* species are all-purpose timbers used for furniture components, cabinet work, flooring and millwork.

Social and environmental values

The region supports one of the oldest and largest tree species in the world, *Fitz-roya cupressoides* – documented at living more than 3600 years (Lara and Villalba 1993). These temperate rain forests have high generic richness, an outstanding percentage of regionally endemic genera (Arroyo *et al.* 1996), and a preponderance of biotic interactions for pollination and dispersal (Riveros 1991; Riveros, Humaña and Lanfranco 1991; Armesto and Rozzi 1989).

In addition to being home to a high diversity of important plant species, the temperate rain forest provides essential habitat for numerous bird species and more than 35 mammal species (other than bats and marine mammals), including the puma and other cats, as well as two species of endangered deer, the huemul (*Hippocamelus bisulcus*) and the pudú (*Pudu pudu*) – one of the smallest deer in the world (Gilroy 1992). The endangered river otter (*Lutra provocax*) has been reduced to remnant populations in Andean streams of the Valdivian rain-forest zone. In addition, habitat destruction in the Valdivian forest is threatening the poorly known marsupials *Dromiciops* and *Rhyncholestes* (Veblen, Schlegel and Oltremari 1983).

Scientific botanical exploration was begun in 1782 by naturalist Juan Ignacio Molina, who published the first works on the Chilean flora to include species from the northern portion of the rain forests. Presently the Universidad de Concepción, Universidad Austral de Chile in Valdivia and Corporación Nacional Forestal (CONAF) are involved in scientific research on this region.

Threats

Historically, exploitation was not prevalent until the mid 16th century when the Spanish arrived. Forests were destroyed as the land was cleared for settlements and agriculture, and trees were cut for construction materials and fuelwood. Severe destruction did not take place until the 19th century – considered one of the most massive deforestations recorded in Latin America (Veblen, Delmastro and Schlatter 1976) – when colonists began occupying the lower slopes and progressively moved higher. Only 25% of Chile's original area of native forests may be intact (Armesto and Smith-Ramírez 1994).

Present threats are logging, clearing, afforestation, fire, agriculture and grazing. The average rate of deforestation in Chile in the 1980s was 500 km² per year, c. 0.7% of the remaining total annually (Ormazábal 1993).

Although fire was at one time the most destructive force, that has lessened considerably. Regeneration of *Fitz-roya* is helped by light burning, but extensive cutting and burning like that which occurred during the colonization of southern Chile inhibits or prevents regeneration. At one time Alerce National Monument preserved more burned and logged forest than primary forest.

Over the past 15 years many of Chile's rain forests have been clear-cut to satisfy foreign demands for timber and woodchips. Today logging of the Valdivian forest for woodchips and fuelwood for industrial and home uses is without doubt the major threat to the integrity of these temperate rain forests. In 1990 native species produced 52% of the woodchips exported and 20% came from natural old-growth forests (Barnett 1992). Wood of short fibres is used to produce the chips, such as from species of *Nothofagus*, *Eucryphia*, *Aextoxicon*, *Weinmannia* and *Laurelia*.

Considered the most valued timber in Chile, *Fitz-roya* has been exploited commercially since the late 16th century (Veblen and Ashton 1982). Due to its durability and resistance to fungal and insect attacks, it has played an important role in Chile's economy. In 1976, the species itself was declared a National Monument, a decree that prohibits the cutting of live trees. Although some logging has stopped, illegal extraction persists (cf. Veblen and Ashton 1982).

Clear-cutting is destructive to the rain forest, and another devastating factor is its replacement with vast monocultural plantations of the exotic *Pinus radiata*. Areas of extensive exploitation are in the provinces Valdivia, Cautín and Malleco, where 60% of the country's timber is produced. To some extent this has relieved pressure on harvesting primary forest; however, the risk of epidemic diseases and soil changes due to rotation of the plantations pose real threats.

The most recent threat to Chilean rain forest is in the southernmost forests, including areas as far south as Tierra del Fuego. Timber companies from Canada (British Columbia) and a joint Chilean-Japanese firm have purchased several large tracts to set up logging operations for woodchips production (Gilroy 1992).

Although 40 national areas are protected legally, many undergo illegal logging and grazing.

Conservation

There are 25 National Parks, 14 National Reserves and one National Monument in the region (WCMC 1992). Several National Parks are quite extensive, such as Laguna San Rafael NP with 17,420 km². One of the important parks is Alerce Andino NP (established in 1982), which protects *Fitz-roya* (alerce) and other important rain-forest species, such as *Nothofagus* spp. The amount of forest included in protected areas throughout Chile is estimated to be less than 15,000 km² (Ormazábal 1992) and possibly as low as just over 10,000 km² (Armesto and Smith-Ramírez 1994).

References

Alaback, P.B. (1991). Comparison of temperate rainforests of the Americas. *Rev. Chilena Hist. Nat.* 64: 399–412.

Armesto, J.J. and Rozzi, R. (1989). Seed dispersal syndromes in the rainforest of Chiloé: evidence for the importance of biotic dispersal in a temperate rainforest. *J. Biogeography* 16: 219–226.

Armesto, J.J. and Smith-Ramírez, C. (eds) (1994). Propuesta de la Sociedad de Ecología de Chile respecto al Proyecto de Ley: "Recuperación del Bosque Nativo y de Fomento Forestal". *Noticiero Biol.* 2(1): 2–8.

Arroyo, M.T. Kalin, Riveros, M., Peñaloza, A., Cavieres, L. and Faggi, A.M. (1996). Phytogeographic relationships and regional richness patterns of the cool temperate rainforest of southern South America. In Lawford, R.G., Alaback, P.B. and Fuentes, E.R. (eds), *High latitude rainforests and associated ecosystems of the west coast of the Americas: climate, hydrology, ecology and conservation.* Springer-Verlag, Heidelberg, Germany. Pp. 134–172.

Axelrod, D.I., Arroyo, M.T. Kalin and Raven, P.H. (1991). Historical development of temperate vegetation in the Americas. *Rev. Chilena Hist. Nat.* 64: 413–446.

Barnett, A. (1992). Vanishing worlds of temperate forests. *New Scientist* 1846: 10.

Bernath, E.L. (1937). Coniferous forest trees of Chile. *Tropical Woods* 52: 19–26.

Dimitri, M.J.L. (ed.) (1972). *La región de los bosques andino-patagónicos: sinópsis general.* Instituto Nacional de Tecnología Agropecuaria (INTA), Col. Cientif. 10. Buenos Aires. 381 pp.

Dimitri, M.J.L. and Correa, H. (1966-1967). La flora andino-patagónica. Estudio filosociológico de una comunidad edáfica entre Puerto Blest y Laguna Frías, del Parque Nacional Nahuelhuapi. *Anales Parques Nacionales* 11: 5–39.

Donoso, C., Escobar, B. and Urrutia, J. (1985). Estructura y estrategias regenerativas de un bosque virgen de Ulmo (*Eucryphia cordifolia* Cav.) - Tepa (*Laurelia philippiana* (Phil.) Looser) en Chiloé, Chile. *Rev. Chilena Hist. Nat.* 58: 171–186.

Donoso, C., Sandoval, V., Grez, R. and Rodríguez, J. (1993). Dynamics of *Fitzroya cupressoides* forests in southern Chile. *J. Vegetation Sci.* 4: 303–312.

Galloway, D.J. (1992). Lichens of Laguna San Rafael, Parque Nacional "Laguna San Rafael", southern Chile: indicators of environmental change. *Global Ecol. Biogeogr. Letters* 2: 37–45.

Gilroy, S. (1992). Disturbing the ancients. *Buzzworm* 4(1): 38–43.

INFOR (1991). *Estadísticas forestales 1990*. Bol. Estadística No. 21. Instituto Forestal (INFOR), Santiago.

Irwin, J.J. (1989). A note on Jurassic and early Cretaceous alkaline magmatism in the Coastal Cordillera of central Chile (30°30'–34°S). *J. South Amer. Earth Sciences* 2: 305–309.

Jarman, S.J., Kantvilas, G. and Brown, M.J. (1991). *Floristic and ecological studies in Tasmanian rainforest*. Tasmanian NRCP Report No. 3. 67 pp.

Kellogg, E. (ed.) (1993). *Coastal temperate rain forests: ecological characteristics, status and distribution worldwide*. A working manuscript, July 1993. Ecotrust and Conservation International, Occasional Paper Series No. 1. Portland, Oregon and Washington, D.C. 64 pp.

Lara, A. and Villalba, R. (1993). A 3620-year temperature record from *Fitzroya cupressoides* tree rings in southern South America. *Science* 260: 1104–1106.

Martínez, O. (1981). Flora and phytosociological description of a relict of *Pilgerodendron uvifera* (D. Don) Florin in the San Pablo de Tregua farm (Valdivia-Chile). *Bosque (Valdivia)* 4: 3–11.

Oberdorfer, E. (1960). *Pflanzensoziologische Studien in Chile*. Tüxen, R. (ed.), Flora et Vegetatio Mundi II. Verlag J. Cramer, Weinheim, Germany. 208 pp.

Ormazábal, C.S. (1992). Bosques naturales en Chile: ¿Cuánto privado? ¿Cuánto estatal? *Ambiente y Desarrollo (Chile)* 8: 38–41.

Ormazábal, C.S. (1993). The conservation of biodiversity in Chile. *Rev. Chilena Hist. Nat.* 66: 383–402.

Pérez, C. and Villagrán, C. (1985). Distribución de abundancias de especies de bosques relictos de la zona mediterránea de Chile. *Rev. Chilena Hist. Nat.* 58: 157–170.

Pisano, E. (1977). Fitogeografía de Fuego-Patagonia chilena. I. Comunidades vegetales entre las latitudes 52° y 56°S. *Anales Instituto Patagonia (Chile)* 8: 121–250.

Ramírez, C. (1968). Die Vegetation del Moore del Cordillera Pelada, Chile. *Bericht Oberhess. Gesellschaft für Natur- und Heilkunde zu Gieben, Neue Folge, Naturwissenschaftliche Abteilung* 36: 95–101.

Ramírez, C., San Martín, C., Figueroa, H., MacDonald, R. and Ferrada, V. (1991). Estudios ecosociológicos en la vegetación de los ñadis de la Décima Región de Chile. *Agro Sur* 19: 34–47.

Riveros, M. (1991). *Biología reproductiva en especies vegetales de dos comunidades de la zona templada del sur de Chile, (40°S)*. Ph.D. thesis, Universidad de Chile, Santiago. 301 pp.

Riveros, M., Humaña, A.M. and Lanfranco, D. (1991). Actividad de los polinizadores en el Parque Nacional Puyehue, X Región, Chile. *Medio Ambiente* 11(2): 5–12.

Ruthsatz, B. and Villagrán, C. (1991). Vegetation pattern and soil nutrients of a Magellanic moorland on the Cordillera de Piuchué, Chiloé Island, Chile. *Rev. Chilena Hist. Nat.* 64: 461–478.

Veblen, T.T. and Ashton, D.H. (1978). Catastrophic influences on the vegetation of the Valdivian Andes, Chile. *Vegetatio* 36: 149–167.

Veblen, T.T. and Ashton, D.H. (1982). The regeneration status of *Fitzroya cupressoides* in the Cordillera Pelada, Chile. *Biol. Conserv.* 23: 141–161.

Veblen, T.T. and Schlegel, F.M. (1982). Ecological review of the southern forests of Chile. *Bosque (Valdivia)* 4: 73–116.

Veblen, T.T., Delmastro, R.J. and Schlatter, J.E. (1976). The conservation of *Fitzroya cupressoides* and its environment in southern Chile. *Environm. Conserv.* 3: 291–301.

Veblen, T.T., Schlegel, F.M. and Oltremari, J.V. (1983). Temperate broad-leaved evergreen forests of South America. In Ovington, J.D. (ed.), *Temperate broad-leaved evergreen forests*. Ecosystems of the World Vol. 10. Elsevier, Amsterdam. Pp. 5–31.

Villagrán, C. (1990). Glacial climates and their effects on the history of the vegetation of Chile: a synthesis based on palynological evidence from Isla de Chiloé. *Rev. Palaeobot. Palynol.* 65: 17–24.

WCMC (World Conservation Monitoring Centre) (1992). *Protected areas of the world: a review of national systems*. Vol. 4: *Nearctic and Neotropical*. IUCN, Gland, Switzerland and Cambridge, U.K. 459 pp.

Authors

This Data Sheet was written by Dra. Mary T. Kalin Arroyo (Universidad de Chile, Facultad de Ciencias, Departamento de Biología, Casilla 653, Santiago, Chile) and Adriana E. Hoffmann-J. (Fundación Claudio Gay, Alvaro Casanova 613 - Peñalolén, Santiago, Chile).

PATAGONIA
Argentina and Chile

Location: Far south-eastern South America between Cordillera de los Andes and Atlantic Ocean, almost entirely in Argentina, between about longitudes 73°–64°W and latitudes 55°–40°S with a narrowing strip extending north to c. 33°S along cordillera's eastern slope.

Area: c. 500,000 km².

Altitude: 0–2000 m.

Vegetation: Xerophytic, with ample predominance of semi-desert, shrub-steppe and grass-steppe; also desert, halophytic steppe, moist meadow.

Flora: Close to 1200 species of vascular plants; nearly 30% endemism, including quasi-endemic family Halophytaceae and 6 endemic genera.

Useful plants: Genetic resources starting to be explored; most natural vegetation used as sheep pasturage; several species were used as crops by aboriginal peoples but presently are not important economically.

Other values: Tourism significant, particularly along Atlantic seaboard; palaeontological interest high – during Jurassic there were large coniferous forests with *Araucaria*; area has been inhabited by Tehuelche and Araucano – latter people have left their cultural mark on whole region.

Threats: Overgrazing, wood-cutting, desertification.

Conservation: Laguna Blanca National Park, Bosques Petrificados Natural National Monument, low eastern portions of several Andean NPs.

Geography

The phytogeographic region of Patagonia (the south-eastern portion of the geopolitical region of Patagonia) extends over 500,000 km² in extreme south-eastern South America (Ragonese 1967; Vila and Bertonatti 1993). A narrow strip in the north between the Andean Cordillera and the monte region widens to the south into a vast tableland from the Andes to the Atlantic Ocean (Map 80) (Cabrera and Willink 1973). The border to the west is with the temperate rain forest (Dimitri 1972) and the Altoandina region (Cabrera 1976), and to the north and east with the monte region (Ruiz Leal 1972). Most of Patagonia lies within the Argentine Republic, with a small portion penetrating Chilean territory in the far south on both sides of the Strait of Magellan.

The region is formed by a Precambrian nucleus – the Patagonian massif, with deposits of eruptive rock (basalt) and terrestrial and marine sediments from the Early Permian to Tertiary. The terrain is a sequence of mesa plains stepping down eastward to the sea, from varied Andean piedmont elevations (2000 m in the north to 700 m in the south). In Chubut Province east of the Andes and longitude 71°W, the precordillera Patagónica (Sierra de Tecka and extensions) rises to 1300–1500 m or more (e.g. Cerro Putrachoique, 1700 m); farther east between longitudes 70°–69°W and the Chubut and Senguerr rivers occur the Patagonian Central Ranges (Sierras Centrales Patagónicas), with elevations to 1690 m (Cerro Boquete) and 1651 m (Cerro Negro) (Volkheimer in Soriano 1983).

The predominant soils are sandy and stony, and poor in organic matter. Along a narrow strip near the Patagonian cordillera down to the Atlantic Coast in the far south are soils richer in organic matter. In restricted moist areas such as bogs, the soils are acidic or neutral. Rivers descend from the cordillera and traverse the region from west to east with scarce flow (being unnavigable); they have little influence on the vegetation.

The climate is characterized by severe dryness, with average annual precipitation below 200 mm, and by low temperatures (annual mean 9°–5°C) and constant drying high winds from the west. The precipitation can vary greatly from year to year (e.g. 157–519 mm at Pilcaniyeu east of San Carlos de Bariloche). Winter lasts 4–5 months (about June–September), with the daily minimum of the coldest month averaging 1°–3°C below freezing, and with no month assuredly frost-free (Walter and Box in Soriano 1983).

Vegetation

The vegetation is xerophytic, with ample predominance of adaptations to the low temperatures and the desiccating effect of the strong winds constantly blowing. Plant cover generally varies from 20–40%, with extremes near 0% in desert areas ("huayquerías" or badlands) and 100% in wet meadows ("vegas" and "mallines"). The dominant types of vegetation are semi-desert (45%), shrub-steppe (30%) and grass-steppe (20%) (Movia, Soriano and León 1987; Soriano 1982, 1983). Up to 28 different communities have been described, distributed among five phytogeographic districts (Soriano 1956; Cabrera 1976). Numerous physiographic units have been recognized (Ambrosetti and Méndez 1983; Beeskow, del Valle and Rostagno 1987; Movia, Soriano and León 1987; Speck *et al.* 1982).

MAP 80. PATAGONIA (CPD SITE SA46)

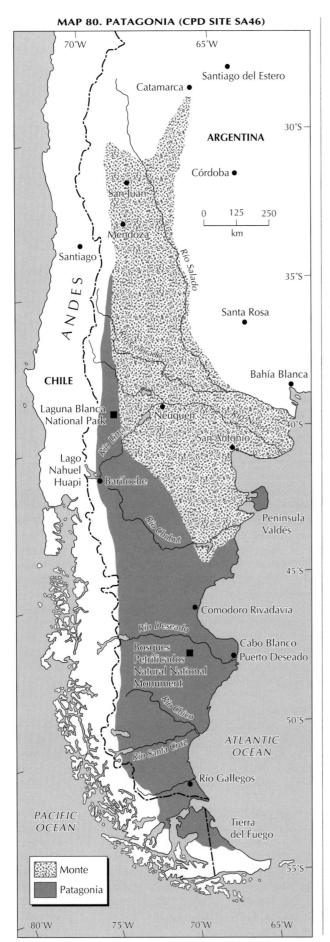

In **deserts** the plant cover is c. 5–15%, with halophytes of Chenopodiaceae predominant (*Atriplex, Nitrophila, Suaeda, Chenopodium*), sometimes accompanied by *Halophyton ameghinoi* (Halophytaceae).

Semi-desert, with c. 40% of the soil surface covered, is the most widely occurring vegetation type (Soriano, Sala and León 1980; Soriano 1983). Dwarf and cushion shrubs are dominant (e.g. *Acantholippia, Benthamiella, Brachyclados, Chuquiraga, Nassauvia, Nardophyllum, Verbena*) and tuft grasses common (*Stipa, Poa*) (Paruelo *et al.* 1992).

Shrub-steppe is characterized by the presence of taller woody shrubs, which in restricted areas can reach 3 m (Soriano 1982). Among the most conspicuous elements are species of *Adesmia, Anarthrophyllum, Berberis, Chuquiraga, Lycium, Mulinum, Schinus* and *Verbena*. In the east around San Jorge Gulf grow thick shrub communities (*Trevoa, Colliguaja*), covering up to 60% or more of the soil surface.

In the west bordering the temperate rain forest, where soils are richer, extends **grass-steppe** dominated by *Festuca pallescens*, which accounts for up to 90% of the vegetation, often accompanied by other grass species (e.g. *Agrostis, Bromus, Deschampsia, Poa, Trisetum*). The grass-steppe may reach 80% cover in non-degraded areas.

Moist meadows occupy valleys and lowlands scattered through the region where water availability is higher. Sedges (*Eleocharis, Scirpus*) and rushes (*Juncus*) are abundant, as are numerous species of grasses of various genera (*Agrostis, Deschampsia, Hordeum, Polypogon*), which grow intermingled with diverse dicotyledons (*Acaena, Caltha, Geranium, Plantago, Ranunculus*). In saline lowlands grow turfs of predominantly halophytic species in genera such as *Distichlis, Frankenia, Nitrophila, Puccinellia* and *Suaeda*.

In some areas intrusions of species characteristic of the monte region are abundant – *Acantholippia seriphioides, Prosopidastrum globosum, Schinus polygamus* and *Stipa tenuis* (Ruiz Leal 1972).

In modified areas numerous adventitious species have established, especially near highways and populated areas. The presence of weeds introduced from the Mediterranean region is not uncommon, and at times very conspicuous (*Bromus, Erodium, Galium, Stellaria, Vulpia*, various crucifers).

In the northern part of Tierra del Fuego two main vegetation types can be recognized: the grass-steppe, dominated by *Festuca gracillima* accompanied by other grasses (e.g. *Agropyron, Hordeum, Festuca, Poa*); and Fuegian meadows. The grass species dominant in the steppe are sometimes replaced by other plants (*Chiliotrichum diffusum, Empetrum rubrum*) that give a distinct physiognomy to the community. The Fuegian meadows occupy low flooded areas; their most important components are grasses (e.g. *Alopecurus, Deschampsia, Hordeum, Phleum*), accompanied by diverse dicotyledons (Moore 1983).

Flora

The flora of the Patagonian region shows clear affinities with the Andean flora (Cabrera 1976; Cabrera and Willink 1973), and is the product of a unique set of extreme climatic conditions of low temperature and aridity. It has an estimated 1200 species of vascular plants (Correa 1969–1988–), almost exclusively angiosperms, with c. 30% endemics. *Ephedra* is the only genus of gymnosperms and very important in some areas. Compositae is the best-represented family (200 spp., 33% endemic), followed by Gramineae (190 spp., 13% endemic) and Leguminosae (120 spp., 60% endemic). Other well-represented families are Cruciferae (80 spp., 30% endemic), Cyperaceae (40 spp., 7% endemic) and Umbelliferae (40 spp., 33% endemic). There is one quasi-endemic family in the region, Halophytaceae; and there are six endemic genera, usually represented by few species: *Philippiella* (Caryophyllaceae), *Neobaclea* (Malvaceae), *Xerodraba* (Cruciferae), *Benthamiella* (Solanaceae), and *Eriachaenium* and *Duseniella* (Compositae) (Soriano 1956).

The absence of deep differences in relief poorly defines the northern boundary of the region, which has a vague delimitation with the Alto-Andean region, and with the "monte" scrubland following approximately the 13°C annual isotherm (Roig 1972; Ruiz Leal 1972). The western limit and the southern limit, with the temperate rain forest (CPD Site SA45), are well defined with clear physiognomic and floristic characteristics (Dimitri 1972).

Useful plants

The knowledge of useful Patagonian plants is very fragmentary. According to information collected by travellers, numerous native species are (or have been) used for food or medicinals. Recent efforts to collect data are far from sufficient to cover the folkloric cultural wealth, which is being lost rapidly. Some species (*Festuca pallescens, F. gracillima*) are important forage in naturally vegetated areas that are grazed, whereas others are toxic to cattle (*Astragalus* spp., *Festuca argentina, Poa huecu*). Recently programmes have been implemented to collect germplasm of native species (Oliva, Montes and Mascó 1993).

Social and environmental values

The Tehuelche, who originally occupied the region, have dwindled to c. 50 persons. The Araucano (from Chile) penetrated the region especially in the 19th century (having begun to immigrate in the 16th century); the 50,000 who live in Argentina are being acculturated and losing their language. Many species were grown, such as maize, "quinoa" (*Chenopodium*), *Amaranthus*, "mango" (*Bromus mango*), potatoes, "oca" (*Oxalis*), "madi" (*Madia sativa*) and "frutilla" or "fresa" (*Frageria chiloensis*) (Parodi 1966). Except for maize, their local cultivation has ceased.

The predominant economic activity in Patagonia since the end of the 19th century has been sheep-farming (INTA 1993; Soriano 1983). The whole region is dependent in its development on the fluctuations in the price of wool. Extensive exploitation is based on heavy grazing of the natural vegetation, which takes place by c. 10,000 ranches of up to several thousand hectares with flock sizes that may reach 10,000 sheep or more (20–60 per km²).

The regional palaeoflora is very rich; some fossils from the Carboniferous have been found. However the most striking specimens, which draw some tourists, are the huge petrified trees of *Araucaria mirabilis* (100 m tall and 3.5 m dbh) in the conifer forests of the Jurassic (Menéndez 1972) – before the rise of the Andes.

Threats

The density of people is low (1 inhabitant per km²), but over-grazing from the sheep is very intense, which (made worse by wood-cutting) has led to serious desertification that threatens all of the region (INTA 1993; Soriano and Movia 1986). For example in 1972, 26% of Chubut Province and 38% of Santa Cruz Province showed eroded areas (Soriano 1983). Within National Parks, the principal problem is persistence of populated areas, although their overall impact is not great.

Conservation

Most important are considerations related to control of the overgrazing (León and Aguiar 1985; Soriano and Movia 1986; Soriano, Sala and León 1980). There is now an official plan to reverse the process of desertification in Patagonia (INTA 1993).

Few protected areas strictly correspond to this phytogeographic unit, but many cordillera National Parks include a representative part of adjacent Patagonian steppe (Erize 1993; WCMC 1992). Three of the protected areas bordering Chile include important sectors of Patagonia in their eastern portions: (1) down to 720 m, Nahuel Huapi National Park (4281 km²) and National Nature Reserve (3300 km²) – the first National Park in South America, begun in 1903; (2.1) down to 400 m, Los Alerces NP (1875 km²) and NNR (755 km²), and (2.2) particularly Perito Francisco P. Moreno NP (851 km²) and NNR (299 km²) down to 900 m – both parks/reserves were established in 1937.

Entirely within this phytogeographic unit in north-western Patagonia (39°30'S, 70°20'W) in the Province Neuquén, is Laguna Blanca NP (82 km²) and adjoining NNR (30 km²), which was created in 1940 (Roquero 1968; Correa-Luna 1977). Bosques Petrificados Natural National Monument (100 km²), created in 1954, is in south-eastern Patagonia (47°40'S, 68°00'W) in the Province Santa Cruz. It protects impressive remnants of the Jurassic petrified forest. There is a proposal to create a park in the east in the Valdés Peninsula Multiple Use Reserve (3600 km²).

References

Ambrosetti, J.A. and Méndez, E. (1983). Los tipos biológicos de Raunkier en las comunidades vegetales de Río Turbio, Provincia de Santa Cruz, Argentina. *Deserta* 7: 12–39.

Beeskow, A.M., del Valle, H.F. and Rostagno, C.M. (1987). *Los sistemas fisiográficos de la región árida y semiárida de la Provincia de Chubut*. Centro Nacional Patagónico, Consejo Nacional de Investigaciones Científicas y Técnicas (CONICET). Puerto Madryn, Argentina. 173 pp.

Cabrera, A.L. (1976). *Regiones fitogeográficas argentinas*. In Parodi, L.R. (ed.), *Enciclopedia argentina de agricultura y jardinería*, 2nd edition. Vol. 2(1). Editorial Acmé, Buenos Aires. Pp. 1–85.

Cabrera, A.L. and Willink, A. (1973). *Biogeografía de América Latina*. Organización de los Estados Americanos (OEA), Serie de Biología, Monogr. No. 13, Washington, D.C. 117 pp.

Correa, M.N. (1969-1988-). *Flora patagónica*. Instituto Nacional de Tecnología Agropecuaria (INTA), Col. Científ. 8. Buenos Aires.

Correa-Luna, H. (1977). *La conservación de la naturaleza: parques nacionales argentinos*. Servicio Nacional de Parques Nacionales, Buenos Aires. 169 pp.

Dimitri, M.J.L. (ed.) (1972). *La región de los bosques andino-patagónicos: sinópsis general*. INTA, Col. Científ. 10. Buenos Aires. 381 pp.

Erize, F. (1993). *Los parques nacionales de la Argentina y otras de sus áreas naturales*, 2nd edition. El Ateneo, Buenos Aires. 238 pp.

INTA (1993). *Proyecto de prevención y control de la desertificación en la Patagonia/Project for prevention and control of desertification in Patagonia*. INTA, Buenos Aires. 19 pp.

León, R.J.C. and Aguiar, M.R. (1985). El deterioro por uso pasturil en estepas herbáceas patagónicas. *Phytocoenology* 13: 181–196.

Menéndez, C.A. (1972). Paleofloras de la Patagonia. In Dimitri, M.J.L. (ed.), *La región de los bosques andino-patagónicos*. INTA, Col. Científ. 10: 129–165.

Moore, D.M. (1983). *Flora of Tierra del Fuego*. A. Nelson, Oswestry, Shropshire, U.K. and Missouri Botanical Garden, St. Louis, Missouri, U.S.A. 396 pp.

Movia, C.P., Soriano, A. and León, R.J.C. (1987). La vegetación de la cuenca del Río Santa Cruz (Provincia de Santa Cruz, Argentina). *Darwiniana* 28: 9–78.

Oliva, G.E., Montes, L. and Mascó, E.M. (1993). Collecting native forage germplasm in Patagonia. FAO/IBPGR, *Plant Genetic Resources Newsletter* 93: 34–37.

Parodi, L.R. (1966). *La agricultura aborigen argentina*. Editorial Universidad de Buenos Aires (EUDEBA), Buenos Aires. 48 pp.

Paruelo, J.M., Aguiar, M.R., Golluscio, R.A. and León, R.J.C. (1992). La Patagonia extrandina: análisis de la estructura y funcionamiento de la vegetación a distintos niveles. *Ecología Austral* 2: 123–136.

Ragonese, A.R. (1967). *Vegetación y ganadería en la República Argentina*. INTA, Col. Científ. 5. Buenos Aires. 218 pp.

Roig, F.A. (1972). Bosquejo fisionómico de la vegetación de la Provincia de Mendoza. *Bol. Soc. Argent. Bot.* 13 (Supl.): 49–80.

Roquero, M.J. (1968). La vegetación del Parque Nacional Laguna Blanca (estudio fitosociológico preliminar). *An. Parques Nac.* 11: 209–223.

Ruiz Leal, A. (1972). Los confines boreal y austral de las provincias patagónica y central respectivamente. *Bol. Soc. Argent. Bot.* 13 (Supl.): 89–118.

Soriano, A. (1956). Los distritos florísticos de la provincia patagónica. *Rev. Invest. Agríc.* 10: 323–347.

Soriano, A. (1982). Patagonia. In Vervoorst, F. (ed.), *Conservación de la vegetación natural en la República Argentina*. Soc. Argent. Bot. – Fundación Miguel Lillo, Serie Conserv. Naturaleza No. 2. Tucumán.

Soriano, A. (1983). Deserts and semi-deserts of Patagonia. In West, N.E. (ed.), *Temperate deserts and semi-deserts*. Ecosystems of the World Vol. 5. Elsevier, Amsterdam. Pp. 423–460.

Soriano, A. and Movia, C.P. (1986). Erosión y desertización en la Patagonia. *Interciencia* 11(2): 77–83.

Soriano, A., Sala, O.E. and León, R.J.C. (1980). Vegetación actual y vegetación potencial en el pastizal de coirón amargo (*Stipa* spp.) del sudoeste del Chubut. *Bol. Soc. Argent. Bot.* 19: 309–314.

Speck, N.H., Sourrouille, E.A., Wijnhoud, S., Munist, E., Monteith, N.H., Volkheimer, W. and Menéndez, J.A. (1982). *Sistemas fisiográficos de la zona Ingeniero Jacobacci-Maquinchao (Provincia de Río Negro)*. INTA, Col. Científ. 19. Buenos Aires. 215 pp.

Vila, A.R. and Bertonatti, C. (1993). *Situación ambiental de la Argentina*. Bol. Técn., Fundación Vida Silvestre Argentina, Buenos Aires. 74 pp.

WCMC (1992). *Protected areas of the world. A review of national systems*. Vol. 4. *Nearctic and Neotropical*. IUCN, Gland, Switzerland and Cambridge, U.K. 459 pp.

Author

This Data Sheet was written by Dr Carlos B. Villamil (Universidad Nacional del Sur, Departamento de Biología, Perú 670, 8000 Bahía Blanca, Argentina).

APPENDIX 1
DEFINITIONS OF SOME TERMS AND CATEGORIES

1. IUCN Conservation (Red Data Book) Categories

Conservation categories are given in the text for some threatened species where this information was readily available. The IUCN categories are given a capital letter to distinguish them from general usage of the terms.

A. Threatened Categories

Extinct (Ex)
Taxa which are no longer known to exist in the wild after repeated searches of their type localities and other known or likely places.

Endangered (E)
Taxa in danger of extinction and whose survival is unlikely if the causal factors continue operating.

Included are taxa whose numbers have been reduced to a critical level or whose habitats have been so drastically reduced that they are deemed to be in immediate danger of extinction.

Vulnerable (V)
Taxa believed likely to move into the Endangered category in the near future if the causal factors continue operating.

Included are taxa of which most or all the populations are *decreasing* because of over-exploitation, extensive habitat or other environmental disturbance; taxa with populations that have been seriously *depleted* and whose ultimate security is not yet assured; and taxa with populations that are still abundant but are *under threat* from serious adverse factors throughout their range.

Rare (R)
Taxa with small world populations that are not at present Endangered or Vulnerable, but are at risk.

These taxa are usually localized within restricted geographical areas or habitats or are thinly scattered over a more extensive range.

Indeterminate (I)
Taxa *known* to be Extinct, Endangered, Vulnerable, or Rare but there is not enough information to say which of the four categories is appropriate.

B. Unknown Categories

Status Unknown (?)
No information is available with which to assign a conservation category. [Note that this category was not used in CPD, but is included here for completeness.]

Insufficiently Known (K)
Taxa that are suspected but not definitely known to belong to any of the above categories, following assessment, because of lack of information.

C. Not Threatened Category

Safe (nt)
Neither rare nor threatened.

2. Categories and Management Objectives of Protected Areas

IUCN Management Categories for protected areas are given in the text where this information was readily available. In some cases, nationally designated areas (such as some National Parks) do not meet the criteria required to be assigned an IUCN Management Category. However, the omission of an IUCN category in the CPD text does not imply that the area concerned has not been assigned a management category.

Category I: Scientific Reserve/Strict Nature Reserve

To protect nature and maintain natural processes in an undisturbed state in order to have ecologically representative examples of the natural environment available for scientific study, environmental monitoring, education, and for the maintenance of genetic resources in a dynamic and evolutionary state.

Category II: National Park

To protect natural and scenic areas of national or international significance for scientific, educational and recreational use.

Category III: Natural Monument/ Natural Landmark

To protect and preserve nationally significant natural features because of their special interest or unique characteristics.

Category IV: Managed Nature Reserve/Wildlife Sanctuary

To assure the natural conditions necessary to protect nationally significant species, groups of species, biotic communities, or physical features of the environment where these require specific human manipulation for their perpetuation.

Category V: Protected Landscape or Seascape

To maintain nationally significant natural landscapes which are characteristic of the harmonious interaction of man and land while providing opportunities for public enjoyment through recreation and tourism within normal life style and economic activity of these areas.

Category VI: Resource Reserve

To protect resources of the area for future use and prevent or contain development activities that could affect the resource pending the establishment of objectives which are based upon appropriate knowledge and planning.

Category VII: Natural Biotic Area/ Anthropological Reserve

To allow the way of life of societies living in harmony with the environment to continue undisturbed by modern technology.

Category VIII: Multiple-Use Management Area/Managed Resource Area

To provide for the sustained production of water, timber, wildlife, pasture, and outdoor recreation, with the conservation of nature primarily oriented to the support of economic activities (although specific zones may also be designed within these areas to achieve specific conservation objectives).

Category IX: Biosphere Reserve

These are part of an international scientific programme, the UNESCO Man and the Biosphere (MAB) Programme, which is aimed at developing a reserve network representative of the world's ecosystems to fulfil a range of objectives, including research, monitoring, training and demonstration, as well as conservation roles. In most cases the human component is vital to the functioning of the Biosphere Reserve.

Category X: World Heritage Site

The *Convention Concerning the Protection of the World Cultural and Natural Heritage* (which was adopted in Paris in 1972 and came into force in December 1975) provides for the designation of areas of "outstanding universal value" as World Heritage Sites, with the principal aim of fostering international cooperation in safeguarding these important sites. Sites, which must be nominated by the signatory nation responsible, are evaluated for their world heritage quality before being declared by the World Heritage Committee. Article 2 of the Convention considers as natural heritage: natural features consisting of physical and biological formations or groups of such formations, which are of outstanding universal value from the aesthetic or scientific point of view; geological or physiographical formations and precisely delineated areas which constitute habitat of threatened species of animals or plants of outstanding universal value; and natural sites or precisely delineated areas of outstanding universal value from the point of view of science, conservation or natural beauty. Criteria for inclusion in the list are published by UNESCO.

The definitions above are abridged from:

IUCN (1990). *1990 United Nations list of national parks and protected areas*. IUCN, Gland, Switzerland and Cambridge, U.K. 284 pp.

McNeely, J.A. and Miller, K.R. (eds) (1984). *National parks, conservation, and development. The role of protected areas in sustaining society*. Smithsonian Institution Press, Washington, D.C. Pp. 47–53.

3. Endemic Bird Areas (EBAs)

Analysis of the patterns of bird distribution by BirdLife International shows that species of birds with restricted ranges tend to occur together in places which are often islands or isolated patches of a particular habitat, especially montane and other tropical forests. Boundaries of these natural groupings have been identified and the areas thus defined have been called Endemic Bird Areas (EBAs). They number 221 and embrace 2484 bird species, which is the vast majority (95%) of all restricted-range birds (which are those species with known breeding ranges below 50,000 km^2). The presence of restricted-range bird species in CPD sites is noted in the Data Sheets.

For more information and a full list of EBAs, the reader is referred to the following:

ICBP (1992). *Putting biodiversity on the map: priority areas for global conservation*. International Council for Bird Preservation, Cambridge, U.K. 90 pp.

Stattersfield, A.J., Crosby, M.J., Long, A.J. and Wege, D.C. (in prep.). *Global directory of Endemic Bird Areas*. BirdLife International, Cambridge, U.K.

AC – Area de Conservación (Costa Rica)

ACEC – Area of Critical Environmental Concern (U.S.A.)

ACOSA – Area de Conservación de la Península de Osa (Costa Rica)

AHE – Asociación Hondureña de Ecología (Honduras)

ANCON – Asociación Nacional para la Conservación de la Naturaleza (Panama)

APA – Área de Proteção Ambiental / Environmental Protection Zone (Brazil)

ARIE – Área de Relevante Interesse Ecológico (Brazil)

BA – Botanical Area (U.S.A)

BCI – Barro Colorado Island (Panama)

BCS – Baja California Sur (Mexico)

BGCI – Botanic Gardens Conservation International

BLM – Bureau of Land Management (U.S.A.)

BOSCOSA – Programa de Manejo y Conservación de Bosques de la Península de Osa (Costa Rica)

BP – (years) Before Present

BR – Biosphere Reserve

CARDER – Corporación Autónoma Regional de Risaralda (Colombia)

CARE – Cooperative for Assistance and Relief

CARL – Conservation and Recreational Lands (U.S.A.)

CATIE – Centro Agronómico Tropical de Investigación y Enseñanza (Costa Rica)

CCAD – Comisión Centroamericana de Ambiente y Desarrollo

CCT – Centro Científico Tropical (Costa Rica)

CDC – Conservation Data Center / Centro de Datos para la Conservación

CDC-UNALM – Centro de Datos para la Conservación – Universidad Nacional Agraria La Molina (Peru)

CECON – Centro de Estudios Conservacionistas, Universidad de San Carlos (Guatemala)

CENARGEN – Centro Nacional de Pesquisas de Recursos Genéticos e Biotecnologia (Brazil)

CEPEC – Centro de Pesquisas do Cacau (Brazil)

CFP – California Floristic Province (U.S.A.)

CI – Conservation International (U.S.A.)

CIIDIR-IPN – Centro Interdisciplinario de Investigación para el Desarrollo Integral Regional, Unidad Durango – Instituto Politécnico Nacional (Mexico)

CITES – Convention on International Trade in Endangered Species of Wild Fauna and Flora

CMC – IUCN Conservation Monitoring Centre (now WCMC)

COA – Corporación Colombiana para la Amazonia (Colombia)

CODEFF – Comité Nacional Pro Defensa de la Fauna y Flora (Chile)

COHDEFOR – Corporación Hondureña de Desarrollo Forestal (Honduras)

CONACYT – Consejo Nacional de Ciencia y Tecnología (Mexico)

CONAF – Corporación Nacional Forestal (Chile)

CONAMA – Comisión Nacional del Medio Ambiente (Guatemala)

CONAP – Consejo Nacional de Areas Protegidas (Guatemala)

CONDEPHAAT – Conselho de Defesa do Patrimônio Histórico, Artístico, Arquitetônico e Turístico (Brazil)

CONIF – Corporación Nacional de Investigaciones Forestales (Colombia)

COPFA – Comité para la Prevención de la Fiebre Aftosa

CPATU – Centro de Pesquisa Agropecuária do Trópico Úmido (Brazil)

CPC – Center for Plant Conservation (U.S.A.)

CPD – Centre of Plant Diversity

CRQ – Corporación Regional de Quindío (Colombia)

CUMAT – Centro de Investigaciones de la Capacidad de Uso Mayor de la Tierra (Bolivia)

CVC – Corporación Valle del Cauca (Colombia)

CVRD – Companhia Vale do Rio Doce (Brazil)

dbh – diameter at breast height

DESFIL – USAID Development Strategies for Fragile Lands Project

DF – Distrito Federal (Brazil)

DFG – Department of Fish and Game (U.S.A.)

DGFC – Dirección General Forestal y de Caza (Peru)

DGFF – Dirección General Forestal y de Fauna (Peru)

DIGEBOS – Dirección General de Bosques y Vida Silvestre (Guatemala)

DIGERENARE – Dirección General de Recursos Naturales Renovables (Honduras)

DINAF – Dirección Nacional Forestal (Ecuador)

DPNVS – Dirección de Parques Nacionales y Vida Silvestre (Paraguay)

EBA – Endemic Bird Area

EC – Commission of the European Communities

ECOCIENCIA – Fundación Ecuatoriana de Estudios Ecológicos (Ecuador)

ECODES – Estrategia de Conservación para el Desarrollo Sostenible (Costa Rica)

EEJI – Estação Ecológica Juréia-Itatins (Brazil)

EMBRAPA – Empresa Brasileira de Pesquisa Agropecuária

ENDESA/BOTROSA – Enchapes Decorativos, S.A. / Bosques Tropicales, S.A. (Ecuador)

FAO – Food and Agriculture Organization of the United Nations

FEDECANAL – Federación Departamental de Campesinos y Nativos de Loreto (Peru)

FEEMA – Fundação Estadual de Engenharia do Meio Ambiente (Brazil)

FFPS – Fauna and Flora Preservation Society (now FFI, Fauna and Flora International) (U.K.)

FMB – Fundación Moisés Bertoni para la Conservación de la Naturaleza (Paraguay)

FN – Fundación Natura (Ecuador)

FNC – Fundación Natura Colombia (Colombia)

FNT – Fundación Neotrópica (Costa Rica)

FOIN – Federación de Organizaciones Indígenas del Napo (Ecuador)

FPCN – Fundación Peruana para la Conservación de la Naturaleza (Peru)

FR – Forest Reserve
FUNATURA – Fundação Pró-Natureza (Brazil)
FUNDECOR – Fundación para el Desarrollo de la Cordillera Volcánica Central (Costa Rica)
FUNDEMABAV – Fundación del Medio Ambiente para Baja Verapaz (Guatemala)
GCTM – Group for the Conservation of the Tropics in Mexico
GDF – Governo do Distrito Federal (Brazil)
GEF – Global Environment Facility
GEOBOL – Servicio Geológico de Bolivia
GIS – Geographic Information System
IAN – Instituto Agronômico do Norte (now CPATU) (Brazil)
IBAMA – Instituto Brasileiro do Meio Ambiente e dos Recursos Naturais Renováveis (Brazil)
IBDF – Instituto Brasileiro de Desenvolvimento Florestal (Brazil)
IBGE – Instituto Brasileiro de Geografia e Estatística (Brazil)
IBP – International Biological Programme
IBPGR – International Board for Plant Genetic Resources (now IPGRI)
ICBP – International Council for Bird Preservation (now BirdLife International)
IDA – Instituto de Desarrollo Agrario (Costa Rica)
IDAEH – Instituto de Antropología e Historia (Guatemala)
IDESP – Institute for Economic and Social Development of Pará (Brazil)
IFEA – Instituto Francés de Estudios Andinos (Peru)
IGAC – Instituto Geográfico "Agustín Codazzi" (Colombia)
IGN – Instituto Geográfico Nacional "Tommy Guardia" (Panama)
IHAH – Instituto Hondureño de Antropología e Historia (Honduras)
IIED – International Institute for Environment and Development
INADE – Instituto Nacional de Desarrollo (Peru)
INAFOR – Instituto Nacional Forestal (Guatemala)
INBio – Instituto Nacional de Biodiversidad (Costa Rica)
INDERENA – Instituto Nacional de los Recursos Naturales Renovables y del Ambiente (Colombia)
INEFAN – Instituto Ecuatoriano Forestal y de Areas Naturales y Vida Silvestre (Ecuador)
INFOR – Instituto Forestal (Chile)
INI – Instituto Nacional Indigenista (Mexico)
INIAA – Instituto Nacional de Investigación Agropecuaria y Agroindustrial (Peru)
INIF – Instituto Nacional de Investigaciones Forestales (Mexico)
INIREB – Instituto Nacional de Investigación sobre Recursos Bióticos (Mexico)
INPA – Instituto Nacional de Pesquisas da Amazônia (Brazil)
INPE – Instituto Nacional de Pesquisas Espaciais (Brazil)
INRENARE – Instituto Nacional de Recursos Naturales Renovables (Panama)
INTA – Instituto Nacional de Tecnología Agropecuaria (Argentina)
IPGRI – International Plant Genetic Resources Institute (formerly IBPGR)
IRENA – Instituto Nicaraguense de Recursos Naturales y del Ambiente (Nicaragua)
ITCZ – Intertropical Convergence Zone
IUCN – International Union for Conservation of Nature and Natural Resources – The World Conservation Union
JBRJ – Jardim Botânico do Rio de Janeiro (Brazil)
JUNAC – Junta del Acuerdo de Cartagena (Colombia)
KFW – German Bank of Reconstruction and Development

LNLJ – Laboratorio Natural Las Joyas de la Sierra de Manantlán (Mexico)
Ma – Million years ago
MAB – Man and the Biosphere Programme (UNESCO)
MAG-PRONAREG – Ministerio de Agricultura y Ganadería, Programa Nacional de Regionalización (Ecuador)
MAYAREMA – USAID Maya Resource Management Project (Guatemala)
MCSE – Minimum Critical Size of Ecosystems Project (Brazil)
MIDEPLAN – Ministerio de Planificación Nacional y Política Económica (Panama)
MIRENEM – Ministerio de Recursos Naturales, Energía y Minas (Costa Rica)
MOPAWI – Mosquitia Pawisa (Honduras)
MRN – Mineração Rio do Norte (Brazil)
NF – National Forest
NGO – non-governmental organization
NP – National Park
NNP – Natural National Park
OAS/OEA – Organization of American States/Organización de Estados Americanos
ODA – Overseas Development Administration (U.K.)
ONERN – Oficina Nacional de Evaluación de Recursos Naturales (Peru)
ORSTOM – Institut Français de Recherche Scientifique pour le Développement en Coopération (formerly Office de la Recherche Scientifique et Technique Outre-Mer) (France)
OTS – Organization for Tropical Studies
PACA – Programa Ambiental para Centroamérica
PAG – IUCN-WWF Plant Advisory Group
PAUT – Proyecto Amazonia Universidad de Turku (Finland/Peru)
PILA – Parque Internacional La Amistad (Costa Rica and Panama)
Ramsar Convention – Convention on Wetlands of International Importance Especially as Waterfowl Habitat
RAP – Rapid Assessment Program (of Conservation International)
RARE – Rare Animal Relief Effort
RENARE – Dirección General de Recursos Naturales Renovables (Honduras)
RI – Resguardo Indígena (Colombia)
RNA – Research Natural Area (U.S.A.)
SAE – Secretariat of Strategic Studies (Brazil)
SAG – Secretaría de Agricultura y Ganadería (Mexico)
SANPES – Sistema de Areas Naturales Protegidas del Estado de Sonora (Mexico)
SARH – Secretaría de Agricultura y Recursos Hidráulicos (Mexico)
SECAB – Secretaría Ejecutiva del Convenio Andrés Bello (Colombia)
SECPLAN – Secretaria de Planejamento e Controle (Brazil)
SECPLAN – Secretaría de Planificación Coordinación y Presupuesto (Honduras)
SEDESOL – Secretaría de Desarrollo Social (Mexico)
SEDUE – Secretaría de Desarrollo Urbano y Ecología (Mexico) (formerly SEDESOL)
SEMARNAP – Secretaría de Medio Ambiente, Recursos Naturales y Pesca (Mexico) (formerly SEDUE and SEDESOL)
SEMATEC – Secretaria do Meio Ambiente, Ciência e Tecnologia (Brazil)
SERBO – Sociedad para el Estudio de Recursos Bióticos de Oaxaca (Mexico)

SFRNR – Subsecretaría Forestal y de Recursos Naturales Renovables (Ecuador)

SIAP – Sistema Integrado de Areas Protegidas de El Petén (Guatemala)

SIAPAZ – Sistema Internacional de Areas Protegidas para la Paz (Sí-A-Paz)

SICAP – Sistema Centroamericano de Areas Protegidas

SIGAP – Sistema de Areas Protegidas de Guatemala

SINAC – Sistema Nacional de Areas de Conservación (Costa Rica)

SINANPE – Sistema Nacional de Areas Naturales Protegidas por el Estado (Peru)

SINAP – Sistema Nacional de Areas Protegidas (Mexico)

SINUC – Sistema Nacional de Unidades de Conservación (Peru)

SNAP – Sistema Nacional de Areas Protegidas (Bolivia)

SNEM – Sistema Nacional para la Erradicación de la Malaria

STP – Secretaría Técnica de Planificación (Paraguay)

SUBIR – USAID Sustainable Use of Biological Resources Program (Ecuador)

SUDAM – Superintendência para o Desenvolvimento da Amazônia (Brazil)

TFAP – Tropical Forestry Action Plan

TNC – The Nature Conservancy (U.S.A.)

UAF – Unidades de Administración Forestal or Unidades de Conservación y Desarrollo Forestal (Mexico)

UN – United Nations

UNAM – Universidad Nacional Autónoma de México (Mexico)

UNCED – United Nations Conference on Environment and Development

UNDP – United Nations Development Programme

UNEP – United Nations Environment Programme

UNESCO – United Nations Educational, Scientific and Cultural Organization

UNMSM – Universidad Nacional Mayor de San Marcos (Peru)

USAID – United States Agency for International Development

USFWS – United States Fish and Wildlife Service

WCMC – World Conservation Monitoring Centre

WRI – World Resources Institute (U.S.A.)

WWF – World Wide Fund for Nature (formerly World Wildlife Fund)

WWF-U.K. – World Wildlife Fund – U.K.

WWF-U.S. – World Wildlife Fund – U.S.

INDEX